Electrophosphorescent Materials and Devices

Electrophosphorescent Materials and Devices

edited by

Mark E. Thompson

Published by

Jenny Stanford Publishing Pte. Ltd.
101 Thomson Road
#06-01, United Square
Singapore 307591

Email: editorial@jennystanford.com
Web: www.jennystanford.com

British Library Cataloguing-in-Publication Data
A catalogue record for this book is available from the British Library.

Electrophosphorescent Materials and Devices

Copyright © 2024 Jenny Stanford Publishing Pte. Ltd.

All rights reserved. This book, or parts thereof, may not be reproduced in any form or by any means, electronic or mechanical, including photocopying, recording or any information storage and retrieval system now known or to be invented, without written permission from the publisher.

For photocopying of material in this volume, please pay a copying fee through the Copyright Clearance Center, Inc., 222 Rosewood Drive, Danvers, MA 01923, USA. In this case permission to photocopy is not required from the publisher.

ISBN 978-981-4877-34-3 (Hardcover)
ISBN 978-1-003-08872-1 (eBook)

Contents

Preface   xxv

1 **Highly Efficient Phosphorescent Emission from Organic Electroluminescent Devices**   1
 Marc A. Baldo, D. F. O'Brien, Y. You, A. Shoustikov, Scott Sibley, Mark E. Thompson, and Stephen R. Forrest

2 **Improved Energy Transfer in Electrophosphorescent Devices**   13
 D. F. O'Brien, Marc A. Baldo, Mark E. Thompson, and Stephen R. Forrest

3 **Efficient, Saturated Red Organic Light Emitting Devices Based on Phosphorescent Platinum(II) Porphyrins**   23
 Raymond C. Kwong, Scott Sibley, Timur Dubovoy, Marc Baldo, Stephen R. Forrest, and Mark E. Thompson
    3.1 Introduction   24
    3.2 Experimental Section   25
       3.2.1 Materials   25
          3.2.1.1 PtDPP   26
          3.2.1.2 PtOX   26
       3.2.2 Photolummescence   26
       3.2.3 Device Fabrication and Testing   27
    3.3 Result and Discussion   28
       3.3.1 Photoluminescence   28
       3.3.2 Electroluminescence   30

4 Excitonic Singlet-Triplet Ratio in a Semiconducting Organic Thin Film   39
Marc A. Baldo, D. F. O'Brien, Mark E. Thompson, and Stephen R. Forrest
    4.1 Introduction   40
    4.2 Theory   40
    4.3 Experiment   43
    4.4 Significance of the Recombination Zone   51
    4.5 Discussion   53
    4.6 Conclusion   56

5 Very High-Efficiency Green Organic Light-Emitting Devices Based on Electrophosphorescence   59
Marc A. Baldo, Sergey Lamansky, Paul E. Burrows, Mark E. Thompson, and Stephen R. Forrest

6 Organic Light-Emitting Devices Based on Phosphorescent Hosts and Dyes   69
Raymond C. Kwong, Sergey Lamansky, and Mark E. Thompson

7 High-Efficiency Organic Electrophosphorescent Devices with tris(2-Phenylpyridine)Iridium Doped into Electron-Transporting Materials   81
Chihaya Adachi, Marc A. Baldo, Stephen R. Forrest, and Mark E. Thompson

8 High-Efficiency Fluorescent Organic Light-Emitting Devices Using a Phosphorescent Sensitizer   91
Marc A. Baldo, Mark E. Thompson, and Stephen R. Forrest

9 Nearly 100% Internal Phosphorescence Efficiency in an Organic Light Emitting Device   103
Chihaya Adachi, Marc A. Baldo, Mark E. Thompson, and Stephen R. Forrest
    9.1 Introduction   104
    9.2 Experimental Method   105
    9.3 Results   107
    9.4 Discussion   108

        9.4.1   Internal Electrophosphorescent Quantum
Efficiency   108
        9.4.2   Exciton Formation process   111
  9.5 Summary   112

10 Endothermic Energy Transfer: A Mechanism for Generating Very Efficient High-Energy Phosphorescent Emission in Organic Materials   115

Chihaya Adachi, Raymond C. Kwong, Peter Djurovich, Vadim Adamovich, Marc A. Baldo, Mark E. Thompson, and Stephen R. Forrest

11 High-Efficiency Yellow Double-Doped Organic Light-Emitting Devices Based on Phosphor-Sensitized Fluorescence   125

Brian W. D'Andrade, Marc A. Baldo, Chihaya Adachi, Jason Brooks, Mark E. Thompson, and Stephen R. Forrest

12 High-Efficiency Red Electrophosphorescence Devices   135

Chihaya Adachi, Marc A. Baldo, Stephen R. Forrest, Sergey Lamansky, Mark E. Thompson, and Raymond C. Kwong

13 Highly Phosphorescent Bis-Cyclometalated Iridium Complexes: Synthesis, Photophysical Characterization, and Use in Organic Light Emitting Diodes   145

Sergey Lamansky, Peter Djurovich, Drew Murphy, Feras Abdel-Razzaq, Hae-Eun Lee, Chihaya Adachi, Paul E. Burrows, Stephen R. Forrest, and Mark E. Thompson

  13.1 Introduction   147
  13.2 Experimental Section   149
        13.2.1   Synthesis   149
        13.2.2   Synthesis of $(C^\wedge N)_2$ Ir(acac) Complexes   149
                13.2.2.1   General Procedure   149
        13.2.3   Optical Measurements   151
        13.2.4   OLED Fabrication and Testing   151
  13.3 Results and Discussion   152

       13.3.1 Synthesis and Characterization of $C^\wedge N_2$ Ir(LX) Complexes   152
       13.3.2 Photophysical Properties of $C^\wedge N_2$Ir(LX) Complexes   155
              13.3.2.1 Correlation of absorption and emission bands   156
              13.3.2.2 $C^\wedge N$ ligand tuning of phosphorescence   160
              13.3.2.3 $C^\wedge N$ vs LX centered emission   162
              13.3.2.4 OLEDs prepared with $C^\wedge N_2$Ir(LX) complexes   163
13.4 Summary   167

14 Synthesis and Characterization of Phosphorescent Cyclometalated Iridium Complexes   171

Sergey Lamansky, Peter Djurovich, Drew Murphy, Feras Abdel-Razzaq, Raymond Kwong, Irina Tsyba, Manfred Bortz, Becky Mui, Robert Bau, and Mark E. Thompson

14.1 Introduction   172
14.2 Experimental Section   174
       14.2.1 Synthesis of $(\underline{C^\wedge N_2}$Ir (acac) Complexes: General Procedure   175
       14.2.2 Method A   177
       14.2.3 Method B   177
       14.2.4 Electrochemistry   178
       14.2.5 Crystallography   178
14.3 Results and Discussion   180
       14.3.1 Synthesis and Characterization of $\underline{C^\wedge N_2}$Ir(LX) Complexes   180
       14.3.2 Photophysical Properties of $\underline{C^\wedge N_2}$Ir(acac) Compounds   184
       14.3.3 Structures of $\underline{C^\wedge N_2}$ Ir(LX) Complexes   187
14.4 Conclusion   191

15 Synthesis and Characterization of Phosphorescent Cyclometalated Platinum Complexes — 197

Jason Brooks, Yelizaveta Babayan, Sergey Lamansky, Peter I. Djurovich, Irina Tsyba, Robert Bau, and Mark E. Thompson

- 15.1 Introduction — 198
- 15.2 Experimental Section — 201
  - 15.2.1 Equipment — 201
  - 15.2.2 Electrochemistry — 202
  - 15.2.3 X-ray Crystallography — 202
  - 15.2.4 Density Functional Calculations — 203
  - 15.2.5 Synthesis of $(C^\wedge N)(O^\wedge O)$ Complexes: General Procedure — 203
  - 15.2.6 Characterization — 205
- 15.3 Results and Discussion — 211
  - 15.3.1 Synthesis and Structure — 211
  - 15.3.2 Crystal Structure — 212
  - 15.3.3 DFT Calculations — 214
  - 15.3.4 Electrochemistry — 215
  - 15.3.5 Electronic Spectroscopy — 219
- 15.4 Conclusions — 227

16 White Light Emission Using Triplet Excimers in Electrophosphorescent Organic Light-Emitting Devices — 233

Brian W. D'Andrade, Jason Brooks, Vadim Adamovich, Mark E. Thompson, and Stephen R. Forrest

17 Electrophosphorescent p–i–n Organic Light-Emitting Devices for Very-High-Efficiency Flat-Panel Displays — 247

Martin Pfeiffer, Stephen R. Forrest, Karl Leo, and Mark E. Thompson

18 Cyclometalated Ir Complexes in Polymer Organic Light-Emitting Devices — 261

Sergey Lamansky, Peter I. Djurovich, Feras Abdel-Razzaq, Simona Garon, Drew L. Murphy, and Mark E. Thompson

- 18.1 Introduction — 262
- 18.2 Experiment — 264

18.3	Results and Discussion	266
18.4	Conclusions	274

19 High Efficiency Single Dopant White Electrophosphorescent Light Emitting Diodes — 279

Vadim Adamovich, Jason Brooks, Arnold Tamayo, Alex M. Alexander, Peter I. Djurovich, Brian W. D'Andrade, Chihaya Adachi, Stephen R. Forrest, and Mark E. Thompson

19.1	Introduction	280
19.2	Experimental	283
	19.2.1 Equipment	283
	19.2.2 Synthesis	283
	19.2.3 Estimation of HOMO and LUMO Energies	285
	19.2.4 OLED Fabrication	285
19.3	Results and Discussion	286
	19.3.1 Optimizing the Emissive Layer	286
	19.3.2 Single Dopant WOLEDs	294
19.4	Conclusion	300

20 High Operational Stability of Electrophosphorescent Devices — 305

Raymond C. Kwong, Matthew R. Nugent, Lech Michalski, Tan Ngo, Kamala Rajan, Yeh-Jiun Tung, Michael S. Weaver, Theodore X. Zhou, Michael Hack, Mark E. Thompson, Stephen R. Forrest, and Julie J. Brown

21 Controlling Exciton Diffusion in Multilayer White Phosphorescent Organic Light Emitting Devices — 315

Brian W. D'Andrade, Mark E. Thompson, and Stephen R. Forrest

22 Blue Organic Electrophosphorescence Using Exothermic Host–Guest Energy Transfer — 327

Russell J. Holmes, Stephen R. Forrest, Yeh-Jiun Tung, Raymond C. Kwong, Julie J. Brown, Simona Garon, and Mark E. Thompson

| 23 | Efficient, Deep-Blue Organic Electrophosphorescence by Guest Charge Trapping | 337 |

Russell J. Holmes, Brian W. D'Andrade, Stephen R. Forrest, Xiaofan Ren, Jian Li, and Mark E. Thompson

| 24 | Synthesis and Characterization of Facial and Meridional Tris-cyclometalated Iridium(III) Complexes | 347 |

Arnold B. Tamayo, Bert D. Alleyne, Peter I. Djurovich, Sergey Lamansky, Irina Tsyba, Nam N. Ho, Robert Bau, and Mark E. Thompson

24.1	Introduction	349
24.2	Experimental Section	351
	24.2.1 Equipment	351
	24.2.2 Electrochemistry	351
	24.2.3 X-ray Crystallography	352
	24.2.4 Density Functional Calculations	352
	24.2.5 Synthesis	355
	24.2.6 Synthesis of fac-Ir$(C^\wedge N)_3$ Complexes: General Procedure	355
	24.2.7 Synthesis of mer-Ir$(C^\wedge N)_3$ Complexes: General Procedure	356
	24.2.8 Characterization	357
	24.2.9 Isomerization of mer-Ir$(C^\wedge N)_3$ to fac-Ir$(C^\wedge N)_3$: Thermal Isomerization	358
	24.2.10 Photochemical Isomerization	359
24.3	Results and Discussion	359
	24.3.1 Synthesis and Structure of Ir$(C^\wedge N)_3$ Complexes	359
	24.3.2 NMR Characterization	361
	24.3.3 X-ray Crystallography	363
	24.3.4 DFT Calculations	368
	24.3.5 Electrochemistry	371
	24.3.6 Electronic Spectroscopy	372
	24.3.7 mer-to-fac Isomerization	376
24.4	Conclusion	378

25 Phosphorescence Quenching by Conjugated Polymers 383
Madhusoodhanan Sudhakar, Peter I. Djurovich,
Thieo E. Hogen-Esch, and Mark E. Thompson

26 Simultaneous Light Emission from a Mixture of Dendrimer Encapsulated Chromophores: A Model for Single-Layer Multichromophoric Organic Light-Emitting Diodes 391
Paul Furuta, Jason Brooks, Mark E. Thompson, and
Jean M. J. Fréchet
   26.1 Introduction 392
   26.2 Results and Discussion 394
       26.2.1 Dendrimer Encapsulated Chromophores 394
       26.2.2 Photo and Device Emission Spectra 395
   26.3 Conclusion 407

27 Ultrahigh Energy Gap Hosts in Deep Blue Organic Electrophosphorescent Devices 413
Xiaofan Ren, Jian Li, Russell J. Holmes, Peter I. Djurovich,
Stephen R. Forrest, and Mark E. Thompson
   27.1 Introduction 414
   27.2 Results and Discussion 416
       27.2.1 UGH Design and Characterization 416
       27.2.2 Electrophosphorescent OLEDs Employing UGH Materials 419
   27.3 Conclusions 422
   27.4 Experimental Section 422

28 Saturated Deep Blue Organic Electrophosphorescence Using a Fluorine-Free Emitter 427
Russell J. Holmes, Stephen R. Forrest, Tissa Sajoto,
Arnold Tamayo, Peter I. Djurovich, Mark E. Thompson,
Jason Brooks, Yeh-Jiun Tung, Brian W. D'Andrade,
Michael S. Weaver, Raymond C. Kwong, and Julie J. Brown

29 Excimer and Electron Transfer Quenching Studies of a Cyclometalated Platinum Complex 437
Biwu Ma, Peter I. Djurovich, and Mark E. Thompson
   29.1 Introduction 438
   29.2 Experimental 440

		29.2.1	Materials	440
		29.2.2	Physical Measurements	440
		29.2.3	Data Analysis for Bimolecular Quenching	442
	29.3	Results and Discussion		443
		29.3.1	Self-quenching	443
		29.3.2	Excited State Properties	449
			29.3.2.1 Redox properties estimated from electrochemistry and spectra	449
			29.3.2.2 Energy transfer quenching	450
			29.3.2.3 Reductive electron transfer quenching	451
			29.3.2.4 Oxidative electron transfer quenching	454
	29.4	Conclusions		455

30 Synthetic Control of Excited-State Properties in Cyclometalated Ir(III) Complexes Using Ancillary Ligands — **461**

Jian Li, Peter I. Djurovich, Bert D. Alleyne, Muhammed Yousufuddin, Nam N. Ho, J. Christopher Thomas, Jonas C. Peters, Robert Bau, and Mark E. Thompson

	30.1	Introduction		463
	30.2	Experimental Section		465
		30.2.1	X-Ray Crystallography	467
		30.2.2	Density Functional Theory Calculation	467
		30.2.3	Electrochemistry	468
		30.2.4	Synthesis	469
	30.3	Results and Discussion		473
		30.3.1	Synthesis and Characterization	473
		30.3.2	DFT Calculations	477
		30.3.3	Electrochemistry	477
		30.3.4	Electronic Spectroscopy	480
		30.3.5	Optical Transition Energies vs Redox Potentials	486
		30.3.6	Ground-State and Excited-State Properties	489
	30.4	Conclusion		499

31 Cationic Bis-cyclometalated Iridium(III) Diimine Complexes and Their Use in Efficient Blue, Green, and Red Electroluminescent Devices — 507

Arnold B. Tamayo, Simona Garon, Tissa Sajoto, Peter I. Djurovich, Irina M. Tsyba, Robert Bau, and Mark E. Thompson

- 31.1 Introduction — 508
- 31.2 Experimental Section — 511
  - 31.2.1 General Procedures — 511
  - 31.2.2 Synthesis of $(C^\wedge N)_2 \text{Ir}(N^\wedge N)^+ PF_6^-$ Complexes: General Procedure — 511
  - 31.2.3 X-ray Crystallography — 515
  - 31.2.4 Density Functional Theory (DFT) Calculations — 515
  - 31.2.5 Electrochemical and Photophysical Characterization — 515
  - 31.2.6 Device Fabrication — 516
- 31.3 Results and Discussion — 517
  - 31.3.1 Synthesis and Structure — 517
  - 31.3.2 DFT Calculations — 519
  - 31.3.3 Electrochemistry — 521
  - 31.3.4 Electronic Spectroscopy — 522
  - 31.3.5 OLED Studies — 529
- 31.4 Conclusion — 532

32 Blue and Near-UV Phosphorescence from Iridium Complexes with Cyclometalated Pyrazolyl or *N*-Heterocyclic Carbene Ligands — 537

Tissa Sajoto, Peter I. Djurovich, Arnold Tamayo, Muhammed Yousufuddin, Robert Bau, Mark E. Thompson, Russell J. Holmes, and Stephen R. Forrest

- 32.1 Introduction — 538
- 32.2 Results and Discussion — 543
  - 32.2.1 Cyclometalated Pyrazolyl-Based Ligands for Blue Phosphorescence — 543
  - 32.2.2 Cyclometalated Carbene Ligands for Near-UV Phosphorescence — 550
- 32.3 Conclusion — 558

32.4 Experimental Section 559
    32.4.1 Synthesis of *fac*-Iridium(III) Tris(1-[2-(9,9-dimethylfluorenyl)]pyrazolyl-$N,C^{3'}$), *fac*-Ir(flz)$_3$ 559
    32.4.2 Synthesis of [(pmi)$_2$IrCl]$_2$ 560
    32.4.3 Synthesis of *fac*-Iridium(III) Tris(1-phenyl-3-methylimidazolin-2-ylidene-$C,C^{2'}$), *fac*-Ir(pmi)$_3$ 561
    32.4.4 Synthesis of *mer*-Iridium(III) Tris(1-phenyl-3-methylimidazolin-2-ylidene-$C,C^{2'}$), *mer*-Ir(pmi)3 562
    32.4.5 Synthesis of [(pmb)$_2$IrCl]$_2$ 562
    32.4.6 Synthesis of Iridium(III) Tris(1-phenyl-3-methylbenzimidazolin-2-ylidene-$C,C^{2'}$), Ir(pmb)$_3$ 563
    32.4.7 X-ray Crystallography 564
    32.4.8 Electrochemical and Photophysical Characterization 564

33 Synthetic Control of Pt···Pt Separation and Photophysics of Binuclear Platinum Complexes 571
Biwu Ma, Jian Li, Peter I. Djurovich, Muhammed Yousufuddin, Robert Bau, and Mark E. Thompson

34 Platinum Binuclear Complexes as Phosphorescent Dopants for Monochromatic and White Organic Light-Emitting Diodes 579
Biwu Ma, Peter I. Djurovich, Simona Garon, Bert Alleyne, and Mark E. Thompson
  34.1 Introduction 580
  34.2 Results and Discussion 582
    34.2.1 Material Properties 582
    34.2.2 Monochromatic OLEDs 585
    34.2.3 Concentration Dependence of the EL of Compound 1 588
    34.2.4 White OLEDs 593
        34.2.4.1 Devices with dual emissive layers 594
        34.2.4.2 Devices with a single emissive layer 596

34.3	Conclusions	597
34.4	Experimental	598

35 Management of Singlet and Triplet Excitons for Efficient White Organic Light-Emitting Devices   **605**
Yiru Sun, Noel C. Giebink, Hiroshi Kanno, Biwu Ma, Mark E. Thompson, and Stephen R. Forrest

36 Highly Efficient, Near-Infrared Electrophosphorescence from a Pt–Metalloporphyrin Complex   **621**
Carsten Borek, Kenneth Hanson, Peter I. Djurovich, Mark E. Thompson, Kristen Aznavour, Robert Bau, Yiru Sun, Stephen R. Forrest, Jason Brooks, Lech Michalski, and Julie Brown

37 Intrinsic Luminance Loss in Phosphorescent Small-Molecule Organic Light Emitting Devices due to Bimolecular Annihilation Reactions   **631**
Noel C. Giebink, Brian W. D'Andrade, Michael S. Weaver, P. B. Mackenzie, Julie J. Brown, Mark E. Thompson, and Stephen R. Forrest

37.1	Introduction	632
37.2	Theory	633
37.3	Experimental	637
37.4	Results	639
37.5	Discussion	641
37.6	Conclusion	652

38 Blue Light Emitting Ir(III) Compounds for OLEDs: New Insights into Ancillary Ligand Effects on the Emitting Triplet State   **655**
Andreas F. Rausch, Mark E. Thompson, and Hartmut Yersin

38.1	Introduction	656
38.2	Experimental Section	658
	38.2.1 Synthesis	658
	38.2.2 Spectroscopy	658
38.3	Results and Discussion	659
	38.3.1 Spectroscopic Introduction	659

	38.3.2	Triplet Substates and Energy Level Diagram of FIracac in CH_2Cl_2 Based on High-Resolution Spectroscopy	660
	38.3.3	Comparison of FIracac and FIrpic: Influence of Ancillary Ligands	664
	38.3.4	Matrix Effects on SOC and on the Triplet State Properties	669
38.4	Summary and Conclusions		670

39 Temperature Dependence of Blue Phosphorescent Cyclometalated Ir(III) Complexes — 675

Tissa Sajoto, Peter I. Djurovich, Arnold B. Tamayo, Jonas Oxgaard, William A. Goddard III, and Mark E. Thompson

39.1	Introduction		676
39.2	Experimental Section		680
	39.2.1	Synthesis	680
	39.2.2	Quantum Yield Measurement	680
	39.2.3	Emission Intensity Measurement	680
	39.2.4	Lifetime Measurement	681
	39.2.5	Theoretical Calculations	681
39.3	Results and Discussion		682
39.4	Conclusion		698

40 Study of Energy Transfer and Triplet Exciton Diffusion in Hole-Transporting Host Materials — 707

Chao Wu, Peter I. Djurovich, and Mark E. Thompson

40.1	Introduction		708
40.2	Results and Discussion		710
	40.2.1	Electrochemistry	711
	40.2.2	Photophysical Properties	712
	40.2.3	Undoped Devices	714
	40.2.4	Doped Devices	716
	40.2.5	Theoretical Analysis	723
40.3	Conclusions		725
40.4	Experimental		726

41 Synthesis and Characterization of Phosphorescent Three-Coordinate Cu(I)–NHC Complexes   733
Valentina A. Krylova, Peter I. Djurovich, Matthew T. Whited, and Mark E. Thompson

42 A Codeposition Route to CuI–Pyridine Coordination Complexes for Organic Light-Emitting Diodes   743
Zhiwei Liu, Munzarin F. Qayyum, Chao Wu, Matthew T. Whited, Peter I. Djurovich, Keith O. Hodgson, Britt Hedman, Edward I. Solomon, and Mark E. Thompson

43 Structural and Photophysical Studies of Phosphorescent Three-Coordinate Copper(I) Complexes Supported by an N-Heterocyclic Carbene Ligand   755
Valentina A. Krylova, Peter I. Djurovich, Jacob W. Aronson, Ralf Haiges, Matthew T. Whited, and Mark E. Thompson

43.1	Introduction		756
43.2	Results and Discussion		758
	43.2.1	Synthesis and X-ray Structures	758
	43.2.2	NMR Characterization	761
	43.2.3	DFT Calculations	767
	43.2.4	Photophysical Properties	769
43.3	Conclusion		775
43.4	Experimental Section		776
	43.4.1	Synthesis	776
		43.4.1.1 Synthesis of (IPr)Cu(fpyro) (1)	777
		43.4.1.2 Synthesis of (IPr)Cu(ppz) (2)	778
		43.4.1.3 Synthesis of (IPr)Cu(fppz) (3)	778
		43.4.1.4 Synthesis of (IPr)Cu(fpta) (4)	779
	43.4.2	X-ray Crystallography	779
		43.4.2.1 $C_{38}H_{41}CuF_6N_4$ (1)	780
		43.4.2.2 $C_{36}H_{41}CuF_3N_5$ (3)	781
		43.4.2.3 C35H40CuF3N6 (4)	781
	43.4.3	Density Functional Calculations	782
	43.4.4	Photophysical Characterization	782

Contents | xix

44 **Phosphorescence versus Thermally Activated Delayed Fluorescence: Controlling Singlet–Triplet Splitting in Brightly Emitting and Sublimable Cu(I) Compounds** — 789
Markus J. Leitl, Valentina A. Krylova, Peter I. Djurovich, Mark E. Thompson, and Hartmut Yersin
44.1 Introduction — 790
44.2 Ambient Temperature Phosphorescence versus TADF — 792
    44.2.1 Compound 2: Typical Triplet Emitter — 794
    44.2.2 Compound 1: Thermally Activated Delayed Fluorescence — 797
44.3 Controlling TADF by Ligand Orientation — 799
44.4 Conclusion — 804
44.5 Experimental Section — 805

45 **Control of Emission Colour with N-Heterocyclic Carbene (NHC) Ligands in Phosphorescent Three-Coordinate Cu(I) Complexes** — 811
Valentina A. Krylova, Peter I. Djurovich, Brian L. Conley, Ralf Haiges, Matthew T. Whited, Travis J. Williams, and Mark E. Thompson

46 **Synthesis and Characterization of Phosphorescent Platinum and Iridium Complexes with Cyclometalated Corannulene** — 823
John W. Facendola, Martin Seifrid, Jay Siegel, Peter I. Djurovich, and Mark E. Thompson
46.1 Introduction — 824
46.2 Results and Discussion — 826
    46.2.1 Crystal Structures — 827
    46.2.2 NMR and Dynamic Behavior — 830
    46.2.3 Electrochemical Properties — 836
    46.2.4 Photophysical Properties — 837
46.3 Conclusion — 842
46.4 Experimental — 843
    46.4.1 Synthesis — 843
        46.4.1.1 (corpy)Pt(dpm) — 844
        46.4.1.2 (corpy)Ir(ppz)$_2$ — 844
        46.4.1.3 (phenpy)Ir(ppz)$_2$ — 845

	46.4.2	Electrochemistry	846
	46.4.3	NMR Measurements	846
	46.4.4	X-ray Crystallography	847
	46.4.5	Photophysical Measurements	847
	46.4.6	Computational Methods	847

47 Understanding and Predicting the Orientation of Heteroleptic Phosphors in Organic Light-Emitting Materials 851

Matthew J. Jurow, Christian Mayr, Tobias D. Schmidt, Thomas Lampe, Peter I. Djurovich, Wolfgang Brütting, and Mark E. Thompson

	47.1	Introduction	852
	47.2	Results	854
	47.3	Discussion	856
	47.4	Conclusion	865
	47.5	Methods	866

48 Deep Blue Phosphorescent Organic Light-Emitting Diodes with Very High Brightness and Efficiency 877

Jaesang Lee, Hsiao-Fan Chen, Thilini Batagoda, Caleb Coburn, Peter I. Djurovich, Mark E. Thompson, and Stephen R. Forrest

	48.1	Introduction		878
	48.2	Results		880
	48.3	Discussion		889
		48.3.1	Implication of the Employed Device Design	889
		48.3.2	Origin of Both Highly Emissive *fac*- and *mer*-Ir(pmp)$_3$	890
		48.3.3	Differences between D_{fac} and D_{mer}	891
	48.4	Conclusions		892
	48.5	Methods		893

49 Hot Excited State Management for Long-Lived Blue Phosphorescent Organic Light-Emitting Diodes 901

Jaesang Lee, Changyeong Jeong, Thilini Batagoda, Caleb Coburn, Mark E. Thompson, and Stephen R. Forrest

| | 49.1 | Introduction | 902 |
| | 49.2 | Results | 904 |

| | | 49.2.1 | Hot Excited State Management to Extend PHOLED Lifetime | 904 |

		49.2.2	Performance of Managed PHOLEDs	907
	49.3	Discussion		913
	49.4	Methods		918
		49.4.1	Exciton Profile Measurement	919
		49.4.2	Mass Spectrometry Measurement	920
		49.4.3	Lifetime Degradation Model	921

50 Eliminating Nonradiative Decay in Cu(I) Emitters: > 99% Quantum Efficiency and Microsecond Lifetime — 927

Rasha Hamze, Jesse L. Peltier, Daniel Sylvinson, Moonchul Jung, Jose Cardenas, Ralf Haiges, Michele Soleilhavoup, Rodolphe Jazzar, Peter I. Djurovich, Guy Bertrand, and Mark E. Thompson

51 Rapid Multiscale Computational Screening for OLED Host Materials — 947

Daniel Sylvinson M. R., Hsiao-Fan Chen, Lauren M. Martin, Patrick J. G. Saris, and Mark E. Thompson

	51.1	Introduction		948
	51.2	Experimental Section		951
		51.2.1	Quantum Mechanics	951
		51.2.2	Molecular Dynamics/Electron Coupling Calculations	952
		51.2.3	Synthesis	954
			51.2.3.1 General procedure for synthesis of 1H-imidazo[4,5-*f*]-phenanthrolines	954
			51.2.3.2 2-(Tert-butyl)-1-phenyl-1H-imidazo[4,5-*f*][1,10] phenanthroline (10d)	955
			51.2.3.3 2-(Tert-butyl)-1-phenyl-1H-imidazo[4,5-*f*][3,8] phenanthroline (10c)	955
			51.2.3.4 2-(Tert-butyl)-1-phenyl-1H-imidazo[4,5-*f*][4,7] phenanthroline (10b)	956

		51.2.4	Equipment	956
		51.2.5	Electrochemistry	956
		51.2.6	OLED Fabrication and Testing	957
	51.3	Results and Discussion		957
		51.3.1	Molecular Search Strategy	957
		51.3.2	*Tier* 1 Selection	959
		51.3.3	*Tier* 2 Selection	961
		51.3.4	Synthesis and Physical Characterization of H2P Host Materials	964
		51.3.5	*Tier* 3 Selection	967
		51.3.6	OLED Fabrication and Testing	973
	51.4	Summary		977

52 "Quick-Silver" from a Systematic Study of Highly Luminescent, Two-Coordinate, d^{10} Coinage Metal Complexes   **987**

Rasha Hamze, Shuyang Shi, Savannah C. Kapper, Daniel Sylvinson Muthiah Ravinson, Laura Estergreen, Moon-Chul Jung, Abegail C. Tadle, Ralf Haiges, Peter I. Djurovich, Jesse L. Peltier, Rodolphe Jazzar, Guy Bertrand, Stephen E. Bradforth, and Mark E. Thompson

	52.1	Introduction		988
	52.2	Experimental Section		992
		52.2.1	Electrochemical Measurements	992
		52.2.2	X-ray Crystallography	992
		52.2.3	Photophysical Measurements	993
		52.2.4	Molecular Modeling	994
	52.3	Results and Discussion		994
	52.4	Conclusion		1009

53 Highly Efficient Photo- and Electroluminescence from Two-Coordinate Cu(I) Complexes Featuring Nonconventional N-Heterocyclic Carbenes   **1019**

Shuyang Shi, Moon Chul Jung, Caleb Coburn, Abegail Tadle, Daniel Sylvinson M. R., Peter I. Djurovich, Stephen R. Forrest, and Mark E. Thompson

	53.1	Introduction	1020
	53.2	Experimental Section	1023

53.2.1 Synthesis 1023
    53.2.1.1 Synthesis of (MAC*)CuCl (1a): 3-chloro-N-(2,6-diisopropylphenyl)-*N'*-((2,6-diisopropylphenylimino)methyl)-2,2-dimethylpropanamide(1c) 1023
    53.2.1.2 1,3-bis(2,6-diisopropylphenyl)-5,5-dimethyl-4-keto-tetrahydropyrimidin-1-ium Chloride (1b) 1024
    53.2.1.3 (*N,N'*-bis(diisopropylphenyl)-5,5-dimethyl-4-keto-tetrahydropyrimidin-2-ylidene)-Cu(I) chloride (MAC*CuCl) (1a) 1024
    53.2.1.4 Synthesis of [(DAC*)Cu]$_2$Cl$_2$ (2a) 1025
    53.2.1.5 Synthesis of Complexes 1–6: General Procedure 1026
    53.2.1.6 (MAC*)Cu(CzCN$_2$) (1) 1026
    53.2.1.7 (MAC*)Cu(CzCN) (2) 1027
    53.2.1.8 (MAC*)Cu(Cz) (3) 1028
    53.2.1.9 (DAC*)Cu(CzCN$_2$) (4) 1028
    53.2.1.10 (DAC*)Cu(CzCN) (5) 1029
    53.2.1.11 (DAC*)Cu(Cz) (6) 1029
53.2.2 X-ray Crystallography 1030
53.2.3 Electrochemical Measurements 1030
53.2.4 Photophysical Characterization 1030
53.2.5 Density Functional Theory (DFT) Calculations 1031
53.2.6 OLED Fabrication and Characterization 1031
53.3 Results and Discussion 1032
  53.3.1 Synthesis and Characterization 1032
  53.3.2 Photophysical Properties 1036
  53.3.3 Electroluminescence 1046
53.4 Conclusion 1048

54 **Platinum-Functionalized Random Copolymers for Use in Solution-Processible, Efficient, Near-White Organic Light-Emitting Diodes**   1061
Paul T. Furuta, Lan Deng, Simona Garon, Mark E. Thompson, and Jean M. J. Fréchet

Index   1069

Preface

The first report from electroluminescence from organic semiconductors appeared in the 1960s and involved relatively large single-crystal samples with evaporated metal contacts on opposite faces.[1] These devices required very high voltages and had short operational lifetimes. The paper that really created the field of organic electroluminescence came from two scientists at the Eastman Kodak Company, Drs Ching Tang and Stephen van Slyke, in 1987.[2] This paper described a bilayer device consisting of hole and electron transporting layers, sandwiched between an indium-tin oxide anode and a magnesium-silver cathode. The layers were 60–75 nm thick and deposited solely by vapor deposition. Even though their efficiency was somewhat modest at 1% (photons/electrons). Two years later, Tang, van Slyke, and Chen demonstrated that they could improve the efficiency of these organic devices by shifting from emitting films made of a single component to ones that incorporated an emissive dopant.[3] This doping approach largely eliminated self-quenching and led to a marked increase in external efficiencies for doped devices. The doping approach had the added benefit that the emission color of the organic LED, or OLED, can be controlled by choosing a fluorescent dye of a desired color. This allowed the community to develop and optimize materials and device structures for carrier transport, recombination, and emission largely independently. The science and engineering communities realized the potential of these OLEDs quickly and a large number of research groups followed the lead of the Kodak researchers. In less than 10 years, the OLED community had prepared devices that covered the entire visible spectrum, prepared broadband, white emissive devices, and pushed the efficiencies to 5–6%.[4]

While the record OLED efficiencies reported in the late 1990s seem low, it is important to stress that these efficiencies are based on only the light that escapes in the forward direction. A substantial fraction of the light generated by electroluminescence is lost due to self-absorption by the organic and cathode materials as well as to waveguiding in the high-index substrates and organic film. These losses prevent 70–80% of the light generated in the OLED from escaping in the forward direction (perpendicular to the substrate). Thus, an external quantum efficiency (EQE) of 5–6%, corresponds to an internal efficiency for the electroluminescent process of roughly 25%. There is a simple reason why the OLED efficiencies seem to stall at 25% in the late 1990s and is tied to the choice of emissive dopants used in OLEDs. The hole and electron that are injected into the OLED have $m_s = \pm 1/2$. A simple spin statistical analysis predicts that on recombination the probability of generating a singlet excited state is 0.25 and that of forming a triplet is 0.75.[5,6] The best OLEDs prepared in the first 10 years after Tang and van Slyke's seminal paper used emissive dopants with the highest possible photoluminescent efficiency, to ensure the highest possible OLED efficiency. The problem is that these were all fluorescent dyes, which have very inefficient intersystem crossing (ISC). These dyes are designed to efficiently emit from a singlet excited state and if they phosphoresce, do so very inefficiently, often with decay lifetimes in the seconds to minutes regime. Thus, these fluorescence-based OLEDs have an internal efficiency limit of 25%.

Starting in the mid-1990s, the groups of Mark Thompson at the University of Southern California and Stephen Forrest, then at Princeton University, set out on a different tract. Rather than focus on fluorescence-based organic electroluminescence, they chose to explore the use of phosphorescent emitters, with a goal of eliminating the spin statistical limitation on OLED efficiency. This approach required a total rethinking of the emitters used in OLEDs. The key issue was to find emitters that could harvest both the singlet and triplet excitons generated in the electroluminescent process. The fluorescent dyes used previously comprised light elements, i.e., C, H, N, O, etc., giving them very poor spin–orbit coupling, markedly limiting their ISC and phosphorescence efficiencies. Thompson and Forrest shifted from simple organic dyes to emitters that contained a heavy metal ion to provide sufficient spin–orbit coupling to give efficient ISC (singlet → triplet) and phosphorescence (emission from the triplet). This gives these heavy metal or phosphorescence-based OLEDs the potential to reach an internal efficiency of 100%. Their work started with platinum porphyrins and organometallic complexes and in 1999 shifted to iridium-based organometallic emitters. Within a few years, the team demonstrated near-unit efficiency for green OLEDs and soon followed with devices emitting across the visible spectrum. It is noteworthy that less than 10 years after the report of these high-efficiency phosphorescent OLEDs, Samsung Display launched their Galaxy line of cellular phones that incorporated iridium-based phosphors as the principal emitters in their active matrix OLED displays.

This book is comprised of a series of chapters that are reproductions of the papers that described the discovery and development of phosphorescence-based OLEDs. This work involved a close collaboration between the Forrest group, working along the physics and engineering axis, with the Thompson group working predominantly on the chemical aspects of the problems associated with efficient singlet and triplet harvesting in OLEDs. The chapters in this book are focused largely on the chemical aspects of the problem, but it is important to stress that a close collaboration between chemical and physics/engineering groups was critical in advancing the field and finding the best solutions to problems, whether they are tied to the materials that comprise the OLED or the device structure. It is also important to stress that a key player in the evolution and ultimate incorporation of phosphorescent OLEDs into commercial displays and solid-state lighting was Universal Display Corporation. They strongly supported the work in the Thompson and Forrest labs in phosphorescence-based OLEDs and developed it into the viable technology it is today.

This book focuses on papers covering the work of the Thompson group from 1996 to 2019. This book is not meant to be a comprehensive review, but a summary of the work from a single, chemistry group working in the field of phosphorescent OLEDs. The reader is encouraged to read other sources to see the important contributions made by other labs and companies that have made OLEDs a viable display and lighting technology that it is today.

The later chapters in the book cover the work of the Thompson group with copper-based emitters. These group 11 complexes emit via thermally assisted delayed fluorescence. We have published a number of papers since 2019 that have expanded our understanding of these potentially important emitters, as well as extended our work to cover their silver- and gold-based analogs.[7–16] The reader is referred to these papers to see the follow-on story from the copper-based TADF emitters covered here.

Mark E. Thompson
Ray R. Irani Chair of Chemistry
The University of Southern California, USA

References

1. Pope, M.; Kallmann, H. P.; Magnante, P., Electroluminescence in organic crystals. *The Journal of Chemical Physics* **1963**, *38* (8), 2042–2043.
2. Tang, C. W.; VanSlyke, S. A., Organic electroluminescent diodes. *Applied Physics Letters* **1987**, *51* (12), 913–915.
3. Tang, C. W.; VanSlyke, S. A.; Chen, C. H., Electroluminescence of doped organic thin films. *Journal of Applied Physics* **1989**, *65* (9), 3610–3616.
4. Shoustikov, A. A.; You, Y.; Thompson, M. E., Electroluminescence color tuning by dye doping in organic light-emitting diodes. *Selected Topics in Quantum Electronics, IEEE Journal of* **1998**, *4* (1), 3–13.
5. Thompson, M. E.; Djurovich, P. E.; Barlow, S.; Marder, S., Organometallic complexes for optoelectronic applications. In *Comprehensive Organometallic Chemistry III*, 1 ed.; O'Hare, D., Ed. Elsevier Ltd.: Oxford, UK, 2007; vol. 12, pp. 102–194.
6. Forrest, S. R., *Organic Electronics: Foundations to Applications*. Oxford University Press: Oxford, England, 2020.
7. Ravinson, D. S. M.; Thompson, M. E., Thermally assisted delayed fluorescence (TADF): fluorescence delayed is fluorescence denied. *Materials Horizons* **2020**, *7* (5), 1210–1217.
8. Hamze, R.; Shi, S.; Kapper, S. C.; Muthiah Ravinson, D. S.; Estergreen, L.; Jung, M.-C.; Tadle, A. C.; Haiges, R.; Djurovich, P. I.; Peltier, J. L.; Jazzar, R.; Bertrand, G.; Bradforth, S. E.; Thompson, M. E., "Quick-silver" from a systematic study of highly luminescent, two-coordinate, d^{10} coinage

metal complexes. *Journal of the American Chemical Society* **2019**, *141* (21), 8616–8626.

9. Shi, S.; Jung, M. C.; Coburn, C.; Tadle, A.; Sylvinson M. R, D.; Djurovich, P. I.; Forrest, S. R.; Thompson, M. E., Highly efficient photo- and electroluminescence from two-coordinate Cu(I) complexes featuring nonconventional N-heterocyclic carbenes. *Journal of the American Chemical Society* **2019**, *141* (8), 3576–3588.

10. Hamze, R.; Idris, M.; Muthiah Ravinson, D. S.; Jung, M. C.; Haiges, R.; Djurovich, P. I.; Thompson, M. E., Highly efficient deep blue luminescence of 2-coordinate coinage metal complexes bearing bulky NHC benzimidazolyl carbene. *Frontiers in Chemistry* **2020**, *8* (401), 1–9.

11. Li, T.-y.; Muthiah Ravinson, D. S.; Haiges, R.; Djurovich, P. I.; Thompson, M. E., Enhancement of the luminescent efficiency in carbene-Au(I)-aryl complexes by the restriction of Renner–Teller distortion and bond rotation. *Journal of the American Chemical Society* **2020**, *142* (13), 6158–6172.

12. Li, T.-y.; Shlian, D. G.; Djurovich, P. I.; Thompson, M. E., A luminescent two-coordinate AuI bimetallic complex with a tandem-carbene structure: a molecular design for the enhancement of TADF radiative decay rate. *Chemistry—A European Journal* **2021**, *27* (20), 6191–6197.

13. Li, T.-y.; Djurovich, P. I.; Thompson, M. E., Phosphorescent monometallic and bimetallic two-coordinate Au(I) complexes with N-heterocyclic carbene and aryl ligands. *Inorganica Chimica Acta* **2021**, *517*, 120188.

14. Ma, J.; Kapper, S. C.; Ponnekanti, A.; Schaab, J.; Djurovich, P. I.; Thompson, M. E., Dynamics of rotation in two-coordinate thiazolyl copper(I) carbazolyl complexes. *Applied Organometallic Chemistry* **2022**, e6728.

15. Li, T.-y.; Schaab, J.; Djurovich, P. I.; Thompson, M. E., Toward rational design of TADF two-coordinate coinage metal complexes: understanding the relationship between natural transition orbital overlap and photophysical properties. *Journal of Materials Chemistry C* **2022**, *10* (12), 4674–4683.

16. Muniz, C. N.; Schaab, J.; Razgoniaev, A.; Djurovich, P. I.; Thompson, M. E., π-Extended ligands in two-coordinate coinage metal complexes. *Journal of American Chemical Society* **2022**, *144* (39), 17916–17928.

Chapter 1

Highly Efficient Phosphorescent Emission from Organic Electroluminescent Devices

M. A. Baldo,[a] D. F. O'Brien,[a] Y. You,[b] A. Shoustikov,[b] S. Sibley,[b,c] M. E. Thompson,[b] and S. R. Forrest[a]

[a]*Center for Photonics and Optoelectronic Materials, Department of Electrical Engineering and the Princeton Materials Institute, Princeton University, Princeton, New Jersey 08544, USA*
[b]*Department of Chemistry, University of Southern California, Los Angeles, California 90089, USA*
[c]*Permanent address: Department of Chemistry, Goucher College, Baltimore, Maryland 21204-2794, USA*
forrest@ee.princeton.edu

The efficiency of electroluminescent organic light-emitting devices [1, 2] can be improved by the introduction [3] of a fluorescent dye. Energy transfer from the host to the dye occurs via excitons, but only the singlet spin states induce fluorescent emission; these represent a small fraction (about 25%) of the total excited-state population (the remainder are triplet states). Phosphorescent dyes, however, offer a means of achieving improved light-emission efficiencies, as emission may result from both singlet and triplet states. Here we

---

Reprinted from *Nature*, **395**, 151–154, 1998.

Electrophosphorescent Materials and Devices
Edited by Mark E. Thompson
Text Copyright © 1998 Macmillan Publishers Ltd.
Layout Copyright © 2024 Jenny Stanford Publishing Pte. Ltd.
ISBN 978-981-4877-34-3 (Hardcover), 978-1-003-08872-1 (eBook)
www.jennystanford.com

report high-efficiency ($\gtrsim$90%) energy transfer from both singlet and triplet states, in a host material doped with the phosphorescent dye 2,3,7,8,12,13,17,18-octaethyl-21H,23H-porphine platinum(II) (PtOEP). Our doped electroluminescent devices generate saturated red emission with peak external and internal quantum efficiencies of 4% and 23%, respectively. The luminescent efficiencies attainable with phosphorescent dyes may lead to new applications for organic materials. Moreover, our work establishes the utility of PtOEP as a probe of triplet behaviour and energy transfer in organic solid-state systems.

When the absorption spectrum of the acceptor (dye) overlaps the emission spectrum of the donor (host), efficient energy transfer from the host to the dye can occur via a singlet-allowed, induced-dipole coupling between the molecular species. Hence, for a fluorescent emitter, the maximum external quantum efficiency (photons extracted in the forward direction per electron injected) is [4, 5]:

$$\phi_{el} = \chi \phi_{fl} \eta_r \eta_e$$

The fraction of charge carrier recombinations in the host resulting in singlet excitons is χ, which from spin statistics is presumed to be $\sim$1/4, ϕ_{fl} is the photoluminescent efficiency of the dye, η_e is the fraction of emitted photons that are coupled out of the device, and η_r is the fraction of injected charge carriers that form excitons. As both the recombination and fluorescent efficiencies can approach unity for an optimized device, the efficiency is primarily limited by coupling losses and a restriction to singlet excitons imposed by spin conservation in the induced-dipole energy-transfer process.

Although the output coupling of photons can be increased by using shaped substrates [6], further efficiency improvements require that both singlet and triplet excited states contribute to luminescence. It has been proposed that intersystem crossing in lanthanide complexes may achieve this with an intramolecular energy transfer from a triplet state of the organic ligand to the $4f$ energy state of the ion [7]. However, a more general and efficient solution to the problem is to use phosphorescent emissive materials.

Phosphorescence is the forbidden relaxation of an excited state with spin symmetry different from the ground state; in organic molecules it typically results from a triplet to a singlet relaxation. Although low-efficiency electroluminescence has been demonstrated [8] at 100 K using the phosphorescent material benzo-phenone doped into poly(methylmethacrylate), no energy transfer from the host material was found. If a phosphorescent dye participates in energy transfer, then triplet behaviour within the host can be directly examined. This may enable the accurate determination of spin statistics and other triplet properties, as well as resulting in very high electroluminescent efficiency devices.

One relatively well studied red emitting phosphorescent dye is PtOEP. Porphine complexes are known to possess long-lived triplet states useful in oxygen detection [9]. The addition of platinum to the porphine ring reduces the phosphorescence lifetime by increasing spin–orbit coupling; the triplet states gain additional singlet character and vice versa. This also enhances the efficiency of intersystem crossing from the first singlet excited state to the triplet excited state. Transient absorption spectrometry gives a singlet lifetime in PtOEP of ~1 ps, and the fluorescence efficiency is extremely weak [10]. In contrast, the room-temperature phosphorescence efficiency of PtOEP in a polystyrene matrix is [11] 0.5 with an observed lifetime of 91 μs.

Induced dipole (or Förster) energy transfer to the triplet state is disallowed by spin conservation. However, energy transfer may occur by the parallel combination of Förster transfer to the singlet state, along with electron exchange (Dexter energy transfer). Dexter transfer is the 'physical', coherent transfer of an exciton from a donor to an acceptor site [12] at a rate proportional to the orbital overlap of the donor and acceptor molecules. Consequently, it is a short-range process, attenuating exponentially with distance. The transfer rate is proportional to the spectral overlap of the two species because the donor exciton energy must closely match that of the acceptor. Dexter processes allow for the transfer of triplet excitons because only the total spin of the donor–acceptor complex is conserved. Thus it is possible for both singlets and triplets to excite phosphorescence, increasing the theoretical efficiency limit from that for fluorescence by a factor of four or larger.

PtOEP shows [13] strong absorption at a wavelengths of 530 nm, corresponding to the peak emission of the electron transport material, tris-(8-hydroxyquinoline) aluminium (Alq$_3$); this makes PtOEP a suitable dopant for Alq$_3$-based organic light-emitting diodes (LEDs). To study the electroluminescent properties of Alq$_3$/PtOEP systems, organic LEDs were produced by high-vacuum (10^{-6} torr) thermal evaporation onto a cleaned glass substrate precoated with conductive, transparent indium tin oxide (ITO). The test structure consists of a 60-Å-thick layer of copper phthalocyanine to aid hole injection from the ITO, a 350-Å-thick layer of the hole transporting material 4,4'-bis[N-(1-napthyl)-N-phenyl-amino] biphenyl (α-NPD), a 400-Å-thick layer of Alq$_3$ doped with varying molar concentrations of PtOEP, and a further 100-Å-thick layer of Alq$_3$ to prevent quenching of PtOEP excitons at the cathode. Finally, a shadow mask with 1-mm-diameter openings was used to define the cathode consisting of a 1,000-Å-thick layer of 25:1 Mg:Ag, with a 500-Å-thick Ag cap. All measurements were performed in air at room temperature except for the photoluminescence measurements which were performed under nitrogen.

The electroluminescence spectra of the devices with three different concentrations of PtOEP are shown in Fig. 1.1. No significant emission is found for the previously identified singlet state, expected at ~580 nm (Ref. [10]), but strong emission is observed from the triplet excited state at 650 nm, with weaker emission seen at the vibronic harmonic overtones at 623 nm, 687 nm and 720 nm. At high drive currents, the emission at 650 nm saturates, and increasing emission is seen from other features, especially the broad Alq$_3$ peak centred at 530 nm. This results in a reversible shift in colour from deep red to orange, corresponding to a shift in the device chromaticity co-ordinates shown in Fig. 1.1 (inset).

Spectral and time-resolved emission measurements were performed with a streak camera on our organic LEDs excited with current pulses. Slow decay of between 10 and 50 μs was observed for the triplet state, confirming the presence of phosphorescence. The luminescent lifetime was also found to decrease with increasing current density, as shown in Fig. 1.2a for a device with a 6% concentration of PtOEP in Alq$_3$.

From the spectral data, the fraction of output power in the red (subtracting the Alq$_3$ emission) can be determined as a function of current to provide the external quantum efficiency (Fig. 1.2b).

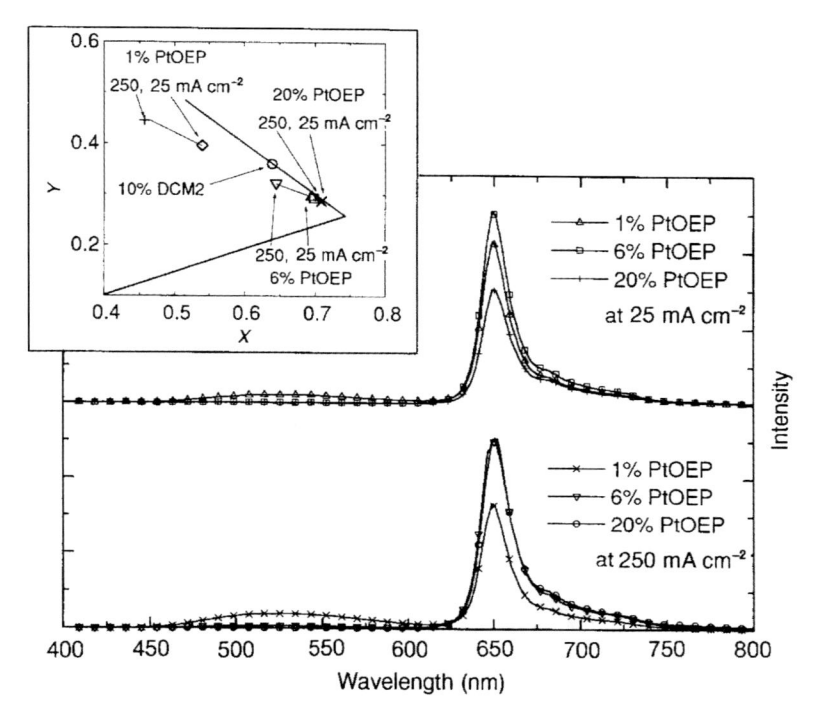

Figure 1.1 Spectra of the organic LEDs with different molar concentrations of PtOEP at different current densities. Upper traces, 1%, 6% and 20% PtOEP in Alq_3 organic LEDs at 25 mA cm^{-2}; lower traces, 1%, 6% and 20% PtOEP in Alq_3 organic LEDs at 250 mA cm^{-2}. We note the increased Alq_3 emission at 530 nm in the 1% PtOEP organic LED. Insert, Commission Internationale de L'Éclairage (CIE) chromaticity coordinates for the devices in the main figure at the specified current densities. Only the red corner of the CIE diagram is shown. We note the trend from saturated red to orange with increasing current. The 6% PtOEP in Alq_3 device at 25 mA cm^{-2} has a luminance of 100 cd m^{-2}. The DCM2 result is taken from Ref. [15].

As expected [12], the trend in phosphorescent lifetime matches the quantum efficiency results: the shorter lifetimes observed at high current densities corresponds to lower quantum efficiencies. The external quantum efficiency reaches a maximum value of 4% at a PtOEP concentration of 6%. To our knowledge, this is considerably higher than the highest external quantum efficiency obtained to date for red-emitting, fluorescent organic LEDs.

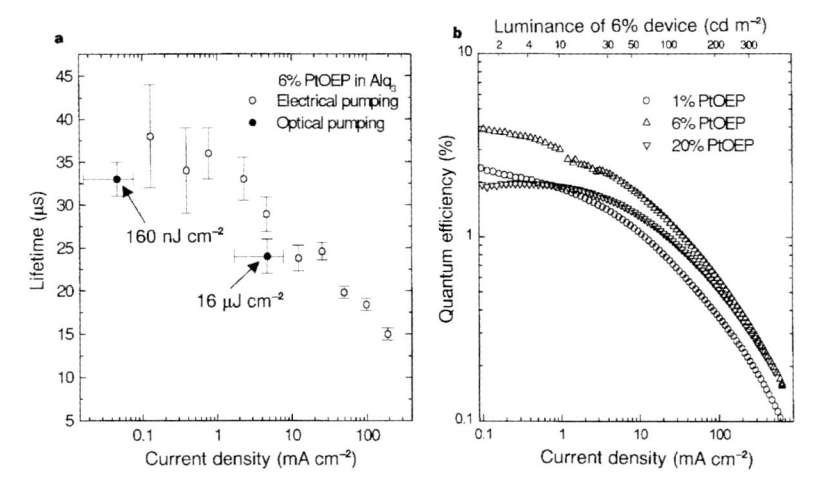

Figure 1.2 Lifetime and quantum efficiency of PtOEP emission. (a) Phosphorescent lifetime of the 6% PtOEP in Alq$_3$ organic LED as function of current density (open circles). To explore the influence of bimolecular processes, PtOEP quenching in the absence of charge carriers was investigated using photoluminescence. This experiment was performed by pumping a 500-Å- thick film of 6% PtOEP in Alq$_3$ with 500-ps pulses from a nitrogen laser at a wavelength of 337 nm. At a pulse energy density of 16 μJ cm^{-2}, we measured a PtOEP lifetime of (24 ± 2) μs. This is similar to the lifetime obtained under electrical pumping at current densities of ∼5 mA cm^{-2}. To check if the optical and electrical decay times have a similar origin, we note that the laser output is absorbed principally by Alq$_3$ with an absorption coefficient [20] of ∼3×10^4 cm^{-1}. Hence, ∼26% of the nitrogen laser pulse is absorbed, giving an exciton density of 1.4×10^{18} cm^{-3}. Assuming 100% transfer of injected carriers to PtOEP excitons, the current density corresponding to this pumping level is shown above (filled circles). When the pulse energy is decreased to ∼160 nJ cm^{-2}, the lifetime increases to (33 ± 2) μs. The trend in the lifetime of optically pumped samples, which has been previously attributed [14] to bimolecular quenching, is consistent with the trend of the electroluminescent lifetime data, suggesting a common physical origin, (b) Quantum efficiency of PtOEP emission as a function of doping concentration and current density. The top axis shows the luminance of the 6% PtOEP in Alq$_3$ device.

Because PtOEP molecules have a long exciton lifetime, it is possible that saturation is responsible for the decreased quantum efficiency at high currents and low dopant concentrations. This can be examined by comparing the total number of PtOEP sites to the maximum number of photons extracted. For a molar concentration of 1%, the number of PtOEP sites in the ~50-Å-thick recombination zone [3] near the α-NPD/PtOEP:Alq$_3$ interface is $\sim 8 \times 10^{12}$ cm^{-2}. After adjusting for the output coupling efficiency [6], the internal emitted photon flux saturates at $\sim 3 \times 10^{16}$ s^{-1} cm^{-2}, yielding a radiative lifetime of ~300 µs, consistent with previous PtOEP phosphorescence efficiency and lifetime measurements [11]. If sites within the recombination zone saturate, then Alq$_3$ excitons are less likely to transfer to PtOEP, corresponding to an increase in the probability for radiative recombination in Alq$_3$. This is supported by the data shown in Fig. 1.1, which shows increased Alq$_3$ emission for devices with low concentrations (~1%) of PtOEP.

Saturation of emissive sites is alleviated by increasing the concentration of PtOEP. Yet from Fig. 1.2b, it is evident that at molar concentrations $\gtrsim 6\%$, the quantum efficiency again decreases. As high densities of porphyrin complexes often show bimolecular quenching [14], it is likely that the PtOEP/Alq$_3$ system is also affected by this process. Hence, the decrease in quantum efficiency with increasing current could be partially accounted for by the decreasing lifetime. As the current increases, the phosphorescent decay becomes bi-exponential, indicating the presence of a quenching process such as bimolecular recombination (see Fig. 1.2a legend).

The doping concentrations required to maximize quantum efficiency are extremely high compared to devices doped with fluorescent dyes and excited by long-range (~40 Å) Förster transfer. This suggests that the shorter-range Dexter process may be dominant in the Alq$_3$:PtOEP system. Calculations of the extent of energy transfer from Alq$_3$ to PtOEP support this hypothesis. Accounting for photons lost via emission through the sides of the device and total internal reflection within the substrate [6], a maximum external efficiency of 4% corresponds to an internal efficiency of 23%. PtOEP doped in polystyrene has a lifetime of 91 µs and a phosphorescent yield of 50% (Ref. [11]). Given a maximum exciton lifetime in the PtOEP/Alq$_3$ devices of 45 µs, we infer a peak phosphorescent yield of $\lesssim$25%. This implies $\gtrsim$90% energy transfer from Alq$_3$ to PtOEP, confirming that both triplets and singlet excitons must participate in energy transfer.

Conclusive evidence of Dexter transfer is obtained by examining the non-normalized spectra of the two devices shown in Fig. 1.3. A 100-Å-thick layer of Alq$_3$ doped with $\sim$1% DCM2 is placed in the recombination zone of both devices. As DCM2 is fluorescent with efficient energy transfer from Alq$_3$ (Ref. [15]), this layer effectively removes singlet excitons from the devices. Remaining singlets eventually recombine in Alq$_3$, yielding the small shoulder in the spectra at $\sim$530 nm. However, in device 2, an additional layer of $\sim$10% PtOEP in Alq$_3$ is introduced 200 Å away from the recombination zone. In this device, emission is seen from PtOEP without any change in the intensity of emission from either DCM2 or Alq$_3$. Hence, PtOEP cannot be an efficient trap, as carriers removed by PtOEP in device 2 would result in a decrease in the DCM2 and Alq$_3$ emission; an effect clearly not observed. As the DCM2 acts as a filter that removes singlet Alq$_3$ excitons, the only possible origin of the PtOEP luminescence is Alq$_3$ triplet states that have diffused through the DCM2 and intervening Alq$_3$ layers.

Given the observed decrease of quantum efficiency with increasing current, it is important that Alq$_3$ devices doped with PtOEP achieve a useful luminosity before significant quenching occurs. Unlike some red dyes [16] with peak emission around 650 nm, PtOEP emission is saturated and does not extend significantly into the infrared, thereby maximizing the luminosity. At a molar concentration of 6%, a quantum efficiency of 1.3% and a power efficiency of 0.15 lm W^{-1} at 100 cd m^{-2} is obtained. In comparison with other compounds with saturated red emission, such as the Eu complexes [17–19], PtOEP possesses quantum and power efficiencies that are superior by at least an order of magnitude. Compounds with unsaturated red emission, such as DCM2 and its variants [16], possess comparable quantum efficiencies but emit closer to orange than PtOEP; further red-shifting by increasing the DCM2 concentration causes the quantum efficiency to decrease, along with a considerable increase in infrared emission [15].

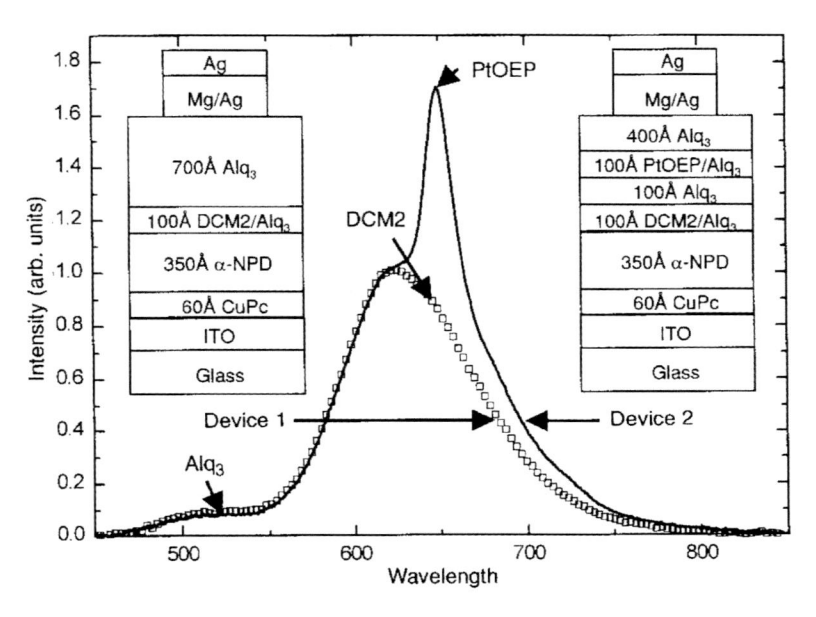

Figure 1.3 Two electroluminescent devices demonstrating that Alq$_3$ triplets are transferred to PtOEP. Each device contains a 100-Å-thick layer of ~1% DCM2 in Alq$_3$ at the recombination zone. This layer acts to remove singlet states. Remaining singlets recombine in Alq$_3$, yielding the shoulder apparent in the spectra at 530 nm. Device 2 contains an additional layer of ~10% PtOEP in Alq$_3$ positioned 200 Å away from the Alq$_3$/α-NPD interface. Strong emission is seen from the PtOEP without a corresponding decrease in emission from DCM2 or Alq$_3$. The spectra were measured at a current density of 6 mA cm^{-2}. (DCM2 is [2-methyl-6-[2-(2,3,6,7-tetrahydro-1H,5Hbenzo[ij]quinolizin-9-yl)ethenyl]-4H-pyran-4-ylidene] propane-dinitrile.)

Although PtOEP demonstrates the efficiency improvements made possible by the participation of triplet excitons in phosphorescence, the long lifetime of PtOEP (>10 μs) and the short-range nature of Dexter energy transfer cause saturation of emission at low dopant concentrations. It is also observed that bimolecular interactions quench emission at high exciton densities, causing a decrease in quantum efficiency for high currents and doping concentrations. But it is possible that other host materials could be used with PtOEP to increase efficiency even beyond the high values reported here. Given the performance of PtOEP, other phosphorescent dyes emitting in the blue and green regions of the spectrum also present an attractive area of study for display applications.

Acknowledgements

We thank V. G. Kozlov for help with the transient measurements, and P. E. Burrows for discussions. This work was supported by Universal Display Corporation, DARPA, AFPSR and NSF.

References

1. VanSlyke, S. A. & Tang, C.W. US Patent No. 4539507 (1985).
2. Burroughes, J. H. et al. Light emitting diodes based on conjugated polymers. *Nature* **347**, 539–541 (1990).
3. Tang, C. W., VanSlyke, S. A. & Chen, C. H. Electroluminescence of doped organic thin films. *J. Appl. Phys.* **65**, 3610–3616 (1989).
4. Tsutsui, T. & Saito, S. *Organic Multilayer-Dye Electroluminescent Diodes—Is There Any Difference with Polymer LED?* (Kluwer Academic, Dordrecht, 1993).
5. Rothberg, L. J. & Lovinger, A. J. Status of and prospects for organic electroluminescence. *J. Mater. Res.* **11**, 3174–3187 (1996).
6. Gu, G. et al. High-external-quantum-efficiency organic light-emitting devices. *Opt. Lett.* **22**, 396–398 (1997).
7. Dirr, S. et al. Vacuum-deposited thin films of lanthanide complexes: spectral properties and applications in organic light-emitting devices. In *SID 97 Digest* (ed. Morreale, J.) 778–781 (Soc. For Information Display, Santa Ana, CA, 1997).
8. Hoshino, S. & Suzuki, H. Electroluminescence from triplet excited states of benzophenone. *Appl. Phys. Lett.* **69**, 224–226 (1996).
9. Mills, A. & Lepre, A. Controlling the response characteristics of luminescent porphyrin plastic film sensors for oxygen. *Anal Chem.* **69**, 4653–4659 (1997).
10. Ponterini, G., Serpone, N., Bergkamp, M. A. & Netzel, T. L. Comparison of radiationless decay processes in osmium and platinum porphyrins. *J. Am. Chem. Soc.* **105**, 4639–4645 (1983).

11. Papkovski, D. B. New oxygen sensors and their applications to biosensing. *Sens. Actuators B* **29**, 213–218 (1995).
12. Turro, N. J. *Modern Molecular Photochemistry* (University Science Books, Mill Valley, CA, 1991).
13. Liu, H.-Y, Switalski, S. C., Coltrain, B. K. & Merkel, P. B. Oxygen permeability of sol-gel coatings. *Appl. Spectrosc.* **46**, 1266–1272 (1992).
14. Rodriguez, J., McDowell, L. & Holten, D. Elucidation of the role of metal to ring charge-transfer states in the deactivation of photo-excited ruthenium porphyrin carbonyl complexes. *Chem. Phys. Lett.* **147**, 235–240 (1988).
15. Bulovic, V. et al. Bright, saturated, red-to-yellow organic light-emitting devices based on polarization- induced spectral shifts. *Chem. Phys. Lett.* **287**, 455–460 (1998).
16. Chen, C. H., Tang, C. W., Shi, J. & Klubeck, K. P. Improved red dopants for organic electroluminescent devices. *Macromol. Symp.* **125**, 49–58 (1997).
17. Tsutsui, T., Takada, N., Saito, S. & Ogino, E. Sharply directed emission in organic electroluminescent diodes with an optical-microcavity structure. *Appl. Phys. Lett.* **65**, 1868–1870 (1994).
18. Kido, J. et al. Bright red light emitting organic electroluminescent devices having a europium complex as an emitter. *Appl. Phys. Lett.* **65**, 2124–2126 (1994).
19. Sano, T. et al. Novel europium complexes for electroluminescent devices with sharp red emission. *Jpn J. Appl. Phys.* **34**, 1883–1887 (1994).
20. Garbuzov, D. Z., Bulovic, V., Burrows, P. E. & Forrest, S. R. Photoluminescence efficiency and absorption of aluminum-tris-quinolate (Alq_3) thin films. *Chem. Phys. Lett.* **249**, 433–437 (1996).

Chapter 2

Improved Energy Transfer in Electrophosphorescent Devices

D. F. O'Brien,[a,c] M. A. Baldo,[a] M. E. Thompson,[b] and S. R. Forrest[a]

[a]*Center for Photonics and Optoelectronic Materials (POEM), Department of Electrical Engineering and the Princeton Materials Institute, Princeton University, Princeton, New Jersey 08544, USA*
[b]*Department of Chemistry, University of Southern California, Los Angeles, California 90089, USA*
[c]*Present address: Department of Physics, Trinity College Dublin, Dublin 2, Ireland*
met@usc.edu

External quantum efficiencies of up to $(5.6 \pm 0.1)\%$ at low brightness and $(2.2 \pm 0.1)\%$ at 100 cd/m^2 are obtained from a red electrophosphorescent device containing the luminescent dye 2,3,7,8,12,13,17,18-octaethyl-21H 23H-phorpine platinum(II) (PtOEP) doped in a 4,4'-N, N'-dicarbazole-biphenyl (CBP) host. Due to weak overlap between excitonic states in PtOEP and CBP, efficiency losses due to nonradiative recombination are low. However, energy transfer between the species is also poor. In compensation, a thin layer of 2,9-dimethyl-4,7 diphenyl-1,10-phenanthroline is used

---

Reprinted from *Appl. Phys. Lett.*, **74**(3), 442–444, 1999.

Electrophosphorescent Materials and Devices
Edited by Mark E. Thompson
Text Copyright © 1999 American Institute of Physics
Layout Copyright © 2024 Jenny Stanford Publishing Pte. Ltd.
ISBN 978-981-4877-34-3 (Hardcover), 978-1-003-08872-1 (eBook)
www.jennystanford.com

as a barrier to exciton diffusion in CBP, improving the energy transfer to PtOEP. This technique may be applied to improve the efficiency of other electrophosphorescent devices.

To date, efficiencies for organic light emitting devices (OLEDs) have been limited by the use of fluorescent emissive materials. This limit arises since fluorescence only involves singlet relaxation, thus eliminating the participation of the triplet exciton population. By employing a phosphorescent dye where both singlet and triplet excited states participate, the OLED *internal* efficiency can, in principle, be increased as high as 100% [1, 2]. Previous work on the red phosphor 2,3,7,8,12,13,17,18-octaethyl-21H,23H-porphine platinum(II) PtOEP in an aluminum tris (8-hydroxyquinoline) (Alq$_3$) host has moved some way towards realizing the potential of electrophosphorescence in OLEDs with internal quantum efficiencies at low brightness as high as 23%, corresponding to an external quantum efficiency of $\eta = 4\%$ [1]. The efficiency of these devices decreased to 1.3% at 100 cd/m^2 and was ultimately limited by the low photoluminescent efficiency ($\sim$25%) of PtOEP in Alq$_3$. In this work, we demonstrate that by altering the host material and the device structure it is possible to reduce nonradiative losses and improve η to as high as (5.6 ± 0.1)% at low brightness and (2.2 ± 0.1)% at 100 cd/m^2.

One notable difference between OLEDs based on fluorescence and phosphorescence is that phosphorescent efficiency decreases rapidly at high current densities. Long phosphorescent lifetimes cause saturation of emissive sites, and triplet-triplet annihilation may also result in significant efficiency losses. Such lifetimes are intrinsic to phosphorescence, but may be reduced by stronger spin state mixing. Another principal difference is that energy transfer of triplets from a conductive host to a luminescent guest molecule is typically slower than that of singlets; the long range dipole–dipole coupling (Förster transfer) which dominates singlet energy transfer is forbidden for triplets by spin conservation. Rather, energy transfer occurs by the diffusion of excitons to neighboring molecules (Dexter transfer), hence significant overlap of donor and acceptor excitonic wavefunctions is critical to energy transfer. Finally, triplet diffusion lengths are long [2] (>1400 Å) compared to singlet diffusion lengths of a few hundred angstroms [3]. In this work, we exploit this latter property to improve energy transfer and thus OLED external quantum efficiency.

The photoluminescent efficiency of PtOEP doped into a host material is approximately proportional [4] to the observed phosphorescent lifetime. Previous measurements of the lifetime of PtOEP triplet excitons have demonstrated that the photoluminescence efficiency of PtOEP in polystyrene [5] is significantly higher than the corresponding efficiency when doped into Alq_3. We speculate that the difference is due to weaker overlap of the excitonic states in PtOEP and polystyrene molecules. This leads to a much lower probability for exciplex formation and reverse energy transfer from the dye to the host, both of which can lead to nonradiative quenching of PtOEP emission. It follows that improved electrophosphorescent efficiencies may result from doping PtOEP into a charge transport material whose excitonic states only weakly overlap with states in PtOEP. Since PtOEP absorbs [6] in the green at a wavelength of $\lambda \sim 530$ nm, blue emitters ($\lambda \sim 400$–450 nm) are likely candidates for efficient host materials.

One such blue emissive material is 4,4'-N, N'-dicarbazole-biphenyl (CBP) [7]. The phosphorescent lifetime in a CBP host is ~ 100 μs (obtained by recording the luminescent transient decay with a streak camera), or approximately twice that of an Alq_3 host. Thus one might expect, provided the effects of reduced energy transfer can be minimized, that the maximum quantum efficiency of a PtOEP/CBP device should be double that of an optimized PTOEP/Alq_3 OLED.

The external quantum efficiencies of several devices where PtOEP is doped at varying concentrations into Alq_3 and CBP are shown in Fig. 2.1, with the device structure presented in the inset. The OLEDs were grown by high vacuum (10^{-6} Torr) thermal evaporation onto a cleaned glass substrate as described elsewhere [8]. Devices were also made with an 80 Å thick 2,9-dimethyl-4,7 diphenyl-1,10-phenanthroline (bathocuproine, or BCP) [9] layer inserted between the luminescent and cap layers to improve device efficiency, as discussed in the following. All measurements were performed in air at room temperature.

Devices using CBP as the host without the BCP layer possess poor external quantum efficiencies: devices with 6% (mol) PtOEP

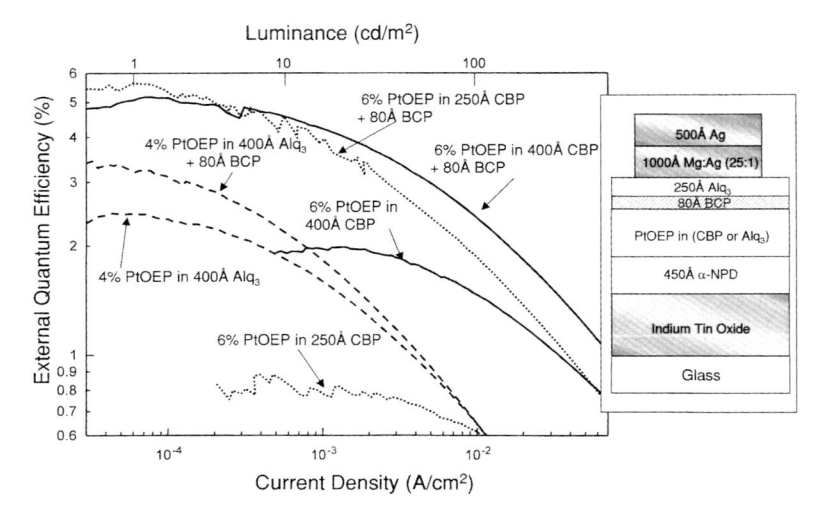

Figure 2.1 External quantum efficiencies of PtOEP/CBP and PtOEP/Alq$_3$ devices as a function of current with and without a BCP blocking layer. The top axis shows the luminance of the device with a 80 Å thick BCP layer and a 400 Å thick CBP luminescent layer doped with 6% PtOEP. Inset: Schematic cross section of the high efficiency OLED, consisting of a 450 Å thick 4,4′-bis[N-(1-naphyl)-N-phenyl-amino] biphenyl (α-NPD) HTL, PtOEP doped ETL codeposited with either Alq$_3$ or CBP acting as the host, an 80 Å thick layer of BCP, and a further 200 Å thick cap layer of Alq$_3$ to prevent nonradiative quenching of PtOEP excitons at the cathode. Finally, a shadow mask with 1 mm diameter openings was used to define the cathode consisting of a 1000 Å thick layer of 25:1 Mg:Ag, with a 500 Å thick Ag cap.

in a 250 Å thick CBP luminescent layer possess $\eta < 1\%$. Here, η is measured by placing the substrate onto the surface of a silicon photodiode and measuring the *red* light scattered into the viewing (i.e., detection) direction. The efficiency improves somewhat as the doped region is extended and devices containing 6% PtOEP within a 400 Å thick CBP layer exhibit a maximum η of approximately 2%. The improvement indicates that the low quantum efficiency of the 250 Å thick CBP devices is most likely due to poor energy transfer in such a thin active layer. It is probable that this and the long lifetime of PtOEP in CBP are manifestations of the same effect: weak overlap between excitonic states in the molecular species leads to inefficient energy transfer.

Thus, high efficiency devices may be obtained by using very thick layers of CBP doped with PtOEP, thereby maximizing the probability that triplets (with their long diffusion lengths) may encounter a PtOEP molecule prior to nonradiative recombination within the film or at the organic/cathode interface. However, since CBP is a relatively poor electron conductor, it is desirable to minimize its thickness to reduce the OLED operating voltage. Alternatively, energy transfer could be improved by increasing the doping concentration of PtOEP, but this also increases nonradiative decay due to PtOEP-PtOEP interactions, i.e., "self-quenching."

By blocking the triplets from leaving the luminescent layer, it should be possible to increase their residence time in this region, thereby increasing the probability for energy transfer from CBP to the phosphorescent center. A material suitable for this purpose is BCP, which has previously been used as a hole blocking layer in OLEDs [9]. When placed between a doped HTL and an Alq_3 ETL, it was found that light emission originated from the HTL. This demonstrates that BCP has a large ionization potential compared to Alq_3, thereby blocking the passage of holes out of the HTL. These results also suggest that the lowest unoccupied molecular orbital level of BCP freely allows the transport of electrons resulting in exciton formation in the HTL. From this, we infer that exciton blocking in our devices can be achieved by capping the PtOEP:CBP layer with 80 Å of BCP.

The current (I)–voltage (V) characteristics for Alq_3 devices with and without the BCP layer are presented in Fig. 2.2. Distinct low and high current regimes of operations are observed as is typical of Alq_3-based OLEDs [8]. Introduction of the BCP cap increases the operating voltage of 10.6 V by only ∼0.7 V at 1 mA/cm^2, indicating the presence of this thin layer minimally affects the electron conduction properties of the device.

The exciton blocking function of BCP is demonstrated by comparison of the spectra in Fig. 2.3 taken for PtOEP:CBP devices with and without BCP. In Fig. 2.3a, the emission spectra of devices without the BCP layer are shown for a range of current densities. At low currents emission is predominantly from PtOEP, with a sharp emission peak at $\lambda = 650$ nm and no measurable luminescence from the PtOEP singlet state at $\lambda = 580$ nm [10]. The red Commission

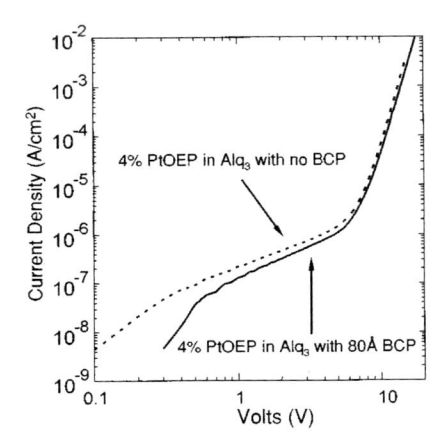

Figure 2.2 Current–voltage characteristics of the devices with and without an 80 Å BCP layer. There is only a ∼0.7 V increase in voltage at 1 mA/cm², for devices with BCP, suggesting that this blocking layer does not significantly alter the conduction properties of the device.

Internationale de L'Eclairage chromaticity coordinates of (0.7, 0.3) are identical to those obtained with an Alq_3 host. As the current increases, green emission characteristic of the Alq_3 cap layer at $\lambda = 520$ nm is observed. The Alq_3 emission is absent in devices containing BCP (Fig. 2.3b) due to the exciton blocking effects of this layer. The lack of Alq_3 emission also suggests that no holes reach the Alq_3 cap. Therefore, BCP may also block holes in addition to confining excitons within the CBP layer.

As expected, the benefits of the BCP barrier are most noticeable for the thinner (250 Å) CBP layers, where energy transfer from CBP to PtOEP is particularly inefficient. It is less useful for devices employing PtOEP in an Alq_3 host where previous studies [1] have shown that under optimum conditions, energy transfer is nearly complete with up to ∼90% of excitons transferred to the dye. Note that the efficiency of the 250 Å thick CBP device begins to decrease at a lower current density than that of the 400 Å thick CBP device (Fig. 2.1) due to the lower number of luminescent PtOEP sites in the thinner device.

The BCP barrier layer substantially improves the quantum efficiency in CBP doped devices to $\eta = (2.2 \pm 0.1)\%$ at 100 cd/m²,

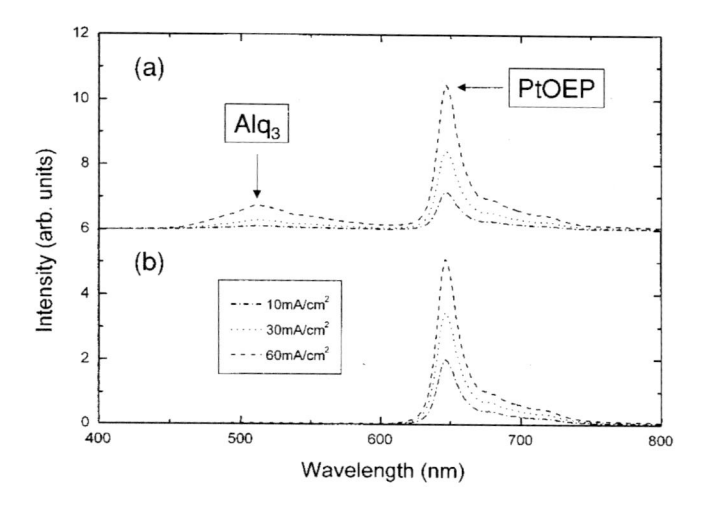

Figure 2.3 Emission spectra of CBP-based electroluminescent devices with and without a BCP exciton blocking layer. The spectra of the device without BCP are shown in (a) to exhibit significant Alq_3 emission at a peak wavelength of $\lambda \sim 520$ nm. The effect of the BCP layer is clearly shown in (b), where no Alq_3 emission is observed.

which compares with a best result of 1.3% in PtOEP/Alq_3 devices [1]. Note that the Alq_3 devices in Fig. 2.1 are more lightly doped with PtOEP and possess $\eta < 1\%$ at 100 cd/m^2. This is due to the weaker energy transfer observed at a lower doping densities, but in this work, poor transfer is required to illustrate the effect of the BCP layer. The peak $\eta = (5.6 \pm 0.1)\%$ is equivalent to an internal quantum efficiency of $\sim 32\%$ [11]. This improvement is less than $\eta \sim 8\%$ implied by the lifetime data, suggesting that some triplet excitons are quenched at the CBP/BCP interface or leak through to the cathode where additional nonradiative recombination occurs. In addition, since the phosphorescent lifetime in CBP is roughly twice that in Alq_3, saturation of PtOEP sites in the CBP devices occurs at a concomitantly lower current density. However, the increased susceptibility of PtOEP in CBP to saturation is overcome by much weaker bimolecular quenching in these devices.

Unlike most red dyes [12] with peak emission around 650 nm, PtOEP emission is color saturated and does not extend significantly into the infrared, thereby maximizing its luminosity. In comparison with saturated reds such as the Eu complexes [13–15], PtOEP possesses quantum efficiencies that are higher by at least an order of magnitude. Unsaturated reds such as DCM2 and its variants [12] possess comparable quantum efficiencies to PtOeP (e.g., 1%–3%) but they typically emit in the orange; and further red-shifting by increasing the DCM2 concentration causes the quantum efficiency to decrease to ($\leq$1%) along with a considerable increase in infrared emission [16].

Evidently, the CBP host reduces bimolecular quenching and lengthens the lifetime of PtOEP triple excitons. Both of these effects are consistent with weaker interactions between CBP and PtOEP. We therefore propose that a general technique for obtaining high efficiency from phosphorescent emitters is to employ a host chosen such that the phosphorescence lifetime is as long as possible and to use a blocking layer such as BCP to keep triplets within the luminescent region.

In conclusion, we have demonstrated saturated red electrophosphorescent devices employing exciton blocking with external efficiencies of (5.6 ± 0.1)% at low brightness ($\sim$1 cd/m^2) and (2.2 ± 0.1)% at 100 cd/m^2. However, significant potential for improvement remains given that the maximum internal quantum efficiency of $\sim$32% is still roughly a factor of three lower than the theoretical limit.

Acknowledgements

This work was funded by Universal Display Corporation, DARPA, AFOSR, and NSF.

References

1. M. A. Baldo, D. F. O'Brien, Y. You, A. Shoustikov, S. Sibley, M. E. Thompson, and S. R. Forrest, *Nature (London)* **395**, 151 (1998).
2. M. A. Baldo, D. F. O'Brien, V. Buloviç, M. E. Thompson, and S. R. Forrest (unpublished).
3. C. W. Tang, S. A. VanSlyke, and C. H. Chen, *J. Appl. Phys.* **65**, 3610 (1989).

4. M. Klessinger and J. Michl, *Excited States and Photochemistry of Organic Molecules* (VCH Publishers, New York, 1995).
5. D. B. Papkovski, *Sens. Actuators B* **29**, 213–218 (1995).
6. H.-Y. Liu, S. C. Switalski, B. K. Coltrain, and P. B. Merkel, *Appl. Spectrosc.* **46**, 1266 (1992).
7. V. G. Kozlov, G. Parthasarathy, P. E. Burrows, S. R. Forrest, Y. You, and M. E. Thompson, *Appl. Phys. Lett.* **72**, 144 (1998).
8. P. E. Burrows, Z. Shen, V. Bulović, D. M. McCarty, S. R. Forrest, J. A. Cronin, and M. E. Thompson, *J. Appl. Phys.* **79**, 7991 (1996).
9. Y. Kijima, in *A Blue Organic Light Emitting Diode*, Materials Research Society Meeting, San Francisco, 1998.
10. G. Ponterini, N. Serpone, M. A. Bergkamp, and T. L. Netzel, *J. Am. Chem. Soc.* **105**, 4639 (1983).
11. G. Gu, D. Z. Garbuzov, P. E. Burrows, S. Venkatesh, S. R. Forrest, and M. E. Thompson, *Opt. Lett.* **22**, 396–398 (1997).
12. C. H. Chen, C. W. Tang, J. Shi, and K. P. Klubeck, *Macromol. Symp.* **125**, 49–58 (1997).
13. T. Tsutsui, N. Takada, S. Saito, and E. Ogino, *Appl. Phys. Lett.* **65**, 1868–1870 (1994).
14. J. Kido, H. Hayese, K. Hongawa, K. Nagai, and K. Okuyama, *Appl. Phys. Lett.* **65**, 2124–2126 (1994).
15. T. Sano, M. Fujita, T. Fujii, Y. Hamada, K. Shibata, and K. Kuroki, *Jpn. J. Appl. Phys.*, Part 1 **34**, 1883–1887 (1994).
16. V. Bulović, A. Shoustikov, M. A. Baldo, E. Bose, V. G. Kozlov, M. E. Thompson, and S. R. Forrest, *Chem. Phys. Lett.* **287**, 455–460 (1998).

Chapter 3

Efficient, Saturated Red Organic Light Emitting Devices Based on Phosphorescent Platinum(II) Porphyrins

Raymond C. Kwong,[a] Scott Sibley,[b] Timur Dubovoy,[b]
Marc Baldo,[c] Stephen R. Forrest,[c] and Mark E. Thompson[a]

[a]*Department of Chemistry, University of Southern California, Los Angeles, California 90089, USA*
[b]*Department of Chemistry, Goucher College, Baltimore, Maryland 21204-2794, USA*
[c]*Department of Electrical Engineering, Princeton University, Princeton, New Jersey 08544, USA*
met@use.edu

Two new platinum(II) porphyrins have been synthesized and their luminescent properties have been studied. The platinum porphyrins exhibited strong phosphorescence in the red with narrow line widths. When they were doped into aluminum(III) tris(8-hydroxyquinolate) (AlQ_3) in the electron-transporting and -emitting layer of an organic light-emitting device, energy transfer occurred between the host AlQ_3 and the platinum porphyrin. Bright saturated red emission with high efficiency at low to moderate current density has been achieved. In the high current regime,

---

Reprinted from *Chem. Mater.*, **11**(12), 3709–3713, 1999.

Electrophosphorescent Materials and Devices
Edited by Mark E. Thompson
Text Copyright © 1999 American Chemical Society
Layout Copyright © 2024 Jenny Stanford Publishing Pte. Ltd.
ISBN 978-981-4877-34-3 (Hardcover), 978-1-003-08872-1 (eBook)
www.jennystanford.com

the electroluminescence efficiency decreased and the perceived emission color blue shifted as a result of mixed emission from the platinum porphyrin and AlQ$_3$. This current dependence was due to the saturation of triplet emissive sites, because of the long-lived phosphorescence state of the platinum porphyrin complex.

3.1 Introduction

Achieving saturated red emission with high quantum and luminous efficiencies remains a challenge in the field of organic electroluminescence. Red organic light emitting diodes (OLEDs) are usually fabricated by doping red dyes into a suitable host [1, 2]. The most commonly used host for this purpose is aluminum(III) tris(8-hydroxyquinolate) (AlQ$_3$). Doping a small amount (∼1%) of a red laser dye such as DCM or DCJ in AlQ$_3$ gives high quantum efficiency (2.3%, photons/electrons), although the emission color appears orange [3]. Increasing the doping concentration to 2% results in red emission due to polarization effects [4], but the efficiency of the devices decreases significantly due to enhanced self-quenching. DCLJ, a sterically bulky derivative of DCJ, has been synthesized to reduce the extent of self-quenching at high level doping [5]. The efficiency of the based DCJT device is relatively high (∼2.5%); however, the red color emission still suffered from a broad bandwidth. Saturated red can also be achieved by doping tetraphenylporphyrin (TPP) into AlQ$_3$, but the quantum efficiency remained low (∼0.1%) [6]. Another approach is to use Eu(III) complexes as dopants [7, 8]. Vapor deposited devices with Eu(III) complexes such as Eu(DBM)$_3$(Phen) [9] show a sharp emission band at 620 nm which is typical of the $^5D_0 \rightarrow {}^7F_2$ transition of the trivalent lanthanide ion [10]. The chromaticity of Eu(III) complex-doped devices is satisfactory, with a quantum efficiency of 0.8%. The most promising candidate for red color display to date is platinum(II) octaethylporphyrin (PtOEP) which has shown exceptionally high efficiency (∼7%) at low injection current [11, 12]. Devices with PtOEP doped into AlQ$_3$ showed narrow emission centered at 650 nm, resulting in a saturated red color. Efficiency as high as 1.3% was achieved, at a luminance of 100 cd/m^2, for

AlQ$_3$-doped OLEDs and 1.9%, in optimized OLEDs [13]. PtOEP has also been used as a phosphorescent dye in polymer LEDs to achieve saturated red emission [14]. Both Eu(III) and Pt(II) complexes are phosphorescent, unlike most other dyes used in OLEDs, which are fluorescent [2]. Using triplet-based emitting centers in OLEDs eliminates the 25% maximum internal quantum efficiency, which is the expected singlet exciton fraction achieved by electrical injection [15, 16]. This is a benefit of all phosphorescence-based OLEDs and gives them the potential of having much higher quantum efficiencies than fluorescence-based ones. The energy of the photophysical transition for lanthanide complexes is determined by the energetics of the metalion [17], while the emission in platinum(II) porphyrins is largely ligand-based [18]. Therefore, by chemical modification of the porphyrin framework, the luminescent properties of the Pt(II) complex can be fine-tuned. To further explore the concept of using phosphorescent dyes in OLEDs, two new platinum(II) porphyrins have been synthesized and studied. We report here our work in utilizing them as red phosphorescent dyes in OLEDs with improved quantum and luminous efficiencies.

3.2 Experimental Section

3.2.1 Materials

All reagents and anhydrous solvents were purchased from Aldrich and used without further purification. AlQ$_3$ [19] and 4,4'-bis[*N*-(l-naphthyl)-*N*-phenylamino]biphenyl (α-NPD) [20] were synthesized according to the literature procedures and purified by vacuum sublimation. 5,15-Diphenylporphyrin (abbreviated here as H$_2$DPP) was synthesized from a condensation–cyclization reaction with 2,2'-dipyrrolemethane and benzaldehyde as described by Therien et al. [21] 2,2'-Dipyrrolemethane was synthesized by a three-step procedure starting with pyrrole and thiophosgene [22]. Etioporphyrin III (abbreviated here as H$_2$OX)was synthesized by reducing the commercially available protoporphyrin IX (isolated from ox hematin) through a one-step hydrogenation–decarboxylation reaction [23]. The platinum complexes PtDPP and PtOX were synthesized by refluxing the corresponding free base in degassed benzonitrile with $PtCl_2$. Typical procedures were as follows: One gram of the free base and $PtCl_2$ (2 equiv) were suspended in 100 mL of anhydrous benzonitrile. The mixture was then purged with N_2 as it was slowly heated to 160°C. It was then brought to reflux under N_2 until there was no free base left as revealed by TLC. The mixture was then cooled to room temperature and the solvent was removed by vacuum distillation. The crude product left behind was then dried completely and purified by sublimation at 300°C at $<10^{-4}$ Torr. Both Pt(II) complexes were characterized spectroscopically and by elemental analysis:

3.2.1.1 PtDPP

^{1}H NMR: δ 10.15 (s, 2H), 9.21 (dd, 4H, $J_1 = 12$ Hz, $J_2 = 7.5$ Hz), 8.93 (dd. 4H, $J_1 = 12$ Hz, $J_2 = 7.5$ Hz), 8.18 (m, 4H), 7.77 (m, 6H). MS (EI): m/z (relative intensity) 655 (M^+, 100), 577 (30), 326 (50), 288 (35). UV–vis (CH_2Cl_2): λ_{max} (nm) (log ϵ) 388 (5.33), 499 (4.15), 530 (4.04). Elemental analysis: calculated C, 58.62; H, 3.07; N, 8.55; Pt, 29.76; found C, 58.35; H, 3.08; N, 8.37. Pt, 29.90.

3.2.1.2 PtOX

^{1}HNMR: δ 9.98 (s, 4H), 3.99 (q, 8H, $J = 8$ Hz), 3.55 (s, 12H), 1.84 (t, 12H, $J = 8$ Hz), MS (EI): m/z (relative intensity) 671 (M^+, 100), 656 (90), 642 (50), 625 (35), 611 (28). UV–vis (CH_2Cl_2): λ_{max} (nm) (log ϵ) 380 (5.30), 500 (4.22), 535 (4.62). Elemental analysis: calculated C, 57.22; H, 5.40; N, 8.34; Pt, 29.04; found C, 57.15; H, 5.20; N, 8.18; Pt, 29.34.

3.2.2 Photolummescence

Quantum efficiency measurements were carried out at room temperature in a toluene–DMF solution (degassed by a freeze–pump–thaw cycle, 20 times) with excitation at 500 nm. All samples were adjusted to give the same absorbance of 0.038 at 500 nm. Estimation was based on ZnTPP in benzene as the reference with a 0.03 quantum efficiency at 298 K [24]. Quantum efficiency measurements were obtained using an Aminco-Bowman Series 2 Luminescence spectrometer. Steady state emission experiments at room temperature were performed on a PTI QuantaMaster Model C-60 Spectrofluorometer. Phosphorescence lifetime measurements were performed on the same fluorimeter equipped with a microsecond Xe flash-lamp. There were three types of samples studied and they were prepared as follows: First, the degassed toluene/DMF solution sample was of the same concentration as that used for the quantum yield measurement. The second sample was prepared by doping the porphyrin into AlQ_3 by vacuum vapor deposition on glass substrates with rates adjusted to give the same doping level as in the emitting layer of the OLED, i.e., 6 mol%. The third sample was prepared by spin-coating a 0.5% porphyrin/polystyrene (w/w) solution in anhydrous toluene on glass slides. The fluorimeter chamber was kept under nitrogen throughout the lifetime measurement to prevent the complication of oxygen quenching. The excitation was set at 390 nm for both porphyrins while the decay was monitored at 648 and 628 nm for PtOX and PtDPP, respectively.

3.2.3 Device Fabrication and Testing

Patterned anode contacts were made from commercially available ITO-coated glass (100 Ω/□) in our laboratory using a photolithographic technique to form circular anodes with 1 mm radii. In a vacuum chamber at pressure $<1.0 \times 10^{-5}$ Torr, 400 Å of α-NPD as the hole-transporting layer; 450 Å of dye-doped AlQ_3 as the electron-transporting and- emitting layer; 50 Å of AlQ_3 as electron-injection layer; and a cathode composed of 600 Å of Mg–Ag (10:1) capped with 300 Å of Ag, were sequentially deposited onto the substrate to give the device structure shown in Scheme 3.1. The platinum porphyrin dye and AlQ_3 were codeposited with rates of ~0.1 and ~1.1 Å/s, respectively, to give a doping level of 6 mol% in the emitting layer. Device current–voltage and light intensity characteristics were measured using a Keithley 2400 SourceMeter/2000 Multimeter coupled to a Newport 1835-C Optical Meter using the LabVIEW program by National Instruments.

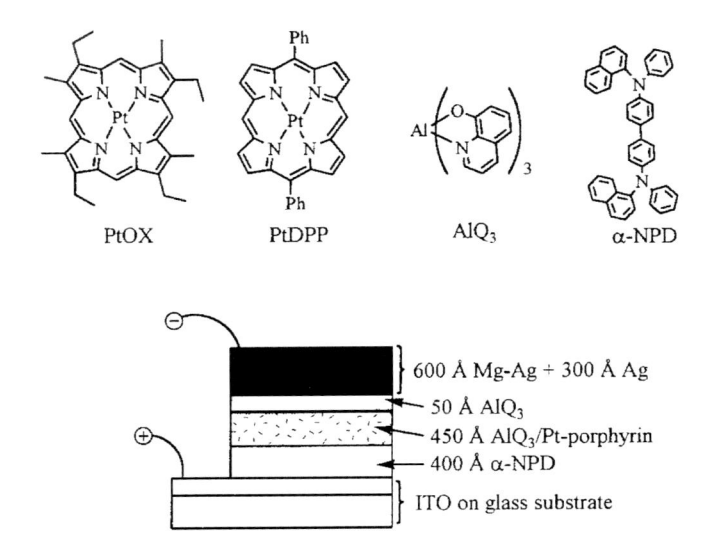

Scheme 3.1 OLED device structure and chemicals structures of the compounds used.

3.3 Result and Discussion

3.3.1 Photoluminescence

The room-temperature photoluminescence spectra of PtOX and PtDPP are shown in Fig. 3.1. A narrow emission peak (26 nm full width at half-maximum intensity) centered at a wavelength of 648 nm was obtained for PtOX, similar to other octaalkylporphyrins [18]. A wider emission centered at 630 nm with a weak second band at 695 nm is observed for PtDPP, characteristic of arylporphyrins [18]. The emission spectra are essentially the same when exciting at either the Soret band (∼390 nm) or the Q-bands (500–540 nm). The luminescence quantum efficiencies measured for the deoxygenated toluene/DMF solutions at room temperature were found to be 0.45, 0.44, and 0.16 for PtOEP, PtOX, and PtDPP, respectively, with error bars of roughly 20%. The quantum efficiency of PtOX is essentially the same as PtOEP. This similarly is not surprising since the only difference in the structures of PtOEP and PtOX is in the alkyl groups at the β-positions of the porphyrin ring. These alkyl groups are not

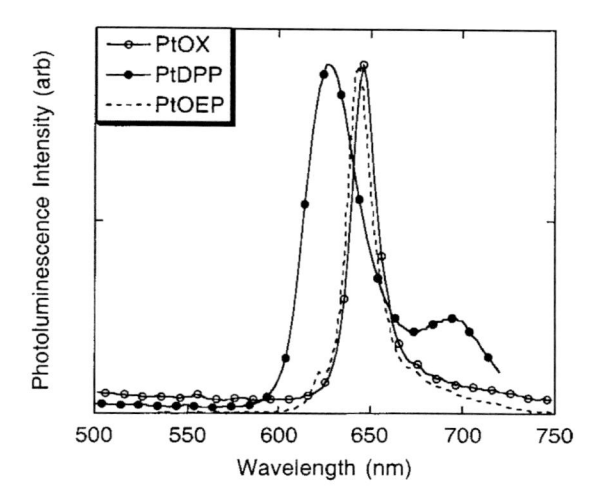

Figure 3.1 Room-temperature photoluminescence spectra of PtOX, PtDPP, and PtOEP in polystyrene matrix, excited at 390 nm.

conjugated to the porphyrin system and consequently have little electronic communication to the π-system. On the other hand, the quantum efficiency of PtDPP is significantly lower. We attribute this to the effect of the phenyl groups at the *meso* positions which are conjugated to the π-system. The rotational of the phenyl groups presumably induces nonradiative decay pathways of the excited states that are not present in either PtOX or PtOEP [25]. Moreover, the *beta* and the *meso* substitution patterns in porphyrins may have different ground state and excited state symmetries [26], which could also affect the radiative and nonradiative decay rates.

Photoluminescence decay studies were carried out to estimate the lifetimes of the excited states. The lifetimes of the toluene/DMF solution samples at room temperature were found to be 83, 76, and 34 μs for PtOEP, PtOX, and PtDPP, respectively. The PtOEP arid PtOX lifetimes are the same within the error limits of the measurement ($\pm$ 3 μs). The long lifetimes of the excited states of these Pt porphyrins clearly indicate their triplet nature. The trend observed in lifetimes is similar to that observed for the quantum yields, suggesting that the radiative rates for the three complexes are very similar. Thin film samples were prepared by spin-coating a solution of the porphyrin and polystyrene onto glass slides. The solution was degassed by thoroughly purging with nitrogen prior to spin-coating, and the measurements were performed in a nitrogen flushed chamber to minimize oxygen quenching. The porphyrin concentration was kept under 1 wt% in the polystyrene matrix to minimize concentration quenching. Lifetimes of 92 and 95 μs were measured for PtOX and PtDPP, respectively. The similarity of the lifetimes for PtOX and PtDPP in polystyrene suggest that the quantum yields for the two films are very similar, in direct contrast to the quantum efficiencies and lifetimes measured in solution. Phenyl group rotation may be responsible for the nonradiative relaxation in PtDPP, which would be active in solution, but may be frozen out in the solid polystyrene thin film. The luminescent lifetimes of PtOX and PtDPP doped into a solid film of AlQ_3 were also measured. The platinum porphyrins were doped into AlQ_3 at 6 mol%, and the films were kept under nitrogen during the lifetime measurement. The measured life times for PtOX and PtDPP in AlQ_3 are 39 and 21 μs, respectively. The lifetime measured for PtOEP under identical conditions was 37 μs [12].

3.3.2 Electroluminescence

The OLED device structure is a multilayer stack deposited on ITO glass, see Scheme 3.1. Similar to the previous study of PtOEP-doped devices [12, 13], significant residual AlQ_3 emission could be observed at low doping concentrations (<3%). Higher platinum porphyrin concentrations (>10%) resulted in a significant drop in quantum efficiency due to self-quenching of the dye. Thus, 6% dye concentration was shown to provide the most saturated color and higher quantum efficiency.

The current–voltage (I–V) profiles in air at room temperature are shown in Fig. 3.2. The turn-on voltages (V_T, defined as the voltage at which the power law dependence of I on V undergoes an abrupt change) for PtOX- and PtDPP-doped devices are 3.3 and 4.5 V, respectively. Bright red emission visible to the naked eye was observable at $V \geq V_T$. The brightness as a function of current density is plotted in Fig. 3.3. In the low current density regime (0.1–10 mA/cm^2), the luminance of both devices shows a linear

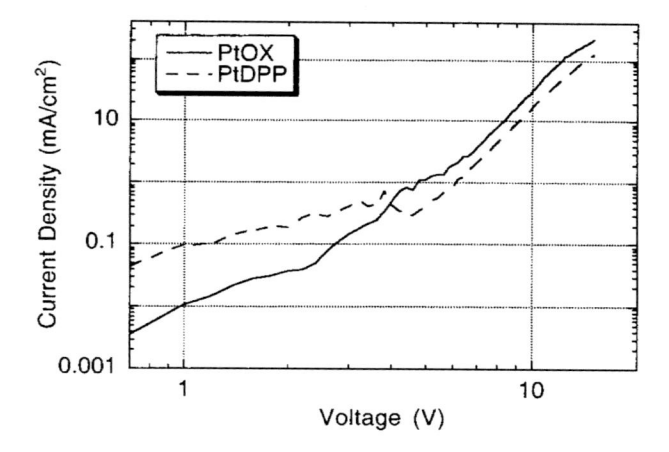

Figure 3.2 Current–voltage profiles of the PtOX and PtDPP OLEDs.

increase with current. In this regime the devices are also the most efficient (Fig. 3.3). For the PtOX device, QE is 1.5% at 3.2 mA/cm^2, corresponding to a luminance of 23 cd/m^2 and a luminous efficiency of 0.3 lm/W. At 100 cd/m^2, the QE drops to 1.1% with a luminous efficiency of 0.17 lm/W which is comparable to the PtOEP device at the same molar concentration of doping which showed QE of 1.3% at 100 cd/m^2 with a luminous efficiency of 0.15 lm/W [11]. For the PtDPP device, the QE reaches a maximum of 0.25% at 3.9 mA/cm^2 with a luminance of 7 cd/m^2 and a luminous efficiency of 0.05 lm/W. At 100 cd/m^2, the QE $= 0.12$% with a luminous efficiency of 0.02 lm/W. The relatively poor performance of PtDPP compared to PtOX or PtOEP is primarily due to lower photoluminescence yield of PtDPP.

The electroluminescent (EL) spectra of the PtOX and PtDPP devices are shown in Fig. 3.4. At low voltages (or current densities), the emission is from the platinum (II) porphyrin, resulting in sharp emission peaks at $\lambda = 650$ and 640 nm for PtOX and PtDPP, respectively. The EL spectra are very similar to their corresponding photoluminescence spectra, shown in Fig. 3.1. No emission from the AlQ$_3$ host is observed, indicating complete energy transfer from the host exciton to the platinum porphyrin. The quality of emission spectra with regard to color and saturation are typically defined

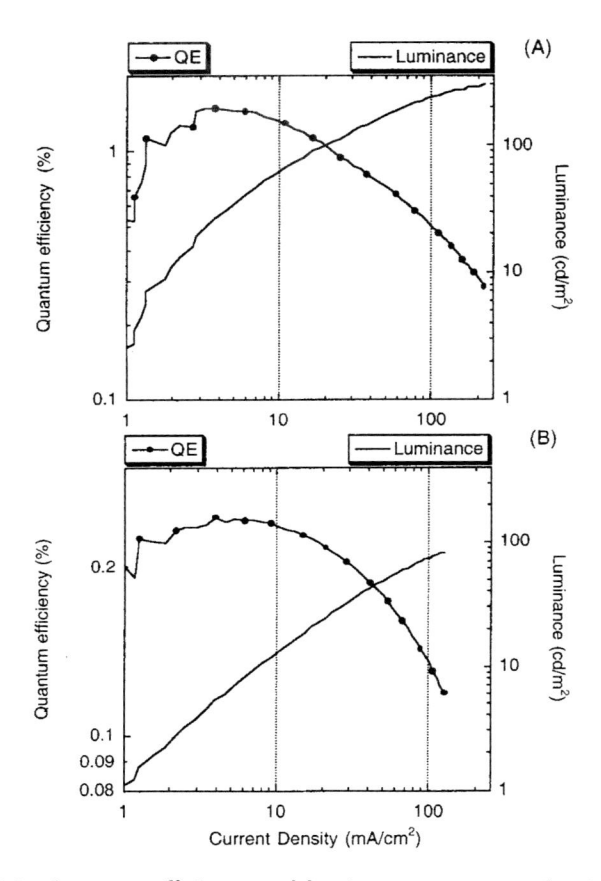

Figure 3.3 Quantum efficiency and luminance vs current density of the PtOX (A) and PtDPP (B) OLEDs.

by their CIE chromaticity coordinates x, y (CIE = Commission Internationale de L'Eclairage) [2, 27, 28]. Acceptable red emitters have $0.55 \leq x \leq 0.74$ and $0.25 \leq y \leq 0.35$. CIE coordinates for devices held at a current density of 2 mA/cm^2 are (0.69, 0.30) and (0.67, 0.31) for PtOX and PtDPP, respectively, as shown in the chromaticity chart in Fig. 3.5. The coordinates remain fairly unchanged up to current densities of 10 mA/cm^2.

At higher current densities ($>$10 mA/cm^2), the luminances of both the PtOX and PtDPP devices increase sublinearly, resulting in a concomitant gradual decrease in QE (Fig. 3.3). The emission color also changes from saturated red to orange due to increased emission between 450 and 610 nm. The EL spectra clearly indicate the emergence of peaks other than the porphyrin emission. These phenomena, occurring at high current densities, were also observed in PtOEP devices [11–13]. There are several factors contributing to the current dependence of the EL spectra. The first involves saturation of the emissive PtOX sites. There is a large difference between the singlet lifetime of AlQ$_3$ (10 ns) [29] and phosphorescence lifetime of the platinum porphyrins in AlQ$_3$ (39 and 21 μs for PtOX and PtDPP, respectively). The phosphorescent emissive sites near the HTL/ETL interface become saturated at medium to high current, i.e., most of the platinum porphyrins are promoted to their triplet excited state, which have a very long lifetime relative to the rate of formation of excitons near this interface. Once the platinum porphyrin dopants are saturated, AlQ$_3$ emission increases since there are no platinum porphyrin molecules in their ground state close enough to effectively transfer energy from the AlQ$_3$ singlet excitons prior to their relaxation (Fig. 3.4). Increasing the platinum

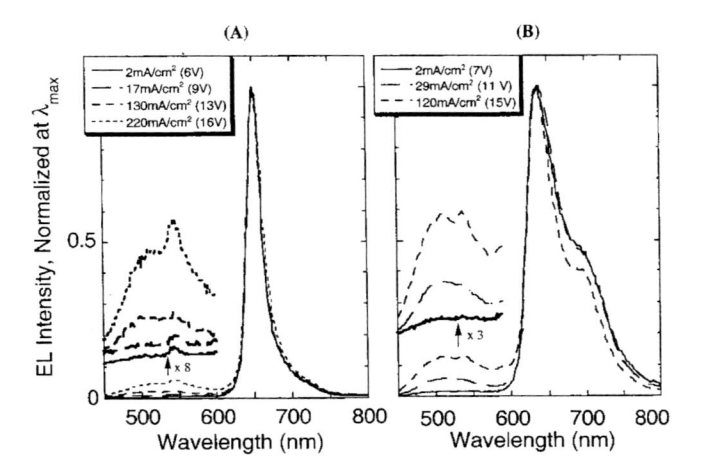

Figure 3.4 Electroluminescent spectra of the PtOX (A) and PtDPP (B) devices at different current densities and the corresponding driving voltages. The gradually increasing AlQ$_3$ emission region of 450–600 nm has been blown up and off-set for clarity. Also note the appearance of the sharp line at 550 nm is due to platinum porphyrin singlet emission.

porphyrin doping level can eliminate the likelihood of saturation of phosphorescent sites. The residual emission from AlQ_3 can be suppressed in such a way, however, the trade-off is the sacrifice of device efficiency due to self-quenching at high platinum porphyrin concentration.

It should be noted that the emission from AlQ_3, although weak compared to the major peak in the red, has a dramatic effect on the CIE coordinates. Since the coordinates are based on the human eye response, any emission near the peak in the photopic response function ($\lambda_{max} = 555$ nm) will have a more pronounced effect on the perceived color of the light source than emission significantly to the blue or red side of the peak. For example, for the PtOX device at 6 V and at 16 V (2 and 220 mA/cm^2, respectively), the spectra only differ by the comparatively weak emission between $\lambda = 450$ nm and $\lambda = 610$ nm seen in the EL spectrum at 16 V. The addition of this weak band leads to a pronounced shift in the CIE coordinates from (0.69, 0.30) at 6 V to (0.55, 0.36) at 16 V. Figure 3.5 shows the CIE coordinates of the electroluminescence at the low, medium, and high current density regimes. The OLED at 6 V appears pure red, while the device at 16 V appears orange to orange-red.

A second factor that contributes to the color change at high current densities is triplet–triplet annihilation ($T_1 + T_1 \rightarrow S_1 + S_0$) [25] taking place at high currents. The appearance and growth of the emission line centered at 550 nm in the electroluminescent spectra as the driving voltage increases (Fig. 3.4) is indicative of this triplet quenching process. This peak does not match the well-studied emission of AlQ_3, but instead coincides with the weak singlet emission (i.e., fluorescence) of platinum porphyrins [18]. This 550-nm emission contributes to the shift of the perceived color of the OLEDs to the orange, as described above for the AlQ_3 contribution. Since triplet–triplet annihilation is a second-order process, the amount of triplet quenching and the resulting PtOX fluorescence are expected to increase significantly as the current is increased. PtOX fluorescence does not come from incomplete intersystem crossing (ISC) from the PtOX singlet to the PtOX triplet. If incomplete ISC were responsible for the 550-nm band, the ratio of the PtOX fluorescence to AlQ_3 emission should decrease as the amount of AlQ_3 fluorescence increases, since the ratio of AlQ_3 to total

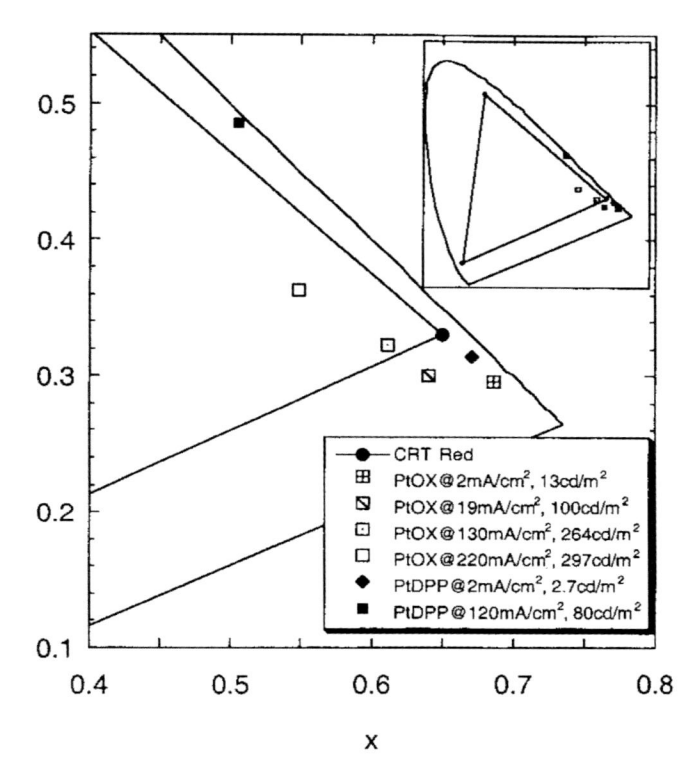

Figure 3.5 CIE coordinates and the corresponding brightness of the emission from the PtOX and PtDPP devices at low and high current regimes. The inset is the full CIE diagram and the RGB coordinates of a standard cathode ray tube display (CRT).

PtOX luminescence is increasing. The PtOX singlet signal increases relative to AlQ$_3$, however, consistent with their origin being from triplet–triplet annihilation.

Further evidence for triplet–triplet annihilation is seen in the small increase in brightness of the PtOX OLED as the current level was raised from 130 mA/cm^2 (red device) to 220 mA/cm^2 (orange device). That is, as the current was increased by 70%, the brightness only increases by 13%. If the color shift on increasing current were due only to saturation of the PtOX molecules, the contribution from PtOX to the emission intensity would be fairly constant and the AlQ$_3$ emission would grow in. The fluorescence from AlQ$_3$ is in the green to yellow part of the spectrum, where the contribution to the luminance on a per-photon basis is roughly 5 times greater than for red photons. This would lead to significantly enhanced luminescence, not a smaller than expected increase. If the contribution of AlQ$_3$ and PtOX fluorescence (450–600 nm) is subtracted from the spectra at 130 and 220 mA/cm^2, the luminance levels drop to 168 and 115 cd/m^2, respectively, corresponding to 64% and 39% of the total measured luminance values, respectively. Note that even though the current has been increased by a factor of 1.7 the amount light coming out of the device from PtOX phosphorescence has decreased by one-third. The fractional contribution of PtOX phosphorescence to the measured luminance decreased substantially as the current level is increased, consistent with enhanced triplet–triplet annihilation at the higher current level. A similar drop is observed in the quantum efficiencies estimated for red light only. If the contribution of the green/yellow part of the EL spectra are removed the quantum efficiencies at 130 and 220 mA/cm^2 drop to 0.39% and 0.24%, respectively.

The PtDPP OLED is not as efficient as the PtOX device, and the contribution to the emission from AlQ$_3$ has a stronger effect on the loss of the red color saturation for PtDPP-based OLEDs. At 130 mA/cm^2, the PtDPP device has become yellow, consistent with a greater AlQ$_3$ emission mixed with the PtDPP emission.

In conclusion, we have demonstrated the use of two new platinum porphyrins as efficient red phosphorescent dyes in small molecule-based OLEDs. Both dyes, at 6% doping level into AlQ$_3$, give saturated red emission at drive voltages <10 V. Higher current densities lead to saturation of phosphorescent emissive sites. Triplet–triplet annihilation, producing platinum porphyrin singlet excitons, was also observed at high current. PtOX-doped devices reach a quantum efficiency of 1.1% at 100 cd/m^2 with a luminous efficiency of 0.17 lm/W.

Acknowledgment

We thank Universal Display Corporation, the Defense Advanced Research Projects Agency, and the National Science Foundation for financial support of this work. The authors also thank Mr. Sergey Lemansky for his assistance with the lifetime measurements. S. Sibley thanks support of the Camille and Henry Dreyfus Foundation and the Beatrice Aitchison '28 fund of Goucher.

References

1. Chen, C. H., Shi, J., Tang C. W. *Macrolmol. Symp.* **1997**, *125*, 1. Sibley, S. P., Thompson, M. E., Burrows, P. E., Forrest, S. R. *Optoelectronic Properties of Inorganic Compounds*; Roundhill, D. M., Fackler, J., eds; Plenum Publishing Co.: New York, 1999; Chapter 2. Rothberg, L. J., Lovinger, A. J. *J. Mater. Res.* **1996**. *11*, 3174.
2. Shoustikov, A. A., You Y., Thompson, M. E. *IEEE J. Selected Top. Quantum Electron.* **1998**, *4*, 3.
3. Tang, C.W., Van Slyke, S. A., Chen, C. H. *J. Appl. Phys.* **1989**, *65*, 3610.
4. Bulovic, V., Shoustikov, A., Baldo, M. A., Bose, E., Kozlov, V. G., Thompson, M. E., Forrest, S. E. *Chem. Phys. Lett.* **1998**, *287*, 455.
5. Tang, C. W. *Dig. Soc. Information Display Int. Symp.* **1996**, *27*, 181.
6. Burrows, P. E., Forrest, S. R., Sibley, S. P., Thompson, M. E. *Appl. Phys. Lett.* **1996**, *69*, 2959.
7. Kido, J., Nagai, K., Okamato, Y., Skotheim, T. *Chem. Lett.* **1991**, 1267.
8. Kido, J., Nagai. K., Ohashi, Y. *Chem. Lett.* **1990**, 657.
9. Kido, J., Hayase, H., Hongawa, K., Nagai, K., Okuyama, K. *Appl. Phys. Lett.* **1994**, *65*, 2124.
10. Sabbatini, N., Guardigli, M., Lehn, J.-M. *Coord. Chem. Rev.* **1993**, *123*, 201.
11. Thompson, M. E., Shoustikov, A., You. Y., Sibley. S., Baldo, M., Koslov, V., Burrows, E. P., Forrest, S. R. *MRS Abstract*, G2.4, Spring Meeting. **1998**.
12. Baldo, M. A., O'Brien, D. F., You, Y., Shoustikov, A., Sibley, S., Thompson, M. E., Forrest. S. R. *Nature* **1998**, *395*, 151.
13. O'Brien, D. F., Baldo, M. A., Thompson, M. E., Forrest, S. R. *Appl. Phys. Lett.* **1999**, *74*, 442.
14. Cleave, V., Yahioglu, G., Barny, P. L., Friend, R. H., Tessler, N. *Adv. Mater.* **1999**, *11*, 285.
15. Rothberg, L. J., Lovinger, A. J. *J. Mater. Res.* **1996**, *11*, 3174.
16. Pope, M., Swenberg, C. E. *Electroinc Processes in Organic Crystals*; Clarendon Press: Oxford, **1982**.

17. Richardson, F. S. *Chem. Rev.* **1982**, *82*, 541.
18. Smith, K. M. *Porphyrins and Metalloporphyrins*; Elsevier Press: New York, 1975.
19. Schmidbauer, H., Lettenbauer, J., Kumberger, O., Lachmann, J., Muller, G. Z. *Naturforsch.* **1991**, *46b*, 1065.
20. Drive, M. S., Hartwig, J. F. *J. Am. Chem. Soc.* **1996**, *118*, 7217. Wolre, J. P., Wagaw, S., Buchwald, S. L. *J. Am. Chem. Soc.* **1996**, *118*, 7215.
21. DiMagno, S. G., Lin, V. S.-Y., Therien. M. *J. Org. Chem.* **1993**, *58*, 5983.
22. Chong, R., Clezy, P. S., Liepa, A. J., Nichol, A. W. *Aust. J. Chem.* **1969**, *22*, 229.
23. Kämpfen, U., Eschenmoser, A. *Hel. Chirn. Acta* **1989**, *72*, 185.
24. Seybold, P. G. and Gouterman, M. *J. Mol. Spectrosc.* **1969**, *31*, 1.
25. Turro, N. J. *Modern Molecular Photochemistry*, The Benjamin/Cummings Press: San Francisco. CA, 1978; p. 170.
26. Shelnutt, J. A., Ortiz, V. *J. Phys. Chem.* **1985**, *89*, 4733.
27. Dartnall, H. J. A., Bowmaker, J. K., Mollon, J. D. *Proc. R. Soc. Lon. B* **1983**, *220*, 115.
28. Gupta, B. D., Goyal, I. C. *J. Photochem.* **1985**, *30*, 173.
29. Ballardini, R., Varani, G., Indelli, M. T., Scandola, F. *Inorg. Chem.* **1986**, *25*, 3858.

Chapter 4

Excitonic Singlet-Triplet Ratio in a Semiconducting Organic Thin Film

M. A. Baldo,[a] D. F. O'Brien,[a,c] M. E. Thompson,[b] and S. R. Forrest[a]

[a] Center for Photonics and Optoelectronic Materials (POEM),
Department of Electrical Engineering and the Princeton Materials Institute,
Princeton University, Princeton, New Jersey 08544, USA
[b] Department of Chemistry, University of Southern California,
Los Angeles, California 90089, USA
[c] Present address: Department of Physics, Trinity College Dublin, Dublin 2, Ireland
met@usc.edu

A technique is presented to determine the spin statistics of excitons formed by electrical injection in a semiconducting organic thin film. With the aid of selective addition of luminescent dyes, we generate either fluorescence or phosphorescence from the archetype organic host material aluminum tris (8-hydroxyquinoline) (Alq_3). Spin statistics are calculated from the ratio of fluorescence to phosphorescence in the films under electrical excitation. After accounting for varying photoluminescent efficiencies, we find a singlet fraction of excitons in Alq_3 of $(22 \pm 3)\%$.

Reprinted from *Phys. Rev. B*, **60**(20), 14422–14428, 1999.

Electrophosphorescent Materials and Devices
Edited by Mark E. Thompson
Text Copyright © 1999 The American Physical Society
Layout Copyright © 2024 Jenny Stanford Publishing Pte. Ltd.
ISBN 978-981-4877-34-3 (Hardcover), 978-1-003-08872-1 (eBook)
www.jennystanford.com

4.1 Introduction

An assumption often employed in the study of electroluminescence (EL) in organic materials is that excitons are formed in the ratio of one singlet to three triplets [1]. However, it is not obvious that this should be the case, especially given that exchange interactions reduce the triplet state energy relative to that of the singlet. Accurate knowledge of spin statistics might therefore provide insight into the poorly understood process of exciton formation by electrical injection in conductive organic materials. Furthermore, since only singlets fluoresce, the singlet fraction (χ_s) is required to calculate the efficiency limit for an increasing diversity of fluorescent organic EL materials [2].

Triplet excitons can be extracted from a semiconducting host material using a phosphorescent dye dopant [3–6]. Since it is well known that singlets can be similarly extracted using a fluorescent dye [7], it follows that χ_s within a host can be determined if efficient energy transfer is possible to both fluorescent and phosphorescent dyes. In this work, we measure χ_s of the archetype host material, aluminum tris (8-hydroxyquinoline) (Alq$_3$) under electrical injection by doping it with either the phosphorescent dye, 2,3,7,812,13,17,18-octaethyl-21H,23H-porphine platinum(II) (PtOEP) [3], or the fluorescent dye [7] [2-methyl-6-[2-(2,3,6,7- tetrahydro-1H,5H-benzo[ij]quinolizin-9-yl)ethenyl]-4H-pyran-4-ylidene] propane-dinitrile (DCM2). We chose Alq$_3$ as the host material since emission by this compound arises from ligand-localized fluorescence [8] (i.e., no triplet emission). Thus undoped Alq$_3$ devices provide a second, independent, measurement of the singlet fraction without the complication of energy transfer. Furthermore, Alq$_3$ is an important organic semiconductor commonly used in organic light emitting devices [9].

4.2 Theory

In guest-host systems such as PtOEP or DCM2 doped into Alq$_3$, excitons formed in the host are transferred to the luminescent dye via a combination of Förster and Dexter energy transfer [10]. Förster transfer is a long range (~40–100 Å), nonradiative, dipole-dipole coupling of donor (D) and acceptor (A) molecules. Since it requires that the transitions from the ground to the excited states be allowed for both D and A species, this mechanism only transfers energy to the singlet state of the acceptor molecule. Dexter transfer is a short-range process where excitons diffuse from D to A sites via intermolecular electron exchange. In contrast to Förster transfer, Dexter processes require only that the total spin of the D–A pair be conserved under the Wigner-Witmer selection rules [10]. Consequently, Dexter transfer permits both singlet-singlet and triplet-triplet transfers. Singlet-triplet and triplet-singlet transfers are also possible if the donor exciton breaks up and reforms on the acceptor via incoherent electron exchange [10]. This latter process is considered to be relatively unlikely as it requires the dissociation of the donor exciton, which in most molecular systems has a binding energy of ~1 eV.

Figure 4.1 summarizes the energy-transfer pathways responsible for guest fluorescence and phosphorescence in a semiconducting host. In Fig. 4.1a, we show the singlet-to-singlet transfer responsible for fluorescence in most doped organic EL devices. Although both Förster and Dexter processes are capable of singlet-to-singlet energy transfer, Förster transfer dominates [11] at low fluorescent dye concentrations because of its long-range nature. Indeed, we find that >99% of the photoluminescent (PL) spectra of DCM2:Alq$_3$ films under modest optical pump intensities ($\lesssim 1$ mW/cm^2) is due to emission only from DCM2. Thus the energy-transfer rate is much faster than either the radiative or non-radiative rates in Alq$_3$. Significantly, this means that singlets are transferred directly after formation, without any preceding nonradiative losses such as intersystem crossing (ISC). On the other hand, a triplet state in DCM2 could be excited by close range and possibly slower, triplet-triplet Dexter transfer from the host, as shown in Fig. 4.1b. Most fluorescent dyes (including DCM2) possess short (~10 ns) radiative lifetimes, and at room temperature, phosphorescence is rarely observed from their triplet states. We assume therefore that all emission from DCM2 ultimately results from singlet states initially formed within the semiconducting host. Figure 4.1c shows singlet singlet transfer

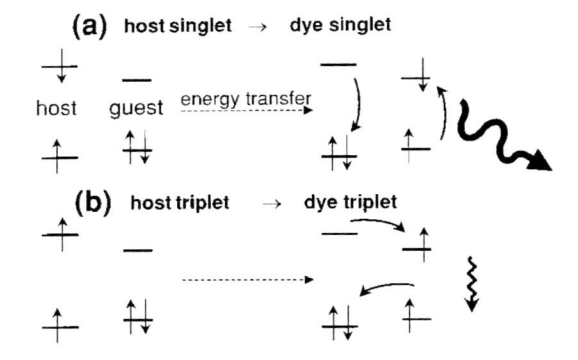

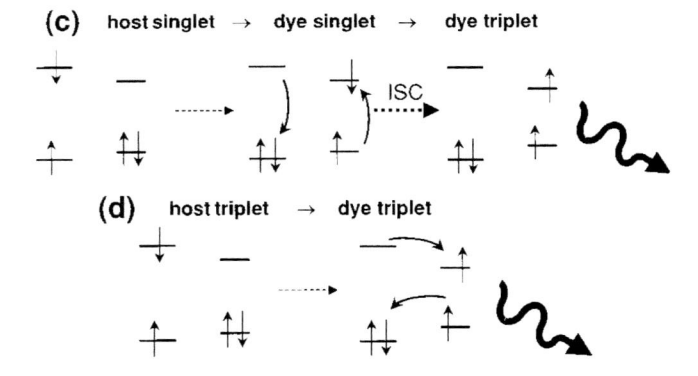

Figure 4.1 Proposed energy-transfer mechanisms in films doped with a fluorescent dye (e.g., DCM2:Alq$_3$) and films doped with a phosphorescent dye (e.g., PtOEP:Alq$_3$). For each molecule we show the ground-state energy level S_0, the excited-state singlet level S_1 and the excited-state triplet level T_1.

between a host and a phosphor such as PtOEP; it is expected that Förster is still the dominant process in this case. The singlet lifetime of PtOEP is [12] $\lesssim$ 10 ps and its fluorescence efficiency is [12] 2 × 10^{-5}. Hence we assume that singlet energy transfer from Alq$_3$ is followed by ISC in PtOEP with near unity efficiency. Finally, as described previously [3], Fig. 4.1d represents direct Dexter transfer between triplet states in the host and the phosphor dopant.

To quantify χ_s in an organic host, we write the external EL quantum efficiency (photons extracted in the forward direction per electron injected) as [13, 14]

$$\Phi_{el} = \left[\chi_s \Phi_{fl} \eta_s + \chi_t \Phi_{ph} \eta_t\right] \eta_r \eta_e \tag{4.1}$$

Here, $\chi_t = (1-\chi_s)$ is the triplet fraction of excitons. Also, Φ_{fl} and Φ_{ph} are the PL efficiencies of fluorescence and phosphorescence of the acceptor, η_s and η_t are the transfer efficiencies of singlet and triplet excitons from D to A, η_r is the fraction of injected charge carriers that form excitons on the donor, and η_e is the fraction of emitted photons that are coupled out of the device.

By quantitatively comparing the EL efficiencies ($\Phi_{el}^{(fl)}$ and $\Phi_{el}^{(ph)}$) of separate devices employing either a fluorescent or phosphorescent dye, we can use Eq. (4.1) to determine χ_s and χ_t provided that η_r and η_e are identical in both devices. Furthermore, if all singlets are transferred in both cases (i.e., $\eta_s^{(fl)} = \eta_s^{(ph)} \sim 1$), we can obtain χ_s from the ratio of fluorescent and phosphorescent efficiencies ($\Phi_{el}^{(fl)}$ and $\Phi_{el}^{(ph)}$) to obtain

$$\chi_s = \frac{\eta_t^{(ph)}}{\left(\frac{\Phi_{el}^{(ph)}}{\Phi_{el}^{(fl)}} \cdot \frac{\Phi_{pl}^{(fl)}}{\Phi_{pl}^{(ph)}}\right) - (1 - \eta_t^{(ph)})}. \tag{4.2}$$

Since the ratios reflect relative measurements, the only absolute result required is the triplet state transfer efficiency from the host to the phosphor (i.e., $\eta_t^{(ph)}$).

4.3 Experiment

To determine the ratio of the EL efficiency of PtOEP and DCM2 in Alq$_3$ (i.e., $\Phi_{el}^{(ph)}/\Phi_{el}^{(fl)}$), a series of devices [9] were made using either of these two materials doped into the Alq$_3$ host (see Fig. 4.2). Organic layers were deposited in a vacuum of $<10^{-6}$ Torr onto a glass substrate precoated with a 1700-Å-thick layer of indium tin oxide (ITO). A 400-Å-thick film of the hole transport material 4,4′-bis [N-(1-napthyl)-N-phenyl-amino] biphenyl (α-NPD) was deposited on the ITO, followed by a thin (100-Å) doped Alq$_3$ layer acting as the emissive region. On top of the doped layer, a thin (80-Å) layer of the

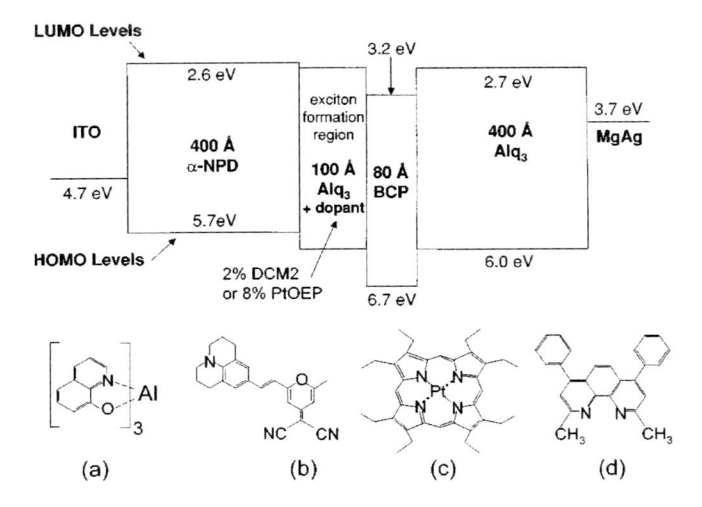

Figure 4.2 Proposed energy level diagram of the electroluminescent devices. The luminescent region is sandwiched between electron blocking α-NPD and hole blocking BCP. Three different luminescent regions were employed: undoped Alq_3, which is fluorescent, 2% DCM2 in Alq_3, which is also fluorescent, and 8% PtOEP in Alq_3, which is phosphorescent. Also shown are the chemical structures of (a) Alq_3, (b) DCM2, (c) PtOEP, and (d) BCP.

hole and exciton blocking material [4] 2,9-dimethyl-4,7-diphenyl-1,10-phenanthroline (bathocuproine, or BCP) [15] was deposited. A 400-Å-thick cap layer of Alq_3 was used as buffer between the emissive region and the Mg:Ag (25:1) cathode. Finally, a 1000-Å-thick cap layer of silver was deposited to protect the cathode from decomposition. For comparison, undoped Alq_3 devices were prepared with and without the BCP layer. Samples were mounted directly onto the surface of a calibrated silicon photodetector [16], and the forward-scattered luminescence was measured.

In Fig. 4.3 we observe that the efficiency of the Alq_3/BCP device has a slight upward trend but that the efficiency of both doped devices decreases with increasing current; a point that we will return to in Section 4.4. The efficiency data therefore were obtained at current densities $\lesssim 10^{-5}$ A/cm^2, where both efficiency curves are relatively flat and there is a low density of excited states. In this

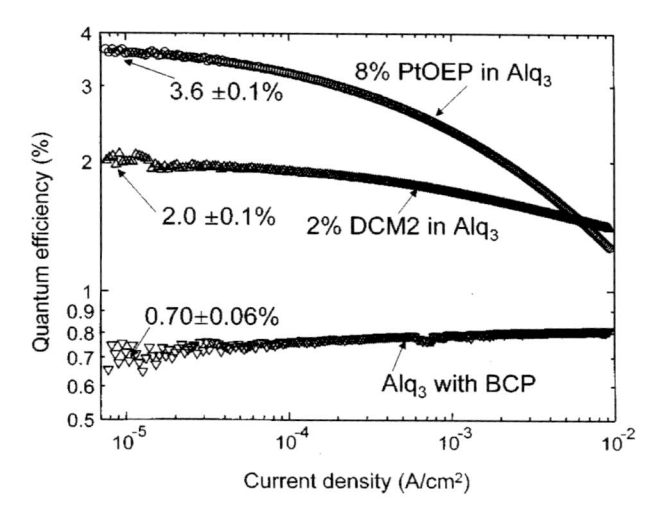

Figure 4.3 EL quantum efficiency of the DCM2, PtOEP, and Alq$_3$-only devices. All curves are constant in the low current regime where multiexciton interactions leading to quenching are negligible. The structure of the devices is shown in Fig. 4.2. All devices contain a BCP blocking layer.

regime, using the data of Fig. 4.3, we get $\Phi_{el}^{(DCM2)}/\Phi_{el}^{(PtOEP)} = 0.56 \pm 0.06$ and $\Phi_{el}^{(Alq3)}/\Phi_{el}^{(PtOEP)} = 0.20 \pm 0.02$.

The values of $\Phi_{pl}^{(PtOEP)}/\Phi_{pl}^{(DCM2)}$, $\Phi_{pl}^{(PtOEP)}/\Phi_{pl}^{(Alq3)}$, $\eta_s^{(DCM2)}$, and $\eta_s^{(PtOEP)}$ were obtained by optically pumping doped films near the absorption peak of Alq$_3$ at a wavelength [16] of $\lambda = 400$ nm using a broad-spectrum mercury xenon lamp and monochromator. Photoluminescence from the sample was coupled into a fiber bundle after being filtered to remove interference from the pump and then analyzed using a second spectrometer. As in the case of electrical pumping, low optical pump intensities of $\lesssim 0.2$ mW/cm^2 were used to minimize bimolecular quenching due to dye exciton interactions [3]. Equating the number of photons absorbed with the number of carriers injected in the EL devices and assuming $\eta_r = 1$, yields an equivalent current density of $\sim 10^{-5}$ A/cm^{-2} for these optical pump intensities. The integrated PL efficiency ratios were $\Phi_{pl}^{(PtOEP)}/\Phi_{pl}^{(DCM2)} = (0.37 \pm 0.03)$ and $\Phi_{pl}^{(PtOEP)}/\Phi_{pl}^{(Alq3)} = (1.2 \pm 0.2)$. The relative PL quantum efficiencies as a function of pump intensity are shown in Fig. 4.4 and the spectra at a pump power of

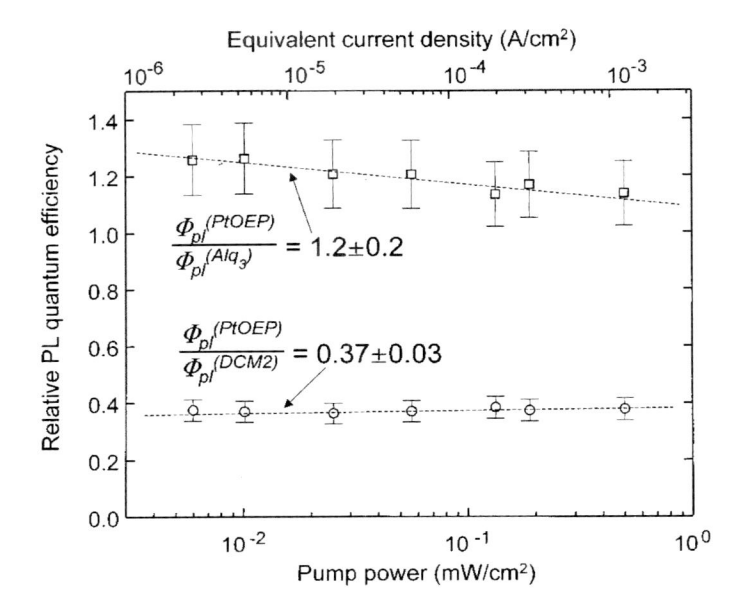

Figure 4.4 PL efficiencies of Alq$_3$ and DCM2:Alq$_3$ relative to PtOEP:Alq$_3$. The efficiencies are plotted versus pump power (at 400 nm) and also, for comparison with the EL data, versus an estimation of the current density equivalent to the pump power. The lines are guides to the eye only.

$\sim$10 µW/cm^2 are shown in Fig. 4.5. The trends in the relative PL data of Fig. 4.4 match those observed in the EL data of Fig. 4.3, indicating that similar quenching phenomena occur in both cases. However, the rapid decrease seen, for example in the PtOEP:Alq$_3$ EL efficiency as the current density increases, is not observed in PL. As discussed in Section 4.4, there are differences between PL and EL processes; and in this work, we attempt to minimize these discrepancies by taking our measurements at very low excitation densities.

No Alq$_3$ emission was observed in the doped films, hence we conclude that $\eta_s^{(fl)} = \eta_s^{(ph)} = 1$. Note that direct absorption at the pump wavelength of $\lambda = 400$ nm by DCM2 and PtOEP molecules is less than 1% and 8% of the total absorption, respectively [11, 16, 17].

To determine $\eta_t^{(ph)}$, we assume that Alq$_3$ triplets either nonradiatively decay or are transferred to PtOEP. The rate of nonradiative

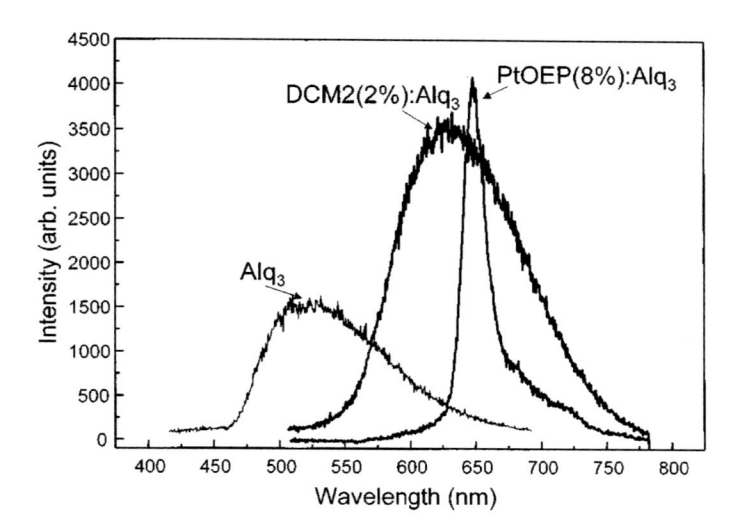

Figure 4.5 The PL spectra of Alq$_3$, DCM2:Alq$_3$, and PtOEP:Alq$_3$ at a pump ($\lambda = 400$ nm) intensity of 10 µW/cm^2.

decay is a function of the triplet diffusion length L_d. Similarly, the rate of triplet transfer to PtOEP is proportional to the transfer length L_t, which depends on dopant concentration. For example, low doping densities result in a reduced likelihood of transfer, and we expect this to be reflected in a larger L_t. If A_T^* is the concentration of Alq$_3$ triplet excited states and x is the distance from the exciton formation zone [7] at the interface between Alq$_3$ and the α-NPD hole transport layer (i.e., at $x = 0$) then

$$\frac{dA_T^*}{dx} = -\frac{A_T^*}{L_t} - \frac{A_T^*}{L_d} = -\frac{A_T^*}{L}. \quad (4.3)$$

Since η_t is the ratio of triplets transferred to the total number of Alq$_3$ triplets formed at the α-NPD interface, then

$$\eta_t(d) = \frac{A}{A^*(0)} \int_0^d \frac{A^*(x)}{L_t} dx = \frac{L_d}{L_t + L_d}(1 - e^{-d/L}) \quad (4.4)$$

where d is the thickness of the doped layer.

To measure L_d, a 100-Å-thick layer of DCM2 doped at 2% molar concentration in Alq$_3$ was deposited in a vacuum of $<10^{-6}$ Torr onto the surface of a previously deposited α-NPD + copper phthalocyanine hole transport layer on an indium tin oxide (ITO) coated glass substrate (see inset, Fig. 4.5a). The DCM2 placed in the exciton formation zone of this otherwise conventional [9] organic light emitting device serves to remove singlets through Förster transfer [3]. The diffusion length of the remaining triplet states is determined by depositing an Alq$_3$ spacer layer between the DCM2 singlet "filter" and a 100-Å-thick layer of PtOEP(10%):Alq$_3$. A second, 300-Å-thick Alq$_3$ layer located between the cathode and the PtOEP layer reduces exciton quenching at the electrode. As the thickness of the spacer between the singlet filter and the PtOEP is increased, triplets must diffuse farther to reach the PtOEP layer, and its luminescence should decrease accordingly. Within experimental error, however, no appreciable decrease in PtOEP emission was observed even at the maximum spacer thickness of 600 Å. From the data in Fig. 4.6a, we estimate therefore that the triplet diffusion length in Alq$_3$ must be $L_d \geq 1400$ Å.

The triplet transfer distance in PtOEP is independently obtained by inserting layers of varying thickness of PtOEP(8%):Alq$_3$ separated from the DCM2 singlet filter by a 100-Å-thick Alq$_3$ spacer layer (inset, Fig. 4.6b). As the thickness of PtOEP increases, the total phosphorescence intensity increases until all Alq$_3$ triplets are either transferred or else nonradiatively decay. Note that extending the PtOEP layer and, consequently, the proximity of PtOEP sites to the cathode, increases the effect of cathode absorption. However, because most luminescence occurs in the first 100 Å of the PtOEP layer at a distance of over 400 Å from the cathode, cathode quenching effects are small. Figure 4.5b shows the spectral intensity as a function of the phosphorescent layer width. From these data, we obtain $L_t = (140 \pm 30)$ Å and thus $L = (1/L_t + 1/L_d)^{-1} = (125 \pm 25)$ Å. Solving Eq. (4.4) for the triplet transfer efficiency in a 400-Å-thick PtOEP(8%):Alq$_3$ layer, we get $\eta_t^{(PtOEP)}$ (400 Å, 8%) = 0.90 $\pm$ 0.05.

This result is supported by a comparison of the peak EL quantum efficiency of PtOEP(8%):Alq$_3$ devices with and without the BCP confinement layer as shown in Fig. 4.7. The peak efficiencies are 3.6% and 3.2%, respectively, and assuming $\eta_t = 1$ in the device employing BCP (circles), then for the device without BCP (squares) $\eta_t^{(PtOEP)}$ (400 Å, 8%) = 3.2/3.6 = 0.9, in agreement with the above calculation. More lightly doped films possess lower transfer efficiencies; for example in Ref. [4] we observed $\eta_t^{(PtOEP)}$(400 Å,

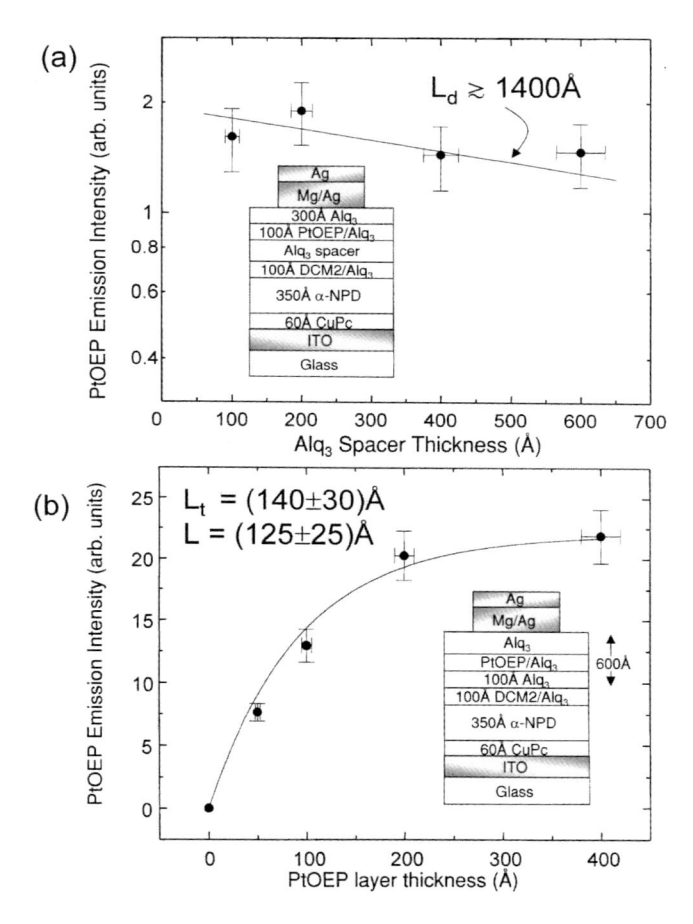

Figure 4.6 (a) To determine the triplet diffusion length of Alq$_3$, we employ the device structure shown in the inset and remove singlet excitons with a 100-Å-thick layer of DCM2(2%):Alq$_3$. The remaining triplets are forced to diffuse through an Alq$_3$ spacer layer before reaching a luminescent PtOEP layer. As the thickness of the spacer layer increases, the rate of decrease in PtOEP luminescence gives a triplet diffusion length of ≥ 1400 Å. (b) The thickness of the luminescent PtOEP layer is increased until all triplets diffusing through the layer either nonradiatively decay or are transferred to PtOEP. By fitting the measured PtOEP emission intensity to Eq. (4.6), the combined diffusion and transfer length of triplets in Alq$_3$ is calculated to be (125 ± 25) Å. From Eq. (4.5), the transfer length is calculated to be (140 ± 30) Å. The total device thickness was kept constant by adjusting the thickness of the Alq$_3$ cap layer adjacent to the cathode. Both the transfer and diffusion measurements were made at 6.5 mA/cm^2.

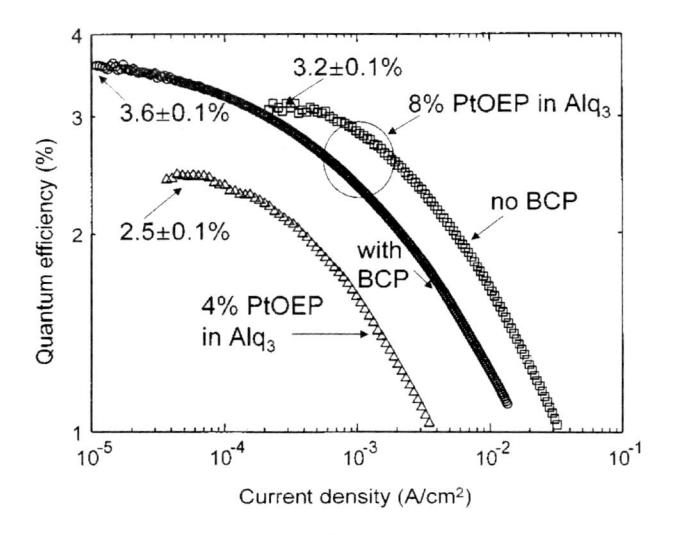

Figure 4.7 Electroluminescent (EL) quantum efficiency of PtOEP:Alq$_3$ devices demonstrating the effect of PtOEP doping density and the BCP exciton blocking layer. The quantum efficiency of a double-heterostructure EL device containing a 100-Å-thick emissive layer of PtOEP(8%):Alq$_3$ and a 80-Å-thick BCP layer is shown (circles). For comparison, the quantum efficiency of a single heterostructure EL device containing a 400-Å-thick emissive layer of PtOEP(8%):Alq$_3$ is also plotted (squares). This later device is predicted to have a triplet transfer efficiency of 0.90 ± 0.05. Hence we expect that the double-heterostructure device has a transfer efficiency of ~100%. Also shown (triangles) is the efficiency of a single heterostructure EL device containing a 400-Å-thick emissive layer of PtOEP(4%):Alq$_3$. The reduction in efficiency may be due to poorer energy transfer in a more diffusely doped device.

4%) ~0.7 (Fig. 4.7, triangles). The complete energy transfer seen in devices employing the blocking layer supports the hypothesis that the principal action of BCP in Alq$_3$-based devices is as a barrier to the diffusion of Alq$_3$ triplet excitons. This finding is also supported by the results in Ref. [5], where phosphorescent efficiencies were observed to increase by an order of magnitude when blocking layers are employed. Hence, we employ BCP in our devices to achieve $\eta_t^{(ph)} \sim 1$.

Given the previous results for the relative EL and PL quantum efficiencies of the devices containing BCP and the calculation of the triplet energy transfer efficiency, we apply Eq. (4.2) to obtain the spin multiplet fractions of $\chi_s = (21 \pm 3)\%$ for the DCM2/PtOEP system, and $\chi_s = (24 \pm 4)\%$ for the Alq$_3$/PtOEP system. Here, the error quoted is the quadrature sum of all measurement errors used in the determination of χ_s, which is justified given that the sources of uncertainty in these measurements are uncorrelated. Taking the weighted average of these results, we obtain an overall value for the singlet fraction in Alq$_3$ of $\chi_s = (22 \pm 3)\%$.

4.4 Significance of the Recombination Zone

The proposed energy level diagram in Fig. 4.2 shows that the Alq$_3$ exciton formation region is surrounded by wide energy-gap materials. Thus the presence of the BCP layer creates a double-heterostructure [18], enforcing identical narrow emission zones in each device, thereby making unbalanced injection unlikely. The disadvantage of employing a heterostructure is that some excitons, notably singlets with their capability for Förster transfer, may be formed in the α-NPD hole transport layer within the energy transfer radius of Alq$_3$ (~32 Å) [19]. However, we consider this possibility to be highly unlikely in our experiments, since even at extremely high injection densities, emission from α-NPD is not observed. Thus the recombination zone, if it exists in α-NPD, must be no wider than a few monolayers.

In principle, it would be preferable to employ single layer Alq$_3$ devices to eliminate the possibility for recombination outside the desired region. However, the location of the recombination zone is uncertain in single layer structures, and experiments such as those by Cao et al. [20]. employing single layer polymer structures do not maintain adequate control over the location of the recombination zone. Hence such measurements must be treated with caution.

The importance of a well-controlled recombination zone is demonstrated by comparing Alq$_3$-only devices with and without BCP. We find that addition of the BCP blocking layer makes little difference to the doped devices but it significantly reduces the slope of the quantum efficiency of the Alq$_3$-only devices (see Fig. 4.8). Furthermore, in the absence of BCP we find that the Alq$_3$-only

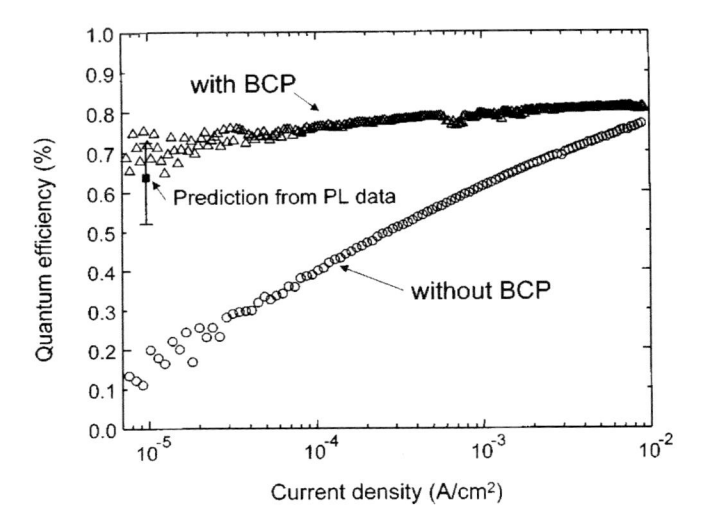

Figure 4.8 EL quantum efficiency of the Alq$_3$-only devices with and without the BCP blocking layer. At $J \sim 10^{-5}$ A/cm^2 the efficiency of the device without BCP is only $\sim$0.2%, which is much less than the expected value of (0.62 ± 0.1)% calculated from the relative PL efficiencies of Alq$_3$ and DCM2:Alq$_3$.

devices cannot be compared to the doped devices. For example, from the relative PL data of DCM2:Alq$_3$ and neat Alq$_3$ we should be able to predict the EL efficiency of an Alq$_3$-only device from the EL efficiency of a DCM2:Alq$_3$ device. Comparing efficiencies at $J \sim 10^{-5}$ A/cm^2, we have $\Phi_{el}^{Alq_3} = \Phi_{el}^{DCM2} \cdot \Phi_{pl}^{Alq_3}/\Phi_{pl}^{DCM2} = (0.62 \pm 0.1)$%. As seen in Fig. 4.8, the efficiency of the Alq$_3$-only device without BCP is significantly lower than predicted (solid square) by the relative PL efficiencies. In contrast, there is much better agreement when the emission zone is restricted by a BCP layer. Calculations for Alq$_3$ heterostructures predict [21] a recombination zone width of $\sim$120 Å. However, the emission zone, as detected by a DCM loped layer [7], is considerably larger ($\sim$400 Å). The difference between the widths of emission and recombination zones is due to exciton diffusion. Hence the discrepancy in efficiency in Fig. 4.8 is due to the diffusion of Alq$_3$ excitons to nonradiative sites near the cathode in the absence of a BCP exciton blocking layer [4]. It is known that doped devices possess a narrow emission zone ($\sim$50 Å) [7], with or without a hole blocking layer. Thus relative intensity experiments are justified only when all devices possess emission regions of similar extent.

4.5 Discussion

Since absolute PL and EL measurements require accounting for every photon [20] and hence are subject to significant systematic error, the technique used here employs relative measurements where possible. For accurate comparisons between different devices, the charge transport layers, injection interfaces and contacts of all devices were fabricated during a single deposition run and hence are identical. In contrast, the measurements of the singlet fraction of excitons in DCM2:Alq$_3$ and Alq$_3$-only devices differ fundamentally by the presence of energy transfer in the DCM2:Alq$_3$ device. For example, the PL efficiency of Alq$_3$ is 0.32 ± 0.2 [16], giving a nonradiative rate approximately twice the radiative rate ($k_{nr} \sim 2\, k_r$). Yet, in a DCM2:Alq$_3$ film, less than 1% of the total emitted photons are observed from Alq$_3$, therefore, given a DCM2:Alq$_3$ PL efficiency of ~80%, we infer an energy transfer rate $k_t \sim 50\, k_{nr}$. Hence energy transfer avoids host quenching processes.

Nevertheless, in relative measurements with PtOEP:Alq$_3$, both Alq$_3$-only and DCM2:Alq$_3$ systems yield approximately the same value for spin statistics. This is because quenching effects are minimized by measuring *relative* efficiencies at low current densities, with the remaining differences in the nonradiative efficiencies quantified by the PL intensities. For example, intersystem crossing [8] in Alq$_3$ should not influence the measurement of spin statistics, since it should equally affect both the EL and PL efficiencies. By taking the ratio of the EL and the PL efficiencies as in Eq. (4.2), the effects of intersystem crossing cancel out.

Thus our measurement of spin statistics cancels the effects of all quenching processes that occur in both EL and PL. However, EL quenching processes which may not be reflected in the PL measurements must also be considered. For example, as shown in Fig. 4.9, there is a 0.5-V increase in the operating voltage of devices containing DCM2 relative to those containing PtOEP. Analysis of energy transfer has shown that charge trapping and

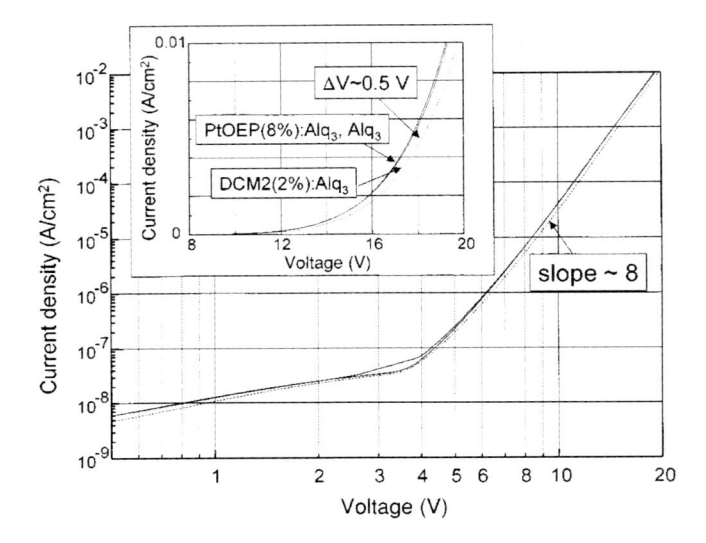

Figure 4.9 The current–voltage characteristics of the devices containing BCP. Devices containing PtOEP overlap the characteristic of the Alq$_3$-only devices, however, DCM2 devices possess a 0.5-V voltage increase, possibly due to trapping on DCM2 molecules.

direct exciton formation is not a major effect in PtOEP:Alq$_3$ films [3]. Thus the discrepancy between the DCM2 and PtOEP device current conduction may imply that a significant fraction of excitons emitted by the DCM2:Alq$_3$ device originated from trapping of carriers on DCM2, rather than transfer from Alq$_3$. But previous work has shown [11] that EL in DCM2:Alq$_3$ is consistent with complete Förster transfer. In addition, we analyzed charge trapping on DCM2 in double heterostructure devices following Tang, VanSlyke, and Tang [7]. That is, the spatial distribution of excitons is probed by moving a DCM2 doped layer within the Alq$_3$ luminescent layer of the double heterostructure. The EL spectra for different DCM2 layer positions are shown in Fig. 4.10. The emission from DCM2 ideally should reflect the density of excitons in that region of the undoped double heterostructure. However, if trapping on the dye is significant, the recombination zone should follow the position of the DCM2 layer [22]. In our experiment, we find that significant Alq$_3$ emission is observed unless the DCM2 layer is positioned immediately adjacent

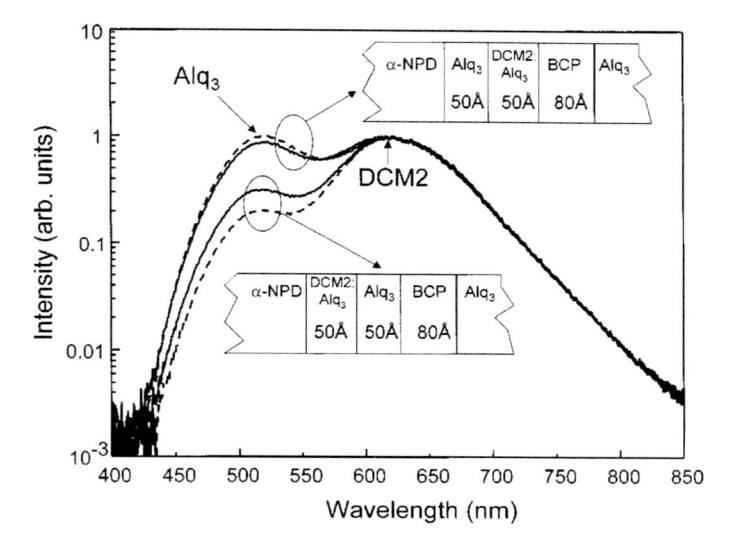

Figure 4.10 A demonstration of the effect of BCP on the location of the recombination zone. Dashed lines indicate the spectra of EL devices without a BCP blocking layer. Relative to devices containing BCP, we observe that the bulk of recombination remains at the α-NPD interface, however, BCP is responsible for a compression of the zone and possibly also a slight shift towards the Alq$_3$/BCP interface. The spectra were recorded at a current density of 60 mA/cm^2. For clarity the spectra have been normalized at the peak of the DCM2 emission (610 nm).

to the α-NPD interface, suggesting that direct charge trapping and exciton formation on DCM2 molecules is probably not significant. This is supported by the equivalence in the singlet fraction obtained from both DCM2:Alq$_3$ and Alq$_3$-only devices.

A final point is the influence of the electric field and polaron formation in EL. Measurements [23] of the magnitude of electric-field quenching of Alq$_3$ luminescence confirm that an external field does indeed lead to exciton dissociation, although the effect is probably not significant at the operating voltages used here. For example, calculations of μE, where μ is the exciton dipole moment and E is the external field, give energies of $\sim$10 meV, much less than the binding energies ($\sim$1 eV) typical of excitons in these materials.

Polaron induced quenching of excitons is also possible, and has been found to have a large effect in some organic materials [24].

However, if polaron quenching were significant in the materials studied here, it is difficult to understand how the Alq$_3$ quantum efficiency could increase with current. Even with a BCP blocking layer and a restricted recombination zone, there remains a slight upward trend in quantum efficiency. Thus we conclude that quenching due to the electric field or polarons has a negligible effect on the calculation of spin statistics.

4.6 Conclusion

An accurate determination of exciton spin statistics requires a thorough understanding of exciton formation and energy transfer. Although we cannot definitively rule out quenching due to polarons or direct exciton formation on the dye molecules, this work has presented a consistent set of data yielding a singlet fraction that agrees within error with the expected value of 25%. Both the Alq3/PtOEP:Alq$_3$ and the DCM2:Alq$_3$/PtOEP:Alq$_3$ systems yield similar results indicating that direct exciton formation on DCM2 molecules is probably not significant. While further work is required to understand exciton formation in doped materials, the techniques presented in this work offer a precise method for the determination of spin statistics. Indeed, it should be possible to extend the techniques introduced in this study to different hosts and different temperatures.

Acknowledgments

This work was funded by NSF and AFOSR. The authors gratefully acknowledge informative discussions with Dr. Paul Burrows, Dr. Vladimir Bulović, and Dr. Scott Sibley.

References

1. A. R. Brown, K. Pichler, N. C. Greenham, D. D. C. Bradley, R. H. Friend, and A. B. Holmes, *Chem. Phys. Lett.*, **210**, 61 (1993).
2. S. R. Forrest, *Chem. Rev.*, **97**, 1793 (1997).

3. M. A. Baldo, D. F. O'Brien, Y. You, A. Shoustikov, S. Sibley, M. E. Thompson, and S. R. Forrest. *Nature (London)*, **395**, 151 (1998).
4. D. F. O'Brien, M. A. Baldo, M. E. Thompson, and S. R. Forrest, *Appl. Phys. Lett.*, **74**, 442 (1999).
5. M. A. Baldo, S. Lamansky, P. E. Burrows, M. E. Thompson, and S. R. Forrest, *Appl. Phys. Lett.*, **75**, 4 (1999).
6. V. Cleave, C. Yahiogiu, P. Le Barny, R. Friend, and N. Tessler, *Adv. Mater.*, **11**, 285 (1999).
7. C. W. Tang, S. A. VanSlyke, and C. H. Chen, *J. Appl. Phys.*, **65**, 3610 (1989).
8. R. Ballardini, G. Varani, M. T. Indelli, and F. Scandola, *Inorg. Chem.*, **25**, 3858 (1986).
9. S. A. VanSlyke and C. W. Tang, U.S. Patent No. 4539507 (1985).
10. M. Klessinger and J. Michl, *Excited States and Photochemistry of Organic Molecules* (VCH Publishers, New York, 1995).
11. V. Bulovic, A. Shoustikov, M. A. Baldo, E. Bose, V. G. Kozlov, M. E. Thompson, and S. R. Forrest, *Chem. Phys. Lett.*, **287**, 455 (1998).
12. G. Ponterini, N. Serpone, M. A. Bergkamp, and T. L. Netzel, *J. Am. Chem. Soc.*, **105**, 4639 (1983).
13. T. Tsutsui and S. Saito, in *Intrinsically Conducting Polymers: An Emerging Technology*, edited by M. Aldissi (Kluwer Academic, Dordrecht, 1993), Vol. 246, p. 123.
14. L. J. Rothberg and A. J. Lovinger, *J. Mater. Res.*, **11**, 3174 (1996).
15. Y. Kijima, Spring Meeting of the Materials Research Society, San Francisco, 1999, Paper G2.1 (in press).
16. D. Z. Garbuzov, V. Bulovic, P. E. Burrows, and S. R. Forrest, *Chem. Phys. Lett*, **249**, 433 (1996).
17. S.-K. Lee and I. Okura, *Anal. Chim. Acta*, **342**, 181 (1997).
18. C. Adachi, T. Tsutsui, and S. Saito, *Appl. Phys. Lett.*, **57**, 531 (1990).
19. R. Deshpande, V. Bulovic, and S. Forrest, *Appl. Phys. Lett.*, **75**, 888 (1999).
20. Y. Cao, I. Parker, G. Yu, Z. Gang, and A. Heeger, *Nature (London)*, **397**, 414 (1999).
21. J. Kalinowski, N. Camaioni, P. Di Marco, V. Fattori, and A. Martelli, *Appl. Phys. Lett.*, **72**, 513 (1998).
22. J. Yang and J. Shen, *J. Appl. Phys.*, **84**, 2105 (1998).
23. W. Stampor, J. Kalinowski, P. D. Marco, and V. Fattori, *Appl. Phys. Lett.*, **70**, 1935 (1997).
24. H. Hieda, K. Tanaka, K. Naito, and N. Gemma, *Thin Solid Films*, **331**, 152 (1998).

Chapter 5

Very High-Efficiency Green Organic Light-Emitting Devices Based on Electrophosphorescence

M. A. Baldo,[a] S. Lamansky,[b] P. E. Burrows,[a] M. E. Thompson,[b] and S. R. Forrest[a]

[a] Center for Photonics and Optoelectronic Materials (POEM),
Department of Electrical Engineering and the Princeton Materials Institute,
Princeton University, Princeton, New Jersey 08544, USA
[b] Department of Chemistry, University of Southern California,
Los Angeles, California 90089, USA
met@usc.edu

We describe the performance of an organic light-emitting device employing the green electrophosphorescent material, *fac* tris(2-phenylpyridine) iridium [Ir(ppy)$_3$] doped into a 4,4'-N,N'-dicarbazole-biphenyl host. These devices exhibit peak external quantum and power efficiencies of 8.0% (28 cd/A) and 31 lm/W, respectively. At 100 cd/m^2, the external quantum and power efficiencies are 7.5% (26 cd/A) and 19 lm/W at an operating voltage of 4.3 V. This performance can be explained by efficient transfer of both singlet and triplet excited states in the host to Ir(ppy)$_3$, leading

---

Reprinted from *Appl. Phys. Lett.*, **75**(1), 4–6, 1999.

Electrophosphorescent Materials and Devices
Edited by Mark E. Thompson
Text Copyright © 1999 American Institute of Physics
Layout Copyright © 2024 Jenny Stanford Publishing Pte. Ltd.
ISBN 978-981-4877-34-3 (Hardcover), 978-1-003-08872-1 (eBook)
www.jennystanford.com

to a high internal efficiency. In addition, the short phosphorescent decay time of Ir(ppy)$_3$ (<1 µs) reduces saturation of the phosphor at high drive currents, yielding a peak luminance of 100 000 cd/m^2.

The recent demonstration [1, 2] of high-efficiency red electrophosphorescence from a platinum porphyrin foreshadowed a breakthrough in organic light-emitting device (OLED) performance. Unlike fluorescence [3], phosphorescence makes use of both singlet and triplet excited states; suggesting [1, 2, 4] the potential for reaching a maximum internal efficiency of 100%. However, at the benchmark luminance of 100 cd/m^2, the porphyrin of the initial studies exhibits an external quantum efficiency of 2.2%, substantially less than its quantum efficiency at low currents (5.6%). Although its efficiency at 100 cd/m^2 is highly competitive with red fluorescent dyes of comparable color saturation [2], that phosphor is hampered by a long decay time, causing saturation of emissive sites and a decrease in efficiency at high drive currents. Consequently, it fails to realize the potential of phosphorescence: external quantum efficiencies of ∼10% with correspondingly high power efficiencies.

In this work, we describe OLEDs employing the green, electrophosphorescent material *fac* tris(2-phenylpyridine) iridium [Ir(ppy)$_3$] [5–8]. The coincidence of a short triplet lifetime and reasonable photoluminescent efficiency allow Ir(ppr)$_3$-based OLEDs to achieve peak quantum and power efficiencies of 8.0% (28 cd/A) and 31 lm/W, respectively. At an applied bias of 4.3 V, the luminance reaches 100 cd/m^2, and the quantum and power efficiencies are 7.5% (26 cd/A) and 19 lm/W, respectively.

Fluorescence is limited to radiative relaxations of organic molecules that conserve spin symmetry. These processes are extremely rapid (∼1 ns), and typically involve transitions between singlet excited and ground states. In contrast, phosphorescence results from "forbidden" transitions where symmetry is not conserved, for example, transitions between triplet excited states and singlet ground states. Under electrical excitation, excitons are formed in both symmetry states; thus, harvesting luminescence from all excitons has the potential to yield significantly higher efficiencies than is possible in purely fluorescent devices.

To maximize performance, electrophosphorescent devices should employ a conductive host material with a phosphorescent guest sufficiently dispersed to avoid "concentration quenching." Although some phosphorescent guests may trap charge and form excitons directly, it is likely that if host emission is to be avoided, then some form of energy transfer to the guest is necessary. Since the triplet energy levels in the host and the absorption of the guest triplet state are frequently unknown and, moreover, are difficult to quantify, optimizing guest–host systems for resonant triplet transfer is problematic. Comparisons of the absorption of the guest and the emission spectrum of the host are relevant only to the transfer of singlet states, but may nevertheless give a general indication of the likelihood for triplet transfer. Based on this assumption, the blue ($\lambda \sim 400$ nm peak) emissive material 4,4'-N,N'-dicarbazole-biphenyl (CBP) [9] was chosen as the host for Ir(ppy)$_3$. In previous work [2], it was found that CBP formed a suitable host for the red phosphor 2,3,7,8,12,13,17,18-ocatethyl-21H,23H-porphine platinum (PtOEP).

A proposed energy level [10] diagram, together with the molecular structural formulas of some of the materials used in the OLEDs, is shown in Fig. 5.1. Organic layers were deposited by high-vacuum (10^{-6} Torr) thermal evaporation onto a cleaned glass substrate precoated with transparent, conductive indium–tin–oxide. A 400-Å-thick layer of 4,4'-bis[N-(1-napthyl)-N-phenyl-amino] biphenyl (α-NPD) is used to transport holes to the luminescent layer consisting of Ir(ppy)$_3$ in CBP. A 200-Å-thick layer of the electron transport material tris-(8-hydroxyquinoline) aluminum (Alq$_3$) is used to transport electrons into the Ir(ppy)$_3$:CBP layer, and to reduce Ir(ppy)$_3$ luminescence absorption at the cathode. A shadow mask with 1-mm-diam openings was used to define the cathode consisting of a 1000-Å-thick layer of 25:1 Mg:Ag, with a 500-Å-thick Ag cap. As previously [2], we found that a thin (60 Å) barrier layer of 2,9-dimethyl-4,7-diphenyl-1,10-phenanthroline (bathocuproine, or BCP) [11] inserted between the CBP and the Alq$_3$ was necessary to confine excitons within the luminescent zone and hence maintain high efficiencies. In Ref. [2], it was argued that this layer prevents triplets from diffusing outside of the doped region. It was also suggested that CBP may readily transport holes and that BCP may be required to force exciton formation within the luminescent layer. Transient studies of triplet exciton diffusion within CBP are

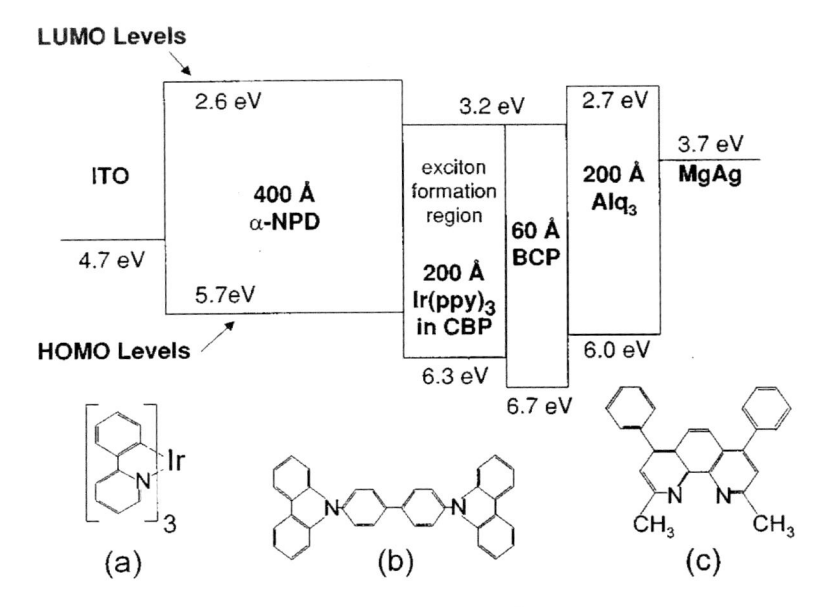

Figure 5.1 Proposed energy level structure of the electrophosphorescent device. The highest occupied molecular orbital (HOMO) energy and the lowest unoccupied molecular orbital (LUMO) energy are also shown (see Ref. [10]). Note that the HOMO and LUMO levels for Ir(ppy)$_3$ are unknown. The inset shows the chemical structural formulas of (a) Ir(ppy)$_3$, (b) CBP, and (c) BCP.

currently in progress in an effort to resolve this question. In either case, the use of BCP in the structure in Fig. 5.1 clearly serves to trap excitons within the luminescent region.

Figure 5.2 shows the external quantum efficiencies of several Ir(ppy)$_3$-based OLEDs. The doped structures exhibit a slow decrease in quantum efficiency with increasing current. Similar to the results for the Alq$_3$:PtOEP system, the doped devices achieve a maximum efficiency ($\sim$8%) for mass ratios of Ir(ppy)$_3$:CBP of approximately 6%–8%. Thus, the energy transfer pathway in Ir(ppy)$_3$:CBP is likely to be similar to that in PtOEP:Alq$_3$ [1, 2], i.e., via short-range Dexter transfer of triplets from the host. At low Ir(ppy)$_3$ concentrations, the lumophores often lie beyond the Dexter transfer radius of an excited Alq$_3$ molecule, while at high concentrations, aggregate quenching is increased. Note that dipole–dipole (Förster) transfer is forbidden for

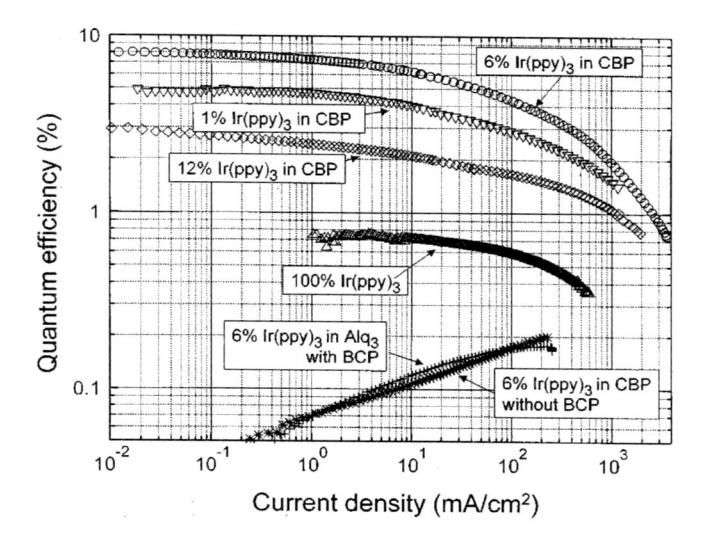

Figure 5.2 The external quantum efficiency of OLEDs using Ir(ppy)$_3$:CBP luminescent layers. Peak efficiencies are observed for mass ratio of 6% Ir(ppy)$_3$:CBP. The 100% Ir(ppy)$_3$ device has a slightly different structure than shown in Fig. 5.1: the Ir(ppy)$_3$ layer is 300 Å thick and there is no BCP blocking layer. The efficiency of a 6% Ir(ppy)$_3$:CBP device grown without a BCP layer is also shown.

triplet transfer, and in the PtOEP:Alq$_3$ system direct charge trapping was not found to be significant.

In addition to the doped device, we fabricated a heterostructure where the luminescent region was a homogeneous film of Ir(ppy)$_3$. The reduction in efficiency (to ~0.8%) of neat Ir(ppy)$_3$ is reflected in the transient decay, which has a lifetime of only ~100 ns, and deviates significantly from mono-exponential behavior. A 6% Ir(ppy)$_3$:CBP device without a BCP barrier layer is also shown together with a 6% Ir(ppy)$_3$:Alq$_3$ device with a BCP barrier layer. Here, very low quantum efficiencies are observed to increase with current. This behavior suggests a saturation of nonradiative sites as excitons migrate into the Alq$_3$, either in the luminescent region or adjacent to the cathode.

In Fig. 5.3, we plot luminance and power efficiency as functions of voltage. The peak power efficiency is 31 lm/W with a quantum efficiency of 8%, (28 cd/A). At 100 cd/m^2, a power efficiency of

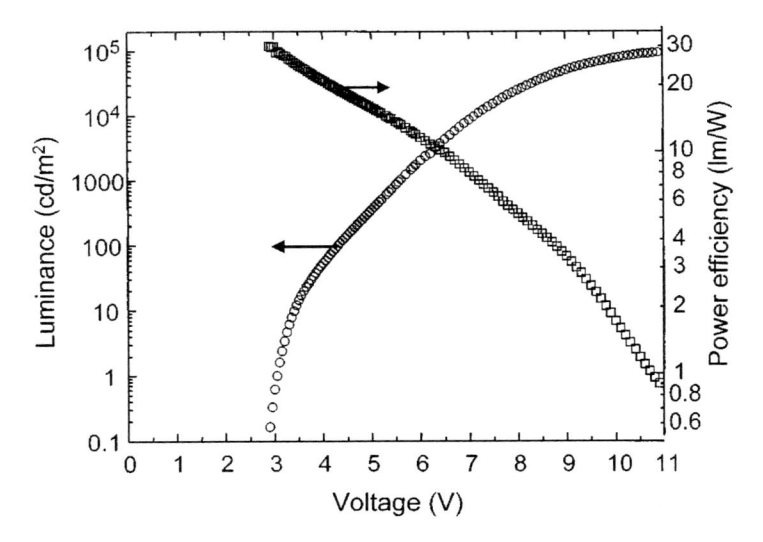

Figure 5.3 The power efficiency and luminance of the 6% Ir(ppy)$_3$:CBP device. At 100 cd/m^2, the device requires 4.3 V and its power efficiency is 19 lm/W.

19 lm/W with a quantum efficiency of 7.5% (26 cd/A) is obtained at a voltage of 4.3 V. The transient response of Ir(ppy)$_3$ in CBP is a mono-exponential phosphorescent decay of ∼500 ns, compared with a measured lifetime [5, 6, 8] of 2 μs in degassed toluene at room temperature. These lifetimes are short and indicative of strong spin orbit coupling, and together with the absence of Ir(ppy)$_3$ fluorescence in the transient response, we expect that Ir(ppy)$_3$ possesses strong intersystem crossing from the singlet to the triplet state. Thus, all emission originates from the long-lived triplet state. Unfortunately, slow triplet relaxation can form a bottleneck in electrophosphorescence and one principal advantage of Ir(ppy)$_3$ is that it possesses a short triplet lifetime. The phosphorescent bottleneck is thereby substantially loosened. This results in only a gradual decrease in efficiency with increasing current, leading to a maximum luminance of ∼100 000 cd/m^2.

In Fig. 5.4, the emission spectrum and Commission Internationale de L'Eclairage (CIE) coordinates of Ir(ppy)$_3$ are shown for the highest efficiency device. The peak wavelength is $\lambda = 510$ nm,

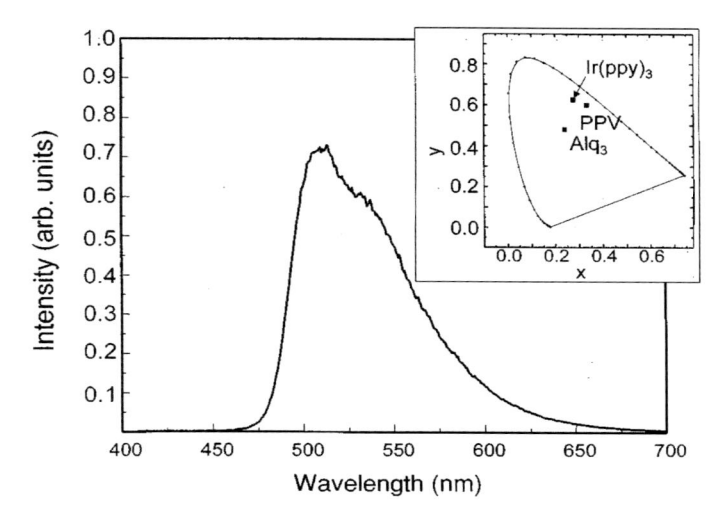

Figure 5.4 The electroluminescent spectrum of 6% Ir(ppy)$_3$:CBP. Inset: the Commission Internationale de L'Eclairage (CIE) chromaticity coordinates of Ir(ppy)$_3$ in CBP are shown relative to the fluorescent green emitters Alq$_3$ and poly(p-phenylenevinylene) (PPV).

and the full width at half maximum is 70 nm. The spectrum and CIE coordinates ($x = 0.27$, $y = 0.63$) are independent of current. Even at very high current densities ($\sim$100 mA/cm^2), blue emission from CBP is negligible—an indication of complete energy transfer.

We note that the device structure has the potential for further optimization. For example, the use of LiF cathodes [12, 13], shaped substrates [14], and novel hole transport materials [15] that result in a reduction in operating voltage or increased quantum efficiency are also applicable to this work. These methods have yielded power efficiencies of $\sim$20 lm/W in fluorescent small molecule devices [15]. The quantum efficiencies in those devices [16] at 100 cd/m^2 is typically $\leq$5%, and hence green-emitting electrophosphorescent devices with power efficiencies of $>$40 lm/W can be expected.

Given the performance advantage inherent to phosphorescence, new phosphors deserve intensive investigation. It has been noted [17] that, since triplet emission is typically red-shifted from singlet emission, it may be difficult to find guest/host systems where the phosphorescent guest emits in the blue or green. Indeed, although fluorescent energy transfer in the blue has been demonstrated using CBP and perylene [18], the additional exchange energy loss must be overcome at some stage in energy transfer to the triplet state of the phosphorescent guest. This may be a problem in the blue, where wide-gap (~3.5 eV) host materials are necessary. However, it is clear from this work that the efficiency improvements offered by phosphorescence outweigh the slight increase in voltage that results from the use of large-energy-gap materials. The alternative is to employ the phosphorescent material as an undoped film, with an attendant loss in efficiency.

Of the few phosphorescent compounds investigated to date, the purely organic materials [19] possess insufficient spin orbit coupling to show strong phosphorescence at room temperature. While one should not rule out the potential of purely organic phosphors, the most promising compounds may be transition-metal complexes with aromatic ligands. The transition metal mixes singlet and triplet states, thereby enhancing intersystem crossing and reducing the lifetime of the triplet excited state. As demonstrated in this work, reasonable photoluminescent efficiencies and lifetimes on the order of 1 μs are sufficient for high-performance devices.

Acknowledgements

This work was funded by the Universal Display Corporation, DARPA, AFOSR, and NSF. The authors also thank Peter Djurovich for helpful suggestions and discussions.

References

1. M. A. Baldo, D. F. O'Brien, Y. You, A. Shoustikov, S. Sibley, M. E. Thompson, and S. R. Forrest, *Nature (London)*, **395**, 151 (1998).
2. D. F. O'Brien, M. A. Baldo, M. E. Thompson, and S. R. Forrest, *Appl. Phys. Lett.*, **74**, 442 (1999).
3. S. A. VanSlyke and C. W. Tang, US Patent No. 4,539,507 (1985).
4. V. Cleave, G. Yahioglu, P. Le Barny, R. Friend, and N. Tessler, *Adv. Mater.*, **11**, 285 (1999).

5. K. A. King, P. J. Spellane, and R. J. Watts, *J. Am. Chem. Soc.*, **107**, 1431 (1985).
6. K. Dedeian, P. I. Djurovich, F. O. Garces, G. Carlson, and R. J. Watts, *Inorg. Chem.*, **30**, 1685 (1991).
7. M. G. Colombo, T. C. Brunold, T. Riedener, H. U. Gudel, M. Fortsch, and H.-B. Burgi, *Inorg. Chem.*, **33**, 545 (1994).
8. E. Vander Donckt, B. Camerman, F. Hendrick, R. Herne, and R. Vandeloise, *Bull. Soc. Chim. Belg.*, **103**, 207 (1994).
9. H. Kanai, S. Ichinosawa, and Y. Sato, International Conference on Electroluminescence of Molecular Materials and Related Phenomena, Fukuoka, Japan, 1997.
10. I. G. Hill and A. Kahn, *J. Appl. Phys.* (in press).
11. Y. Kijima, Spring MRS, San Francisco, 1998.
12. L. S. Hung, C. W. Tang, and M. G. Mason, *Appl. Phys. Lett.*, **70**, 152 (1997).
13. G. E. Jabbour, Y. Kawabe, S. E. Shaheen, J. F. Wang, M. M. Morrell, B. Kippelen, and N. Peyghambarian, *Appl. Phys. Lett.*, **71**, 1762 (1997).
14. G. Gu, D. Z. Garbuzov, P. E. Burrows, S. Venkatesh, S. R. Forrest, and M. E. Thompson, *Opt. Lett.*, **22**, 396 (1997).
15. B. Kippelen, G. E. Jabbour, S. E. Shaheen, M. M. Morrell, J. D. Anderson, P. Lee, E. Bellmann, S. Thayumanavan, S. Barlow, R. H. Grubbs, S. R. Marder, N. R. Armstrong, and N. Peyghambarian, Spring MRS, San Francisco, California, 1999.
16. J. Kido and Y. Iizumi, *Appl. Phys. Lett.*, **73**, 2721 (1998).
17. R. H. Friend, R. W. Gymer, A. B. Holmes, J. H. Burroughes, R. N. Marks, C. Taliani, D. D. C. Bradley, D. A. Dos Santos, J. L. Bredas, M. Logdlund, and W. R. Salaneck, *Nature (London)*, **397**, 121 (1999).
18. V. G. Kozlov, G. Parthasarathy, P. E. Burrows, and S. Forrest, *Appl. Phys. Lett.*, **72**, 144 (1998).
19. S. Hoshino and H. Suzuki, *Appl. Phys. Lett.*, **69**, 224 (1996).

Chapter 6

Organic Light-Emitting Devices Based on Phosphorescent Hosts and Dyes

Raymond C. Kwong, Sergey Lamansky, and Mark E. Thompson
Department of Chemistry, University of Southern California,
Los Angeles, California 90089, USA
met@usc.edu

Dye doping in organic light-emitting diodes (OLED) offers the ability of color tuning and efficiency improvement [1, 2]. Fluorescent dyes with high luminescence efficiencies have been extensively used for this purpose [2]. However, the quantum efficiency of an electrofluorescence device is limited by the low theoretical ratio of singlet excitons (25%) compared to triplet excitons (75%) upon electron–hole recombination from electrical excitation [3, 4]. Experimental results also concurred with this low singlet to triplet exciton ratio [5]. To investigate the concept of utilizing both singlet and triplet excitons, phosphorescent dye doping was proposed and several phosphorescent OLED systems have been studied so far [6–9], with the most impressive results being the highly efficient devices based on phosphorescent dyes such as platinum(II) porphyrins (maximum quantum efficiency, $QE_{\max} = 5\%$) [10–12]

Reprinted from *Adv. Mater.*, **12**(15), 1134–1138, 2000.

Electrophosphorescent Materials and Devices
Edited by Mark E. Thompson
Text Copyright © 2000 WILEY-VCH Verlag GmbH
Layout Copyright © 2024 Jenny Stanford Publishing Pte. Ltd.
ISBN 978-981-4877-34-3 (Hardcover), 978-1-003-08872-1 (eBook)
www.jennystanford.com

and iridium(III) tris(2-phenylpyridine) [Ir(ppy)$_3$] (QE_{max} = 8%) [13].

A general phenomenon among OLEDs with phosphorescent dyes doped into fluorescent host materials is the rapid decrease of efficiency at high current densities. It is postulated that the effects of triplet–triplet annihilation and saturation of emissive phosphorescent dopants at high triplet exciton population contribute to the decrease [12]. These two effects are both affected by the lifetime of the dopant. In a (fluorescent host)–(phosphorescent dye) OLED, the host material has a short singlet exciton lifetime compared to the phosphorescent dopant. Saturation of triplet sites takes place at high current densities. Because the singlet excitons of the host are short lived, they relax before they can diffuse to a dopant molecule in its ground state, leading to host emission. The emission spectrum of the host is usually very different from the dye, leading to a mixed or impure device color. The outcomes of saturation of emissive phosphorescent sites and triplet–triplet annihilation are best exemplified in the device with platinum(II) 2,8,12,17-tetraethyl-3,7,13,18-tetramethylporphyrin (PtOX, lifetime = 39 µs) doped into tris(8-hydroxyquinoline) aluminum (Alq$_3$, lifetime = 13 ns) [14]. In the reported device, the doped Alq$_3$ layer is both the electron transporting and emitting layer. The maximum quantum efficiency is 1.5% at 3.2 mA/cm^2, but the efficiency drops to 1.3% at 10 mA/cm^2, and further to only 0.5% at 100 mA/cm^2 [12]. The emission color from the device changes from saturated red at low current densities to orange at high currents due to increasing Alq$_3$ emission. The extent of saturation can be reduced by increasing the doping level. However, at high doping levels, dopant concentration quenching and triplet–triplet exciton annihilation (at high current densities) significantly decrease the device efficiency. As a result, optimization of electrophosphorescent devices relies on balancing the saturation problem at low doping levels with the concentration quenching or triplet–triplet annihilation, which occurs at high doping levels. It has been shown that 6% doping in PtOX in Alq$_3$ offers the best device efficiency and color purity [10, 12, 13].

If the lifetime difference between the host and the dye exciton is reduced, either by using a shorter lived dye or a longer lived host, it is possible to reduce the extent of the saturation of the dopant triplet emissive sites. The former solution has been realized in the case where Ir(ppy)$_3$ with a relatively short triplet lifetime (500 ns in 4,4'-biscarbazolybiphenyl, CBP) is used as a dopant [13]. The latter strategy can be investigated by using long lived phosphorescent hosts. Triplet–triplet annihilation would be a problem for a thin film of this phosphorescent host if it were used alone as the emitting material, however, a lower energy dopant will effectively quench the host triplet. The quenching process will lead to a significantly deceased lifetime for the doped host material, largely eliminating the annihilation problem, as observed for Ir(ppy)$_3$ in CBP. The lifetime of this doped phosphorescent host will still be long relative to a fluorescent host, so the saturation problem should be lessened. Here we report the employment of phosphorescent host materials doped with phosphorescent emitters. To date, the only reported example of phosphorescent host in OLEDs was the use of benzophenone derivatives as phosphorescent hosts for a Eu[III] diketonate as the dopant [15]. Owing to the low phosphorescence efficiency of the benzophenone hosts, triplet energy transfer could not be studied unless the device was cooled to 77 K. Herein we report OLEDs with room temperature phosphorescent hosts, so that direct triplet energy transfer can be observed and studied under ambient conditions.

Three types of organic light-emitting devices were fabricated as the following:

- Device 1. Indium tin oxide (ITO)α-NPD(400 Å)/Irq$_3$-PtOX(450 Å)/Irq$_3$(50 Å)/Mg:Ag/Ag,
- Device 2. ITO/α-NPD(400 Å)/Ir(ppy)$_3$-PtOX(450 Å)/Ir(ppy)$_3$(50 Å)/Mg:Ag/Ag,
- Device 3. ITO/α-NPD(400 Å)/Ir(ppy)$_3$-PtOX/BCP (60/120 Å)/Alq$_3$(200 Å)/Mg:Ag/Ag (BCP: 2,9-dimethyl-4,7-diphenyl-1,10-phenanthroline).

Figure 6.1 shows the band diagram for Device 3 and the structures of the compounds used. All devices consist of a 400 Å thick hole-transporting layer with 4,4'-bis[N-(1-naphthyl)-N-phenyl-amino]biphenyl (α-NPD) on top of the ITO anode, and a cathode composed of 600 Å of Mg:Ag (10:1) capped with 300 Å of Ag. The difference between the devices is the intervening layers.

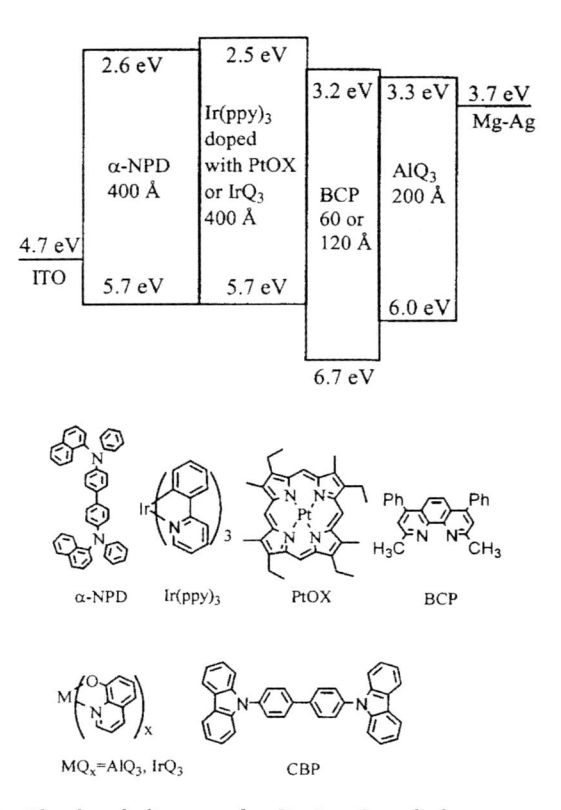

Figure 6.1 The band diagram for Device 3 and the structures of the compounds used.

The doping level of the phosphorescent dye PtOX is 6% for all devices. In Device 1, a 450 Å thick layer of electron transporting and emissive layer consisting of Irq$_3$ (tris(8-hydroxyquinoline) iridium) between the cathode and the hole transporting layer. Alq$_3$ has been widely used as an electron transporting material [16], while Irq$_3$ has not been used in OLEDs. Alq$_3$ shows a quasi-reversible reduction peak at -1.8 V vs saturated calomel electrode (SCE) in cyclic voltammetric scan [17]. Irq$_3$ is isostructural with Alq$_3$, and shows a reversible reduction wave at -1.75 V [14]. The lowest unoccupied molecular orbital (LUMO) levels of metal 8-hydroxyquinolate complexes are localized on the quinolate ligand [18], consistent with the similarity in the reduction potentials for the two complexes. Irq3 is phosphorescent at room temperature in solution with a luminescence yield of 0.009 (λ_{max} = 660 nm) [14]. Irq3 is also phosphorescent in a neat thin film, with λ_{max} of emission slightly blue shifted to 650 nm.

In order to investigate the use of a phosphorescent electron transport layer (ETL) OLEDs were prepared with an Irq3 electron transporting and emissive layer (Device 1). Both undoped and PtOX doped Irq3 films were used. The emission spectra of Irq3 and PtOX are similar. Energy transfer between materials with similar emission spectra has been observed previously for fluorescence based systems (e.g., coumarin-6 doped into Alq3), [19] but has not been examined for phosphorescence based OLEDs. The hope is that energy transfer from Irq3 to PtOX will be observed, which would substantially increase the quantum efficiency of the device, due to the marked increase in the photoluminescence yield of PtOX relative to Irq3 (ϕ_{PL} = 0.44 [12] and 0.009, respectively). The device current–voltage profiles in both the undoped and 6 mol.-% PtOX doped devices are similar to that of the well-studied α-NPD/Alq3 device, with turn-on voltages ~4 V. No emission from α-NPD was detected, suggesting exclusive emission from the heavy metal quinolate. The quantum efficiencies are low (~0.002%), consistent with the intrinsically low phosphorescence yield of Irq3. The electroluminescence spectra (Fig. 6.2) of the both undoped and doped devices are identical. No emission is observed from the phosphorescent dopant, PtOX, which exhibits very narrow electroluminescence centered at 648 nm (full width at half maximum, FWHM = 18 nm). Apparently, the Irq3 exciton energy is too low to effectively transfer to the PtOX dopant, leading to exclusive emission from Irq3 and identical quantum efficiencies for the doped and undoped devices. It is interesting to note that the quantum efficiencies of these Irq3 based devices do not decrease significantly as the current density is raised. Platinum(II) 2,3,7,8,12,13,17,18-octaethylporphyrin (PtOEP) and PtOX based OLEDs typically show a sharp decrease in their quantum efficiencies as the current is raised, due to increased quenching via triplet–triplet annihilation at higher current densities [10–12]. This is not observed in the α-NPD/Irq3 device, which shows a constant quantum efficiency at 0.002% from 10 mA/cm^2 to 450 mA/cm^2. The short lifetime of the Irq3 exciton makes triplet–triplet

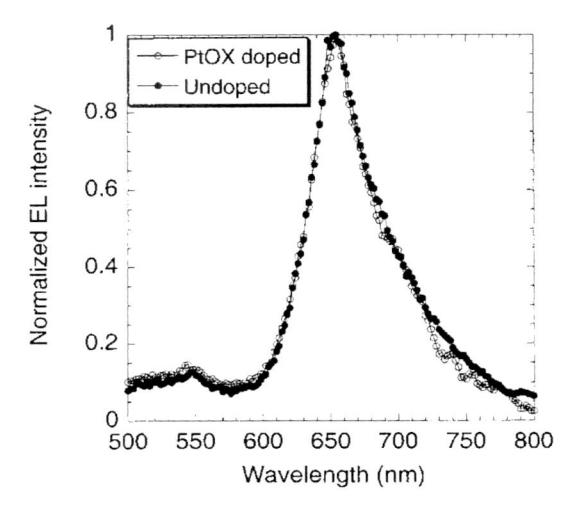

Figure 6.2 Electroluminescence spectra of the devices α-NPD(400 Å)/Irq$_3$–PtOX(450 Å)/Irq$_3$(50 Å) and α-NPD(400 Å)/Irq$_3$(450 Å)/Irq$_3$(50 Å) devices.

annihilation an inefficient quenching process, even at comparatively high currents.

In order to facilitate triplet energy transfer to PtOX from the host, the host triplet energy should be higher than that of PtOX. Device 2 consists of the green phosphorescent Ir(ppy)$_3$ as the host. A device with the structure ITO/α-NPD/Ir(ppy)$_3$/MgAg has been studied previously, and it showed exclusive Ir(ppy)$_3$ triplet emission centered at 510 nm [13]. PtOX was doped at 6 mol.-% into Ir(ppy)$_3$. The resulting electroluminescent color was red, with the majority of the emission from PtOX and a smaller amount of emission from Ir(ppy)$_3$, as expected. The quantum efficiency of this device is low, $QE_{max} = 0.02\%$, see Fig. 6.3. For comparison, a device with the structure ITO/α-NPD/Alq$_3$-PtOX/MgAg has a maximum quantum efficiency of 1.5% at the same current density. The low efficiency may be due to hole leakage in the NPD/Ir(ppy)$_3$ device. The oxidation potential of Ir(ppy)$_3$ is 0.7 V [20]. The highest occupied molecular orbital (HOMO) level of Ir(ppy)$_3$ aligns well with that of α-NPD. Holes may be conducted through the Ir(ppy)$_3$ layer to the cathode, leading to a carrier imbalance, and thus a very low charge recombination efficiency.

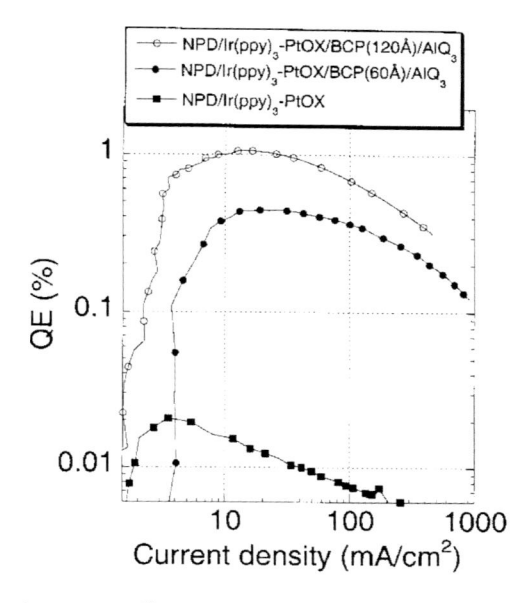

Figure 6.3 Quantum efficiency vs. current density of the devices α-NPD(400 Å)/Ir(ppy)$_3$–PtOX/Ir(ppy)$_3$(50 Å) and α-NPD(400 Å)/Ir(ppy)$_3$–PtOX(400 Å)/BCP(60/120 Å)/Alq$_3$(200 Å).

The value of confining exciton formation and diffusion has been shown to significantly enhance OLED efficiency with phosphorescent dopants [11, 13]. This strategy is also employed in the present studies. The difference between Device 2 and Device 3 is that the latter contains a 60 or 120 Å layer of BCP and a 200 Å layer of Alq$_3$ on top of the doped Ir(ppy)$_3$ layer. The low HOMO level of BCP is well below that of Ir(ppy)$_3$, preventing hole injection from the Ir(ppy)$_3$ into the Alq$_3$ layer. Electrons are injected into the Ir(ppy)$_3$ layer, leading to charge recombination in the Ir(ppy)$_3$ layer. The excitons formed are confined in this layer, since the energy of the Ir(ppy)$_3$ exciton is significantly lower than the energy gaps of α-NPD and BCP. Thus, the charge recombination efficiency is increased and exciton quenching is reduced. The turn-on voltage of the device (∼5 V) is slightly higher than when BCP is absent (∼4 V) (Fig. 6.4a). The quantum efficiency of this device with 120 Å thick of BCP is 1.1%, 50 times higher than Device 2 (Fig. 6.3). The electroluminescent spectra (Fig. 6.4b) show dominant PtOX emission with narrow linewidth

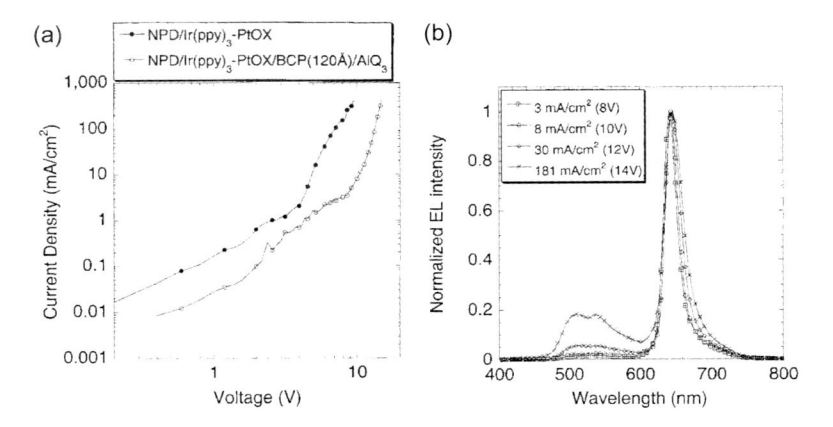

Figure 6.4 (a) Current–voltage profiles of the Ir(ppy)₃ based devices α-NPD (400 Å)/Ir(ppy)₃–PtOX/Ir(ppy)₃(50 Å) and α-NPD(400 Å)/Ir(ppy)3–PtOX(400 Å)/BCP(60/120 Å)/Alq₃(200 Å). (b) Electroluminescene spectra of the Ir(ppy)₃ based devices of the devices α-NPD(400 Å)/Ir(ppy)₃–PtOX(400 Å)/BCP(120 Å)/Alq₃(200 Å).

centered at 648 nm. Residual emissions from at 480–560 nm, grow as the current increases. This is attributed to the emission from Ir(ppy)₃. Obviously, the saturation of the long lived PtOX emissive sites in the present Ir(ppy)₃–PtOX system still occurs. It would be of interest to compare the effect of host exciton lifetime on the saturation of emissive sites in the Ir(ppy)₃–PtOX and Alq₃–PtOX systems The lifetime for neat Ir(ppy)₃ film is 100 ns while PtOX has a lifetime of 68 μs when doped in Ir(ppy)₃. The lifetimes of Alq₃ and PtOX doped into Alq₃ are 13 ns and 39 μs respectively. Similar to the Alq₃–PtOX system, the amount of Ir(ppy)₃ host emission increases with increasing current density. Although the host Ir(ppy)₃ is a much longer lived phosphorescent material than the fluorescent Alq₃, it is still not long lived enough to reduce the saturation of PtOX emissive sites.

In conclusion, we have fabricated efficient organic light-emitting devices with phosphorescent host materials. Triplet energy transfer from the host Ir(ppy)₃ to the red phosphorescent dopant PtOX can be observed directly. Similar to the Alq₃ based devices in which Alq₃ is a singlet short lived emitter, the PtOX doped Ir(ppy)₃ device also shows saturation of the PtOX triplet emissive sites, since the lifetime of Ir(ppy)$_3$ triplet exciton is still far shorter than the exciton lifetime of the phosphorescent PtOX. Further studies on phosphorescent host-guest systems in which the host and dye exciton lifetimes are more closely matched are necessary to further examine the issue of saturation of triplet emissive sites. Nevertheless, our demonstration of efficient phosphorescent host–guest system has opened up new opportunities in OLED research.

Experimental

Materials: Alq$_3$ [21], Irq$_3$ [14], Ir(ppy)$_3$ [20], PtOX [12], and α-NPD [22] and CBP were synthesized according to the literature procedures and purified by vacuum sublimation. BCP was purchased from Aldrich and used as received.

Device Fabrication and Testing: Patterned anode contacts were made from commercially available ITO coated glass (10 Ω/□) in our laboratory using a photolithographic technique to form circular anodes with 1 mm radii. The OLEDs were prepared on the prepatterned substrates as previously described [12]. Organic film depositions were carried out at ~1.5 Å/s. For doped films, the dopant was sublimed from an independently controlled source at a rate of ~0.1 Å/s at the substrate, to give a doping level of 6 mol.-%. Device current–voltage and light intensity characteristics were measured with a Keithley 2400 SourceMeter/2000 Multimeter coupled to a Newport 1835-C Optical Meter using the LabVIEW program by National Instruments. Both photo- and electroluminescence spectra were recorded at room temperature on a PTI QuantaMaster Model C-60 Spectrofluorometer. Lifetime measurements were carried out as previously described [12].

Acknowledgements

We thank Universal Display Corporation, the Defense Advanced Research Projects Agency and the National Science Foundation for financial support of this work. We also thank Prof. Stephen Forrest and Marc Bald of or helpful discussion.

References

1. C. H. Chen, J. Shi, C. W. Tang, *Macromol. Symp.* 1997, **125**, 1. S. P. Sibley, M. E. Thompson, P. E. Burrows, S. R. Forrest, *Optoelectronic Properties of Inorganic Compounds* (eds: D. M. Roundhill, J. Fackler), Plenum, New York, 1999, Ch. 2.
2. A. A. Shoustikov, Y. You, M. E. Thompson, *IEEE J. Selected Topics Quantum Electron.*, 1998, **4**, 3.
3. L. J. Rothberg, A. J. Lovinger, *J. Mater. Res.*, 1996, **11**, 3174.
4. M. Pope, C. E. Swenberg, *Electroinc Processes in Organic Crystals*, Clarendon, Oxford, 1982.
5. M. A. Baldo, D. E. O'Brien, V. Bulovic, M. E. Thompson, S. R. Forrest, *Phys. Rev. Lett.*, in press.
6. S. Hoshino, H. Suzuki, *Appl. Phys. Lett.*, 1996, **69**, 224.
7. Y. Ma, H. Zhang, J. Shen, C. Che, *Synth. Met.*, 1998, **94**, 245.
8. Y. Ma, X. Zhou, J. Shen, H.-Y Chao, C.-M. Che, *Appl. Phys. Lett.*, 1999, **74**, 1361.
9. Y. Ma, C.-M. Che, H.-Y. Chao, X. Zhou, W.-H. Chan, J. Shen, *Adv. Mater.*, 1999, **11**, 852.
10. M. A. Baldo, D. F. O'Brien, Y. You, A. Shoustikov, S. Sibley, M. E. Thompson, S. R. Forrest, *Nature*, 1998, **395**, 151.
11. D. F. O'Brien, M. A. Baldo, M. E. Thompson, S. R. Forrest, *Appl. Phys. Lett.*, 1999, **74**, 442.
12. R. C. Kwong, S. Sibley, T. Dubovoy, M. A. Baldo, S. R. Forrest, M. E. Thompson, *Chem. Mater.*, 1999, **11**, 3709.
13. M. A. Baldo, S. Lamansky, P. E. Burrows, M. E. Thompson, S. R. Forrest, *Appl. Phys. Lett.*, 1999, **75**, 4.
14. R. Ballardini, G. Varani, M. T. Indelli, F. Scandola, *Inorg. Chem.*, 1986, **25**, 3858.
15. Y. Miyamoto, M. Uekawa, H. Ikeda, K. Kaifu, *J. Lumin.*, 1999, **81**, 159.
16. M. Takeuchi, H. Masui, I. Kikuma, M. Masui, T. Muranoi, T. Wada, *Jpn. J. Appl. Phys.*, 1992, **31**, L498. D. J. Fatemi, H. Murata, C. D. Merritt, Z. H. Kafafi, *Synth. Met.*, 1997, **85**, 1225. A. Shoustikov, Y. You, P. E. Burrows, M. E. Thomspon, S. R. Forrest, *Synth. Met.*, 1997, **91**, 217.
17. J. D. Anderson, E. M. McDonald, P. A. Lee, M. L. Anderson, E. L. Ritchie, H. K. Hall, T. Hopkins, E. A. Mash, J. Wang, A. Padias, S. Thayumanavan, S. Barlow, S. R. Marder, G. E. Jabbour, S. Shaheen, B. Kippelen, N. Peyghambarian, R. M. Wightman, *J. Am. Chem. Soc.*, 1998, **120**, 9646.

18. A. Curioni, M. Boero, W. Andreoni, *Chem. Phys. Lett.*, 1998, **294**, 263. P. E. Burrows, Z. Shen, V. Bulovic, D. M. McCarty, S. R. Forrest, J. Cronin, M. E. Thompson, *J. Appl. Phys.*, 1996, **79**, 7991.
19. C. W. Tang, S. A. VanSlyke, C. H. Chen, *J. Appl. Phys.*, 1989, **65**, 3610.
20. K. Dedeian, P. I. Djurovich, F. O. Garces, G. Carlson, R. J. Watts, *Inorg. Chem.*, 1991, **30**, 1685.
21. H. Schmidbauer, J. Lettenbauer, O. Kumberger, J. Lachmann, G. Muller, *Z. Naturforsch.*, 1991, **46b**, 1065.
22. M. S. Drive, J. F. Hartwig, *J. Am. Chem. Soc.*, 1996, **118**, 7217. J. P. Wolre, S. Wagaw, S. L. Buchwald, *J. Am. Chem. Soc.*, 1996, **118**, 7215.

Chapter 7

High-Efficiency Organic Electrophosphorescent Devices with tris(2-Phenylpyridine)Iridium Doped into Electron-Transporting Materials

Chihaya Adachi,[a] Marc A. Baldo,[a] Stephen R. Forrest,[a] and Mark E. Thompson[b]

[a] *Center for Photonics and Optoelectronic Materials (POEM), Department of Electrical Engineering, Princeton University, Princeton, New Jersey 08544, USA*
[b] *Department of Chemistry, University of Southern California, Los Angeles, California 90089, USA*
met@usc.edu

We demonstrate high-efficiency organic light-emitting devices employing the green electrophosphorescent molecule, *fac* tris(2-phenylpyridine)iridium [Ir(ppy)$_3$], doped into various electron-transport layer (ETL) hosts. Using 3-phenyl-4-(1'-naphthyl)-5-phenyl-1,2,4-triazole as the host, a maximum external quantum efficiency (η_{ext}) of 15.4 ± 0.2% and a luminous power efficiency of 40 ± 2 lm/W are achieved. We show that very high internal quantum efficiencies (approaching 100%) are achieved for organic phosphors

---

Reprinted from *Appl. Phys. Lett.*, **77**(6), 904–906, 2000.

Electrophosphorescent Materials and Devices
Edited by Mark E. Thompson
Text Copyright © 2000 American Institute of Physics
Layout Copyright © 2024 Jenny Stanford Publishing Pte. Ltd.
ISBN 978-981-4877-34-3 (Hardcover), 978-1-003-08872-1 (eBook)
www.jennystanford.com

with low photoluminescence efficiencies due to fundamental differences in the relationship between electroluminescence from triplet and singlet excitons. Based on the performance characteristics of single and double heterostructures, we conclude that exciton formation in Ir(ppy)$_3$ occurs within close proximity to the hole-transport layer/ETL:Ir(ppy)$_3$ interface.

The recent demonstration [1, 2] of high-efficiency organic light-emitting devices (OLEDs) using the electrophosphorescent molecules 2,3,7,8,12,13,17,18-octaethyl-21H,23H-porphine platinum (PtOEP) and *fac*tris(2-phenylpyridine)iridium (Ir(ppy)$_3$) led to the prospect of obtaining devices with internal quantum efficiencies (η_{int}) of 100% through radiative recombination of both singlet and triplet excitons. In previous studies [2, 3], an external quantum efficiency (η_{ext}) of 8% (corresponding to $\eta_{ext} \sim 40\%$) using a light-emitting layer (EML) comprised of Ir(ppy)$_3$ doped into a 4,4'-N,N'-dicarbazole-biphenyl (CBP) host was reported. This remarkable result was ascribed [4] to bipolar carrier transport in CBP along with a favorable triplet energy level alignment between the host and the dopant which promotes efficient energy transfer between species. In this study, we describe high-efficiency electrophosphorescent OLEDs employing an electron-transport layer (ETL) as a host. The ETL materials are 2,9-dimethyl-4,7-diphenyl-phenanthroline (BCP) [5], 1,3-bis(N,N-t-butyl-phenyl)-1,3,4-oxadiazole (OXD7) [6], and 3-phenyl-4-(1'-naphthyl)-5-phenyl-1,2,4-triazole (TAZ) [7]. It has been previously established that these materials possess good electron-transport characteristics while also serving to block hole and exciton transport [5–7]. Our experiments are consistent with η_{int} approaching 100%, suggesting that future work on increasing efficiency will realize the largest gains by focusing on improving light out-coupling from the OLED structure.

Organic layers were deposited by high-vacuum (10^{-6} Torr) thermal evaporation onto a clean glass substrate precoated with an indium–tin–oxide (ITO) layer with a sheet resistance of $\sim$20 $\Omega/\square$. A 60-nm-thick film of 4,4'-bis[N,N'-(3-tolyl)amino]-3,3'-dimethylbiphenyl [8] (HMTPD) served as the hole-transport layer (HTL). Next, a 25-nm-thick EML consisting of 6%–8% Ir(ppy)$_3$ was doped into various electron-transporting hosts via thermal codeposition. A 50-nm-thick layer of tris-(8-hydroxyquinoline)aluminum

(Alq$_3$) was used to transport and inject electrons into the EML. A shadow mask with 1-mm-diam openings was used to define the cathode consisting of a 150-nm-thick Mg–Ag layer, with a 20-nm-thick Ag cap. Alternatively, the cathode consisted of a 100-nm-thick layer of Al–0.56 wt% Li.

Current density (J) versus voltage (V) measurements were obtained using a semiconductor parameter analyzer, with the luminance obtained by placing the OLEDs directly onto the surface of a large-area calibrated silicon photodiode, thus avoiding corrections needed to account for non-Lambertian spatial emission patterns [3]. The photoluminescence (PL) and electroluminescence (EL) transient decays were characterized using a streak camera following excitation by a nitrogen laser at a wavelength of $\lambda = 337$ nm and a pulse width of $\sim$500 ps for PL, and by a pulse generator for EL.

A maximum $\eta_{ext} = 15.4 \pm 0.2\%$ and power efficiency of 40 $\pm$ 2 lm/W of the 7%-Ir(ppy)$_3$-doped TAZ device using an Al–Li cathode was obtained (see Fig. 7.1), corresponding to $\eta_{int} \cong 80\%$. The value of η_{ext} is almost double compared with that previously reported for Ir(ppy)$_3$ devices [2] with η_{int} now approaching 100%. The device exhibits a gradual decrease in quantum efficiency with increasing current, characteristic of triplet–triplet annihilation observed in all electrophosphorescent devices [11]. Nevertheless, a high optical output power of 2.5 mW/cm^2 (corresponding to a luminance of $\sim$4000 cd/m^2) with $\eta_{ext} = 10.0\%$ was maintained even at $J = 10$ mA/cm^2. At Ir(ppy)$_3$ concentrations less than 2%, we observed a decrease to $\eta_{ext} \sim 3\%$ along with additional blue host emission ($\lambda \sim 440$ nm), while at high Ir(ppy)$_3$ concentrations, a significant decrease in η_{ext} was also observed due to aggregate-induced quenching.

We summarize the performance of Ir(ppy)$_3$-doped OLEDs with four hosts and cathode metals, including previously reported data for CBP as a host [2] in the inset of Fig. 7.1. The use of an Al–Li cathode in TAZ-based devices slightly enhances η_{ext}, reflecting the improved electron-injection efficiency of Li over that of Mg due to the comparatively low work function of Li. We observed that both TAZ and OXD7 showed comparably high values of η_{ext}, while devices with BCP and CBP hosts exhibited $\sim$30% lower efficiencies. As shown Fig. 7.2, the room-temperature transient phosphorescent

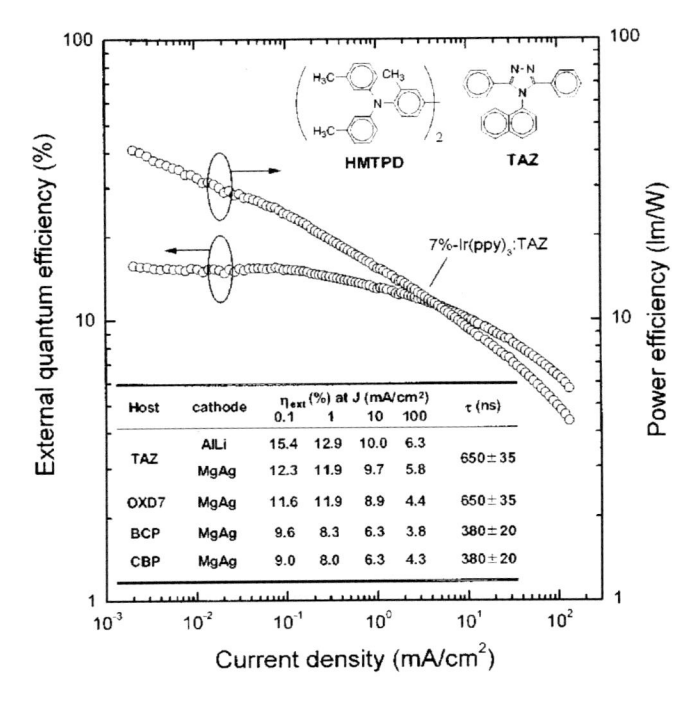

Host	cathode	η_{ext} (%) at J (mA/cm^2)				τ (ns)
		0.1	1	10	100	
TAZ	Al/Li	15.4	12.9	10.0	6.3	650±35
	MgAg	12.3	11.9	9.7	5.8	
OXD7	MgAg	11.6	11.9	8.9	4.4	650±35
BCP	MgAg	9.6	8.3	6.3	3.8	380±20
CBP	MgAg	9.0	8.0	6.3	4.3	380±20

Figure 7.1 External quantum and power efficiencies of an ITO/HMTPD(60 nm)/7%-Ir(ppy)$_3$:TAZ(25 nm)/Alq$_3$(50 nm)/Al–Li(100 nm) OLED. A maximum external quantum efficiency $\eta = 15.4\%$ and power efficiency of 40 lm/W were obtained. Insets: The chemical structures of HMTPD and TAZ. External quantum efficiencies (η_{ext}) and photoluminescence decay times (τ) for Ir(ppy)$_3$ doped into various hosts.

lifetime of 7%-Ir(ppy)$_3$:BCP in TAZ and OXD7 is $\tau = 650 \pm 35$ ns, compared with $\tau = 380 \pm 20$ ns in 7%-Ir(ppy)$_3$:BCP and 7%-Ir(ppy)$_3$:CBP. Since the phosphorescence efficiency is approximately proportional to the lifetime [9], the longer lifetime in TAZ and OXD7 is consistent with the higher EL efficiencies of these devices. We note that all of the Ir(ppy)$_3$-doped hosts showed one order of magnitude longer lifetime compared to that of a neat film in which Ir(ppy)$_3$–Ir(ppy)$_3$ exciton interactions, i.e., "self-quenching," increase the probability for nonradiative decay.

To elucidate the emission mechanisms leading to high efficiency, we varied the EML thickness from 2.5 to 30 nm while maintaining

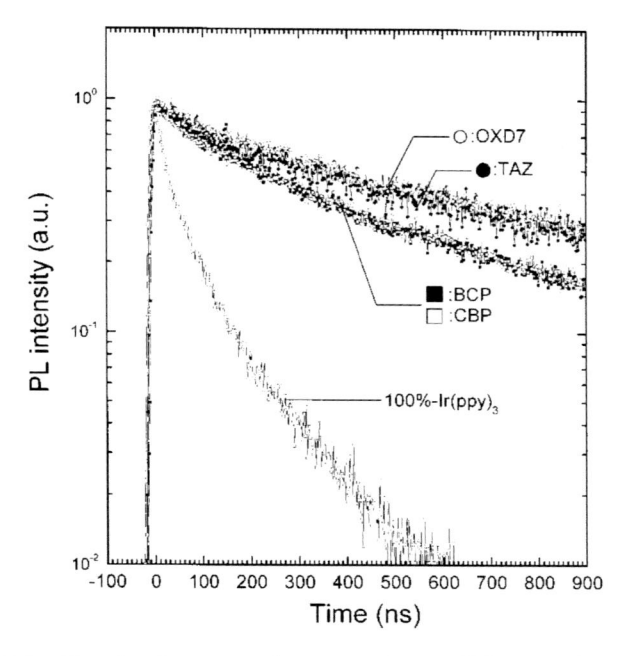

Figure 7.2 Photoluminescence decay transients of 7%-Ir(ppy)$_3$:OXD7(50 nm), 7%-Ir(ppy)$_3$:TAZ(50 nm), 7%-Ir(ppy)$_3$:BCP(50 nm), 7%-Ir(ppy)$_3$:CBP(50 nm), and 100%-Ir(ppy)$_3$(50 nm). Excitation power density is ~5 × 10^{16} cm^3.

both the HTL and ETL thicknesses at 50 and 40 nm, respectively. This device is a single heterostructure (SH) due to the barrier to charge carriers and excitons only at the HTL/EML interface. These devices were compared with a double heterostructure [10] (DH) comprised of an EML sandwiched between the HTL (HMTPD) and a 10-nm-thick neat ETL, as shown in the inset of Fig. 7.3. Figure 7.3 shows the thickness dependence of η_{ext} at a fixed current density of 0.1 mA/cm^2 for both SH and DH devices using BCP as a host. At an EML thickness of <15 nm, a significant decrease of η_{ext} was observed in the SH device, while η_{ext} = 9% was retained even with a 2.5-nm-thick EML in the DH device. This suggests the confinement of both charge carriers and triplet excitons within the very thin EML, characteristic of the DH architecture.

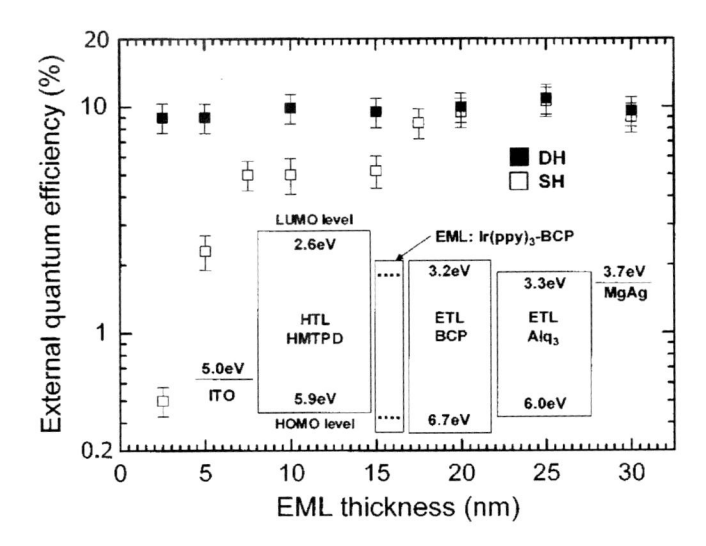

Figure 7.3 External quantum efficiency (η) of single (SH) and double heterostructure (DH) OLEDs as a function of light-emitting-layer thickness. The SH and DH structures are composed of ITO/HMTPD(50 nm)/7%-Ir(ppy)$_3$:BCP(variable)/Alq(40 nm)/MgAg(150 nm)/Ag(20 nm) and ITO/HMTPD(50 nm)/7%-Ir(ppy)$_3$:BCP(variable)/BCP(10 nm)/Alq(40 nm)/MgAg(150 nm)/Ag(20 nm), respectively. In the DH structure, a high η was retained even with a light-emitting-layer thickness of only 2.5 nm. Inset: Energy levels of the constituent materials used in a DH OLED as referenced to vacuum. The highest occupied molecular orbital (HOMO) and the lowest unoccupied molecular orbital (LUMO) energies are indicated (see Ref. [19]), and are unknown for Ir(ppy)$_3$ (dashed lines).

In both the SH and DH devices, the EL spectral shapes due to Ir(ppy)$_3$ triplet emission (with a peak wavelength of $\lambda_{max} = 515$ nm) are independent of the EML thickness. Even in the SH structure with a 2.5-nm-thick EML, the spectral shape is identical to that of the neat Ir(ppy)$_3$ PL spectrum, with a negligible contribution from Alq$_3$.

The EL transient decays of the three devices are shown in Fig. 7.4. The DH device with a 2.5-nm-thick EML (curve i) shows an EL transient decay time ($\tau = 320 \pm 15$ ns) comparable with that of the SH device ($\tau = 380 \pm 20$ ns) with a 30-nm-thick EML (curve ii). The slightly reduced lifetime of the DH device may be due to a high density of triplet excitons confined within a very narrow

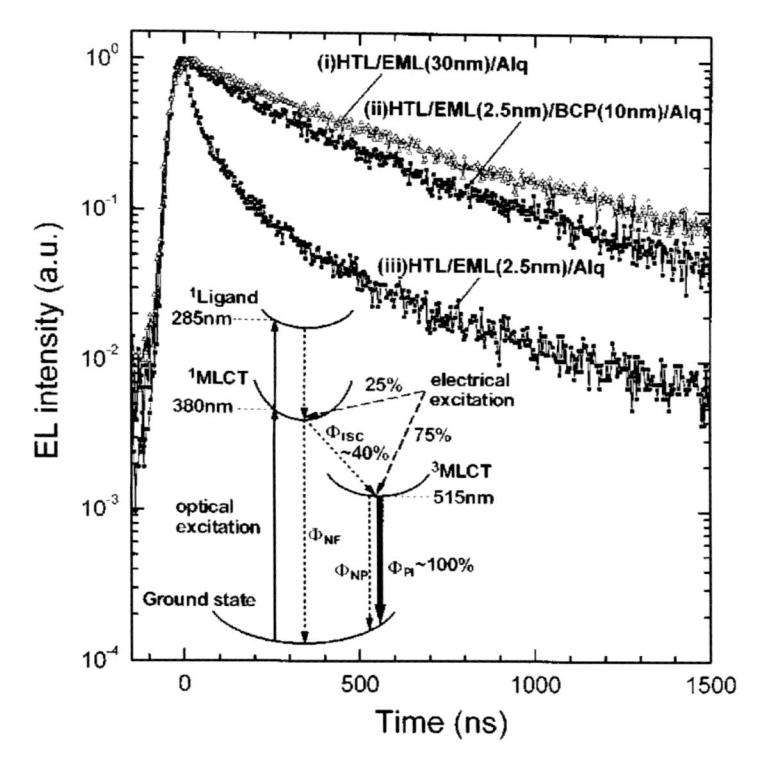

Figure 7.4 Transient electroluminescence decay of the following: (i) single heterostructure (SH): ITO/HMTPD(50 nm)/7%-Ir(ppy)$_3$:BCP(30 nm)/Alq(40 nm)/MgAg(150 nm)/Ag(20 nm), (ii) double heterostructure (DH): ITO/HMTPD(50 nm)/7%-Ir(ppy)$_3$:BCP(2.5 nm)/BCP(10 nm)/Alq(40 nm)/MgAg(150nm)/Ag(20 nm), and (iii) SH: ITO/HMTPD(50 nm)/7%-Ir(ppy)$_3$:BCP(2.5 nm)/Alq(40 nm)/MgAg(150 nm)/Ag(20 nm) under a 100 ns, 9 V pulse excitation. Inset: Energy-level diagram of Ir(ppy)$_3$. The ligand singlet state (^{1}Ligand) and metal-to-ligand charge-transfer singlet state (^{1}MLCT) were determined by the absorption peaks in toluene solution (10^{-5} M). Also, the triplet MLCT state (^{3}MLCT) was estimated from the phosphorescence peak. Φ_{NF}, Φ_{ISC}, Φ_{PI}, and Φ_{NP} are quantum yields for nonemissive transitions from ^{1}MLCT, intersystem crossing, intrinsic phosphorescent transitions, and nonemissive transitions from ^{3}MLCT, respectively.

EML, leading to enhanced triplet–triplet annihilation [11]. On the other hand, the SH device with a 2.5-nm-thick EML (curve iii) shows a very short lifetime ($\tau = 50 \pm 3$ ns), suggesting the presence of significant dissipative nonradiative transitions. We note, also, that in the thinnest EML SH structure, the Ir(ppy)$_3$ triplets are strongly quenched by the adjacent Alq$_3$ layer with its significantly less energetic, nonradiative triplet state. The DH structure consisting of HMTPD and BCP double blocking layers, in contrast, confines both charge carriers and triplet excitons even within a 2.5-nm-thick EML. The lack of blue fluorescence from either HMTPD or BCP suggests that leakage of electrons or holes into these adjacent layers is negligible. Finally, the time-resolved EL spectrum of this latter device is similar to neat Ir(ppy)$_3$, even during the initial $\sim$20 ns after excitation. This, too, indicates that exciton formation occurs entirely within the EML, with little or no carrier recombination in the adjacent Alq$_3$ layer.

Two possible mechanisms lead to light emission: direct charge trapping by Ir(ppy)$_3$ or energy transfer from BCP to Ir(ppy)$_3$ followed by carrier recombination by BCP. Since BCP serves as both an electron-transporting and hole-blocking material [5], holes can be directly injected from the HTL into the Ir(ppy)$_3$ highest occupied molecular orbital and trapped, where subsequently they combine with electrons which are transported across the BCP layer leading to direct exciton formation on Ir(ppy)$_3$. Alternatively, exciton formation may occur first on BCP, and then are subsequently transferred to Ir(ppy)$_3$ through Förster or Dexter processes, leading to efficient Ir(ppy)$_3$ emission. We are unable to determine which process is dominant since both occur within close proximity to the HTL/EML interface.

Finally, we discuss strategies for further improvement of EL efficiency of Ir(ppy)$_3$-doped devices. The inset of Fig. 7.4 shows the energy levels of an Ir(ppy)$_3$ molecule based on its absorption and emission spectra. We demonstrated here an internal efficiency of 80% by assuming that 20% of the emitted light is extracted from the structure [12]. However, the PL efficiency of Ir(ppy)$_3$ in a dilute solution is only 40% $\pm$ 10% [13]. In photoexcitation, phosphorescence takes place via intersystem crossing (ISC) from the metal to ligand charge-transfer singlet state (^{1}MLCT), since direct excitation to the triplet state (^{3}MLCT) is prohibited. Thus, the phosphorescence quantum yield (Φ_P) follows:

$$\Phi_P = \Phi_{ISC} \frac{\kappa_P}{\kappa_P + \kappa_{NP}}, \tag{7.1}$$

where Φ_{ISC} represents the probability for ISC, κ_P is the phosphorescence emission rate, and κ_{NP} that of nonemissive triplet decay. Under electrical excitation, both singlet and triplet excitons are directly created on either the guest or host molecules with a statistical splitting of $\chi \sim 25\%$ singlets and $(1-\chi) \sim 75\%$ triplets [14]. Thus, η_{int} follows:

$$\eta_{int} = [(1-\chi) + \chi \Phi_{ISC}] \frac{\kappa_P}{\kappa_P + \kappa_{NP}}. \tag{7.2}$$

For consistent interpretation of both the EL and PL data, we conclude for Ir(ppy)$_3$ that $\kappa_P \gg \kappa_{NP}$ and $\Phi_{ISC} \cong 40\%$. Since the intrinsic phosphorescence efficiency [$\kappa_P/(\kappa_P + \kappa_{NP})$] is already near to its maximum value, further efficiency enhancement is limited even if we have $\Phi_{ISC} = 100\%$. Hence, to further increase OLED efficiency beyond that obtained via electrophosphorescence, we must focus on schemes to increase light out-coupling by incorporating microcavities [15], shaped substrates [12, 16], or an index-matching medium [17, 18]. Also, phosphorescent materials with low-PL efficiencies appear to be useful in EL devices provided that the intrinsic phosphorescence efficiency is high.

Acknowledgements

The authors thank the Universal Display Corporation and the Defense Advanced Research Projects Agency for their support of this research. The authors thank M. Ohta at Ricoh Co. Ltd. for providing the hole-transport material.

References

1. M. A. Baldo, D. F. O'Brien, Y. You, A. Shoustikov, S. Sibley, M. E. Thompson, and S. R. Forrest, *Nature (London)*, **395**, 151 (1998).

2. M. A. Baldo, S. Lamansky, P. E. Burrows, M. E. Thompson, and S. R. Forrest, *Appl. Phys. Lett.*, **75**, 4 (1999).
3. T. Tsutsui, M.-J. Yang, M. Yahiro, K. Nakamura, T. Watanabe, T. Tsuji, Y. Fukuda, T. Wakimoto, and S. Miyaguchi, *Jpn. J. Appl. Phys.*, Part 2 **38**, L1502 (1999).
4. H. Kanai, S. Ichinosawa, and Y. Sato, *Synth. Met.*, **91**, 195 (1997).
5. H. Nakada, S. Kawami, K. Nagayama, Y. Yonemoto, R. Murayama, J. Funaki, T. Wakimoto, and K. Imai, *Polym. Prepr. Jpn.*, **35**, 2450 (1994).
6. Y. Hamada, C. Adachi, T. Tsutsui, and S. Saito, *Jpn. J. Appl. Phys.*, Part 1, **31**, 1812 (1992).
7. J. Kido, K. Hongawa, K. Okuyama, and K. Nagai, *Jpn. J. Appl. Phys.*, Part 2, **32**, L917 (1993).
8. C. Adachi, K. Nagai, and N. Tamoto, *Appl. Phys. Lett.*, **66**, 2679 (1995).
9. M. Klessinger and J. Michl, *Excited States and Photochemistry of Organic Molecules* (VCH, New York, 1995).
10. C. Adachi, T. Tsutsui, and S. Saito, *Appl. Phys. Lett.*, **57**, 531 (1990).
11. M. A. Baldo, C. Adachi, and S. R. Forrest (unpublished).
12. G. Gu, D. Z. Garbuzov, P. E. Burrows, S. Venkatesh, and S. R. Forrest, *Opt. Lett.*, **22**, 396 (1997).
13. K. A. King, P. J. Spellane, and R. J. Watts, *J. Am. Chem. Soc.*, **107**, 1431 (1985).
14. W. Helfrich and W. G. Schneider, *J. Chem. Phys.*, **44**, 2902 (1966).
15. V. Bulovic, V. V. Khalfin, G. Gu, P. E. Burrows, D. Z. Garbuzov, and S. R. Forrest, *Phys. Rev. B*, **58**, 3730 (1998).
16. C. F. Madigan, M.-H. Lu, and J. C. Sturm, *Appl. Phys. Lett.*, **76**, 1650 (2000).
17. M. Borodilsky, T. F. Kraus, R. Coccioli, R. Vrijen, R. Bhat, and E. Yablonovitch, *Appl. Phys. Lett.*, **75**, 1036 (1999).
18. T. Yamazaki, K. Sumioka, and T. Tsutsui, *Appl. Phys. Lett.*, **76**, 1243 (2000).
19. I. G. Hill and A. Kahn, *J. Appl. Phys.*, **85**, 6589 (1999).

Chapter 8

High-Efficiency Fluorescent Organic Light-Emitting Devices Using a Phosphorescent Sensitizer

M. A. Baldo,[a] M. E. Thompson,[b] and S. R. Forrest[a]

[a]*Center for Photonics and Optoelectronic Materials (POEM), Department of Electrical Engineering and the Princeton Materials Institute, Princeton University, Princeton, New Jersey 08544, USA*
[b]*Department of Chemistry, University of Southern California, Los Angeles, California 90089, USA*
forrest@ee.princeton.edu

To obtain the maximum luminous efficiency from an organic material, it is necessary to harness both the spin-symmetric and anti-symmetric molecular excitations (bound electron–hole pairs, or excitons) that result from electrical pumping. This is possible if the material is phosphorescent, and high efficiencies have been observed in phosphorescent [1, 2] organic light-emitting devices [3]. However, phosphorescence in organic molecules is rare at room temperature. The alternative radiative process of fluorescence is more common, but it is approximately 75% less efficient, due to the requirement of spin-symmetry conservation

---

Reprinted from *Nature*, 403(6771), 750–753, 2000.

Electrophosphorescent Materials and Devices
Edited by Mark E. Thompson
Text Copyright © 2000 Macmillan Magazines Ltd.
Layout Copyright © 2024 Jenny Stanford Publishing Pte. Ltd.
ISBN 978-981-4877-34-3 (Hardcover), 978-1-003-08872-1 (eBook)
www.jennystanford.com

[4]. Here, we demonstrate that this deficiency can be overcome by using a phosphorescent sensitizer to excite a fluorescent dye. The mechanism for energetic coupling between phosphorescent and fluorescent molecular species is a long-range, non-radiative energy transfer: the internal efficiency of fluorescence can be as high as 100%. As an example, we use this approach to nearly quadruple the efficiency of a fluorescent red organic light-emitting device.

Light is generated in organic materials from the decay of molecular excited states, also known as excitons. Understanding the properties and interactions of excitons is crucial to the design of efficient organic devices for use in displays, lasers and other illumination applications. For example, the spin-symmetry of an exciton determines its probability of radiative recombination and also its multiplicity. Spin-symmetric excitons with a total spin of $S = 1$ have a multiplicity of three and are known as triplets. Spin-antisymmetric excitons ($S = 0$) have a multiplicity of one and are known as singlets. During electrical excitation approximately one singlet exciton is created for every three triplet excitons [4], but because the ground state is typically also spin-antisymmetric, only relaxations of singlet excitons conserve spin and generate fluorescence. Usually the energy in triplet excitons is wasted; however, given some perturbation in symmetry, triplets may slowly radiatively decay, producing the delayed luminescence known as phosphorescence. Although it is often inefficient, phosphorescence may be enhanced if spin–orbit coupling mixes the singlet and triplet states, an effect often promoted by the presence of a heavy metal atom [5]. Indeed, phosphorescent dyes with these properties have demonstrated very high-efficiency electroluminescence [1, 2].

Very few organic materials have been found to be capable of efficient room-temperature phosphorescence from triplets [1, 2, 6]. In contrast, many organic molecules exhibit fluorescence [7, 8] and fluorescence is also unaffected by triplet–triplet annihilation, which degrades phosphorescent emission efficiency at high excitation densities [1]. Consequently, fluorescent materials are suited to many electroluminescent applications, particularly those such as passive matrix displays that require high excitation densities.

It is desirable therefore to find a process whereby triplets that are formed after electrical excitation are not wasted, but are instead transferred to the singlet excited state of a fluorescent dye. There are two mechanisms for triplet–singlet energy transfer from a donor molecule (D) to an acceptor (A). In Dexter transport [5], the exciton hops directly between molecules. This is a short-range process dependent on the overlap of molecular orbitals of neighbouring molecules. It also preserves the symmetry of the donor and acceptor pair [5]. Thus, a triplet–singlet energy transfer is not possible by a Dexter mechanism. A change in spin-symmetry is possible if the donor exciton breaks up and reforms on the acceptor by incoherent electron exchange [5]. However, this process is considered to be relatively unlikely as it requires the dissociation of the donor exciton, which in most molecular systems has a binding energy of $\sim$1 eV.

The alternative mechanism is Förster energy transfer [5]. Here, molecular transition dipoles couple and exchange energy. The efficiency of energy transfer (η_{ET}) is:

$$\eta_{ET} = \frac{k_{ET}}{k_{ET} + k_R + k_{NR}} \quad (8.1)$$

Here k_{ET} is the rate of Förster energy transfer from D to A and k_R and k_{NR} are the radiative and non-radiative rates on the donor, respectively. From Eq. (8.1), energy transfer is efficient if $k_{ET} \cdot k_R + k_{NR}$; however, in Förster's theory k_{ET} is proportional to the oscillator strength of the donor transition [5], as is k_R. Thus, η_{ET} is approximately independent of oscillator strength if $k_R \gg k_{NR}$, that is, if the donor is efficiently phosphorescent then it is possible to obtain triplet–singlet energy transfer by the Förster mechanism. We note that a 'pure' triplet has an infinite lifetime and no probability of Förster energy transfer because the oscillator strength of its decay is zero. However, the slightest perturbation (that is, some overlap between the triplet and ground state on the donor) can counteract the presence of non-radiative modes and yield efficient energy transfer. Such triplet–singlet energy transfers were predicted by Förster [9] and confirmed by Ermolaev and Sveshnikova [10], who detected the energy transfer using a range of phosphorescent donors and fluorescent acceptors in rigid media at 77 K or 90 K. Large transfer distances were observed; for example with triphenylamine as the donor and chrysoidine as the acceptor, the interaction range is 52 Å.

Unfortunately, when a fluorescent acceptor is doped directly into a phosphorescent donor material, the close proximity of the donor and acceptor increases the likelihood of Dexter transfer between the donor and the acceptor triplets. Once excitons reach the acceptor triplet state, they are effectively lost since these fluorescent dyes typically exhibit extremely inefficient phosphorescence (that is, $k_{NR} \gg k_R$).

Another technique is to dope both the phosphorescent material and the fluorescent acceptor into a conductive organic host. Ideally, the phosphor then sensitizes the energy transfer from the host, now acting as the donor, to the fluorescent acceptor. Cascade Förster energy transfer of singlets has been demonstrated for fluorescent materials [11]; however, here all energy is ideally transferred into the triplet state of the sensitizer, where it is then transferred to the singlet state of the fluorescent dye, that is:

$$^1D^* + ^1X \rightarrow {}^1D + {}^1X^*$$
$$^1X^* \rightarrow {}^3X^*$$
$$^3X^* + {}^1A \rightarrow {}^1X + {}^1A^*$$
$$^1A^* \rightarrow {}^1A + h\nu \qquad (8.2)$$

and

$$^3D^* + {}^1X \rightarrow {}^1D + {}^3X^*$$
$$^3X^* + {}^1A \rightarrow {}^1X + {}^1A^*$$
$$^1A^* \rightarrow {}^1A + h\nu \qquad (8.3)$$

Here, the photon energy is $h\nu$, and the donor, sensitizer and fluorescent acceptor are represented by D, X and A, respectively. Triplet and singlet states are signified by a superscript 3 or 1, respectively, and excited states are marked by asterisks. The multiple-stage energy transfer is described schematically in Fig. 8.1. Dexter transfers are indicated by dotted arrows and Förster transfers by solid arrows. Processes resulting in a loss in efficiency are marked with a cross. In addition to the energy transfer paths shown in Fig. 8.1, direct electron–hole recombination is possible on the phosphorescent and fluorescent dopants as well as the host. Triplet exciton formation after charge recombination on the fluorescent dye is another potential loss mechanism.

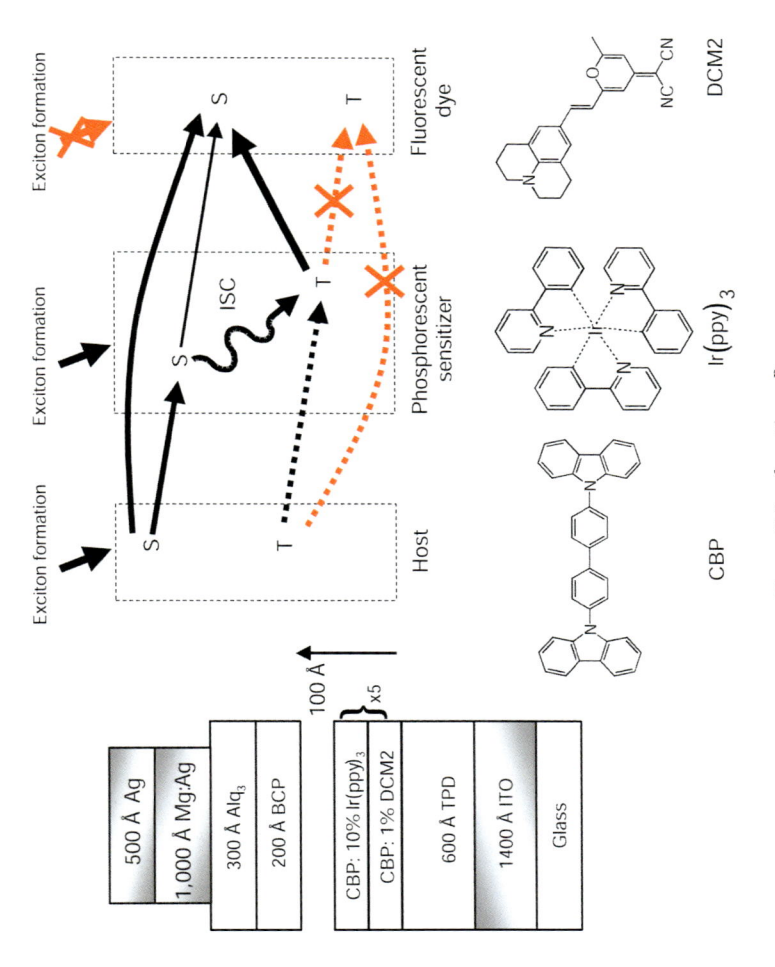

Figure 8.1 *(continued)*

Figure 8.1 (*continued*) Proposed energy transfer mechanisms in the sensitized system. In principle, all excitons are transferred to the singlet state of the fluorescent dye, as triplets in the dye non-radiatively recombine. Förster transfers are represented by solid lines and Dexter transfers by dotted lines. Electron–hole recombination creates singlet (*S*) and triplet (*T*) excitons in the host material, although as indicated, charge trapping may be responsible for exciton formation in the other materials as well. There is a probability of direct transfer into the singlet state of the fluorescent dye by a Förster process, or Dexter transfer into the triplet state. This is a source of loss and is indicated by a cross. Singlet excitons in the phosphor are subject to intersystem crossing (ISC) and transfer to the triplet state. From this state, the triplets may either dipole–dipole couple with the singlet state of the fluorescent dye or, in another loss mechanism, they may Dexter transfer to the triplet state. Direct formation of triplets on the fluorescent dye is an additional path to loss. Inset, the structure of the principal materials and electroluminescent devices fabricated in this work. We use the green emitting phosphor [2] *fac* tris(2-phenylpyridine) iridium (Ir(ppy)$_3$) and the red fluorescent dye [17] [2-methyl-6-[2-(2,3,6,7-tetrahydro-1H,5H-benzo[ij]quinolizin-9-yl) ethenyl]-4H-pyran-4-ylidene] propane-dinitrile (DCM2). DCM2 absorbs in the green and it emits at wavelengths between $\lambda = 570$ nm and $\lambda = 650$ nm (Ref. [14]). To demonstrate the multiple-stage transfer, we used 4,4'-N,N'-dicarbazole-biphenyl (CBP) as the donor and host material [18]. Organic layers were deposited by high vacuum (10^{-6} Torr) thermal evaporation onto a clean glass substrate pre-coated with a layer 1,400 Å in thickness of transparent and conductive indium tin oxide. A layer 600 Å in thickness of N,N'-diphenyl-N,N'-bis(3-methylphenyl)-[1,1'-biphenyl]-4,4'-diamine (TPD) is used for hole transport. We used an alternating series of layers 10 Å in thickness of 10% Ir(ppy)$_3$/CBP and 1% DCM2/CBP. In total, 10 doped layers were grown with a total thickness of 100 Å. Excitons were confined within the luminescent region by a layer 200 Å in thickness of the wide-energy-gap material 2,9-dimethyl-4,7-diphenyl-1,10-phenanthroline (bathocuproine, or BCP). A layer 300 Å in thickness of tris-(8-hydroxyquinoline) aluminium (Alq$_3$) is used to transport electrons to the luminescent region and to reduce absorption at the cathode. Metal deposition through a shadow mask with openings 1 mm in diameter defined cathodes consisting of a layer 1000 Å in thickness of 25:1 Mg/Ag, with an Ag cap 500 Å in thickness.

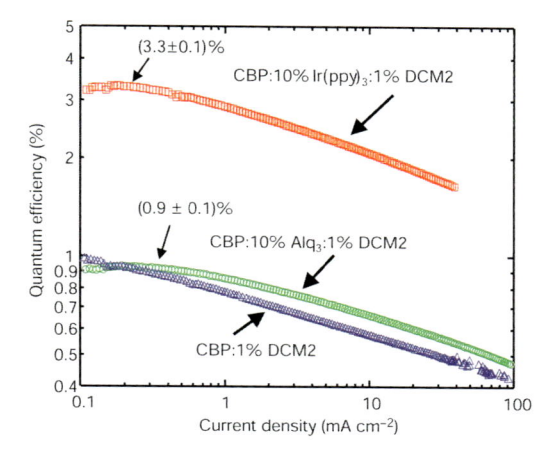

Figure 8.2 The external quantum efficiencies of DCM2 emission in the three devices. The sensitizing action of Ir(ppy)$_3$ improves the efficiency. We also note that the presence of Alq$_3$ in the all-fluorescent devices makes little or no difference. The CBP/10% Ir(ppy)$_3$/1% DCM2 device reaches a brightness of 100 cd m^{-2} at 9.9 V (3 lm/W).

The structure of the organic devices and the principal materials used to demonstrate the sensitized energy transfer are shown in the inset of Fig. 8.1. We used CBP as the host and exciton donor, the green phosphor Ir(ppy)$_3$ as the sensitizer and the red fluorescent dye DCM2 as the acceptor. As a control, two other organic light-emitting device (OLED) structures were made. In the first, Ir(ppy)$_3$ was replaced by Alq$_3$, which has similar emission and absorption spectra, but no observable phosphorescence at room temperature. From the spectral overlap with DCM2 and the photoluminescent efficiencies [12, 13] of Alq$_3$ and Ir(ppy)$_3$, the Förster transfer radii for Alq$_3$ to DCM2 and Ir(ppy)$_3$ to DCM2 are both calculated to be ~40 Å (Ref. [14]). In the second device, the sensitizer was completely omitted in order to examine direct energy transfer from the donor, CBP, to the acceptor, DCM2.

The external quantum efficiency (photons per electron) of the DCM2 portion of the emission spectrum of these three devices is shown as a function of injected current density in Fig. 8.2. The DCM2 emission efficiency of the device containing the phosphorescent sensitizer is significantly higher than its fluorescent analogue.

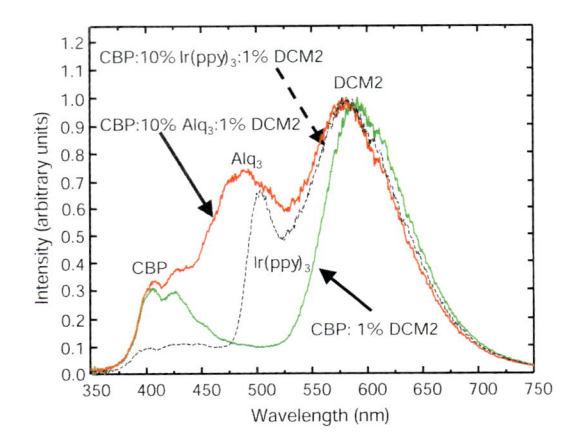

Figure 8.3 The spectra of the three electroluminescent devices fabricated in this work. Characteristic peaks are observed for CBP (at wavelength $\lambda \approx 400$ nm), TPD ($\lambda \approx 430$ nm), Alq$_3$ ($\lambda \approx 490$ nm), Ir(ppy)$_3$ ($\lambda \approx 500$ nm) and DCM2 ($\lambda \approx 590$ nm). Approximately 80% of the photons in the Ir(ppy)$_3$ device are emitted by DCM2. All spectra were recorded at a current density of ~ 1 mA cm^{-2}. Note that the peak intensities of each of the spectra are normalized for comparison.

Indeed, the peak efficiency of $(3.3 \sim 0.1)\%$ is significantly higher than the best result of $\sim 2\%$ observed for DCM2-based OLEDs in previous studies [15], suggesting that host triplets are transferred to the fluorescent singlet state. We also note that addition of Alq$_3$ to the unsensitized CBP/DCM2 device makes no significant difference to the maximum efficiency of $(0.9 \sim 0.1)\%$ of DCM2 emission.

The emission spectra of the OLEDs are shown in Fig. 8.3. All devices show energy transfer to the fluorescent dye. By calculating the area under the various spectral peaks, we find that approximately 80% of photons are emitted by DCM2 in the device containing the Ir(ppy)$_3$ sensitizer. The remainder contribute to CBP luminescence at $\lambda \approx 400$ nm, TPD luminescence at $\lambda \approx 430$ nm and Ir(ppy)$_3$ luminescence at $\lambda \approx 500$ nm. In the device doped with 10% Alq$_3$, an emission peak is also observed at $\lambda \approx 490$ nm. This is consistent with earlier observations [16] of Alq$_3$ emission in a nonpolar host such as CBP.

Conclusive evidence of the energy transfer of Eqs. (8.2) and (8.3) is provided by examining the transient behaviour of the DCM2 and

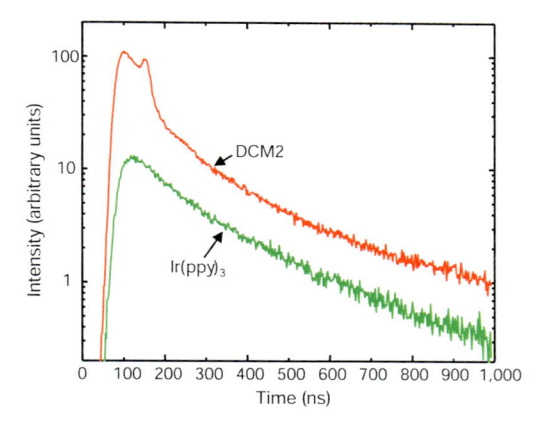

Figure 8.4 The transient response of the DCM2 and Ir(ppy)$_3$ spectral components in the CBP/10% Ir(ppy)$_3$/1% DCM2 device. The transient lifetime of DCM2 is ∼1 ns, thus in the case of energy transfer from Ir(ppy)$_3$, the response of DCM2 should be governed by the transient lifetime of Ir(ppy)$_3$. After the initial 100-ns-wide electrical excitation pulse, this is clearly the case, demonstrating that energy is transferred from the triplet state in Ir(ppy)$_3$ to the singlet state in DCM2. However, during the excitation pulse, singlet transfer to DCM2 is observed, resulting in the ripples in the transient response. These ripples are due to fluctuations in the current density and the discharge of traps at the falling edge of the pulse. We note that the trends in the DCM2 and Ir(ppy)$_3$ transient responses eventually diverge. This is caused by charge trapped on DCM2 molecules recombining and causing luminescence.

Ir(ppy)$_3$ components of the emission spectra. These data are shown in Fig. 8.4, and were obtained by applying a ∼100 ns electrical pulse to the electroluminescent device. The resulting emission was measured using a streak camera. If a fraction of the DCM2 emission originates by transfer from Ir(ppy)$_3$ triplets, then the proposed energy transfer must yield delayed DCM2 fluorescence. Furthermore, since the radiative lifetime of DCM2 is much shorter than that of Ir(ppy)$_3$, the transient decay of DCM2 should match that of Ir(ppy)$_3$. After an initial peak, most probably due to CBP/DCM2 singlet–singlet transfer and exciton formation on DCM2, the DCM2 decay does indeed trace the Ir(ppy)$_3$ decay. The transient lifetime of Ir(ppy)$_3$ in this system is ∼100 ns, compared to a lifetime of ∼500 ns in the absence of DCM2. Since these lifetimes are inversely proportional to the rates of exciton transfer and relaxation, respectively, this confirms an energy transfer of ∼80%. The decrease in the triplet lifetime as a result of energy transfer to the fluorescent acceptor is advantageous. Not only does it increase the transient response of the system, but it should also reduce the probability of triplet–triplet annihilation. Thus, it is expected that this multi-stage energy transfer will reduce the quenching of triplet states, thereby further enhancing the potential efficiency of sensitized fluorescence.

The transient response of the sensitized system may be further examined to determine the relative efficiencies of the singlet-to-singlet and triplet-to-singlet energy transfer pathways. By subtracting the transient decay of Ir(ppy)$_3$ from the DCM2 transient, the ratio of instantaneous to delayed DCM2 fluorescence is found to be ∼1:2. Consequently, we expect that the sensitized OLED should have a quantum efficiency roughly triple that of an OLED that does not exhibit delayed fluorescence. Indeed, from the data in Fig. 8.2, the improvement in peak quantum efficiencies is $(3.7 \sim 0.5)$%. Ideally the ratio of instantaneous to delayed fluorescence should reflect the ratio of singlets to triplets (1:3), or even be weighted towards triplets if intersystem crossing on Ir(ppy)$_3$ is significant. It is likely that direct triplet exciton formation on DCM2 contributes to some loss. In addition, the likelihood of triplet transfer from CBP to DCM2, CBP to Ir(ppy)$_3$, or even Ir(ppy)$_3$ to CBP is unknown, so it is possible that some triplets are transferred directly to DCM2, bypassing the sensitizer.

In conclusion, we have demonstrated a general technique for improving the efficiency of fluorescence in OLEDs. Further improvement in energy transfer efficiency could be expected by mixing the host, phosphorescent sensitizer and fluorescent dye rather than doping in alternating thin layers as was done here. However, a mixture may possess increased losses from Dexter transfer from the sensitizer to the triplet state of the fluorescent dye. To reduce this loss, an ideal system may incorporate low concentrations of a sterically hindered fluorescent dye. For example, adding spacer groups to the DCM2 molecule could decrease the probability of Dexter transfer to the dye while minimally affecting its participation in Förster transfer or its luminescence efficiency. As Dexter transfer can be understood as the simultaneous transfer of an electron and a hole, steric hindrance may also reduce the likelihood of charge trapping on the fluorescent dye. Similar efforts have already reduced non-radiative excimer formation in a DCM2 variant [15].

Acknowledgements

This work was supported by Universal Display Corporation, the Defense Advanced Research Projects Agency, the Air Force Office of Scientific Research and the National Science Foundation.

References

1. Baldo, M. A. et al. Highly efficient phosphorescent emission from organic electroluminescent devices. *Nature* **395**, 151–154 (1998).
2. Baldo, M. A., Lamansky, S., Burrows, P. E., Thompson, M. E. and Forrest, S. R. Very high-efficiency green organic light-emitting devices based on electrophosphorescence. *Appl. Phys. Lett.* **75**, 4–6 (1999).
3. VanSlyke, S. A. and Tang, C. W. Organic electroluminescent devices having improved power conversion efficiencies (US Patent No. 4539507, 1985).
4. Baldo, M. A., O'Brien, D. F., Thompson, M. E. and Forrest, S. R. The excitonic singlet-triplet ratio in a semiconducting organic thin film. *Phys. Rev. B* **60**, 14422–14428 (1999).
5. Klessinger, M. and Michl, J. *Excited States and Photochemistry of Organic Molecules* (VCH Publishers, New York, 1995).
6. Cleave, V., Yahioglu, G., Le Barny, P., Friend, R. and Tessler, N. Harvesting singlet and triplet energy in polymer LEDs. *Adv. Mater.* **11**, 285–288 (1999).
7. Chen, C. H., Shi, J. and Tang, C. W. Recent developments in molecular organic electroluminescent materials. *Macromol. Symp.* **125**, 1–48 (1997).
8. Brackmann, U. *Lambdachrome Laser Dyes* (Lambda Physik, Gottingen, 1997).
9. Förster, T. Transfer mechanisms of electronic excitation. *Discuss. Faraday Soc.* **27**, 7–17 (1959).

10. Ermolaev, V. L. and Sveshnikova, E. B. Inductive-resonance transfer of energy from aromatic molecules in the triplet state. *Dokl. Akad. Nauk* **149**, 1295–1298 (1963).
11. Berggren, M., Dodabalapur, A., Slusher, R. E. and Bao, Z. Light amplification in organic thin films using cascade energy transfer. *Nature* **389**, 466–469 (1997).
12. Garbuzov, D. Z., Bulovic, V., Burrows, P. E. and Forrest, S. R. Photoluminescence efficiency and absorption of aluminium-tris-quinolate (Alq_3) thin films. *Chem. Phys. Lett.* **249**, 433–437 (1996).
13. Vander Donckt, E., Camerman, B., Hendrick, F., Herne, R. and Vandeloise, R. Polystyrene immobilized Ir(III) complex as a new material for oxygen sensing. *Bull. Soc. Chim. Belg.* **103**, 207–211 (1994).
14. Bulovic, V. et al. Bright, saturated, red-to-yellow organic light-emitting devices based on polarization induced spectral shifts. *Chem. Phys. Lett.* **287**, 455–460 (1998).
15. Chen, C. H., Tang, C. W., Shi, J. and Klubeck, K. P. Improved red dopants for organic electroluminescent devices. *Macromol. Symp.* **125**, 49–58 (1997).
16. Bulovic, V., Deshpande, R., Thompson, M. E. and Forrest, S. R. Tuning the color emission of thin film molecular organic light emitting devices by the solid state solvation effect. *Chem. Phys. Lett.* **308**, 317 (1999).
17. Tang, C.W., VanSlyke, S. A. and Chen, C. H. Electroluminescence of doped organic thin films. *J. Appl. Phys.* **65**, 3610–3616 (1989).
18. O'Brien, D. F., Baldo, M. A., Thompson, M. E. and Forrest, S. R. Improved energy transfer in electrophosphorescent devices. *Appl. Phys. Lett.* **74**, 442–444 (1999).

Chapter 9

Nearly 100% Internal Phosphorescence Efficiency in an Organic Light Emitting Device

Chihaya Adachi, Marc A. Baldo, Mark E. Thompson,[*] and Stephen R. Forrest

Center for Photonics and Optoelectronic Materials (POEM), Department of Electrical Engineering, Princeton University, Princeton, New Jersey 08544, USA
[*]*Present address: Department of Chemistry, University of Southern California, Los Angeles, CA 90089, USA*
met@usc.edu

We demonstrate very high efficiency electrophosphorescence in organic light-emitting devices employing a phosphorescent molecule doped into a wide energy gap host. Using bis(2-phenylpyridine)iridium(III) acetylacetonate [(ppy)$_2$Ir(acac)] doped into 3-phenyl-4-(1'-naphthyl)-5-phenyl-1,2,4-triazole, a maximum external quantum efficiency of (19.0 ± 1.0)% and luminous power efficiency of (60 ± 5) lm/W are achieved. The calculated internal quantum efficiency of (87 ± 7)% is supported by the observed absence of thermally activated nonradiative loss in the photoluminescent efficiency of (ppy)$_2$Ir(acac). Thus, very high

---

Reprinted from *J. Appl. Phys.*, **90**(10), 5048–5051, 2001.

Electrophosphorescent Materials and Devices
Edited by Mark E. Thompson
Text Copyright © 2001 American Institute of Physics
Layout Copyright © 2024 Jenny Stanford Publishing Pte. Ltd.
ISBN 978-981-4877-34-3 (Hardcover), 978-1-003-08872-1 (eBook)
www.jennystanford.com

external quantum efficiencies are due to the nearly 100% internal phosphorescence efficiency of (ppy)$_2$Ir(acac) coupled with balanced hole and electron injection, and triplet exciton confinement within the light-emitting layer.

9.1 Introduction

Inorganic, direct band gap semiconductor heterostructure light-emitting devices and lasers comprised of III–V compounds based on GaAs and InP, are known to have nearly 100% internal quantum efficiency (η_{int}) [1]. Since holes and electrons in the valence and conduction bands, respectively, are considered as free particles, radiative and nonradiative transitions in these direct band gap materials do not involve the intermediate formation of excitons. In contrast, many optical processes in organic thin films are mediated by excitons which can ultimately lead to a reduction in electroluminescence efficiency in polymer and molecular organic light emitting devices (OLEDs). Besides nonradiative pathways due to strong exiton–phonon coupling, the fraction of singlet excitons (χ) under electrical excitation is 0.25 in a molecular solid such as aluminum tris(8-hydroxyquinoline) (Alq$_3$) [2], limits η_{int} to only 25% when fluorescence-based light emitting molecules are employed. The OLED external quantum efficiency (η_{ext}) follows:

$$\eta_{ext} = \eta_{int}\eta_{ph} = \gamma\eta_{ex}\phi_p\eta_{ph}, \tag{9.1}$$

where η_{ph} is the light out-coupling efficiency, η_{ex} is the fraction of total excitons formed which result in radiative transitions ($\eta_{ex} \sim 1/4$ for fluorescent molecular dyes, and 1 for phosphorescent materials), γ is the ratio of electrons to holes (or vice versa, to maintain $\gamma \leq 1$) injected from opposite contacts (the electron–hole charge-balance factor), and ϕ_p is the intrinsic quantum efficiency for radiative decay (including both fluorescence and phosphorescence). If only singlets are radiative in fluorescent materials, η_{ext} is limited to ~5%, assuming [3] $\eta_{ph} \sim 1/2n^2 \sim 20\%$ for a glass substrate with index of refraction $n = 1.5$. In contrast, by using high efficiency phosphorescent materials which harvest both singlet and triplet excitons, η_{int} can approach 100%, in which case we can anticipate $\eta_{ext} \sim 20\%$.

The recent fabrication of extremely efficient electrophosphorescent OLEDs employing Pt and Ir complexes suggests that devices with internal quantum efficiencies of 100% are achievable through radiative recombination of both singlet and triplet excitons [4–13]. Attempts to observe electrophosphorescence (EP) were reported using keto-coumarin [14] and benzophenone [15] derivatives. The efficiency of these materials was low, $\eta_{ext} < 1\%$, even at 77 K, because the rate of phosphorescent light emission was comparable to the excited state nonradiative decay rate. However, organometallic compounds which introduce spin–orbit coupling due to the central heavy metal atom show a relatively high ligand based phosphorescence efficiency ($\phi_p > 20\%$) even at room temperature because of the strong radiative transition moment of $n - \pi^*$, $\pi - \pi^*$ and the metal-to-ligand charge transfer states [12]. Using this approach, demonstrations of high-efficiency ($\eta_{ext} > 5\%$) OLEDs were made possible using 2,3,7,8,12,13,17,18-octaethyl-21H,23H−porphine platinum(II) [4, 5] and *fac* tris(2-phenylpyridine)iridium [Ir(ppy)$_3$] and its derivatives [6–13].

Here, we demonstrate very high efficiency EP-OLEDs employing the green electrophosphorescent molecule, bis(2-phenylpyridine)iridium(III)acetylacetonate [(ppy)$_2$Ir(acac)], with a maximum $\eta_{ext} = (19.0 \pm 0.5)\%$ and luminous power efficiency of $\eta_p = (60 \pm 5)$ lm/W [16]. We show that these values correspond to $\eta_{int} = (87 \pm 7)\%$. We also demonstrate that the very high internal quantum efficiency is due to direct exciton formation at the guest phosphor, along with subsequent exciton confinement within the emissive layer.

9.2 Experimental Method

Organic layers were deposited by high-vacuum (10^{-6}Torr) thermal evaporation onto a clean glass substrate pre-coated with an indium tin oxide (ITO) layer (160 mm thick) with a sheet resistance of ~ 20 Ω/sq. Prior to use, the substrate was degreased with solvents and cleaned in a UV-ozone chamber before it was loaded into the evaporation system. First, a 60 nm thick hole transport layer of 4,4'-

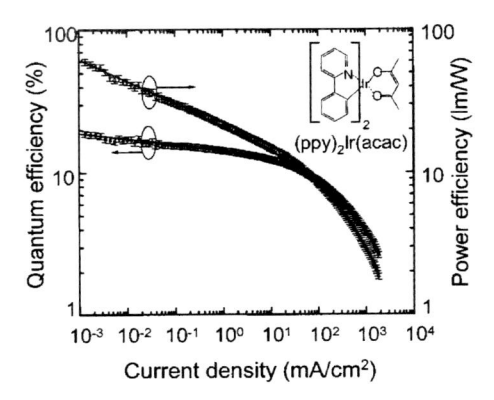

Figure 9.1 The external quantum and power efficiencies of an ITO/HMTPD (60 nm)/12%-(ppy)$_2$Ir(acac):TAZ (25 nm)/Alq$_3$ (50 nm)/Mg:Ag/Ag OLED. A maximum external quantum efficiency of $\eta_{ext} = (19.0 \pm 0.5)$% and power efficiency of $\eta_p = (60 \pm 5)$ lm/W were obtained. Inset: Molecular structure of (ppy)$_2$Ir(acac).

bis[N,N'-(3-tolyl)amino]-3,3'-diemthyl biphenyl (HMTPD) [8] was deposited, followed by a 25 nm thick light-emitting layer (EML) consisting of [17] (ppy)$_2$Ir(acac) codeposited with [8] a 3-phenyl-4-(1'-naphthyl)-5-phenyl-1,2,4-triazole (TAZ) electron-transport host layer. A 50 nm thick layer of Alq$_3$ was then deposited onto the EML surface to transport and inject electrons into the EML. A shadow mask with 1 mm diameter openings was used to define the cathode consisting of a 150 nm thick Mg:Ag (10:1) layer, with a 20 nm thick Ag cap. Similar results to those reported here were also obtained employing a LiF/Al cathode.

Current density (J)–voltage (V)–luminance (L) characteristics were measured using an HP4145B semiconductor parameter analyzer, with the quantum efficiency directly obtained by placing the OLED approximately 3 mm above the center of a large diameter (1.13 cm) calibrated Si photodiode. Since almost all of the emitted light from the OLED substrate surface is detected by the photodiode, this method avoids systematic errors introduced by corrections needed to account for non-Lambertian spatial emission patterns [7]. Furthermore, this direct measurement of the quantum efficiency also eliminates errors often introduced by first measuring

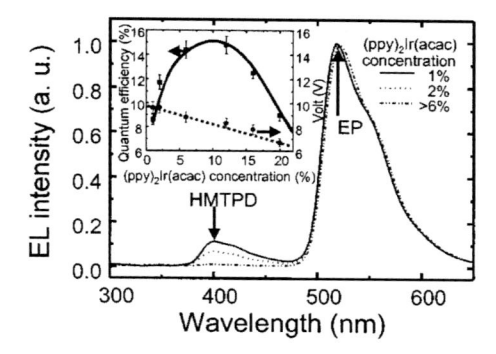

Figure 9.2 Concentration dependence of EL spectra on (ppy)$_2$Ir(acac) of an ITO/HMTPD (60 nm)/1%, 2%, and 6%-(ppy)$_2$Ir(acac):TAZ (25 nm)/Alq$_3$ (50 nm)/Mg:Ag/Ag OLED. Inset: Concentration dependence of the external quantum efficiency (η_{ext}) and driving voltage of a (ppy)$_2$Ir(acac) device with a structure: ITO/HMTPD (60 nm)/X%-(ppy)$_2$Ir(acac):TAZ (25 nm)/Alq$_3$ (50 nm) Mg:Ag/Ag ($X = 1$% to 20%).

device luminance and then converting the data to efficiency after determination of the OLED emission spectrum [18].

9.3 Results

Figure 9.1 shows the EP-OLED efficiency with 12%-(ppy)$_2$Ir(acac) doped into TAZ. A maximum $\eta_{ext} = (19.0 \pm 0.5)$ % and $\eta_p = (60 \pm 5)$ lm/W are obtained at a current of $J = 1.5$ µA/cm^2 and luminance of 1.3 cd/m^2. The device exhibits a gradual decrease in η_{ext} with increasing current at $J > 10$ mA/cm^2 due to triplet–triplet annihilation [19, 20]. Even at a luminance of ~1000 cd/m^2 corresponding to $J = 2.1$ mA/cm^2, $\eta_{ext} = (13.7 \pm 0.5)$% was observed.

Figure 9.2 shows the dependence of the (ppy)$_2$Ir(acac) EP spectrum on doping concentration in TAZ. At concentrations >6%, we only observe the electroluminescence (EL) component centered at a wavelength of $\lambda = 520$ nm due to the (ppy)$_2$Ir(acac) phosphorescence. At (ppy)$_2$Ir(acac) concentrations less than 2%, an additional deep blue emission ($\lambda = 395$ nm) due to HMTPD fluorescence is also observed. The inset of Fig. 9.2 shows (ppy)$_2$Ir(acac) concentration

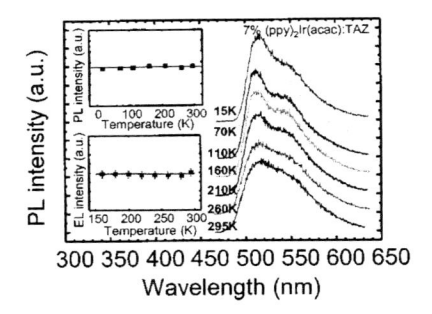

Figure 9.3 Temperature dependence of the PL spectrum of a 12%-(ppy)$_2$Ir(acac):TAZ film from $T = 275$ to 15 K. Inset: Temperature dependence of the PL intensity of a 12%-(ppy)$_2$Ir(acac):TAZ film and EP intensity of the OLED at $J = 0.1$ mA/cm^2.

dependences of η_{ext} and drive voltage at 1 mA/cm^2. A maximum η_{ext} was observed at (ppy)$_2$Ir(acac) concentrations from 5% to 12%, while a significant decrease in η_{ext} was observed at both higher and lower concentrations. In addition, a gradual decrease of the drive voltage was observed with an increase in (ppy)$_2$Ir(acac) concentration.

We also measured the temperature dependence of the EP and photoluminescence (PL) quantum efficiencies of the EML from $T = 15$ to 300 K. Figure 9.3 shows the temperature dependence of the PL spectra in a (ppy)$_2$Ir(acac) doped TAZ film. The temperature dependencies of the integrated PL and EP intensities under a constant current density of $J = 0.1$ mA/cm^2 are shown in the insets of Fig. 9.3. While the spectral width decreased slightly with temperature, both the PL and EP intensities are temperature independent.

9.4 Discussion

9.4.1 Internal Electrophosphorescent Quantum Efficiency

To determine the internal quantum efficiency of the EP-OLED, we begin by calculating the radiative modes in the OLED optical microcavity [21, 22] employing the theory of Chance et al. [23].

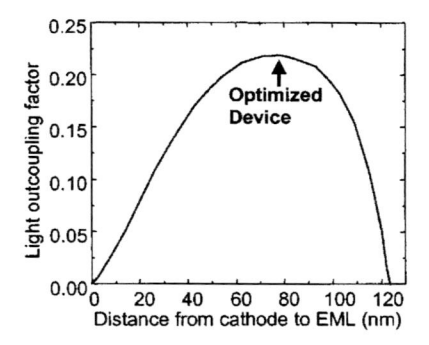

Figure 9.4 Calculated light out-coupling factor for glass ($n = 1.5$)/ITO (160 nm)/organic layer (125 nm)/Mg:Ag/Ag. The abscissa shows the distance of the light emitting layer from the Mg:Ag cathode.

Dyadic Green's functions are used to compute the radiative decay rates, allowing us to consider arbitrarily complex structures. This treatment also accounts for nonradiative losses due to dipole coupling with surface plasmon modes at the metalorganic cathode interface. We model the OLED in Fig. 9.1 as a four-layer structure: the Mg:Ag cathode is considered semi-infinite with a refractive index of $n_{Mg} = 0.25 + i4.36$ [24]; the organic layers are represented by a single layer of thickness 125 nm with refractive index $n = 1.7$ [21]; the ITO has a thickness of 160 nm and a refractive index of $n = 1.9 + i0.01$ [25], where the absorption was calculated from transmission measurements; and finally the semi-infinite glass layer has a refractive index of $n = 1.5$. The glass substrate is approximately 1 mm thick, allowing for the use of ray optics to calculate the angular emission pattern within the glass, and consequently the coupling into air. For luminescence at the HMTPD (ppy)$_2$Ir(acac):TAZ interface 75 nm from the cathode, we calculate an output coupling efficiency of $\eta_p = (22 \pm 2)\%$ (Fig. 9.4). Hence, for $\eta_{ext} = (19.0 \pm 0.5)\%$, we obtain $\eta_{int} = (87 \pm 7)\%$.

With our estimate of η_{int}, we can infer the efficiencies of the molecular transitions leading to EP using Fig. 9.5, which shows the absorption and emission spectra of (ppy)$_2$Ir(acac) identifying transitions from several excited state manifolds. The energy level scheme inferred from these spectra is shown in the inset. Phosphorescence proceeds via either direct injection into the

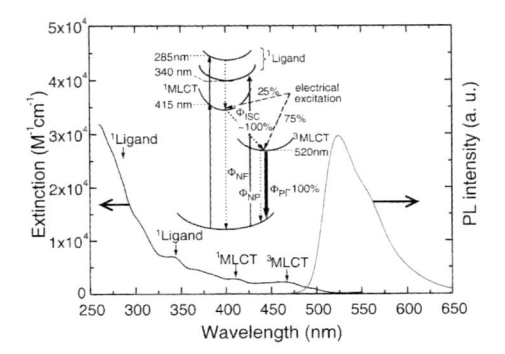

Figure 9.5 Absorption and emission spectra of (ppy)$_2$Ir(acac). Inset: Energy level diagram of (ppy)$_2$Ir(acac). The ligand singlet (^{1}Ligand) and triplet (^{3}Ligand) states, (^{1}MLCT) and (^{3}MCLT) were determined from the absorption and emission spectra. Φ_{NF}, Φ_{ISC}, Φ_{PI}, and Φ_{NP} are quantum yields for nonemissive transitions from ^{1}MLCT intersystem crossing, intrinsic phosphorescent transitions, and nonemissive transitions from ^{3}MCLT, respectively.

triplet metal ligand charge transfer state (^{3}MLCT), or via intersystem crossing (ISC) from the singlet charge transfer state (^{1}MLCT). Now, the phosphorescence quantum yield (ϕ_p) follows:

$$\phi_p = \kappa_p/[\kappa_p + \kappa_{NP}], \quad (9.2)$$

where κ_p is the phosphorescence emission rate and κ_{NP} that of nonemissive triplet decay. Under electrical excitation, on the other hand, both singlet and triplet excitons are directly created on either the guest or host molecules with a statistical splitting of $\chi = 25\%$ singlets and $(1 - \chi) = 75\%$ triplets [2]. Thus, η_{int} follows [c.f. Eqs. (9.1) and (9.2)]:

$$\eta_{int} = \gamma \phi_p \eta_{ex} = \gamma \kappa_p/[\kappa_p + \kappa_{NP}][(1 - \chi) + \chi \phi_{ISC}]. \quad (9.3)$$

From the temperature independence of the PL and EP efficiencies (see inset of Fig. 9.3), we infer that the temperature-dependent nonradiative pathways are almost negligible in a (ppy)$_2$Ir(acac):TAZ solid-state film even at room temperature. Note, however, that some nonradiative processes (e.g., triplet–triplet and triplet–polaron annihilation, and field-dependent exciton dissociation, etc.) do not depend strongly on temperature. From our measurements of η_{int}

$= (87 \pm 7)\%$, Eq. (9.3) implies that the $(13 \pm 7)\%$ loss in total efficiency must arise either from residual nonradiative processes in (ppy)$_2$Ir(acac) or because ϕ_{ISC} or γ is less than 1. However, it is unlikely that $\phi_{ISC} < 1$ since, in that case, we would anticipate fluorescent emission from the ^{1}MCLT to ground state on a time scale of ~10 ns, but this emission is not observed at any temperature or pumping intensity used in these studies. Hence, we conclude that both the photoluminescent efficiency of (ppy)$_2$Ir(acac) and γ are at least ~0.9.

9.4.2 Exciton Formation process

The EP spectral characteristics in Fig. 9.2 suggest that at (ppy)$_2$Ir(acac) concentrations less than 2%, hole injection from the HMTPD highest occupied molecular orbital (HOMO) into the TAZ HOMO is energetically unfavorable, and carrier recombination partly occurs within HMTPD (c.f. energy level diagram in Fig. 9.6a). The large energy difference between the HOMO level of HMTPD and TAZ of ~1.0 eV prevents direct hole injection from HMTPD into TAZ. The accumulated holes at the interface then recombine with electrons injected from TAZ layer, leading to blue HMTPD emission in addition to exciton formation at (ppy)$_2$Ir(acac). This analysis is consistent with the dominant electron transport characteristics of TAZ. At (ppy)$_2$Ir(acac) concentrations higher than 6%, on the other hand, there is no HMTPD blue emission (Fig. 9.2). Thus, we conclude that (ppy)$_2$Ir(acac) exciton formation occurs directly on (ppy)$_2$Ir(acac) from holes injected from HMTPD and electrons primarily transported at the TAZ lowest unoccupied molecular orbital (LUMO) energy (Fig. 9.6b). A decrease of the driving voltage with increasing phosphor concentration is evident for the direct hole injection process. Since the HOMO levels of HMTPD and (ppy)$_2$Ir(acac) are aligned at 5.6 eV, direct hole injection at the HMTPD/EML interface should reduce the drive voltage, as observed.

Recall that the electron–hole charge-balance factor, γ 0.9. Since most holes injected into the (ppy)$_2$Ir(acac) HOMO level recombine with electrons at this interface, and since hole transport in the doped TAZ layer is probably more likely than electron transport through the HMTPD HTL, we speculate that the hole density is slightly higher

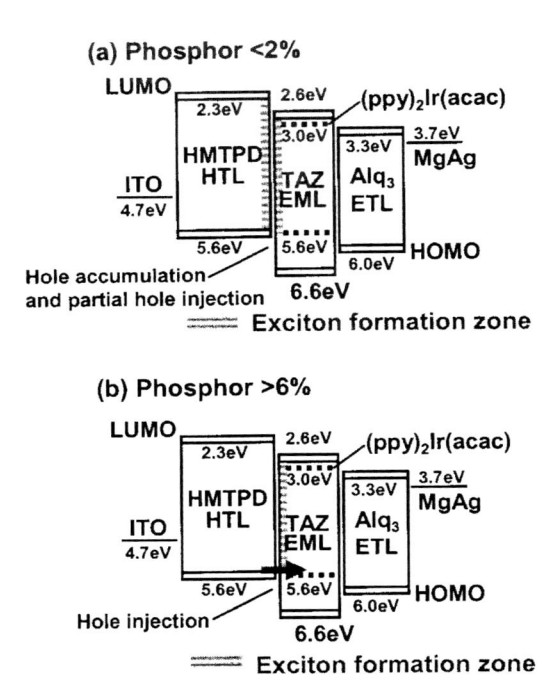

Figure 9.6 Energy diagrams of the ITO/HMTPD/(((ppy)$_2$Ir(acac):TAZ/Alq$_3$/MgAg/Ag EP-OLED showing the relative positions of the HOMO and LUMO levels of the various organic layers, corresponding to dopant concentrations of (a) <2% (ppy)$_2$Ir(acac) and (b) >6% (ppy)$_2$Ir(acac) in a TAZ host.

than that of electrons, possibly leading to deviation in γ from its ideal value of 1. After direct exciton formation on (ppy)$_2$Ir(acac), the exciton radiatively decays due to the wide energy gap of a TAZ host which confines triplet excitons on the guest molecule. No host fluorescence and phosphorescence even at a low temperature thus ensures good exciton confinement on the phosphor guest.

9.5 Summary

In conclusion, we demonstrated a very high-efficiency EP-OLED approaching 100% internal quantum efficiency. The high internal phosphorescence efficiency and charge balance in the structure are responsible for the high efficiency. From these results, we find that further increases in OLED efficiency will only be obtained by developing schemes for increasing light out-coupling by incorporating microcavities [21, 26], shaped substrates [27, 28], or an index matching medium [29, 30] in combination with the use of phosphorescent molecular dyes.

Acknowledgments

The authors are grateful to Universal Display Corporation, the National Science Foundation MRSEC program, and the Defense Advanced Research Projects Agency for their support of this research.

References

1. L. A. Coldren and S. W. Corzine, *Diode Lasers and Photonic Integrated Circuits* (Wiley, New York, 1995), p. 54.
2. M. A. Baldo, D. F. O'Brien, M. E. Thompson, and S. R. Forrest, *Phys. Rev. B* **60**, 14422 (1999).
3. N. C. Greenham, R. H. Friend, and D. D. C. Bradley, *Adv. Mater.* **6**, 491 (1994).
4. M. A. Baldo, D. F. O'Brien, Y. You, A. Shoustikov, S. Sibley, M. E. Thompson, and S. R. Forrest, *Nature (London)* **395**, 151 (1998).
5. D. F. O'Brien, M. A. Baldo, M. E. Thompson, and S. R. Forrest, *Appl. Phys. Lett.* **74**, 442 (1999).
6. M. A. Baldo, S. Lamansky, P. E. Burrows, M. E. Thompson, and S. R. Forrest, *Appl. Phys. Lett.* **75**, 4 (1999).
7. T. Tsutsui, M. J. Yang, M. Yahiro, K. Nakamura, T. Watanabe, T. Tsuji, Y. Fukuda, T. Wakimoto, and S. Miyaguchi, *Jpn. J. Appl. Phys.*, Part 2 **38**, L1502 (1999).
8. C. Adachi, M. A. Baldo, and S. R. Forrest, *Appl. Phys. Lett.* **77**, 904 (2000).
9. C.-L. Lee, K. B. Lee, and J.-J. Kim, *Appl. Phys. Lett.* **77**, 2280 (2000).
10. C. Adachi, M. A. Baldo, S. R. Forrest, S. Lamansky, M. E. Thompson, and R. C. Kwong, *Appl. Phys. Lett.* **78**, 1622 (2001).

11. C. Adachi, M. A. Baldo, M. E. Thompson, and S. R. Forrest, *Bull. Am. Phys. Soc.* **46**, 863 (2001).
12. S. Lamansky, P. Djurovich, D. Murphy, F. Abdel-Razzaq, C. Adachi, P. E. Burrows, S. R. Forrest, and M. E. Thompson, *J. Am. Chem. Soc.* **123**, 4304 (2001).
13. C. Adachi, R. C. Kwong, and S. R. Forrest, *Org. Electron.* **2**, 37 (2001).
14. M. Morikawa, C. Adachi, T. Tsutsui, and S. Saito, 51st Fall Meeting, *Jpn. Soc. Appl. Phys.*, Paper 28a-PB-8 (1990).
15. S. Hoshino and H. Suzuki, *Appl. Phys. Lett.* 69, 224 (1996).
16. These results are similar to those recently reported for OLEDs employing the analogous phosphor, Ir(ppy)$_3$ doped into a hole-transporting host by M. Ikai, S. Tokito, Y. Sakamoto, T. Suzuki, and Y. Taga, *Appl. Phys. Lett.* **79**, 156 (2001).
17. S. Lamansky, P. Djurovich, D. Murphy, F. Abdel-Razaq, R. Kwong, I. Tsyba, M. Bortz, B. Mui, R. Bau, and M. E. Thompson, *Inorg. Chem.* **40**, 1704 (2001).
18. J. Kido and Y. Iizumi, *Appl. Phys. Lett.* **73**, 2721 (1998).
19. C. Adachi, M. A. Baldo, and S. R. Forrest, *J. Appl. Phys.* **87**, 8049 (2000).
20. M. A. Baldo, C. Adachi, and S. R. Forrest, *Phys. Rev. B* **62**, 10967 (2000).
21. V. Bulovic, V. B. Khalfin, G. Gu, P. E. Burrows, D. Z. Garbuzov, and S. R. Forrest, *Phys. Rev. B* **58**, 3730 (1998).
22. J.-S. Kim, P. K. H. Ho, N. C. Greenham, and R. H. Friend, *J. Appl. Phys.* **88**, 1073 (2000).
23. R. R. Chance, A. prock, and R. Sibley, *Adv. Chem. Phys.* **37**, 1 (1978).
24. E. D. Palik, *Handbook of Optical Constants of Solids* (Academic, Orlando, FL, 1985).
25. J. S. Kim, P. K. H. Ho, N. C. Greenham, and R. H. Friend, *J. Appl. Phys.* **88**, 1073 (2000).
26. R. H. Jordan, L. J. Rothberg, A. Dodabalapur, and R. E. Slusher, *Appl. Phys. Lett.* **69**, 1997 (1996).
27. G. Gu, D. Z. Garbuzov, P. E. Burrows, S. Venkatesh, and S. R. Forrest, *Opt. Lett.* **22**, 396 (1997).
28. C. F. Madigan, M. H. Lu, and J. C. Sturm, *Appl. Phys. Lett.* **76**, 1650 (2000).
29. M. Borodilsky, T. F. Kraus, R. Coccioli, R. Vrijen, R. Bhat, and E. Yablonovitch, *Appl. Phys. Lett.* **75**, 1036 (1999).
30. T. Yamazaki, K. Sumioka, and T. Tsutsui, *Appl. Phys. Lett.* **76**, 1243 (2000).

Chapter 10

Endothermic Energy Transfer: A Mechanism for Generating Very Efficient High-Energy Phosphorescent Emission in Organic Materials

Chihaya Adachi,[a] Raymond C. Kwong,[b] Peter Djurovich,[b] Vadim Adamovich,[c] Marc A. Baldo,[a] Mark E. Thompson,[c] and Stephen R. Forrest[a]

[a]*Center for Photonics and Optoelectronic Materials, Department of Electrical Engineering, Princeton University, Princeton, New Jersey 08544, USA*
[b]*Universals Display Corporation, 375 Phillips Blvd., Ewing, New Jersey 08618, USA*
[c]*Department of Chemistry, University of Southern California, Los Angeles, California 90089, USA*
forrest@princeton.edu

Intermolecular energy transfer processes typically involve an exothermic transfer of energy from a donor site to a molecule with a substantially lower-energy excited state (trap). Here, we demonstrate that an endothermic energy transfer from a molecular organic host (donor) to an organometallic phosphor (trap) can lead to highly efficient blue electroluminescence. This demonstration of

---

Reprinted from *Appl. Phys. Lett.*, **79**(13), 2082–2084, 2001.

Electrophosphorescent Materials and Devices
Edited by Mark E. Thompson
Text Copyright © 2001 American Institute of Physics
Layout Copyright © 2024 Jenny Stanford Publishing Pte. Ltd.
ISBN 978-981-4877-34-3 (Hardcover), 978-1-003-08872-1 (eBook)
www.jennystanford.com

endothermic transfer employs iridium(III)bis(4,6-di-fluorophenyl)-pyridinato-$N,C^{2'}$)picolinate as the phosphor. Due to the comparable energy of the phosphor triplet state relative to that of the 4,4'-N, N'-dicarbazole-biphenyl conductive host molecule into which it is doped, the rapid exothermic transfer of energy from phosphor to host, and subsequent slow endothermic transfer from host back to phosphor, is clearly observed. Using this unique triplet energy transfer process, we force emission from the higher-energy, blue triplet state of the phosphor (peak wavelength of 470 nm), obtaining a very high maximum organic light-emitting device external quantum efficiency of (5.7 ± 0.3)% and a luminous power efficiency of (6.3 ± 0.3)lm/W.

Energy transfer from a conductive host to a luminescent dopant can result in high external quantum efficiencies in organic thin-film light-emitting devices. For example, we have recently demonstrated high-efficiency green and red organic electrophosphorescent devices which harvested both singlet and triplet excitons, leading to internal quantum efficiencies (η_{int}) approaching 100% [1–5]. In these cases, high efficiencies were obtained by energy transfer from both the host singlet and triplet states to the phosphor triplet, or via direct trapping of charge on the phosphor, thereby harvesting up to 100% of the excited states. These transfers entail a resonant, exothermic process. As the triplet energy of the phosphor increases, it becomes less likely to find an appropriate host with a suitably high-energy triplet state. The very large excitonic energies required of the host also suggest that this material layer may not have appropriate energy-level alignments with other materials used in an OLED structure, hence, resulting in a further reduction in efficiency. To eliminate this competition between the conductive and energy transfer properties of the host, a route to efficient blue electrophosphorescence may involve the endothermic energy transfer from a near-resonant excited state of the host to the higher triplet energy of the phosphor [6, 7]. Provided that the energy required in the transfer is not significantly greater than the thermal energy, this process can be very efficient.

Here, we demonstrate blue electrophosphorescence using energy transfer from a conductive organic host to the iridium complex: iridium(III)bis[4,6-di-fluorophenyl)-pyridinato-$N,C^{2'}$] picolinate

(FIrpic) [8]. The introduction of the electron withdrawing fluorine complex results in an increase of the triplet exciton energy and, hence, a blueshift of the phosphorescence compared with that of Ir(ppy)$_3$. We obtained a maximum external quantum electroluminescent (EL) efficiency (η_{ext}) of (5.7 ± 0.3)% and a luminous power efficiency (η_p) of (6.3 ± 0.3)lm/W, representing a significant improvement of the efficiencies compared with the blue fluorescent emitters reported to date [9–11].

Figure 10.1a shows photoluminescent (PL) spectra of three different iridium-based phosphors, bis(2-phenyl-pyridinato-N,C^2)' iridium(acetylacetonate) [ppy$_2$Ir(acac)], bis[4,6-di-fluorophenyl)-pyridinato-$N,C^{2'}$]iridium(acetylacetonate) [FIr(acac)], and FIrpic, demonstrating a spectral shift with ligand modification. The presence of the heavy metal iridium results in strong spin-orbit coupling and metal ligand charge transfer, allowing for rapid intersystem crossing of excitons into the radiative triplet manifold of the ligand [8]. All three complexes give high photoluminescent efficiencies of $\Phi_{pl} = 0.5$–0.6 in fluid solution. With introduction of fluorine atoms into the 4,6-positions in 2-phenylpyridine, the triplet excited state experiences a blue-shift of ∼40 nm in the PL peak FIr(acac) as compared with the green emitting ppy$_2$Ir(acac). Furthermore, replacement of the acetylacetonate ligand of FIr(acac) with picolinate (i.e., FIrpic) resulted in an additional ∼20 nm blueshift.

Organic light-emitting devices were grown on a glass substrate precoated with a ∼130-nm-thick indium–tin–oxide (ITO) layer with a sheet resistance of ∼20 Ω/□. Prior to organic layer deposition, the substrate was degreased with solvents and cleaned for 5 min by exposure to an UV–ozone ambient, after which it was immediately loaded into the evaporation system. With a base pressure of ∼4 × 10^{-8} Torr, the organic and metal cathode layers were grown successively without breaking vacuum using an *in vacuo* mask exchange mechanism. First, a 10-nm-thick copper phthalocyanine (CuPc) hole injection layer followed by a 30-nm-thick 4,4'-bis[N-(1-naphthyl)-N-phenyl-amino]biphenyl (α-NPD) hole transport layer (HTL) were deposited. Next, a 30-nm-thick light-emitting layer (EML) consisting of 6% FIrpic doped into a 4,4'-N,N'-dicarbazole-biphenyl (CBP) host was prepared via thermal codeposition. Finally, a 30-nm-thick layer

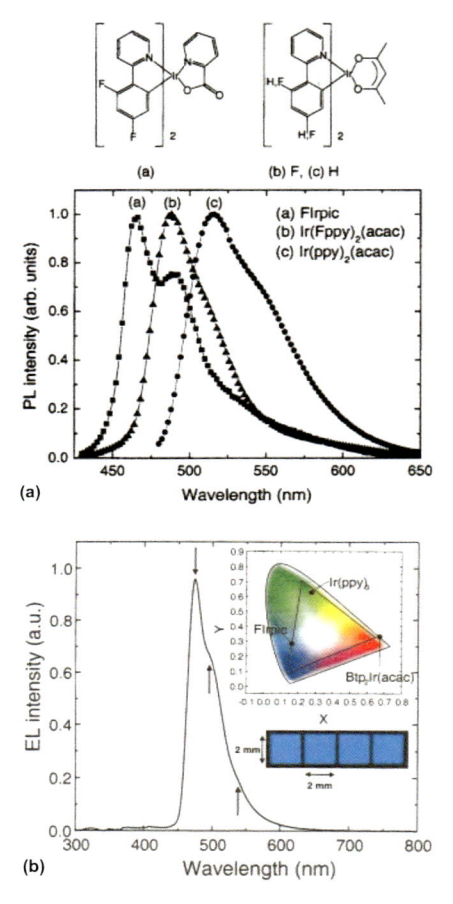

Figure 10.1 (a) Molecular structures of the iridium complexes: ppy$_2$Ir(acac), FIr(acac), and FIrpic, with their photoluminescence spectra in a dilute (10^{-5} M) chloroform solution, (b) Electroluminescence spectra of the following OLED structure: ITO/CuPc (10 nm)/α-NPD(30 nm)/CBP host doped with 6% FIrpic (30 nm)/BAlq (30 nm)/LiF (1 nm)/Al (100 nm). The EL spectrum has a maximum at the peak wavelength of $\lambda_{max} = 475$ nm and additional subpeaks at $\lambda_{sub} = 495$ and 540 nm (arrows), which agrees with the PL spectral shape. (Inset) CIE coordinates of FIrpic ($x = 0.16, y = 0.29$), Ir(ppy)$_3$ ($x = 0.28, y = 0.62$), and btp$_2$Ir(acac) ($x = 0.67, y = 0.33$), and a color photograph of an array of four FIrpic OLEDs.

of 4-biphenyloxolato aluminum(III)bis(2-methyl-8-quinolinato)4-phenylphenolate (BAlq) was used to transport and inject electrons into the EML. A shadow mask with rectangular 2 mm × 2 mm openings was used to define the cathode consisting of a 1-nm-thick LiF layer, followed by a 100-nm-thick Al layer. After deposition, the device was encapsulated using an UV-epoxy resin under a nitrogen atmosphere with <1 ppm oxygen and water. Given that the peak CBP triplet wavelength [9] is $\lambda = 484$ nm [(2.56 ± 0.10) eV], compared to $\lambda = 475$ nm [(2.62 ± 0.10) eV] for FIrpic (see spectra in Fig. 10.3), endothermic transfer may be interrupted by nonradiative defect states of intermediate energy. Introduction of oxygen or water may be the source of such defects. Indeed, we have found that breaking vacuum at any point in the fabrication process and exposure to air or purified oxygen (<1 ppm oxygen and water) results in a decrease in efficiency of at least a factor of 2 below the values reported here. A similar ambient sensitivity is not observed for green and red electrophosphorescence OLEDs employing conventional exothermic energy transfer mechanisms.

Figure 10.1b shows the EL spectrum with a maximum at the peak wavelength of $\lambda_{max} = 475$ nm and additional sub-peaks at $\lambda_{sub} = 495$ and 540 nm (arrows), which generally agrees with the PL spectral shape. The Commission Internationale de L'Eclairage (CIE) coordinates of $(x = 0.16, y = 0.29)$ for a FIrpic OLED is shown in the inset of Fig. 10.1b along with the coordinates of green [Ir(ppy)$_3$] and red [Btp$_2$Ir(acac)] electrophosphorescence devices.

Figure 10.2 shows η_{ext} and η_p as functions of current density. A maximum $\eta_{ext} = (5.7 ± 0.3)\%$ and a luminous power efficiency (η_p) of $(6.3 ± 0.3)$lm/W are achieved at $J = 5$ and 0.1 mA/cm^2, respectively. While the device shows a gradual decrease in η_{ext} with increasing current which has previously been attributed to triplet–triplet annihilation [12], a maximum luminance of 6400 cd/m^2 with $\eta_{ext} = 3.0\%$ was obtained even at a high current of $J = 100$ mA/cm^2. These values compare favorably with $\eta_{ext} = 2.4\%$ for fluorescent devices with a similar blue color emission spectrum. Since the triplet energy level of a CBP host $(2.56 ± 0.10)$ eV is slightly less than that of FIrpic at $(2.62 ± 0.10)$ eV (inset of Fig. 10.2), exothermic energy transfer from FIrpic to CBP is inferred. The pronounced roll off at small J is indicative of the sensitivity of backward energy transfer to

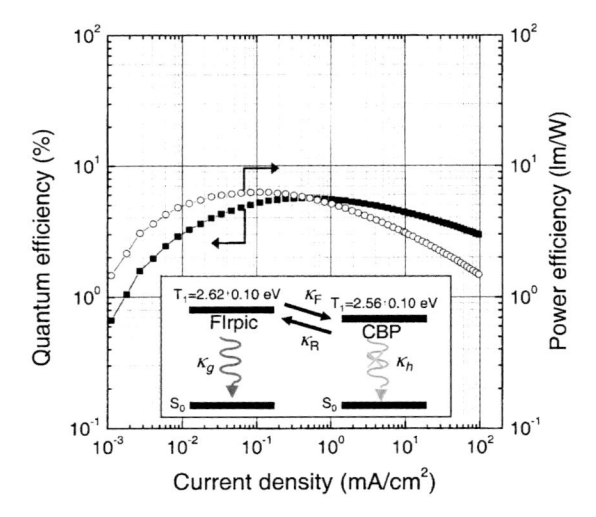

Figure 10.2 External electroluminescent quantum (η_{ext}: filled squares) and power (η_p : open circles) efficiencies of the following OLED structure: ITO/CuPc (10 nm)/α-NPD (30 nm)/CBP host doped with 6% FIrpic (30 nm)/BAlq (30 nm)/LiF (1 nm)/Al (100 nm). (Inset) Energy-level diagram of triplet levels of a CBP host and a FIrpic guest. Due to the energy lineup of CBP and FIrpic triplet levels, both exothermic and endothermic transfer is possible. Here, κ_g and κ_h are the radiative decay rates of triplets on the guest (phosphor) and host molecules, and the rates of exothermic (forward) (κ_F) and endothermic (reverse) (κ_R) energy transfers between CBP and FIrpic are also indicated.

the presence of energy dissipative pathways, reducing the efficiency via nonradiative triplet recombination when the density of triplets is too low to saturate these parasitic mechanisms.

Figure 10.3 shows a streak image of the transient decay of a 6% FIrpic:CBP film at $T = 100$ K with two time-resolved emission spectra. In addition to the prompt phosphorescence of FIrpic, we observe an extremely long decay component lasting for $\tau \sim 10$ ms, which follows the CBP triplet lifetime. Since the PL spectrum of the slow component coincides with that of FIrpic PL, this supports the conclusion that exothermic energy transfer from FIrpic to CBP occurs. The triplet state then migrates through the CBP host molecules, and finally, is endothermally transferred back to FIrpic, resulting in the delayed phosphorescence observed. Due to the

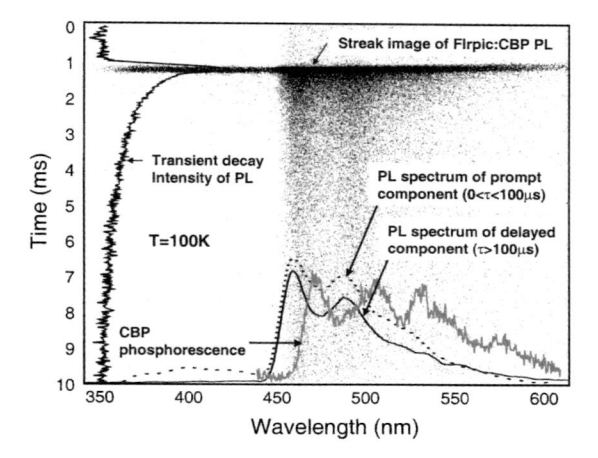

Figure 10.3 Streak image of a 6% FIrpic:CBP film (100 nm thick) on a Si substrate under nitrogen pulse excitation (∼500 ps) at $T = 100$ K. Two distinct decay processes, prompt and delayed phosphorescence, are demonstrated along with their photoluminescent spectra: dashed line = delayed and solid line = prompt. Also shown is the CBP phosphorescence spectrum obtained at 10 K.

significant difference of lifetimes of the excited states, $\kappa_h \ll \kappa_g$, (κ_h and κ_g are the radiative decay rates of the triplets on the host and guest molecules, respectively), the triplet exciton decay originates from FIrpic, as desired. The blue emission centered at $\lambda_{max} = 400$ nm in the prompt emission spectrum is due to fluorescence of CBP, with a transient lifetime ≪100 ns, which is significantly shorter than the decay of FIrpic.

Figure 10.4 shows the temperature dependence of the transient decay and the relative PL efficiency (η_{PL}) of FIrpic doped into CBP. After a slight enhancement of η_{PL} as the temperature is increased from 50 to 200 K, it once again decreases at yet higher temperatures. The transient decay characteristics are also temperature dependent. In particular, a significant decrease in the nonexponential decay time was observed at $T = 50$ and 100 K. The increase of η_{PL} from $T = 300$ to 200 K is due to the suppression of nonradiative decay of FIrpic. The decrease below $T \sim 200$ K, however, is a signature of retardation of the endothermic process of energy transfer from CBP to FIrpic, leading to loss of the radiative triplet excitons. Since

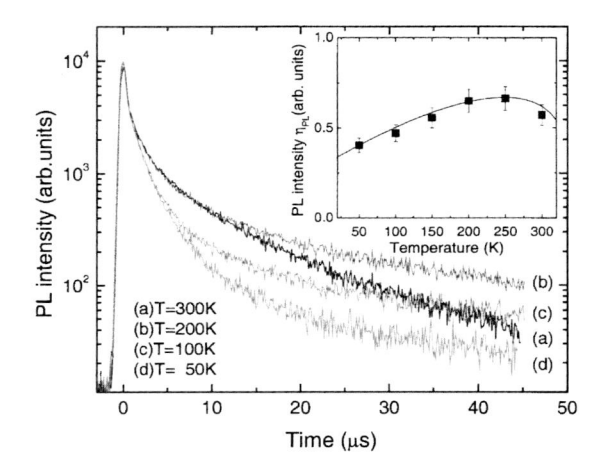

Figure 10.4 Transient photoluminescence decay characteristics of a 6% FIrpic:CBP film (100 nm thick) on a Si substrate under nitrogen pulse excitation (~500 ps) at $T = 50, 100, 200$, and 300 K. (Inset) Temperature dependences of the relative photoluminescence efficiency (η_{PL}) of the film.

we observe no delayed component at $T = 300$ K, energy transfer from CBP to FIrpic is very efficient with thermal assistance. In contrast, the PL intensity of Ir(ppy)$_3$:CBP shows no temperature dependence along with no evidence for such a slow component at low temperature, suggesting the absence of backward energy transfer in that system.

In summary, we demonstrated efficient blue electrophosphorescence using FIrpic as the phosphor molecule. The transient phosphorescence decay suggests the presence of endothermic energy transfer between the phosphor and the conductive CBP host.

This work was funded by the Universal Display Corporation, the Defense Advanced Research Projects Agency, and the Air Force Office of Scientific Research.

References

1. M. A. Baldo, D. F. O'Brien, Y. You, A. Shoustikov, S. Sibley, M. E. Thompson, and S. R. Forrest, *Nature (London)* **395**, 151 (1998).

2. M. A. Baldo, S. Lamansky, P. E. Burrows, M. E. Thompson, and S. R. Forrest, *Appl. Phys. Lett.* **75**, 4 (1999).
3. C. Adachi, M. A. Baldo, and S. R. Forrest, *Appl. Phys. Lett.* **77**, 904 (2000).
4. C. Adachi, S. Lamansky, M. A. Baldo, R. C. Kwong, M. E. Thompson, and S. R. Forrest, *Appl. Phys. Lett.* **78**, 1622 (2001).
5. C. Adachi, M. A. Baldo, M. E. Thompson, and S. R. Forrest, *Bull. Am. Phys. Soc.* **46**, 863 (2001).
6. M. A. Baldo and S. R. Forrest, *Phys. Rev. B* **62**, 10958 (2000).
7. W. E. Ford and M. A. J. Rogers, *J. Phys. Chem.* **96**, 2917 (1992).
8. S. Lamansky, P. Djurovich, D. Murphy, F. Abdel-Razzaq, C. Adachi, P. E. Burrows, S. R. Forrest, and M. E. Thompson, *J. Am. Chem. Soc.* **123**, 4304 (2001).
9. A. W. Grice, D. D. C. Bradley, M. T. Bernius, M. Inbasekaran, W. W. Wu, and E. P. Woo, *Appl. Phys. Lett.* **73**, 629 (1998).
10. C. Hosokawa, H. Higashi, H. Nakamura, and T. Kusumoto, *Appl. Phys. Lett.* **67**, 3853 (1995).
11. C. Hosokawa, M. Eida, M. Matsuura, K. Fukuoka, H. Nakamura, and T. Kusumoto, *Synth. Met.* **91**, 3 (1997).
12. M. A. Baldo, C. Adachi, and S. R. Forrest, *Phys. Rev. B* **62**, 10967 (2000).

Chapter 11

High-Efficiency Yellow Double-Doped Organic Light-Emitting Devices Based on Phosphor-Sensitized Fluorescence

Brian W. D'Andrade,[a] Marc A. Baldo,[a] Chihaya Adachi,[a] Jason Brooks,[b] Mark E. Thompson,[b] and Stephen R. Forrest[a]

[a] *Center for Photonics and Optoelectronic Materials,*
Department of Electrical Engineering and the Princeton Materials Institute,
Princeton University, Princeton, New Jersey 08544, USA
[b] *Department of Chemistry, University of Southern California,*
Los Angeles, California 90089, USA
forrest@princeton.edu

We demonstrate high-efficiency yellow organic light-emitting devices (OLEDs) employing [2-methyl-6-[2,3,6,7-tetrahydro-1H,5H-benzo[ij]quinolizin-9-yl]ethenyl]-4H-pyran-4-ylidene] propanedinitrile (DCM2) as a fluorescent lumophore, with a green electrophosphorescent sensitizer, fac tris(2-phenylpyridine) iridium [Ir(ppy)$_3$] co-doped into a 4,4'-N,N'-dicarbazole-biphenyl host. The devices exhibit peak external fluorescent quantum and power efficiencies of 9% ± 1% (25 cd/A) and 17 ± 2 lm/W at 0.01 mA/cm^2,

---

Reprinted from *Appl. Phys. Lett.*, **79**(7), 1045–1047, 2001.

Electrophosphorescent Materials and Devices
Edited by Mark E. Thompson
Text Copyright © 2001 American Institute of Physics
Layout Copyright © 2024 Jenny Stanford Publishing Pte. Ltd.
ISBN 978-981-4877-34-3 (Hardcover), 978-1-003-08872-1 (eBook)
www.jennystanford.com

respectively. At 10 mA/cm², the efficiencies are 4.1% ± 0.5% (11 cd/A) and 3.1 ± 0.3 lm/W. We show that this exceptionally high performance for a fluorescent dye is due to the ∼100% efficient transfer of both singlet and triplet excited states in the doubly doped host to the fluorescent material using Ir(*ppy*)₃ as a sensitizing agent. These results suggest that 100% internal quantum efficiency fluorescent OLEDs employing this sensitization process are within reach.

Over the last two decades, advances in organic light-emitting device (OLED) efficiencies have been made through the synthesis of efficient lumophores, optimization of the OLED structure, and doping of OLEDs with highly emissive phosphorescent and fluorescent materials [1, 3, 4]. In particular, the highest external quantum efficiencies [4, 5] (η_{ext}) of ∼18% reported thus far employ the electrophosphorescent molecule fac tris(2-phenylpyridine) iridium [Ir(*ppy*)₃] doped into hosts such as 4,4'-*N,N*'-dicarbazole-biphenyl (CBP). The large η_{ext} of these devices is attributed to emissive triplet excitons that set an upper limit to the internal quantum efficiency (η_{int}) of 100%. In contrast, given a statistical limit of 1:3 for the singlet-to-triplet exciton ratio in molecular organic materials [6], the maximum η_{ext} for fluorescent materials is only ∼5%, ultimately limited by the output coupling efficiency [2, 7] of ∼20% for OLEDs with flat glass substrates.

The recent demonstration of phosphor sensitization [8] provides a means for fluorescent OLEDs to approach efficiencies similar to those of electrophosphorescent devices through resonant energy transfer between triplet excitons in the phosphor and singlets in the fluorophore. Phosphor sensitization works as follows: By doping a phosphor at high concentrations (∼5–10 wt%) into a conductive host, both singlet and triplet excitons can transfer onto the phosphor molecule. If the phosphor contains a heavy metal atom, spin-orbit coupling transfers all excited states on the phosphor to the radiative triplet manifold. These radiative states can then be readily transferred via the dipole–dipole Förster [9] process to the radiative singlet state of the fluorophore co-doped with both the host and phosphor molecules. By lightly doping (∼1%) the fluorophore, hopping from the host triplets to the nonradiative triplet state of the fluorescent molecule is discouraged.

In principle, therefore, phosphor sensitization can lead to 100% internal quantum efficiency of OLEDs radiating from the singlet manifold of the fluorescent dopant molecules.

Here, we describe high-efficiency fluorescent OLEDs using a phosphorescent sensitizer. These OLEDs were improved and simplified from previous work [8] by uniformly co-doping the two dyes into the host material instead of using thin layers of the host alternately doped with the phosphorescent sensitizer and the fluorophore. At low current densities, we obtain a maximum external quantum efficiency of 9% ± 1% (corresponding to an internal efficiency of $\sim$50%) and a power efficiency of 17 ± 2 lm/W for fluorescent materials. Moreover, η_{ext} is found to be equal to that obtained in control devices containing only the phosphor, suggesting that the exciton transfer efficiency from phosphor to fluorophore is $\sim$100%. This result indicates that phosphor-sensitized fluorescence provides a realistic route for obtaining 100% internal electroluminescence in fluorescent OLEDs.

As previously [8], the host material was CBP, and the dopants were Ir(*ppy*)$_3$ and the yellow-red fluorescent dye [10] [2-methyl-6-[2,3,6,7-tetrahydro-1H,5H-benzo[ij]quinolizin-9-yl)ethenyl]-4H-pyran-4-ylidene]propane-dinitrile (DCM2). Organic layers were deposited by high-vacuum (10^{-7} Torr) thermal evaporation onto a precleaned and uv–ozone [11] treated glass substrate pre-coated with an indium–tin–oxide (ITO) anode having a sheet resistance of $\sim$20 Ω/sq. A 40-nm-thick film of *N,N*-diphenyl-*N,N*-bis(3-methylphenyl)-[1,1-biphenyl]-4,4-diamine (TPD) served as the hole-transport layer (HTL). Next, a 30-nm-thick light-emitting layer (EML) consisting of the CBP host was deposited while being simultaneously doped with 4–12 wt% Ir(*ppy*)$_3$, and 0.1–1.3 wt% DCM2 via thermal co-deposition from three independent source boats. A 12-nm-thick layer of bathocuproine (BCP) was used to confine excitons [12] in the EML. By trapping the excitons, their residence time and recombination probability in the EML were increased, leading to a concomitant increase in OLED efficiency. A 30-nm-thick layer of tris-(8-hydroxy-quinoline)aluminum (Alq$_3$) was used to transport and inject electrons into the EML. A shadow mask with 1-mm-diam openings was used to define the cathode consisting of a 100-nm-thick 10:1 layer with a 30-nm-thick silver

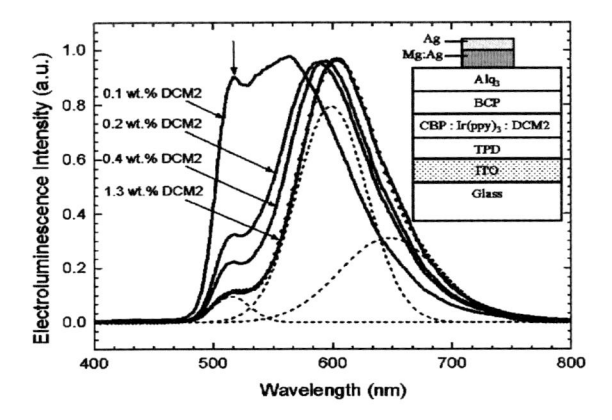

Figure 11.1 Electroluminescence spectrum of devices doped with 0.1, 0.2, 0.4, and 1.3% DCM2 and 8 wt% Ir(ppy)$_3$ in CBP. The intensity of Ir(ppy)$_3$ emission, indicated by the arrow, decreases, and the DCM2 peak emission redshifts with increasing DCM2 concentration. Gaussian fits of the spectral components of the device doped with 1.3 wt% DCM2 are shown as dashed lines. Inset: Schematic cross section of a phosphor-sensitized fluorescent OLED.

cap. A schematic cross section of the device structure is shown in the inset of Fig. 11.1.

Electroluminescence (EL) was only observed from Ir(ppy)$_3$ and DCM2 in all double-doped devices, indicating efficient transfer of excitons from the CBP host to the dopants, in addition to direct formation of excitons on Ir(ppy)$_3$. Figure 11.1 (dashed lines) shows the emission of a device doped with 8 wt% Ir(ppy)$_3$ and 1.3 wt% DCM2, fit using three Gaussian curves by an iterative chi-square minimization routine with peaks at wavelengths of $\lambda = 520$, 600, and 650 nm. The summation of all three Gaussians is shown by the dotted line overlapping the actual spectrum. From the ratio of the areas of the two DCM2 Gaussian spectral components to the total area under the curve, we can calculate the percentage of photons emitted from DCM2. Nearly 100% DCM2 emission was observed with 1.3 wt% DCM2, and ~86% with a doping concentration of 0.1 wt% DCM2. Figure 11.1 also shows the reduction in the residual Ir(ppy)$_3$ emission intensity (whose peak, indicated by the vertical arrow, is at $\lambda = 520$ nm) for devices doped with an increasing

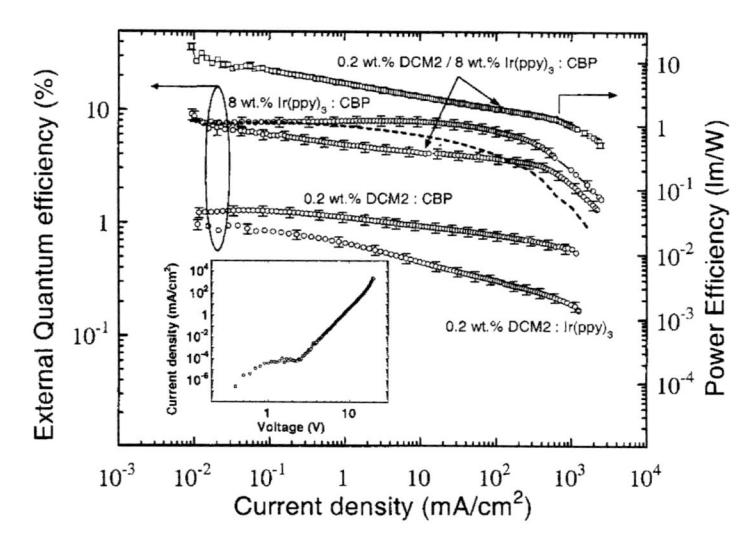

Figure 11.2 External quantum and power efficiencies of single- and double-doped OLEDs vs current density. A maximum external DCM2 quantum efficiency of $9 \pm 1\%$ (25 cd/A) and power efficiency of 17 ± 2 lm/W are obtained for the 8 wt% Ir$(ppy)_3$/0.2 wt% DCM2:CBP device. Dashed line shows the convolution between the 8 wt% Ir$(ppy)_3$:CBP and the 0.2 wt % DCM2:CBP emission curves. Inset: Current–voltage characteristic of an 8 wt% Ir$(ppy)_3$/0.2 wt% DCM2:CBP device.

concentration of DCM2. Furthermore, a red spectral shift of the DCM2 peak from $\lambda = 560$ nm to $\lambda = 600$ nm is observed with increasing DCM2 concentration. Since DCM2 is a polar molecule, higher doping concentrations lead to a redshift in the emission [13] along with a substantial increase in aggregate-induced quenching, which lowers the quantum efficiency. There was no significant variation in the percentage of photons emitted by DCM2 between devices with different Ir$(ppy)_3$ concentrations, provided that the DCM2 concentration was held fixed. This indicates efficient energy transfer from Ir$(ppy)_3$ to DCM2 at all concentration combinations employed.

The maximum DCM2 external quantum efficiency at low current densities ($J \leq 10^{-2}$ mA/cm^2) was $\eta_{\text{ext}} = 9 \pm 1\%$ (25 cd/A) for an 8 wt% Ir$(ppy)_3$/0.2 wt% DCM2-doped device, corresponding to a maximum power efficiency of $\eta_p = 17 \pm 2$ lm/W (see Fig. 11.2). The

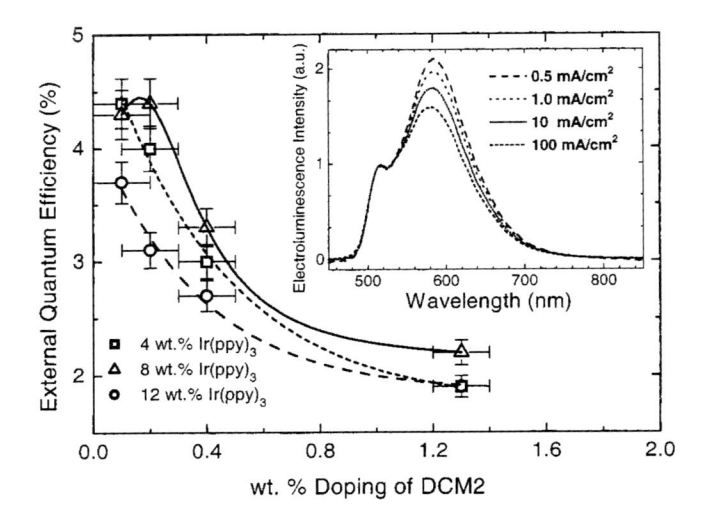

Figure 11.3 External quantum efficiencies at 10 mA/cm² of devices co-doped with various concentrations of Ir(ppy)₃ and DCM2. Inset: electroluminescent emission intensity as a function of wavelength for an 8 wt% Ir(ppy)₃/0.2 wt% DCM2 OLED at several current densities. DCM2 emission at 600 nm decreases from 96% at 0.5 mA/cm² to 93% at 100 mA/cm².

device had a luminance of 100 cd/m² at 8.3 V (corresponding to ∼1 mA/cm²), calculated using current–voltage characteristics shown in the inset of Fig. 11.2. The variation in the hue was small over several orders of magnitude in current density, as inferred from the EL spectra shown in Fig. 11.3, inset. The value of η_{ext} is double that of previously reported [8] sensitized devices between current densities of 0.1 and 30 mA/cm², and nearly three times higher at current densities below 0.1 mA/cm², where η_{ext} is comparable to a purely phosphorescent Ir(ppy)³ device, also shown in Fig. 11.2. This equivalence in maximum η_{ext} for the phosphor and the phosphor-sensitized fluorescent devices indicates that the exciton transfer efficiency between Ir(ppy)³ and DCM2 is ∼100%. Given that only radiative triplet states produced by spin-orbit coupling on the Ir(ppy)₃ molecule can transfer by Förster processes to DCM2, this suggests that the phosphor has successfully transferred all excited states from the host to the singlet manifold of the fluorescent dye.

A slow decrease in η_{ext} between 0.01 mA/cm² at 4.7 V and 300 mA/cm² at 17 V is observed for all double-doped systems. It is also observed in the purely fluorescent DCM2:CBP OLED and the device using Ir(*ppy*)₃ as the host doped with 0.2 wt% DCM2. Figure 11.2 shows that the decrease in η_{ext} with current of the 0.2 wt% DCM2:Ir(*ppy*)₃ device is faster than that of the 0.2 wt% DCM2:CBP device, indicative of additional nonradiative energy pathways available between Ir(*ppy*)₃ and DCM2, which are eliminated in the double-doped system. Furthermore, aggregate quenching of Ir(*ppy*)₃ excitons substantially lowers the maximum η_{ext} of the 0.2 wt% DCM2:Ir(*ppy*)₃ device to <1%.

Efficiencies of double-doped devices roll off sharply at $J \geq$ 300 mA/cm², similar to the 8 wt% Ir(*ppy*)₃:CBP device. This roll off is due to triplet–triplet $(T-T)$ annihilation [14] between the Ir(*ppy*)₃ and host triplets. The product of the efficiency of the 8 wt% Ir(*ppy*)₃:CBP and that from the 0.2 wt% DCM2:CBP devices results in the dashed curve shown in Fig. 11.2. The roll off in η_{ext} at high currents of these structures precedes that in the double-doped device, indicating that $T-T$ annihilation in Ir(*ppy*)₃ is reduced by the presence of the DCM2. This reduced annihilation is most probably due to rapid Förster transfer of the radiative Ir(*ppy*)₃ triplet to the singlet manifold of DCM2. Hence, phosphor-sensitized fluorescence provides a mechanism for substantially reducing $T-T$ quenching [4, 5].

We attribute the exceptionally high efficiencies observed in DCM2 to the mixing of the two dopants with the host afforded by the co-deposition process. We can estimate the concentrations of dopants required to maximize η_{ext} by considering the Förster radii and hopping distances between different molecules in the EML layer. Hopping occurs between nearest-neighbor molecules that have separation distances of ≤5 Å since intermolecular orbital overlap determines the transfer probability. In contrast, long-range Förster singlet–singlet transfer radii and radiative triplet–singlet transfer radii extend over a considerably longer range. These radii have been calculated [9] for the several material combinations employed, with the results provided in Table 11.1.

Given the density of thin films of ∼1.3 g/cm³, the weight ratio doping of the EML of 8 wt% Ir(*ppy*)₃/0.2 wt% DCM2 and the

Table 11.1 Förster transfer radii and the internal photoluminescence efficiencies (η^0) of the different molecules studied

Material combination	Förster transfer radii/Å	η^0
CBP–DCM	30 ± 3	0.53 ± 0.05
CBP–Ir(ppy)$_3$	29 ± 3	0.66 ± 0.06
Ir(ppy)$_3$–DCM2	40 ± 4	0.66 ± 0.06

molecular masses of the organic materials, we find that each DCM2 molecule lies in a spherical volume of radius 38 Å, close to a calculated Förster radius of 40 ± 4 Å. At >1 wt% DCM2 doping, there should be little improvement in the transfer efficiency and a concomitant increase in the concentration quenching [10] of DCM2 emission, consistent with the results in Fig. 11.3.

By independently varying concentrations of the dopants, we optimized the efficiency of Ir(ppy)$_3$ and DCM2 OLEDs, with results shown in Fig. 11.3. The highest η_{ext} of similarly doped DCM2 OLEDs were observed for devices containing 8 wt% Ir(ppy)$_3$. Hopping between CBP and Ir(ppy)$_3$ is less likely at lower doping concentrations since fewer CBP molecules have Ir(ppy)$_3$ as a nearest neighbor. At higher doping levels, concentration quenching of Ir(ppy)$_3$ emission is dominant [6]. Thus, we expect lower η_{ext} for devices doped with 4 and 12 wt% Ir(ppy)$_3$, as observed. Also, there is a noticeable decrease in the efficiency of the devices as the concentration of DCM2 is increased due to concentration quenching of singlets transferred to that molecule.

In conclusion, we have demonstrated that co-doping of the conductive host molecule CBP with <1 wt% DCM2 and an Ir(ppy)$_3$ sensitizer considerably improves the external fluorescence quantum efficiency of OLEDs. Optimized devices exhibit $\eta_{ext} = 9 \pm 1\%$ (25 cd/A) and $\eta_p = 17 \pm 2$ lm/W at 0.01 mA/cm^2. The results show that Förster transfer between Ir(ppy)$_3$ and DCM2 can be made ~100% efficient leading to fluorescent OLEDs whose efficiency is equal to that of purely electrophosphorescent OLEDs. In addition, phosphor-sensitized fluorescent OLEDs exhibit reduced $T-T$ annihilation as compared with purely phosphor-doped devices due to the rapid and efficient transfer of radiative triplet states to the singlet manifold of the fluorescent molecular dopant.

The authors thank the Universal Display Corporation the Defense Advanced Research Projects Agency, and the National Science Foundation's Materials Research Science and Engineering Center for their support.

References

1. C. W. Tang and S. A. VanSlyke, *Appl. Phys. Lett.* **51**, 913 (1987).
2. V. Bulovic, V. B. Khalfin, G. Gu, and P. E. Burrows, *Phys. Rev. B* **58**, 3730 (1998).
3. M. A. Baldo, D. F. O'Brien, Y. You, A. Shoustikov, S. Sibley, M. E. Thompson, and S. R. Thompson, *Nature (London)* **395**, 151 (1998).
4. M. A. Baldo, S. Lamansky, P. E. Burrows, M. E. Thompson, and S. R. Forrest, *Appl. Phys. Lett.* **75**, 4 (1999).
5. C. Adachi, M. A. Baldo, S. R. Forrest, and M. E. Thompson, *Appl. Phys. Lett.* **77**, 904 (2000).
6. M. A. Baldo, D. F. O'Brien, M. E. Thompson, and S. R. Forrest, *Phys. Rev. B* **60**, 14422 (1999).
7. G. Gu, D. Z. Garbuzov, P. E. Burrows, S. Venkatesh, S. R. Forrest, and M. E. Thompson, *Opt. Lett.* **22**, 396 (1997).
8. M. A. Baldo, M. E. Thompson, and S. R. Forrest, *Nature (London)* **403**, 750 (2000).
9. T. Förster, *Discuss. Faraday Soc.* **27**, 7 (1959).
10. C. W. Tang, S. A. VanSlyke, and C. H. Chen, *J. Appl. Phys.* **65**, 3610 (1989).
11. D. J. Milliron, I. G. Hill, C. Shen, A. Kahn, and J. Schwartz, *J. Appl. Phys.* **87**, 572 (2000).
12. D. F. O'Brien, M. A. Baldo, M. E. Thompson, and S. R. Forrest, *Appl. Phys. Lett.* **74**, 442 (1999).
13. V. Bulovic, A. Shoustikov, M. A. Baldo, E. Bose, V. G. Kozlov, M. E. Thomspon, and S. R. Forrest, *Chem. Phys. Lett.* **287**, 455 (1998).
14. M. A. Baldo, C. Adachi, and S. R. Forrest, *Phys. Rev. B* **62**, 10967 (2000).

Chapter 12

High-Efficiency Red Electrophosphorescence Devices

Chihaya Adachi,[a] Marc A. Baldo,[a] Stephen R. Forrest,[a]
Sergey Lamansky,[b] Mark E. Thompson,[b] and
Raymond C. Kwong[c]

[a] *Center for Photonics and Optoelectronic Materials,*
Department of Electrical Engineering, Princeton University,
Princeton, New Jersey 08544, USA
[b] *Department of Chemistry, University of Southern California,*
Los Angeles, California 90089, USA
[c] *Universal Display Corporation, 375 Phillips Boulevard, Ewing,*
New Jersey 08618, USA
forrest@ee.princeton.edu

We demonstrate high-efficiency red electrophosphorescent organic light-emitting devices employing *bis*(2-(2'-benzo[4,5-a]thienyl)pyridinato-N,C$^{3'}$) iridium(acetylacetonate) [Btp$_2$Ir(acac)] as a red phosphor. A maximum external quantum efficiency of $\eta_{ext} = (7.0 \pm 0.5)\%$ and power efficiency of $\eta_p = (4.6 \pm 0.5)$ lm/W are achieved at a current density of $J = 0.01$ mA/cm^2. At a higher current density of $J = 100$ mA/cm^2, $\eta_{ext} = (2.5 \pm 0.3)\%$ and $\eta_p = (0.56 \pm 0.05)$ lm/W are obtained. The electroluminescent spectrum has a maximum at a wavelength of $\lambda_{max} = 616$ nm

---

Reprinted from *Appl. Phys. Lett.*, **78**(11), 1622–1624, 2001.

Electrophosphorescent Materials and Devices
Edited by Mark E. Thompson
Text Copyright © 2001 American Institute of Physics
Layout Copyright © 2024 Jenny Stanford Publishing Pte. Ltd.
ISBN 978-981-4877-34-3 (Hardcover), 978-1-003-08872-1 (eBook)
www.jennystanford.com

with additional intensity peaks at λ_{sub} = 670 and 745 nm. The Commission Internationale de L'Eclairage coordinates of (x = 0.68, y = 0.32) are close to meeting video display standards. The short phosphorescence lifetime (~4 µs) of Btp$_2$Ir(acac) leads to a significant improvement in η_{ext} at high currents as compared to the previously reported red phosphor, 2,3,7,8,12,13,17,18-octaethyl-12H, 23H-prophine platinum (II) PtOEP with a lifetime of ~50 µs.

Heavy-metal complexes [1, 2], where strong spin-orbit coupling leads to singlet–triplet state mixing, can result in high-efficiency electrophosphorescence in organic light-emitting devices (OLEDs) [3–6]. For example, OLEDs employing the phosphor, *fac* tris(2-phenylpyridine)iridium [Ir(ppy)$_3$], exhibit green emission with an external quantum efficiency (η_{ext}) of ~15% [6]. By designing appropriate ligands for heavy-metal complexes, high-efficiency electrophosphorescence at other emission colors is anticipated for high-performance full-color display applications [7]. Both red-emitting fluorescent and phosphorescent dopants have shown promise for use in OLEDs. While fluorescent dyes, including 4-dicyanomethylene-2-methyl-6-[2-(2,3,6,7-tetra-hydro-1H, 5Hbenzo[ij]quinolizin-8-yl)-4H-pyran (DCM2) [8, 9] and porphyrin derivatives (TPP [10], ZnTPP [11], and TPC [12]), have been investigated, their maximum external quantum efficiencies are limited to less than 1%. Furthermore, the red phosphorescent europium complexes (EuL$_3$) [13, 14] and 2,3,7,8,12,13,17,18-octaethyl-12H,23Hporphine platinum (II) (PtOEP) have been studied [3, 4]. Although EuL$_3$ complexes show very strong red Eu^{3+} ion emission (with an internal quantum efficiency η_{int} > 80%) [15], the long lifetime of the Eu^{3+} excited state (350 µs) results in pronounced triplet–triplet (T–T) annihilation at high current [14]. In addition, the high triplet energy of the ligands causes backward energy transfer to the host molecule, leading to η_{ext} < 1.4% [14]. The highest η_{ext} = 5.6% for a red phosphor has been achieved with PtOEP doped into 4,4'-N,N'-dicarbazole-biphenyl (CBP) [4]. However, the relatively long phosphorescence lifetime (~80 µs) again results in T–T annihilation at high current [16].

In this study, we demonstrate red organometallic phosphors characterized by a high quantum efficiency and a short radiative lifetime [17]. The metal complexes contain cyclometalated benzothienylpyridine ligands, i.e., bis(2-(2'-benzo[4,5-α]thienyl) pyridinato-N,C$^{3'}$) iridium (acetyl-acetonate) [btp$_2$Ir(acac)] [7] and (2-(2'-benzo[4,5-α]thienyl)pyridinato-N,C$^{3'}$) platinum (acetylacetonate) [btpt(acac)], shown in Fig. 12.1. The synthesis of btpPt(acac) was prepared by a method analogous to that used [7] for btp$_2$Ir(acac). The small $\pi-\pi^*$ transition energy of the btp ligand relative to other ligands used in heavy-metal phosphors leads to a low-energy triplet excited state, giving strong red phosphorescence. OLEDs employing these phosphors were grown by high vacuum (10^{-6} Torr) thermal evaporation onto pre-cleaned glass substrates as described elsewhere [6]. Device I is comprised of a 20 $\Omega/\square$ indium tin oxide (ITO) anode, a 50-nm-thick 4,4'-bis[N-(1-naphthyl)-N-phenyl-amino]biphenyl hole transport layer (HTL), a 20-nm-thick light-emitting layer (EML) consisting of a conductive CBP host doped with ~7% of the guest phosphor, a 10-nm-thick 2,9-dimethyl-4,7-diphenyl-phenanthroline hole and exciton blocking layer [4], a 65-nm-thick tris(8-hydroxyquinoline)aluminum (Alq$_3$) electron transport layer (ETL), and a cathode comprised of a 100-nm-thick (10:1) MgAg layer, with a further 20 nm Ag deposited as a protective cap (Fig. 12.1). Device II has the ITO anode, a 60-nm-thick 4,4'-bis[N,N'-(3-tolyl)amino]-3,3'-dimethylbiphenyl [6] HTL, a 25-nm-thick 2,2',2"-(1,3,5-benzenetriyl)tris[1-phenyl-1H-benzimidazole] (TPBI) [18] EML doped with ~7% phosphor, a 50-nm-thick Alq$_3$ ETL, and the same cathode as in device I.

Figure 12.2 shows η_{ext} and the power efficiency (η_p) as functions of current density for device I with a btp$_2$Ir(acac) guest. A maximum $\eta_{ext} = (7.0 \pm 0.5)\%$ and $\eta_p = (4.6 \pm 0.5)$ lm/W, and a luminance of 6.5 cd/m^2 were obtained at $J = 0.1$ mA/cm^2. The device showed a gradual decrease of η_{ext} with increasing current, which has been previously attributed to T–T annihilation [14, 19]. Nevertheless, at $J = 100$ mA/cm^2, the device efficiency remained high at $\eta_{ext} = (2.5 \pm 0.3)\%$ and $\eta_p = (0.56 \pm 0.05)$ lm/W. A maximum luminance of 6800 cd/m^2 was obtained at $J = 690$ mA/m^2.

Table 12.1 summarizes the electroluminescence characteristics of btp$_2$Ir(acac), btpPt(acac), PtOEP and Eu(TTA)$_3$phen (TTA = thenoyltrifluoroacetone, phen = 1,10-phenanthroline) guests in devices I and II. A maximum $\eta_{ext} \sim 7\%$ was obtained with btp$_2$Ir(acac) in both devices I and II, and PtOEP in device II. Also,

Table 12.1 Red electrophosphorescent OLED characteristics. External quantum efficiency (η_{ext}), and power efficiency (η_p) are given as functions of current density. Peak wavelength (λ_{max}) in electroluminescent spectrum. Transient electrophosphorescent time (τ) under electrical pulse excitation with pulse width of 500 ns. The characteristic current (J_0) of triplet–triplet annihilation

		η_{ext} (%) [η_p (lm/W)] at J/(mA/cm^2)						λ_{max} (nm)	τ (μs)	J_0 (mA/cm^2)
		$J = 0.01$	0.1	1	10	100	1000			
Btp$_2$Ir(acac)	Device I	7.0 (4.6)	6.7 (3.1)	6.0 (2.1)	4.4 (1.3)	2.5 (0.56)	0.69 (0.16)	616	6.5	27.4
	Device II	6.9 (5.7)	6.8 (3.6)	5.9 (2.5)	3.7 (1.4)	1.6 (0.57)	0.53 (0.14)		4.0	11.0
BtpPt(acac)	Device I	2.7 (2.5)	2.7 (1.4)	2.6 (0.93)	1.9 (0.54)	1.0 (0.25)	0.37 (0.083)	610	9.3	37.7
	Device II	2.2 (1.7)	2.2 (1.1)	2.1 (0.72)	1.3 (0.34)	0.57 (0.12)	0.02 (0.036)		5.6	14.5
PtOEP	Device I	5.2 (1.3)	5.2 (1.1)	4.1 (0.64)	2.1 (0.25)	0.67 (0.066)	0.18 (0.015)	650	86.5	6.4
	Device II	6.9 (1.4)	6.9 (1.0)	4.3 (0.53)	1.9 (0.17)	0.5 (0.039)	0.10 (0.010)		49.6	2.6
Eu(TTA)$_3$ phen	Device I	1.4 (1.2)	1.3 (0.94)	1.2 (0.53)	0.68 (0.21)	0.21 (0.08)	...	614	350	3.6

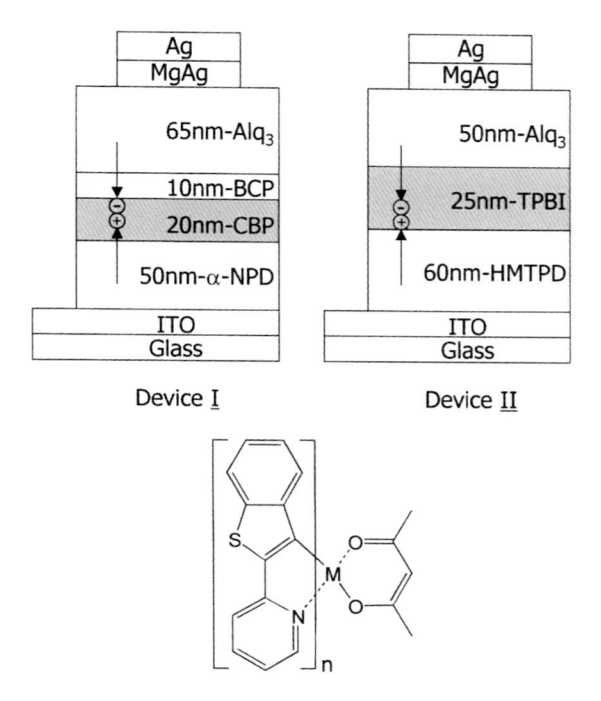

Figure 12.1 Device structures I and II. Molecular structure of the btp$_2$M(acac) phosphors (here $M =$ Ir, $n = 2$ and $M =$ Pt, $n = 1$).

the maximum η_p of btp$_2$Ir(acac) was (4.6 ± 0.5) lm/W which is significantly higher than that of PtOEP, since the spectral peak of the latter compound is ∼30 nm redshifted relative to btp$_2$Ir(acac). A pronounced improvement in η_{ext} of btp$_2$Ir(acac) was observed at high current. At $J = 100$ mA/cm^2, the btp$_2$Ir(acac) doped device I showed a relatively high $\eta_{ext} = (2.5 ± 0.5)\%$, compared with PtOEP and Eu(TTA)$_3$phen doped devices I with $\eta_{ext} = (0.61 ± 0.05)\%$ and $\eta_{ext} = (0.21 ± 0.05)\%$, respectively. The roll off in η_{ext} with J is consistent with the electrophosphorescent lifetimes, τ, of the several materials employed. As shown previously, the characteristic current (J_0) at which η_{ext} falls to 50% its peak value due to T–T annihilation is inversely proportional to τ^2 [14, 19]. This figure of merit of a phosphor is listed in Table 12.1 along with τ for the devices tested. The btp$_2$Ir(acac) lifetime is ∼12 times smaller than that of PtOEP, leading to considerably improved high-current performance of the

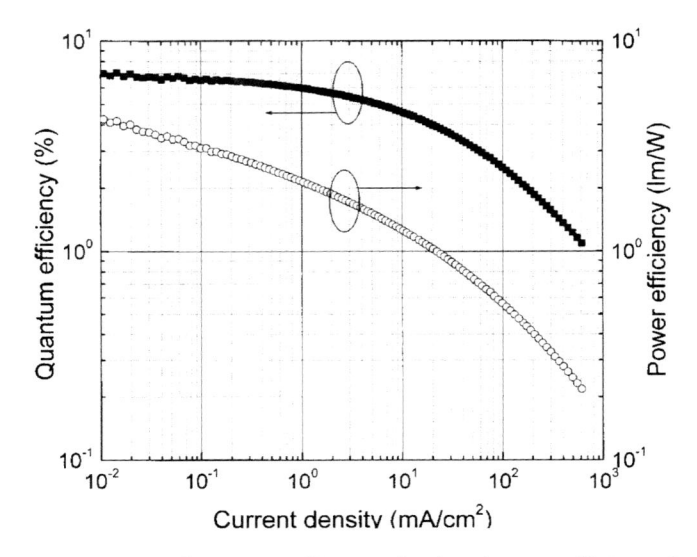

Figure 12.2 External quantum efficiency (η_{ext}) and power efficiency (η_p) vs current density for device I with a btp$_2$Ir(acac) guest.

Ir-based compound. The low η_{ext} of btpPt(acac) is consistent with its low photoluminescence efficiency [(8 ± 2)%] compared with that of btp$_2$Ir(acac) [(21 ± 5)%] and PtOEP [(40 ± 10)%].

The EL spectra originating from the triplet-excited states of the phosphors shown in Fig. 12.3 are coincident with those of the phosphors in a dilute solution. The Commission Internationale de L'Eclairage (CIE) coordinates for the four devices are compared with a fluorescent 2%-DCM2:Alq3 device ($x = 0.61, 0.39$) (inset of Fig. 12.2). Similar to the Eu(TTA)$_3$phen of ($x = 0.68, y = 0.32$), the btp$_2$Ir(acac) and btpPt(acac) doped devices demonstrate a saturated red emission ($x = 0.67, y = 0.33$) which is close to the National Television Standards Committee recommended red for a video display. Furthermore, the EL spectra and CIE coordinates of btp$_2$Ir(acac) in devices I and II are independent of current (Fig. 12.4). Even at $J > 100$ mA/cm^2, blue emission from either the CBP or TPBI host is negligible, indicating complete energy transfer under the excitation conditions used: a direct consequence of the short phosphor lifetime of this compound [20].

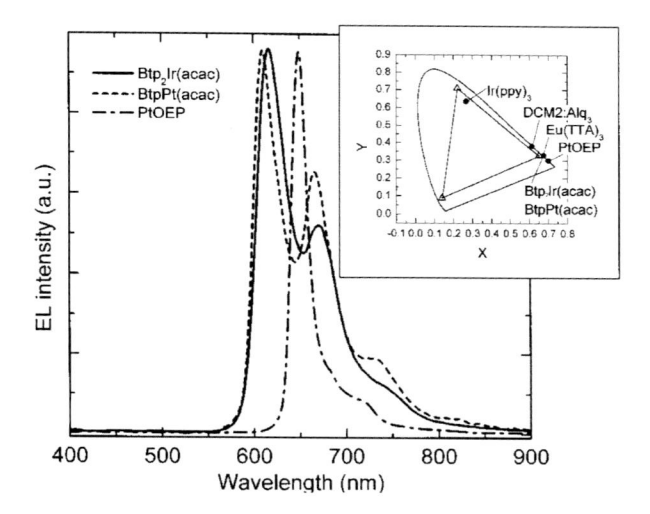

Figure 12.3 Electroluminescent spectra of device II employing btp$_2$Ir(acac), PtOEP or btpPt(acac) at a current density of $J = 1.3$ mA/cm^2. Inset: CIE coordinates for btp$_2$Ir(acac) ($x = 0.67, y = 0.33$), btpPt(acac) ($x = 0.67, y = 0.33$), PtOEP ($x = 0.70, y = 0.30$) and Eu(TTA)$_3$phen ($x = 0.68, y = 0.32$) doped devices and a 2%-DCM2: Alq$_3$ ($x = 0.61, y = 0.39$) device. The triangles show the NTSC recommended blue, green, and red coordinates.

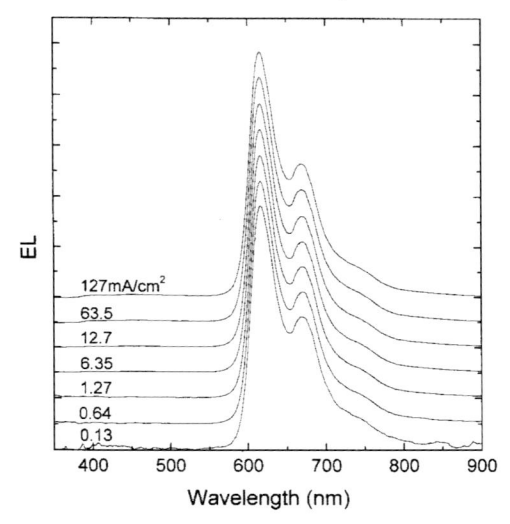

Figure 12.4 Electroluminescent spectra of device II with a btp$_2$(acac) guest dopant as a function of OLED drive current density.

In summary, we demonstrated high-efficiency, high-brightness red phosphorescent OLEDs employing benzothienylpyridine (btp) as a ligand in iridium and platinum complexes. Significant improvements in η_{ext} as compared with PtOEP were achieved due to the short phosphorescence lifetimes of < 10 μs of the new compounds studied, thereby minimizing T–T annihilation and saturation of the ligand excited state.

This work was funded by Universal Display Corporation, the Defense Advanced Research Projects Agency, and the Air Force Office of Scientific Research.

References

1. J.-M. Lehn, *Supermolecular Chemistry* (VCH, Weinheim, Germany, 1995).
2. C. A. Bignozzi, J. R. Schoonover, and F. Scandola, *Prog. Inorg. Chem.* **44**, 1 (1997).
3. M. A. Baldo, D. F. O'Brien, Y. You, A. Shoustikov, S. Sibley, M. E. Thompson, and S. R. Forrest, *Nature (London)* **395**, 151 (1998).
4. D. F. O'Brien, M. A. Baldo, M. E. Thompson, and S. R. Forrest, *Appl. Phys. Lett.* **74**, 442 (1999).
5. M. A. Baldo, S. Lamansky, P. E. Burrows, M. E. Thompson, and S. R. Forrest, *Appl. Phys. Lett.* **75**, 4 (1999).
6. C. Adachi, M. A. Baldo, and S. R. Forrest, *Appl. Phys. Lett.* **77**, 904 (2000).
7. S. Lamansky, P. Djurovich, D. Murphy, F. Abdel-Razzaq, H.-E. Lee, C. Adachi, P. E. Burrows, S. R. Forrest, and M. E. Thompson, *J. Am. Chem. Soc.* (submitted).
8. C. W. Tang, *SID Digest* **96**, 181 (1996).
9. Y. Hamada, H. Kanno, T. Tsujioka, H. Takahashi, and T. Usuki, *Appl. Phys. Lett.* **75**, 1682 (1999).
10. P. E. Burrows, S. R. Forrest, S. P. Sibley, and M. E. Thompson, *Appl. Phys. Lett.* **69**, 2959 (1996).
11. Y. Hamada, *IEEE Trans. Electron Devices* **44**, 1208 (1997).
12. Y. Sakakibara, S. Okutsu, T. Enokida, and T. Tani, *Appl. Phys. Lett.* **74**, 2587 (1999).
13. J. Kido, H. Hayase, K. Hongawa, K. Nagai, and K. Okamoto, *Appl. Phys. Lett.* **65**, 2124 (1994).

14. C. Adachi, M. A. Baldo, and S. R. Forrest, *J. Appl. Phys.* **87**, 8049 (2000).
15. G. L. J. A. Rikken, *Phys. Rev. A* **51**, 4906 (1995).
16. M. A. Baldo, M. E. Thompson, and S. R. Forrest, *Nature (London)* **403**, 750 (2000).
17. B. W. D'Andrade, M. A. Baldo, C. Adachi, M. E. Thompson, and S. R. Forrest (unpublished).
18. Y. T. Tao, E. Balasubramaniam, A. Danel, B. Jarosz, and P. Tomasik, *Appl. Phys. Lett.* **77**, 1575 (2000).
19. M. A. Baldo, C. Adachi, and S. R. Forrest, *Phys. Rev. B* **62**, 10967 (2000).
20. R. C. Kwong, S. Sibley, T. Dubovoy, M. A. Baldo, S. R. Forrest, and M. E. Thompson, *Chem. Mater.* **11**, 3709 (1999).

Chapter 13

Highly Phosphorescent Bis-Cyclometalated Iridium Complexes: Synthesis, Photophysical Characterization, and Use in Organic Light Emitting Diodes

Sergey Lamansky,[a] Peter Djurovich,[a] Drew Murphy,[a] Feras Abdel-Razzaq,[a] Hae-Eun Lee,[a] Chihaya Adachi,[b] Paul E. Burrows,[b,c] Stephen R. Forrest,[b] and Mark E. Thompson[a]

[a]*Department of Chemistry, University of Southern California, Los Angeles, California 90089, USA*
[b]*Department of Electrical Engineering, Princeton University, Princeton, New Jersey 08544, USA*
[c]*Current address: Pacific Northwest National Laboratory, K3-59 PO Box 999, Richland, Washington 99352, USA*
forrest@ee.princeton.edu, met@usc.edu

The synthesis and photophysical study of a family of cyclometalated iridium(III) complexes are reported. The iridium complexes have two cyclometalated ($C^\wedge N$) ligands and a single monoanionic, bidentate ancillary ligand (LX), i.e., $C^\wedge N_2$Ir(LX). The $C^\wedge N$ ligands

---

Reprinted from *J. Am. Chem. Soc.*, **123**, 4304–4312, 2001.

Electrophosphorescent Materials and Devices
Edited by Mark E. Thompson
Text Copyright © 2001 American Chemical Society
Layout Copyright © 2024 Jenny Stanford Publishing Pte. Ltd.
ISBN 978-981-4877-34-3 (Hardcover), 978-1-003-08872-1 (eBook)
www.jennystanford.com

can be any of a wide variety of organometallic ligands. The LX ligands used for this study were all β-diketonates, with the major emphasis placed on acetylacetonate (acac) complexes. The majority of the $C^{\wedge}N_2$Ir(acac) complexes phosphoresce with high quantum efficiencies (solution quantum yields, 0.2–1.0), and microsecond lifetimes (e.g., 1–14 μs). The strongly allowed phosphorescence in these complexes is the result of significant spin–orbit coupling of the Ir center. The lowest energy (emissive) excited state in these $C^{\wedge}N_2$Ir(acac) complexes is a mixture of ^{3}MLCT and $^3(\pi - \pi*)$ states. By choosing the appropriate $C^{\wedge}N$ ligand, $C^{\wedge}N_2$Ir(acac) complexes can be prepared which emit in any color from green to red. Simple, systematic changes in the $C^{\wedge}N$ ligands, which lead to bathochromic shifts of the free ligands, lead to similar bathochromic shifts in the Ir complexes of the same ligands, consistent with "$C^{\wedge}N_2$-Ir"-centered emission. Three of the $C^{\wedge}N_2$Ir(acac) complexes were used as dopants for organic light emitting diodes (OLEDs). The three Ir complexes, i.e., bis(2-phenylpyridinato-N,$C^{2'}$)iridium(acetylacetonate) [ppy_2Ir-(acac)], bis(2-phenyl benzothiozolato-N,$C^{2'}$) iridium(acetylacetonate) [bt_2Ir(acac)], and bis(2-(2'-benzothienyl)-pyridinato-N,$C^{3'}$)iridium(acetylacetonate) [btp_2Ir(acac)], were doped into the emissive region of multilayer, vapor-deposited OLEDs. The ppy_2Ir(acac)-, bt_2Ir(acac)-, and btp_2Ir(acac)-based OLEDs give green, yellow, and red electroluminescence, respectively, with very similar current–voltage characteristics. The OLEDs give high external quantum efficiencies, ranging from 6 to 12.3%, with the ppy_2Ir(acac) giving the highest efficiency (12.3%, 38 lm/W, >50 Cd/A). The btp_2Ir(acac)-based device gives saturated red emission with a quantum efficiency of 6.5% and a luminance efficiency of 2.2 lm/W. These $C^{\wedge}N_2$Ir(acac)-doped OLEDs show some of the highest efficiencies reported for organic light emitting diodes. The high efficiencies result from efficient trapping and radiative relaxation of the singlet and triplet excitons formed in the electroluminescent process.

13.1 Introduction

The photophysics of octahedral $4d^6$ and $5d^6$ complexes has been studied extensively [1]. These complexes, particularly those prepared with Ru and Os, have been used in a variety of photonic applications, including photocatalysis and photoelectrochemistry [2]. These d^6 complexes are attractive in photochemical applications, because they generally have long-lived excited states and high luminescence efficiencies, increasing the likelihood of either energy or electron transfer occurring prior to radiative or nonradiative relaxation. Strong spin–orbit coupling of the 4d or 5d ion leads to efficient intersystem crossing of the singlet excited states to the triplet manifold [3]. The long lifetimes of these excited states are due to the triplet character of this state. Mixing of the singlet and triplet excited states, via spin–orbit coupling, removes the spin-forbidden nature of the radiative relaxation of the triplet state, leading to high phosphorescence efficiencies.

Researchers have directed their attention to the photochemistry and photophysics of Ru^{2+} and Os^{2+} complexes, with the majority of the effort being focused on metal–diimine complexes, such as the bipyridine and phenanthroline complexes [4]. More recently, researchers have investigated the photophysics of isoelectronic Rh^{3+} and Ir^{3+} complexes, with both diimine and cyclometalated ligands, such as 2-phenylpyidinato-C^2,N (*ppy*) [5]. The cyclometalated ligands are formally monoanionic and can thus be used to prepare neutral tris-ligand complexes, which are isoelectronic with the cationic trisdiimine complexes of Ru and Os, e.g., *fac*-M(*ppy*)$_3$ [6], *fac*-M(2-(α-thiopheneyl)-pyridine)$_3$ (*fac* = facial) [7]. The d^6 Ir complexes show intense phosphorescence at room temperature, while the Rh complexes give measurable emission only at low temperatures, consistent with stronger spin–orbit coupling of Ir relative to Rh. The electronic transitions responsible for luminescence in these complexes have been assigned to a mixture of metal-to-ligand charge-transfer (MLCT) and $^3(\pi-\pi^*)$ ligand states [8]. We have recently found that highly emissive Ir complexes can be formed with two cyclometalated ligands (abbreviated hereafter as $C^\wedge N$) and a single monoanionic, bidentate ancillary ligand

(LX) [9]. The emission colors from these complexes are strongly dependent on the choice of cyclometalating ligand, ranging from green to red, with room-temperature lifetimes from 1 to 14 µs. The photophysical properties of these $C^\wedge N_2$Ir(LX) complexes are similar to those observed for the tris-cyclometalated complexes, and will be discussed below.

Heavy metal complexes, particularly those containing Pt and Ir, can serve as efficient phosphors in organic light emitting devices [10]. In these devices, holes and electrons are injected into opposite surfaces of a planar multilayer organic thin film. The holes and electrons migrate through the thin film, to a material interface, where they recombine to form radiative excited states, or excitons. This electrically generated exciton can be either a singlet or a triplet. Both theoretical predictions and experimental measurements give a singlet/triplet ratio for these excitons of 1 to 3 [11]. Fluorescent materials typically used to fabricate organic light emitting diodes (OLEDs) do not give detectable triplet emission (i.e., phosphorescence), nor is there evidence for significant intersystem crossing between the triplet and singlet manifolds at room temperature. The singlet/triplet ratio thus implies a limitation of 25% for the internal quantum efficiency for OLEDs based on fluorescence. By doping OLEDs with heavy metal phosphors, we have shown that the singlet–triplet limitation can be eliminated [10]. The excited states generated by electron–hole recombination are trapped at the phosphor, where strong spin–orbit coupling leads to singlet–triplet state mixing and, hence, efficient phosphorescent emission at room temperature. Both singlet and triplet excited states can be trapped at the phosphor. OLEDs prepared with these heavy metal complexes are the most efficient OLEDs reported to date, with internal quantum efficiencies exceeding 75% (photons/electrons) (>15% external efficiency) [12]. Furthermore, OLEDs have been prepared with $C^\wedge N_2$Ir(LX) phosphor dopants, giving efficient green, yellow, or red emission. The external quantum efficiencies for these devices vary from 5% to nearly 20%. In this work, we explore the photophysical and electroluminescent properties of a series of Ir complexes used as efficient phosphorescent dopants in OLEDs. We demonstrate that, by optimizing the molecular structure of $C^\wedge N_2$Ir(LX) dopants and the energy-transfer process, exceedingly high external and power efficiencies can be obtained in the green to red spectral region.

13.2 Experimental Section

13.2.1 Synthesis

All synthetic procedures involving $Ircl_3 \cdot H_2O$ and other Ir(III) species were carried out in inert gas atmosphere despite the air stability of the compounds, the main concern being the oxidative stability of intermediate complexes at the high temperatures used in the reactions. NMR spectra were recorded on Bruker AMX 360- or 500-MHz instruments. High-resolution mass spectrometry was carried out by the mass spectroscopy facility at the Frick Chemistry Laboratory, Princeton University, Elemental analyses (C, H, N) were carried out by standard combustion analysis by the Microanalysis Laboratory at the University of Illinois, Urbana–Champagne.

Cyclometalated Ir(III) μ-chloro-bridged dimers of a general formula $C^\wedge N_2 Ir(\mu\text{-}Cl)_2 IrC^\wedge N_2$ were synthesized according to the Nonoyama route, by refluxing $IrCl_3 \cdot nH_2O$ (Next Chimica) with 2–2.5 equiv of cyclometalating ligand in a 3:1 mixture of 2-ethoxyethanol (Aldrich Sigma) and water [13].

$C^\wedge N_2 Ir(acac)$, $C^\wedge N = ppy, tpy, bzq, bt, \alpha bsn$, and pq were prepared as described previously [9].

13.2.2 Synthesis of $(C^\wedge N)_2$ Ir(acac) Complexes

13.2.2.1 General Procedure

The chloro-bridged dimer complex (0.08 mmo1), 0.2 mmo1of acetyl acetone, and 85–90 mg of sodium carbonate were refluxed in an inert atmosphere in 2-ethoxyethanol for 12–15 h. After cooling to room temperature, a colored precipitate was filtered off and washed with water, hexane, and ether. The crude product was flash chromatographed on a silica column with dichloromethane mobile phase to yield ∼75–90% of the pure $C^\wedge N_2 Ir(acac)$, after solvent evaporation and drying.

*thp*₂Ir(acac): Iridium(III) bis(2-(2'-thienyl)pyridinato-N,C$^{3'}$) (acetyl acetonate) (yield 83%). ^{1}H NMR (360 MHz, acetone-d_6): δ, ppm 8.41 (d, 2H, J 5.8 Hz), 7.79 (td, 2H, J 7.9, 1.6 Hz), 7.56 (d, 2H, J 7.9 Hz), 7.22 (d, 2H, J 4.7Hz), 7.11 (td, 2H, J 6.3, 1.6 Hz), 6.09 (d, 2H, J 4.7 Hz), 5.29 (s, 1H), 1,72 (s, 6H), Anal. Found C 45.33, H 3.00, N 4.81. Calcd C 45.16, H 3.13, N 4.58.

*btp*₂Ir(acac): Iridium(III) bis(2-(2'-benzothienyl)pyridinato-N,C$^{3'}$) (acetylacetonate) (yield 72%). ^{1}H NMR (360 MHz, acdtone-d_6): δ, ppm 8.39 (d, 2H, J 5.9Hz), 7.80 (t, 2H, J 7.9Hz), 7.77 (d, 2H, J 8.0 Hz), 7.68 (d, 2H, J 8.0 Hz), 7.25 (t, 2H, J 7.0 Hz), 7.10 (t, 2H, J 7.1 Hz), 6.82 (t, 2H, J 8.0 Hz), 6.40 (d, 2H, J 7.3 Hz), 5.70 (s, 1H), 1.90 (s, 6H). Anal. Found C 52.51, H 3.29, N 4.01. Calcd C 52.30, H 3.26, N 3.94.

*dpo*₂Ir(acac): Iridium(III) bis(2,4-diphenyloxazolato-1,3-N,C$^{2'}$) (acetyl acetonate) (yield 93%). ^{1}H NMR (360 MHz, CDCl$_3$): δ, ppm 7.79 (d, 4H, 7.4 Hz), 7.53 (d, 2H, J 7.9 Hz), 7.49 (m, 6H), 7.45 (s, 2H), 7.40 (t, 2H, J 7.4 Hz), 6.84 (t, 2H, J 7.4 Hz), 6.76 (t, 2H, J 7.4 Hz), 6.62 (d, 2H, J 7.9 Hz), 5.25 (s, 1H), 1.86 (s, 6H). Anal. Found C 56.35, H 3.67, N 3.89. Calcd C 57.44, H 3.72, N. 3.83.

*C6*₂Ir(acac): Iridium(III) bis(3-(2-benzothiazolyl)-7-(diethylamino)-2H-1-benzopyran-2-onalo-N',C^4) (acetyl acetonate) (yield 59%). ^{1}H NMR (360 MHz, CDCl$_3$): δ, ppm 7.86 (d, 2H, J 8.0 Hz), 7.58 (d, 2H, J 8.0 Hz). 7.29 (t, 2H, J 8.0 Hz), 7.21 (t, 2H, J 7.4 Hz), 6.29 (s, 2H), 6.05 (d, 2H, J 9.7 Hz), 5.83 (d, 2H, J 9.1 Hz), 5.27 (s, 1H), 3.20 (m, 8H), 1.68 (s, 6H), 1.03 (t, 12H, J 7.4 Hz). Anal. Found C 52.45, H 4.33, N 5.33. Calcd C 54.58, H 4.17, N 5.66.

*bon*₂Ir(acac): Iridium(III) bis(2-(1-naphthyl)benzooxazolato-N,C$^{2'}$) (acetyl acetonate) (yield 70%). ^{1}H NMR (360 MHz, CDCl$_3$): δ, ppm 1.82 (s, 6H), 5.24 (s, 1H), 6.70 (d, 2H, J 8.5 Hz), 7.07 (d, 2H, J 8.2 Hz), 7.28 (t, 2H, J 7.5 Hz), 7.41 (t, 2H, J 7.5 Hz), 7.47 (d, 2H, J 8.2 Hz), 7.55 (m, 6H), 7.80 (d, 2H, J 7.50 Hz), 8.89 (d, 2H, J 8.5 Hz). High-resolution MS: calculated M$^+$ 780.1600; observed M$^+$ 780.1592.

β*bsn*₂Ir(acac): Iridium(III) bis(2-(2-naphthyl)benzothiazolato-N,C$^{2'}$) (acetyl acetonate) (yield 70%). ^{1}H NMR (360 MHz, CD$_2$Cl$_2$): δ, ppm 8.25 (s, 2 H), 8.16 (d, 2 H, J 8 Hz), 8.05 (d, 2 H, J 8 Hz), 7.71 (t, 2 H, J 4.5 Hz), 7.52 (m, 4 H), 7.18 (m, 6 H), 6.74 (s, 2 H), 5.24 (s, 1 H), 1.79 (s, 6 H). Anal. Found C 56.68, H 3.22, N 3.59. Calcd C 56.56, H 3.51, N3.03.

op_2Ir(acac): Iridium(III) bis(2-phenyl oxazolinato-N,$C^{2'}$ (acetyl acetonate) (yield 70%). ^{1}H NMR (360 MHz, CD_2Cl_2): δ, ppm 7.26 (d, 2H, J 7.8 Hz), 6.86 (m, 4 H), 6.8 (t, 2H, J 6.8 Hz), 5.22 (s, 1 H), 4.98 (dd, 2H, J 8.6, 8.3 Hz), 4.87 (dd, 2H, J 8.6, 8.3 Hz), 4.04 (dd, 2H, J 12.7, 7.7 Hz), 3.82 (dd, 2H, J 12.7, 7.7 Hz), 1.80 (s, 6H). High-resolution MS: calculated M^+ 584.1287; observed M^+ 584.1295.

13.2.3 Optical Measurements

Absorption spectra were recorded on AVIV model 14DS UV–visible–IR spectrophotometer (re-engineered Cary 14) and corrected for background due to solvent absorption. Emission spectra were recorded on PTI QuantaMaster model C-60SE spectrofluorometer with 928 PMT detector and corrected for detector sensitivity inhomogeneity. Emission quantum yields were determined using fac-Ir(ppy)$_3$ as a reference [14]. Emission lifetimes were obtained by exponential fit of emission decay [15] curves recorded on a PTI TimeMaster model C-60SE spectrofluorometer.

13.2.4 OLED Fabrication and Testing

The OLED structure employed in (this study is shown in the inset of Fig. 13.6. Organic layers were fabricated by high-vacuum (10^{-6} Torr) thermal evaporation onto a glass substrate precoated with an indium–tin–oxide (ITO) layer with a sheet resistance of 20 $\Omega/\square$, Prior to use, the ITO surface was ultrasonicated in a detergent solution followed by a deionized water rinse, dipped into acetone, trichloroethylene, and 2-propanol, and then degreased in 2-propanol vapor. After degreasing, the substrate was oxidized and cleaned in a UV-ozone chamber before it was loaded into an evaporator, A 50-nm-thick film of 4,4'-bis[N-(naphthyl-N-phenylamino) biphenyl (α-NPD) served as the hole transport layer (HTL). The light emitting layer was prepared by coevaporating a 4,4'-N,N'-dicarbazolebiphenyl (CBP) host and a phosphorescent dopant, with both deposition rates being controlled with two independent quartz crystal oscillators. ppy_2-Ir(acac), bt_2Ir(acac), and btp_2Ir(acac) with a dopant concentration of $\sim$7% were utilized to promote short-range Dexter transfer of the triplet excitons without concentration quenching [1]. Next, a 10-nin-thick 2,9-dimethyl-4,7-diphenyl-1,10-phenanthroline (BCP) as a hole and exciton blocking layer (HBL) and 40-nm-thick tris(8-hydroxyquinoline)-aluminum (Alq$_3$) as an electron transport layer were deposited on the emitter layer. Finally, a shadow mask with 1-min-diameter openings was used to define the cathode consisting of a 100-nm-thick Mg–Ag layer, with a 20-nm-mick Ag cap. Current density–voltage–luminance (J–V–L) measurements were obtained using a semiconductor parameter analyzer and a calibrated silicon photodiode.

13.3 Results and Discussion

13.3.1 Synthesis and Characterization of $C^\wedge N_2$ Ir(LX) Complexes

$C^\wedge N_2$Ir(LX) complexes have been prepared, with several different $C^\wedge N$ and LX ligands, Fig. 13.1 and Table 13.1. The synthetic procedure used to prepare these complexes involves two steps [9]. In the first step, IrCl$_3$ · nH$_2$O is reacted with an excess of the desired $C^\wedge N$ ligand to give a chloride-bridged dimer, i.e., $C^\wedge N_2$Ir(μ-Cl)$_2$Ir$C^\wedge N_2$, Eq. 13.1. The NMR spectra of these complexes

$$2\text{IrCl}_3 \cdot n\text{H}_2\text{O} + 2C^\wedge N\text{H} \rightarrow C^\wedge N_2\text{Ir}(\mu - \text{Cl})_2\text{Ir}C^\wedge N_2 + 2\text{HCl} + 2\text{Cl} \tag{13.1}$$

$$C^\wedge N_2\text{Ir}(\mu - \text{Cl})_2\text{Ir}C^\wedge N2 + 2\text{LXH} \rightarrow 2C^\wedge N_2\text{Ir}(\text{LX}) + 2\text{HCl} \tag{13.2}$$

are consistent with the heterocyclic rings of the $C^\wedge N$ ligands being in a trans disposition, as shown in Fig. 13.1b. The chloride-bridged dimers can be readily converted to emissive, monomeric complexes by replacing the bridging chlorides with bidentate, monoanionic β-diketonate ligands (LX), Eq. 13.2. These reactions give $C^\wedge N_2$Ir(LX) with a yield of typically >80%.

X-ray crystallographic studies have been carried out for two $C^\wedge N_2$Ir(LX) complexes, i.e., $(ppy)_2$Ir(acac) and $(tpy)_2$Ir(acac) (acac = acetylacetonate) [9]. The Ir in both of these $C^\wedge N_2$Ir(acac) complexes is octahedrally coordinated by the three chelating ligands, with the pyridyl groups in a trans disposition, as shown schematically in Fig. 13.1b. The coordination geometries of the "$C^\wedge N_2$Ir" fragment

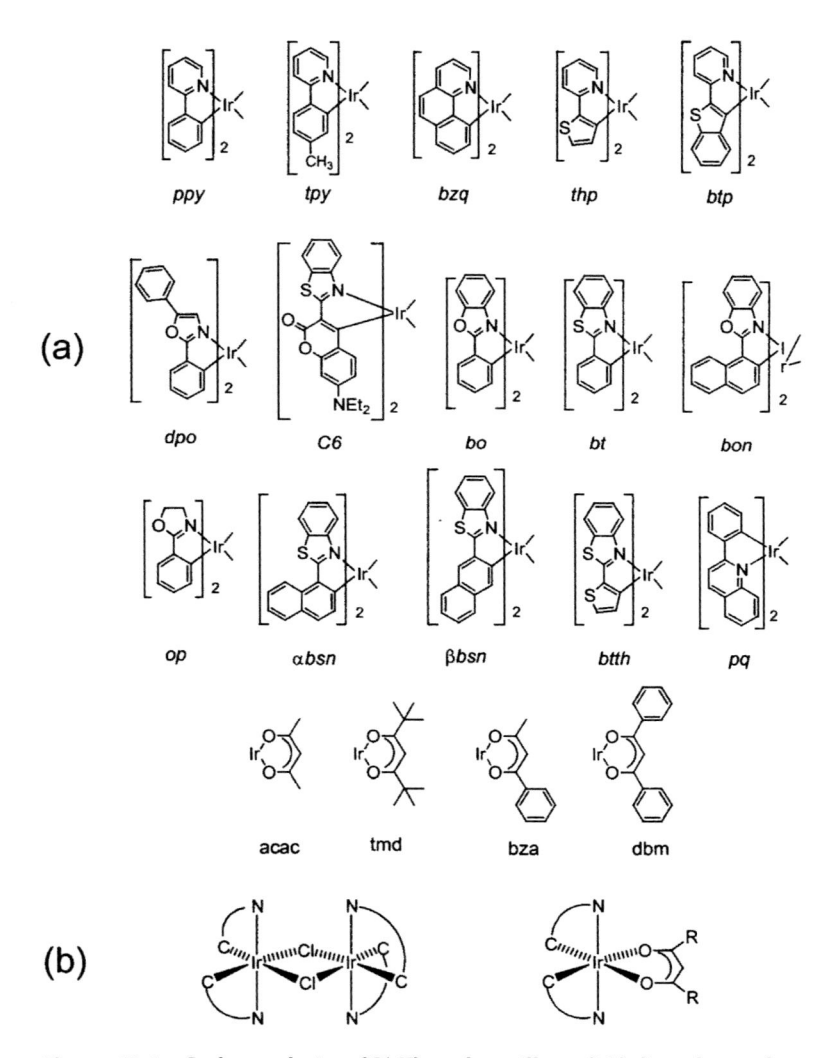

Figure 13.1 Cyclometalating ($C^{\wedge}N$) and ancillary (LX) ligands used to prepare $C^{\wedge}N_2\mathrm{Ir(LX)}$ complexes, (a). The abbreviations used throughout the paper for each ligand are listed below the "$C^{\wedge}N_2\mathrm{Ir}$" or "Ir(LX)" fragment. The coordination geometries of the chloride-bridged dimer and $C^{\wedge}N_2\mathrm{Ir(LX)}$ complexes are shown in (b).

Table 13.1 Photophysical data for $C^{\wedge}N_2\text{Ir(acac)}$ complexes[a]

$C^{\wedge}N$ ligand	Absorbance λ (log ϵ)	Emission λ_{max} (nm)	Lifetime (μs) 298 K	Quantum yield[b]
bo	262 (4.7), 269 (4.6), 298 (4.6), 310 (4.5), 343 (4.0), 383 (3.8), 425 (3.7), 462 (3.6), 510 (2.6)	525	1.1	0.25 0.55
bon	266 (4.8), 298 (4.7), 326 (4.6), 360 (4.4), 410 (4.1), 458 (4.0), 491 (4.0)	586	1.3	0.11 0.24
bzq[9]	260 (4.6), 360 (3.9), 470 (3.3), 500 (3.2)	548	4.5	0.27 0.59
α-bsn	274 (4.7), 300 (4.6), 345 (4.5), 427 (4.0), 476 (4.0), 506 (3.9)	606	1.8	0.16 0.35
β-bsn	328 (4.7), 348 (4.6), 420 (3.8), 496 (3.5)	594	2.2	0.21 0.46
btth	286 (4.4), 327 (4.4), 405 (4.0), 437 (4.0), 478 (3.9)	593	3.6	0.21 0.46
bt[9]	269 (4.6), 313 (4.4), 327 (4.5), 408 (3.8), 447 (3.8), 493 (3.4), 540 (3.0)	557	1.8	0.26 0.57
btp	286 (4.5), 340 (4.1), 355 (3.9), 495 (3.7)	612	5.8	0.21 0.46
C6	444 (4.8), 472 (4.8)	585	14	0.6 >0.95
op	258 (4.4), 294 (4.2), 336 (4.1), 372 (3.8), 456 (3.5)	520	2.3	0.14 0.31
dpo	297 (4.8), 369 (4.0), 420 (3.9), 443 (3.9)	550	3.0	0.1 0.22
ppy[9]	260 (4.5), 345 (3.8), 412 (3.4), 460 (3.3), 497 (3.0)	516	1.6	0.34 0.75
pq[9]	268 (5.0), 349 (4.4), 433 (3.9), 467 (3.9), 553 (3.6)	597	2	0.1 0.22
thp	302 (4.4), 336 (4.1), 387 (3.8), 453 (3.5)	562	5.3	0.12 0.26
tpy[9]	270 (4.5), 370 (3.7), 410 (3.5), 460 (3.4), 495 (3.0)	512	3.1	0.31 0.68

[a] All of the data were collected for 2-methyltetrahydrofuran solutions.
[b] The quantum yield reported in reference 14 for the Ir(ppy)$_3$ reference compound was 0.4. It was later discovered that this number was not correct and the photoluminescence quantum yield for this complex in solution is 0.73 in toluene and (W. Holzer, et al., Chem. Phys., 2005, **308**, 93–102, DOI: 10.1016/j.chemphys.2004.07.051) and 0.9 in THF (Sajoto, T., et al., J. Am. Chem. Soc., 2009, **131**, 9813–9822, DOI: 10.1021/ja903317w). Thus, the quantum yield values previously published have been increased by a factor of 2.2 (0.9/0.4) from the published values. The published values have been struck out.

in both the *tpy* and *ppy* iridium acac complexes are the same as that reported for the chloride-bridged dimers and $C^\wedge N_2\mathrm{Ir(bpy)}^+$ complexes of the same $C^\wedge N$ ligands [16]. The $C^\wedge N_2\mathrm{Ir(LX)}$ complexes are stable in air and can be sublimed in a vacuum without decomposition.

13.3.2 Photophysical Properties of $C^\wedge N_2$ Ir(LX) Complexes

In order for the $C^\wedge N_2\mathrm{Ir(LX)}$ complexes to be useful as phosphors in organic light emitting diodes, strong spin–orbit coupling must be present to efficiently mix the singlet and triplet excited states Clear evidence for significant mixing of the singlet and triplet excited states is seen in both the absorption and emission spectra of these complexes. All of the $C^\wedge N_2\mathrm{Ir(LX)}$ complexes show intense absorption from $C^\wedge N$ ligand $\pi-\pi^*$ and MLCT transitions. The absorption spectra for the *ppy*, *bzq*, *thp*, and *btp* complexes are shown in Fig. 13.2. The extinction coefficients for these bands are in the ranges expected for $\pi-\pi^*$ ligand-centered and MLCT bands, about 10 000–35 000 and 2000–6000 M^{-1} cm^{-1}, respectively. The $\pi-\pi^*$ absorption bands for these complexes fall in the ultraviolet and closely resemble the spectra of the free $C^\wedge N$ ligands. Both ^{1}MLCT and ^{3}MLCT bands are typically observed for these complexes. The high degree of spin–orbit coupling is evident in comparing the oscillator strengths for the two MLCT bands. For example, the singlet and triplet MLCT bands for *ppy*$_2$Ir(acac) fall at 410 and 460 nm, respectively (see Fig. 13.2 and Table 13.1), with less than a factor of 2 difference in their extinction coefficients. Strong spin–orbit coupling on Ir gives the formally spin-forbidden ^{3}MLCT an intensity comparable to the allowed ^{1}MLCT. The energies of these singlet and triplet MLCT absorptions are very similar to those reported for *ppy*$_2$Ir(bpy)$^+$ [17], *ppy*$_2$Ir(H$_2$O)$_2^+$ [18], and *ppy*$_2$Ir(μ-Cl)$_2$Ir*ppy*$_2$ [19]. The *ppy*, *bzq*, *thp*, and *btp* complexes have very similar ^{3}MLCT energies, with λ_{max} values ranging from 440 to 490 nm (top inset of Fig. 13.2 and Table 13. 1). The similarity of MLCT energies for these complexes is not surprising, since all four complexes have MLCT states involving very similar (pyridyl) acceptors.

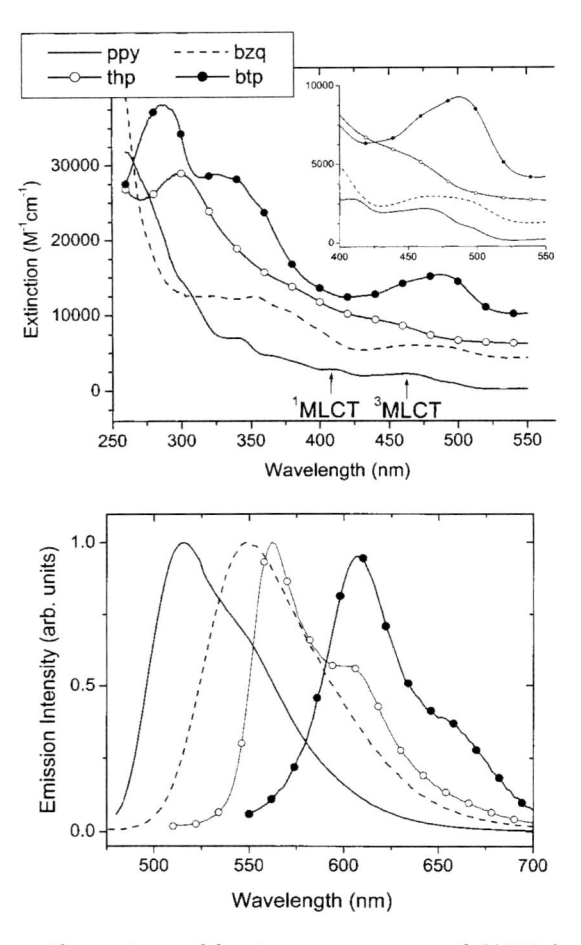

Figure 13.2 Absorption and luminescence spectra of $C^\wedge N_2\text{Ir(acac)}$ complexes. (top) Absorption spectra for $C^\wedge N$ = ppy, bzq, thp, and btp complexes. The spectra have been offset for clarity. All four complexes have extinction coefficients of ~0.0 at 550 nm. (bottom) Photoluminescence spectra for the same complexes.

13.3.2.1 Correlation of absorption and emission bands

In addition to ^{3}MLCT absorption bands with high oscillator strengths, strong spin–orbit coupling leads to efficient phosphorescence in the majority of the $C^\wedge N_2\text{Ir(LX)}$ complexes reported here. The room-temperature (solution) quantum yields of these complexes range from 0.2 to >0.95 (Table 13.1), and their luminescent lifetimes fall between 1 and 14 µs, consistent with emission from a triplet excited state. The positions of the maximums in the excitation spectra for these complexes are very similar to those in their absorption spectra. Pumping either ligand-based or MLCT transitions efficiently gives the same phosphorescent excited state, as illustrated for $C6_2$Ir(acac) and dpo_2Ir(acac) in Fig. 13.3.

Note that $C6$ and dpo are common fluorescent laser dyes. Solutions of $C6$ give green fluorescence, while solutions of dpo give UV/violet fluorescence [20]. Both of these laser dyes give high quantum efficiencies for fluorescence at room temperature and no observable phosphorescence. Further, $C6$ and dpo have the requisite structures to make them suitable for cyclometalation reactions with Ir and have been used to make $C^\wedge N_2$Ir(acac) complexes. Coordination to Ir shifts the emission maximum of $C6$ from 500 nm in the free dye to 570 nm for $C6_2$Ir(acac), as shown in Fig. 13.3, providing striking evidence for intersystem crossing induced by the proximity of the heavy Ir atom. This Ir complex has a high quantum yield for emission of >0.95 and the lifetime is 14 µs. The excitation spectrum for $C6_2$Ir(acac) is very similar to that of pure $C6$ and gives a good match to the absorption spectrum of the Ir complex. A more marked red shift is observed on metalation of dpo, whose λ_{max} for emission shifts from 365 nm for fluorescence from the free dye to 550 nm for phosphorescence from dpo_2Ir(acac). The excitation spectrum matches the absorption spectrum of dpo_2Ir(acac), with a maximum efficiency at an energy close to the MLCT band. The quantum efficiency and lifetime for the dpo Ir complex are 0.1 and 3 µs, as expected for Ir-promoted phosphorescence from the dpo ligand. The quantum yields for these phosphorescent transitions are high, especially $C6_2$Ir(acac), which has a quantum yield nearly as high as those of the fluorescent transition $C6$ alone. In both $C6$- and dpo-based complexes, the Ir center facilitates intersystem crossing into the triplet levels of the laser dye and efficient phosphorescence from these low-energy triplet states. Efficient, ligand-based phosphorescence is common in $C^\wedge N_2$Ir(LX) complexes, making the design of new phosphors straightforward, as described below.

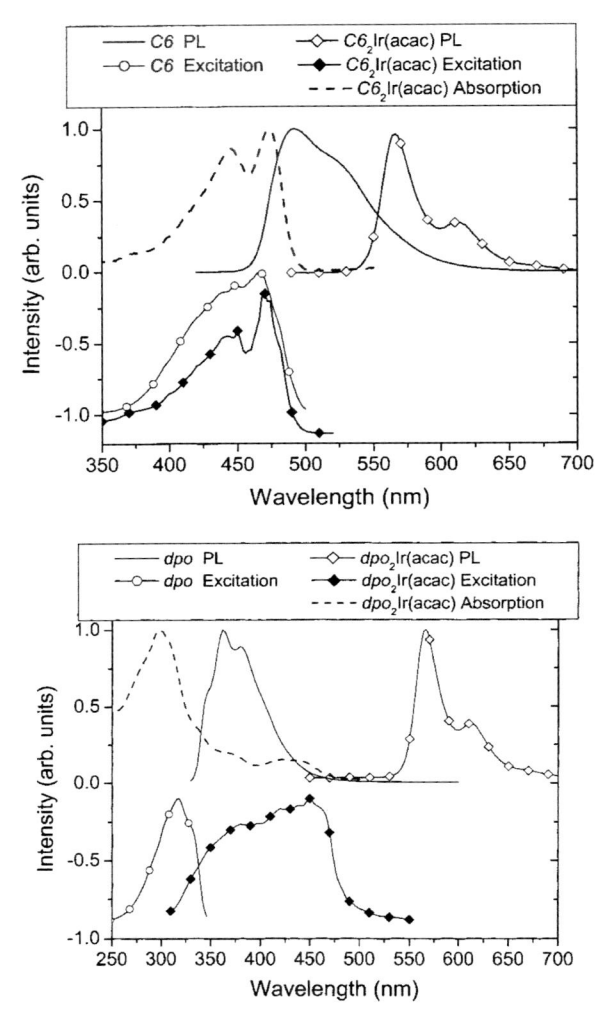

Figure 13.3 Solution photoluminescence, excitation, and absorption spectra of $C6_2$Ir(acac) and $C6$ (top) and dpo_2Ir(acac) and dpo (bottom).

Although the *ppy, bzq, thp,* and *btp* complexes have similar MLCT absorption energies, their emission spectra differ markedly, Fig. 13.2 (bottom). The *ppy* and *bzq* complexes exhibit a small Stokes shift between the ^{3}MLCT absorption and emission bands, while the *thp* and *btp* complexes give larger shifts. The energy differences between

λ_{max} for the ^{3}MLCT absorption and emission spectra of ppy_2Ir(acac) and bzq_2Ir(acac) are 55 and 58 nm, respectively (~2200 cm^{-1}), whereas significantly larger shifts are observed for the *thp* and *btp* complexes (105 and 120 nm, respectively, ~4100 cm^{-1}). It has been shown that the emission spectra of *fac*-Irppy_3 and *fac*-Irthp_3 complexes result from mixtures of MLCT and $^3(\pi-\pi^*)$ transitions [21]. *fac*-Irppy_3 emits from an excited state that is predominantly due to MLCT, while *fac*-Irthp_3 emits from a largely ligand-based $^3(\pi-\pi^*)$ excited state. The $^3(\pi-\pi^*)$ level for phenylpyridine *(ppy)* has an energy of 460 nm [22], which puts it at a sufficiently high energy such that the ^{3}MLCT becomes the lowest energy excited state. In contrast, the $^3(\pi-\pi^*)$ transition for thienylpyridine (*thp*) is at $\lambda_{max} = 564$ nm, which is at an energy below the ^{3}MLCT energy, so a predominantly ligand-based state is the lowest energy excited state in *fac*-Irthp_3. The Stokes shifts observed for the $C^\wedge N_2$Ir(acac) complexes support similar assignments here. Emission from a predominantly ^{3}MLCT state would be expected to have a small Stokes shift between the ^{3}MLCT absorption and emission bands, as seen for *ppy* and *bzq* complexes. Emission from a predominantly ligand-based excited state, however, should give a large Stokes shift between the ^{3}MLCT absorption and emission bands, as observed for the *thp* and *btp* complexes.

The line shapes of the phosphorescence spectra of these complexes also support the hypothesis that the *ppy* and *bzq* complexes emit primarily from a MLCT state, while the other complexes emit from $^3(\pi-\pi^*)$ $C^\wedge N$ states. The emission spectra for thp_2Ir(acac) closely resemble the phosphorescence spectrum reported for the free organic ligand [7, 17, 23], supporting the assignment of the lowest energy excited state to be predominantly ligand based [24]. Vibronic fine structure is clearly observed for the *thp* and *btp* ligands and is absent for the *ppy* and *bzq* complexes. Emission bands from MLCT states are generally broad and featureless, while $^3(\pi-\pi^*)$ states typically give highly structured emission [25]. Significant vibronic fine structure is also observed in the PL spectra of both the *C6* and *dpo* complexes, consistent with emission from ligand-based excited states in these complexes as well.

13.3.2.2 $C^\wedge N$ ligand tuning of phosphorescence

Further evidence for significant ligand character in the emission spectra can be seen in a series of $C^\wedge N_2\text{Ir}(acac)$ complexes prepared with *bo, bt, bon*, and *αbsn* ligands. The photoluminescence spectra of these complexes show a pronounced red shift, Fig. 13.4.

All four complexes show vibronic fine structure in their emission spectra, as expected for ligand-based transitions. The highest energy emission is observed for the benzoxazole (*bo*) complex. Substitution of S for O in a chromophore (*bo* → *bt*) leads to a 30-nm red shift, due the higher polarizibility and basicity of sulfur relative to oxygen [26], in this ligand-based excited state. Increasing the size of the ligand π system is expected to bathochromically shift electronic transitions, as is observed in converting a phenyl group to a naphthyl group (*bo* → *bon*), which leads to a 60-nm red shift. The effects of the naphthyl and sulfur substitutions are nearly additive, leading to an 80-nm red shift when comparing *bo* to *αbsn* complexes. Unfortunately, we have been unable to detect phosphorescence spectra from the free

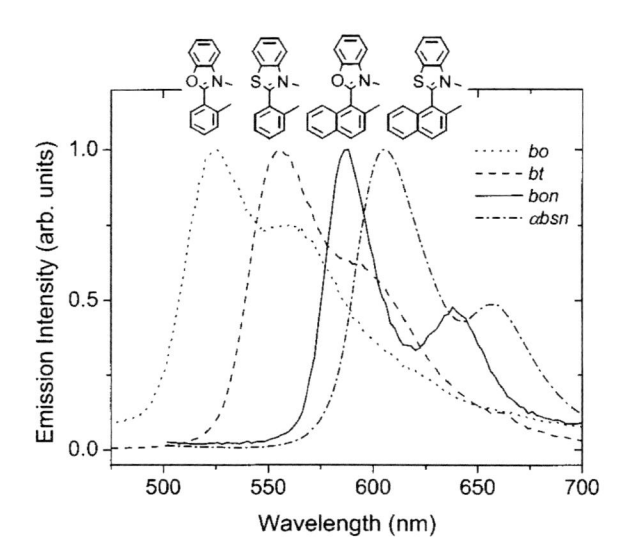

Figure 13.4 Solution photoluminesccence spectra of $bo_2\text{Ir}(acac)$, $bt_2\text{Ir}(acac)$, $bon_2\text{Ir}(acac)$ and $\alpha bsn_2\text{Ir}(acac)$. The structures of the individual $C^\wedge N$ ligands are shown above the corresponding spectrum.

ligands, even at temperatures below 10 K, so we cannot directly correlate the phosphorescence spectra of the Ir complexes with those of the $^3(\pi-\pi^*)$ of the free ligands. The luminescence energies of the four Ir complexes do follow the same trend seen in the fluorescence spectra of the free ligands, however, which have λ_{max} values of 350, 360, 380 and 400 nm, respectively.

We also attempted to decrease the size of the ligand π system, in the hope of achieving a shift form green to blue emission. The first approach we chose was to replace the metalated phenyl group of the *ppy* ligand with a smaller group, such as a vinyl or cycloalkene (e.g., $C^{\wedge}N$ = 2-vinylpyridine, vinylbenzoxazole, and 1-cyclohexenylbenzoxazole). Unfortunately, we have been unable to cyclometalate ligands in which the phenyl has been replaced with olefinic groups. We have succeeded in decreasing the size of the π system in the benzoxazole portion of the ligand, however, i.e., *op* (2-(1-phenyl)oxazole, Fig. 13.1). In *op*, the phenylene group of the *bo* benzoxazole group has been replaced with a vinylene group. The emission spectrum of op_2Ir(acac) has its λ_{max} at 520 nm, only 5-nm blue shifted from bo_2Ir(acac), Table 13.1. The π^* state in both of these complexes is presumably localized in the oxazole group and does not extend significantly into the added phenyl ring of the benzoxazole group, leading to only a modest shift in the phosphorescence energy on removal of the phenyl ring.

Two different isomers of the naphthylbenzothiazole (αbsn and βbsn) have been used to prepare cyclometalated Ir complexes. The only chemical difference between these complexes is the substitution at the naphthyl groups. The emission efficiencies, lifetimes, and emission spectral line shapes of the two complexes are similar. The principal difference in the photophysical properties of αbsn and βbsn complexes are blue shifts in the phosphorescence and absorption spectra of βbsn relative to the αbsn complex (e.g., emission λ_{max} = 593 and 606 nm, respectively). Both phosphorescence spectra show vibronic fine structure, consistent with ligand-based emission. Unfortunately, we have been unable to observe phosphorescence from frozen solutions of the αbsn and βbsn free ligands, so we cannot determine if the blue shift for the β-isomer is due the electronic structure of the two ligands. On the basis of the photophysics of related naphthyl complexes, the βbsn

blue shift is most likely not related to the Ir substitution. Similar blue shifts are observed in other acceptor substituted naphthalene derivatives, substituted in the 1 (α)- and 2 (β)-positions. For example, the phosphorescence energy of 2-nitronaphthalene is blue shifted relative to that of 1-nitronaphthalene by 16 nm [27].

13.3.2.3 $C^{\wedge}N$ vs LX centered emission

Three *fac*-Ir$C^{\wedge}N_3$ complexes have been reported, i.e., *fac*-Ir(*ppy*)$_3$ [28], *fac*-Ir(*bzq*)$_3$ [9], and *fac*-Ir(*thp*)$_3$ [29]. All three tris-chelate complexes have emission spectra that are nearly identical to the $C^{\wedge}N_2$Ir(acac) complexes with the same ligands [9]. The phosphorescence quantum yields and lifetimes for the two classes of complex for a given ligand are also the same. The similarity of the two classes of complexes is consistent with emission predominantly from the "$C^{\wedge}N_2$Ir" fragment in $C^{\wedge}N_2$Ir(acac) complexes.

For all of the complexes discussed thus far, the triplet levels of the LX ligand (acac) lie well above the energies of the $C^{\wedge}N$ ligand and MLCT excited states. Thus, the luminescence is dominated by $C^{\wedge}N$ and MLCT transitions, leading to efficient phosphorescence. If the triplet state energy of the LX ligand is lower in energy than the $C^{\wedge}N^3(\pi-\pi^*)$ or ^{3}MLCT, however, a triplet LX level will be the lowest energy excited state. A switch from "$C^{\wedge}N_2$Ir" to LX based emission can be seen in a series of complexes prepared with different β-diketonates (i.e., acac, tmd, bza, and dbm in Fig. 13.1). The phosphorescence quantum efficiencies for these complexes are provided in the bar graph of Fig. 13.5. The emission spectra and phosphorescence efficiencies for the tmd complexes of *ppy, bzq, thp*, and *btp* are identical to those of the same complexes prepared with an acac LX ligand. All of these complexes emit from states within the "$C^{\wedge}N_2$Ir" fragment as discussed above. When the LX ligand is bza, emission from the *ppy, bzq*, and *thp* complexes are largely quenched ($\phi_{phos} < 0.01$). In contrast, the *btp* complex (*btp*$_2$Ir(bza)) gives a quantum efficiency of 0.1 and an emission spectrum identical to that of *btp*$_2$Ir(acac). For the $C^{\wedge}N_2$Ir(dbm) complexes, all four $C^{\wedge}N$ ligands give very weak phosphorescence ($\phi_{phos} < 0.01$). The triplet levels of these three β-diketonates fall in the order tmd > bza > dbm. Apparently, only the ligand with the lowest triplet

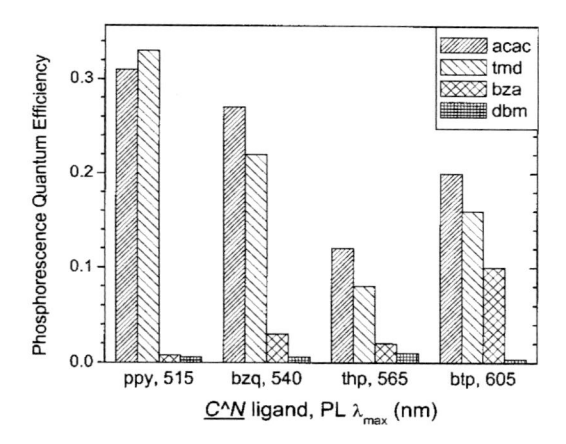

Figure 13.5 Solution phosphorescence efficiencies of $C^\wedge N_2\text{Ir}(LX)$ as a function of the LX ligand for acac, tmb, bza, and dbm (see Fig. 13.1). As discussed above, quantum yields here are low by a factor of 2.2.

energy, i.e., $C^\wedge N = btp$, has a triplet level lower than dbm. Thus, only the btp complex gives emission from an excited state on the "$C^\wedge N_2\text{Ir}$" fragment for bza complexes. The other three complexes have their excited states localized predominantly on the on the bza ligand, leading to weak phosphorescence. All four $C^\wedge N$ ligands give "$C^\wedge N_2\text{Ir}$" fragments with triplet levels higher than dbm, so the emission results from a dbm-based excited state, giving rise to very inefficient phosphorescence for all four $C^\wedge N$ ligands.

13.3.2.4 OLEDs prepared with $C^\wedge N_2\text{Ir}(LX)$ complexes

Using the phosphors discussed above it is possible to prepare efficient OLEDs, which emit in a variety of colors. The OLED structure used is the same one previously developed for *fac* Ir(ppy)$_3$-based OLEDs, which gave green electroluminescence (EL), at an external efficiency of 9% (η_{ext}, photons/electrons) [30]. In the $C^\wedge N_2\text{Ir}(\text{acac})$-based devices, the Ir phosphor was doped into the emissive layer of the OLED, at a concentration of 7 wt%. Figure 13.6 (top) shows external quantum efficiency (η_{ext}) and power efficiency as functions of current density for OLEDs with a ppy$_2$Ir(acac) dopant. A maximum external quantum efficiency (η_{ext}) of 12.3 ±

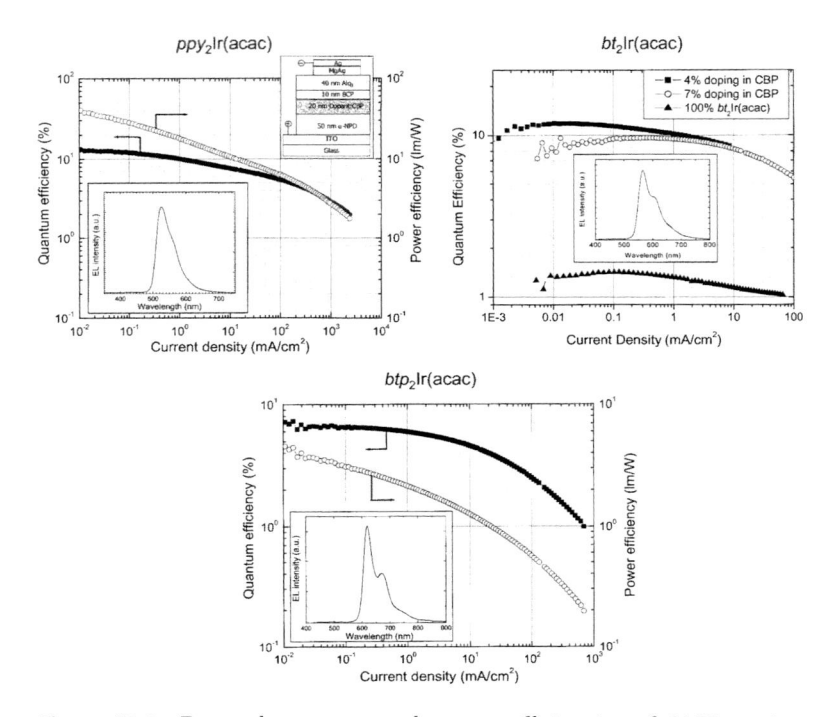

Figure 13.6 External quantum and power efficiencies of OLEDs using ppy_2Ir(acac):CBP (top), bt_2Ir(acac):CBP and bt_2Ir(acac) (middle), and btp_2Ir(acac):CBP (bottom) emissive layers. The inset to the top plot shows the OLED structure used in this study. The EL spectrum of each device is shown as an inset on the relevant plots. α-NPD = 4,4'-bis-[N-(naphthyl-N−phenylamino)biphenyl, CBP = 4,4'-N, N'-dicarbazolyl-biphenyl, BCP = 2,9- dimethyl-4,7-diphenyl-1,10-phenanthroline (bathocuproine), and Alq_3 = tris(8-hydroxyquinoline)aluminum.

0.3% and a power efficiency of 38 ± 3 lm/W [31] were obtained at a current density 0.01 mA/cm². The device showed a gradual decrease in η_{ext} with increasing current density, which is attributed to increasing triplet–triplet annihilation of the phosphor-bound excitons [10–12]. A maximum optical output of 32 500 cd/m² was obtained at $J = 2.4$ A/cm².

OLEDs prepared with btp_2Ir(acac) gave a peak quantum efficiency of 6.6% and a power efficiency of 2.2 lm/W (at 1 mA/cm²). The external quantum efficiency and power efficiency as functions

Table 13.2 OLED performance of ppy_2Ir($acac$)-, bt_2Ir(acac)-, and btp_2Ir(acac)-based OLEDs[a]

	Dopant phosphor		
	ppy_2Ir(acac)	bt_2Ir(acac)	btp_2Ir(acac)
EL color	green	yellow	red
peak wavelength (nm)	525	565	617
CIE-x	0.31	0.51	0.68
CIE-y	0.64	0.49	0.33
luminance			
at 10 mA/cm^2 (cd/m^2)	3.300	2500	470
drive voltage (V)			
at 10 mA/cm^2	9.3	9.5	11.6
ext quantum effic (%)			
at 1 mA/cm^2	10.0	9.7	6.6
at 10 mA/cm^2	7.6	8.3	6.0
at 100 mA/cm^2	5.4	5.5	4.6
power effic (lm/W)			
at 1 mA/cm^2	18	11	2.2

[a] Device structure, ITO/α-NPD (500 Å)/doped CBP (300 Å)/BCP (100 Å)/Alq$_3$ (400 Å)/Mg–Ag. All three devices were prepared with the dopant at a 7% loading in the CBP layer.

of current density for btp_2Ir(acac) based OLEDs are shown in Fig. 13.6 (bottom). These efficiency values are the highest values reported for a red emissive OLED. Table 13.2 summarizes the EL performance with ppy_2Ir(acac), bt_2Ir(acac), and btp_2Ir(acac) doped OLEDs. These devices give high performance with EL emission colors in green, yellow, and red, respectively. The device structures and layer thickness used here were identical to those reported for fac-Ir(ppy)$_3$-based OLEDs [10(b)]. They have not been optimized for either low-voltage or high-efficiency operation, so the quantum and power efficiency values are lower limits. For example, lowering the doping concentration of bt_2Ir(acac) from 7 to 4% increases the maximum η_{ext} from 9.5 to 11.9%, Fig. 13.6 (middle). Data are also shown for an OLED in which the bt_2Ir(acac)-doped CBP film has been replaced with a pure bt_2Ir(acac) luminescent film. Increasing the concentration of bt_2Ir(acac) to 100% lowers the device efficiency to 1.5%, due to enhanced self-quenching in the

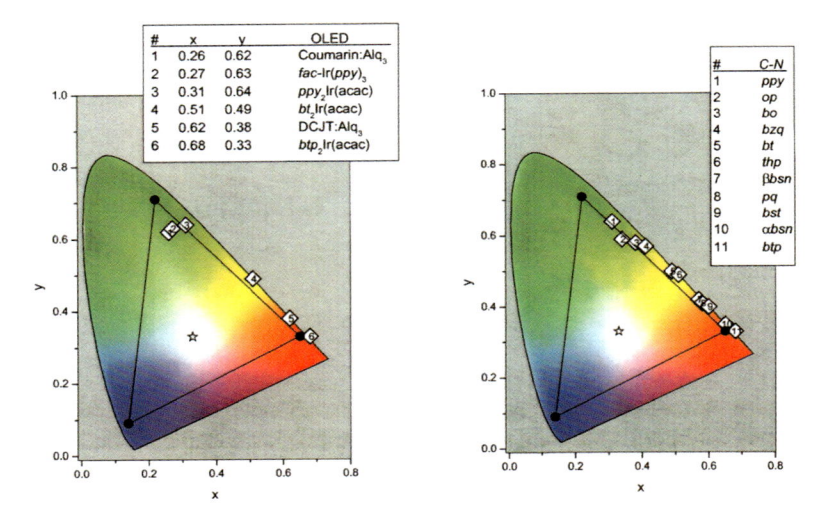

#	x	y	OLED
1	0.26	0.62	Coumarin:Alq$_3$
2	0.27	0.63	fac-Ir(ppy)$_3$
3	0.31	0.64	ppy$_2$Ir(acac)
4	0.51	0.49	bt$_2$Ir(acac)
5	0.62	0.38	DCJT:Alq$_3$
6	0.68	0.33	btp$_2$Ir(acac)

#	C-N
1	ppy
2	op
3	bo
4	bzq
5	bt
6	thp
7	βbsn
8	pq
9	bst
10	αbsn
11	btp

Figure 13.7 The Commission Internationale de L'Eclairage (CIE) chromaticity coordinates of OLEDs and phosphorescence spectra of $C\wedge N_2\text{Ir}(LX)$ complexes. The CIE coordinates for OLEDs with $ppy_2\text{Ir}(acac)$:CBP, $bt_2\text{Ir}(acac)$:CBP, and $btp_2\text{Ir}(acac)\text{Ir}$:CBP are shown relative to the fluorescence-based devices, coumarine6:Alq$_3$ and DCJT:Alq$_3$ on the left. The CIE coordinates of the phosphorescence spectra of many of the $C\wedge N_2\text{Ir}(LX)$ complexes prepared here are shown to the right. The NTSC standard coordinates for the red, green, and blue subpixels of a CRT are at the corners of the black triangle.

neat $bt_2\text{Ir}(acac)$ film. The device performances and electrical properties of the three $C\wedge N_2\text{Ir}(LX)$-based OLEDs are very similar, which is consistent with similar mechanisms for exciton formation and trapping at the phosphorescent centers. The lower efficiency of the $btp_2\text{Ir}(acac)$-based OLED relative to the $ppy_2\text{Ir}(acac)$ and $btp_2\text{Ir}(acac)$ may be due to a lower phosphorescence efficiency of the btp complex relative to the ppy and bt complexes.

The EL spectra of the $C\wedge N_2\text{Ir}(acac)$-based devices match those of the same phosphors in a dilute solution. Thus, all EL emission originates from the triplet excited states of the phosphors. The Commission Internationale de L'Eclairage (CIE) coordinates for the three OLEDs are shown in Fig. 13.7 (left). The CIE system is the standard for evaluating color quality for visual applications [32]. The $btp_2\text{Ir}(acac)$-doped OLED gives a saturated red emission that has CIE coordinates close to the National Television Standards Committee (NTSC) recommended red for a cathode ray tube (CRT). Green emission from ppy_2Ir(acac) is very similar to Ir(ppy)$_3$-based OLEDs and a common fluorescence-based green OLED (Coumarin6:Alq$_3$). The three complexes used to fabricate OLEDs were chosen to be representative of the family of $C^\wedge N_2$Ir(LX) phosphors. Figure 13.7 (right) shows the CIE coordinates of the solution phosphorescent spectra of all of the $C^\wedge N_2$Ir(acac) complexes reported here. All of the $C^\wedge N_2$Ir(acac) complexes are expected to give OLEDs with efficiencies similar to those reported for the *ppy*, *bt*, and *btp* complexes. We expect that the EL and photoluminescence spectra of these phosphors are similar, so the coordinates of the phosphorescence spectra shown in Fig. 13.7 (right) will likely be the same for OLEDs prepared with these phosphors. The $C^\wedge N_2$Ir(acac) are clearly a broad and widely tunable class of OLED phosphors.

13.4 Summary

It has been reported that phosphorescence in iridium tris-cyclometalated complexes comes from a mixture of ligand centered, $^3(\pi-\pi^*)$, and ^{3}MLCT excited states [7, 33]. While phosphorescence results from a mixture of the two triplet states (and he ^{1}MLCT via spin–orbit coupling), the emissive excited state is predominantly $^3(\pi-\pi)$ or ^{3}MLCT, depending on the energies of the two excited states. The same situation is present in the $C^\wedge N_2$Ir(LX) complexes, ppy_2Ir(acac), tpy_2Ir(acac), and bzq_2Ir(acac) have high-energy $^3(\pi-\pi^*)$ states and, thus, give unstructured, predominantly ^{3}MLCT emission. The other complexes reported here give phosphorescence spectra with a reasonable degree of vibronic fine structure and significant Stokes shifts, consistent with predominantly $^3(\pi-\pi^*)$ $C^\wedge N$ based emission. By changing the $C^\wedge N$ ligands in cyclometalated Ir complexes, we demonstrated green to red electrophosphorescence with high η_{ext}. High phosphorescence efficiencies and lifetimes less than 10 μs resulted in record high-performance OLEDs operating from the green to the red. We note that the device structure has the potential for further optimization. For example, the use of Li based cathodes [12, 34], optimization of dopant concentration and thickness of organic layers, and other device configurations [11] can result in a reduction of operating voltage and increased quantum efficiency.

Acknowledgment

The authors thank Universal Display Corp., the Defense Advanced Research Projects Agency, and the National Science Foundation for financial support of this work.

Supporting Information Available

Synthetic and spectroscopic data for all of the $C^\wedge N_2 \text{Ir}(LX)$ complexes where LX = tmd, bza, and dbm reported in this chapter. This material is available free of charge via the Internet at http://pubs.acs.org.

References

1. (a) Balzani, V., Scandola, F. *Supramolecular Photochemistry*, Ellis Horwood: Chichester, U.K., 1991. (b) Balzani, V., Credi, A., Scandola, F. In *Transition Metals in Supramolecular Chemistry*, Fabbrizzi, L., Poggi, A., Eds., Kluwer: Dordrecht, The Netherlands, 1994; p1. (c) Lehn, J.-M. *Supramolecular Chemistry–Concepts and Properties*; VCH: Weinheim. Germany, 1995. (d) Bignozzi, C. A., Schoonover, J. R., Scandola, F. *Prog. Inorg. Chem.* 1997, **44**, 1.

2. (a) Kalyanasundaran, K. *Coord. Chem. Rev.* 1982, **46**, 159–244. (b) Chin, K.-F., Cheung, K.-K., Yip, H.-K., Mak, T. C. W., Che, C. M. *J. Chem. Soc., Dalton Trans.* 1995, **4**, 657–665. (c) Sonoyama, N., Karasawa O., Kaizu. Y. *J. Chem. Soc., Faraday Trans.* 1995, **91**, 437–443. (d) Tan-Sien-Hee. L., Mesmaeker, A. K–D. *J. Chem. Soc., Dalton Trans.* 1994, **24**, 3651–3658. (e) Kalyanasundaram, K., Gratzel, M. *Coord. Chem. Rev.* 1998, **177**, 347–414.

3. (a) Baldo, M. A., O'Brien, D. F., You Y., Shousrikov, A., Sibley, S., Thompson, M. E., Forrest, S. R. *Nature* 1998, **395**, 151–154. (b) Baldo, M. A., Lamansky, S., Burrows, P. E., Thompson, M. E., Forrest, S. R. *Appl. Phys. Lett.* 1999, **75**, 4–6. (c) Thompson, M. E., Burrows, P. E., Forrest, S. R. *Curr. Opin. Solid State Mater. Sci.* 1999, **4**, 369.

4. (a) Anderson, P. A., Anderson, R. F., Fume, M., Junk, P. C., Keene, F. R., Patterson, B. I., Yeomans. B. D. *Inorg. Chem.* 2000, **39**, 2721–2728. (b) Li, C., Hoffman, M. Z. *Inorg. Chem.* 1998, **37**, 830–832. (c) Berg-Brennan, C., Subramanian, P., Absi, M., Stern, C., Hupp, J. T. *Inorg. Chem.* 1996, **35**, 3719–3722. (d) Kawanishi, Y., Kitamura, N., Tazuke, S. *Inorg. Chem.* 1989, **28**, 2968–2975.
5. (a) Balzani, V., Juris, A., Venturi, M., Campagna, S., Serroni, S. *Chem. Rev.* 1996, **96**, 759–834. (b) Shaw, J. R., Sadler, G. S., Wacholtz, W. F., Ryu, C. K., Schmehl, R. H. *New J. Chem.* 1996, **20**, 749.
6. (a) Garces, F. O., King, K. A., Watts, R. J. *Inorg. Chem.* 1988, **27**, 3464–3471. (b) Dedian, K., Djurovich, P. I., Garces, F. O., Carlson, G., Watts, R. J. *Inorg. Chem.* 1991, **30**, 1685–1687. (c) Schild. B., Garces. F. O., Watts, R. J. *Inorg. Chem.* 1994, **33**, 9–14. (d) King, K. A., Spellane, P. J., Watts. R. J. *J. Am. Chem. Soc.* 1985, **107**, 1432–1433.
7. Colombo, M. G., Brunold, T. C., Riedener, T., Güdel, H. U; Förtsch, M., Bürgi, H.-B. *Inorg. Chem.* 1994, **33**, 545–550.
8. (a) Wilde, A. P., King. K. A., Watts, R. J. *J. Phys. Chem.* 1991, **95**, 629–634. (b) Sprouse, S., King, K. A., Spellane, P. J., Watts, R. J. *J. Am. Chem. Soc.* 1984, **106**, 6647–6653. (c) Crosby, G. A. *J. Chim. Phys.* 1967, **64**, 160. (d) Columbo, M. C., Hauser, A., Güdel, H. U. *Top. Curr. Chem.* 1994, **171**, 143.
9. Lamansky, S., Djurovich, P., Murphy, D., Abdel-Razaq, F., Kwong, R., Tsyba, I., Bortz, M., Mui, B., Bau, R., Thompson. M. E. *Inorg. Chem.* 2001, **40**, 1704–1711.
10. (a) Baldo, M. A., O'Brien, D. F., You, Y., Shoustikov, A., Sibley, S., Thompson, M. E., Forrest, S. R. *Nature* 1998, **395**, 151–154. (b) Baldo, M. A., Lamansky, S., Burrows, P. E., Thompson, M. E., Forrest, S. R. *Appl Phys. Lett.* 1999, **75**, 4–6. (c) Thompson, M. E., Burrows, P. E., Forrest, S. R. *Curr. Opin. Solid State Mater. Sci.* 1999, **4**, 369.
11. Baldo, M. A., O'Brien, D. F., Thompson, M. E., Forrest, S. R. *Phys. Rev. B* 1999, **60**, 14422.
12. Adachi, C., Baldo, M. A., Forrest, S. R., Thompson, M. E. *Appl. Phys. Lett.* 2000, **78**, 1704.
13. Nonoyama, M. *Bull. Chem. Soc. Jpn.* 1974, **47**, 767–768.
14. King, K. A., Spellane, P. J., Watts, R. J. *J. Am. Chem. Soc.* 1985, **107**, 1432–1433.
15. O'Connor, D. V., Phillips, D. *Time-Correlated Single Photon Counting*; Academic Press: London, 1984; p. 287.
16. Carlson, G. A., Djurovich, P. I., Watts, R. J. *Inorg. Chem.* 1993, **32**, 4483–4484.

17. (a) Ohsawa, Y., Sprouse, S., King, K. A., DeArmond, M. K., Hanck, K. W., Watts, R. J. *J. Phys. Chem.* 1987, **91**, 1047–1058. (b) Colombo, M. G., Hauser, A., Güdel, H. U. *Inorg. Chem.* 1993, **32**, 3088–3092.
18. Schmid, B., Garces, F. O., Watts, R. J. *Inorg. Chem.* 1994, **33**, 9–14.
19. Sprouse, S., King, K. A., Spellane, P. J., Watts, R. J. *J. Am. Chem. Soc.* 1984, **106**, 6647–6653.
20. Brackman, U. *Lambdachrome Laser Dyes*, 2nd ed., Lambda Physik Ink., Göttingen, Germany, 1997.
21. (a) Colombo, M. G., Brunold, T. C., Riedener, T., Güdel, H. U., Förtsch, M., Bürgi, H.-B. *Inorg. Chem.* 1994, **33**, 545–550. (b) Colombo, M. G., Hauser, A., Güdel, H. U. *Inorg. Chem.* 1993, **32**, 3088–3092.
22. Sarkar, A., Sankar, C. *J. Lumin.* 1995, **65**, 163–168.
23. Colombo, M. G., Hauser, A., Güdel, H. U. *Top. Curr. Chem.* 1994, **171**, 143–171 (Electronic and Vibronic Spectra of Transition Metal Complexes I).
24. Sandrini, D., Maestri, M., Ciano, M., Balzani, V., Lueoend, R., Deuschel-Cornioley. C., Chassot, L., von Zelewsky, A. *Gaz. Chim. Ital.* 1988, **118**, 661.
25. Sandrini, D., Maestri, M., Ciano, M., Balzani, V., Deuschel-Cornioley, C., von Zelewsky, A., Jolliet, P. *Helv. Chim. Acta* 1988, **71**, 1053. Balton, C. B., Murtaza, Z., Shaver, R. J., Rillema, D. P. *Inorg. Chem.* 1992, **31**, 3230.
26. Zollinger, H. *Color Chemistry: Syntheses, Properties and Applications of Organic Dyes and Pigments*, 2nd ed., VSH: Weinheim, Germany, 1991.
27. Rusakowicz, R., Testa, A. C. *Spectochim. Acta, Part A* 1971, **27**, 787–792.
28. (a) Garces, F. O., King, K. A., Watts, R. J. *Inorg. Chem.* 1988, **27**, 3464–3471. (b) Dedian. K., Djurovich, P. I., Garces, F. O., Carlson. G., Watts, R. J. *Inorg. Chem.* 1991, **30**, 1685–1687.
29. Colombo, M. G., Brunold, T. C., Riedener, T., Güdel, H. U., Förtsch, M., Bürgi, H.-B. *Inorg. Chem.* 1994, **33**, 545–550.
30. Baldo, M. A., Lamansky. S., Burrows, P. E., Thompson. M. E., Forrest, S. R. *Appl. Phys. Lett.* 1999, **75**, 4–6.
31. The lumen is a unit of optical power that is weighted by the human eye response. For the EL spectrum of ppy_2Ir(acac), there are roughly 500 lumens/optical watt.
32. Whitaker. J., *Electronic Displays: Technology, Design, and Applications*; McGraw-Hill: New York, 1994; p. 92.
33. Colombo, M. G., Hauser, A., Güdel, H. U. *Inorg. Chem.* 1993, **32**, 3088–3092.
34. Hung, L. S., Tang, C.W., Mason, M. G. *Appl. Phys. Lett.* 1997, **70**, 152.

Chapter 14

Synthesis and Characterization of Phosphorescent Cyclometalated Iridium Complexes

Sergey Lamansky, Peter Djurovich, Drew Murphy,
Feras Abdel-Razzaq, Raymond Kwong, Irina Tsyba,
Manfred Bortz, Becky Mui, Robert Bau, and
Mark E. Thompson

*Department of Chemistry, University of Southern California,
Los Angeles, California 90089, USA*
met@usc.edu

The preparation, photophysics, and solid state structures of octahedral organometallic Ir complexes with several different cyclometalated ligands are reported. $IrCl_3 \cdot nH_2O$ cleanly cyclometalates a number of different compounds (i.e., 2-phenylpyridine, 2-(*p*-tolyl)pyridine, benzoquinoline, 2-phenylbenzothiazole, 2-(1-naphthyl)benzothiazole, and 2-phenylquinoline), forming the corresponding chloride-bridged dimers, $C^{\wedge}N_2Ir(\mu\text{-Cl})_2Ir\ C^{\wedge}N_2$ ($C^{\wedge}N$ is a cyclometalated ligand) in good yield. These chloride-bridged dimers react with acetyl acetone (acacH) and other bidentate, monoanionic

---

Reprinted from *Inorg. Chem.*, **40**, 1704–1711, 2001.

Electrophosphorescent Materials and Devices
Edited by Mark E. Thompson
Text Copyright © 2001 American Chemical Society
Layout Copyright © 2024 Jenny Stanford Publishing Pte. Ltd.
ISBN 978-981-4877-34-3 (Hardcover), 978-1-003-08872-1 (eBook)
www.jennystanford.com

ligands such as picolinic acid (picH) and *N*-methylsalicylimine (salH), to give monomeric *C^N*₂Ir(LX) complexes (LX = acac, pic, sal). The emission spectra of these complexes are largely governed by the nature of the cyclometalating ligand, leading to λ_{max} values from 510 to 606 nm for the complexes reported here. The strong spin–orbit coupling of iridium mixes the formally forbidden ^{3}MLCT and $^3\pi$–π* transitions with the allowed ^{1}MLCT, leading to a strong phosphorescence with good quantum efficiencies (0.2–0.8) and room temperature lifetimes in the microsecond regime. The emission spectra of the *C^N*₂Ir(LX) complexes are surprisingly similar to the *fac*-Ir *C^N*₃ complex of the same ligand, even though the structures of the two complexes are markedly different. The crystal structures of two of the *C^N*₂Ir(acac) complexes (i.e., *C^N* = *ppy* and *tpy*) have been determined. Both complexes show *cis*-C,C′, *trans*-N,N′ disposition of the two cyclometalated ligands, similar to the structures reported for other complexes with a "*C^N*₂Ir" fragment. NMR data (^{1}H arid ^{13}C) support a similar structure for all of the *C^N*₂Ir(LX) complexes. Close intermolecular contacts in both (*ppy*)₂Ir(acac) and (*tpy*)₂Ir(acac) lead to significantly red shifted emission spectra for crystalline samples of the *ppy* and *tpy* complexes relative to their solution spectra.

14.1 Introduction

The photophysics of octahedral 4d^6 and 5d^6 complexes has been studied extensively [1]. These complexes, particularly those prepared with Ru^{2+} and Os^{2+}, have been used extensively in photocatalysis and photoelectrochemistry [2]. The reason these d^6 complexes are attractive in photochemical applications is that they have long-lived excited states and good photoluminescence efficiencies. The majority of the studies have been focused on the photophysical properties of octahedral metal–diimine complexes Ru^{2+} and Os^{2+}, with ligands such as the bipyridine and phenanthroline [3]. More recently a number of groups have investigated the isoelectronic Rh^{3+} and Ir^{3+} complexes [4]. Tris-chelate complexes of Rh and Ir have been prepared with diamine ligands as well as cyclometalated ligands, such as 2-phenyl-pyidinato-C^2,N (*ppy*) [5]. Several tris-cyclometalated complexes of Rh and Ir have been reported, e.g., fac-M$(ppy)_3$, fac-M(2-(2-thienyl)pyridine)$_3$ [6]. The cyclometalated ligands are mono-anionic, thus forming neutral metal tris-chelate complexes, which are isoelectronic to the cationic Ru and Os complexes. When both cyclometalated and diimine ligands are used, monocationic complexes can be obtained, e.g., $(ppy)_2$M(L')$^+$ (L' = 2,2'-bipyridine, phenanthroline) [7].

Both neutral and cationic tris-chelate complexes of Rh and Ir show excited state lifetimes in the microsecond regime, as expected for a high-spin excited state. The Ir complexes show intense phosphorescence at room temperature, while the Rh complexes give measurable emission only at low temperatures. The stronger spin–orbit coupling expected for Ir relative to Rh significantly mixes singlet and triplet excited states for Ir, largely removing the spin-forbidden nature of the phosphorescent transitions, leading to efficient phosphorescence for these complexes [e.g., ϕ_{phos}(fac-Ir$(ppy)_3$) = 0.4 [9]. The phosphorescent lifetimes for these Ir complexes are relatively long ($\tau \approx \mu s$) [4–7], compared to fluorescent lifetimes, which are typically in the nanosecond regime [8]. However, the Ir phosphor lifetimes are significantly shorter than phosphorescent lifetimes of common organic luminophores, which can range from many milliseconds to minutes (common for a spin-forbidden transition) [8]. The significant decrease in the lifetime of the triplet excited state for the Ir complexes relative to the triplet excited state of organic molecules is a direct consequence of the strong spin–orbit coupling of the Ir center. The electronic transitions responsible for luminescence in these cyclometalated complexes have been assigned to mixture of MLCT and ligand-centered transitions [7b, 9]. An interesting application for these organometallic Ir complexes is their use as phosphors in organic light emitting diodes (OLEDs). The excited states generated in electroluminescence are trapped at the phosphor, where strong spin–orbit coupling leads to efficient phosphorescence at room temperature and efficient utilization of these excited states [10]. OLEDs prepared with these heavy metal complexes are the most efficient OLEDs reported to date [10], with efficiencies of greater than 80% (photons/electrons) reported for Ir$(ppy)_3$ based OLEDs [11].

Unfortunately, many of the cyclometalating ligands that could be incorporated into Ir-based phosphors do not give tris-ligand complexes by the reported synthetic methods. Herein we describe the synthesis as well as spectroscopic and structural characterization of a class of highly phosphorescent Ir complexes. These complexes have two cyclometalated ligands and a single bidentate, monoanionic ancillary ligand, making the complexes neutral and sublimable. The emission color from the complex is dependent on the choice of cyclometalating ligand, ranging from green to red. The structures of two of these complexes have been determined and strongly resemble that of the chloride-bridged dimers prepared with the same cyclometalated ligands, i.e., $\underline{C^\wedge N_2}\text{IrIr}(\mu\text{-Cl})_2\text{Ir}\ \underline{C^\wedge N_2}$.

14.2 Experimental Section

UV–visible spectra were measured on an Aviv model 14DS spectrophotometer (a re-engineered Cary 14 spectrophotometer). Photoluminescent spectra were measured with a Photon Technology International fluorimeter. Quantum efficiency measurements were carried out at room temperature in a 2-MeTHF solution (degassed by-several freeze–pump–thaw cycles). Degassed solutions of *fac*-Ir(*ppy*)$_3$ were used as a reference. Steady state emission experiments at room temperature were performed on a PTI Quanta-Master model C-60 spectrofluorimeter. Phosphorescence lifetime measurements were performed on the same fluorimeter equipped with a microsecond Xe flash lamp. NMR spectra were recorded on Bruker AC 250 MHz, AM 360 MHz, or AMX 500 MHz instruments. Solid probe MS spectra were taken with Hewlett-Packard GC/MS instrument with electron impact ionization and model 5873 mass sensitive detector. High-resolution mass spectrometry was done at Frick Chemical Laboratory at Princeton University. The Microanalysis Laboratory at. the University of Illinois, Urbana-Campaign, performed all elemental analyses.

All procedures involving IrCl$_3$ · H$_2$O or any other Ir(III) species were carried out in inert gas atmosphere despite the air stability of the compounds, the main concern being their oxidative and thermal stability of intermediate complexes at the high temperatures used in the reactions. Cyclometalated Ir(III) μ-chloro-bridged dimers of general formula $C^{\wedge}N_2$Ir(μ-Cl)Ir $C^{\wedge}N_2$ were synthesized by the method reported by Nonoyama, which involves refluxing $IrCl_3 \cdot nH_2O$ (Next Chimica) with 2–2.5 equiv of cyclometalating ligand in a 3:1 mixture of 2-methoxy-ethanol (Aldrich Sigma) and water [15].

14.2.1 Synthesis of ($C^{\wedge}N_2$Ir (acac) Complexes: General Procedure

[($C^{\wedge}N$)$_2$IrCl]$_2$ complex (0.078 mmol), 0.2 mmol of 2.4-pentanedione (Aldrich Sigma), and 85–90 mg of sodium carbonate were refluxed under inert gas atmosphere in 2-ethoxyethanol for 12–15 h. After cooling to room temperature, the colored precipitate was filtered off and was washed with water, followed by 2 portions of ether and hexane. The crude product was flash chromatographed using a silica/dichloromethane column to yield ca. 75–90% of the pure ($C^{\wedge}N$)$_2$Ir(acac) after solvent, evaporation and drying,

(ppy)$_2$Ir(acac): iridium(III) bis(2-phenylpyridiriato-N, $C^{2'}$) acetylacetonate (yield 83%). ^{1}H NMR (360 MHz, acetone-d_6), ppm: 8.55 (d, 2H, J 5.8 Hz), 8.07 (d, 2H, J 7.9 Hz), 7.91 (t, 2H, J 7.4 Hz), 7.63 (d, 2H, J 7.9 Hz), 7.32 (t, 2H, J 7.4 Hz), 6.74 (t, 2H, J 7.4 Hz), 6.59 (t, 2H, J 5.8Hz), 6.21 (d, 2H, J 7.4 Hz), 5.26 (s, 1H), 1.69 (s, 6H), Anal. Found: C 54.20, H 3.92, N 4.71. Calcd: C 54.08, H3.87, N4.67.

(tpy)$_2$Ir(acac): iridium(III) bis(2-(4-tolyl)pyridinato-$N,C^{2'}$) acetylacetonate (yield 75%). ^{1}H NMR (500 MHz, CDCl3), δ, ppm: 8.45 (d, 2H, J 5.1 Hz), 7.77 (d, 2H, J 8.3 Hz), 7.68 (dt, 2H, 7.4, 1.4 Hz), 7.42 (d, 2H, J 8.3 Hz), 7.07 (ddd, 2H, J 6.9, 6.5, 1.4 Hz), 6.61 (d, 2H, J 7.8 Hz), 6.04 (s, 2H), 5.17 (s, 1H), 1.54 (s, 6H). Anal. Found: C 55.35, H 4.40, N 4.52. Calcd: C 55.49, H 4.43, N 4.46.

(bzq)$_2$Ir(acac): iridium(III) bis(7,8-benzoquinolinato-$N,C^{3'}$) acetylacetonate (yield 90%). ^{1}H NMR (360 MHz, acetone-d_6), ppm: 8.96 (d, 2H, J 4.8 Hz), 8.49 (d, 2H, J 7.9 Hz), 7.80 (6H, 6, 8, m), 7.26 (d, 2H, J 7.9 Hz), 6.89 (t, 2H, J 7.4 Hz), 6.21 (t, 2H, J 7.9 Hz), 5.33 (s, 1H), 1.69 (s, 6H). Anal. Found: C 58.15, H 3.89, N 4.37. Calcd: C 58.08, H 3.81, N 4.23.

(bt)₂Ir(acac): iridium(III) bis(2-phenylbenzothiozoiato-$N,C^{2'}$) acetvlacetonate. ^{1}H NMR (360 MHz, DMSO-d_6), ppm: 8.25 (m, 2H), 7.93 (m, 2H), 7.74 (dd, 2H, J 8.0, 2.0 Hz), 7.55 (m, 4H), 6.85 (td, 2H, J 7.6, 1.0 Hz), 6.60 (td, 2H, J 7.6, 1.0 Hz), 6.20 (dt, 2H, 7.5, 0.6 Hz), 5.12 (s, 1H), 1.71 (s, 6H). Anal. Found: C 52.08, H 3.30, N 3.47. Calcd: C 52.23, H 3.39, N 3.93.

(bt)₂Ir(pico): iridium(III) bis(2-phenylbenzothiozolato-N,C^2) picolinate, ^{1}H NMR (360 MHz, DMSO-d_6), ppm: 8.24 (m, 3H), 8.09 (td, 1H, J 7.5, 1.7 Hz), 7.99 (d, 1H, J 7.7 Hz), 7.91 (dd, 1H, J 7.8, 0.8 Hz), 7.83 (m, 2H), 7.70 (m, 1H), 7.49 (m, 2H), 7.43 (t, 1H, J 4.5 Hz), 7.13 (t, 1H, J 7.9 Hz), 7.00 (td, 1H, J 7.7, 1.0 Hz), 6.95 (td, 1H, J 7.3, 1.2 Hz), 6.78 (td, 1H, J 7.3, 0.9 Hz), 6.73 (td, 1H, J 7.5, 1.2 Hz), 6.41 (d, 2H, J 7.7 Hz), 6.15 (d, 1H, J 7.7 Hz), 6.00 (d, 1H, J 8.3 Hz). Anal. Found: C 52.44, H 2.72, N 5.35. Calcd: C 52.30, H 2.74, N 5.72.

(bt)₂Ir(sal): iridium(III) bis(2-phenylbenzothiozolato-N,C^2) (N-methylsalicylimine-N,O). ^{1}H NMR (360 MHz, acetone-d_6), ppm: 8.60 (d, 1H, J 8.4 Hz), 8.15 (d, 1H, J 8.4 Hz), 8.09 (m, 1H), 7.86 (d, 1H, J 8.0 Hz), 7.78 (d, 1H, J 7.6 Hz), 7.70 (d, 1H, J 7.6 Hz), 7.41–7.52 (m, 3H), 7.31 (t, 1H, J 8.4 Hz), 6.99–7.05 (m, 2H), 6.83–6.89 (m, 2H), 6.61–6.67 (m, 2H), 6.43–6.49 (m, 2H), 6.27 (d, 1H, J 7.6 Hz), 6.18 (t, 1H, J 7.6 Hz), 5.61 (s, 1H), 3.13 (s, 3H). Anal. Found: C 53.62, H 3.44, N 4.85. Calcd: C 54.67, H 3.24, N 5.63.

(bsn)₂Ir(acac): iridium (III) bis(2-(1-naphthyl)benzothiazolato-$N,C^{2'}$) acetylacetonate. ^{1}H NMR (250 MHz, CDCl₃), ppm: 8.56 (d, 2H, J 7.8 Hz), 8.09 (d, 2H, J 7.9 Hz), 7.99 (d, 2H, J 7.5 Hz), 7.60 (m, 4H), 7.43 (m, 4H), 7.31 (t, 2H, J 7.8 Hz), 6.99 (d, 2H, J 8.2 Hz), 6.55 (d, 2H, J 8.5 Hz), 5.07 (s, 1H), 1.72 (s, 6H). Anal. Found: C 57.69, H 3.35, N 3.45. Calcd: C 56.56, H 3.51, N 3.03.

(pq)₂Ir(acac): iridium(III) bis(2-phenylquinolyi-$N,C^{2'}$) acetylacetonate (yield ca. 95%). ^{1}H NMR (360 MHz, DMSO-d_6), ppm: 8.51 (d, 2H, 8 Hz), 8.37 (m, 4H), 8.01 (m, 4H), 7.55 (m, 4H), 6.88 (tm, 2H, 5 Hz), 6.53 (td, 2H, 7 Hz, 1 Hz), 6.29 (dd, 2H, 8 Hz. 1 Hz), 4.70 (s, 1H), 2.08 (s, 6H).

***fac*-Ir(ppy)₃:** iridium(III) *fac*-tris(2-phenylpyridinato-$N,C^{2'}$). Iridium-(III) bis(2-phenylpyridinato-$N,C^{2'}$) acetylacetonate (0.20 mmol) and phenylpyridine (0.5 mmol) were refluxed for 15 h in

10 mL of glycerol to give a yellow suspension upon cooling. The crude product was isolated by filtration, washed with hexane and ether, dried in a vacuum, and zone sublimed (sample sublimed in a temperature gradient of 300–270°C) to give a green-yellow product (yield 85%). Analytical data matches that reported for fac-Ir(ppy)$_3$ [5].

Ir(*bzq*)$_3$: iridium(III) *fac*-tris(7,8-benzoquinolinate-$N,C^{3'}$). This complex was prepared by two different methods, method A and method B.

14.2.2 Method A

This complex was prepared from Ir(acac)$_3$ by a procedure analogous to the one reported for Ir(*ppy*)$_3$ [5]. Ir(acac)$_3$ (50 mg) and 7,8-benzoquinoline (113 mg) were dissolved in 5 mL, of glycerol and refluxed for 12–15 h under an inert atmosphere. After cooling, the product was filtered off and washed with several 10 mL portions of hexanes, ether, and acetone. The crude product was sublimed to give a 32% yield of orange Ir(*bzq*)$_3$. The Ir(*bzq*)$_3$ product from this reaction is a mixture of *fac*- and *mer*-isomers. Several wash cycles (acetone and dichloromethane) cause significant enrichment of the mixture in *fac*-product but still does not allow isolation of a pure facial complex. ^{1}H NMR (500 MHz, CD$_2$Cl$_2$), ppm: 8.31 (d, J 8.1 Hz), 8.19 (d, J 8.0 Hz), 8.12 (d, J 10.0 Hz), 8.03 (m), 7.9 (m), 7.6 (m), 7.47 (t, J 7.5 Hz), 7.39 (t, J 7.4 Hz), 7.22 (m), 7.14 (m), 7.07 (m), 6.96 (d, J 8.0 Hz), 6.80 (d, J 8.1 Hz), 6.57 (d, J 7.9 Hz). Anal. Found: C 63.55, H 3.51, N 5.32. Calcd: C 64.45, H 3.33. N 5.78.

14.2.3 Method B

(*bzq*)$_2$Ir(acac) (0.16 mmol) and 7,8-benzoquinoline (0.5 mmol) were refluxed for 15 h in 10 mL of glycerol to give a yellow-orange suspension upon cooling. The crude product was isolated by filtration, washed with hexane and ether, dried in a vacuum, and zone sublimed (sample sublimed in a temperature gradient of 300–270°C) to give an orange product, a mixture of *fac*- and *mer*-isomers (total yield 75%). ^{1}H NMR and elemental analysis characterization match those listed for Ir(*bzq*)$_3$ prepared by method A.

14.2.4 Electrochemistry

Cyclic voltammograms (CV) were recorded on an EG&G potentiostat/galvanostat model 283 in degassed dichloromethane solutions using 0.1 M tetra(*n*-butyl)ammonium hexafluoro-phosphate as a supporting electrolyte and a Ag/AgCl reference electrode. The Cp_2Fe/Cp_2Fe^+ redox couple was used as a secondary, internal reference. All studied complexes showed reversible oxidation waves with half-wave potentials of 870, 820, 860, 930, and 1000 mV for $(ppy)_2$Ir(acac), $(tpy)_2$Ir(acac), $(bzq)_2$Ir(acac), $(bsn)_2$Ir(acac), and $(bt)_2$Ir(acac) vs Ag/AgCl, respectively.

14.2.5 Crystallography

Diffraction data for $(tpy)_2$Ir(acac) and $(ppy)_2$Ir(acac) were collected at $T = -100°C$ on a Siemens P4 diffractometer with Mo Kα radiation ($\lambda = 0.71073$ Å). The cell dimensions and orientation matrix for data collection were obtained from a least-squares refinement of the setting angles of at least 25 accurately centered reflections. Data collection in the 2θ scan mode, cell refinement, and data reduction were carried out with the use of the program XSCANS [12] and resulted in data sets of 2176 reflections for $(ppy)_2$Ir(acac) and 4663 reflections for $(tpy)_2$Ir(acac). Pertinent crystal and unit cell parameters are given in Table 14.1. The $(ppy)_2$Ir(acac) crystal used in the X-ray diffraction study was of lower quality than the $(tpy)_2$Ir(acac) crystal. We attempted to grow better crystals of $(ppy)_2$Ir(acac), but could not get better crystals than the one used here. As a result the $(ppy)_2$Ir(acac) crystal only gave useful data out to $2\theta = 45°$, while the $(tpy)_2$Ir(acac) diffracted strongly to well beyond the 50° cutoff used here. The lower quality of the $(ppy)_2$Ir(acac) crystal is reflected in the comparatively higher R factor for the $(ppy)_2$Ir(acac) structure.

The structures were solved by direct methods using the SHELXS [13] package of computer programs. Both structures were refined by full-matrix least-squares methods based on F^2 using SHELX93 [13] and corrected for absorption by the program DIFABS [14]. Calculated hydrogen positions were input and refined in a riding manner along with their attached carbons. Due to the limited data

Table 14.1 Crystal data and summary of intensity data collection and structure refinement

	(ppy)$_2$Ir(acac)	(tpy)$_2$Ir(acac)
empirical formula	C$_{27}$H$_{23}$IrN$_2$O$_2$	C$_{29}$H$_{26}$IrN$_2$O$_2$
fw	599.67	626.72
temp, K	173	173
wavelength, Å	0.71073	0.71073
cryst syst	Orthorhombic	triclinic
space group	Pbcn (No. 60)	$P\bar{1}$ (No. 2)
unit cell dimens		
a(Å)	13.171(3)	9.697(2)
b(Å)	10.086(3)	11.616(1)
c(Å)	16.613(3)	12.371(1)
α (deg)	90	64.340(6)
β (deg)	90	76.153(9)
γ (deg)	90	87.211(10)
vol, Å^3	2206.9(8)	1217.2(4)
Z	4	2
density (calcd), g/cm^3	1.805	1.710
abs coeff, mm^{-1}	6.08	5.51
$F(000)$	1168	614
θ range for data collection, deg	2.5–22.5	2.0–25.0
index ranges	$0 \leq h \leq 14$,	$-9 \leq h \leq 9$,
	$0 \leq k \leq 10$,	$-12 \leq k \leq 12$,
	$0 \leq l \leq 17$	$0 \leq l \leq 14$
reflns collected	2176	4663
indep reflns	1410 [R(int) 0.074]	3745 [R(int) 0.0233]
refinement meth	full-matrix least-squares on F^2	full-matrix least-squares on F^2
data/restraints/params	1409/0/71	3745/0/309
GOF on F^2	1.129	0.725
final R index [$I > 2\sigma(I)$]	0.0678	0.0270
R index (all data)	0.0870	0.0336

set for (ppy)$_2$Ir(acac), the Ir was refined anisotropically and all other atom positions were refined isotropically. For (tpy)$_2$Ir(acac), the non-hydrogen atom positions were refined anisotropically and the hydrogen positions were refined isotropically. A summary of the refinement details and the resulting agreement factors is given in Table 14.1. In both cases the Ir centers are octahedrally coordinated by the three chelating ligands, with the N atoms in a trans configuration, and a 2-fold rotation axis passes through the Ir atom and the acac ligand.

14.3 Results and Discussion

14.3.1 Synthesis and Characterization of $C^\wedge N_2$Ir(LX) Complexes

Iridium chloride reacts with phenylpyridine in refluxing 2-ethoxyethanol to give a cyclometalated complex [15]. The product of this reaction is a chloride-bridged dimer, $(ppy)_2$Ir(μ-Cl)$_2$Ir$(ppy)_2$, ppy = 2-phenylpyridinato-C^2,N, Eq. 14.1. This cyclometallation reaction leads to a similar chloride-bridged dimer for other substituted pyridyl or heterocyclic ligands as well, i.e., 2-phenylbenzothiazole (bt), benzoquinoline (bzq), and donor- or acceptor-substituted phenylpyridines [9b, 16]. Crystallographic studies of the ppy complex show that the Ir center in $(ppy)_2$Ir$(\mu$-Cl$)_2$Ir$(ppy)_2$ is octahedrally coordinated by the two bridging chlorides and two cyclometalated ppy ligands, with the heterocyclic rings of the ligands in a trans configuration [16a, 17], as illustrated in Fig. 14.1b. Crystal structures have not been reported for the other chloride-bridged dimers, but NMR data suggests a structure similar to that of $(ppy)_2$Ir$(\mu$-Cl$)_2$Ir$(ppy)_2$ [16b, 18].

$$IrCl_3 \cdot nH_2O + 4\,[ppy] \longrightarrow [(ppy)_2 Ir(\mu\text{-}Cl)_2 Ir(ppy)_2] + 4\,HCl + n\,H_2O \tag{14.1}$$

We have found this cyclometalation reaction to be a general one, giving good yields of the corresponding $C^\wedge N_2$Ir$(\mu$-Cl$)_2$Ir $C^\wedge N_2$ dimers ($C^\wedge N$ = cyclometalated ligand) for a number of ligands. The ligands examined here are shown in Fig. 14.1. The reactions to form the dimers proceed under fairly mild conditions, giving

(a) ppy   tpy   bzq

bt   bsn   pq

(b)

(c) $\genfrac{}{}{0pt}{}{L}{X}\Big)\Longrightarrow$

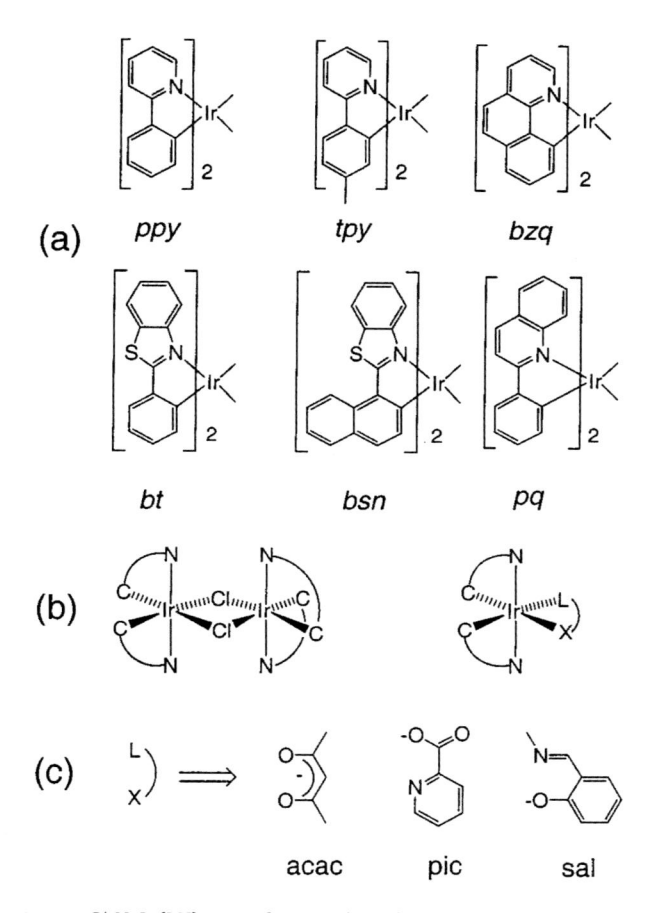

acac   pic   sal

Figure 14.1 $C^{\wedge}N_2$Ir(LX) complexes. The $C^{\wedge}N_2$Ir fragments are shown in part a, with the acronym used for each $C^{\wedge}N$ ligand listed under the fragment. The coordination geometries of $C^{\wedge}N_2$Ir(μ-Cl)$_2$Ir $C^{\wedge}N_2$ and $C^{\wedge}N_2$Ir(acac) complexes are shown in part b. The identities of the LX ligands are shown in part c.

good yields of the organometallic complexes, typically greater than 75%. All of the ligands used here have a heterocyclic nitrogen and an sp^2-hybridized carbon positioned such that the cyclometallation reaction leads to a five-membered ring. The NMR spectra of these complexes are consistent with the heterocyclic rings of the C,N ligands being in a trans disposition, as shown in Fig. 14.1b. The chloride-bridged complexes form a racemic mixture of $\Delta\Delta/\Lambda\Lambda$ dimers that are characterized by pronounced downfield shifts in the ^{1}H NMR resonance of heterocyclic protons nearest to the bridging chlorides [19].

$C^{\wedge}N_2$Ir(μ-Cl)$_2$Ir $C^{\wedge}N_2$ complexes show strong MLCT- and ligand-based absorption bands. These dimers are not strongly emissive at room temperature; however, they phosphoresce strongly in frozen glasses at 77 K. When the dimer is dissociated into monomeric Ir complexes, either by preparing an Ir$C^{\wedge}N_3$ complex or a cationic species, e.g., $(ppy)_2$Ir$(bpy)^+$ (bpy = 2,2'-bipyridine), the phosphorescence efficiency increases markedly. The weak phosphorescence of the chloride-bridged dimers has been attributed to a monomer–dinner equilibrium that occurs in solution [16c]. The tris-chelate complexes are attractive in a number of applications due to their high phosphorescence efficiencies and microsecond lifetimes; however, we have found it very difficult to make the complexes for many of the cyclometalated ligands of interest. While we were able to prepare the chloride bridged dimers for all of the $C^{\wedge}N$ ligands shown in Fig. 14.1, the preparation of tris-chelates, i.e., Ir $C^{\wedge}N_3$, has only been possible for ppy, tpy, and bzq ligands.

The chloride-bridged dimers can be converted to emissive, monomeric complexes by treating the dimer with a bidentate, anionic ancillary ligand (LX), such as a β-diketonate, 2-picolinic acid, or an N-alkylsalicylimine, Eq. 14.2. These reactions give $C^{\wedge}N_2$Ir(LX) in high yield, typically greater than 80%. The majority of the $C^{\wedge}N_2$Ir(LX) complexes prepared are emissive at room temperature, with λ_{max} values between 510 and 610 nm, Table 14.2. The complexes are stable in air and can be sublimed in a vacuum without decomposition.

$$C^{\wedge}N_2\text{Ir}(\mu - \text{Cl})_2\text{Ir}C^{\wedge}N_2 + 2\text{LXH} \rightarrow C^{\wedge}N_2\text{Ir}(\text{LX}) + 2\text{HCl} \quad (14.2)$$

Analysis by cyclic voltammetry shows that the $C^{\wedge}N_2$Ir(acac) complexes all undergo a reversible one-electron oxidation; however, no reduction processes were observed in CH_2Cl_2. The oxidation potentials of these complexes vary between 0.8 and 1.0 V (vs Ag/AgCl), Table 14.2. The complexes with pyridyl type ligands (i.e., ppy, tpy, and bzq) are easier to oxidize than the complexes with benzothiazolyl based ligands (i.e., bt and bsn). The oxidation

Table 14.2 Photophysical and electrochemical data for $C\wedge N_2\text{Ir}(LX)$ complexes

$C\wedge N$	LX	Absorbance λ (log ϵ)	Emission λ_{max} (nm)	Lifetime (μsec)[a] 77 K	Lifetime (μsec)[a] 298 K	Quantum yield[a,b]	Oxidation potential mV vs Ag/AgCl
ppy	acac	260 (4.5), 345 (3.8), 412 (3.4), 460 (3.3), 497 (3.0)	516	3.2	1.6	0.34 0.8	870
tpy	acac	270 (4.5), 370 (3.7), 410 (3.5), 460 (3.4), 495 (3.0)	512	4.5	3.1	0.31 0.7	820
bzq	acac	260 (4.6), 360 (3.9), 470 (3.3), 500 (3.2)	548	23.3	4.5	0.27 0.6	860
bt	acac	269 (4.6), 313 (4.4), 327 (4.5), 408 (3.8), 447 (3.8), 493 (3.4), 540 (3.0)	557	4.4	1.8	0.26 0.6	1000
bt	pic	206 (4.6), 326 (4.6), 355 (4.1), 388 (3.9), 437 (3.8), 475 (3.7)	541	3.6	2.3	0.37 0.8	
bt	sal	276 (4.6), 325 (4.3), 450 (3.7), 540 (2.9)	562	3.1	1.4	0.22 0.5	
bsn	acac	274 (4.7), 300 (4.6), 345 (4.5), 427 (4.0), 476 (4.0), 506 (3.9)	606	2.5	1.8	0.22 0.5	930
pq	acac	268 (5.0), 349 (4.4), 433 (3.9), 467 (3.9), 553 (3.6)	597		2.0	0.10 0.2	

[a] Lifetime and quantum yield measurements were made in 2-MeTHF, except for the 77 K lifetime of $(ppy)_2\text{Ir}(acac)$, which was carried out in a CH_2Cl_2 frozen glass. The lifetimes and quantum yields have error bars of have ±10%.

[b] The quantum yield reported in reference 9 for the $\text{Ir}(ppy)_3$ reference was 0.4. It was later discovered that this number was not correct and the photoluminescence quantum yield for this complex in solution is 0.73 in toluene and (W. Holzer, et al., *Chem. Phys.* 2005, **308**, 93–102, DOI: 10.1016/j.chemphys.2004.07.051) and 0.9 in THF (Sajoto, T, et al., *J. Am. Chem. Soc.*, 2009, **131**, 9813–9822, DOI: 10.1021/ja903317w). Thus, the quantum yield values previously published have been increased by a factor of 2.2 (0.9/0.4) from the published values. The published values have been struck out.

potentials of (*ppy*)$_2$Ir(acac) and (*tpy*)$_2$Ir(acac) are ca. 300 mV less positive than the values reported for the corresponding cationic (*ppy*)$_2$Ir(L')$^+$ and (*tpy*)$_2$Ir(L')$^+$ complexes, where L' = 2,2'-bipyridine or 1,10-phenanthroline [7a, 16a]. The oxidation potentials of the *ppy and tpy* complexes are comparable to those found in Ir C^N$_3$ complexes with the same ligands, showing that the metal localized HOMOs are at very similar energies.

The C^N$_2$Ir(acac) complexes can be used to prepare Ir C^N$_3$. Treating (*ppy*)$_2$Ir(acac) with phenylpyridine in refluxing glycerol gives Ir(*ppy*)$_3$ in greater than 75% yield. The overall yield of Ir(*ppy*)$_3$ (based on IrCl$_3$ · nH$_2$O) by this route is greater than 60%, compared to less than 20% when Ir(acac)$_3$ is used as the intermediate [20]. Similar yields have been achieved for the synthesis of Ir(*bzq*)$_3$ and Ir(2-(2-thienyl)pyridine)$_3$ via the C^N$_2$Ir(acac) complexes. Only C^N$_2$Ir(acac) complexes of phenyl pyridines, benzoquinoline, and thienylpyridine form tris complexes when reacted with excess C^N ligand. To the best of our knowledge this is the first report of a tris-cyclometalated complex of benzoquinoline, Ir(*bzq*)$_3$. Based on molecular modeling there are no obvious steric interactions that would prevent the formation tris complexes from *bt, bsn,* or *pq* ligands. However, other factors such as imine basicity could play a controlling role in stabilizing intermediates in the mechanism of tris complex formation.

14.3.2 Photophysical Properties of C^N$_2$Ir(acac) Compounds

The absorption spectra for (*tpy*)$_2$Ir(acac) and (*bzq*)$_2$Ir(acac) are characteristic of C ^ N$_2$Ir(LX) complexes, Fig. 14.2. Intense bands are observed in the ultraviolet part of the spectra, between 250 and 350 nm, which can be assigned to the allowed $^1(\pi-\pi^*)$ transitions of the C^N ligands. The energies and extinction coefficients for these bands correlate well with the transitions observed for the free hydrocarbons, i.e., tolylpyridine and benzoquinoline [6, 16]. Somewhat weaker bands are observed at lower energies ($\lambda_{max} > 400$ nm). in other cyclometalated complexes these bands have been assigned to singlet, and triplet MLCT transitions, and the same assignment is likely here [6, 16]. In the spectra for (*tpy*)$_2$Ir(acac),

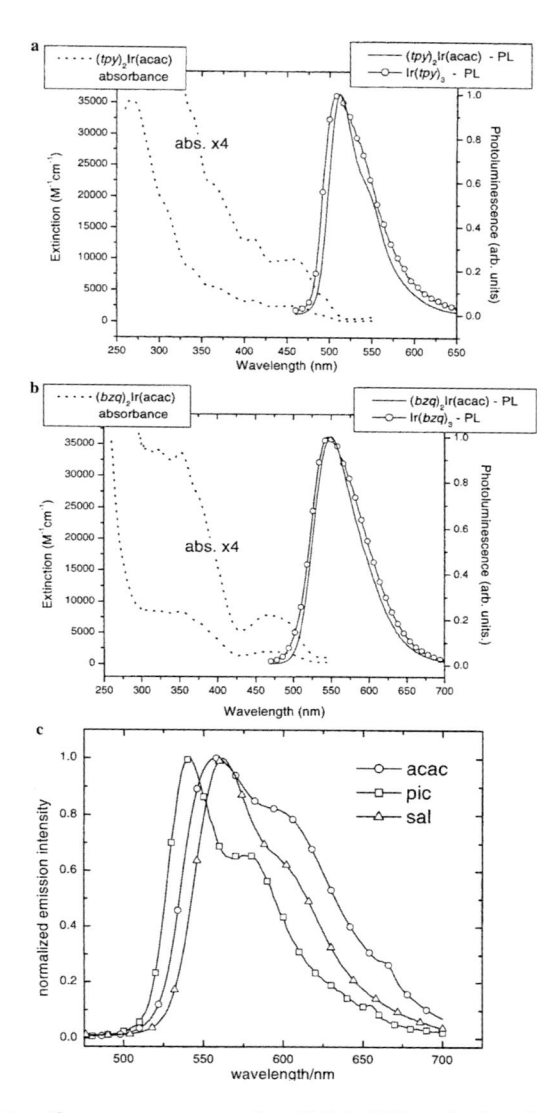

Figure 14.2 Absorption spectra for $C^{\wedge}N_2$Ir(LX) and photoluminescence spectra for $C^{\wedge}N_2$Ir(LX) and Ir $C^{\wedge}N_3$ complexes. Spectra for $(tpy)_2$Ir(acac) and fac-Ir$(ppy)_3$ are shown in part a. The absorption and emission spectra for $(tpy)_2$Ir(acac) and $(ppy)_2$Ir(acac) are identical. The absorption spectrum for $(bzq)_2$Ir(acac) as well as the emission spectra for $(bzq)_2$Ir(acac) and Ir$(bzq)_3$ are shown in part b. Spectra for $(bt)_2$Ir(LX), LX = acac, pic, sal, are shown in part c. Samples were in degassed 2-MeTHF solutions, measured at room temperature.

two MLCT bands are clearly resolved at 410 and 460 nm, with extinction coefficients of 3500 and 2500 M^{-1} cm^{-1}, respectively. The formally spin forbidden ^{3}MLCT at 460 nm gains intensity by mixing with the higher lying ^{1}MLCT transition through the strong spin–orbit coupling of Ir. This mixing is strong enough in these Ir complexes that the formally spin forbidden ^{3}MLCT has an extinction coefficient that is more than half that of the spin-allowed ^{1}MLCT. In some cases, the ^{1}MLCT and ^{3}MLCT are not clearly resolved, as seen in (bzq)$_2$Ir(acac), where a broad band is observed between 430 and 520 nm. The spectra observed for $C^{\wedge}N_2$Ir(LX) complexes are very similar to those of the $C^{\wedge}N_2$Ir(μ-Cl)$_2$Ir $C^{\wedge}N_2$ and $C^{\wedge}N_2$Ir(bpy)$^+$ complexes with the same $C^{\wedge}N$ ligand [7a, 16b], suggesting that the dominant absorptions in the $C^{\wedge}N_2$Ir(LX) are due to the "$C^{\wedge}N_2$Ir" fragment, which has the same structure in all three types of complexes, *vide infra*.

All of the $C^{\wedge}N_2$Ir(LX) complexes reported here give phosphorescence quantum yields between 0.2 and 0.8, and luminescence lifetimes between 1 and 5 µs for degassed solutions at room temperature, Fig. 14.2. These photophysical parameters are very similar to those of the *fac*-Ir $C^{\wedge}N_3$ complexes. The emission spectra of the $C^{\wedge}N_2$Ir(acac) complexes match those of the *fac*-Ir $C^{\wedge}N_3$ complexes prepared with the same $C^{\wedge}N$ ligands for *ppy*-, *tpy*-, and *bzq*-based complexes, Fig. 14.2. It has been shown that emission from Ir(*ppy*)$_3$ comes from MLCT transitions [6, 21]. Considering that the absorption spectra of the *ppy* and *bzq* complexes reported here are very similar to their Ir$C^{\wedge}N_3$ and $C^{\wedge}N_2$Ir(L')$^+$ analogues, it is not surprising that the emission spectra of $C^{\wedge}N_2$Ir(acac) and Ir $C^{\wedge}N_3$ complexes are similar, since all of the *ppy*, *tpy*, and *bzq* complexes presumably emit from MLCT states of the same energy.

Changes in the $C^{\wedge}N$ ligand in a $C^{\wedge}N_2$Ir(LX) complex typically have a marked effect on the phosphorescence spectrum, while changing the LX ligand leads to a relatively minor shift. *ppy* and *tpy* ligands give green emission, *bzq* and *bt* ligands give yellow emission, *pq* ligands give orange emission, and the *bsn* complex gives red emission. If the pyridyl group of *ppy* is replaced with a 2-quirioliriyl group, the emission line is shifted by 77 nm. A similar red shift is observed if the phenyl group of the *bt* ligand is replaced with an α-naphthyl group (*bsn*). The spectra of $C^{\wedge}N_2$Ir(LX) complexes show very little LX ligand character in the excited state. For example, $(bt)_2$Ir(LX) (LX = acac, pic, and sal) give very similar spectra, Fig. 14.2c, with only minor shifts as a function of the LX ligand. Vibronic fine structure is observed in the $(bt)_2$Ir(LX) emission spectra, consistent with a significant ligand $^3(\pi-\pi^*)$ contribution to the phosphorescence, as observed in related Ir $\underline{C^\wedge N}_3$ complexes [6]

14.3.3 Structures of $\underline{C^\wedge N}_2$ Ir(LX) Complexes

Molecular plots of $(ppy)_2$Ir(acac) and $(tpy)_2$Ir(acac) are shown in Figs. 14.3a and b, respectively. Crystallographic data are given in Table 14.1, and selected bond lengths (Å) and angles (deg) for these complexes are presented in the Table 14.3. Crystals of $(ppy)_2$Ir(acac) were of lower quality than those of $(tpy)_2$Ir(acac), which is reflected in the poorer agreement factor for the $(ppy)_2$Ir(acac) complex. The molecular structures and solid state packing determined for the two complexes are very similar. The *tpy* and *ppy* complexes have an octahedral coordination geometry around Ir, retaining the *cis*-C,C *trans*-N,N chelate disposition of the chloride-bridged precursor complex, i.e., $(tpy)_2$Ir(μ-Cl)$_2$Ir$(tpy)_2$. The Ir–C bonds of these complexes (Ir–C$_{av}$ = 1.993(8) Å) are shorter than the Ir–N bonds (Ir–N$_{av}$ = 2.021(7) Å). These Ir–C bond lengths are similar to values of 2.02(2)–2.13(6) Å reported for the $(tpy)_2$Ir(μ-Cl)$_2$Ir$(tpy)_2$ and *fac*-Ir$(tpy)_3$ [17] complexes as well those given for other mononuclear complexes with the "ppy_2Ir" fragment [22, 23]. The Ir–N bond lengths also fall within the range of values given for complexes with a "ppy_2Ir" fragment and a *trans*-N,N' disposition of pyridyl ligands (1.98(1)–2.07(1) Å). The Ir–N bonds are shorter than the value of 2.132 Å observed in *fac*-Ir$(tpy)_3$, which has all pyridyl groups trans to phenyl groups, consistent with the strong trans influence of phenyl groups. The Ir–O bond lengths of 2.146(6) and 2.161(4) Å are longer than the mean Ir–O value of 2.088 Å reported in the Cambridge Crystallographic Database [24] and reflect the large trans influence of the phenyl groups. All other bond lengths and angles within the chelate ligands are typical for cyclometalated *ppy* [17, 22, 23] and acac ligands bound to Ir(III) [25–27].

The crystal packing of $(tpy)_2$Ir(acac) shows a nearest neighbor molecule related by a $C2$ rotation that positions the pyridyl rings

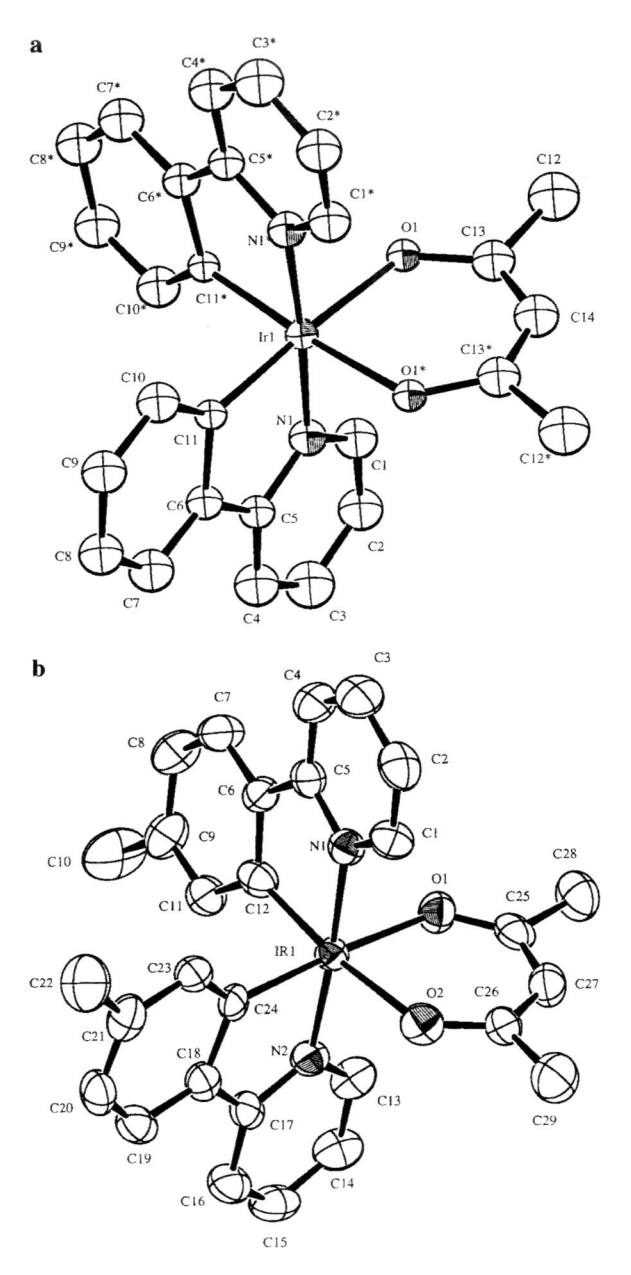

Figure 14.3 ORTEP drawings of $(ppy)Ir_2(acac)$ (a) and $(tpy)Ir_2(acac)$ (b). The thermal ellipsoids for both images represent a 50% probability limit.

Table 14.3 Selected bond distances (Å) and angles (deg) for $(ppy)_2\text{Ir(acac)}$ and $(ppy)_2\text{Ir(acac)}$

	$(ppy)_2\text{Ir(acac)}$	$(tpy)_2\text{Ir(acac)}$
Bond distances (Å)		
Ir(1)-N(1)	2.010(9)	2.040(5)
Ir(1)-N(1)*/Ir(1)-N(2)	2.010(9)	2.023(5)
Ir(1)-C(11)/Ir(1)-C(12)	2.003(9)	1.985(7)
Ir(1)-C(11)*/Ir(1)-C(24)	2.003(9)	1.982(6)
Ir(1)-O(1)/Ir(1)-O(2)	2.146(6)	2.161(4)
Ir(1)-O(1)*/Ir(1)-O(1)	2.146(6)	2.136(4)
Angles (deg)		
N(1)-Ir(1)-N(1)*/N(1)-Ir(1)-N(2)	176.3(4)	176.2(2)
N(1)-Ir(1)-C(11)/N(1)-Ir(1)-C(12)	81.7(4)	80.4(2)
N(1)*-Ir(1)-C(11)*/N(2)-Ir(1)-C(24)	81.7(4)	80.3(2)
N(1)-Ir(1)-O(1) /N(1)-Ir(1)-O(2)	94.5(3)	94.2(2)
C(11)*-Ir(1)- O(1) /C(24)-Ir(1)- O(2)	87.5(3)	91.3(2)
O(1)-Ir(1)-O(1)* /O(12)-Ir(1)-O(1)	90.0(3)	88.2(2)

over the aryl rings of adjacent molecules, as shown in Fig. 14.4 (top). The aromatic ring systems of the dyads have a significant degree of $\pi-\pi$ overlap with a ca. 3.4 Å face–face separation between the adjacent ring planes. A similar dyad formation also occurs in the crystal packing of $(ppy)_2\text{Ir(acac)}$, albeit with a lesser degree of ppy ligand overlap relative to tpy ligands and a separation of ca. 3.5 Å between the aromatic ring systems, Fig. 14.4 (bottom). On the basis of the higher degree of spatial overlap and closer face–face spacing of the tpy ligands, relative to the ppy ligands, we expect the $\pi-\pi$ interaction to be greater for $(tpy)_2\text{Ir(acac)}$.

The distance separating the aromatic rings in the solid state dimers is well within range for electronic interaction between the adjacent π systems [28]. Consequently, despite having nearly identical solution absorption and emission properties, the two complexes exhibit pronounced differences in their solid state luminescence spectra (Fig. 14.5). Powdered samples of both complexes are an indistinguishable yellow color. Powdered $(ppy)_2\text{Ir(acac)}$ luminesces yellow-green with a spectrum that is broader and red-shifted (λ_{max} = 540 nm) relative to the solution spectrum, while powdered $(tpy)_2\text{Ir(acac)}$ exhibits an orange emission that is more structured

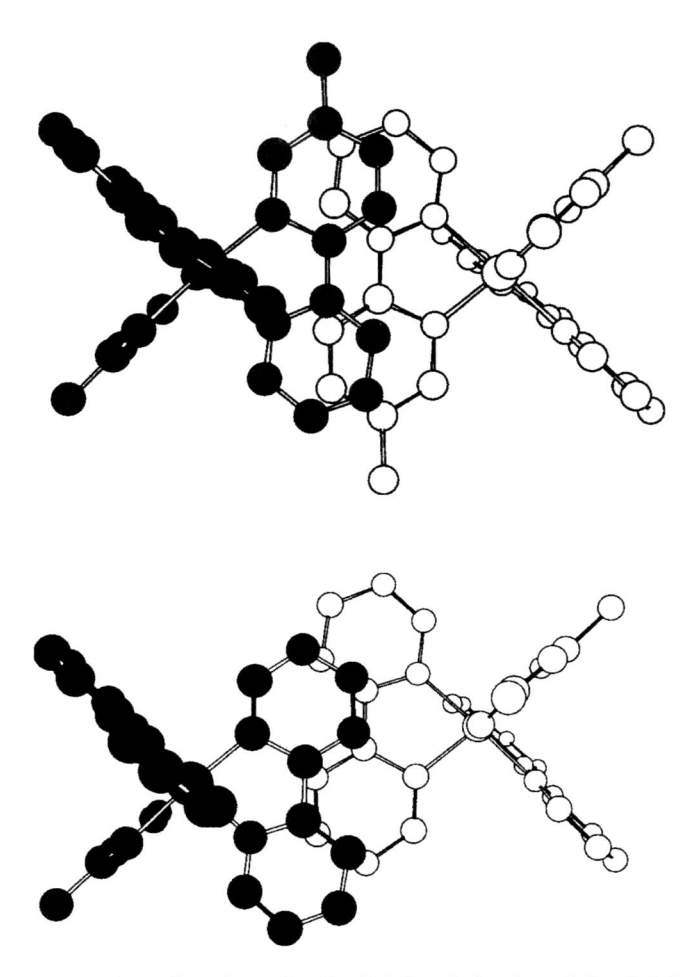

Figure 14.4 Crystal packing for $(tpy)_2$Ir(acac) (top) and $(ppy)_2$Ir(acac) (bottom). The molecule in the foreground is shown in black. The $\pi-\pi$ spacing between the arylpyridine (*tpy* or *ppy*) ligands of adjacent molecules is 3.4 and 3.5 Å for the *tpy and ppy* complexes, respectively.

than that of $(tpy)_2$Ir(acac) and has a maximum at 575 nm. The shift to lower emission energy for solid $(tpy)_2$Ir(acac) versus solid $(ppy)_2$Ir(acac) correlates with the greater degree of π overlap present in the dimeric units of the former complex [29]. Face to face $\pi-\pi$ interactions between aromatic molecules in the solid state

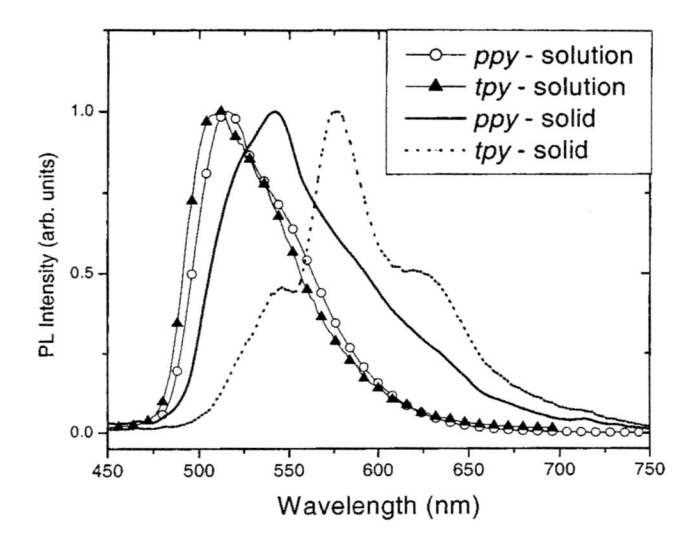

Figure 14.5 Solution (CH_2Cl_2) and solid state photoluminescence spectra of $(ppy)Ir_2(acac)$ and $(tpy)Ir_2(acac)$. The spectra were recorded at room temperature; however, cooling the samples to 77 K does not significantly alter the spectra.

often lead to a broad and featureless luminescence that is red-shifted from the emission that occurs in dilute solution and is typically ascribed to excimer formation [30]. Excimer formation has been reported in several examples of square planar Pt(diimine) complexes [31–33]; however, the influence of π stacking on the emission properties of octahedral complexes is less well documented [34]. While the luminescence from solid $(ppy)_2Ir(acac)$ is characteristic of excimer emission, the structured luminescence from $(tpy)_2Ir(acac)$ suggests a different mechanism, perhaps originating from ground state dimers which act as luminescent traps.

14.4 Conclusion

Iridium tris-chelates have high phosphorescence efficiency and microsecond lifetimes, which make them ideal for OLED applications [10]. However, the synthesis of a large variety of these complexes is difficult due to steric and electronic effects of the ligands.

Here it has been shown that a facile route to a variety of phosphorescent compounds exists through chloride bridged dimers. These complexes have the general formula $C^\wedge N_2$Ir(LX), where $C^\wedge N$ is a cyclometalating ligand and LX is a monoanionic bidentate ligand. These complexes give good phosphorescence efficiencies at room temperature and have luminescent lifetimes of 1–5 μs. The absorption and emission spectra of the $C^\wedge N_2$Ir(LX) complexes are very similar to the spectra of the chloride bridged dimers and mononuclear cationic ($C^\wedge N_2$Ir(bpy)$^+$) complexes leading to the conclusion that the emission is mainly due to the "$C^\wedge N_2$Ir" fragment. We have recently prepared organic LEDs with these complexes as emissive dopants and achieved quantum efficiencies comparable to those of devices with Ir(ppy)$_3$ dopants. This LED work will be reported elsewhere [35].

Acknowledgment

We thank Universal Display Corporation, the Defense Advanced Research Projects Agency, and the National Science Foundation for financial support of this work.

Supporting Information Available

Crystal data, structure refinement, atomic coordinates, bond distances, bond angles, and anisotropic displacement parameters ((*tpy*)$_2$Ir(acac) only) for (*ppy*)$_2$Ir(acac) and (*tpy*)$_2$Ir(acac) and the indexed powder patterns for (*ppy*)$_2$Ir(acac) and (*tpy*)$_2$Ir(acac). This material is available free of charge via the Internet at http://pubs.acs.org.

References

1. (a) Balzani, V., Scandola, F. *Supramolecular Photochemistry*, Ellis Horwood: Chichester, U.K., 1991. (b) Balzani, V., Credi, A., Scandola, F. In *Transition Metals in Supramolecular Chemistry*, Fabbrizzi, L., Poggi,

A., Eds., Kluwer: Dordrecht, The Netherlands, 1994; p1. (c) Lehn, J.-M. *Supramolecular Chemistry—Concepts and Properties*, VCH: Weinheim, Germany, 1995. (d) Bignozzi, C. A., Schoonover, J. R., Scandola, F. *Prog. Inorg. Chem.* 1997, **44**, 1.

2. (a) Kalyanasundaran, K. *Coord. Chem. Rev.* 1982, **46**, 159. (b) Chin, K.-F., Cheung, K.-K., Yip, H.-K., Mak, T. C. W., Che, C. M. *J. Chem. Soc. Dalton Trans.* 1995, **4**, 657–665. (c) Sonoyama, N., Karasawa, O., Kaizu, Y. *J. Chem. Soc. Faraday Trans.* 1995, **91**, 437. (d) Tan-Sien-Hee, L., Mesmaeker, A. K.-D. *J. Chem. Soc., Dalton Trans.* 1994, **24**, 3651–3658. (e) Kalyanasundaram, K., Gratzel, M. *Coord. Chem. Rev.* 1998, **177**, 347–414.

3. (a) Anderson, P. A., Anderson, R. F., Fume, M., Junk, P. C., Keene, F. R., Patterson, B. T., Yeomans, B. D. *Inorg. Chem.* 2000, **39**, 2721–2728. (b) Li, C., Hoffman, M. Z. *Inorg. Chem.* 1998, **37**, 830–832. (c) Berg-Brennan, C., Subramanian, P., Absi, M., Stern, C., Hupp, J. T. *Inorg. Chem.* 1996, **35**, 3719–3722. (d) Kawanishi, Y., Kitamura, N., Tazuke, S. *Inorg. Chem.* 1989, **28**, 2968–2975.

4. (a) Balzani, V., Juris. A., Venturi, M., Campagna, S., Serroni, S. *Chem. Rev.* 1996, **96**, 759. (b) Shaw, J. R., Sadler, G. S., Wacholtz, W. F., Ryu, C. K., Schmehl, R. H. *New J. Chem.* 1996, **20**, 749.

5. Dedeian, K., Djurovich, P. I., Garces, F. O., Carlson, G., Watts, R. J. *Inorg. Chem.* 1991, **30**, 1685–1687.

6. Colombo, M. G., Brunold, T. C., Riedener, T., Gudel, H. U. *Inorg. Chem.* 1994, **33**, 545–550.

7. (a) Ohsawa, Y., Sprouse, S., King, K. A., DeArmond, M. K., Hanck, K. W., Watts, R. J. *J. Phys. Chem.* 1987, **91**, 1047–1058. (b) Wilde, A. P., King, K. A., Watts, R. J. *J. Phys. Chem.* 1991, **95**, 629–634.

8. Turro, N. J. *Modern Molecular Photochemistry*, The Benjamin/Cummings Publishing Co., Inc., Menlo Park. California, 1978. Murov, S. L., Carmichael, I., Hug, G. L. *Handbook of Photochemistry*, Marcel Dekker: New York, 1993.

9. (a) Sprouse, S., King, K. A., Spellane, P. J., Watts, R. J. *J. Am. Chem. Soc.* 1984, **106**, 6647–6653. (b) Crosby, G. A. *J. Chem. Phys.* 1967, **64**, 160.

10. (a) Baldo, M. A., O'Brien, D. F., You, Y., Shoustikov, A., Sibley, S., Thompson, M. E., Forrest, S. R. *Nature* 1998, **395**, 151. (b) Baldo, M. A., Lamansky, S., Burrows, P. E., Thompson, M. E., Forrest, S. R. *Appl. Phys. Lett.* 1999, **75**, 4. (c) Thompson, M. E., Burrows, P. E., Forrest, S. R. *Curr. Opin. Solid State Mater. Sci.* 1999, **4**, 369.

11. Adachi, C., Forrest, S. R., Thompson, M. E. *Appl. Phys. Lett.* in press.

12. XSCANS system of data collection solution software: Siemens Analytical Instruments, Madison, Wisconsin.
13. G. M. Sheldrick, SHELX system of programs for crystallographic analysis, University of Göttingen, Germany.
14. Program DIFABS for empirical absorption corrections: Walker, N., Stuart, D. *Acta Crystallogr*. 1983, **A39**, 158.
15. Nonoyama, M. *Bull. Chem. Soc. Jpn*. 1974, **47**, 767.
16. (a) Garces, F. O., King, K. A., Watts, R. J. *Inorg. Chem*. 1988, **27**, 3464–3471. (b) Carlson, G. A., Djurovich, P. I., Watts, R. J. *Inorg. Chem*. 1993, **32**, 4483–4484.
17. Garces, F. O., Dedian, K., Keder, N. L., Watts, R. J. *Acta Crystallogr*. 1993, **C49**, 1117.
18. (a) Yin, C. C., Deeming, A. J. *J. Chem. Soc., Dalton Trans*. 1975, 2091. (b) Nord, G., Hazell, A. C., Hazell, R. G., Farver, O. *Inorg. Chem*. 1983, **22**, 3429–3434. (c) Spellane, P. J., Watts, R. J., Curtis, C. J. *Inorg. Chem*. 1983, **22**, 4060–4062.
19. Garces, F. O., Watts, R. J. *J. Magn. Reson. Chem*. 1993, **31**, 529.
20. The reported yield of Ir(acac)$_3$ from IrCl$_3 \cdot n$H$_2$O is 35% (Collins, J. E., Castellani, M. P., Rheingold, A. L., Miller, E. J., Geiger, W. E., Rieger, A. L., Rieger, P. H. *Organometallics* 1995, **14**, 1232–1238), and the yield of Ir $C^\wedge N_3$ from Ir(acac)$_3$ is 40% [5].
21. Colombo, M. G., Hauser, A., Güdel, H. U. *Inorg. Chem*. 1993, **32**, 3088–3092.
22. Urban, R., Krämer, R., Mihan, S., Polborn, K., Wagner, B., Beck, W. *J. Organomet. Chem*. 1996, **517**, 191.
23. Neve, F., Crispini, A. *Eur. J. Inorg. Chem*. 2000, 1039.
24. Allen, F. H., Davies, J. E., Galloy, J. J., Johnson, O., Kennard, O., Macrae, C. F., Mitchell, E. M., Mitchell, G. F., Smith, J. M., Watson, D. G. *J. Chem. Inf. Comput. Sci*. 1991, **31**, 187.
25. Esteruelas, M. A., Lahoz, F. J., Oñate, E., Oro, L. A., Rodríguez, L. *Organometallics* 1996, **15**, 823.
26. Papenfuhs, B., Mahr, N., Werner, H. *Organometallics* 1993, **12**, 4244.
27. Bezman, S. A., Bird, P. H., Fraser, A. R., Osborn, J. A. *Inorg. Chem*. 1980, **19**, 3755.
28. Hunter, C. A., Sander, J. K. M. *J. Am. Chem. Soc*. 1990, **112**, 5525.
29. An alternate explanation involves the formation of aggregates in the powered sample, which are not present in solution or in the crystals examined by X-ray diffraction. The powder X-ray diffraction patterns

for the *ppy* and *tpy* complexes exclude this possibility. Both complexes give well-resolved powder patterns, which can be indexed to the same unit cell found for single crystals of the same complexes. The measured powder patterns and the fits to the single-crystal unit cell parameters are given in the Supporting Information. Spectroscopic analysis of solutions of the complexes (NMR, UV–vis, and emission) match those expected for $(ppy)_2$Ir(acac) and $(tpy)_2$Ir(acac).

30. Birks, J. B. *Photophysics of Aromatic Molecules*; Wiley-Interscience: New York, 1970.
31. Lai, S.-W., Chan, M. C. W., Cheung, K.-K., Che, C.-M. *Inorg. Chem.* 1999, **38**, 4262.
32. Bailey, J. A., Hill, M. G., Marsh, R. E., Miskowski, V. M., Schaefer, W. P., Gray, H. B. *Inorg. Chem.* 1989, **28**, 1529.
33. Miskowski, V. M., Houlding, V. H. *Inorg. Chem.* 1989, **28**, 1529.
34. Alcock, N. W., Barker, P. R., Haider, J. M., Hannon, M. J., Painting, C. L., Pikramenou, Z., Plummer, E. A., Rissanen, K., Saarenketo, P. *J. Chem. Soc., Dalton Trans.* 2000, 1447.
35. Lamansky, S., Djurovich, P., Murphy, D., Abdel-Razzaq, F., Adachi, C., Burrows, P. E., Forrest, S. R., Thompson, M. E. Submitted.

Chapter 15

Synthesis and Characterization of Phosphorescent Cyclometalated Platinum Complexes

Jason Brooks, Yelizaveta Babayan, Sergey Lamansky,
Peter I. Djurovich, Irina Tsyba, Robert Bau, and
Mark E. Thompson

*Department of Chemistry, University of Southern California,
Los Angeles, California 90089-0744, USA*
met@usc.edu

The synthesis, electrochemistry, and photophysics of a series of square planar Pt(II) complexes are reported. The complexes have the general structure $C^\wedge NPt(O^\wedge O)$ where $C^\wedge N$ is a monoanionic cyclometalating ligand (e.g., 2-phenylpyridyl, 2-(2'-thienyl)pyridyl, 2-(4,6-difluorophenyl)pyridyl, etc.) and $O^\wedge O$ is a β-diketonato ligand. Reaction of K_2PtCl_4 with a $HC^\wedge N$ ligand precursor forms the chloride-bridged dimer, $C^\wedge NPt(\mu\text{-Cl})_2PtC^\wedge N$, which is cleaved with β-diketones such as acetyl acetone (acacH) and dipivaloylmethane (dpmH) to give the corresponding monomeric $C^\wedge NPt(O^\wedge O)$ complex. The *thpy*Pt(dpm) (*thpy* = 2-(2'-thienyl)pyridyl) complex has been characterized using X-ray crystallography. The bond lengths

---

Reprinted from *Inorg. Chem.*, **41**, 3055–3066, 2002.

Electrophosphorescent Materials and Devices
Edited by Mark E. Thompson
Text Copyright © 2002 American Chemical Society
Layout Copyright © 2024 Jenny Stanford Publishing Pte. Ltd.
ISBN 978-981-4877-34-3 (Hardcover), 978-1-003-08872-1 (eBook)
www.jennystanford.com

and angles for this complex are similar to those of related cyclometalated Pt complexes. There are two independent molecular dimers in the asymmetric unit, with intermolecular spacings of 3.45 and 3.56 Å, consistent with moderate $\pi-\pi$ interactions and no evident Pt–Pt interactions. Most of the $C^\wedge N Pt(O^\wedge O)$ complexes display a single reversible reduction wave between -1.9 and -2.6 V (vs Cp_2Fe/Cp_2Fe^+), assigned to largely $C^\wedge N$ ligand based reduction, and an irreversible oxidation, assigned to predominantly Pt based oxidation. DFT calculations were carried out on both the ground (singlet) and excited (triplet) states of these complexes. The HOMO levels are a mixture of Pt and ligand orbitals, while the LUMO is predominantly $C^\wedge N$ ligand based. The emission characteristics of these complexes are governed by the nature of the organometallic cyclometalating ligand allowing the emission to be tuned throughout the visible spectrum. Twenty-three different $C^\wedge N$ ligands have been examined, which gave emission λ_{max} values ranging from 456 to 600 nm. Well-resolved vibronic fine structure is observed in all of the emission spectra (room temperature and 77 K). Strong spin–orbit coupling of the platinum atom allows for the formally forbidden mixing of the ^{1}MLCT with the ^{3}MCLT and $^3\pi-\pi^*$ states. This mixing leads to high emission quantum efficiencies (0.02–0.25) and lifetimes on the order of microseconds for the platinum complexes.

15.1 Introduction

A significant research effort has focused on the photo-physical properties of luminescent square planar platinum complexes. These complexes have been investigated as emissive probes for DNA [1], photosensitizers for hydrogen production [2], singlet oxygen sensitizers [3], photooxidants [4], liquid crystal optical storage materials [5], and photochemical devices for the conversion of light to chemical energy through photoinduced charge separation [6]. Recently, platinum complexes have been used as luminescent centers in organic light-emitting diodes (OLEDs) [7, 8]. A recent breakthrough using phosphorescent luminophores has demonstrated the ability to make highly efficient electroluminescent devices [7, 9]. The strong spin–orbit coupling of the heavy metal atom allows for efficient intersystem crossing (ISC) between singlet and triplet, states, which can lead to a high quantum yield of emission from the triplet state. Thus, OLEDs can be fabricated that utilize all of the electrogenerated singlet and triplet excitons, thereby approaching an internal efficiency of 100% [10].

While the chemistry and photophysics of coordinatively unsaturated square planar platinum complexes have been extensively explored, it is interesting to note that surprisingly few of them are reported to be emissive in room temperature solution [11]. Only Pt(II) species chelated with aromatic ligands, such as bipyridine, phenanthroline, 2-phenylpyridine, or similar derivatives, emit in fluid solution from excited states localized in the aromatic π systems. These Pt(II) complexes benefit from having relatively high energy metal centered (MC) or ligand field excited states when compared with, for example, their palladium analogues [12]. For the Pt complexes, the MC states lie at higher energy than the intraligand and MLCT states, a situation advantageous for enhancing luminous efficiency. If the emitting state (either intraligand or MLCT) and MC states lie too close in energy, they can thermally equilibrate, thereby quenching the emission through fast radiationless decay through the MC states. The energy gap between the MC states and lowest energy emitting excited state may be considered to be one of the limiting factors for emission efficiency (i.e., a small energy gap between emitting and MC states leads to poor emission efficiency) [13]. Additionally, open coordination sites of the square planar complex can allow for other deactivating pathways to occur through metal interactions with the environment [11].

A number of early studies of these emissive complexes were focused on the photophysics of Pt(II) diimine dithiolate compounds [11, 14]. Since then, similar diimine-based complexes have been reported with other ancillary ligands, such as acetylide [15], aryl [16], and pyrazole [17]. Studies on the photophysical properties of tridentate terpyridine [13, 18] and phenylbipyridine [19] derivatives have also been reported. The nature and energy of the excited state in these complexes varies depending on the ligand [14a, 17]. For the dithiolate and tridentate complexes, the lowest energy excited state has been attributed to either a triplet metal-to-ligand charge transfer state (^{3}MLCT) or a ligand-based state (either a ligand–ligand charge transfer, LLCT, or single ligand centered triplet excited states, ^{3}LC).

Several cyclometalated organometallic platinum complexes, based on $C^\wedge N$ aromatic chelate derivatives (e.g., 2-arylpyridine), have been reported in the literature. Among the most basic and thoroughly studied complexes are those based on 2-phenylpyridine (*ppy*) and 2-(2'-thienyl)pyridine (*thpy*). Examples of these include homoleptic complexes, $(C^\wedge N)_2$ [20], heteroleptic complexes, $(C^\wedge N)(C^\wedge N)$ [21], and complexes with a single cyclometalating ligand $(C^\wedge N)$ and a non-cyclometalating, ancillary ligand (LX) [22]. The strong ligand field influence of the aromatic carbon atom, along with the added stabilization of π donation from the metal into the aromatic ring, helps to make these types of chelates very stable, while concurrently increasing the energy gap of the MC excited states [22a]. These cyclometalated complexes are known to be strongly emissive arid have long luminescent lifetimes (microseconds) in fluid solution indicative of emission from the triplet excited state. The photoluminescent spectra for many of these complexes are red-shifted from those observed for the uncoordinated $(HC^\wedge N)$ ligand precursor. The emission spectra are highly structured with vibrational progressions decreasing in intensity to lower energy. On the basis of the red shift, structured emission pattern, and long luminescent lifetime, i.e., microseconds, the emission is thought to arise from an excited state that is predominantly a ^{3}LC state, mixed with ^{1}MLCT character by the strong spin–orbit coupling [23].

Heteroleptic $C^\wedge N$PtLX complexes offer several advantages over the bis-$C^\wedge N$ derivatives. Physical properties such as the charge and solubility of the complexes can be independently adjusted through proper choice of the ancillary ligand. It has also been shown that the nature of the ancillary ligand can have profound effects on the lowest emitting excited state [22a, c]. The electron-donating or -withdrawing character of the LX ligand can increase or decrease the amount of electron density at the metal center. The amount of ground state electron density on the metal will subsequently affect the amount of MLCT character mixed into the lowest energy transition, thus altering both the color of emission arid radiative lifetime of the excited state. A series of mixed cyclomctalatcs based on 2-phenylpyridine and 2-thienylpyridine have been reported where LX is an ancillary bidentate ligand based on unsaturated α, α''-diimines and saturated diamines, such as 2,2- bipyridine (bpy) or 1,2-diaminoethane (en), respectively [22a]. It was shown that emission from complexes with unsaturated (α, α''-diimine ligands originates from states on the diimine, rather than those of the cyclometalate. For complexes with saturated LX ligands, the emitting state is strongly governed by the ^{3}LC transition from the cyclometalate, similar to bis-homoleptic analogues.

We report here a series of cyclometalated platinum complexes of the general structure, $C^\wedge NPt(O^\wedge O)$, where $O^\wedge Q$ is a β-diketonate ligand, i.e., acetyl acetonate (acac) or dipivolylmethane (dpm). All of these complexes are strongly emissive in frozen glass solutions, and several are strongly emissive in degassed room temperature solution, with life-times on the order of microseconds. We observe that the nature of the cyclometalating ligand strongly affects the energy of the emissive state and have been able to tune the emission color throughout the visible spectrum either by modifying the 2-phenylpyridyl ligand with electron-donating and -withdrawing substituents or by varying the character of the cyclometalate. The emission for all of these complexes is attributed to a mixed ^{3}LC–MLCT state.

15.2 Experimental Section

15.2.1 Equipment

The UV–visible spectra were recorded on an Aviv model 14DS spectrophotometer (a re-engineered Cary 14 spectrophotometer). Photoluminescent spectra were measured using a Photon Technology International fluorimeter. Quantum efficiency measurements were carried out at room temperature in a 2-meth-yltetrahydrofuran solution that, was distilled over sodium. Before spectra were measured, the solution was degassed by several freeze–pump–thaw cycles using a diffusion pump. Solutions of coumarin 47 in ethanol ($\Phi = 0.60$) or degassed fac-Ir$(ppy)_3$ in 2-MeTHF ($\Phi = 0.40$) were used as a reference. The equation

$$\Phi_s = \Phi_r \left(\frac{\eta_s^2 A_r I_s}{\eta_r^2 A_s I_r} \right)$$

was used to calculate quantum yields where Φ_s is the quantum yield of the sample, Φ_r is the quantum yield of the reference, η is the refractive index of the solvent, A_s and A_r are the absorbance of the sample and the reference at the wavelength of excitation, and I_s and I_r are the integrated areas of emission bands [22d]. Steady state emission experiments at room temperature were performed on a PTI QuantaMaster model C-60 spectrofluorimeter. Phosphorescence lifetime measurements were performed on the same fluorimeter equipped with a microsecond Xe flash lamp and were limited to lifetimes >2 µs. NMR spectra were recorded on Bruker AC 250 MHz, AM 360 MHz, or AMX 500 MHz instruments. Solid probe MS spectra were taken with Hewlett-Packard GC/MS instrument with electron impact ionization and model 5873 mass selective detector. The Microanalysis Laboratory at the University of Illinois, Urbana-Champaign, performed all elemental analyses.

15.2.2 Electrochemistry

Cyclic voltammetry and differential pulsed voltammetry were performed using an EG&G potentiostat/galvanostat model 283, Anhydrous DMF (Aldrich) was used as the solvent under a nitrogen atmosphere, and 0.1 M tetra(*n*-butyl)-ammonium hexafluorophosphate was used as the supporting electrolyte. A Ag wire was used as the pseudoreference electrode, and a Pt wire was used as the counter electrode. The working electrode was glassy carbon. The redox potentials are based on values measured from differential pulsed voltammetry and are reported relative to a ferrocene/ferrocenium (Cp_2Fe/Cp_2Fe^+) redox couple used as an internal reference (0.45 V vs SCE) [24] while electrochemical reversibility was determined using cyclic voltammetry.

15.2.3 X-ray Crystallography

Diffraction data for *thpy*Pt(dpm) was collected at room temperature ($T = 23°C$) on a Bruker SMART APEX CCD diffractometer with graphite-monochromated Mo Kα radiation ($\lambda = 0.71073$ Å). The cell parameters for the Pt complex were obtained from the least-squares refinement of the spots (from 60 collected frames) using the SMART program, A hemisphere of the crystal data was collected up to a resolution of 0.75 Å, and the intensity data was processed using the Saint Plus program. All calculations for structure determination were carried out using the SHELXTL package (version 5.1) [25]. Initial atomic positions were located by Patterson methods using XS, and the structure was refined by least-squares methods using SHELX with 6983 independent reflections arid within the range of (Φ1.38 $-$24.71° (completeness 98.8%). Absorption corrections were applied by using SADABS [26]. Calculated hydrogen positions were input and refined in a riding manner along with the attached carbons. A summary of the refinement details and the resulting factors are given in Table 15.1.

15.2.4 Density Functional Calculations

DFT calculations were performed using Titan software package (Wavefunction, Inc.) at the B3LYP/LACVP** level. The HOMO and LUMO energies were determined using minimized singlet geometries to approximate the ground state. The minimized singlet geometries were used to calculate the triplet molecular orbitals and approximate the triplet HSOMO (HSOMO = highest singly occupied molecular orbital).

15.2.5 Synthesis of ($C^\wedge N$)($O^\wedge O$) Complexes: General Procedure

The compounds 2-bromo-4-methoxypyridine [27], 2-bromo-4-di-methyarninopyridirie [28], 2-(1,5-dimethylpyrrol-2-yl)pyridine [29], and 2-(2'-benzothienyl)pyridine [30] were prepared following literature procedures. 2-Phenylpyridine and *p*-tolylpyridine were purchased from Aldrich Chemical Co. Other phenylpyridine derivatives were prepared by Suzuki coupling reaction with commercially available boronic acids and the appropriate bromopyridines [31]. All other ligands and materials were purchased from either Aldrich Chemical Co. or Frontier Scientific and used without further purification.

Table 15.1 Crystal data and structure refinement for *Thpy*Pt (dpm)

empirical formula	$C_{20}H_{25}NO_2PtS$
fw	538.56
temp	296(2) K
wavelength	0.71073 Å
cryst syst	triclinic
space group	$P\,1$
unit cell dimens	
a	10.7376(17) Å
b	13.804(2) Å
c	15.995(3) Å
α	110.487(2)°
β	91.959(3)°
γ	108.925(3)°
vol	2071.5(6) Å^3
Z	4
density (calcd)	1.727 g/cm^3
abs coeff	6.886 mm^{-1}
$F(000)$	1048
cryst size	0.35 × 0.13 × 0.05 mm^3
Θ range for data collection	1.38–24.71°
index ranges	$-12 \leq h \leq 12, -16 \leq k \leq 15, -16 \leq l \leq 18$
reflns collected	10636
indep reflns	6983 [R(int)) 0.0212]
completeness (to $\Theta = 24.71°$)	98.8%
transm factors	min/max ratio 0.663952
refinement meth	full-matrix least squares on F^2
data/restraints/params	6983/0/399
GOF on F^2	0.986
final R indices [$I > 2\sigma(I)$]	R1 = 0.0455, wR2 = 0.1118
R indices (all data)	R1 = 0.0637, wR2 = 0.1215
largest diff peak and hole	1.464 and -0.699 e.Å^{-3}

All procedures were carried out in inert gas atmosphere despite the air stability of the complexes, the main concern being the oxidative and thermal stability of intermediates at the high temperatures in the reactions. The Pt(II) μ-dichloro-bridged dimers were prepared by a modified method of Lewis [32]. This involves heating the K_2PtCl_4 salt with 2–2.5 equiv of cyclometalating ligand in a 3:1 mixture of 2-ethoxyethanol (Aldrich) and water to 80 °C for 16 h. The dimers were isolated in water and were subsequently reacted with 3 equiv of the chelating diketone derivative and 10 equiv of Na_2CO_3 in 2-ethoxyethanol at 100 °C for 16 h. The solvent was removed under reduced pressure, and the compound was purified by flash chromatography using dichloromethane. The product was recrystallized with dichloromethane/methanol.

15.2.6 Characterization

***ppy*Pt(acac):** platinum(II) (2-phenylpyridinato-$N,C^{2'}$) (2,4-pentanedionato-O,O). Yield: 36%. ^{1}H NMR (360 MHz, acetone-d_6), ppm: 9.00 (d, 1H, J = 5.8 Hz), 8.02 (dd, 1H, J = 1.6, 7.4 Hz), 7.89 (d, 1H, J = 7.9 Hz), 7.57 (dd, 1H, J =1.6, 7.4 Hz), 7.51 (dd, 1H, J = 1.6, 7.4 Hz), 7.32 (dd, 1H, J = 1.6, 6.8 Hz), 7.11 (ddd, 1H, J = 1.6, 7.9 Hz), 7.04 (dd, 1H, J = 1.6, 7.4 Hz), 5.55 (s, 1H), 1.98 (s, 3H), 1.95 (s, 3H). Anal. for $C_{16}H_{15}NO_2Pt$: found C 43.13, H 3.41, N 3.42, calcd C 42.86, 3.37, 3.12.

***ppy*Pt(dpm):** platinum(II) (2-phenylpyridinato-$N,C^{2'}$) (2,2,6,6-tetramethyl-3,5-heptanedionato-O,O). Yield: 30%, ^{1}H NMR (250 MHz, CDCl$_3$), ppm: 8.98 (d, 1H, J = 6.2 Hz), 7.77 (dd, 1H, J = 8.4, 9.5, 1.8 Hz), 7.62 (m, 2H), 7.42 (dd, 1H, J = 7.7, 1.1 Hz), 7.19 (ddd, 1H, J = 7.7, 8.8, 1.5 Hz), 7.08 (m, 2H), 5.79 (s, 1H), 1.26 (s, 9H), 1.25 (s, 9H). Anal. for $C_{22}H_{27}NO_2Pt$: found C 49.94, H 5.11, N 2.69, calcd C 49.62, H 5.11, N 2.63.

***tpy*Pt(acac):** platinum (II) (2-(*p*-tolyl)pyridinato-$N,C^{2'}$) (2,4-pentanedionato-O,O), Yield: 42%. ^{1}H NMR(360 MHz, CDCl$_3$), ppm: 8.94 (d, 1H, J = 5.9 Hz), 7.74 (d, 1H, J = 6.8 Hz), 7.53 (d, 1H, J = 7.8 Hz), 7.39 (s, 1H), 7.30 (d, 1H, J = 7.8 Hz), 7.04 (d, 1H, J = 6.8 Hz), 6.88 (d, 1H, J = 7.8 Hz), 5.45 (s, 1H), 2.00 (s, 3H), 1.98 (s, 3H), 1.95 (s, 3H). Anal. for $C_{17}H_{17}NO_2Pt$: found C 44.10, H 3.89, N 3.32, calcd C 44.16; H 3.71; N 3.03.

***tpy*Pt(dpm):** platinum(II) (2-(*p*-tolyl)pyridinato-$N,C^{2'}$) (2,2,6,6-tetramethyl-3,5-heptanedionato-O,O). Yield; 40%. ^{1}H NMR (250 MHz, CDCl$_3$), ppm: 8.94 (d, 1H, J = 5,9 Hz), 7.74 (dd, 1H, J = 9.1, 8.4, 1.5 Hz), 7.54 (d, 1H, J = 8.04 Hz), 7.45 (s, 1H), 7.31 (d, 1H, J = 7.7

Hz), 7.04 (dd, 1H, $J = 5.9, 8.4, 1.1$ Hz), 6.89 (d, 1H, $J = 7.7$ Hz), 5.78 (s, 1H), 2.38 (s, 3H), 1.27 (s, 9H), 1.25 (s, 9H). Anal. for $C_{23}H_{29}NO_2Pt$: found C 49.18, H 5.13, N 2.64, calcd C 50.54, H 5.35, N 2.56.

***otpy*Pt(acac):** platinum(II) (2-(*o*-tolyl)pyridinato-$N,C^{2'}$) (2,4-pentanedionato-O,O). Yield: 29%. ^{1}H NMR (250 MHz, CDCl$_3$), ppm: 9.41 (d, 1H, $J = 5.9$ Hz), 7.81 (m, 2H), 7.56 (d, 1H, $J = 7.7$ Hz), 7.06 (m, 2H), 6.87 (d, 1H, $J = 7.3$ Hz), 5.44 (s, 1H), 2.64 (s, 3H), 1.98 (s, 3H), 1.97 (s, 3H). Anal. for $C_{17}H_{17}NO_2Pt$: found C 44.05, H 3.55, N 3.06, calcd C 44.16, H 3.71, N 3.03.

***6Fppy*Pt(dpm):** platinum(II) (2-(6'-fluorophenyl)pyridiriato-$N,C^{2'}$) (2,2,6,6-tetramethyl-3,5-heptanedionato-O,O). Yield: 25%, ^{1}H NMR (250 MHz, CDCl$_3$), ppm: 9.04 (d, 1H, $J = 5.8$ Hz), 8.03 (d, 1H, $J = 8.2$Hz), 7.81 (dd, 1H, $J = 8.2, 8.2$ Hz), 7.45 (d, 1H, $J = 7.2$ Hz), 7.15 (m, 2H), 6.77 (ddd, 1H, $J = 12.3, 8.2, 1.0$ Hz), 5.80 (s, 1H), 1.26 (s, 9H), 1.25 (s, 9H), Anal. for $C_{22}H_{26}FNO_2Pt$: found C 48.07, H 4.83, N 2.68, calcd C 48.00, H 4.76, N 2.54.

***6tfmppy*Pt(dpm):** platinum(II) (2-(6'-trifluoromethylphenyl)-pyridinato-$N,C^{2'}$) (2,2,6,6-tetramethyl-3,5-heptanedionato-O,O). Yield: 30%, ^{1}H NMR (250 MHz, CDCl$_3$), ppm: 9.20 (d, 1H, $J = 6.1$ Hz), 8.08 (d, 1H, $J = 8.5$ Hz), 7.99 (d, 1H, $J = 7.5$ Hz), 7.84 (ddd, 1H, $J = 8.8, 7.5, 1.7$ Hz), 7.47 (d, 1H, $J = 7.8$ Hz), 7.12 (m, 2H), 5.81 (s, 1H), 1.27 (s, 9H), 1.26 (s, 9H). Anal. for $C_{23}H_{26}F_3NO_2Pt$: found C 45.86, H 4.32, N 2.38, calcd C 46.00, H 4.36, N 2.33.

***45dfppy*Pt(acac):** platinum(II) (2-(4',5'-difluorophenyl)pyridinato-$N,C^{2'}$)(2,4-pentanedionato-O,O). Yield: 37%, ^{1}H NMR (360 MHz, acetone-d_6), ppm: 8.99 (d, 1H, $J = 5.7$ Hz), 8.06 (dd, 1H, $J = 2.3, 8.0$ Hz), 7.90 (d, 1H, $J = 8.0$ Hz), 7.62–7.68 (m, 1H), 7.37 (ddd, 1H, $J = 1.7, 5.7$ Hz), 7.20–7.25 (m, 1H), 5.58 (s, 1H), 1.99 (s, 3H), 1.98 (s, 3H). Anal. for $C_{16}H_{13}F_2NO_2Pt$: found C 39.40, 2.91, 2.61, calcd C 39.68, H 2.71, N 2.89.

***46dfppy*Pt(acac):** platinum(II) (2-(4',6'-difluorophenyi)pyridinato-$N,C^{2'}$) (2,4-pentanedionato-O,O) Yield: 47%, ^{1}H NMR (250 MHz, CDCl$_3$), ppm: 8.95(d, 1H, $J = 5.8$ Hz), 7.91 (d, 1H, $J = 8.2$ Hz), 7.79 (m, 2H), 7.08 (m, 2H), 6.54 (ddd, 1H, $J = 11.6, 9.3, 2.4$ Hz), 5.45 (s, 1H), 1.99 (s, 3H), 1.98 (s, 3H), Anal. for $C_{16}H_{13}F_2NO_2Pt$: found C 39.52, H 2,56, N 2.87, calcd C 39.68, H 2,71, N 2.89.

***46dfppy*Pt(dpm):** platinum(II) (2-(4',6'-difluorophenyl)pyridinato-$N,C^{2'}$) (2,2,6,6-tetramethyl-3,5-heptanedionato-O,O) Yield: 38%. ^{1}H NMR (250 MHz, CDCl$_3$), ppm: 9.00 (d, 1H, J = 6.1 Hz), 7.96 (d, 1H, J = 8.5 Hz), 7.81 (dd, 1H, J = 7.5, 7.5 Hz), 7.11 (m, 2H), 6.55 (ddd, 1H, J = 11.7, 9.4, 2.4 Hz), 5.81 (s, 1H), 1.26 (s, 9H), 1.25 (s, 9H). Anal. for C$_{22}$H$_{25}$F$_2$NO$_2$Pt: found C 46.33, H 4.35, N 2.55, calcd C 46.48, H 4.43, N 2.46.

***46dfp-3Mepy*Pt(dpm):** platinum(II) (2-(4',6'-difluoropbenyl)-3-methylpyridinato-$N,C^{2'}$) (2,2,6,6-tetramethyl-3,5-heptanedionato-O,O). Yield: 25%. ^{1}H NMR (250 MHz, CDCl$_3$), ppm: 8.98 (d, 1H, J = 5.8 Hz), 7.65 (d, 1H, J = 7.9 Hz), 7.17 (dd, 1H, J = 8.5, 2.7 Hz), 7.07 (dd, 1H, J = 7.8, 5.8 Hz), 6.53 (m, 1H), 5.80 (s, 1H), 2.53 (d, 3H, J = 11.6 Hz), 1.25 (s, 9H), 1.24 (s, 9H). Anal. for C$_{23}$H$_{27}$F$_2$NO$_2$Pt: found C 47.24, H 4.57, N 2.51, calcd C 47.42, H 4.67, N 2.46.

***46dfp-4Mepy*Pt(dpm):** platinum(II) (2-(4',6'-difluorophenyl)-4-methylpyridinato-$N,C^{2'}$) (2,2,6,6-tetramethyl-3,5-heptanedionato-O,O). Yield: 30%. ^{1}H NMR (250 MHz, CDCl$_3$), ppm: 8.80 (d, 1H, J = 5.4 Hz), 7.77 (s, 1H), 7.10 (dd, 1H, J = 8.2, 2.4 Hz), 6.96 (d, 1H, J = 6.8 Hz), 6.55 (ddd, 1H, J = 12.2, 9.2, 2.4 Hz), 5.80 (s, 1H), 2.43 (s, 3H), 1.25 (s, 9H), 1.24 (s, 9H). Anal. for C$_{23}$H$_{27}$F$_2$NO$_2$Pt: found C 47.22, H 4.05, N 2.41, calcd C 47.42, H 4.67, N 2.40.

***46dfp-5Mepy*pt(dpm):** platinum(II) (2-(4',6'-difiuoropheny])-5-methylpyridinato-$N,C^{2'}$) (2,2,6,6-tetramethyl-3,5-heptanedionato-O,O). Yield: 31%, ^{1}H NMR (250 MHz, CDCl$_3$), ppm: 8.84 (s, 1H), 7.83 (d, 1H, J = 8, 5 Hz), 7.62 (d, 1H, J = 8.5 Hz), 7.09 (dd, 1H, J = 8.8, 2.7 Hz), 6.54 (ddd, 1H, J = 11.9, 9.5, 2.7 Hz), 5.81 (s, 1H), 2.38 (s, 3H), 1.26 (s, 9H), 1.25 (s, 9H), Anal. for C$_{23}$H$_{27}$F$_2$NO$_2$Pt: found C 47.40, H 4.64, N 2.49, calcd C 47.42, H 4,67, N 2.40.

***46dfp-6Mepy*Pt(dpm):** platinum(II) (2-(4',6'-difluorophenyl)-6-metbylpyridinato-$N,C^{2'}$) (2,2,6,6-tetramethyl-3,5-heptanedionato-O,O). Yield: 5%. ^{1}H NMR (250 MHz, CDCl$_3$), ppm: 7.86 (d, 1H, J = 7.2 Hz), 7.64 (dd, 1H, J = 8.2, 8.2 Hz), 7.20 (dd, 1H, J = 9.6, 2.7 Hz), 6.96 (d, 1H, J = 8.2 Hz), 6.55 (ddd, 1H, J = 11.9, 8.9, 2.4 Hz), 5.85 (s, 1H), 3.04 (s, 3H), 1.26 (s, 9H), 1.19 (s, 9H). Anal. for C$_{23}$H$_{27}$F$_2$NO$_2$Pt: found C 47.01, H 4.26, N 2.44, calcd C 47.42, H 4.67, N 2.40.

***46dFp-4meopy*Pt(acac):** platinum(II) (2-(4',6'-difluorophenyl)-4-methoxypyridinato-$N,C^{2'}$) (2,4-pentanedionato-O,O). Yield: 21%, ^{1}H NMR (250 MHz, CDCl$_3$), ppm: 8.70 (d, 1H, J = 7.0 Hz), 7.45 (dd, 1H, J = 2.9Hz), 7.06 (dd, 1H, J = 8.8, 2.6Hz), 6.68 (dd, 1H, J = 6.6, 2.9 Hz), 6.53 (ddd, 1H, J = 11.7, 9.1, 2.6 Hz), 5.45 (s, 1H), 1.98 (s, 3H), 1.97 (s, 3H). Anal. for C$_{17}$H$_{15}$F$_2$NO$_3$Pt: found C 39.16, H 2.39, N 2.61, calcd C 39.69, H 2.94, N 2.72.

***46dFp-4meopy*Pt(dpm):** platinum(II)(2-(4',6'-difluoropheriyl)-4-methoxypyridinato-$N,C^{2'}$) (2,2,6,6-tetrametbyl-3,5-heptanedionato-O,O). Yield: 22%. ^{1}H NMR (250 MHz, CDCl$_3$), ppm: 8.71 (d, 1H, J = 6.8 Hz), 7.47 (dd, 1H, J = 2.7, 2.7 Hz), 7.07 (dd, 1H, J = 8.2, 2.1 Hz), 6.69 (dd, 1H, J = 6.8, 2.7 Hz), 6.54 (ddd, 1H, J = 11.9, 9.2, 2.4 Hz), 5.46 (s, 1H), 3.93 (s, 3H), 1.98 (s, 3H), 1.97 (s, 3H). Anal. for C$_{23}$H$_{27}$F$_2$NO$_3$Pt: found C 45.94, H 4.36, N 2.45, calcd C 46.15, H 4.55, N 2.34.

***46dFp-4dmapy*Pt(dpm):** platinum(II) (2-(4',6'-difluoropheriyl)-4-dimethylaminopyridinato-$N,C^{2'}$) (2,2,6,6-tetramethyl-3,5-heptanediona-O,O). Yield: 20%. ^{1}H NMR (250 MHz, CDCl$_3$), ppm: 8.42 (d, 1H, J = 6.6 Hz), 7.17 (dd, 1H, J = 2.94, 2.94 Hz), 7.08 (dd, 1H, J = 8.8, 2.9 Hz), 6.51 (ddd, 1H, J = 11.7, 8.83, 2.9 Hz), 6.36 (dd, 1H, J = 6.6, 2.9 Hz), 5.76 (s, 1H), 3.10 (s, 6H), 1.24 (s, 9H), 1.23 (s, 9H). Anal. for C$_{24}$H$_{30}$F$_2$N$_2$O$_2$Pt: found C 46.97, H 4.78, N 4.57, calcd C 47.13, H 4.94, N 4.58.

***4meoppy*Pt(dpm):** platinum(II) (2-(4'-methoxypbenyl)pyridinato-$N,C^{2'}$) (2,2,6,6-tetramethyl-3,5-heptanedionato-O,O). Yield: 28%. ^{1}H NMR (250 MHz, CDCl$_3$), ppm: 8.88 (d, 1H J = 4.9Hz), 7.71 (d, 1H, J = 9.2, 9.2, 1.8 Hz), 7.45 (d, 1H, J = 7.9 Hz), 7.35 (d, 1H, J = 8.5 Hz), 7.20 (d, 1H, J = 2.4 Hz), 6.98 (m, 1H), 6.65 (dd, 1H, J = 8.5, 3.1 Hz), 5.79 (s, 1H), 3.86 (s, 3H), 1.26 (s, 9H), 1.25 (s, 9H). Anal. for C$_{23}$H$_{29}$NO$_3$Pt: found C 47.24, H 4.57, N 2.51, calcd C 47.42, H 4.67, N 2.46.

***5meoppy*Pt(dpm):** platinum(II) (2-(5'-methoxyphenyl)pyridinato-$N,C^{2'}$) (2,2,6,6-tetramethyl-3,5-heptanedionato-O,O). Yield: 25%. ^{1}H NMR (250 MHz, CDCl$_3$), ppm: 8.97 (d, 1H, J = 5.5 Hz), 7.78 (ddd, 1H, J = 8.1, 9.1, 1.8 Hz), 7.57 (s, 1H), 7.51 (d, 1H, J = 8.4 Hz), 7.09 (ddd, 1H, J = 8.8, 5.9, 1.1 Hz), 7.01 (d, 1H, J = 2.6 Hz), 6.90 (dd,

1H, $J = 8.4, 2.6$ Hz), 5.77 (s, 1H), 1.25 (s, 9H), 1.24 (s, 9H) Anal. for $C_{23}H_{29}NO_3Pt$: found C 48.90, H 4.94, N 2.62, calcd C 49.11, H 5.20, N 2.49.

pqPt(dpm): platinum(II) (2-phenylbenzothiozolato-$N,C^{2'}$) (2,2,6,6-tetramethyl-3,5-heptanedionato-O,O). Yield: 19% ^{1}H NMR (250 MHz, CDCl$_3$), ppm: 9.65 (d, 1H, $J = 8.8$ Hz), 8.23 (d, 1H, $J = 8.3$ Hz), 7.77 (m, 4H), 7.54 (m, 2H), 7.17 (m, 2H), 5.89 (s, 1H), 1.30 (s, 9H), 1.27 (s, 9H). Anal. for $C_{20}H_{27}NO_2Pt$: found C 52.95, H 4.74, N 2.53, calcd C 53.60, H 5.02, N 2.40.

btPt(dpm): platinum(II) bis(2-phenylbenzothiozolato-$N,C^{2'}$) (2,2,6.6-tetramethyl-3,5-heptanedionato-O,O). Yield: 23%. ^{1}H NMR (250 MHz, CDCl$_3$), ppm: 9.34 (d, 1H, $J = 8.2$), 7.78 (m, 2H), 7.52 (m, 2H), 7.41 (m, 1H), 7.20 (m, 1H), 7.09 (m, 1H), 5.89 (s, 1H), 1.33 (s, 9H), 1.29 (s, 9H). Anal. for $C_{26}H_{29}NO_2Pt$: found C 48.05, H 4.16, N 2.41, calcd C 48.97, H 4.62, N 2.38.

btpPt(aeac): platinum(II) (2–2'-(4',5'-benzo)thienyl)pyridinato-$N,C^{3'}$) (2,4-pentanedionato-O,O) Yield: 20%. ^{1}H NMR (360 MHz, CDCl$_3$), ppm: 8.90 (d, 1H, $J = 5.9$ Hz), 8.75–9.79 (m, 1H), 7.77–7.81 (m, 1H), 7.71 (dd, 1H, $J = 1.5, 7.8$ Hz), 7.27–7.34 (m, 3H), 6.95 (dd, 1H, $J = 1.5, 6.8$ Hz), 5.54 (s, 1H), 2.08 (s, 3H), 2.01 (s, 3H). Anal. for $C_{18}H_{15}NO_2PtS$: found C 42.51, H 3.53, N 2.56, calcd C 42.86, 3.00, 2.78.

c6Pt(acac): platinum(II) (3-(2-benzothiazolyl)-7-(diethylamino)-2H-1-benzopyran-2-onato-N', C^4) (2,4-pentanedionato-O,O). Yield: 10%. ^{1}H NMR (250 MHz, CDCl$_3$), ppm: 8.93 (d, 1H, $J = 8.8$ Hz), 7.73 (d, 1H, $J = 7.5$ Hz), 7.33 (m, 2H), 6.49 (d, 1H, $J = 9.5$ Hz), 6.35 (s, 1H), 5.50 (s, 1H), 3.43 (q, 4H, $J = 7.16$ Hz), 1.99 (s, 3H), 1.91 (s, 3H), 1.23 (t, 6H, $J = 7.2$ Hz). Anal. for $C_{25}H_{24}N_2O_4PtS$: found C 46.36, H 3.53, N 4.17, calcd C 46.65, H 3.76, N 4.35.

dpoPt(dpm): platinum(II) (2,4-diphenyloxazolato-1,3-$N,C^{2'}$) (2,2,6,6-tetramethyl-3,5-heptanedionato-O,O). Yield: 6%. ^{1}H NMR (250 MHz, CDCl$_3$), ppm: 7.71 (d, 2H, $J = 8.5$ Hz), 7.66 (d, 1H, $J = 7.9$ Hz), 7.45 (m, 5H), 7.20 (dd, 1H, $J = 7.5, 1.4$ Hz), 7.08 (ddd, 1H, $J = 8.5, 7.5, 1.0$ Hz), 5.81 (s, 1H), 1.27 (s, 9H), 1.26 (s, 9H), Anal. for $C_{26}H_{29}NO_3Pt$: found C 51.03, H 5.13, N 2.21, calcd C 52.17, H 4.88, N 2.34.

***thpy*Pt(acac):** platinum(II) (2-(2'-thienyl)pyridinato-$N,C^{3'}$) (2,4-pentanedionato-O,O). Yield: 20%. ^{1}H NMR (500 MHz, CDCl$_3$), ppm: 8.78 (d, 1H), 7.67 (m, 1H), 7.46 (d, 1H), 7.26 (d, 1H), 7.17 (d, 1H), 6.86 (m, 1H), 5.46 (s, 1H), 1.98 (s, 3H), 1.95 (s, 3H). Anal. for C$_{14}$H$_{13}$NO$_2$PtS: found C 38.10, H 3.02, N 3.21, calcd C 37.00, H 2.88, N 3.08.

***thpy*Pt(dpm):** platinum(II) (2-(2'-thienyl)pyridinato-$N,C^{3'}$) (2,2,6,6-tetramethyl-3,5-heptanedionato-O,O). Yield: 22%. ^{1}H NMR (250 MHz, CDCl$_3$), ppm: 8.79 (d, 1H, J = 5.8 Hz), 7.65 (ddd, 1H, J = 9.2, 7.5, 1.7 Hz), 7.47 (d, 1H, J = 4.8 Hz), 7.27 (d, 1H, J = 8.5 Hz), 7.18 (d, 1H, J = 4.8 Hz), 6.90 (ddd, 1H, J = 7.2, 5.8, 1.4 Hz), 5.79 (s, 1H), 1.25 (s, 9H), 1.23 (s, 9H). Anal. for C$_{20}$H$_{25}$NO$_2$PtS: found C 44.57, H 4.70, N 2.74, calcd C 44.60, H 4.68, N 2.60.

***pyrpy*Pt(acac):** platinum(II) (2-(1',5'-dimethyl-pyrrol-2'-yl)pyridinato-$N,C^{2'}$) (2,4-pentanedioiiato-O,O). Yield: 16%. ^{1}H NMR (250 MHz, CDCl$_3$), ppm: 8.70 (d, 1H, J = 6.3 Hz), 7.44 (m, 1H), 7.05 (d, 1H, J = 8.5 Hz), 6.58 (m, 1H), 5.92 (s, 1H), 5.40 (s, 1H), 3.71 (s, 3H), 2.23 (s, 3H), 1.92 (s, 3H), 1.89 (s, 3H). Anal. for C$_{16}$H$_{18}$N$_2$O$_2$Pt: found C 41.48, H 3.90, N 5.53, calcd C 41.29, H 3.90, N 6.02.

***pyrpy*Pt(dpm):** platinum (II) (2-(1',5'-dimethyl-pyrrol-2'-yl)pyridinato-$N,C^{2'}$) (2,2,6,6-tetramethyl-3,5-heptanedionato-O,O). Yield: 20%. ^{1}H NMR (250 MHz, CDCl$_3$), ppm: 8.70 (d, 1H, J = 6.1 Hz), 7.45 (ddd, 1H, J = 8.8, 7.5, 1.7 Hz), 7.06 (d, 1H, J = 8.2 Hz), 6.59 (ddd, 1H, J = 7.2, 5.8, 1.4 Hz), 5.90 (s, 1H), 5.72 (s, 1H), 3.71 (s, 3H), 2.23 (s, 3H), 1.22 (s, 9H), 1.20 (s, 9H). Anal. for C$_{22}$H$_{30}$N$_2$O$_2$Pt: found C 48.39, H 5.60, N 4.91, calcd C 48.08, H 5.50, N 5.10.

***bzq*Pt(acac):** platinum (II) (7,8-benzoquinolinato-$N,C^{3'}$) (2,4-pentanedionato-O,O). Yield: 27%. ^{1}H NMR (360 MHz, acetone d_6), ppm: 9.13 (d, 1H, J = 5.4 Hz), 8.25 (d, 1H, J = 8.3 Hz), 7.75 (m, 2H), 7.50–7.57 (m, 3H), 7.44 (dd, 1H, J = 5.4, 5.4 Hz), 5.52 (s, 1H), 2.04 (s, 6H). Anal. for C$_{18}$H$_{15}$NO$_2$Pt: found C 45.60, H 3.11, N 2.93, calcd C 45.77; H 3.20; N 2.97.

***bzq*Pt(dpm):** platinum(II) (7,8-benzoquinolinato-$N,C^{3'}$) (2,2,6,6-tetramethyl-3,5-heptanedionato-O,O). Yield: 31%. ^{1}H NMR (250 MHz, CDCl$_3$), ppm: 9.12 (d, 1H, J = 5.5 Hz), 8.25 (dd, 1H, J = 8.0, 1.5 Hz), 7.77 (m, 2H), 7.51 (m, 5H), 5.85 (s, 1H), 1.31 (s, 9H), 1.30

(s, 9H). Anal. for $C_{24}H_{27}NO_2Pt$: found C 51.19, H 4.76, N 2.59, calcd C 51.79, H 4.89, N 2.52.

15.3 Results and Discussion

15.3.1 Synthesis and Structure

Dichloride-bridged dimers of the general structure $C^\wedge NPt(\mu\text{-}Cl)_2PtC^\wedge N$ were prepared by the reaction of potassium tetrachloroplatinate with a cyclometalating ligand precursor ($HC^\wedge N$) in 2-ethoxyethanol (Eq. 15.1). The structures of the $C^\wedge N$ ligands used are shown in Fig. 15.1.

$$K_2PtCl_4 + HC^\wedge N \rightarrow C^\wedge NPt(\mu - Cl)_2PtC^\wedge N \qquad (15.1)$$

The dimers were isolated as solids that varied in color from gray to red, depending on the purity of the dimer complex and identity of the chelating ligand. Related square planar dimers have been well characterized in the literature [22b, 33]. In most instances, the dimers were not further characterized and were used directly in subsequent reactions. The dimers were dissociated in the presence of a base and either acetyl acetone (acacH) or dipivaloylmethane (dpmH) (Eq. 15.2).

$$C^\wedge NPt(\mu - Cl)_2PtC^\wedge N + 2(\beta - \text{diketone}) \xrightarrow{Na_2CO_3} C^\wedge NPtO^\wedge O$$
$$(15.2)$$

The isolated yields for these complexes, based on K_2PtCl_4, varied between 5% and 50%. The complexes have the general structure $C^\wedge NPt(O^\wedge O)$, as shown in Fig. 15.1, are air stable and sublimable, and varied in color between yellow and red. Changing the ancillary β-diketonate ligand from acac to dpm increases the solubility of the complexes and greatly simplifies purification by column chromatography. The increased solubility also allows for photophysical measurements to be made in solvents having a broad range of polarities, such as THF or hexanes. Our synthesis is quite general, and complexes can be easily prepared with a variety of functional groups on the $C^\wedge N$ ligand. The synthesis of the biscyclometalated analogues, in contrast, requires use of more reactive organolithium intermediates that are intolerant of many of the functional groups

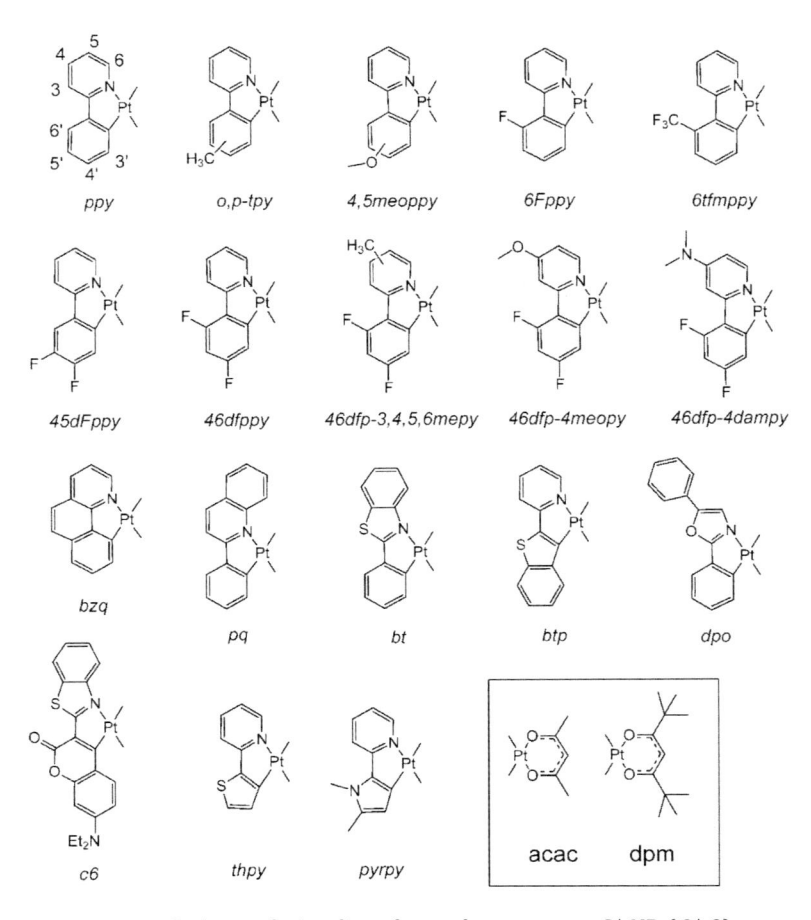

Figure 15.1 Cyclometalating ligands used to prepare $C^\wedge N\text{Pt}(O^\wedge O)$ complexes. Abbreviations used throughout the paper are listed below the $C^\wedge N$ or $O^\wedge O$ fragment.

used here [20a,c]. In addition, our synthesis avoids the use of thallium acetyl acetonate, which has been previously employed to prepare Pt(II) acac complexes [32, 34].

15.3.2 Crystal Structure

Single crystals of *thpy*Pt(dpm) were grown from dichloromethane/ methanol solution and characterized using X-ray crystallography.

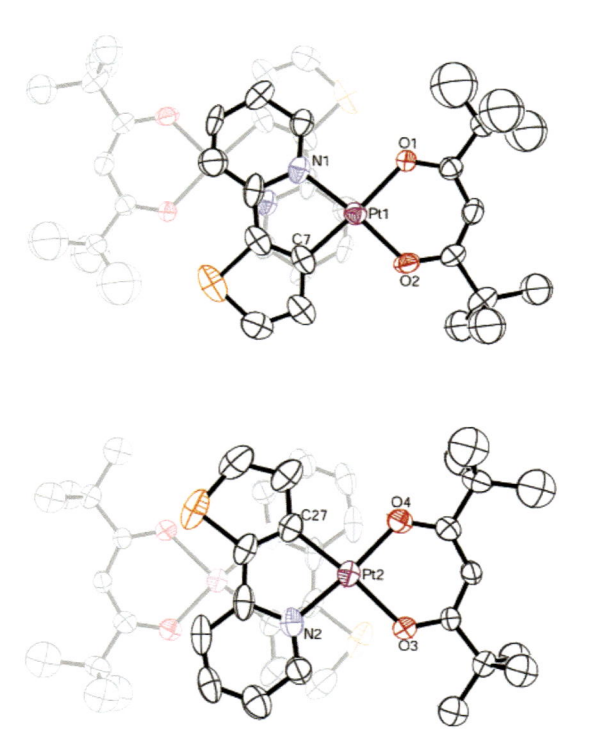

Figure 15.2 ORTEP diagram of two *thpy*Pt(dpm) dimers. One of the *thpy*Pt(dpm) molecules in each dimer has been lightened in color for clarity.

There are two unique molecules in the asymmetric unit. The molecules pack as head-to-tail dimers, each molecule of the dimer related to the other by a center of inversion (Fig. 15.2). The dimers have a plane-to-plane separation of 3.45 Å for Pt(1) and 3.56 Å for Pt(2) indicative of moderate $\pi-\pi$ interactions (Fig. 15.3). There are no metal–metal interactions, the closest Pt–Pt distance being 4.92 Å. Each molecule has a distorted square planar geometry. The Pt(1)–C(7) (1.953(7) Å), Pt(2)–C(27) (1.983(7) Å) bond lengths are similar to the mean value reported for the cis-(*thpy*)$_2$Pt complex (1.989(6) Å) [35]. The Pt(1)–N(1) (1.984(5) Å), Pt(2)–N(2) (2.006(6) Å) bond lengths are slightly larger and comparable to mean values of 2.05 Å from other (*thpy*)Pt derivatives where N is opposite a ligand of weak trans influence [36]. The Pt(1)–O(1)

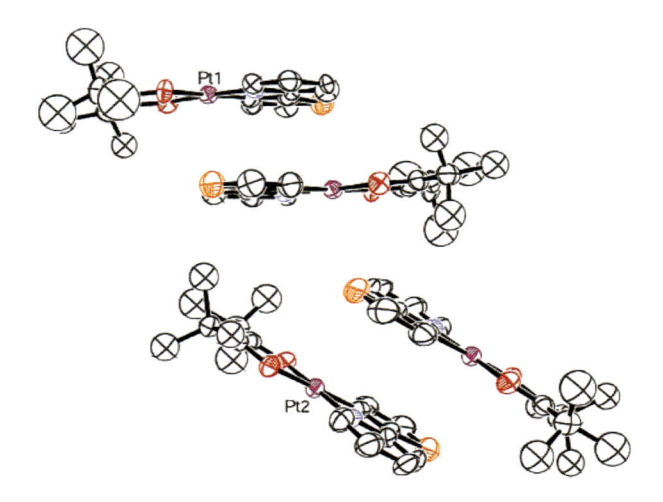

Figure 15.3 ORTEP diagram showing π stacking of *thpy*Pt(dpm).

(2.064(4) Å), Pt(1)–O(2) (1.998(4) Å) and Pt(2)–O(3) (2.083(4) Å), Pt(2)–O(4) (1.990(4) Å) bond lengths are within the 1.985(6)–2.156(15) Å range of other cyclometalated Pt(β-diketonato) derivatives [34] with the longer distances corresponding to oxygen trans to carbon. The C(7)–Pt(1)–N(1) (81.7(3)°), C(27)–Pt(2)–N(2) (80.9(3)°) and O(1)–Pt(1)–O(2) (92.26(16)°), O(3)–Pt(2)–O(4) (92.28(17)°) bond angles are typical for cyclometalates and β-diketonate derivatives of Pt [34, 37]. There is very little distortion away from the square plane, and the C–Pt–N and O–Pt–O chelate planes are only slight bowed and subtend angles of 4.57° for Pt(1) and 3.36° for Pt(2).

15.3.3 DFT Calculations

B3LYP density functional theory (DFT) calculations were carried out on several of the $C^\wedge N$Pt(acac) complexes using a LACVP** basis set. A similar approach has been used to investigate the ground and excited state properties of related Ir complexes (i.e., Ir(*ppy*)$_3$ and (*ppy*)$_2$Ir(acac)) [38]. We will focus our discussion here on the results for *ppy*Pt(acac); however, all of the $C^\wedge N$Pt(acac) complexes we examined gave a similar picture of the HOMO and

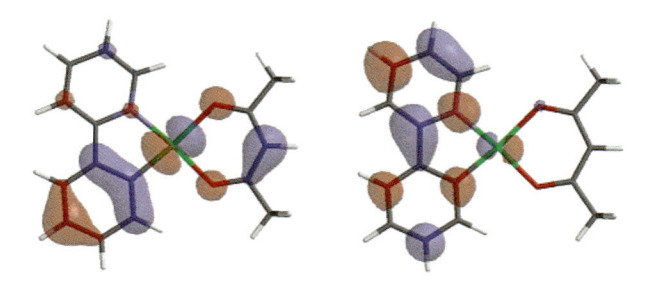

Figure 15.4 Density functional theory calculation (DFT) of HOMO (left) and LUMO (right) for *ppy*Pt(acac).

LUMO orbitals. The calculated values for Pt–C (1.98 Å), Pt–N (2.03 Å), Pt–O (2.04 Å), and Pt–O (2.15 Å) bond lengths and C–Pt–N (81.2°) and O–Pt–O (90.9°) chelate angles compare favorably to the corresponding experimental values determined in the X-ray structure of *thpy*Pt(dpm) and related cyclometalates and β-diketonate derivatives of Pt (vide supra). The HOMO and LUMO orbitals are shown in Fig. 15.4. The HOMO and LUMO levels displayed in the figure have π symmetry, with opposite phases above and below the molecular plane. The HOMO (−5.41 eV) consists of a mixture of phenyl, Pt, and acac orbitals while the LUMO (−1.60 eV) is predominantly 2-phenylpyridyl in character. The triplet HSOMO (−2.64 eV) has a phase and spatial relation nearly identical to those of the singlet LUMO. The energy of the triplet state was estimated as the difference between the singlet ground state (HOMO) and triplet (HSOMO) energies. The theoretical triplet energy of 2.77 eV (448 nm) is comparable to lowest energy absorption band found for *ppy*Pt(dpm) (2.88 eV, 430 nm) (vide infra). The valence orbital picture is consistent with combined LC and MLCT components in the lowest energy optical transition. The results of calculations for other $C^\wedge N$Pt(acac) will be discussed in the following text as they apply to spectral interpretation.

15.3.4 Electrochemistry

The electrochemical properties of the complexes were examined using cyclic voltammetry, and the redox data are given in Table 15.2.

Table 15.2 Absorption, 77 K emission, and redox properties of $C^\wedge NPt(O^\wedge O)$ complexes

$C^\wedge N$	LX	abs, λ_{max}^a (nm) {$\epsilon(10^{-3}$ cm^{-1} M^{-1})}	Emission at 77K[b] λ_{max} (nm)	τ (μs)	Redox (V)[c] $E_{1/2}^{Red}$
ppy	dpm	249 (28), 286 (19), 322 (8.8), 333(8.9), 381 (6.6), 406 (2.8), 430 sh (1.8)	477	8.9	−2.41
ppy	acac	269 (32), 292 (28), 310 (22), 360 (11), 389 (6.0), 410 sh (3.7)	480	9.0	−2.39
p-tpy	acac	252 (28), 281 (22), 317 (11), 330 (9.7), 360 (6.6), 398 sh (3.1)	480	10.9	−2.34
o-tpy	acac	254 (31), 281 (19), 316 (11), 328 (11), 368 (5.7), 404 sh (2.7)	480	8.6	−2.39
6fppy	dpm	246 (33), 284 (18), 317 (10), 329 (12), 363 (4.7), 381 (6.9), 410 (2.9)	468	7.2	−2.37
6tfmppy	dpm	240 (24), 257 (27), 281 (22), 305 (11), 326 sh (9.4), 395 (7.0), 420 sh (3.7)	487	7.7	−2.14
46dfppy	dpm	280 (20), 315 (11), 327 (12), 358 (5.7), 373 (8.0), 394 (3.1)	458	8.1	−2.31
46dfppy	acac	252 (30), 273 sh (19), 308 (11), 321 (11), 359 (6.9), 390 sh (2.4)	458	9.1	−2.29
45dfppy	acac	274 (22), 312 (10), 323 (8.9), 361 (6.0) 400 sh (2.3)	476	10.5	−2.27
46dfp-3mepy	dpm	247 (28), 280 (17), 317 (11), 330 (10), 377 (6.0), 400 (2.9)	467	9.1	−2.38
46dfp-4mepy	dpm	246 (34), 279 (20), 312 (11), 324 (11), 369 (8.1), 390 (3.2)	456	7.7	−2.42
46dfp-5mepy	dpm	246 (28), 281 (16), 315 (9.0), 329 (10), 369 (6.1), 395 (2.1)	466	10.0	−2.32
46dfp-6mepy	dpm	246 (21), 276 (16), 315 (8.1), 330 (9.7), 373 (5.4), 396 (2.8)	464	11.5	−2.31
46dfp-4meopy	dpm	211 (23), 246 (30), 276 (20), 306 (9.0), 349 (6.1), 367 (7.7)	438	7.0	−2.51
46dfp-4dmapy	dpm	214 (9.2), 244 (15), 277 (16), 348 (4.5), 365 (6.0)	440	6.1	−2.60
4meoppy	dpm	249 (25), 270 (21), 298 (20), 322 (14),370 (8.0), 398 (4.3), 419 (3.5)	480	13.9	−2.50[d]
5meoppy	dpm	250 (31), 288 (22), 327(9.6), 358 (7.3), 384 (7.2), 426 (3.6), 450 (3.0)	525	13.3	−2.49[e]
bt	dpm	216 (38), 258 (25), 321 (19), 340 (11), 379 (5.8), 401 (8.8), 443 (3.0)	530	7.7	−2.16
pq	dpm	250 (28), 298 (27), 325(8.8), 341 (8.8), 359 (9.3), 422 (6.8), 455 (3.6)	555	10.3	−2.00

c6	acac	300 (20), 388 (23), 469 (41), 496 (51)	590	54.1	−1.94
dpo	dpm	236 (24), 303 (27), 320 (22), 363 (10) 386 (8.8), 409 (6.7)	538	6.8	−2.40
btp	acac	265 (20), 318 (20), 344 sh (14), 427 (6.7), 444 (6.8)	600	11.3	−2.25
thpy	dpm	254 (13), 288 (17), 322 (15), 340 (17), 376 (8.2), 414 (4.6), 440 (4.6)	550	20.4	−2.41
pyrpy	dpm	237 (17), 300 (14), 331 (17), 354 (15), 370 (13), 460 (4.5), 487 (3.3)	580	16.1	−2.60[e]
bzq	dpm	217 (38), 250 (36), 281 (9.3), 325 (11), 344 (9.0), 391 (5.8), 442 (2.4)	493	122	−2.22[e]

[a]Absorption measurements of dpm complexes were carried out in hexanes; acac complexes were measured in CH_2Cl_2.
[b]77 K emission and lifetime measurements were carried out in 2-methyltetrahydrofuran.
[c]Redox measurements were carried out in DMF solution; values are reported relative to Cp_2Fe/Cp_2Fe^+.
[d]Quasi-reversible.
[e]Irreversible.

All of the electrochemical potentials reported here were measured relative to an internal ferrocene reference (Cp_2Fe/Cp_2Fe^+) or were adjusted to a ferrocene reference (for literature data only). Most complexes described here show a single reversible reduction wave between −1.9 and −2.6 V and irreversible oxidation. For example, *ppy*Pt(dpm) has a reversible reduction wave at −2.41 V in DMF and an irreversible oxidation near 1.0 V. Similarly, the heteroleptic complex *ppy*Pt(en)$^+$ is also reported to have a single reversible reduction wave at −2.6 V and an irreversible oxidation wave at +0.51 V in DMF solution [22a]. Likewise, the Pt(*ppy*)$_2$ complex is reported to have a reversible reduction at −2.5 V [39] and an irreversible oxidation at 0.1 V. Therefore, for these cyclometalated complexes, it is generally considered that reduction is localized on the $C^\wedge N$ ligand while oxidation occurs at the metal center. However, since square planar Pt(I) and Pt(III) metal centers are susceptible to nucleophilic attack by solvents, the Pt(II) redox processes are usually irreversible [22a]. This electrochemical behavior is consistent with a description of ligand-localized LUMO states and a HOMO with substantial metal character, as seen in our DFT calculations. In addition, since Pt(II) complexes with *thpy* ligands have reversible reductions at the same potential as those with *ppy* ligands, it has been suggested that reduction occurs primarily on the more electron accepting pyridyl portion of the $C^\wedge N$ ligand [40].

Substituents on the 2-phenylpyridyl ligand strongly affect the redox properties of these complexes. Incorporating electron-withdrawing fluorine atoms on the phenyl ring gives a marked decrease in the reduction potential. The monoflu-orinated complex, *6fppy*Pt(dpm), has a reduction at −2.37 V, making it 40 mV easier to reduce than the *ppy* derivative, while the difluorinated derivatives, *46dfppy*Pt(dpm) and *45dfppy*Pt(acac), are an additional 100 mV easier to reduce with reversible reduction waves at −2.31 and −2.27 V, respectively. Introduction of the weakly σ-donating methyl group on the pyridyl ring makes the *46dfp-4mepy*Pt(dpm) complex 100 mV harder to reduce, relative to *46dfppy*Pt(dpm). Furthermore, substituting stronger electron donating groups on the pyridyl ring, such as methoxy and dimethyl amino, increases the reduction potential of *46dfp-4meopy*Pt(dpm) to −2.51 V and *46dfp-*

4dmapyPt(dpm) to -2.60 V. Similarly, substitution of electron-donating methoxy groups onto the phenyl ring leads to an equivalent cathodic shift. The 4meoppyPt(dpm) complex shows a quasi-reversible reduction potential at -2.50 V, and the 5meoppyPt(dpm) is irreversibly reduced at -2.49 V.

As the conjugated π system of the $C^\wedge N$ ligand is increased, complexes such as btPt(dpm), pqPt(dpm), and btpPt(acac) show a corresponding decrease in the reduction potential. For example, pqPt(dpm) has a reversible reduction wave at -2.00 V, and a second irreversible reduction is seen at -2.91 V. The decrease in reduction potential for these complexes is likely due to greater stabilization of the negative charge on the more delocalized π-orbital system. Interestingly, the bzqPt(dpm) complex is irreversibly reduced at -2.2 V. Similar behavior has been previously observed in related bzqPtLX systems and has been attributed to a weaker metal–ligand interaction due to the rigid planarity and enhanced conjugation of the benzoquinolyl ligand [40].

15.3.5 Electronic Spectroscopy

Absorption and low-temperature emission spectra were recorded for all of the complexes. The data are given in Table 15.2, and spectra for ppyPt(dpm), 46dfppyPt(dpm), and 46dfp-4dmapyPt(dpm) are shown in Fig. 15.5. The dpm derivatives have nearly identical photophysical properties (dilute solution and frozen glasses) as compared to their acac counterparts. The energies and extinction coefficients of these spectra are similar to those of other $C^\wedge N$Pt complexes reported in the literature [20, 22]. Low-energy transitions in the range of 350–450 nm, with extinctions between 2000 and 6000 $M^{-1}cm^{-1}$, are assigned as metal to ligand charge transfer (MLCT) transitions. These bands are solvatochromatic [41] and are not observed in the free $HC^\wedge N$ ligand precursors. The higher energy, more intense absorption bands are assigned to $\pi-\pi^*$ ligand-centered (LC) transitions. These transitions are shifted from the free ligand absorption due to perturbation from the metal, but are not strongly solvatochromic. Metal-centered, d–d, transitions are not observed for these complexes. It is believed that the strong ligand field of the $C^\wedge N$ ligands shifts the d–d transitions to high energy,

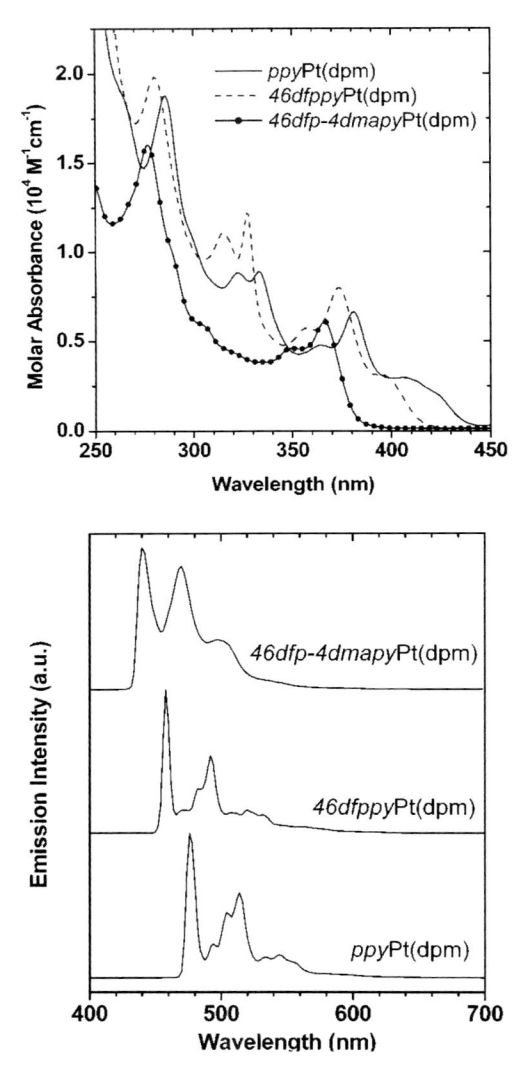

Figure 15.5 Absorption (top) and emission (bottom) spectra for *ppy*Pt(dpm), *46dfppy*Pt(dpm), and *46dpf-4dmapy*Pt(dpm) complexes. The absorption spectra were measured in hexanes at room temperature, and the emission spectra were measured at 77 K in a 2-methyltetrahydrofuran glass.

putting them under the more intense LC and MLCT transitions [20b,c].

All of the complexes are intensely emissive in low temperature glasses (77 K), and many are luminescent in fluid solution (298 K) as well. Most of the complexes show a small rigidochromic blue shift of 5–10 nm on cooling of the solution sample to 77 K. The emission bands for all of the complexes are red shifted from the phosphorescence of the corresponding protonated $C^\wedge N$ ligand and have shorter radiative lifetimes. For example, at 77 K, the emission maximum and lifetime of the *ppy*Pt(dpm) complex (λ_{max} = 480 nm, τ = 9.0 μs) can be compared to the phosphorescence emission maximum and lifetime of 2-phenylpyridine (λ_{max} = 430 nm, τ > 100 ms) [22b]. The highly structured emission spectra display vibronic progressions of ca. 1400 cm^{-1}, which are typical for breathing modes on the aromatic ring. The structured luminescence and microsecond radiative lifetimes at 77 K are consistent with emission originating from a mixed ^{3}LC–MLCT excited state [23b]. Longer lifetimes for some of the other complexes are indicative of greater ^{3}LC character in the excited state. The β-diketonato ligand does not significantly perturb the excited state since the luminescent properties of the *ppy* and *thpy*Pt($O^\wedge O$) complexes are comparable in wavelength, efficiency, and lifetime relative to *ppy*Pt(en)$^+$ and *thpy*Pt(en)$^+$ complexes reported in the literature [22a,c].

Substituent effects can be used to tune the energy of the emissive state for these *ppy*Pt(β-diketonate) complexes over a wide range (Table 15.2). Incorporating an electronegative atom such as fluorine onto the phenyl ring hypsochromically shifts the emission spectrum. A single fluoride substituent in the 6'-position leads to a 12 nm blue shift in the emission of *6fppy*Pt(dpm) (λ_{max} = 468 nm), relative to *ppy*Pt(dpm). Difluoro substitution at the 4',6'-positions on the phenyl ring gives a more pronounced blue shift in *46dfppy*Pt(dpm) (λ_{max} = 458 nm). However, difluoro substitution at the 4',5'-positions causes a much smaller shift in *45dfppy*Pt(acac) (λ_{max} = 476 nm). On the basis of the DFT calculations, it appears that a large amount of electron density in the HOMO is centered at the 5'-position of the phenyl ring and nodes exist at the 4' and 6'-positions (the HOMO orbitals for difluoro *ppy*Pt complexes look very similar to the one shown for *ppy*Pt(acac) in Fig. 15.4). Therefore, for

*45dfppy*Pt(acac), weak π donation into this molecular orbital from the 5'-fluoro group raises the HOMO level and offsets the electron withdrawing effect from the 4'-fluoro group. Similarly, substitution with a more strongly electron donating methoxy group in the 5'-position gives a pronounced red shift in *5meoppy*Pt(dpm) (λ_{max} = 525 nm) relative to the 4'-position in *4meoppy*Pt(dpm) (λ_{max} = 480 nm). This sensitivity of transition energy to the substitution position on 2-phenyl-pyridyl ligands has been previously observed in related tris-cyclometalated iridium complexes [42].

In order to further increase the energy gap of the emitting excited state, electron-donating groups were incorporated onto the pyridyl ring since substitution on this ring should raise the LUMO energy and thereby increase the HOMO–LUMO energy gap. Consistent with this argument, methyl and dimethylamino donors in the 4-position of the respective pyridyl rings of *46dpf-4mepy*Pt(dpm) and *46dpf-4dmapy*Pt(dpm) clearly increase the reduction potential relative to *46dpfppy*Pt(dpm) (vide supra). Consequently, the lowest energy MLCT absorption bands for the *46dfppy*, *46dfp-4mepy*, and *46dfp-4dmapy* Pt(dpm) complexes decrease in series 394, 390, and 365 nm, respectively. Higher energy $\pi-\pi^*$ transitions between 250 and 300 nm are not as strongly perturbed, but also show hypsochromic shifts in wavelength such that *46dfppy* > *46dfp-4mepy* > *46dfp-4dmapy*. A similar trend is seen in the emission spectra; a weakly σ-donating methyl group in the 4-position on the pyridyl ring results in a slight blue shift of 2 nm for *46dfp-4mepy*Pt(dpm) (λ_{max} = 456 nm) while the stronger dimethylamino donor imparts a more substantial shift for *46dfp-4dmapy*Pt(dpm) (λ_{max} = 440 nm).

The emission characteristics of the methoxy-substituted complexes display some distinct features that require separate consideration. A comparison of the room temperature and 77 K emission spectra for *46dfp-4meopy*Pt(acac) in 2-MeTHF is shown in Fig. 15.6. At room temperature the structured spectrum has a maximum at 456 nm with a vibronic spacing of 1400 cm^{-1}. Interestingly, at 77 K the luminescence spectrum displays two emission bands as separate vibronic progressions (1475 cm^{-1}) separated by 600 cm^{-1}. The intensity of the higher energy progression is greatly attenuated in 3-methylpentane glass at 77 K. The same emissive behavior is observed with the dpm derivative (i.e., *46dfp-4meopy*Pt(dpm)) in

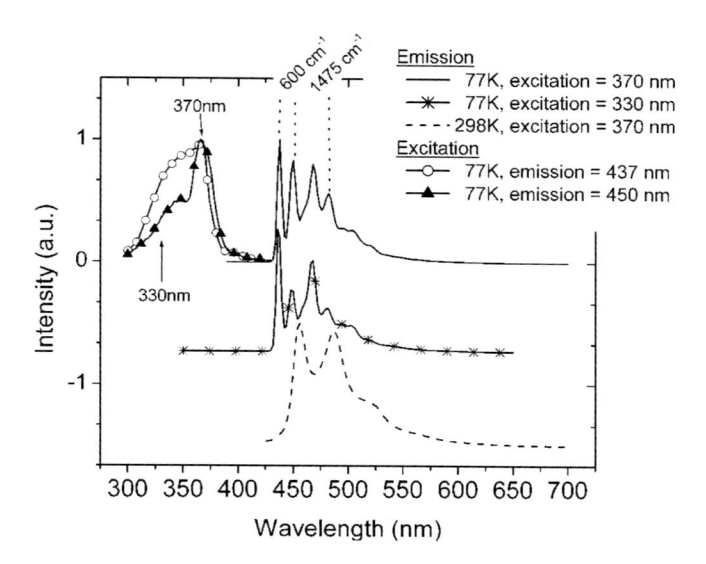

Figure 15.6 Room temperature and 77 K emission and excitation spectra for *46dtp-4meopy*Pt(acac) in 2-methyltetrahydrofuran solution.

2-MeTHF. A similar effect is also observed for the *5meoppy*Pt(dpm) complex in which two sets of vibronic progressions are observed. These two progressions are also separated from each other by roughly 600 cm^{-1} with the higher energy transition of much lower intensity. While the spectra could be explained as being due to two vibronic progressions (600 and 1450 cm^{-1}) on a single electronic transition, the excitation and emission spectra suggest that this is not the case. The excitation spectra for the two different progressions (λ_{max} = 437 and 450 nm) are different, and the ratio of the higher energy to the lower energy progression is strongly affected by the excitation wavelength (Fig. 15.6). Had the spectral line shape been caused by two vibronic progressions on a single electronic transition, the ratio of the two progressions should not depend on the excitation energy, which they clearly do. The best explanation for the low-temperature spectra of this complex is the appearance of two different emitting species in the low-temperature glass. All of the methoxy derivatives were carefully examined by NMR spectroscopy and found to consist of a single species. This analysis is further substantiated by the fact that the ratio of the two states is dependent on the coordinating nature of the solvent, as seen from the differences in emission for 3-methylpentane and 2-methyltetrahydrofuran solutions. The source of the two emitting species is unknown at present but could be due to conformational isomerism of the methoxy group with respect to the pyridyl ring.

The nature of the $C^\wedge N$ ring system can also have profound effects on the color of emission. Complexes with softer, more polarizable atoms such as sulfur and nitrogen incorporated in the ring system are much easier to oxidize and have significantly lower transition energies. Therefore, the *thpy*Pt(dpm) and *pyrpy*Pt(dpm) complexes display orange-red emission with a λ_{max} up to 600 nm. Extending the size of the conjugated π system for $C^\wedge N$ ligands compared to 2-phenylpyridyl expectedly lowers the energy of the ^{3}LC transition by simultaneously raising the HOMO and lowering the LUMO energy. Thus, the complexes *bt*Pt(dpm), *pq*Pt(dpm), and *btp*Pt(acac) emit yellow to red compared to the green of *ppy*Pt(dpm) (Fig. 15.7).

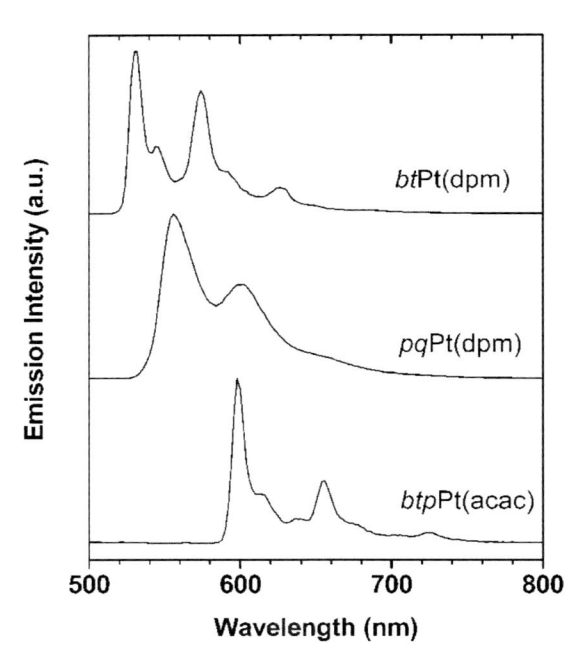

Figure 15.7 77 K emission spectra for *bt*Pt(dpm), *pq*Pt(dpm), and *btp*Pt(acac) complexes in 2-methyltetrahydrofuran.

The *bzq*Pt(dpm) complex is unique because of its exceptionally long lifetime of 120 µs in 77 K solution. A similar long lifetime for the related *bzq*Pt(en)$^+$ complex ($\tau = 217$ µs at 77 K)[40] has been attributed to the increased conjugation within the ligand π system due to the rigid planarity of the benzoquinolyl chelate. This leads to both a decrease in the energy of the ^{3}LC excited state and a weaker metal–ligand interaction, thus increasing the ^{3}LC character of the excited state.

The compounds *dpo* and *c6* are commercially available laser dyes with fluorescence maxima at 365 and 500 nm, respectively. It

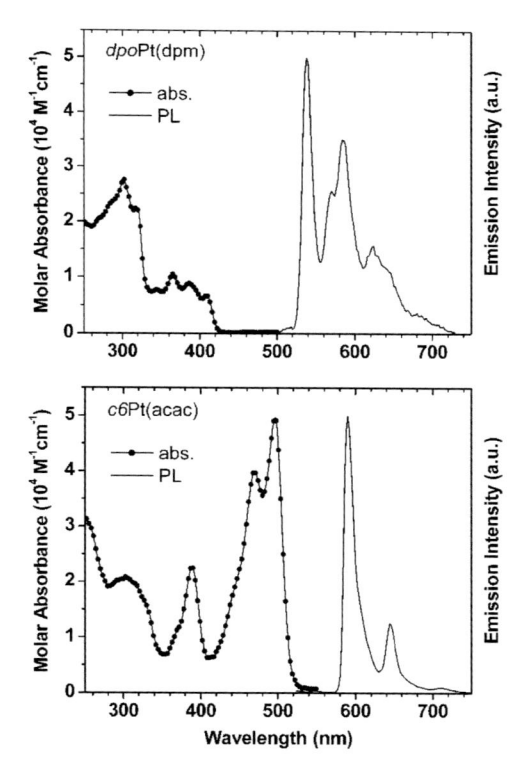

Figure 15.8 Absorption and emission spectra for *dpo*Pt(dpm) (top) and *c6*Pt(acac) (bottom) complexes. The absorption spectrum of *dpo*Pt(dpm) was measured in hexanes, and *c6*Pt(acac) in CH_2Cl_2, both at room temperature. The emission spectra were measured at 77 K in a 2-methyltetrahydrofuran glass.

Table 15.3 Emission properties of $C^\wedge NPt(O^\wedge O)$ complexes at room temperature[a]

$C^\wedge N$	$O^\wedge O$	λ_{max}, nm	τ_{298}, μs	Φ_{PL}
ppy	acac	486	2.6	0.15
p-tpy	acac	485	4.5	0.22
6fppy	dpm	476	<1.0	0.06
45dfppy	acac	484	3.0	0.22
46dfppy	acac	466	<1.0	0.02
46dfppy	dpm	466	<1.0	0.02
46dfp-5mepy	dpm	472	<1.0	0.05
46dfp-4meopy	dpm	456	<1.0	
46dfp-4dmapy	dpm	447	<1.0	
4meoppy	dpm	490	7.4	0.20
c6	acac	589	27.9	0.25
btp	acac	612	3.4	0.08
thpy	acac	575	4.5	0.11
pyrpy	acac	603	7.14	0.02

[a] 2-Methyltetrahydrofuran solutions.

has been recently demonstrated that these dyes can cyclometalate onto indium [9a]. Laser dyes, in general, have very high extinction coefficients and fluoresce very efficiently, whereas emission from the triplet state may only be observed at low temperature, if at all. However, the Ir cyclometalated dyes only display emission from ligand-centered triplet states. Likewise, the *dpo*Pt(dpm) and *c6*Pt(acac) complexes emit exclusively from triplet excited states (*dpo*Pt(dpm), $\lambda_{max} = 538$ nm, $\tau = 6.8$ μs; *c6*Pt(acac), $\lambda_{max} = 590$ nm, $\tau = 28$ μs). Hence, cyclometalating a laser dye to the heavy Pt metal atom also allows for facile intersystem crossing between the singlet and triplet manifolds of the dye.

Many of these complexes are also highly emissive in degassed room temperature solution with lifetimes in the microsecond regime; data for several of the complexes is shown in Table 15.3. As the energy of emission increases, a corresponding decrease in room temperature quantum efficiency and lifetime is observed. The luminescent quantum yield of the *ppy* complex is 15% ($\tau = 2.6$ μs), whereas the *46dfppy* complex has a quantum yield of only 2% ($\tau < 2$ μs) and the quantum yields of the *46dfp-4meopy* and *46dfp-4dmapy* complexes are <0.1%. It appears that a new luminescent quenching mechanism comes into play as the energy of the ^{3}LC excited state increases. The nature of this process is unknown at this time, but may involve a self-quenching mechanism or perhaps thermal activation to either a MC state or a competing MLCT state on the ancillary β-diketonato ligand. Further work is underway in order to elucidate the origin of this phenomenon.

15.4 Conclusions

The different $C^\wedge N\text{Pt}(O^\wedge O)$ complexes presented here demonstrate that these complexes can be readily tuned to emit with high efficiency throughout the visible spectrum. The emission characteristics of these complexes are governed by the nature of the cydometalating ligand, $C^\wedge N$, with the spectra resembling the ligand phosphorescence spectra. Strong spin–orbit coupling of the heavy platinum atom allows for mixing of ^{1}MLCT with ligand-based $^3\pi-\pi^*$, making the formally forbidden radiative relaxation of these states an efficient process, with quantum efficiencies as high as 0.25 at room temperature. All of these complexes are strongly emissive in the solid state and at 77 K with lifetimes on the order of microseconds. Many of these complexes are also emissive in fluid solution. These complexes are easily prepared, neutral, stable in air, and sublimable, making them good candidates for use in optoelectronic devices. The wide range of excited state energies and redox potentials reported here make this series of Pt complexes attractive as highly tunable photooxidants and photoreductants.

We have recently incorporated one of these Pt complexes into an organic light emitting diode: which emitted white light very efficiently (Q.E. = 4% (photons/electrons)) [43]. We are currently examining a range of Pt complexes for both monochrome and white OLEDs. These results of these studies will be reported in future publications.

Acknowledgment

The authors thank Universal Display Corp., the Defense Advanced Research Projects Agency, and the National Science Foundation for financial support of this work. We would also like to thank Dr. Stephen Bradforth and Amy Germaine for helpful discussions.

Supporting Information Available

Crystal data, atomic coordinates, bond distances, bond angles, anisotropic displacement parameters, and ORTEP diagrams for *thpy*Pt(dpm). This material is available free of charge via the Internet at http://pubs.acs.org.

References

1. (a) Peyratout, C. S., Aldridge, T. K., Crites, D. K., McMillin, D. R. *Inorg. Chem.* 1995, **34**, 4484–4489. (b) Wang, A. H.-J., Nathans, J., van der Marel, G. A., van Boom, J. H., Rich, A. *Nature (London)* 1978, **276**, 471–474.
2. (a) Houlding, V. H., Frank, A. J. *Inorg. Chem.* 1985, **24**, 3664–3668. (b) Maruyama, T., Yamamoto, T. *J. Phys. Chem. B* 1997, **101**, 3806–3810.
3. Anbalagan, V., Srivastava, T. S. *J. Photochem. PhotobioL, A: Chem.* 1995, **89**, 113–119.
4. Connick, W. B., Gray, H. B. *J. Am. Chem. Soc.* 1997, **119**, 11620–11627.
5. Buey, J., Diez, L., Espinet, P., Kitzerow, H.-S., Miguel, J. A. *Chem. Mater.* 1996, **8**, 2375–2381.
6. (a) Hissler, M., McGarrah, J. E., Connick, W. B., Geiger, D. K., Cummings, S. D., Eisenberg, R. *Coord. Chem. Rev.* 2000, **208**, 115–137. (b) McGarrah, J. E., Kim, Y.-J., Hissler, M., Eisenberg, R. *Inorg. Chem.* 2001, **40**, 4510–4511.
7. (a) Baldo, M. A., O'Brien, D. F., You. Y., Shoustikov, A., Sibley, S., Thompson, M. E., Forrest, S. R. *Nature* 1998, **395**, 151. (b) Kwong, R. C., Sibley, S., Dubovoy, T., Baldo, M. A., Forrest, S. R., Thompson, M. E. *Chem. Mater.* 1999, **11**, 3709–3713. (c) O'Brien, D. F., Baldo, M. A., Thompson, M. E., Forrest, S. R. *Appl. Phys. Lett.* 1999, **74**, 442–444. (d) Adachi, C., Baldo, M. A., Forrest, S. R., Lamansky, S., Thompson, M. E., Kwong, R. C. *Appl. Phys. Lett.* 2001, **78**, 1622–1624. (e) Cleave, V., Yahioglu, G., Barny, P. L., Friend R. H., Tessler, N. *Adv. Mater.* 1999, **11**, 285.
8. (a) Kunugi, Y., Mann, K. R., Miller, L. L., Exstrom, C. L. *J. Am. Chem. Soc.* 199, **120**, 589–590. (b) Chan, S.-C., Chan, M. C. W., Wang, Y., Che, C.-M., Cheung, K. K., Zhu, N. *Chem. Eur. J.* 2001, **7**, 4180–4190. (c) Lu, W., Mi,

B.-X., Chan, M. C. W., Hui, Z., Zhu, N., Lee, S.-T., Che, C.-M. *Chem. Commun.* 2002, 206–207.

9. (a) Lamansky, S., Djurovich, P., Murphy, D., Abdel-Razzaq, F., Lee, H.-E., Adachi, C., Burrows, P. E., Forrest, S. R., Thompson, M. E. *J. Am. Chem. Soc.* 2001, **123**, 4304–4312. (b) Lamanksy, S., Djurovich, P., Murphy, D., Abdel-Razzaq, F., Kwong, R., Tsyba, I., Bortz, M., Mui, B., Bau, R., Thompson, M. E. *Inorg. Chem.* 2001, **40**, 1704–1711.

10. (a) Baldo, M. A., Lamansky, S., Burrows, P. E., Thompson, M. E., Forrest, S. R. *Appl. Phys. Lett.* 1999, **75**, 4–6. (b)Adachi, C., Baldo, M. A., Forrest, S. R., Thompson, M. E. *Appl. Phys. Lett.* 2000, **77**, 904–906. Ikai, M., Tokito, S., Sakamoto, Y., Suzuki, T., Taga, Y. *Appl. Phys. Lett.* 2001, **79**, 156–158.

11. Bevilacqua, J. M., Eisenberg, R. *Inorg. Chem.* 1994, **33**, 2913–2923.

12. Barigelletti, F., Sandrini, D., Maestri, M., Balzani, V., von Zelewsky, A., Chassot, L., Jolliet, P., Maeder, U. *Inorg. Chem.* 1988, **27**, 3644–3647.

13. Aldridge, T. K., Stacy, E. M., McMillin, D. R. *Inorg. Chem.* 1994, **33**, 722–727.

14. (a) Cummings, S. D., Eisenberg, R. *J. Am. Chem. Soc.* 1996, **118**, 1949–1960. (b) Zuleta, J. A., Chesta, C. A., Eisenberg, R. *J. Am. Chem. Soc.* 1989, **111**, 8916–8917.

15. Whittle, C. E., Weinstein, J. A., George, M. W., Schanze, K. S. *Inorg. Chem.* 2001, **40**, 4053–4062. Hissler,M., Connick, W. B., Geiger, D. K., McGarrah, J. E., Lipa, D., Lachicotte, R. J., Eisenberg, R. *Inorg. Chem.* 2000, **39**, 447. Connick, W. B., Geiger, D., Eisenberg, R. *Inorg. Chem.* 1999, **38**, 3264. Chan, C. W., Cheng, L. K., Che, C. M. *Coord. Chem. Rev.* 1994, **132**, 87.

16. Dungey, K. E., Thompson, B. D., Kane-Maguire, N. A. P., Wright, L. L. *Inorg. Chem.* 2000, **39**, 5192–5196.

17. Connick, W. B., Miskowski, V. M., Houlding, V. H., Gray, H. B. *Inorg. Chem.* 2000, **39**, 2585–2592.

18. Lai, S.-W., Chan, M. C. W., Cheung, K.-K., Che, C.-M. *Inorg. Chem.* 1999, **38**, 4262–4267.

19. (a) Cheung, T. C., Cheung, K. K., Peng, S. M., Che, C. M. *J. Chem. Soc., Dalton Trans.* 1996, 1645–1653. (b) Lai, S.-W., Chan, C.-W., Cheung, K.-K., Che, C.-M. *Organometallics* 1999, **18**, 3327–3336.

20. (a) Chassot, L., Müller, E., von Zelewsky, A. *Inorg. Chem.* 1984, **23**, 4249–4253. (b) Maestri, M., Sandrini, D., Balzani, V., Chasson, L., Jolliet, P., von Zelewsky, A. *Chem. Phys. Lett.* 1985, **122**, 375–379. (c) Chassot, L., von Zelewsky, A. *Inorg. Chem.* 1987, **26**, 2814–2818. (d) Jolliet, P., Gianini, M., von Zelewsky, A., Bernardinelii, G., Stoeckli-Evans, H. *Inorg. Chem.* 1996, **35**, 4883–4888.

21. Deuschel-Cornioley, C., Lüönd, R., von Zelewsky, A. *Helv. Chim. Act.* 1989, **72**, 377–382.
22. (a) Kvam, P.-I., Puzyk, M. V., Balashev, K. P., Songstad, J. *Acta Chem. Scand.* 1995, **49**, 335–343. (b) Mdleleni, M. M., Bridgewater, J. S., Watts, R. J., Ford, P. C. *Inorg. Chem.* 1995, **34**, 2334–2342. (c) Balashev, K. P., Puzyk, M. V., Kotlyar, V. S., Kulikova, M. V. *Coord. Chem. Rev.* 1997, **159**, 109–120. (d) DePriest, J., Zheng, G. Y., Goswami, N., Eichhorn, D. M., Woods, C., Rillema, D. P. *Inorg. Chem.* 2000, **39**, 1955–1963.
23. (a) Juris, A., Balzani, V., Barigelletni, F., Campagna, S., Belser, P., von Zelewsky, A. *Coord. Chem. Rev.* 1988, **84**, 85–277. (b) Crosby, G. A. *Acc. Chem. Res.* 1975, **8**, 231–238.
24. (a) Gagne, R. R., Koval, C. A., Lisensky, G. C. *Inorg. Chem.* 1980, **19**, 2854–2855. (b) Sawyer, D. I., Sobkowiak, A., Roberts, J. L., Jr. *Electrochemistry for Chemists*, 2nd ed.; John Wiley and Sons: New York, 1995; pp. 467.
25. Sheldrick, G. M. *SHELXTL*, versions5.1; Bruker Analytical X-ray System, Inc., Madison, WI, 1997.
26. Blessing, R. H. *Acta Crystallogr.* 1995, **A51**, 33–38.
27. Effenberger, F., Krebs, A., Willrett, P. *Chem. Ber.* 1992, **125**, 1131–1140.
28. Sammakia, T., Hurley, T. B. *J. Org. Chem.* 1999, **64**, 4652–4664.
29. Savoia, D., Concialini, V., Tarsi, L. *J. Org. Chem.* 1991, **56**, 1822–1827.
30. (a) Chippendale, K. E., Iddon, B., Suschitzky, H. *J. Chem. Soc., Perkin Trans.* 1972, 2023–2030. (b) Gilman, H., Shirley, D. A. *J. Am. Chem. Soc.* 1949, **90**, 1871.
31. Lohse, O., Thevenin, P., Waldvogel, E. *Synlett* 1999, **1**, 45–48.
32. Cockburn, B. N., Howe, D. V., Keating, T., Johnson, B. F. G., Lewis, J. *J. Chem. Soc., Dalton Trans.* 1973, 404–410.
33. (a) Balashev, K. P., Simon, J., Ford, P. C. *Inorg. Chem.* 1991, **30**, 859–861. (b) Pregosin, P. S., Wombacher, F., Albinati, A., Lianza, F. *J. Organamet. Chem.* 1991, **418**, 249–267. (c) Cope, A. C., Siekman, R. W. *J. Am. Chem. Soc.* 1965, **87**, 3272–3273. (d) Cave, G. W. V., Fanizzi, F. P., Deeth, R. J., Errington, W., Rourke, J. P. *Organometallics* 2000, **19**, 1355–1364. (e) Nonoyama, M., Takayanagi, H. *Transition Met. Chem.* 1976, **1**, 10–13.
34. Ghedini, M., Pucci, D., Crispini, A., Barberio, G. *Organometallics* 1999, **18**, 2116–2124.
35. Breu, J., Range, K.-J., von Zelewsky, A., Yersin, H. *Acta Crystallogr.* 1997, **C53**, 562–565.
36. (a) Stückl, A. C., Klement, U., Range, K.-J. *Z. Kristallogr.* 1993, **208**, 297–298. (b) Giordano, T. J., Rasmussen, P. G. *Inorg. Chem.* 1975, **14**, 1628–1634.

37. Katoh, H., Miki, K., Kai, Y., Tanaka, N., Kasai, N. *Bull. Chem. Soc. Jpn.* 1981, **54**, 611–612.
38. Hay, P. J. *J. Phys. Chem. A* 2002, **106**, 1634.
39. Cornioley-Deuschel, C., von Zelewsky, A. *Inorg. Chem.* 1987, **26**, 3354–3358.
40. Kulikova, M. V., Balashev, K. P., Kvam, P.-I., Songstad, J. *Russ. J. Gen. Chem.* 2000, **70**, 163–170.
41. *ppy*Pt(dbm) is representative of the complexes examined here. The band at 381 nm in hexanes shifts to higher energies as the polarity of the solvent increases. This band comes at 374, 371, 366, and 360 nm in toluene, THF, acetone, and acetonitrile, respectively. The solvent polarity indexes for hexanes, toluene, THF, acetone, and acetonitrile are 0.1, 2.4, 4.0, 5.1, and 5.8, respectively. Polarity indexes were taken from the following: Snyder, L. R. *J. Chromatogr. Sci.* 1978, **16**, 223–234. Snyder, L. R. *J. Chromatogr.* 1974, **92**, 223–230.
42. (a) Dedeian, K., Djurovich, P. I., Garces, F. O., Carlson, G., Watts, R. J. *Inorg. Chem.* 1991, **30**, 1685–1687. (b) Grushin, V. V., Herrron, N., LeCloux, D. D., Marshall. W. J., Petrov, V. A., Wang, Y. *Chem. Commun.* 2001, 1494–1495.
43. D'Andrade, B. W., Brooks, J., Adamovich, V., Thompson, M. E., Forrest, S. R. *Adv. Mater.*, in press.

Chapter 16

White Light Emission Using Triplet Excimers in Electrophosphorescent Organic Light-Emitting Devices

Brian W. D'Andrade,[a] Jason Brooks,[b] Vadim Adamovich,[b] Mark E. Thompson,[b] and Stephen R. Forrest[a]

[a] Center for Photonics and Optoelectronic Materials (POEM),
Princeton Materials Institute (PMI), Department of Electrical Engineering,
Princeton University, Princeton, New Jersey 08544, USA
[b] Department of Chemistry, University of Southern California,
Los Angeles, California 90089, USA
forrest@ee.princeton.edu

A general solution to the problem of efficient white light generation in organic light-emitting devices is demonstrated using the novel process of phosphor excimer emission. These electrophosphorescent organic light-emitting devices solve the problem of interactions between multiple dopants for the generation of white light by using a single phosphor for blue mission, as well as for forming excimers that emit from green to red. Energy transfer between dopants (leading to current-dependent color changes) is significantly reduced since excimers lack a bound ground state,

---

Reprinted from *Adv. Mater.*, **14**(15), 1032–1036, 2002.

Electrophosphorescent Materials and Devices
Edited by Mark E. Thompson
Text Copyright © 2002 WILEY-VCH Verlag GmbH
Layout Copyright © 2024 Jenny Stanford Publishing Pte. Ltd.
ISBN 978-981-4877-34-3 (Hardcover), 978-1-003-08872-1 (eBook)
www.jennystanford.com

thus offering a simplified means to achieve broad spectral emission from organic light-emitting devices. This approach is well suited to low-cost white light emitters, which eventually might replace existing, illumination sources. Using two phosphor dopants, devices with a high color rendering index of 78, a maximum external quantum efficiency of $\eta_{ext} = 4.0 \pm 0.4\%$ corresponding to 9.2 ± 0.9 cd A^{-1}, and a maximum luminance of 31000 ± 3000 cd m^{-2} at 16.6 V, demonstrate that efficient and bright electrophosphorescent white organic light-emitting devices based on excimer emission can significantly exceed efficiencies of fluorescent white light emitters based on this same principle. Furthermore, efficient white light emitters are demonstrated using only a single phosphor in a highly simplified device architecture.

The pace of research in the search for new white light sources has recently seen a marked increase in response to the need for more efficient and environmentally friendly solutions to the impending energy shortage. One particularly attractive means for producing energy efficient lighting is the use of organic light-emitting devices (OLEDs), which operate at very low voltages (~5 V) and with high luminance efficiency [1]. Among methods for producing white light, electrophosphorescence [2] stands out as the most effective mechanism for OLED emission due to its demonstrated potential for achieving 100 % internal emission efficiency [3]. While electrophosphorescence, typically achieved by doping an organometallic phosphor into a conductive host, has successfully been used to generate the primary colors necessary for display applications [3–5], efficient generation of the broad spectral emission required of a white light source has remained elusive.

Several different approaches have been used to generate white light in OLEDs [6–9]. One involves the combination (either in mixed or separate layers) of three different dopants located within the emission zone of the device. This approach has been demonstrated to obtain a high color-rendering index (CRI) necessary for lighting applications. However, this triple dopant strategy can be problematic due to the different efficiencies of energy transfer to each dopant in the mix, leading to an imbalance in the white color, or pronounced variations in color with brightness. Nevertheless, we have recently demonstrated that separating two or three phosphors into different "bands" in the emitting zone can generate high-efficiency (5.2%) white light emission [6]. While that technique may ultimately be practical, devising yet more simplified solutions may be desirable to lower the cost and increase the yield of electrophosphorescent white light sources.

One particularly promising approach to reducing the number of dopants and structural heterogeneities inherent in the multiple color-band architecture is to employ a lumophore that forms a broadly emitting exciplex (i.e., a state whose wavefunction overlaps a neighboring, dissimilar molecule) in its excited state. Recently, fluorescent exciplex OLEDs have been demonstrated with Commission Internationale de L'Eclairage (CIE) coordinates close to the ideal white light source of (x, y) (0.33, 0.33). Some of the best results from various groups report an external quantum efficiency [10] of $\eta_{ext} = 0.3\%$, a luminance efficiency [7] of $\eta_p = 0.58$ lm W^{-1} and a maximum luminance [7] of 2000 cd m^{-2}. These performance values lie well below those needed in practical lighting applications.

In contrast to this past work on fluorescent excimers [11], we report here on the use of phosphorescent excimer emission [12, 13] from the phosphor, platinum(II)(2-(4',6'-difluorophenyl)pyridinato-N, C$^{2'}$)(2,4-pentanedionato) (FPt1), coupled with blue monomer emission from either iridium-*bis*(4,6,-difluorophenyl-pyridinato-N, C^2)-picolinate (FIrpic) [5] or from platinum(n)(2-(4',6'-difluorophenyl)pyridinato-N, C$^{2'}$)(6-methyl-2,4-heptanedionato-O,O) (FPt2) to achieve broadly white light-emitting, efficient electrophosphorescent OLEDs (or WOLEDs). Molecular formulae for FIrpic, FPt1, and FPt2 are shown in Scheme 16.1. In contrast to an exciplex, an excimer is an emissive excited state whose wavefunction overlaps two adjacent molecules of like composition. Both excimers and exciplexes lack a bound ground state, and hence provide a unique solution to achieving efficient energy transfer from the host to the light-emitting centers. Indeed, the use of an excimer prevents the cascade of energy from the host material and from higher energy (blue) dopants to the lower energy (excimer-orange) dopant since the excimer has no bound ground state. Complicated intermolecular interactions, which make color balancing using multiple dopants problematic, are thus eliminated. The excimers used here have the broad emission spectrum desired, spanning wavelengths from

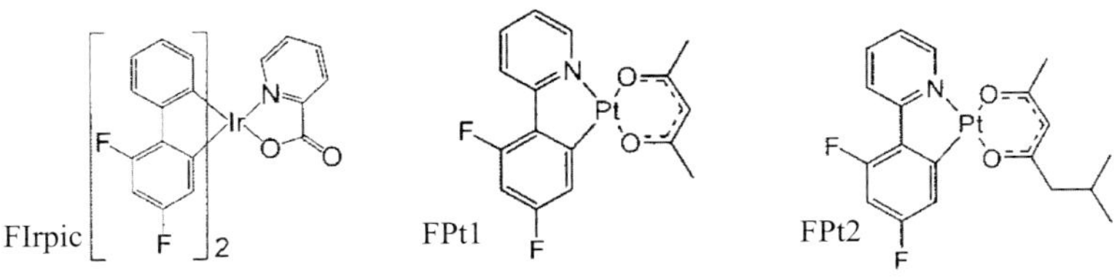

Scheme 16.1 Chemical structures of FIrpic, FPt1, and FPt2.

$\lambda = 450$ nm to 800 nm. Hence, using only one, or at most two dopants, coverage of the entire visible spectrum is achieved with an electrophosphorescent OLED.

To achieve efficient excimer emission from a doped layer, control of the dopant concentration is crucial. Excimers typically form when planar molecules are in close proximity to each other [12]. Square-planar Pt complexes are known to form excimers in concentrated solutions and thin films [14–16]. For example, FPt1 shows monomer emission in 10^{-6} M solutions of dichloromethane and excimer emission in more concentrated 10^{-3} M solutions. Neat films of FPt1 show no additional absorption peaks characteristic of dimer emission. Further, the photoluminescence excitation (PLE) spectra of the monomer and excimer peaks are the same, suggesting they have a common excitation pathway, which would not be the case for a different (e.g., dimer) species.

We begin the design of the WOLED by examining the photoluminescence emission (PL) and PLE spectra of the materials used in the device emissive region. Figure 16.1 shows the PL (solid lines) and PLE (open circles) spectra of the films examined. Film 1 shows the 4,4'-N,N'-dicarbazole-biphenyl (CBP) PL spectrum with a peak at $\lambda = 390$ nm and the corresponding PLE at $\lambda = 220$–370 nm. For emission at $\lambda = 390$ nm, the PLE of CBP has a shoulder at $\lambda = 300$ nm and a main peak at $\lambda = 350$ nm. The CBP PLE peaks correspond to the absorption peaks at $\lambda = 300$ nm and 350 nm (arrows, inset of Fig. 16.2). These two CBP features appear in the PLE spectra of all films where CBP was used as a host; therefore, it is the main absorbing species in all the films, and energy must be transferring efficiently from CBP to both FPt1 and FIrpic for emission from these molecules to occur.

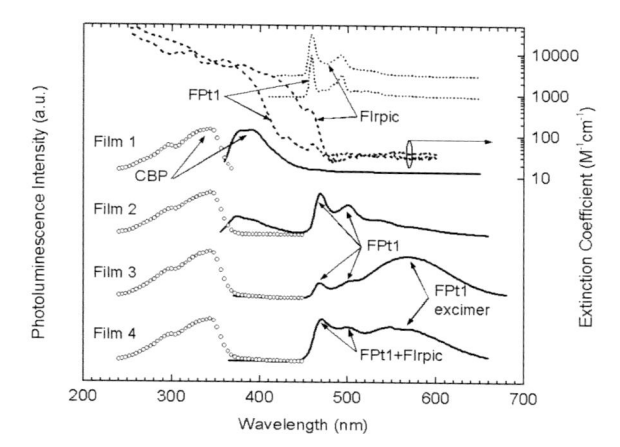

Figure 16.1 Photoluminescence emission (solid lines) and excitation spectra (open circles) of films 1–4. Film 1 is neat CBP. Film 2 is CBP doped with <1 wt.% FPt1. Film 3 is CBP doped with 7 wt.% FPt1 and Film 4 is CBP doped with both FIrpic and FPt1 at 6 wt.%. There is an increase in the excimer emission at higher concentrations of the FPt1 dopant, and balanced white emission with CBP emission totally quenched is achieved in Film 4. Dotted lines: Photoluminescence of FIrpic and FPt1 in 2-methyltetrahydrofuran at 77 K shows that the triplet energy levels of FIrpic and FPt1 are very similar due to good spectral overlap. Both phosphors had $\lambda_{max} = 458$ nm indicating the ligand-based nature of the lowest energy transition. Dashed lines: Molar extinction coefficients for FIrpic and FPt1 in dichloromethane.

The PL spectrum of Film 2 (<1 wt.% FPt1 in CBP) shows bands consistent with CBP and FPt1 monomer emission only. CBP emission is at $\lambda = 390$ nm, and FPt1 monomer emission has peaks at $\lambda = 470$ nm and $\lambda = 500$ nm. The spectrum observed for FPt1 in CBP is very similar to the same molecule in dilute solution. At <1 wt.%, the randomly distributed FPt1 molecules are, on average, separated by 30 Å, precluding significant excimer formation. We note that the lack of a broad, long wavelength peak in Film 2 suggests that exciplexes do not form between CBP and FPt1. That is, if exciplexes form between these moieties, exciplex emission from a FPt1–CBP complex would be present even in the most lightly doped samples.

As the FPt1 doping concentration in CBP is increased to ∼7 wt.% (Film 3), we observe strong excimer emission with an orange–

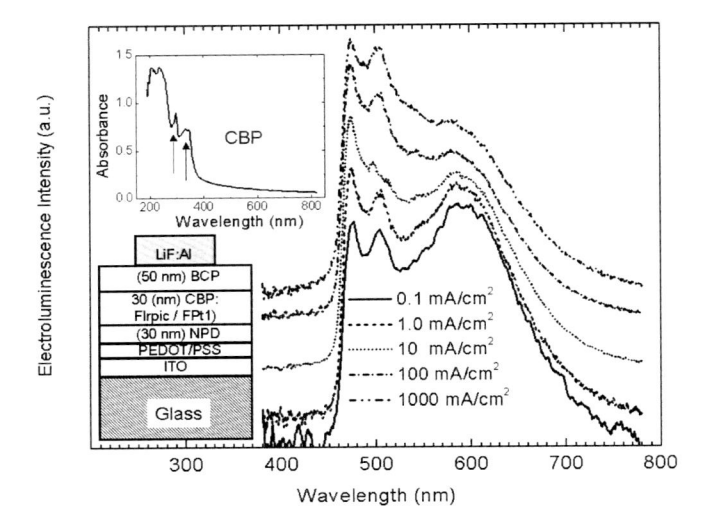

Figure 16.2 Normalized electroluminescence spectra of the WOLED at several current densities, vertically spaced for clarity. The excimer emission is shown to decrease relative to the blue emission at higher current densities. Upper inset: Absorbance versus wavelength of a 1000 Å thick CBP film on quartz. Arrows correspond to PLE peaks for neat CBP in Fig. 16.1. Lower inset: WOLED structure showing the organic layers that were sequentially deposited in vacuum.

red peak at $\lambda = 570$ nm along with the characteristic monomer emission at $\lambda = 470$ nm and $\lambda = 500$ nm. The higher doping level leads to complete quenching of the CBP fluorescence. For Film 3, the measured lifetime of $\tau = 7.2$ μs of the FPt1 emission at $\lambda = 570$ nm, compared to 8.3 μs at $\lambda = 470$ nm, is consistent with phosphorescent excimer formation on FPt1 complexes.

Film 4 consists of CBP doped with 6 wt.% FIrpic and 6 wt.% FPt1. Here, CBP emission is absent in the PL, but the PLE spectrum still indicates that it is the main absorbing species, and that energy is efficiently transferred to both FIrpic and FPt1. The combined FIrpic (monomer) and FPt1 (excimer) emissions give the co-doped film a white photoluminescence with CIE coordinates (0.33, 0.44).

We can understand the energy transfer process in the double-doped system by referring to the highest occupied molecular orbital (HOMO) and lowest unoccupied molecular orbital (LUMO) energies

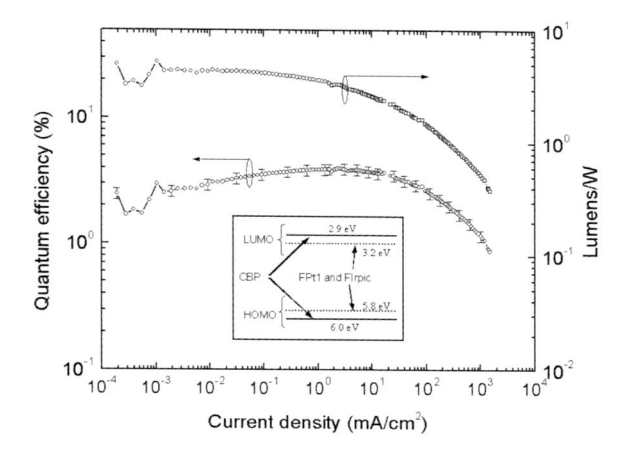

Figure 16.3 External quantum and power efficiencies versus current density of the double-doped electrophosphorescent white organic light-emitting device. The emissive layer consists of 6 wt.% FIrpic and 6 wt.% FPt1 doped into CBP. Inset: Energy level diagram of CBP (solid lines) doped with FIrpic and FPt1 (dashed lines). Here, HOMO indicates the position of the highest occupied molecular orbital and LUMO indicates the position of the lowest unoccupied molecular orbital. These levels are inferred from the reduction potentials of FIrpic (−2.30 V) and FPt1 (−2.29 V), measured in dimethylformamide relative to ferrocene. The HOMO levels of the two compounds are similar based on the reduction potential values given above, and on the extinction coefficients, ε, of both FIrpic and FPt1, which sharply increase at $\lambda = 450$ nm as shown in Fig. 16.1.

of the three organic components (inset, Fig. 16.3). Triplet energy transfer from CBP to FIrpic occurs via an endothermic process described elsewhere [5]. FIrpic and FPt1 have similar absorption and emission spectra for their monomer states, as apparent in Fig. 16.1, suggesting similar electronic structures. Assuming the same positions of the HOMO (5.8 ± 0.1 eV), and the LUMO (3.2 ± 0.1 eV) energy levels of both FPt1 and FIrpic, a similar endothermic triplet energy transfer pathway can be expected between CBP and FPt1.

The FPt1 excimer is formed after the exciton is localized on the molecule (via energy transfer from CBP to the FPt1 monomer or hole–electron trapping at FPt1). Resonant energy transfer between the triplet levels of the two dopants is also likely since they are present at high concentration in the CBP matrix. However, direct energy transfer from FIrpic to the excimer does not occur because the excimer has zero ground state absorption, preventing the cascade of energy from the blue to the orange emission centers. This essentially decouples the excited states of these molecules, allowing for simple optimization of the doping in order to achieve the desired color balance. In terms of Dexter, or direct-hopping energy transfer, the transfer rate is zero if the overlap integral between the normalized donor emission and acceptor absorption spectra is zero, which is the case for FIrpic and the FPt1 excimer [12].

We measured (in the forward viewing direction) the light output power from the WOLED using a calibrated silicon photodiode, and then calculated η_{ext} using current density-voltage characteristics shown in Fig. 16.4. A lambertian intensity profile is assumed to calculate η_p (Fig. 16.3) and luminance (Fig. 16.4). Here, $\eta_{ext} \geq 3.0\%$ between $J = 1 \times 10^{-3}$ mA cm^{-2} and 10 mA cm^{-2}. The roll-off at $J > 300$ mA cm^{-2} is attributed to sample heating and triplet–triplet annihilation [17, 18]. The WOLED has a maximum $\eta_{ext} = 4.0 \pm 0.4\%$ corresponding to 9.2 ± 0.9 cd A^{-1}, a luminance of 31000 ± 3000 cd m^{-2} at 16.6 V, $\eta_p = 4.4 \pm 0.4$ lm W^{-1} and a CRI of 78.

A neat film of FPt1 has measured lifetimes of $\tau = 4.8$ μs and 5.2 μs at $\lambda = 470$ nm and $\lambda = 600$ nm, respectively. Hence, as the current density increases, the FPt1 excimer states may become saturated compared to the monomer and FIrpic, leading to the increased blue

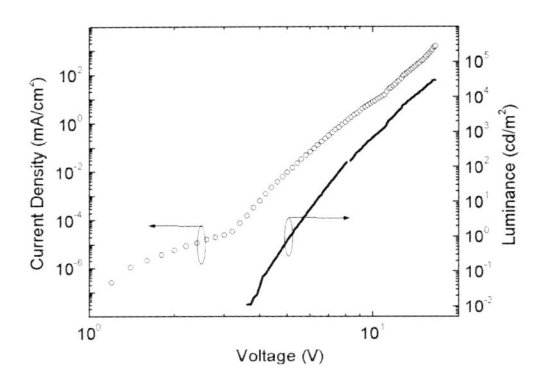

Figure 16.4 Current density and luminance versus voltage characteristics of the WOLED shown in Fig. 16.2.

emission shown in Fig. 16.2. The spectral changes reflect small changes in CIE coordinates from (0.40, 0.44) to (0.35, 0.43).

To achieve well-balanced monomer and excimer emission from a *single dopant*, excimer formation is possible only at high doping levels (thereby allowing for monomer phosphorescence to simultaneously occur). Increasing the steric bulk of the β-diketonate ligand can hinder excimer formation, leading to an optimal monomer–excimer balance at higher doping levels than for FPt1. Indeed, the bulky alkyl substituent incorporated into the β-diketonate ligands of FPt2 does not change the solution (monomer) PL spectra of the complex relative to FPt1, but gives FPt2 balanced monomer and excimer emission at doping levels between 10 and 12 wt.%.

White light from a balance of monomer and excimer emission is observed for FPt1 doped between 3 and 4 wt.%. CBP emission is not totally quenched for these doping concentrations; hence devices are less efficient than for higher dopant concentrations. There is a concurrent increase in the intensity of the excimer emission at higher doping concentrations of FPt1, requiring addition of FIrpic to balance emission to obtain white. No excimer emission is observed from FIrpic.

The EL spectra for a single dopant FPt2 WOLED over a range of luminance values are shown in Fig. 16.5. The PL spectrum of a neat film of CBP–8 wt.% FPt2 is shown in the inset in Fig. 16.5. This WOLED gives white emission from a single dopant emitting simultaneously from both its monomer and excimer states. The spectrum has a contribution from NPD emission at 435 nm, which becomes significant only at luminances > 1000 cd m^{-2}, resulting from poor carrier confinement in the luminescent region at high currents. This leads to a slightly more blue spectrum than observed in the double doped WOLEDs. An electron-blocking layer inserted between the emitting layer and the hole-transport layer (HTL) completely eliminates this spectral feature, as will be reported elsewhere. This limits the peak brightness for this device to only 5500 cd m^{-2} (12 V). The ratio of excimer to monomer emission gradually becomes richer in monomer emission as the brightness is increased, analogous to the FIrpic-FPt1 WOLEDs. The CIE coordinates of approximately (0.34, 0.35) are not severely affected by the shift in the monomer-excimer ratio, remaining nearly

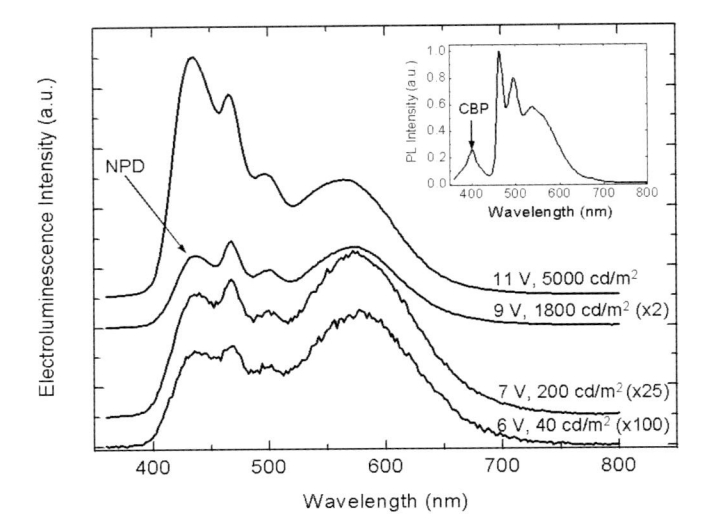

Figure 16.5 Electroluminescence spectra of a single doped CBP-8 wt.% FPt2 WOLED. The FPt2 dopant emits from both the monomer and excimer states in the luminescent region of the device. NPD emission degrades the device's match to CIE white at higher voltages, and also reduces device efficiency since triplet excitons are lost to a fluorescent emitter. Inset: Photoluminescence of a neat film of CBP-8 wt.% FPt2. The CBP emission is still present, and the excimer emission is not as significant as for FPt1 at similar doping concentrations in CBP.

ideal white up to $\sim$2000 cd m^{-2}, with a CRI value at 75 or above for nearly all of the spectra considered. If we ignore the NPD emission, the CIE coordinates are slightly shifted to (0.33, 0.40). The CBP–FPt2 WOLEDs give $\eta_{ext} = 1.9 \pm 0.2$% at a brightness of 100 cd m^{-2} ($J = 2$ mA cm^{-2}). The power efficiency at this brightness is 2.5 lm W^{-1}.

We have demonstrated that control of the molecular and electronic structures of an organic phosphor can lead to surprisingly efficient white light emission from an OLED containing only a single emissive dopant. We expect that these device structures will have several benefits over stacked or multiply doped single emissive layer devices [6]. The simple device structures make them well suited to low cost lighting applications. Indeed, a serious problem in multiple doped systems for white emission is differential aging. If one of the dopants degrades at a different rate from the others, the color of the device will change over time. This should not be the case for the single dopant monomer–excimer device. Degradation that results from charging or other reactions of the ground state molecules will equally affect the monomer and excimer emission. The photostability of these phosphorescent dopants is typically quite high, and device failure generally occurs through charged, rather than excitonic states [19].

Experimental

The three phosphorescent dopants (FIrpic, FPt1, and FPt2) and NPD were prepared by literature procedures [20, 21]. BCP and Alq$_3$ were purchased from Aldrich Chemical Company. PEDOTS-PSS was purchased from Bayer Chemical. All of the molecular materials used to prepare WOLEDs were purified by thermal gradient sublimation before use.

Photoluminescent, photoluminescence excitation, and absorbance spectroscopy were performed on thin-films grown in vacuum ($<1 \times 10^{-6}$ torr) by thermal evaporation on solvent-cleaned quartz substrates. The films were 1000 Å thick and were degassed with nitrogen during testing. Five films were grown: neat CBP, two films of CBP doped at <1 wt.% and ~7 wt.% with FPt1, CBP doped with both FIrpic and FPt1 at 6 wt.%, and CBP doped with 8 wt.% of FPt2.

Electrophosphorescent double-doped monomer-excimer WOLEDs were grown with a 6 wt.% FIrpic–6 wt.% FPt1 doped CBP luminescent layer. The devices were fabricated on a glass substrate pre-coated with indium tin oxide (ITO, sheet resistance = 20 Ω/sq). The substrate cleaning and device fabrication were carried out by procedures described previously [6]. The device structure was ITO/PEDOT:PSS (400 Å)/NPD (300 Å)/CBP–FIrpic–FPt1 (300 Å)/BCP (500 Å)/LiF (5 Å)/Al (700 Å) (PEDOT:PSS=poly (ethylene-dioxythiophene): poly(styrene sulfonic acid), NPD= N, N'-diphenyl-N,N'-bis(α-napthyl)-4,4'-biphenyl, BCP = bathocuproine). A cross section of the device structure is shown in Fig. 16.2. Devices were grown with areas as large as 2.7 cm^2 without loss in efficiency. The samples were exposed to air during tests. The device structure of the single dopant WOLEDs was ITO/NPD (400 Å)/CBP-8 wt.% FPt2 (300 Å)/BCP (150 Å)/Alq$_3$ (200 Å)/LiF–Al (1000 Å).

Acknowledgment

The authors thank Universal Display Corporation, the Defense Advanced Research Projects Agency, and the National Science Foundation's Materials Research Science and Engineering Center for their support.

References

1. L. S. Hung, M. G. Mason, *Appl. Phys. Lett.* 2001, **78**, 3732.
2. M. A. Baldo, D. F. O'Brien, Y. You, A. Shoustikov, S. Sibley, M. E. Thompson, S. R. Forrest, *Nature* 1998, **395**, 151.
3. C. Adachi, M. A. Baldo, M. E. Thompson, S. R. Forrest, *J. Appl. Phys.* 2001, **90**, 5048.
4. C. Adachi, M. A. Baldo, S. R. Forrest, S. Lamansky,M. E. Thompson, R. C. Kwong, *Appl. Phys. Lett.* 2001, **78**, 1622.
5. C. Adachi, R. C. Kwong, P. Djurovich, V. Adamovich, M. A. Baldo, M. E. Thompson, S. R. Forrest, *Appl. Phys. Lett.* 2001, **79**, 2082.
6. B. W. D'Andrade, M. E. Thompson, S. R. Forrest, *Adv. Mater.* 2001, **13**, 147.
7. J. Feng, F. Li, W. Gao, S. Liu, Y. Liu, Y. Wang, *Appl. Phys. Lett.* 2001, **78**, 3947.
8. X. Jiang, Z. Zhang, W. Zhao, W. Zhu, B. Zhang, S. Xu, *J. Phys. D: Appl. Phys.* 2000, **33**, 473.
9. J. Kido, H. Shionoya, K. Nagai, *Appl. Phys. Lett.* 1995, **67**, 2281.
10. M. Berggren, G. Gustafsson, O. Inganäs, M. R. Andersson, T. Hjertberg, O. Wennerström, *J. Appl. Phys.* 1994, **76**, 7530.
11. S. A. Jenekhe, J. A. Osaheni, *Science* 1994, **265**, 765.
12. N. J. Turro, *Modern Molecular Photochemistry,* University Science Books, Mill Valley, CA 1991, Ch. 5.
13. M. Pope, C. E. Swenberg, *Electronic Processes in Organic Crystals,* Oxford University Press, New York 1982, Ch. 1.

14. S. W. Lai, M. C. W. Chan, K. K. Cheung, C. M. Che, *Inorg. Chem.* 1999, **38**, 4262.
15. S. W. Lai, M. C. W. Chan, T. C. Cheung, S. M. Peng, C. M. Che, *Inorg. Chem.* 1999, **38**, 4046.
16. P. I. Kvam, M. V. Puzyk, K. P. Balashev, J. Songstad, *Acta Chem. Scand.* 1995, **49**, 335.
17. M. A. Baldo, C. Adachi, S. R. Forrest, *Phys. Rev. B* 2000, **62**, 10967.
18. N. Tessler, N. T. Harrison, D. S. Thomas, R. H. Friend, *Appl. Phys. Lett.* 1998, **73**, 732.
19. Z. Popovic, H. Aziz, N. Hu, A. Ioannidis, P. d. Anjos, *J. Appl. Phys.* 2001, **89**, 4673.
20. J. Brooks, Y. Babayan, S. Lamansky, P. I. Djurovich, I. Tsyba, R. Bau, M. E. Thompson, *Inorg. Chem.* 2002, in press.
21. S. Lamansky, P. I. Djurovich, D. Murphy, F. Abdel-Razzaq, R. Kwong, I. Tsyba, M. Bortz, R. Bau, M. E. Thompson, *Inorg. Chem.* 2001, **40**, 1704.

Chapter 17

Electrophosphorescent p–i–n Organic Light-Emitting Devices for Very-High-Efficiency Flat-Panel Displays

Martin Pfeiffer,[a,b] Stephen R. Forrest,[b] Karl Leo,[a] and Mark E. Thompson[c]

[a] *Institut für Angewandte Photophysik, Technische Universität Dresden, George-Baehr, Str. 1, Dresden D-01062, Germany*
[b] *Department of Electrical Engineering, Princeton University, Princeton, New Jersey 08544, USA*
[c] *Department of Chemistry, University of Southern California, Los Angeles, California 90089, USA*
forrest@ee.princeton.edu, pfeiffer@iapp.de

Displays are ubiquitous in our everyday world, providing the primary interface with computers, appliances, and test instruments. Some critical properties of all displays used in mobile applications are mechanical robustness, reliability, full color, light weight, and high efficiency to extend battery life. Current candidates for these advanced, mobile displays are based on back-lit liquid crystals, field-emission displays, and organic light-emitting devices (OLEDs), based on either polymers or small molecular weight materials. Until recently, the highest luminance efficiencies of devices based on

---

Reprinted from *Adv. Mater.*, **14**(22), 1633–1636, 2002.

Electrophosphorescent Materials and Devices
Edited by Mark E. Thompson
Text Copyright © 2002 WILEY-VCH Verlag GmbH
Layout Copyright © 2024 Jenny Stanford Publishing Pte. Ltd.
ISBN 978-981-4877-34-3 (Hardcover), 978-1-003-08872-1 (eBook)
www.jennystanford.com

these technologies have been in the range of 4–6 lm W^{-1} at display operating brightnesses of $\sim$100–300 cd m^{-2}. This situation improved dramatically, achieving green-emitting efficiencies exceeding 15 lm W^{-1} with the demonstration of electrophosphorescent OLEDs (PHOLEDs) based on small molecular weight organic materials [1–4]. Still, the efficiency of small molecular weight materials is limited by the voltage drop across the highly resistive semiconducting layers comprising the devices. Thus, as is apparent in Table 17.1, the voltage drop across an OLED is considerably higher than the minimum potential (V_λ) needed to generate a photon of wavelength λ. Recently, however, it has been shown [5–9] that increasing the conductivity of organic semiconductor layers by doping with either donors (for electron conducting materials) or acceptors (for hole-transport materials) can significantly reduce the voltage drop across organic thin films, and in particular, fluorescent OLEDs. This approach has recently led to the demonstration of a fluorescent p–i–n type OLED with an operating voltage of 2.55 V for 100 cd m^{-2}, i.e., close to the minimum drive voltage limit of V_λ [9]. In this work, we extend this approach to small molecular weight electrophosphorescent light-emitting devices, thereby achieving very high efficiency OLEDs that perform in a manner similar to LEDs employing conventional semiconductor materials. Indeed, we obtain significant light emission at applied voltages $V \leq V_\lambda$. This advance represents a significant step in realizing very high brightness displays with efficiencies unmatched by any current competing technology. Our results also suggest that white-emitting PHOLEDs will have applications in very high efficiency room lighting in the future, as well as in high current devices such as bipolar transistors and lasers.

The doped PHOLEDs use a p–i–n heterostructure configuration, as shown in the proposed equilibrium energy level diagram in the inset of Fig. 17.1. Here the 50 nm thick 4,4',4"-tris-(3-methylphenylphenylamino)triphenylamine (m-MTDATA) hole-transport layer (HTL) doped with 2 mol.% of 2,3,5,6-tetrafluoro-7,7,8,8-tetracyanoquinodimethane (F$_4$-TCNQ) is deposited on an indium tin oxide (ITO) coated glass substrate. The dopant, F$_4$-TCNQ, has previously been shown to be an effective acceptor, whose electron affinity [5, 10] approximately coincides with that of m-MTDATA

Table 17.1 Comparison of the efficiencies of several different phosphor/host combinations reported for PHOLEDs

Phosphor	Host	Φ [lm/W]	$J = 1\,\mu A/cm^2$				$J = 1\,mA/cm^2$			
			η_P [lm/W]	η_{Qext}	η_{Qint}	V_1/V	η_P [lm/W]	η_{Qext}	η_{Qint}	V_1/V
ppy$_2$Ir(acac) [3]	TAZ	53	60	0.19	0.87	0.60	20	0.15	0.68	0.25
btpIr(acac) [28, 29]	CBP	17	4	0.13	0.65	0.34	2.2	0.10	0.44	0.22
Firpic [30]	CBP	26	1.3	0.06	0.27	0.83	5.0	0.057	0.23	0.34
Ir(ppy)$_3$ p-i-n	CBP	53	22	0.05	0.25	1.0	29	0.09	0.45	0.86

Note: Here, Φ is the theoretical maximum power efficiency given the emission spectrum of the phosphor, η_P is the measured power efficiency at current density J, η_{Qint} is the internal quantum efficiency, η_{Qext} is the external efficiency, V_λ is the photon energy divided by the elementary charge at the peak of the emission spectrum, and V is the applied voltage at J.

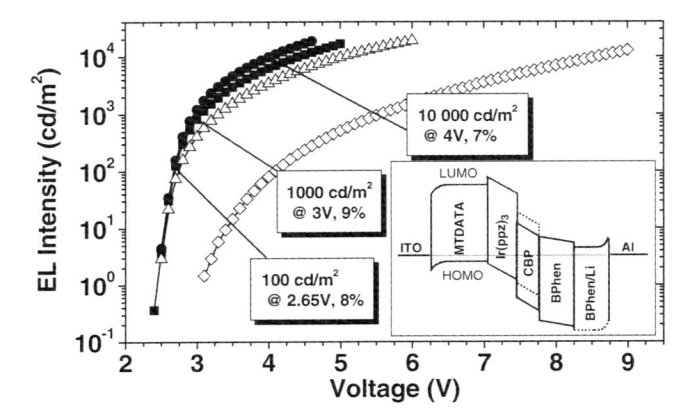

Figure 17.1 Electroluminescence (EL) intensity versus applied voltage for CBP:Ir(ppy)$_3$ based PHOLEDs. Shown are data for p–i–n PHOLED samples A (triangles), B (squares), and C (circles), compared with an undoped, conventional PHOLED (diamonds). The percentages shown in the boxes are the external quantum efficiencies measured at the corresponding voltages. Inset: Proposed equilibrium energy level diagram of the p–i–n PHOLED.

[11, 12], thereby providing a high density ($\sim 10^{17}$ cm^{-3}) of holes to the HTL [13, 14]. To prevent excitons generated in the emission layer (EML) from drifting into the heavily doped and highly conducting HTL where non-radiative quenching might occur, an electron blocking buffer layer of 10 nm thick tris(phenylpyrazole)iridium (Ir(ppz)$_3$) is deposited, followed by a emission layer of 5 nm thick N, N'-dicarbazolbiphenyl (CBP) doped with 7 wt.% of the green-emitting phosphor, tris(phenylpyridine)iridium (Ir(ppy)$_3$). Next, a 4,7-diphenyl-1,10-phenanthroline (Bphen) hole and exciton-blocking buffer layer is deposited, followed by a Li-doped BPhen electron-transport layer (ETL) obtained by co-deposition of pure Li, contributing a free electron concentration of $\sim 10^{18}$ cm^{-3} [15]. The total thickness of the BPhen layers is kept constant at 55 nm, while the thickness of the Li-doped layer varies between 0 and 30 nm. For the device without Li doping, a 0.5 nm thick layer of LiF is deposited prior to the deposition of the Al cathode, which has also been shown to result in n-type conductivity in organic materials such as Bphen [16, 17].

Conductivity-doped PHOLEDs are fundamentally different from undoped, or adventitiously doped organic electronic devices. One difference is that the electron- and hole-confining barriers in doped PHOLEDs are due, at least at low current injection, to a combination of built-in potential as well as the energy-level offsets between the adjacent layers. In conventional organic electronic devices, these barriers are solely determined by the energy-level offsets. While at high injection levels this added energy barrier vanishes by application of an external bias, at low levels of injection they can improve charge and carrier confinement, and hence the device efficiency. Furthermore, the very low resistance of the doped layers (typically $< 50\ \Omega\text{cm}^{-2}$) allows for very thick transport layers to be employed. Whereas a typical small-molecule-based device thickness should not exceed 100 nm for the organic material layers, doped devices can be made up to five times thicker without appreciably increasing the voltage drop across the layers. This provides a significant advantage in device yield when large displays are mass produced, due to the elimination of many parasitic shorts between electrodes often encountered in thinner, undoped devices.

The effect of doping the charge-transport layers on the performance of the PHOLEDs is shown in the luminance–voltage (L–V) characteristics in Fig. 17.1. Here, the L–V characteristics of p–i–n PHOLEDs with LiF-doped BPhen (sample A), and a 15 nm and 30 nm thick Li-doped BPhen (samples B and C, respectively) are compared with a conventional, undoped PHOLED structure. The undoped sample reaches 100 cd m^{-2} at 4 V, with the luminance increase flattening out at higher voltages due to ohmic losses in the transport layers or at injection barriers. Accordingly, a luminance of 10^4 cd m^{-2} is achieved at 8.5 V. In comparison, the intentionally doped p–i–n PHOLEDs operate at a significantly lower voltage, with 100 cd m^{-2} obtained at 2.65 V for samples B and C, increasing to only 4 V at 10^4 cd m^{-2}. A slightly increased voltage is obtained for sample A with a LiF interlayer instead of bulk Li-doping of BPhen, due to the limited diffusion of Li into the BPhen layers from the Al/LiF cathode. The external quantum efficiencies, measured with a calibrated Si detector capturing only light emitted in the forward (viewing) direction, is also shown at several luminances.

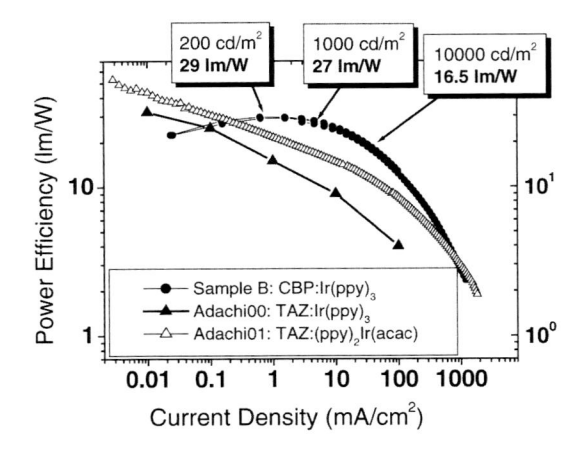

Figure 17.2 Luminous power efficiency versus current density for a p–i–n PHOLED sample B, compared to undoped PHOLEDs using a similar CBP:Ir-complex emissive region reported in the literature [2, 3].

While the external quantum efficiencies are typical of conventional CBP:Ir(ppy)$_3$ PHOLEDs reported previously [1], they remain ∼50% that of optimized devices, where external efficiencies have been shown to approach ∼20% [3, 4]. Nevertheless, their luminous power efficiencies exceed that of optimized conventional devices. This improvement, shown in Fig. 17.2, is due to the lower voltage operation of the doped structures. Here, we plot the power efficiency versus current density for sample B, along with those previously reported for optimized, conventional ultrahigh efficiency green-emitting PHOLEDs. While the efficiency of the conventional PHOLEDs approaches 70 lm W^{-1} at very low brightness, the rapid increase in voltage with current (and hence brightness) results in a concomitant rapid decrease in luminance efficiency. Hence, while the power efficiencies of both the doped and conventional PHOLEDs are approximately equal at 100 cd m^{-2}, the doped PHOLED efficiency is much higher as the brightness continues to increase. For example, at 1000 cd m^{-2}, the p–i–n PHOLED has a power efficiency of 27 lm W^{-1}, approximately equal to its efficiency at 100 cdm^{-2}, falling to 16.5 lm/W at 10^4 cd m^{-2}. This is compared with 29 lm W^{-1} at 100 cd m^{-2}, dropping to 20 lm W^{-1} at 1000 cd m^{-2}, and 13 lm W^{-1} at 10^4 cd m^{-2} for the best conventional, undoped PHOLEDs reported to date [3].

Two principle reasons for the low operation voltage in p–i–n-type OLEDs are the efficient injection from contacts into doped transport layers [6, 9, 18, 19], and the low ohmic losses in these high conductivity layers [20]. However, these advantages can only be exploited when using resistive buffer and emission layers with minimal thicknesses. Here, we succeeded in reducing the thickness of the emission layer to only 5 nm without losses in the quantum efficiency. As shown in the inset of Fig. 17.1, a reduction of the EML thickness is made possible using Ir(ppz)$_3$ as a buffer between the p-doped HTL and the EML. The high LUMO energy of Ir(ppz)$_3$, combined with the alignment (as determined from its ionization potential) of its highest occupied molecular orbital (HOMO) with that of m-MTDATA [12, 21], make this material suitable for blocking electron- and exciton-transport, while allowing for efficient hole injection into the EML. Indeed, using Ir(ppz)$_3$ as an electron blocking layer, we observe only Ir(ppy)$_3$ electrophosphorescence even with an EML thickness of only 5 nm. Furthermore, the triplet energy of BPhen also lies well above that of Ir(ppy)$_3$. Combined with its deep HOMO energy of 6.7 eV [1], both BPhen and Ir(ppz)$_3$ provide nearly complete isolation of the EML from the conductive dopants, as well as good confinement of both carriers and excitons in this region, allowing for high efficiencies to be obtained with thin, low resistance EMLs [1, 2].

The external quantum efficiency as a function of EML thickness is shown in Fig. 17.3. Here, it is clear that the highest efficiencies (9.5 ± 0.5%) at high current densities are obtained for an EML thickness of 5 nm. The decrease in efficiency at low currents is attributed to quenching by impurities such as Li atoms having diffused through the buffer layer during the deposition process. In contrast, the efficiencies obtained using the hole-transport material, N, N'-di(naphthalene-2-yl)-N, N'-diphenylbenzidine (α-NPD), in place of Ir(ppz)$_3$ are considerably lower. This reduction can be reversed to a limited degree if the EML is made thicker (e.g., 20 nm), although this increases the operating voltage of the device, and hence significantly decreases the power efficiency. This is due to the shallow LUMO of α-NPD (approximately 2.6 eV) [1], which allows generation of excitons

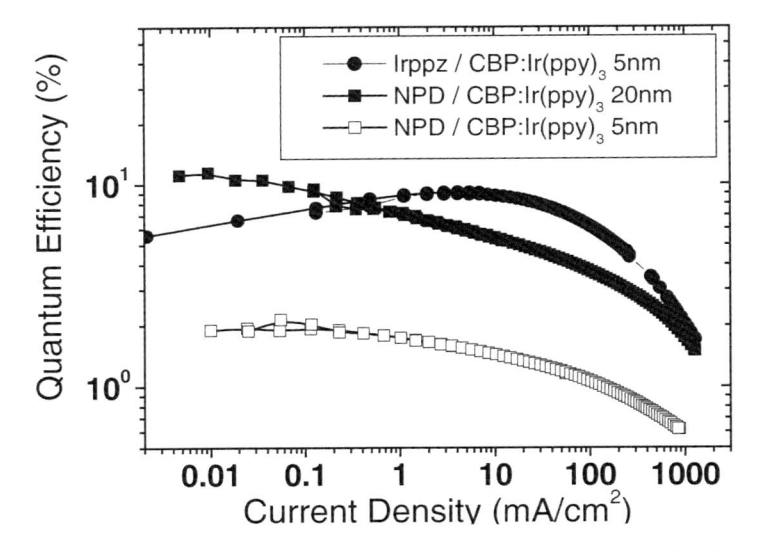

Figure 17.3 External quantum efficiency versus current density for the p–i–n CBP:Ir(ppy)$_3$ PHOLEDs with different emissive-layer thickness (shown in the legend), compared to the efficiency of a thin (5 nm) emissive layer combined with an Ir(ppz)$_3$ electron and exciton blocking layer.

on α-NPD from carrier pairs at the α-NPD/EML interface. Moreover, we found that the photoluminescence of Ir(ppy)$_3$ mixed into α-NPD is strongly quenched due to the lower triplet exciton energy of α-NPD as compared to Ir(ppy)3. Accordingly, Dexter transfer of triplet excitons from the EML to α-NPD is another possible loss mechanism whenever the exciton generation occurs near to the α-NPD interface.

To understand the conduction mechanism in p–i–n PHOLEDs, we fit the current–density vs. voltage (j–V) characteristics of the doped samples to the Shockley model for a p–n junction:

$$j(V) = j_s \exp\left[\frac{e(V - (j/j_T)^{1/m})}{nk_B T}\right] \quad (17.1)$$

where j_s is the saturation current density, k_B is the Boltzmann constant, T is the absolute temperature, e is the electron charge, and n is the ideality factor. Here, the second term in the exponential accounts for voltage drops across the EML and undoped buffer

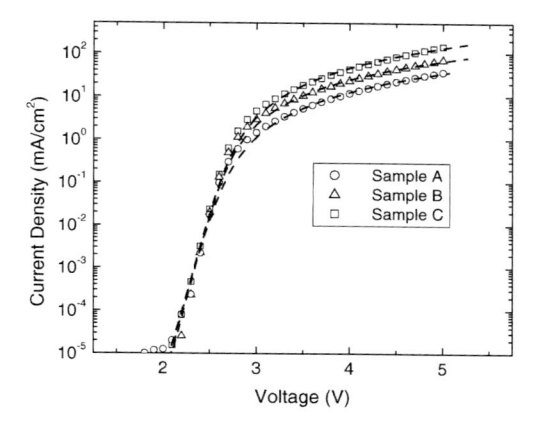

Figure 17.4 Current vs. voltage characteristics of three p–i–n CBP:Ir(ppy)$_3$ PHOLED structures. Also shown are fits to the data assuming a theory combining the Shockley equation [27] and trap-free space-charge limited currents as discussed in text.

layers of the device, which are expected to be space-charge limited, where $V_T = (j/j_T)^{1/m}$ is the voltage drop across these layers. For trap-free space-charge-limited currents, $m = 2$ and $j_T = 9/8[\varepsilon\mu/d^3]$, where ε is the permittivity of the layer, d is its thickness, and μ is the charge mobility. Note that $m > 2$ in the presence of traps [22, 23] or for injection-rather than bulk-limited transport [24].

Least-square fits to the forward-biased j–V characteristics for samples A–C in Fig. 17.4 show a reasonable fit to Eq. 17.1 over seven orders of magnitude in current density. The fits yield $n = 2.1 \pm 0.1$, a value typical for inorganic double-heterojunction LEDs [25]. This is characteristic of recombination of charge carriers within the intrinsic (i) region due to traps, energy wells formed by interfaces, or other recombination mechanisms to be considered in detail elsewhere. Further, we find that $m = 1.7 \pm 0.1$ for these devices, which is close to the value expected for trap-free space-charge-limited transport. The differences in the high current behavior of the three samples is mainly caused by different values for j_τ, 10 mA cm^{-2}, 20 mA cm^{-2}, and 42 mA cm^{-2}, for samples A, B, and C, respectively, with the increase corresponding to a decrease in total thicknesses of the undoped region for these samples. The significantly lower value for sample A indicates that free Li from the LiF leads to only a thin doped injection layer while the majority of the BPhen remains essentially undoped. To check the fitting parameters obtained from the high current regime, we infer a charge carrier mobility of $\mu \sim 10^{-5}$ cm^2 V^{-1}s^{-1}, which is consistent with values obtained for most organic amorphous materials. However, this represents a mean value, as the undoped region comprises an electron-transporting, a hole-transporting and a bipolar layer with different mobilities. Using 3×10^{-7} S cm^{-1} for the conductivity of doped m-MTDATA [13], the voltage drop over a 50 nm thick layer at $\sim$10 mA cm^{-2} is < 0.2 V, and thus is small compared to the total voltage (3 V) applied.

The applicability of the ideal-diode and space-charge-limited model over the device operating range spanning seven orders of magnitude in current, strongly suggests that the limiting mechanisms for carrier transport in doped organic devices are similar to those governing p–i–n junctions employing conventional inorganic semiconductors. That is, carrier diffusion to overcome the built-in energy barrier and subsequent recombination at low voltages, followed by space-charge-limited current due to transport across undoped layers at higher voltages, is similar to the characteristics of a Mott diode. This is in striking contrast to an injection-limited transport characteristic of the undoped layers typically comprising organic thin film devices such as OLEDs [24, 26].

In conclusion, we have demonstrated very-high-efficiency, conductivity doped p–i–n green-emitting electrophosphorescent OLEDs. By doping the hole- and electron-transport layers, low voltage ($\sim$3 V) operation is achieved, leading to a luminous power efficiency of > 20 lm W^{-1} at very high brightness. The doped PHOLEDs exhibit nearly ideal diode characteristics up to a current density that generates a brightness of $\sim$1000 cd m^{-2} at 3 V. At higher brightness, we find that the current is space-charge limited. These results suggest a new architecture for organic electronic devices, where charge confinement is provided by appropriate orbital offsets at hetero-junctions, along with the built-in potential due to the Fermi energy offsets of the p- and n-doped regions, and the total undoped region can be kept extremely thin. Nevertheless, doping allows for the use of thick, low resistance layers, with the potential for increasing yields in large-area organic electronic applications.

Experimental

Here the 50 nm thick (m-MTDATA) hole-transport layer (HTL) doped with 2 mol.% 2,3,5,6-tetrafluoro-7,7,8,8-tetracyano-quinodimethane (F_4-TCNQ) is first deposited following procedures described previously [3]. All films are deposited on a pre-cleaned indium tin oxide (ITO) coated glass substrate by vacuum thermal evaporation in a chamber with a base pressure of 10^{-7} torr. Doping is by co-evaporation from independent sources, whose effusion rates are monitored using quartz crystal microbalances. The ITO serves as a transparent anode contact with a sheet resistance of 20 Ωsq^{-1}.

The materials used in the devices are as follows: m-MTDATA = 4,4′,4″-tris-(3-methylphenylphenylamino)triphenylamine; F_4-TCNQ = 2,3,5,6-tetrafluoro-7,7,8,8-tetracyanoquinodimethane; Ir(ppz)$_3$ = tris(phenylpyrazole)iridium; CBP = N, N'-dicarbazolbiphenyl; Ir(ppy)$_3$ = tris(phenylpyridine)iridium; Bphen = 4,7-diphenyl-1,10-phenanthroline; α-NPD = N, N'-di(naphthalene-2-yl)-N, N'-diphenylbenzidine.

Acknowledgment

The authors thank the Air Force Office of Scientific Research, DARPA, Universal Display Corp., and the German Research Foundation, DFG, for support of this work.

References

1. M. A. Baldo, S. Lamansky, P. E. Burrows, M. E. Thompson, S. R. Forrest, *Appl. Phys. Lett.* 1999, **75**, 4.
2. C. Adachi, M. A. Baldo, S. R. Forrest, M. E. Thompson, *Appl. Phys. Lett.* 2000, **77**, 904.
3. C. Adachi, M. A. Baldo, M. E. Thompson, S. R. Forrest, *J. Appl. Phys.* 2001, **90**, 5048.
4. M. Ikai, S. Tokito, Y. Sakamoto, T. Suzuki, Y. Taga, *Appl. Phys. Lett.* 2001, **79**, 156.

5. M. Pfeiffer, A. Beyer, T. Fritz, K. Leo, *Appl. Phys. Lett.* 1998, **73**, 3202.
6. J. Blochwitz, M. Pfeiffer, T. Fritz, K. Leo, *Appl. Phys. Lett.* 1998, **73**, 729.
7. M. Pfeiffer, T. Fritz, J. Blochwitz, A. Nollau, B. Plönnigs, A. Beyer, K. Leo, *Adv. Solid State Phys.* 1999, **39**, 77.
8. X. Zhou, M. Pfeiffer, J. Blochwitz, A. Werner, A. Nollau, T. Fritz, K. Leo, *Appl. Phys. Lett.* 2001, **78**, 410.
9. J. Huang, M. Pfeiffer, A. Werner, J. Blochwitz, S. Liu, K. Leo, *Appl. Phys. Lett.* 2002, **80**, 139.
10. W. Gao, A. Kahn, *Appl. Phys. Lett.* 2001, **79**, 4040.
11. Y. Shirota, *Proc. SPIE—Int. Soc. Opt. Eng.* 1997, **3148**, 186.
12. Y. Shirota, *J. Mater. Chem.* 2000, **10**, 1.
13. J. Drechsel, M. Pfeiffer, X. Zhou, A. Nollau, K. Leo, *Synth. Met.* 2002, **127**, 201.
14. The hole density is estimated from the time of flight mobility of undoped m-MTDATA reported by Shirota et al. [12], and the conductivity of doped m-MTDATA reported by Drechsel et al. [13].
15. G. Parthasarathy, C. Shen, A. Kahn, S. R. Forrest, *J. Appl. Phys.* 2001, **89**, 4986.
16. C. Ganzorig, M. Fujihira, *Jpn. J. Appl. Phys.*, Part 2, 1999, **38**, L1348.
17. H. Heil, J. Steiger, S. Karg, M. Gastel, H. Ortner, H. van Seggern, *J. Appl. Phys.* 2001, **89**, 420.
18. J. Kido, T. Matsumoto, *Appl. Phys. Lett.* 1998, **73**, 2866.
19. G. Parthasarathy, C. Adachi, P. E. Burrows, S. R. Forrest, *Appl. Phys. Lett.* 2000, **76**, 2128.
20. X. Zhou, J. Blochwitz, M. Pfeiffer, A. Nollau, T. Fritz, K. Leo, *Adv. Funct. Mater.* 2001, **11**, 310.
21. M. E. Thompson, C. Adachi, unpublished.
22. M. A. Lampert, *Rep. Prog. Phys.* 1964, **27**, 329.
23. P. E. Burrows, Z. Shen, V. Bulovic, D. M. McCarty, S. R. Forrest, J. A. Cronin, M. E. Thompson, *J. Appl. Phys.* 1996, **79**, 7991.
24. M. A. Baldo, S. R. Forrest, *Phys. Rev. B* 2001, **64**, 5201.
25. L. A. Coldren, S. W. Corzine, *Diode Lasers and Photonic Integrated Circuits*, Wiley, New York 1995.
26. S. Barth, P. Müller, H. Riel, P. E Seidler, W. Riess, H. Vestweber, U. Wolf, H. Bässler, *Synth. Met.* 2000, **111**, 327.

27. The Shockley equation holds for both themionic emission theory and diffusion theory.
28. C. Adachi, S. Lamansky, M. A. Baldo, R. Kwong, M. E. Thompson, S. R. Forrest, *Appl. Phys. Lett.* 2001, **78**, 1622.
29. J. J. Brown, personal communication.
30. C. Adachi, R. C. Kwong, P. Djurovich, V. Adamovich, M. A. Baldo, M. E. Thompson, S. R. Forrest, *Appl. Phys. Lett.* 2001, **79**, 2082.

Chapter 18

Cyclometalated Ir Complexes in Polymer Organic Light-Emitting Devices

Sergey Lamansky, Peter I. Djurovich, Feras Abdel-Razzaq, Simona Garon, Drew L. Murphy, and Mark E. Thompson

Department of Chemistry, University of Southern California, Los Angeles, California 90089-0074, USA
met@usc.edu

Several new iridium based cyclometalated complexes were investigated as phosphorescent dopants for molecularly doped polymeric organic light-emitting diodes. Specifically, the complexes used in this study were iridium (III) bis(2-phenylpyridinato-$N,C^{2'}$) (acetylacetonate) [**ppy**], iridium (III) bis(7,8-benzoquinolinato-$N,C^{3'}$) (acetylacetonate) [**bzq**], iridium (III) bis(2-phenylbenzothiazolato-$N,C^{2'}$) (acetylacetonate) [**bt**], iridium (III) bis(2-(2'-naphthyl)benzothiazolato-$N,C^{2'}$) (acetylacetonate) [**bsn**] and iridium (III) bis(2-(2'-benzo[4,5-a]thienyl)pyridinato-$N,C^{3'}$) (acetylacetonate) [**btp**]. Single layer devices of doped polyvinylcarbazole: 2-(4-biphenyl)-5-(4-tert-butylphenyt)-1,3,4-oxadiazole give maximum external quantum efficiencies that varied from 3.5% for the **ppy** dopant to 0.4% for the **btp** dopant. Several different device

---

Reprinted from *J. Appl. Phys.*, **92**(3), 1633–1636, 2002.

heterostructure architectures were explored, and the best quantum efficiency of the devices reached 4.2% for the heterostructures.

18.1 Introduction

Organic light-emitting diodes (OLEDs) are a promising new technology for both thin-film display and lighting applications. Phosphorescent based OLEDs have been shown to perform significantly better than purely fluorescent based devices. The reason for such an enhanced performance involves the efficient utilization of both singlet and triplet excitons in the radiative processes in phosphorescence based OLEDs. By effectively harvesting all of the excitons formed on hole–electron recombination, it has been possible to prepare OLEDs with quantum efficiencies approaching 100% [1]. Studies on phosphorescence based OLEDs have been carried out with either vacuum-deposited small molecule based devices [2] or with phosphorescent doped polymer-based devices [3]. The polymer OLEDs prepared with phosphorescent dopants show enhanced efficiencies over their fluorescence based counterparts, however, there is still ample room for improvement in these phosphorescence based polymer OLEDs.

The best efficiencies reported for phosphorescent polymer based devices are for molecularly doped polymer OLEDs (mdpLEDs) using nonconjugated polymers [4]. The mdpLEDs are convenient devices to use in the study of phosphorescent light-emitting devices (LEDs) due to their relative ease of preparation and the ready availability of carrier transporting polymeric and molecular materials [5, 6]. These devices incorporate some of the best features of both small molecule and polymer-based devices in that the high degree of electronic variability of the small molecule building blocks is combined with the ease of fabrication of polymer OLEDs. Moreover, the high triplet energies available in small molecules can be replicated using nonconjugated polymers (e.g., polyvinylcarbazole, PVK) as a host matrix, making it possible to fabricate phosphorescent mdpLED with emission wavelengths that cover the entire visible spectrum. In mdpLEDs, the inert or carrier transporting polymer is doped with molecular materials that are chosen to show efficient emission and/or carrier transport. Electron transporting molecules are either dispersed in mdpLED polymer matrix to form a single-layer device or vacuum-deposited on top of the polymer film to form an organic/polymer heterostructure. PVK is one of the most commonly used polymers in mdpLEDs due to its excellent film-forming properties, high glass transition temperature, wide energy gap (emission is observed in the blue spectral region) and good hole mobility of $\sim 10^{-5}$ cm^2 V^{-1} s^{-1}, at the electric fields typical for OLEDs (10^6 V cm^{-1}) [7]. The best examples of both single-layer and heterostructured fluorescent based mdpLEDs show peak external quantum efficiency of 1.1%–1.8% [8], whereas examples of PVK-based mdpLEDs with phosphorescent dopants, i.e., Os [9], Au, and Cu (Ref. [10]) complexes, give electroluminescence efficiencies of 0.1% or less, values that are lower than would be expected from their photoluminescence quantum yields.

One key to the success of phosphorescent complexes in mdpLEDs is not only utilizing the triplet character of emission, but also properly matching the orbital energies of the dopants with those of the host compound(s). In order to achieve high efficiency for doped PVK-based devices, the phosphorescent complexes must not only serve as singlet and triplet energy traps, but also effectively trap holes or/and electrons. Proper device architecture is also critical in order to both improve the charge balance and control the exciton recombination zone with a high degree of fidelity. With proper dopant and materials choices, it has proven possible to make efficient white mdpLEDs [11]. Recently, efficient multilayer devices with *fac*-Ir(ppy)$_3$ (ppy = 2-phenylpyridinato-$N,C^{2'}$) doped into a hole transporting polymer layer have been reported [12]. As part of our continuing studies of electrophosphorescence we have also described molecularly doped polymer OLEDs incorporating phosphorescent Pt and Ir complexes and have achieved a peak efficiency of 2.8% [13]. This article discusses our efforts to expand the number of potential phosphorescent dopants for polymer OLEDs and to develop a greater insight into which parameters are important in optimizing the performance of these devices. Herein we describe on our current investigations of newly developed organometallic phosphorescent complexes as dopants in mdpLEDs [14]. These same dopants have proven to be very effective at improving the efficiency of small molecule based OLEDs [15]. By examining the effect of different architectures on the device performance and comparing heterostructures to single-layer polymer devices, we demonstrate that the efficiency of the phosphorescent polymer OLEDs can be substantially improved by building a heterostructure polymer device rather than a single-layer device, and that excitons can be efficiently confined within a desired emissive area through proper use of charge and exciton blocking materials.

18.2 Experiment

All solvents, poly(N-vinylcarbazole) (PVK), and 2-(4-biphenyl)-5-(4-*tert*-butylphenyl)-1,3,4-oxadiazole (PBD) were purchased from Sigma-Aldrich and used as received. Aluminum(III) *tris*(8- hydroxyquinolinate) (Alq$_3$) and bathocuproine (BCP) were purchased from Sigma-Aldrich and sublimed prior to use. The $C^\wedge N_2$Ir(acac) complexes studied here are shown in Fig. 18.1; the five complexes were synthesized and purified as described elsewhere [14, 15]. The PVK solutions were prepared by dissolving 100 mg of PVK, 40 mg of PBD, and 3–4 mg of a phosphorescent dopant in 7.5 mL of dry chloroform. Doped PVK thin films were spin coated from a chloroform solution onto patterned, solvent cleaned, and oxygen plasma treated indium tin oxide (ITO) coated glass substrates. The sheet resistance of the ITO substrates was 40 Ω/sq. Immediately prior to spinning, the solutions were filtered through a 0.2 µm Teflon filter. 3,4-polyethylenedioxythiophene-polystyrene sulfonate (PEDOT-PSS) films were deposited by spin coating from solution at 8000 rpm for 40 s followed by drying at 110°C for 15 min. For the PVK:PBD:dopant films, the chosen spin-coating conditions (3000 rpm, 40 s) led to 1300 $\pm$ 20 Å thick films (film thicknesses were measured ellipsometrically with a 633 nm HeNe laser, using a Rudolph Instruments model AutoEL II Ellipsometer). The polymer coated substrates were then loaded into the vacuum deposition system. The base pressure for both molecular and metal thin-film deposition was $< 10^{-5}$ Torr. Alq$_3$ and BCP films were prepared by resistively heating source boats and subliming the materials on to substrates held at room temperature. After the polymer and

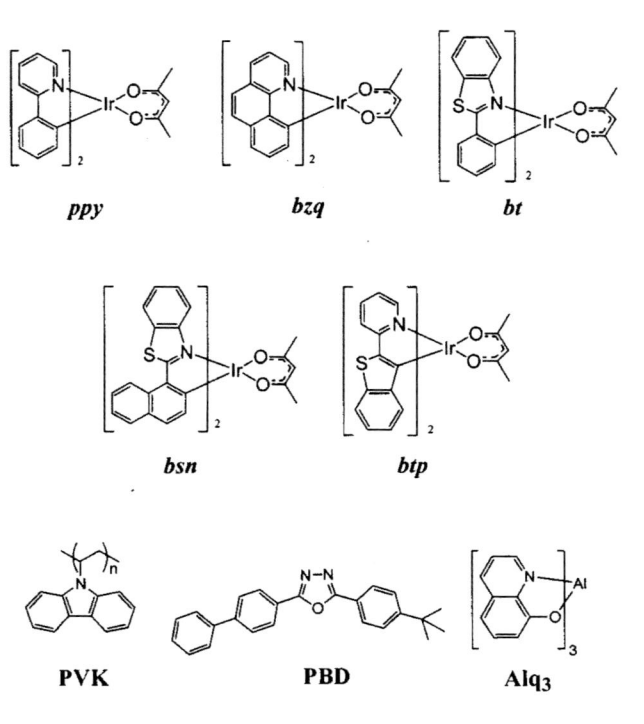

Figure 18.1 Chemical structures of compounds used in the mdpLEDs.

evaporated, molecular layers were deposited, 500 Å Mg:Ag (10:1 weight ratio), and cathodes were vacuum deposited onto the organic films followed by a 1000 Å thick film of Ag.

All device measurements were carried out in air at room temperature. The total time needed to perform the measurements was typically 30 min. While the devices showed marked degradation in air after several hours, there was no detectable change in performance over the 30 min. It took to collect the data. Device current–voltage and light intensity characteristics were measured using the LABVIEW™ program by National Instruments with a Keithley 2400 SourceMeter/2000 Multimeter coupled to a Newport 1835-C Optical Meter, equipped with a Si photocathode. Only light emitting from the front face of the device was collected and used in subsequent efficiency calculations. Electroluminescence spectra were recorded on a PTI QuantaMaster™ Model C-60SE spectrofluorometer.

18.3 Results and Discussion

Five different cyclometalated Ir complexes, $C^\wedge N_2\text{Ir(acac)}$, were investigated as dopants in polymer OLEDs. The $C^\wedge N_2\text{Ir(acac)}$ complexes used in this study were iridium (III) bis(2-phenylpyridinato-$N,C^{2'}$) (acetylacetonate) [**ppy**], iridium (III) bis(7,8-benzoquinolinato-$N,C^{3'}$) (acetylacetonate) [**bzq**], iridium (III) bis(2-phenylbenzothiazolato-$N,C^{2'}$) (acetylacetonate) [**bt**], iridium (III) bis[2-(2'-naphthyl)benzothiazolato-$N,C^{2'}$] (acetylacetonate) [**bsn**], and iridium (III) bis(2-(2'-benzo [4,5-a]thienyl) pyridinato-$N,C^{3'}$) (acetylacetonate) [**btp**], shown in Fig. 18.1. These five complexes have emission $\lambda_{\max}$ values that range from 516 to 612 nm (green to red in color, Table 18.1) and make good candidates for study as emissive dopants in polymer OLEDs due their high photoluminescent (PL) quantum yields and their greater solubility than the related tris-cyclometalated derivatives. Moreover, three of these complexes, **ppy**, **bt**, and **btp**, have been used as emissive dopants in vacuum-deposited OLEDs to give high efficiency green, yellow, and red electroluminescence (EL) respectively [15].

The molecularly doped polymer LEDs used here consisted of a hole conducting polymer matrix (PVK) doped with a phosphorescent $C^\wedge N_2\text{Ir(acac)}$ complex as an emitter. A well known electron transporting oxidiazole, 2-(4-biphenyl)-5-(4-tert-butylphenyl)-1,3,4-oxadiazole, PBD, was also added to the PVK in order to facilitate electron injection and transport in the doped PVK film. This combination of materials has been shown to lead to highly efficient devices using either fluorescent [6] and phosphorescent dopants [13]. In our previous studies using phosphorescent dopants, we found that the optimized structure and composition for this single-layer device is a 1300 Å thick film composed of PVK (67% by weight), PBD (30% by weight), and the phosphor at 2%–3% by weight, sandwiched between an ITO anode and a Mg–Ag (10:1) cathode [6]. This single-layer structure will be referred to as a type **I** device, Fig. 18.2. EL spectra from these type **I** mdpLEDs

Table 18.1 Characteristics of the single-layer ITO/PVK-PBD-$C^\wedge N_2$Ir(acac)/Mg-Ag/Ag LEDs with **ppy**, **bzq**, **bt**, **bsn**, and **btp** as emissive dopants

Dopant	Solution λ_{max} (nm), Φ_{PL} (%)	Turn-on voltage (V)	Peak extΦ_{EL} (%)	Peak current efficiency (Cd/A)	CIE coordinates X, Y	Maximum luminance $(Cd/cm)^2$
ppy	516, 35	8	3.4 ± 0.2	12.3 ± 0.5	0.28, 0.65	10^4
bzq	548, 27	8	1.9 ± 0.1	6.9 ± 0.3	0.41, 0.57	4500
bt	557, 26	9	1.6 ± 0.2	4.5 ± 0.4	0.50, 0.49	3100
bsn	606, 22	8	0.7 ± 0.1	1.0 ± 0.2	0.62, 0.34	1100
btp	612, 21	10	0.4 ± 0.1	0.5 ± 0.1	0.66, 0.33	500

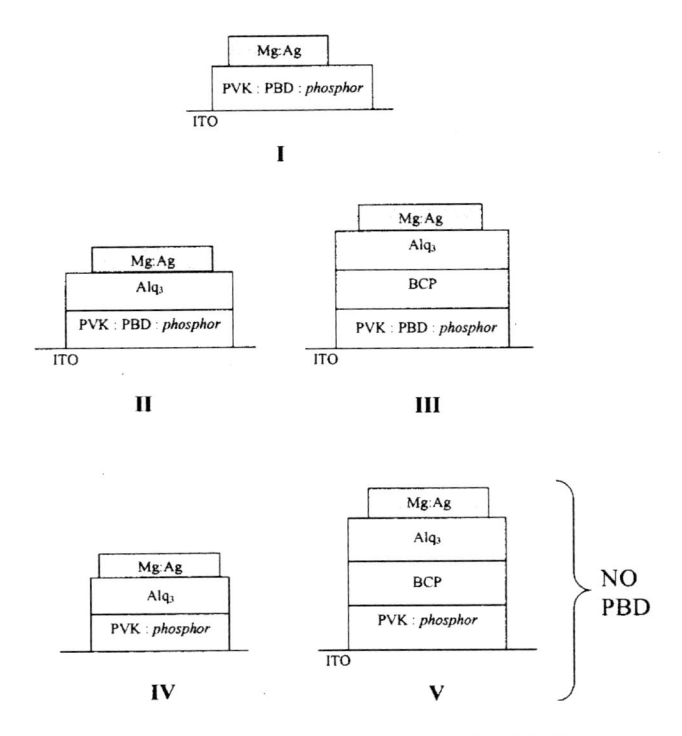

Figure 18.2 Schematic structures of mdpLEDs.

show no significant contribution from either the PVK host or PBD and closely match solution emission spectra of the corresponding $C^\wedge N_2$Ir(acac) complexes (Fig. 18.3). The absence of either PVK or PBD emission indicates that the excitons are exclusively localized on the phosphorescent dopant molecules prior to relaxation. The CIE (Commission Internationale d'Eclairage) coordinates for these devices fall between the National Television Standards Committee (NTSC) green and red standards [16], Table 18.1.

The performance parameters for these $C^\wedge N_2$Ir(acac) doped devices are summarized in Table 18.1. The maximum external quantum efficiency of **ppy** based devices ($\sim$3.3%) exceeds our previously reported efficiencies for single-layer PVK devices [13]. The mpdLEDs utilizing **bzq** and **bt** also give good external quantum efficiencies of 1.9% and 1.5%, respectively, while those with **bsn** and **btp** dopants give efficiencies below 1%. Turn-on voltages for these

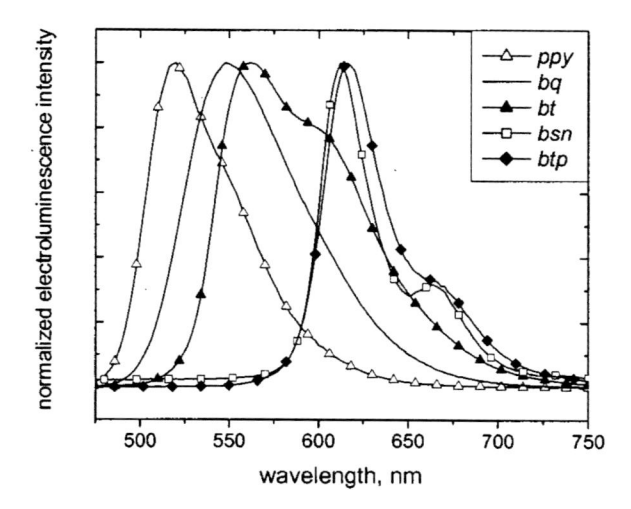

Figure 18.3 EL spectra of the single-layer ITO/PVK-PBD-$C^\wedge N_2$Ir(acac)/Mg–Ag/Ag LEDs with **ppy**, **bzq**, **bt**, **bsn**, and **btp** as phosphorescent dopants.

devices are typical for mdpLEDs, falling in the 7–10 V range. The turn-on voltages reported here correspond to the voltage required to produce a light output of 10^{-7} W/cm^2 (1 Cd/m^2 for **ppy**) from the device.

The quantum efficiencies of these devices decrease as the emission color shifts from green to red, i.e., η_{ext}, **ppy** > **bzq** > **bt** > **bsn** > **btp**. This trend may partly reflect the decreasing luminance quantum efficiency for the dopants since the trend parallels the solution PL yields, however, the drop in EL efficiency in going from green to red is greater than that seen for the PL efficiencies of the same complexes. For example, the **btp** doped device has an EL efficiency nearly ten times lower than a type **I ppy** device yet the difference in PL efficiencies between **btp** and **ppy** is only 1.5. Clearly, there is a loss process active in the longer wavelength type **I** mdpLEDs that is not active in the green devices.

Charge imbalance caused by either carrier leakage and/or charge trapping near the cathode has been invoked as one possible quenching mechanism in single-layer polymeric LEDs since the simple structure provides no barrier for either carrier or exciton confinement within the organic layer [17]. Consequently, the use of additional charge transporting layers in order to create a heterostructure has been shown to significantly enhance the efficiency of both small molecule [18] and polymer OLEDs [19]. Heterostructure OLEDs with polymer emissive layers exhibit higher EL efficiencies than single-layer devices, a result attributed to the enhanced charge injection for the heterostructures [19]. Therefore, several heterostructure OLEDs incorporating a PVK-$C^\wedge N_2$Ir(acac) emissive layer were prepared to determine if the device properties could be improved by carrier and exciton confinement. The heterostructures studied are shown schematically in Fig. 18.2 (**II–V**). The influence of PBD in these devices was also probed by examining devices with and without PBD (**IV** and **V**).

Figure 18.4 shows external quantum efficiency as a function of current density for single-layer PVK-PBD-**btp** mdpLED (type **I**) and heterostructured PVK-**btp**-PBD/Alq$_3$ (type **II**) and PVK-**btp**-PBD BCP/Alq$_3$ (type **III**) devices. The type **III** device was investigated because the use of a bathcuproine (BCP) layer to serve as an exciton and hole blocking barrier has been found to significantly improve the performance of small molecule based devices [20]. Both type **II** and **III** structures give exclusive **btp** emission. The type **II** device shows a significant improvement in the EL efficiency over the type **I** mdpLEDs, i.e., 2.7% and 0.4%, respectively (Tables 18.1 and 18.2). For the type **III** device, the addition of a BCP layer between the PVK

Table 18.2 Performance characteristics of the heterostructure mdpLEDs ITO/PVK-PBD-$C^\wedge N_2$Ir(acac)/Alq$_3$/Mg-Ag/Ag LEDs and ITO/PVK–PBD$C^\wedge N_2$Ir(acac)/BCP/Alq$_3$/Mg-Ag/Ag LEDs with **ppy**, **bt**, **bsn**, and **btp** as emitters

Dopant	Device type	Turn-on voltage (V)	Peak ext Φ_{EL} (%)	Peak current efficiency (Cd/A)
ppy	II	8	4.2 ± 0.2	15.4 ± 0.5
ppy	III	7	4.4 ± 0.2	16.2 ± 0.5
bt	II	8	2.6 ± 0.2	7.8 ± 0.5
bt	III	7	3.2 ± 0.3	9.5 ± 0.7
bsn	II	8	1.4 ± 0.2	2.0 ± 0.4
bsn	III	8	1.5 ± 0.2	2.1 ± 0.4
btp	II	8	2.7 ± 0.2	3.4 ± 0.5
btp	III	7	2.8 ± 0.2	3.5 ± 0.5

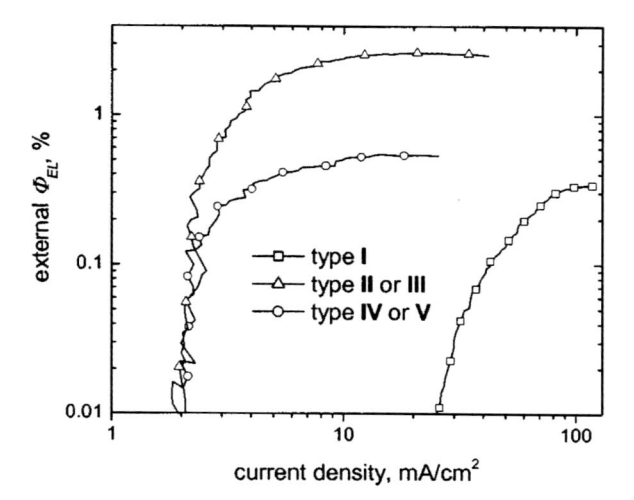

Figure 18.4 External quantum efficiency as a function of current density for single-layer PVK-PBD-**btp** LED (type **I** device) and heterostructures PVK-**btp**-PBD/(BCP)/Alq$_3$ (type **II** or **III** device) and PVK-**btp**/(BCP)/Alq$_3$ (type **IV** or **V** device).

and Alq$_3$ layers does not significantly alter the device properties since the efficiency versus current density plot for this device is identical to the one shown for the type **II** device (Fig. 18.4). The added Alq$_3$ or BCP/Alq$_3$ layers facilitate electron injection into the PVK layer, leading to improved carrier balance and thus higher quantum efficiency. The ratio of the EL efficiencies for **btp** and **ppy** doped heterostructure devices (**II** or **III**) of 2.7%:4.2% is close that of the PL efficiencies for the same dopants (21%:35%). This suggests that the losses in the two mdpLEDs are similar and not dependent on the nature of the phosphorescent dopant. The turn-on voltage is decreased for the heterostructure device consistent with more efficient electron injection into Alq$_3$ than into the PBD doped PVK film. Similarly, for the type **III** device, the turn-on voltage decreases only slightly whereas the quantum efficiency is unchanged (Table 18.2), thus indicating that both type **II** and **III** structures confine carriers and excitons to the emissive region.

All of the other dopants examined here also show significantly improved performance for the heterostructure devices (type **II** of

III) versus the single layer type I structure, albeit to a lesser extent than with **btp**. In all cases, the EL spectrum is due solely to the organometallic dopant. Devices based on **bt** are the only ones that show a difference in the efficiencies of type **II** and **III** devices. In these **bt** mdpLEDs, the type **III** device gives a higher efficiency than the type **II** device (3.2% and 2.6%, respectively). The hole blocking function of BCP may be responsible for the enhanced performance for **bt** based mdpLEDs. Since the highest occupied molecular orbital (HOMO) level for BCP (6.7 eV) [21] is 0.8 eV lower than that of Alq_3 (5.9 eV) [22] and its optical gap is nearly 0.8 eV higher than Alq_3, the added BCP layer of the type **III** device can provide a greater degree of hole and exciton blocking in this device than can occur in the Alq_3 only device (type **II**). We have previously shown that holes are trapped at phosphorescent dopants in doped PVK devices [13]. The oxidation potential for PVK (1.1 V versus Ag/AgCl) [23] is greater than for any of the $C^\wedge N_2 Ir(acac)$ complexes used here, suggesting that the HOMO level of PVK is below those of all of the dopants, leading to hole trapping at the dopant sites. Of the $C^\wedge N_2 Ir(acac)$ complexes used in this study, **bt** has the highest oxidation potential (1.0 V versus Ag/AgCl) [14] and will therefore have the lowest lying HOMO and best likelihood of injecting holes into Alq_3. Thus, the BCP layer in **bt** based devices is needed to prevent hole leakage into the Alq_3 layer. The other dopants used here have lower oxidation potentials than **bt** and presumably trap holes more efficiently, preventing significant hole leakage into the Alq_3 layer.

Heterostructured mdpLED devices were also prepared without PBD, i.e., the PVK layer had only added **btp** (type **IV** and **V** devices in Fig. 18.2), in order to examine the extent of carrier transport by the phosphor dopant. It has been previously observed that single-layer type **I** devices without PBD have both very high turn-on voltages and low efficiencies. In contrast, the heterostructured devices with no PBD have a relatively low turn-on voltages—comparable to the type **II** and **III** devices—and quantum efficiencies close to values obtained for a type **I** mdpLED (Fig. 18.4). The reason for the significant decrease in quantum efficiency relative to the type **II** and **III** devices can be seen in the spectra of the EL devices, Fig. 18.5. The EL spectrum of the type **IV** device displays emission bands originating

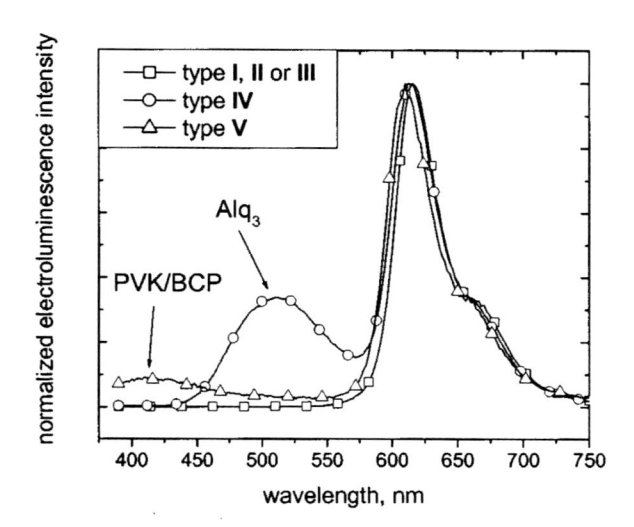

Figure 18.5 EL spectra of the heterostructures PVK-**btp**-PBD/Alq$_3$ (type II), PVK-**btp**-PBD/BCP/Alq$_3$ (type III), and PVK-**btp**/(BCP)/Alq$_3$ (type IV or V device).

from both **btp** and Alq$_3$, while the spectrum of the type **V** device shows a contribution from BCP. The type **IV** and **V** structures behave as simple hetero-structure devices with holes transporting through the doped PVK layer, electrons through the Alq$_3$ or BCP/Alq$_3$ layers and recombination occurring at or near the interface between the hole and electron transporting layers. For both devices the absence of PBD decreases both the efficiency of electron injection into, and electron conduction within, the PVK film. The result is that a significant fraction of the hole–electron recombination takes place in the Alq$_3$ or BCP layers. When this occurs, the benefit of using the phosphorescent dopant is lost and the device efficiency drops to a level comparable to mdpLEDs prepared with fluorescent dopants. It is important to note that if the electron transporting moiety (i.e., PBD) is present, excitons do not leak into the electron injecting Alq$_3$ layer even though the triplet energy of Alq$_3$ is lower than the triplet energies of all the other $C^\wedge N_2$Ir(acac) dopants [24].

The mpdLEDs have turn-on voltages that are higher than those observed for small molecule based devices prepared with the same phosphorescent dopants. High turn-on voltages are often observed for PVK based devices, most likely due to a significant barrier for injecting holes from the ITO anode into the PVK film [6]. One method to decrease the injection barrier is to coat the ITO electrode with a hole injecting film. The most commonly used hole injecting polymer is PEDOT-PSS. This material makes a good hole injecting contact for both polymer and small molecule based devices. Since PEDOTPSS is a water soluble material, the PVK solution does not dissolve or disrupt the PEDOT-PSS film, making it useful for preparing the multilayer type of devices described here. The PEDOT-PSS film is deposited and annealed on the ITO substrate, followed by spin coating of PVK film from an organic solvent. When a type **II btp** device was prepared on a PEDOT-PSS coated ITO substrate (i.e., ITO/PEDOT-PSS (500 Å)/PVK-**btp**-PBD/Alq$_3$/Mg–Ag) the turn-on voltage decreased by roughly 2 V while the quantum efficiency was not significantly changed. This has been observed for other mpdLEDs as well [11]. The quantum efficiency was not affected by the presence of the PEDOT-PSS film, indicating that the type **II** heterostructure leads to good carrier balance, even without the PEDOT-PSS injection layer.

18.4 Conclusions

Molecularly doped polymer LEDs based on phosphorescence show high efficiency when efficient triplet emitters such as the $C^{\wedge}N_2$Ir(acac) complexes are used and a proper device architecture is provided. Generally, multilayer PVK-dopants ETL devices perform more efficiently than the corresponding single-layer PVK-dopants LEDs and give external quantum efficiencies of 2.0%–4.4% with the most of the phosphors studied here. To maximize the efficiency of such devices electron-transporting materials should be used not only in the form of a thin vacuum deposited layer (forming the heterostructure), but also as a component blended in the PVK thin film. Moreover, the hole-blocking ability of BCP in the PVK-based electrophosphorescent devices allows for further LED architecture modifications and improvements. While PVK has been examined extensively in mdpLED structures, it may not be the optimal polymer for this application. A useful approach to improve the performance of mdpLEDs is to explore other polymeric host materials with phosphorescent dopants. Toward that end, we are currently investigating a number of different conjugated and nonconjugated polymers as host polymers for mdpLEDs.

Acknowledgments

This work was supported by the Universal Display Corporation, Defense Advanced Research Projects Agency, the Airforce Office of Scientific Research through Multidisciplinary University Research Institution (MURI) program, and the National Science Foundation.

References

1. C. Adachi, M. A. Baldo, S. R. Forrest, and M. E. Thompson, *Appl. Phys. Lett.* **77**, 904 (2000); *J. Appl. Phys.* **90**, 5048 (2001).
2. M. A. Baldo, D. F. O'Brien, Y. You, A. Shoustikov, S. Sibley, M. E. Thompson, and S. R. Forrest, *Nature (London)* **395**, 151 (1998); M. A. Baldo, S. Lamansky, P. E. Burrows, M. E. Thompson, and S. R. Forrest, *Appl. Phys. Lett.* **75**, 4 (1999); Z. Hong, C, Liang, R. Li, W. Li, D. Zhao, D. Fan, D. Wang, B. Chu, F. Zang, L.-S. Hong, and S.-T. Lee, *Adv. Mater.* **13**, 1241 (2001); H. Z. Xie, M. W. Liu, O. Y. Wang, X. H. Zhang, C. S. Lee, L. S. Hung, S. T. Lee, P. F. Teng, H. L. Kwong, H. Zheng, and C.-M. Che, *Adv. Mater.* **13**, 1245 (2001).
3. V. Cleave, G. Yahioglu, P. Le Barny, R. H. Friend, and N. Tessler, *Adv. Mater.* **11**, 285 (1999); M. D. McGehee, T. Bergstedt, T. Zhang, A. P. Saab, M. B. O'Regan, G. C. Bazan, V. I. Srdanov, and A. J. Heeger, *Adv. Mater.* **11**, 1349 (1999); A. Wu, D. Yoo, J. K. Lee, and M. F. Rubner, *J. Am. Chem. Soc.* **121**, 4883 (1999); T.-F. Guo, S.-C. Chang, Y. Yang, R. C. Kwong, and M. E. Thompson, *Org. Electronics* **1**, 15 (2000); D. F. O'Brien, C. Giebler, R. B. Fletcher, A. J. Cadby, L. C. Palilis, D. G. Lidzey, P. A. Lane, D. D. C. Bradley, and W. Blau, *Synth. Met.* **116**, 379 (2001); S. Lamansky, R. C. Kwong, M. Nugent, P. I. Djurovich, and M. E. Thompson, *Org. Electron.* **2**, 53 (2001); R. W. T. Higgins, A. P. Monkman, H.-G. Nothofer, and U. Scherf, *J. Appl. Phys.* **91**, 99 (2002); Y. Kawamura, S. Yanagida, and S. R. Forrest (unpublished).

4. X. Jiang, A. K.-Y Jen, D. Huang, G. D. Phelan, T. M. Londergan, and L. R. Dalton, *Synth. Met.* **125**, 331 (2002); N. A. H. Male, O. V. Salata, and V. Christou, *ibid* **126**, 7 (2002).
5. J. Kido, M. Kohda, K. Okuyama, and K. Nagai, *Appl. Phys. Lett.* **61**, 761 (1992).
6. C. Wu, J. C. Sturm, R. A. Register, J. Tian, E. P. Dana, and M. E. Thompson, *IEEE Trans. Electron Devices* **44**, 1269 (1997).
7. W. D. Gill, *J. Appl. Phys.* **43**, 5033 (1972).
8. J. S. Huang, H. F. Zhang, W. J. Tian, J. Y. Hou, Y. G. Ma, J. C. Shen, and S. Y. Liu, *Synth. Met.* **87**, 105 (1997); R. Zhang, H. Zheng, and J. Shen, *ibid.* **105**, 49 (1999); S. E. Shaheen, B. Kippelen, N. Peyghambarian, J.-F. Wang, J. D. Anderson, E. A. Mash, P. A. Lee, N. R. Armstrong, and Y. Kawabe, *J. Appl. Phys.* **85**, 7939 (1999).
9. Y. Ma, H. Zhang, J. Shen, and C.-M. Che, *Synth. Met.* **94**, 245 (1998).
10. Y. Ma, C.-M. Che, H. Chao, X. Zhou, W. Chan, and J. Shen, *Adv. Mater.* **11**, 852 (1999).
11. Y. Kawamura, S. Yanagida, and S. R. Forrest, *J. Appl. Phys.* **92**, 87 (2002).
12. M. J. Yang and T. Tsutsui, *Jpn. J. Appl. Phys.*, Part 2 **39**, L828 (2000); C. L. Lee, K. B. Lee, and J. J. Kim, *Appl. Phys. Lett.* **77**, 2280 (2000).
13. S. Lamansky, R. C. Kwong, M. Nugent, P. I. Djurovich, and M. E. Thompson, *Org. Electron.* **2**, 53 (2001).
14. S. Lamansky, P. Djurovich, D. Murphy, F. Abdel-Razzaq, M. Bortz, I. Tsyba, B. Mui, R. Bau, and M. E. Thompson, *Inorg. Chem.* **40**, 1704 (2001).
15. S. Lamansky, P. Djurovich, D. Murphy, F. Abdel-Razzaq, H.-E. Lee, C. Adachi, P. E. Burrows, S. R. Forrest, and M. E. Thompson, *J. Am. Chem. Soc.* **123**, 4304 (2001).
16. J. Whitaker, *Electronic Displays: Technology, Design, and Applications* (McGraw–Hill, New York, 1994), p. 92.
17. R. H. Friend, R. W. Gymer, A. B. Holmes, J. H. Burroughes, R. N. Marks, C. Taliani, D. D. C. Bradley, D. A. Dos Santos, J. L. Brédas, M. Lögdlund, and W. R. Salaneck, *Nature (London)* **397**, 121 (1999).
18. C. Adachi, T. Tsutsui, and S. Saito, *Appl. Phys. Lett.* **57**, 531 (1990); Z. Y. Xie, L. S. Hung, and S. T. Lee, *ibid.* **79**, 1048 (2001).
19. P. K. H. Ho, J.-S. Kim, J. H. Burroughes, H. Becker, S. F. Y. Li, T. M. Brown, F. Cacialli, and R. H. Friend, *Nature (London)* **404**, 481 (2000).
20. D. F. O'Brien, M. A. Baldo, M. E. Thompson, and S. R. Forrest, *Appl. Phys. Lett.* **74**, 442 (1999).

21. Y. Kijima, N. Asai, and S. Tamura, *Jpn. J. Appl. Phys.*, Part 1 **38**, 5274 (1999).
22. A. Schmidt, M. L. Anderson, and N. R. Armstrong, *J. Appl. Phys.* **78**, 5619 (1995).
23. J. D. Anderson, E. M. McDonald, P. A. Lee, M. L. Anderson, E. L. Ritchie, H. K. Hall, T. Hopkins, E. A. Mash, J. Wang, A. Padias, S. Thayumanavan, S. Barlow, S. R. Marder, G. E. Jabbour, S. Shaheen, B. Kippelen, N. Peyghambarian, R. M. Wightman, and N. R. Armstrong, *J. Am. Chem. Soc.* **120**, 9646 (1998).
24. R. J. Curry and W. P. Gillin, *J. Appl. Phys.* **88**, 781 (2000).

Chapter 19

High Efficiency Single Dopant White Electrophosphorescent Light Emitting Diodes

Vadim Adamovich,[a] Jason Brooks,[a] Arnold Tamayo,[a] Alex M. Alexander,[a] Peter I. Djurovich,[a] Brian W. D'Andrade,[b] Chihaya Adachi,[c] Stephen R. Forrest,[b] and Mark E. Thompson[a]

[a] Department of Chemistry, University of Southern California, Los Angeles, California 90089, USA
[b] Center for Photonics and Optoelectronic Materials (POEM), Princeton Materials Institute (PMI), Department of Electrical Engineering, Princeton University, Princeton, New Jersey 08544, USA
[c] Department of Photonics Materials Science, Chitose Institute of Science & Technology, 758-65 Bibi, Chitose 066-8655, Japan
met@usc.edu, forrest@princeton.edu

Efficient white electrophosphorescence has been achieved with a single emissive dopant. The dopant in these white organic light emitting diodes (WOLEDs) emits simultaneously from monomer and aggregate states, leading to a broad spectrum and high quality white emission. The dopant molecules are based on a series of platinum(II) [2-(4,6-difluorophenyl)pyridinato-$N,C^{2'}$] β-diketonates. All of the dopant complexes described herein have identical photophysics in

---

Reprinted from *New J. Chem.*, **26**, 1171–1178, 2002.

Electrophosphorescent Materials and Devices
Edited by Mark E. Thompson
Text Copyright © 2002 The Royal Society of Chemistry and the Centre National de la Recherche Scientifique
Layout Copyright © 2024 Jenny Stanford Publishing Pte. Ltd.
ISBN 978-981-4877-34-3 (Hardcover), 978-1-003-08872-1 (eBook)
www.jennystanford.com

dilute solution with structured blue monomer emission (λ_{max} = 468, 500, 540 nm). A broad orange aggregate emission ($\lambda_{max} \approx$ 580 nm) is also observed, when doped into OLED host materials. The intensity of the orange band increases relative to the blue monomer emission, as the doping level is increased. The ratio of monomer to aggregate emission can be controlled by the doping concentration, the degree of steric bulk on the dopant and by the choice of the host material. A doping concentration for which the monomer and excimer bands are approximately equal gives an emission spectrum closest to standard white illumination sources. WOLEDs have been fabricated with doped CBP and mCP luminescent layers (CBP = N,N'-dicarbazolyl-4,4'-biphenyl, mCP = N,N'-dicarbazolyl-3,5-benzene). The best efficiencies and color stabilities were achieved when an electron/exciton blocking layer (EBL) is inserted into the structure, between the hole transporting layer and doped CBP or mCP layer. The material used for an EBL in these devices was *fac*-tris(1-phenylpyrazolato-$N,C^{2'}$)iridium(III). The EBL material effectively prevents electrons and excitons from passing through the emissive layer into the hole transporting NPD layer. CBP based devices gave a peak external quantum efficiency of 3.3 ± 0.3% (7.3 ± 0.7 lm W^{-1}) at 1 cd m^{-2}, and 2.3 ± 0.2% (5.2 ± 0.3 lm W^{-1}) at 500 cd m^{-2}. mCP based devices gave a peak external quantum efficiency of 6.4% (12.2 lm W^{-1}, 17.0 cd A^{-1}), CIE coordinates of 0.36, 0.44 and a CRI of 67 at 1 cd m^{-2} (CIE = Commission Internationale de l'Eclairage, CRI = color rendering index). The efficiency of the mCP based device drops to 4.3 ± 0.5% (8.1 ± 0.6 lm W^{-1}, 11.3 cd A^{-1}) at 500 cd m^{-2}, however, the CIE coordinates and CRI remain unchanged.

19.1 Introduction

The efficiencies and color purities of monochromatic organic light emitting diodes (OLEDs) have improved markedly over the last five years, leading to devices with close to 100% internal quantum efficiencies [1]. While the progress toward achieving high efficiency in white OLEDs (WOLEDs) has been slower, there have nonetheless been significant advances in this area as well. The motivation for improving WOLEDs is the need for novel lighting sources that are less expensive and more efficient alternatives to conventional incandescent and fluorescent illumination sources [2]. WOLEDs make attractive candidates as future illumination sources for several reasons, including compact size, the suitability for fabrication on flexible substrates [3], low operating voltages and good power efficiencies.

Most WOLEDs utilize emission from several different colored emitters, such that the combined output covers the visible spectrum uniformly. While WOLEDs with less than three distinct emitters have been reported, the most common approach in WOLEDs is to use three separate emitters, that is, blue, green and red. It has been demonstrated that three emitters can be mixed together in a single layer to achieve the desired white emission [4]. However, this approach is problematic because energy readily transfers from the higher energy blue dye to the green dye and from the green dye to the red dye. Therefore, careful adjustment of the concentration of each dye is required to achieve a well-balanced emission color. A solution to this energy transfer problem is to segregate the dyes into different layers. Efficient WOLEDs have been prepared using this stacked concept with both fluorescent [5] and phosphorescent emitters [6]. More simplified structures have also been described, which use dual component fluorescent blue and orange emitters doped into separate layers [5c, 7]. While stacking the emitters eliminates these energy transfer problems, the device architecture can become significantly more complicated due to difficulties in achieving balanced carrier recombination and exciton localization in each of the separate emitting layers.

A further simplification of the device structure can be achieved by using an exciplex as the emitting species. An exciplex is a metastable complex formed by associative excited state interactions between two different molecules. Exciplexes are known to emit over a wide spectral range and have been used in several WOLED architectures [8–10]. Although the use of exciplex emission simplifies the structure of white OLEDs, the efficiency of these devices remains very low, typically no more than 0.6 lm W^{-1}, due to the inherent low luminescence efficiency of the exciplexes described in the literature [8–10].

Recently, we have reported a new approach to the fabrication of WOLEDs that combines the monomer and excimer phosphorescence of two emitters co-doped into a single emissive layer [11]. This approach has led to a significant simplification of the device structure without a loss in efficiency. Continuing this study, we now report the achievement of well balanced white emission from an emissive layer with only a single luminescent dopant, which emits simultaneously from monomer and aggregate states. Single dopant WOLEDs give voltage independent white emission with external efficiencies as high as 6.4% (*ca.* 12 lm W^{-1}). In addition to reducing the complexity of the device structure, single dopant WOLEDs may also solve the problems associated with differential dopant aging. WOLEDs utilizing multiple dopants may change color over time, due to differences in the degradation rates of each dopant. It is anticipated that a single dopant WOLED would not suffer from this drawback.

The first section in this paper details the photoluminescent analysis of doped thin films, so as to optimize the emission ratios of the monomer and aggregate states in the emissive layer. It is important to note that doping concentrations of > 5 wt% are typically required to efficiently quench the host luminescence and achieve good carrier transport in phosphorescent OLEDs [12–14]. Therefore, one requirement for a single dopant WOLED is to have the proper monomer/aggregate emission ratio for white emission at a high enough doping concentration that will also give a good photon-to-electron quantum efficiency. This optimization was achieved by modifying the steric bulk of the dopant molecule and by changing the host matrix material. Both approaches affected the degree of association of the dopant in the emissive layer and hence the ratio of the emissive states. The second section in this paper demonstrates efficient single dopant WOLEDs. An architecture is used that utilizes a novel electron blocking material, *fac*-tris(1-phenylpyrazole)iridium(III), to confine exciton recombination to the doped luminescent layer. The performance characteristics of these devices are described.

19.2 Experimental

19.2.1 Equipment

Absorption spectra were recorded on an AVIV Model 14DS-UV-Vis-IR spectrophotometer (re-engineered Cary 14) and corrected for background due to solvent absorption. Emission spectra (photoluminescence and electroluminescence) were recorded on a PTI QuantaMaster™ Model C-60SE spectrofluorometer, equipped with a 928 PMT detector and corrected for detector response. Phosphorescence lifetime measurements were performed on the same fluorimeter equipped with a microsecond Xe flash lamp and were limited to lifetimes > 2 ms. NMR spectra were recorded on Bruker AC 250 MHz or AM 360 MHz instruments. Solid probe MS spectra were taken with a Hewlett Packard GC/MS instrument with electron impact ionization and model 5873 mass selective detector. Elemental analyses were performed by the Microanalysis Laboratory at the University of Illinois, Urbana-Champaign.

19.2.2 Synthesis

Solvents and reagents were purchased form Aldrich Chemical Company. The reagents were of the highest purity available and used as received.

The Pt complexes (**1–4**) were prepared by a procedure that is detailed elsewhere [15]. A general description is given here. The ligand 2-(2,4-difluorophenyl)pyridine (F_2ppy) was prepared by Suzuki coupling of 2,4-difluorophenylboronic acid and 2-bromopyridine (Aldrich) [16]. The Pt(II) μ-dichloro-bridged dimer [(F_2ppy)$_2$Pt(m-Cl)$_2$Pt(F_2ppy)$_2$] was prepared by a modified method of Lewis [17]. The dimer was treated with 3 equiv. of the chelating diketone ligand and 10 equiv. of Na_2CO_3. 2,6-Dimethyl-3,5-heptanedione and 6-methyl-2,4-heptanedione were purchased from TCI. 3-Ethyl-2,4-pentandione was purchased from Aldrich. The solvent was removed under reduced pressure and the compound purified chromatographically. The product was recrystallized from dichloromethane–methanol and then sublimed. The characterization data (NMR and mass spectra, as well as CHN analysis) for **1** matched those reported

[15]. The characterization data for compounds **2**, **3**, and **4** are given below. Based on NMR spectra, compound **4** was determined to be a mixture of 2 isomers in an approximate 1:1 ratio due to the asymmetry of the diketonate ligand.

fac-Ir(ppz)$_3$ was prepared from Ir(acac)$_3$ by a procedure analogous to the one reported for *fac*-Ir(ppy)$_3$ [18]. Ir(acac)$_3$ (3.0 g) and 1-phenylpyrazole (3.1 g) were dissolved in 100 ml glycerol and refluxed for 12 h under an inert atmosphere. After cooling the product was isolated by filtration and washed with several portions of distilled water, methanol, ether and hexanes and then vacuum dried. The crude product was then sublimed in a temperature gradient of 220–250°C to give a pale yellow product (yield 58%).

The synthesis of mCP was based on a known literature procedure using palladium-catalyzed cross coupling of aryl halides and arylamines [19].

[2-(4′,6′-Difluorophenyl)pyridinato-*N*,*C*$^{2'}$]platinum(II) (2,6-dimethyl-3,5-heptanedionato-*O*,*O*), 2. ^{1}H NMR (250 MHz, CDCl$_3$): ppm 9.01 (d, J = 5.8 Hz, 1H), 7.96 (d, J = 7.8 Hz, 1H), 7.82 (dd, J = 7.5, 6.8 Hz, 1H), 7.13 (m, 2H), 6.56 (ddd, J = 12.3, 9.2, 2.7 Hz, 1H), 5.50 (s, 1H), 2.55 (hep, J = 7.2 Hz, 1H), 2.54 (hep, J = 6.8 Hz, 1H), 1.20 (d, J = 6.8 Hz, 6H), 1.19 (d, J = 6.8 Hz, 6H). Anal. calcd for C$_{20}$H$_{21}$F$_2$NO$_2$Pt: C 44.45, H 3.92, N 2.59; found: C 44.39, H 3.87, N 2.68.

[2-(4′,6′-Difluorophenyl)pyridinato-*N*,*C*$^{2'}$]platinum(II) (3-ethyl-2,4-pentanedionato-*O*,*O*), 3. ^{1}H NMR (250 MHz, CDCl$_3$): ppm 8.95 (d, J = 5.1 Hz, 1H), 7.94 (d, J = 8.5 Hz, 1H), 7.80 (dd ,J = 8.2, 7.2 Hz, 1H), 7.1 (m, 2H), 6.54 (ddd, J = 11.9, 9.2, 2.4 Hz, 1H), 2.38 (q,J = 6.8 Hz, 2H), 2.14 (s, 3H), 2.13 (s, 3H), 1.07 (t, J = 7.5 Hz, 3H). Anal. calcd for C$_{18}$H$_{17}$F$_2$NO$_2$Pt: C 42.19, H 3.34, N 2.73; found: C 42.15, H 3.26, N 2.77.

[2-(4′,6′-Difluorophenyl)pyridinato-*N*,*C*$^{2'}$]platinum(II) (6-methyl-2,4-heptanedionato-*O*,*O*), 4. ^{1}H NMR (250 MHz, CDCl$_3$): ppm 8.98 (two overlapped d, J = 5.8 Hz, 1H), 7.95 (d ,J = 7.8 Hz, 1H), 7.81 (dd, J = 7.5, 7.5 Hz, 1H), 7.10 (m, 2H), 6.56 (ddd, J = 11.9, 9.2, 2.0 Hz, 1H), 5.45 (s, 1H), 2.14 (s, 2H), 2.12 (m, 1H), 2.00 (two overlapped s, 3H), 0.97, 0.96 (two overlapped d, J = 6.5 Hz,

6H). Anal. calcd for $C_{19}H_{19}F_2NO_2Pt$: C 43.35, H 3.64, N 2.66; found: C 43.41, H 3.65, N 2.77.

fac-tris(1-Phenylpyrazolato-*N,C2'*)iridium(III), (Irppz). ^{1}H NMR (360 MHz, CDCl$_3$): ppm 7.94 (d, J = 2.4 Hz, 3H), 7.18 (dd, J = 7.8, 1.0 Hz, 3H), 6.97 (d, J = 2.0 Hz, 3H), 6.91 (dd, J = 7.8, 7.3, 1.5 Hz, 3H), 6.85 (ddd, J = 7.8, 1.9, 1.5 Hz, 3H), 6.77 (dd, J = 7.3, 1.5 Hz, 3H), 6.35 (t, J = 2.4 Hz, 3H). Anal. calcd for $C_{27}H_{21}N_6Ir$: C 52.16, H 3.40, N 13.52; found: C 52.04, H 3.39, N 13.52.

3,5-Bis(*N*-carbazolyl)benzene (mCP). ^{1}H NMR (250 MHz, CDCl$_3$): ppm 8.16 (d, J = 7.8 Hz, 4H), 7.85 (dd, J = 8.6, 7.5 Hz, 1H), 7.82 (t, J = 1.8 Hz, 1H), 7.70 (ddd, J = 10.1, 7.9, 2.0 Hz, 2H), 7.54 (d, J = 8.4 Hz, 4H), 7.44 (t, J = 7.6 Hz, 4H), 7.31 (t, J = 7.7 Hz, 4H). Anal. calcd for $C_{30}H_{20}N_2$: C 88.21, H 4.93, N 6.86; found: C 88.3, H 4.91, N 7.01.

19.2.3 Estimation of HOMO and LUMO Energies

The HOMO energies for NPD, CBP, Irppz, BCP, Alq$_3$ and mCP were determined by photoelectron spectroscopy, using an AC-1 (Riken Keiki Co., Japan) UV photoelectron spectrometer. The values determined here are consistent with literature values for NPD, CBP, BCP and Alq$_3$ [20]. The LUMO energies were estimated by using the optical energy gap of each material to approximate its carrier gap. The low energy edge of the absorption spectra for Irppz and mCP (370 and 350 nm, respectively) were used as their optical gaps. The HOMO and LUMO values for **1** were taken from the literature [11.

19.2.4 OLED Fabrication

Prior to device fabrication, ITO on glass was patterned as 2 mm wide stripes with a resistivity of 20 Ω $\square^{-1}$. The substrates were cleaned by sonication in soap solution, rinsed with deionized water, boiled in trichloroethylene, acetone and ethanol for 3–4 min in each solvent and dried with nitrogen. Finally, the substrates were treated with UV ozone for 10 min.

Organic layers were deposited sequentially by thermal evaporation from resistively heated tantalum boats onto the substrate at a rate of 2.5 Å s^{-1}. The base pressure at room temperature was 3–4 × 10^{-6} torr. The rate for single component layers was controlled using one crystal monitor that was located near the substrate. A second crystal monitor located near the evaporation source of the dopant was used to control the rate of dopant molecule incorporation into the host matrix. The additional monitor was screened from the host evaporation, allowing for increased precision of the dopant concentration.

After organic film deposition, the chamber was vented and a shadow mask with a 2 mm wide stripe was put onto the substrate perpendicular to the ITO stripes. A cathode consisting of 10 Å LiF followed by 1000 Å of aluminum was deposited at a rate of 0.3–0.4 Å s^{-1} for LiF and 3–4 Å s^{-1} for aluminum. OLEDs were formed at the 2 × 2 mm squares where the ITO (anode) and Al (cathode) stripes intersected.

The devices were tested in air within 2 h of fabrication. Device current–voltage and light intensity characteristics were measured using the LabVIEW™ program by National Instruments with a Keithley 2400 SourceMeter/2000 Multimeter coupled to a Newport 1835-C Optical Meter, equipped with a UV-818 Si photocathode. Only light emitting from the front face of the WOLED was collected and used in subsequent efficiency calculations. Electroluminescence spectra were recorded on a PTI QuantaMaster™ Model C-60SE spectrofluorometer and corrected for detector response.

19.3 Results and Discussion

19.3.1 Optimizing the Emissive Layer

The solid state and solution photophysics of luminescent square planar platinum(II) complexes, chelated with aromatic ligands, have been studied in detail [15, 21–26]. In dilute solution, these complexes generally emit as isolated molecules or monomers from a mixed MLCT/^{3}LC excited state (MLCT = metal to ligand charge transfer, LC = ligand centered). However, as the concentration of the Pt complexes in solution is raised, monomer emission decreases and a broad, lower energy emission band is observed. The low energy emission is typically the result of intermolecular stacking interactions, leading to the formation of either excimers or metal-metal bound oligomers. Excimers involve the formation of excited state dimers [23, 25, 26]. An excimer is only bound in the excited state and rapidly dissociates to two discrete molecules after relaxation to the ground state. In contrast, metal-metal bound oligomers are stable in the ground state, typically involving the formation of weak Pt $\cdots$ Pt bonds. Oligomers of this type have been observed both in solution [22, 23, 26] and in the solid state [22–24, 26, 27]. The oligomeric structures seen in crystallographic studies can range in length from dimers to continuous chains [22, 23, 26–28]. Emission from these oligomeric structures is attributed to a $^3[\pi^* \rightarrow d\sigma^*]$ (MMLCT: metal-metal to ligand charge transfer) transition. The emission spectra observed from both excimer and oligomer states are typically broad and unstructured, falling at lower energy than emission from the monomeric species.

Determining the electronic origin for a low energy transition in the solid state for a single compound can often be achieved by correlating the solid state photophysical behavior with the crystal structure. Unfortunately, differentiating between excimer and oligomer excited states in solution or doped thin films is problematic. The oscillator strengths of the MMLCT absorption transitions are typically very low, making them difficult to resolve from the more intense MLCT transitions. The phosphorescent Pt dopants used here may well be involved in both $\pi-\pi$ stacking and metal-metal interactions in the doped films, leading to contributions from both excimeric and oligomeric excited states in the emission spectra. For the present study we will not differentiate between excimer and oligomer states, since we do not have conclusive evidence to show if either transition is more important for the dopants examined here. Hereafter the term "aggregate" will be used to describe both excited state (excimer) and ground state (oligomer) aggregated species.

Modifying the Pt dopant. The synthesis and characterization of the $(C^\wedge N)Pt(O^\wedge O)$ dopants used in this study have been described recently [15]. It has been shown that the alteration of the alkyl groups on the β-diketonate ligand $(O^\wedge O)$ does not affect the solution photophysics of the complex. We examined the role that alkyl substituents on the β-diketonate ligand play in controlling the monomer/aggregate emission ratio for two reasons. First, the separation between the square-planar emitting molecules can be adjusted by increasing the degree of steric bulk of the Pt complex. A Pt complex more sterically encumbered than the planar derivative, **1**, should be less prone to form aggregates that are able to electronically interact. Hence, these complexes should require a higher doping concentration to achieve balanced monomer/aggregate emission. Second, larger alkyl groups should increase the "solubility" of the complex, thereby allowing for more uniform dispersal in the host matrix. The series of Pt complexes (**1–4**, Fig. 19.1) was examined to find the best compound for balanced emission.

Compounds **2–4** have greater steric bulk than compound **1**. Complexes **2** and **3** are symmetric, whereas the β-diketonate ligand of **4** is asymmetric. The resulting Pt complex is a mixture of the two inseparable isomers, determined by ^{1}H NMR to be in an approximate 1:1 ratio, only one of which is shown in Fig. 19.1. The alternate isomer has the β-diketonate ligand reversed (i.e., methyl group *trans* to the pyridyl group). The narrow emission linewidths observed for dilute solutions of **4** indicate that the two isomers have identical excited state energies.

Most of our previous work with electrophosphorescent red, yellow, green and blue OLEDs has involved the use of carbazole biphenyl (CBP) as the host material for the doped luminescent layer (Fig. 19.1) [12–14]. CBP has a number of important properties as a matrix material, such as a high triplet energy of 2.56 eV (484 nm) [29] and ambipolar charge transporting properties [30], that make it an excellent host for phosphorescent dopants. Therefore, it was an obvious choice to begin with in the doping studies described below.

Thin films of CBP doped with 1–30 wt% of Pt complexes **1–4** were prepared by co-depositing the two materials onto a glass substrate. The photoluminescent excitation spectra for all of the doped films are identical to the excitation spectra of an undoped CBP film, however, emission comes primarily from the dopant, demonstrating that energy transfer from CBP to dopant is an efficient process at >3 wt% doping levels. The peak positions of the monomer (λ_{max} = 468, 500, and 540 nm) and aggregate

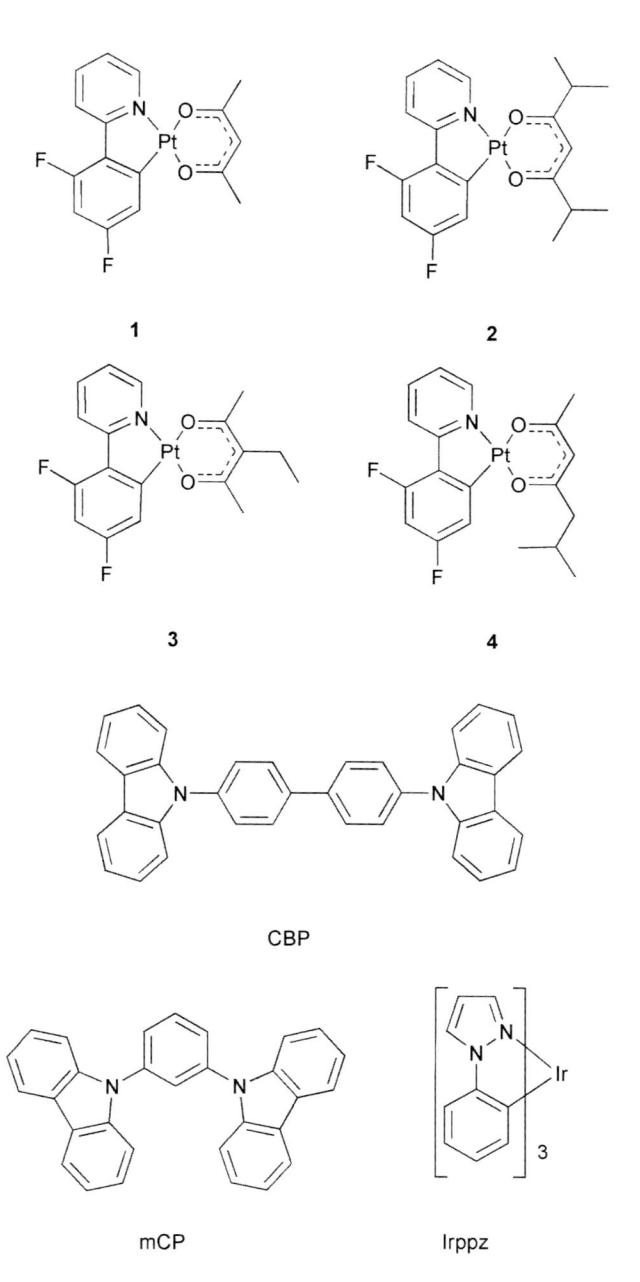

Figure 19.1 The structures of the Pt phosphors and OLED materials used in this study.

($\lambda_{max} \sim 580$ nm) transitions are the same for all four dopants and do not shift in energy with increasing dopant concentration. The spectra of **1** doped into CBP at a range of different concentrations are shown in Fig. 19.2. At a 1.5 wt% doping level, emission from **1** is that of the monomer, closely resembling the spectrum observed for **1** in dilute fluid solution. Fluorescence from CBP is also observed ($\lambda_{max} = 406$ nm), since this doping level is too low to effectively quench all the CBP emission. As the doping level is increased, the band at 580 nm grows in, due to the aggregate emission. The aggregate emission ultimately dominates the spectrum at doping levels of 8 wt% and higher. At intermediate doping levels (3–5 wt%), both monomer and aggregate emission are observed from a single doped film. The effect of increasing the size of the alkyl groups can be seen by comparing the spectra of 8 wt% doped **1** to that of 20 wt% doped **2** in CBP. The 8 wt% doped film of **1** shows nearly complete aggregate emission, while the 20 wt% doped film of **2** shows nearly exclusive monomer emission. Hence, changing the methyl groups in **1** to *i*-propyl groups in **2** strongly suppresses aggregate formation in doped CBP.

In order to achieve balanced monomer/aggregate emission at >10 wt% doping levels, a molecule is needed that has intermediate steric bulk between that of **1** and **2**. To that end, films of compounds **3** and **4** were examined. At low doping levels of either **3** or **4**, weak CBP fluorescence is observed, as is the case for compounds **1** and **2**. Based on the photoluminescence studies, compound **3** appears to have greater steric bulk than **1**, giving balanced monomer/aggregate emission at ~8 wt% doping levels (see Electronic supplementary information; ESI). The emission spectra of **4** doped CBP shows the expected transition from monomer to aggregate emission as the concentration is increased, however, for this derivative, balanced monomer/aggregate emission is observed at doping levels approaching 10 wt% (Fig. 19.2). Thus, compound **4** has the appropriate amount of steric bulk to give a balance of monomer and aggregate emission at concentrations necessary for WOLED fabrication and, therefore, was the best dopant option for use with the CBP host matrix material.

Modified host matrices. The second approach taken to vary the monomer/aggregate ratio involved modifying the host matrix

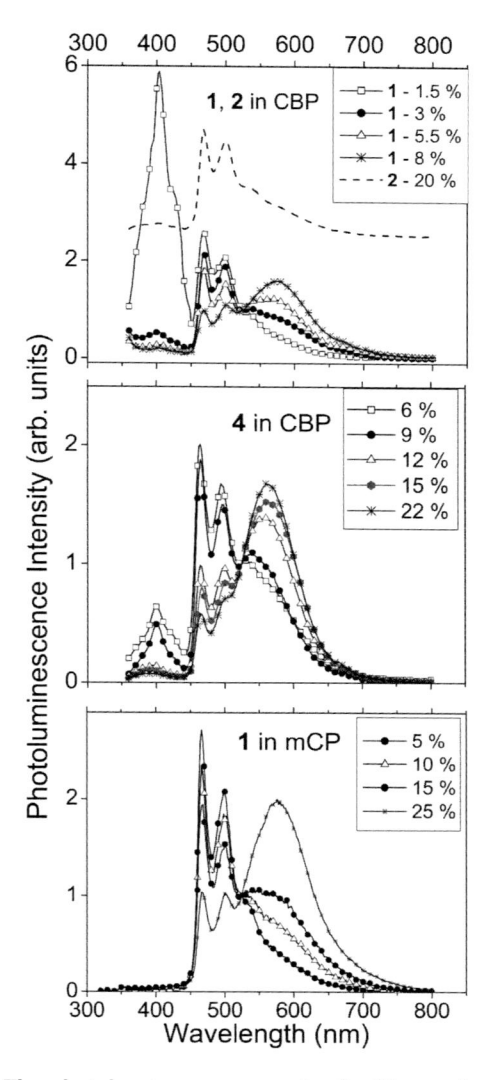

Figure 19.2 The photoluminescence spectra for Pt complexes doped into either CBP or mCP thin films are shown. The doping levels for each spectrum are indicated in the legend to each plot. (top): Spectra of **1** and **2** in CBP. The spectrum of **2** doped into CBP at 20% has been vertically shifted for clarity. (middle): Spectra of **4** in CBP. (bottom): Photoluminescence spectra of **1** in mCP and (inset to bottom plot) the CIE coordinate plot for **1** in mCP. The spectra were measured by exciting the film at the excitation maximum of the matrix material (340 nm for CBP and 300 nm for mCP).

material. During the growth of doped films, there are competing processes between the aggregation of dopants and their dispersion in the host matrix. If the host acts as a good solvent, the dopants will be more evenly dispersed in the film, favoring monomeric species. A poorly solvating host matrix will not disperse the monomer dopant efficiently, leading to dopant aggregation. Two different materials (CBP and mCP, Fig. 19.1) have been examined, which give different degrees of aggregation of **1**.

The spectra of **1** doped in mCP, at a range of concentrations, are shown in Fig. 19.2. The wavelengths of the emission maxima for the monomer and aggregate states of **1** doped into mCP are the same as those of **1** in CBP. Balanced monomer/aggregate emission is observed at a doping level of approximately 15 wt%, roughly three times the concentration required to achieve an equivalent monomer/aggregate emission ratio from **1** doped CBP films. This suggests that mCP is a better solvent for **1**, leading to fewer **1** ⋯ **1** interactions in the doped mCP film, at a given concentration.

In contrast to the CBP doped films, no host emission is observed in the photoluminescence spectra of lightly doped mCP films (< 1 wt% **1**), indicating that energy transfer from mCP to **1** is more efficient than from CBP to **1**. Despite the high triplet energy of CBP (phosphorescence $\lambda_{max} = 460$ nm) [29], energy transfer from CBP to blue phosphorescent dopants, such as the Pt complexes used here, is an endothermic process [11, 14]. In contrast, mCP has a phosphorescence spectrum peaked at 410 nm [31] (see ESI), making energy transfer from mCP to the Pt complex dopants a more efficient, exothermic process. A more efficient energy transfer from the host to the dopant will affect the amount of dopant necessary to quench emission, as observed.

Both CBP and mCP have low dipole moments (*ca.* 0.5 D), so electrostatic interactions between the dopants and host materials are expected to be similar. This is consistent with the observation that the spectra of monomer and aggregate states for doped mCP and CBP films are the same. Our best explanation for the differences between CBP and mCP, which give rise to differing dopant solubilities, is related to their molecular structures. Planar molecules tend to have high association energies, which promote crystallization and hinder glass formation [31]. CBP is expected to be largely planar in the solid state. This is consistent with our observation that undoped CBP thin films rapidly crystallize when deposited directly on glass or ITO substrates. The high CBP association energy may tend to exclude monomer dopant, leading to aggregate formation at moderate doping levels [32]. mCP readily forms a stable glass when deposited on either inorganic or organic substrates, suggesting it has a nonplanar ground state structure [31]. The glass transition temperature for mCP is 65°C. Steric interactions between adjacent carbazole groups and the phenyl ring lead to a prediction that both CBP and mCP should have nonplanar ground state structures, as seen in the geometry of the energy minimized structures in Fig. 19.3 [33]. While the minimized structure of CBP appears somewhat nonplanar, it is important to

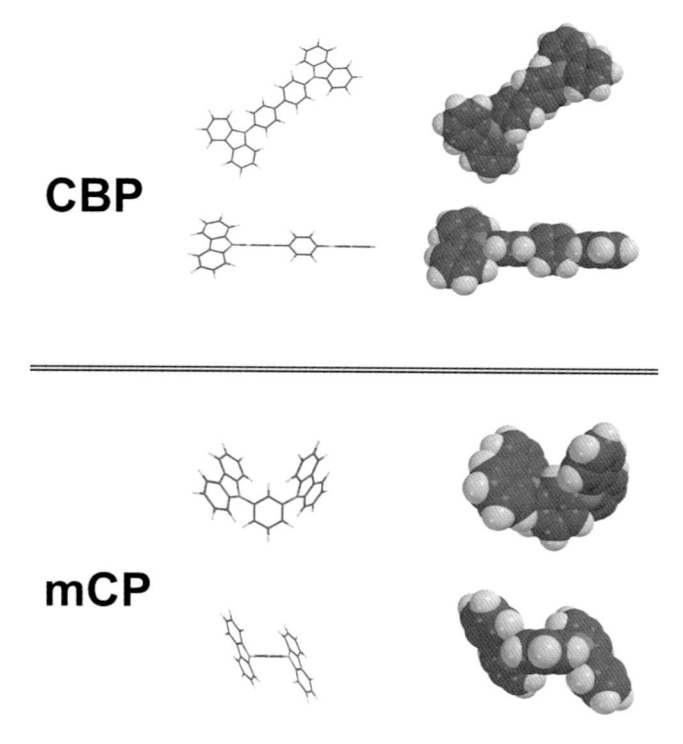

Figure 19.3 The energy minimized structures of the CBP and mCP host molecules [33].

Table 19.1 CIE coordinates and color rendering indices for photoluminescence spectra of **1** doped mCP films at a range of different dopant concentrations

Concentration/wt%	CIE x	CIE y	CRI
0.1	0.15	0.28	–
0.5	0.19	0.32	–
4.5	0.21	0.35	44.5
10	0.27	0.39	59.7
15	0.32	0.41	68.3
20	0.32	0.39	73.2
25	0.41	0.46	64.1
30	0.41	0.45	69.2

note that the calculated energy difference between the structure shown and the planar conformer is only 18 kJ mol^{-1}. In contrast, the energy cost to planarize mCP is 35 kJ mol^{-1}. The principal cause of the large barrier to flatten mCP is H $\cdots$ H repulsions between adjacent carbazoles, interactions that are absent in CBP. Based on the structural differences, we expect the degree of solvation of a square planar Pt dopant by mCP to be very different from that of CBP. This change significantly affects the monomer/aggregate ratio at a given doping level in CBP vs. mCP.

The CIE coordinates and the color rendering index (CRI) for the photoluminescence spectra of **1** doped into mCP are given in Table 19.1. Concentrations between 4–10 wt% gave the CIE coordinates closest to white (0.33, 0.33) while the maximum CRI was observed for concentrations ranging between 15–20 wt%. At the higher concentrations, the CIE coordinates are close to those found in incandescent lamps (*ca.* 0.41, 0.41). Therefore, the 10–20 wt% concentration range for **1** doped mCP was chosen to be optimal for use in WOLEDs.

19.3.2 Single Dopant WOLEDs

OLEDs of the general structure ITO/NPD/CBP:dopant/BCP/Alq$_3$/LiF/Al [NPD = N,N'-diphenyl-N,N'-bis(1-naphthyl)benzidine, BCP = bathocuproine, Alq$_3$ = aluminum tris(8-hyrdroxyquinolate)] have been fabricated previously and have proven to be very effective for red-to-green phosphorescent OLEDs [1, 6, 12, 13]. The doped CBP luminescent layer is sandwiched between the hole transporting layer (HTL, i.e., NPD) and the hole blocking layer (HBL, i.e., BCP). The HBL is used to confine both carriers and excitons to the luminescent layer, preventing hole and exciton leakage into the electron transporting layer (ETL, i.e., Alq$_3$). OLEDs of the general structure ITO/NPD/CBP:4/BCP/Alq$_3$/LiF/Al have been shown to emit white light from a single dopant luminescent layer (i.e., CBP doped with 4) [11]. Unfortunately, while this device gave the desired white emission (CIE = 0.33, 0.31; CRI = 86 at 11 V), the electroluminescence (EL) spectrum had a significant contribution from NPD emission [11]. As the bias was increased, the NPD emission band (λ_{max} = 430 nm) grew relative to the monomer/aggregate features, dominating the EL spectrum at biases of 10 V and above. The cause of this NPD emission was either electron or exciton leakage from the luminescent layer into the NPD layer. This is a problem for these devices, because the dopant is a high energy (blue) phosphorescent emitter. High energy phosphorescent dopants tend to have high energy LUMO levels, approaching those of the transport and host materials [14]. If the dopant LUMO level approaches the LUMO energy of the NPD, electrons can leak into the NPD layer. Likewise, exciton leakage into the HTL layer can occur as the emission energy of the dopant approaches the absorption energy of NPD. The voltage dependent NPD emission in the WOLED described above is indicative of poor charge confinement, which may decrease OLED efficiency. The energy level diagram for this device, shown in Fig. 19.4, illustrates that the barrier for migration of electrons from the dopant/CBP LUMO levels to the NPD LUMO may be comparable to the hole injection barrier from NPD into the emissive layer. Eliminating electron/exciton leakage into the HTL should improve both the WOLED efficiency and color stability. Therefore, introduction of an electron/exciton blocking layer (EBL) between the HTL and luminescent layer was deemed necessary to improve the device characteristics.

An efficient EBL material needs to fulfill several criteria. It must have a wide energy gap to prevent exciton leakage into the HTL, a

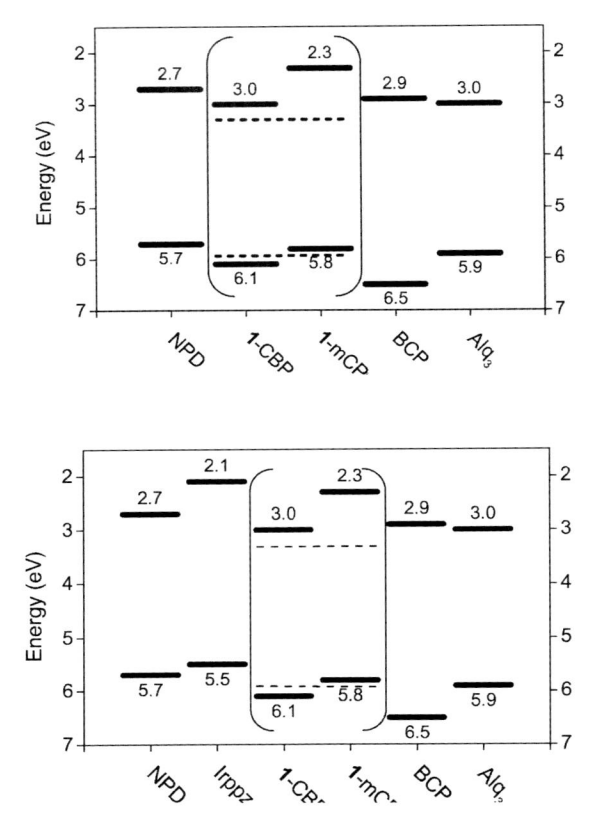

Figure 19.4 Energy level diagrams showing the HOMO and LUMO levels for the OLED materials investigated here. The energy for each orbital is listed below (HOMOs) or above (LUMOs) the appropriate bar. The HOMO and LUMO levels for the emissive dopant **1** is shown as a dashed line in each of the plots. The doped luminescent layers (CBP or mCP) are enclosed in brackets. Each device had either a CBP or an mCP layer, not both. The top plot shows the diagram for a four-layer OLED (no electron blocking layer), and the bottom plot shows a similar OLED with an Irppz EBL.

high LUMO level to block electrons, and a HOMO level above that of the HTL. We have found that the *fac*-tris(1-phenylpyrazolato-$N,C^{2'}$)iridium(III) (Irppz, Fig. 19.1) complex satisfies these requirements. The Irppz complex emits exclusively from a phosphorescent excited state (λ_{max} = 414 nm at 77 K, τ = 15 μsec). The optical gap for this complex was taken as the low energy edge of the absorption spectrum, at 370 nm (3.4 eV). This estimate of the optical gap represents a lower limit for the carrier gap. Irppz shows a reversible oxidation in fluid solution at 0.38 V (vs. ferrocene/ferrocenium), but no reduction wave occurs out to −3.0 V in DMF, consistent with a carrier gap of > 3.4 eV. The HOMO energy for Irppz was measured by ultraviolet photoelectron spectroscopy (UPS) and found to be 5.5 eV. Using the Irppz optical gap to approximate the carrier gap, we estimate the Irppz LUMO is 2.1 eV, well above both the CPB and dopant LUMOs. The energy scheme of Fig. 19.4 suggests that Irppz should make an excellent EBL.

The spectra and CIE coordinates (inset) of a single dopant WOLEDs (using 8, 10 and 12 wt% **4** in CBP), with an Irppz EBL, are shown in Fig. 19.5. The devices give an EL spectrum consistent with only dopant emission, that is no NPD emission is observed at any bias level. The ratio of monomer/aggregate contributions in the EL spectrum is also invariant with applied bias, leading to a voltage independent, high quality white emission (0.36, 0.44 and CRI of 67 for the 10% doped device). The peak brightness of the 10% doped device was 8000 cd m^{-2} and the maximum quantum efficiency was 3.3 ± 0.3% (7.3 ± 0.7 lm W^{-1}) at 0.5 cd m^{-2}, dropping to 2.3 ± 0.2% (5.2 ± 0.3 lm W^{-1}) at 500 cd m^{-2}. The quantum efficiency of the device with an Irppz blocking layer is nearly double that of the device with no EBL (peak efficiency = 1.9%) [11]. It is also apparent from the current–voltage and quantum efficiency plots that increasing the dopant concentration improves the performance of the devices (higher quantum efficiency and lower leakage current at a given bias). However, the higher doping levels also increase the amount of aggregate emission in the spectrum, leading to a shift in the color of the device from white to yellow.

The use of the mCP host in place of CBP significantly improves the device performance. A device was fabricated with the structure NPD (400 Å)/Irppz (200 Å)/mCP:**1** (16%, 300 Å)/BCP (150 Å)/Alq$_3$ (200 Å)/LiF (10 Å)/Al (1000 Å). The efficiency, current–voltage characteristics, and spectra of the device are shown in Fig. 19.6. The higher doping concentrations and improved energy transfer from mCP to the dopant gave a maximum quantum efficiency of 6.4 ± 0.6% (12.2 ± 1.4 lm W^{-1}, 17.0 cd A^{-1}) at low brightness

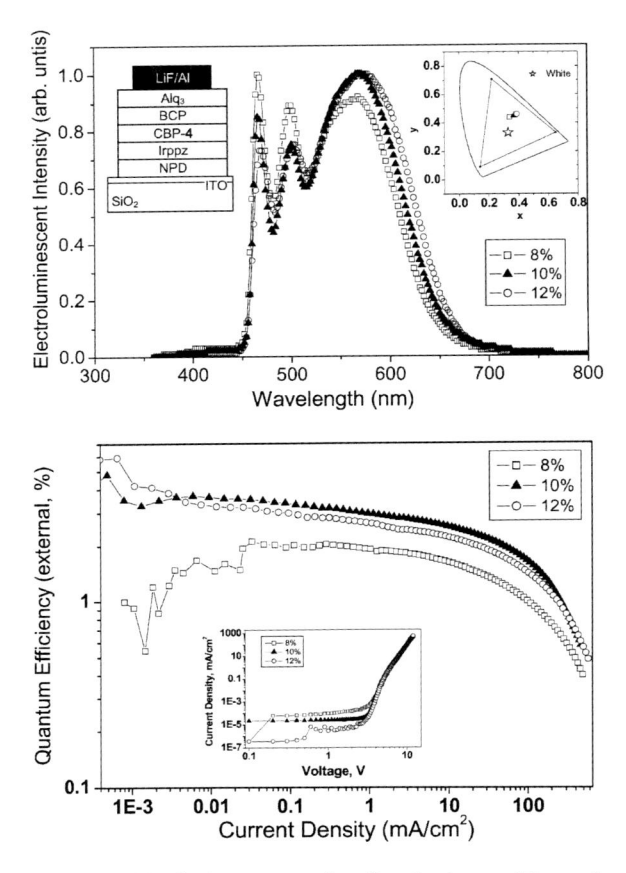

Figure 19.5 WOLED device properties for devices with an Irppz EBL [ITO/NPD (400 Å)/Irppz (200 Å)/CBP:**4** (300 Å)/BCP (150 Å)/Alq$_3$ (200 Å)/ LiF–Al]. A schematic drawing of the device is shown as an inset to the top plot. Data for devices doped at 8, 10, 12% are shown. The spectra and CIE coordinates (inset) are shown in the top plot and the quantum efficiency vs. current density and current–voltage characteristics (inset) are shown in the bottom plot.

levels (1 cd m^{-2}) and 4.3 ± 0.5% (8.1 ± 0.6 lm W^{-1}, 11.3 cd A^{-1}) at 500 cd m^{-2}. The quantum efficiencies demonstrated by these mCP · **1** WOLEDs are the highest reported efficiencies for a WOLED [4–11]. The quantum efficiency decreases with increasing current density, as observed for other devices [34], however, the

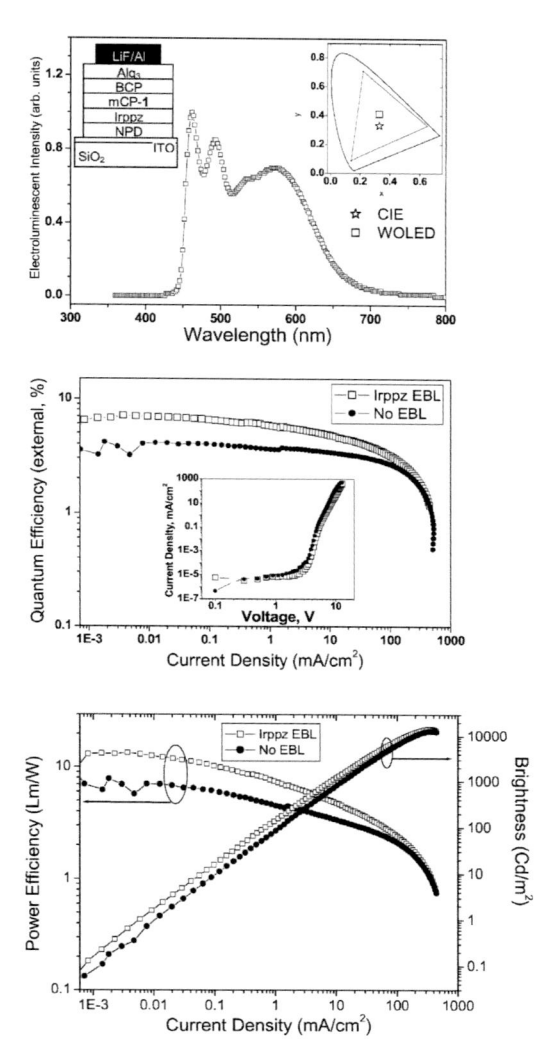

Figure 19.6 WOLED device properties for a mCP based WOLED [ITO/NPD (400 Å)/Irppz (200 Å)/mCP:**1** (doping level 16%, 300 Å)/BCP (150 Å)/Alq$_3$ (200 Å)/ LiF–Al]. A schematic drawing of the device with the Irppz EBL is shown as an inset to the top plot. The spectra and CIE coordinates (inset) are shown in the top plot and the quantum efficiency vs. current density and current–voltage characteristics (inset) are shown in the middle plot. Lumens per watt and brightness vs. current density plots for the WOLED and the related structure without the Irppz EBL are shown in the bottom plot.

Figure 19.7 A color bar illuminated only by the singly doped white OLED described in Fig. 19.6. A picture of the WOLEDs in room light is shown to the lower left.

decrease is less severe than in most other electrophosphorescent devices. If the Irppz EBL is omitted (i.e., NPD/mCP-**1**/BCP/Alq$_3$), the EL spectrum again has a significant contribution from NPD and quantum efficiency of the devices drops by roughly a factor of two (see Fig. 19.6). Overall, the Irppz EBL increases the OLED efficiency, removes NPD emission from the spectra and makes the spectrum independent of voltage.

The high color rendering index of a single dopant OLED is demonstrated in Fig. 19.7. The OLEDs used for this figure are a mCP:**1** based device. The CIE diagram is illuminated *only* by the mCP · **1** OLEDs. All of the colors on the diagram are readily distinguished. The spectral characteristics of these single dopant WOLEDs clearly make them suitable for use as white light illumination sources.

19.4 Conclusion

We have demonstrated the most efficient WOLEDs reported [4–11]. These devices emit from a single doped luminescent layer, containing only one emissive dopant. In order to accomplish this it was important to control both the emission character, by tuning the degree of dopant-dopant and dopant-host interactions, and the carrier/exciton confinement in the device. The structure reported here utilizes an electron blocking layer (Irppz) to confine carriers and excitons to the desired luminescent layer. This structure may be useful for monochromatic electrophosphorescent OLEDs as well as for WOLEDs. As the dopant energies are increased toward the blue end of the spectrum, the dopant energy gap can exceed that of the available transport and host materials, leading to poor confinement of charge and energy in the desired luminescent layer. The approach developed here to eliminate HTL (NPD) emission and improve the quantum efficiency may work equally well for blue electrophosphorescent OLEDs, by effectively controlling the recombination and emission zones of the device. Studies in this area are ongoing.

The use of only a single dopant in WOLEDs significantly simplifies the fabrication of WOLEDs relative to other approaches to white organic electroluminescence. It may also solve the problems associated with differential dopant aging. The lifetimes of monochromatic OLEDs, prepared with different dopants, vary over a wide range, due to different chemical and electrochemical stabilities of the various dopants and host materials that are used. While there have been no reports of the lifetimes or color stabilities of WOLEDs, it is expected that WOLEDs utilizing multiple dopants will show different characteristic aging times for each of the dopants. Differential aging of the dopants would change the color of the WOLED over time, as the dopant emission ratio changes. It is expected that a single dopant WOLED will not suffer from these limitations, since the two emission bands (monomer and aggregate) come from the same dopant. Experiments are currently underway with the single dopant WOLEDs reported here to examine their lifetimes and verify that they are color stable over the life of the WOLED.

Acknowledgements

The authors wish to thank the Universal Display Corporation and DARPA for their financial support of this work.

Supporting Information

Electronic supplementary information (ESI) available: emission spectra as a function of doping concentration for **3** in CBP, as well as the absorption and emission spectra of Irppz, CBP and mCP. See http://www.rsc.org/suppdata/nj/b2/b204301g/.

References

1. (a) C. Adachi, M. A. Baldo, S. R. Forrest and M. E. Thompson, *J. Appl. Phys.*, 2001, **90**, 4058; (b) M. Ikai, S. Tokito, Y. Sakamoto, T. Suzuki and Y. Taga, *Appl. Phys. Lett.*, 2001, **79**, 156–158.
2. T. Justel, H. Nikol and C. Ronda, *Angew. Chem., Int. Ed.*, 1998, **37**, 3084–3113.
3. J. Mahon, J. Brown, T. Zhou, P. Burrows and S. Forrest, *Annu. Tech. Conf.Proc. - Soc. Vac. Coaters*, 1999, **42**, 456–459.
4. (a) J. Kido, H. Shionoya and K. Nagai, *Appl. Phys. Lett.*, 1995, **67**, 2281–2283; (b) S. Tasch, E. J. W. List, O. Ekström, W. Graupner, G. Leising, P. Schlichting, U. Rohr, Y. Geerts, U. Scherf and K. Müllen, *Appl. Phys. Lett.*, 1997, **71**, 2883–2885; (c) Y. Kawamura, S. Yanagida and S. R. Forrest, *J. Appl. Phys.*, 2002, **92**, 87–93.
5. (a) Y. S. Huang, J. H. Jou, W. K. Weng and J. M. Liu, *Appl. Phys. Lett.*, 2002, **80**, 2782–2784; (b) C. Ko and Y. Tao, *Appl. Phys. Lett.*, 2001, **79**, 4234–4236; (c) X. Jiang, Z. Zhang, W. Zhao, W. Zhu, B. Zhang and S. Xu, *J. Phys. D: Appl. Phys.*, 2000, **33**, 473–476; (d) S. Liu, J. Huang, Z. Xie, Y. Wang and B. Chen, *Thin Solid Films*, 2000, **363**, 294; (e) J. Kido, M. Kimura and K. Nagai, *Science*, 1995, **267**, 1332–1334.
6. B. D'Andrade, M. Thompson and S. Forrest, *Adv. Mater.*, 2002, **14**, 147–151.
7. J. Yang, Y. Jin, P. Heremans, R. Hoefnagels, P. Dieltiens, F. Blockhuys, H. Geise, M. Van der Auweraer and G. Borghs, *Chem. Phys. Lett.*, 2000, **325**, 251–256.
8. J. Thompson, R. Blyth, M. Mazzeo, M. Ani, G. Gigli and R. Cingolani, *Appl. Phys. Lett.*, 2001, **79**, 560–562.
9. C.-I. Chao and S.-A. Chen, *Appl. Phys. Lett.*, 1998, **73**, 426–428.
10. J. Feng, F. Li, W. Gao and S. Liu, *Appl. Phys. Lett.*, 2001, **78**, 3947–3949.

11. B. D'Andrade, J. Brooks, V. Adamovich, M. E. Thompson and S. R. Forrest, *Adv. Mater.*, 2002, **14**, 1032–1036.
12. D. O'Brien, M. Baldo, M. E. Thompson and S. R. Forrest, *Appl. Phys. Lett.*, 1999, **74**, 442–444.
13. S. Lamansky, P. Djurovich, D. Murphy, F. Abdel-Razzaq, H.-E. Lee, C. Adachi, P. E. Burrows, S. R. Forrest and M. E. Thompson, *J. Am. Chem. Soc.*, 2001, **123**, 4304–4312.
14. C. Adachi, R. Kwong, P. Djurovich, V. Adamovich, M. Baldo, Mark E. Thomson and S. R. Forrest, *Appl. Phys. Lett.*, 2001, **79**, 2082–2084.
15. J. Brooks, Y. Babayan, S. Lamansky, P. I. Djurovich, I. Tsyba, R. Bau and M. E. Thompson, *Inorg. Chem.*, 2002, **41**, 3055–3066.
16. O. Lohse, P. Thevenin and E. Waldvogel, *Synlett*, 1999, **1**, 45–48.
17. B. N. Cockburn, D. V. Howe, T. Keating, B. F. G. Johnson and J. Lewis, *J. Chem. Soc., Dalton Trans.*, 1973, 404–410.
18. K. Dedeian, P. I. Djurovich, F. O. Garces, G. Carlson and R. J. Watts, *Inorg. Chem.*, 1991, **30**, 1685–1687.
19. T. Yamamoto, M. Nishiyama and Y. Koie, *Tetrahedron Lett.*, 1998, **39**, 2367–2370.
20. (a) H. Ishii, K. Sugiyama, E. Ito and K. Seki, *Adv. Mater.*, 1999, **11**, 605–625; (b) I. Hill and A. Kahn, *J. Appl. Phys.*, 1998, **84**, 5583
21. (a) V. M. Miskowski and V. M. Houlding, *Inorg. Chem.*, 1989, **28**, 1529–1533; (b) V. H. Houlding and V. M. Miskowski, *Coord. Chem. Rev.*, 1991, **111**, 145–152; (c) V. M. Miskowski and V. H. Houlding, *Inorg. Chem.*, 1991, **30**, 4446–4452; (d) V. M. Miskowski, V. M. Houlding, C.-M. Che and Y. Wang, *Inorg. Chem.*, 1993, **32**, 2518–2524.
22. J. A. Bailey, M. G. Hill, R. E. Marsh, V. M. Miskowski, W. P. Schaefer and H. B. Gray, *Inorg. Chem.*, 1995, **34**, 4591–4599.
23. T.-C. Cheung, K.-K. Cheung, S.-M. Peng and C.-M. Che, *J. Chem. Soc., Dalton Trans.*, 1996, 1645–1651.
24. G. Y. Zheng and D. P. Rillema, *Inorg. Chem.*, 1998, **37**, 1392–1397.
25. (a) S.-W. Lai, M. C. W. Chan, K.-K. Cheung, S.-M. Peng and C.-M. Che, *Organometallics*, 1999, **18**, 3991–3997; (b) S.-W. Lai, M. C. W. Chan, K.-K. Cheung and C.-M. Che, *Inorg. Chem.*, 1999, **38**, 4262–4267; (c) W. Lu, M. C. W. Chan, K.-K. Cheung and C.-M. Che, *Organometallics*, 2001, **20**, 2477–2486.
26. (a) R. Büchner, C. T. Cunningham, J. S. Field, R. J. Haines, D. R. McMillan and G. C. Summerton, *J. Chem. Soc., Dalton Trans.*, 1999, 711–717; (b) W. Lu, N. Zhu and C.-M. Che, *Chem. Commun.*, 2002, 900–901; (c) S. W. Lai,

H. W. Lam, W. Lu, K. K. Cheung and C. M. Che, *Organometallics.*, 2002, **21**, 226–234.
27. J. P. H. Charmant, J. Forniés, J. Gómez, E. Lalinde, R. I. Merino, M. T. Moreno and A. G. Orpen, *Organometallics*, 1999, **18**, 3353–3358.
28. W. B. Connick, R. E. Marsh, W. P. Schaefer and H. B. Gray, *Inorg. Chem.*, 1997, **36**, 913–922.
29. M. A. Baldo and S. R. Forrest, *Phys. Rev. B*, 2000, **62**, 10 958–10 966.
30. C. Adachi, R. Kwong and S. R. Forrest, *Org. Electron.*, 2001, **2**, 37–43.
31. K. Naito and A. Miura, *J. Phys. Chem.*, 1993, **97**, 6240.
32. " Aggregate formation" in this context could mean either the formation of Pt–Pt bonded oligomers or the formation of regions of the film with a high local concentration of dopant that will readily form excimers on excitation.
33. The molecular modeling and energy minimization was carried out at PM3 level using the MacSpartan Pro v. 1.02 software package, Wavefunction Inc, Irvine, CA 92612.
34. M. A. Baldo and S. R. Forrest, *Phys. Rev. B*, 2000, **62**, 10 958–10 966.

Chapter 20

High Operational Stability of Electrophosphorescent Devices

Raymond C. Kwong,[a] Matthew R. Nugent,[a] Lech Michalski,[a]
Tan Ngo,[a] Kamala Rajan,[a] Yeh-Jiun Tung,[a] Michael S. Weaver,[a]
Theodore X. Zhou,[a] Michael Hack,[a] Mark E. Thompson,[b]
Stephen R. Forrest,[c] and Julie J. Brown[a]

[a]*Universal Display Corporation, Ewing, New Jersey 08618, USA*
[b]*Department of Chemistry, University of Southern California,
Los Angeles, California 90089, USA*
[c]*Department of Electrical Engineering, Princeton University,
Princeton, New Jersey 08544, USA*
rkwong@universaldisplay.com

lectrophosphorescent devices with *fac*-tris(2-phenylpyridine) iridium as the green emitting dopant have been fabricated with a variety of hole and exciton blocking materials. A device with aluminum(III)bis(2-methyl-8-quinolinato)4-phenylphenolate (BAlq) demonstrates an efficiency of 19 cd/A with a projected operational lifetime of 10 000 h, operated at an initial brightness of 500 cd/m^2; or 50,000 h normalized to 100 cd/m^2. An orange-red electrophosphorescent device with iridium(III) bis(2-phenylquinolyl-$N,C^{2'}$) acetylacetonate as the dopant emitter and

---

Reprinted from *Appl. Phys. Lett.*, **81**(1), 162–164, 2002.

Electrophosphorescent Materials and Devices
Edited by Mark E. Thompson
Text Copyright © 2002 American Institute of Physics
Layout Copyright © 2024 Jenny Stanford Publishing Pte. Ltd.
ISBN 978-981-4877-34-3 (Hardcover), 978-1-003-08872-1 (eBook)
www.jennystanford.com

BAlq as the hole blocker demonstrates a maximum efficiency of 17.6 cd/A with a projected operational lifetime of 5000 h at an initial brightness of 300 cd/m^2; or 15,000 h normalized to 100 cd/m^2. The average voltage increase for both devices is < 0.3 mV/h. The device operational lifetime is found to be inversely proportional to the initial brightness, typical of fluorescent organic light emitting devices.

Since the initial report of efficient fluorescent small molecule OLEDs in 1987 [1], both the efficiency and lifetime of OLEDs have increased dramatically. For example, the external quantum efficiency of green OLEDs has been improved from $\sim 1\%$ for undoped tris(8-hydroxyquinolinato) aluminum (Alq$_3$) devices with operational lifetimes of several thousand hours to 2.5% for fluorescent dye doped devices with lifetimes of more than 10,000 h.^{2} Recently, phosphorescent organic light emitting diodes (PHOLED^{TM}) have shown very high efficiencies and lifetimes [3–7]. For example, OLEDs employing phosphorescent organometallic iridium complexes exhibited green emission with maximum external quantum efficiencies of 19% [4, 5].

Device lifetimes are strongly determined by the materials used. The most efficient devices are composed of multiple layers serving different roles, i.e., hole transporting, light emitting, charge, or exciton blocking, and electron transporting functions. Green emitting *fac*-tris(2-phenylpyridine)iridium [Ir(ppy)$_3$] doped OLEDs employing a 2,9-dimethyl-4,7-diphenyl-1,10-phenanthroline (BCP) hole blocking layer result in low stability, with $T_{1/2} < 700$ h at an initial luminance (L_0) of 600–1200 cd/m^2, whereas devices employing an aluminum(III)bis(2-methyl-8-quinolinato)4-phenylphenolate (BAlq) hole blocking layer have $T_{1/2} \sim 4000$ h at $L_0 \sim 570$ cd/m^2 [8]. Half-life, $T_{1/2}$, is defined as the time for the luminance to decay to $0.5L_0$. In the present study we examine a number of different blocking materials in an effort to improve the device lifetime.

All devices were fabricated by high vacuum (10^{-8} Torr) thermal evaporation. The devices consisted of an indium tin oxide anode, a 100 Å copper phthalocyanine (CuPc) hole injection layer, a 300 Å 4,4'-bis[N-(1-naphthyl)-N-phenylamino]biphenyl (α-NPD) hole transporting layer, a 300 Å 4,4'-bis(N-carbazolyl)biphenyl

(CBP) doped with 6 wt% of Ir(ppy)$_3$ emissive layer, a 100 Å hole blocking layer, and a 400 Å Alq$_3$ electron transporting layer. The 5 mm^2 cathode was comprised of 10-Å-thick lithium fluoride followed by 1000 Å aluminum. All devices were encapsulated under nitrogen (<1 ppm of H$_2$O and O$_2$). A calcium oxide moisture getter was incorporated inside the package. Device operational stabilities, measured on several identically fabricated devices, are determined to be within ± 10%.

Table 20.1 lists the structures of the hole blockers, along with the OLED initial luminous efficiency (η_L), external quantum

Table 20.1 Performance of Ir(ppy)$_3$ devices $L_0 = 500$ cd/m^2 with different hole blockers

Hole blocker	η_L (cd/A)	η_{ext} (%)	$T_{1/2}$ (h)
TPBi	25.3	7.5	70
	15.6	4.7	640
SAlq	18.0	5.3	1075
PAlq	19.0	5.6	10 000[a]
BAlq			

[a] Projected lifetimes.

efficiency (η_{ext}), and $T_{1/2}$. The efficiencies were recorded at $L_0 = 500$ cd/m^2. The device luminous efficiencies range from 15.6 to 25.3 cd/A, corresponding to external quantum efficiencies of 4.5%–7.5%. The operational lifetimes under constant dc drive at room temperature for the various hole blocking groups differ by more than two orders of magnitude. The hole blocking material 2,2',2"-(1,3,5- benzenetriyl) tris -[1-phenyl -1H-benzimidazole] (TPBi), is an electron deficient heterocyclic compound capable of electron transport [9]. Although the TPBi device shows the highest efficiency, it has a $T_{1/2} = 70$ h. Electron deficient heterocyclic compounds such as triazoles [3], triazines [10], oxadiazole [11], and imidazoles [9] have also been widely used as electron transporting and hole blocking materials. However, devices employing these materials exhibit $T_{1/2}$ typically less than 100 h at $L_0 = 500$ cd/m$_2$ [12]. Other hole blockers in Table 20.1 are based on bis(8-hydroxy-2-methylquinolato)aluminum chelates, yielding devices that exhibit significantly longer $T_{1/2}$ compared to those with TPBi, BCP, or other heterocyclic compounds. Devices with aluminum(III)bis(2-methyl-8-quinolinato) [SAlq] triphenylsilanolate [13] and aluminum(III) bis(2-methyl-8-quinolinato)4-phenolate [9] [PAlq] as the hole blocker give $T_{1/2} = 640$ and 1075 h, respectively. Devices with BAlq show a further improvement, with a projected $T_{1/2} = 10,000$ h ($L_0 = 500$ cd/m^2). These results are a significant improvement over that reported by Watanabe et al. of $T_{1/2} \sim 4000$ h at $L_0 = 570$ cd/m^2 [8]. We attribute this improvement to the effect of Ir(ppy)$_3$ doping concentration in the emissive layer. In Watanabe's device, the Ir(ppy)$_3$ doping concentration was 1.4 wt%. With this doping concentration, we obtain $T_{1/2} = 5000$ h at $L_0 = 500$ cd/m^2, or half the lifetime of the 6 wt% doped devices. The electroluminescent spectrum of the 1.4 wt% doped device reveals, in addition to the strong Ir(ppy)$_3$ green emission, a weak blue emission at wavelengths between 430 and 480 nm. The resulting Commission Internationale de L'Eclairage (CIE) coordinates of (0.29, 0.62) of the 1.4 wt% doped device compared with (0.30, 0.63) of the 6 wt% doped device also indicates this slight spectral difference. The blue luminescence is due to the emission of NPD when the electron-hole recombination occurs near the NPD/CBP:Ir(ppy)$_3$ interface. Ir(ppy)$_3$ is a hole transporting material [14], and as the doping

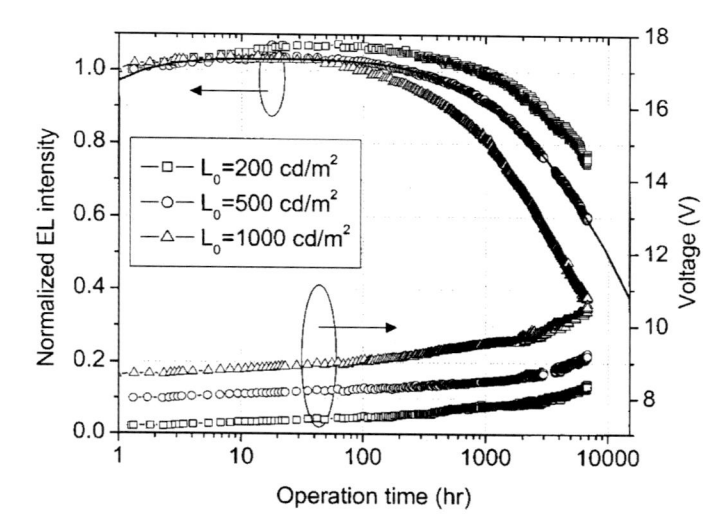

Figure 20.1 Luminance (normalized) and operating voltage evolution for a CuPc (100 Å)/NPD (300 Å)/CBP:Ir(ppy)$_3$ (300 Å, 6%-doped)/BAlq (100 Å)/Alq$_3$ (400 Å)/LiF (10 Å)/Al (1000 Å) organic light emitting device operated at initial luminance $L_0 = 200, 500,$ and 1000 cd/m^2 under constant dc drive. The polynomial fit (black line) projects an operational lifetime of $T_{1/2} = 10,000$ h at $L_0 = 500$ cd/m^2.

concentration increases, the recombination zone moves away from the NPD/CBP:Ir(ppy)$_3$ interface and into the CBP:Ir(ppy)$_3$ layer, eliminating the residual NPD emission. The difference in $T_{1/2}$ between the 1.4 and 6 wt% doped devices demonstrates that device stability is sensitive to the location of the electron-hole recombination.

The stability of the 6 wt% Ir(ppy)$_3$ devices using a BAlq hole blocker was further examined at $L_0 = 200, 500,$ and 1000 cd/m^2 under constant currents of 1.2, 2.7, and 5.2 mA/cm^2, corresponding to initial voltages of 7.2, 8.0, and 8.6 V, respectively. The normalized luminance decay curves and the operating voltages as functions of time are shown in Fig. 20.1. After 3000 h of operation, the device luminances dropped to 88%, 75%, and 57% of their initial values, and the voltages increased to 8.0, 8.7, and 9.9 V, respectively. The voltage increase rates are 0.26, 0.24, and 0.41 mV/h. Extrapolating the lifetime at $L_0 = 500$ cd/m^2, this green Ir(ppy)$_3$ PHOLED has $T_{1/2}$ $\sim 10,000$ h.

The use of BAlq also results in high stability for red electrophosphorescent devices. Based on the same structure of the Ir(ppy)$_3$ device, an orange-red PHOLED was fabricated with 6 wt% iridium(III) bis(2-phenylquinolyl-$N,C^{2'}$) acetylacetonate [PQ$_2$Ir(acac)] doped in CBP as the emissive layer. The thicknesses of the NPD, BAlq, and Alq$_3$ layers are 500, 150, and 500 Å, respectively. The electroluminescence spectrum is identical to the photoluminescence spectrum reported previously [15], with CIE coordinates of (0.61, 0.38). The maximum luminous efficiency, obtained at 600 cd/m^2, is 17.6 cd/A, and corresponds to an external quantum efficiency of 10.3%. The device lifetests were performed at $L_0 = 100$, 300, and 500 cd/m^2, under constant currents of 0.6, 1.7, and 2.8 mA/cm^2, corresponding to initial voltages of 8.1, 9.0, and 9.6 V, respectively. The normalized luminance decay curves and the operating voltages as functions of time are shown in Fig. 20.2. After 3000 h of operation, the luminances dropped to 79%, 59%, and 50% of their initial values, and the voltages increased to 8.6, 9.8, and 10.7 V, respectively. The voltage increase rates were 0.16, 0.26, and 0.36 mV/h. Extrapolating the lifetime at $L_0 = 300$ cd/m^2, this orange-red PQ$_2$Ir(acac) PHOLED has $T_{1/2} \sim 5000$ h.

It is important to establish a correlation between the rate of device degradation and the initial brightness to provide insights into the origin of the degradation. In addition, it allows accelerated lifetime tests at high initial brightness to reliably estimate device lifetimes at low to moderate initial brightnesses more typical of display applications. A scaling relationship of $T_{1/2} \propto 1/L_0$ for fluorescent OLEDs has been previously suggested [16]. The data presented earlier can be used to examine this relationship in PHOLEDs. To correlate luminance decay against initial brightness, $T_{1/2}$ is typically employed as a figure of merit. However, since most of the Ir(ppy)$_3$ and PQIr devices in this case are still operating at $>0.5L_0$ at the end of our tests, we find it convenient to use $T_{0.8}$, corresponding to the time for the luminance to decay to $0.8L_0$. $T_{0.8}$ of the Ir(ppy)$_3$ device at $L_0 = 200$, 500, and 1000 cd/m^2 and the PQIr devices at $L_0 = 100$, 300, and 500 cd/m^2 is plotted versus $1/L_0$ in Fig. 20.3. In both devices, linear relationships result,

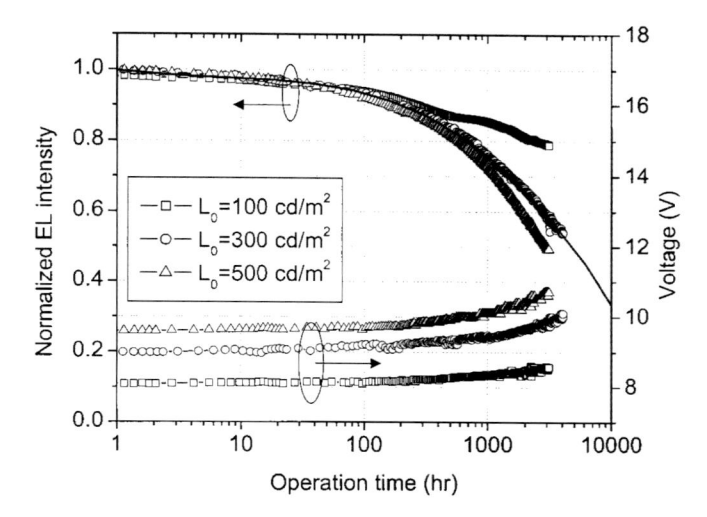

Figure 20.2 Luminance (normalized) and operating voltage evolution for a CuPc (100 Å)/NPD (500 Å)/CBP:PQ$_2$Ir(acac) (300 Å, 6%-doped)/BAlq (100 Å)/Alq$_3$ (500 Å)/LiF (10 Å)/Al (1000 Å) organic light emitting device operated initial luminance L_0 = 100, 300, and 500 cd/m^2 under constant dc drive. The polynomial fit (black line) projects an operational lifetime of $T_{1/2}$ = 5000 hours at L_0 = 300 cd/m^2.

indicating the operational lifetime is inversely proportional to the initial luminance in PHOLEDs, similar to that found for fluorescent OLEDs. Since the brightness is linearly proportional to the current density in the regime used for the lifetest (~1–10 mA/cm^2), the lifetime is also inversely proportional to the current density in PHOLEDs. Despite the difference between PHOLED and fluorescent OLED emission mechanisms which gives rise to the much higher efficiency of PHOLEDs as compared with fluorescent OLEDs, our stability studies show that the same lifetime scaling law applies in both cases, suggesting that electrophosphorescence is not subject to additional failure mechanisms beyond those commonly observed in fluorescent devices.

The best OLED lifetimes reported are $T_{1/2}$ = 7500 h at L_0 = 1450 cd/m^2 for a 7.3 cd/A, DMQA doped green emitting device [17], and $T_{1/2}$ = 8000 h at L_0 = 560 cd/m^2 for a 2.8 cd/A, α-NPD-rubrene-DCJTB triple-doped red emitting device [18]. Normalizing

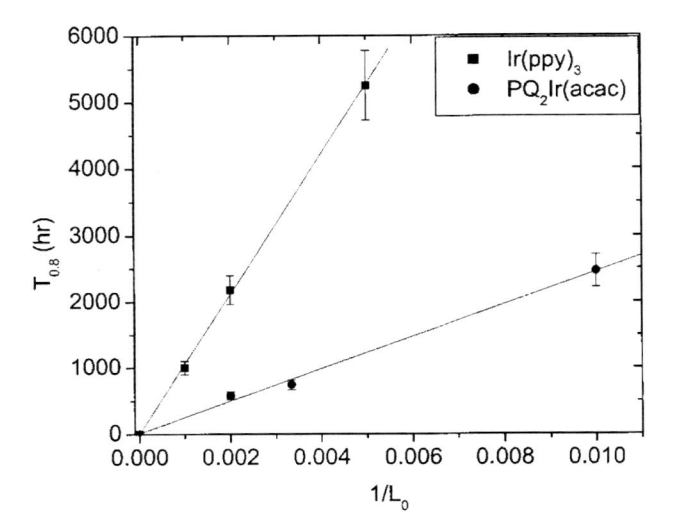

Figure 20.3 The time for the luminance to decay to $0.8L_0$, $T_{0.8}$, vs $1/L_0$, for a green Ir(ppy)$_3$ PHOLED and an orange-red PQ$_2$Ir(acac) PHOLED. Linear relationships are obtained.

to $L_0 = 100$ cd/m^2 using $T_{1/2} \propto 1/L_0$, these values correspond to $T_{1/2} = 100{,}000$ and 45,000 h, respectively; whereas the phosphorescent Ir(ppy)$_3$ green and PQ$_2$Ir(acac) red devices have $T_{1/2} = 50{,}000$ and 15,000 h, respectively. While the phosphorescent device lifetimes are shorter than the fluorescent device lifetimes, their stability has significantly improved to commercially viable values.

The voltage increase rates of both the Ir(ppy)$_3$ and PQ$_2$Ir(acac) devices are low, <0.3 mV/h during 3000 h of operation at $L_0 = 500$ and 300 cd/m^2, respectively, and comparable to those reported for fluorescent devices [17]. The high operational stability and the low voltage increase rate indicate that the integrity of the device structure is largely maintained during operation, suggesting the absence of material degradation due to decomposition from current-driven oxidation and reduction.

In conclusion, high-stability and high-efficiency electrophosphorescence in green and orange-red has been demonstrated, utilizing a multilayer architecture with BAlq as the hole blocking material. The efficiencies of the Ir(ppy)$_3$ green and PQ$_2$Ir(acac) orange-red devices are 19 and 17.6 cd/A with projected half-lives of 10,000 and 5000 h at L_0 of 500 and 300 cd/m^2, respectively. A linear degradation relationship of $T_{0.8} \propto 1/L_0$ is obtained, indicating that electrophosphorescence, despite its significantly higher efficiency compared with fluorescent OLEDs, follows the same lifetime scaling relationship.

Acknowledgements

This work was funded in part by the Defense Advanced Research Projects Agency and the US Army. The authors thank Dr. C. Adachi of Chitose Institute of Technology; and colleagues at PPG Industries for their assistance in this work.

References

1. C. W. Tang and S. A. VanSlyke, *Appl. Phys. Lett.* **51**, 913 (1987).
2. C. W. Tang, *SID International Sym. Digest of Technical Papers* (SID, Santa Ana, CA, 1996), p. 181.
3. C. Adachi, M. A. Baldo, M. E. Thompson, and S. R. Forrest, *J. Appl. Phys.* **90**, 5048 (2001).
4. M. Ikai, S. Tokito, Y. Sakamoto, T. Suzuki, and Y. Taga, *Appl. Phys. Lett.* **79**, 156 (2001).
5. M. A. Baldo, D. F. O'Brien, Y. You, A. Shoustikov, S. Sibley, M. E. Thompson, and S. R. Forrest, *Nature (London)* **395**, 151 (1998).
6. M. A. Baldo, S. Lamansky, P. E. Burrows, M. E. Thompson, and S. R. Forrest, *Appl. Phys. Lett.* **75**, 4 (1999).
7. P. E. Burrows, S. R. Forrest, T. X. Zhou, and L. Michalski, *Appl. Phys. Lett.* **76**, 2493 (2000).
8. T. Watanabe, K. Nakamura, S. Kawami, Y. Fukuda, T. Tsuji, T. Wakimoto, and S. Miyahuchi, *Proc. SPIE* **4105**, 175 (2000).
9. C. H. Chen, J. Shi, and C. W. Tang, *Macromol. Symp.* **125**, 1 (1997).
10. R. Fink, Y. Heischkel, M. Thelakkat, and H.-W. Schmidt, *Chem. Mater.* **10**, 3620 (1998).
11. M.-J. Yang and T. Tsutsui, *Jpn. J. Appl. Phys.*, Part 2 **39**, L828 (2000).
12. Unpublished results.

13. Y. Sato, T. Ogata, S. Ichinosawa, and Y. Murata, *Synth. Met.* **91**, 103 (1997).
14. C. Adachi, R. C. Kwong, and S. R. Forrest, *Organic Electronics* **2**, 37 (2001).
15. S. Lamansky, P. Djurovich, D. Murphy, F. Abdel-Razaq, R. C. Kwong, I. Tsyba, M. Bortz, B. Mui, R. Bau, and M. E. Thompson, *Inorg. Chem.* **40**, 1704 (2001).
16. S. A. Van Slyke, C. H. Chen, and C. W. Tang, *Appl. Phys. Lett.* **69**, 2160 (1996).
17. J. Shi and C. W. Tang, *Appl. Phys. Lett.* **70**, 1665 (1997).
18. T. K. Hatwar, G. Rajeswaran, J. Shi, Y. Hamada, H. Kanno, and H. Takahashi, *Proceedings of the 10th International Workshop on Inorganic and Organic Electroluminescence*, 2000, p. 31.

Chapter 21

Controlling Exciton Diffusion in Multilayer White Phosphorescent Organic Light Emitting Devices

Brian W. D'Andrade,[a] Mark E. Thompson,[b] and Stephen R. Forrest[a]

[a] *Center for Photonics and Optoelectronic Materials (POEM),
Princeton Materials Institute (PMI), Department of Electrical Engineering,
Princeton University, Princeton, New Jersey 08544, USA*
[b] *Department of Chemistry, University of Southern California,
Los Angeles, California 90089, USA*
forrest@ee.princeton.edu

White organic light-emitting devices (WOLEDs) are of interest because they offer low-cost alternatives for backlights in flat-panel displays, and may eventually find use in lighting. White light emission can be obtained from multilayer OLED structures [1, 2] n which different layers emit different parts of the visible spectrum, from single layer polymer blends [3, 4], or from hybrid organic/inorganic structures, white light emitting materials, or exciplexes [5–7]. Most of these approaches rely on the use of a combination of several emitting organic molecules to fully span

---

Reprinted from *Adv. Mater.*, **14**(2), 147–151, 2002.

Electrophosphorescent Materials and Devices
Edited by Mark E. Thompson
Text Copyright © 2002 WILEY-VCH Verlag GmbH
Layout Copyright © 2024 Jenny Stanford Publishing Pte. Ltd.
ISBN 978-981-4877-34-3 (Hardcover), 978-1-003-08872-1 (eBook)
www.jennystanford.com

the entire visible spectrum. Here we report on the use of blue, yellow, and red phosphor doped emissive regions combined in two multilayer OLEDs to efficiently produce white light. One device had Commission Internationale de l'Eclairage (CIE) coordinates of (0.35, 0.36), a maximum external quantum efficiency $\eta_{ext} = 3.8 \pm 0.4\%$ corresponding to 6.1 ± 0.6 cd/A, a maximum luminance of 29,000 $\pm$ 3000 cd/m^2 at 13.4 V, and a color rendering index (CRI) of 50; the second device had CIE coordinates (0.37, 0.40), $\eta_{ext} = 5.2 \pm 0.5\%$ corresponding to 11 ± 1 cd/A, a maximum luminance of 31,000 $\pm$ 3000 cd/m^2 at 14 V, and a CRI of 83.

As defined by the CIE chromaticity coordinate system, an ideal white light source has coordinates of (0.33, 0.33), and is obtained by balancing the emission from each of the colors employed in the WOLED. Additionally, the CRI of a white light source is a measure of the color shift that an object undergoes when illuminated by the light source as compared with the color of the same object when illuminated by a reference source of comparable color temperature [8]. The values of CRI range from 0 to 100, with 100 representing no shift in color. White light sources are referenced to daylight, with fluorescent bulbs having ratings between 60 and 99, mercury lamps near 50, and high-pressure sodium lamps can have a CRI of 20.

Typical luminous power efficiencies for white light sources are 15 lm/W for the incandescent light bulb, and 80 lm/W for a fluorescent lamp. Over the last several years, the power (η_p) and external quantum (η_{ext}) efficiencies of WOLEDs have been steadily improving. Some of the best results published report a maximum brightness of 20,000 cd/m^2 [2], an η_{ext} of 1.2% [3], and a η_p of 1.4 lm/W [1]. To our knowledge, there are presently no reports on the CRI of organic white devices.

Electrophosphorescent OLEDs have been shown to have very high η_{ext} when used for monochromatic light emission [9, 10], so their incorporation into a white emitting device should lead to similarly high efficiency WOLEDs. The devices reported here exhibit $\eta_{ext} = 5.2 \pm 0.5\%$, $\eta_p = 6.4 \pm 0.6$ lm/W, a maximum brightness of $31,000 \pm 3000$ cd/m^2, CIE coordinates (0.37, 0.40), and a CRI of 83.

Two white emitting devices were grown with structures listed in Table 21.1. The WOLEDs were grown on a glass substrate precoated

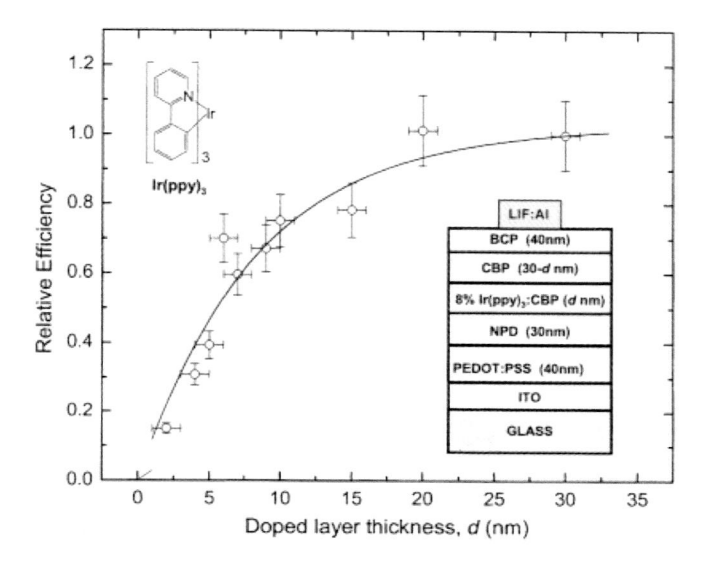

Figure 21.1 Relative efficiencies of devices with 300 - d Å of CBP doped with Ir(ppy)$_3$ compared to a device with 300 Å of CBP doped with Ir(ppy)$_3$. The solid line is a chi-square fit of the points to Eq. 21.1 in text. A CBP triplet transfer length of 83 ± 10 Å was obtained from the fit. Inset: Device structure used to probe the triplet exciton concentration in CBP and the molecular structural formula of Ir(ppy)$_3$.

with an indium tin oxide (ITO) layer having a sheet resistance of 20 Ω/sq, see Experimental section. Molecular structural formulae of the phosphors used [10] are shown in the inset of Fig. 21.1 and on the right hand side of Fig. 21.2.

The control of the diffusion of triplet excitons provides a means for obtaining the desired color balance. Triplets have lifetimes that are several orders of magnitude longer than singlets, hence they have longer diffusion lengths, allowing emissive layers to be >10 nm thick [11]. Hence, to achieve a desired emission color, the thickness of each layer doped with a different phosphor can be adjusted to serve as a recombination zone of the appropriate fraction of excitons initially formed at the hole transport layer/emissive region (HTL/EMR) interface.

To design such a structure, the 4,4'-N,N'-dicarbazole-biphenyl (CBP) triplet exciton transfer length, L_T, must first be determined.

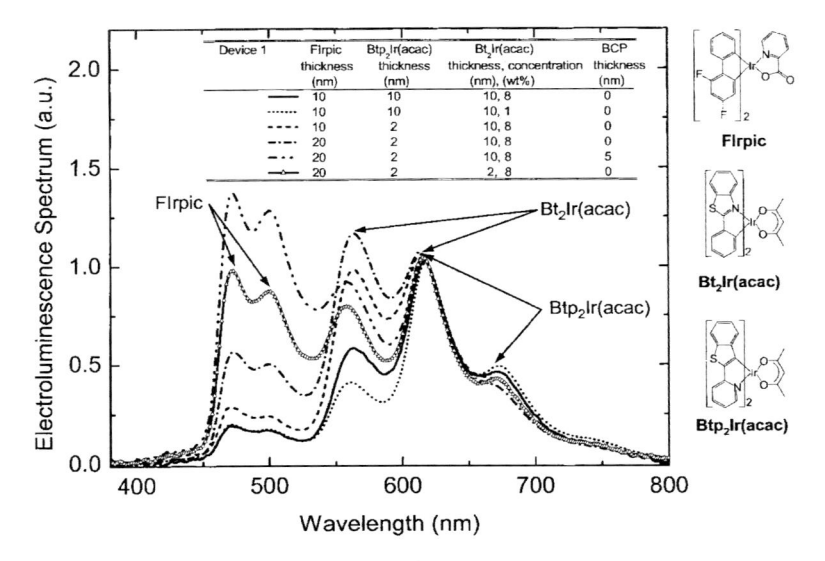

Figure 21.2 Variation, at 10 mA/cm², in the electroluminescence spectra with layer thickness, dopant concentration, and the insertion of an exciton/hole blocking layer between the FIrpic and Btp$_2$Ir(acac) doped layers for the Device 1. Right: The molecular structural formulae of Btp$_2$Ir(acac), FIrpic, and Bt$_2$Ir(acac).

Device 1	FIrpic thickness (nm)	Btp$_2$Ir(acac) thickness (nm)	Bt$_2$Ir(acac) thickness, concentration (nm, wt%)	BCP thickness (nm)
	10	10	10, 8	0
	10	10	10, 1	0
	10	2	10, 8	0
	20	2	10, 8	0
	20	2	10, 8	5
	20	2	2, 8	0

This is done by varying the thickness (d) of a *fac* tris(2-phenylpyridine)iridium (Ir(ppy)$_3$) phosphor doped region within a 30 nm thick CBP layer in the structure shown in Fig. 21.1. The efficiencies of the various devices are then compared to the efficiency of a device with the entire CBP region doped with Ir(ppy)$_3$. That is, assuming a linear relationship between the exciton density between position x and $x + \Delta x$, and the amount of light emitted from that region, the CBP exciton density is then related to the green Ir(ppy)$_3$ emission via:

$$\eta(x) = \frac{\eta(30)(1 - e^{-d/L_T})}{(1 - e^{-d_0/L_T})} \tag{21.1}$$

where $\eta(d)$ is the efficiency of a device with a doped CBP layer of thickness d, and $\eta(30)$ is the efficiency of the device with $d_0 = 30$ nm of CBP doped with Ir(ppy)$_3$. This technique, along with assumptions made, are similar to those discussed previously by Baldo et al. [12].

Table 21.1 WOLED structures employed in this study

Device 1 (Three phosphor structure)	Device 2 (Blocking layer structure)
Cathode [LiF/Al]	Cathode [LiF/Al]
BCP (40 nm)	BCP (40 nm)
Yellow EMR [8 wt.-% Bt$_2$Ir(acac):CBP (2 nm)]	Red EMR [8 wt.-% Btp$_2$IR(acac):CBP (10 nm)]
Red EMR [8 wt.-% Btp$_2$IR(acac):CBP (2 nm)]	Hole/Exciton blocker [BCP (3 nm)]
Blue EMR [6 wt.-% FIrpic:CBP (20 nm)]	Blue EMR [6 wt.-% FIrpic:CBP (20 nm)]
HTL [NPD (30 nm)]	HTL [NPD (30 nm)]
PEDOT:PSS (40 nm)	PEDOT:PSS (40 nm)
ITO	ITO
Substrate [Glass]	Substrate [Glass]

We obtain a CBP triplet transfer length of 8.3 ± 1 nm using a chi-square fit (solid line) of the data in Fig. 21.1 to Eq. 21.1.

By varying the concentration of the dopants, the location of the different color regions with respect to the HTL interface where exciton formation occurs, the thicknesses of each of the layers, and by inserting an exciton blocking layer between emissive layers, the CIE coordinates of the OLED emission can be tuned over a wide range. However, we note that the phosphors with lower triplet energy, which trap excitons most readily, should be positioned farthest from the exciton formation region. This ensures that the excitons can diffuse throughout the luminescent region, producing the desired output color balance. Figure 21.2 shows the dependence of electrophosphorescent spectrum of Device 1 on the layer thickness, phosphor doping concentration, and the insertion of a blocking layer between the iridium(III)bis(4,6-di-fluorophenyl)-pyridinato-N,C^2) picolinate (FIrpic) and bis(2-(2'-benzo[4,5-a]thienyl)pyridinato-N,C^3) iridium(acetylacetonate) (Btp$_2$Ir(acac)) doped regions. All spectra were recorded at 10 mA/cm^2, corresponding to luminances ranging between 400 and 800 cd/m^2.

FIrpic emission peaks at $\lambda = 472$ nm and at $\lambda = 500$ nm, shown in Fig. 21.2, increase relative to the Btp$_2$Ir(acac) emission at

$\lambda = 620$ nm when the thickness of the bis(2-phenyl benzothiozolato-$N,C^{2'}$) iridium(acetylacetonate) (Bt$_2$Ir(acac)) and Btp$_2$Ir(acac) doped layers are reduced to 2 nm, and when the thickness of the FIrpic layer is increased to 20 nm, because a larger fraction of the total number of excitons diffuse into the FIrpic layer and hence are available for emission from this somewhat less efficient dopant. However, FIrpic emission does not increase relative to Btp$_2$Ir(acac) emission for FIrpic doped layer thicknesses exceeding 30 nm. This suggests that the exciton formation zone is not at the HTL/EMR interface as seen in the diffusion profile of CBP triplets measured using Ir(ppy)$_3$, since FIrpic emission should continue to increase relative to all other phosphor emission with increasing FIrpic layer thickness. The precise location of the exciton formation zone in the WOLED is difficult to establish since it may shift depending on the several variables considered for color balancing.

The color balance (particularly enhancement of blue emission) can be improved by inserting a thin BCP, hole/exciton blocking layer between the FIrpic and Btp$_2$Ir(acac) doped layers in Device 2. This layer retards the flow of holes from the FIrpic doped layer towards the cathode and thereby forces more excitons to form in the FIrpic layer, and it prevents excitons from diffusing towards the cathode after forming in the FIrpic doped layer. These two effects increase FIrpic emission relative to Btp$_2$Ir(acac).

The main Bt$_2$Ir(acac) emission peak at $\lambda = 563$ nm is easily discernable from its sub-peak at $\lambda = 600$ nm which overlaps the main Btp$_2$Ir(acac) peak (Fig. 21.2). The peak at $\lambda = 563$ nm decreases when the doping concentration of Bt$_2$Ir(acac) is reduced from 8 wt.% to 1 wt.% and when the layer thickness is reduced from 10 nm to 2 nm. At 1 wt.% Bt$_2$Ir(acac), the transfer of triplets between the host and guest molecules is hindered because fewer guest molecules are within the Förster transfer radius (~30 Å) of the host, decreasing proportionately the fraction of Bt$_2$Ir(acac) emission. For the exciton concentration profile described by Eq. 21.1, $\eta(2) < \eta(10)$ because the dopant can capture more of the CBP triplets for thicker doped regions. Hence, the emission from a 2 nm thick Bt$_2$Ir(acac) doped layer should be lower than for a 10 nm thick layer because it captures fewer CBP excitons.

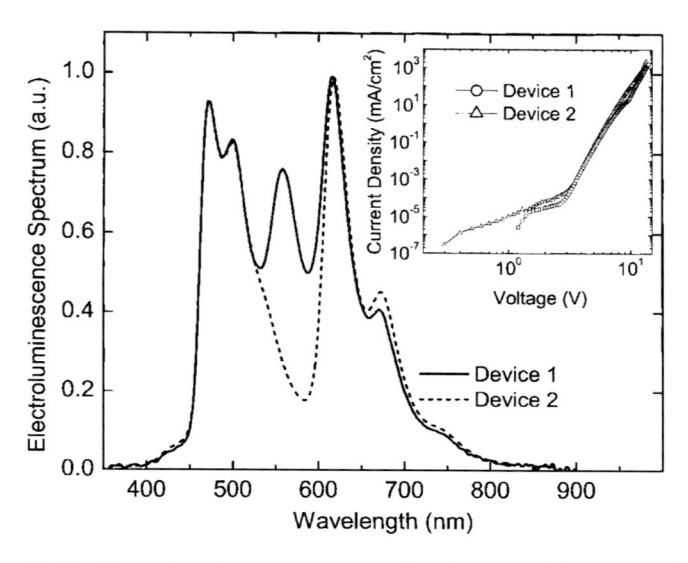

Figure 21.3 Electroluminescent spectra of Devices 1 and 2 at 10 mA/cm^2. Inset: Current density versus voltage characteristics of Device 1 and 2.

Device 1 and 2 electroluminescent spectra are compared in Fig. 21.3. For Device 2, there is almost no emission between $\lambda = 520$ nm and $\lambda = 600$ nm, whereas Device 1 has significantly more emission from Bt$_2$Ir(acac) in this region. The enhancement of the yellow region of the spectrum for Device 1 increases the CRI from 50 to 83 and shifts the CIE from (0.35, 0.36) to (0.37, 0.40) relative to Device 2. The x and y CIE coordinates of all the devices varied by <10% between 1 mA/cm^2 and 500 mA/cm^2, corresponding to luminances in the range from 60 to 20 000 cd/m^2.

Device 2 is useful for flat-panel displays since the human perception of white from the display will be unaffected by the lack of emission in the yellow region of the spectrum. In theory, the best white that can be made with FIrpic and Btp$_2$Ir(acac) doped into CBP is at (0.33, 0.32), close to that of Device 2 of (0.35, 0.36). With a CRI of 83, Device 1 can be used in flat-panel displays, but it can also be used as an illumination source, since at this high a CRI value, objects will appear as they would under daylight conditions. The CRI of Device 2 can be theoretically improved to a maximum value of 88, however, the CIE of such an optimized device is (0.47, 0.40). The

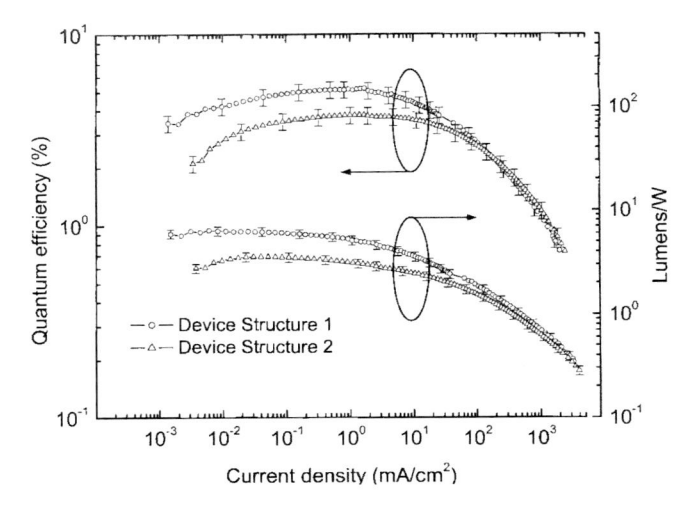

Figure 21.4 Power and external quantum efficiency versus current density for Devices 1 and 2.

additional doped layer also improves the efficiency of Device 2 as compared to Device 1 by boosting the yellow emission where the human eye has the highest photopic response efficiency, and by the use of Bt$_2$Ir(acac) [10], which has a higher η_{ext} than FIrpic.

We assume lambertian intensity profiles and calculate the η_{ext}, η_p (shown in Fig. 21.4) and luminance using the current density–voltage characteristics shown in the inset of Fig. 21.3. Table 21.2 shows the results for both Devices 1 and 2. Here, $\eta_{ext} \geq 3.0\%$ over three orders of magnitude in current density, and η_{ext} is found to increase to a maximum value and then roll-off at higher-current densities. The initial low η_{ext} is possibly due to current leakage; whereas at high current densities ($J > 10$ mA/cm^2), the roll off has been previously ascribed to triplet–triplet annihilation [13]. Device 1 attains a maximum luminance of 31,000 cd/m^2 at 14 V, and Device 2 emits 30,000 cd/m^2 at 13.4 V.

The break in vacuum necessary to define the cathode region limits the efficiency of the all-phosphor WOLED because of the introduction of non-radiative defect states due to exposure to atmosphere. Exciton transfer between CBP and FIrpic is especially sensitive to defects due to the endothermic process characteristic

Table 21.2 Chromaticity and efficiency values for WOLEDs

Device structure	Max η_{ext} [%]	Max efficiency [cd/A]	Max η_p [lm/W]	Max luminance [cd/m^2]	CRI	CIE (x,y) at 10 mA/cm^2
1	5.2 ± 0.5	11 ± 1	6.4 ± 0.6	31,000 at 13.4 V	83	(0.37,0.40)
2	3.8 ± 0.4	6.1 ± 0.6	3.6 ± 0.4	30,000 at 13.4 V	50	(0.35,0.36)

of this material system [14]. Higher efficiency WOLEDs, therefore, can be expected if the entire device is grown under high vacuum conditions.

In summary, we have demonstrated that multi-emissive layer fully electrophosphorescent WOLEDs can take advantage of the diffusion of triplets to produce bright white devices with high power and quantum efficiencies. The device color can be tuned by varying the thickness and the dopant concentrations in each layer, and by introducing exciton blocking layers between emissive layers.

Experimental

Prior to organic layer deposition, the substrates were degreased in ultrasonic solvent baths and then treated with an oxygen plasma [15] for 8 min at 20 W and 150 mtorr. Poly(3,4-ethylenedioxythiophene):poly(styrene sulfonic acid) (PEDOT:PSS), used to decrease OLED leakage current [16] and to increase OLED fabrication yield [17], was spun onto the ITO at 4000 rpm for 40 s and then baked for 15 min at 120°C, attaining an approximate thickness of 40 nm. All molecular organic layers were sequentially deposited without breaking vacuum by thermal evaporation at a base pressure of $<3 \times 10$ torr.

Deposition began with a 30 nm-thick 4,4'-bis[N-(1-napthyl)-N-phenyl-amino]biphenyl (α-NPD) hole transport layer. For Device 1, an emissive region was grown consisting of a 20 nm thick layer of the primarily electron conducting host 4,4'-N,N'-dicarbazole-biphenyl doped with 6 wt.% of the blue emitting phosphor

[14], iridium(III)bis(4,6-di-fluorophenyl)-pyridinato-N,C^2) picolinate, followed by a 2 nm-thick CBP layer doped at 8 wt.% with the red phosphor; bis(2-(2'-benzo[4,5-a]thienyl)pyridinato-N,C^3) iridium(acetylacetonate), and a 2 nm-thick CBP layer doped at 8 wt.% with the yellow phosphor; bis(2-phenyl benzothiozolato-$N,C^{2'}$) iridium(acetylacetonate).

For Device 2, the EMR consisted of a 20 nm thick layer of CBP doped with 6 wt.% of FIrpic, followed by a 3 nm-thick 2,9-dimethyl-4,7-diphenyl-1,10-phenanthroline (BCP) exciton blocking layer, and a 10 nm-thick CBP layer doped with 8 wt.% Btp$_2$Ir(acac). BCP was the final organic layer deposited on all devices and served as both a hole/exciton blocker and an electron transport layer.

After deposition of the organic layers, the samples were transferred from the evaporation chamber into a N$_2$ filled glove box containing $\leq$ 1 ppm of H$_2$O and O$_2$. After affixing masks with 1 mm diameter openings to the samples, they were transferred into a second vacuum chamber ($<10^{-7}$ torr) where a cathode was deposited through the masks. The cathode consisted of 5 Å of LiF followed by 100 nm of Al. The samples were only exposed to air while being tested.

Acknowledgements

The authors thank the Universal Display Corporation, the Defense Advanced Research Projects Agency, and the National Science Foundation's Materials Research Science and Engineering Center for their support.

References

1. X. Jiang, Z. Zhang, W. Zhao, W. Zhu, B. Zhang, S. Xu, *J. Phys. D: Appl. Phys.* 2000, **33**, 473.
2. S. Liu, J. Huang, Z. Xie, Y. Wang, B. Chen, *Thin Solid Films* 2000, **363**, 294.
3. S. Tasch, E. J. W. List, O. Ekström, W. Graupner, G. Leising, P. Schlichting, U. Rohr, Y. Geerts, U. Scherf, K. Müllen, *Appl. Phys. Lett.* 1997, **71**, 2883.
4. J. Kido, H. Shionoya, K. Nagai, *Appl. Phys. Lett.* 1995, **67**, 2281.

5. J. Feng, F. Li, W. Gao, S. Liu, Y. Liu, Y. Wang, *Appl. Phys. Lett.* 2001, **78**, 3947.
6. F. Hide, P. Kozodoy, S. P. DenBaars, A. J. Heeger, *Appl. Phys. Lett.* 1997, **70**, 2664.
7. Y. Hamada, T. Sano, H. Fujii, Y. Nishio, *Jpn. J. Appl. Phys.* 1996, **35**, 1339.
8. *Method of Measuring and Specifying Colour Rendering Properties of Light Sources*, Commission Internationale de L'Éclairage (CIE), Paris 1974.
9. C. Adachi, M. A. Baldo, M. E. Thompson, R. C. Kwong, M. E. Thompson, S. R. Forrest, *Appl. Phys. Lett.* 2001, **78**, 1622.
10. S. Lamansky, P. Djurovich, D. Murphy, F. Abdel-Razzaq, H.-E. Lee, C. Adachi, P. E. Burrows, S. R. Forrest, M. E. Thompson, *J. Am. Chem. Soc.* 2001, **123**, 4303.
11. M. A. Baldo, S. R. Forrest, *Phys. Rev. B* 2000, **62**, 10958.
12. M. A. Baldo, D. F. O'Brien, M. E. Thompson, S. R. Forrest, *Phys. Rev. B* 1999, **60**, 14 422.
13. M. A. Baldo, C. Adachi, S. R. Forrest, *Phys. Rev. B* 2000, **62**, 10967.
14. C. Adachi, R. C. Kwong, P. Djurovich, V. Adamovich, M. A. Baldo, M. E. Thompson, S. R. Forrest, *Appl. Phys. Lett.* 2001, **19**, 2082.
15. D. J. Milliron, I. G. Hill, C. Shen, A. Kahn, J. Schwartz, *J. Appl. Phys.* 2000, **87**, 572.
16. T. M. Brown, F. Cacialli, *IEE Proc.: Optoelectron.* 2001, **148**, 74.
17. P. Peumans, S. R. Forrest, *Appl. Phys. Lett.* 2001, **79**, 126.

Chapter 22

Blue Organic Electrophosphorescence Using Exothermic Host–Guest Energy Transfer

R. J. Holmes,[a] S. R. Forrest,[a] Y.-J. Tung,[b] R. C. Kwong,[b] J. J. Brown,[b] S. Garon,[c] and M. E. Thompson[c]

[a] Department of Electrical Engineering, Princeton University,
Princeton, New Jersey 08544, USA
[b] Universal Display Corporation,
375 Phillips Boulevard, Ewing, New Jersey 08618, USA
[c] Department of Chemistry, University of Southern California,
Los Angeles, California 90089, USA
forrest@ee.princeton.edu

We demonstrate efficient blue electrophosphorescence using exothermic energy transfer from a host consisting of N,N'-dicarbazolyl-3,5-benzene (mCP) to the phosphorescent iridium complex iridium(III)bis[(4,6-difluorophenyl)-pyridinato-$N,C^{2'}$] picolinate (FIrpic). By examining the temperature dependence of the radiative lifetime and the photoluminescence of a film of mCP doped with FIrpic, we confirm the existence of exothermic energy transfer in contrast to the endothermic transfer characteristic of the N,N'-dicarbazolyl-4-4'-biphenyl and FIrpic system. In employing

---

Reprinted from *Appl. Phys. Lett.*, **82**(15), 2422–2424, 2003.

Electrophosphorescent Materials and Devices
Edited by Mark E. Thompson
Text Copyright © 2003 American Institute of Physics
Layout Copyright © 2024 Jenny Stanford Publishing Pte. Ltd.
ISBN 978-981-4877-34-3 (Hardcover), 978-1-003-08872-1 (eBook)
www.jennystanford.com

exothermic energy transfer between mCP and FIrpic, a maximum external electroluminescent quantum efficiency of (7.5 ± 0.8)% and a luminous power efficiency of (8.9 ± 0.9)lm/W are obtained, representing a significant increase in performance over previous endothermic blue electrophosphorescent devices.

The design of an efficient, electrophosphorescent organic light-emitting device (OLED) often begins with the selection of a materials combination that allows for exothermic energy transfer between conductive host and phosphorescent guest molecules. Exothermic energy transfer occurs between a host excited state and a lower-energy guest unoccupied orbital, resulting in an energetically favorable excited state transition between molecules. Unlike the situation for red and green electrophosphorescent OLEDs [1, 2], the challenge to achieving exothermic energy transfer in the blue lies in the lack of suitable high-energy hosts (i.e., with a triplet level emitting at wavelengths $\lambda \leq 450$ nm). Until now, the only means demonstrated to achieve efficient blue electrophosphorescence has been through endothermic energy transfer in which the energy of the host triplet lies below that of the guest triplet state [3]. This allows for the generation of efficient high-energy electrophosphorescence provided that the energy required for the endothermic energy transfer is not significantly greater than the thermal energy. Unfortunately, the endothermic transfer efficiency is strongly affected by defect levels in the host that limit device efficiency, operating temperature, and long-term operational stability [3].

Here, we demonstrate efficient blue electrophosphorescence using exothermic energy transfer from the conductive host [4] N,N'-dicarbazolyl-3,5-benzene (mCP) to the phosphorescent iridium complex [5] iridium(III)bis[(4,6-difluorophenyl)-pyridinato-$N,C^{2'}$]picolinate (FIrpic). Device performance for an mCP-based structure is demonstrated to be superior to that employing the endothermic host–guest combination consisting of N,N'-dicarbazolyl-4-4'-biphenyl (CBP) [6, 7] and FIrpic. We obtain a maximum external electroluminescent quantum efficiency of $\eta_{\text{EXT}} = (7.5 \pm 0.8)\%$ and a luminous power efficiency of $\eta_\text{P} = (8.9 \pm 0.9)$lm/W for devices based on mCP and FIrpic. These results represent a significant increase in efficiency over a similar structure based on endothermic energy transfer from CBP to FIrpic, where $\eta_{EXT} = (6.1 \pm 0.6)\%$ and $\eta_P = (7.7 \pm 0.8)$lm/W.

OLEDs were grown on a glass substrate precoated with a ~100 nm thick layer of indium-tinoxide having a sheet resistance of ~20Ω/□. Substrates were degreased with solvents and then cleaned by exposure to oxygen plasma and UV-ozone ambient. All organic and cathode metal layers were grown in succession without breaking the vacuum (~2 × 10^{-8} Torr) by using an *in vacuo* mask exchange system. Two device structures were grown, differing only in the choice of host material. First, a 10-nm-thick copper phthalocyanine (CuPc) hole injection layer and a 30-nm-thick 4-4'-bis[N-(1-naphthyl)-N-phenyl-amino]biphenyl (α-NPD) hole transport layer were deposited. Next, 6% FIrpic (by weight) was codeposited with either mCP or CBP to form the 30-nm-thick emissive layer. Finally, a 40-nm-thick electron transport layer consisting of 4-biphenyloxolato aluminum(III)bis(2-methyl-8-quinolinato)4-phenylphenolate (BAlq) was deposited and used to block holes and confine excitons in the emissive zone. Literature procedures were used to prepare all materials [4, 5, 8, 9] except CuPc which was commercially obtained [10]. All materials were purified by repeated temperature gradient vacuum sublimation. Cathodes consisting of a 1-nm-thick layer of LiF followed by a 100-nm-thick layer of Al were patterned using a shadow mask with an array of 2 mm × 2 mm openings. Devices were encapsulated in a nitrogen ambient immediately after deposition, using a glass coverslip sealed to the substrate with a UV-curable epoxy.

Figure 22.1 depicts the chemical structures and thin-film phosphorescence spectra of the materials employed in the device emissive layers. The thin film, blue phosphorescence of FIrpic is visible at room temperature while weak phosphorescence from mCP and CBP is only observed at low temperature ($T = 10$ K). The phosphorescence of FIrpic, shown in Fig. 22.1, was collected under steady-state excitation by a filtered Hg lamp ($\lambda = 334.1$ nm) at $T = 10$ K. The phosphorescence spectrum of mCP was obtained ~1 ms after excitation by a nitrogen laser ($\lambda = 337$ nm, 50 Hz, and 1 ns pulses). The CBP phosphorescence in Fig. 22.1 is from Ref. [3]. The triplet emission from mCP is blueshifted from that of CBP, giving an estimated mCP triplet energy (arrows in Fig. 22.1) of (2.90 ± 0.10)

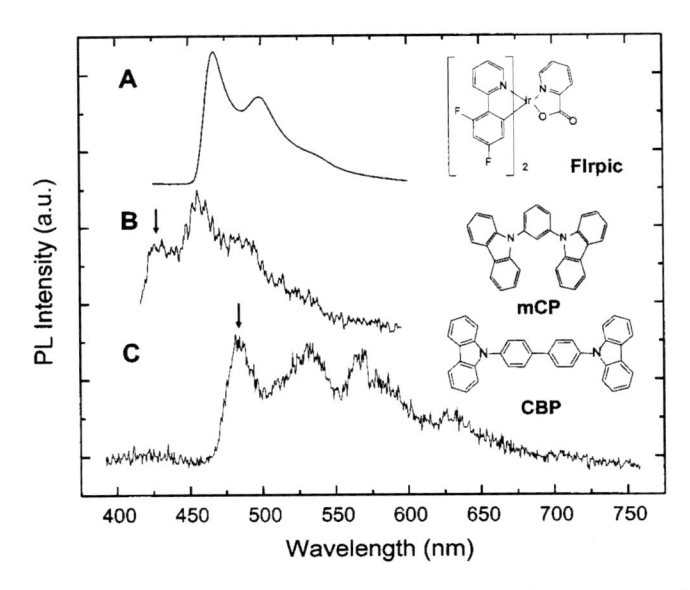

Figure 22.1 Phosphorescence spectra and chemical structures of the emissive materials used in this study measured at 10 K. Curves A, B, and C are collected for FIrpic, mCP, and CBP, respectively. (The phosphorescence spectrum of CBP (line C) is reproduced from Ref. [3].) The arrows indicate the inferred triplet level positions of CBP and mCP.

eV, compared to (2.56 ± 0.10) eV for CBP [3]. With a triplet energy of FIrpic of (2.65 ± 0.10) eV, we infer that energy transfer from mCP to FIrpic is exothermic.

Figure 22.2 compares the external efficiencies obtained for both CBP:6% FIrpic (circles) and mCP:6% FIrpic (squares) devices. A maximum quantum efficiency of (6.1 ± 0.6)% and a luminous power efficiency of (7.7 ± 0.8)lm/W are realized at current densities of 0.2 and 0.02 mA/cm^2, respectively, for the CBP-based devices. A peak luminance of 6400 cd/m^2 is achieved at a current density of 100 mA/cm^2 with a quantum efficiency of 2.9%. Energy transfer in this host–guest materials system has been previously demonstrated to be endothermic [3]. For mCP:6% FIrpic devices, a maximum quantum efficiency of (7.5 ± 0.8)% and a luminous power efficiency of (8.9 ± 0.9)lm/W are obtained at current densities of 0.2 and 0.03 mA/cm^2, respectively. A peak luminance of 9500 cd/m^2 is achieved at a current density of 100 mA/cm^2 with a quantum efficiency of

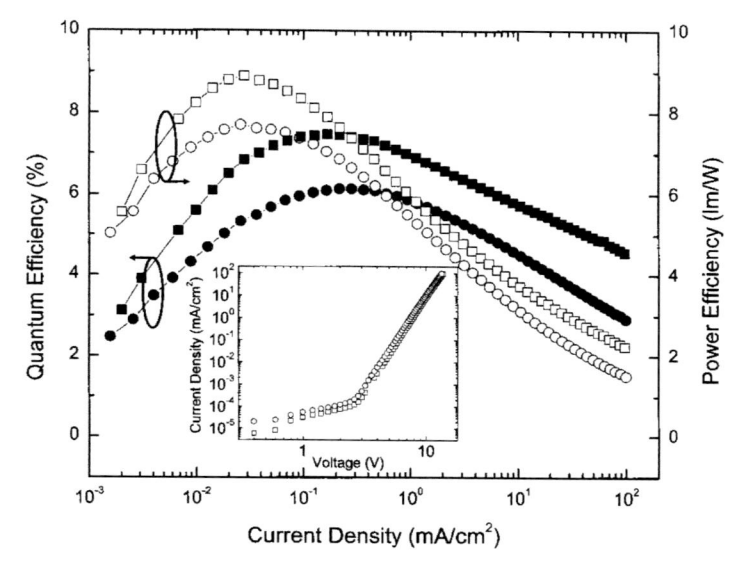

Figure 22.2 Efficiency vs drive current density for ITO/CuPc (10 nm)/α-NPD (30 nm)/CBP (circles) and mCP (squares):6% FIrpic (30 nm)/BAlq (40 nm)/LiF (1 nm)/Al (100 nm). The quantum and power efficiencies increase significantly in devices that employ the exothermic materials combination of mCP:FIrpic. Inset: Current–voltage characteristics for devices consisting of both CBP:FIrpic (circles) and mCP:FIrpic (squares) emissive layers.

4.6%. At these high current densities, the mCP:6% FIrpic device has > 50% higher quantum efficiency than the CBP:6% FIrpic device.

Several competing factors can yield a higher efficiency. For example, improved charge injection or an increase in carrier mobility in mCP versus CBP can lead to such an increase. However, this would lead to a corresponding change in the current–voltage characteristics of the device. From Fig. 22.2, inset, the charge injection and transport in CBP and mCP based devices are nearly identical, with the mCP device having a slightly larger drive voltage consistent with its larger energy gap, suggesting that the higher efficiencies measured for mCP based devices are not the result of improved carrier injection. Alternatively, the increase in quantum efficiency can result from an increase in the photoluminescence (PL) efficiency for FIrpic in mCP versus CBP. In fact, the PL efficiency of FIrpic in mCP relative to FIrpic in CBP is 40%–60% more efficient over a range of excitation wavelengths from $\lambda = 250$ nm to 350 nm. This is a further indication of the improved energy transfer between mCP and FIrpic as well as a reduced susceptibility to traps that would serve to limit the PL efficiency of FIrpic in an endothermic system.

The most reliable means for determining the energetics of the transfer process is to examine the temperature dependence of the guest luminescence efficiency and its radiative transient response. In Fig. 22.3, we find that the temperature dependence of the PL intensity from a film of mCP:6% FIrpic (open symbols) is much weaker than that observed for CBP:6% FIrpic (closed symbols) over a range of temperature from 10 K to 300 K. For CBP:FIrpic, an initial increase in PL intensity is observed with decreasing temperature as non-radiative pathways via traps in the guest and host are "frozen out" as the temperature is decreased. Below 200 K, a further reduction in temperature leads to a rapid decrease in the PL efficiency of the film as endothermic energy transfer between the CBP host molecules and FIrpic guest molecules becomes increasingly inefficient. Due to the energetically unfavorable transfer process, the PL efficiency is strongly influenced by traps deep within the host and/or guest energy gaps. As the temperature is decreased, energy transfer from CBP to FIrpic competes less effectively with nonradiative recombination at defects, leading to a decrease in the PL efficiency. In contrast, the mCP:FIrpic film shows almost no temperature dependence except at very low temperatures, characteristic of efficient exothermic energy transfer [1].

The reduction in PL efficiency at low temperatures is due to the decreased probability of exciton diffusion from mCP to FIrpic [11]. While the mCP emission increases by a factor of 7.5 due to exciton trapping followed by radiative recombination as the temperature is lowered to 10 K, the FIrpic emission decreases by only 40%. To understand this apparent discrepancy, the neat film PL efficiencies of mCP and FIrpic were measured to increase by factors of 2 and 6.5, respectively, as the temperature was reduced from 300 K to 10 K. Hence, the 40% change in the PL intensity from FIrpic in mCP implies that the number of excitons that are trapped on mCP increases by $0.4 \times 6.5 = 2.6$ at 10 K. Further, since the neat film PL efficiency of mCP doubles in going from $T = 300$ K to $T = 10$ K, a

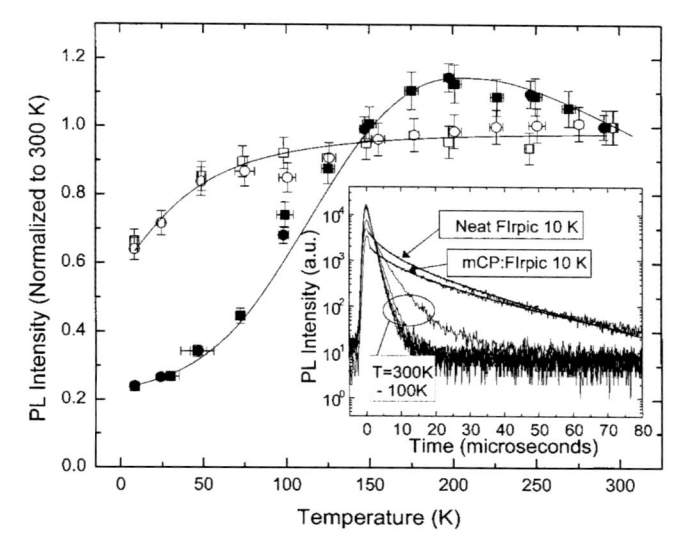

Figure 22.3 PL intensity originating from films of CBP:6% FIrpic (closed symbols) and mCP:6% FIrpic (open symbols) as a function of temperature. Data were collected first as the temperature was ramped down (squares) followed by ramping up to room temperature (circles). The excitation source for both films was a filtered Hg lamp ($\lambda = 334.1$ nm). Inset: The transient decay of mCP:6% FIrpic PL under excitation by a nitrogen laser ($\lambda = 337$ nm, 50 Hz, and 1 ns pulses). No change in the decay transient is observed above $T = 50$ K. At $T = 50$ K and below, the decay time increases while the relative PL intensity decreases. Depicted for comparison is the transient decay of a neat film of FIrpic at 10 K. (The solid lines are fits to the transients based on the theory of Ref. [12].)

total increase in mCP PL intensity of $2.6 \times 2 = 5.2$ is expected for the doped film, which is consistent with the increase of ∼7.5 observed for mCP fluorescence from a doped film of mCP and FIrpic.

The transient response of CBP:FIrpic PL reported previously [3] indicated that the radiative lifetime is longest at $T \sim 200$ K, where the PL efficiency is maximum, decreasing monotonically as temperature is further reduced to 10 K. Also indicative of endothermic energy transfer is the observation of delayed phosphorescence lasting for milliseconds beyond the natural FIrpic lifetime (∼23 μs at 10 K and ∼1 μs at 300 K in neat film as in Fig. 22.3). Delayed phosphorescence implies that triplet excitons are trapped on the long-lived CBP triplet state prior to energy transfer back to the FIrpic triplet where they radiatively recombine.

From the foregoing, the significantly reduced temperature dependence of the PL from an mCP:FIrpic film is strong evidence for exothermic energy transfer. This conclusion was confirmed by measuring the transient decay of FIrpic phosphorescence in mCP as a function of temperature (Fig. 22.3, inset). In contrast to the CBP:FIrpic system, the transient lifetime is again nearly temperature independent until $T = 50$ K, at which point it increases monotonically as the temperature is further decreased. No delayed phosphorescence is observed in a doped film of mCP:6% FIrpic, indicating that the mCP triplet transfers efficiently to FIrpic. The transient decay of a neat film of FIrpic at 10 K is also shown in Fig. 22.3, inset. Fits to both the FIrpic and mCP:FIrpic data, following Ref. [12], yield phosphorescent lifetimes of (22.3 $\pm$ 0.3 μs) and (27.8 $\pm$ 0.5 μs), respectively. As in the case of the PL intensity at the lowest temperatures, we expect that the radiative decay time is ultimately determined by thermally activated diffusion of excitons from host to guest, giving rise to the increased lifetime associated with the doped film of mCP:FIrpic at low temperatures.

We have demonstrated efficient blue electrophosphorescence by employing an emissive exothermic host–guest material system consisting of mCP and FIrpic. The favorable relative energies of the triplet states of these two materials leads to improved host–guest energy transfer, allowing for significant increases in device performance over structures employing the endothermic host–guest pair of CBP and FIrpic.

Acknowledgements

This work was partially supported by the Defense Advanced Research Projects Agency/Air Force Research Labs (Dr. R. Tulis and Dr. D. Hopper) and Universal Display Corporation.

References

1. C. Adachi, M. A. Baldo, M. E. Thompson, and S. R. Forrest, *J. Appl. Phys.* **90**, 5048 (2001).
2. C. Adachi, M. A. Baldo, R. C. Kwong, S. Lamansky, M. E. Thompson, and S. R. Forrest, *Appl. Phys. Lett.* **78**, 1622 (2001).
3. C. Adachi, R. C. Kwong, P. Djurovich, V. Adamovich, M. A. Baldo, M. E. Thompson, and S. R. Forrest, *Appl. Phys. Lett.* **79**, 2082 (2001).
4. V. Adamovich, J. Brooks, A. Tamayo, A. M. Alexander, P. Djurovich, B. W. D'Andrade, C. Adachi, S. R. Forrest, and M. E. Thompson, *New J. Chem.* **25**, 1171 (2002).
5. S. Lamansky, P. Djurovich, D. Murphy, F. Abdel-Razzaq, R. C. Kwong, I. Tsyba, M. Bortz, B. Mui, R. Bau, and M. E. Thompson, *Inorg. Chem.* **40**, 1704 (2001).
6. D. F. O'Brien, M. A. Baldo, M. E. Thompson, and S. R. Forrest, *Appl. Phys. Lett.* **74**, 442 (1999).
7. M. A. Baldo, S. Lamansky, P. E. Burrows, M. E. Thompson, and S. R. Forrest, *Appl. Phys. Lett.* **75**, 4 (1999).
8. D. E. Loy, B. E. Koene, M. E. Thompson, P. E. Burrows, and S. R. Forrest, *Chem. Mater.* **10**, 2235 (1998).
9. High-purity BAlq is obtained from PPG Industries, Pittsburgh, PA 15272.
10. Copper(II) phthalocyanine, Cat. 54,668-2, Sigma-Aldrich Corp., St. Louis, Missouri, 63178.
11. R. Priestley, A. D. Wasler, and R. Dorsinville, *Opt. Commun.* **158**, 93 (1998).
12. M. A. Baldo, C. Adachi, and S. R. Forrest, *Phys. Rev. B* **62**, 10967 (2000).

Chapter 23

Efficient, Deep-Blue Organic Electrophosphorescence by Guest Charge Trapping

R. J. Holmes,[a] B. W. D'Andrade,[a] S. R. Forrest,[a] X. Ren,[b] J. Li,[b] and M. E. Thompson[b]

[a] Center for Photonics and Optoelectronic Materials (POEM),
Department of Electrical Engineering and the Princeton Materials Institute,
Princeton University, Princeton, New Jersey 08544, USA
[b] Department of Chemistry, University of Southern California,
Los Angeles, California 90089, USA
forrest@ee.princeton.edu

We demonstrate efficient, deep-blue organic electrophosphorescence using a charge-trapping phosphorescent guest, iridium(III) bis(4',6'-difluorophenylpyridinato)tetrakis(1-pyrazolyl)borate (FIr6) doped in the wide-energy-gap hosts, diphenyldi(o-tolyl)silane (UGH1) and *p*-bis(triphenylsilyly)benzene (UGH2), where exciton formation occurs directly on the guest molecules. Charge trapping on the guest is confirmed by the dependence of the drive voltage and electroluminescence spectrum on guest concentration. Ultraviolet photoemission spectroscopy measurements establish the relative

---

Reprinted from *Appl. Phys. Lett.*, **83**(18), 3818–3820, 2003.

Electrophosphorescent Materials and Devices
Edited by Mark E. Thompson
Text Copyright © 2003 American Institute of Physics
Layout Copyright © 2024 Jenny Stanford Publishing Pte. Ltd.
ISBN 978-981-4877-34-3 (Hardcover), 978-1-003-08872-1 (eBook)
www.jennystanford.com

highest occupied molecular orbital positions of FIr6 in UGH1 and UGH2. Peak quantum and power efficiencies of $(8.8 \pm 0.9)\%$ and (11.0 ± 1.1) lm/W in UGH1 and $(11.6 \pm 1.2)\%$ and (13.9 ± 1.4) lm/W in UGH2 are obtained, while the emission in both cases is from FIr6 and is characterized by Commission Internationale de l'Eclairage coordinates of $(x = 0.16, y = 0.26)$ in UGH2.

Recently, efficient blue electrophosphorescence in organic light-emitting devices (OLEDs) has been generated by energy transfer from a fluorescent host to a phosphorescent guest molecule [1–3]. However, there are obstacles to using energy transfer to excite deeper blue guest phosphors. As the energy gap of the guest becomes large, successively larger host energy gaps are required to maintain efficient exothermic energy transfer from host to guest [2]. While finding such hosts is possible, electrically injecting carriers and forming excitons on wide-gap materials is often difficult, requiring the traversing of large energy barriers between carrier transport layers and the highest occupied molecular orbital (HOMO) and lowest unoccupied molecular orbital (LUMO) of the wide-gap host. To alleviate this difficulty, we employ host molecules that act as an inert matrix, leaving the guest molecules to both conduct and trap charge, allowing for direct exciton formation on the guest phosphor. This eliminates the need to electrically excite the host, while allowing for efficient carrier collection, exciton formation, and recombination at the guest molecular sites. For efficient emission, however, an exothermic path from host to guest must be maintained to eliminate the transfer of excitons formed on guest molecules back to the lower energy host molecules [1, 2].

A number of studies have examined means to distinguish between the two primary mechanisms of exciton formation and emission in OLEDs: host–guest energy transfer and direct charge trapping followed by exciton formation at the guest. Distinctions between these mechanisms can be made from the dependence of drive voltage or changes in output spectrum on guest concentration, as will be shown subsequently [4–6].

OLEDs were grown on a glass substrate precoated with a $\sim$100-nm-thick layer of indium tin oxide (ITO) having a sheet resistance of $\sim$20 $\Omega/\square$. Substrates were degreased with solvents and then cleaned by exposure to a UV-ozone ambient. All organic layers were grown in succession without breaking vacuum ($\sim 1 \times 10^{-7}$ Torr). Masking for cathode deposition after the layers were deposited was completed in a glove box with <1 ppm water and oxygen. The device structure consisted of a 40-nm-thick 4-4'-bis [N-(1-naphthyl)-N-phenyl-amino]biphenyl (NPD) hole transport layer, followed by a 15-nm-thick layer of N,N'-dicarbazolyl-3,5-benzene (mCP) [7] that confines excitons to the OLED emissive layer (EML). The 25-nm-thick blue EML consists of bis(4',6'-difluorophenylpyridinato)tetrakis(1-pyrazolyl)borate [8] (FIr6) codeposited with either diphenyldi(o-tolyl)silane (UGH1) or p-bis(triphenylsilyly)benzene (UGH2). Finally, a 40-nm-thick electron transporting and hole blocking layer consisting of 2,9-dimethyl-4,7-diphenyl-1,10-phenanthroline (bathocuproine, BCP) was deposited. Literature procedures were used to prepare NPD [9], mCP [7], UGH1 [10, 11], UGH2 [12], and FIr6 [8], while BCP was purchased from the Aldrich Chemical Company [13]. All materials were purified by repeated temperature gradient vacuum sublimation. Cathodes consisting of a 0.5-nm-thick layer of LiF followed by a 50-nm-thick layer of Al were patterned using a shadow mask with an array of 1-mm-diameter circular openings. Current–voltage and external efficiency measurements were carried out using standard methods [14].

Figure 23.1 shows the chemical structures and the emission and excitation spectra measured for the host and guest materials used. The room-temperature emission spectra for UGH1, UGH2, and FIr6 were collected under illumination by a Hg lamp monochromated to wavelengths of $\lambda = 239$, 230, and 325 nm, respectively. The excitation spectra of UGH1, UGH2, and FIr6 were collected for emission wavelengths of 303, 320, and 458 nm, respectively. The blue phosphorescence of FIr6 is readily visible at room temperature, allowing for assignment of the triplet state to the peak at $\lambda = 457$ nm. The phosphorescence of each fluorescent host is measured in 2-methyltetrahydrofuran solution at 77 K to be $\lambda = 393$ nm for UGH1 and $\lambda = 390$ nm for UGH2. From these energies, we expect the energy transfer to FIr6 from both UGH1 and UGH2 to be exothermic.

Figure 23.2 depicts the performance of two phosphorescent OLEDs, one with UGH1 (closed symbols) and the other with UGH2 (open symbols) codeposited with FIr6. Peak external electroluminescent (EL) quantum and power efficiencies of $\eta_{EXT} = (8.8 \pm 0.9)\%$

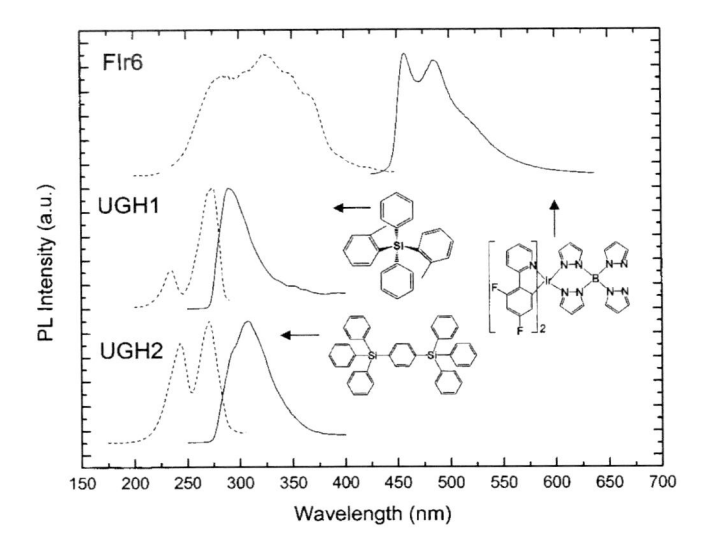

Figure 23.1 Room-temperature emission and excitation spectra and the chemical structures of the emissive materials used in this study. Emission spectra (solid lines) were collected under excitation at $\lambda = 239$, 230, and 325 nm, for UGH1, UGH2, and FIr6, respectively, while excitation spectra (broken lines) were collected at $\lambda = 303$, 320, and 458 nm.

and $\eta_P = (11.0 \pm 1.1)$ lm/W for a UGH1:10% FIr6 (by weight) OLED are achieved at current densities of 22.5 and 6.2 µA/cm², respectively. With a UGH2 host also doped with 10% FIr6, $\eta_{EXT} = (11.6 \pm 1.2)\%$ and $\eta_P = (13.9 \pm 1.4)$ lm/W are obtained at current densities of 9.9 and 5.6 µA/cm², respectively. Devices using UGH1 and UGH2 have luminances of 12,600 cd/m² (320 mA/cm²) and 11,800 cd/m² (156 mA/cm²), respectively, at 15 V. The difference in efficiency between UHG1 and UGH2 devices is possibly due to the enhanced dopant aggregation in UGH1, leading to increased exciton quenching in this device. The tendency to aggregate is inferred from the strong peak in quantum efficiency for UGH1:FIr6 devices between 5% and 10% FIr6, compared to the nearly concentration-independent efficiency of UGH2:FIr6 devices between 5% and 20% dopant concentrations.

The inset of Fig. 23.2 shows the EL spectra at 10 mA/cm² originating from FIr6, with Commission Internationale de L'Eclairage

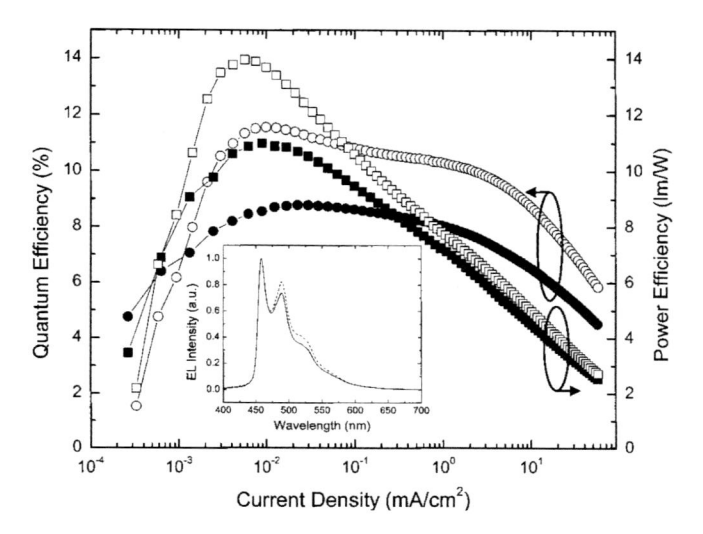

Figure 23.2 Quantum (circles) and power (squares) efficiency versus drive current density for ITO/NPD (40 nm)/mCP (15 nm)/UGH1 (closed symbols) or UGH2 (open symbols): 10% FIr6 (25 nm)/BCP (40 nm)/LiF (0.5 nm)/Al (50 nm) OLEDs. Inset: EL at 10 mA/cm^2 originating solely from FIr6 in UGH1 (broken line) and UGH2 (solid line).

(CIE) coordinates of (0.16, 0.28) and (0.16, 0.26) for UGH1 and UGH2 devices, respectively. The EL spectra were independent of drive current density from 1 to 100 mA/cm^2.

In all devices, a thin layer of mCP is inserted between the NPD layer and the EML. Without the mCP, the EL spectrum shows a significant contribution from NPD, and the device efficiency drops by an order of magnitude. Here, mCP efficiently prevents the leakage of both excitons and electrons from the EML into NPD, increasing color purity and carrier balance in the emissive layer.

To determine whether charge transport is influenced by the presence of the guest, we fabricated an OLED consisting of a neat FIr6 EML, whose quantum and power efficiencies are shown in Fig. 23.3. The peak efficiencies are $\eta_{EXT} = (5.0 \pm 0.5)\%$ and $\eta_P = (7.0 \pm 0.7)$ lm/W, comparable to other highly efficient devices whose emissive zone consists of a neat layer of phosphorescent material [15]. The CIE coordinates at 10 mA/cm^2 are (0.19, 0.34), with a slight blueshift with increasing drive current density. The high efficiencies

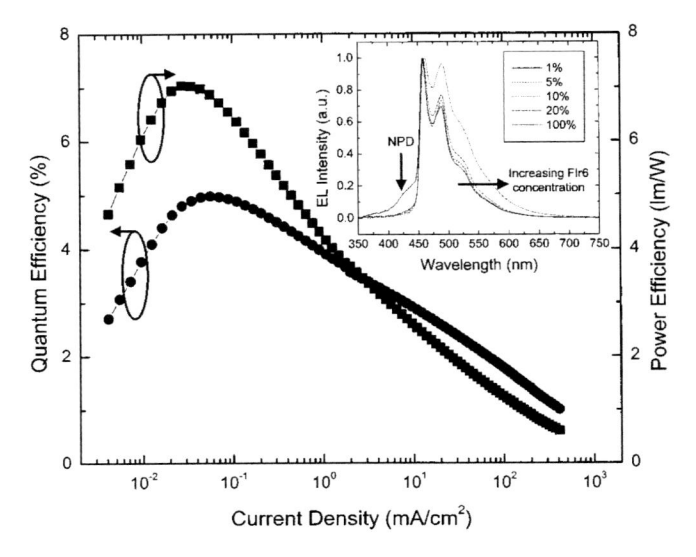

Figure 23.3 Quantum (circles) and power (squares) efficiency versus drive current density for ITO/NPD (40 nm)/mCP (15 nm)/FIr6 (25 nm)/BCP (40 nm)/LiF (0.5 nm)/Al (50 nm) OLEDs. Inset: Dependence of the EL spectrum on guest concentration at 10 mA/cm^2 for various concentrations of FIr6 in UGH2. For FIr6 concentrations less than 5%, incomplete electron trapping by FIr6 is evidenced by the observation at $\lambda = 435$ nm of NPD emission.

observed for devices with a neat layer of FIr6 suggest that this dopant can efficiently transport charge without the need for a host.

We also fabricated devices consisting only of a neat layer of UGHx. We find that $\eta_{EXT} = (0.9 \pm 0.1)\%$ and $\eta_P = (0.4 \pm 0.1)$ lm/W for both UGH1 and UGH2, with emission originating almost entirely from the NPD hole transport layer. This suggests that UGH1 and UGH2 are effective hole blocking materials, although both molecules transport electrons, allowing for exciton formation to occur on NPD, as observed.

Figure 23.4 shows the current-density–voltage $(J-V)$ characteristics of a UGH2:FIr6 device as a function of FIr6 concentration, as well as a proposed energy level scheme for the device. The HOMO energies for each material were measured using ultraviolet photoemission spectroscopy, except those of BCP and NPD, which were taken from Ref. [16]. The LUMO energies were estimated by

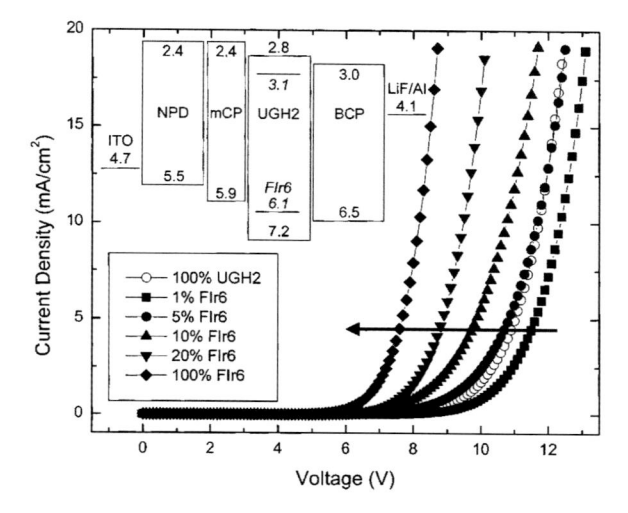

Figure 23.4 $J-V$ characteristics for ITO/NPD (40 nm)/mCP (15 nm)/ UGH2:x% FIr6 (25 nm)/BCP (40 nm)/LiF (0.5 nm)/Al (50 nm) OLEDs. The arrow points in the direction of increasing FIr6 concentration. Inset: Proposed energy level scheme for the device under zero applied bias.

adding the energy corresponding to the onset of optical absorption to the HOMO energy. While this yields an underestimate of the carrier transport gap, the accuracy of the relative energy level placements for all of the organic materials considered is sufficient to understand device performance [16]. From Fig. 23.4, we infer a barrier to hole injection into UGH2 from mCP of 1.3 eV (consistent with the observation of NPD emission in a neat UGH2 device), while hole injection into FIr6 is energetically favorable with only a 0.2 eV barrier. This is supported in the $J-V$ characteristics of doped device. The increase in drive voltage with a low (1%) concentration of FIr6 results since FIr6 acts as an electron trap in the EML [6]. In a device consisting of undoped UGH2, current is carried by electrons in the EML, allowing for exciton formation and recombination directly on NPD. The addition of FIr6 traps reduces the electron mobility while also increasing hole injection from NPD/mCP into the EML. This is inferred from the inset of Fig. 23.3, where a significant component of the EL spectrum ($\lambda \leq 450$ nm) from a UGH2:1% FIr6 device originates from NPD. As the concentration of FIr6 is increased, the hole injection efficiency into the EML increases, while electron leakage into NPD decreases as electrons are trapped by the guest molecules. A further increase in FIr6 concentration beyond 5% leads to a decreasing drive voltage, with the minimum drive voltage at 100% FIr6. Thus, for guest concentrations >5%, both electrons and holes are transported by FIr6, with UGH2 acting as an inert host matrix.

The same trends are also observed for UGH1:FIr6-based devices. However, for UGH1, the increase in voltage at 1% FIr6 is larger, remaining larger than for neat UGH1 even at 5% FIr6. Increases in concentration beyond 5% FIr6 lead to drive voltages similar to those for UGH2 in Fig. 23.4. This behavior is consistent with the LUMO level difference for UGH1:FIr6 being larger than for UGH2:FIr6. For UGH1, the HOMO is at (7.2 ± 0.2) eV, and the LUMO energy is at (2.6 ± 0.4) eV, yielding a LUMO difference with FIr6 of 0.5 eV, reduced to only 0.3 eV for UGH1. Hence, FIr6 is a more effective electron trap in UGH1.

We have demonstrated highly efficient blue electrophosphorescence by charge trapping on guest molecules of FIr6 doped into the wide-energy-gap hosts UGH1 and UGH2. The EL of FIr6 in a host–guest system is noticeably "bluer" in color at (0.16, 0.26) than the previously studied phosphorescent blue emitter iridium(III)bis[(4,6-difluorophenyl)pyridinato-$N,C^{2'}$]picolinate (FIrpic), which has CIE coordinates of (0.16, 0.37) [2]. In Ref. [2], efficient blue electrophosphorescence from FIrpic is demonstrated using exothermic energy transfer from a wide-gap host to the FIrpic guest, although this mechanism is more difficult to achieve for deep-blue emitters. The deep HOMO levels of UGH1 and UGH2 make hole trapping on FIr6 the dominant means for hole injection into the EML. The direct formation of excitons on FIr6 avoids electrical excitation of the wide-gap host, while eliminating exothermic back transfer from guest to host.

Acknowledgements

This work was partially supported by the Defense Advanced Research Projects Agency/Air Force Research Labs (Dr. R. Tulis and

Dr. D. Hopper) and Universal Display Corporation. We would like to thank Dr. David Knowles and Dr. Jie-Yi Tsai for helpful discussions.

References

1. C. Adachi, R. C. Kwong, P. Djurovich, V. Adamovich, M. A. Baldo, M. E. Thompson, and S. R. Forrest, *Appl. Phys. Lett.* **79**, 2082 (2001).
2. R. J. Holmes, S. R. Forrest, Y.-J. Tung, R. C. Kwong, J. J. Brown, S. Garon, and M. E. Thompson, *Appl. Phys. Lett.* **82**, 2422 (2003).
3. S. Tokito, T. Iijima, Y. Suzuri, H. Kita, T. Tsuzuki, and F. Sato, *Appl. Phys. Lett.* **83**, 569 (2003).
4. M. Uchida, C. Adachi, T. Koyama, and Y. Taniguchi, *J. Appl. Phys.* **86**, 1680 (1999).
5. X. J. Wang, M. R. Andersson, M. E. Thompson, and O. Inganas, *Synth. Met.* **137**, 1019 (2003).
6. H. Murata, C. D. Merritt, and Z. H. Kafafi, *IEEE J. Sel. Top. Quantum Electron.* **4**, 119 (1998).
7. V. Adamovich, J. Brooks, A. Tamayo, A. M. Alexander, P. Djurovich, B. W. D'Andrade, C. Adachi, S. R. Forrest, and M. E. Thompson, *New J. Chem.* **26**, 1171 (2002).
8. J. Li, P. I. Djurovich, B. D. Alleyne, I. Tsyba, N. N. Ho, R. Bau, and M. E. Thompson (unpublished).
9. D. E. Loy, B. E. Koene, M. E. Thompson, P. E. Burrows, and S. R. Forrest, *Chem. Mater.* **10**, 2235 (1998).
10. H. Gilman and G. N. R. Smart, *J. Org. Chem.* **15**, 720 (1950).
11. M. Charisse, A. Zickgraf, H. Stenger, E. Brau, C. Besmarquet, M. Drager, S. Gerstmann, D. Dakternieks, and J. Hook, *Polyhedron* **17**, 4497 (1998).
12. H. Gilman and G. D. Lichtenwalte, *J. Am. Chem. Soc.* **80**, 608 (1958).
13. Bathocuproine, Cat. 14,091-0, Sigma-Aldrich Corp., St. Louis, Missouri.
14. S. R. Forrest, D. D. C. Bradley, and M. E. Thompson, *Adv. Mater.* (Weinheim, Ger.) **15**, 1043 (2003).
15. Y. Wang, N. Herron, V. V. Grushin, D. LeCloux, and V. Petrov, *Appl. Phys. Lett.* **79**, 449 (2001).
16. I. G. Hill and A. Kahn, *J. Appl. Phys.* **86**, 4515 (1999).

Chapter 24

Synthesis and Characterization of Facial and Meridional Tris-cyclometalated Iridium(III) Complexes

Arnold B. Tamayo, Bert D. Alleyne, Peter I. Djurovich, Sergey Lamansky, Irina Tsyba, Nam N. Ho, Robert Bau, and Mark E. Thompson

Contribution from the Department of Chemistry, University of Southern California, Los Angeles, California 90089-0744, USA
met@usc.edu

The synthesis, structures, electrochemistry, and photophysics of a series of facial (*fac*) and meridional (*mer*) tris-cyclometalated Ir(III) complexes are reported. The complexes have the general formula Ir$(C^\wedge N)_3$ [where $C^\wedge N$ is a monoanionic cyclometalating ligand; 2-phenylpyridyl (*ppy*), 2-(*p*-tolyl)pyridyl (*tpy*), 2-(4,6-difluorophenyl)pyridyl (*46dfppy*), 1-phenylpyrazolyl (*ppz*), 1-(4,6-difluorophenyl)pyrazolyl (*46dfppz*), or 1-(4-trifluoromethylphenyl)pyrazolyl (*tfmppz*)]. Reaction of the dichloro-bridged dimers [$(C^\wedge N)_2$Ir(μ-Cl)$_2$Ir$(C^\wedge N)_2$] with 2 equiv of H$C^\wedge N$ at 140–150 °C forms the corresponding meridional isomer, while higher reaction temperatures give predominantly the facial isomer. Both facial and

---

Reprinted from *J. Am. Chem. Soc.*, **125**, 7377–7387, 2003.

Electrophosphorescent Materials and Devices
Edited by Mark E. Thompson
Text Copyright © 2003 American Chemical Society
Layout Copyright © 2024 Jenny Stanford Publishing Pte. Ltd.
ISBN 978-981-4877-34-3 (Hardcover), 978-1-003-08872-1 (eBook)
www.jennystanford.com

meridional isomers can be obtained in good yield (>70%). The meridional isomer of Ir(*tpy*)$_3$ and facial and meridional isomers of Ir(*ppz*)$_3$ and Ir(*tfmppz*)$_3$ have been structurally characterized using X-ray crystallography. The facial isomers have near identical bond lengths (av Ir–C = 2.018 Å, av Ir–N = 2.123 Å) and angles. The three meridional isomers have the expected bond length alternations for the differing trans influences of phenyl and pyridyl/ pyrazolyl ligands. Bonds that are trans to phenyl groups are longer (Ir–C av = 2.071 Å, Ir–N av = 2.031 Å) than when they are trans to heterocyclic groups. The Ir–C and Ir–N bonds with trans N and C, respectively, have bond lengths very similar to those observed for the corresponding facial isomers. DFT calculations of both the singlet (ground) and the triplet states of the compounds suggest that the HOMO levels are a mixture of Ir and ligand orbitals, while the LUMO is predominantly ligand-based. All of the complexes show reversible oxidation between 0.3 and 0.8 V, versus Fc/Fc$^+$. The meridional isomers are easier to oxidize by ca. 50–100 mV. The phenylpyridyl-based complexes have reduction potentials between −2.5 and −2.8 V, whereas the phenylpyrazolyl-based complexes exhibit no reduction up to the solvent limit of −3.0 V. All of the compounds have intense absorption bands in the UV region assigned into $^1(\pi \rightarrow \pi^*)$ transitions and weaker MLCT (metal-to-ligand charge transfer) transitions that extend to the visible region. The MLCT transitions of the pyrazolyl-based complexes are hypsochromically shifted relative to those of the pyridyl-based compounds. The phenylpyridyl-based Ir(III) tris-cyclometalates exhibit intense emission both at room temperature and at 77 K, whereas the phenylpyrazolyl-based derivatives emit strongly only at 77 K. The emission energies and lifetimes of the phenylpyridyl-based complexes (450–550 nm, 2–6 μs) and phenylpyrazolyl-based compounds (390–440 nm, 14–33 μs) are characteristic for a mixed ligand-centered/MLCT excited state. The meridional isomers for both pyridyl and pyrazolyl-based cyclometalates show markedly different spectroscopic properties than do the facial forms. Isolated samples of *mer*-Ir($C^\wedge N$)$_3$ complexes can be thermally and photochemically converted to facial forms, indicating that the meridional isomers are kinetically favored products. The lower thermodynamic stabilities of the meridional isomers are likely related to structural features of these complexes; that is, the meridional configuration places strongly trans influencing phenyl groups opposite each other, whereas all three phenyl groups are opposite pyridyl or pyrazolyl groups in the facial complexes. The strong trans influence of the phenyl groups in the meridional isomers leads to the observation that they are easier to oxidize, exhibit broad, red-shifted emission, and have lower quantum efficiencies than their facial counterparts.

24.1 Introduction

A significant research effort has focused on the synthesis and photophysical characterization of octahedral $4d^6$ and $5d^6$ metal complexes [1]. These d^6 complexes, particularly the diimine (i.e., bipyridine, phenanthroline) chelates of Ru(II) and Os(II), have been widely used in a variety of photonic applications, including photocatalysis and photoelectrochemistry [2, 3]. The attraction of these d^6 complexes for such applications comes from their long excited-state lifetimes and high luminescent efficiencies. These properties increase the likelihood of either an energy or an electron-transfer process occurring prior to a radiative or nonradiative relaxation. The photophysics of related tris-chelate Ir(III) complexes have also been investigated [4–6]. These Ir(III) complexes have been prepared with either diimine ligands or cyclometalated ligands, such as 2-phenylpyidinato-C^2,N (*ppy*) [5, 6] and 2,2-thienylpyridinato-C^2,N (*thpy*) [6]. Complexes with the formally monoanionic cyclometalating ligands are isoelectronic with the cationic tris-diimine complexes of Ru(II) and Os(II). As compared to Ru(II) and Os(II) complexes, however, the d^6 Ir(III) complexes exhibit longer excited-state lifetimes, typically in the order of microseconds, and higher luminescence efficiencies [e.g., ϕ_{phos} (*fac*-Ir(*ppy*)$_3$) = 0.4] [7] in fluid solutions. These properties are due to efficient intersystem crossing between the singlet and triplet excited states brought about by the strong spin-orbit coupling of the Ir(III) metal ion.

The photophysical properties of bis- and tris-cyclometalated complexes of Ir(III) make them very useful for several photonic applications [4, 5a, 8–10]. These compounds can be employed as sensitizers for outer-sphere electron-transfer reactions [11, 12], photocatalysts for CO_2 reduction [13], photooxidants, and singlet oxygen sensitizers [14]. A recent application for these compounds is in the field of organic light-emitting devices (OLEDs), where they have been used as phosphorescent dopants in the emitting layer [8b]. The singlet and triplet excited states that are created during charge recombination are trapped at the phosphor, where the effective intersystem crossing leads to efficient electrophosphorescence at room temperature [15]. Recently, it has also been shown that a phenylpyrazolyl-based Ir(III) tris-cyclometalate can be used as an electron blocking material (preventing electron leakage to the hole-transporting layer) to make highly efficient single-dopant white OLEDs [16].

Metal d^6 tris-complexes with asymmetric chelate ligands can have either a facial (*fac*) or a meridional (*mer*) configuration. The photophysical and electrochemical properties of facial and meridional isomers have been investigated for a number of Ru(II) complexes having asymmetric diimine ligands [17–21]. However, for these cationic Ru(II) diimine complexes, it has been difficult to achieve control over which isomer is obtained from the synthesis, as well as to isolate and purify a specific isomer [17–20]. Moreover, the class of ligands that have been used (e.g., pyridyl-pyrazoles) gave only minor differences in the photophysical properties between the facial and meridional isomers because the two types of coordinating ligand, that is, pyridyl and pyrazolyl, have very similar electronic and coordinating characteristics [17, 18]. On the other hand, tris-chelates with cyclometalating ligands such as 2-phenylpyridyl should have pronounced differences between the facial and meridional isomers due to the marked disparity between the formally anionic phenyl ligand and neutral pyridyl ligand. To date, all investigations into the synthesis and photophysics of tris-cyclometalated metal d^6 complexes have been reported on the facial isomers [5b,h, 6, 9a, 10], whereas the chemistry of the meridional analogues has remained unexplored. Herein, we describe new synthetic routes which allow for isolation of either facial or meridional isomers of tris-cyclometalates of Ir(III). The meridional isomers have been prepared in good yield (ca. >70%) as kinetically favored products. The structures of these complexes have been determined by X-ray diffraction and NMR methods. The photophysical properties of both facial and meridional complexes have also been examined and show distinct differences between the two forms. The meridional isomers display red-shifted emission and decreased quantum efficiencies relative to their facial analogues. We also report on the conversion of the meridional isomers to facial forms by both thermal and photochemical routes.

24.2 Experimental Section

24.2.1 Equipment

UV–visible spectra were measured on an AVIV model 14DS UV–vis–IR (a reengineered Cary 14) or a Hewlett-Packard 4853 diode array spectrophotomer. Steady-state emission spectra were measured with a Photon Technology International QuantaMaster model C-60 spectrofluorimeter. Phosphorescence lifetime measurements (>2 μs) were performed on the same fluorimeter equipped with a microsecond xenon flash lamp or using an IBH Fluorocube fluorimeter equipped with a blue LED (λ_{max} = 373 nm). The quantum efficiency (QE) measurements were carried out at room temperature in degassed 2-methyltetrahydrofuran solutions using the optically dilute method [22]. Solutions of Ir(ppy)$_3$ in 2MeTHF were used as reference. NMR spectra were recorded on Bruker AMX 360 and 500 MHz instruments. Mass spectra were taken with a Hewlett-Packard GC/MS instrument with electron impact ionization and model 5873 mass sensitive detector. Elemental analyses (CHN) were performed at the Microanalysis Laboratory at the University of Illinois, Urbana-Champaign.

24.2.2 Electrochemistry

Cyclic voltammetry and differential pulsed voltammetry were performed using an EG&G potentiostat/galvanostat model 283. Anhydrous DMF (Aldrich) was used as the solvent under inert atmosphere, and 0.1 M tetra(n-butyl)ammonium hexafluorophosphate was used as the supporting electrolyte. A glassy carbon rod was used as the working electrode, a platinum wire was used as the counter electrode, and a silver wire was used as a pseudoreference electrode. The redox potentials are based on values measured from differential pulsed voltammetry and are reported relative to a ferrocene/ferrocenium (Cp_2Fe/Cp_2Fe^+) redox couple used as an internal reference [23], while electrochemical reversibility was determined using cyclic voltammetry.

24.2.3 X-ray Crystallography

Diffraction data for mer-Ir(ppz)$_3$, fac-Ir(tfmppz)$_3$, and mer-Ir(tfmppz)$_3$ were collected at room temperature ($T = 23°C$), while data for mer-Ir(tpy)$_3$ and fac-Ir(ppz)$_3$ were taken at -50 and $0°C$, respectively. The data sets were collected on a Bruker SMART APEX CCD diffractometer with graphite monochromated Mo Kα radiation ($\lambda = 0.71073$ Å). The cell parameters for the Ir complexes were obtained from a least-squares refinement of the spots (from 60 collected frames) using the SMART program. One hemisphere of crystal data for each compound was collected up to a resolution of 0.80 Å, and the intensity data were processed using the Saint Plus program. All of the calculations for the structure determination were carried out using the SHELXTL package (version 5.1) [24]. Absorption corrections were applied by using SADABS [25]. In most cases, hydrogen positions were input and refined in a riding manner along with the attached carbons. All of the structural analyses proceeded smoothly except that of mer-Ir(ppz)$_3$, which was complicated by a slight packing disorder of one of the three ppz ligands. This necessitated several data sets to be collected for mer-Ir(ppz)$_3$ until a crystal was found that yielded a satisfactory result. Anisotropic refinement of this data set led to unreasonable bond distances so only the isotropically refined data were used in the subsequent analysis. A summary of the refinement details and the resulting factors for mer-Ir(tpy)$_3$, fac-Ir(ppz)$_3$, mer-Ir(ppz)$_3$, Ir(tfmppz)$_3$, and mer-Ir(tfmppz)$_3$ are given in Table 24.1.

24.2.4 Density Functional Calculations

DFT calculations were performed using the Titan software package (Wavefunction, Inc.) at the B3LYP/LACVP** level. The HOMO

Table 24.1 Crystallographic data for mer-Ir(tpy)$_3$, fac-Ir(ppz)$_3$, mer-Ir(ppz)$_3$, fac-Ir(tfmppz)$_3$, and mer-Ir(tfmppz)$_3$

	mer-Ir(tpy)$_3$·CH$_2$Cl$_2$	fac-Ir(ppz)$_3$	mer-Ir(ppz)$_3$	fac-Ir(tfmppz)$_3$	mer-Ir(tfmppz)$_3$
Empirical formula	C$_{36}$H$_{30}$IrN$_3$·CH$_2$Cl$_2$	C$_{27}$H$_{21}$IrN$_6$	C$_{27}$H$_{21}$IrN$_6$	C$_{30}$H$_{18}$F$_9$IrN$_6$	C$_{30}$H$_{18}$F$_9$IrN$_6$
Formula weight	781.76	621.70	621.70	825.70	825.70
Temperature, K	223(2)	273(2)	296(2)	296(2)	296(2)
Wavelength (Å)	0.71073	0.71073	0.71073	0.71073	0.71073
Crystal system	monoclinic	tetragonal	monoclinic	orthorhombic	monoclinic
Space group	P2(1)/n	P-42(1)c	P2(1)/c	Pbca	P2(1)/n
Unit cell dimensions					
a (Å)	15.113(2)	23.125(11)	15.2907(13)	17.5088(13)	11.5923(10)
b (Å)	9.9696(15)	23.125(11)	14.8863(11)	17.0833(12)	33.878(3)
c (Å)	21.093(3)	8.854(8)	10.1503(9)	40.472(3)	15.5667(14)
α (deg)	90	90	90	90	90
β (deg)	91.383(3)	90	97.607(5)	90	103.030(2)
γ (deg)	90	90	90	90	90
Volume (Å^3)	3177.2(8)	4735(5)	2290.1(3)	12,105.4(15)	5956.0(9)
Z	4	8	4	16	8
Density, calcd (g/cm^3)	1.634	1.744	1.803	1.812	1.842
Absorption coefficient (mm^{-1})	4.401	5.667	5.858	4.499	4.572
F(000)	1544	2416	1208	6368	3184

(Continued)

Table 24.1 (Continued)

	mer-Ir(tpy)₃·CH₂Cl₂	fac-Ir(ppz)₃	mer-Ir(ppz)₃	fac-Ir(tfmppz)₃	mer-Ir(tfmppz)₃
θ range for data collection (deg)	1.64–24.71	1.25–27.56	1.34–28.42	1.01–24.71	1.47–24.71
Reflections collected	24,547	27,943	13,642	59,022	30,328
Independent reflections	5422	5315	5263	10,321	10,165
	[R(int) = 0.0489]	[R(int) = 0.0484]	[R(int) = 0.0262]	[R(int) = 0.0459]	[R(int) = 0.0313]
Refinement method	full-matrix least-squares on F^2	full-matrix least-squares on F^2	full-matrix least-squares on F^2	full-matrix least-squares on F^2	full-matrix least-squares on F^2
Data/restraints/parameters	5219/0/353	5315/0/308	5263/0/143	10,321/108/786	10,165/108/786
Goodness-of-fit on F^2	1.066	1.054	1.004	1.037	1.023
Final R indices [$I > 2\sigma(I)$]	0.0521	0.0366	0.0480	0.0495	0.0468
R indices (all data)	0.0686	0.0477	0.0706	0.0731	0.0595

and LUMO energies were determined using minimized singlet geometries to approximate the ground state. The minimized singlet geometries were used to calculate the triplet molecular orbitals and approximate the triplet HSOMO (HSOMO = highest singly occupied molecular orbital).

24.2.5 Synthesis

The compounds 2-(4,6-difluorophenyl)pyridine [26], 1-(4,6-difluorophenyl)pyrazole [27], and 1-(4-trifluoromethylphenyl)pyrazole [27] were prepared following literature procedures. Ir$(acac)_3$ was purchased from Strem Chemical Co., IrCl$_3 \cdot n$H$_2$O was from Next Chimica, and all other chemicals were purchased from Aldrich Chemical Co. and used as received.

All experiments involving IrCl$_3 \cdot x$H$_2$O or any other Ir(III) species were carried out in inert atmosphere despite the stability of the compounds in air, the main concern being their oxidative and thermal stability of intermediate complexes at the high temperatures used in the reactions. Cyclometalated Ir(III) μ-chloro-bridged dimers of general formula $(C^\wedge N)_2$Ir$(\mu$-Cl$)_2$Ir$(C^\wedge N)_2$ (where $C^\wedge N$ represents a cyclometalating ligand) were synthesized by the method reported by Nonoyama [28], which involves heating IrCl$_3 \cdot$H$_2$O to 110°C with 2–2.5 equiv of cyclometalating ligand in a 3:1 mixture of 2-ethoxyethanol and deionized water. $(C^\wedge N)_2$Ir$(O^\wedge O)$ (where $O^\wedge O$ represents 2,2,6,6-tetramethyl-3,5-heptanedione, dipivaloylmethane - dpm) was prepared by reacting the dimers with 2–2.5 equiv of the chelating diketone and an equivalent amount of K$_2$CO$_3$ in 1,2-dichloroethane at 90°C for 24 h [8].

24.2.6 Synthesis of *fac* -Ir$(C^\wedge N)_3$ Complexes: General Procedure

fac-Ir$(ppz)_3$, *fac*-Ir$(46dfppz)_3$, *fac*-Ir$(tfmppz)_3$. **Method A.** This method involves treating Ir$(acac)_3$ ($acac$ = acetylacetonate) with 3–3.5 equiv of the appropriate cyclometalating ligand in refluxing glycerol, as previously described (see Eq. 24.1) [5h].

***fac*-Ir(*46dfppy*)₃, *fac*-Ir(*46dfppz*)₃, *fac*-Ir(*tfmppz*)₃. Method B.**
The $(C^{\wedge}N)_2\text{Ir}(O^{\wedge}O)$ complex and 1.2–1.5 equiv of the appropriate cyclometalating ligand were refluxed under inert gas atmosphere in 10 mL of glycerol for 20–24 h. After the mixture was cooled to room temperature, 20 mL of 5% HCl solution was added, and the product was thrice extracted with 25 mL of CH_2Cl_2. The organic extracts were combined and then dried with anhydrous $MgSO_4$, after which the solvent was removed in vacuo. The crude material was then flash chromatographed on a silica column using dichloromethane to yield 60–80% product (see Eq. 24.2).

***fac*-Ir(*ppy*)₃, *fac*-Ir(*tpy*)₃, *fac*-Ir(*ppz*)₃. Method C.** $[(C^{\wedge}N)_2\text{IrCl}]_2$ complex, 2–2.5 equiv of the appropriate cyclometalating ligand, and 5–10 equiv of K_2CO_3 were heated to $\sim 200°C$ under inert atmosphere in 10 mL of glycerol for 20–24 h. After the mixture was cooled to room temperature, 20 mL of deionized H_2O was added, and the resulting precipitate was filtered off, washed with two portions of methanol, followed by ether and hexanes. The crude product was then flash chromatographed on a silica column using dichloromethane to yield 65–80% pure *fac*-Ir$(C^{\wedge}N)_3$ (see Eq. 24.3).

24.2.7 Synthesis of *mer*-Ir($C^{\wedge}N$)₃ Complexes: General Procedure

***mer*-Ir(*ppy*)₃, *mer*-Ir(*tpy*)₃, *mer*-Ir(*ppz*)₃, *mer*-Ir(*46dfppz*)₃, *mer*-Ir(*tfmppz*)₃.** All of the meridional isomers, except *mer*-Ir(*46dfppy*)₃, were prepared by using a modified version of Method C. $[(C^{\wedge}N)_2\text{IrCl}]_2$ complex, 2–2.5 equiv of the appropriate cyclometalating ligand, and 5–10 equiv of K_2CO_3 were heated to 140–145°C under inert atmosphere in 10 mL of glycerol for 20–24 h. After the mixture was cooled to room temperature, distilled water was added, and the resulting precipitate was filtered off, washed with two more portions of distilled water, and air-dried. The crude product was then flash chromatographed on a silica column using dichloromethane to give 68–80% (based on the starting dichloro-bridged dimer) of the pure meridional tris-cyclometalated complex.

***mer*-Ir(*46dfppy*)₃.** This complex was prepared by dissolving the corresponding dimer, 2 equiv of ligand, and K_2CO_3 in

2-ethoxyethanol and heated to 120°C for overnight. The reaction mixture was then allowed to cool to room temperature, and the solvent was removed in vacuo. The crude product mixture was then flash chromatographed on a silica column using dichloromethane to give 74% pure *mer*-Ir(*46dfppy*)$_3$.

24.2.8 Characterization

^{1}H and ^{13}C NMR chemical shifts and coupling constants for all of the complexes listed below are given in the Supporting Information.

fac-Ir(*ppy*)$_3$ **(Method C):** *fac*-tris(2−(phenyl)pyridinato,$N,C^{2'}$) iridium(III). Yield: 79%. ^{1}H and ^{13}C NMR spectroscopy and C, H, N analysis match that reported for *fac*-Ir(ppy)$_3$ [5h].

fac-Ir(*tpy*)$_3$ **(Method C):** *fac*-tris(2-(*p*-tolyl)pyridinato,$N,C^{2'}$) iridium(III). Yield: 81%. MS: *m/z* calcd 696.9; found 697. Anal. Calcd for $C_{36}H_{30}N_3Ir$: C, 62.05; H, 4.34; N, 6.03. Found: C, 61.67; H, 4.26; N, 6.05.

fac-Ir(*46dfppy*)$_3$ **(Method B):** *fac*-tris(2-(4,6-difluorophenyl) pyridinato,$N,C^{2'}$)iridium(III). Yield: 74%. MS: *m/z* calcd 762.7; found 763. Anal. Calcd for $C_{33}H_{18}F_6N_3Ir$: C, 51.97; H, 2.38; N, 5.51. Found: C, 51.92; H, 2.36; N, 5.51.

fac-Ir(*ppz*)$_3$ **(Method C):** *fac*-tris(1-phenylpyrazolato,$N,C^{2'}$) iridium(III). Yield: 84%. MS: *m/z* calcd 621.7; found 622. Anal. Calcd for $C_{27}H_{21}N_6Ir$: C, 52.16; H, 3.40; N, 13.52. Found: C, 52.04; H, 3.39; N, 13.52.

fac-Ir(*dfppz*)$_3$ **(Method B):** *fac*-tris(1-(4,6-difluoro-phenylpyrazolato,$N,C^{2'}$)iridium(III). Yield: 78%. MS: *m/z* calcd 729.7; found 730. Anal. Calcd for $C_{27}H_{15}F_6N_6Ir$: C, 44.44; H, 2.07; N, 11.52. Found: C, 44.23; H, 1.89; N, 11.28.

fac-Ir(*tfmppz*)$_3$ **(Method B):** *fac*-tris(1-(4-trifluoromethylphenyl-pyrazolato,$N,C^{2'}$)iridium(III). Yield: 72%. MS: *m/z* calcd 789.7; found 790. Anal. Calcd for $C_{27}H_{18}F_9N_6Ir$: C, 43.64; H, 2.20; N, 10.18. Found: C, 43.68; H, 2.01; N, 9.90.

mer-Ir(*ppy*)$_3$: *mer*-tris(2-(phenyl)pyridinato,$N,C^{2'}$)iridium(III). Yield: 75%. MS: *m/z* calcd 654.8; found 655. Anal. Calcd for

$C_{36}H_{30}N_3Ir$: C, 60.53; H, 3.69; N, 6.42. Found: C, 60.25; H, 3.59; N, 6.46.

***mer*-Ir(*tpy*)$_3$:** *mer*-tris(2-(*p*-tolyl)pyridinato,$N,C^{2'}$)iridium(III). Yield: 78%. MS: *m/z* calcd 696.9; found 697. Anal. Calcd for $C_{36}H_{30}N_3Ir$: C, 62.05; H, 4.34; N, 6.03. Found: C, 61.59; H, 4.23; N, 6.04.

***mer*-Ir(*46dfppy*)$_3$:** *mer*-tris(2-(4,6-difluorophenyl)pyridinato, $N,C^{2'}$)iridium(III). Yield: 74%. MS: *m/z* calcd 762.7; found 763. Anal. Calcd for $C_{33}H_{18}F_6N_3Ir$: C, 51.97; H, 2.38; N, 5.51. Found: C, 50.25; H, 2.35; N, 5.31.

***mer*-Ir(*ppz*)$_3$:** *mer*-tris(1-phenylpyrazolato,$N,C^{2'}$)iridium(III). Yield: 80%. MS: *m/z* calcd 621.71; found 622. Anal. Calcd for $C_{27}H_{21}N_6Ir$: C, 52.16; H, 3.40; N, 13.52. Found: C, 51.74; H, 3.13; N, 13.28.

***mer*-Ir(*dfppz*)$_3$:** *mer*-tris(1-(4,6-difluorophenyl)pyrazolato,$N,C^{2'}$) iridium(III). Yield: 86%. MS: *m/z* calcd 729.7; found 730. Anal. Calcd for $C_{27}H_{15}F_6N_6Ir$: C, 44.44; H, 2.07; N, 11.5. Found: C, 44.32; H, 1.95; N, 11.27.

***mer*-Ir(*tfmppz*)$_3$:** *mer*-tris(1-(4-trifluoromethylphenylpyrazolato, $N,C^{2'}$)iridium(III). Yield: 69%. MS: *m/z* calcd 789.7; found 790. Anal. Calcd for $C_{27}H_{18}F_9N_6Ir$: C, 43.64; H, 2.20; N, 10.18. Found: C, 44.0; H, 2.12; N, 9.95.

24.2.9 Isomerization of *mer*-Ir($C^\wedge N$)$_3$ to *fac*-Ir($C^\wedge N$)$_3$: Thermal Isomerization

Samples of *mer*-Ir($C^\wedge N$)$_3$ (100 mg) were refluxed in 10 mL of glycerol under inert atmosphere for 24 h. After the mixture was cooled to room temperature, 100 mL of deionized H_2O was added, and the mixture was filtered off and washed with several portions of water and allowed to dry. The crude product was flashed chromatographed using a silica/dichloromethane column.

24.2.10 Photochemical Isomerization

Samples of mer-Ir($C^\wedge N$)$_3$ (10-15 mg) were dissolved in 0.5 mL of d_6-DMSO in Young's NMR tube (Wilmad Co.). The samples were degassed, and the ^{1}H NMR spectrum was taken before and after exposure to UV light. Handheld TLC lamps (UVP model UVGL-25) positioned face-to-face and covered with aluminum foil were the UV source. After 2 h, >95% conversion was observed as verified by ^{1}H NMR spectroscopy.

24.3 Results and Discussion

24.3.1 Synthesis and Structure of Ir($C^\wedge N$)$_3$ Complexes

Tris-cyclometalated Ir(III) complexes can be prepared by three different synthetic routes (Eqs. 24.1–24.3, Scheme 24.1) using either 2-phenylpyridine (ppyH) or 1-phenylpyrazole (ppzH) as the cyclometalating ligand precursors (Fig. 24.1). The first method involves treating Ir(acac)$_3$ with 3 equiv of the free ligand in glycerol, at refluxing temperatures (Method A, Eq. 24.2) [5h]. The tris-cyclometalated complexes can also be prepared from the appropriate β-diketonate derivative [($C^\wedge N$)$_2$Ir($O^\wedge O$), $O^\wedge O$ = 2,2,6,6-tetramethyl-3,5-heptanedione (dpm)] (Method B) or dichloro-bridged dimer [($C^\wedge N$)$_2$Ir(μ-Cl)$_2$Ir($C^\wedge N$)$_2$] (Method C), by heating the Ir complex with a 2–3-fold excess of cyclometalating ligand in glycerol. These syntheses work equally well for other pyridine-type ligands (e.g., tpyH, 46dfppyH), as well as for phenylpyrazoles (e.g., 46dfppzH, tfmppzH).

Methods B and C have several advantages over Method A. The dichloro-bridged dimers [28] and ($C^\wedge N$)$_2$Ir($O^\wedge O$) [8] compounds are easily prepared in high yield from IrCl$_3$·H$_2$O, a starting material less expensive than Ir(acac)$_3$. In addition, Methods B and C give higher yields than does Method A. For example, reactions using Method C and either ppyH or ppzH give yields between 80 and 85% (based on the starting dimer, reaction temperature ~200°C) versus 45–60% using Method A [5h, 16]. Previous routes to make tris-cyclometalated Ir(III) complexes also utilize either IrCl$_3$·H$_2$O or ($C^\wedge N$)$_2$Ir(μ-Cl)$_2$Ir($C^\wedge N$)$_2$ complexes as starting materials [6, 9a].

However, these methods employ a large excess of the cyclometalating precursor ligands as solvent, making it necessary to prepare the desired $HC^\wedge N$ compounds on a relatively large scale.

METHOD A

$$Ir(acac)_3 + 3C^\wedge NH \xrightarrow[\Delta,\ N_2]{glycerol} Ir(C^\wedge N)_3 + 3\ acacH \quad (24.1)$$

METHOD B

$$(C^\wedge N)_2Ir(O^\wedge O) + HC^\wedge N \xrightarrow[\Delta,\ N_2]{glycerol} Ir(C^\wedge N)_3 + O^\wedge OH \quad (24.2)$$

METHOD C

$$(C^\wedge N)_2Ir(\mu\text{-Cl})_2Ir(C^\wedge N)_2 + 2HC^\wedge N \xrightarrow[\Delta,\ N_2]{glycerol,\ K_2CO_3} Ir(C^\wedge N)_3 + 2KCl + H_2O + CO_2 \quad (24.3)$$

Scheme 24.1

The reaction temperature and nature of the cyclometalating ligand strongly affect the facial/meridional product ratios of the reactions. For the nonfluorinated ligands *ppy*H, *tpy*H, and *ppz*H, both Methods B and C give the facial isomer as the predominant product when reaction temperatures are >200°C. A small amount of the meridional isomer (typically 1–3%) is sometimes also present in the crude reaction mixture. The meridional impurity can be readily removed by recrystallization or column chromatography. Higher yields of the *mer*-Ir$(C^\wedge N)_3$ complexes (68–80%) are obtained using Method C at a lower temperature, that is, 140°C. Interestingly, different results are obtained with the fluorine-substituted ligands, *46dfppz*H, *tfmppz*H, and *46dfppy*H. All of these ligands give the facial isomer as the major product when Methods A or B are used at high temperature (>200°C). However, when using Method C, we found that *46dfppz*H and *tfmppz*H give meridional isomers as principal products, whereas *46dfppy*H gives mixtures of *fac/mer*-isomers, along with other unidentified products, even at 140°C. Therefore, pure samples of *mer*-Ir(*46dfppy*)$_3$ were prepared by Method C using 2-ethoxyethanol as solvent at 120°C.

The coordination geometry of the precursor complexes used in Method C [$(C^\wedge N)_2$Ir(μ-Cl)$_2$Ir$(C^\wedge N)_2$] has the Ir–N bonds in a mutually trans disposition. This suggests that cyclometalation by the

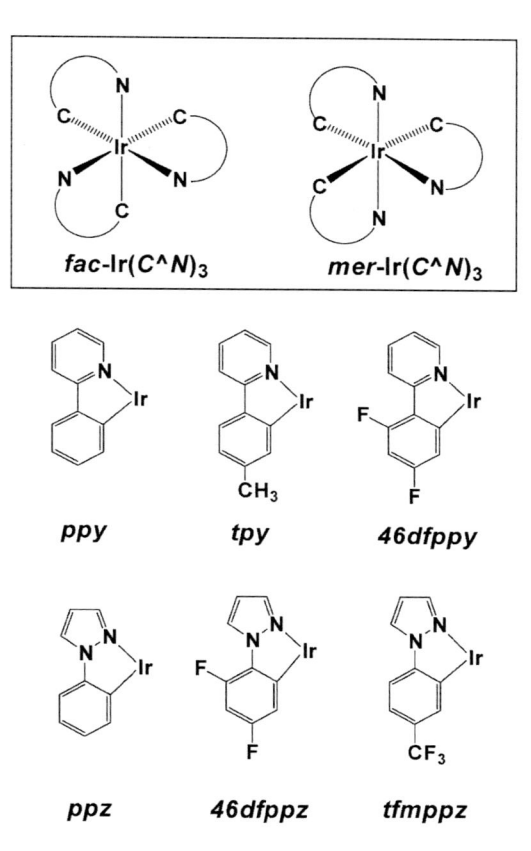

Figure 24.1 Cyclometalating ligands used to prepare Ir(C^N)₃. Abbreviations used throughout the paper are listed below the $C^\wedge N$ fragment.

third $C^\wedge N$ ligand leads directly to the formation of the meridional isomer, as observed when the syntheses are carried out at a lower temperature [5b]. When the reaction temperature for Method C is raised to >200°C, however, isomerization of either the starting materials or the formed meridional isomers needs to occur to form the observed facial products (*vide infra*).

24.3.2 NMR Characterization

The solution structures of the complexes were established using ^{1}H, ^{13}C, and, where applicable, ^{19}F NMR spectroscopy. The ^{1}H and

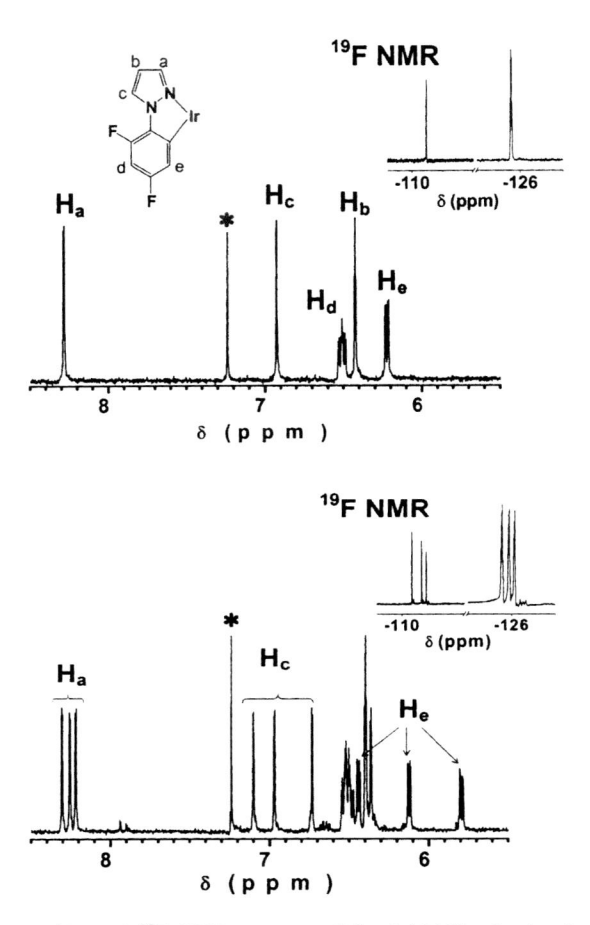

Figure 24.2 ^{1}H and ^{19}F NMR spectra of *fac*-Ir(*46dfppz*)$_3$ (top) and *mer*-Ir(*46dfppz*)$_3$ (bottom). The solvent peak is indicated by *. The ^{19}F spectra (C$_6$F$_6$ was used as an external reference δ(ppm): −164 ppm) are shown as insets to the ^{1}H spectra. The ^{19}F spectrum of the meridional isomer is shown with a split scale for clarity.

^{19}F NMR spectra of *fac*- and *mer*-Ir(*46dfppz*)$_3$ are presented in Fig. 24.2; the ^{1}H NMR spectra of all of the other complexes are given in the Supporting Information. The facial and meridional isomers of Ir(C^N)$_3$ complexes are readily distinguished by NMR spectroscopy, as all of the derivatives display similar spectral characteristics. In the facial tris-cyclometalates, the three ligands surrounding the iridium atom are magnetically equivalent due to the inherent C_3 symmetry of the complexes. This gives rise to first-order NMR spectra and makes spectral interpretation for these isomers relatively straightforward because the total number of resonances in the complex are equal to the number of resonances in a single anionic $C^\wedge N$ ligand. In the ^{1}H NMR spectrum of *fac*-Ir(*46dfppz*)$_3$, for example, five distinct aromatic proton resonances are displayed, each peak integrating to three equivalent H's that correspond to an individual aromatic proton in the *46dfppz* ligand (Fig. 24.2). In contrast, the C_1 symmetry of the *mer*-Ir($C^\wedge N$)$_3$ complexes gives rise to non-first-order ^{1}H NMR spectra with the total of number of resonances equal to the total number of aromatic protons in the complex. Hence, the ^{1}H NMR spectrum of the *mer*-Ir(*46dfppz*)$_3$ is more complicated than the facial isomer with 15 distinct aromatic proton resonances appearing in the spectrum, each integrating as a single proton. For example, three aromatic resonances in the facial isomer, labeled H$_a$, H$_c$, and H$_e$ ($\delta = 8.30$, 6.9, 6.2 ppm, respectively), appear as nine separate resonances in the meridional isomer. Similarly, the ^{1}H-decoupled ^{13}C NMR spectrum of *fac*-Ir(*46dfppz*)$_3$ displays only nine inequivalent aromatic carbon resonances, whereas 27 carbon resonances are observed for the meridional isomer. The ^{19}F NMR spectroscopy also clearly distinguishes between the isomers of Ir(*dfppz*)$_3$ (see insets to Fig. 24.2). The facial isomer shows two distinct fluorine resonances, while the meridional isomer displays six distinct fluorine resonances.

24.3.3 X-ray Crystallography

Single crystals of *mer*-Ir(*tpy*)$_3$, as well as the facial and meridional isomers of Ir(*ppz*)$_3$ and Ir(*tfmppz*)$_3$, were grown from methanol/dichloromethane solution and characterized using X-ray crystallography. The crystal data are given in Table 24.1. Tables of atomic coordinates, bond lengths, and angles for each complex are given in the Supporting Information. All of the complexes examined here (both *fac*- and *mer*-isomers) have the three cyclometalating ligands in a pseudooctahedral coordination geometry around the metal center. The C–C and C–N intraligand bond lengths and angles are within normal ranges expected for cyclometalated Ir(III)

Table 24.2 Comparison of selected bond distances (Å) for fac-Ir(tpy)$_3$, (tpy)$_2$Ir(acac) [8a], and mer-Ir(tpy)$_3$

Bond type	Bond distances (Å)		
	fac-Ir(tpy)$_3$	mer-Ir(tpy)$_3$	(tpy)$_2$Ir(acac)
Ir–N1		2.151(9)	
Ir–N2	2.132(5)	2.044(8)	2.023(5)
Ir–N3		2.065(8)	2.040(5)
Ir–C1		2.076(10)	
Ir–C2	2.024(6)	2.086(12)	1.985(7)
Ir–C3		2.020(8)	1.982(6)

complexes and are similar to values reported for the (C^N)$_2$Ir-(μ-Cl)$_2$Ir(C^N)$_2$) [29] and fac-Ir(C^N)$_3$ [9a, 29] complexes, as well as to reported values for other mononuclear complexes with the (C^N)$_2$Ir fragment [30, 31].

Figure 24.3 gives molecular plots for fac-Ir(tpy)$_3$, mer-Ir(tpy)$_3$, and (tpy)$_2$Ir(acac). The Ir–C and Ir–N bond lengths for each of the complexes are given in Table 24.2. The structural data for fac-Ir(tpy)$_3$ [29] and (tpy)$_2$Ir(acac) [8a] were taken from literature references. The facial isomer sits on a three-fold axis, leading to identical Ir–C and Ir–N bond lengths of 2.024(6) and 2.132(5) Å, respectively. The bond lengths in the meridional isomer of tpy differ markedly from those of the facial isomer. The Ir–C bond trans to a pyridyl group (Ir–C3 = 2.020(8) Å) shares an electronic environment and, thus, bond length similar to those of the Ir–C bonds of the facial isomer. Likewise, the Ir–N bond trans to the phenyl group (Ir–N1 = 2.151(9) Å) is nearly the same length as the Ir–N bonds of the facial isomer. The Ir–N bonds of the mutually trans pyridyl groups in the meridional complex (Ir–N2 = 2.044(8) Å and Ir–N3 = 2.065(8) Å) are significantly shorter than Ir–N bonds of the facial isomer. This is consistent with the weaker trans influence of a pyridyl group relative to a phenyl ligand [32]. In contrast, the Ir–C bonds trans to phenyl groups (Ir–C1 = 2.076(10) Å and Ir–C2 = 2.086(12) Å) have lengths markedly longer than the Ir–C bonds of the facial isomer, consistent with the significant trans influence of phenyl groups on each other. It is also interesting to compare mer-Ir(tpy)$_3$ and (tpy)$_2$Ir(acac) structures. The bis-cyclometalated

fac-Ir(*tpy*)₃   *mer*-Ir(*tpy*)₃   (*tpy*)₂Ir(*acac*)

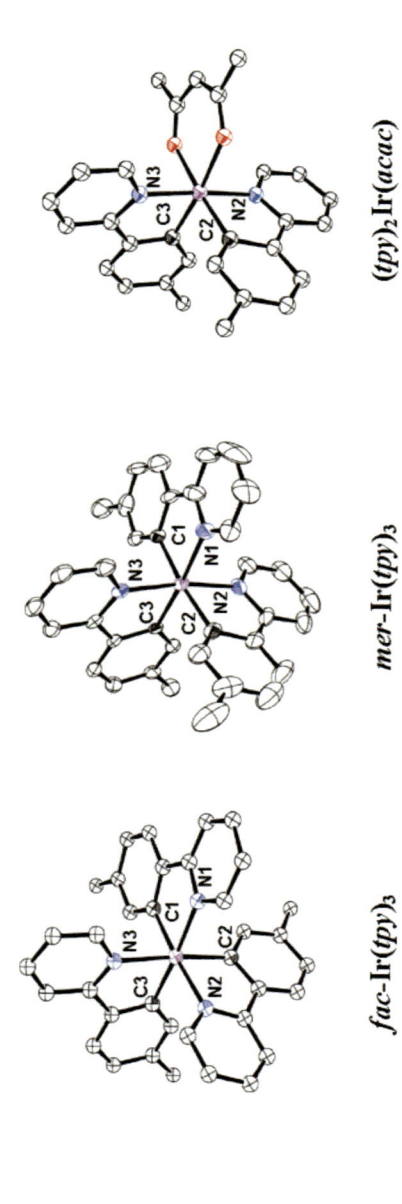

Figure 24.3 ORTEP drawings of *fac*-Ir(*tpy*)₃, *mer*-Ir(*tpy*)₃, and (*tpy*)₂Ir(*acac*). The thermal ellipsoids for the image represent 25% probability limit. The hydrogen atoms have been omitted for clarity.

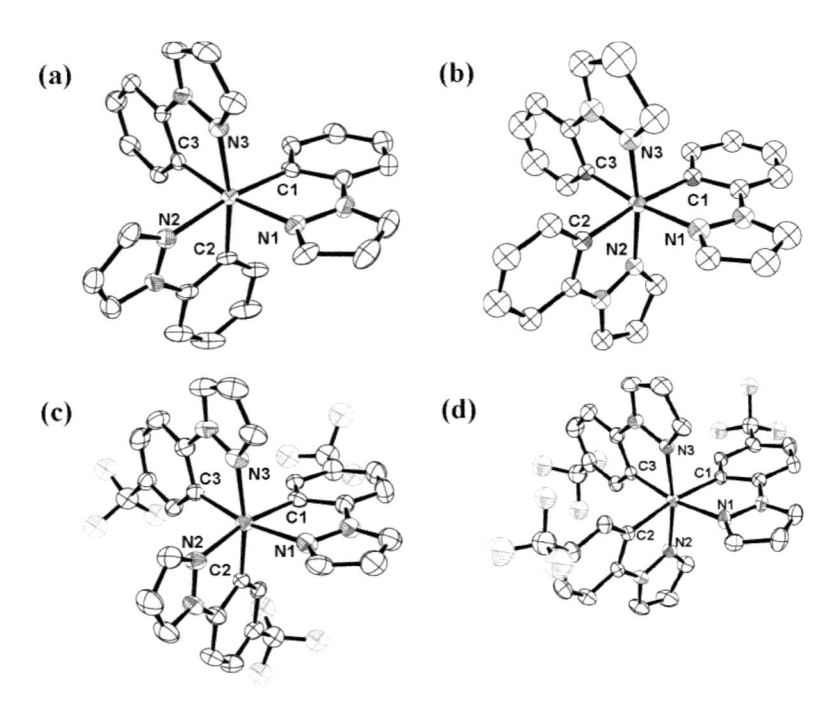

Figure 24.4 ORTEP drawings of (a) *fac*-Ir(*ppz*)$_3$, (b) *mer*-Ir(*ppz*)$_3$, (c) *fac*-Ir(*tfmppz*)$_3$, and (d) *mer*-Ir(*tfmppz*)$_3$.

fragment of (*tpy*)$_2$Ir(*acac*) has the same disposition of *tpy* ligands as found in *mer*-Ir(tpy)$_3$, and the mutually trans disposed Ir–N bonds in both complexes lengths have similar lengths. On the other hand, the weak trans influence of the acetylacetonate ligand leads to shorter Ir–C bonds (av = 1.984(6) Å) for the (*tpy*)$_2$Ir(*acac*) complex than those observed in either the meridional or the facial Ir(*tpy*)$_3$ complexes.

The structures of the facial and meridional isomers of Ir(*ppz*)$_3$ are shown in Fig. 24.4. The Ir–C and Ir–N bond lengths for the two complexes are given in Table 24.3. The average Ir–C (2.021(6) Å) and Ir–N (2.124(5) Å) bond lengths of *fac*-Ir(*ppz*)$_3$ are very similar to those of *fac*-Ir(*tpy*)$_3$, suggesting a similar trans influence for pyridyl and pyrazolyl groups. In the meridional isomer, the mutually trans Ir–C bond lengths (Ir–C1 and Ir–C2, av = 2.054(2) Å) are similar to those of *mer*-Ir(*tpy*)$_3$ and reflect an equivalent degree of trans influence for both aryl groups in the two chelate systems. The mutually trans Ir–N bond lengths (Ir–N1 and Ir–N2, av = 2.019(2) Å) are also similar to values for the corresponding bonds in mer-Ir(tpy)$_3$ consistent with pyrazole having a trans influence comparable to that of pyridine. However, the bond lengths for Ir–C trans to pyrazolyl (1.993(2) Å) and Ir–N trans to phenyl (2.053 Å) are shorter than the corresponding pair in mer-Ir(tpy)$_3$, as well as to the Ir–C and Ir–N bonds in the facial isomer. This decrease in bond distance is most likely due to less steric repulsion between the ppz ligands when the complex is in a meridional configuration.

The structures of the facial and meridional isomers of Ir(tfmppz)$_3$ are shown in Fig. 24.4, and the Ir–C and Ir–N bond lengths are given in Table 24.3. The electron-accepting trifluoromethyl group of the tfmppz ligand is in a meta disposition relative to Ir and, electronically, is thus weakly coupled to the metal center. Hence, the average Ir–C (2.015(8) Å) and Ir–N (2.114(7) Å) bond lengths for the facial Ir(tfmppz)$_3$ are the same as those observed for the facial isomers of both Ir(tpy)$_3$ and Ir(ppz)$_3$. For the meridional isomer of Ir(tfmppz)$_3$, the mutually trans Ir–C bonds (av = 2.078(5) Å) and Ir–N bonds (av = 2.020(5) Å) are similar in length to the equivalent bonds in mer-Ir(tpy)$_3$ and mer-Ir(ppz)$_3$. The Ir–C trans to pyrazolyl (1.991(5) Å) and Ir–N trans to phenyl (2.095(4) Å) bond lengths are also similar to values for the corresponding bonds in the meridional

Table 24.3 Selected bond distances (Å) and angles (deg) for fac-Ir(ppz)$_3$, mer-Ir(ppz)$_3$, fac-Ir(tfmppz)$_3$, and mer-Ir(tfmppz)$_3$

Bond type	Ir(ppz)$_3$Ir(tfmppz)$_3$		Ir(tfmppz)$_3$	
	Facial	Meridional	Facial	Meridional
Bond distances (Å)				
Ir–N1	2.117(5)	2.053(2)	2.114(7)	2.095(4)
Ir–N2	2.135(5)	2.026(2)	2.113(7)	2.024(5)
Ir–N3	2.120(6)	2.013(2)	2.116(8)	2.016(5)
Ir–C1	2.015(7)	2.051(2)	2.016(8)	2.073(5)
Ir–C2	2.027(6)	2.057(2)	2.016(8)	2.083(5)
Ir–C3	2.021(6)	1.993(2)	2.013(8)	1.991(5)
Bond angles (deg)				
N2-Ir-N3	94.4(2)	171.54(8)	95.1(3)	171.26(17)
C1-Ir-C2	93.7(3)	172.67(9)	93.7(3)	170.9(2)

tpy and *ppz* complexes. Likewise, the bond distances for the trans disposed phenyl and pyrazolyl groups are relatively unperturbed by the CF_3 substituent.

24.3.4 DFT Calculations

B3LYP density functional theory (DFT) calculations were carried out on all of the $Ir(C^\wedge N)_3$ complexes using the Titan software package (Wavefunction, Inc.) with a LACVP** basis set. A similar approach has been used to investigate the ground- and excited-state properties of cyclometalated Ir and Pt compounds [33, 34]. The HOMO and LUMO surfaces for *fac-* and *mer-*Ir(*ppz*)$_3$ are illustrated in the Supporting Information. The discussion here will focus on the results for *fac-* and *mer-*Ir(*ppz*)$_3$; however, all of the examined facial and meridional isomers give a similar picture for the HOMO and LUMO orbitals. The calculated values for the Ir–C (2.17 Å) and Ir–N (2.04 Å) bond distances and N-Ir–N (96.3°) and C-Ir-C (95.2°) bond angles are comparable to the experimental values determined in the X-ray structure of *fac-*Ir(*ppz*)$_3$ (see Table 24.3). Likewise, the calculated Ir–C and Ir–N bond lengths of *mer-*Ir(*ppz*)$_3$ have the same length alternations as those observed in the X-ray structure. The calculated HOMO energies for *fac-* and *mer-*Ir(*ppz*)$_3$ are −5.02 and −4.81 eV, respectively. The HOMO pictures for both isomers consist of a mixture of phenyl and Ir orbitals. The HOMO in the *fac-*isomer is distributed equally among the three *ppz* ligands due to the C_3 symmetry of the complex. The HOMO of the *mer-*isomer is localized primarily on the two *ppz* ligands with the transoid disposition of nitrogen. Similarly, the LUMO - while predominantly phenylpyrazolyl in character - is delocalized among the three ligands in the *fac-*isomer (−0.57 eV) as opposed to being localized on a single ligand in the *mer-*isomer (−0.60 eV). The triplet HSOMOs (HSOMO = highest singly occupied molecular orbital) are calculated to be −1.60 and −1.53 eV for the *fac-* and *mer-*isomer, respectively. The triplet state energy can then be estimated as the difference between the ground-state singlet (HOMO) and triplet (HSOMO) energies [33, 34]. The values obtained for the theoretical triplet energy are 3.42 eV (362 nm) for *fac-*Ir(*ppz*)$_3$ and 3.28 eV (378 nm) for the meridional isomer. These values are in close agreement with the data obtained from

Table 24.4 Photophysical and electrochemical properties of $Ir(C^\wedge N)_3$ complexes

Complex	Absorption[a]	Emission at 77 K[b]		Emission at 298 K		Redox (V)[c]	
$Ir(C^\wedge N)_3$	$\lambda(nm)$ {ϵ, 10^3 L mol^{-1} cm^{-1}}	λ_{max}	τ (μs)	λ_{max}	τ (μs)	$E^{ox}_{1/2}$	$E^{red}_{1/2}$
ppy fac	244 (45.5), 283 (44.8), 341 (9.2), 377 (12.0), 405 (8.1), 455 (2.8), 488 (1.6)	492	3.6	510	1.9	0.31	−2.70, −3.00
mer	246 (47.3), 276 (51.0), 339 (9.2), 382 (10.7), 410 (7.2), 457 (3.4), 488 (1.4)	493	4.2	512	0.15	0.25	−2.63, −2.82
tpy							
fac	248 (41.4), 287 (44.7), 347 (10.6), 374 (11.8), 410 (7.0), 450 (2.9), 485 (1.4)	492	3.0	510	2.0	0.30	−2.78, −3.09
mer	276 (53.6), 336 (13.4), 383 (8.4), 420 (5.3), 451 (3.7), 485 (1.5)	530[d]	4.8	550	0.26	0.18	−2.73, −3.04
46dfppy							
fac	240 (50.2), 274 (44.7), 292 (25.5), 346 (11.3), 379 (7.1), 427 (1.6), 457 (0.3)	450	2.5	468	1.6	0.78	−2.51, −2.81
mer	264 (50.7), 312 (15.5), 353 (8.8), 388 (7.8), 428 (2.5), 456 (0.9)	460	5.4	482	0.21	0.69	−2.50, −2.86
ppz							
fac	244 (49.1), 261 (41.1), 292 (16.5), 321 (13.5), 366 (4.2)	414	14			0.39	

(Continued)

Table 24.4 (Continued)

Complex	Absorption[a]	Emission at 77 K[b]		Emission at 298 K	Redox (V)[c]
mer	228 (44.8), 246 (44.0), 293 (17.3), 320 (9.6), 349 (5.1)	427[d]	28		0.28
46dfppz fac	246 (55.3), 254 (49.9), 283 (23.1), 316 (10.0)	390[d]	27		0.80
mer	244 (48.0), 278 (24.0), 322 (8.9)	402[d]	33		0.72
tfmppz					
fac	243 (45.3), 251 (40.8), 261 (37.0), 289 (12.2), 323 (10.6), 364 (3.6)	422	17	428 0.05	0.73
mer	234 (42.3), 247 (46.6), 287 (19.4), 344 (7.7), 372 (3.2)	430[d]	32		0.62

[a] Absorption measurements of complexes were taken in CH_2Cl_2.
[b] 77 K emission and lifetime measurements were carried out in 2-methyltetrahydrofuran.
[c] Redox measurements were carried out in anhydrous DMF solution; values are reported relative to Cp_2Fe/Cp_2Fe^+.
[d] The λ_{max} values correspond to the highest energy peak in the spectrum. See Supporting Information.

photophysical measurements (vide infra). The calculation results for other tris-cyclometalated Ir complexes will be discussed in the following text as they pertain to electrochemical and spectral interpretation.

24.3.5 Electrochemistry

The electrochemical properties of the tris-cyclometalated iridium complexes were examined by cyclic voltammetry. A summary of the redox potentials, measured relative to an internal ferrocene reference ($Cp_2Fe/Cp_2Fe^+ = 0.45$ V vs SCE in DMF solvent) [23], is given in Table 24.4. All of the complexes show reversible oxidation, with potentials of 0.30–0.80 V. The phenylpyridyl-based derivatives exhibit reversible reduction, in the range from -2.51 to -2.78 V. However, no reduction is observed for the phenylpyrazolyl-based complexes out to -3.0 V (the solvent limit in DMF). The DFT calculations suggest that the reductive process is largely localized on the heterocyclic portion of the cyclometalating ligands. The reductive electrochemistry of the $Ir(C^\wedge N)_3$ complexes is thus consistent with the pyrazolyl being significantly more difficult to reduce than a pyridyl group. The absence of reductive processes in pyrazolyl ligated complexes is a common occurrence noted by other research groups [35–38]. For example, neither Pt(II) nor Rh(III) phenylpyrazolyl cyclometalates undergo measurable reduction, unlike the phenylpyridyl analogues [6, 34].

Replacing the pyridyl moiety with a pyrazolyl group also affects the oxidation potentials of the metal complexes. The oxidation potential of *fac*-Ir(*ppz*)$_3$ (0.39 V) is shifted to a slightly higher potential than that of *fac*-Ir(*ppy*)$_3$ (0.31 V). The DFT calculations show that the HOMOs of all of the tris-cyclometalated complexes examined here are composed of Ir-d and phenyl-π orbitals, similar to related studies [33]. While the pyridyl and pyrazolyl ligands do not contribute markedly to the HOMO, they do affect the energies of the metal orbitals through π-back-bonding. Previous studies with neutral pyrazole ligands and deprotonated pyrazoles have indicated that the pyrazolyl ligands are weak π-acceptors similar to pyridyl ligands [19, 39]. A similar observation has been reported for analogous Pt complexes; Pt(*phpz*)$_2$ has a higher oxidation potential than Pt(*ppy*)$_2$ (0.49 and 0.26 V, respectively) [38]. In accord with other studies, fluoro and/or trifluoromethyl substituents also increase the oxidation potential of the complexes while simultaneously making the phenylpyridyl-based derivatives easier to reduce [5h, 9, 34].

The meridional tris-cyclometalates have oxidation potentials ca. 50–100 mV less positive than their facial isomers, while the reduction potentials for the *mer*-phenylpyridyl-based complexes are only slightly less negative than those in the *fac*-analogues. As mentioned above, the oxidation processes involve the Ir-phenyl center, while reduction occurs primarily on the heterocyclic portion of the $C^\wedge N$ ligands [40]. The difference in electrochemical behavior between the meridional and facial isomers can be explained by the presence of mutually trans phenyl ligands in the *mer*-isomers. This configuration leads to a lengthening of the transoid Ir–C bonds and, consequently, destabilizes the HOMO to a significant extent, while only slightly stabilizing the LUMO. The electrochemistry is in accordance with results from DFT calculations that suggest that the HOMO energies of the meridional isomers are higher than those of the facial forms, while the LUMO energies are roughly the same. For example, the calculated HOMO energy for *mer*-Ir(*tpy*)$_3$ is -4.67 versus -4.78 eV for *fac*-Ir(*tpy*)$_3$, while the LUMO energies are less strongly perturbed (-1.18 versus -1.12 eV, respectively). Therefore, on the basis of these calculations, we predicted that it should be easier to both oxidize and reduce *mer*-Ir(*tpy*)$_3$ than would be the case for *fac*-Ir(*tpy*)$_3$, as is indeed observed.

24.3.6 Electronic Spectroscopy

The absorption and emission spectra were recorded for all of the complexes. These data are summarized in Table 24.4, and spectra for *fac*/*mer*-Ir(*tpy*)$_3$ and *fac*/*mer*-Ir(*ppz*)$_3$ are given in Fig. 24.5. The photophysical properties of the facial isomers of the phenylpyridyl-based Ir(III) cyclometalates have been examined by a number of research groups [5, 6, 9]. The absorption spectra of these compounds show intense bands appearing in the ultraviolet part of the spectrum between 240 and 350 nm. These bands have been assigned to the spin-allowed $^1(\pi \rightarrow \pi^*)$ transitions of the phenylpyridyl ligand. The

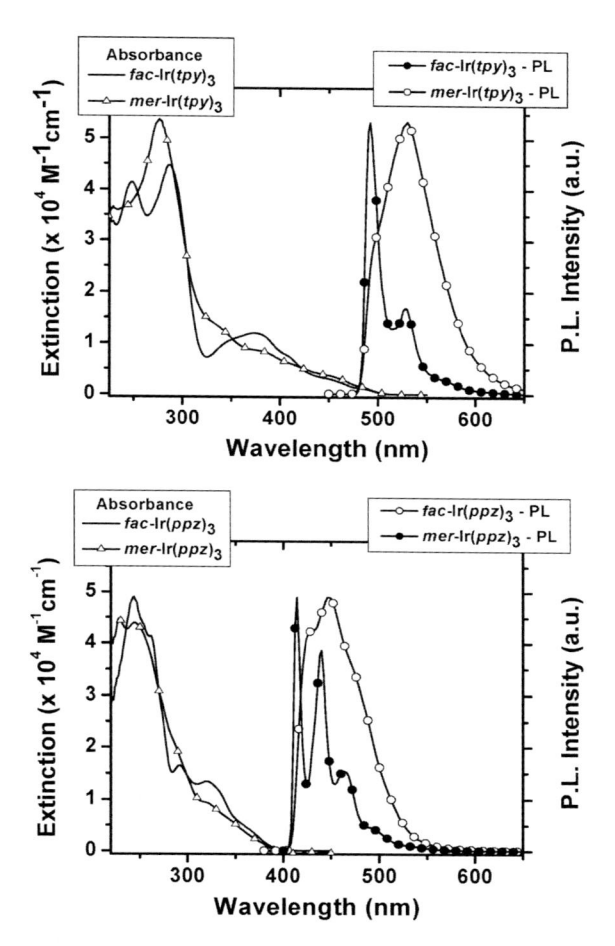

Figure 24.5 Absorption and emission spectra for facial and meridional isomers of Ir(*tpy*)$_3$ (top) and Ir(*ppz*)$_3$ (bottom). The absorption spectra were measured in CH$_2$Cl$_2$ at room temperature, and the emission spectra were measured at 77 K in 2-methyltetrahydrofuran (2-MeTHF) glass.

$^1(\pi \rightarrow \pi^*)$ bands are accompanied by weaker, lower energy features extending into the visible region from 350 to 450 nm that have been assigned to both allowed and spin-forbidden MLCT transitions. The high intensity of these MLCT bands has been attributed to an effective mixing of these charge-transfer transitions with higher lying spin-allowed transitions on the cyclometalating ligand [6].

This mixing is facilitated by the strong spin–orbit coupling of the Ir(III) center. The absorption spectra for all of the phenylpyrazolyl-based Ir complexes investigated here also show intense bands appearing in the ultraviolet part of the spectrum between 240 and 350 nm. The measured energies and extinction coefficients are comparable to those of the free ligands, that is, *ppz*H, *46dfppz*H, and *tfmppz*H. Thus, these features are similarly assigned to the allowed $^1(\pi \rightarrow \pi^*)$ transitions of the phenylpyrazolyl-based ligands [41]. These bands, as in the phenylpyridyl-based analogues, are also accompanied by weaker, low-lying charge-transfer transitions at 350–380 nm. However, the MLCT transitions of the phenylpyrazolyl-based complexes are significantly higher in energy than those of phenylpyridyl analogues. A similar hypsochromic shift has also been observed in the absorption spectra of phenylpyrazolyl-based Rh [36, 37] and Pt [35, 38] cyclometalates, as well as in pyridylpyrazolyl-based Ru tris-chelates [18, 19]. The shift to higher energy indicates an increase in energy separation between the metal d- and π^*-orbitals and is principally due to the fact that the π^*-orbitals of pyrazole are higher in energy than the π^*-orbitals of pyridine [18, 19, 39].

The facial isomers of the phenylpyridyl-based complexes are all intensely luminescent, both at 77 K and at room temperature (Table 24.4), with emission characteristic of phosphorescence from a mixed-ligand-centered-MLCT (LC-MLCT) triplet state [6, 8, 34]. The photophysical properties of *fac*-Ir(*tpy*)$_3$ at 77 K ($\lambda_{max} = 492$ nm, $\tau = 3.0$ μs) are nearly the same as those of *fac*-Ir(*ppy*)$_3$, while the difluoro-substituted analogue, *fac*-Ir(*46dfppy*)$_3$, has a blue-shifted emission ($\lambda_{max} = 450$ nm, $\tau = 2.5$ μs). The large hypsochromic shift caused by 4,6-difluoro substitution is consistent with behavior seen in Pt(II) cyclometalates [34]. Substitution of the phenyl hydrogens with inductively electron-withdrawing fluorine atoms, particularly on the 4′- and 6′-positions, stabilizes the HOMO more than the LUMO, thus increasing the triplet energy gap [34].

In contrast to the phenylpyridyl-based complexes, the facial isomers of the phenylpyrazolyl-based compounds are all very weak emitters at room temperature ($\Phi < 0.1\%$), but are intensely luminescent at 77 K. Several related *ppz* cyclometalated Pt and Rh complexes have also been reported to be poorly emissive in fluid solution but highly emissive in glassy matrixes at 77 K [35, 36]. The highly structured emissions of the facial phenylpyrazolyl-based complexes occur at higher energies and have longer lifetimes than the phenylpyridyl-based analogues. For example, at 77 K, the emission for *fac*-Ir(*ppz*)$_3$ has a $\lambda_{max} = 414$ nm and $\tau = 14$ µs, while for *fac*-Ir(*ppy*)$_3$, the values are $\lambda_{max} = 492$ nm and $\tau = 3.6$ µs. The behavior is consistent with a strongly perturbed ligand-centered transition in the phenylpyrazolyl derivatives. The triplet energy of the $C^\wedge N$ ligand strongly influences the phosphorescence energy of the corresponding cyclometalate because the excited state has both MLCT and intraligand ($\pi \rightarrow \pi^*$) triplet character [6, 8]. The triplet energy of phenylpyrazole (378 nm, 26,500 cm^{-1}) [36, 41] is greater than that of phenylpyridine (430 nm, 23,300 cm^{-1}) [42]; therefore, a phenylpyrazolyl-based cyclometalate is expected to have a higher emission energy than a related phenylpyridyl-based complex. Also, as seen with the phenylpyridyl-based complexes [5, 34], addition of electron-withdrawing fluorine substituents on the phenyl ring of *ppz* blue-shifts the emission for the corresponding cyclometalated complexes; thus, the triplet energy of *fac*-Ir(*46dfppz*)$_3$ is 390 nm. To our knowledge, this is the first Ir(III) tris-cyclometalate that emits in the ultraviolet region (below 400 nm).

The meridional isomers of the tris-cyclometalates exhibit photophysical characteristics different from those of their facial analogues. For example, the $^1(\pi \rightarrow \pi^*)$ absorption in *mer*-Ir(*tpy*)$_3$ (Fig. 24.5a) occurs as a single intense band at 276 nm, whereas in *fac*-Ir(*tpy*)$_3$ it appears as two features at 248 and 287 nm. Likewise, the MLCT band shapes in the meridional isomer are less sharply defined and have lower extinctions than those in the facial isomer. Greater differences exist in the luminescent behavior. Unlike the highly structured emission displayed by *fac*-isomers, the *mer*-isomers display broader, red-shifted luminescence. At room temperature, the *mer*-isomers of the phenylpyridyl derivatives also have much lower luminescent efficiencies and, concomitantly, shorter emission decay lifetimes than their facial counterparts. Assuming that the emitting state of a complex is formed with unit efficiency, one can calculate the radiative (k_r) and nonradiative (k_{nr}) rate constants using the relationships $k_r = \Phi_{PL}/\tau$ and $\Phi_{PL} = k_r/(k_r + k_{nr})$. The calculated radiative and nonradiative rate constants for the phenylpyridyl-based complexes are listed in Table 24.5. The *fac*- and *mer*-isomers have similar radiative rate constants; however, the nonradiative rate constants for the *mer*-isomers are more than an order of magnitude larger than those of the *fac*-isomers. The nonradiative rate constant is a sum of rates for several processes that quench emission. One of these processes could involve bond dissociation in the excited state. The mutually trans Ir–C_1 and Ir–C_2 bonds in the *mer*-Ir(*tpy*)$_3$ are already significantly longer than those of the facial analogue as a result of the strong trans influence of phenyl groups. The broadened emission of the *mer*-isomers indicates that the excited-state geometry is further distorted from that of the ground state. Photolytic cleavage of either an Ir–C or an Ir–N bond can then lead to subsequent rearrangement of the complex. This sort of bond breaking in the excited state is most likely responsible for the photoisomerization process described below.

24.3.7 *mer*-to-*fac* Isomerization

The fact that the meridional Ir($C^\wedge N$)$_3$ isomers can be obtained in good yield at low temperature, via synthetic Method C, suggests that the meridional isomers are kinetically favored products, while the facial isomers, obtained at higher temperatures, are thermodynamically favored. To test this hypothesis, the thermal conversion of meridional to facial isomers was examined. A pure sample of *mer*-Ir(*tpy*)$_3$ was dissolved in glycerol and refluxed for 24 h. Subsequent purification by column chromatography, under conditions where the facial and meridional isomers are cleanly separated, gave only a single colored band, and pure *fac*-Ir(*tpy*)$_3$ was isolated in >70% yield. This result supports the hypothesis that the meridional isomer is the kinetically favored product and converts to the facial isomer during high-temperature synthesis. The *mer*-Ir(*ppy*)$_3$, *mer*-Ir(*ppz*)$_3$, *mer*-Ir(*46dfppz*)$_3$, and *mer*-Ir(*tfmppz*)$_3$ complexes also isomerize to their facial forms using the same reaction conditions. However, the *mer*-Ir(*46dfppy*)$_3$ complex did not isomerize under these conditions, indicating that a larger kinetic barrier needs be overcome to isomerize this derivative.

The large nonradiative rate constants determined for some of the *mer*-isomers suggest that a bond rupture process may occur

Table 24.5 Luminescent quantum efficiencies, lifetimes, and the radiative/nonradiative decay rates for Ir$(C^\wedge N)_3$ complexes at room temperature

Complex	$\Phi_{PL}{}^a$	τ (μs)	k_r	k_{nr}
fac-Ir(ppy)$_3$	~~0.40~~ 0.9	1.9	2.1×10^5	3.2×10^5
mer-Ir(ppy)$_3$	~~0.036~~ 0.08	0.15	2.4×10^5	6.4×10^6
fac-Ir(tpy)$_3$	~~0.50~~ 1.0	2.0	2.5×10^5	2.5×10^5
mer-Ir(tpy)$_3$	~~0.05~~ 0.1	0.26	2.0×10^5	3.6×10^6
fac-Ir(46dfppy)$_3$	~~0.43~~ 0.9	1.6	2.7×10^5	3.6×10^5
mer-Ir(46dfppy)$_3$	~~0.053~~ 0.1	0.21	2.5×10^5	4.5×10^6

[a]The quantum yield reported in reference 5b for the Ir(ppy)$_3$ reference was 0.4. It was later discovered that this number was not correct and the photoluminescence quantum yield for this complex in solution is 0.73 in toluene and (W. Holzer, et al., Chem. Phys. 2005, **308**, 93–102, DOI: 10.1016/j.chemphys.2004.07.051) and 0.9 in THF (Sajoto, T., et al., J. Am. Chem. Soc. 2009, **131**, 9813–9822, DOI: 10.1021/ja903317w). Thus, the quantum yield values previously published have been increased by a factor of 2.2 (0.9/0.4) from the published values. The published values have been struck out.

in the excited state; hence, the photostability of mer-Ir$(C^\wedge N)_3$ complexes was also examined. The ^{1}H NMR spectra of the samples of the meridional complexes in degassed DMSO-d_6 were taken before and after irradiation of the sample with a handheld UV lamp. Nearly complete conversion (>95%) to the corresponding facial isomers was found to occur after less than 2 h of irradiation. This photochemical mer-to-fac conversion was observed for all of the Ir$(C^\wedge N)_3$ complexes reported here. The photoisomerization process is slower in less-coordinating solvents; thus, a sample of mer-Ir(tpy)$_3$ in toluene-d_8 took 72 h to undergo a similar degree of conversion to the fac-isomer. No intermediate species were observed when the NMR spectrum of this sample was examined at various time intervals during the photolysis (see Supporting Information). A clean mer-to-fac isomerization process is also indicated from the observation of isobestic points in the absorption spectra of mer-Ir(tpy)$_3$, taken at various stages of the photolysis in 2-methyltetrahydrofuran (see Supporting Information).

The greater thermodynamic stability of the fac-Ir$(C^\wedge N)_3$ complexes relative to the mer-isomers can be contrasted to the greater stability of mer-Ru(btpz)$_3$ [btpz = 1-(2'-(4',5'-benzothiazolyl) pyrazolyl)] [19] and mer-Al(8-hydroxyquinolyl)$_3$ [43] relative to their respective facial analogues. This has been attributed to the relief of steric interactions in the latter *mer*-compounds. The differing stability between *fac*- and *mer*-isomers of $N^\wedge N$ and $C^\wedge N$ tris-chelates may be brought about by the preference for maintaining the three strong trans influence phenyl groups of the cyclometalates on the same face of the molecule, trans to the heterocyclic groups. The thermodynamic instability of the meridional configuration is supported by the DFT calculations which show a $\sim$30 kJ/mol stabilization of total energy in favor of the *fac*-isomers.

24.4 Conclusion

The preparative methods for tris-cyclometalates reported here demonstrate that controlling the reaction conditions can impart significant control in the product configuration. For all of the cyclometalating ligands studied here, there is a preference to form the facial isomers at high temperatures ($>200^\circ$C) and the meridional isomers at lower temperatures ($<150^\circ$C). At high temperatures, the meridional isomer can be efficiently converted to the facial form, demonstrating the facial isomer is the thermodynamic product and the meridional form is the kinetic product. The differences in the ligand configuration of these complexes result in significantly different electrochemical and photophysical properties. The meridional isomers are easier to oxidize, and the emission is broad and red-shifted relative to the facial forms. The solution photoluminescent quantum efficiencies of the meridional isomers and their emission lifetimes are significantly lower than the facial isomers of the same cyclometalating ligands. The large difference between the quantum efficiencies of the meridional and facial isomers can be explained by an efficient bond breaking process for the meridional excited state, acting as an effective quenching pathway and giving subsequent isomerization to the facial form.

    The photochemical isomerization of meridional $Ir(C^\wedge N)_3$ complexes provides a new route to prepare the facial isomers, at temperatures much lower than conditions previously employed to make these complexes. This synthetic procedure can then lead to more efficient utilization of the iridium precursor materials because high reaction temperatures can promote undesired side reactions, decreasing product yields. Also, many potentially interesting and useful cyclometalating ligands have substituents that are unstable in the harsh reaction conditions required by the prior synthetic methods. Therefore, a greater variety of compounds can now be considered as potential ligands for facial tris-cyclometalated iridium complexes. In addition to the synthetic utility of these reactions, the mechanisms of the thermal and photochemical *mer*-to-*fac* isomerizations are of interest. Investigations into the mechanisms of these isomerization processes are currently being pursued in our laboratories.

Acknowledgment

The authors thank Universal Display Corp. and the Defense Advanced Research Projects Agency for financial support of this work.

Supporting Information Available

^{1}H NMR spectra, ^{1}H and ^{13}C NMR chemical shifts and coupling constants, UV-vis absorption and emission spectra, and table of DFT-calculated orbital energies for all of the facial and meridional Ir$(C^\wedge N)_3$ compounds examined here, as well as HOMO and LUMO plots for *fac*- and *mer*-isomers of Ir$(ppz)_3$ and a series of absorption and ^{1}H NMR spectra showing the photoconversion of *mer*-Ir$(tpy)_3$ to *fac*-Ir$(tpy)_3$ (PDF). Also included are tables of crystal data, atomic coordinates, bond distances, bond angles, and anisotropic displacement parameters for *mer*-Ir$(tpy)_3$, *fac*- and *mer*-isomers of Ir$(ppz)_3$, and *fac*- and *mer*-isomers of Ir$(tfmppz)_3$, as well as the corresponding CIF files. This material is available free of charge via the Internet at http://pubs.acs.org.

References

1. (a) Balzani, V., Scandola, F. *Supramolecular Photochemistry*; Ellis Horwood: Chichester, U.K., 1991. (b) Balzani, V., Credi, A., Scandola, F. In *Transition Metals in Supramolecular Chemistry*; Fabbrizzi, L., Poggi, A., Eds., Kluwer: Dordrecht, The Netherlands, 1994; p 1. (c) Lehn, J.-M. *Supramolecular Chemistry-Concepts and Properties*; VCH: Weinheim, Germany, 1995. (d) Bignozzi, C. A., Schoonover, J. R., Scandola, F. *Prog. Inorg. Chem.* 1997, **44**, 11.

2. (a) Kalyanasundaran, K. *Coord. Chem. Rev.* 1982, **46**, 159. (b) Chin, K.-F., Cheung, K.-K., Yip, H.-K., Mak, T. C. W., Che, C. M. *J. Chem. Soc., Dalton Trans.* 1995, **4**, 657. (c) Sonoyama, N., Karasawa, O., Kaizu, Y. *J. Chem. Soc., Faraday Trans.* 1995, **91**, 437. (d) Tan-Sien-Hee, L., Mesmaeker, A. K.-D. *J. Chem. Soc., Dalton Trans.* 1994, **24**, 3651. (e) Kalyanasundaram, K., Gratzel, M. *Coord. Chem. Rev.* 1998, **177**, 347.

3. (a) Anderson, P. A., Anderson, R. F., Furue, M., Junk, P. C., Keene, F. R., Patterson, B. T., Yeomans, B. D. *Inorg. Chem.* 2000, **39**, 2721. (b) Li, C., Hoffman, M. Z. *Inorg. Chem.* 1998, **37**, 830. (c) Berg-Brennan, C., Subramanian, P., Absi, M., Stern, C., Hupp, J. T. *Inorg. Chem.* 1996, **35**, 3719. (d) Kawanishi, Y., Kitamura, N., Tazuke, S. *Inorg. Chem.* 1989, **28**, 2968.

4. (a) Balzani, V., Juris, A., Venturi, M., Campagna, S., Serroni, S. *Chem. Rev.* 1996, **96**, 759. (b) Shaw, J. R., Sadler, G. S., Wacholtz, W. F., Ryu, C. K., Schmehl, R. H. *New J. Chem.* 1996, **20**, 749.

5. (a) Sprouse, S., King, K. A., Spellane, P. J., Watts, R. J. *J. Am. Chem. Soc.* 1984, **106**, 6647. (b) King, K, A., Spellane, P. J., Watts, R. J. *J. Am. Chem. Soc.* 1985, **107**, 1432. (c) Ohsawa, Y., Sprouse, S., King, K. A., DeArmond, M. K., Hanck, K. W., Watts, R. J. *J. Phys. Chem.* 1987, **91**, 1047. (d) Ichimura, K., Kobayashi, T., King, K. A., Watts, R. J. *J. Phys. Chem.* 1987, **91**, 6104. (e) Garces, F. O., King, K. A., Watts, R. J. *Inorg. Chem.* 1988, **27**, 3464. (f) Garces, F. O., Watts, R. J. *Inorg. Chem.* 1990, **29**, 582. (g) Wilde, A. P., King, K. A., Watts, R. J. *J. Phys. Chem.* 1991, **95**, 629. (h) Dedeian, K., Djurovich, P. I., Garces, F. O., Carlson, G., Watts, R. J. *Inorg. Chem.* 1991, **30**, 1685.

6. (a) Colombo, M. G., Hauser, A., Gudel, H. U. *Inorg. Chem.* 1993, **32**, 3088. (b) Colombo, M. G., Brunold, T. C., Riedener, T., Gudel, H. U. *Inorg. Chem.* 1994, **33**, 545.

7. (a) Sprouse, S., King, K. A., Spellane, P. J., Watts, R. J. *J. Am. Chem. Soc.* 1984, **106**, 6647. (b) Crosby, G. A. *J. Chim. Phys.* 1967, **64**, 160.

8. (a) Lamansky, S., Djurovich, P., Murphy, D., Abdel-Razzaq, F., Kwong, R., Tsyba, I., Bortz, M., Mui, B., Bau, R., Thompson, M. E. *Inorg. Chem.* 2001,

40, 1704. (b) Lamansky, S., Djurovich, P., Murphy, D., Abdel-Razzaq, F., Lee, H., Adachi, C., Burrows, P. E., Forrest, S. R., Thompson, M. E. *Inorg. J. Am. Chem. Soc.* 2001, **40**, 1704.

9. (a) Grushin, V. V., Herron, N., LeCloux, D. D., Marshall, W. J., Petrov, V. A., Wang, Y. *Chem. Commun.* 2001, 1494. (b) Wang Y., Herron, N., Grushin, V. V., LeCloux, D., Petrov, V. *Appl. Phys. Lett.* 2001, **79**, 479.

10. Ostrowski, J. C., Robinson, M. R., Heeger, A. J., Bazan, G. C. *Chem. Commun.* 2002, 784.

11. (a) Sutin, N. *Acc. Chem. Res.* 1968, **1**, 225. (b) Meyer, T. J. *Acc. Chem. Res.* 1978, **11**, 94.

12. Schmid, B., Garces, F. O., Watts, R. J. *Inorg. Chem.* 1994, **32**, 9.

13. (a) Belmore, K. A., Vanderpool, R. A., Tsai, J.-C., Khan, M. A., Nicholas, K. M. *J. Am. Chem. Soc.* 1988, **110**, 2004. (b) Silaware, N. D., Goldman, A. S., Ritter, R., Tyler, D. R. *Inorg. Chem.* 1989, **28**, 1231.

14. (a) Demas, J. N., Harris, E. W., McBride, R. P. *J. Am. Chem. Soc.* 1977, **99**, 3547. (b) Demas, J. N., Harris, E. W., Flynn, C. M., Diemente, D. *J. Am. Chem. Soc.* 1975, **97**, 3838. (c) Gao, R., Ho, D. G., Hernandez, B., Selke, M., Murphy, D., Djurovich, P. I., Thompson, M. E. *J. Am. Chem. Soc.* 2002, **124**, 14828.

15. (a) Baldo, M. A., O'Brien, D. F., You, Y., Shoustikov, A., Sibley, S., Thompson, M. E., Forrest, S. R. *Nature* 1998, **395**, 151. (b) Baldo, M. A., Lamansky, S., Burrows, P. E., Thompson, M. E., Forrest, S. R. *Appl. Phys. Lett.* 1999, **75**, 4. (c) Thompson, M. E., Burrows, P. E., Forrest, S. R. *Curr. Opin. Solid State Mater. Sci.* 1999, **4**, 369.

16. Adamovich, V., Brooks, J., Tamayo, A., Djurovich, P., Alexander, A., Thompson, M. E. *New J. Chem.* 2002, 1171.

17. Steel, P. J., Lahousse, F., Lerner, D., Marzin, C. *Inorg. Chem.* 1983, **22**, 1488.

18. Steel, P. J., Constable, E. C. *J. Chem. Soc., Dalton Trans.* 1990, 1389.

19. Luo, Y., Potvin, P. G., Tse, Y., Lever, A. B. P. *Inorg. Chem.* 1996, **35**, 5445.

20. Fletcher, N. C., Nieuwenhuyzen, M., Rainey, S. *J. Chem. Soc., Dalton Trans.* 2001, 2641.

21. Fletcher, N. C., Nieuwenhuyzen, M., Prabaharan, R., Wilson, A. *Chem. Commun.* 2002, 1188.

22. (a) Demas, J. N., Crosby, G. A. *J. Phys. Chem.* 1978, **82**, 991. (b) DePriest, J., Zheng, G. Y., Goswami, N., Eichhorn, D. M., Woods, C., Rillema, D. P. *Inorg. Chem.* 2000, **39**, 1955.

23. (a) Gagne, R. R., Koval, C. A., Lisensky, G. C. *Inorg. Chem.* 1980, **19**, 2854. (b) Sawyer, D. T., Sobkowiak, A., Roberts, J. L., Jr. *Electrochemistry for Chemists*, 2nd ed., John Wiley and Sons: New York, 1995; p 467.

24. Sheldrick, G. M. *SHELXTL*, version 5.1; Bruker Analytical X-ray System, Inc.: Madison, WI, 1997.
25. Blessing, R. H. *Acta Crystallogr*. 1995, **A51**, 33.
26. Lohse, O., Thevenin, P., Waldvogel, E. *Synlett* 1999, **1**, 45.
27. Finar, I. L., Rackham, D. M. *J. Chem. Soc. B* 1968, 211.
28. Nonoyama, M. *Bull. Chem. Soc. Jpn*. 1974, **47**, 767.
29. Garces, F. O., Dedian, K., Keder, N. L., Watts, R. J. *Acta Crystallogr*. 1993, **C49**, 1117.
30. Urban, R., Kramer, R., Mihan, S., Polborn, K., Wagner, B., Beck, W. *J. Organomet. Chem*. 1996, **517**, 191.
31. Neve, F., Crispini, A. *Eur. J. Inorg. Chem*. 2000, 1039.
32. Douglas, B., McDaniel, D., Alexander, J. *Concepts and Models in Inorganic Chemistry*, 3rd ed., John Wiley & Sons, Inc.: New York, 1994.
33. Hay, P. J. *J. Phys. Chem. A*. 2002, **106**, 1634.
34. Brooks, J., Babayan, Y., Lamansky, S., Djurovich, P. I., Tsyba, I., Bau, R., Thompson, M. E. *Inorg. Chem*. 2002, **41**, 3055.
35. Sandrini, D., Maestri, M., Ciano, M., Balzani, V., Lueoend, R., Deuschel-Cornioley, C., Chassot, L., Von Zelewsky, A. *Gazz. Chim. Acta* 1988, **118**, 661.
36. Maeder, U., Stoeckli-Evans, H., von Zelewsky, A. *Helv. Chim. Acta* 1992, **73**, 1321.
37. Sandrini, D., Maestri, M., Ciano, M., Maeder, U., von Zelewsky, A. *Helv. Chim. Acta* 1990, **73**, 1307.
38. Chassot, L., von Zelewsky, A. *Inorg. Chem*. 1987, **26**, 2814.
39. Sullivan, P., Salmon, D. J., Meyer, T. J., Peeding, J. *Inorg. Chem*. 1979, **18**, 3369.
40. Kulikova, M., Balashev, K. P., Kvam, P. I., Songstad, J. *Russ. J. Gen. Chem*. 2000, **70**, 163.
41. (a) Pavlik, J. W., Connors, R. E., Burns, D. S., Kurzweil, E. M. *J. Am. Chem. Soc*. 1993, **115**, 7645. (b) Cativiela, C., Laureiro, J. I. G., Elguero, J., Elguero, E. *Gazz. Chim. Ital*. 1991, **121**, 477. (c) Cativiela, C., Laureiro, J. I. G., Elguero, J., Elguero, E. *Gazz. Chim. Ital*. 1989, **119**, 41. (d) Cativiela, C., Laureiro, J. I. G., Elguero, J., Elguero, E. *Gazz. Chim. Ital*. 1986, **116**, 119.
42. Murov, S. L., Carmichael, I., Hug, G. L. *Handbook of Photochemistry*; Marcel Dekker, Inc.: New York, 1993.
43. Utz, M., Chen, C., Morton, M., Papadimitrakopoulos, F. *J. Am. Chem. Soc*. 2003, **125**, 1371.

Chapter 25

Phosphorescence Quenching by Conjugated Polymers

Madhusoodhanan Sudhakar, Peter I. Djurovich,
Thieo E. Hogen-Esch, and Mark E. Thompson
*Department of Chemistry, University of Southern California,
Los Angeles, California 90089, USA*
met@usc.edu

Efficient organic light-emitting diodes (OLEDs) have been fabricated with low-molecular weight materials [1]. Hole–electron recombination in these OLEDs leads to the formation of both singlet and triplet excited states (excitons) within the molecular thin film [2]. For most compounds, only the singlet state is emissive, leading to a significant limitation in the OLED efficiency. Incorporation of phosphorescent compounds into the OLED gives a substantial improvement in device efficiency, since both the singlet and triplet excitons are trapped at the phosphor. This approach has led to OLEDs with external quantum efficiencies (photon/electron) of roughly 20% which correspond to internal efficiencies of nearly 100% [3].

Reprinted from *J. Am. Chem. Soc.*, **125**, 7796–7797, 2003.

Electrophosphorescent Materials and Devices
Edited by Mark E. Thompson
Text Copyright © 2003 American Chemical Society
Layout Copyright © 2024 Jenny Stanford Publishing Pte. Ltd.
ISBN 978-981-4877-34-3 (Hardcover), 978-1-003-08872-1 (eBook)
www.jennystanford.com

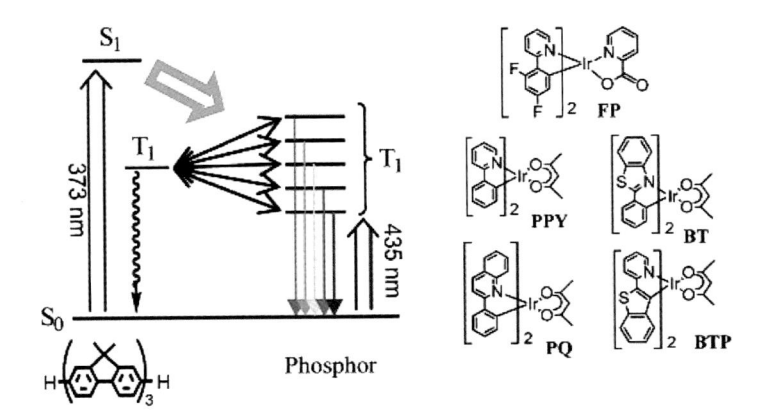

Figure 25.1 Schematic energy level diagram showing energy transfer between F_3 (left) and the iridium complexes (right).

Conjugated polymers have also been used to prepare OLEDs, e.g. poly(phenylenevinylenes) [4a], polyfluorenes [4b], and poly(p-phenylenes) [4c]. While these devices can have good power efficiencies (lum/W or W_{opt}/W_{elect}), the polymer-based OLEDs tend to give external quantum efficiencies of less than 5%. Several groups have attempted to increase the quantum efficiencies of conjugated polymer-based OLEDs by incorporating phosphorescent dopants [5]. However, while the efficiencies of conjugated polymer OLEDs are improved by phosphor doping, values are still markedly lower than those of small-molecule-based devices (<5%) [5]. In contrast, phosphor-doped OLEDs fabricated with nonconjugated polymers (e.g. poly(vinylcarbazole), PVK) can have good external quantum efficiencies [6], with some as high as 8% [7].

To understand why phosphor-doped OLEDs give low quantum efficiencies with conjugated polymers, we have carried out a luminescent quenching study of phosphorescent emission using a model polyfluorene oligomer as a triplet energy quencher (F_3 shown in Fig. 25.1). The phosphorescent cyclometalated Ir complexes used in this study had a range of emission energies, spanning from blue to red (Fig. 25.2) [8]. A fluorene trimer, rather than a polymer, was chosen as a quencher for several reasons. F_3 is estimated to have a triplet energy somewhat higher than that of the polymer,

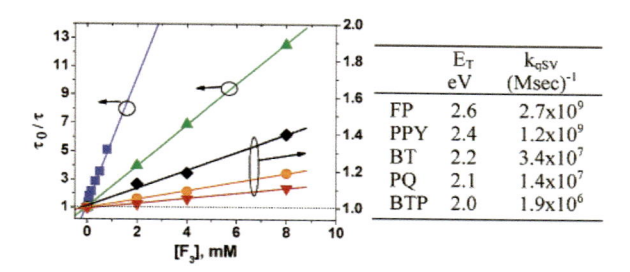

	E_T eV	$k_q SV$ $(Msec)^{-1}$
FP	2.6	2.7×10^9
PPY	2.4	1.2×10^9
BT	2.2	3.4×10^7
PQ	2.1	1.4×10^7
BTP	2.0	1.9×10^6

Figure 25.2 Stern–Volmer plot for quenching of the five phosphors by F_3. (Legend: ■ FP, ▲ PPY, ♦ BT, ● PQ, ▼ BTP). Table: triplet energy [8, 10] of phosphors and their Stern–Volmer quenching constants [11]. The estimated errors in the k_qSV values are ±10%.

roughly about 540–580 nm [9]. In addition, unlike polyfluorenes, F_3 has negligible absorption in the region where the phosphors absorb strongly (i.e., 435 nm), allowing the direct excitation of the phosphor. Also, on the basis of solution electrochemistry of F_3 and the complexes, the HOMO and LUMO orbitals of each of the phosphors are expected to fall within the HOMO/LUMO gap of F_3 [10]. This precludes exciplex formation or excited-state electron transfer as the origin of luminescent quenching. Last, F_3 is more soluble than a polymer and has a discrete molecular weight, making for straightforward Stern–Volmer analysis of the quenching phenomena.

A schematic energy level diagram for F_3 and the Ir phosphors is shown in Fig. 25.1. The singlet state of F_3 is significantly higher in energy than any of the phosphor triplet states, but the F_3 triplet energy is expected to fall somewhere in the middle of the phosphor energy range. Hence, phosphorescence quenching should occur by energy transfer between the excited phosphor and the triplet state of F_3.

The luminescent lifetimes were recorded in 2-methyltetra-hydrofuran as a function of F_3 concentration and plotted using standard Stern–Volmer analysis (i.e., τ_0/τ versus [F_3], Fig. 25.2) [10, 11]. Quenching rate constants were derived from the slope of linear fits to the data for each complex. The Stern–Volmer quenching rates given in Fig. 25.2 were measured using direct excitation of the complexes with 435 nm light; however, similar rates are obtained when the solutions are excited using 373 nm light and energy is transferred from the F_3 singlet to the triplet levels of the complex, as illustrated by the gray arrow in Fig. 25.1. It can be seen from the data in Fig. 25.2 that quenching of FP and PPY emission by F_3 is an exothermic process, occurring at near diffusion controlled rates, while quenching of the yellow, orange, and red emission (from BT, PQ, and BTP, respectively) is an endothermic process, occurring at rates well below those of FP or PPY. This indicates that the triplet energy of the trimer is less than 2.3 eV, as predicted from Bässler's data [9].

Upon excitation of Ir phosphor/F_3 mixtures, the Ir complex can either relax directly to the ground state (via radiative or nonradiative pathways) or transfer energy to the F_3 triplet state. Similar behavior in the type of thin films used in phosphorescent OLEDs, where high dopant concentrations (5–10 wt%) are employed, should lead to pseudo-first-order reaction conditions for energy transfer from phosphor to quencher. It is therefore worth noting that emission from BTP is efficiently quenched by F_3 (k_qSV) $1.9 \times 10^6 M^{-1} s^{-1}$) despite having a lower triplet energy. Using this k_qSV value, the half-life for BTP emission quenching in a doped thin film of F_3 (ca. 1 M) is estimated to be 0.4 µs. Since the phosphorescent half-life of an excited BTP molecule is 4 µs (based on a luminescent lifetime of 5.8 µs) [8], near complete phosphorescence quenching is predicted for a BTP-doped F_3 film. Quite the contrary, BTP emission is observed for 5 wt% doped polyfluorene thin films [5e]. This discrepancy comes about because of significant differences between concentrated solid films and the dilute solution mixtures used for our studies. First, the doping levels of phosphors in the solid films are considerably higher than the µM to mM concentrations used in the quenching experiments. The high doping concentrations promote phosphor aggregation [12]; consequently, the reduced intermolecular contact between F_3 molecules and the dopant complexes decreases the phosphorescent quenching rate. Second, energy transfer between the triplets of the phosphor and F_3 is most likely an electron exchange or "Dexter" transfer process. Dexter transfer requires a good intermolecular overlap between the pertinent molecular orbitals of the donor and acceptor [11]. Since molecular motion is inhibited in amorphous solid films, a dopant-F_3 configuration that gives poor Dexter energy transfer cannot readily reorient to a configuration appropriate for efficient energy transfer. Last, in fluid solution, the F_3 and phosphor can physically separate after energy transfer. This physical diffusion enhances quenching by suppressing reverse energy transfer (F_3 to phosphor), especially for those phosphors (e.g., BTP and PQ) where energy transfer to the F_3 triplet state is an endothermic process. In a rigid matrix, such as a doped thin film, physical diffusion cannot occur. In this case, separating excited F_3 and the phosphor involves energy migration from the excited F_3 to an adjacent F_3, a slower process than the competing back energy transfer. These three factors together lead to k_qSV values in doped thin films that are lower than those measured in dilute fluid solutions.

The results obtained from our quenching studies highlight an important criterion that needs to be considered when designing polymer-based phosphorescent OLEDs. Energy transfer from either singlet or triplet levels of a conjugated polymer to a phosphorescent dopant certainly looks appealing at the outset. However, one needs also to consider the emission quenching brought about by the low-energy triplet states of the polymer since even endothermic transfer can effectively quench phosphorescence. High-molecular weight polyfluorenes have conjugation lengths that are greater than that of F_3; thus, triplet-state energies are lower for the polymers (2.1 eV) than for the oligomer (2.3 eV) [9]. Hence, one can expect that phosphorescence quenching will be more efficient in polyfluorenes than in F_3. With nonconjugated polymers such as PVK, however, phosphorescence quenching is less favorable due to the absence of any low-energy triplet states [6, 7]. Although doped OLEDs with moderate efficiencies have been prepared by blending conjugated polymers with red phosphors [5c–e], it becomes increasingly difficult to use this strategy for devices emitting at higher energy since it requires conjugated host polymers with high triplet-state energies. A solution to this dilemma is either to use nonconjugated polymers, such as PVK, or to design new conjugated polymers with higher triplet energies. Such new polymer systems may then allow fabrication of highly efficient (>10% external) polymer-based phosphorescent OLEDs.

Acknowledgment

We thank The Universal Display Corporation (M.E.T.) and the NSF-DMR (T.H.E.) for financial support of this work.

Supporting Information Available

Phosphorescence spectra from the phosphors, synthesis and characterization of F_3, electrochemical properties of the compounds, and data for the Stern–Volmer analysis (PDF). This material is available via the Internet at http://pubs.acs.org.

References

1. (a) Mitscheke, U., Bauerle, P. *J. Mater. Chem.* 2000, **10**, 1471–1507. (b) Tsutsui, T., Fujita, K. *Adv. Mater.* 2002, **14**, 949–952.
2. Baldo, M. A., O'Brien, D. F., Thompson, M. E., Forrest, S. R. *Phys. Rev. B* 1999, **60**, 14422–14428.
3. (a) Adachi, C., Baldo, M. A., Thompson, M. E., Forrest, S. R. *J. Appl. Phys.* 2001, **90**, 5048–5051. (b) Ikai, M., Tokito, S., Sakamoto, Y., Suzuki, T., Taga, Y. *Appl. Phys. Lett.* 2001, **79**, 156–158.
4. (a) Friend, R. H., Gymer, R. W., Holmes, A. B., Burroughes, J. H., Marks, R. N., Taliani, C., Bradley, D. D. C., Dos Santos, D. A., Bredas, J. L., Salaneck, W. R. *Nature* 1999, **397**, 121–128. (b) Millard, I. S. *Synth. Met.* 2000, **111–112**, 119–123. (c) Wohlegenannt, M., Tandon, K., Mazumdar, S., Ramashesha, S., Vardeny, Z. V. *Nature* 2001, **409**, 494–497.
5. (a) Gross, M., Nuller, D. C., Nothofer, H.-G., Scherf, U., Neher, D., Brauchle, C., Meerholz, K. *Nature* 2000, **405**, 661–665. (b) Chen, X., Liao, J.-L., Liang, Y., Ahmed, M. O., Tseng, H.-E., Chen, S.-C. *J. Am. Chem. Soc.* 2003, **125**, 636–637. (c) Higgins, R. W. T., Monkman, A. P., Nothofer, H.-G., Scherf, U. *J. Appl. Phys.* 2002, **91**, 99–105. (d) Zhu, W., Liu, C., Su, L., Yang, W., Yuan, M., Cao, Y. *J. Mater. Chem.* 2003, **13**, 50–55. (e) Chen, F.-C., Yang, Y., Thompson, M. E., Kido, J. *Appl. Phys. Lett.* 2002, **80**, 2308–2310.
6. (a) Yang, M.-J., Tsutsui, T. *Jpn. J. Appl. Phys., Part 2* 2000, **39**, L828–L829. (b) Kawamura, Y., Yanagida, S., Forrest, S. R. *J. Appl. Phys.* 2002, **92**, 87–93.

7. Vaeth, K. M., Tang, C. W. *J. Appl. Phys.* 2002, **92**, 3447–3453.
8. Lamansky, S., Djurovich, P., Murphy, D., Abdel-Razzaq, F., Lee, H.-E., Adachi, C., Burrows, P. E., Forrest, S. R., Thompson, M. E. *J. Am. Chem. Soc.* 2001, **123**, 4304–4312.
9. Hertel, D., Romanovskii, Y. V., Schweitzer, B., Scherf, U., Bässler, H. *Macromol. Symp.* 2001, **175**, 141–150.
10. Refer to the Supporting Information.
11. Turro, N. J. *Modern Molecular Photochemistry*; The Benjamin/Cummings Publishing Co., Inc., Menlo Park, California, 1978.
12. Noh, Y.-Y., Lee, C.-L., Kim, J.-J., Yase, K., *J. Chem. Phys.* 2003, **118**, 2853–2864.

Chapter 26

Simultaneous Light Emission from a Mixture of Dendrimer Encapsulated Chromophores: A Model for Single-Layer Multichromophoric Organic Light-Emitting Diodes

Paul Furuta,[a] Jason Brooks,[b] Mark E. Thompson,[b] and Jean M. J. Fréchet[a]

Contribution from the Department of Chemistry, University of California, Berkeley, California 94720-1460, USA, and Materials Science Division, Lawrence Berkeley National Laboratory, Berkeley, California 94720, USA, and Department of Chemistry, University of Southern California, Los Angeles, California 90089-0744, USA
[a]*Department of Chemistry, University of California and Materials Science Division, Lawrence Berkeley National Laboratory*
[b]*Department of Chemistry, University of Southern California*
frechet@cchem.berkeley.edu, mthompso@chem1.usc.edu

The site isolation of two dyes capable of electronic interaction via Forster energy transfer has been studied with the two dyes coumarin 343 and pentathiophene encapsulated by dendrons containing both solubilizing and electroactive moieties.

Reprinted from *J. Am. Chem. Soc.*, **125**, 13165–13172, 2003.

Electrophosphorescent Materials and Devices
Edited by Mark E. Thompson
Text Copyright © 2003 American Chemical Society
Layout Copyright © 2024 Jenny Stanford Publishing Pte. Ltd.
ISBN 978-981-4877-34-3 (Hardcover), 978-1-003-08872-1 (eBook)
www.jennystanford.com

Photoluminescence studies of mixtures of the dendritic dyes show that at high dendron generation, significant site isolation is achieved with relative emission characteristics influenced by both the degree of site isolation and the emission quantum yield of the dyes. Electroluminescence studies carried out in organic light emitting diode devices confirm that color tuning may be achieved by mixing the two encapsulated dyes in a single layer. However, selective carrier trapping by one of the core component dyes can dramatically influence the effectiveness of other components in the device.

26.1 Introduction

It is well-known that the environment of a chromophore is able to affect its photophysical behavior including its absorption and emission characteristics. Nature has been able to perfect a variety of microenvironments to enhance performance by spatially arranging multiple chromophores with respect to one another in a highly ordered fashion thereby also maximizing synergism between various species [1]. Although proteins are nature's main building blocks for the precise assembly of most functional moieties, including chromophores, synthetic systems do not benefit from the structural precision achievable with proteins. Recently, dendrimers have been used in an attempt to build multichromophoric systems that mimic some of the primary events of photosynthesis. Therefore, using a dendritic framework, multiple peripheral chromophores can be assembled around a single core unit with which they can "communicate" via an energy transfer interaction to form a light-harvesting antenna [2]. In such an antenna, the energy harvested at the periphery of the dendrimer is funneled to the core where it is reprocessed, for example through amplified core emission. In this instance, the dendritic framework facilitates the communication between the chromophores by keeping them at an appropriate average distance—within the Forster radius—of the core. In other instances, nature uses proteins to accomplish site isolation of a reactive site, thereby preventing destructive interaction. This is the case with a variety of catalytic enzymes such as the Cytochromes for which isolation of the catalytic hemes is essential to derive function.

Once again, dendrimers have been used to perform analogous site isolation, for example, bulky dendrons assembled around a central moiety have been used to prevent the self-aggregation of dye molecules in the solid state [2o]. Encapsulating individual chromophores greatly enhances their optical properties due to reduced self-quenching [3]. As described below, a challenging test of the ability of dendrimers to afford site isolation is afforded by light emitting diodes containing multiple emitters in a single layer.

In an OLED application [4], it is difficult to tune the color of emission by mixing several light emitting dyes in a single layer because energy transfer between the dyes lead to emission solely from the dye with the lowest HOMO–LUMO band gap [5]. In addition, the emitting fluorophores can self-quench via excimer formation when doped at high concentrations. These problems can, at least in principle, be overcome by the use of dendrimers. For example, a fluorophore surrounded by a dendritic shell of sufficient size can isolate the emitting moieties from one another, thereby enabling simultaneous emission from several dyes, as site-isolation should inhibit both energy transfer between dyes and self-quenching [3]. Color tuning in a single layer device would be achieved by the proper selection and mixing of dendrimers with different site-isolated fluorescent dyes as their cores. Moreover, the solubility and processing behavior will be greatly affected by the dendritic shell, allowing for a wide range of dyes to be encapsulated within the dendrimer while peripheral structural changes can be used to optimize and control process parameters [6].

Recently, a variety of OLEDs [7] in which dendrimers or dendronized polymers used as both hole transporting (HT) [8] and electron transporting (ET) [9] components, as well as unimolecular emitters [10] have been reported. Building upon an early design of Devadoss et al. [2b], Freeman et al. [11] successfully utilized triarylamine (TAA) HT-labeled poly(benzyl ether) dendrimers possessing Coumarin 343 (C343) and pentathiophene (T5) core dyes to give blue and green OLEDs, respectively. Initial attempts to achieve synergistic emission from the two types of dendrimers by blending them in a single layer were only moderately successful due to detrimental energy transfer from C343 to the smaller band-gap T5. Due to the large Förster radius of these dyes, energy transfer is an unfortuitous consequence of the small size and flexibility (i.e., poor shielding effect) of the surrounding dendrons that were attached at a single point to the lumophores. Attempts to synthesize larger dendrimers were unsuccessful due to the poor solubility and increased crystallinity of the triarylamine dendrons beyond generation 2, which resulted in very poor device performance. In the accompanying report (this issue), we describe the design and synthesis of novel dendrimers containing not only triarylamine groups, but also solubilizing alkyl ether groups at the periphery of the dendrimers to solve both the solubility and crystallinity problems, enabling the synthesis of larger dendrimers for a fundamental investigation of the concept of color tuning via site isolation.

26.2 Results and Discussion

26.2.1 Dendrimer Encapsulated Chromophores

Both coumarin 343 (C343) and pentathiophene (T5) labeled dendrimers were utilized to effect the site isolation of light emitting lumophores in an OLED configuration. The two core dyes were chosen as models primarily on the basis of the overlap of their absorbances with the emission of the peripheral TAA hole transporting moieties at 425 nm. The dendritic structure we tested in a preliminary study [11] has now been completely redesigned with a periphery containing a combination of alkyl ether groups [12] and TAA donors (Fig. 26.1). This combination has enabled the construction of much larger dendrimers for a more realistic testing of dye encapsulation while removing the issues of solubility and crystallization that earlier plagued film coating processes, affected device construction, and limited performance. For example a coumarin 343 laser dye could be encapsulated with 3 generation three [G-3] or generation four [G-4] dendrons to form [G-3]3-C343, **5**, and [G-4]3-C343, **6**, as shown in Scheme 26.1. Given the elongated structure of the pentathiophene chromophore and the availability of new, highly soluble electroactive dendrons, we targeted better encapsulation by capping both ends of the T5 with high generation dendrons (Scheme 26.2). An efficient synthesis

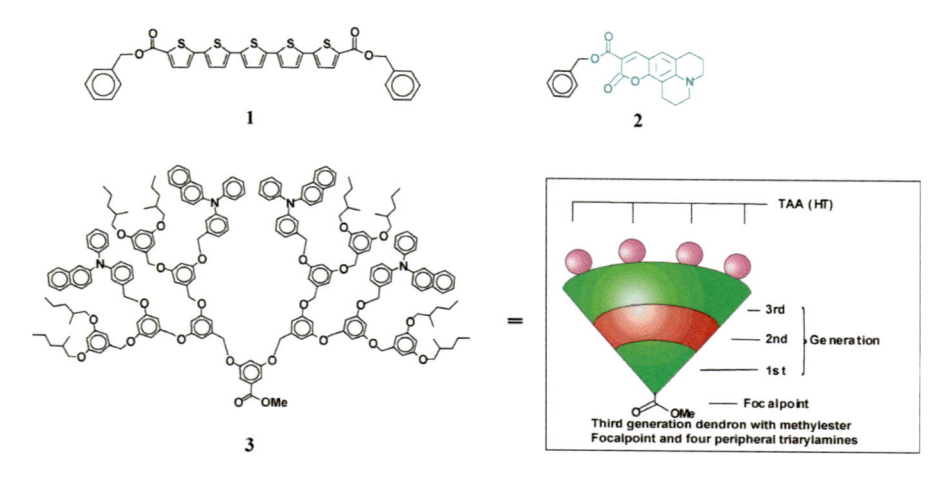

Figure 26.1 Fluorescent dyes, Coumarin 343 (C343) **1** and pentathiophene dicarboxylic ester (T5) **2**, used as cores for the dendrimers of this study. The surrounding dendrons architecture and periphery are shown with a third generation dendron **3** that features hole transporting (TAA) and solubilizing (alkyl ether) end-groups.

was chosen, which simultaneously forms the T5 core and links properly functionalized dendrimers, forming the target bis [G-3], [G-4], and [G-5] T5 "dumbbell-shaped" dendrimers (**7**, **8**, and **9**, respectively). As a result of the dendron design with its peripheral energy donating/hole transporting TAA and solubilizing alkyl ether groups, all the C343 and T5 dendrimers were fully soluble in common organic solvents and formed homogeneous uniform films suitable for OLED's.

26.2.2 Photo and Device Emission Spectra

As the main objective of this study was a determination of the site isolation provided by our synthetic blueprint, solid-state thin film photoluminescence (PL) and device electroluminescence (EL) studies of the dendrimers were conducted utilizing different mixtures of C343 to T5 dendrimers of varying sizes and used in different ratios as outlined in Table 26.1. The C343 has an absorption λ_{max} at 446 nm with an extinction coefficient of 44,000 cm^{-1}M^{-1}, and T5 absorbs at 425 nm with an extinction coefficient of 49,000 cm^{-1}M^{-1}. As mentioned earlier these absorptions overlap well with the emission maximum of TAA at 425 nm (Fig. 26.2). The dyes emit at ~470 nm for C343, and at ~525 nm for T5, via Förster resonance energy transfer [13]. PL emission is produced by selective excitation of the peripheral TAA groups at 350 nm and energy is efficiently transferred into both the C343 and T5 cores. The Förster radius between the C343 and T5 dyes was calculated to be about 3.8 nm, due to a significant overlap of the T5 absorption band with the emission of the C343. It is the prevailing occurrence of such long-range Coulombic interactions that makes large dendrimers necessary to spatially separate the dyes and prevent energy transfer from C343 to T5. Dendrimers films with film thicknesses ranging between 1100 and 1300 Å were spun onto glass substrate using chloroform as the solvent. Upon inspection under a microscope, the films appeared to be consistent and uniform without any observable morphological features over the area of the substrate. The alkyl surface groups interspersed between the TAA moieties in our new dendron design provide the desired solubility enhancement, preventing crystallization, and leading to good properties for the spin-coated dendrimer films. Thin film photoluminescence experiments conducted with the various mixtures of C343 and T5 dendrimers of Table 26.1 clearly demonstrate that site-isolation increases as the size of the dendrons surrounding the cores increases. The PL spectra of dendrimer mixtures 1 and 2 (Fig. 26.3) display emission at ~525 nm corresponding primarily to the T5 with a very small contribution from C343 at ~470 nm, which increases slightly as the molar ratio of **5** to **7** increases. As expected, the smaller generation 3 (G-3) dendrons provide only modest isolation and significant energy transfer occurs from C343 to T5. Therefore, a second series of films were prepared using mixtures of the larger dendrimers **6** and **9** in the same 1:1 and 5:1 molar ratios (mixtures 7 and 9, Table 26.1). With the higher generation dendrimers, the increase in the degree of site isolation can be seen from the noticeable increase of C343 emission compared to that observed for the corresponding third generation dendrimers. In a 1:1 molar ratio of **6** and **9**, the observed emission for the higher energy C343 chromophore far exceeded that of a 5:1 molar ratio of **5** and **7** giving a strong indication that there is increased average spatial separation between the two dendronized dyes.

Scheme 26.1

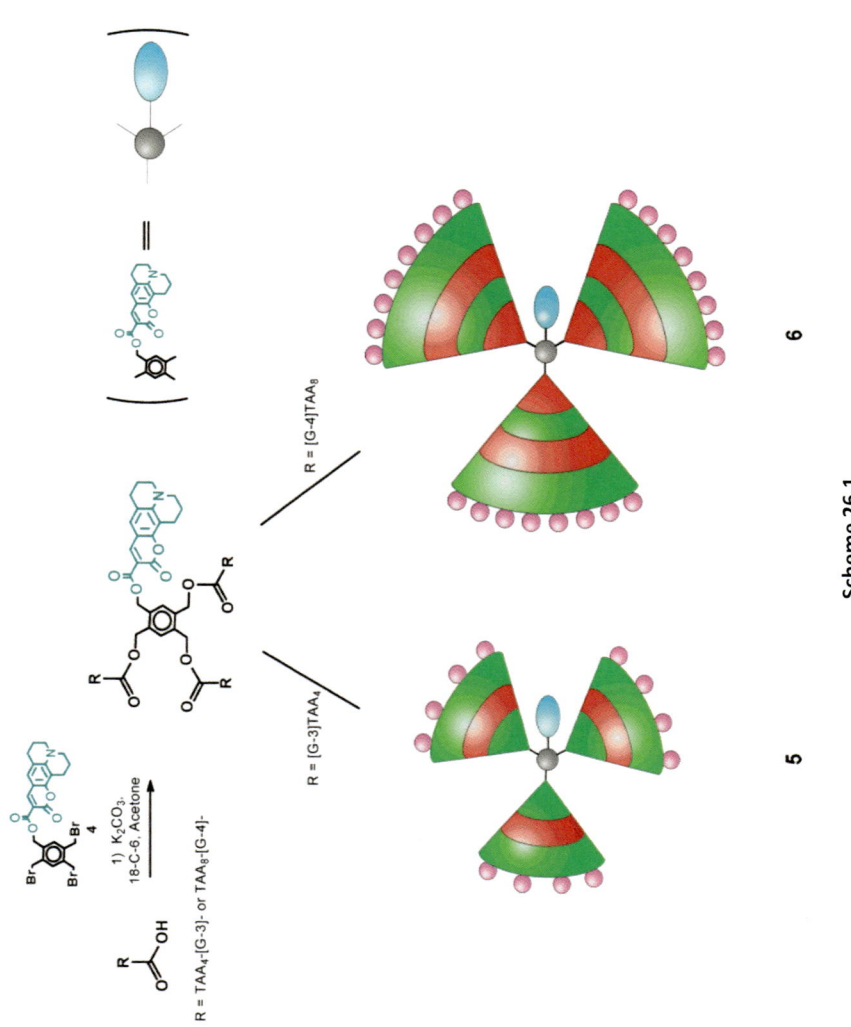

Scheme 26.2

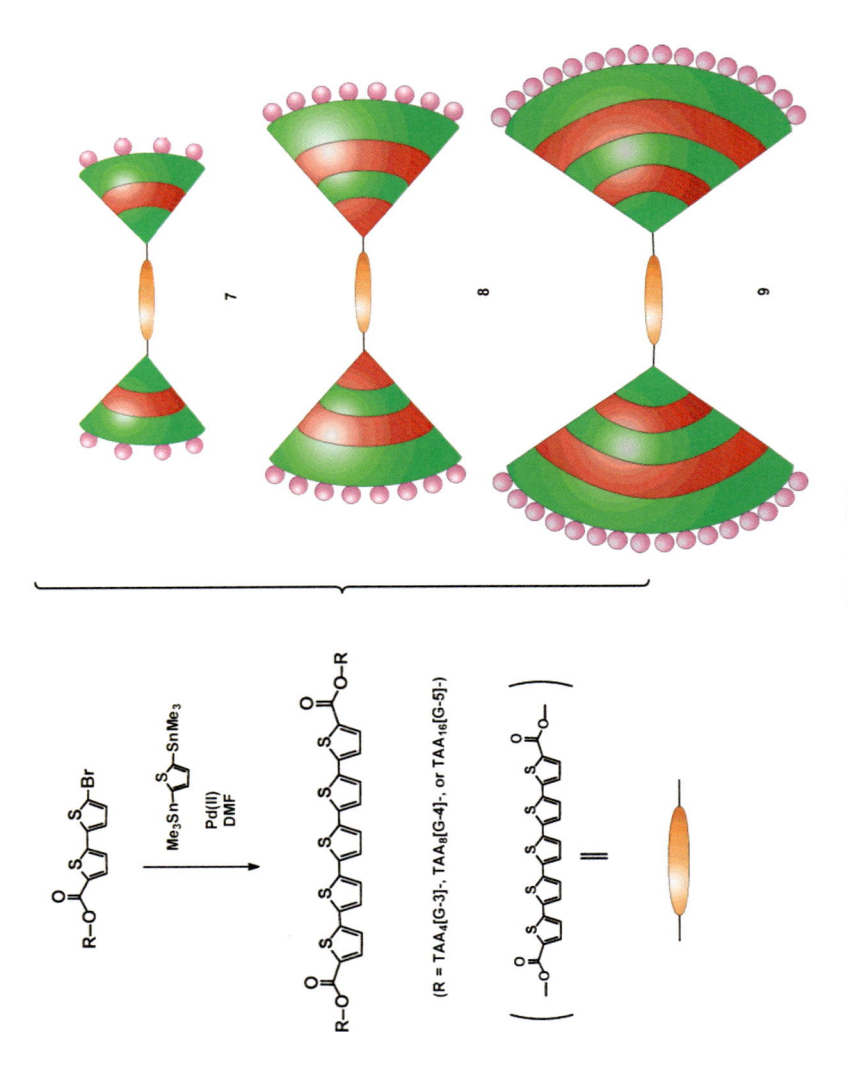

(R = TAA$_4$[G-3]-, TAA$_8$[G-4]-, or TAA$_{16}$[G-5]-)

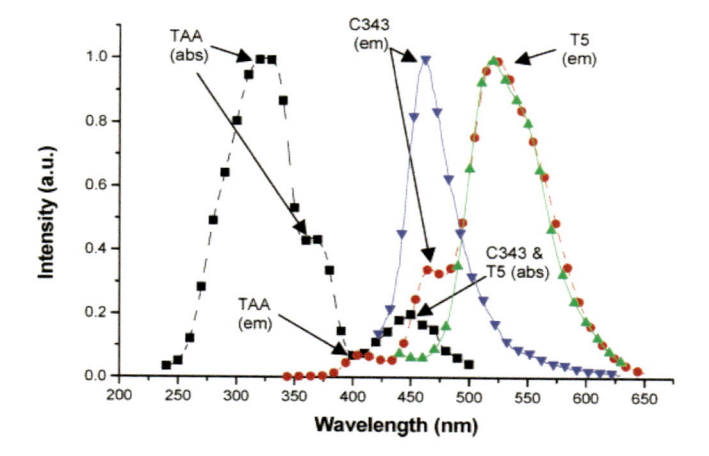

Figure 26.2 Normalized PL emission of **6** (blue downward-pointing triangle) and **9** (green upward-pointing triangle). Dendrimer mixture: PL absorbance (black square) and emission (red circle) spectra of a single layer mixture of **6** and **9** in a 1:1 molar ratio of C343: T5 dye cores.

Table 26.1 Mixtures of C343 and T5 dendrimers

Mixture	Composition		C343:T5
	C343	T5	
1	**5** ([G-3])	**7** ([G-3])	1:1
2	**5**	**7**	5:1
3	**5**	**8** ([G-4])	5:1
4	**5**	**9** ([G-4])	5:1
5	**6** ([G-5])	**7**	5:1
6	**6**	**8**	5:1
7	**6**	**9**	1:1
8	**6**	**9**	3:1
9	**6**	**9**	5:1

A further set of experiments was done in order to investigate the effect of changing dendrimer size at a constant 5:1 molar ratio of C343 to T5. Figure 26.4 shows the PL spectra of the third and fourth generation C343 dendrimers (**5**, **6**) with third, fourth, and fifth generation T5 dendrimers (**7**, **8**, and **9**). These results clearly demonstrate the effect of increasing isolation via an increase in

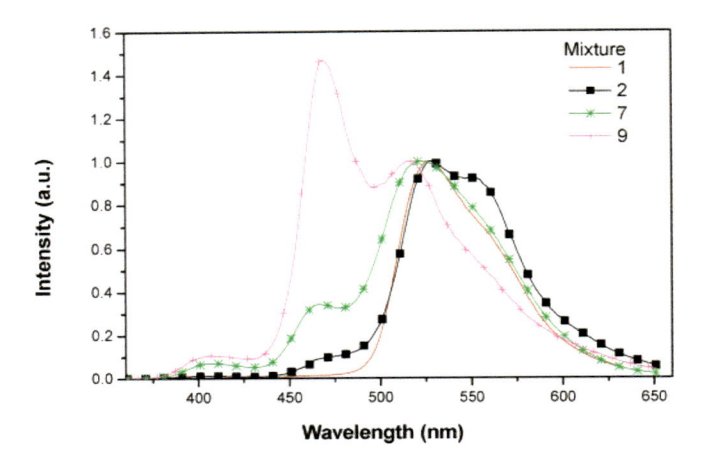

Figure 26.3 PL emission of mixtures of "small" dendrimers **5** and **7** (mixtures 1 and 2, Table 26.1), and "large" dendrimers **6** and **9** (mixtures 7 and 9, Table 26.1), in both 1:1 and 5:1 molar ratios of C343: T5 dye cores.

dendrimer size. In the cases where the C343 is changed from third to fourth generation with each size thiophene (mixtures 2 and 5, 3 and 6, 4 and 9), a large increase in the C343 emission is observed, an indication of increased site-isolation for the fourth generation C343 (resulting in a decrease in the C343 to T5 energy transfer). A similar trend is also seen for increased generations of thiophene dendrimers, but in this case only a very small increase in C343 emission is seen between the third and fourth generation T5's (mixtures 2 and 3, 5 and 6), whereas a much larger increase in C343 emission is measured between the fourth and fifth generations (mixtures 3 and 4, 6 and 9).

It follows that T5 dendrimers only begin to have adequate site-isolation after the fifth generation while an appreciable amount of isolation is already observed for fourth generation C343. This finding is not unexpected as the elongated shape of the T5 moiety makes encapsulation difficult, even when both ends of the T5 are capped in a "dumbbell" arrangement. Therefore, with G-3 or G-4 dendrons the T5 core remains exposed, whereas the more compact C343 core is relatively well shielded by three G-4 dendrons.

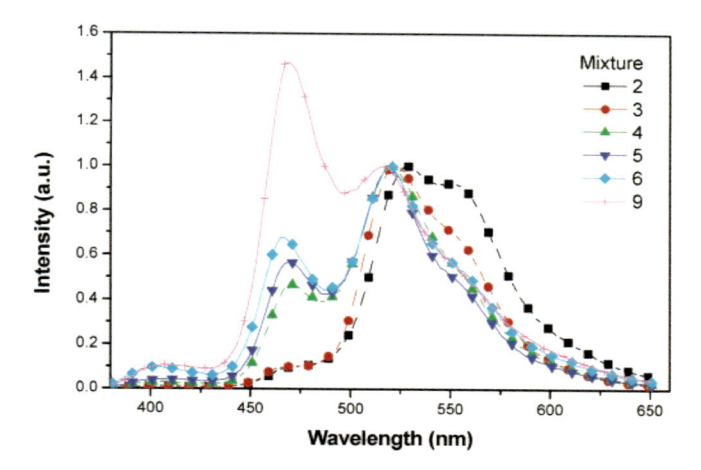

Figure 26.4 PL emission of different size mixtures of C343 and T5 dendrimers at 5:1 molar ratios of C343: T5 dye cores, normalized to T5 emission. (See Table 26.1 for the compositions of the various dendrimer mixtures).

To measure the spectrum of a dendrimer mixture in the absence of interdendrimer energy transfer, we have also examined the PL spectra of dendrimer mixtures diluted into a glassy polystyrene matrix. All of the polystyrene samples studied contained a 1:1 molar mixture of **6** and **9**. Figure 26.5 shows the PL spectra of samples ranging in concentration from 0.35 to 4.2%, as well as a neat film of a mixture of dendrimers **6** and **9**. As the polystyrene solution is made more dilute, thereby reducing interdendrimer interactions, the T5 contribution to the spectrum decreases. At the highest dilution (0.36% dendrimer by weight in PS), the intensity of the C343 emission is about twice that of the T5 emission, whereas it is roughly a third of the T5 emission in the neat dendrimer film. The ratio of **6** to **9** emission is not thought to involve different absorption characteristics for the two, since the principal absorber for the dendrimers are the TAA groups, which are present in relatively similar numbers in the two dendrimers (24 TAA groups for **6** compared to 32 for **9**). The imbalance in C343 and T5 in the most dilute sample can be attributed to differences in the quantum yields of emission for the coumarin and thiophene. The ϕ_{em} of C343 is 0.63

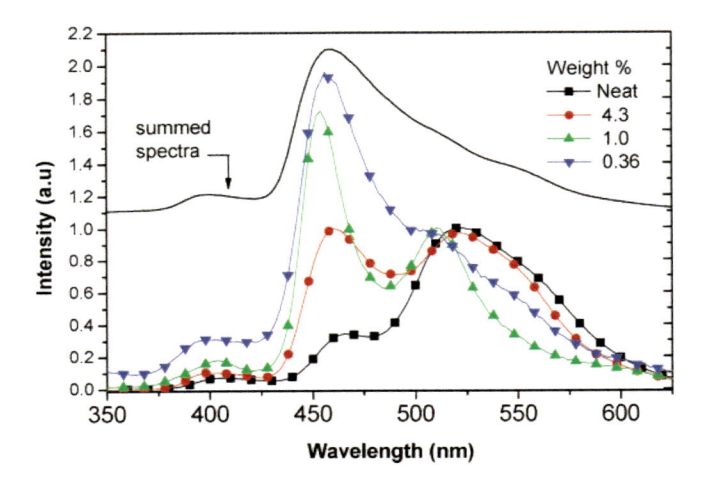

Figure 26.5 PL emission of **6** and **9** at 1:1 molar ratios neat and diluted with polystyrene (showing the % dendrimer by weight), normalized to T5 emission. The offset (upper) spectrum was generated by summing the spectra of pure samples of **6** and **9**, weighted by their PL quantum efficiencies (i.e., 0.63 × **6** + 0.15 × **9**).

[14], whereas that for T5, estimated from related pentathiophenes, is approximately $\phi_{em} = 0.15$ [15]. To provide a perspective for the PL measurements made with the 0.36% mixed dendrimer film, a simulated PL spectrum was created by summing the emission spectra of pure **6** and **9**, weighted by their relative quantum yields. Although photoinduced electron transfer between TAAs and the C343 [16] could be a possible reason for the reduced C343 emission in the mixtures, the absence of noticeable quenching [16a] of the C343 in the most dilute samples compared to the weighted relative quantum yields suggests that electron transfer is not a significant factor for the dendrimers of this study. As shown in Fig. 26.5, the weighted simulated spectra very closely match that of the highly diluted PS film, showing that a spectrum rich in C343 emission is expected for a sample with complete site isolation. Hence, alhtough not completely isolating the dyes from energy transfer in neat films, these higher generation dendrimers do provide a considerable degree of site-isolation when compared to the free dye [11] or to the

lower generation dendrimers (mixtures of dendrimers **5** and **7** give mostly T5 emission).

OLED devices were fabricated using the same molar ratios and spin casting parameters used for the previously described thin film PL spectra. ITO substrates were pretreated with a layer of poly(3,4-ethylenedioxythiophene):polystyrene sulfonic acid (PEDOT:PSS) which is known to facilitate hole injection and provides for planarization of the substrate [17]. Additionally, an electron transporter, 2-(4-biphenylyl)-5-(4-*tert*-butylphenyl)-1,3,4-oxadiazole (PBD), was incorporated in the dendrimer solutions using 1:1 molar ratio of PBD to peripheral TAA moiety. The purpose of the PBD is to transport electrons, which should help to balance electron–hole transport within the device [11, 18]. The holes are expected to be transported by the easily oxidized TAA groups, and exciton recombination is expected to occur on the dye cores. Furthermore, a thin 100 Å thick film of 2,9-dimethyl-4,7-diphenyl-1,10-phenanthroline (BCP) was subsequently vapor deposited as a hole blocker to confine exciton recombination, and prevent the loss of the faster moving holes to the cathode [6, 19]. This was followed by a 150 Å electron injecting layer of aluminum-tris(8-hydroxyquinolate) (AlQ_3) and a LiF/Al cathode. It was found that devices with BCP/AlQ_3 layers were an order of magnitude more efficient than those without these blocking and injecting layers, demonstrating their importance in the device.

Figure 26.6 shows the electroluminescent (EL) emission spectra of devices with only dendrimers **6** and **9** (as well as varying mixtures of the two), compared to a 5:1 mixture of the smaller sized dendrimers **5** and **7**. These measurements are in qualitative agreement with those for PL spectra as site isolation is also significantly enhanced in the EL spectra for dendrimers in which the dyes are shielded by the larger dendrons. Interestingly, a greater amount of T5 emission relative to C343 emission is observed in the EL compared to the PL. Because the films used for PL and EL were spin cast from the same stock solutions, the observed differences in intensities are unlikely to be related to differences in composition. It is believed that the difference between EL and PL characteristics may be due to direct hole or electron (carrier) trapping on the T5 core. Hence, if carrier recombination were to

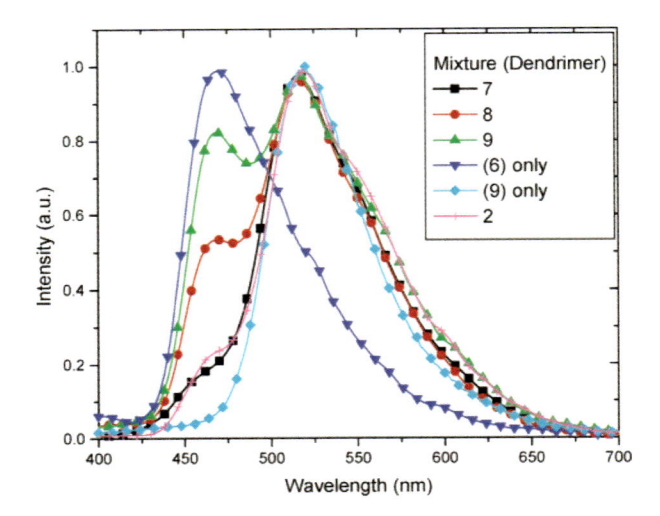

Figure 26.6 Device EL emission of **6** and **9** alone and in 1:1, 3:1, and 5:1 molar ratio mixtures. EL emission of 5:1 molar ratio mixture of **5** and **7** shown for comparison. (See Table 26.1 for the compositions of the various dendrimer mixtures).

occur more efficiently at the T5 than at C343, it could result in preferential exciton formation at T5. This carrier trapping process gives T5 an added pathway for exciton formation that may not be present for C343, thereby shifting the spectra further toward the lower energy T5 emitter. This effect results in nearly equal emission intensities of both T5 and C343 at a 5:1 molar ratio (mixture 9) whereas in the PL, the same mixture gives a much higher relative intensity for the C343.

The device performance data for a film obtained from mixture 9 with its 5:1 ratio of C343 to T5, which affords balanced EL intensities for the two chromophores, are shown in Fig. 26.7. The maximum external quantum efficiency of the mixed [G4]$_3$C-343: [G-5]$_2$T5 device was 0.2% with a power efficiency of 0.1 lm/W. The device had a turn-on voltage of approximately 12 V and a maximum luminescence of only 22 cd/m^2 at 20 V. Device performance data for an OLED prepared from the [G-5]$_2$T5 (**9**) alone is shown in Fig. 26.8. Interestingly, while the turn-on voltages are comparable, the T5-only device had much better performance characteristics. A maximum

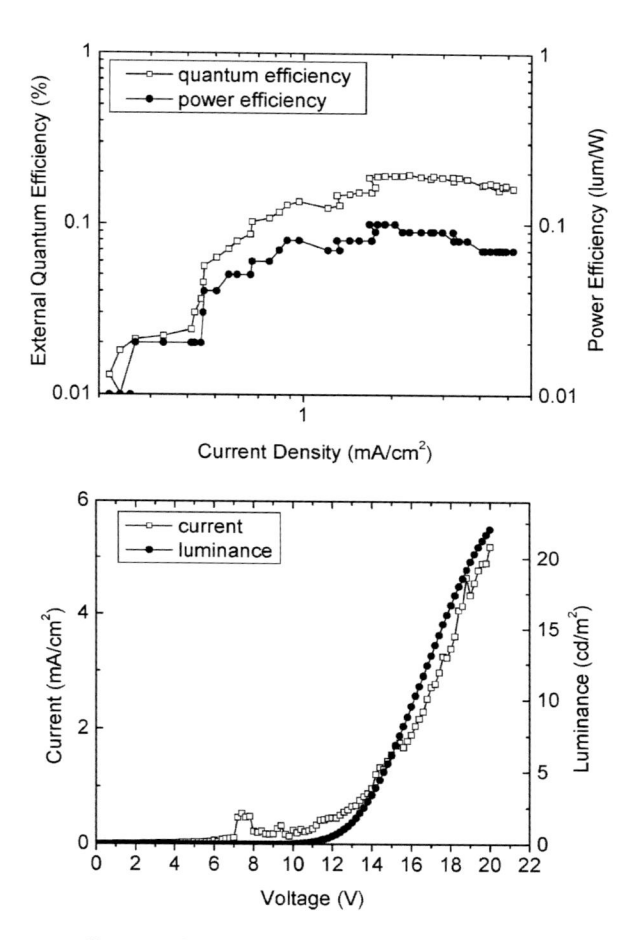

Figure 26.7 Efficiency (top) and IV/luminescence characteristics (bottom) for ITO/PEDOT/G4C:G5T(5:1)-PBD/BCP/AlQ$_3$/LiF/Al device.

luminance of 420 cd/m^2 was recorded at 20 V. The device had a maximum efficiency of 0.59 lm/W (0.76% photon/electron) at a brightness of 100 cd/m^2.

When the PL and EL spectra of a material are significantly different, as is seen for the 5:1 mixture of **6:9**, the cause can generally be attributed to carrier trapping affects. If one or both of the carriers were preferentially trapped at the T5 core of **9**, then a higher than statistical amount of hole-electron

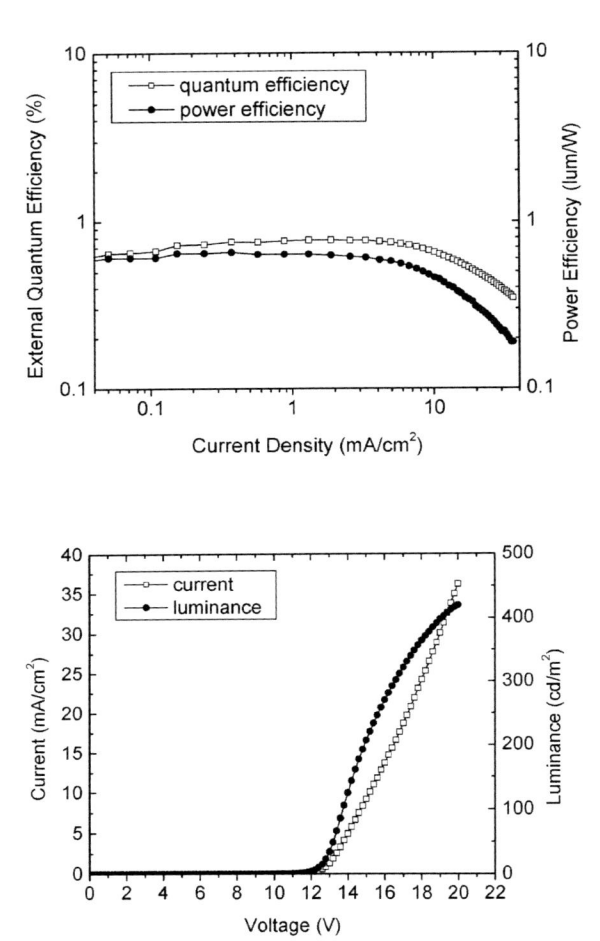

Figure 26.8 Caption: Efficiency (top) and IV/luminescence characteristics (bottom) for ITO/PEDOT/G5T-PBD/BCP/AlQ$_3$/LiF/Al device.

recombination will occur at **9**, leading to an EL spectrum which is richer in T5 emission than expected, as is in fact observed. To determine if a trapping assignment makes sense, a series of solution electrochemical measurements were made to estimate the hole and electron levels of the individual components. For this study, model T5 **1** and C343 **2** (Fig. 26.1) compounds were capped with benzyl esters to simulate the cores, and a G3 TAA methyl ester dendron **3** was used to simulate the characteristics of the hole transporting periphery. Electrochemical measurements were recorded in dimethylformamide (DMF) solution and potentials are given relative to an internal ferrocene reference (Fc/Fc$^+$). In addition, all literature values were adjusted to Fc/Fc$^+$ for clarity. The T5 core had two reversible oxidations at 0.62 V and 0.76 V and a reversible reduction at -1.79 V. These oxidation values are consistent with two similar T5 containing polymers that were previously measured to have single reversible oxidations in the solid state at 0.61 and 0.51 V [15]. The reduction potential of the PBD electron transport material is reported to be -2.34 [20]. Thus, it is expected that the T5 core will trap electrons, from the electron transporting PBD molecules. Unfortunately, the oxidation and reduction potentials of both the C343 core and TAA dendron were irreversible, so we cannot estimate their HOMO and LUMO energies reliably. The irreversible oxidation of the TAA dendron was observed at 0.49 V, which is greater than the 0.3 V oxidation potential reported for the corresponding benzidine derivative (*N,N'*-diphenyl-*N,N'*-bis-1-naphthyl-4,4'-biphenyl, NPD), a commonly used hole transport material [21]. This sort of difference is expected when comparing a monoamine to a conjugated diamine. The irreversible oxidation for C343 is between 0.49 and 0.59 V, thus we do not expect either C343 or T5 to effectively trap holes from the TAA groups. Hence, the likely explanation for the enhanced performance of T5 involves the trapping of electrons at the T5 core, followed by hole-electron recombination at that emitter. This gives an additional pathway for exciton localization at T5, other than energy transfer from C343 or TAA, leading to enhanced T5 emission in EL, relative to the PL measurements.

26.3 Conclusion

The use of dendrons to site isolate chromophores capable of electronic "communication" has been explored through their use in organic light emitting diodes. The novel design of the peripheral moieties of the dendrons allowed tuning of properties for the casting of films that did not show a tendency to crystallize thereby enabling both photoluminescence and electroluminescence studies to be carried out. These studies confirm that the size of the dendritic shell is critical to site isolation. Although energy transfer between dendrimers still occurs to some extent, even with the largest dendrimers we have prepared, several important findings have emerged. First, it is clear that the film forming properties of the dendrimers are critical and the addition of peripheral alkyl groups to the electroactive but low solubility TAA moieties helps with the formation of clear films and solves the issues associated with the inherent crystallinity of the TAA dendrons at higher generations. Furthermore, moderately efficient devices can be fabricated using these dendrimers along with BCP/AlQ$_3$ hole blocking/electron injecting films. Best results based on these unoptimized dendrimer systems were an external quantum efficiency of 0.2% for mixed films containing two chromophores in the same layer, and 0.76% for a fifth generation thiophene-only device.

These results also demonstrate a potential problem that may occur in many dendrimer based OLEDs. Carrier trapping at the dendrimer core can lead to significant changes in the properties of the devices. This is evidenced by the different device performances observed for the mixed dendrimer devices and the T5-only device, as well as differences in PL and EL emission intensity for mixed films. Therefore, care must be taken when choosing dye cores for such dendrimer-based devices. Selective carrier trapping by one of the core components can dramatically influence the effectiveness of other components in the device. We believe that the high voltages required to run these dendrimer based OLEDs was due, at least in part, to carrier trapping at the T5 dye cores. When choosing dye cores for dendrimers of this type it is important to select dyes whose HOMO levels are below the hole transporting material, and whose LUMO levels are above the electron transporting materials, to prevent charge trapping within the dendrimer. In addition, even at large generations, complete site-isolation remains a difficult problem for chromophores with large Förster radii. It is believed that once these issues are understood, dendritic systems such as these could provide access to color tunable solution processed large area displays.

Acknowledgment

Financial support of this research by the Air Force Office of Scientific Research and by the U.S. Department of Energy under Contract No. DE-AC03-76SF00098 is acknowledged with thanks.

Supporting Information Available

OLED fabrication procedures and equipment: a full Experimental Section (PDF) is provided. This material is available free of charge via the Internet at http://pubs.acs.org.

References

1. Glazer, A. N. *Annu. Rev. Biophys. Biophys. Chem.* 1995, **14**, 47.
2. (a) Stewart, G. M., Fox, M. A. *J. Am. Chem. Soc.* 1996, **118**, 4354. (b) Devadoss, C., Bharati, P., Moore, J. S. *J. Am. Chem. Soc.* 1996, **118**, 9635. (c) Wang, P.-W., Liu, Y.-J., Devadoss, C., Bharathi, P., Moore, J. S. *Adv. Mater.* 1996, **3**, 237. (d) Shortreed, M. R., Swallen, S. F., Shi, Z.-Y., Tan, W., Xu, Z., Devadoss, C., Moore, J. S., Kopelman, R. *J. Phys. Chem. B.* 1997, **101**, 6318. (e) Jiang, D.-L., Aida, T. *Nature* 1997, **388**, 454. (f) Aida, T., Jiang, D.-L. *J. Am. Chem. Soc.* 1998, **120**, 10895. (g) Sato, T., Jiang, D.-L., Aida, T. *J. Am. Chem. Soc.* 1999, **121**, 10658. (h) Balzani, V., Campagna, S., Denti, G., Juris, A., Serroni, S., Venturi, M. *Acc. Chem. Res.* 1998, **31**, 26. (i) Maruo, N., Uchiyama, M., Kato, T., Arai, T., Akisada, H., Nishino, N. *Chem. Commun.* 1999, 2057. (j) Plevoets, M., Vögtle, F., Cola, L. D., Balzani, V. *New J. Chem.* 1999, **23**, 63. (k) Adronov, A., Fréchet, J. M. J. *Chem. Commun.* 2000, **18**, 1701. (l) Peng, Z., Pan, Y., Xu, B., Zhang, J. *J. Am. Chem. Soc.* 2000, **122**, 6619. (m) Weil, T., Wiesler, U. M., Herrmann, A., Bauer, R., Hofkens, J., De Schryver, F. C., Müllen, K. *J. Am. Chem. Soc.* 2001, **123**, 8101. (n) Balzani, V., Ceroni, P., Juris, A., Venturi, M., Campagna, S., Puntoriero, F., Serroni, S. *Coord. Chem. Rev.* 2001, **219**, 545. (o) Herrmann, A., Weil, T., Sinigersky, V., Wiesler, U.-M., Vosch, T., Hofkens, J., De Schryver, F. C., Müllen, K. *Chem. Eur. J.* 2001, **7**, 4844–4853. (p) Campagna, S., Di Pietro, C., Loiseau, F., Maubert, B., McClenaghan, N., Passalacqua, R., Puntoriero, F., Ricevuto, V., Serroni, S. *Coord. Chem. Rev.* 2002, **229**, 67. (q) Hahn, U., Gorka, M., Vögtle, F., Vicinelli, V., Ceroni, P., Maestri, M., Balzani, V. *Angew. Chem.*,

Int. Ed. 2002, **41**, 3595. (r) Weil, T., Reuther, E., Müllen, K. *Angew. Chem., Int. Ed.* 2002, **41**, 1900–1904.

3. Kawa, M., Fréchet, J. M. J. *Chem. Mater.* 1998, **10**, 286. Zhu, L., Tong, X., Li, M., Wang, E. *J. Phys. Chem. B.* 2001, **105**, 2461. Chou, C.-H., Shu, C.-F. *Macromolecules* 2002, **35**, 9673.
4. Tang, C. W., VanSlyke, S. A. *Appl. Phys. Lett.* 1987, **51**, 913.
5. Shoustikov, A., You, Y., Thompson, M. E. *IEEE J. Sel. Top. Quant.* 1998, **4**, 3.
6. Halim, M., Samuel, I. D. W., Pillow, J. N. G., Monkam, A. P., Burn, P. L. *Synth. Met.* 1999, **102**, 1571–1574.
7. (a) Halim, M., Pillow, J. N. G., Samuel, I. D. W., Burn, P. L. *Synth. Met.* 1999, **102**, 922–923. (b) Halim, M., Samuel, I. D. W., Pillow, Burn, P. L. *Synth. Met.* 1999, **102**, 1113–1114. (c) Lo, S.-C. L., Male, N. A. H., Markham, J. P. J., Magennis, S. W., Burn, P. L., Salata, O. V., Samuel, D. W. *Adv. Mater.* 2002, **14**, 975–979.
8. Kuwabara, Y., Ogawa, H., Inada, H., Noma, N., Shirota, Y. *Adv. Mater.* 1994, **6**, 677. (b) Satoh, N., Cho, J., Higuchi, M., Yamamoto, K. *J. Am. Chem. Soc.* 2003, **125**, 8104.
9. Bettenhausen, J., Greczmiel, M., Jandke, M., Strohiegl, P. *Synth. Met.* 1997, **91**, 223.
10. (a) Shirshendu, K. D., Maddux, T. M., Yu, L. *J. Am. Chem. Soc.* 1997, **119**, 9079. (b) Pogantsch, A., Wenzl, F. P., List, E. J. W., Leising, G., Grimsdale, A. C., Müllen, K. *Adv. Mater.* 2002, **14**, 1061. (c) Anthopoulos, T. D., Markham, J. P. J., Namdas, E. B., Samuel, I. D. W., Lo, S., Burn, P. L. *Appl. Phys. Lett.* 2003, **82**, 4824.
11. (a) Freeman, A. W., Fréchet, J. M. J., Koene, S. C., Thompson, M. E. *Polym. Prepr.* 1999, **40**, 1246 (b) Freeman, A. W., Koene, S. C., Malenfant, P. R. L., Thompson, M. E., Fréchet, J. M. J. *J. Am. Chem. Soc.* 2000, **122**, 12385.
12. Robinson, M. R., O'Regan, M. B., Bazan, G. C. *Chem. Com.* 2000, 1645.
13. Turro, N. J. *Modern Molecular Photochemistry*; University Science Books: New York, 1991, p 302.
14. Adronov, A., Gilat, S. L., Fréchet, J. M. J., Ohta, K., Neuwahl, F. V. R., Fleming, G. R. *J. Am. Chem. Soc.* 2000, **122**, 1175–1185.
15. Donat-Bouillud, A., Mazerolle, L., Gagnon, P., Goldenberg, L., Petty, M. C., Leclerc, M. *Chem. Mater.* 1997, **9**, 2815–2821.
16. (a) Nad, S., Pal, H. *J. Photochem. Photobiol. A: Chem.* 2000, **134**, 9–15. (b) Lor, M., Thielemans, J., Viaene, L., Cotlet, M., Hofkens, J., Weil, T., Hampel, C., Müllen, K., Verhoeven, J. W., Van der Auweraer, M., De Schryver, F. C. *J. Am. Chem. Soc.* 2002, **124**, 9918–9925.

17. Lamansky, S., Djurovich, P. I., Abdel-Razzaq, F., Garon, S., Murphy, D. L., Thompson, M. E. *J. Appl. Phys.* 2002, **92**, 1570–1575.
18. Kido, J., Shionoya, H., Nagai, K. *Appl. Phys. Lett.* 1995, **67**(16), 2281–2283.
19. (a) Baldo, M. A., Lamansky, S., Burrows, P. E., Forrest, S. R., Thompson, M. E. *Appl. Phys. Lett.* 1999, **75**, 4. (b) Tsutsui, T., Yang, M. J., Masayuki, M., Nakamura, Y., Wanabe, T., Tsuji, T., Fukuda, Y., Wakimoto, T., Miyaguchi, S. *Jpn. J. Appl. Phys.* 1999, **38**, L1502.
20. Wu, C.-C., Sturm, J. C., Register, R. A., Tian, J., Dana, E. P., Thompson, M. E. *IEEE Trans. Elec. Dev.* 1997, **44**, 1269–1281.
21. Koene, B., Loy, D., Thompson, M. E. *Chem. Mater.* 1998, **10**, 2235–2250.

Chapter 27

Ultrahigh Energy Gap Hosts in Deep Blue Organic Electrophosphorescent Devices

Xiaofan Ren,[a] Jian Li,[a] Russell J. Holmes,[b] Peter I. Djurovich,[a] Stephen R. Forrest,[b] and Mark E. Thompson[a]

[a] Department of Chemistry, University of Southern California, Los Angeles, California 90089, USA
[b] Department of Electrical Engineering, Princeton University, Princeton, New Jersey 08544, USA
met@usc.edu

Four ultrahigh energy gap organosilicon compounds [diphenyldi(o-tolyl)silane (UGH1), p-bis(triphenylsilyl)benzene (UGH2), m-bis(triphenylsilyl)benzene (UGH3), and 9,9'-spirobisilaanthracene (UGH4)] were employed as host materials in the emissive layer of electrophosphorescent organic light-emitting diodes (OLEDs). The high singlet (~4.5 eV) and triplet (~3.5 eV) energies associated with these materials effectively suppress both the electron and energy transfer quenching pathways between the emissive dopant and the host material, leading to deep blue phosphorescent devices with high (~10%) external quantum efficiencies. Furthermore, by direct charge injection from the adjacent hole and electron transport layers onto the phosphor doped into the UGH matrix, exciton formation

---

Reprinted from *Chem. Mater.*, **16**, 4743–4747, 2004.

Electrophosphorescent Materials and Devices
Edited by Mark E. Thompson
Text Copyright © 2004 American Chemical Society
Layout Copyright © 2024 Jenny Stanford Publishing Pte. Ltd.
ISBN 978-981-4877-34-3 (Hardcover), 978-1-003-08872-1 (eBook)
www.jennystanford.com

occurs directly on the dopant, thereby eliminating exchange energy losses characteristic of guest–host energy transfer. We discuss the material design, and present device data for OLEDs employing UGHs. Among the four host materials, UGH2 and UGH3 have higher quantum efficiencies than UGH1 when used in OLEDs. Rapid device degradation was observed for the UGH4-based device due to electro- and/or photooxidation of the diphenylmethane moiety in UGH4. In addition to showing that UGH materials can be used to fabricate efficient blue OLEDs, we demonstrate that very high device efficiencies can be achieved in structures where the dopant transports both charge and excitons.

27.1 Introduction

Phosphorescent electroluminescent materials and devices are a prime focus of organic light-emitting device (OLED) research due to their ability to efficiently utilize both singlet and triplet excitons [1]. The most common design for phosphorescence-based OLEDs involves a doped emissive region, where the emissive dopant is either an Ir or a Pt complex. The large spin–orbit coupling for these heavy metals leads to efficient intersystem crossing and, hence, short emissive lifetime phosphorescence. To achieve the maximum efficiency in these host–dopant OLEDs, the triplet level of the host must not quench dopant phosphorescence. Such host–guest systems have been realized in green and red organic electrophosphorescent devices, leading to internal quantum efficiencies approaching 100% [2]. Recently, the applicability of host materials with high triplet energies has been demonstrated in blue electrophosphorescent devices, which utilize emission from a bis[2-(4',6'-difluorophenyl)pyridinato-$N,C^{2'}$]iridium(III) picolinate (FIrpic) dopant [3]. Efficient blue electrophosphorescence (η_{ext} = 7.5%) was achieved when using an N,N'-dicarbazolyl-3,5-benzene (mCP) host, whose triplet level (E_T = 2.9 eV) is nearly 0.3 eV above that of FIrpic (E_T = 2.62 eV). On the other hand, when an N,N'-dicarbazolyl-4,4'-biphenyl (CBP) host with a triplet energy of 2.56 eV, ca. 0.06 eV below that of the FIrpic dopant, is used, an efficiency of only 5% is observed. Clearly, the higher triplet energy of mCP is beneficial in achieving the higher efficiency.

To achieve a close match to the National Television Standards Committee recommended blue for video displays, the phosphors used in OLEDs generally need to have triplet energies near 2.8–2.9 eV (emitting at $\lambda \approx 440$ nm). However, as the triplet energy of the phosphor is shifted into the deep blue, it becomes increasingly difficult to find a host with a suitably high energy triplet state. The most common hosts used in phosphorescent OLEDs, carbazole-based materials [4], have triplet energies of 2.9 eV or less [5], putting them close to the desired dopant energies. An added complication of deeper blue phosphors is their deep highest occupied molecular orbital (HOMO) levels that typically fall near or below the HOMO levels of the carbazole-based host materials; e.g., mCP and FIr6 (vide infra) have HOMO energies of 5.9 and 6.1 eV, respectively. Having host triplet energies that are close to the phosphor energy and similar host–dopant HOMO levels leads to both energy and electron transfer quenching pathways, which ultimately limit device efficiency [6].

To circumvent the limitations of carbazole-based hosts, we have studied a series of ultrawide energy gap hosts (UGHs).The materials have large HOMO to lowest unoccupied molecular orbital (LUMO) energy gaps (in the range of 4.5–5.0 eV), and triplet energies greater than that of carbazole (i.e., 3.0 eV). These UGHs are compatible with the organometallic phosphors used in efficient electrophosphorescent devices. It is desirable for the UGH HOMO and LUMO energies to be large enough to lie below and above of the same levels in the dopant, respectively. The guest energy levels are then nested between the HOMO–LUMO gap of the host, thereby eliminating the potential for exciplex formation between the dopant and host. Hence, the guest acts as the primary site for electron and hole conduction within the emissive layer (EML), as well as the trap site for excitons. One example of such a UGH material is polystyrene (PS), which has been used as a host for small-molecule chromophores dispersed by solution processing [7]. However, these PS devices had a single-layer structure, leading to poor charge carrier balance and confinement in the EML, and thus poor efficiency.

In this work, we present a class of small-molecule-based UGH materials applicable to vacuum-deposited OLEDs. We have previously shown that by using ultrahigh energy gap materials it is possible to make highly efficient deep blue phosphorescence devices. By this method, a maximum external quantum efficiency of $\eta_{ext} = 11.2\%$ [8] was achieved using the deep blue phosphorescent iridium complex bis[2-(4′,6′-difluorophenyl)pyridinato-$N,C^{2'}$]tetrakis(1-pyrazolyl)borate (FIr6) as the dopant [9]. Here we discuss the material design, and present device data for OLEDs employing four different UGHs. In addition, we show that high device efficiencies can be achieved in structures where the dopant conducts both holes and electrons in the EML.

27.2 Results and Discussion

27.2.1 UGH Design and Characterization

Figure 27.1 lists the chemical structures of diphenyldi(o-tolyl)silane (UGH1), p-bis(triphenylsilyl)benzene (UGH2), m-bis(triphenylsilyl)benzene (UGH3), and 9,9′-spirobisilaanthracene (UGH4). The structures of 5,5′-spirobi(dibenzosilole) (Si(bph)$_2$), mCP, and FIr6 are also shown. Tetraarylsilane compounds have been chosen for two reasons: (i) tetraarylsilicon compounds can be synthesized and modified readily under mild conditions compatible with the demands of organic electronic devices and (ii) tetraarylsilicon compounds can be sublimed without decomposition [10]. The approach taken to maximize the HOMO–LUMO gaps in these materials is to electronically isolate each phenyl ring in the structure, avoiding any conjugating substituents. Direct phenyl–phenyl linkages must be avoided since biphenyl has a triplet energy of 2.8 eV. For example, we have also prepared Si(bph)$_2$, which, due to the biphenyl moieties in the structure, has a triplet energy of only 2.72 eV (Fig. 27.2), making it an unsuitable host for blue electrophosphorescent OLEDs. That is, a blue OLED, having the same architecture as the high-efficiency UGH device [8], using FIr6 ($E_T = 2.72$ eV, 457 nm) as the dopant and (Si(bph)$_2$) as a host had a maximum external quantum efficiency of only $\eta_{ext} = 1\%$. Nevertheless, Si(bph)$_2$ is an

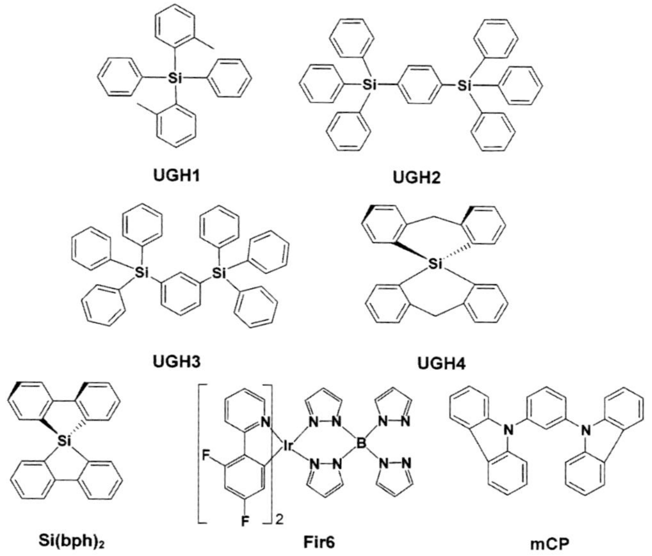

Figure 27.1 Chemical structures of UGH1, UGH2, UGH3, mCP, and FIr6.

efficient host for green phosphorescent OLEDs [11]. Two OLEDs prepared using tris[2-(phenyl)pyridinato-$N,C^{2'}$]iridium(III) (Irppy; $E_T = 2.46$ eV, 505 nm) as the dopant, one with a doped Si(bph)$_2$ host layer and the other with a doped CBP host layer [12], gave similar current and luminescence vs voltage characteristics (see the Supporting Information). For example, the Irppy doped, Si(bph)$_2$ host OLED has $\eta_{ext} = 4.5\%$ at a luminance of 10,000 cd/m^2.

The singlet and triplet energies along with the absolute HOMO–LUMO energies of the UGH materials were determined through a combination of optical and ultraviolet photoemission (UPS) spectroscopies, as described in Section 27.4. UGH1, UGH2, UGH3, and freshly prepared UGH4 all have nearly identical absorption, room temperature emission, and 77 K photoluminescence (PL) spectra (see Fig. 27.2, showing only the UGH1 spectra as an example). This similarity indicates that little or no conjugation exists between the triarylsilicon groups in UGH2 and UGH3, and between the methylene-linked phenyl rings in UGH4. The singlet energy gaps are estimated to be 4.4 ± 0.1 eV (corresponding to a

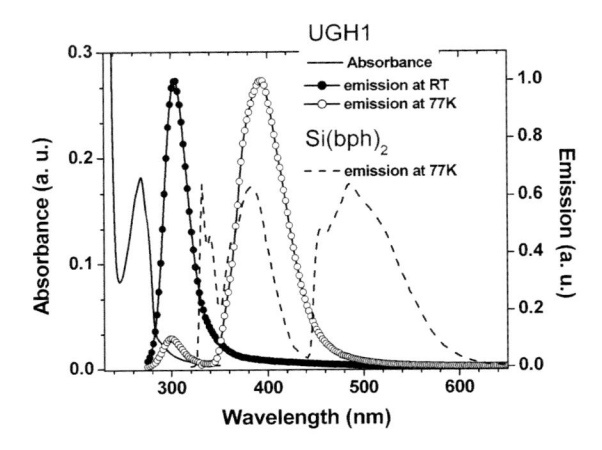

Figure 27.2 Absorption and emission spectra of UGH1 at room temperature and 77 K, and emission spectrum of Si(bph)$_2$ at 77 K.

long-wavelength absorption cutoff of $\lambda = 280$ nm) for each compound, while the triplet energies of the four UGHs are estimated to be approximately 3.5 eV ($\lambda = 360$ nm). Since the redox potentials of the UGH compounds fall outside the range for convenient electrochemical analysis, UPS was used to measure their HOMO energies relative to those in a vacuum. These measurements indicate the HOMO levels of the four UGH materials are at 7.2 ± 0.1 eV [8]. By subtracting the optical energy gap from the HOMO energy, the LUMO levels are estimated at 2.8 eV. The HOMO energies are well below those of other, more conventional materials used in phosphorescent OLEDs, suggesting that UGHs may also block hole injection from the adjacent layer. Furthermore, the LUMO energies of the UGHs fall very near those of the common electron-injecting materials, such as 2,9-dimethyl-4,7-diphenyl-1,10-phenanthroline (BCP) and aluminum 8-hydroxyquinoline (Alq$_3$), allowing for efficient electron injection into the EML (Fig. 27.4).

Two approaches have been adopted to improve the film morphological stability: (i) using Si compounds with multiple arylsilane groups, as in UGH2 and UGH3, or (ii) using nonconjugated linking groups, to prevent ring rotation, as in UGH4. Both approaches lead to an increase in the glass transition temperatures (T_g), thus enhancing the likelihood that the materials will form morphologically stable amorphous films [13]. Glass transition temperatures for UGH3 and UGH4 are 46 and 53°C, respectively, compared to 26°C for UGH1, whereas UGH2 shows no detectable T_g, with a melting point of 345°C.

The quality of neat thin films of the UGH1 prepared by thermal deposition was investigated by optical microscopy. When UGH1 is deposited directly onto a silicon wafer surface or onto a substrate precoated with an organic film (e.g., N,N'-diphenyl-N,N'-bis(1-naphthyl)benzidine, NPD), the surface texture appears rough, suggesting the UGH layer has crystallized, as expected for a material with a T_g near room temperature. Codeposition of 92% UGH1 with 8% (by weight) FIr6 inhibits crystallization, such that freshly prepared films are uniform, as expected for an amorphous thin film, although significant surface roughness develops after the film stands for 4–5 h. The surface roughness may be due to the formation of crystalline domains within the doped thin film, which is expected to lead to poor OLED operational stability [14]. Optical microscopy showed that UGH2, UGH3, and UGH4 all give smooth, pinhole-free films, whether they are deposited alone or with a phosphorescent dopant.

27.2.2 Electrophosphorescent OLEDs Employing UGH Materials

Blue emitting OLEDs employing the phosphor FIr6 in the EML were fabricated with the structure indium tin oxide (ITO)/NPD (400 Å)/mCP (100 Å)/10% FIr6:UGH (250 Å)/BCP (150 Å)/Alq$_3$ (250 Å)/LiF/Al. Due to the deep HOMO levels of the three UGH materials, it is expected that holes are injected directly onto and carried by the dopant. Consistent with the proposal that the dopant carries charge is the observation that low dopant concentrations in UGH host materials lead to high OLED drive voltages (due to carrier trapping at the dopant), while increasing the doping level gives a marked decrease in drive voltage [8]. Similar dopant-based carrier trapping/transport has been proposed for OLEDs with related wide gap host materials [15]. To increase OLED efficiency, a thin layer of mCP was inserted between the NPD hole transport layer and the EML to reduce the 0.7 eV energy barrier between the HOMO levels

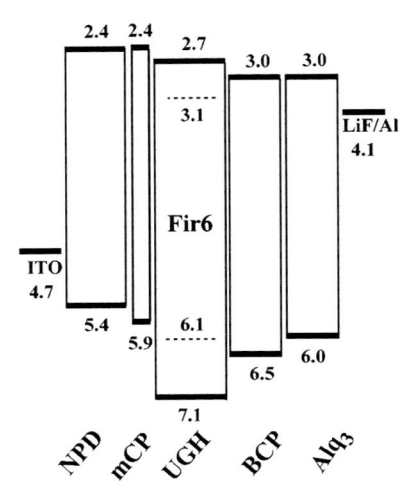

Figure 27.3 Proposed energy level diagram for the following device: ITO/NPD (400 Å)/mCP (100 Å)/FIr6:UGH (10%, 250 Å)/BCP (150 Å)/Alq$_3$ (250 Å)/LiF/Al.

of NPD and FIr6, thereby facilitating resonant hole injection [8] into the dopant (see the proposed OLED energy level scheme, Fig. 27.3) [16]. In addition, the higher triplet energy of mCP compared to that of NPD (E_T = 2.3 eV [17]) confines excitons within the emissive layer. The external quantum efficiency and current–voltage characteristics of the three UGH devices are shown in Fig. 27.4. All devices have electroluminescence (EL) spectra that match the dopant photoluminescence (PL) spectrum, and at voltages of ~5 V achieve a luminance of 1 cd/m^2. The maximum quantum efficiencies of the three devices are 7.0% for UGH1, 9.1% for UGH2, and 8.8% for UGH3. One reason for the relatively low η_{ext} of the UGH1 device may be that UGH2 and UGH3 are better solvents for the FIr6 dopant, leading to reduced dopant aggregation, and a concomitantly lower exciton quenching in the EML.

When UGH4 was used as the host, device degradation occurred rapidly during operation and, unlike the behavior of the other UGH devices, an additional broad emission band appeared at λ = 545 nm at voltages higher than 8 V (see Fig. 27.4b), changing the emission

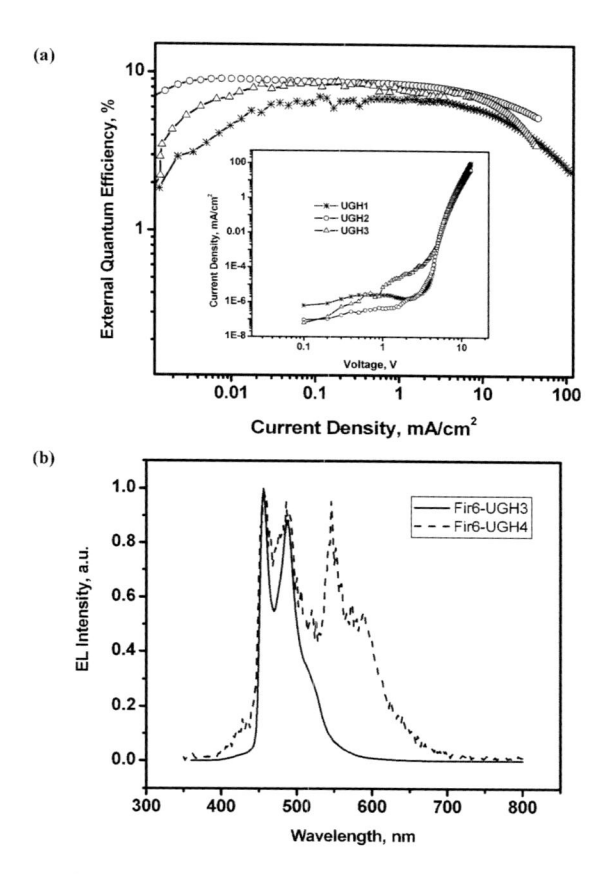

Figure 27.4 (a) External quantum efficiency vs current density for the device NPD (400 Å)/mCP (100 Å)/FIr6:UGH (10%, 250 Å)/BCP (150 Å)/Alq$_3$ (250 Å)/LiF/Al. Inset: current density vs voltage for the same device. (b) EL spectra of FIr6–UGH3- and FIr6–UGH4-based OLEDs, after several minutes of operation. Devices prepared with UGH1 and UGH3 give spectra identical to the one shown for the UGH3-based device. The UGH1, UGH2, and UGH3 devices give the same EL spectrum for extended periods, due to only FIr6 emission. The UGH4-based device initially gives the spectrum shown for the UGH3-based device, but changes to the spectrum shown here occur within a few minutes.

color from blue to yellow-white. The position and shape of the additional peak, together with the fact that device operation was carried out in air, suggest that the rapid spectral changes of the UGH4 device are related to electro- and/or photooxidation of the fluorene moiety, resulting in ketonic defects (i.e., formation of fluorenone groups) [18]. This is consistent with the observation that solution PL of UGH4 films exposed to air for a few days also displays a broad peak in the green ($\lambda_{max} = 515$ nm). We anticipate that the chemical stability of the compound can be significantly improved by replacing the hydrogens in the 9-position of UGH4 with alkyl or aryl groups.

27.3 Conclusions

In summary, we have demonstrated that the use of UGHs can lead to high-efficiency deep blue phosphorescent OLEDs. The ultrahigh energy gaps inherent in these hosts effectively reduce dopant quenching by the host matrix. Furthermore, we demonstrate that ultrahigh energy gap hosts can be "inert", thereby relying on the phosphor dopant to transport both charge and excitons across the EML, resulting in improved blue electrophosphorescent OLED efficiency compared with that of conventional systems relying on host-to-guest energy transfer.

27.4 Experimental Section

UGH1 [19], UGH2 [20], UGH3 [21], UGH4 [22], FIr6 [9], NPD [23], mCP [5], and Alq$_3$ [24] were prepared according to literature procedures. BCP was purchased from the Aldrich Chemical Co. All materials were purified by temperature gradient vacuum sublimation. Absorption spectra were recorded on an AVIV model 14DS-UV-Vis-IR spectrophotometer and corrected for background due to solvent absorption. Emission spectra (photoluminescence and electroluminescence) were recorded on a PTI QuantaMaster model C-60SE spectrofluorometer, equipped with a 928 PMT detector and corrected for detector response. Solution photoluminescence was taken in 2-methyltetrahydrofuran at both room temperature and 77 K. The HOMO–LUMO gap energy was defined as the point at which the absorption and fluorescence spectra intersect, while the triplet-state energy was determined from the highest energy feature of the low-energy emission band in the 77 K photoluminescence spectra. The HOMO energies for UGH1, UGH2, UGH3, Alq3, mCP, and FIr6 were determined by photoelectron spectroscopy with a He I UV source that has a photon energy of 21.2 eV (Thermo VG Scientific.). The HOMO energies for BCP and NPD were taken from the literature [25].

Prior to device fabrication, ITO on glass was patterned as 2 mm wide stripes. The substrates were cleaned by sonication in soap solution, followed by boiling for 5 min in trichloroethylene, acetone, and ethanol in succession for each solvent. Then the substrates were treated with UV ozone for 10 min. Organic layers were deposited sequentially by thermal evaporation from resistively heated tantalum boats at a rate of 2–2.5 Å s^{-1}. After organic film deposition, the chamber was vented and a shadow mask with a 2 mm wide stripe was put on the substrate perpendicular to the ITO stripes. Cathodes consisted of a 10 Å thick layer of LiF followed by a 1000 Å thick layer of Al.

The devices were tested in air within 2 h of fabrication. Device current–voltage and light-intensity characteristics were measured using the LabVIEW program by National Instruments with a Keithley 2400 SourceMeter/2000 multimeter coupled to a Newport 1835-C optical meter, equipped with a UV-818 Si photocathode. Only light emitting from the front face of the devices was collected and used in subsequent calculations.

Acknowledgment

We thank Dr. David Knowles and Dr. Jie-Yi Tsai for helpful discussions. Financial support was provided by the Universal Display Corp., DARPA, and NSF.

Supporting Information Available

Current vs voltage and brightness vs voltage plots for Irppy-doped Si(bph)$_2$ and FIr6-doped mCP-based OLEDs (PDF). This material is available free of charge via the Internet at http://pubs.acs.org.

References

1. (a) Baldo, M. A., O'Brien, D. F., You, Y., Shoustikov, A., Sibley, S., Thompson, M. E., Forrest, S. R. *Nature (London)* 1998, **395**, 151. (b) Cleave, V., Yahioglu, G., Le Barny, P., Friend, R., Tessler, N. *Adv. Mater.* 1999, **11**, 285.
2. (a) Adachi, C., Baldo, M. A., Forrest, S. R., Thompson, M. E. *J. Appl. Phys.* 2001, **90**, 4058. (b) Ikai, M., Tokito, S., Sakamoto, Y., Suzuki, T., Taga, Y. *Appl. Phys. Lett.* 2001, **79**, 156. (c) Adachi, C., Lamansky, S., Baldo, M. A., Kwong, R. C., Thompson, M. E., Forrest, S. R. *Appl. Phys. Lett.* 2001, **78**, 1622.
3. Holmes, R. J., Forrest, S. R., Tung, Y.-J., Kwong, R. C., Brown, J. J., Garon, S., Thompson, M. E. *Appl. Phys. Lett.* 2003, **82**, 2422.
4. (a) O'Brien, D. F., Baldo, M. A., Thompson, M. E., Forrest, S. R. *Appl. Phys. Lett.* 1999, **74**, 442. (b) Burrows, P. E., Forrest, S. R., Zhou, T. X., Michalski, L. *Appl. Phys. Lett.* 2000, **76**, 2493.
5. (a) Adamovich, V., Brooks, J., Tamayo, A., Alexander, A. M., Djurovich, P. I., D'Andrade, B. W., Adachi, C., Forrest, S. R., Thompson, M. E. *New J. Chem.* 2002, **26**, 1171. (b) Thoms, T., Okada, S., Chen, J.-P., Furugori, M. *Thin Solid Films* 2003, **436**, 264.
6. An mCP–FIr6-based device is a good illustration of the problems associated with having host triplet energies that are close to the phosphor energy and closely matched host–dopant HOMO levels. Optimized devices with mCP–FIr6 emissive layers give peak efficiencies of only 2.9%, compared to nearly 11% for devices that have both high host triplet energy and a HOMO energy roughly 1 eV below that of FIr6 (see Ref. [8]). The device data for the mCP–FIr6-based devices are given in the Supporting Information.
7. (a) Kang, T.-S., Harrison, B. S., Foley, T. J., Knefely, A. S., Boncella, J. M., Reynolds, J. R., Schanze, K. S. *Adv. Mater.* 2003, **15**, 1093. (b) Goes, M., Verhoeven, J. W., Hofstraat, H., Brunner, K. *Chem. Phys. Chem.* 2003, **4**, 349.

8. Holmes, R. J., D'Andrade, B. W., Forrest, S. R., Ren, X., Li, J., Thompson, M. E. *Appl. Phys. Lett.* 2003, **83**, 3818.
9. Li, J., Djurovich, P. I., Alleyne, B. D., Tsyba, I., Ho, N. N., Bau, R., Thompson, M. E. *Polyhedron* 2004, **23**, 419–428.
10. Uchida, M., Izumizawa, T., Nakano, T., Yamaguchi, S., Tamao, K., Furukawa, K. *Chem. Mater.* 2001, **13**, 2680. (b) Chen, L. H., Lee, R. H., Hsieh, C. F., Yeh, H. C., Chen, C. T. *J. Am. Chem. Soc.* 2002, **124**, 6469.
11. A recent theoretical study suggested that Si(bph)$_2$ would be a poor host material for blue phosphorescent OLEDs and a good host for green devices. Matsushita, T., Uchida, M. *J. Photopolym. Sci. Technol.* 2003, **16**, 315–316.
12. The device structures used were ITO/NPD (400 Å)/Irppy$_3$:HOST (8%, 250 Å)/BCP (150 Å)/Alq$_3$ (250 Å)/LiF/Al, where NPD = N,N'-diphenyl-N,N'-bis(1-naphthyl)benzidine, Irppy$_3$ = tris(2-phenylpyridine)iridium, HOST = either Si(bph)$_2$ or CBP, BCP = bathocuproine, and Alq$_3$ = aluminium tris(8-hydroxyquinolate).
13. D'Andrade, B. W., Forrest, S. R., Chwang, A. B. *Appl. Phys. Lett.* 2003, **83**, 3858.
14. Tokito, S., Tanaka, H., Noda, K., Okada, A., Taga, T. *Appl. Phys. Lett.* 1997, **70**, 1929.
15. Lamansky, S., Kwong, R. C., Nugent, M., Djurovich, P. I., Thompson, M. E. *Org. Electron.* 2001, **2**, 53. Lamansky, S., Djurovich, P. I., Abdel-Razzaq, F., Garon, S., Murphy, D., Thompson, M. E. *J. Appl. Phys.* 2002, **92**, 1570. He, G., Li, Y., Liu, J., Yang, Y. *Appl. Phys. Lett.* 2002, **80**, 4247. Mwaura, J. K., Knefely, A. S., Schanze, K. S., Boncella, J. M., Reynolds, J. R. *Polym. Prepr. (Am. Chem. Soc., Div. Polym. Chem.)* 2004, **45**, 907.
16. UGH devices without the mCP interface layer have about 1% external quantum efficiency, with maximum light output at 450 Cd/m^2.
17. Shizuka, H., Yamaji, M. *Bull. Chem. Soc. Jpn.* 2000, **73**, 267.
18. Bliznyuk, V. N., Carter, S. A., Scott, J. C., Klarner, G., Miller, R. D., Miller, D. C. *Macromolecules* 1999, **32**, 361.
19. Gilman, H., Smart, R. G. N. *J. Org. Chem.* 1950, **15**, 720.
20. Wittenberg, D., Wu, T. C., Gilman, H. *J. Org. Chem.* 1958, **23**, 1898.
21. El-Attar, A. A., Cerny, M. *Collect. Czech. Chem. Commun.* **1975**, **40**, 2806.
22. (a) Van Beelen, D. C., Van Rijn, J., Heringa, K. D., Wolters, J., De Vos, D. *Main Group Met. Chem.* 1997, **20**, 37. (b) Bickelhaupt, F., Jongsma, C., De Koe, P., Lourens, R., Mast, N. R., Van Mourik, G. L. *Tetrahedron* 1976, **32**, 1921.

23. (a) Drive, M. S., Hartwig, J. F. *J. Am. Chem. Soc.* 1996, **118**, 7217. (b) Wolre, J. P., Wagaw, S., Buchwald, S. L. *J. Am. Chem. Soc.* 1996, **118**, 7215.
24. Schmidbauer, H., Letterbauer, J., Kumberger, O., Lachmann, L., Muller, Z. *Z. Naturforsch.* 1991, **466**, 1065.
25. Hill, I. G., Kahn, A. *J. Appl. Phys.* 1999, **86**, 4515.

Chapter 28

Saturated Deep Blue Organic Electrophosphorescence Using a Fluorine-Free Emitter

R. J. Holmes,[a] S. R. Forrest,[a] T. Sajoto,[b] A. Tamayo,[b]
P. I. Djurovich,[b] M. E. Thompson,[b] J. Brooks,[c] Y.-J. Tung,[c]
B. W. D'Andrade,[c] M. S. Weaver,[c] R. C. Kwong,[c] and J. J. Brown[c]

[a] Princeton Institute for the Science and Technology of Materials (PRISM),
Department of Electrical Engineering, Princeton University,
Princeton, New Jersey 08544, USA
[b] Department of Chemistry, University of Southern California,
Los Angeles, California 90089, USA
[c] Universal Display Corporation, 375 Phillips Boulevard, Ewing, New Jersey 08618, USA
forrest@princeton.edu

We demonstrate saturated, deep blue organic electrophosphorescence using the *facial-* and *meridianal-*isomers of the fluorine-free emitter *tris*(phenyl-methyl-benzimidazolyl)iridium(III) (f-Ir(pmb)$_3$ and m-Ir(pmb)$_3$, respectively) doped into the wide energy gap host, p-*bis*(triphenylsilyly)benzene (UGH2). The highest energy electrophosphorescent transition occurs at a wavelength of $\lambda = 389$ nm for the *fac*-isomer and $\lambda = 395$ nm for the *mer*-isomer. The emission chromaticity is characterized by Commission Internationale

---

Reprinted from *Appl. Phys. Lett.*, **87**, 243507, 2005.

Electrophosphorescent Materials and Devices
Edited by Mark E. Thompson
Text Copyright © 2005 American Institute of Physics
Layout Copyright © 2024 Jenny Stanford Publishing Pte. Ltd.
ISBN 978-981-4877-34-3 (Hardcover), 978-1-003-08872-1 (eBook)
www.jennystanford.com

de l'Eclairage coordinates of ($x = 0.17$, $y = 0.06$) for both isomers. Peak quantum and power efficiencies of $(2.6 \pm 0.3)\%$ and (0.5 ± 0.1)lm/W and $(5.8 \pm 0.6)\%$ and (1.7 ± 0.2)lm/W are obtained using f-Ir(pmb)$_3$ and m-Ir(pmb)$_3$ respectively. This work represents a departure from previously explored, fluorinated blue phosphors, and demonstrates an efficient deep blue/near ultraviolet electrophosphorescent device.

The study of blue organic electrophosphorescence (EP) has focused predominantly on the use of electron-withdrawing fluorine atoms to shift the molecular triplet state to the higher energies required [1, 2]. For example, external quantum efficiencies exceeding 10% have been demonstrated using fluorinated phenyl-pyridine complexes [3–6]. There are drawbacks to using this technique, namely that the saturated blue phosphorescence required for many display applications may not be achievable through fluorination. In addition, the large electronegativity of the fluorine atom may destabilize the molecule, making it electrochemically reactive, leading to potentially short device operational lifetimes. Both of these challenges underscore the need for fluorine-free, deep blue emitting phosphors.

To achieve deep blue EP, the pyridine ring common to blue phosphors such as iridium(III)bis[(4,6-difluorophenyl)pyridinato-$N,C^{2'}$]picolinate (FIrpic) [3–5, 7] and iridium(III)bis(4',6'-difluorophenylpyridinato)tetrakis(1-pyrazolyl) borate (FIr6) [6] can be replaced with a methyl-benzimidazolyl group. The resulting phenyl-methyl-benzimidazole ligand (pmb) has a higher triplet energy than difluorophenyl-pyridine, leading to a high phosphorescence energy for the tris-cyclometalated Ir complexes (i.e. fac- and mer-Ir(pmb)$_3$) [8]. This substitution pushes the triplet energy of the molecule toward the blue (with photoluminescence (PL) at a peak wavelength of $\lambda < 400$ nm for fac- and mer-Ir(pmb)$_3$) without adding potentially unstable fluorine atoms [9]. Here we study such compounds that are co-deposited with the wide energy gap host, p-bis(triphenylsilyly)benzene (UGH2) to encourage the direct trapping of holes on the emissive phosphor [6]. The use of a wide gap host such as UGH2 also ensures that the high energy triplet excitons remain confined to the phosphor dopant, eliminating guest-to-host back transfer of triplet excitons [1, 7].

Organic light emitting devices (OLEDs) were grown on a glass substrate precoated with a ~150-nm-thick layer of indium tin oxide having a sheet resistance of ~Ω/sq. Substrates were degreased with solvents and then exposed to a UV-ozone ambient. All organic layers were grown in succession by thermal sublimation without breaking vacuum (base pressure ~1×10^{-7} Torr). Masking for cathode deposition was completed in a glovebox with <1 ppm water and oxygen. The devices consisted of a 30-nm-thick 4-4'-*bis*[*N*-(1-naphthyl)-*N*-phenyl-amino]biphenyl (NPD) hole transport layer, followed by a 10-nm-thick exciton/electron blocking layer (EBL) of either *fac-tris*(1-phenylpyrazolato, *N*, $C^{2'}$)iridium(III) [Ir(ppz)$_3$] [10] or 4, 4', 4"-*tris*(carbazol-9-yl)-triphenylamine (TCTA) [11, 12], that confine electrons and possibly excitons in the OLED emissive layer (EML). The 25-nm-thick EML consists of either *fac*- or *mer-tris*(phenyl-methyl-benzimidazolyl)iridium(III), (*f*-Ir(pmb)$_3$ or *m*-Ir(pmb)$_3$, respectively) [9] co-deposited with UGH2 [6, 13]. Finally, a 35-nm-thick electron transporting and hole blocking layer consisting of 2,9-dimethyl-4,7-diphenyl-1,10-phenanthroline (bathocuproine, BCP) was deposited. Literature procedures were used to prepare NPD [14], Ir(ppz)$_3$ [10], *f*-Ir(pmb)$_3$ and *m*-Ir(pmb)$_3$ [9], and UGH2 [13, 15]. Also, BCP [16] and TCTA [17] were obtained from commercial suppliers. Prior to deposition, all materials were purified by temperature gradient vacuum sublimation except TCTA, which was used as supplied. Cathodes consisting of a 0.5-nm-thick layer of LiF followed by a 50-nm-thick layer of Al were patterned by deposition through a shadow mask with an array of 1-mm-diam circular openings. Current-voltage and external efficiency measurements were obtained using standard methods [18].

Figure 28.1a, inset, shows the molecular structure of the blue phosphor Ir(pmb)$_3$, with EP spectra collected at 10 mA/ cm^2 for OLEDs using either Ir(ppz)$_3$ or TCTA as an EBL shown in Fig. 28.1. The OLED EML consists of UGH2 doped with 10 wt% of either the *fac*- or *mer*-isomer of Ir(pmb)$_3$. For Ir(ppz)$_3$-based devices, the emission originates solely from the triplet state of the phosphor and is independent of current density from 1 to 100 mA/ cm^2. Devices using TCTA show pure dopant emission at 1 mA/ cm^2, with increasing amounts of NPD emission visible between 10 and 100 mA/ cm^2. Both isomers have significant EP in the ultra-violet

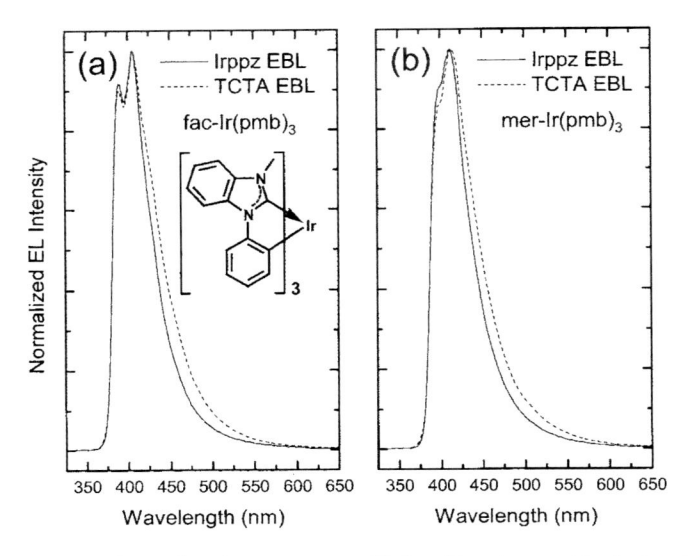

Figure 28.1 The molecular structure of Ir(pmb)$_3$, along with electrophosphorescent spectra from devices with emissive layers of either UGH2:10 wt% (a) *fac*-Ir(pmb)$_3$ or (b) *mer*-Ir(pmb)$_3$. The spectra are obtained at 10 mA/ cm^2 for devices whose exciton/electron blocking layers consist of either Ir(ppz)$_3$ (solid lines) or TCTA (broken lines). Inset: Structural formula of Ir(pmb)$_3$.

($\lambda < 400$ nm) region of the spectrum. The first phosphorescent transition of the *fac*- isomer at $\lambda = 389$ nm (3.2 eV) is slightly higher in energy than that of the *mer*-isomer at $\lambda = 395$ nm (3.1 eV). The emission chromaticity is characterized by Commission Internationale de l'Eclairage (CIE) co-ordinates of ($x = 0.17, y = 0.06$) and ($x = 0.17, y = 0.08$) for f-Ir(pmb)$_3$, and ($x = 0.17, y = 0.06$) and ($x = 0.17, y = 0.08$) for m-Ir(pmb)$_3$, for Ir(ppz)$_3$ and TCTA EBLs, respectively.

Figure 28.2 shows the external quantum and power efficiencies of phosphorescent OLEDs using f-Ir(pmb)$_3$ (top) and m-Ir(pmb)$_3$ (bottom), co-deposited at 10 wt% with UGH2. Both Ir(ppz)$_3$ (squares) and TCTA (circles) were used as EBLs, with the TCTA blocked devices yielding higher efficiencies for both isomers. With the *fac*-isomer, peak external quantum (η_{EXT}) and power (η_P) efficiencies of $\eta_{EXT} = (1.8 \pm 0.2)\%$ and $\eta_P = (0.3 \pm 0.1)$lm/ W

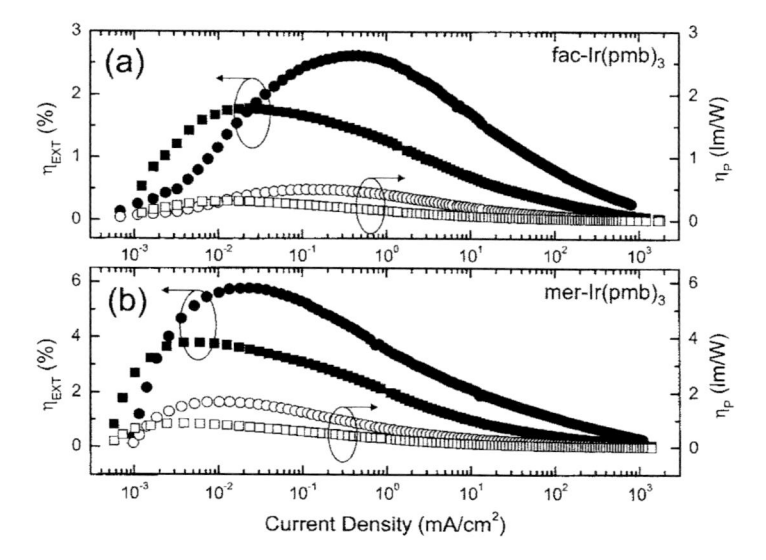

Figure 28.2 External quantum (η_{EXT}, closed symbols) and power (η_P, open symbols) efficiencies for devices with emissive layers of either UGH2:10 wt% *fac*-Ir(pmb)$_3$ (top) or UGH2:10 wt% *mer*-Ir(pmb)$_3$ (bottom) with exciton/electron blocking layers of either Ir(ppz)$_3$ (squares) or TCTA (circles).

are obtained using Ir(ppz)$_3$, while $\eta_{EXT} = (2.6 \pm 0.3)\%$ and $\eta_P = (0.5 \pm 0.1)$lm/W are obtained with TCTA. For UGH2:10 wt% *m*-Ir(pmb)$_3$ devices, efficiencies of $\eta_{EXT} = (3.8 \pm 0.4)\%$ and $\eta_P = (0.9 \pm 0.1)$lm/W are obtained using Ir(ppz)$_3$, while $\eta_{EXT} = (5.8 \pm 0.6)\%$ and $\eta_P = (1.7 \pm 0.2)$lm/W are obtained with TCTA. The relatively low power efficiencies compared to previously reported phosphor dopants reflect the limited overlap between the deep blue EP spectra of these devices and the photopic response of the human eye [19].

While OLEDs prepared with the meridianal isomer of Ir(pmb)$_3$ give higher EP efficiencies than those with the facial isomer by roughly a factor of 2, the PL efficiencies of the two complexes in fluid solution differ by more than an order of magnitude, with the meridianal giving the lower efficiency ($\Phi_{PL} = 0.04$, 0.002 for *f*-Ir(pmb)$_3$ and *m*-Ir(pmb)$_3$, respectively) [9]. The two complexes have similar radiative rates in solution [($k_r = (1.8$ and $1.4) \times 10^5$ s^{-1}, respectively)], as well as similar luminescent lifetimes in the solid

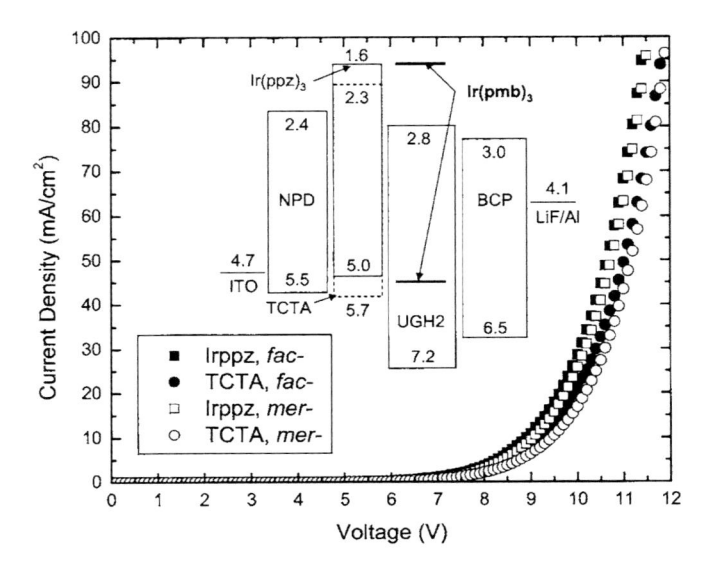

Figure 28.3 Current-voltage characteristics for the devices of Figs. 28.1 and 28.2. The lower voltage observed for Ir(ppz)$_3$ devices reflects the ease of hole injection from NPD into Ir(ppz)$_3$ relative to TCTA. Inset: Proposed equilibrium energy level diagram for the devices. The HOMO and LUMO levels of Ir(ppz)$_3$ and TCTA are (5.0, 1.6 eV) and (5.7, 2.3 eV), respectively. The energy levels of the phosphors are not pictured but lie at (5.1, 1.8 eV) and (4.8, 1.4 eV) for the *fac-* and *mer-*isomers, respectively.

state ($\tau = 1$ μs for both complexes) [9], suggesting that the facial and meridianal isomers should have similar luminescent efficiencies in the solid state. Thus, the difference in device efficiencies for OLEDs made with the two dopants is most likely due to differences in charge conduction, trapping, or recombination as a result of the different highest occupied molecular orbital (HOMO) and triplet energies of the two species.

The inset of Fig. 28.3 shows the proposed energy level diagram for the devices under zero bias. The HOMO energies were measured using ultraviolet photoelectron spectroscopy, except that of BCP which was taken from Ref. [20]. The HOMO levels of the phosphors are at (5.1 ± 0.1) eV and (4.8 ± 0.1) eV referenced to vacuum for the *fac-* and *mer-*isomers, respectively. Lowest unoccupied molecular orbital (LUMO) energies were estimated by adding the energies corresponding to the onset of optical absorption to the HOMO energies, giving LUMO energies of 1.8 and 1.4 eV, respectively [20]. However, the use of the absorption edge often underestimates LUMO energies given in Fig. 28.3, particularly for poorly conjugated materials such as UGH2 [21, 22]. Hole injection into the HOMO of UGH2 is expected to be inefficient given the large energy barrier between the Ir(ppz)$_3$ (or TCTA) HOMO and the UGH2 HOMO [6]. As a result, holes are transported primarily by the phosphorescent dopant after direct, and nearly resonant injection from the injection layer onto the dopant. Electron injection onto the phosphor is energetically unfavorable with the LUMO levels of the *fac*- and *mer*- isomers. Hence, excitons may be formed by the injection of electrons directly into the triplet level of the phosphor from the LUMO of either BCP or UGH2.

Figure 28.3 shows the current density-voltage (J-V) characteristics for the devices under study. Both isomers have nearly identical J-V characteristics, consistent with their comparable HOMO and LUMO energies. The Ir(ppz)$_3$-based devices have a slightly lower operating voltage than those that use TCTA due to the closer alignment of the Ir(ppz)$_3$ HOMO level to that of NPD that leads to a higher injection efficiency.

In summary, we have demonstrated efficient, saturated blue electrophosphorescence using the fluorine-free *fac*- and *mer*- isomers of Ir(pmb)$_3$ doped into the wide gap host UGH2. The electrophosphorescent spectrum of Ir(pmb)$_3$ with CIE coordinates of ($x = 0.17$, $y = 0.06$), is dramatically "bluer" than previously explored blue phosphors such as FIrpic ($x = 0.16$, $y = 0.37$) [7] and FIr6 ($x = 0.16, y = 0.26$) [6]. Although these former, fluorinated molecules exhibit high quantum efficiencies, they lack the desired deep blue color purity [3–7].

Acknowledgment

The authors thank PPG Industries for providing some of the materials used, and the Department of Energy Office of Basic Energy Sciences for partial support of this work.

References

1. C. Adachi, R. C. Kwong, P. Djurovich, V. Adamovich, M. A. Baldo, M. E. Thompson, and S. R. Forrest, *Appl. Phys. Lett.* **79**, 2082 (2001).
2. J. Li, P. I. Djurovich, B. D. Alleyne, M. Yousufuddin, N. N. Ho, J. C. Thomas, J. C. Peters, R. Bau, and M. E. Thompson, *Inorg. Chem.* **44**, 1713 (2005).
3. S.-J. Yeh, M.-F. Wu, C.-T. Chen, Y.-H. Song, Y. Chi, M.-H. Ho, S.-F. Hsu, and C. H. Chen, *Adv. Mater. (Weinheim, Ger.*) **17**, 285 (2005).
4. S. Tokito, T. Iijima, Y. Suzuri, H. Kita, T. Tsuzuki, and F. Sato, *Appl. Phys. Lett.* **83**, 569 (2003).
5. S.-C. Lo, G. J. Richards, J. P. J. Markham, E. B. Namdas, S. Sharma, P. L. Burn, and I. D. W. Samuel, *Adv. Funct. Mater.* **15**, 1451 (2005).
6. R. J. Holmes, B. W. D'Andrade, S. R. Forrest, X. Ren, J. Li, and M. E. Thompson, *Appl. Phys. Lett.* **83**, 3818 (2003).
7. R. J. Holmes, S. R. Forrest, Y.-J. Tung, R. C. Kwong, J. J. Brown, S. Garon, and M. E. Thompson, *Appl. Phys. Lett.* **82**, 2422 (2003).
8. M. E. Thompson, J. Li, A. Tamayo, T. Sajoto, P. I. Djurovich, S. R. Forrest, R. J. Holmes, and J. J. Brown, *SID Proc.* 1058 (2005).
9. T. Sajoto, P. I. Djurovich, A. Tamayo, M. Yousufuddin, R. Bau, M. E. Thompson, R. J. Holmes, and S. R. Forrest, *Inorg. Chem.* **44**, 7992 (2005).
10. V. Adamovich, J. Brooks, A. Tamayo, A. M. Alexander, P. Djurovich, B. W. D'Andrade, C. Adachi, S. R. Forrest, and M. E. Thompson, *New J. Chem.* **26**, 1171 (2002).
11. Y. Kuwabara, H. Ogawa, H. Inada, N. Noma, and Y. Shirota, *Adv. Mater. (Weinheim, Ger.*) **6**, 677 (1994).
12. B. W. D'Andrade, R. J. Holmes, and S. R. Forrest, *Adv. Mater.* (*Weinheim, Ger.*) **16**, 624 (2004).
13. X. Ren, J. Li, R. J. Holmes, P. I. Djurovich, S. R. Forrest, and M. E. Thompson, *Chem. Mater.* **16**, 4743 (2004).
14. D. E. Loy, B. E. Koene, M. E. Thompson, P. E. Burrows, and S. R. Forrest, *Chem. Mater.* **10**, 2235 (1998).
15. H. Gilman and G. D. Lichtenwalte, *J. Am. Chem. Soc.* **80**, 608 (1958).
16. Bathocuproine, Cat. 14,091-0, Sigma-Aldrich Corp., St. Louis, Missouri.
17. 4,4',4''-tris(carbazol-9-yl)-triphenylamine, Cat. LT-E207, Luminescence Technology Corp., Taiwan.
18. S. R. Forrest, D. D. C. Bradley, and M. E. Thompson, *Adv. Mater.* (*Weinheim, Ger.*) **15**, 1043 (2003).

19. M. Born and E. Wolf, *Principles of Optics*, 7th ed. (Cambridge University Press, Cambridge, 1999).
20. I. G. Hill and A. Kahn, *J. Appl. Phys.* **86**, 4515 (1999).
21. R. S. Ruoff, K. M. Kadish, P. Boulas, and E. C. M. Chen, *J. Phys. Chem.* **99**, 8843 (1995).
22. X. Zhan, C. Risko, F. Amy, C. Chan, W. Zhao, S. Barlow, A. Kahn, J.-L. Bredas, and S. R. Marder, *J. Am. Chem. Soc.* **127**, 9021 (2005).

Chapter 29

Excimer and Electron Transfer Quenching Studies of a Cyclometalated Platinum Complex

Biwu Ma,[a] Peter I. Djurovich,[b] and Mark E. Thompson[a,b]

[a]*University of Southern California, Department of Materials Science, Los Angeles, California 90089, USA*
[b]*University of Southern California, Department of Chemistry, 840 Downey Way, Los Angeles, California 90089, USA*
met@usc.edu

Luminescence quenching studies of a cyclometalated complex, platinum(II) (2-(4′,6′-difluorophenyl)pyridinato-$N,C^{2'}$)(2,4-pentanedionato-O,O) (**FPt**), in solution at room temperature are reported. The **FPt** complex undergoes efficient self-quenching in solution at room temperature that can be successfully modeled by a monomer/excimer phosphorescence mechanism with a diffusion limited rate constant ($4.2 \pm 0.3 \times 10^9$ M^{-1} s^{-1}). The emission lifetimes for **FPt** monomer and excimer in 2-methyltetrahydrofuran at room temperature are 330 ns ($\pm$ 15 ns) and 135 ns ($\pm$ 10 ns), respectively. The excited state properties of **FPt** were also investigated. A triplet energy E_T of 2.8 eV and excited-state reduction potential

---

Reprinted from *Coord. Chem. Rev.*, **249**, 1501–1510, 2005.

Electrophosphorescent Materials and Devices
Edited by Mark E. Thompson
Text Copyright © 2005 Elsevier B.V.
Layout Copyright © 2024 Jenny Stanford Publishing Pte. Ltd.
ISBN 978-981-4877-34-3 (Hardcover), 978-1-003-08872-1 (eBook)
www.jennystanford.com

$E(\text{FPt}^{*/-})$ of 0.81 V versus SCE were determined from quenching studies in agreement with values estimated from emission spectra and a thermochemical cycle. The excited-state oxidation potential $E(\text{FPt}^{*/+})$ cannot be determined from electrochemical data since **FPt** undergoes irreversible oxidation. However, a value of $E(\text{FPt}^{*/+})= -1.41$ V versus SCE was established from an electron transfer quenching study and thus, a ground state oxidation potential for **FPt** can be estimated to be 1.30 V versus SCE.

29.1 Introduction

Square planar platinum complexes have attracted considerable interest for their potential use in a wide range of applications such as optical chemosensors [1], photocatalysis [2], and molecular photochemical devices for solar energy conversion [3]. Platinum complexes have also been applied successfully as phosphorescent emitters in organic light-emitting diodes (OLEDs) [4]. OLEDs fabricated with platinum complexes can exhibit high emission efficiency, as the strong spin-orbital coupling of platinum allows for efficient phosphorescence at room temperature [5]. Efficient phosphorescent dopants are important components in OLEDs since the hole–electron recombination process leads to a mixture of singlet and triplet excited states in the device [6]. Recently, white OLEDs have been prepared with an organometallic Pt complex, platinum(II) (2-(4′,6′-difluorophenyl)pyridinato-$N,C^{2'}$)(2,4-pentanedionato-O,O) (**FPt**), with emission that originates from both isolated Pt complexes and excimers or aggregates of the same complex [7]. The devices utilized a very simple device structure since only a single dopant is used in the emissive layer and were highly efficient. In order to better understand the excited state properties of **FPt**, we have carried out an extensive photophysical quenching study of this complex and report it herein.

Self-quenching of platinum complexes has been well known since Che and co-workers reported concentration dependence of the emission intensity for the complex Pt(5,5′-dimethyl-2,2′-bipyridine)(CN)$_2$ [8a]. Other researchers have reported similar self-quenching for related square planar platinum complexes, as well as the observation of weak excimer emission at high platinum complex concentrations [8, 9]. The self-quenching mechanism used to explain this concentration dependent quenching process involves initial formation of the excited complex, M*, followed by association with a ground state complex to give a weakly emissive excimer, [M, M]*. The latter process is clearly dependent on the concentration of the metal complex. The model is identical to one used to describe the monomer/excimer fluorescence kinetics of aromatic hydrocarbons [10]. Using this model, the diffusion controlled self-quenching rate constants have been determined for a number of platinum complexes by monitoring the self-quenching reactions of the Pt complex as a function of concentration. Direct characterization of these platinum excimers, however, has been less documented due to limited solubility of the complexes, as well as weak emission and short lifetimes for the excimers [9].

Metal complexes in their excited states may be potent oxidizing and reducing agents [11], making them potentially useful as sensitizers for photochemical energy conversion (i.e. solar energy conversion). In this context, it is important to know the redox properties of excited metal complex, in order to design the optimal combination of donor and acceptor materials to interact with the sensitizer in a solar cell configuration. Likewise, in organic light-emitting diodes, a good understanding of the excited state redox properties is also important, since such knowledge will enable the best materials to be chosen in order to eliminate luminescent quenching by electron transfer between the excited dopant and the other materials in the device, processes that lead to a decreased OLED efficiency.

In this chapter, we present a number of photophysical studies of **FPt** in solution at room temperature. The concentration dependent luminescent behavior of **FPt** was studied by static and transient spectroscopy. The **FPt** complex was found to undergo luminescent self-quenching at near diffusion controlled rates from monomer/excimer phosphorescence kinetic analysis. Energy transfer and electron transfer quenching studies of the complex **FPt** were also carried out to estimate the redox properties of the **FPt** ground and excited state [12].

29.2 Experimental

29.2.1 Materials

All of the neutral organic quenchers used here were commercially available and used as received. The pyridinium salts were prepared by refluxing the corresponding substituted pyridine with appropriate alkylating reagent in an acetone–methanol mixture overnight, followed by metathesis in water with NH_4PF_6. The solvent 2-methyltetrahydrofuran (2-MeTHF) was distilled under N_2 after being refluxed over sodium prior to use. The acetonitrile used was EM Science DriSolv solvent and used as received. **FPt** was prepared by a literature procedure [5].

29.2.2 Physical Measurements

The UV–vis spectra were recorded on an Aviv model 14DS spectrophotometer (a re-engineered Cary 14 spectrophotometer). Photoluminescent spectra were measured using a Photon Technology International fluorimeter. Emission lifetime measurements were performed using time-correlated single photon counting on an IBH Fluorocube instrument. Samples were excited with 405 nm pulsed diode laser having a pulse duration of ca. 1.2 ns and an energy of 500 nJ/pulse. The energy per laser pulse was adequate to provide a transient signal yet low enough to prevent formation of high concentrations of excited state species, which would lead to quenching by triplet–triplet annihilation. The measured lifetimes have an error of 5%.

Quantum efficiency measurements were carried out at room temperature in a 2-methyltetrahydrofuran solution. Before spectra were measured, the solution was degassed by several freeze–pump–thaw cycles using a diffusion pump. Solutions of coumarin-47 in ethanol ($\Phi = 0.60$) were used as a reference. The equation $\Phi_s = \Phi_r(\eta_s^2 A_r I_s / \eta_r^2 A_s I_r)$ was used to calculate quantum yields, where Φ_s is the quantum yield of the sample, Φ_r is the quantum yield of the reference, η is the refractive index of the solvent, A_s and A_r are the absorbance of the sample and the reference at the wavelength of excitation, and I_s and I_r are the integrated areas of emission bands.

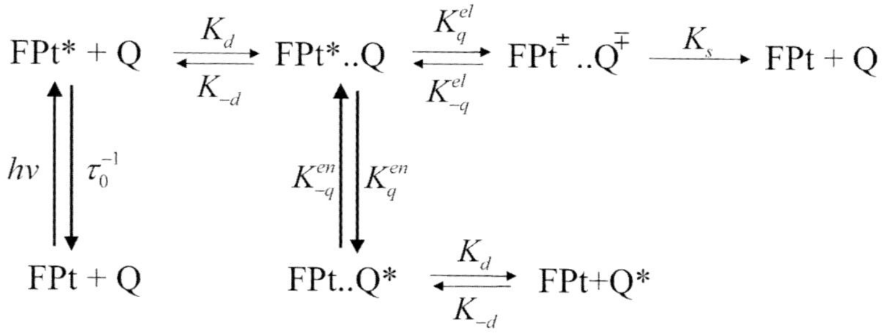

$$\text{FPt}^* + Q \underset{K_{-d}}{\overset{K_d}{\rightleftarrows}} \text{FPt}^*..Q \underset{K^{rel}_{-q}}{\overset{K^{rel}_q}{\rightleftarrows}} \text{FPt}^{\pm}..Q^{\mp} \overset{K_s}{\longrightarrow} \text{FPt} + Q$$

$$hv \Big\Updownarrow \tau_0^{-1} \qquad\qquad K^{en}_{-q}\Big\Updownarrow K^{en}_q$$

$$\text{FPt} + Q \qquad\qquad \text{FPt}..Q^* \underset{K_{-d}}{\overset{K_d}{\rightleftarrows}} \text{FPt}+Q^*$$

Scheme 29.1

Cyclic voltammetry and differential pulsed voltammetry were performed using an EG&G potentiostat/galvanostat model 283. Anhydrous DMF (Aldrich) was used as the solvent under a nitrogen atmosphere, and 0.1 M tetra(*n*-butyl)ammonium hexafluorophosphate was used as the supporting electrolyte. A Pt wire acted as the counter electrode, an Ag wire was used as the pseudo reference electrode, and the working electrode was glassy carbon. The redox potentials are based on values measured from differential pulsed voltammetry and are reported relative to an internal ferrocenium/ferrocene (Cp_2Fe^+/Cp_2Fe, 0.45 V versus SCE) [13] reference. Electrochemical reversibility was determined using cyclic voltammetry.

For the bimolecular quenching experiments, **FPt** concentration was set as 0.01 mM and the concentrations of quenchers were from 0 to 50 mM. Before each lifetime measurement, the solution samples were degassed by several freeze–pump–thaw cycles using a diffusion pump. In the case of oxidative quenching studies, the samples in acetonitrile were bubble degassed with dry argon for 15–30 min. A Stern–Volmer quenching analysis, as in Eq. (29.1), was applied to determine the bimolecular quenching rate constants, where I_0 and τ_0 are the emission intensity and excited state lifetime in absence of quenchers, I and τ are the corresponding values with the quenchers present, K_q is the experimental quenching rate constant and [Q] is the molar concentration of the quencher. For the self-quenching studies, the emission lifetimes in different concentrations were measured and the self-quenching rate K_q was determined using Eq. (29.2), where [FPt] is the total concentration of **FPt**.

$$\frac{I_0}{I} = \frac{\tau_0}{\tau} = 1 + K_q\tau_0[Q] \qquad (29.1)$$

$$\frac{\tau_0}{\tau} = 1 + K_q\tau_0\,[\mathrm{FPt}] \qquad (29.2)$$

29.2.3 Data Analysis for Bimolecular Quenching

The kinetics for electron transfer or energy transfer (shown in Scheme 29.1) has been modeled and can be fit with the following equations [12].

$$K_q^{el} = \frac{K_d}{1 + (K_{-d}/K_s)\exp(\Delta G_{el}/RT)} + (K_{-d}/K_{el}^0)\exp(\Delta G_{el}^{\neq}/RT)$$
(electron transfer quenching) (29.3)

$$K_q^{en} = \frac{K_d}{1 + \exp(\Delta G_{en}/RT)} + (K_{-d}/K_{en}^0)\exp(\Delta G_{en}^{\neq}/RT)$$
(energy transfer quenching) (29.4)

where $\Delta G_{en(el)}^{\neq} = \Delta G_{en(el)} + (\Delta G_{en(el)}^{\neq}(0)/\ln 2)\ln[1 + \exp(-\Delta G_{en(el)} \ln 2/\Delta G_{en(el)}^{\neq}(0))]$ (the free energy of activation: Agmon–Levine relationship) [14].

In these equations, the diffusion rate constant K_d and dissociation rate constant K_{-d} were determined as $K_d = 1.3 \times 10^{10}$ s^{-1} for the solvent 2-Me-THF and $K_d = 1.9 \times 10^{10}$ M^{-1} s^{-1} and $K_{-d} = 2.1 \times 10^{10}$ M^{-1} s^{-1} for the solvent acetonitrile at room temperature [15]. $K_{en(el)}^0$, $\Delta G_{en(el)}^{\neq}(0)$, and K_s are the pre-exponential factor, the reorganizational intrinsic barrier and rate constant for back electron transfer to form the reactants in their ground state, respectively. These values can be determined based on quenching studies done with closely related compounds and were chosen as $K_{en(el)}^0 \approx 1.4 \times 10^{10}$ s^{-1} and $\Delta G_{en(el)}^{\neq}(0) \approx 0.1$ eV to achieve the best fits to the experimental results and fall close to the values reported for closely related Pt complexes [12]. In electron transfer quenching, K_s was taken to equal K_{-d} [16]. The standard free energy change for energy transfer, G_{en}, is given by the following equation: $G_{en} = -\{E^{00}([\mathrm{Pt}]^*, [\mathrm{Pt}]) - E^{00}(Q^*, Q)\}$, where E^{00} is the energy of the excited state derived from spectra for **FPt** and quenchers. The standard free energy change for electron transfer, G_{el}, is given by $G_{el} = (E_{ox} - E_{red}) - E^{00} + W_p - W_r$, in which E_{ox} is the oxidation potential of the electron donor, E_{red} is the reduction potential of the electron acceptor, W_p and W_r are the work terms, which take the value of $W_p - W_r = 0.15$ eV for reductive electron transfer quenching [17] and have a negligible value for oxidative electron transfer quenching [25].

29.3 Results and Discussion

29.3.1 Self-quenching

The absorption spectrum along with the room temperature luminescence spectra of **FPt** in 2-MeTHF at three different concentrations is shown in Fig. 29.1. Low energy transitions in the absorption spectrum (300–450 nm) are assigned as metal-to-ligand charge transfer (MLCT) transitions, while higher energy transitions (230–300 nm) are assigned to $\pi - \pi^*$ ligand centered (LC) transitions. At low concentration, a vibronically structured luminescence ($\lambda_{max} = 465$ nm) is observed from a single emissive species. This emission has been previously assigned to phosphorescence from a triplet ligand centered (^{3}LC) state of the cyclometalated ligand that is strongly perturbed by mixing with a higher energy, singlet metal-to-ligand charge transfer (^{1}MLCT) state [5]. As the **FPt** concentration increases, a broad, low energy emission band ($\lambda_{max} = 650$ nm) grows into the spectrum, whereas dilution of the sample eliminates the low energy emission band. The absorption spectra of the dilute and concentrated samples are the same, having a lowest energy triplet absorption transition at 460 nm ($\varepsilon = 40$ M^{-1} cm^{-1}) at room temperature. Excitation spectra recorded by monitoring emission wavelengths of 465 and 620 nm are identical (Fig. 29.1), and match the low energy absorption bands for **FPt**. Upon cooling a concentrated solution to 77 K, the 650 nm emission disappears and the higher energy emission increases in intensity and undergoes a small rigidochromic blue shift to 458 nm [5]. A vibronic progression is observed in the spectra at 77 K with a spacing of 1470 cm^{-1}.

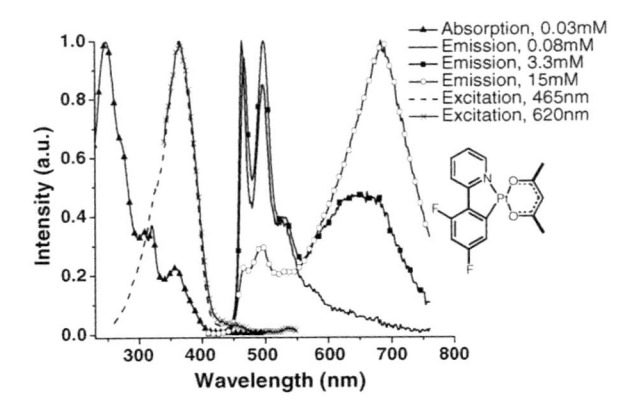

Figure 29.1 Absorption, emission and excitation spectra of **FPt** at different concentrations in 2-MeTHF. The inset shows the molecular structure of **FPt**.

On the basis of this spectroscopic data, the structured high energy emission can be assigned to the **FPt** monomer and the featureless low energy emission band can be assigned to an **FPt** excimer. The time dependence of the emission intensity from the **FPt** monomer and excimer were measured by monitoring the luminescence decay at 465 and 650 nm, respectively (Fig. 29.2). Whereas the monomer signal shows emission decay commencing immediately after the excitation pulse, the excimer emission initially increases and then

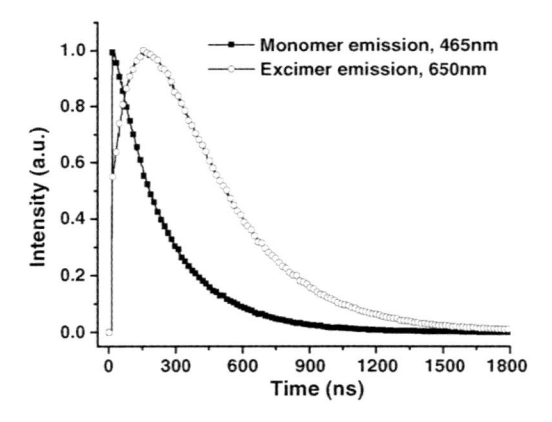

Figure 29.2 Excited state decay curves of 0.25 mM **FPt** in 2-MeTHF monitored at 465 and 650 nm.

decays, consistent with initial formation of the excimer followed by its return back to the ground state.

The kinetics for the photophysics of an **FPt** solution can be treated as shown in Scheme 29.2 [9, 10]. Here, K_q^M represents the monomer self-quenching rate constant (also the rate constant for excimer formation), K_d^M represents the decay rate for the monomer, and K_d^E is the decay rate for the excimer. A similar model has been successfully applied to treat monomer/excimer emission from both pyrene [10] and square planar platinum complexes [9]. The monomer emission intensity $i_{M(t)}$ and the excimer emission intensity $i_E(t)$ can then be described using Eqs. (29.5) and (29.6) respectively, where λ_1 and λ_2 are decay constants that are related to the rate parameters in the manner shown in Eq. (29.7). In this model, the large energy difference between the monomer and excimer (roughly 0.76 eV) makes the excimer extremely stable towards dissociation back to **FPt*** and **FPt**. Therefore, it can be assumed that the rate constant for excimer dissociation, K_s^E, is significantly smaller than K_q^M and thus, K_s^E can be neglected. This assumption leads to $(K_d^E + K_s^E - K_d^M - K_q^M[\text{FPt}]) \gg \sqrt{4K_q^M K_s^E[\text{FPt}]}$ at low to modest concentrations solutions ($\leq$3 mM). The decay constant, λ_1, can then be approximated as $\lambda_1 = K_d^M + K_q^M[\text{FPt}]$, i.e. the observed monomer decay rate at a given solution concentration, and the decay constant λ_2 can be treated approximately as $\lambda_2 = K_d^E$, i.e. the excimer decay rate.

$$i_M(t) \propto e^{-\lambda_1 t} + A e^{-\lambda_2 t} \tag{29.5}$$

$$i_E(t) \propto e^{-\lambda_1 t} - e^{-\lambda_2 t} \tag{29.6}$$

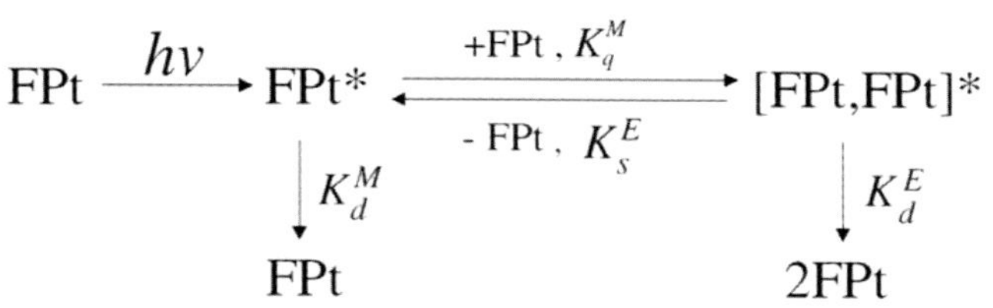

$$\text{FPt} \xrightarrow{h\nu} \text{FPt*} \underset{-\text{FPt}, K_s^E}{\overset{+\text{FPt}, K_q^M}{\rightleftharpoons}} [\text{FPt,FPt}]^*$$

$$\downarrow K_d^M \qquad\qquad\qquad\qquad \downarrow K_d^E$$

$$\text{FPt} \qquad\qquad\qquad\qquad 2\text{FPt}$$

Scheme 29.2

where $A = (K_d^M + K_q^M[\text{FPt}] - \lambda_1)/(\lambda_2 - K_d^M + K_s^E[\text{FPt}])$, and

$$\lambda_{1,2} = \frac{1}{2}(K_d^M + K_q^M[\text{FPt}] + K_d^E + K_s^E$$
$$\mp \sqrt{(K_d^E + K_s^E - K_d^M - K_q^M[\text{FPt}])^2 + 4K_q^M K_s^E[\text{FPt}]}) \quad (29.7)$$

The luminescent decay of **FPt** monomer at $\lambda_{max} = 465$ nm was examined at a number of different concentrations (Fig. 29.3). According to the model shown above, monomer emission decay should exhibit bi-exponential kinetics. However, our experimental results display a simple mono-exponential decay for the monomer at all concentrations studied (Fig. 29.3, inset), which can be simply explained as due to the pre-exponential factor, A, being $\ll 1$ (i.e. $K_s^E \ll K_q^M$). A plot of the measured emission decay rate versus the **FPt** concentration is linear (Fig. 29.4), matching the expected dependence, i.e. $K_{obs} = K_d^M + K_q^M[\text{FPt}]$. The slope of the linear fit to this data gives a large, near diffusion controlled self-quenching rate constant of $4.2 \pm 0.3 \times 10^9$ M^{-1} s$^-$ for excimer formation. This finding supports the assumption that K_s^E is quite small relative to K_q^M. The intercept of the linear fit to the data gives the emission rate in the absence of self-quenching (i.e. infinite dilution limit [FPt] $\sim$0) and thus, an estimate of the intrinsic lifetime (τ_{rad}) for **FPt**. The intercept of the best-fit line ($K_d^M = 3.0 \times 10^6$ s^{-1}) corresponds to

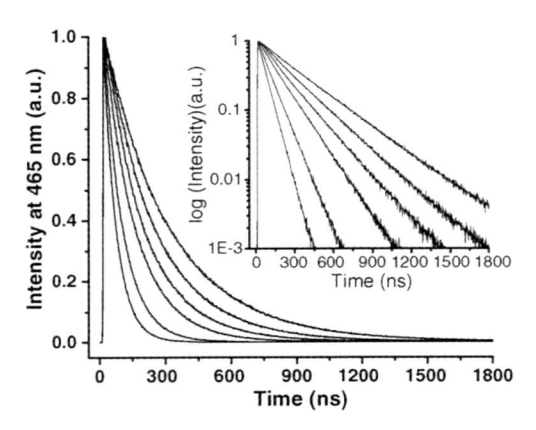

Figure 29.3 Monomer luminescence decay curves for **FPt** in 2-MeTHF at concentrations listed in inset of Fig. 29.4. The inset shows the decay curves in a semilog plot.

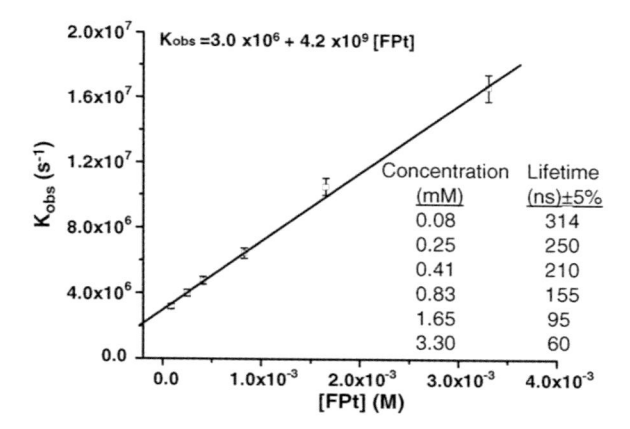

Figure 29.4 Modified Stern–Volmer plot of self-quenching (squares) for **FPt**. The inset is a table of monomer lifetimes and concentrations.

an **FPt** monomer lifetime of 330 ns (±15 ns) at infinite dilution. A quantum efficiency of 0.6% was measured for **FPt** emission for a dilute sample (0.01 mM) where only monomer emission is observed. The radiative and non-radiative rate constants are then calculated to be 1.8×10^4 s^{-1} ($\tau_{rad} = 55$ μs) and 3.0×10^6 s^{-1}, respectively [18]. The radiative rate constants observed for **FPt** are similar to values reported for related square planar platinum complexes [9].

Several groups have described excimer formation during self-quenching of platinum complexes [9]. Although the intrinsic monomer lifetimes and the self-quenching rate constants have been determined for most of the complexes through analysis of monomer luminescent self-quenching, few studies have determined the excimer emission decay rate directly [9b, f, i]. Fortunately, the **FPt** complex has sufficient solubility and high enough excimer luminescent efficiency to allow for direct measurement of the excimer formation and decay kinetics. The rise and fall of the excimer emission intensity in a 0.25 mM **FPt** solution was monitored at 650 nm. The excimer decay can be well fit with a bi-exponential function, representing the formation and decay of the excimer (Fig. 29.2). According to discussion above, the decay constant (λ_2) at such a concentration can be treated as equal to the excimer decay rate, K_d^E. Application of Eq. (29.6) to the excimer signal at

650 nm then gives K_d^E a value of $7.5 \pm 0.5 \times 10^6$ s^{-1} and thus, an excimer lifetime at ca. 135 ± 10 ns. Therefore, the **FPt** complex has comparable monomer (330 ± 15 ns) and excimer (135 ± 10 ns) lifetimes, which causes the pronounced rise and decay signals for the excimer emission intensity (Fig. 29.2). The monomer and excimer lifetimes for **FPt** are similar to those of other Pt complexes reported in the literature, although there is a wide spread in excimer lifetimes (40–3000 ns) [9i]. Some platinum complexes, e.g. Pt(4,4′-di-*tert*-butylbipyridine)(CN)$_2$, have monomer and excimer lifetimes that differ by a factor of 100 ($\tau_{monomer} = 2900$ ns; $\tau_{excimer} \approx 40$ ns) [9b]. Unfortunately, it is not possible to determine a value for the quantum efficiency for excimer emission, so the radiative and non-radiative rates for the two types of emissive species cannot be compared.

While we have focused on the solution photophysical properties of **FPt**, D'Andrade and Forrest have examined the excimer formation and decay kinetics for **FPt** in the solid state in both neat and doped thin films [7d]. Both **FPt** monomer and excimer emission were observed under these conditions at room temperature and 10 K, with the proportion of emission from the monomer increasing with decreasing temperature. The behavior is consistent with only a small fraction of the **FPt** molecules being in the appropriate disposition to generate excimer emission upon excitation of individual **FPt** molecules. Energy from an **FPt*** molecule in the solid state migrates via exciton diffusion through the lattice by intermolecular energy transfer before being trapped by an **FPt** "excimeric" dimer, as opposed to the situation in fluid solution, where excimer formation occurs by physical diffusion of an **FPt*** molecule. In a neat **FPt** film, the lifetime of monomeric emission ($\lambda = 470$ nm) decreases from $\tau = 8.7$ μs at 5 K to $\tau = 250$ ns at 220 K [7d]. Exciton quenching of **FPt*** at a relatively small number of excimer sites presumably contributes to the decrease in lifetime with increasing temperature in the solid state. At higher temperatures, the **FPt*** exciton has a larger diffusion radius (due to thermally activated intermolecular energy transfer) and is more likely to encounter an excimer formation site prior to unimolecular decay. Since the processes for generating **FPt** excimer emission in solution and the solid state are very different, it is not proper to compare the parameters determined for the rates of excimer formation in the two systems. It is, however, useful to compare monomer and excimer decay rates in the two systems. In fluid solution, the room temperature lifetime of **FPt*** decreases from 330 ns at infinite dilution to 60 ns at a concentration of 3 mM. The low barrier for physical diffusion of **FPt*** in fluid solution leads to **FPt*** quenching by excimer formation at a markedly faster rate than does **FPt*** exciton trapping by "excimeric" dimers in the solid state. The lifetime for **FPt** excimer emission in the neat solid (corrected to eliminate the weak monomer contribution) is nearly temperature independent, decreasing from 1.7 μs at 100 K to 1.3 μs at 220 K. On the other hand, the **FPt** excimer lifetime measured in 0.25 mM solution is shorter (135 ns) and consistent with non-radiative decay channels being present in fluid solution that are inaccessible in the neat solid.

29.3.2 Excited State Properties

29.3.2.1 Redox properties estimated from electrochemistry and spectra

Cyclic voltammetry of **FPt** shows a single reversible reduction at -2.29 V versus Cp_2Fe^+/Cp_2Fe (-1.84 V versus SCE) and an irreversible oxidative wave at 0.52 V versus Cp_2Fe^+/Cp_2Fe (0.97 V versus SCE) [5]. The reduction is considered to be localized on the cyclometalated ligand while the oxidation is thought to occur predominantly at the metal center [5]. Oxidation of square planar Pt^{II} complexes is typically irreversible due to rapid solvolysis of the resultant Pt^{III} species [19]. While differential solvation of the cationic and anionic forms of **FPt** may affect the potentials measured here, the difference in oxidation and reduction potentials, 2.81 eV, can be used as an estimate of the HOMO–LUMO energy gap. An alternate method for evaluating the HOMO–LUMO separation is to use low energy edge of the ^{1}MLCT transition in the absorption spectrum of **FPt** (Fig. 29.1), which occurs at 3.1 eV (∼400 nm). This spectroscopic method for estimating the HOMO–LUMO gap may also give an inaccurate value, due to solvation, correlation and reorganization effects, thus both electrochemical and spectroscopic methods should be treated as estimates [20]. The difference of ca. 0.3 eV between the two estimates of the HOMO–LUMO energy gap for **FPt**, combined with the irreversible nature of the electrochemical oxidation process, creates considerable uncertainty in the value of the thermodynamic oxidation potential of **FPt**. Fortunately, a thermochemical cycle (Scheme 29.3) can be constructed to estimate the ground state redox potentials from the excited state redox properties and the luminescence data. Therefore, in order to gain a more quantitative picture of the ground and excited state energetics, a number of bimolecular quenching studies were carried out with the **FPt** complex.

29.3.2.2 Energy transfer quenching

A range of organic molecules with different triplet energies were used as triplet energy transfer quenchers in this study (Table 29.1). The redox properties of the chosen quenchers make electron transfer unlikely, based on a preliminary estimate of the excited state redox properties of **FPt**. The rate constants for energy transfer quenching by each of the quenchers were determined from the slopes of the Stern–Volmer plots and are listed in Table 29.1. A semilog plot of K_q^{en} versus E_T (the triplet energy of a given quencher) shows a marked dependence of the quenching rate on

Table 29.1 Rate constants for energy transfer quenching of **FPt**

Triplet quenchers	$E(D^+/D)^a$	$E(A^+/A)^{0\,b}$	E_T (eV)^c	ΔG (eV)^d	K_q (M^{-1} s^{-1})^e
Anthracene	1.16	−1.93	1.85	−0.86	7.22×10^9
trans-Stilbene		−2.26	2.12	−0.59	5.75×10^9
1-Cyanonaphthalene		−1.98	2.49	−0.22	5.96×10^9
2-Methoxynaphthalene	1.42		2.62	−0.09	3.72×10^9
Naphthalene	1.60	−2.29	2.64	−0.07	4.23×10^9
Phenanthrene	1.58	−2.20	2.69	−0.02	2.66×10^9
Triphenylene	1.64	−2.22	2.89	0.18	6.66×10^7
Fluorene	1.55		2.94	0.23	2.99×10^7
Methyl-p-cyanobenzoate		−1.76	3.12	0.41	$<10^6$

^{a}Oxidation potential (vs. SCE, values from [12a]).
^{b}Reduction potential (vs. SCE, values from [12a]).
^{c}Triplet energy (values from [12a]).
^{d}Free energy change for energy transfer (values calculated using $E^{00} = 2.71$ eV).
^{e}Bimolecular quenching rate constant.

the triplet energy of the quencher (Fig. 29.5a). Using the kinetic analysis discussed in the experimental section, the best fit to the experimental energy transfer quenching data gives an E^{00} value of 2.80 eV. The triplet energy of **FPt** obtained in this way is very close to the E^{00} value of 2.71 eV determined using the highest energy emission peak in a 2-MeTHF solution at 77 K (458 nm) [21]. This study also emphasizes the importance of using high triplet energy host materials in order to prevent luminescent quenching by energy transfer and thereby maximize efficiencies in phosphor-doped OLEDs [22].

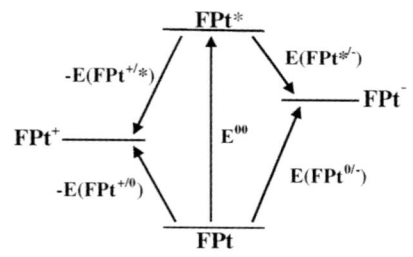

$$\text{FPt}^*$$

$$-E(\text{FPt}^{+/*}) \quad E(\text{FPt}^{*/-})$$

$$\text{FPt}^+ \quad\quad\quad E^{00} \quad\quad\quad \text{FPt}^-$$

$$-E(\text{FPt}^{+/0}) \quad E(\text{FPt}^{0/-})$$

$$\text{FPt}$$

Excited state reduction potential: $E(\text{FPt}^{*/-}) = E^{00} + E(\text{FPt}^{0/-})$
Excited state oxidation potential: $-E(\text{FPt}^{+/*}) = E^{00} - E(\text{FPt}^{+/0})$

Scheme 29.3

29.3.2.3 Reductive electron transfer quenching

Amines have proven to be excellent reductive electron transfer quenchers in the studies of Pt complex excited states [23]. The high triplet energies of these amines, as well as their high reduction potentials (Table 29.2), make both energy transfer and oxidative electron transfer quenching of the **FPt** excited state unfavorable. The reductive quenching rate constants (from Stern–Volmer analysis) for each of the amines are given in Table 29.2 and plotted in Fig. 29.5b versus the oxidation potential of each quencher. Increasing the oxidation potential of the quencher leads to a pronounced decrease in the reductive quenching rate. Applying the model for electron transfer quenching discussed in the experimental section, the best fit to the experimental data for the excited-state reduction potential is

Table 29.2 Rate constants for reductive electron transfer quenching of **FPt**

Reductive quenchers	$E(D^+/D)$ (V)[a]	E_T (eV)[b]	K_q (M^{-1} s^{-1})[c]
N,N,N',N'-Tetramethyl-1,4-benzenediamine	0.12		10.6×10^9
N,N,N',N'-Tetramethylbenzidine	0.43	2.73	8.21×10^9
1,4-Diazabicyclo[2.2.2]octane	0.60	>3.9	1.16×10^9
N,N,N',N'-Tetramethyldiaminoethane	0.67	>3.9	4.28×10^8
N,N-Dimethyl-p-toluidene	0.71	3.1	2.19×10^8
Triethylamine	0.78	>3.9	2.21×10^7

[a] Potential vs. SCE (values from [23]).
[b] Triplet energy (values from [12a]).
[c] Bimolecular quenching rate constant.

Table 29.3 Rate constants for oxidative electron transfer quenching of **FPt**

Oxidative quenchers[a]	$E(A^+/A)$ (V)[b]	K_q (M^{-1} s^{-1})[c]
N,N'-Dimethyl-4,4'-pyridinium	−0.45	5.42×10^9
4-Amido-N-ethylpyridinium	−0.93	4.56×10^9
3-Amido-N-methylpyridinium	−1.14	1.63×10^9
N-Ethylpyridinium	−1.37	2.28×10^7

[a] Hexafluorophosphate salts.
[b] Potential vs. SCE (values from [25]).
[c] Bimolecular quenching rate constant.

0.81 V versus SCE. This value is very close to one estimated from the E^{00} energy and the ground state reduction potential using Scheme 29.3 (0.87 V versus SCE). This is to be expected, since the reduction is completely reversible and the phosphorescence energy is similar to the triplet energy determined in the quenching studies.

The effects of both energy transfer and electron transfer quenching need to be considered when employing **FPt** as the dopant in OLEDs. Since **FPt** has a high triplet energy and a high excited-state oxidation potential, an appropriate host material should exceed both parameters to maximize the luminescent efficiency of a device. A good demonstration of the problem is seen for mixtures of **FPt** and N,N'-diphenyl-N,N'-di(3-toly)benzidine (TPD), a common hole transporter for OLEDs. TPD has a triplet energy of 2.3 eV and an

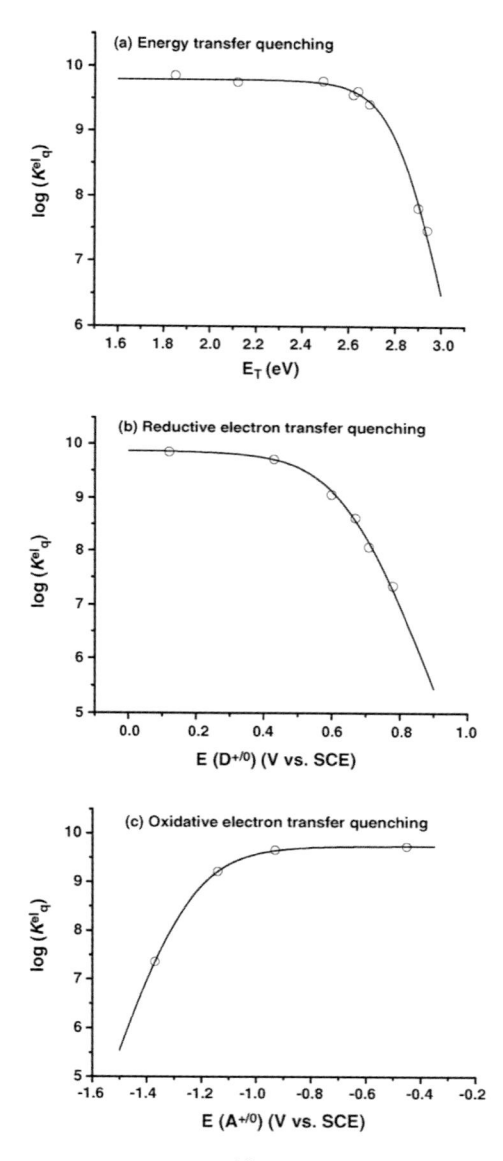

Figure 29.5 Plots of the log($K_q^{en(el)}$) vs. E_T (Q*, Q), $E_{1/2}^{Ox}$ and $E(A^{+/0})$ for energy transfer quenching (a), reductive electron transfer quenching (b) and oxidative electron transfer quenching (c) of **FPt***, respectively, by organic molecules in 2-MeTHF at room temperature. Circles represent experimental data, solid lines are best fitting curves.

oxidation potential of 0.73 V versus SCE. As shown in Fig. 29.6, thin films of TPD doped with 1 wt.% **FPt** do not show any monomer **FPt** emission, rather TPD fluorescence and **FPt** excimer-like emission are observed instead [24]. This quenching of the **FPt** monomer emission is to be expected when one considers the differences in electrochemical potentials and triplet energies between the two host materials. Reductive electron transfer reaction of **FPt** is spontaneous with a 0.08 V driving force, whereas energy transfer to TPD is exergonic by nearly 0.4 eV. On the other hand, the **FPt** excimer-like state in the doped TPD film is not quenched. The lower triplet energy, and consequently less positive excited-state oxidation potential, of the excimer are endogonic with respect to the TPD triplet state energy and reduction potentials. Hence, if **FPt** is doped into a material with a higher oxidation potential and higher triplet energy, such as CBP (CBP = 4,4′-N,N-dicarbazole-biphenyl, E_{oxid} = 0.98 V versus SCE, E_T = 2.6 eV), both **FPt** monomer and excimer-like luminescence are observed from doped thin films (see Fig. 29.6).

29.3.2.4 Oxidative electron transfer quenching

Pyridinium acceptors have previously been used for oxidative quenching experiments [25]. These compounds have both high

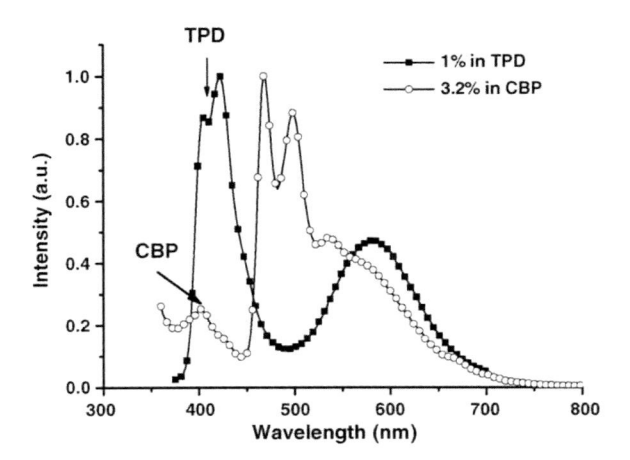

Figure 29.6 Photoluminescence spectra of ∼1% **FPt** doped into TPD and 3.2% **FPt** doped into CBP.

triplet energies and oxidation potentials [26], precluding energy transfer or reductive quenching processes from occurring with most donor compounds. The quenching rate constants of **FPt*** by these pyridinium quenchers are presented in Table 29.3. A plot of log K_q^{el} versus $E(A^{+/0})$ for the oxidative quenching is shown in Fig. 29.5c. The curve shown in Fig. 29.5c is the best fit to the experimental data using Eq. (29.3) and gives an excited-state oxidation potential of -1.41 V versus SCE. In the fitting process used here, the only parameter being adjusted to achieve the best fit is the excited-state oxidation potential (see previous discussion). Application of the thermochemical cycle in Scheme 29.3 gives a value of 1.30 V versus SCE for the ground state oxidation potential of **FPt**. This value calculated from the excited-state reduction potential, when combined with the reversible reduction potential of -1.84 V versus SCE determined by cyclic voltammetry, gives a HOMO–LUMO energy gap in 3.14 eV. The energy of the HOMO–LUMO gap (395 nm) obtained in this manner closely corresponds to the onset of the ^{1}MLCT transition in the absorption spectrum. Therefore, we can conclude that thermodynamic ground state oxidation potential for **FPt** is nearer 1.30 V versus SCE, rather than the value of 0.97 V versus SCE obtained using electrochemical methods.

29.4 Conclusions

We have conducted a series of luminescent quenching studies of the molecule **FPt** in solution phase at room temperature. The formation of red shifted emissive phosphorescent triplet excimers from the self-quenched triplet excited state monomers has been observed in 2-MeTHF solution at room temperature. A rate constant of $4.2 \pm 0.3 \times 10^9$ M^{-1} s^{-1} for the excimer formation has been determined from the monomer self-quenching behavior. The intrinsic lifetime for the excited state of the monomer was determined to be 330 ns (± 15 ns) at infinite dilution, while the excimer lifetime was found to be ca. 135 ns (± 10 ns) for a 0.25 mM solution.

Energy transfer and electron transfer quenching studies were carried out to determine the photoredox properties of excited state of the **FPt** complex. Values obtained for the triplet energy

(2.80 eV) and the excited-state reduction potential (0.81 V versus SCE) from these studies are in accord with values estimated using the E^{00} energy from the emission spectra and a thermodynamic cycle (Scheme 29.3) based on the electrochemical and spectral properties of the **FPt** complex. An excited-state reduction potential of -1.41 V versus SCE was also obtained by an oxidative electron transfer quenching study. This value does not concur with the value predicted from electrochemical and spectroscopic measurements due to the irreversibility of the electrochemical oxidation. However, using the thermodynamic cycle and the E^{00} energy from the emission spectrum, the ground state oxidation potential for **FPt** is estimated to be near 1.30 V versus SCE. This value for the oxidation potential of **FPt** is a consistent with the value calculated using the onset of the ^{1}MLCT transition and the reversible reduction potential of the complex. With a good picture of the excited state energy and redox properties in hand, we expect to be able to design OLEDs with higher efficiencies, as well as high efficiency solar cells, using **FPt** as the light emitter or absorber, respectively.

Acknowledgements

This work was supported by Universal Display Corp. and the National Science Foundation. We thank Dr. Jason Brooks, Dr. Bert Alleyne, Jian Li, and Arnold Tamayo for helpful discussions. Biwu Ma would like to thank USC School of Engineering for Dean's Fellowship support.

References

1. (a) C.S. Peyratout, T.K. Aldridge, D.K. Crites, D.R. McMillin, *Inorg. Chem.* **34** (1995) 4484; (b) A.H.-J. Wang, J. Nathans, G.A. Van Der Marel, J.H. Van Boom, A. Rich, *Nature* **276** (1978) 471; (c) V.H. Houlding, A.J. Frank, *Inorg. Chem.* **24** (1985) 3664; (d) T. Maruyama, T. Yamamoto, *J. Phys. Chem. B* **101** (1997) 3806; (e) V. Anbalagan, T.S. Srivastava, *J. Photochem. Photobiol. A: Chem.* **89** (1995) 113.

2. (a) D.M. Roundhill, H.B. Gray, C.M. Che, *Acc. Chem. Res.* **22** (1989) 55; (b) D.G. Nocera, *Acc. Chem. Res.* **28** (1995) 209; (c) W. Paw, S.D. Cummings, M.A. Mansour, W.B. Connick, D.K. Geiger, R. Eisenberg, *Coord. Chem. Rev.* **171** (1998) 125.
3. (a) M. Hissler, J.E. McGarrah, W.B. Connick, D.K. Geiger, S.D. Cummings, R. Eisenberg, *Coord. Chem. Rev.* **208** (2000) 115; (b) J.E. McGarrah, Y.-J. Kim, M. Hissler, R. Eisenberg, *Inorg. Chem.* **40** (2001) 4510; (c) E.M. James, R. Eisenberg, *Inorg. Chem.* **42** (2003) 4355.
4. (a) M.A. Baldo, D.F. O'Brien, Y. You, A. Shoustikov, S. Sibley, M.E. Thompson, S.R. Forrest, *Nature* **395** (1998) 151; (b) R.C. Kwong, S. Sibley, T. Dubovoy, M.A. Baldo, S.R. Forrest, M.E. Thompson, *Chem. Mater.* **11** (1999) 3709; (c) D.R. O'Brien, M.A. Baldo, M.E. Thompson, S.R. Forrest, *Appl. Phys. Lett.* **74** (1999) 442; (d) C. Adachi, M.A. Baldo, S.R. Forrest, S. Lamansky, M.E. Thompson, R.C. Kwong, *Appl. Phys. Lett.* **78** (2001) 1622; (e) V. Cleave, G. Yahioglu, P.L. Barny, R.H. Friend, N. Tessler, *Adv. Mater.* **11** (1999) 285; (f) Y. Kunugi, K.R. Mann, L.L. Miller, C.L. Exstrom, *J. Am. Chem. Soc.* **120** (1998) 589; (g) S.-C. Chan, M.C. Chan, Y. Wang, C.M. Che, K.K. Cheung, N. Zhu, *Chem. Eur. J.* **7** (2001) 4180; (h) W. Lu, B.X. Mi, M.C.W. Chan, Z. Hui, N. Zhu, S.T. Lee, C.M. Che, *Chem. Commun.* (2002) 206.
5. J. Brooks, Y. Babayan, S. Lamansky, P.I. Djurovich, I. Tsyba, R. Bau, M.E. Thompson, *Inorg. Chem.* **41** (2002) 3055.
6. (a) M.A. Baldo, D.F. O'Brien, M.E. Thompson, S.R. Forrest, *Phys. Rev. B* **60** (1999) 14422; (b) M.A. Baldo, S. Lamansky, P.E. Burrows, M.E. Thompson, S.R. Forrest, *Appl. Phys. Lett.* **75** (1999) 4; (c) C. Adachi, M.A. Baldo, S.R. Forrest, M.E. Thompson, *Appl. Phys. Lett.* **77** (2000) 904; (d) M. Ikai, S. Tokito, Y. Sakamoto, T. Suzuki, Y. Taga, *Appl. Phys. Lett.* **79** (2001) 156.
7. (a) B. D'Andrade, M.E. Thompson, S.R. Forrest, *Adv. Mater.* **14** (2002) 147; (b) B. D'Andrade, J. Brook, V. Adamovich, M.E. Thompson, S.R. Forrest, *Adv. Mater.* **14** (2002) 1032; (c) V. Adamovich, J. Brooks, A. Tamayo, A.M. Alexander, P.I. Djurovich, B.D. Andrade, C. Adachi, S.R. Forrest, M.E. Thompson, *New J. Chem.* **9** (2002) 1171; (d) B. D'Andrade, S.R. Forrest, *J. Chem. Phys.* **286** (2003) 321.
8. (a) C.M. Che, K.T. Wan, L.Y. He, C.K. Poon, V.W.-W. Yam, *J. Chem. Soc., Chem. Commun.* (1989) 943; (b) C.W. Chan, C.M. Che, M.C. Cheng, Y. Wang, *Inorg. Chem.* **31** (1992) 4874; (c) W.B. Connick, H.B. Gray, *J. Am. Chem. Soc.* **119** (1997) 11620.
9. (a) H. Kunkely, V. Vogler, *J. Am. Chem. Soc.* **112** (1990) 5625; (b) K.-T. Wan, C.M. Che, K.C. Cho, *J. Chem. Soc., Dalton Trans.* (1991) 1077; (c) C.W.

Chan, T.F. Lai, C.M. Che, S.M. Peng, *J. Am. Chem. Soc.* **115** (1993) 11245;
(d) H.K. Yip, K.K. Cheng, S.M. Peng, C.M. Che, *J. Chem. Soc., Dalton Trans.*
(1993) 2933; (e) C.N. Pettijohn, E.B. Jochnowitz, B. Chuong, J.K. Nagle,
A. Vogler, *Coord. Chem. Rev.* **171** (1998) 85; (f) W.B. Connick, D. Geiger,
R. Eisenberg, *Inorg. Chem.* **38** (1999) 3264; (g) M. Hissler, W.B. Connick,
D.K. Geiger, J.E. McGarrah, D. Lipa, R.J. Lachicotte, R. Eisenberg, *Inorg.
Chem.* **39** (2000) 447; (h) M. Hissler, J.E. McGarrah, W.B. Connick, D.K.
Geiger, S.D. Cummings, R. Eisenberg, *Coord. Chem. Rev.* **208** (2000) 115;
(i) W.L. Fleeman, W.B. Connick, Comments *Inorg Chem.* **23** (2002) 205;
(j) J.A.G. Williams, A. Beeby, E.S. Davies, J.A. Weinstein, C. Wilson, *Inorg.
Chem.* **42** (2003) 8609.

10. (a) J.B. Birks, D.J. Dyson, I.H. Munro, *Proc. R. Soc. A* **275** (1963) 575; (b) J.B. Birks, *Photophysics of Aromatic Molecules*, University of Manchester, 1970.

11. D.M. Roundhill, *Photochemistry and Photophysics of Metal Complexes*, Plenum Press, New York, 1994, p. 17.

12. (a) J.K. George, N.J. Turro, *Chem. Rev.* **86** (1986) 401, and reference therein; (b) V. Balzani, F. Bolletta, F. Scandola, *J. Am. Chem. Soc.* **102** (1980) 2152; (c) B. Marciniak, G.L. Hug, *Coord. Chem. Rev.* **159** (1997) 55, and reference therein.

13. (a) R.R. Gagne, C.A. Koval, G.C. Lisensky, *Inorg. Chem.* **19** (1980) 2854; (b) D.T. Sawyer, A. Sobkowiak, J.L. Roberts, *Electrochemistry for Chemists*, second ed., Wiley, New York, 1995, p. 467.

14. (a) N. Agmon, R.D. Levine, *Chem. Phys. Lett.* **52** (1977) 197; (b) N. Agmon, *J. Chem. Soc., Faraday Trans.* **74** (1978) 388.

15. (a) J. Saltiel, B.W. Atwater, in: D.H. Volman, G.S. Hammond, K. Gollnick (Eds.), *Advances in Photochemistry*, vol. 14, Wiley, New York, 1988 ($K_d = 8000RT/3\eta$; Debye equation); (b) M. Eigen, *Z. Phys. Chem. (Leipzig)* **203** (1954) 176 ($K_{-d} = K_d (3000/4\pi R^3 N_0)$, η is the solvent viscosity, R is the reaction radius (taken as 7×10^{-8} cm) and N_0 is the Avogadro number).

16. (a) G.L. Hug, B. Marciniak, *J. Phys. Chem.* **98** (1994) 7523; (b) F. Wilkinson, C. Tsiamis, *J. Am. Chem. Soc.* **105** (1983) 767; (c) B. Marciniak, H.G. Lohmannsroben, *Chem. Phys. Lett.* **148** (1988) 29.

17. (a) H. Knibbe, D. Rehm, A. Weller, B. Bunsenges, *Phys. Chem.* **72** (1968) 257; (b) A.Z. Weller, *Phys. Chem. (Wiesbaden)* **133** (1982) 93.

18. The radiative (k_r) and non-radiative (k_{nr}) rate constants are determined from the emission lifetime (τ) and quantum efficiency (ϕ_{em}) using the following equations: $k_r = \phi_{em} \times \tau^{-1}$ and $k_{nr} = k_r(\phi_{em}^{-1} - 1)$.

19. P.I. Kvam, M.V. Puzyk, K.P. Balashev, J. Songstad, *Acta Chem. Scand.* **49** (1995) 335.
20. A.B.P. Lever, E.S. Dodsworth, in: E.I. Solomon, A.B.P. Lever (Eds.), *Inorganic Electronic Structure and Spectroscopy*, vol. 2, Wiley, New York, 1999, p. 227.
21. N.J. Turro, *Modern Molecular Photochemistry*, University Science Books, Sausalito, 1991.
22. (a) M. Sudhakar, P.I. Djurovich, T.E. Hogen-Esch, M.E. Thompson, *J. Am. Chem. Soc.* **125** (2003) 7796; (b) M.A. Baldo, S.R. Forrest, *Phys. Rev. B* **62** (2000) 10958.
23. (a) J.M. Bevilacqua, R. Eisenberg, *Inorg. Chem.* **33** (1994) 1886; (b) N. Kitamura, H.B. Kim, S. Okano, S. Tazuke, *J. Phys. Chem.* **93** (1989) 5750; (c) D. Sandrini, M. Maestri, P. Belser, A. Von Zelewsky, V. Balzani, *J. Phys. Chem.* **89** (1985) 3675; (d) C.R. Bock, J.A. Connor, A.R. Gutierrez, T.J. Meyer, D.G. Whitten, B.P. Sullivan, J.K. Nagle, *J. Am. Chem. Soc.* **101** (1979) 4815; (e) B. Ballardini, G. Varani, M.T. Indelli, F. Scandola, V. Balzani, *J. Am. Chem. Soc.* **100** (1978) 7219; (f) W. Hub, S. Schneider, F. Doerr, J.D. Oxman, F.D. Lewis, *J. Am. Chem. Soc.* **106** (1984) 701.
24. J. Brooks, Ph.D. thesis, University of Southern California, 2002.
25. (a) J.L. Marshall, S.R. Stobart, H.B. Gray, *J. Am. Chem. Soc.* **106** (1984) 3027; (b) V.W. Yam, K.K. Lo, K.K. Cheung, *Inorg. Chem.* **35** (1996) 3459; (c) C.M. Che, H.Y. Chao, V.M. Miskowski, Y. Li, K.K. Cheung, *J. Am. Chem. Soc.* **123** (2001) 4985; (d) H.Y. Chao, W. Lu, Y. Li, C.W. Chan, C.M. Che, K.K. Cheung, N. Zhu, *J. Am. Chem. Soc.* **124** (2002) 14696.
26. A. Ledwith, *Acc. Chem. Res.* **5** (1972) 133.

Chapter 30

Synthetic Control of Excited-State Properties in Cyclometalated Ir(III) Complexes Using Ancillary Ligands

Jian Li,[a] Peter I. Djurovich,[a] Bert D. Alleyne,[a]
Muhammed Yousufuddin,[a] Nam N. Ho,[a]
J. Christopher Thomas,[b] Jonas C. Peters,[b]
Robert Bau,[a] and Mark E. Thompson[a]

[a]*University of Southern California, Department of Chemistry,
Los Angeles, California 90089, USA*
[b]*Division of Chemistry and Chemical Engineering,
Arnold and Mabel Beckman Laboratories of Chemical Synthesis,
California Institute of Technology, Pasadena, California 91125, USA*
met@usc.edu

The synthesis and photophysical characterization of a series of $(N,C^{2'}$-(2-*para*-tolylpyridyl$))_2$Ir(LL′) [$(tpy)_2$Ir(LL′)] (LL′ = 2,4-pentanedionato (acac), bis(pyrazolyl)borate ligands and their analogues, diphosphine chelates and *tert*-butylisocyanide (CN-*t*-Bu)) are reported. A smaller series of [$(dfppy)_2$Ir(LL′)] ($dfppy$ = N, $C^{2'}$-2-(4′,6′-difluorophenyl)pyridyl) complexes were also examined

---

Reprinted from *Inorg. Chem.*, **44**(6), 1713–1727, 2005.

Electrophosphorescent Materials and Devices
Edited by Mark E. Thompson
Text Copyright © 2005 American Chemical Society
Layout Copyright © 2024 Jenny Stanford Publishing Pte. Ltd.
ISBN 978-981-4877-34-3 (Hardcover), 978-1-003-08872-1 (eBook)
www.jennystanford.com

along with two previously reported compounds, $(ppy)_2Ir(CN)_2^-$ and $(ppy)_2Ir(NCS)_2^-$ ($ppy = N,C^{2'}$-2-phenylpyridyl). The $(tpy)_2Ir$ $(PPh_2CH_2)2BPh_2$ and $[(tpy)_2Ir(CN\text{-}t\text{-}Bu)_2](CF_3SO_3)$ complexes have been structurally characterized by X-ray crystallography. The Ir−C$_{aryl}$ bond lengths in $(tpy)_2Ir(CN\text{-}t\text{-}Bu)_2^+$ (2.047(5) and 2.072(5) Å) and $(tpy)_2Ir(PPh_2CH_2)_2BPh_2$ (2.047(9) and 2.057(9) Å) are longer than their counterparts in $(tpy)_2Ir(acac)$ (1.982(6) and 1.985(7) Å). Density functional theory calculations carried out on $(ppy)_2Ir(CN\text{-}Me)_2^+$ show that the highest occupied molecular orbital (HOMO) consists of a mixture of phenyl-π and Ir-d orbitals, while the lowest unoccupied molecular orbital is localized primarily on the pyridyl-π orbitals. Electrochemical analysis of the $(tpy)_2Ir(LL')$ complexes shows that the reduction potentials are largely unaffected by variation in the ancillary ligand, whereas the oxidation potentials vary over a much wider range (as much as 400 mV between two different LL' ligands). Spectroscopic analysis of the cyclometalated Ir complexes reveals that the lowest energy excited state (T_1) is a triplet ligand-centered state (^{3}LC) on the cyclometalating ligand admixed with ^{1}MLCT (MLCT = metal-to-ligand charge-transfer) character. The different ancillary ligands alter the ^{1}MLCT state energy mainly by changing the HOMO energy. Destabilization of the ^{1}MLCT state results in less ^{1}MLCT character mixed into the T_1 state, which in turn leads to an increase in the emission energy. The increase in emission energy leads to a linear decrease in $\ln(k_{nr})$ (k_{nr} = nonradiative decay rate). Decreased ^{1}MLCT character in the T_1 state also increases the Huang–Rhys factors in the emission spectra, decreases the extinction coefficient of the T_1 transition, and consequently decreases the radiative decay rates (k_r). Overall, the luminescence quantum yields decline with increasing emission energies. A linear dependence of the radiative decay rate (k_r) or extinction coefficient (ϵ) on $(1/\Delta E)^2$ has been demonstrated, where ΔE is the energy difference between the ^{1}MLCT and ^{3}LC transitions. A value of 200 cm^{-1} for the spin–orbital coupling matrix element $\langle ^3LC|H_{SO}|^1MLCT\rangle$ of the $(tpy)_2Ir(LL')$ complexes can be deduced from this linear relationship. The $(fppy)_2Ir(LL')$ complexes with corresponding ancillary ligands display similar trends in excited-state properties.

30.1 Introduction

During the last two decades, luminescent cyclometalated Ir(III) complexes have exhibited an enormous potential for a range of photonic applications. For example, these Ir complexes can be used as emissive dopants in organic light emitting devices (OLEDs) [1, 2], sensitizers for outer-sphere electron-transfer reactions [3, 4], photocatalysts for CO_2 reduction [5, 6], photooxidants and singlet oxygen sensitizers [7], and biological labeling reagents [8]. Since the optical properties and related uses of the cyclometalated Ir complexes are strongly dependent on the characteristics of their ground and lowest excited states, it becomes desirable to better understand the interactions between these states and thus determine how to systematically alter the photophysical properties by appropriate ligand design.

Several research groups have established that luminescence from cyclometalated Ir(III) and Rh(III) complexes originates from a lowest triplet excited (T_1) state that is ligand centered (^{3}LC) with singlet metal-to-ligand charge-transfer (^{1}MLCT) character mixed in through spin–orbit coupling [9–13]. The admixture of ^{1}MLCT character into what is principally a ^{3}LC state has dramatic effects on the optical properties of those complexes [9], including a large decrease in the luminescence lifetimes and the appearance of metal–ligand vibrational sidebands in absorption and luminescence spectra [13, 14]. In addition, since the lowest excited state is ^{3}LC-dominant, employing different cyclometalating ligands enables the excited-state energy of Ir complexes to be varied over a wide spectral range. Thus, through careful selection of ligands, it is possible to "tune" the emission color from red to blue [15–22]. Interesting questions still remain regarding the luminescent properties of the cyclometalated Ir complexes; e.g., to what extent can the admixture of ^{3}LC and ^{1}MLCT states be controlled, and how does a change in the admixture influence the excited-state properties of Ir complexes? Several studies have investigated the use of electron donating and withdrawing groups on the cyclometalating ligand in order to raise or lower the MLCT emission energy [14]. However, the relative positions between ^{1}MLCT and ^{3}LC state energies will not systematically vary by modifying the cyclometalating ligand, since altering the cyclometalate will influence both state energies simultaneously [14].

To investigate the interactions between ^{1}MLCT and ^{3}LC states on the excited-state properties of Ir cyclometalates, we have prepared a series of bis-cyclometalated Ir(III) complexes having the same cyclometalating ligand, either $N,C^{2'}$-(2'-para-tolylpyridyl) (*tpy*) or $N,C^{2'}$-2-(4',6'-difluorophenyl)pyridyl (*dfppy*), and different ancillary ligands (LL'). Two recently reported anionic complexes with $N,C^{2'}$-2-phenylpyridyl (*ppy*) ligands were also examined [18]. The structures and abbreviations for the cyclometalated complexes reported here, numbered **1–12** for (*tpy*)$_2$Ir(LL') and **13–14** for (*ppy*)$_2$Ir(LL'), are listed in Fig. 30.1, where $C^\wedge N$ is a general abbreviation for both cyclometalating ligands and LL' is the ancillary ligand. Some selected (*dfppy*)$_2$Ir(LL') complexes were also synthesized with corresponding ancillary ligands. The ancillary ligands form air stable complexes [23–27] and were also chosen to be "nonchromophoric", i.e., to have sufficiently high singlet and triplet energies such that the excited-state properties are dominated by the "$(C^\wedge N)_2$Ir" fragment. Therefore, the energy of the $_3$LC state is expected to be relatively constant for all related $(C^\wedge N)_2$Ir(LL') complexes while the energy of $_1$MLCT states can be altered by varying the electron withdrawing/donating effects of the ancillary ligand. Although the influence of various nonchromophoric ligands on the excited states of polypyridyl complexes of Ru(II) and Os(II) has been previously explored [27, 28], a similar study has not been conducted on cyclometalated Ir(III) derivatives. The degree of metal participation in the ground and excited states of d^6 metal complexes varies greatly depending on whether the metal is from group 8 or 9, e.g., Os(II) and Ir(III), and consequently has a strong influence on the photophysical properties of the complexes [29–31]. For example, luminescence from [Os(*bpy*)$_3$]$^{2+}$ (*bpy* = 2,2'-bipyridyl) originates from a low-energy MLCT state that, despite having a high radiative rate constant ($\sim$10^5 s^{-1}), is efficiently deactivated by metal–ligand nonradiative transitions [32]. In contrast, emission from [Ir(*bpy*)$_3$]$^{3+}$ occurs from a high-energy, ligand-centered state that is effectively quenched because of the low radiative rate for the emissive state ($\sim$10^3 s^{-1}) [33]. In addition, replacing the neutral *bpy* ligand with the formally anionic *ppy* ligand increases the electron density on the metal center, which enhances the MLCT character in the excited states of Ir complexes [6, 14, 34]. These differences between the metal (Os vs Ir) and ligand (*bpy* vs *ppy*) imply that systematic variation of the nonchromophoric ligand in the $(C^\wedge N)_2 Ir(LL')$ derivatives will be essential to generate new insights into the excited-state properties of cyclometalated Ir complexes.

The electrochemical and photophysical properties of the $(tpy)_2 Ir(LL')$ complexes are discussed in detail. The lowest excited state of cyclometalated Ir complexes is identified as a dominant ligand-centered $^3\pi\text{-}\pi^*$ state with minor to significant ^{1}MLCT character [14, 21, 29]. The electrochemical studies of all $(tpy)_2 Ir(LL')$ complexes demonstrate that the ancillary ligands increase the absorption and emission energies of $(tpy)_2 Ir(LL')$ complexes by stabilizing the metal-based highest occupied molecular orbital (HOMO), leaving the lowest unoccupied molecular orbital (LUMO) largely unchanged. Besides increasing the emission energy, the lower HOMO energies increase the energy separation between the ^{1}MLCT and ^{3}LC states, which in turn modify the excited-state properties of the Ir complexes primarily by decreasing the radiative rates.

30.2 Experimental Section

The UV–visible spectra were recorded on a Hewlett-Packard 4853 diode array spectrometer. The IR spectra were obtained on a Perkin-Elmer FTIR spectrometer (model Spectrum 2000). Steady-state emission experiments at room temperature and 77 K were performed on a PTI QuantaMaster model C-60 spectrometer. Quantum efficiency measurements were carried out at room temperature in a 2-methyltetrahydrofuran (2-MeTHF) solution that was distilled over sodium. Before emission spectra were measured, the solutions were degassed by several freeze–pump–thaw cycles using a high-vacuum line equipped with a diffusion pump. Solutions of coumarin 47 (coumarin 1) in ethanol ($\Phi = 0.73$) [35] were used as a reference. The equation $\Phi_s = \Phi_r (\eta_s^2 A_r I_s / \eta_r^2 A_s I_r)$ was used to calculate the quantum yields where Φ_s is the quantum yield of the sample, Φ_r

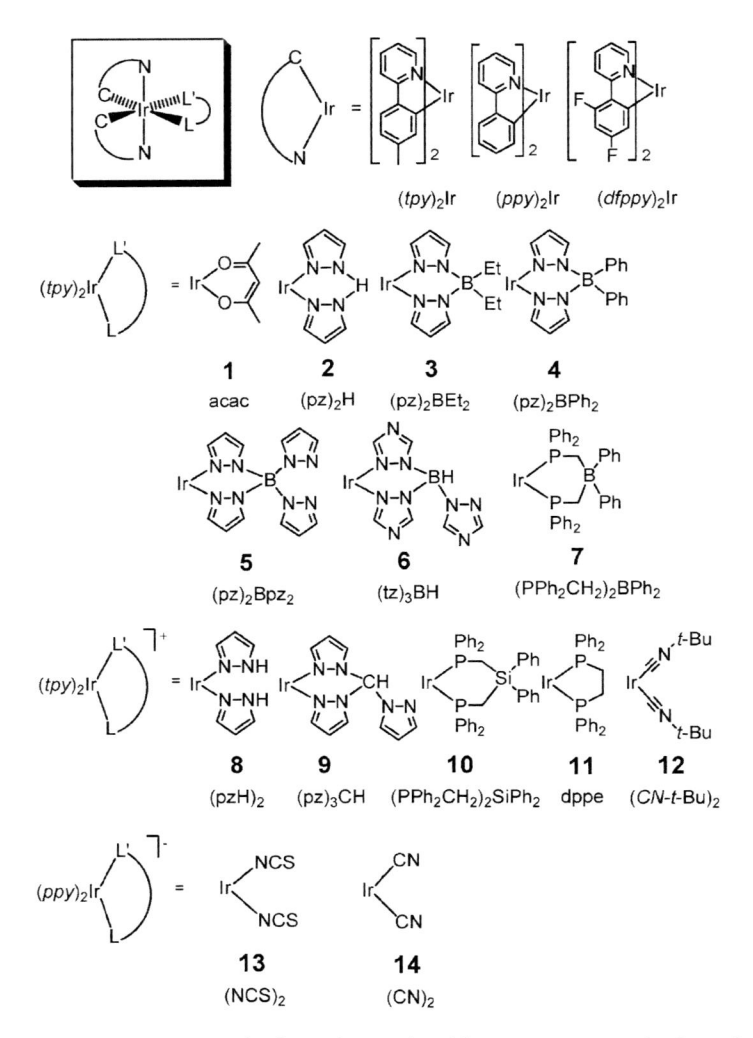

Figure 30.1 Structural formula and abbreviations used for the $(C^\wedge N)_2 \text{Ir}(LL')$ complexes.

is the quantum yield of the reference, η is the refractive index of the solvent, A_s and A_r are the absorbance of the sample and the reference at the wavelength of excitation, and I_s and I_r are the integrated areas of emission bands [36]. Phosphorescence lifetime measurements were performed on an IBH Fluorocube fluorimeter by a time-correlated single photon counting method using either a 373 nm or a 403 nm LED excitation source. NMR spectra were recorded on Bruker AM 360 MHz and AMX 500 MHz instruments, and chemical shifts were referenced to residual protiated solvent. The Microanalysis Laboratory at the University of Illinois, Urbana–Champaign, performed all elemental analysis.

30.2.1 X-Ray Crystallography

X-ray diffraction data were collected on a Bruker SMART APEX CCD diffractometer with graphite-monochromated Mo Kα radiation ($\lambda =$ 0.71073 Å) at 298(2) K for $(tpy)_2$Ir(PPh$_2$CH$_2$)$_2$BPh$_2 \cdot$ H$_2$O and 143(2) K for [$(tpy)_2$Ir(CN-t−Bu)$_2$](CF$_3$SO$_3$)$\cdot$ CHCl$_3$. The cell parameters for the Ir complexes were obtained from the least-squares refinement of spots (from 60 collected frames) using the SMART program. A hemisphere of the crystal data was collected up to a resolution of 0.75 Å, and the intensity data were processed using the Saint Plus program. All calculations for the structure determination were carried out using the SHELXTL package (version 5.1) [37]. Initial atomic positions were located by Patterson methods using XS, and the structure of $(tpy)_2$Ir(PPh$_2$CH$_2$)$_2$BPh$_2 \cdot$ H$_2$O was refined by least-squares methods using SHELX93 with 6983 independent reflections within the range of $\theta = 1.30$–$27.50°$ (completeness 95.2%). The data for [$(tpy)_2$Ir(CN-t-Bu)$_2$](CF$_3$SO$_3$)$\cdot$ CHCl$_3$ were $\theta = 1.63$–$27.47°$ (completeness 96.7%). Absorption corrections were applied by using SADABS [38]. Calculated hydrogen positions were input and refined in a riding manner along with the attached carbons. A summary of the refinement details and resulting factors are given in Table 30.1.

30.2.2 Density Functional Theory Calculation

Density functional theory (DFT) calculations were performed using the Titan software package (Wavefunction, Inc.) at the B3LYP/LACVP** level. The HOMO and LUMO energies were determined using a minimized singlet geometry to approximate the ground state.

Table 30.1 Crystal data and summary of intensity data collection and structure refinement for $(tpy)_2Ir(PPh_2CH_2)_2BPh_2 \cdot H_2O$ and $(tpy)_2Ir(CN\text{-}t\text{-}Bu)_2(CF_3SO_3) \cdot CHCl_3$

	$(tpy)_2Ir(PPh_2CH_2)_2$ $BPh_2 \bullet H_2O$	$[(tpy)_2Ir(CN\text{-}t\text{-}Bu)_2](CF_3SO_3) \bullet CHCl_3$
empirical formula	$C_{62}H_{56}BIrN_2OP_2$	$C_{36}H_{39}Cl_3F_3IrN_4O_3S$
formula weight	1110.10	963.32
temperature, K	298(2)	143(2)
wavelength, Å	0.71073	0.71073
crystal system	Monoclinic	Monoclinic
space group	$P2(1)/n$	$P2(1)/n$
unit cell dimensions		
a (Å)	17.6248(14)	13.655(3)
b (Å)	13.6990(11)	13.262(3)
c (Å)	23.8155(19)	22.981(5)
α (deg)	90	90
β (deg)	104.034(2)	98.624(4)
γ (deg)	90	90
volume, Å^3	5578.4(8)	4114.6(15)
Z	4	4
d_{calcd}, Mg/m^3	1.365	1.555
abs coeff, mm^{-1}	2.494	3.542
$F(000)$	2316	1912
θ range for data collection, deg	1.30–27.50	1.63–27.47
reflns collected	32905	24489
indep reflns	12187 [R(int) = 0.0579]	9126 [R(int)) 0.0532]
refinement method	full-matrix least-squares on F^2	full-matrix least-squares on F^2
data/restraints/params	12187/0/625	9126/0/468
goodness-of-fit on F^2	0.943	1.046
final R indices [$I > 2\sigma(I)$]	0.0652	0.0433
R indices (all data)	0.0882	0.0588

30.2.3 Electrochemistry

Cyclic voltammetry and differential pulsed voltammetry were performed using an EG&G potentiostat/galvanostat model 283. Anhydrous DMF (Aldrich) was used as the solvent under a nitrogen atmosphere, and 0.1 M tetra(n-butyl)ammonium hexafluorophosphate was used as the supporting electrolyte. A silver wire was used as the *pseudo*-reference electrode, a Pt wire was used as the counter electrode, and glassy carbon was used as the working electrode. The redox potentials are based on the values measured from differential pulsed voltammetry and are reported relative to a ferrocenium/ferrocene (Fc^+/Fc) redox couple used as an internal reference (0.45 V vs saturated calomel electrode (SCE)) [39]. The reversibility of reduction or oxidation was determined using cyclic voltammetry [40]. As defined, if peak anodic and peak cathodic currents have an equal magnitude under the conditions of fast scan (100 mV/s or above) and slow scan (50 mV/s), then the process is *reversible*; if the magnitudes in peak anodic and peak cathodic currents are the same in fast scan but slightly different in slow scan, the process is defined as *quasi-reversible*; otherwise, the process is defined as *irreversible*.

30.2.4 Synthesis

The $(tpy)_2$Ir(LL′) complexes $(tpy)_2$Ir(acac) (**1**) [15], $(tpy)_2$Ir(pz)$_2$H (**2**), [$(tpy)_2$Ir(pzH)$_2$](CF$_3$SO$_3$) (**8**), $(tpy)_2$Ir(pz)$_2$BEt$_2$ (**3**), $(tpy)_2$Ir(pz)$_2$BPh$_2$ (**4**), $(tpy)_2$Ir(pz)$_2$Bpz$_2$ (**5**), $(dfppy)_2$Ir(acac) (**1**), $(dfppy)_2$Ir(pz)$_2$H (**2**), and $(dfppy)_2$Ir(pz)$_2$Bpz$_2$ (**5**) were prepared according to previous reported synthetic methods [41]. (Bu$_4$N)(PPh$_2$-CH$_2$)$_2$BPh$_2$ [42], (PPh$_2$CH$_2$)$_2$SiPh$_2$ [42], K(tz)$_3$BH [43], [$(tpy)_2$Ir(H$_2$O)$_2$](CF$_3$-SO$_3$) [4, 41], (Bu$_4$N)[$(ppy)_2$Ir(NCS)$_2$] (**13**) and (Bu$_4$N)[$(ppy)_2$Ir(CN)$_2$] (**14**) [18] were prepared following published procedures. Hydrotris(pyrazolyl)methane (pz$_3$CH) was purchased from STREM Chemical Co. Silver trifluoromethylsulfonate (AgOTf), *tert*-butyl isocyanide (CN-*t*-Bu), bis(diphenylphosphino)ethane, and all other materials were purchased from Aldrich Chemical Co. and used without further purification.

(a) Synthesis of $(tpy)_2$Ir(tz)$_3$BH (6): Iridium(III) Bis(2′-para-tolylpyridinato-*N,C$^{2'}$*)η^2-hydrotris(triazolyl)borate [$(tpy)_2$Ir(H$_2$O)$_2$](CF$_3$SO$_3$) (0.15 g, 0.2 mmol) was dissolved in 20 mL CH$_3$CN, and 2 equiv K(tz)$_3$BH (0.1 g, 0.4 mmol) was added to the solution. The solution was refluxed under N$_2$ overnight. After cooling to room temperature, the reaction mixture was evaporated to dryness under reduced pressure. The raw product was crystallized in methanol/hexane followed by recrystallization in CH_2Cl_2/hexane. The yellow compound was obtained in a yield of 50%. ^{1}H NMR (500 MHz, $CDCl_3$), ppm: 8.20 (s, 1H), 8.04 (s, 1H), 7.81 (d, $J = 8.4$ Hz, 1H), 7.77 (d, $J = 8.4$ Hz, 1H), 7.77 (vT, $J = 7.5$ Hz, 1H), 7.49 (m, 3H), 7.36 (s, 1H), 7.30 (d, $J = 5.6$ Hz, 1H), 7.21 (s, 1H), 6.94 (vT, $J = 6.6$ Hz, 1H), 6.80 (d, $J = 8.0$ Hz, 1H), 6.77 (d, $J = 8.0$ Hz, 1H), 6.66 (vT, $J = 6.6$ Hz, 1H), 6.05 (s, 1 H), 5.98 (s, 1H), 5.2–4.2 (br, 1H), 2.09 (s, 3H), 2.05 (s, 3H). Anal. for (*tpy*)$_2$Ir(tz$_3$-BH) · 0.5CH_2Cl_2: found: C 46.96, H 3.31, N 19.87; calcd: C 46.54, H 3.59, N 19.57.

(b) Synthesis of [(*tpy*)$_2$Ir(pz)$_3$CH](CF$_3$SO$_3$) (9): Iridium(III) Bis(2′-*para*-tolylpyridinato-*N*,*C*$^{2′}$)[η^2-hydrotris(pyrazolyl) methane] Trifluoromethylsulfonate. (*tpy*)$_2$Ir(H$_2$O)$_2$(CF$_3$SO$_3$) (0.15 g, 0.2 mmol) was dissolved in 20 mL CH_3CN, and 2 equiv pz$_3$CH (0.09 g, 0.4 mmol) was added to the solution. The solution was refluxed under N$_2$ overnight. After cooling to room temperature, the reaction mixture was evaporated to dryness under reduced pressure. The raw product was crystallized in acetone/hexane followed by recrystallization in CH_2Cl_2/hexane. The yellow-brownish compound was obtained in a yield of 35%. ^{1}H NMR (500 MHz, acetone-d_6), ppm: 8.20 (s, 1H), 8.04 (s, 1H), 7.81 (d, $J = 8.4$ Hz, 1H), 7.77 (d, $J = 8.4$ Hz, 1H), 7.77 (vT, $J = 7.5$ Hz, 1H), 7.49 (m, 3H), 7.36 (s, 1 H), 7.30 (d, $J = 5.6$ Hz, 1H), 7.21 (s, 1H), 6.94 (vT, $J = 6.6$ Hz, 1H), 6.80 (d, $J = 8.0$ Hz, 1H), 6.77 (d, $J = 8.0$ Hz, 1H), 6.66 (vT, $J = 6.6$ Hz, 1H), 6.05 (s, 1H), 5.98 (s, 1H), 5.2–4.2 (br, 1H), 2.09 (s, 3H), 2.05 (s, 3H). Anal. for [(*tpy*)$_2$Ir(pz$_3$CH)](CF$_3$SO$_3$)· 0.25CH_2Cl_2: found: C 46.57, H 3.33, N 11.99; calcd: C 46.36, H 3.37, N 12.27.

(c) Synthesis of (*tpy*)$_2$Ir(PPh$_2$CH$_2$)2BPh$_2$ (7): Iridium(III) Bis(2′-*para*-tolylpyridinato-*N*,*C*$^{2′}$) Bis(diphenylphosphinomethylene)diphenylborate. [(*tpy*)$_2$Ir(H$_2$O)$_2$](CF$_3$SO$_3$) (0.15 g, 0.2 mmol) was dissolved in 25 mL CH_3CN, and 2 equiv (Bu$_4$N) [(PPh$_2$CH$_2$)$_2$BPh$_2$] (0.28 g, 0.4 mmol) was added to the solution. The solution was refluxed under N$_2$ for 18 h. After cooling to room temperature, the reaction mixture was evaporated to dryness under reduced pressure. The yellow crystalline product (yield 50%) was obtained from column chromatography on silica using a CH_2Cl_2 mobile phase. ^{1}H NMR (500 MHz, $CDCl_3$), ppm: 8.72 (d, $J = 2.8$ Hz,

2H), 7.48–7.39 (m, 6H), 7.20 (dd, $J = 8.0, 8.0$ Hz, 4H), 7.12 (d, $J = 4.2$ Hz, 2H), 7.09–6.94 (m, 10H), 6.80–6.70 (m, 8H), 6.56 (d, $J = 3.8$ Hz, 2H), 6.45 (dd, $J = 7.5, 7.5$ Hz, 4H), 6.30 (dd, $J = 7.0, 7.0$ Hz, 4H), 6.15 (s, 2H), 2.30 (m, 2H), 2.21 (m, 2H), 2.05 (s, 6H). Anal. for $(tpy)_2Ir[(PPh_2CH_2)_2BPh_2]$: found: C 68.07, H 4.80, N 2.70; calcd: C 68.19, H 4.98, N 2.57.

(d) Synthesis of [(tpy)$_2$Ir(PPh$_2$CH$_2$)2SiPh$_2$](CF$_3$SO$_3$) (10): Iridium(III) Bis(2′-para-tolylpyridinato-N,C$^{2'}$)[bis(diphenylphosphinomethylene)diphenylsilane] Trifluoromethylsulfonate. $[(tpy)_2Ir(H_2O)_2](CF_3SO_3)$ (0.15 g, 0.2 mmol) was dissolved in 25 mL CH$_2$Cl$_2$, and 2 equiv Ph$_2$Si(CH$_2$PPh$_2$)$_2$ (0.23 g, 0.4 mmol) was added to the solution. The solution was stirred at room temperature overnight and was evaporated to dryness under reduced pressure. The yellow product (yield 60%) was obtained by recrystallization in CH$_2$Cl$_2$/hexane. ^{1}H NMR (500 MHz, CDCl$_3$), ppm: 8.59 (d, $J = 5.8$ Hz, 2H), 7.73 (dd, $J = 8.4, 8.4$ Hz, 2H), 7.47 (m, 6H), 7.39 (dd, $J = 7.0, 7.0$ Hz, 4H), 7.36–7.26 (m, 8H), 6.98 (m, 8H), 6.91 (dd, $J = 7.0, 7.0$ Hz, 2H), 6.62 (dd, $J = 7.8, 7.8$ Hz, 6H), 6.21 (d, $J = 8.4$, Hz, 4H), 5.98 (s, 2H), 2.62–2.46 (m, 4H), 2.04 (s, 6H). Anal. for $(tpy)_2Ir[(PPh_2CH_2)_2SiPh_2](CF_3SO_3)\cdot 0.5CH_2Cl_2$: found: C 58.96, H 4.24, N 2.30; calcd: C 58.63, H 4.26, N 2.15.

(e) Synthesis of [(tpy)$_2$Ir(dppe)](CF$_3$SO$_3$) (11): Iridium(III) Bis(2′-para-tolylpyridinato-N,C$^{2'}$)[bis(diphenylphosphino)ethane] Trifluoromethylsulfonate. $[(tpy)_2Ir(H_2O)_2](CF_3SO_3)$ (0.15 g, 0.2 mmol) was dissolved in 25 mL dicholoroethane, and 2 equiv bis(diphenylphosphino)ethane (0.16 g, 0.4 mmol) was added to the solution. The solution was refluxed under N$_2$ overnight and was evaporated to dryness under reduced pressure. The raw product was crystallized in CH$_2$Cl$_2$/hexane followed by recrystallization in CH$_2$Cl$_2$/ether. The yellowish compound was obtained in a yield of 30%. ^{1}H NMR (500 MHz, acetone-d_6), ppm: 7.90 (d, $J = 5.7$ Hz, 2H), 7.84 (vT, $J = 8.0$ Hz, 6H), 7.72 (d, $J = 8.0$ Hz, 2H), 7.62 (vT, $J = 8.0$ Hz, 2H), 7.47 (vT, $J = 7.5$ Hz, 2H), 7.37 (vT, $J = 7.5$ Hz, 4H), 7.06 (vT, $J = 7.0$ Hz, 2H), 6.96–6.87 (m, 6H), 6.78 (vT, $J = 8.9$ Hz, 4H), 6.44 (vT, $J = 7.1$ Hz, 2H), 6.22 (s, 2H), 4.12–3.93 (m, 2H), 3.06 (d, $J = 9.0$ Hz, 2H), 2.10 (s, 6H). Anal. for [(tpy)₂Ir(dppe)](CF₃SO₃)· H₂O: found: C 55.23, H 4.01, N 2.07; calcd: C 55.98, H 4.24, N 2.56.

(f) **Synthesis of [(tpy)₂Ir(CN-t-Bu)₂](CF₃SO₃) (12): Iridium(III) Bis(2′-para-tolylpyridinato-N,C²′) Bis(tert-butyl isocyanide) Trifluoromethylsulfonate.** [(tpy)₂Ir(H₂O)₂](CF₃SO₃) (0.15 g, 0.2 mmol) and 0.2 g t-BuNC were added to 20 mL of CH₂Cl₂. The solution was stirred for 2 days at room temperature. The reaction mixture was evaporated to dryness under reduced pressure, and the yellow product (yield 35%) was obtained by successive recrystallization in CH₂Cl₂/hexane. ^{1}H NMR (500 MHz, CDCl₃), ppm: 8.93 (d, J = 5.6 Hz, 2H), 7.97 (dd, J = 8.0, 8.0 Hz, 2H), 7.90 (d, J = 8.0 Hz, 2H), 7.50 (d, J = 8.0 Hz, 2H), 7.41 (vT, J = 6.4 Hz, 2H), 6.77 (d, J = 7.6 Hz, 2H), 6.05 (s, 1H), 5.91 (s, 2H), 2.04 (s, 6H), 1.31 (s, 18H). IR: 2192, 2168 cm^{-1} (terminal C≡N stretch). Anal. for [(tpy)₂Ir(CN-t-Bu)₂](CF₃SO₃): found: C 48.96, H 4.38, N 6.48; calcd: C 49.81, H 4.54, N 6.64.

(g) **Synthesis of (dfppy)₂Ir(PPh₂CH₂)₂BPh₂ (7): Iridium(III) Bis(4′,6′-difluorophenylpyridinato-N,C²′) Bis(diphenylphosphinomethylene)diphenylborate.** This yellow compound was prepared analogous to the synthesis of (tpy)₂Ir[Ph₂B(CH₂PPh₂)₂] and was obtained after chromatography on a silica/CH₂Cl₂ column in 55% yield. ^{1}H NMR (500 MHz, CDCl₃), ppm: 8.84 (d, J = 2.8 Hz, 2H), 7.61 (dd, J = 4.2, 3.8 Hz, 2H), 7.53 (dd, J = 8.0, 8.0 Hz, 2H), 7.42 (d, J = 3.8 Hz, 4H), 7.18–7.04 (m, 10H), 6.99 (dd, J = 7.0, 7.0 Hz, 2H), 6.90–6.76 (m, 8H), 6.65 (dd, J = 7.5, 7.5 Hz, 4H), 6.40–6.26 (m, 6H), 5.84 (m, 2H), 2.29 (m, 2H), 2.21 (m, 2H). Anal. for (dfppy)₂Ir[(PPh₂CH₂)₂BPh₂]: found: C 62.75, H 3.92, N 2.68; calcd: C 63.44, H 4.08, N 2.47.

(h) **Synthesis of [(dfppy)₂Ir(CN-t-Bu)₂](CF₃SO₃) (12): Iridium(III) Bis(4′,6′-difluorophenylpyridinato-N,C²′)bis(tert-butyl isocyanide) Trifluoromethylsulfate.** This yellow compound was prepared analogous to the synthesis of (tpy)₂Ir(CN-t−Bu)₂(CF₃SO₃) and was obtained by further recrystallization in CH₂Cl₂/hexane twice in 40% yield. ^{1}H NMR (360 MHz, CDCl₃), ppm: 9.05 (d, J = 5.9 Hz, 2H), 8.35 (d, J = 8.3 Hz, 2H), 8.03 (vT, J = 7.3 Hz, 2H), 7.52 (vT, J = 7.3 Hz, 2H), 6.49 (m, 2H), 5.55 (dd, J = 7.8, 2.4 Hz, 2H), 1.39 (s, 18H). IR: 2204, 2183 cm^{-1} (terminal C≡N stretch). Anal.

for (*dfppy*)$_2$Ir(CN-*t*—Bu)$_2$(CF$_3$SO$_3$): found: C 44.17, H 3.18, N 6.01; calcd: C 44.64, H 3.41, N 6.31.

30.3 Results and Discussion

30.3.1 Synthesis and Characterization

The (*tpy*)$_2$Ir(LL′) complexes in Fig. 30.1 were prepared from the chloride-bridged Ir(III) dimer, [(*tpy*)$_2$Ir(μ-Cl)]$_2$, using three different routes. The *tpy*, as opposed to the parent 2-phenylpyridyl (*ppy*), ligand was chosen in order to increase solubility, and thereby ease synthesis, and also to simplify characterization by NMR spectroscopy and X-ray crystallography. Only minor perturbation is caused by the methyl substituent in the photophysical properties of the (*tpy*)$_2$Ir(LL′) complexes relative to the *ppy* analogues [15]. We have previously shown that the Ir dimer can be readily converted to emissive, monomeric complex by treating the dimer with acetylacetone and base (Eq. 30.1) [15]. Likewise, the synthesis and characterization of (*tpy*)$_2$Ir(LL*) (LL* = pyrazolyl or pyrazolyl borate) complexes have been discussed in an earlier report [41]. Reaction of the dichloro-bridged dimer with excess bis(pyrazolyl)borate ligand leads only to formation of a protonated-dipyrazolyl Ir complex, (*tpy*)$_2$Ir(pz)$_2$H (Eq. 30.2) [41]. Therefore, the chloride-free Ir complex, [(*tpy*)$_2$Ir(H$_2$O)$_2$](CF$_3$SO$_3$), was prepared by chloride abstraction with CF$_3$SO$_3$Ag and used in the syntheses of Ir complexes with bis(pyrazolyl)borate ligands, e.g., (*tpy*)$_2$Ir(pz)$_2$Bpz$_2$ (Eq. 30.3). Similarly, [(*tpy*)$_2$Ir(H$_2$O)$_2$](CF$_3$SO$_3$) readily reacts with the anionic ligands of bidentate borates, e.g., (tz)$_3$BH$^-$, and (PPh$_2$CH$_2$)2BPh$_2^-$, and their neutral analogues to form the corresponding (*tpy*)$_2$Ir(LL′) complexes.

$$[(tpy)_2\text{Ir}(\mu - \text{Cl})]_2 + \text{excess acacH} \xrightarrow[\text{C}_2\text{H}_4\text{Cl}_2, \text{reflux}]{\text{Na}_2\text{CO}_3} (tpy)_2\text{Ir}(\text{acac}) \quad (30.1)$$

$$[(tpy)_2\text{Ir}(\mu - \text{Cl})]_2 + \text{excess KBpz}_4 \xrightarrow[\text{reflux}]{\text{CH}_3\text{CN}} (tpy)_2\text{Ir}(\text{pz})_2\text{H} \quad (30.2)$$

$$[(tpy)_2\text{Ir}(\mu-\text{Cl})]_2 \xrightarrow[\text{CH}_4\text{Cl}_2/\text{MeOH}]{\text{AgOTf}} [(tpy)_2\text{Ir}(\text{H}_2\text{O})_2]$$

$$(\text{CF}_3\text{SO}_3) \xrightarrow[\text{CH}_3\text{CN, reflux}]{\text{excess KBpz}_4} (tpy)_2\text{Ir}(pz)_2\text{Bpz}_2 \qquad (30.3)$$

Single crystals of $(tpy)_2\text{Ir}[(\text{PPh}_2\text{CH}_2)_2\text{BPh}_2] \cdot \text{H}_2\text{O}$ were prepared by slow evaporation of a chloroform solution, while single crystals of $[(tpy)_2\text{Ir}(\text{CN-}t\text{-Bu})_2](\text{CF}_3\text{SO}_3) \cdot \text{CHCl}_3$ were grown from chloroform/hexane. Molecular plots of $(tpy)_2\text{Ir}[(\text{PPh}_2\text{CH}_2)_2\text{BPh}_2]$ and $(tpy)_2\text{Ir}(\text{CN-}t\text{-Bu})_2^+$ are shown in Fig. 30.2; crystallographic data are given in Table 30.1. The Ir-P-C-B-C-P ring of $(tpy)_2$Ir$[(\text{PPh}_2\text{CH}_2)_2\text{BPh}_2]$ adopts a twisted-boat conformation (Fig. 30.2a, inset) similar in structure to that reported in other complexes of the same ligand, e.g., [Me$_2$Pt(PPh$_2$CH$_2)_2$BPh$_2$] [ASN] (ASN = 5-azonia-spiro[4.4]nonane)) [42a,b], or in the Li salt, [(PPh$_2$CH$_2)_2$BPh$_2$][Li(TMEDA)$_2$] (TMEDA = N,N,N',N'-tetramethylethylene-1,2-diamine) [42c]. The CN-t-Bu ligands in $(tpy)_2$Ir(CN-t-Bu)$_2^+$ are slightly distorted (average Ir–C–N = 170°, average C–N–C = 167°). Similar deviations from linearity have been reported for the CN-t-Bu ligands in other Rh or Ir complexes such as $[(\eta^2\text{-acac})_2\text{Rh}(\mu\text{-CPh}_2)_2\text{Rh}(\text{CN-}t\text{-Bu})_2]$ [44], [(t-BuNC)$_2$(Cl)Ir(μ-pz)$_2$Ir(η^1-CH$_2$Ph)(CN-t-Bu)$_2$] [45], and [Cp*Ir(CN-t-Bu)(μ-S)$_2$Ir(CN-t-Bu)Cp*] [46].

Selected bond lengths for seven Ir complexes with the same "tpy_2Ir" fragment, including previously reported $(tpy)_2$Ir(acac) [15], $(tpy)_2$Ir(pz)$_2$H [41], $[(tpy)_2$Ir(pzH)$_2](\text{CF}_3\text{SO}_3)$ [41], $(tpy)_2$Ir(pz)$_2$Bpz$_2$ [41], and $(tpy)_2$Ir(pz)$_2$BEt$_2$ [41], are provided in Table 30.2. All of these complexes have an octahedral coordination geometry around Ir, retaining the cis-C,C $trans$-N,N chelate disposition of the chloride-bridged precursor complex, $[(tpy)_2$Ir(μ-Cl)]$_2$. The N–Ir–N angles for the two $trans$-N,N atoms in these complexes are between 169 and 175°. The Ir–C$_{aryl}$ bond lengths of $(tpy)_2$Ir(PPh$_2$CH$_2)_2$BPh$_2$ (2.047(9), 2.057(9) Å) and $[(tpy)_2$Ir(CN-t-Bu)$_2]^+$ (2.047(5), 2.072(5) Å) are longer than their counterparts in $(tpy)_2$Ir(acac) (1.982(6), 1.985(7) Å). However, the Ir–C$_{aryl}$ bond lengths of the $(tpy)_2$Ir(LL') [41] complexes with pyrazolyl ligands (range = 1.995(12)—2.020(13) Å) are not significantly different than those found in $(tpy)_2$Ir(acac).

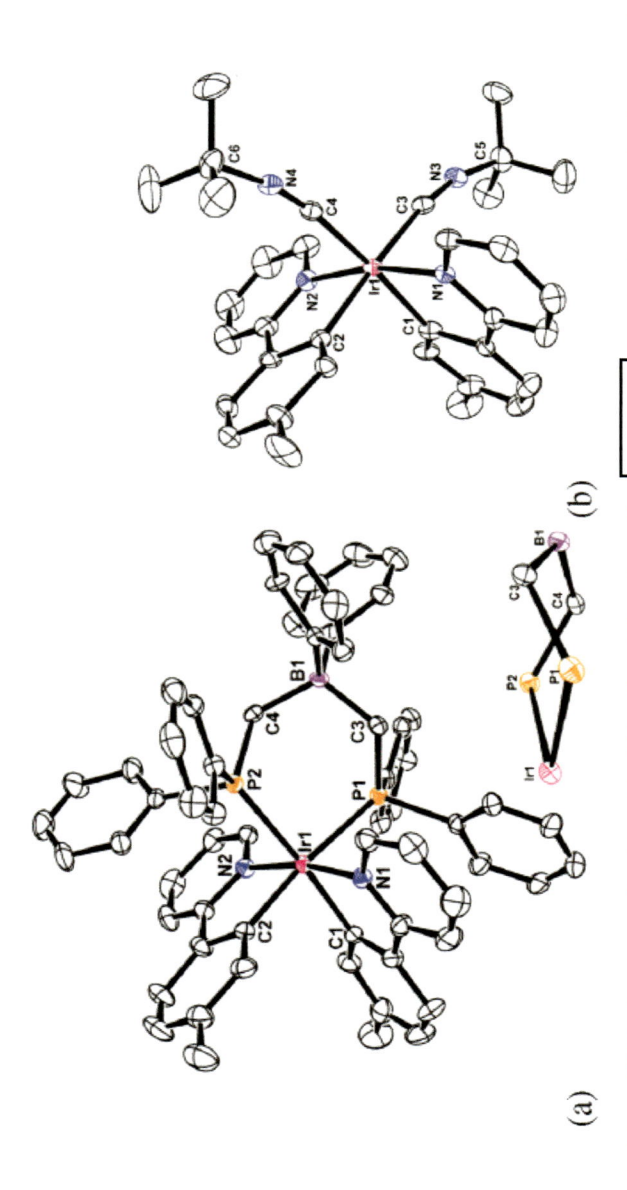

Figure 30.2 ORTEP drawings of (a) $(tpy)_2Ir(PPh_2CH_2)_2BPh_2 \cdot H_2O$ (the Ir-P-C-B-C-P ring is shown in the inset) and (b) $(tpy)_2Ir(CN\text{-}t\text{-}Bu)_2(CF_3SO_3) \cdot CHCl_3$. The thermal ellipsoids for the image represent 25% probability limit. The hydrogen atoms, counteranion, and solvent are omitted for clarity.

Table 30.2 Selected bond distances (Å) for $(tpy)_2Ir(acac)$, $(tpy)_2Ir(pz)_2H$, $[(tpy)_2Ir(pzH)_2](CF_3SO_3)$, $(tpy)_2Ir(pz)_2BEt_2$, $(tpy)_2Ir(pz)_2Bpz_2$, $(tpy)_2Ir(PPh_2CH_2)BPh_2$, and $(tpy)_2Ir(CN-t-Bu)_2(CF_3SO_3)$

$(tpy)_2Ir(X_1X_2)$	Ir-C$_1$	Ir-C$_2$	Ir-N$_1$	Ir-N$_2$	Ir-X$_1$	Ir-X$_2$
1 $(tpy)_2Ir(acac)$	1.982(6)	1.985(7)	2.023(5)	2.040(5)	2.161(4)	2.136(4)
2 $(tpy)_2Ir(pz)_2H$	2.005(12)	2.020(13)	2.031(10)	2.042(9)	2.155(9)	2.181(10)
8 $[(tpy)_2Ir(pzH)_2](CF_3SO_3)$	1.995(12)	1.995(12)	2.045(10)	2.045(10)	2.180(9)	2.180(10)
3 $(tpy)_2Ir(pz)_2BEt_2$	2.005(4)	2.007(4)	2.030(3)	2.033(3)	2.141(3)	2.137(3)
5 $(tpy)_2Ir(pz)_2Bpz_2$	2.009(5)	2.007(5)	2.044(4)	2.039(4)	2.147(4)	2.137(4)
7 $(tpy)_2Ir(PPh_2CH_2)BPh_2$	2.047(9)	2.057(9)	2.082(7)	2.083(7)	2.420(2)	2.431(2)
11 $[(tpy)_2Ir(CN-t-Bu)_2](CF_3SO_3)$	2.072(5)	2.047(5)	2.059(4)	2.061(4)	2.004(6)	2.018(6)

30.3.2 DFT Calculations

B3LYP/LACVP** DFT calculations were carried out on the [(ppy)$_2$Ir(CN-Me)$_2$]$^+$ complex to ascertain the influence of isocyanide ligands on the bis-cyclometalated Ir complexes. A similar theoretical approach has been used to investigate the ground- and excited-state properties of related cyclometalated Ir and Pt compounds [16, 19, 47]. The calculated metric parameters of [(ppy)$_2$Ir(CN-Me)$_2$]$^+$ (Ir–C$_{aryl}$ (2.08 Å), Ir–N (2.10 Å), Ir–C$_{CN-Me}$ (2.04 Å)) are similar to the values found in [(tpy)$_2$Ir(CN-t-Bu)$_2$]$^+$ (Table 30.2). The calculated singlet state energy [48] of [(ppy)$_2$Ir(CN-Me)$_2$]$^+$ is 4.24 eV. The HOMO and LUMO surfaces for [(ppy)$_2$Ir(CN-Me)$_2$]$^+$ are illustrated in Fig. 30.3. The orbital contours correspond quite closely to the HOMO and LUMO surfaces reported for (ppy)$_2$Ir(acac) [47]. The HOMO consists principally of a mixture of phenyl-π and Ir-d orbitals, whereas the LUMO is localized largely on the ppy orbitals with very little metal orbital character. Interestingly, the two ancillary ligands appear to contribute very little electron density to either the HOMO or LUMO. The HOMO surfaces for the complex suggest a possible overlap between the metal-phenyl and isocyanide orbitals. However, no such clear interaction exists between the ppy and isocyanide orbitals because the LUMO is predominantly distributed on the pyridyl-π orbitals, which are orthogonal to the isocyanide orbitals. Thus, the isocyanides can only directly interact with the cyclometalating ligands in the HOMO via the metal orbitals, and consequently this leaves the LUMO energy largely unperturbed, as confirmed by electrochemical analysis (vide infra). This electronic model, where the ancillary ligand interacts strongly only with the HOMO, is believed to apply to all the other (tpy)$_2$Ir(LL′) complexes as well.

30.3.3 Electrochemistry

The electrochemical properties of the (tpy)$_2$Ir(LL′) complexes, as well as for (Bu$_4$N)[(ppy)$_2$Ir(NCS)$_2$] (**13**) and (Bu$_4$N)[(ppy)$_2$Ir(CN)$_2$] (**14**) [49], were examined using cyclic voltammetry, and the values of redox potentials were determined using differential pulsed voltammetry (Table 30.3). All of the electrochemical data reported

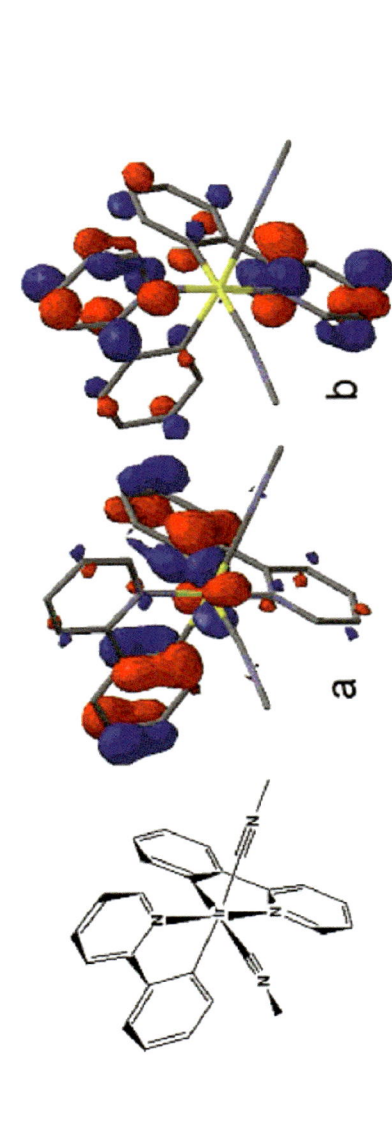

Figure 30.3 The HOMO (a) and LUMO (b) surface of $(ppy)_2\text{Ir}(\text{CN-Me})_2^+$ from DFT calculations. The HOMO orbital consists mainly of a mixture of phenyl-π and Ir-d orbitals, while the LUMO orbital is largely localized on the pyridyl moiety.

Table 30.3 Redox properties of $(tpy)_2\text{Ir}(LL')$ complexes[e]

$(C \wedge N)_2\text{Ir}(LL')$	$E^{Ox}_{1/2}$ (V)	$E^{Red}_{1/2}$ (V)	$\Delta E_{1/2}$ (V)	E_{1MLCT} (V)
1 $(tpy)_2\text{Ir}(acac)$	0.41	−2.68	3.09	3.02
2 $(tpy)_2\text{Ir}(pz)_2\text{H}$	0.55,[d] 0.77[d]	−2.65[d]	3.20	3.16
3 $(tpy)_2\text{Ir}(pz)_2\text{BEt}_2$	0.63,[d] 0.77[d]	−2.66[c]	3.29	3.18
4 $(tpy)_2\text{Ir}(pz)_2\text{BPh}_2$	0.67[d]	−2.64	3.31	3.17
5 $(tpy)_2\text{Ir}(pz)_2\text{Bpz}_2$	0.72[d]	−2.64	3.36	3.24
6 $(tpy)_2\text{Ir}(tz)_3\text{BH}$	0.82[a]	−2.56	3.38	3.25
7 $(tpy)_2\text{Ir}(PPh_2\text{CH}_2)_2\text{BPh}_2$	0.81[a,d]	−2.62	3.43	3.35
8 $[(tpy)_2\text{Ir}(pzH)_2]\text{CF}_3\text{SO}_3$	0.75[a]	−2.40[d]	3.15	3.27
9 $[(tpy)_2\text{Ir}(pz)_3\text{CH}](\text{CF}_3\text{SO}_3)$	0.89[a]	−2.43[d]	3.32	3.29
10 $[(tpy)_2\text{Ir}(PPh_2\text{CH}_2)_2\text{SiPh}_2](\text{CF}_3\text{SO}_3)$	1.10[d]	−2.42[c]	3.52	3.44
11 $[(tpy)_2\text{Ir}(dppe)](\text{CF}_3\text{SO}_3)$	1.03[b,d]	−2.46	3.49	3.46
12 $[(tpy)_2\text{Ir}(CN\text{-}t\text{-Bu})_2](\text{CF}_3\text{SO}_3)$	1.23[b,d]	−2.39[d]	3.62	3.56
13 $(NBu_4)[(ppy)_2\text{Ir}(NCS)_2]$	0.42[d]	−2.71[d]	3.13	3.12
14 $(NBu_4)[(ppy)_2\text{Ir}(CN)_2]$	0.50[d]	−2.78	3.28	3.25

[e]Redox measurements were carried out in DMF solution unless noted: (a) in CH_2Cl_2; (b) in CH_3CN. The redox values are reported relative to Fc^+/Fc. The electrochemical process is reversible unless noted: (c) quasi-reversible; (d) irreversible.

here were measured relative to an internal ferrocenium/ferrocene reference (Fc^+/Fc). The electrochemistry of related tris cyclometalated Ir(III) complexes, e.g., *fac*-Ir$(ppy)_3$, has been thoroughly studied [16, 50]. Oxidation is considered to be a metal-aryl centered process, whereas reduction is localized mainly on the pyridyl rings of the cyclometalating ligands [50]. Thus, if no other competitive oxidation or reduction process occurs in the ancillary ligand, each $(tpy)_2\text{Ir}(LL')$ complex should only display a single one-electron oxidation (occurring at the $(tpy)_2\text{Ir}$-based HOMO) and two one-electron reductions (occurring at each pyridyl ring of the *tpy* ligands). Under the current experimental conditions, i.e., anhydrous and nitrogen-purged DMF solution, one reduction is clearly observed for all $(tpy)_2\text{Ir}(LL')$ complexes. The ancillary ligands themselves are difficult to reduce, so the first reduction process can be assigned to one of the pyridyl rings of the "$(tpy)_2\text{Ir}$" fragment [47, 50]. The reduction potentials of the neutral $(tpy)_2\text{Ir}(LL')$ complexes stay in a narrow range (−2.56 to −2.68 V), which indicates that the LUMO is little affected by the nature of the LL' ligand.

Contrary to the relatively invariant reduction potentials of neutral $(tpy)_2$Ir(LL′) complexes, the corresponding oxidation potentials span a wider range (0.41–0.82 V). Increasing the ligand field strength of the ancillary ligands leads to higher oxidation potentials for the $(tpy)_2$Ir(LL′) complexes. For example, $(tpy)_2$Ir(pz)$_2$Bpz$_2$ has the highest oxidation potential among all neutral $(tpy)_2$Ir(LL*) (LL* = pyrazolyl ligands) complexes due to the relatively high ligand field strength of Bpz$_4^-$ [51]. Coordination of stronger π-acid ancillary ligands, e.g., diphosphines, onto the $(tpy)_2$Ir(LL′) complexes leads to further increases in the oxidation potential. Although the DFT calculations suggest that the ancillary ligands do not contribute significant electron density to the HOMO of the $(tpy)_2$Ir(LL′) compounds, they still can influence the HOMO energy by interacting with the Ir d-orbitals. Thus, ancillary ligands with stronger ligand field strength stabilize the HOMO. In addition, the $(tpy)_2$Ir(pz)$_2$BEt$_2$ and $(tpy)_2$Ir(pz)$_2$H complexes display a second oxidation process that can be ascribed to oxidation processes involving the ancillary ligands. For $(tpy)_2$Ir(pz)$_2$BEt$_2$, the electron-rich boron atom and comparably weak B–C bonds may be sites of oxidation [52], as evidenced by the fact that Na(pz)$_2$BEt$_2$ has an irreversible oxidation (ca. 0.4 V vs Fc$^+$/Fc) in a dichloroethane solution. For $(tpy)_2$Ir(pz)$_2$H, on the other hand, the acidic proton in the "(pz)$_2$H" ligand can undergo solvent-induced deprotonation in the oxidized complex to generate additional electroactive species [53].

The redox potentials of cationic $(tpy)_2$Ir(LL′) complexes are shifted by roughly 200–300 mV to more positive values relative to potentials found in the neutral analogues. The anodic shifts are due to the overall positive charge of the complexes. Regardless, the redox potentials display trends similar to those observed in the neutral complexes; the reduction potentials for cationic complexes fall in a narrow range (-2.39 to -2.43 V), whereas the oxidation potentials vary more widely (0.75–1.23 V).

30.3.4 Electronic Spectroscopy

The room-temperature (RT) absorption and emission spectra and low-temperature (77 K) spectra were recorded for all $(tpy)_2$Ir(LL′) complexes (Table 30.4). A typical example of absorption and

Table 30.4 Photophysical properties of $(tpy)_2\text{Ir}(LL')$ complexes[a]

$(C \wedge N)_2\text{Ir}(LL')$	abs, λ_{max} (nm) $\{E, 10^3 \text{ cm}^{-1} \text{ M}^{-1}\}$	Emission at RT					Emission at 77 K	
		λ_{max} (nm)	τ (μs)	Φ_{PL}	k_r 10^5 s^{-1}	k_{nr} 10^5 s^{-1}	λ_{max} (nm)	τ (μs)
1 $(tpy)_2\text{Ir}(\text{acac})$	269 (40.5), 406 (4.1), 451 (2.8), 488 (1.0)	512	1.4	0.38	2.7	4.4	500	3.8
2 $(tpy)_2\text{Ir}(\text{pz})_2\text{H}$	268 (37.1), 392 (4.8), 478 (0.42)	490	2.2	0.44	2.0	2.6	480	4.2
8 $[(tpy)_2\text{Ir}(\text{pzH})_2]$ (CF_3SO_3)	267 (40.0), 382 (4.5), 472 (0.17)	490	2.4	0.26	1.1	3.1	480	3.7
3 $(tpy)_2\text{Ir}(\text{pz})_2\text{BEt}_2$	266 (36.0), 390 (5.1), 477 (0.35)	484	2.6	0.52	2.0	1.8	480	3.3
4 $(tpy)_2\text{Ir}(\text{pz})_2\text{BPh}_2$	266 (40.0), 391 (4.8), 475 (0.31)	484	3.4	0.49	1.4	1.5	474	4.1
5 $(tpy)_2\text{Ir}(\text{pz})_2\text{Bpz}_2$	263 (40.0), 383 (5.1), 472 (0.25)	480	3.5	0.52	1.5	1.4	473	4.2
6 $(tpy)_2\text{Ir}(\text{tz})_3\text{BH}$	259 (38.0), 382 (4.5), 471 (0.16)	478	4.1	0.55	1.3	1.1	470	5.1
9 $[(tpy)_2\text{Ir}(\text{pz})_3\text{CH}]$ (CF_3SO_3)	258 (40.2), 377 (4.9), 469 (0.16)	476	3.0	0.33	1.1	2.2	469	4.2
7 $(tpy)_2\text{Ir}(\text{PPh}_2\text{CH}_2)_2\text{BPh}_2$	250 (41.5), 370 (4.5), 463 (0.04)	468	4.7	0.038	0.08	2.0	462	23.5

(Continued)

Table 30.4 (Continued)

$(C \wedge N)_2\text{Ir}(LL')$	abs, λ_{max} λ (nm) $\{E, 10^3\text{ cm}^{-1}\text{ M}^{-1}\}$	Emission at RT					Emission at 77 K	
		λ_{max} (nm)	τ (μs)	Φ_{PL}	k_r 10^5 s^{-1}	k_{nr} 10^5 s^{-1}	λ_{max} (nm)	τ (μs)
10 [(tpy)$_2$Ir(PPh$_2$CH$_2$)$_2$SiPh$_2$](CF$_3$SO$_3$)	260 (39.0), 322 (14.0), 360 (7.4), 464 (0.04)	468	1.1	0.006	0.05	9.0	462	26.2
11 [(tpy)$_2$Ir(dppe)](CF$_3$SO$_3$)	250 (34.3), 267 (26.3), 322 (8.6), 358 (5.7), 425 (0.06), 456(0.03)	458	2.1	0.005	0.03	4.7	452	32.3
12 [(tpy)$_2$Ir(CN-t-Bu)$_2$](CF$_3$SO$_3$)	260 (33.5), 315 (15.8), 348 (12.1), 422 (0.07), 452 (0.03)	458	35.6	0.28	0.08	0.20	454	45.4
13 (NBu$_4$)[(ppy)$_2$Ir(NCS)$_2$]	263 (42.3), 336 (7.4), 398 (3.8), 444 (2.3), 482 (0.59)	508	1.5	0.40	2.7	4.0	486	3.0
14 (NBu$_4$)[(ppy)$_2$Ir(CN)$_2$]	258 (36.3), 344 (6.0.9), 381 (4.9), 462 (0.25)	476	2.6	0.48	1.8	2.0	458	4.8

[a]The absorption spectra were measured in CH$_2$Cl$_2$, and the emission spectra were measured in 2-MeTHF solution.

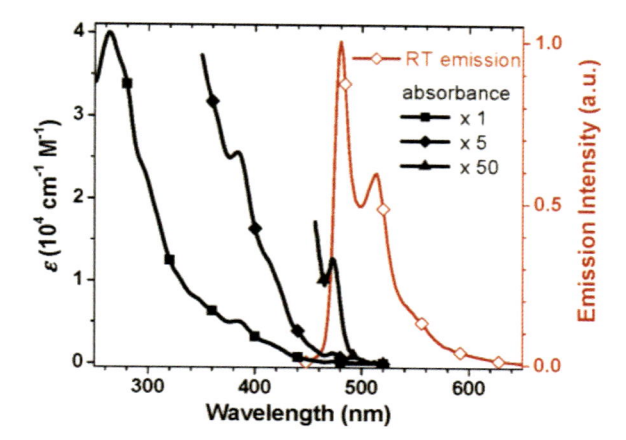

Figure 30.4 Room-temperature absorption and emission spectra of $(tpy)_2Ir(pz)_2Bpz_2$. The absorption spectrum (filled symbols) was measured in CH_2Cl_2, and the emission spectrum (empty symbols) was measured in 2-methyltetrahydrofuran (2-MeTHF).

emission spectra at RT displayed by the $(tpy)_2Ir(LL')$ complexes is shown in Fig. 30.4 for $(tpy)_2Ir(pz)_2Bpz_2$. Three characteristic types of well-resolved absorption bands are observed. High-energy, intense absorption bands (250–270 nm, $\epsilon \approx 4.0 \times 10^4$ cm^{-1} M^{-1}) can be assigned to allowed $^1(\pi-\pi^*)$ transitions of the *tpy* ligand [15, 54–56]. The energies and extinction coefficients of these bands correlate well with similar absorption features observed in 2'-*para*-tolylpyridine (*tpy*H). Weaker bands located at longer wavelength (350–440 nm, ϵ = 1000–8000 cm^{-1} M^{-1}) can be assigned to Ir ⟶ *tpy* charge-transfer transitions [15, 18]. The energies of these MLCT transitions are only weakly solvatochromic, undergoing a 6 nm red-shift in nonpolar (hexanes) solvent, indicating a ground state that is more polar than the excited state. The weak, lowest energy absorption band (472 nm, ϵ = 250 cm^{-1} M^{-1}) can be identified as a triplet transition (T_1) on the basis of the small energy shift (350 cm^{-1}) between absorption and emission at room temperature.

A comparison of lowest energy absorption features among $(tpy)_2$Ir(acac), $(tpy)_2Ir(pz)_2Bpz_2$, and $(tpy)_2Ir(PPh_2CH_2)_2BPh_2$ is shown in Fig. 30.5. For $(tpy)_2Ir(acac)$, a series of strong overlapping absorption bands diminish in energy down to 500 nm. The

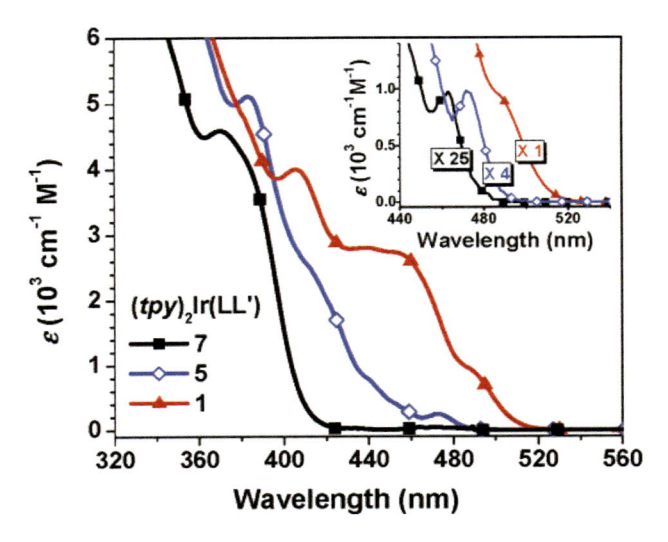

Figure 30.5 ^{1}MLCT absorption spectra of $(tpy)_2$Ir(PPh$_2$CH$_2$)$_2$BPh$_2$ (**7**), $(tpy)_2$Ir(pz)$_2$Bpz$_2$ (**5**), and $(tpy)_2$Ir(acac) (**1**) complexes in CH$_2$Cl$_2$. The T$_1$ absorption transitions are shown in the inset.

absorption bands between 420 and 500 nm can be assigned to a combination of singlet and triplet MLCT transitions involving both the *tpy* and acac ligands on the basis of time-dependent DFT calculations of the $(ppy)_2$Ir analogue [47]. The peak at 406 nm (3.06 eV) coincides in energy with the difference in electrochemical oxidation and reduction potentials (3.09 V) and can therefore be assigned to a ^{1}MLCT transition between the HOMO and LUMO. For both $(tpy)_2$Ir(pz)$_2$Bpz$_2$ and $(tpy)_2$Ir(PPh$_2$CH$_2$)$_2$BPh$_2$ complexes, the intensities of the mixed singlet and triplet MLCT transitions at wavelengths greater than 420 nm are markedly attenuated, whereas strong ^{1}MLCT transitions still occur between 350 and 410 nm. The ^{1}MLCT and T$_1$ transition energies increase with change in the ancillary ligand (acac$^-$ < Bpz$_4^-$ < (PPh$_2$CH$_2$)$_2$BPh$_2^-$). However, while the ^{1}MLCT energies increase with only minor changes in intensity, the increase in the T$_1$ absorption energy coincides with a pronounced decrease in the extinction coefficient (Fig. 30.5, inset).

The room-temperature and 77 K emission spectra of $(tpy)_2$Ir(PPh$_2$CH$_2$)$_2$BPh$_2$, $(tpy)_2$Ir(pz)$_2$Bpz$_2$, $(tpy)_2$Ir(pz)$_2$H, and $(tpy)_2$

Ir(acac) are shown in Fig. 30.6. Most of the $(tpy)_2$Ir(LL′) complexes are strongly luminescent (quantum yields $(\Phi) = 0.26$–0.55) and have short luminescence lifetimes ($\tau = 2$–5 μs) at room temperature, similar to values reported for fac-Ir$(tpy)_3$. Exceptions are complexes with diphosphine ($\Phi < 0.04$) and isocyanide ($\tau = 35$ μs) ligands. All of the $(tpy)_2$Ir(LL′) complexes are intensely emissive at low temperature (77 K), and most have short luminescence lifetimes ($\tau = 3$–5 μs). Again, the complexes with diphosphine and isocyanide ligands are exceptions with $\tau > 23$ μs. The radiative (k_r) and nonradiative decay (k_{nr}) rates can be calculated from the room-temperature Φ and τ data [57]. The k_r values of the $(tpy)_2$Ir(LL′) complexes range between 2.7×10^5 and 3×10^3 s^{-1} with the lowest values for complexes with the highest emission energy. The k_{nr} values span a narrower range of values (2.2×10^4 to 9.0×10^5 s^{-1}) and tend to decrease as the emission energy increases.

The structured luminescent spectra display vibronic progressions that become more highly resolved with increasing emission energy. Two prominent vibronic features are present in the emission spectra (Fig. 30.6). The vibrational fine-structure observed in emission spectra is often the result of several overlapping satellites belonging to different vibronic transitions [58]. Here we analyze the emission spectra qualitatively and simply assume that the vibrational progression is only due to the dominant vibrational stretch. The intensity ratio of this first major vibrational transition to the highest energy peak ($E_{em}(0\text{-}0)$) is a measure of vibronic coupling between the ground and excited state (Huang–Rhys factor, S_M) and is proportional to the degree of structural distortion that occurs in the excited state relative to the ground state [28, 32, 59–62]. The dominant vibrational mode associated with the excited-state distortion ($\hbar\omega_M$) can be obtained from the energy difference (in cm^{-1}) of these vibronic transitions at 77 K, whereas the S_M value can be estimated from the peak heights [60]. Table 30.5 lists some parameters of the emission spectra (77 K) including $E_{em}(0\text{-}0)$, S_M, $\hbar\omega_M$, and selected IR absorption data for several $(tpy)_2$Ir(LL′) complexes and tpyH. Although all of the Ir complexes have the same "$(tpy)_2$Ir" fragment, $E_{em}(0\text{-}0)$ shifts to higher energy in complexes with stronger ligand field strength ancillary ligands such as $Ph_2B(CH_2PPh_2)_2^-$. The Huang–Rhys factors also increase

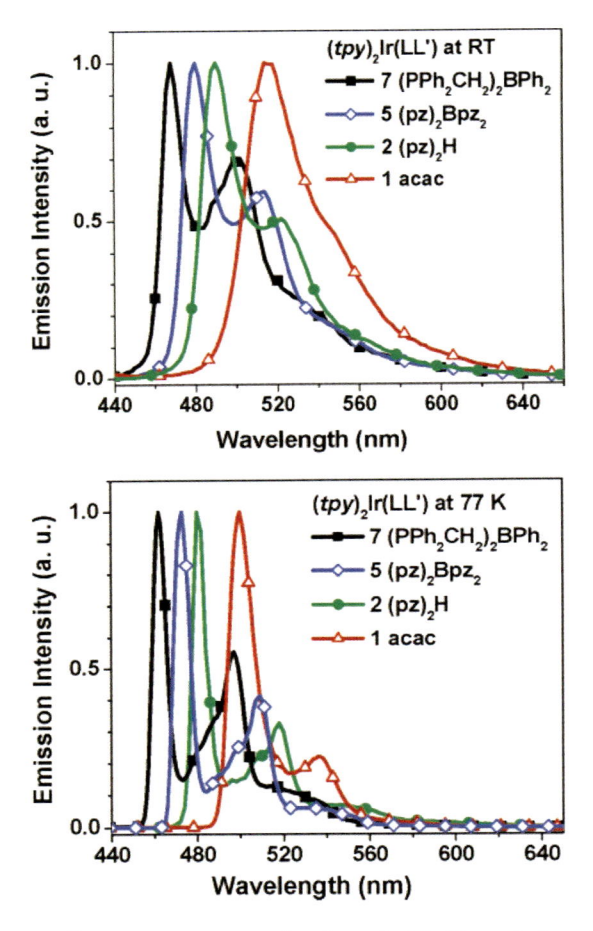

Figure 30.6 Room-temperature (top) and 77 K emission (bottom) spectra of $(tpy)_2\text{Ir}(\text{PPh}_2\text{CH}_2)_2\text{BPh}_2$, $(tpy)_2\text{Ir}(pz)_2\text{Bpz}_2$, $(tpy)_2\text{Ir}(pz)_2\text{H}$, and $(tpy)_2\text{Ir}(acac)$ complexes in 2-MeTHF.

monotonically with increasing $E_{em}(0-0)$, and the S_M value of *tpy*H is larger than values found for any of the $(tpy)_2\text{Ir}(LL')$ complexes.

30.3.5 Optical Transition Energies vs Redox Potentials

Previous investigations of diimine complexes of Ru(II) and Os(II), e.g., $\text{Ru}(bpy)_2\text{L}_2^{2+}$, have shown that absorption and emission energies

Table 30.5 Excited state properties and IR absorption of selected (tpy)2Ir(LL′) complexes and free tpyH ligand[a]

(tpy)$_2$Ir(LL′)	E_{em} (0-0) (10^6 cm^{-1})	$\hbar\omega_M$ (cm^{-1})	S_M	IR absorption (1400-1550 cm^{-1})
1 (tpy)$_2$Ir(acac)	2.00	1413	0.22	1514, 1476, 1428, 1401
2 (tpy)$_2$Ir(pz)$_2$H	2.08	1528	0.33	1476, 1428, 1406
5 (tpy)$_2$Ir(pz)$_2$Bpz$_2$	2.11	1495	0.41	1505, 1478, 1466, 1429, 1411, 1402
7 (tpy)$_2$Ir(PPh$_2$CH$_2$)$_2$BPh$_2$	2.16	1524	0.55	1480, 1433
12 [(tpy)$_2$Ir(CN-t-Bu)$_2$] (CF$_3$SO$_3$)	2.20	1535	0.64	1508, 1482, 1466, 1432, 1402
Free tpy ligand	2.29	1522	1.53	1514, 1467, 1433

[a]The emission energy, E_{em}(0-0), was obtained from the maximum emission wavelength at 77 K. The Huang–Rhys factor, S_M, was estimated from the peak heights of the first two features of the 77 K emission spectra. The energy of $\hbar\omega_M$ was obtained from the energy difference (in cm^{-1}) of the first two emission peaks.

increase linearly with an increase in the electrochemical gap, $\Delta E_{1/2}$ (the energy difference between the first oxidation potential and first reduction potential of the parent complex) [28]. The linear relationship is consistent with the proposed model that describes formation of the MLCT state from an electronic transition between a metal-centered HOMO to ligand-localized LUMO [27, 28]. For the (tpy)$_2$Ir(LL′) complexes, there is also a close correspondence between the $\Delta E_{1/2}$ values and the singlet (E_{1MLCT}) absorption energies (Table 30.3). The small differences between the $\Delta E_{1/2}$ and E_{1MLCT} values (<0.15 eV) indicate that Franck–Condon factors contribute little to the intra- and intermolecular reorganization energies upon optical excitation. A linear correlation is obtained when the E_{1MLCT} and triplet (E_{T1}) absorption energies are plotted versus $\Delta E_{1/2}$ (Fig. 30.7). The near unity of the slope of E_{1MLCT} versus $\Delta E_{1/2}$ (0.84) provides further support for the assignment of E_{1MLCT} to a charge-transfer transition.

To illustrate the influence the various ancillary ligands have on the HOMO and LUMO energies, the reduction (E_{red}) and oxidation potentials (E_{ox}) of the neutral (tpy)$_2$Ir(LL′) complexes are plotted separately versus E_{T1} in Fig. 30.8. It is apparent from the plots that

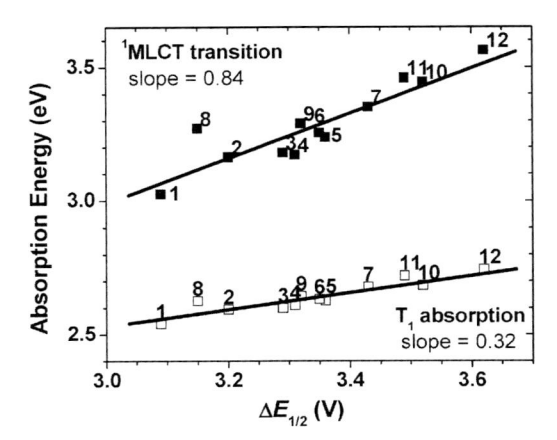

Figure 30.7 Plot of ^{1}MLCT absorption energy ($E_{1\text{MLCT}}$, filled squares) and T$_1$ absorption energy (E_{T1}, empty squares) vs electrochemical gap ($\Delta E_{1/2}$) of $(tpy)_2$Ir(LL′) complexes. The linear fits are also shown in the graph The ancillary ligands (LL′) are numbered as in Fig. 30.1.

nearly all the variation in triplet energy occurs from changes in the E_{ox}, not E_{red}. Likewise, the redox potentials of all the cationic $(tpy)_2$Ir(LL′) complexes (triangles in Fig. 30.8) follow the same trends as their neutral analogues. The experimental data concur with the DFT calculation results: a LUMO localized predominantly on the pyridyl orbitals and a HOMO that is largely metal-aryl in character, in which only the HOMO is affected by the ancillary ligand. Thus, the ancillary ligands increase the optical energy gap, i.e., the $E_{1\text{MLCT}}$ and T$_1$ absorption energies, by lowering the HOMO energy while leaving the LUMO energy relatively unchanged.

It is noteworthy that the slope of E_{T1} versus $\Delta E_{1/2}$ (0.32) in Fig. 30.7 is much lower than that of $E_{1\text{MLCT}}$ versus $\Delta E_{1/2}$ (0.84). The difference between $E_{1\text{MLCT}}$ and E_{T1} (ΔE_{ST}) is related to the exchange energy of a complex. Large exchange energies occur when the orbitals involved with a triplet state have a significant spatial overlap with those of a singlet state, whereas small exchange energies occur when the orbital overlap between the two states is poor. For example, a decrease in ΔE_{ST} is observed in linear conjugated polymers as a function of chain length in going from monomer to polymer [63]. The decrease in ΔE_{ST} with increasing chain length

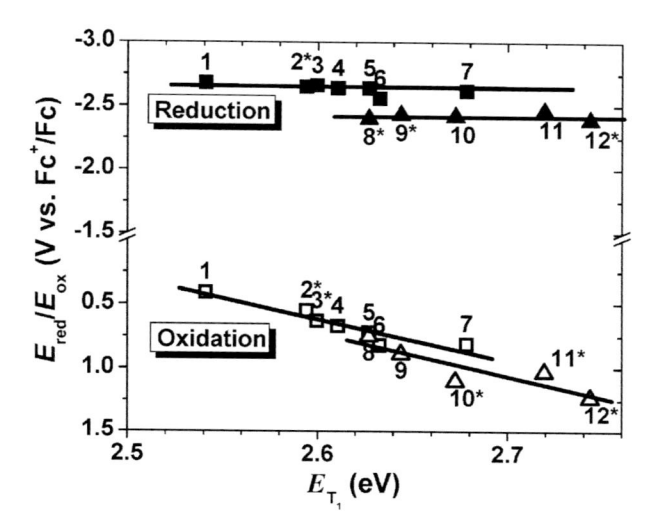

Figure 30.8 Plot of red/ox potentials vs emission energy, E_{em}(RT), of $(tpy)_2$Ir(LL′) complexes. The ancillary ligands (LL′) are numbered as in Fig. 30.1. Neutral $(tpy)_2$Ir(LL′) complexes (squares) and cationic $(tpy)_2$Ir(LL′) complexes (triangles); reduction potentials (filled) and oxidation potentials (empty). The asterisk (*) indicates an irreversible oxidation or reduction process; otherwise the electrochemical process is reversible or *quasi-*reversible.

in the conjugated systems has been attributed to the singlet state becoming more delocalized compared to the relatively localized triplet state. By analogy, the decrease in ΔE_{ST} with decreasing $\Delta E_{1/2}$ in the $(tpy)_2$Ir(LL′) complexes implies that the ^{1}MLCT state acquires a more delocalized character with decreasing energy than does the T_1 state. Similar reductions in ΔE_{ST} with decreasing $\Delta E_{1/2}$ also occur in M$(bpy)_2$L$_2^{2+}$ complexes (M = Ru, Os) [27, 64]. The small slope of T_1 vs $\Delta E_{1/2}$ and consequent decrease in ΔE_{ST} may be due to localization of the triplet state onto a single cyclometalated ligand in the $(tpy)_2$Ir(LL′) complexes [65].

30.3.6 Ground-State and Excited-State Properties

The lowest energy excited state of 4d^6 and 5d^6 complexes with π-accepting cyclometalating ligands can be characterized as a dominant ligand-centered $^3\pi-\pi^*$ state with some ^{1}MLCT character mixed in by spin–orbit coupling (Fig. 30.9) [14, 29, 66]. By applying first-order perturbation theory, the following formula can be used to define the lowest excited state (Eq. 30.4):

$$\Psi_{T_1} = \sqrt{1-\alpha^2}|^3\text{LC}\rangle + \alpha|^1\text{MLCT}\rangle \quad (30.4)$$

where Ψ_{T_1} is the wave function of the lowest excited state and α is a coefficient that gives an estimate of the degree of singlet character mixed into the unperturbed triplet state (^{3}LC) [14]. The value of α can be approximated with the formula

$$\alpha = \frac{\langle^3\text{LC}|H_{SO}|^1\text{MLCT}\rangle}{\Delta E} \quad (30.5)$$

where $\langle^3\text{LC}|H_{SO}|^1\text{MLCT}\rangle$ is the spin–orbital coupling matrix element, characterizing the strength of spin–orbital coupling between ^{3}LC and ^{1}MLCT, and ΔE is the energy difference between the ^{3}LC and ^{1}MLCT transitions [14]. Equations 30.4 and 30.5 have been used to correlate α values with the luminescent properties of diimine and cyclometalated Rh(III) and Ir(III) complexes [14]. Only small amounts of ^{1}MLCT character need be mixed into the lowest excited state to significantly increase the ^{3}LC oscillator strength and radiative decay rate in luminescent metal complexes. For example, Güdel and co-workers have estimated that α is 0.085 in the strongly luminescent complex $(ppy)_2\text{Ir(bpy)}^+$ [9, 14]. Moreover, since α is inversely proportional to ΔE, the ΔE value can be used to evaluate the properties of the lowest excited state, provided that $\langle^3\text{LC}|H_{SO}|^1\text{MLCT}\rangle$ remains invariant in the series of metal complexes. For all the $(tpy)_2\text{Ir(LL')}$ complexes, the energy level of the unperturbed ^{3}LC state can, to a first approximation, be assumed to be constant since all species have the same cyclometalating ligand. However, the energy of the ^{1}MLCT state, and likewise ΔE, will increase with increasing stability of the HOMO. Increasing ΔE decreases the amount of ^{1}MLCT character mixed into the T_1 state (Fig. 30.9b), which in turn is expected to decrease the oscillator strength for this transition [14], a premise that is strongly supported by our study. Ancillary ligands that increase the T_1 transition energy of the $(tpy)_2\text{Ir(LL')}$ complexes also decrease the extinction coefficient (Fig. 30.5, Table 30.4). For example, the value of the T_1 extinction coefficient for $(tpy)_2\text{Ir(Ph}_2\text{B(CH}_2\text{PPh}_2)_2)$ is 4% that of $(tpy)_2\text{Ir(acac)}$. However, the ^{1}MLCT extinction coefficients in the

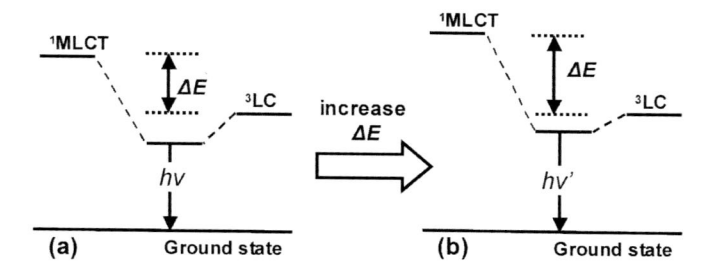

Figure 30.9 Schematic energy level diagram for state mixing in cyclometalated Ir(III) complexes with different ΔE: (a) small ΔE, large admixture of ^{1}MLCT and ^{3}LC, low emission energy; (b) large ΔE, small admixture of ^{1}MLCT and ^{3}LC, high emission energy.

corresponding $(tpy)_2$Ir(LL') complexes do not change significantly with increasing transition energy. The decrease in oscillator strength with increasing absorption energy for the T_1 transition is consistent with a lowest excited state that has less ^{1}MLCT admixture and thus, more spin-forbidden character.

For all $(tpy)_2$Ir(LL') complexes, the increase of emission energies (E_{em}) is accompanied by an increase in the Huang–Rhys factor, S_M (shown in Table 30.5). The values of the vibronic spacing for all the selected $(tpy)_2$Ir(LL') complexes and tpyH are the same within experimental error ($\hbar\omega_M \approx 1480 \pm 90$ cm^{-1}). The dominant vibronic transition that appears in the emission spectra can be correlated with vibrational features between 1400 and 1520 cm^{-1} in the IR spectra of the corresponding $(tpy)_2$Ir(LL') complexes and tpyH. The IR transitions can be assigned to C–C inter-ring stretching modes by comparison to similar vibrations identified in $[Ru(bpy)_3]^{2+}$ by normal coordinate analysis [21, 67]. The S_M value quantifies the degree of electron-vibrational coupling, and large S_M values indicate strong coupling between the dominant ligand-localized vibrations in the excited and ground states [28, 59]. Therefore, the increase in S_M with increasing emission energies is consistent with a T_1 state that becomes more ligand-localized (^{3}LC) in character. An alternative explanation is that an increasing ^{1}MLCT admixture expands the electronic spatial extensions, resulting in a decreased charge density per cyclometalating ligand and weaker coupling between vibrational levels in the ground and excited T_1

state [13]. Thus, this explanantion also suggests that the increase of S_M values indicates less ^{1}MLCT character in the T_1 state, i.e., the lowest excited state of the $(tpy)_2$Ir(LL') complexes has more ^{3}LC (less delocalized) character as the T_1 energy increases.

The polarity of a molecule in an excited state is correlated to the rigidochromic shift, i.e., the energy difference of $E_{em}(0-0)$ for dilute solutions at room temperature and 77 K [20, 68]. A molecule in the excited state can reach a fully relaxed geometry upon solvent reorientation in a low-viscosity medium at room temperature, whereas the molecular excited state cannot fully relax in highly viscous, frozen media at 77 K. Thus, emission spectra from molecules with luminescent charge-transfer transitions typically display hypsochromic shifts upon going from room-temperature fluid solution to 77 K glass. A greater ^{1}MLCT character in the lowest excited state of the Ir complexes is expected to lead to greater change in the dipole moment upon excitation, resulting in larger rigidochromic shifts [20, 68]. A plot of the rigidochromic shift (ν_{max}(77 K) – ν_{max}(RT)) versus E_{em}(RT) for the $(tpy)_2$Ir(LL') complexes in 2-MeTHF (Fig. S4, see also Table 30.4) shows an increase with decreasing emission energies, consistent with the proposed increase in ^{1}MLCT character with decreasing T_1 state energy.

The ancillary ligands of the $(tpy)_2$Ir(LL') complexes significantly affect the luminescent quantum yield. Since Φ is dependent on both the radiative and nonradiative decay rates [69], the k_r and k_{nr} values need to be considered individually in order to determine the overall influence on Φ brought about by the ancillary ligands. Previous investigations of Ru(II) and Os(II) polypyridyl complexes have shown that the Φ values for these species are dictated by the variation in k_{nr} with energy. The k_{nr} values for a large number of luminescent diimine transition complexes have also been shown to follow the energy gap law for radiationless transitions [28, 59–62]. A linear decrease between ln(k_{nr}) with increasing emission energy is deduced from the energy gap law if the dominant radiationless process is assigned to the same vibrational states in closely related systems. For the $(tpy)_2$Ir(LL') complexes, the k_{nr} values decrease with increasing E_{em}(RT). A linear relationship of ln(k_{nr}) vs E_{em}(RT) is obtained for the series of neutral $(tpy)_2$Ir(LL*)

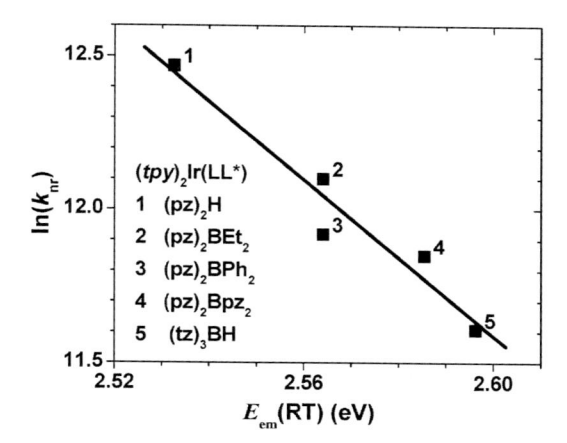

Figure 30.10 Plot of nonradiative decay rate ($\ln(k_{nr})$) vs emission energy ($E_{em}(RT)$) for neutral $(tpy)_2Ir(LL^*)$ (LL^* = pyrazolyl ligands) complexes.

complexes (Fig. 30.10), where the Ir complexes considered here retain a similar "$(tpy)_2Ir(pz)_2$" core. Since the k_{nr} values for the $(tpy)_2Ir(LL')$ complexes also appear to follow the energy gap law, any decrease in Φ with increasing emission energy cannot be attributed to an increase in nonradiative decay rate. In particular, the low Φ values of the $(tpy)_2Ir(LL')$ complexes with diphosphine ligands are mostly due to their small respective k_r values, since their k_{nr} values are similar to rates found in other $(tpy)_2Ir(LL')$ complexes with much larger values of Φ.

The radiative rates of the $(tpy)_2Ir(LL')$ complexes decrease with increasing emission energies (Table 30.4) [70]. The k_r values for $(tpy)_2Ir(LL')$ complexes with $E_{em}(RT) < 2.65$ eV are ca. 10^5 s^{-1}, comparable to values found in Os(II) polypyridyl complexes [61], whereas k_r values decrease to $< 10^4$ s^{-1} for the $(tpy)_2Ir(LL')$ complexes with $E_{em}(RT) > 2.65$ eV. The radiative decay rate can be related to the emission energy using Eq. 30.6:

$$k_r = \frac{4E_{em}^3}{3\hbar}|\langle\psi_e|\xrightarrow{d}|\psi_g\rangle|^2 \tag{30.6}$$

where $|\langle\psi_e|\xrightarrow{d}|\psi_g\rangle|^2$ describes the probability for an excited-to-ground-state transition [28, 71, 72]. In an investigation of Os(II) polypyridyl complexes, Meyer et al. assumed a constant value of

$|\langle\psi_e|\xrightarrow{d}|\psi_g\rangle|^2$ for all complexes studied and noted a roughly linear increase in k_r with an increase in E_{em}^3 [61]. However, since k_r decreases with increasing $E_{em}(RT)$ in the $(tpy)_2Ir(LL')$ complexes, the transition probability for these cyclometalated species must be decreasing with increasing $E_{em}(RT)$.

According to classical theory [61, 72], the radiative decay rate for emission can be related to the oscillator strength (f) for absorption, provided that the absorption and emission transitions involve the same initial and final states. The oscillator strength is proportional to the width of the absorption band at $^{1/2}\epsilon_{max}(\Delta\nu_{1/2})$ for transitions with a Gaussian line shape. Equation 30.7 can then be used to calculate k_r using the intensity of the corresponding absorption band:

$$k_r = 3 \times 10^{-9}\nu_0^2 \int \epsilon d\nu \approx 3 \times 10^{-9}\nu_0^2 \epsilon \Delta\nu_{1/2} \tag{30.7}$$

where ν_0 is the energy (in cm^{-1}) corresponding to the maximum wavelength of absorption and $\int \epsilon d\nu$ is the area of the molecular extinction coefficient [72]. The distinctive T_1 absorption bands displayed by the $(tpy)_2Ir(LL')$ complexes (Fig. 30.5, inset) are in contrast to the broad, ill-defined bands that are observed for similar transitions in Os(II) and Ru(II) polypyridyl complexes [28]. The well-resolved T_1 absorption transition allows for an accurate assessment of ϵ. However, for $(tpy)_2Ir(acac)$, the T_1 absorption band is overlapped with higher energy MLCT transitions, which precludes accurate identification of ϵ; therefore, this complex was not included in the following analysis. A plot of k_r versus $\nu_0^2\epsilon$ for complexes **2-14** is shown in Fig. 30.11. The good linear fit between k_r and $\nu_0^2\epsilon_{T_1}$ is consistent with emission originating from a common state, i.e., the "$(tpy)_2Ir$" fragment. A $\Delta\nu_{1/2}$ value of 410 cm^{-1} is derived from the slope of the plot, which is comparable to the line width reported for the lowest triplet transition in [Os(bpy)$_3$]$^{2+}$ (300 cm^{-1} at 5 K) [73]. Therefore, the absorption properties of the T_1 band reflect the luminescent characteristics and can be used to evaluate the emission properties of the respective $(tpy)_2Ir(LL')$ complexes.

The oscillator strength and consequently k_r are dependent on the amount of ^{1}MLCT character that is mixed into the T_1 state which, in turn, depends on ΔE. It is possible to calculate k_r as a function of

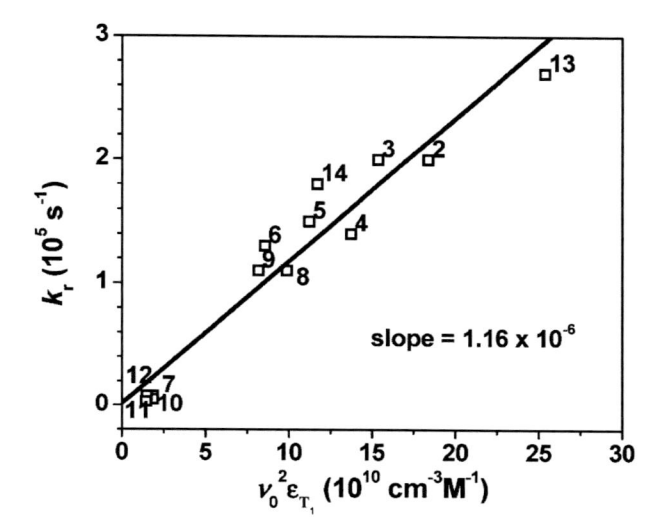

Figure 30.11 Plot of k_r vs $\nu_0^2 \epsilon_{T1}$ for $(tpy)_2\mathrm{Ir}(LL')$ complexes. The ancillary ligands (LL′) are numbered as in Fig. 30.1 (k_r = rate of radiative decay, ν_0 and ϵ_{T1} = T_1 absorption energy and extinction coefficient, respectively).

ΔE using Eq. 30.8 [58, 74].

$$k_r = k_r(1\mathrm{MLCT}) \left(\frac{\langle 3\mathrm{LC}|H_{SO}|1\mathrm{MLCT}\rangle}{\Delta E} \right)^2 \left(\frac{\nu_{T_1}}{\nu_{1\mathrm{MLCT}}} \right)^3 \quad (30.8)$$

Here, $k_r(^1\mathrm{MLCT})$ is the radiative rate of the perturbing state while ν_{T1} and $\nu_{1\mathrm{MLCT}}$ are the respective absorption transition energies. The $k_r(^1\mathrm{MLCT})$ value can be considered essentially constant since the magnitude is proportional to the oscillator strength of the $^1\mathrm{MLCT}$ transition, which is roughly the same for all the complexes. Values for ΔE can be obtained from the $\nu_{1\mathrm{MLCT}}$ data and by using the emission energy of $(ppy)_2\mathrm{Pt}(CH_2Cl)Cl$ (444 nm, 22500 cm^{-1}, $k_r = 10_3$ s^{-1}) as the $^3\mathrm{LC}$ energy for an unperturbed "$(tpy)_2\mathrm{Ir}$" fragment, since the Pt(IV) complex has a ligand-centered excited state perturbed only by metalation to the heavy atom [85]. On the basis of Eq. 30.8, k_r values should be linearly proportional to $(1/\Delta E)^2 (\nu_{T1}/\nu_{1\mathrm{MLCT}})^3$, provided that $\langle ^3\mathrm{LC}|H_{SO}|^1\mathrm{MLCT}\rangle$ is constant (Fig. 30.12a). Alternatively, Eqs. 30.7 and 30.8 can be combined and rearranged to give Eq. 30.9.

$$\epsilon = k_\text{r}(1\text{MLCT})\left(\frac{1}{3\times 10^{-9}\Delta v_{1/2}}\right)$$
$$\times \left(\frac{\langle ^3\text{LC}|H_\text{SO}|^1\text{MLCT}\rangle}{\Delta E}\right)^2 \left(\frac{v_{T_1}}{v_{^1\text{MLCT}^3}}\right) \quad (30.9)$$

The derivation used to generate Eq. 30.9 is similar to the approach used by Demas and Crosby to correlate ^{1}MLCT and T_1 energies of Ru and Os polypyridyl complexes to extinction coefficients and radiative rates [75]. The analysis has also been shown by Watts to be applicable to both cyclo-metalated and polypyridyl Ir complexes [4, 76, 77]. Equation 30.9 has the advantage of using only absorption data to generate a plot. Thus, on the basis of Eq. 30.9, a linear fit of ϵ vs $(1/\Delta E)^2 (v_{T1}/v_{^1\text{MLCT}}^3)$ can also be used to derive a value of $\langle ^3\text{LC}|H_\text{SO}|^1\text{MLCT}\rangle$ for the $(tpy)_2\text{Ir}(LL')$ complexes (Fig. 30.12b). If one assumes that $k_\text{r}(^1\text{MLCT})$ has the same value as for [Ru$(bpy)_3$]$^{2+}$ ($\sim 10^8$ s^{-1}) [78], a value of 200 cm^{-1} is obtained for $\langle ^3\text{LC}|H_\text{SO}|^1\text{MLCT}\rangle$ from the slopes of Fig. 30.12a,b (using a value of 410 cm^{-1} for $\Delta v_{1/2}$ in Eq. 30.9). Interestingly, the $\langle ^3\text{LC}|H_\text{SO}|^1\text{MLCT}\rangle$ value estimated from such an analysis of room-temperature data is intermediate between $\langle ^3\text{LC}|H_\text{SO}|^1\text{MLCT}\rangle$ values determined for the cyclometalated complexes $(ppy)_2\text{Ir}(bpy)^+$ (147 cm^{-1}) and $(thpy)_2\text{Ir}(bpy)^+$ (237 cm^{-1}) ($thpy$ = 2-(2-thienyl)pyridyl) using low-temperature, high-resolution spectroscopy [14]. A close inspection of the data (Table S1) also reveals that variation in k_r and ϵ in Fig. 30.12a,b, respectively, is most strongly correlated by the nearly 5-fold change in the ΔE^{-2} term. Therefore, the decrease in the k_r (and ϵ) values with increasing emission energy can be ascribed to an increase in the ^{1}MLCT transition energy, and likewise ΔE, brought about by a decrease in the HOMO energy, which leads to less mixing of ^{1}MLCT character into the lowest T_1 state.

A large decrease in k_r values in both the CN-t-Bu and (PPh$_2$CH$_2$)$_2$BPh$_2^-$ based $(tpy)_2\text{Ir}(LL')$ complexes can also be correlated with an increase in the Ir–C$_\text{aryl}$ bond lengths. The Ir–C$_\text{aryl}$ bonds are longer than the corresponding bonds in $(tpy)_2\text{Ir}(acac)$ due to the relatively stronger *trans* influence of the two types of ancillary ligands. The longer Ir–C$_\text{aryl}$ bonds are expected to influence the metal–ligand stretching vibrations, which also play a role in the vibronic coupling between the excited and ground states (Herzberg–

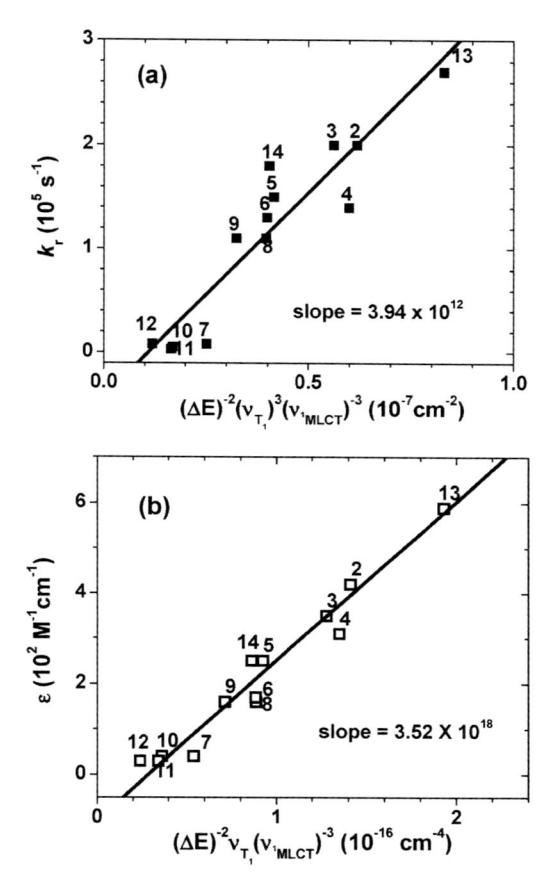

Figure 30.12 (a) Plot of radiative decay rate (k_r) vs $(1/\Delta E)^2(\nu_{T_1}/\nu_{1_{MLCT}})^3$ for $(tpy)_2$Ir(LL′) complexes. (b) Plot of ϵ vs $(1/\Delta E)^2(\nu_{T_1}/\nu^3_{1_{MLCT}})$. The ancillary ligands (LL′) are numbered as in Fig. 30.1.

Teller coupling) [12]. Thus, in addition to the decline in k_r caused by an increase in ΔE, the isocyanide and phosphino ligands may by weakening the Ir–C$_{aryl}$ bonds, resulting in a further decrease the coupling between the ground and excited states decrease in k_r values.

A smaller series of $(dfppy)_2$Ir(LL′) complexes were examined using a related set of ancillary ligands (Table 30.6). Substituting fluorine in the 4′,6′-positions of the 2-phenylpyridyl ligand leads

Table 30.6 Photophysical properties of $(dfppy)_2Ir(LL')$ complexes[a]

(C^N)₂Ir(LX)	abs, λ_{max} (nm) {ϵ, 10^3 cm^{-1} M^{-1}}	Emission at RT λ_{max} (nm)	τ (μs)	Φ_{PL}	k_r 10^5 s^{-1}	k_{nr} 10^5 s^{-1}	Emission at 77 K λ_{max} (nm)	τ (μs)
1 $(dfppy)_2Ir(acac)$	254 (47.8), 387 (5.0), 461 (0.9)	482	1.2	0.62	5.2	3.2	469	2.8
2 $(dfppy)_2Ir(pz)_2H$	253 (42.3), 323 (11.8), 375 (5.3), 456 (0.4)	466	1.5	0.62	4.1	2.5	458	2.4
5 $(dfppy)_2Ir(pz)_2Bpz_2$	252 (39.8), 367 (4.9), 420 (0.52), 451(0.22)	456	3.7	0.73	2.0	0.73	450	4.0
7 $(dfppy)_2Ir(PPh_2CH_2)_2BPh_2$	251 (51.5), 312 (13.8), 413 (0.1), 442 (0.04)	448	8.4	0.19	0.22	1.0	443	19.7
12 $[(dfppy)_2Ir(CN-t-Bu)_2](CF_3SO_3)$	252 (44.9), 308 (20.1), 412 (0.1), 440 (0.06)	444 (sh), 468	6.2	0.16	0.26	1.4	442	7.8

[a]The absorption spectra were measured in CH_2Cl_2 and the emission spectra were measured in 2-MeTHF solution.

to a hypsochromic shift in the emission spectra relative to their (tpy)$_2$Ir(LL′) analogues, as expected from results obtained with related tris-cyclometalated complexes [16, 19]. The higher triplet energy of *dfppy* relative to *tpy* leads to both a decrease in k_{nr} (energy gap law) and an increase in k_r (Eq. 30.6) resulting in higher Φ values for the (*dfppy*)$_2$Ir(LL′) derivatives. The (*dfppy*)$_2$Ir(LL′) complexes display a decrease in k_r with increasing emission energy similar to that observed in the (*tpy*)$_2$Ir(LL′) complexes, suggesting that the effects of ancillary ligands on the excited-state properties of Ir complexes are independent of the choice of cyclometalating ligand. The fact that the same ancillary ligands decrease k_r values in both types of cyclometalated species supports the proposal that the low k_r values are caused by a decrease in the HOMO energy and consequently an increase in ΔE, that decreases the amount of ^{1}MLCT character mixed into the T$_1$ state.

30.4 Conclusion

The photophysical and electrochemical properties of a series of (*tpy*)$_2$Ir(LL′) complexes have been examined. The studies presented here demonstrate that it is possible to tune the properties of the lowest excited state chemically by only employing different ancillary ligands. The ancillary ligands increase the optical energy gap of Ir complexes by lowering the HOMO (related to the oxidation of metal-centered orbitals) and leave the LUMO (related to the reduction of ligand-localized orbitals) unchanged. The destabilization of the ^{1}MLCT state results in a decreased ^{1}MLCT character in the lowest excited state, thereby increasing the lowest excited state energy. In addition, the reduced ^{1}MLCT character within the lowest excited state has pronounced effects on the photophysical properties of the Ir complexes: the excited states become more ligand-localized with stronger coupling with the dominant vibrational mode, the oscillator strength of the T$_1$ transition decreases, and the decreased ^{1}MLCT admixture leads to a decrease in radiative decay rates.

Most of the neutral Ir complexes reported here are highly emissive at room temperature and stable to sublimation, making them ideal for use as emissive dopants in blue and white phosphorescent OLEDs. Devices fabricated using films doped with (dfppy)$_2$Ir(pz)$_2$Bpz$_2$ have external quantum efficiencies of > 11% for blue phosphorescent OLEDs [79] and > 12% for white phosphorescent OLEDs [80]. We are examining other related complexes as emissive dopants in monochromatic and white OLEDs. The tetrakis(pyrazolyl)borate ligand also offers the possibility of expanded coordination [81, 82], since the noncoordinated pyrazoles of (tpy)$_2$Ir(pz)$_2$Bpz$_2$ can be used to bind a second metal center. Related dinuclear organometallic complexes have been reported for Pd and Ru complexes but not, thus far, for photoactive metal fragments such as the Ir complexes reported here. The ability to make homo-metallic and heterometallic complexes of this type may lead to further applications for cyclometalated Ir complexes, such as photosensitizers [83] and donor-acceptor dyads [84].

Acknowledgment

The authors thank the Universal Display Corporation, the Defense Advanced Research Projects Agency, and the National Science Foundation for their financial support.

Supporting Information Available

The UV–vis absorption and emission spectra for selected (dfppy)$_2$Ir(LL′) complexes (Figs. S5–S7); the absorption spectra of (tpy)$_2$Ir(pz)$_2$Bpz$_2$ in different solvents at room temperature (Fig. S2); the absorption, excitation (77 K), and emission spectra (77 K) of (tpy)$_2$Ir(CN-t-Bu)$_2$(CF$_3$SO$_3$) in 2-MeTHF (Fig. S3); plot of rigidochromic shifts vs emission energies for (tpy)$_2$Ir(LL′) complexes in 2-MeTHF (Fig. S4); the electrochemical and photophysical properties of all (dfppy)$_2$Ir(LL′) complexes examined here (Tables S2–S4). Also included are tables of crystal data, atomic coordinates, bond distances, bond angles, and anisotropic displacement paramenters for (tpy)$_2$Ir(PPh$_2$CH$_2$)$_2$BPh$_2$· H$_2$O and (tpy)$_2$Ir(CN-t-Bu)$_2$(CF$_3$SO$_3$)· CHCl$_3$, as well as the corresponding CIF files. This material is available free of charge via the Internet at http://pubs.acs.org.

References

1. (a) Baldo, M. A., O'Brien, D. F., You, Y., Shoustikov, A., Sibley, S., Thompson, M. E., Forrest, S. R. *Nature* 1998, **395**, 151. (b) Baldo, M. A., Lamansky, S., Burrows, P. E., Thompson, M. E., Forrest, S. R. *Appl. Phys. Lett.* 1999, **75**, 4. (c) Thompson, M. E., Burrows, P. E., Forrest, S. R. *Curr. Opin. Solid State Mater. Sci.* 1999, **4**, 369. (d) Baldo, M. A., Thompson, M. E., Forrest, S. R. *Nature* 2000, **403**, 750.
2. (a) Lamansky, S., Djurovich, P. I., Abdel-Razzaq, F., Garon, S., Murphy, D. L., Thompson, M. E. *J. Appl. Phys.* 2002, **92**, 1570. (b) Chen, F. C., Yang, Y., Thompson, M. E., Kido, J. *Appl. Phys. Lett.* 2002, **80**, 2308. (c) Markham, J. P. J., Lo, S.-C., Magennis, S. W., Burn, P. L., Samuel, I. D. W. *Appl. Phys. Lett.* 2002, **80**, 2645. (d) Zhu, W., Mo, Y., Yuan, M., Yang, W., Cao, Y. *Appl. Phys. Lett.* 2002, **80**, 2045.
3. (a) Sutin, N. *Acc. Chem. Res.* 1968, **1**, 225. (b) Meyer, T. J. *Acc. Chem. Res.* 1978, **11**, 94.
4. Schmid, B., Garces, F. O., Watts, R. J. *Inorg. Chem.* 1994, **32**, 9.
5. (a) Belmore, K. A., Vanderpool, R. A., Tsai, J. C., Khan, M. A., Nicholas, K. M. *J. Am. Chem. Soc.* 1988, **110**, 2004. (b) Silavwe, N. D., Goldman, A. S., Ritter, R., Tyler, D. R. *Inorg. Chem.* 1989, **28**, 1231.
6. King, K. A., Spellane, P. J., Watts, R. J. *J. Am. Chem. Soc.* 1985, **107**, 1431.
7. (a) Demas, J. N., Harris, E. W., McBride, R. P. *J. Am. Chem. Soc.* 1977, **99**, 3547. (b) Demas, J. N., Harris, E. W., Flynn, C. M., Diemente, J. D. *J. Am. Chem. Soc.* 1975, **97**, 3838. (c) Gao, R., Ho, D. G., Hernandez, B., Selke, M., Murphy, D., Djurovich, P. I., Thompson, M. E. *J. Am. Chem. Soc.* 2002, **124**, 14828.
8. Lo, K. K.-W., Chung, C.-K., Lee, T. K.-M., Lui, L.-K., Tsang, K. H.-K., Zhu, N. *Inorg. Chem.* 2003, **42**, 6886.
9. Colombo, M. G., Güdel, H. U. *Inorg. Chem.* 1993, **32**, 3081.
10. Strouse, G. F., Güdel, H. U., Bertolasi, V., Ferretti, V. *Inorg. Chem.* 1995, **34**, 5578.
11. Lever, A. P. B. *Inorganic Electronic Spectroscopy*, 2nd ed.: Elsevier: New York, 1984; pp. 174–178.
12. (a) Wiedenhofer, H., Schützenmeier, S., von Zelewsky, A., Yersin, H. *J. Phys. Chem.* 1995, **99**, 13385. (b) Schmidt, J., Wiedenhofer, H., von Zelewsky, A., Yersin, H. *J. Phys. Chem.* 1995, **99**, 226.
13. Yersin, H., Donges, D. *Top. Curr. Chem.* 2001, **214**, 81–186.

14. Vanhelmont, F. W. M., Güdel, H. U., Förtsch, M., Bürgi, H.-B. *Inorg. Chem.* 1997, **36**, 5512.
15. Lamansky, S., Djurovich, P., Murphy, D., Abdel-Razzaq, F., Kwong, R., Tsyba, I., Bortz, M., Mui, B., Bau, R., Thompson, M. E. *Inorg. Chem.* 2001, **40**, 1704.
16. Tamayo, A. B., Alleyne, B. D., Djurovich, P. I., Lamansky, S., Tsyba, I., Ho, N. H., Bau, R., Thompson, M. E. *J. Am. Chem. Soc.* 2003, **125**, 7377.
17. (a) Grushin, V. V., Herron, N., LeCloux, D. D., Marshall, W. J., Petrov, V. A., Wang, Y. *Chem. Commun.* 2001, (16), 1494. (b) Wang, Y., Herron, N., Grushin, V. V., LeCloux, D., Petrov, V. *Appl. Phys. Lett.* 2001, **79**, 449.
18. Nazeeruddin, Md. K., Humphry-Baker, R., Berner, D., Rivier, S., Zuppiroli, L., Graetzel, M. *J. Am. Chem. Soc.* 2003, **125**, 8790.
19. Brooks, J., Babayan, Y., Lamansky, S., Djurovich, P. I., Tsyba, I., Bau, R., Thompson, M. E. *Inorg. Chem.* 2002, **41**, 3055.
20. Tsuboyama, A., Iwawaki, H., Furugori, M., Mukaide, T., Kamatani, J., Igawa, S., Moriyama, T., Miura, S., Takiguchi, T., Okada, S., Hoshino, M., Ueno, K. *J. Am. Chem. Soc.* 2003, **125**, 12971.
21. (a) Finkenzeller, W. J., Yersin, H. *Chem. Phys. Lett.* 2003, **377**, 299. (b) Finkenzeller, W. J., Stoessel, P., Yersin, H. *Chem. Phys. Lett.* 2004, **397**, 289.
22. Adachi, C., Kwong, R. C., Djurovich, P., Admovich, V., Baldo, M. A., Thompson, M. E., Forrest, S. R. *Appl. Phys. Lett.* 2001, **79**, 2082.
23. Trofimenko, S. *J. Am. Chem. Soc.* 1967, **89**, 3170.
24. (a) Breakell, K. R., Patmore, D. J., Storr, A. *J. Chem. Soc., Dalton Trans.* 1975, 749. (b) Komorowski, L., Maringgele, W., Meller, A., Niedenzu, K., Serwatowski, J. *Inorg. Chem.* 1990, **29**, 3845.
25. (a) Tellers, D. M., Bergman, R. G. *J. Am. Chem. Soc.* 2001, **123**, 11508. (b) Tellers, D. M., Bergman, R. G. *J. Am. Chem. Soc.* 2000, **122**, 954. (c) Wiley, J. S., Heinekey, D. M. *Inorg. Chem.* 2002, **41**, 4961. (d) Gutierrez-Puebla, E., Monge, A., Paneque, M., Poveda, M. L., Salazar, V., Carmona, E. *J. Am. Chem. Soc.* 1999, **121**, 248. (e) Tellers, D. M., Bergman, R. G. *Organometallics* 2001, **20**, 4819.
26. (a) Trofimenko, S. *Chem. Rev.* 1993, **93**, 943. (b) Vicente, J., Chicote, M. T., Guerrero, R., Herber, U., Bautista, D. *Inorg. Chem.* 2002, **41**, 1870. (c) Paulo, A., Domingos, A., Santos, I. *Inorg. Chem.* 1996, **35**, 1798. (d) Kitano, T., Sohrin, Y., Hata, Y., Kawakami, H., Hori, T., Ueda, K. *J. Chem. Soc., Dalton Trans.* 2001, 3564. (e) Shi, X., Ishihara, T., Yamanaka, H., Gupton, J. T. *Tetrahedron Lett.* 1995, **36**, 1527. (f) Yamanaka, H., Takekawa, T., Morita, K., Ishihara, T., Gupton, J. T. *Tetrahedron Lett.* 1996, **37**, 1829.

27. Kober, E. M., Marshall, J. L., Dressick, W. J., Sullivan, B. P., Casper, J. V., Meyer, T. J. *Inorg. Chem.* 1985, **24**, 2755.
28. Caspar, J. V., Meyer, T. J. *Inorg. Chem.* 1983, **22**, 2444.
29. Yersin, H., Humbs, W. *Inorg. Chem.* 1999, **38**, 5820.
30. Zheng, K., Wang, J., Shen, Y., Kuang, D., Yun, F. *J. Phys. Chem. A* 2001, **105**, 7248.
31. Yersin, H., Humbs, W., Strasser, J. *Coord. Chem. Rev.* 1997, **159**, 325.
32. Kober, E. M., Caspar, J. V., Lumpkin, R. S., Meyer, T. J. *J. Phys. Chem.* 1986, **90**, 3722.
33. Flynn, C. M., Jr., Demas, J. N. *J. Am. Chem. Soc.* 1974, **96**, 1959.
34. Watts, R. J. *Comments Inorg. Chem.* 1991, **11**, 303.
35. Jones, G., II; Jackson, W. R., Choi, C.-Y., Bergmark, W. R. *J. Phys. Chem.* 1985, **89**, 294.
36. DePriest, J., Zheng, G. Y., Goswami, N., Eichhorn, D. M., Woods, C., Rillema, D. P. *Inorg. Chem.* 2000, **39**, 1955.
37. Sheldrick, G. M. *SHELXTL*, version 5.1; Bruker Analytical X-ray System, Inc.: Madison, WI, 1997.
38. Blessing, R. H. *Acta Crystallogr.* 1995, **A51**, 33.
39. Connelly, N. G., Geiger, W. E. *Chem. Rev.* 1996, **96**, 877.
40. Harris, D. C. *Quantitative Chemical Analysis*, 6th ed., W. H. Freeman: New York, 2003; pp. 394–396.
41. Li, J., Djurovich, P. I., Alleyne, B. D., Tsyba, I., Ho, N. N., Bau, R., Thompson, M. E. *Polyhedron*, 2004, **23**, 419.
42. (a) Thomas, J. C., Peters, J. C. *J. Am. Chem. Soc.* 2001, **123**, 5100. (b) Thomas, J. C., Peters, J. C. *J. Am. Chem. Soc.* 2003, **125**, 8870. (c) Thomas, J. C., Peters, J. C. *Inorg. Chem.* 2003, **42**, 5055.
43. Shiu, K.-B., Lee, J. Y., Wang, Y., Cheng, M.-C., Wang S.-L., Liao, F.-L. *J. Organomet. Chem.* 1993, **453**, 211.
44. Herber, U., Pechmann, T., Weberndorfer, B., Ilg, K., Werner, H. *Chem.— Eur. J.* 2002, **8**(1), 309.
45. (a) Tejel, C., Ciriano, M. A., Lopez, J. A., Lahoz, F. J., Oro, L. A. *Organometallics* 1998, **17**, 1449. (b) Tejel, C., Ciriano, M. A., Lopez, J. A., Lahoz, F. J., Oro, L. A. *Organometallics* 2000, **19**, 4977.
46. Dobbs, D. A., Bergman, R. G. *Inorg. Chem.* 1994, **33**, 5329.
47. Hay, P. J. *J. Phys. Chem. A* 2002, **106**, 1634.
48. The singlet state energy is estimated from the difference between the calculated HOMO and LUMO energies.

49. The redox potentials we measure for compounds **13** and **14** are considerably different from those reported in Ref. [18]. In particular, the reduction potentials we observe in DMF lie outside the range of accessible potentials for the CH_2Cl_2 solvent (used in Ref. [18]).
50. Ohsawa, Y., Sprouse, S., King, K. A., DeArmond, M. K., Hanck, K. W., Watts, R. J. *J. Phys. Chem.* 1987, **91**, 1047.
51. Sohrin, Y., Kokusen, H., Matsui, M. *Inorg. Chem.* 1995, **34**, 3928.
52. Pal, P. K., Chowdhury, S., Drew, M. G. B., Datta, D. *New J. Chem.* 2002, **26**, 367.
53. Haga, M. *Inorg. Chim. Acta* 1983, **75**, 29.
54. Garces, F. O., King, K. A., Watts, R. J. *Inorg. Chem.* 1988, **27**, 3464.
55. Colombo, M. G., Brunold, T. C., Riedener, T., Güdel, H. U., Förtsch, M., Bürgi, H.-B. *Inorg. Chem.* 1994, **33**, 545.
56. Carlson, G. A., Djurovich, P. I., Watts, R. J. *Inorg. Chem.* 1993, **32**, 4483.
57. The equations of $k_r = \Phi/\tau$ and $k_{nr} = (1 - \Phi)/\tau$ were used to calculate the rates of radiative and nonradiative decay, where Φ is the quantum efficiency and τ is the luminescence lifetime of the sample at room temperature.
58. (a) Humbs, W., Yersin, H. *Inorg. Chim. Acta* 1997, **265**, 139. (b) Yersin, H., Schuetzenmeier, S., Wiedenhofer, H., von Zelewsky, A. *J. Phys. Chem.* 1993, **97**, 13496.
59. Caspar, J. V., Westmoreland, T. D., Allen, G. H., Bradley, P. G., Meyer, T. J., Woodruff, W. H. *J. Am. Chem. Soc.* 1984, **106**, 3492.
60. Rillema, D. P., Blanton, C. B., Shaver, R. J., Jackman, D. C., Boldaji, M., Bundy, S., Worl, L. A., Meyer, T. J. *Inorg. Chem.* 1992, **31**, 1600.
61. Allen, G. H., White, R. P., Rillema, D. P., Meyer, T. J. *J. Am. Chem. Soc.* 1984, **106**, 2613.
62. Damrauer, N. L., Boussie, T. R., Devenney, M., McCusker, J. K. *J. Am. Chem. Soc.* 1997, **119**, 8253.
63. (a) Köhler, A., Beljonne, D. *Adv. Funct. Mater.* 2004, **14**, 11. (b) Liu, Y., Jiang, S., Glusac, K., Powell, D. H., Anderson, D. F., Schanze, K. S. *J. Am. Chem. Soc.* 2002, **124**, 12412.
64. Vlcek, A. A., Dodsworth, E. S., Pietro, W. J., Lever, A. B. P. *Inorg. Chem.* 1995, **34**, 1906.
65. (a) Vacha, M., Koide, Y., Kotani, M., Sato, H. *J. Luminescence* 2004, **107**, 51. (b) Yeh, A. T., Shank, C. V., McCusker, J. K. *Science* 2000, **289**, 935.
66. Komada, Y., Yamauchi, S., Hirota, N. *J. Phys. Chem.* 1986, **90**, 6425.

67. (a) Finkenzeller, W., Stoessel, P., Kulikova, M., Yersin, H. *Proceedings of SPIE "Optical Science and Technology"*, San Diego (USA), August 2003, (*Conference 5214*). (b) Strommen, D. P., Mallick, P. K., Danzer, G. D., Lumpkin, R. S., Kincaid, J. R. *J. Phys. Chem.* 1990, **94**, 1357.
68. Cummings, S. D., Eisenberg, R. *J. Am. Chem. Soc.* 1996, **118**, 1949.
69. $\Phi = k_r/(k_r + k_{nr})$.
70. For comparative purposes the radiative decay rates (k_r) of compounds **13** and **14** given in Table 30.4 are from data measured in degassed 2-MeTHF solution. The absorption spectra and extinction coefficients of **13** and **14** (in CH_2Cl_2) and room-temperature luminescence lifetimes ($\tau = 1.5$ μs for **13** and $\tau = 2.9$ μs for **14** in degassed CH_3CN) are comparable to the reported data in Ref. [18]. However, the quantum efficiencies we measure for these Ir complexes are significantly lower than the reference data.
71. Herzberg, G. *Molecular Spectra and Molecular Structure*; Van Nostrand: New York, 1950; Vol. 1, Chapter 4.
72. Turro, N. J. *Modern Molecular Photochemistry*; University Science Books: Mill Valley, CA, 1978; pp. 86–88.
73. (a) Felix, F., Ferguson, J., Güdel, H. U., Ludi, A. *Chem. Phys. Lett.* 1979, **62**, 153. (b) Decurtins, S., Felix, F., Ferguson, J., Güdel, H. U., Ludi, A. *J. Am. Chem. Soc.* 1980, **102**, 4102.
74. McGlynn, S. P., Azumi, T., Kinoshita, M. *Molecular Spectroscopy of the Triplet State*; Prentice Hall: Englewood Cliffs, NJ, 1969; Chapter 5.
75. Demas, J. N., Crosby, G. A. *J. Am. Chem. Soc.* 1971, **93**, 2841.
76. Watts, R. J., Crosby, G. A. *J. Am. Chem. Soc.* 1972, **94,** 2606.
77. Watts, R. J., Crosby, G. A., Sansregret, J. L. *Inorg. Chem.* 1972, **11**, 1474.
78. Yamauchi, S., Komada, Y., Hirota, N. *Chem. Phys. Lett.* 1986, **129**, 197.
79. (a) Holmes, R. J., D'Andrade, B. W., Forrest, S. R., Ren, X., Li, J., Thompson, M. E., *Appl. Phys. Lett.* 2003, **83**, 3818. (b) Ren, X., Li, J., Holmes, R. J., Djurovich, P. I., Forrest, S. R., Thompson, M. E. *Chem. Mater.* 2004, **16**, 4743.
80. D'Andrade, B. W., Holmes, R. J., Forrest, S. R. *Adv. Mater.* 2004, **16**, 624.
81. Ruiz, J., Florenciano, F., Rodríguez, V., de Haro, C., Lopóz, G., Pérez, J. *Eur. J. Inorg. Chem.* 2002, 2736.
82. Huang, L., Seward, K. J., Sullivan, B. P., Jones, W. E., Mecholsky, J. J., Dressick, W. J. *Inorg. Chim. Acta* 2000, **310**, 227.

83. (a) Subhan, M. A., Suzuki, T., Kaizaki, S. *J. Chem. Soc., Dalton Trans.* 2002, 1416. (b) Sanada, T., Suzuki, T., Yoshida, T., Kaizaki, S. *Inorg. Chem.* 1998, **37**, 4712. (c) Klink, S. I., Keizer, H., van Veggel, F. C. J. M. *Angew. Chem., Int. Ed.* 2000, **39**(23), 4319.
84. Kercher, M., Konig, B., Zieg, H., De Cola, L. *J. Am. Chem. Soc.* 2002, **124**, 1154.
85. Chassot, L., von Zelewsky, A., Sandrini, D., Maestri, M., Balzani, V. *J. Am. Chem. Soc.* 1986, **108**, 6084.

Chapter 31

Cationic Bis-cyclometalated Iridium(III) Diimine Complexes and Their Use in Efficient Blue, Green, and Red Electroluminescent Devices

Arnold B. Tamayo, Simona Garon, Tissa Sajoto, Peter I. Djurovich, Irina M. Tsyba, Robert Bau, and Mark E. Thompson

Department of Chemistry, University of Southern California, Los Angeles, California 90089, USA
met@usc.edu

A series of cationic Ir(III) complexes with the general formula $(C^\wedge N)_2 Ir(N^\wedge N)^+ PF_6^-$ featuring bis-cyclometalated 1-phenylpyrazolyl-$N,C^{2'}$ $(C^\wedge N)$ and neutral diimine ($N^\wedge N$, e.g., 2,2'-bipyridyl) ligands were synthesized and their electrochemical, photophysical, and electroluminescent properties studied. Density functional theory calculations indicate that the highest occupied molecular orbital of the compounds is comprised of a mixture of Ir d and phenylpyrazolyl-based orbitals, while the lowest unoccupied molecular orbital has predominantly diimine character. The oxidation and

---

Reprinted from *Inorg. Chem.*, **44**(24), 8723–8732, 2005.

Electrophosphorescent Materials and Devices
Edited by Mark E. Thompson
Text Copyright © 2005 American Chemical Society
Layout Copyright © 2024 Jenny Stanford Publishing Pte. Ltd.
ISBN 978-981-4877-34-3 (Hardcover), 978-1-003-08872-1 (eBook)
www.jennystanford.com

reduction potentials of the complexes can be independently varied by systematic modification of either the $C^{\wedge}N$ or $N^{\wedge}N$ ligands with donor or acceptor substituents. The electrochemical redox gaps ($E_{ox}-E_{red}$) were adjusted to span a range between 2.39 and 3.08 V. All of the compounds have intense absorption bands in the UV region assigned to $^1(\pi-\pi^*)$ transitions and weaker charge-transfer (CT) transitions that extend to the visible region. The complexes display intense luminescence both in fluid solution and as neat solids at 298 K that is assigned to emission from a triplet metal–ligand-to-ligand CT (^{3}MLLCT) excited state. The energy of the ^{3}MLLCT state varies in nearly direct proportion to the size of the electrochemical redox gap, which leads to emission colors that vary from red to blue. Three of the $(C^{\wedge}N)_2Ir(N^{\wedge}N)^+PF_6^-$ complexes were used as active materials in single-layer light-emitting electrochemical cells (LECs). Single-layer electroluminescent devices were fabricated by spin-coating the Ir complexes onto an ITO–PEDOT/PSS substrate followed by deposition of aluminum contacts onto the organic film. Devices were prepared that give blue, green, and red electroluminescence spectra ($\lambda_{max} = 492, 542$, and 635 nm, respectively), which are nearly identical with the photoluminescence spectra of thin films of the same materials. The single-layer LECs give peak external quantum efficiencies of 4.7, 6.9, and 7.4% for the blue, green, and red emissive devices, respectively.

31.1 Introduction

Light-emitting electrochemical cells (LECs) offer a number of advantages over traditional organic light-emitting diodes (OLEDs). Typical OLEDs require a multilayered structure for charge injection, transport, and light emission, as well as a low-work-function metal cathode and a high-work-function metal or metal oxide anode to provide efficient charge injection [1–4]. LECs require only a single layer of organic semiconductor, which is processed directly from solution. LECs generally give low turn-on voltages, close to the photon energy, which are largely independent of the thickness of the active layer. Furthermore, charge injection in an LEC is insensitive to the work function of the electrode material, thus permitting the use of a wide variety of metals as cathode materials, including those with moderate to high work functions, such as gold [5–7].

The first example of a solid-state LEC was demonstrated by Pei et al. [8]. The LEC was based on a spin-cast polymer blend sandwiched between two electrodes. The polymer blend was comprised of a mixture of a semiconducting and luminescent polymer (MEH/PPV), a salt ($Li^+CF_3SO_3^-$), and an ion-conducting polymer [poly(ethylene oxide), PEO] in which the salt dissolves. When an external bias is applied, the polymer is oxidized (p-type doping) at the anode and reduced at the cathode (n-type doping). Ions redistribute within the organic film to compensate for the charge buildup due to polymer doping. Li^+ ions concentrate near the cathode and $CF_3SO_3^-$ near the anode. This ion redistribution leads to a significant delay between the time a bias is applied and light emission is observed because electron–hole recombination occurs at the newly formed p–n junction. This process is reversible, such that when the bias is removed, the p and n dopings are lost and the ions return to a homogeneous distribution within the film. The function of the PEO is to facilitate the transport of charge-compensating ions to the p- and n-doped regions. Luminance values of 200 cd m^{-2} at 4 V and external quantum efficiencies (EQEs) as high as 2% have been reported for this system [8–10].

Recently, it has been demonstrated that cationic chelated complexes containing 4d^6 and 5d^6 metal ions, e.g., Ru^{2+} and Os^{2+}, can be used as active materials in solid-state LECs [11–14] and have several advantages over the polymer-based LECs. These metal complexes can be synthesized and purified with relative ease (compared to semiconducting polymers). They show excellent electrochemical, photochemical, and thermal stability, good charge-transport properties, long-lived excited states, and good photoluminescence (PL) efficiencies. The positive charges on the complexes are compensated for by negatively charged counterions (e.g., PF_6^- or ClO_4^-). Thus, LECs based on charged metal complexes do not require the presence of an added electrolyte or an ion-conducting polymer, making the device architecture simpler than that of polymer-based LECs. Similar to the situation for polymer-based LECs, the anions migrate into the p-doped region; however, the n-doping process involves reduction of the metal complex, so cation migration into the n-doped region is not required. This simplified redistribution process results in a short turn-on time in these devices, which ranges from several seconds to minutes [7, 11–14].

Most of the single-layer electroluminescent devices based on transition-metal complexes have employed bipyridyl complexes of Ru(II) and Os(II). While devices fabricated using these metal complexes have reached EQEs of 5.5% [11c] and 1.1% [14], respectively, only orange–red emission has been reported thus far. Recently, charged complexes containing the bis–cyclometalated Ir(III) moiety have been found to be more versatile in terms of color tunability [15–17] and device efficiency [18, 19]. The cationic complexes that were examined in LECs have been limited to those with a phenylpyridine-based cyclometalating ligand, e.g., (ppy)$_2$Ir(bpy)$^+$ [ppy = 2-phenylpyridinato-$N,C^{2'}$; bpy = 2,2'-bipyridine]. However, there is some ambiguity concerning the nature of the lowest excited state in these complexes because emission can occur from triplet states localized on either the ppy or bpy ligands [20]. The uncertainty comes about from the fact that the triplet energies of phenylpyridyl and bipyridyl ligands are nearly identical (ppy, $^3\pi - \pi = 430$ nm; bpy, $^3\pi - \pi = 436$ nm) [21]. Replacing ppy with a higher energy $C^\wedge N$ ligand such as 1-phenylpyrazolyl (ppz, $^3\pi - \pi = 380$ nm) [22] simplifies spectral characterization, such that the lowest triplet state can be unambiguously assigned to the "Ir(bpy)" fragment of (ppz)$_2$Ir(bpy)$^+$. Cyclometalated compounds of Ir-containing ppz are known to have emission energies higher than those of ppy–based derivatives [23–25]. To date, however, a systematic study on the effect that changes in the ppz or bpy ligands have on the electrochemical and photophysical properties of these compounds has not been reported, nor electroluminescent devices.

In this chapter, we report on the synthesis and photophysical characterization of a series of luminescent $(C^\wedge N)_2$Ir$(N^\wedge N)^+$ compounds containing ppz-based cyclometalating ligands. We demonstrate the ability to tune the emission color by varying the electrochemical gap of the parent compound, (ppz)$_2$Ir(bpy)$^+$PF$_6^-$, which emits yellow–green. Blue emission is achieved by substituting the 4' and 6' positions on the ppz ligand with electron-withdrawing F atoms, while red emission is achieved by changing the diimine ligand from bpy to 2,2′-biquinolyl (biq). We also report on the device characteristics of single-layer electroluminescent devices fabricated using these compounds. The LEC devices cover a larger color gamut and exhibit higher EQEs than values reported for similar single-layer LEC devices made with ruthenium(II) and osmium(II) diimine complexes.

31.2 Experimental Section

31.2.1 General Procedures

Solvents and reagents for synthesis were purchased from Aldrich, Matrix Scientific, and EM Science and used without further purification. Acetonitrile (EM Science, anhydrous, 99.8%) and tetra-*n*-butylammonium hexafluorophosphate (TBAH; Fluka, electrochemistry grade) were used for spectroscopic and electrochemical measurements. $IrCl_3 \cdot nH_2O$ was purchased from Next Chimica. The compounds 1-(2′,4′-difluorophenyl)pyrazole, 1-(4′-*tert*-butylphenyl)pyrazole, and 1-(5′-methoxyphenyl)pyrazole were prepared following literature procedures [26]. All experiments involving $IrCl_3 \cdot nH_2O$ or any other Ir(III) species were carried out in an inert atmosphere despite the stability of the compounds in air, with the main concern being the oxidative and thermal stability of the intermediate complexes at the high temperatures used in the reactions. Cyclometalated Ir(III) dichloro-bridged dimers of the general formula $(C^{\wedge}N)_2 Ir(\mu\text{-}Cl)_2 Ir(C^{\wedge}N)_2$ (where $C^{\wedge}N$ represents a cyclometalating ligand) were synthesized by the method reported by Nonoyama [27], which involves the heating of $IrCl_3 \cdot nH_2O$ with 2–2.5 equiv of a cyclometalating ligand in a 3:1 mixture of 2-ethoxyethanol and deionized water to 110 °C.

31.2.2 Synthesis of $(C^{\wedge}N)_2 Ir(N^{\wedge}N)^+ PF_6^-$ Complexes: General Procedure

In a round-bottomed flask, 0.100 g of dichloro-bridged iridium dimer and 2.0 equiv of the diimine are mixed together in 25 mL of methanol. The solution was then refluxed overnight under an inert atmosphere. After cooling to room temperature, counterion exchange from Cl$^-$ to PF$_6^-$ was accomplished via a metathesis reaction in which complexes **1–9** were precipitated from the methanol solution with an excess of NH$_4$PF$_6$, washed with water and methanol, and dried.

(i) Iridium(III) Bis[1′-phenylpyrazolato-N,$C^{2'}$]-2,2′-bipyridine Hexafluorophosphate (1). Yield: 83.2%. ^{1}H NMR (360 MHz, CD$_2$Cl$_2$): δ 8.49 (d, J = 8.1 Hz, 2H), 8.19 (dd, J = 5.4 and 1.5 Hz, 2H), 8.17 (d, J = 2.7 Hz, 2H), 8.15 (ddd, J = 8.1, 8.1, and 1.7 Hz, 2H), 7.48 (ddd, J = 8.1, 5.4, and 1.0 Hz, 2H), 7.36 (dd, J = 8.1 and 1.2 Hz, 2H), 7.09 (ddd, J = 7.6, 7.3, and 1.2 Hz, 2H), 6.9 (dt, J = 7.8, 7.8, and 1.5 Hz, 2H), 6.87 (dt, J = 2.2 Hz, 2H), 6.56 (dd, J = 3.2 and 2.4 Hz, 2H), 6.33 (dd, J = 7.3 and 1.2 Hz, 2H). ^{13}C NMR (90.55 MHz, CD$_2$Cl$_2$): δ156.8, 151.6, 143.3, 140.0, 138.5, 133.7, 132.2, 128.3, 127.6, 127.4 124.7, 123.9, 112.9, 108.7. Anal. Calcd for C$_{28}$H$_{22}$F$_6$IrN$_6$P: C, 43.13; H, 2.84; N, 10.78. Found: C, 42.86; H, 2.55; N, 10.34.

(ii) Iridium(III) Bis[1′-phenylpyrazolato-N,$C^{2'}$]-4,4′-di-*tert*-butyl-2,2′-bipyridine Hexafluorophosphate (2). Yield: 78.5%. ^{1}H NMR (360 MHz, CD$_3$CN): δ 8.32 (d, J = 2.9 Hz, 2H), 8.05 (d, J = 2.9 Hz, 2H), 7.97 (d, J = 5.9 Hz, 2H), 7.38 (dd, J = 5.6 and 1.2 Hz, 2H), 7.01 (dd, J = 8.8 and 7.8 Hz, 2H), 6.94 (d, J = 2.2 Hz, 2H), 6.83 (dd, J = 8.1 and 7.6 Hz, 2H), 6.54 (dd, J = 2.9 and 2.44 Hz, 2H), 6.29 (d, J = 7.6 Hz, 2H), 1.42 (s, 18H). ^{13}C NMR (90.55 MHz, CD$_3$CN): δ 165.1, 157.2, 151.4, 144.2, 139.3, 134.0, 133.7, 128.8, 127.5, 125.8, 124.1, 122.6, 112.8, 109.6, 36.5, 30.5. Anal. Calcd for C$_{36}$H$_{38}$F$_6$IrN$_6$P: C, 48.48; H, 4.29; N, 9.42. Found: C, 48.48; H, 4.29; N, 9.06.

(iii) Iridium(III) Bis[1′-phenylpyrazolato-N,$C^{2'}$]-4,4′-dimethoxy-2,2′-bipyridine Hexafluorophosphate (3). Yield: 90.2%. ^{1}H NMR (360 MHz, CD$_3$CN): δ 7.80 (dd, J = 3.1 and 0.4 Hz, 2H), 7.42 (d, J = 2.89 Hz, 2H), 7.32 (d, J = 8.0 and 1.2 Hz, 2H), 6.44–6.55 (m, 6H), 6.30 (ddd, J = 7.8, 7.3, and 1.2 Hz, 2H), 6.03 (dd, J = 2.8 and 2.3 Hz, 2H) 5.72 (dd, J = 7.5 and 1.4 Hz, 2H), 3.45 (s, 6H). ^{13}C NMR (90.55 MHz, CD$_3$CN): δ 168.9, 158.9, 152.8, 144.4, 139.4, 134.1, 133.6, 128.8, 127.5, 124.0, 114.4, 112.9, 112.1, 109.1, 57.7.

Anal. Calcd for $C_{30}H_{26}F_6IrN_6O_2P$: C, 42.91; H, 3.12; N, 10.01. Found: C, 43.01; H, 3.08; N, 9.98.

(iv) Iridium(III) Bis[1′-phenylpyrazolato-N,$C^{2'}$]-4,4′-(dicarboxylic acid diethyl ester)-2,2′-bipyridine Hexafluorophosphate (4). Yield: 85.1%. ^{1}H NMR (360 MHz, CD$_3$CN): δ 9.04 (s, 2H), 8.35 (d, J = 3.0 Hz, 2H), 8.3 (d, J = 6.7 Hz, 2H), 7.96 (dd, J = 3.9 and 1.5 Hz, 2H), 7.47 (d, J = 8.1 Hz, 2H), 7.08 (dd, J = 8.4 and 7.8 Hz, 2H), 7.01 (d, J = 2.3 Hz, 2H), 6.89 (dd, J = 8.5 and 6.4 Hz, 2H), 6.55 (dd, J = 3.2 and 2.5 Hz, 2H), 6.24 (d, J = 7.4 Hz, 2H), 4.46 (q, J = 7.1 Hz, 4H) 1.41 (dd, J = 7.1 Hz, 6H). ^{13}C NMR (90.55 MHz, CDCl$_3$): δ 162.8, 156.7, 157.7, 142.4, 140.6, 138.7, 133.1, 130.9, 127.5, 127.0, 126.9, 124.0, 123.7, 111.6, 108.7, 63.1, 14.2. Anal. Calcd for $C_{34}H_{30}F_6IrN_6O_4P$: C, 44.20; H, 3.27; N, 9.10. Found: C, 44.16; H, 3.23; N, 9.05.

(v) Iridium(III) Bis[1′-phenylpyrazolato-N,$C^{2'}$]-2,2′-biquinoline Hexafluorophosphate (5). Yield: 89.0%. ^{1}H NMR (360 MHz, CD$_3$CN): δ8.67−8.76 (m, 2H), 8.35 (d, J = 2.9 Hz, 1H), 8.01 (dd, J = 7.8 and 2.9 Hz, 1H), 7.97 (d, J = 8.8 Hz, 2H), 7.58 (ddd, J = 8.3, 6.83, and 0.97 Hz, 1H), 7.38 (dd, J = 7.8 and 0.98 Hz, 2H), 7.26 (ddd, 8.8, 6.8, and 1.66 Hz, 1H), 6.97 (dt, J = 7.81, 7.82, and 0.98 Hz, 1H), 6.96 (d, J = 2.44 Hz, 1H), 6.76 (ddd, J = 7.31, 6.32, and 0.98 Hz, 1H), 6.55 (dd, J = 3.42 and 2.93 Hz, 1H), 6.05 (dd, J = 7.33 and 1.46 Hz, 1H). ^{13}C (90.55 MHz, CD$_3$CN): δ 160.8, 149.4, 142.0, 140.9, 140.9, 134.0, 132.2, 132.2, 129.9, 129.7, 129.4, 129.2, 129.0, 128.8, 127, 2, 124.1, 122.2, 113.0. Anal. Calcd for $C_{36}H_{26}F_6IrN_6P$: C, 49.15; H, 2.98; N, 9.55. Found: C, 48.95; H, 2.90; N, 9.52.

(vi) Iridium(III) Bis[1′-(4′-*tert*-butylphenyl)pyrazolato-N,$C^{2'}$]-2,2′-biquinoline Hexafluorophosphate (6). Yield: 93.2%. ^{1}H NMR (360 MHz, CDCl$_3$): δ 8.88 (d, J = 8.8 Hz, 2H), 8.72 (d, J = 8.8 Hz, 2H), 8.06 (d, J = 2.9 Hz, 2H), 7.98 (d, J = 8.8 Hz, 2H), 7.91 (d, J = 8.3 Hz, 2H), 7.50 (dd, J = 8.8 and 7.3 Hz, 2H), 7.16 (ddd, J = 9.3, 6.8, and 1.2 Hz, 2H), 7.09 (d, J = 8.6 Hz, 2H), 6.93 (dd, 8.3 and 2.0 Hz, 2H), 6.75 (d, 2.4 Hz, 2H), 6.52 (dd, J = 2.9 and 2.4 Hz, 2H), 5.92 (d, J = 1.7 Hz, 2H), 1.03 (s, 18H). ^{13}C (90.55 MHz, CDCl$_3$): δ141.4, 139.5, 139.0, 130.9, 130.4, 129.8, 129.6, 128.7, 128.3, 126.2, 121.9, 119.7,

110.7, 108.3, 34.1, 31.1. Anal. Calcd for $C_{44}H_{42}F_6IrN_6P$: C, 53.27; H, 4.27; N, 8.47. Found: C, 53.05; H, 4.27; N, 8.25.

(vii) Iridium(III) Bis[1′-(5′-methoxyphenyl)pyrazolato-$N,C^{2'}$]-4,4′-di-*tert*-butyl-2,2′-bipyridine Hexafluorophosphate (7).
Yield: 61.3%. ^{1}H NMR (360 MHz, CD$_3$CN): δ7.91 (d, $J = 1.5$ Hz, 2H), 7.81 (d, $J = 2.9$ Hz, 2H), 7.48 (d, $J = 5.9$ Hz, 2H), 6.96 (dd, $J = 7.3$ and 2.0 Hz, 2H), 6.58 (d, $J = 2.5$ Hz, 2H), 6.38 (d, $J = 2.0$ Hz, 2H), 6.02 (dd, $J = 4.9$ and 2.4 Hz, 2H), 6.00 (d, $J = 2.4$ Hz, 2H), 5.57 (d, $J = 8.3$ Hz, 2H), 3.22 (s, 6H), 1.44 (s, 18H). ^{13}C NMR (90.55 MHz, CDCl$_3$): δ164.0, 156.6, 156.3, 150.3, 142.9, 138.3, 133.1, 126.8, 124.7, 121.2, 121.0, 112.5, 108.3, 98.9, 55.5, 35.7, 30.2. Anal. Calcd for $C_{38}H_{42}F_6IrN_6O_2P$: C, 47.94; H, 4.45; N, 8.83. Found: C, 47.6; H, 4.3; N, 8.34.

(viii) Iridium(III) Bis[1′-(3′-biphenylyl)pyrazolato-$N,C^{2'}$)]-4,4′-di-*tert*-butyl-2,2′-bipyridine Hexafluorophosphate (8).
Yield: 91.1%. ^{1}H NMR (360 MHz, CD$_3$CN): δ 8.51 (d, $J = 2.7$ Hz, 2H), 8.49 (d, $J = 1.7$ Hz, 2H), 8.07 (d, $J = 5.9$ Hz, 2H), 7.78 (d, $J = 1.7$ Hz, 2H), 7.6 (dd, $J = 7.3$ and 1.2 Hz, 4H), 7.51 (dd, $J = 6.1$ and 1.7 Hz, 2H), 7.43 (m, 4H), 7.32 (dt, 7.3 and 1.7 Hz, 2H), 7.05 (d, $J = 2.2$ Hz, 2H), 6.6 (dd, $J = 3.2$ and 2.4 Hz, 2H), 6.38 (d, $J = 7.6$ Hz, 2H), 1.42 (s, 18H). ^{13}C NMR (90.55 MHz, CDCl$_3$): δ 164.0, 156.1, 150.2, 143.4., 140.6, 138.6, 136.5, 128.7, 128.7, 126.9, 126.8, 126.5, 126.5, 125.6 124.9, 120.9, 110.0, 108.6, 35.6, 30.2. Anal. Calcd for $C_{48}H_{46}F_6IrN_6P$: C, 55.22; H, 4.44; N, 8.05. Found: C, 54.92; H, 4.13; N, 7.91.

(ix) Iridium(III) Bis[1′-(4′,6′-difluorophenyl)pyrazolato-$N,C^{2'}$]-4,4′-di-*tert*-butyl-2,2′-bipyridine Hexafluorophosphate (9).
Yield: 65.6%. ^{1}H NMR (360 MHz, CDCl$_3$): δ 8.40 (d, $J = 1.7$ Hz, 2H), 8.33 (d, $J = 2.7$ Hz, 2H), 7.95 (d, $J = 5.9$ Hz, 2H), 7.46 (dd, $J = 5.9$ and 1.7 Hz, 2H), 7.01 (d, $J = 2.2$ Hz, 2H), 6.66 (ddd, $J = 12.5$, 8.8, and 2.2 Hz, 2H), 6.59 (dd, $J = 3.2$ and 2.7 Hz, 2H), 5.72 (dd, $J = 8.1$ and 2.4 Hz, 2H), 1.45 (s, 18H). ^{13}C NMR (90.55 MHz, CDCl$_3$): δ 164.8, 160.3, 156.2, 149.7, 148.9, 138.7, 137.1, 131.4, 131.2, 126.9, 125.2, 121.9, 109.2, 115.1, 109.2, 99.4, 35.8, 30.1. Anal. Calcd for $C_{36}H_{34}F_{10}IrN_6P$: C, 45.86; H, 3.56; N, 8.72. Found: C, 45.62; H, 3.63; N, 8.45.

31.2.3 X-ray Crystallography

Diffraction data for compound **2'''** w ere collected at $T = 296$ K [2]. The data set was collected on a Bruker SMART APEX CCD diffractometer with graphite monochromated Mo Kα radiation ($\lambda = 0.710\ 73$ Å). The cell parameters for the iridium complex were obtained from a least-squares refinement of the spots (from 60 collected frames) using the SMART program. One hemisphere of crystal data for the compound was collected up to a resolution of 0.86 Å, and the intensity data were processed using the Saint Plus program. All of the calculations for the structure determination were carried out using the SHELXTL package (version 5.1) [28]. Initial atomic positions were located by Patterson methods using X-ray spectroscopy, and the structure of **2** was refined by a least-squares method using SHELX93 with 17 203 independent reflections within the range of $\theta = 0.86\text{--}27.57°$. Absorption corrections were applied by using SADABS [29]. In most cases, hydrogen positions were input and refined in a riding manner along with the attached carbons. A summary of the refinement details and the resulting factors for the complex are given in the Supporting Information.

31.2.4 Density Functional Theory (DFT) Calculations

DFT calculations were performed using the Titan software package (Wavefunction, Inc.) at the B3LYP/LACVP** level. The HOMO (highest occupied molecular orbital) and LUMO (lowest unoccupied molecular orbital) energies were determined using minimized singlet geometries to approximate the ground state. The minimized triplet geometries were used to calculate the triplet molecular orbitals and to approximate the triplet HSOMO (where HSOMO = highest singly occupied molecular orbital) as well as the spin-density surface.

31.2.5 Electrochemical and Photophysical Characterization

Cyclic voltammetry and differential pulse voltammetry were performed using an EG&G potentiostat/galvanostat model 283.

Anhydrous acetonitrile was used as the solvent under an inert atmosphere, and 0.1 M TBAH was used as the supporting electrolyte. A glassy carbon rod was used as the working electrode, a platinum wire was used as the counter electrode, and a silver wire was used as a pseudo reference electrode. The redox potentials are based on values measured from differential pulse voltammetry and are reported relative to a ferrocenium/ferrocene (Cp_2Fe^+/Cp_2Fe) redox couple used as an internal reference [30], while electrochemical reversibility was determined using cyclic voltammetry.

The UV–visible spectra were recorded on a Hewlett-Packard 4853 diode array spectrophotometer. Steady-state emission experiments at room temperature and 77 K were performed on a Photon Technology International QuantaMaster model C-60SE spectrofluorimeter. Phosphorescence lifetime measurements were performed on the same fluorimeter equipped with a microsecond xenoflash lamp or on an IBH Fluorocube lifetime instrument by a time-correlated single-photon counting method using either a 331-, 373-, or 405-nm LED excitation source. Quantum efficiency (QE) measurements were carried out at room temperature in degassed 2-MeTHF solutions using the optically dilute method [31]. Solutions of Coumarin 47 in ethanol ($\Phi = 0.73$) [32] were used as the reference. NMR spectra were recorded on Bruker AM 360-MHz instrument, and chemical shifts were referenced to a residual protiated solvent. Elemental analyses (CHN) were performed at the Microanalysis Laboratory at the University of Illinois, Urbana-Champaign, IL.

31.2.6 Device Fabrication

Prior to device fabrication, indium/tin oxide (ITO) on glass was patterned as 2-mm-wide stripes with a resistivity of 20 Ω $\square^{-1}$. The substrates were cleaned by sonication in a soap solution, rinsed with deionized water, boiled in trichloroethylene, acetone, and ethanol for 5–6 min in each solvent, and dried with nitrogen. Finally, the substrates were treated with UV ozone for 10 min. Poly(3,4-ethylenedioxythiophene)/poly(styrene sulfonate) (PEDOT/PSS), a widely used hole-injecting material, was filtered and spin-coated onto the ITO substrates at a rate of 8000 rpm for 60 s. The substrates were then baked in the oven under vacuum for 40 min at

$T = 90–100\,°C$ to remove the residual solvent. Films of the iridium complexes were then spin-coated from an acetonitrile solution (30 mg mL^{-1}) onto the cooled substrates. The thicknesses of the films were between 70 and 90 nm, as measured by ellipsometry. Aluminum cathodes (1200 Å) were vapor-deposited onto the substrates in a high-vacuum chamber through a shadow mask that defined four devices per substrate with a 2-mm^2 active area each. The devices were tested within 3h of device fabrication. The electrical and optical intensity characteristics of the devices were measured with a Keithly 2400 source/meter/2000 multimeter coupled to a Newport 1835-C optical meter, equipped with a UV-818 Si photodetector. Only light emitting from the front face of the device was collected and used in subsequent efficiency calculations. The electroluminescence (EL) spectra were measured on a PTI QuantaMaster model C-60SE spectrofluorimeter, equipped with a 928 PMT detector and corrected for detector response [33]. The emission was found to be uniform throughout the area of each device.

31.3 Results and Discussion

31.3.1 Synthesis and Structure

Complexes **1–9** (see Fig. 31.1) were synthesized from the dichloro-bridged Ir(III) dimer and 2.0 equiv of the diimine ligand in refluxing methanol, followed by metathesis with NH$_4$PF$_6$. The complexes were characterized using spectroscopic methods in addition to elemental analysis. ^{1}H and ^{13}C NMR spectroscopies indicate that the complexes are C_2-symmetric stereoisomers (one set of proton signals for the cyclometalating ligand). The ligand configuration at the Ir(III) center, with cis–metalated phenyl groups and *trans*-pyrazolyl moieties, is illustrated schematically in Fig. 31.1 and confirmed by X-ray crystallography.

Single crystals of complex **2** were grown from an acetone solution and characterized by X-ray crystallography. The crystal data, atomic coordinates, and a complete listing of bond lengths and angles are given in the Supporting Information; selected bond lengths and

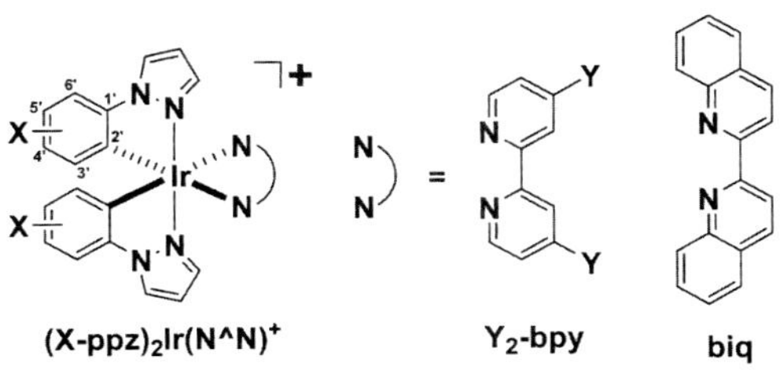

Compound	ppz, X =	N^N, Y =
1	H	bpy, H
2	H	bpy, *t*-butyl
3	H	bpy, OCH$_3$
4	H	bpy, CO$_2$Et
5	H	biq
6	5'-*t*-butyl	biq
7	4'-methoxy	bpy, *t*-butyl
8	4'-phenyl	bpy, *t*-butyl
9	4',6'-difluoro	bpy, *t*-butyl

Figure 31.1 Structure of the $(C^\wedge N)_2 Ir(N^\wedge N)^+ PF_6^-$ complexes.

angles are given in Table 31.1. A perspective view of the cation is given in Fig. 31.2. For comparison, structural data for the related compounds *mer*-Ir(ppz)$_3$ [23] and Ir(bpy)$_3^{3+}$ [34] are also given in Table 31.1. The C–C and C–N bond lengths and angles are within the normal ranges expected for those of organic fragments. The ligands are arranged in a pseudo-octahedral geometry around the metal center with the expected trans configuration of pyrazolyl groups. The measured bond angle between the *trans*-pyrazolyl groups is 173.4(4)°, comparable to the value of 171.54(8)° reported for *mer*-Ir(ppz)$_3$. The average Ir–C bond length and C$_{pz}$–Ir–N$_{pz}$ angle are 2.019(11) Å and 80.7(4)°, respectively. These values are also consistent with the values 2.025(2) Å and 79.15(8)° reported for *mer*-Ir(ppz)$_3$. The strong trans influence of the phenyl groups results in slightly longer Ir–N(bpy) bond lengths in **2** [Ir-N$_{ave}$ = 2.136(9) Å] as compared to those of Ir(bpy)$_3^{3+}$, where the average Ir–N bond length is 2.021(6) Å. Likewise, the average N$_{pyr}$–Ir–N$_{pyr}$ angle for

Table 31.1 Selected bond distances and angles for **2**, mer-Ir(ppz)$_3$,[a] and Ir(bpy)$_3^{3+b}$

Complex	2	mer-Ir(ppz)$_3$	Ir(bpy)$_3^{3+}$
Ir–C	2.017(11)	2.057 (2) (cis)	
	2.022(10)	1.993 (2) (trans)	
		2.051 (2) (trans)	
Ir–N	2.005(9) (pz)	2.026(2) (trans)	2.021(6)
	2.015(9) (pz)	2.013(2) (trans)	
	2.140(9), 2.132(9)	2.053 (2)	
N$_{pz}$–Ir–N$_{pz}$	173.4(4)	171.54(8)	
C$_{pz}$–Ir–N$_{pz}$	80.1(4)	78.57(8)	
	81.3(4)	79.73(8)	
N$_{bpy}$–Ir–N$_{bpy}$	76.4(3)		78.7(8)[c]

[a]Reference [23]. [b]Reference [34]. [c]Average.

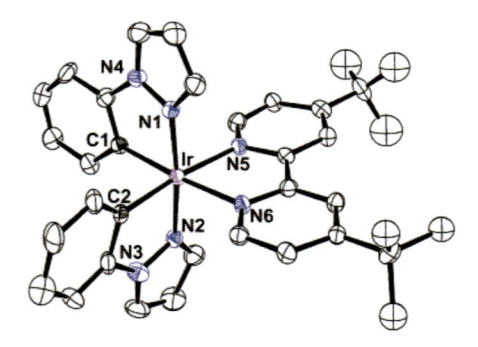

Figure 31.2 ORTEP drawing of **2**. The thermal ellipsoids for the image represent a 25% probability limit. The hydrogen atoms, counterion, and solvent are omitted for clarity.

2 is measured to be 76.4(3)°, comparable to the average value of 78.7(8)° reported for Ir(bpy)$_3^{3+}$.

31.3.2 DFT Calculations

DFT calculations were carried out on all of the complexes at a B3LYP level using a LACVP** basis set to get some insight into the electronic structure of the complexes. A similar approach has been shown to be effective at reproducing the ground- and

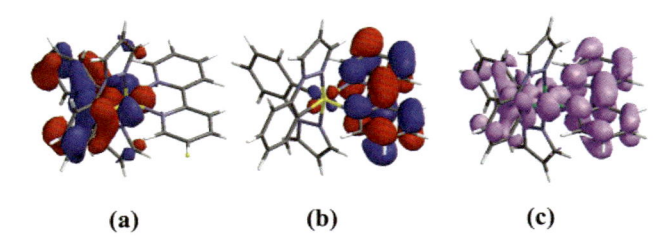

Figure 31.3 (a) HOMO, (b) LUMO, and (c) triplet spin-density surfaces of $(ppz)_2Ir(bpy)^+$ determined using DFT. The HOMO consists mainly of a mixture of Ir and phenyl π orbitals, while the LUMO has predominantly bipyridyl character.

excited-state properties of related cyclometalated Ir(III) and Pt(II) complexes [16, 35]. The discussion here will focus on the results for **1**; however, all of the examined $(C^\wedge N)_2Ir(N^\wedge N)^+$ complexes gave a similar picture for the HOMO and LUMO surfaces. The calculated Ir–C_{ave}(2.05 Å), Ir–$N_{pz(ave)}$ (2.06 Å), and Ir–N_{pyr}(2.18 Å) bond distances and N_{pz}–Ir–N_{pz} (170.93°), C_{pz}–Ir–N_{pz} (79.56°), and N_{pyr}–Ir–N_{pyr} (75.783°) chelate angles for $(ppz)_2Ir(bpy)^+$ compare favorably to the corresponding experimental values determined in the X-ray structure of **2** as well as to literature values of similar bis-cyclometalated iridium(III) diimine complexes [25]. The singlet HOMO and LUMO and triplet spin-density surfaces of $(ppz)_2Ir(bpy)^+$ are illustrated in Fig. 31.3. The calculated HOMO and LUMO energies for $(ppz)_2Ir(bpy)^+$ in the ground state are -8.01 and -5.10 eV, respectively. The HOMO is principally composed of a mixture of Ir d and phenyl π orbitals distributed equally among the two phenylpyrazolyl ligands, whereas the LUMO is predominantly localized on the bipyridyl ligand. The HOMO and LUMO orbitals are orthogonal to each other, and thus, there is little electronic overlap between them. The orthogonal relationship between the HOMO and LUMO suggests that their energies are amenable to independent variation by simple substitution of either the cyclometalate or the diimine, respectively. The energy calculated for the HSOMO in the geometry-optimized triplet state is -5.50 eV. The difference in energy between the singlet ground-state HOMO and triplet HSOMO was then used to estimate the singlet–triplet transition energy. The calculated transition energy of 2.51 eV (494 nm) is comparable to the energy of the emission band found for (ppz)$_2$Ir(bpy)$^+$ (487 nm) in 2-MeTHF at 77 K (vide infra). The location of the unpaired spins in the frontier orbitals of the triplet state is illustrated by the spin-density surface (Fig. 31.3c). The spin-density surface shares the same spatial extent as the singlet HOMO and LUMO surfaces, which leads to a description of the lowest energy excited state as having metal-ligand-to-ligand charge-transfer (MLLCT) character. The DFT results will be further used in the text as they pertain to the electrochemical and spectral interpretation.

31.3.3 Electrochemistry

The electrochemical properties of the bis-cyclometalated iridium diimine complexes in acetonitrile were examined using cyclic voltammetry. A summary of the redox potentials, measured relative to an internal ferrocene reference (Cp$_2$Fe$^+$/Cp$_2$Fe = +0.35 V vs SCE in acetonitrile), is listed in Table 31.2. All complexes exhibit reversible oxidation and reduction processes. For example, (ppz)$_2$Ir(bpy)$^+$ (**1**) has a reversible oxidation wave at 0.95 V and a reversible reduction wave at -1.80 V. For comparison, neutral tris-cyclometalated (ppz)Ir(III) complexes also undergo reversible oxidation at potentials near 0.40 V but are extremely difficult to reduce (reduction potentials <-3.0) [23]. The electrochemical gap ($\Delta E_{1/2}^{\text{redox}} = E_{1/2}^{\text{ox}} - E_{1/2}^{\text{red}}$) found for complex **1** (-2.75 V) is

Table 31.2 Redox properties of compounds **1–9**[a]

Complex	$E_{1/2}^{\text{ox}}$ (V)	$E_{1/2}^{\text{red}}$ (V)	$\Delta E_{1/2}$ (V)
1	0.95	-1.80	2.75
2	0.95	-1.89	2.84
3	0.99	-1.90	2.89
4	0.97	-1.42	2.39
5	1.04	$-1.36, -1.96$	2.40
6	1.01	$-1.44, -2.04$	2.45
7	0.62	-1.90	2.52
8	0.84	-1.88	2.72
9	1.25	-1.83	3.08

[a]Redox measurements were carried out in an anhydrous CH$_3$CN solution, and values are reported relative to Fc$^+$/Fc.

similar to the difference in the singlet HOMO and LUMO energies obtained from DFT calculations (−2.91 eV). Therefore, it can be presumed that oxidation is associated with the bis-cyclometalated phenyl-Ir moiety while reduction occurs on the bipyridyl ligand. It follows from the results of the DFT calculations that the oxidation and reduction potentials of the $(C^\wedge N)_2 Ir(N^\wedge N)^+$ complexes can be independently adjusted by modification of the phenyl and diimine ligands, respectively, with donor or acceptor substituents. For example, complexes **1–5**, which differ only by the identity of the diimine ligand, have reduction potentials that vary over 0.53 V (from −1.89 V for **2**, with electron-donating 4,4′-di-*tert*-butyl groups, to −1.36 V for **5**, which has a π-extended 2,2′-biquinolyl ligand). However, the corresponding oxidation potentials differ over a much smaller range, 0.09 V (from 0.95 V for **2** to 1.04 V for **5**). Similarly, complexes **2** and **7–9**, each with an identical diimine ligand and modified ppz ligands, have oxidation potentials that vary over 0.63 V (from 0.62 V for **7**, with electron-donating 5′-methoxy substituents, to 1.25 V for **9**, with electron-withdrawing 4′,6′-difluoro substituents), whereas the reduction potentials fall within a narrow range of 0.07 V. The ability to independently adjust the oxidation and reduction potentials of these complexes over a wide range leads to $\Delta E_{1/2}^{\text{redox}}$ values that vary from 2.39 V for **4** to 3.08 V for **9**.

31.3.4 Electronic Spectroscopy

Absorption and emission data are summarized in Table 31.3, while spectra for complexes **2**, **6**, **7**, and **9** are presented in Fig. 31.4a. Spectra for the other complexes are given in the Supporting Information. The absorption spectra of the compounds in acetonitrile show intense bands ($\epsilon > 10^4$ M^{-1} cm^{-1}) in the ultraviolet part of the spectrum between 200 and 300 nm. These bands are assigned to spin-allowed π–π^* ligand-centered (LC) transitions in both $C^\wedge N$ and $N^\wedge N$ ligands. The less intense, lower energy absorption features from 300 to 600 nm are assigned to both spin-allowed and spin-forbidden charge-transfer (CT) transitions. Two types of CT transitions can be distinguished in the spectra, bands between 300 and 400 nm of moderate intensity [$\epsilon \approx (1–2)$

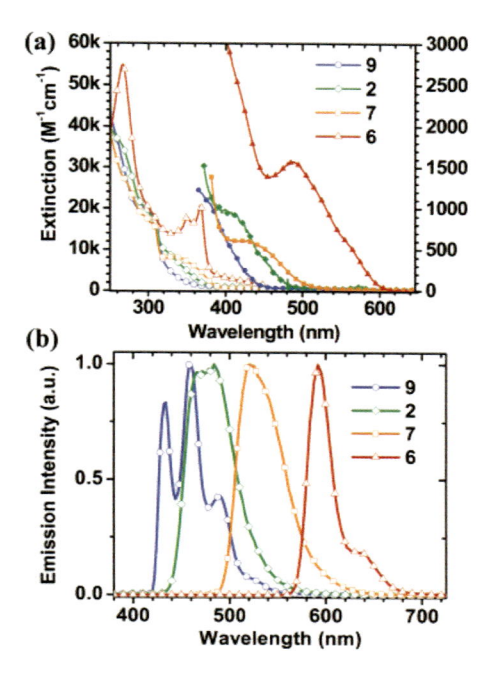

Figure 31.4 (a) Room-temperature absorption (in CH_3CN) and (b) 77 K emission (in 2-MeTHF) spectra of compounds **2** (◊), **6** (△), **7** (□), and **9** (○). The filled symbols in the absorption spectra correspond to the right axis.

× 10^4 M^{-1} cm^{-1}] and transitions in the visible region with much weaker intensity ($\epsilon \leq 10^3$ M^{-1} cm^{-1}). The higher energy bands are assigned to metal-to-ligand CT (MLCT) transitions involving either the cyclo-metalated or diimine ligand, in analogy to features seen at similar energies in related tris-cyclometalated or diimine Ir complexes [20, 25, 36]. The lower energy absorption bands are assigned to $(ppz)_2Ir \rightarrow$ diimine MLLCT transitions. There is a direct correlation between the absorption energy of these MLLCT bands and the $\Delta E_{1/2}^{redox}$ values of the respective complexes (Fig. 31.5). The visible absorption bands have low extinction coefficients because of the poor spatial overlap between the molecular orbitals involved in the MLLCT transitions (see Fig. 31.3a,b) [36]. The lowest energy absorption shoulders on these bands are assigned to spin-forbidden components of the MLLCT transitions.

Table 31.3 Photophysical properties of compounds **1–9**[a]

Complex	Absorbance λ (nm) $(\epsilon, \times 10^4 \text{M}^{-1}\text{cm}^{-1})$	Emission at room temperature λ_{max} (solvent) (nm)	$\tau_{solution}$ (μs)	τ_{powder} (μs)	Φ_{PL}	k_r (10^4 s^{-1})	k_{nr} (10^6 s^{-1})	Emission at 77 K λ (nm)	τ (μs)
1	216 (5.21), 244 (4.37), 261 (3.81), 298 (1.39), 308 (1.26), 345 (0.97), 436 (0.13)	563 (CH$_3$CN), 554 (CH$_2$Cl$_2$), 565 (2-MeTHF), 550 (neat)	0.48	0.37	0.17	35	2	487	4.8
2	207 (5.70), 248 (3.95), 266 (3.53), 296 (2.04), 308 (1.76), 328 (0.89), 420 (0.09)	555 (CH$_3$CN), 538 (CH$_2$Cl$_2$), 550 (2-MeTHF), 535 (neat)	0.72	0.76	0.076	11	1	465	4.3
3	221 (5.98), 247 (4.50), 267 (3.60), 300 (1.47), 325 (0.92), 417 (0.07)	554 (CH$_3$CN), 541 (CH$_2$Cl$_2$), 550 (2-MeTHF), 532 (neat)	0.53	0.69	0.032	6.0	2	440	6.9
4	212 (5.49), 243 (4.07), 298 (2.10), 325 (1.60), 369 (0.57), 482 (0.10)	628 (CH$_3$CN), 620 (CH$_2$Cl$_2$), 625 (2-MeTHF), 615 (neat)	0.10	0.38	0.0012	1.2	10	545	4.4
5	215 (5.48), 258 (6.02), 277 (4.77), 300 (2.23), 323 (1.87), 351 (1.55), 367 (1.88), 417 (0.27), 489 (0.17), 578 (0.06)	616 (CH$_3$CN), 610 (CH$_2$Cl$_2$), 614 (2-MeTHF), 613 (neat)	0.86	0.26	0.017	2.0	1	585	7.7

6	215 (5.39), 266 (5.48), 304 (1.92), 338 (1.44), 350 (1.79), 368 (2.08), 403 (0.32), 426 (0.23), 491 (0.16), 568 (0.05)	627 (CH$_3$CN), 620 (CH$_2$Cl$_2$), 624 (2-MeTHF), 622 (neat)	0.56	0.62	0.0068	1.2	2	592	8.5
7	206 (5.51), 223 (4.85), 244 (3.98), 273 (2.67), 297 (1.82), 307 (1.67), 342 (0.85), 454 (0.05)	618 (CH$_3$CN), 598 (CH$_2$Cl$_2$), 615 (2-MeTHF), 570 (neat)	0.031	0.12	0.0007	2.2	32	522	2.3
8	297 (8.00), 256 (7.81), 297 (3.98), 307 (3.70), 426 (0.09)	570 (CH$_3$CN), 551 (CH$_2$Cl$_2$), 570 (2-MeTHF), 560 (neat)	0.32	0.37	0.010	3.1	3	480	2.4
9	210 (5.72), 246 (4.32), 295 (2.05), 310 (1.68), 386 (0.10)	495 (CH$_3$CN), 484 (CH$_2$Cl$_2$), 493 (2-MeTHF), 487 (neat)	1.34	3.48	0.40	30	0.4	433	3.9

[a]The absorption spectra were measured in CH$_3$CN. The solution PL efficiencies and lifetimes were measured in degassed 2-MeTHF.

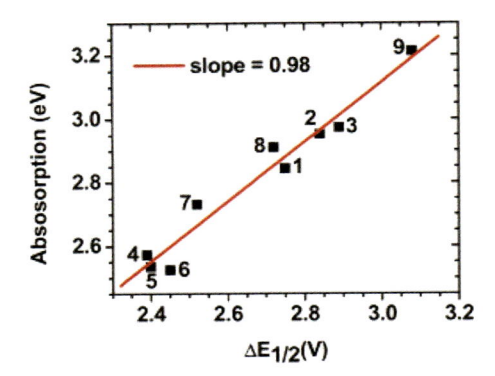

Figure 31.5 Plot of the absorption energy (in CH_3CN) of lowest CT band vs the redox gap, $\Delta E_{1/2}$, for complexes **1–9**.

The Ir complexes all display broad, featureless emission spectra in a fluid solution or a neat solid at room temperature, with maxima ranging from 485 to 630 nm and luminescence decay lifetimes that fall between 0.031 and 1.34 μs (Table 31.1). The emission spectra exhibit large Stokes shifts (ca. 4000–6000 cm^{-1}) from the lowest energy absorption bands. The emission properties at room temperature are consistent with luminescence originating from a triplet MLLCT state. The complexes all undergo large rigidochromic blue shifts upon cooling of the solutions to 77 K. For example, the emission maximum of **8** in 2-MeTHF at room temperature is 570 nm, while at 77 K, the maximum is centered at 480 nm. The hypsochromic shifts are due to solvent reorganization in a fluid solution at room temperature, which stabilize the CT states prior to emission. This process is significantly impeded in a rigid matrix at 77 K, and thus emission occurs at higher energy. The emission spectra remain featureless in a 2-MeTHF solution at 77 K except for spectra of complexes **5**, **6**, and **9**, which show a series of vibronic transitions (Fig. 31.4b). The structured emission from complexes **5**, **6**, and **9** implies that a considerable LC character develops in these three species at low temperature. In particular, the highly structured emission for complex **9** at 77 K has a highest energy peak (433 nm) that is comparable to what is found for $Ir(bpy)_3^{3+}$ (430 nm) [37], a complex with emission assigned to a ^{3}LC transition. However, the emission lifetime for **9** at 77 K ($\tau = 3.9$ μs) indicates that the MLLCT state still exerts a strong perturbation on the ^{3}LC transition since the value is much shorter than that found for Ir(bpy)$_3^{3+}$ ($\tau = 80$ μs).

The radiative (k_r) and nonradiative (k_{nr}) decay rates calculated from luminescence quantum yield data are given in Table 31.3. The luminescent quantum yields for the complexes in solution range between 0.0007 for **8** and 0.40 for **9**. The difference in the quantum yields can be attributed to a decline in the nonradiative rates with increasing emission energy because variation in k_{nr} is 10-fold greater than that in k_r. The trend in the nonradiative rates is consistent with energy gap law behavior. The energy gap law states that a series of compounds with similar ground- and excited-state vibrational modes will show a linear correlation between $\ln(k_{nr})$ and the emission energy. A plot $\ln(k_{nr})$ vs emission energy for the $(C^{\wedge}N)_2 \text{Ir}(N^{\wedge}N)^+$ complexes is given in Fig. 31.6. The nonradiative rates for complexes with bipyridyl-type $N^{\wedge}N$ ligands can be linearly correlated with the emission energy, whereas the nonradiative rates for complexes **5** and **6** with biquinolyl ligands are lower than those of complexes with similar emission energies such as **4** and **7**. This difference indicates that the vibrational modes responsible for excited-state deactivation are less effectively coupled to the ground state in the biquinolyl ligands than in the bipyridyl ligands, which leads to more efficient emission from complexes **5** and **6** than from **4** and **7**.

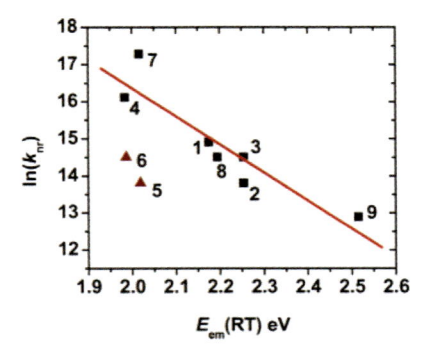

Figure 31.6 Plot of the nonradiative decay rate [$\ln(k_{nr})$] vs emission energy [$E_{em}(\text{RT})$] of complexes **1–9**. The line shown is the linear fit for complexes with bipyridyl–type ligands.

Table 31.4 Summary of device data for LECs prepared with compounds 2, 6, and 9[a]

Compound	λ_{max} (nm)	CIE coordinates	Turn-on time[b] (min)	Peak EQE (%)	Brightness at peak EQE (cd m^{-2})
2	542	x = 0.37 y = 0.59	35	6.9	11500
6	635	x = 0.67 y = 0.32	32	7.4	7500
9	492	x = 0.20 y = 0.41	30	4.6	1700

[a] All devices were run at a constant bias of 3 V. [b] Time required to reach 1.0 cd m^{-2}.

The luminescence properties for neat sample compounds **1–9** are also listed in Table 31.3. The emission spectra for the neat solids are similar to those observed in a CH_2Cl_2 solution at room temperature. The luminescent lifetimes measured for most of the compounds in the solid state are longer than those for samples in degassed solutions. Significant quenching occurs for most of the compounds in a fluid solution. Thus, the luminescent quantum yield of the compounds measured in a solution is not an ideal parameter for evaluating their utility as emissive materials for single-layer electroluminescent devices. However, the lifetimes exhibited by powdered samples of compounds **1** and **5** are shorter than the values in degassed solutions, which indicates that solid-state interactions can lead to significant self-quenching in these species. Increasing the steric bulk of either the bpy or ppz ligands by the addition of *tert*-butyl substituents (i.e., **2** and **6**), leads to diminished self-quenching, such that the powdered lifetimes for these derivatives are longer than the fluid solution values. This strategy of using bulky alkyl groups to inhibit self-quenching in the solid state has likewise been shown to be effective for $Ru(bpy)_3^{2+}$ complexes, where single-layer LEC devices made with *tert*-butyl-substituted $Ru(bpy)_3^{2+}$ derivatives show improved device performance (Table 31.4) [11c, 13a]. Interestingly, the phosphorescent lifetime of **9** in a degassed solution (1.34 μs) is shorter than the value found in the powdered sample (3.5 μs); however, both values are markedly larger than those of compounds **1–8**. The τ value for the powdered sample of **9** is similar to the lifetime measured in a solid solution at 77 K.

Assuming a quantum yield of near unity at 77 K, the PL quantum yield of **9** in the neat solid at room temperature is also near unity.

31.3.5 OLED Studies

Blue, green, and red phosphorescent LEC devices were fabricated with compounds **9**, **2**, and **6**, respectively, to test the utility of these Ir complexes in LEC structures. ITO substrates coated with PEDOT/PSS were used to prepare the LECs. The use of PEDOT/PSS is essential for the formation of uniform thin films of the $(C^{\wedge}N)_2Ir(N^{\wedge}N)^+PF_6^-$ salt complexes. Films of the Ir(III) salts spin-coated onto bare ITO crystallize rapidly and are prone to the formation of pinholes, whereas PEDOT/PSS-treated substrates give films that have low pinhole densities and appear uniform, even at high magnification (1000°). Thus, the device structure investigated here was ITO-PEDOT/PSS (400 Å)/Ir complex (700–900 Å)/Al (1200 Å). LECs prepared with the three emitters display excellent reproducibility in both their fabrication and electroluminescent behavior. The EL spectra have Commission Internationale de L'Eclairage (CIE) coordinates, which correspond to blue ($x = 0.20$, $y = 0.41$), green–yellow ($x = 0.37$, $y = 0.59$), and red ($x = 0.67$, $y = 0.32$) emission for devices made with complexes **9**, **2**, and **6**, respectively. All three LECs give EL spectra with line widths similar to those of their thin-film PL spectra but show red shifts of 5–10 nm (Fig. 31.7). The EL spectra are independent of the applied voltage and invariant when tested under forward and reverse bias (forward is ITO = anode). This behavior is in contrast to the bias-dependent EL spectral shifts reported for LEC devices made with (ppy)$_2$Ir(4,4′-di-*tert*-butylbipyridine)$^+$PF$_6^-$ [19].

The time dependence of the brightness and QE for each of the three devices is shown in Fig. 31.8. These devices exhibit the characteristic delay in light output observed for LECs and only begin to give visible EL (0.1 cd m^{-2}) after 20–30 min when held at a constant bias of 3 V. The formation of a stable p–n junction by the slow migration of the PF$_6^-$ ions in the thin film toward the anode is responsible for the delay in the LEC response. As the voltage is increased, the delay decreases. For example, at a bias of 5 V, devices show visible emission after only 3–5 min. While this delay in light output is long relative to that of OLEDs, it is shorter than delays

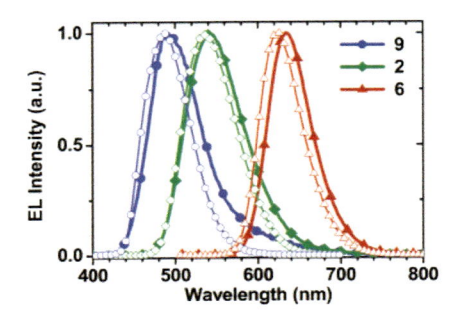

Figure 31.7 EL and thin-film PL spectra for compounds **2** (◇), **6** (△), and **9** (◯). EL spectra are illustrated with solid symbols and PL spectra with open symbols.

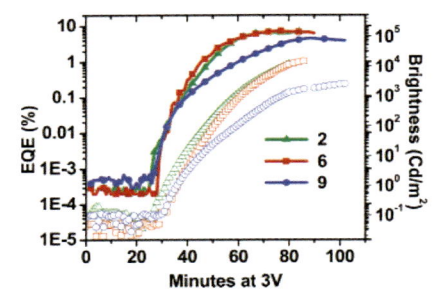

Figure 31.8 Plot of the light output and device efficiency as a function of time for LECs based on compounds **2** (△), **6** (◇), and **9** (◯), held at a constant bias of 3 V. The open symbols show the brightness (cd m^{-2}), and the solid symbols correspond to the EQE (%) values.

reported for other LECs, which can be on the order of hours [11–14, 19]. The EQE values for these Ir-based LECs increased steadily after the devices were turned on. The EQE values of LECs made with either complex **2** or **6** take ca. 70 min to reach peak EQE values of 6.9 and 7.4%, respectively, when held at 3 V. In contrast, devices made with the blue emissive complex **9** reach a lower peak EQE of 4.6% after 55 min. The LECs reported here give high power efficiencies. Devices with compounds **2**, **6**, and **9** give peak power efficiencies of 25, 10, and 11 lm W^{-1}, respectively, when biased at 3 V.

Current–voltage (J–V) plots for the LEC using complex **2** were recorded at four different times (Fig. 31.9). The J–V characteristics

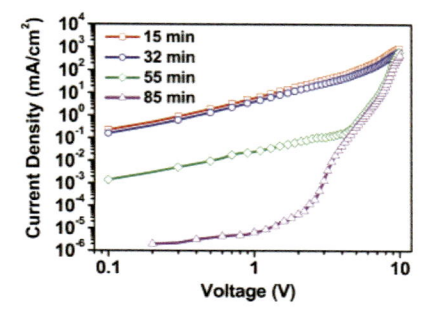

Figure 31.9 Current–voltage plots measured at different times for an LEC using complex **2**.

of the device gradually evolve and ultimately begin to resemble those commonly observed in OLEDs [38]. The initial J–V plot, as well as the one recorded just after emission is observed (15 and 32 min, respectively), shows a high current at low bias that is consistent with a high degree of carrier leakage in the device. This is not surprising because at short operating times the unpolarized device is expected to show a J–V response that is dominated by conduction of a single charge carrier. As the device continues to operate, a p–n junction is formed and the current at low bias decreases significantly. The QE for the device increases as a result of the improved carrier balance. The changes in the current–voltage plots observed for the LEC provide a clear signature of a single-layer device evolving into a single-heterostructure device, in which charge recombination occurs within the organic film, thereby minimizing the losses associated with carrier leakage to either electrode.

The EQE values reported here are some of the highest reported for LEC devices yet still fall short of the values that are common for electrophosphorescent OLEDs ($> 10\%$). Because the heterostructure of an LEC device is formed under bias and not in place at the outset of testing, charge recombination is somewhat more complicated than that which occurs in an OLED. Nevertheless, the EQEs should increase with improved carrier balance, and the power efficiencies of LECs could then approach values similar to those of the best p- and n-doped phosphorescent OLEDs [39].

The reported device lifetimes for LECs have been comparatively short (minutes to hours to half-brightness) [11–13, 40, 41]. Unfortunately, the methods used to evaluate the lifetimes of LECs are not directly comparable to those of OLEDs. LECs have been tested at constant bias, and OLEDs are typically tested at constant current. In constant-current mode, the device brightness is directly proportional to the QE, so the time to go to half-brightness is a good estimate of the half-life of the device. When the LEC is held at a constant bias, the brightness and QE evolve, somewhat independently. The EQE increases over time, as illustrated in Fig. 31.8; however, the current at constant bias decreases over time (Fig. 31.9), so the device brightness at a given bias may actually decrease even though the EQE is increasing. When we tried to evaluate the lifetimes of LECs prepared with **2** at constant current (after first prebiasing to form the p–n junction), the device was not stable toward loss of the p–n junction. The formation/loss of the p–n junction makes lifetime measurements for LECs a problem. This problem could be eliminated by immobilizing the counterions after the formation of the p–n junction. We are currently examining methods to positionally stabilize the counterions after the p–n junction is formed, similar to the approaches reported for polymer-based LECs [5, 8–10, 42]. We believe that continual ion diffusion may be related to the short lifetimes typically observed for LECs. Thus, with "frozen" anion LECs, we will be able to carry out lifetime measurements at constant current, analogous to the methods used for OLEDs.

31.4 Conclusion

A series of cationic phenylpyrazole-based iridium diimine complexes have been synthesized and their electrochemical, spectroscopic, and electroluminescent properties examined. By simple modification of either the pyrazolyl or diamine ligands, the electrochemical gap and the energy of the lowest emissive state can be varied over a wide range. The HOMO and LUMO energies in the complexes can be independently altered by simple ligand design. Substitution at the diimine ligand led to reduction potentials spanning 0.53 V, while substitution on the ppz ligand led to complexes whose oxidation potentials spanned 0.63 V. The $(ppz)_2Ir(N^\wedge N)^+$ complexes can be tuned to emit throughout the visible spectrum from a MLLCT state. These materials have high phosphorescence efficiencies, which make them ideal active materials for single-layer LECs. Single-layer LEC devices fabricated with these materials exhibit blue, green, and red EL and have EQEs as high as 7.4%.

Acknowledgment

The authors thank Universal Display Corp. and the Department of Energy for financial support of this work.

Supporting Information Available

UV–vis absorption spectra, 77 K emission spectra, and table of DFT-calculated HOMO–LUMO energies for all of compounds examined here. Also included are tables of crystal data, atomic coordinates, bond distances, bond angles, and anisotropic displacement parameters for $(ppz)_2Ir(^tBu\text{-}bpy)^+PF_6^-$ as well as the corresponding CIF files. This material is available free of charge via the Internet at http://pubs.acs.org.

References

1. Tang, C. W., Van Slyke, S. A. *Appl. Phys. Lett.* 1987, **51**, 913–915.
2. Parker, I. D. *J. Appl. Phys.* 1994, **75**, 1656–1666.
3. Hughes, G., Bryce, M. R. *J. Mater. Chem.* 2005, **15**(1), 94–107.
4. Forrest, S. R. *Nature* 2004, **428**, 911–918.
5. Armstrong, N. R., Wightman, R. M., Gross, E. M. *Annu. Rev. Phys. Chem.* 2001, **52**, 391–422.
6. Edman, L., Summers, M. A., Buratto, S. K., Heeger, A. *J. Phys. Rev. B* 2004, **70**, 115212–115217.

7. Slinker, J., Bernards, D., Houston, P. L., Abruña, H. D., Bernhard, S., Malliaras, G. G. *Chem. Commun.* 2003, 2392–2399.
8. (a) Pei, Q., Yu, G., Zhang, C., Yang, Y., Heeger, A. J. *Science* 1995, **269**, 1086–1088. (b) Pei, Q., Yu, G., Zhang, C., Yang, Y., Heeger, A. J. *J. Am. Chem. Soc.* 1996, **118**, 3922–3929.
9. Pei, Q., Yang, Y., Yu, G., Cao, Y., Heeger, A. J. *Synth. Met.* 1997, **85**, 1229–1232.
10. Pei, Q., Yang, Y. *Synth. Met.* 1996, **80**, 131–136.
11. (a) Lee, J., Yoo, D., Rubner, M. F. *Chem. Mater.* 1997, **9**, 1710–1712. (b) Handy, E. S., Pal, A. J., Rubner, M. F. *J. Am. Chem. Soc.* 1999, **121**, 3525–3528. (c) Rudmann, H., Shimada, S., Rubner, M. F. *J. Am Chem. Soc.* 2002, **124**, 4918–4921. (d) Takane, N., Gaynor, W., Rubner, M. F. *Polym. Prepr. (Am. Chem. Soc., Div. Polym. Chem.)* 2004, **45**, 349. (e) Rudmann, H., Shimada, S., Rubner, M. F. *J. Appl. Phys.* 2003, **94**(1), 115–122. (f) Rudmann, H., Rubner, M. F. *J. Appl. Phys.* 2001, **90**, 4338–4345. (g) Rudmann, H., Shimada, S., Rubner, M. F., Oblas, D. W., Whitten, J. E. *J. Appl. Phys.* 2002, **92**, 1576–1581.
12. (a) Kalyuzhny, G., Buda, M., McNeill, J., Barbara, P., Bard, A. J. *J. Am. Chem. Soc.* 2003, **125**, 6272–6283. (b) Gao, F. G., Bard, A. J. *Chem. Mater.* 2002, **14**, 3465–3470. (c) Buda, M., Kalyuzhny, G., Bard, A. J. *J. Am. Chem. Soc.* 2002, **124**, 6090–6098. (d) Gao, F. G., Bard, A. J. *J. Am. Chem. Soc.* 2000, **122**, 7426–7427. (e) Fan, F. F., Bard, A. J. *J. Phys. Chem. B* 2003, **107**, 1781–1787. (f) Liu, C., Bard, A. J. *J. Am. Chem. Soc.* 2002, **124**, 4190–4191.
13. (a) Bernhard, S., Barron, J. A., Houston, P. L., Abruña, H. D., Ruglovsky, J. L., Gao, X., Malliaras, G. G. *J. Am. Chem. Soc.* 2002, **124**, 13624–13628.
14. Bernhard, S., Gao, X., Malliaras, G. G., Abruna, H. D. *Adv. Mater.* 2002, **14**(6), 433–436.
15. Nazeeruddin, M. K., Humphry-Baker, R., Berner, D., Rivier, S., Zuppiroli, L., Graetzel, M. *J. Am. Chem. Soc.* 2003, **125**, 8790–8797.
16. Lowry, M. S., Hudson, W. R., Pascal, R. A., Jr., Bernhard, S. *J. Am. Chem. Soc.* 2004, **126**, 14129–14135.
17. (a) Lo, K. K., Chung, C., Lee, T. K., Lui, L., Tsing, K. H., Zhu, N. *Inorg. Chem.* 2003, **42**, 6886. (b) Lo, K. K., Chan, J. S., Chung, C., Lui, L. *Organometallics* 2004, **23**, 3108–3116. (c) Lepeltier, M., Lee, T. K., Lo, K. K., Toupet, L., Le Bozec, H., Guerchais, V. *Eur. J. Inorg. Chem.* 2005, 110–117.
18. Slinker, J., Koh, C. Y., Malliaras, G. G. *Appl. Phys. Lett.* 2005, **8**, 1735061–1735063.

19. Slinker, J. D., Gorodetsky, A. A., Lowry, M. S., Wang, J., Parker, S., Rohl, R., Bernhard, S., Malliaras, G. G. *J. Am. Chem. Soc.* 2004, **126**, 2763–2767.
20. (a) Columbo, M. G., Güdel, H. U. *Top. Curr. Chem.* 1994, **171**, 143–171. (b) King, J. A., Watts, R. J. *J. Am. Chem. Soc.* 1987, **109**, 1589–1590. (c) Garces, F. O., King, J. A., Watts, R. J. *Inorg. Chem.* 1988, **27**, 3464–3471.
21. Maestri, M., Sandrini, D., Balzani, V., Maeder, U., von Zelewsky, A. *Inorg. Chem.* 1987, **26**, 1323–1327.
22. Pavlik, J. W., Connors, R. E., Burns, D. S., Kurzwell, E. M. *J. Am. Chem. Soc.* 1993, **115**, 7645–7652.
23. Tamayo, A. B., Alleyne, B. D., Djurovich, P. I., Lamansky, S., Tsyba, I., Ho, N. N., Bau, R., Thompson, M. E. *J. Am. Chem. Soc.* 2003, **125**, 7377–7387.
24. Dedeian, K., Shi, J., Nigel, S., Forsythe, E., Morton, D. *Inorg. Chem.* 2005, **44**, 4445–4447.
25. Lo, K. K., Chan, J. S., Chung, C., Tsang, V. W., Zhu, N. *Inorg. Chim. Acta* 2004, **357**, 3109–3118.
26. (a) Finar, I. L., Rackham, D. M. *J. Chem. Soc. B* 1968, 211–214. (b) Finar, I. L., Godfrey, K. E. *J. Chem. Soc.* 1954, 2293–2298.
27. Nonoyama, M. *Bull. Chem. Soc. Jpn.* 1974, **47**, 767–768.
28. Sheldrich, G. M. *SHELXTL*, version 5.1; Bruker Analytical X-ray System, Inc.: Madison, WI, 1997.
29. Blessing, R. H. *Acta Crystallogr.* 1995, **A51**, 33–38.
30. (a) Gagne, R. R., Koval, C. A., Lisenski, G. C. *Inorg. Chem.* 1980, **19**, 2854–2855. (b) Sawyer, D. T., Sobkowiak, A., Roberts, J. L., Jr. *Electrochemistry for Chemists*, 2nd ed., John Wiley and Sons: New York, 1995; p. 467.
31. (a) Demas, J. N., Crosby, G. A. *J. Phys. Chem.* 1978, **82**, 991–1024. (b) Crosby, G. A., Demas, J. N. *J. Am. Chem. Soc.* 1971, **93**, 2841–2847. (c) Depriest, J., Zheng, G. Y., Goswami, N., Eichhorn, D. M., Woods, C., Rillema, D. P. *Inorg. Chem.* 2000, **39**, 1955–1963.
32. Jones, G., Jackson, W. R., Choi, C. Y., Bergmark, W. R. *J. Phys. Chem.* 1985, **89**, 294–300.
33. Forrest, S. R., Bradley, D. D. C., Thompson, M. E. *Adv. Mater.* 2002, **15**(13), 1043–1048.
34. Hazell, A. C., Hazell, R. G. *Acta Crystallogr.* 1984, **C40**, 806–811.
35. (a) Hay, P. J. *J. Phys. Chem. A* 2002, **106**, 1634–1641. (b) Brooks, J., Babayan, Y., Lamansky, S., Djurovich, P. I., Tsyna, I., Bau, R., Thompson, M. E. *Inorg. Chem.* 2002, **41**, 3055–3066. (c) Zheng, K., Wang, J., Shen, Y., Kung, D., Yun, F. *J. Phys. Chem. A* 2001, **105**, 7248–7253.

36. Neve, F., La Deda, M., Crispini, A., Bellusci, A., Puntoriero, F., Campagna, S. *Organometallics* 2004, **23**, 5856–5863.
37. Flynn, C. M., Jr., Demas, J. N. *J. Am. Chem. Soc.* 1974, **96**, 1959–1960.
38. Segal, M., Baldo, M. A., Holmes, R. J., Forrest, S. R., Soos, Z. G. *Phys. Rev. B* 2003, **68**(7), 075211–075214.
39. He, G., Pfeiffer, M., Leo, K., Hofmann, M., Birnstock, J., Pudzich, R., Salbeck, J. *Appl. Phys. Lett.* 2004, **85**, 3911–3913.
40. Parker, S. T., Slinker, J. D., Lowry, M. S., Cox, M. P., Bernhard, S., Malliaras, G. G. *Chem. Mater.* 2005, **17**, 3187–3190.
41. (a) Maness, K. M., Terrill, R. H., Meyer, T. J., Murray, R. W., Wightman, R. M. *J. Am. Chem. Soc.* 1996, **118**, 10609–10616. (b) Maness, K. M., Masui, H., Wightman, R. M., Murray, R. W. *J. Am. Chem. Soc.* 1997, **118**, 3987–3993.
42. Gao, J., Yu, G., Heeger, A. *J. Appl. Phys. Lett.* 1997, **71**, 1293–1295.

Chapter 32

Blue and Near-UV Phosphorescence from Iridium Complexes with Cyclometalated Pyrazolyl or *N*-Heterocyclic Carbene Ligands

Tissa Sajoto,[a] Peter I. Djurovich,[a] Arnold Tamayo,[a] Muhammed Yousufuddin,[a] Robert Bau,[a] Mark E. Thompson,[a] Russell J. Holmes,[b] and Stephen R. Forrest[b]

[a]*Department of Chemistry, University of Southern California, Los Angeles, California 90089-0744, USA*
[b]*Department of Electrical Engineering, Princeton University, Princeton, New Jersey 08544, USA*
met@usc.edu, forrest@princeton.edu

Two approaches are reported to achieve efficient blue to near-UV emission from triscyclometalated iridium(III) materials related to the previously reported complex, *fac*-Ir(ppz)$_3$ (ppz = 1-phenylpyrazolyl-$N,C^{2'}$). The first involves replacement of the phenyl group of the ppz ligand with a 9,9-dimethyl-2-fluorenyl group, i.e., *fac*-tris(1-[(9,9-dimethyl- 2-fluorenyl)]pyrazolyl-$N,C^{2'}$)iridium(III), abbreviated as *fac*-Ir(flz)$_3$. Crystallographic analysis reveals that

---

Reprinted from *Inorg. Chem.*, **44**(22), 7992–8003, 2005.

Electrophosphorescent Materials and Devices
Edited by Mark E. Thompson
Text Copyright © 2005 American Chemical Society
Layout Copyright © 2024 Jenny Stanford Publishing Pte. Ltd.
ISBN 978-981-4877-34-3 (Hardcover), 978-1-003-08872-1 (eBook)
www.jennystanford.com

both *fac*-Ir(flz)$_3$ and *fac*-Ir(ppz)$_3$ have a similar coordination environment around the Ir center. The absorption and emission spectra of *fac*-Ir(flz)$_3$ are red shifted from those of *fac*-Ir(ppz)$_3$. The *fac*-Ir(flz)$_3$ complex gives blue photoluminescence (PL) with a high efficiency ($\lambda_{max} = 480$ nm, $\phi_{PL} = 0.9$) at room temperature. The lifetime and quantum efficiency were used to determine the radiative and nonradiative rates (3.0×10^4 and 3.0×10^3 s^{-1}, respectively). The second approach utilizes *N*-heterocyclic carbene (NHC) ligands to form triscyclometalated Ir complexes. Complexes with two different NHC ligands, i.e., iridium tris(1-phenyl-3-methylimidazolin-2-ylidene-$C,C^{2'}$), abbreviated as Ir (pmi)$_3$, and iridium tris(1-phenyl-3-methylbenzimidazolin-2-ylidene-$C,C^{2'}$), abbreviated as Ir(pmb)$_3$, were both isolated as facial and meridianal isomers. Comparison of the crystallographic structures of the *fac*- and *mer*-isomers of Ir(pmb)$_3$ with the corresponding Ir(ppz)$_3$ isomers indicates that the imidazolyl-carbene ligand has a stronger trans influence than pyrazolyl and, thus, imparts a greater ligand field strength. Both *fac*-Ir(pmi)$_3$ and *fac*-Ir(pmb)$_3$ complexes display strong metal-to-ligand-charge-transfer absorption transitions in the UV ($\lambda = 270$–350 nm) and phosphoresce in the near-UV region ($E_{0-0} = 380$ nm) at room temperature with ϕ_{PL} values of 0.02 and 0.04, respectively. The radiative decay rates for *fac*-Ir(pmi)$_3$ and *fac*-Ir(pmb)$_3$ (5×10^4 s^{-1} and 18×10^4 s^{-1}, respectively) are somewhat higher than that of *fac*-Ir(flz)$_3$, but the nonradiative rates are two orders of magnitude faster (i.e., $(2-4) \times 10^6$ s^{-1}).

32.1 Introduction

Luminescent Ir(III) complexes have exhibited enormous potential in a range of photonic applications. For example, these Ir complexes can be used as emissive dopants in organic light emitting devices (OLEDs) [1, 2], sensitizers for outer-sphere electron-transfer reactions [3, 4], photocatalysts for CO_2 reduction [5], photoreductants [6] and singlet oxygen sensitizers [7], as well as biological labeling reagents [8]. In particular, cyclometalated iridium complexes with emission colors that vary from blue to red have received a great deal of attention recently for their application to light emitting diodes [2, 9]. Since the optical properties and related uses of the cyclometalated Ir complexes are strongly dependent on the characteristics of their ground and lowest excited states, it has become desirable to better understand the interactions between these states and thus determine how to systematically alter the photophysical properties by appropriate ligand or complex design. Moreover, the formation of a high energy emitting species would be beneficial for many of the proposed applications for these materials. To these ends, we report a study of the synthesis and photophysical properties of cyclometalated Ir complexes, some of which emit in the near-UV part of the spectrum, with *N*-pyrazolyl- or imidazolyl-type carbene ligands.

The emission energies of luminescent cyclometalated Ir complexes are principally determined by the triplet energy of the cyclometalating ligand (C^N). To understand the strategies used to alter the emission energies by C^N ligand modification, the *fac*-Ir(ppy)$_3$ complex (ppy = 2-phenylpyridyl, Fig. 32.1) can be used as a prototypical cyclometalated phosphor [6]. Both ligand- and metal-based orbitals are involved in the excited states of this complex. The phenyl group of ppy carries a formal negative charge; the highest occupied molecular orbital (HOMO) is principally composed of π orbitals of the phenyl ring and the metal d orbitals. The pyridine is formally neutral and the principal contributor to the lowest unoccupied molecular orbital (LUMO). The absorption spectra of the complex display strong metal-to-ligand-charge-transfer (MLCT) transitions at energies lower than the ligand π–π* transitions. The MLCT absorption bands have been assigned to transitions between the metal-ligand HOMO and ligand-localized LUMO orbitals [10]. However, the low temperature emission spectrum resembles simple ligand phosphorescence, suggesting decay from a ligand-localized (^{3}LC, π–π*) lowest excited state [11]. The radiative lifetime for the complex is ca. 2 µs at room temperature, which is indicative of a strong perturbation of the ^{3}LC state by participation of the singlet MLCT state through spin–orbit coupling. Thus, the lowest excited state of the complex is best described as an admixture of ^{3}LC and singlet metal-to-ligand-charge-transfer (^{1}MLCT) states [12].

The Ir complexes with ppy ligands emit green light, λ_{max} = 515 nm, for both *fac*-Ir(ppy)$_3$ and (ppy)$_2$Ir(acac) (acac = 2,4-

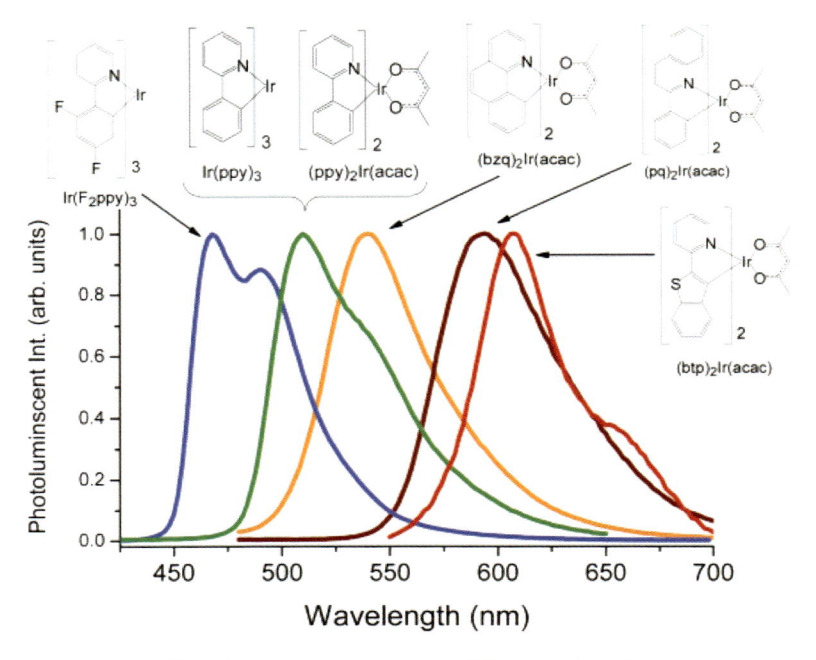

Figure 32.1 Photoluminescence spectra of Ir phosphors (room temperature, 2-MeTHF solutions).

pentanedionato-O,O, Fig. 32.1). If the C^N π-system of the phenyl and pyridyl ligands is expanded by a bridging vinyl group, i.e., (bzq)$_2$Ir(acac), a bathochromic shift is observed. If the π-system of the pyridyl fragment is enlarged using a 2-quinolyl moiety, e.g., (pq)$_2$Ir(acac) in Fig. 32.1, the emission color red shifts further [2h]. A similar red shift occurs when the phenyl fragment of ppy is replaced with a benzothiophene group ((btp)$_2$Ir(acac), Fig. 32.1). On the other hand, attempts to blue shift the emission spectrum by decreasing the size of the C^N π-system, e.g., replacing the phenyl ring with a vinyl group, also lead to a red shifted emission [13]. The decrease in emission energy in this case is understandable when one considers that the electron–electron repulsion in the triplet state increases with reduction in the size of the π-system, thereby lowering the triplet energy through increased singlet–triplet splitting (greater exchange energy) [14]. For example, the triplet energies of vinyl substituted benzene derivatives are ca. 5 kcal mol^{-1} (0.2 eV) [15] lower than that of biphenyl ($E_{0-0} = 436$ nm, 2.84 eV) [16]. Therefore, alternative methods are required to increase the triplet energies of Ir complexes with cyclometalated aromatic ligands.

A number of approaches to increase the emission energy of cyclometalated Ir complexes have focused on methods to decrease the HOMO energy while keeping the LUMO energy relatively unchanged. The addition of electron withdrawing groups to the phenyl ring has been used as one way to achieve this goal [17, 18a]. The most common electron withdrawing group used for this purpose is fluoride, and a typical example of this blue shifting is seen for *fac*-Ir(F$_2$ppy)$_3$ in Fig. 32.1 [18a]. An alternate approach to lower the HOMO energy involves the use of ancillary ligands to tune the HOMO energies of biscyclometalated derivatives. We have recently reported a detailed study of ancillary ligand effects on the emission energies of (C^N)$_2$Ir(L^X) complexes [19]. The emission energy can be significantly increased by judicious choice of the ancillary ligand; however, increasing the emission energy of the ppy$_2$Ir(L^X) complexes to generate a saturated blue color leads to a pronounced decrease in luminance efficiency. The decline in quantum efficiency is due to a significant decrease in the radiative rates, relative to the nonradiative rates, which comes about from an increased separation between the ^{1}MLCT and ^{3}LC energies as a result of the decreased HOMO energy [19].

In contrast to methods which rely on altering the HOMO energy to blue shift the phosphorescence of cyclometalated Ir complexes, the strategy described herein involves replacing the heterocyclic fragment of the C^N ligands with moieties that destabilize the LUMO relative to that of a cyclometalated ppy ligand. We show that replacing the pyridyl ring with either an *N*-pyrazolyl or *N*-heterocyclic carbene-based group leads to a significant increase in the LUMO energy and, consequently, increases the emission energy of complexes coordinated to these ligands. It, thus, becomes possible to observe efficient blue or near-UV phosphorescence at room temperature from Ir complexes that have cyclometalated *N*-pyrazolyl- or carbene-based ligands.

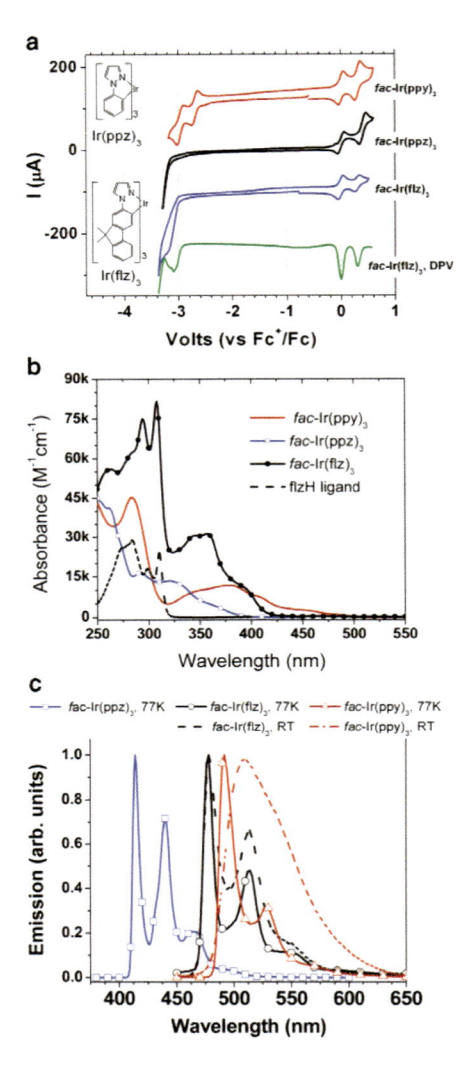

Figure 32.2 (a) Cyclic voltammetric (CV) traces, (b) absorption spectra, and (c) emission spectra for fac-Ir(ppy)$_3$, fac-Ir(ppz)$_3$, and fac-Ir(flz)$_3$. The DPV trace for fac-Ir(flz)$_3$ is also shown. The wave at 0 V in each of the CV traces is the ferrocene reference; the fac-Ir(ppy)$_3$ and fac-Ir(flz)$_3$ traces are offset in the current axis for clarity. Absorption spectra were recorded at room temperature, and emission spectra were recorded in 2-MeTHF solutions at either 77 K or room temperature (see legend).

32.2 Results and Discussion

32.2.1 Cyclometalated Pyrazolyl-Based Ligands for Blue Phosphorescence

The effect that an increase in the LUMO energy has on triscyclometalated Ir complexes can be illustrated by comparing the electrochemical and photophysical properties of *fac*-Ir(ppy)$_3$ with those of the analogous complex that has a ligand with a high triplet energy, *fac*-Ir(ppz)$_3$ (ppz = 1-phenylpyrazolyl) [18, 20]. Both Ir complexes have similar HOMO energies; however, the LUMO energy of *fac*-Ir(ppz)$_3$ is shifted to a considerably higher energy relative to that of *fac*-Ir(ppy)$_3$. This difference in electronic energy is reflected in the redox properties of the two species (Fig. 32.2). The oxidation potentials for the two complexes are nearly identical (*fac*-Ir(ppz)$_3$, $E_{1/2}^{ox} = 0.41$ V; *fac*-Ir(ppy)$_3$, $E_{1/2}^{ox} = 0.31$ V), while the reduction potentials are markedly different; that is, reduction for *fac*-Ir(ppz)$_3$ is not observed prior to solvent reduction (-3.2 V vs Fc$^+$/Fc), whereas *fac*-Ir(ppy)$_3$ displays two reduction waves under the same conditions ($E_{1/2}^{red} = -2.70, -3.00$ V) [18a]. The higher LUMO energy of *fac*-Ir(ppz)$_3$ relative to that of *fac*-Ir(ppy)$_3$ is also manifested in the photophysical properties of the complexes. The MLCT absorption transitions of *fac*-Ir(ppz)$_3$ ($\lambda_{max} = 321$ nm, 3.86 eV) are much higher in energy than those of *fac*-Ir(ppy)$_3$ ($\lambda_{max} = 377$ nm, 3.29 eV) (Fig. 32.2) [18a]. Likewise, the emission energy of *fac*-Ir(ppz)$_3$ ($E_{0-0} = 414$ nm, 3.00 eV) is greater than that of *fac*-Ir(ppy)$_3$ ($E_{0-0} = 494$ nm, 2.51 eV) (Fig. 32.2). The photophysical properties of the Ir complexes are mirrored by those of the ligands themselves. The triplet energy of the ppzH ligand (380 nm, 3.26 eV) [21] is much higher than that of ppyH (430 nm, 2.88 eV) [22]. Notably, whereas the photoluminescent (PL) efficiency of *fac*-Ir(ppy)$_3$ is quite high and invariant with temperature (when dispersed in a solid matrix) [23], the PL efficiency of *fac*-Ir(ppz)$_3$ is strongly temperature dependent. No detectable emission from *fac*-Ir(ppz)$_3$ is observed in fluid or solid solutions at room temperature [24]. Weak emission from *fac*-Ir(ppz)$_3$ is only observed in fluid solutions at low temperature (<230 K), while intense emission occurs at 77 K.

Evidently, *fac*-Ir(ppz)$_3$ decays nonradiatively by thermal population of a higher lying nonemissive excited state at room temperature.

Considerable work has been done to understand the thermally activated decay processes in luminescent transition metal complexes [25]. In particular, Ru(II) and Os(II) tris-diimines have been extensively studied in this regard, since the photophysical properties of these materials exhibit strong temperature-dependent behavior. It has been found that the temperature-dependent luminescence in Ru(II) diimine complexes is characteristic of thermal population to ligand field (dissociative) states, whereas the Os(II) analogues have kinetic parameters more consistent with deactivation through higher energy MLCT states [26]. The difference between the Os and Ru complexes is due to the larger ligand field splitting observed for 5d over 4d elements, which destabilizes the ligand field states in the Os analogues to energies that are thermally inaccessible from the emissive triplet state. The cyclometalated Ir(III) complexes reported here are isoelectronic with the Ru(II) and Os(II) tris-diimine materials and are expected to mimic the photophysics of their Os(II) counterparts [27]. Thus, the *fac*-Ir(ppz)$_3$ complex could decay nonradiatively through higher energy MLCT states [20b]. However, the triplet energy of *fac*-Ir(ppz)$_3$ is very high (3.0 eV, 70 kcal mol^{-1}) and comparable to the Ir-phenyl bond strength [28]. Therefore, thermal population to accessible ligand field states can also be considered to be a possible nonradiative luminescent decay mechanism. Another possible decay process could be through a ligand localized, $n-\pi^*$ state. A schematic representation of the emission and potential thermally activated nonradiative decay processes of cyclometalated Ir complexes is given in Fig. 32.3a. The absence of luminescence at room temperature requires that the rate of non-radiative decay (k_{nr}) from the nonemissive excited state (NR state in Fig. 32.3) be significantly greater than the radiative rate (k_r). In addition, the energy difference between the emissive triplet (T_1) and the NR states must be small enough that the higher energy nonemissive state is thermally accessible.

To achieve efficient luminescence from cyclometalated ppz-based Ir complexes at room temperature, it is necessary to retard or eliminate nonradiative processes that thermally deactivate the excited state. The pyrazolyl moiety itself is not the origin

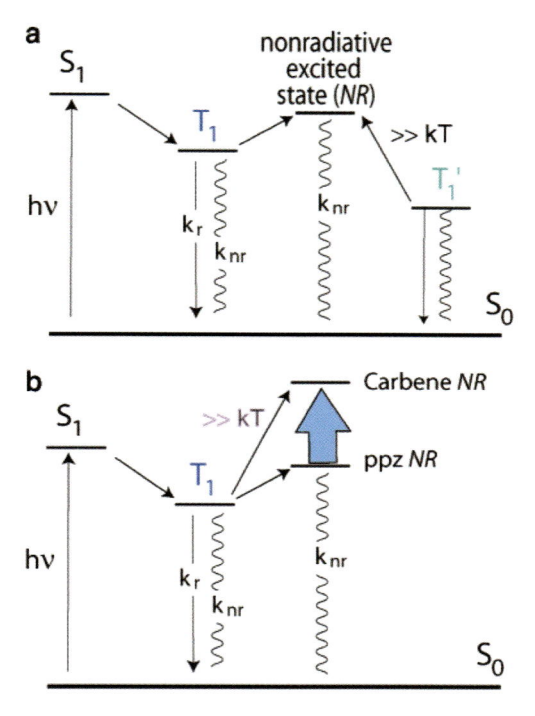

Figure 32.3 (a) Energy level scheme for emissions from fac-Ir(ppz)$_3$ (T$_1$) and fac-Ir(flz)$_3$ (T$_1'$). The two compounds are expected to have nonradiative excited states (NR) with similar energies. (b) Energy level scheme for emissions from fac-Ir(ppz)$_3$ and fac-Ir(C^C:)$_3$ compounds.

of emission decay, since several complexes with pyrazolyl-based ancillary ligands have very high PL efficiencies at room temperature [17c, 29, 30]. One way to suppress the luminescent deactivation in a complex is to lower the energy of its triplet state to such a value that the NR state is no longer thermally accessible at room temperature. This concept was applied in the design of fac-Ir(flz)$_3$ (flz = 1-[(9,9-dimethyl-2-fluorenyl)]pyrazolyl), an analogue of fac-Ir(ppz)$_3$ in which the π-system of the ppz ligand is extended using a fluorenyl moiety. The incorporation of fluorenyl groups in cyclometalated Ir complexes has been previously reported and found to lower the triplet energy of Ir(ppy)$_3$ type derivatives [31]. In these earlier examples, however, the decreased triplet energy of the fluorenyl

Table 32.1 Selected bond distances (Å) and angles (deg) for *fac-* and *mer-*Ir(ppz)$_3$ [18a], *fac-*Ir(flz)$_3$, and *fac-* and *mer-*Ir(pmb)$_3$

Bond	Ir(ppz)$_3$ [18a] Facial	Ir(flz)$_3$ Meridianal	Ir(pmb)$_3$ Facial	Facial	Meridianal
		Selected Bond Distances (Å)			
Ir1–N1	2.117(5)	2.053(2)	2.115(6)		
Ir1–N2	2.135(5)	2.026(2)	2.095(7)		
Ir1–N3	2.120(6)	2.013(2)	2.121(6)		
Ir1–C1	2.015(7)	2.051(2)	2.023(7)	2.077(7)	2.099(4)
Ir1–C2	2.027(6)	2.057(2)	2.019(7)	2.071(7)	2.078(4)
Ir1–C3	2.021(6)	1.993(2)	2.024(8)	2.094(7)	2.086(4)
Ir1–C4				2.035(7)	2.043(4)
Ir1–C5				2.022(7)	2.019(4)
Ir1–C6				2.022(8)	2.032(4)
		Bond Angles (deg)			
N2–Ir1–N3	94.4(2)	171.54(8)	91.7(2)		
C1–Ir1–C2	93.7(3)	172.67(9)	92.2(3)	91.0(3)	90.91(16)
C4–Ir1–C6				100.2(3)	93.96(16)

analogues had little impact on the luminescence efficiency of the complexes, since the *fac-*Ir(ppy)$_3$ complex is already highly emissive at room temperature.

The *fac-*Ir(flz)$_3$ complex was prepared by a two step procedure analogous to what we have reported for *fac-*Ir(ppz)$_3$ [18a]. Single crystals of *fac-*Ir(flz)$_3$ were analyzed by X-ray crystallography. The *fac-*Ir(flz)$_3$ complex is expected to have a similar electronic configuration around the iridium center as *fac-*Ir(ppz)$_3$, and a comparison of metric parameters for both compounds supports a high degree of structural similarity between the two species. Selected bond lengths and bond angles of the two complexes are listed in Table 32.1, and molecular plots are shown in Fig. 32.4. Both *fac-*Ir(flz)$_3$ and *fac-*Ir(ppz)$_3$ have a pseudo-octahedral coordination geometry of three pyrazolyl nitrogens and three phenyl carbons in a facial arrangement. The Ir–N and Ir–C bond lengths for the two structures are statistically equivalent. The average Ir–N (2.110(6) Å) and Ir–C (2.022(7) Å) bond lengths of *fac-*Ir(flz)$_3$ are within one sigma of the corresponding averages for *fac-*Ir(ppz)$_3$ (average Ir–N = 2.124(5) Å, average Ir–C = 2.021(6) Å). The structural data

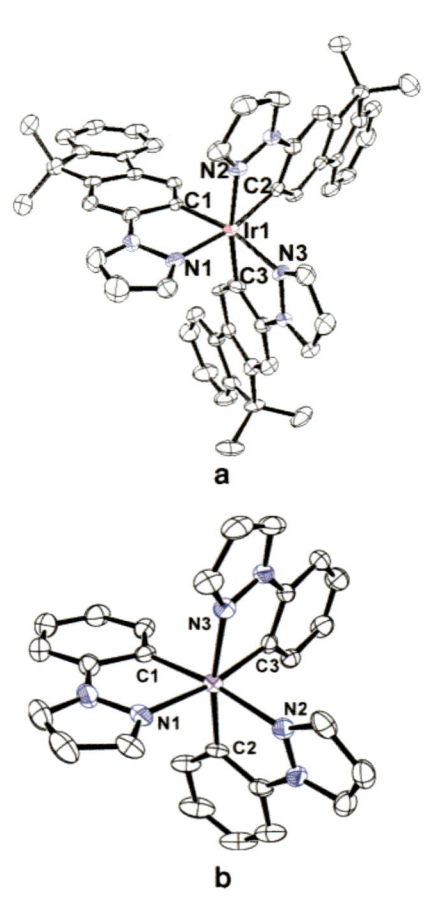

Figure 32.4 Thermal ellipsoid (ORTEP) plots of *fac*-Ir(flz)$_3$ (a) and *fac*-Ir(ppz)$_3$ [18a] (b). The hydrogen atoms have been omitted for clarity. The atom numbering used in Table 32.1 is shown.

suggest a nearly identical coordination environment around the Ir center for the two complexes, and thus, both are expected to have ligand field states with similar energies.

The influence of the extended π-system in *fac*-Ir(flz)$_3$ is apparent in the electrochemical properties of the complex. While the oxidation potential of *fac*-Ir(flz)$_3$ ($E^{ox}_{1/2} = 0.31$ V) is similar to that of *fac*-Ir(ppz)$_3$, a quasi-reversible reduction can now be discerned in the envelope of the solvent reduction wave (Fig. 32.2).

A distinct reduction process ($E_{1/2}^{red} = -3.1$ V) can be identified using differential pulse voltammetric (DPV) analysis. The expanded aromatic ring system of *fac*-Ir(flz)$_3$ clearly serves to lower the LUMO energy of the complex.

The introduction of the fluorenyl chromophore has a marked impact on the absorption spectra of *fac*-Ir(flz)$_3$ (Fig. 32.2). The complex displays a series of intense, high energy absorption bands ($\lambda < 320$ nm, $\epsilon > 5 \times 10^4$ M^{-1} cm^{-1}), which can be assigned to ligand localized π–π^* transitions by comparison to the absorption spectrum of the flzH ligand (Fig. 32.2). The extinction coefficients for these bands are roughly three times greater than those for the same transitions in the free flzH compound, as expected for a metal complex with three such ligands. Strong absorption bands at lower energy ($\lambda = 320$–420 nm, $\epsilon > 1 \times 10^4$ M^{-1} cm^{-1}) are assigned to a combination of singlet and triplet MLCT transitions, since their energy corresponds quite closely to a redox gap measured by solution electrochemistry ($\lambda_{max} = 357$ nm, 3.47 eV; $\Delta E_{1/2}^{redox} = E_{1/2}^{ox} - E_{1/2}^{red} = 3.41$ V). The intensities of these MLCT transitions are higher than their counterparts in either the *fac*-Ir(ppz)$_3$ or *fac*-Ir(ppy)$_3$ spectra, most likely due to the high oscillator strength of the fluorenyl chromophore. Much weaker absorption bands at still lower energy ($\lambda = 420$–485 nm, $\epsilon < 100$ M^{-1} cm^{-1}) can be observed in concentrated solutions of the complex [32] and are assigned to ^{3}LC transitions that have enhanced intensity due to mixing with the ^{1}MLCT states.

The *fac*-Ir(flz)$_3$ complex displays an intense, structured emission in solution at 77 K ($E_{0-0} = 480$ nm, 2.58 eV) with a lifetime of 50 µs (Fig. 32.2). The well-defined vibronic fine structure in the emission spectrum of *fac*-Ir(flz)$_3$ is indicative of a predominant ^{3}LC character in the excited state. The difference in the emission energy of *fac*-Ir(flz)$_3$ relative to *fac*-Ir(ppz)$_3$ ($E_{0-0}^{Ir(ppz)3} - E_{0-0}^{Ir(flz)3} = 0.42$ eV) closely corresponds to the difference in triplet energy between their respective ligands (flzH ligand, $E_{0-0} = 445$ nm, 2.79 eV [32]; $E_{0-0}^{ppzH} - E_{0-0}^{flzH} = 0.47$ eV). More importantly, the PL efficiency of *fac*-Ir(flz)$_3$ at room temperature is high ($\Phi = 0.38$) and the measured lifetime is 37 µs. The radiative (k_r) and nonradiative (k_{nr}) decay rates estimated from the lifetime and PL efficiency [33] are 3.0×10^4 s^{-1} and 3.0×10^3 s^{-1}, respectively. These rates are an order of magnitude lower than those reported for both blue and green emissive cyclometalated Ir complexes, such as those illustrated in Fig. 32.1 [2, 12, 18a]. The low radiative rate for fac-Ir(flz)$_3$ is consistent with a low level of ^{1}MLCT character in the lowest excited state and is due to the large separation between the ^{1}MLCT and ligand triplet energies ($\Delta E_{1/2}^{\text{redox}} - E_{0-0}^{\text{flzH}} = 0.62$ eV). Note that the corresponding energy separation for fac-Ir(ppy)$_3$ is small (0.13 eV), and thus, the radiative rate for this species is high ($k_r = 7.5 \times 10^5$ s^{-1}) [23b]. A small ^{1}MLCT character in the lowest energy excited state of fac-Ir(flz)$_3$ relative to fac-Ir(ppz)$_3$ is also expected, since the fluorenyl substituent markedly lowers the ^{3}LC energy without decreasing the ^{1}MLCT level to the same extent. A correlation between the radiative rate and the separation of ^{1}MLCT and ligand triplet state energies was similarly observed for a series of related (C^N)$_2$Ir(L^X) complexes [19].

The fac-Ir(flz)$_3$ complex shows a negligible rigidochromic shift, in contrast to what is observed when comparing the room temperature and 77 K PL spectra of fac-Ir(ppy)$_3$ (Fig. 32.2). The magnitude of the rigidochromic shift is also related to the degree of ^{1}MLCT character in the excited state [31f, 34]. Molecular reorientation of the solvent dipoles enables a complex to reach a fully relaxed geometry in the excited state at room temperature, whereas the solvent dipoles are unperturbed in the rigid matrix at low temperature. Since the formation of MLCT excited states involves substantial charge redistribution, such transitions typically display hypsochromic shifts upon going from a room temperature fluid solution to a 77 K glass. In contrast, excited states that are principally ligand based, e.g., $\pi-\pi^*$ states, generally have low levels of charge redistribution and, thus, give very small rigidochromic shifts. Hence, a greater MLCT character in the excited state of the Ir complexes is expected to lead to a large rigidochromic shift. For example, a clear correlation between the degree of MLCT character in the excited state and the magnitude of the rigidochromic shift was demonstrated for (C^N)$_2$Ir(L^X) complexes [19]. Therefore, the absence of a rigidochromic shift for fac-Ir(flz)$_3$ provides further support that the excited state is largely ^{3}LC in character.

32.2.2 Cyclometalated Carbene Ligands for Near-UV Phosphorescence

Although modification of *fac*-Ir(ppz)$_3$ with the fluorenyl group leads to a complex that is highly emissive at room temperature, the emission energy of *fac*-Ir(flz)$_3$ is lower than that of compounds such as *fac*-Ir(F$_2$ppy)$_3$ (Fig. 32.1). For some applications, complexes with higher emission energy are highly desirable. Thus, a potentially more attractive approach to achieving efficient blue phosphorescence from cyclometalated Ir complexes at room temperature would be to increase the energy of the nonemissive excited state so that thermal population to it is inhibited, as shown in Fig. 32.3b. The key to this approach is to design cyclometalating ligands that destabilize the NR state. If this state is a metal-localized, ligand field state, strengthening the metal-ligand bonds will destabilize its energy, since the state is comprised of antibonding counterparts to the metal–ligand bonding orbitals. Of the number of potential ligand choices that could have stronger ligand fields than pyrazolyl, the ligand type used here is an *N*-heterocyclic carbene (NHC). The syntheses of NHC ligand precursors are well-known [35], making access to the complexes straightforward. The NHC ligands form very strong bonds with transition metals [36], which will shift the metal–carbene antibonding orbitals to high energy, thereby decreasing or eliminating decay through the ligand field state. Specifically, the two NHC ligands studied here are 1-phenyl-3-methylimidazolin-2-ylidene (pmi) and 1-phenyl-3-methylbenzimidazolin-2-ylidene (pmb) (see Scheme 32.1). The carbene moiety is a neutral, two electron donor ligand, which makes the cyclo-metalated ligand a bidentate monoanionic ligand (C^C: is used here as a general abbreviation for a cyclometalated carbene ligand), similar to the C^N ligands that have been used to make stable iridium trischelates. An earlier example of triscyclometalated carbene complexes, Ir(C^C:)$_3$, was reported in 1980 by Lappert et. al. (see Ir–Lappert in Scheme 32.1) [37]. While an X-ray structure was given for this particular complex, to our knowledge, no photophysical data have been reported for these types of species [38].

The Ir(C^C:)$_3$ complexes were prepared by a modification of the method used to prepare the Ir(C^N)$_3$ analogues [39]. A

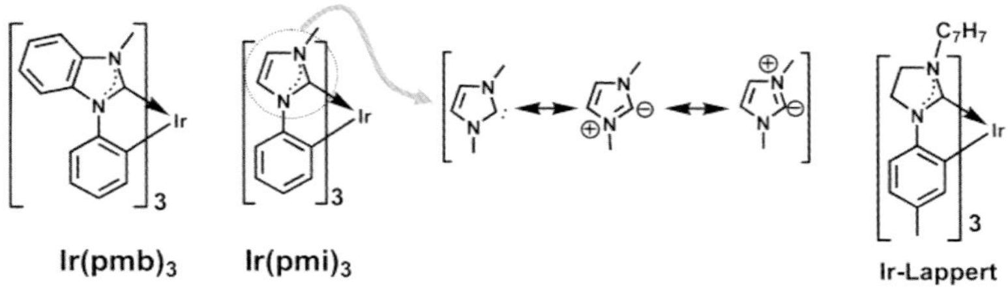

Ir(pmb)₃   Ir(pmi)₃   Ir-Lappert

Scheme 32.1

stoichiometric amount of imidazolium (pmiH$^+$) or benzimidazolium (pmbH$^+$) iodide salts, silver(I) oxide, and iridium(III) chloride hydrate were refluxed in a 2-ethoxyethanol solution to give a mixture of *fac*- and *mer*-Ir(C^C:)$_3$ complexes in low yield (< 10%) along with a product formulated as [(C^C:)$_2$IrCl]$_2$ (Eq. 32.1) [40]. The [(C^C:)$_2$IrCl]$_2$

$$3\text{pmiH}^+\text{I}^- \left(\text{or pmbH}^+\text{I}^-\right) + 3\text{Ag}_2\text{O} + \text{IrCl}\cdot x\text{H}_2\text{O}$$
$$\longrightarrow \text{Ir}\left(C^\wedge C:\right)_3 + [(C^\wedge C:)_2\text{IrCl}]_2 \tag{32.1}$$

complex can then be treated with additional pmiH$^+$ or pmbH$^+$ and silver(I) oxide in 1,2-dichloroethane to form a mixture of *fac*- and *mer*-Ir(C^C:)$_3$ complexes (Eq. 32.2). The

$$\text{pmiH}^+\text{I}^- \left(\text{or pmbH}^+\text{I}^-\right) + \text{Ag}_2\text{O} + [(C^\wedge C:)_2\text{IrCl}]_2 \longrightarrow \text{Ir}\left(C^\wedge C:\right)_3 \tag{32.2}$$

silver(I) oxide has several potential roles in these reactions: to deprotonate the imidazolium salt and stabilize the resultant carbene ligand, to abstract the chloride ligand from the Ir salt, and to serve as a transmetalating agent [41, 17a]. To date, we have been unable to optimize Eq. 32.1 to give a high yield of any particular cyclometalated product. The synthesis using Eq. 32.2 also gives a mixture of the facial or meridianal isomers of the Ir(C^C:)$_3$ complex (see Section 32.4). The *mer*-isomers of the Ir(C^C:)$_3$ complexes do not convert to their facial isomers upon photolysis but, instead, eventually decompose under long term UV exposure. Likewise, the facial isomers also decompose during extensive irradiation; hence, it is not possible to determine which configuration of the Ir(C^C:)$_3$ complexes is the thermodynamically favored isomer. Pure samples of *fac*- and *mer*-Ir(C^C:)$_3$ isomers were, therefore, isolated by either column chromatography or selective precipitation/crystallization, a procedure aided by the lower solubility of the *fac*-isomers. These synthetic difficulties are in contrast to the case of the previously developed high yield synthesis for Ir(C^N)$_3$ complexes using the chloride bridged dimer, (C^N)$_2$Ir(μ-Cl)$_2$Ir(C^N)$_2$ [18a]. The *mer*-isomers of the Ir(C^N)$_3$ complexes are the kinetically favored species and can be readily converted to the facial isomers, either by thermal or photochemical means, with minimal decomposition. We presume that the failure to interconvert the *fac*- and *mer*-isomers of the Ir(C^C:)$_3$ complexes is principally due to Ir–C$_{carbene}$ bonds being stronger than Ir–N bonds, making the partial ligand dissociation that is necessary for isomerization more difficult for the Ir(C^C:)$_3$ complexes than for their C^N counterparts.

Single crystals of *fac*-Ir(pmb)$_3$ and *mer*-Ir(pmb)$_3$ were grown from a methanol/dichloromethane solution and characterized using X-ray crystallography. Molecular structures of *fac*-Ir(pmb)$_3$ and *mer*-Ir(pmb)$_3$ are shown in Fig. 32.5. Selected bond distances and angles of the complexes are provided in Table 32.1, along with those of *mer*-Ir(ppz)$_3$, for which structural data were taken from the literature [18a]. All of the complexes analyzed here have the three cyclometalating ligands in a pseudo-octahedral coordination geometry around the iridium center. The C^C: ligands in both *fac*- and *mer*-Ir(pmb)$_3$ are twisted from planarity around the bridging C$_{phenyl}$–N$_{carbene}$ bond. The distortion is presumably due to steric repulsion between adjacent hydrogen atoms on the phenyl and benzimidazolyl moieties. The twist is smaller in the *fac*-isomer (dihedral angles are between 1 and 10°) than in the *mer*-isomer (dihedral angles vary between 5 and 27°). This variation in ligand distortion is most likely due to crystal packing effects.

It is instructive to compare the Ir coordination environment of the facial isomer of Ir(pmb)$_3$ to that of the same isomer for Ir(ppz)$_3$. In *fac*-Ir(pmb)$_3$, the average Ir–C$_{carbene}$ distance (2.026(7) Å) is significantly shorter than the average Ir–N distance in *fac*-Ir(ppz)$_3$ (2.124(5) Å). The short distance in *fac*-Ir(pmb)$_3$ suggests that the carbene moiety is more strongly bound to the Ir than the pyrazolyl ligand. Moreover, the average Ir–C$_{phenyl}$ distance (2.081(7) Å) in *fac*-Ir(pmb)$_3$ is longer than the average Ir–C$_{phenyl}$ distance in *fac*-Ir(ppz)$_3$ (2.021(6) Å), suggesting that the carbene is a stronger field ligand than the pyrazolyl. The bonding parameters found for the meridianal isomer of the complex support this view as well. In mer-Ir(pmb)$_3$, the bond length of Ir–C$_{aryl}$ trans to benzimidazolyl (Ir–C2 = 2.078(4) Å) is greater than the bond length of Ir–C$_{aryl}$ trans to the pyrazolyl group in mer-Ir(ppz)$_3$ (Ir–C3 = 1.993(2) Å), illustrating the stronger trans influence of the carbene ligand over that of pyrazolyl. The lengths of the mutually trans Ir–C$_{aryl}$ bond (Ir–C1 and Ir–C3, average (av) = 2.093(4) Å) in mer-Ir(pmb)$_3$ are slightly longer than those in mer-Ir(ppz)$_3$ (av = 2.040(2) Å), indicating greater electron donation from the carbene ligand than from the pyrazolyl moiety. The structures of both isomers of Ir(pmb)$_3$ are consistent with a strong trans influence of a formally neutral carbene ligand. The bond length differences suggest that the cyclometalated carbenes are stronger field ligands than their pyrazolyl or pyridyl counterparts, and therefore, the Ir(C$^\wedge$C:)$_3$ complexes should have high energy ligand field states.

The Ir(pmi)$_3$ and Ir(pmb)$_3$ complexes display electrochemical and absorption characteristics comparable to those of fac-Ir(ppz)$_3$. The data are consistent with the Ir(C$^\wedge$C:)$_3$ complexes having both low HOMO and high LUMO energies. The fac-isomers undergo reversible oxidation (fac-Ir(pmi)$_3$, $E_{1/2}{}^{ox}$ = 0.22 V; fac-Ir(pmb)$_3$, $E_{1/2}{}^{ox}$ = 0.48 V) and no observable reduction within the accessible solvent window, similar to the data shown in Fig. 32.2 for fac-Ir(ppz)$_3$. The mer-isomers are easier to oxidize than their fac counterparts (mer-Ir(pmi)$_3$, $E_{1/2}{}^{ox}$ = 0.14 V; mer-Ir(pmb)$_3$, $E_{1/2}{}^{ox}$ = 0.31 V), and an irreversible reduction can be observed in mer-Ir(pmb)$_3$ at –3.2 V. The absorption spectra show strong bands at high energy ($\lambda < 270$ nm, $\epsilon > 2 \times 10^4$ M^{-1} cm^{-1}) that are assigned to ligand π–π* transitions of the cyclometalating ligands, whereas weaker bands at lower energy ($\lambda = 270$–360 nm, $\epsilon \sim 10^4$ M^{-1} cm^{-1}) are ascribed to MLCT transitions (Fig. 32.6). A weak, low energy band is also present in both complexes ($\lambda = 380$ nm, $\epsilon \sim 100$ M^{-1} cm^{-1}) and is assigned to a perturbed ^{3}LC transition. The fac- and mer-isomers have similar absorption band shapes in each of the respective complexes. An interesting feature of fac-Ir(pmi)$_3$ and fac-Ir(pmb)$_3$ complexes is the small difference in their MLCT absorption energies. This similarity is surprising in view of the large decrease in the MLCT absorption energy (ca. 0.3 eV) that occurs when the

π-system of the pyridyl ring in (ppy)$_2$Ir(acac) is expanded by adding a fused phenyl ring to form (pq)$_2$Ir(acac) [2]. Apparently, the minor variance of the MLCT energy upon extention of the π-system of the imidazolyl–carbene moiety is due to poor conjugation of the phenyl and benzimidazolyl fragments, such that the phenyl and carbene moieties behave as independent chromophores.

All of the Ir(C^C:)$_3$ complexes display intense emission at 77 K in the near-UV spectrum and, notably, also luminesce at room temperature in fluid solution (Fig. 32.6). The quantum efficiencies and lifetimes for the Ir(C^C:)$_3$ complexes are given in Table 32.2. The excited state properties of the Ir(C^C:)$_3$ complexes are related to those of the Ir(C^N)$_3$ analogues in that both types of species emit from perturbed ^{3}LC states. The emission spectra at 77 K are highly structured (vibronic spacings of ca. 1300 cm^{-1}) and have luminescent lifetimes between 2 and 7 µs. The emission spectra of the *fac*- and *mer*-isomers are very similar in appearance (Fig. 32.6), unlike the distinctly different line shapes seen for the emission spectra of the *fac*- and *mer*-isomers of Ir(C^N)$_3$ complexes [18a]. The Ir(pmi)$_3$ and Ir(pmb)$_3$ complexes have high emission energies that are nearly identical in value (E_{0-0} = 380 nm, 3.26 eV). These energies approach the value for the triplet state of *N,N*-diethylaniline (E_{0-0} = 370 nm, 3.35 eV) [42]. The high emission energies of the Ir(C^C:)$_3$ complexes in conjunction with their low oxidation potentials imply that these materials have extremely high reduction potentials when in the excited state ($E_{1/2}^{+/0*} \approx E_{1/2}^{ox} - E_{0-0}$; *fac*-Ir(pmi)$_3$, $E_{1/2}^{+/0*}$ = −3.04 V; *fac*-Ir(pmb)$_3$, $E_{1/2}^{+/0*}$ = −2.78 V). The potent reducing ability of the Ir(C^C:)$_3$ complexes in their excited states could make these materials interesting candidates for use in photoinduced electron-transfer studies.

The PL efficiencies at room temperature for the Ir(C^C:)$_3$ complexes are low (0.002–0.05) but, nevertheless, higher than those for their pyrazolyl counterparts. The radiative and nonradiative decay rates of *fac*-Ir(pmi)$_3$ (k_r = 5 × 10^4 s^{-1}, k_{nr} = 2 × 10^6 s^{-1}) are comparable to those of *fac*-Ir(pmb)$_3$ (k_r = 1.8 × 10^5 s^{-1}, k_{nr} = 4 × 10^6 s^{-1}). The *mer*-isomer of Ir(pmb)$_3$ has a lower PL efficiency and a higher nonradiative decay rate than its facial analogue, which is similar to what was found for the corresponding properties in the *fac*- and *mer*-isomers of Ir(C^N)$_3$ complexes [18]. The radiative

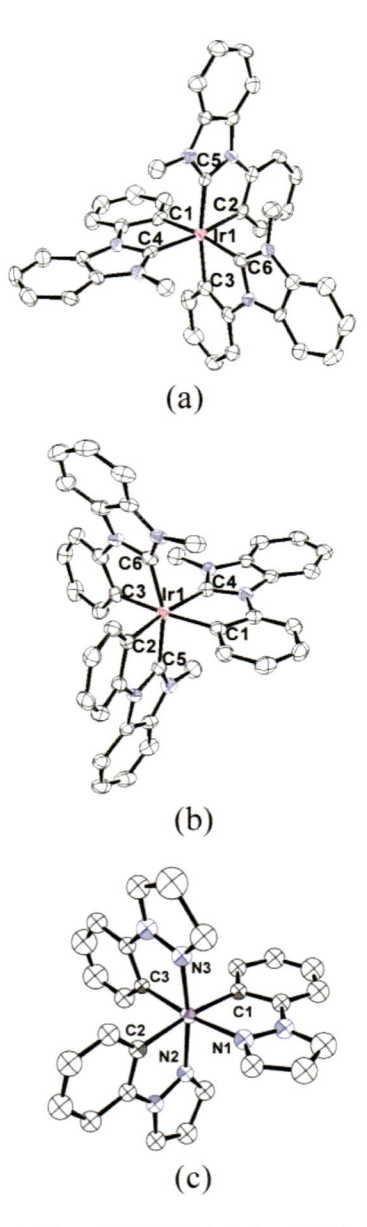

Figure 32.5 Thermal ellipsoid (ORTEP) plots of (a) *fac*-Ir(pmb)$_3$, (b) *mer*-Ir(pmb)$_3$, and (c) *mer*-Ir(ppz)$_3$ [18a]. The hydrogen atoms have been omitted for clarity. The atom numbering used in Table 32.1 is shown.

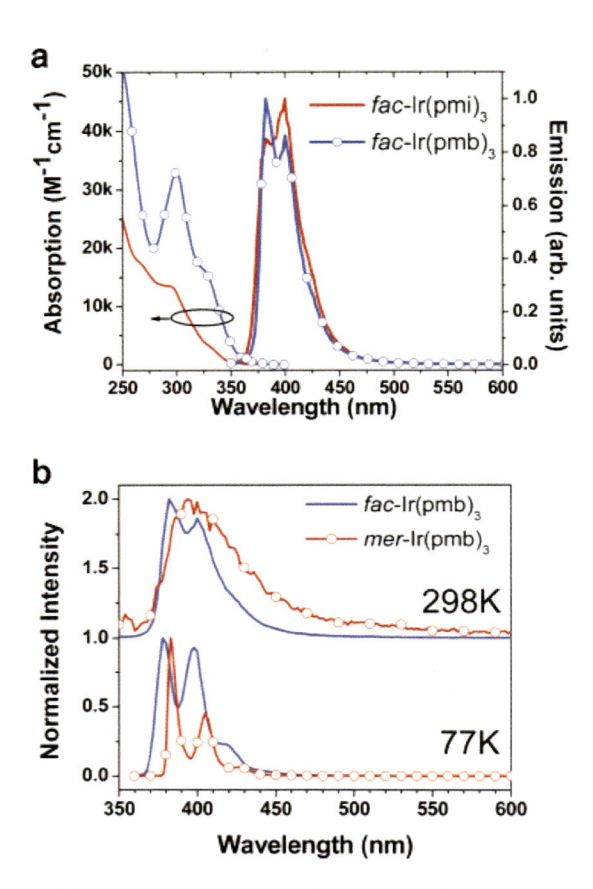

Figure 32.6 (a) Absorption and emission spectra for *fac*-Ir(pmi)$_3$ and *fac*-Ir(pmb)$_3$ in CH$_2$Cl$_2$ at room temperature and (b) emission spectra of *fac*- and *mer*-Ir(pmb)$_3$ in 2-MeTHF at room temperature and 77 K.

rates for the Ir(C^C:)$_3$ complexes are higher than that for *fac*-Ir(flz)$_3$ but less than that for *fac*-Ir(ppy)$_3$, consistent with an intermediate degree of ^{1}MLCT character in the excited states of the Ir(C^C:)$_3$ complexes. The moderate rigidochromic shifts for the Ir(C^C:)$_3$ complexes (see Fig. 32.6) also suggest that their excited states have more MLCT character than those of *fac*-Ir(flz)$_3$ (largely ^{3}LC based) but less than those of *fac*-Ir(ppy)$_3$. The greater ^{1}MLCT character in the excited states of the Ir(C^C:)$_3$ complexes relative to *fac*-Ir(flz)$_3$ is

Results and Discussion

Table 32.2 Photophysical parameters for Ir(C^N)$_3$ and Ir(C^C:)$_3$ complexes[a]

	ϕ_{PL}[b]	τ (μs) 300 K	k_r (10^4 s^{-1})	k_{nr} (10^6 s^{-1})	τ (μs) 77 K	τ (μs) 77 K (polystyrene)
fac-Ir(flz)$_3$	~~0.38~~ 0.9	37	1.0	0.020	50	48
fac-Ir(pmi)$_3$	~~0.02~~ 0.4	0.40	5	2	6.8	
mer-Ir(pmi)$_3$	~~0.05~~ 0.1	0.62	8	2	2.4	3.4
fac-Ir(pmb)$_3$	~~0.04~~ 0.09	0.22	18	4	3.1	1.1
mer-Ir(pmb)$_3$	~~0.002~~ 0.004	0.015	13	65	2.4	1.0

[a]The values listed are for 2-MeTHF solutions at room temperature, unless otherwise noted.
[b]The quantum yield reported in reference 6 for the Ir(ppy)$_3$ reference was 0.4. It was later discovered that this number was not correct and the photoluminescence quantum yield for this complex in solution is 0.73 in toluene and (W. Holzer, et al., Chem. Phys. 2005, **308**, 93–102, DOI: 10.1016/j.chemphys.2004.07.051) and 0.9 in THF (Sajoto, T., et al., J. Am. Chem. Soc. 2009, **131**, 9813–9822, DOI: 10.1021/ja903317w. Thus, the quantum yield values previously published have been increased by a factor of 2.2 (0.9/0.4) from the published values. The published values have been struck out.

presumably due to the ^{1}MLCT and ^{3}LC levels being closer in energy and, thus, more effectively mixed in the Ir(C^C:)$_3$ complexes than in fac-Ir(flz)$_3$.

The fact that near-UV emission is observed at room temperature from the Ir(C^C:)$_3$ complexes suggests that the nonemissive excited state for these complexes has been significantly destabilized relative to the similar state in fac-Ir(ppz)$_3$. However, the Ir(C^C:)$_3$ complexes still undergo nonradiative decay to a greater extent than fac-Ir(flz)$_3$, as evidenced by their high nonradiative rates and low PL efficiencies. The high nonradiative decay rates for the Ir(C^C:)$_3$ complexes indicate that a thermally accessible NR state is still present in these species. Thermal activation to this state can be largely eliminated by cooling the compounds to low temperature, and the measured lifetimes under these conditions are close to the radiative lifetimes. An alternative means to suppress the nonradiative decay processes at room temperature is to immobilize the complexes in a rigid matrix. Accordingly, when the Ir(C^C:)$_3$ complexes are dispersed in polystyrene, the lifetimes increase by nearly an order of magnitude at room temperature (see Table 32.2). A similar increase in radiative lifetime for Ru(II) and Os(II) tris-diimine complexes in polymeric matrices was previously reported and has been attributed to the inhibition of ligand field decay pathways in the rigid media [25, 43]. The long lifetimes obtained in polystyrene for the Ir(C^C:)$_3$ complexes imply that the luminance efficiencies at room temperature are much higher in the polymer matrix ($\Phi > 0.1$–0.2) than values measured in fluid solution. This behavior is noteworthy with regards to the use of the Ir(C^C:)$_3$ complexes as phosphorescent dopants in applications such as OLEDs, since it suggests that the PL efficiencies in doped films are similarly quite high in value. If the luminance efficiencies were to remain low in the solid state, the Ir(C^C:)$_3$ complexes would be of little use as phosphorescent emitters. Consequently, we have fabricated OLEDs using the Ir(C^C:)$_3$ complexes as dopants and have obtained high external quantum efficiencies ($> 5\%$), consistent with values of $\Phi > 0.2$. A detailed study of these devices is currently underway and will be presented in a future report [44].

32.3 Conclusion

The approach to tuning the photophysical properties of cyclometalated Ir complexes investigated here involved the control of the relative energies of emissive and nonemissive excited states. The starting point for this work was a high triplet energy Ir complex, i.e., *fac*-Ir(ppz)$_3$, that gives no measurable emission at room temperature. By increasing the energy separation between emissive and nonemissive states, nonradiative decay rates could be decreased, leading to substantially improved luminescent quantum yields. This was demonstrated either by lowering the energy of the emitting state, as in *fac*-Ir(flz)$_3$, or by increasing the energies of the non-emissive states, as in the Ir(C^C:)$_3$ complexes. While the *fac*-Ir(flz)$_3$ complex has a markedly lower triplet state energy than the *fac*-Ir(flz)$_3$ complex, the Ir(C^C:)$_3$ complexes have higher triplet energies and display near-UV luminescence at room temperature. The next step in this study is to examine the temperature dependence of the photophysical properties of these materials. The temperature dependence of the emission lifetime and efficiency can be used to determine the relative energies and decay rates for both the emissive and nonemissive excited states of the NHC- and pyrazolyl-based materials. With this information in hand, we will be able to evaluate the nature of the nonradiative decay processes and design ligands and complexes that more effectively destabilize the non-emissive states, leading to higher luminance efficiencies. A detailed temperature dependent study is currently in progress.

32.4 Experimental Section

Solvents and reagents for synthesis were purchased from Aldrich, Matrix Scientific, and EM Science and were used without further purification. $IrCl_3 \cdot nH_2O$ was purchased from Next Chimica. N,N-Dimethylformamide (EM Science, anhydrous, 99.8%) and tetra-n-butylammonium hexafluorophosphate (TBAH) (Fluka, electrochemistry grade) were used for spectroscopic and electrochemical measurements. The syntheses of 1-[2-(9,9-dimethylfluorenyl)]pyrazole, 1-phenyl-3-methylimidazolium iodide, and 1-phenyl-3- methylbenzimidazolium iodide are given in the Supporting Information. The synthesis and characterization of *fac*-Ir(flz)$_3$, *fac*- and *mer*-Ir(pmi)$_3$, and *fac*- and *mer*-Ir(pmb)$_3$ are given below. All syntheses involving $IrCl_3 \cdot nH_2O$ or any other Ir(III) species were carried out under an inert atmosphere.

32.4.1 Synthesis of *fac*-Iridium(III) Tris(1-[2-(9,9-dimethylfluorenyl)]pyrazolyl-N,$C^{3'}$), *fac*-Ir(flz)$_3$

The *fac*-Ir(flz)$_3$ complex was prepared using the same two step procedure as that used for *fac*-Ir(ppz)$_3$ [18a]. A fluorenylpyrazolyl-based Ir(III) μ-dichloro-bridged dimer, [(flz)$_2$IrCl]$_2$, was synthesized in a 100 mL round-bottomed flask that was charged with iridium trichloride hydrate (0.250 g, 0.837 mmol), 1-[2-(9,9-dimethylfluorenyl)]pyrazole (0.500 g, 1.92 mmol), and 60 mL of a 2-ethoxyethanol–water mixture (3:1). The reaction mixture was stirred and heated with an oil bath at 110 °C for 16 h under nitrogen and then allowed to cool to ambient temperature. Addition of water gave the [(flz)$_2$IrCl]$_2$ complex as a yellowish-white solid (0.415 g, 66%), which was used without further purification in the next step to prepare the diketonate complex.

In a 100 mL round-bottomed flask, the [(flz)$_2$IrCl]$_2$ complex (0.250 g, 0.168 mmol), 2,4-pentanedione (0.036 mL, 0.352 mmol), and excess K$_2$CO$_3$ were combined with 60 mL of 1,2-dichloroethane. The reaction mixture was stirred and heated with an oil bath at 90°C for 16 h under nitrogen. After being cooled to ambient temperature, the reaction mixture was filtered to remove the insoluble salts and the solvent was removed from the filtrate under reduced pressure. Addition of methanol gave 0.248 g of a light yellow solid which was found to be (flz)$_2$Ir(acac) (91% yield). ^{1}H NMR (360 MHz, CDCl$_3$), ppm: 8.16 (dd, $J = 2.9, 1.0$ Hz, 2H), 7.69 (dd, $J = 2.2, 0.73$ Hz, 2H), 7.25–7.29 (m, 2H), 7.31–7.35 (m, 2H), 7.19 (s, 2H), 7.09-7.18 (m, 4H), 6.71 (dd, $J = 2.9, 2.2$ Hz, 2H), 6.54 (s, 2H), 5.23 (s, 1H), 1.40 (s, 6H), 1.38 (s, 12H).

A 50 mL round-bottomed flask was charged with the (flz)$_2$Ir(acac) complex (0.100 g, 0.125 mmol), 1-[2-(9,9-dimethyl-fluorenyl)]pyrazole (0.036 g, 0.138 mmol), and 25 mL of glycerol. The reaction mixture was stirred and heated in an oil bath at 210–220°C for 24 h under a nitrogen atmosphere. After being cooled to ambient temperature, distilled water was added to the reaction mixture and the precipitate was vacuum filtered and dried. The crude product was purified by flash column chromatography on silica gel using dichloromethane as the eluent to give pure *fac*-Ir(flz)$_3$ (0.0895 g, 75%). ^{1}H NMR (360 MHz, CDCl$_3$), ppm: 8.1 (dd, $J = 2.8, 0.5$ Hz, 3H), 7.36 (dd, $J = 6.7, 1.3$ Hz, 3H), 7.29 (s, 3H), 7.28 (s, 3H), 7.26 (dd, $J = 6.3, 1.5$ Hz, 3H), 7.14 (ddd, $J = 7.5, 7.4, 1.5$ Hz, 3H), 7.10 (ddd, $J = 7.4, 7.2, 1.3$ Hz, 3H), 7.03 (dd, $J = 2.1, 0.4$ Hz, 3H), 6.41 (dd, $J = 2.8, 2.2$ Hz, 3H), 1.56 (s, 9H), 1.46 (s, 9H). ^{13}C (90.55 MHz, CDCl$_3$), ppm: 153.74, 146.77, 143.75, 140.02, 138.47, 136.74, 136.55, 128.86, 126.54, 125.73, 124.87, 121.98, 119.88, 106.53, 105.55, 46.23, 28.03, 27.19. CHN Analysis Calcd: C, 66.85; H, 4.68; N, 8.66. Observed: C, 66.22; H, 4.52; N, 8.55.

32.4.2 Synthesis of [(pmi)$_2$IrCl]$_2$

A 100 mL round-bottomed flask was charged with silver(I) oxide (0.428 g, 1.85 mmol), 1-phenyl-3-methylimidazolate iodide (0.946 g, 3.31 mmol), iridium trichloride hydrate (0.301 g, 1.01 mmol), and 60 mL of 2-ethoxyethanol. The reaction mixture was stirred and heated with an oil bath at 120°C for 15 h under nitrogen while protected from light with aluminum foil. The reaction mixture was cooled to ambient temperature, and the solvent was removed under reduced pressure. The black mixture was extracted with ca. 20 mL of dichloromethane, and the extract was reduced to a volume of ca. 2 mL. Addition of methanol gave an off-white solid, which was formulated as the dichlorobridged dimer on the basis of its relatively simple ^{1}H NMR spectrum, along with some unreacted ligand. ^{1}H NMR (250 MHz, CDCl$_3$), ppm: 7.50 (d, $J = 2.0$ Hz, 4H), 7.07 (d, $J = 2.0$ Hz, 4H), 6.09 (dd, $J = 7.8, 1.0$ Hz, 4H), 6.63 (ddd, $J = 7.5, 7.5, 1.4$ Hz, 4H), 6.15 (ddd, $J = 7.5, 7.5, 1.4$ Hz, 4H), 3.87 (s, 12H). This product was used without further purification for the synthesis of the *fac*-Ir(pmi)$_3$.

32.4.3 Synthesis of *fac*-Iridium(III) Tris(1-phenyl-3-methylimidazolin-2-ylidene-$C,C^{2'}$), *fac*-Ir(pmi)$_3$

A 50 mL round-bottomed flask was charged with silver(I) oxide (0.278 g, 1.20 mmol), 1-phenyl-3-methylimidazolium iodide (0.080 g, 0.280 mmol), [(pmi)$_2$IrCl]$_2$ (0.108 g, 0.091 mmol), and 25 mL of 1,2-dichloroethane. The reaction mixture was stirred and heated with an oil bath at 77°C for 15 h under nitrogen while protected from light with aluminum foil. The reaction mixture was cooled to ambient temperature and concentrated under reduced pressure. The dark residue was taken up in dichloromethane, and the silver(I) salts were removed by filtration through Celite. The resultant light brown solution was further purified by flash column chromatography on silica gel using dichloromethane as the eluent and was then reduced in volume to ca. 2 mL. Addition of methanol gave 0.010 g (8% yield) of the iridium complex as an off-white solid. 1H NMR (360 MHz, CD$_2$Cl$_2$), ppm: 7.43 (d, $J = 2.0$ Hz, 3H), 7.16 (dd, $J = 7.7, 1.0$ Hz, 3H), 6.89 (ddd, $J = 8.0, 7.3, 1.5$ Hz, 3H), 6.78 (d, $J = 2.0$ Hz, 3H), 6.71 (ddd, $J = 8.0, 7.3, 1.5$ Hz, 3H), 6.60 (dd, $J = 7.2, 1.4$ Hz, 3H). ^{13}C (90.55 MHz, CD$_2$Cl$_2$), ppm: 176.77, 150.12, 148.05, 137.80, 125.02, 120.73, 120.56, 114.71, 110.42, 36.95. CHN Analysis Calcd: C, 54.28; H, 4.10; N, 12.66. Observed: C, 54.09; H, 3.83; N, 12.47.

32.4.4 Synthesis of *mer*-Iridium(III) Tris(1-phenyl-3-methylimidazolin-2-ylidene-*C,C*$^{2'}$), *mer*-Ir(pmi)3

A 50 mL round-bottomed flask was charged with silver(I) oxide (0.076 g, 0.328 mmol), 1-phenyl-3-methylimidazolium iodide (0.109 g, 0.381 mmol), iridium trichloride hydrate (0.029 g, 0.097 mmol), and 20 mL of 2-ethoxyethanol. The reaction mixture was stirred and heated with an oil bath at 120°C for 15 h under nitrogen while protected from light with aluminum foil. The reaction mixture was cooled to ambient temperature and concentrated under reduced pressure. Filtration through Celite, using dichloromethane as the eluent, was performed to remove the silver(I) salts. A white solid was obtained after removing the solvent in vacuo and was washed with methanol to give 0.016 g (24% yield) of the meridianal tris-iridium complex as a white solid. ^{1}H NMR (360 MHz, CDCl$_3$), ppm: 7.42 (d, J = 2.0 Hz, 1H), 7.37 (d, J = 2.0 Hz, 1H), 7.28 (d, J = 2.0 Hz, 1H), 6.9–7.2 (m, 3H), 6.78–6.85 (m, 3H), 6.5–6.75 (m, 3H), 3.04 (s, 1H), 3.02 (s, 1H), 2.95 (s, 1H). ^{13}C (90.55 MHz, CDCl$_3$), ppm: 175.08, 173.62, 172.46, 151.43, 150.09, 148.87, 148.07, 147.44, 146.69, 139.76, 139.54, 137.27, 124.76, 124.71, 124.50, 124.33, 120.57, 120.40, 119.98, 119.66, 119.44, 119.39, 114.13, 114.11, 114.08, 110.07, 109.48, 36.99, 36.93, 35.87. CHN Analysis Calcd: C, 54.28; H, 4.10; N, 12.66. Observed: C, 54.25; H, 3.77; N, 12.44.

32.4.5 Synthesis of [(pmb)$_2$IrCl]$_2$

A 100 mL round-bottomed flask was charged with silver(I) oxide (5.590 g, 24.1 mmol), 1-phenyl-3-methylbenzimidazolium iodide (6.756 g, 20.1 mmol), iridium trichloride hydrate (1.50 g, 5.02 mmol), and 50 mL of 2-ethoxyethanol. The reaction mixture was stirred and heated with an oil bath at 120°C for 24 h under nitrogen while protected from light with aluminum foil. The reaction mixture was cooled to ambient temperature and concentrated under reduced pressure. Flash column chromatography on Celite using dichloromethane as the eluent was performed to remove the silver(I) salts. A brown oil was obtained, and addition of ethanol gave a light brown solid. The brownish solid was filtered and washed with ethanol. It was further purified by flash column chromatography on silica gel using dichloromethane as the eluent to give a yellowish solid product (0.412 g, 12.7% yield), which was formulated as [(pmb)$_2$IrCl]$_2$ based on ^{1}H NMR spectral features that are similar to those for [(pmi)$_2$IrCl]$_2$. ^{1}H NMR (360 MHz, CDCl$_3$), ppm: 8.18 (m, 4H), 7.62 (dd, J = 7.9, 1.2 Hz, 4H), 7.52 (m, 4H), 7.30 (m, 4H), 6.78 (ddd, J = 8.3, 7.8, 1.3 Hz, 4H), 6.38 (ddd, J = 8.2, 7.6, 1.1 Hz, 4H), 6.24 (dd, J = 7.6, 1.3 Hz, 4H), 3.78 (s, 12H).

32.4.6 Synthesis of Iridium(III) Tris(1-phenyl-3-methylbenzimidazolin-2-ylidene-C,C$^{2'}$), Ir(pmb)$_3$

A 50 mL round-bottomed flask was charged with silver(I) oxide (0.0886 g, 0.382 mmol), 1-phenyl-3-methylbenzimidazolium iodide (0.225 g, 0.669 mmol), [(pmb)$_2$IrCl]$_2$ (0.412 g, 0.319 mmol), and 25 mL of 1,2-dichloroethane. The reaction mixture was stirred and heated with an oil bath at 95°C for 24 h under nitrogen while protected from light with aluminum foil. The reaction mixture was cooled to ambient temperature and concentrated under reduced pressure. Flash column chromatography on 50:50 Celite and silica gel using dichloromethane as the eluent was performed to give 0.514 g of a mixture of meridianal and facial Ir(pmb)$_3$ (3:1) as an-off white solid. Column chromatography using ethyl acetate/hexanes (20:80) as the eluent was performed to give 0.400 g of predominantly *mer*-Ir(pmb)$_3$ with some facial impurity (77% yield). The remaining *fac*-Ir(pmb)$_3$ was eluted using ethyl acetate/hexanes (40:60) to give 0.110 g (21% yield) of pure *fac*-Ir(pmb)$_3$. Repeated column chromatography was required to obtain *mer*-Ir(pmb)3 free of the fac impurity.

(A) *fac*-Ir(pmb)$_3$. ^{1}H NMR (360 MHz, CDCl$_3$), ppm: 8.08 (d, J = 8.2 Hz, 3H), 7.86 (d, J = 7.8 Hz, 3H), 7.24 (ddd, J = 8.8, 6.9, 1.3 Hz, 3H), 7.15 (ddd, J = 7.7, 8.06, 0.85 Hz, 3H), 7.12 (d, J = 1.1 Hz, 2H), 7.09 (d, J = 1.1 Hz, 1H), 7.05 (ddd, J = 8.2, 7.0, 1.6 Hz, 3H), 6.72 (ddd, J = 7.7, 7.1, 1.1 Hz, 3H), 6.65 (dd, J = 7.3 Hz, 3H) 3.22 (s, 9H). ^{13}C (90.55 MHz, CDCl$_3$), ppm: 189.58, 148.75, 148.65, 137.02, 136.31, 132.63, 124.64, 122.59, 121.68, 120.86, 111.98, 111.125, 109.49, 33.42. CHN Analysis Calcd: C, 61.97; H, 4.09; N, 10.32. Observed: C, 62.16; H, 3.77; N, 10.31.

(B) *mer*-Ir(pmb)$_3$. ^{1}H NMR (360 MHz, CDCl$_3$), ppm: 8.16 (d, $J = 8.7$ Hz, 1H), 8.14 (d, $J = 8.2$ Hz, 1H), 8.05 (d, $J = 7.8$ Hz, 1H), 7.85 (d, $J = 7.5$ Hz, 1H), 7.82 (d, $J = 7.8$ Hz, 1H), 7.75 (d, $J = 7.3$ Hz, 1H), 6.46–7.47 (m, 18H), 3.25 (s, 3H), 3.18 (s, 3H), 3.17 (s, 3H). ^{13}C (90.55 MHz, CDCl$_3$), ppm: 188.22, 185.95, 184.89, 150.84, 149.64, 149.27, 148.79, 147.86, 147.84, 139.10, 138.93, 136.67, 136.60, 136.22, 132.49, 132.52, 132.48, 124.75, 124.53, 124.34, 122.69, 122.61, 121.86, 121.75, 121.43, 120.61, 120.21, 120.23, 112.35, 111.77, 111.13, 111.09, 111.05, 109.64, 109.56, 109.39, 33.40, 33.39, 33.33, 32.73. CHN Analysis Calcd: C, 61.97; H, 4.09; N, 10.32. Observed: C, 61.88; H, 3.69; N, 10.21.

32.4.7 X-ray Crystallography

Diffraction data for *fac*-Ir(flz)$_3$, *fac*-Ir(pmb)$_3$, and *mer*-Ir(pmb)$_3$ were collected at $T = 153(2)$, 143(2), and 143(2) K, respectively. The data sets were collected on a Bruker SMART APEX CCD diffractometer with graphite monochromated Mo Kα radiation ($\lambda = 0.71073$ Å). The cell parameters for the iridium complexes were obtained from a least-squares refinement of the spots (from 60 collected frames) using the SMART program. Intensity data were processed using the Saint Plus program. All the calculations for the structure determination were carried out using the SHELXTL package (version 5.1) [45]. Initial atomic positions were located by Patterson methods using XS, and the structures of the compounds were refined by the least-squares method using SHELX97. Absorption corrections were applied by using SAD-ABS [46]. In most cases, hydrogen positions were input and refined in a riding manner along with the attached carbons. A summary of the refinement details and the resulting factors for the complexes are given in the Supporting Information.

32.4.8 Electrochemical and Photophysical Characterization

Cyclic voltammetry and differential pulsed voltammetry were performed using an EG&G potentiostat/galvanostat model 283. Anhydrous DMF was used as the solvent under an inert atmosphere, and 0.1 M tetra-n-butylammonium hexafluorophosphate was used as the supporting electrolyte. A glassy carbon rod was used as the working electrode, a platinum wire was used as the counter electrode, and a silver wire was used as a pseudoreference electrode. The redox potentials are based on values measured from differential pulsed voltammetry and are reported relative to either a ferrocenium/ferrocene (Cp_2Fe^+/Cp_2Fe) redox couple or a decamethylferrocenium/decamethylferrocene ($Me_5Cp_2Fe^+/Me_5Cp_2Fe$) redox couple, used as an internal reference [47].

The UV–visible spectra were recorded on a Hewlett-Packard 4853 diode array spectrophotometer. Steady state emission spectra at room temperature and 77 K were determined using a Photon Technology International QuantaMaster model C-60SE spectrofluorimeter. Phosphorescence lifetime measurements (> 2 μs) were performed on the same fluorimeter, equipped with a microsecond xenon flashlamp, or on an IBH Fluorocube lifetime instrument by a time correlated single photon counting method using either a 331 or 373 nm LED excitation source. Quantum efficiency (QE) measurements were carried out at room temperature in degassed toluene solutions using the optically dilute method [48]. Solutions of *fac*-Ir(ppy)$_3$ ($\Phi = 0.4$)* [6, 18a] were used as reference. ^{1}H and ^{13}C NMR spectra were recorded on a Bruker AM 360 MHz instrument, and chemical shifts were referenced to residual protiated solvent. Elemental analyses (CHN) were performed at the Microanalysis Laboratory at the University of Illinois, Urbana-Champaign.

Acknowledgment

We would like to acknowledge Universal Display Corporation and the Department of Energy for their support of this work.

Supporting Information Available

The syntheses of 1-[2-(9,9-dimethylfluorenyl)]pyrazole, 1-phenyl-3-methylimidazolium iodide, and 1-phenyl-3-methylbenzimidazolium iodide are described. The emission spectrum of flzH at 77 K, cyclic voltammetric traces for *fac*- and *mer*-Ir(pmi)$_3$ and Ir(pmb)$_3$, absorption spectra of *mer*-Ir(pmi)$_3$ and *mer*-Ir(pmb)$_3$, and emission spectra of *fac*- and *mer*-Ir(pmi)$_3$ at 77 K, as well as crystallographic data for *fac*-Ir(flz)$_3$ and *fac*- and *mer*-isomers of Ir(pmb)$_3$ (i.e., tables of bond lengths and angles, crystal data, atomic coordinates, bond distances, bond angles, and anisotropic displacement parameters) are also given. The crystallographic information files (.cif) are also given for each compound.

References

1. (a) Baldo, M. A., O'Brien, D. F., You, Y., Shoustikov, A., Sibley, S., Thompson, M. E., Forrest, S. R. *Nature* 1998, **395**, 151. (b) Baldo, M. A., Lamansky, S., Burrows, P. E., Thompson, M. E., Forrest, S. R. *Appl. Phys. Lett.* 1999, **75**, 4. (c) Thompson, M. E., Burrows, P. E., Forrest, S. R. *Curr. Opin. Solid State Mater. Sci.* 1999, **4**, 369. (d) Baldo, M. A., Thompson, M. E., Forrest, S. R. *Nature* 2000, **403**, 750.

2. (a) Lamansky, S., Djurovich, P. I., Abdel-Razzaq, F., Garon, S., Murphy, D. L., Thompson, M. E. *J. Appl. Phys.* 2002, **92**, 1570. (b) Chen, F. C., Yang, Y., Thompson, M. E., Kido, J. *Appl. Phys. Lett.* 2002, **80**, 2308. (c) Markham, J. P. J., Lo, S.-C., Magennis, S. W., Burn, P. L., Samuel, I. D. W. *Appl. Phys. Lett.* 2002, **80**, 2645. (d) Zhu, W., Mo, Y., Yuan, M., Yang, W., Cao, Y. *Appl. Phys. Lett.* 2002, **80**, 2045. (e) Adachi, C., Baldo, M. A., Forrest, S. R., Thompson, M. E. *J. Appl. Phys.* 2001, **90**, 4058. (f) Ikai, M., Tokito, S., Sakamoto, Y., Suzuki, T., Taga, Y. *Appl. Phys. Lett.* 2001, **79**, 156. (g) Adachi, C., Lamansky, S., Baldo, M. A., Kwong, R. C., Thompson, M. E., Forrest, S. R. *Appl. Phys. Lett.* 2001, **78**, 1622. (h) Lamansky, S., Djurovich, P. I., Murphy, D., Abdel-Razzaq, F., Lee, H. E., Adachi, C., Burrows, P. E., Forrest, S. R., Thompson, M. E. *J. Am. Chem. Soc.* 2001, **123**, 4304.

3. (a) Sutin, N. *Acc. Chem. Res.* 1968, **1**, 225. (b) Meyer, T. J. *Acc. Chem. Res.* 1978, **11**, 94.

4. Schmid, B., Garces, F. O., Watts, R. J. *Inorg. Chem.* 1994, **33**, 9.

5. (a) Belmore, K. A., Vanderpool, R. A., Tsai, J. C., Khan, M. A., Nicholas, K. M. *J. Am. Chem. Soc.* 1988, **110**, 2004. (b) Silavwe, N. D., Goldman, A. S., Ritter, R., Tyler, D. R. *Inorg. Chem.* 1989, **28**, 1231.

6. King, K. A., Spellane, P. J., Watts, R. J. *J. Am. Chem. Soc.* 1985, **107**, 1431.

7. (a) Demas, J. N., Harris, E. W., McBride, R. P. *J. Am. Chem. Soc.* 1977, **99**, 3547. (b) Demas, J. N., Harris, E. W., Flynn, C. M., Diemente, J. D. *J. Am.*

Chem. Soc. 1975, **97**, 3838. (c) Gao, R., Ho, D. G., Hernandez, B., Selke, M., Murphy, D., Djurovich, P. I., Thompson, M. E. *J. Am. Chem. Soc.* 2002, **124**, 14828.

8. Lo, K. K.-W., Chung, C.-K., Lee, T. K.-M., Lui, L.-K., Tsang, K. H.-K., Zhu, N. *Inorg. Chem.* 2003, **42**, 6886.

9. (a) Li, J., Djurovich, P. I., Alleyne, B. D., Tsyba, I., Ho, N. N., Bau, R., Thompson, M. E. *Polyhedron* 2004, **23**, 419. (b) Holmes, R. J., Forrest, S. R., Tung, Y. J., Kwong, R. C., Brown, J. J., Garon, S., Thompson, M. E. *Appl. Phys. Lett.* 2003, **82**, 2422. (c) Holmes, R. J., D'Andrade, B. W., Forrest, S. R., Ren, X., Thompson, M. E. *Appl. Phys. Lett.* 2003, **83**, 3818. (d) Ren, X., Li, J., Holmes, R. J., Djurovich, P. I., Forrest, S. R., Thompson, M. E. *Chem. Mater.* 2004, **16**, 4743.

10. (a) Stampor, W., Mȩzyk, J., Kalinowski, J. *Chem. Phys.* 2004, **300**, 189–195. (b) Hay, P. J. *J. Phys. Chem. A* 2002, **106**, 1634.

11. Finkelstein, W., Yersin, H. *Chem. Phys. Lett.* 2004, 300.

12. (a) Colombo, M. G., Güdel, H. *Inorg. Chem.* 1993, **32**, 3081. (b) Strouse, G. F., Güdel, H. U., Bertolasi, V., Ferretti, V. *Inorg. Chem.* 1995, **34**, 5578. (c) Lever, A. P. B. *Inorganic Electronic Spectroscopy*, 2nd ed., Elsevier: New York, 1984; p. 174. (d) Wiedenhofer, H., Schützenmeier, S., von Zelewsky, A., Yersin, H. *J. Phys. Chem.* 1995, **99**, 13385. (e) Yersin, H., Donges, D. *Top. Curr. Chem.* 2001, **214**, 81–186. (f) Vanhelmont, F. W. M., Guḋel, H. U., Förtsch, M., Bürgi, H.-B. *Inorg. Chem.* 1997, **36**, 5512.

13. Paulose, B. M. J. S., Rayabarapu, D. K., Duan, J.-P. Cheng, C.-H. *Adv. Mater.* 2004, **16**, 2003.

14. Turro, N. J. *Modern Molecular Photochemistry*; Benjamin/Cummings: Menlo Park, 1978; pp. 30–32.

15. (a) Ni, T., Caldwell, R. A., Melton, L. A. *J. Am. Chem. Soc.* 1989, **111**, 457–464. (b) Ramamurthy, V., Caspar, J. V., Eaton, D. F., Kuo, E. W., Corbin, D. R. *J. Am. Chem. Soc.* 1992, **114**, 3882–3892.

16. Taylor, H. V., Allred, A. L., Hoffman, B. M. *J. Am. Chem. Soc.* 1973, **95**, 3215–3219.

17. (a) Grushin, V. V., Herron, N., LeCloux, D. D., Marshall, W. J., Petrov, V. A., Wang, Y. *J. Chem. Soc., Chem. Commun.* 2001, 1494–1495. (b) Coppo, P., Plummer, E. A., De Cola, L. *J. Chem. Soc., Chem. Commun.* 2004, 1774–1775. (c) Dedeian, K., Shi, J., Nigel, S., Forsythe, E., Morton, D. *Inorg. Chem.* 2005, **44**, 4445–4447.

18. (a) Tamayo, A., Alleyne, B., Djurovich, P. I., Lamansky, S., Tsyba, I., Ho, N., Bau, R., Thompson, M. E. *J. Am. Chem. Soc.* 2003, **125**, 7377. (b) Adamovich, V., Brooks, J., Tamayo, A., Alexander, A. M., Djurovich, P. I.,

D'Andrade, B. W., Adachi, C., Forrest, S. R., Thompson, M. E. *New J. Chem.* 2002, **26**, 1171.

19. Li, J., Djurovich, P. I., Alleyne, B. D., Yousufuddin, M., Ho, N. N., Thomas, J. C., Peters, J. C., Bau, R., Thompson, M. E. *Inorg. Chem.* 2005, **44**, 1713.

20. (a) Nam, E. J., Kim, J. H., Kim, B., Kim, S. M., Park, N. G. *Bull. Chem. Soc. Jpn.* 2004, **77**, 751–755. (b) Choi, G. C., Lee, J. E., Park, N. G., Kim, Y. S. *Mol. Cryst. Liq. Cryst.* 2004, **424**, 173–185.

21. Pavlik, J. W., Connors, R. E., Burns, D. S., Kurzwell, E. M. *J. Am. Chem. Soc.* 1993, **115**, 7465.

22. Maestri, M., Sandrini, D., Balzani, V., Maeder, U., von Zelewsky, A. *Inorg. Chem.* 1987, **26**, 1323.

23. Kawamura, Y., Goushi, K., Brooks, J., Brown, J. J., Sasabe, H., Adachi, C. *Appl. Phys. Lett.* 2004. Kawamura, Y., Sasabe, H., Adachi, C. *Jpn. J. Appl. Phys.* 2004, **43**, 7729.

24. The authors in Ref. [20a] report a green emission at room temperature from solid state samples of *fac*-Ir(ppz)$_3$. We have also observed a similar emission for neat solids of *fac*-Ir(ppz)$_3$ and attribute it to luminescence from an impurity in the material. Extensive purification of *fac*-Ir(ppz)$_3$ leads to weakly or nonemissive samples. Analysis of the green luminescence in the solid state by time-resolved spectroscopy shows that the emission displays a rise time of ca. 2 μs at 77 K before decay. This temporal behavior is consistent with trap emission from an impurity in the solid. Moreover, the emission is only observed for neat solids. Samples of *fac*-Ir(ppz)$_3$ doped into polystyrene or other high triplet energy host materials are nonemissive at room temperature and show only the near-UV emission illustrated in Fig. 32.2 at temperatures below 200 K.

25. (a) Forster, L. S. *Coord. Chem. Rev.* 2002, **227**, 59 and references therein. (b) Brennaman, M. K., Meyer, T. J., Papanikolas, J. M. *J. Phys. Chem. A* 2004, **108**, 9938. (c) Wang, X.-Y., Del Guerzo, A., Schmehl, R. H. *J. Photochem. Photobiol. C* 2004, **5**, 55 and references therein.

26. Lumpkin, R. S., Kober, E. M., Worl, L. A., Murtaza, Z., Meyer, T. J. *J. Phys. Chem.* 1990, **94**, 239.

27. Dixon, I. M., Collin, J.-P., Sauvage, J.-P., Flamigni, L., Encinas, S., Barigelletti, F. *Chem. Soc. Rev.* 2000, **6**, 385–391.

28. Nolan, S. P., Hoff, C. D., Stoutland, P. O., Newman, L. J., Buchanan, J. M., Bergman, R. G., Yang, G. K., Peters, K. S. *J. Am. Chem. Soc.* 1987, **109**, 3143.

29. (a) Holmes, R. J., D'Andrade, B. W., Forrest, S. R., Ren, X., Thompson, M. E. *Appl. Phys. Lett.* 2003, **83**, 3818. (b) Ren, X., Li, J., Holmes, R. J., Djurovich, P. I., Forrest, S. R., Thompson, M. E. *Chem. Mater.* 2004, **16**, 4743.

30. (a) Lo, K. K., Chung, C., Lee, T. K., Lui, L., Tsing, K. H., Zhu, N. *Inorg. Chem.* 2003, **42**, 6886. (b) Lo, K. K., Chan, J. S., Chung, C., Lui, L. *Organometallics* 2004, **23**, 3108. (c) Kwon, T.-H., Cho, H. S., Kim, M. K., Kim, J.-W., Kim, J.-J., Lee, K. H., Park, S. J., Shin, I.-S., Kim, H., Shin, D. M., Chung, Y. K., Hong, J.-I. *Organometallics* 2005, **24**, 1578–1585.
31. (a) Gong, X., Robinson, M. R., Ostrowski, J. C., Moses, D., Bazan, G. C., Heeger, A. J. *Adv. Mater.* 2002, **14**, 581. (b) Ostrowski, J. C., Robinson, M. R., Heeger, A. J., Bazan, G. C. *Chem. Commun.* 2002, **7**, 784. (c) Gong, X., Robinson, M. R., Ostrowski, J. C., Moses, D., Bazan, G. C., Heeger, A. J., Liu, M. S., Jen, A. K. *Adv. Mater.* 2003, **15**, 45. (d) Gong, X., Robinson, M. R., Ostrowski, J. C., Moses, D., Bazan, G. C., Heeger, A. J. *Appl. Phys. Lett.* 2005, **86**, 171108. (e) Wu, F., Su, H., Shu, C., Luo, L., Diau, W., Cheng, C., Duan, J., Lee, G. *J. Mater. Chem.* 2005, **15**, 1035. (f) Tsuboyama, A., Iwawaki, H., Furugori, M., Mukaide, T., Kamatani, J., Igawa, S., Moriyama, T., Miura, S., Takiguchi, T., Okada, S., Hoshino, M., Ueno, K. *J. Am. Chem. Soc.* 2003, **125**, 12971.
32. See the Supporting Information.
33. The equations $k_r = \Phi/\tau$ and $k_{nr} = (1 - \Phi)/\tau$ were used to calculate the radiative (k_r) and nonradiative (k_{nr}) decay rates, where Φ is the quantum efficiency and τ is the luminescence lifetime of the sample at room temperature.
34. Cummings, S. D., Eisenberg, R. *J. Am. Chem. Soc.* 1996, **118**, 1949.
35. (a) Bourissou, D., Guerret, O., Gabbaï, F. P., Bertrand, G. *Chem. Rev.* 2000, **100**, 39. (b) Klapars, A., Antilla, J. C., Huang, X., Buchwald, S. L. *J. Am. Chem. Soc.* 2001, **123**, 7727–7729.
36. Nemcsok, D., Wichmann, K., Frenking, G. *Organometallics* 2004, **23**, 3640.
37. Hitchcock, P. B., Lappert; M. F., Terreros, P. *J. Organomet. Chem.* 1982, **239**, C26.
38. For a recent example of a related luminescent Ru complex with a noncyclometalated NHC ligand, see: Son, S. U., Park, K. H., Lee, Y.-S., Kim, B. Y., Choi, C. H., Lah, M. S., Jang, Y. H., Jang, D.-J., Chung, Y. K. *Inorg. Chem.* 2004, **43**, 6896–6898.
39. Lamansky, S., Djurovich, P. I., Murphy, D., Abdel-Razaq, F., Kwong, R., Tsyba, I., Bortz, M., Mui, B., Bau, R., Thompson, M. E. *Inorg. Chem.* 2001, **40**, 1704–1711.
40. The formulation of the [(C^C:)$_2$IrCl]$_2$ complexes as μ-dichlorobridged dimers is based on their relatively simple ^{1}H NMR spectra and by analogy to the [(C^N)$_2$IrCl]$_2$ complexes. In addition, the [(C^C:)$_2$IrCl]$_2$

complexes can be treated with 4,4'-di-*tert*-butyl-2,2'-bipyridine to form cationic complexes with the general formula $[(C^\wedge C:)_2Ir(N^\wedge N)]^+$ (unpublished results).

41. (a) Wang, H. M. J., Lin, I. J. B. *Organometallics* 1998, **17**, 972–975. (b) Chianese, A. R., Li, X., Janzen, M. C., Faller, J. W., Crabtree, R. H. *Organometallics* 2003, **22**, 1663–1667.
42. Murov, S. L., Carmichael, I., Hug, G. L. *Handbook of Photochemistry*, 2nd ed., Marcel Dekker: New York, 1993.
43. Draxler, S. *J. Phys. Chem. A* 1999, **103**, 4719.
44. Holmes, R., Forrest, S. R., Sajoto, T., Tamayo, A., Djurovich, P. I., Thompson, M. E., Brooks, J., Tung, Y.-J., D'Andrade, B. W., Weaver, M. S., Kwong, R. C., Brown, J. J. *Appl. Phys. Lett.,* submitted for publication.
45. Sheldrick, G. M. *SHELXTL*, version 5.1; Bruker Analytical X-ray System, Inc.: Madison, WI, 1997.
46. Blessing, R. H. *Acta Crystallogr*. 1995, **A51**, 33.
47. (a) Gagne, R. R., Koval, C. A., Lisensky, G. C. *Inorg. Chem.* 1980, **19**, 2854. (b) Sawyer, D. T., Sobkowiak, A., Roberts, J. L., Jr. *Electrochemistry for Chemists*, 2nd ed., John Wiley and Sons: New York, 1995; p. 467.
48. (a) Demas, J. N., Crosby, G. A. *J. Phys. Chem.* 1978, **82**, 991. (b) DePriest, J., Zheng, G. Y., Goswami, N., Eichhorn, D. M., Woods, C., Rillema, D. P. *Inorg. Chem.* 2000, **39**, 1955.

Chapter 33

Synthetic Control of Pt···Pt Separation and Photophysics of Binuclear Platinum Complexes

Biwu Ma, Jian Li, Peter I. Djurovich, Muhammed Yousufuddin, Robert Bau, and Mark E. Thompson

Departments of Chemistry and Materials Science, University of Southern California, Los Angeles, California 90089, USA
met@usc.edu

Luminescent square-planar platinum(II) complexes have attracted a great deal of interest because of their potential applications in many fields, such as chemosensors [1], photocatalysts [2], light emitting diodes (LEDs) [3], and photovoltaic devices [4]. Emission from an isolated platinum(II) complex is typically assigned to ligand centered (LC) and/or metal-to-ligand charge transfer (MLCT) states. In addition, the square-planar Pt complexes have a rich excimer and aggregate/dimer photophysics [5], leading to marked red shifts relative to the mononuclear emission spectra. These transitions are denoted as either metal–metal-to-ligand charge transfer (MMLCT) or excimeric ligand-to-ligand charge transfer. The MMLCT transition involves charge transfer between a filled Pt–Pt antibonding orbital

---

Reprinted from *J. Am. Chem. Soc.*, **127**, 28–29, 2005.

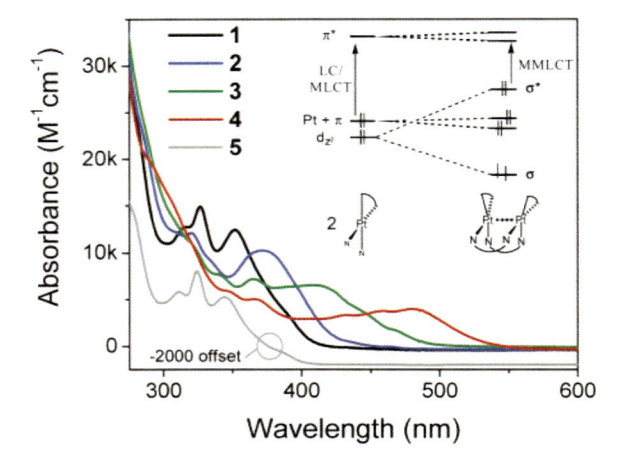

Figure 33.1 Absorption spectra for **1–5** (CH_2Cl_2 solvent, 298 K) are shown. Qualitative MO schemes for **1–4** and a mononuclear analogue are shown in the inset.

and a vacant, ligand-based π^* orbital ($\sigma^* \rightarrow \pi^*$; see Fig. 33.1 inset), often luminescing in the visible part of the spectrum [6]. The energy of the σ^* orbital, and thus the MMLCT excited state, shows a strong dependence on the metal–metal distance; the transition energy decreases with decreasing Pt–Pt separation.

In this communication, we report a series of pyrazolate-bridged cyclometalated platinum(II) complexes that have a boatlike conformation (Fig. 33.2) [7]. The bridging pyrazolate controls the degree of metal–metal interaction and, thus, the nature of the excited state. The complexes have the general formula $C^\wedge NPt(\mu\text{-}pz')_2PtC^\wedge N$ (where $C^\wedge N$ = 2-(2,4-difluorophenyl)pyridyl, $pz' = \mu$-pyrazolate (**1**), 3,5-dimethylpyrazolate (**2**), 3-methyl-5-tert-butylpyrazolate (**3**), and 3,5-bis(tert-butyl)pyrazolate (**4**)) (see Supporting Information for syntheses). We have also prepared $C^\wedge NPt(pz)_2BEt_2$, **5**, which has an isolated Pt center (Fig. 33.2).

The single-crystal X-ray structure of each complex has been determined. Complexes **1–4** have similar geometries, consisting of two square-planar Pt moieties, bridged by two μ-pyrazolate ligands in an exo-bidentate fashion (Fig. 33.2) [7]. The two $C^\wedge N$ ligands are oriented in a C_s symmetric manner for compounds **1**, **2**, and

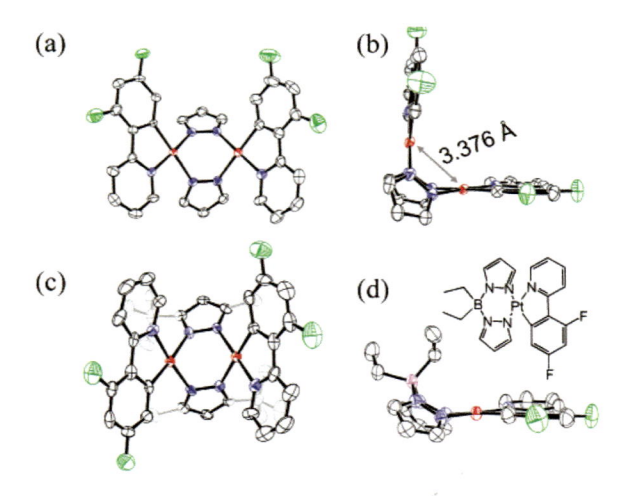

Figure 33.2 ORTEP views of **1**: (a) and (b), **3**: (c) and **5**: (d).

4 (i.e., pyridyl group at the bottom in Fig. 33.2a), while the $C^\wedge N$ ligands have a C_2 symmetric, opposed orientation in **3** (Fig. 33.2c). The structural difference for **3** is most likely due to the asymmetric substitution of the pyrazolate, which leads to a steric preference for the opposed $C^\wedge N$ orientation. Substituting bulky groups to the 3- and 5-positions of the pyrazolate bridges forces the two $C^\wedge N$ Pt moieties closer together, decreasing the Pt–Pt distance. The Pt–Pt spacings are **1** = 3.3763(7) Å, **2** = 3.1914(9) Å, **3** = 3.0457(7) Å, and **4** = 2.8343(6) Å. A related μ-pyrazolate bridged binuclear Pt complex with thienylpyridine cyclometallating ligands ($C^\wedge N$) has a Pt–Pt spacing of 3.4863(6) Å [7a], slightly longer than that observed for **1**. Despite the short Pt–Pt distance in **4**, the separation between the $C^\wedge N$ ligands ranges from 3.2 Å near the Pt atom to more than 5.1 Å at the periphery of the $C^\wedge N$ ligand; these values are outside of the distances expected for any π–π interaction. Thus, we expect the principal effect of the increased steric bulk of the μ-pyrazolate in **2–4** is to enhance Pt-Pt interaction (stabilizing the Pt–Pt σ and destabilizing the σ^* bonding orbitals) [8]. A number of platinum complexes exhibiting a range of Pt–Pt distances (and thus different photophysical properties) have been reported [5, 6];

however, simple synthetic control of Pt–Pt separation, as illustrated here, has not been previously achieved.

The complexes range in color, from yellow for **1** and **2**, to orange for **3**, and red for **4**. The UV–visible absorption spectra of **1**–**4** (Fig. 33.1) show bands between 300 and 400 nm that are very similar to analogous bands in the related mononuclear species **5**; these are assigned to MLCT transitions [9]. Compound **5** has an extinction coefficient roughly half that of **1**, consistent with the lower molecular weight for **5**. For complexes **3** and **4**, a lower energy band is also observed between 400 and 550 nm ($\epsilon \approx 6000$ $M^{-1}cm^{-1}$); these bands are assigned to MMLCT transitions. The broad absorption bands red-shift with decreasing Pt–Pt distance, indicating that the ^{1}MMLCT is strongly dependent on the Pt–Pt separation.

All four complexes emit in fluid solution, in the solid state (dispersed in polystyrene (PS)), and in 2-MeTHF glass at 77 K (Fig. 33.3). Single-crystal emission spectra closely resemble the same spectra in PS (see Supporting Information). There is a large variation in the emission λ_{max} as a function of Pt–Pt distance for **1**–**4** in both glassy 2-MeTHF solution and dispersed in PS. The luminescent lifetimes (τ) for **1**–**4** are in the microsecond regime, indicating that the emission is phosphorescence. For example, $\tau = 2.4$ μs for **1** and $\tau = 1.4$ μs for **4**, in PS at room temperature. The emission spectra for **1**–**4** in 2-MeTHF at 77 K narrow considerably, and the λ_{max} blue-shift to 458, 462, 515, and 570 nm, respectively. The emission line shapes of **1** and **2** at 77 K have distinct vibronic features, similar to the mononuclear analogue **5** and have $\tau = 8.5$ and 7.4 μs, respectively. In contrast, complexes **3** and **4** display broad, featureless emission spectra at 77 K, with $\tau = 4.1$ and 5.4 μs, respectively, consistent with an assignment to a triplet metal–metal-to-ligand charge transfer (^{3}MMLCT) transition. The photophysical properties of **1** and **2** in fluid solution differ from those in the solid state. Complexes **1** and **2** display broad featureless spectra similar to **3** in fluid solution. Complex **1** also shows additional emission features between 450 and 525 nm, consistent with mononuclear emission. In contrast, the mononuclear analogue, **5**, shows only blue structured emission in all solution media.

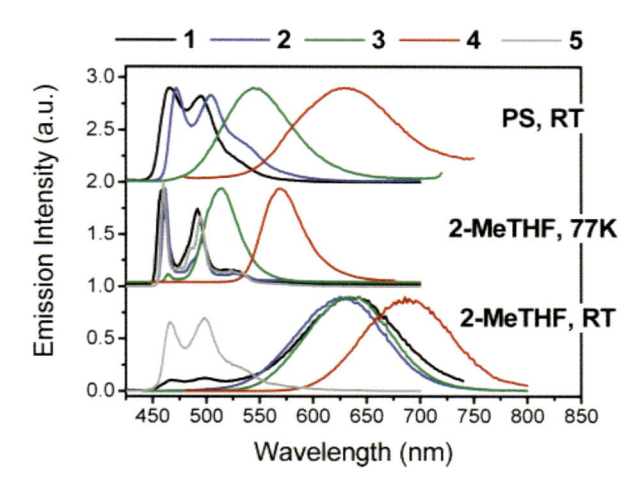

Figure 33.3 Normalized emission spectra of complexes **1–5** in 2-MeTHF solution (bottom and middle) and complexes **1–4** dispersed in polystryene at room temperature (top).

For complexes **1** and **2** in the solid state, the close similarity of the emission spectra to that of **5** indicates luminescence originating from a mixed ligand center triplet/metal-to-ligand charge transfer (^{3}LC/MLCT) excited state, consistent with little-to-no metal–metal ground-state interaction under these conditions. On the other hand, for complexes **3** and **4** in the same rigid media, only ^{3}MMLCT emission is observed, indicating that a strong intramolecular Pt–Pt ground-state interaction exists in both of these complexes. The complex with the shortest Pt–Pt spacing, **4**, displays red emission, while the complex with an intermediate Pt–Pt spacing, **3**, gives green emission. In fluid solution, all four binuclear complexes readily collapse into the ^{3}MMLCT excited state, giving a broad featureless, red emission. In the ^{3}MMLCT excited state, the σ^* orbital is depopulated, which shortens the Pt–Pt bond. A similar bond shortening process has been experimentally observed for other d^8–d^8 Pt binuclear complexes [10]. The enhanced Pt–Pt interactions destabilize the σ^* orbital and, in fluid solution, lead to luminescence that is red-shifted relative to emission from complexes with static, ground-state structures, as occurs in the PS and 77 K samples. Whereas all four complexes can relax into the ^{3}MMLCT excited state in fluid solution, the samples in PS (and at 77 K) retain their equilibrium ground-state structures and the two which have a σ^* HOMO, **3** and **4**, continue to display MMLCT absorption and emission.

The color control and bright luminescence of **1–4** in the solid state make these complexes materials of interest for application as phosphorescent emitters in organic LEDs; these device results will be reported elsewhere. It has also not escaped our notice that the shortening of the Pt–Pt distance in the excited state of these binuclear Pt complexes makes them ideal candidates for photoactuated nanohinges and nanotweezers.

Acknowledgment

We thank the Universal Display Corporation for financial support of this work.

Supporting Information Available

X-ray crystallographic data in CIF format, synthetic procedures, characterization data, and photophysical data are given for **1–5** (PDF). This material is available free of charge via the Internet at http://pubs.acs.org.

References

1. (a) Peyratout, C. S., Aldridge, T. K., Crites, D. K., McMillin, D. R. *Inorg. Chem.* 1995, **34**, 4484–4489. (b) Houlding, V. H., Frank, A. J. *Inorg. Chem.* 1985, **24**, 3664–3668. (c) Kunugi, Y., Mann, K. R., Miller, L. L., Exstrom, C. L. *J. Am. Chem. Soc.* 1998, **120**, 589.
2. (a) Zhang, D., Wu, L.-Z., Zhou, L., Han, X., Yang, Q.-Z., Zhang, L.-P., Tung, C.-H. *J. Am. Chem. Soc.* 2004, **126**, 3440–3441. (b) Connick, W. B., Gray, H. B. *J. Am. Chem. Soc.* 1997, **119**, 11620–11627. (c) Hissler, M., McGarrah, J. E., Connick, W. B., Geiger, D. K., Cummings, S. D., Eisenberg, R. *Coord. Chem. Rev.* 2000, **208**, 115.

3. (a) Adamovich, V., Brooks, J., Tamayo, A., Alexander, A. M., Djurovich, P. I., D'Andrade, B. W., Adachi, C., Forrest, S. R., Thompson, M. E. *New J. Chem.* 2002, **26**, 1171–1178. (b) Lin, Y. Y., Chan, S. C., Chan, M. C. W., Hou, Y. J., Zhu, N., Che, C. M., Liu, Y., Wang, Y. *Chem.-Eur. J.* 2003, **9**, 1263–1272. (c) Lu, W., Mi, B. X., Chan, M. C. W., Hui, Z., Che, C. M., Zhu, N., Lee, S. T. *J. Am. Chem. Soc.* 2004, **126**, 4958–4971.

4. (a) Islam, A., Sugihara, H., Hara, K., Singh, L. P., Katoh, R., Yanagida, M., Takahashi, Y., Murata, S., Arakawa, H., Fujihashi, G. *Inorg. Chem.* 2001, **40**, 5371–5380. (b) McGarrah, J. E., Kim, Y. J., Hissler, M., Eisenberg, R. *Inorg. Chem.* 2001, **40**, 4510–4511. (c) McGarrah, J. E., Eisenberg, R. *Inorg. Chem.* 2003, **42**, 4355–4365.

5. Lu, W., Chan, M. C. W., Zhu, N., Che, C.-M., Li, C., Hui, Z. *J. Am. Chem. Soc.* 2004, **126**, 7639–7651 and references therein.

6. (a) Yam, V. W.-W., Wong, K. M.-C., Zhu, N. *J. Am. Chem. Soc.* 2002, **124**, 6506–6507. (b) Lai, S.-W., Lam, H.-W., Lu, W., Cheung, K.-K., Che, C.-M. *Organometallics* 2002, **21**, 226–234. (c) Yersin, H., Dinges, D., Humbs, W., Strasser, J., Sitters, R., Glasbeek, M. *Inorg. Chem.* 2002, **41**, 4915. (d) Miskowski, V. M., Houlding, V. H. *Inorg. Chem.* 1989, **28**, 1529–1533. (e) Connick, W. B., Marsh, R. E., Schaefer, W. P., Gray, H. B. *Inorg. Chem.* 1997, **36**, 913–922. (f) Bailey, J. A., Hill, M. G., Marsh, R. E., Miskowski, V. M., Schaefer, W. P., Gray, H. B. *Inorg. Chem.* 1995, **34**, 4591–4599. (g) Buchner, R., Cunningham, C. T., Field, J. S., Haines, R. J., McMillin, D. R., Summerton, G. C. *J. Chem. Soc., Dalton Trans.* 1999, 711–717. (h) Lai, S.-W., Chan, M. C.-W., Cheung, T.-C., Peng, S.-M., Che, C.-M. *Inorg. Chem.* 1999, **38**, 4046.

7. Similar structures have been previously reported for binuclear Pt(II) and Ir(I) complexes. (a) Lai, S.-W., Chan, M. C. W., Cheung, K.-K., Peng, S.-M., Che, C.-M. *Organometallics* 1999, **18**, 3991–3997. (b) Jain, V. K., Kannan, S., Tiekink, E. R. T. *J. Chem. Soc., Dalton Trans.* 1993, 3625. (c) Coleman, A. W., Eadie, D. T., Stodart, S. R., Zaworotko, M. J., Atwood, J. L. *J. Am. Chem. Soc.* 1982, **104**, 922.

8. Adding bulky groups to the pyrazole bridge also changes the angle between the two Pt square planes and thus, the dz^2 orbitals. The angle between the two Pt square planes in **1** is 85°, while for **4** the angle is 55°. The smaller angle in **4** increases dz^2 overlap, enhancing the Pt–Pt interaction in the same way the decreased separation does.

9. Brooks, J., Babayan, Y., Lamansky, S., Djurovich, P. I., Tsyba, I., Bau, R., Thompson, M. E. *Inorg. Chem.* 2002, **41**, 3055–3066.

10. (a) Kim, C. D., Pillet, S., Wu, G., Fullagar, W. K., Coppens, P. *Acta Crystallogr., Sect. A* 2002, **58**, 133–137. (b) Novozhilova, I. V., Volkov, A. V., Coppens, P. *J. Am. Chem. Soc.* 2003, **125**, 1079–1087. (c) Ozawa, Y., Terashima, M., Mitsumi, M., Toriumi, K., Yasuda, N., Uekusa, H., Ohashi, Y. *Chem. Lett.* 2003, **32**, 62. (d) Rice, S. F., Gray, H. B. *J. Am. Chem. Soc.* 1983, **105**, 4571–4575.

Chapter 34

Platinum Binuclear Complexes as Phosphorescent Dopants for Monochromatic and White Organic Light-Emitting Diodes

Biwu Ma,[a,b] Peter I. Djurovich,[a] Simona Garon,[a] Bert Alleyne,[a,c] and Mark E. Thompson[a]

[a] *Department of Chemical Engineering and Materials Science, and Department of Chemistry, University of Southern California Los Angeles, California 90089, USA*
[b] *Current address: Materials Science Division, Lawrence Berkeley National Laboratory, Berkeley, California 94720, USA*
[c] *Current address: Universal Display Corporation, Ewing, New Jersey 08618, USA*
met@usc.edu

Efficient blue-, green-, and red-light-emitting organic diodes are fabricated using binuclear platinum complexes as phosphorescent dopants. The series of complexes used here have pyrazolate bridging ligands and the general formula C^NPt(μ-pz)$_2$PtC^N (where C^N = 2-(4′,6′-difluorophenyl)pyridinato-$N,C^{2'}$, pz = pyrazole (**1**), 3-methyl-5-*tert*-butylpyrazole (**2**), and 3,5-bis(*tert*-butyl)pyrazole (**3**)). The Pt–Pt distance in the complexes, which decreases in the

---

Reprinted from *Adv. Funct. Mater.*, **16**, 2438–2446, 2006.

Electrophosphorescent Materials and Devices
Edited by Mark E. Thompson
Text Copyright © 2006 WILEY-VCH Verlag GmbH & Co. KGaA, Weinheim
Layout Copyright © 2024 Jenny Stanford Publishing Pte. Ltd.
ISBN 978-981-4877-34-3 (Hardcover), 978-1-003-08872-1 (eBook)
www.jennystanford.com

order **1** > **2** > **3**, solely determines the electroluminescence color of the organic light-emitting diodes (OLEDs). Blue OLEDs fabricated using 8% **1** doped into a 3,5-bis(*N*-carbazolyl)benzene (mCP) host have a quantum efficiency of 4.3% at 120 Cd m^{-2}, a brightness of 3900 Cd m^{-2} at 12 V, and Commission Internationale de L'Eclairage (CIE) coordinates of (0.11, 0.24). Green and red OLEDs fabricated with **2** and **3**, respectively, also give high quantum efficiencies (−6.7%), with CIE coordinates of (0.31, 0.63) and (0.59, 0.46), respectively. The current-density–voltage characteristics of devices made using dopants **2** and **3** indicate that hole trapping is enhanced by short Pt–Pt distances (< 3.1 Å). Blue electrophosphorescence is achieved by taking advantage of the binuclear molecular geometry in order to suppress dopant intermolecular interactions. No evidence of low-energy emission from aggregate states is observed in OLEDs made with 50% **1** doped into mCP. OLEDs made using 100% **1** as an emissive layer display red luminescence, which is believed to originate from distorted complexes with compressed Pt–Pt separations located in defect sites within the neat film. White OLEDs are fabricated using **1** and **3** in three different device architectures, either with one or two dopants in dual emissive layers or both dopants in a single emissive layer. All the white OLEDs have high quantum efficiency (−5%) and brightness (−600 Cd m^{-2} at 10 V).

34.1 Introduction

Organic light-emitting diodes (OLEDs) have attracted a great deal of attention due to their potential applications in both flat-panel displays and lighting sources [1–3]. A recent advance has demonstrated the ability to make highly efficient electroluminescent devices using phosphorescent emitters based on heavy transition metals (e.g., Ir, Pt, Os) [4–16]. The strong spin-orbit coupling of the heavy-metal atom allows for efficient intersystem crossing (ISC) between singlet and triplet states, which can lead to a high quantum yield of emission from the triplet state. Thus, OLEDs can be fabricated that utilize all of the electrogenerated singlet and triplet excitons, thereby approaching an internal efficiency of 100% [13,

17, 18]. Spectral tuning of the luminescent complexes has been successfully achieved by changing the organic ligands coordinated to the heavy-metal center, leading to electroluminescence (EL) colors ranging from red to blue [8, 10, 19].

Platinum(II) complexes have been used as phosphorescent emitters in small-molecule OLEDs [4, 12, 20–25]. Since the first phosphorescent OLED was reported with 2,3,7,8,12,13,17,18-octaethyl-$21H,23H$-porphine platinum(II) (PtOEP) as a red emissive dopant [4], platinum(II) complexes have been used to prepare OLEDs that give green, red, and even white EL with external quantum efficiencies as high as 16.5% [15]. However, to date no blue OLEDs with high efficiency using platinum-complex emitters have been reported, which may be due to the fact that luminescence from platinum(II) complexes show significant spectral shifts upon dopant aggregation, leading to broad, red-shifted emission. This color-shift caused by intermolecular interaction has been exploited to prepare a single-dopant white OLED with platinum (2-(4′,6′-difluorophenyl)pyridinato-$N,C^{2'}$) (2,4-pentanedionato-O,O) (FPt1), in which emission comes simultaneously from the blue-light-emitting monomer and the red-light-emitting aggregated states [24, 26, 27].

Herein, we report efficient monochromatic and white OLEDs (WOLEDs) by applying a series of color-tunable pyrazolate bridged platinum(II) binuclear complexes as the phosphorescent dopants. Three different phosphorescent dopants were used to prepare the OLEDs, compounds **1–3**, which emit blue, green, and red light, respectively. The general structures for these binuclear complexes are shown in Scheme 34.1; the photophysical properties for these materials are listed in Table 34.1. Also included in Scheme 34.1 and Table 34.1 are structures and data for the reference compounds FPt1 and [(2-(4′,6′-difluorophenyl)pyridinato-$N,C^{2'}$)Pt(2-thiopyridyl)]$_2$ (FPt2). FPt2 is analogous to a binuclear complex with cyclometallated 2-phenylpyridyl (ppy) ligands recently prepared and used as a red dopant in OLEDs [28, 29]. Synthesis and characterization data for FPt2 are given in the Experimental, and crystallographic data are given in the Supporting Information. The synthesis and photophysics of compounds **1–3** have been reported elsewhere [30]. The emission colors of **1–3** are governed by the

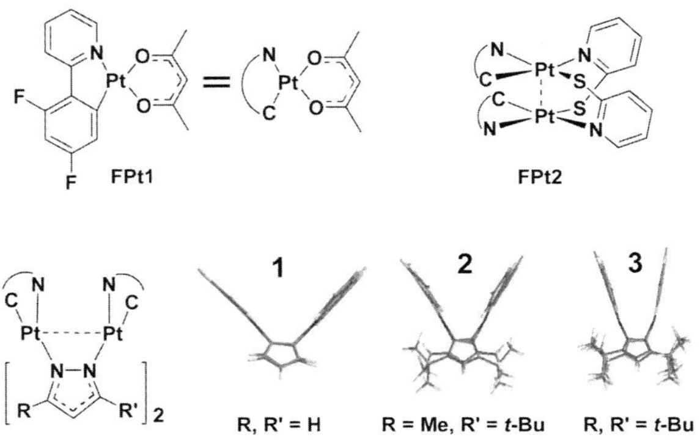

Scheme 34.1 The structures of the binuclear complexes used as phosphorescent dopants to prepare the OLEDs in this work. Compounds **1–3** emit blue, green, and red light, respectively. The structures of the reference compounds, FPt1 and FPt2 are also shown.

steric bulk at the 3,5-positions of the pyrazolate bridging ligands, which in turn alters the Pt–Pt distance within the complex. As the Pt–Pt separation decreases, the emissive triplet state changes from one with a mixed ligand/ MLCT character similar to that in FPt1 [31], to a lower energy, metal-metal-to-ligand charge-transfer (MMLCT) state involving the shortened Pt–Pt bond. The use of these binuclear species as phosphors has enabled us to examine the role that Pt–Pt interactions can play in controlling the color and electrical characteristics of OLED devices. This information can be extended to include the behavior of other square-planar metal complexes that are capable of forming close metal–metal contacts when used as electrically active components in organic devices.

34.2 Results and Discussion

34.2.1 Material Properties

An energy-level diagram is shown in Fig. 34.1 for the materials used to make the OLED devices. The highest occupied molecular orbital (HOMO) energies for 4,4′-bis[N-(1-naphthyl)

Table 34.1 Photophysical and electrochemical properties of the platinum complexes used in this work

Compound	R, R′	Pt–Pt [Å]	Emission max. λ_{max} [nm] [a]	Reduction potential [V] [b]	Oxidation potential [V] [b]	HOMO energy [eV]
FPt1	—	—	462	−2.29	0.51	5.9 [c]
1	H, H	3.376	466	−2.28, −2.35	0.52	5.9 [d]
2	Me, t-Bu	3.064	546	−2.30, −2.51	0.44	5.6 [d]
3	t-Bu, t-Bu	2.834	630	−2.31, −2.60	0.37	5.4 [d]
FPt2	—	2.87	610	−2.23, −2.53	0.30	5.5 [c]

[a] Measurement carried out in a doped polystyrene thin film at room temperature. [b] Values are versus a ferrocenium/ferrocene (Fc$^+$/Fc) couple and were determined using differential pulse voltammetry (see Supporting Information). [c] Measured made using UV photoelectron spectroscopy (UPS). [d] Estimated value (see text).

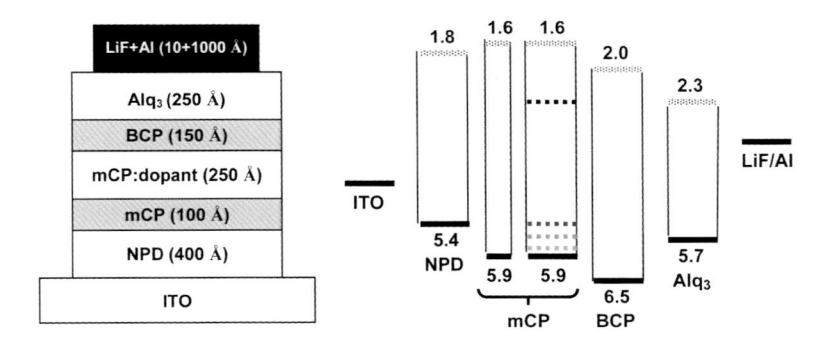

Figure 34.1 Device architecture for OLEDs (left) along with proposed energy-level diagram (right). HOMO energies are shown with filled lines, LUMO energies with shaded lines, and dopant energies with dashed lines.

N-phenylamino]biphenyl (NPD), 3,5-bis(N-carbazolyl)benzene (mCP), 2,9-dimethyl-4,7-diphenyl-1,10-phenanthroline (BCP), and tris(8-hydroxyquinolinato)aluminum(III) (Alq$_3$) are literature values obtained from UV photoelectron spectroscopy (UPS) data [32]. The lowest unoccupied molecular orbital (LUMO) energies were estimated from the electrochemical reduction potentials of the respective compounds using a recently described method, and the LUMO energy of Alq$_3$ was measured using inverse photoelectron spectroscopy [33]. The optical and electrochemical data for complexes **1–3** given in Table 34.1 allow us to estimate the HOMO and LUMO energies for the platinum dopants. Compounds **1–3** have nearly identical reduction potentials, comparable to that of Alq$_3$, and thus are expected to have similar LUMO energies. In contrast, the oxidation potentials for the three complexes are substantially different ($E_{1/2}^{ox}$ **1** > **2** > **3**), indicating that the HOMO energies differ in the same order. Unfortunately, electrochemical oxidation of Pt complexes is typically an irreversible process caused by solvent coordination to the open coordination site on the metal center and subsequent ligand rearrangement. Therefore, the $E_{1/2}^{ox}$ values do not necessarily represent the true thermodynamic potential for removing an electron from the complex, particularly for complex **1**. However, we can estimate the HOMO energy for **1–3** using comparative data obtained for the related FPt1 and FPt2 complexes. While a ppy-based analog of FPt2 has been reported, the F$_2$ppy-based complex reported here is new. The MLCT and MMLCT state energies for the Pt complexes are effectively governed by changes in the HOMO energy since the moiety that determines the LUMO energy, the 4′,6′-F$_2$ppy ligand, is the same in all species. The HOMO energy for FPt1 measured using UPS (5.9 eV) corresponds to a thermodynamic $E_{1/2}^{ox}$ value of −1.0 V (vs. Fc$^+$/Fc). This HOMO energy also agrees with the oxidation potential for FPt1 determined from excited-state quenching experiments [34]. Since **1** and FPt1 have similar emission energies and $E_{1/2}^{ox}$ values, both can be expected to have comparable HOMO energies. The increased metal–metal interaction in **2** and **3** caused by the decreased Pt–Pt separation raises the HOMO and consequently decreases the emission energy. The emission energies of **2** and **3** differ from **1** by values of 0.28 and 0.53 eV, respectively; therefore, the HOMO energies for the former complexes can be expected to vary by a comparable amount. This analysis is supported by the HOMO energy measured for FPt2, a related complex that has both a Pt–Pt separation and oxidation potential similar to **3**.

34.2.2 Monochromatic OLEDs

The OLEDs were fabricated by high-vacuum thermal evaporation of the OLED materials onto ITO-coated (ITO: indium tin oxide) glass that is used as the anode. All of the devices have the configuration shown in the left of Fig. 34.1. NPD (40 nm) acts as a hole-transport layer (HTL), mCP (10 nm) acts as an electron-blocking layer and host, BCP (15 nm) and Alq_3 (25 nm) work as hole-blocking and electron-transport layers (HBLs and ETLs), respectively. Platinum complexes were doped into mCP at a level of 8 wt%. The deep HOMO of the BCP layer prevents holes from leaking into the Alq_3 layer and the high LUMO of the mCP layer hinders electron leakage into the NPD layer. The mCP layer also prevents recombination from occurring at the NPD interface, which improves the efficiency and color stability of the devices.

All of the devices exhibited bright emission when a positive bias was applied between the electrodes. Figure 34.2 shows the voltage-independent EL spectra of the three devices, the device performance data is summarized in Table 34.2. No emission from the mCP host is observed, indicating complete energy transfer from the host exciton to the platinum complexes. Also, there is no NPD emission observed

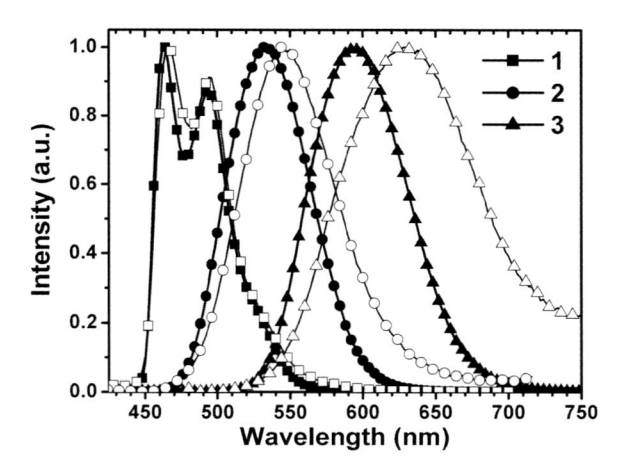

Figure 34.2 EL spectra (filled symbols) for OLEDs using dopants **1**, **2**, and **3**. Photoluminescence (PL) spectra (empty symbols) recorded in polystyrene are also shown for comparison.

Table 34.2 Performance characteristics of the monochromatic OLEDs

Dopant	Current density [mA cm^{-2}]	Voltage [V]	External quantum efficiency [%]	Brightness [cd m^{-2}]	CIE coordinates (x, y)
1 (blue)	1.0	7.3	3.8	59	0.11, 0.24
	10.0	9.2	3.8	592	
	31.3	10.3	3.2	1544	
2 (green)	1.0	8.5	6.6	272	0.31, 0.63
	10.5	10.7	6.0	2517	
	30.8	11.9	5.1	6415	
3 (red)	1.0	9.0	6.6	144	0.59, 0.46
	10.8	11.5	6.3	1435	
	30.1	12.7	5.6	3553	

at any bias level, indicating that electrons do not leak into the HTL layer. In these devices, the OLED structure and composition were not changed. The only difference in the devices is the Pt–Pt intramolecular distance of the bimetallic dopant. Thus, the emission color of the OLEDs is determined solely by the molecular geometry of the luminescent platinum complexes.

The EL spectra are nearly coincident with the photoluminescence (PL) spectra of the corresponding complexes in a polystyrene matrix at room temperature. The blue OLED utilizing compound **1** has Commission Internationale de L'Eclairage (CIE) coordinates of (0.11, 0.24), which is slightly more saturated than the previously studied phosphorescent deep-blue emitter bis(2-(4′,6′-difluorophenyl)pyridinato-$N,C^{2'}$)iridium(III) tetrakis(1-pyrazolyl)borate, (FIr6, CIE = (0.16, 0.26)) [35]. Emission from aggregate states is not observed from devices using **1**; the EL spectrum is similar to that found for devices that use FPt1 at low concentration (1–2 wt%) [24]. In contrast, a device with 8 wt% FPt1 doping in CBP gives roughly equal amounts of blue and red (excimer/dimer) emission (see the Supporting Information for a direct comparison). The absence of red EL from **1** can be explained by the fact that the boatlike geometry of the binuclear complex inhibits the intermolecular interactions responsible for the lower-energy emission. This aspect of **1** will be explored in greater detail in the next section. Devices using dopants **2** and **3** display lower-energy

EL as a direct result of the decreased Pt–Pt separation in these two complexes. In addition, the dependence of emission energy on Pt–Pt distance provides a sensitive means to indirectly probe the matrix environment surrounding the dopant molecules. The green and orange-red EL spectra from devices using dopants **2** and **3**, respectively, are at higher energy and have narrower lineshapes than the corresponding PL in polystyrene. These differences indicate that complexes **2** and **3** are located in rigid sites in the mCP matrix. The rigid surroundings prevent the dopants from relaxing their molecular geometry and shortening the Pt–Pt separation upon forming the MMLCT excited state [30].

The electrical characteristics of the devices are displayed in the current density–voltage (J–V) and luminance-voltage characteristics of the OLEDs (Fig. 34.3). The turn-on voltages were all near 4 V, and the luminance reached 4000, 6800 and 2200 Cd m^{-2} for the respective blue-, green-, and red-light-emitting devices at 12 V. From the J–V characteristics, it can be seen that the devices using dopants **2** and **3** are more resistive than those using **1**. This behavior

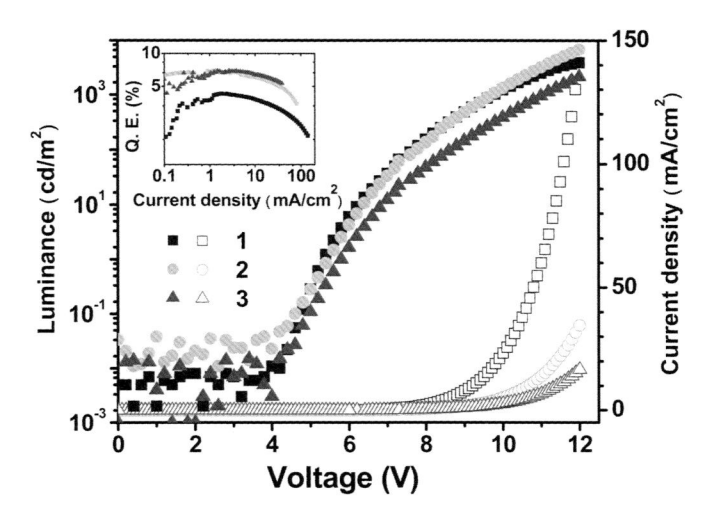

Figure 34.3 Applied voltage–luminance (solid symbols) and applied voltage–current density (open symbols) characteristics for the monochromatic OLEDs. Inset shows the external quantum efficiency (Q.E.) as a function of current density.

can be rationalized by considering the HOMO-LUMO levels of the three dopants. If the dopant has a HOMO energy close to that of the host material, holes are less likely to be trapped by the dopant. When the HOMO energy of dopant is less than that of the host, hole trapping by the dopant is expected to increase. One can thus expect the greatest amount of hole trapping to occur with **3**, followed by **2**, whereas **1** should have the least amount of trapping. This effect is clearly reflected in the J–V characteristics, which show that the current levels at any given voltage fall in the order $\mathbf{1} > \mathbf{2} > \mathbf{3}$. The inset of Fig. 34.3 shows the external quantum efficiency (Q.E.) as a function of current density for all three devices. All three devices show a gradual decrease of the external quantum efficiency with an increase of current density, an effect that is likely due to triplet-triplet and triplet-polaron annihilation. The maximum external quantum efficiency was 4.3% for blue, 7.0% for green, and 7.0% for red. The higher quantum efficiencies of devices using dopants **2** and **3**, relative to **1**, are likely a consequence of the greater hole trapping ability of the former dopants. Exciton formation caused by direct charge recombination on the dopant is a characteristic EL mechanism in OLEDs that incorporate phosphorescent dopants.

34.2.3 Concentration Dependence of the EL of Compound 1

The photophysical properties of luminescent compounds typically undergo large changes upon going from an isolated molecule to the neat solid state. Intermolecular interactions often lead to emission from a neat solid that is red-shifted in comparison to what is observed from the isolated molecule. This phenomenon is particularly pronounced for platinum(II) complexes coordinated to ligands with conjugated π-systems as the square-planar geometry of the molecule easily accommodates π-stacking interactions in the solid state [20, 23, 36]. To probe the role that intermolecular interactions caused by dopant aggregation have in controlling the EL spectrum, an investigation was carried out on the concentration-dependent EL of the blue-emissive compound **1**. The same OLED architecture shown in Fig. 34.1 was used for these devices. The EL

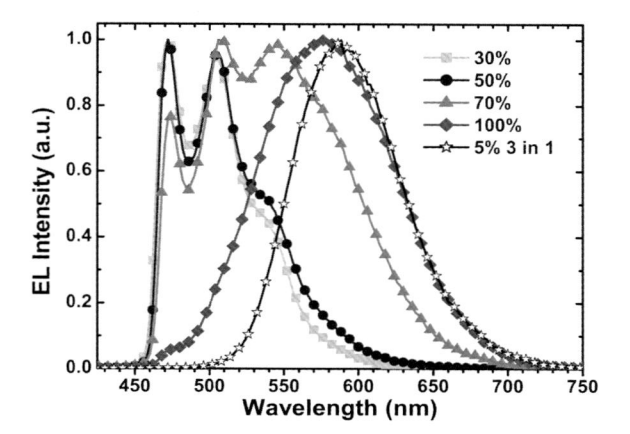

Figure 34.4 Concentration dependence of EL for devices using compound **1** either as a phosphorescent dopant in mCP or as a host for **3**.

spectra for devices with dopant concentration of 30, 50, 70, and 100 wt% in mCP are shown in Fig. 34.4. It is noteworthy that OLEDs with 30 and 50 wt% **1** display the same blue EL spectra associated with the parent dopant in the 8 wt% device. At these high loading levels there should be numerous close contacts between individual dopant molecules; however, no red emission can be discerned in the EL spectrum. Upon increasing the concentration to higher levels, new emission bands appear at lower energy. For the device with 70 wt% **1**, the EL shows three peaks, the first two being vibronic transitions from the parent dopant, while a new emission band appears near 550 nm. At the highest concentration, 100%, the EL spectrum from **1** is predominantly orange, consisting of a broad band centered at 578 nm and a minor emission component at high energy (< 500 nm). The broad, low-energy EL from the neat layer of **1** is similar in appearance to the EL from a monochromatic device made using compound **3** as a dopant in a host of mCP. The low-energy band is likewise akin to the excimer band that is observed in the PL from concentrated solutions of mononuclear complex FPt1, as well as the EL from highly doped and neat films of FPt1. The similarity in the luminescence spectra under all these conditions suggest emission originating from species that have a common excited state with close Pt–Pt contacts (< 3.2 Å). It should be pointed out that no equivalent low-energy emission can be observed in the PL from crystalline samples of **1**, despite the fact that analysis using X-ray crystallography reveals packing motifs that display extensive intermolecular π–π overlaps (see Supporting Information). It is therefore unlikely that excited states stabilized by π–π interactions are responsible for the low-energy emission in the sublimed films. Due to the steric constraints of this complex, it is also unlikely to form intermolecular dimers in amorphous thin films or as dopants in thin films. In compounds **1–3**, the pyrazole groups extend well above the platinum square plane (see Supporting Information), which prevents the close approach of platinum atoms in adjacent molecules. This close approach is needed to form emissive intermolecular dimers. An alternative possibility is that some individual binuclear complexes with an intramolecular Pt–Pt distance of 3.2 Å or less are present in the 70 wt% and neat film of **1**. We propose that these distorted complexes are located in disordered sites that allow the intramolecular Pt–Pt separation to compress to values that are comparable to those that occur in complexes **2** (3.06 Å) and **3** (2.83 Å). The presence of related dimeric species with close intermolecular Pt–Pt contacts has been proposed to be responsible for red emission in neat films of FPt1 [27]. The species with short Pt–Pt separations will act as both luminescent and hole-trapping sites in the amorphous films. In devices prepared with **1** at levels of 70% and higher, the concentration of dimer species with a short Pt–Pt distance is high enough to trap excitons and carriers, markedly affecting the EL spectrum. On the basis of the transition energies displayed in the EL spectra, the HOMO energy of the Pt–Pt trap species can be as high as that of compound **3**.

All the devices have turn-on voltages near 4 V. They also have high brightness and external quantum efficiencies greater than 3.0% (Fig. 34.5). For example, the device using a neat film of **1** had a luminance of 8000 Cd m^{-2} at 11.5 V and an external quantum efficiency of 4.8% at 6.1 V. The data presented in Fig. 34.5 show that the device with 50 wt% dopant supports a higher current density than a 30 wt% device at the same driving potential. However, it is also interesting to observe that the conductivity for the device with

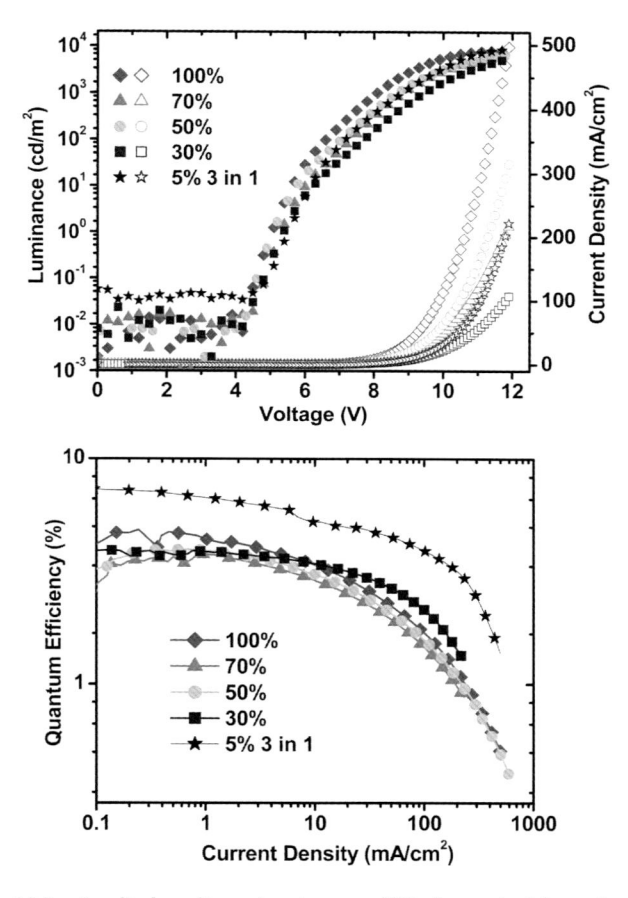

Figure 34.5 Applied voltage–luminance (filled symbols) and applied voltage–current density (open symbols) characteristics (top) and external quantum efficiency as a function of current density for OLEDs (bottom) for devices using compound 1 as either dopant or host.

70 wt% dopant is lower than the one with 50 wt% dopant. On the basis of their nearly identical blue EL spectra, emission from the 30 and 50 wt% devices comes from isolated molecules of compound **1**. On the other hand, for the device with 70 wt% dopant, the presence of species with short Pt–Pt separations, like those in compound **2**, leads to an increased amount of hole trapping. The results are consistent with the observation that hole trapping also decreases

the current density in the film of mCP doped with **3** (see Fig. 34.3). The J–V characteristics show that the device made with a neat film of **1** has the highest conductivity: at 10 V, the current density reached 113 mA cm^{-2}, versus 23.3 mA cm^{-2} for the device using 8 wt% dopant **1**. As both **1** and mCP have comparable HOMO energies, the J–V data indicates that compound **1** has a greater propensity to transport electrons than does mCP. This behavior can be explained by the fact that the LUMO energy of compound **1** is much lower than that of mCP, and this reduction in the barrier height at the interface facilitates electron injection from BCP layer.

Besides acting as a dopant, compound **1** can also be used as the host for the phosphorescent host–guest type devices. A related type of emissive layer was used in devices where the phosphorescent host material *fac*-Ir(ppy)$_3$ was doped with the red-emissive PtOEP complex [37]. The EL spectrum of a device that uses an emissive layer composed of complex **1** doped with 5 wt% **3** is shown in Fig. 34.4. Devices using either **1** or mCP as the host material and dopant **3** display nearly identical EL spectra, indicating that efficient energy transfer occurs from the host **1** to the red phosphorescent dopant **3**. Devices using the two different host materials also have similar external quantum efficiencies (near 5.4% at 8 V). A decrease in quantum efficiency with increasing current density is more pronounced for the device using **1** as a host than for the mCP host, possibly due to a greater contribution from triplet–triplet annihilation in the former matrix. The brightness and J–V characteristics for the device using **1** as a host were improved relative to the mCP host. At an applied bias of 10 V, the device using host **1** had a luminance of 3300 Cd m^{-2} at 31 mA cm^{-2}, compared to 390 Cd m^{-2} at 2.6 mA cm^{-2} for the device with the mCP host. This is consistent with our finding that compound 1 has better electron-transport properties than mCP. Figure 34.4 also shows that the Pt–Pt dimer compound **3** is effective at trapping holes as the current density is significantly lower than that found for the device using a neat film of **1**. Moreover, the higher conductivity of the device using a neat film of **1**, plus the absence of any EL at short wavelengths (< 500 nm) in the doped device, indicates that a neat film of **1** has a rather low (< 5 wt%) concentration of trap states comprised of disordered species with short Pt–Pt distances.

34.2.4 White OLEDs

White-light emission requires a mixture of either two complementary colors or three primary colors. Various methods for generating white light from OLEDs have been reported [3, 24, 26, 38–41]. One approach to create well-balanced white EL is to use separate emissive layers that have blue- and red-luminescent dopants. Alternatively, the red and blue dopants can be combined into a single emissive layer. The EL lineshapes from monochromatic devices using dopants **1** and **3**, or a neat film of compound **1**, are therefore appropriate for use in generating white emission.

Herein, a preliminary evaluation is given for WOLEDs that use these platinum binuclear complexes in the device architectures shown in Scheme 34.2. The basic device configuration is similar

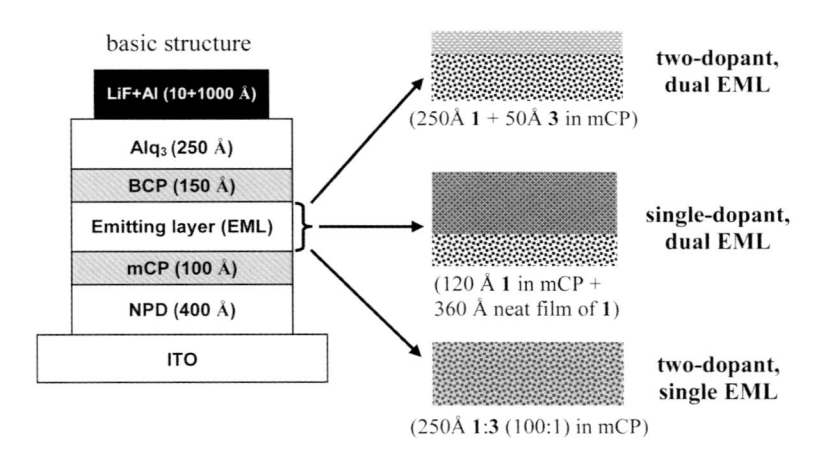

Scheme 34.2 Device architectures of the WOLEDs using the platinum binuclear complexes **1** and **3**.

to the one shown in the Fig. 34.1 with NPD, mCP, BCP, and Alq_3 as HTL, electron-blocking layer (and host, if necessary), HBL, and ETL, respectively. The devices differ only in the composition of the emitting layer (EML); using an either dual EML with two dopants, a dual EML with a single dopant, or a single EML with two dopants.

34.2.4.1 Devices with dual emissive layers

In devices using a dual EML, blue and red emission are generated by two discrete emitting regions. Blue EL is produced in a layer of mCP doped with 8 wt% compound **1**. Two methods were used to generate red EL; either from a layer of mCP doped with 8 wt% compound **3** or by using a layer of a neat compound **1**. The EL spectra for devices with a dual EML are strongly dependent on the thickness of each emissive layer; thus, to create white emission, a proper balance of blue and red colors was achieved by optimizing the thickness of each EML.

A device using a dual EML consisting of 8 wt% **1** (25 nm) and 8 wt% **3** (5 nm) produced white emission. The turn-on voltage for this device (ca. 4 V) is almost the same as the monochromatic devices. The EL spectrum displays three emission peaks at 465, 500, and 580 nm; the first two peaks are from vibronic transitions of compound **1** while the third peak corresponds to EL from compound **3**. The white EL from this device is voltage dependent (Fig. 34.6a); increasing the applied voltage increases the proportion of blue emission (CIE = (0.42, 0.46) at 9 V and CIE = (0.39, 0.45) at 13 V). Since compound **1** has deeper HOMO than compound **3**, holes are trapped in the red-emissive layer at low voltage and build up at the BCP interface. At higher voltages, more electrons are injected into the layer doped with compound **1** and this leads to a recombination zone that is more evenly distributed in both emitting layers. The device has similar luminance-efficiency–voltage characteristics as the monochromatic mCP-based red-light-emitting device (see Supporting Information), that is, a maximum luminance of 4300 Cd m^{-2} occurs at 13 V and the maximum quantum efficiency is 7.7% at 8.5 V.

As discussed in the previous section, devices highly doped with compound **1** display concentration-dependent EL, that is, low concentrations produce blue emission, whereas a neat film produces orange-red emission. Therefore, it is possible to get white emission using the combined emission from a dual EML of 8 wt% **1** doped into mCP and a neat film of **1**. The sequence of emissive layers is the same to that of the previous device with a dual EML, except that a neat layer of compound **1** is used in place of the doped layer of the

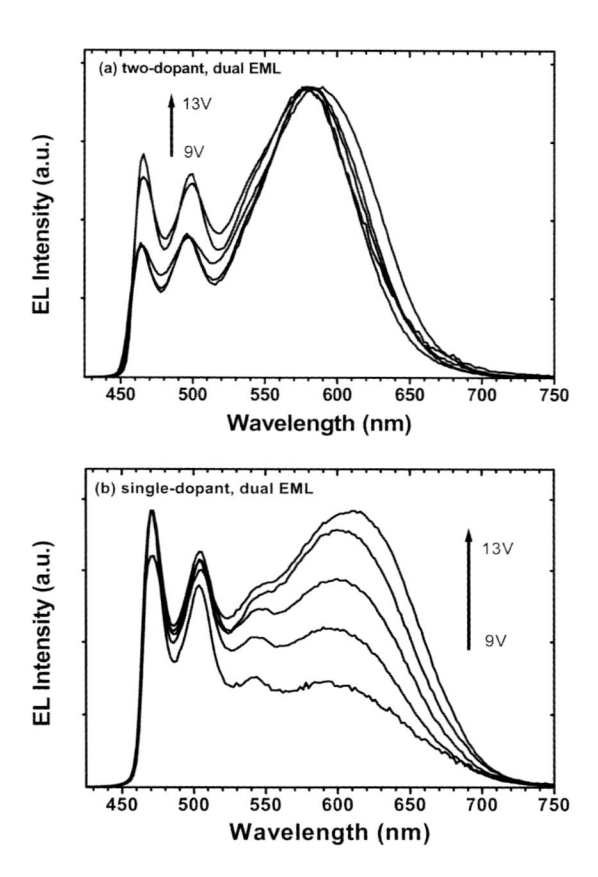

Figure 34.6 The EL spectra for WOLEDs with dual emissive layers: (a) two-dopant, dual EML; (b) single-dopant, dual EML.

red compound **3**. To get balanced white emission, the thickness of the doped layer was kept constant while the thickness of the neat film was varied. It was found that a device with a 12 nm layer of mCP doped with 8% **1** and a 36 nm layer of neat **1** gives white EL, as shown in Fig. 34.6b. The device luminance reached 650 Cd m^{-2} at an applied bias of 10 V and the maximum quantum efficiency was 4.2% at 10.7 V (see Supporting Information).

The EL spectrum of the single-dopant, dual EML device is also voltage-dependent. However, the spectral change is opposite to that of the former device; higher voltages lead to increasing amounts of red emission (CIE = (0.33, 0.42) at 9 V; CIE = (0.42, 0.44) at 13 V). As shown earlier (Fig. 34.5), the conductivity of a neat film of **1** is higher than that of a layer of mCP doped with a low concentration of **1**, a characteristic attributed to the good electron-transport properties of **1**. The charge-transport properties of the various EML layers follow the order: neat film of **1** > 8% **1**-doped mCP > 8% **3**-doped mCP. Therefore, the improved electron-transport ability of the neat film of **1** (relative to a mCP film doped with **3**) leads to efficient electron injection into the mCP film doped with **1**. With increasing voltage, the proportion of red emission increases as more holes are injected into the neat film of **1**.

34.2.4.2 Devices with a single emissive layer

For this single EML device, blue and red emission was generated by co-doping compounds **1** and **3** into a mCP host. Typically, thermal evaporation of two different molecules with disparate volatility requires the use of separate deposition sources in order to precisely control the doping ratio. Fortunately, complexes **1** and **3** have the same sublimation temperature, most likely a result of the structural similarity between the two species. This feature enabled us to control the doping ratio during sublimation by simply mixing the two complexes into a single source boat prior to deposition. A related co-deposition approach from a single sublimation source has been used to fabricate WOLEDs with multiple fluorescent dopants [42]. The co-deposition of two dopants from a single source significantly simplified the device fabrication procedure. For the two-dopant single EML device, the EL is strongly influenced by the ratio of the two dopants. We determined through systematic variation that a device using 8 wt% dopant having a **1:3** (blue/red) ratio of 100:1 gives the white EL shown in Fig. 34.7.

The turn-on voltage for this device (4 V) is the same as the previous WOLEDs. A maximum luminance of 6820 Cd m^{-2} is achieved at 13 V, and the maximum external quantum efficiency is near 5.4% at 8.8 V. As with the previous devices, the EL is voltage-dependent, albeit less so (CIE = (0.36, 0.48) at 9 V; CIE = (0.32, 0.48) at 13 V). The improved color stability with applied bias of this device can be attributed to a uniform concentration of the two dopants in

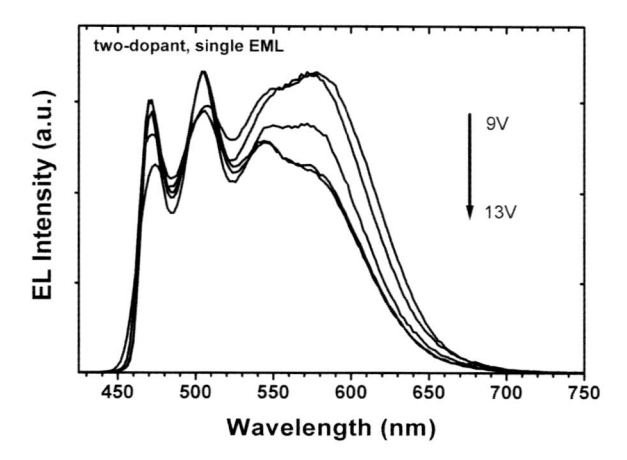

Figure 34.7 EL spectra for a WOLED with a two dopant, single emissive layer.

the EML. The EL spectrum is similar to that seen in WOLEDs that use FPt1 as a single dopant [24]. The related performance with the two-dopant, single EML device suggests that when FPt1 is used to make a single-dopant WOLED, the orange–red portion of the EL originates from platinum "dimer" trap states that have short intermolecular Pt–Pt distances comparable to **3**. The data also imply that the concentration of these dimer traps need be only as low as 1% of the original doping concentration in order to add a significant red component in the EL spectrum. The proportion of red dopant needed to produce balanced emission in the device can be this low due to the combined effects of efficient energy transfer from the blue phosphor and exciton formation by charge trapping on the red dopant.

34.3 Conclusions

The present study demonstrates the ability to control OLED properties through simple molecular structure manipulation. The Pt–Pt spacing in a series of pyrazolate-bridged platinum(II) binuclear complexes has a profound effect on the performance of devices that use these compounds as luminescent dopants. Efficient blue, green, and red OLEDs were fabricated with the binuclear complexes. The emission color is determined only by modifying the pyrazolate bridging ligand used to control the intramolecular Pt–Pt separation. The nonplanar geometry of the binuclear complexes also suppresses unwanted intermolecular dopant interactions that lead to concentration dependent spectral changes. The absence of such dopant aggregation effects has enabled us to prepare blue OLEDs with high efficiency and brightness using one of these Pt-based compounds. An efficient, Pt-based host–guest system has also been achieved by employing the phosphorescent blue emitter as host for red OLEDs. Taking these two device concepts together, WOLEDs have been fabricated by using different device structures and the platinum binuclear complexes. All the WOLEDs have high brightness and quantum efficiencies; however, their color quality shows a marked voltage dependence.

In addition to controlling the emission color, the adjustable Pt–Pt separation makes it possible to systematically determine the influence that metal–metal interactions have on the OLED performance. A smaller Pt–Pt distance raises the HOMO energy for the dopant, which leads to an increased resistivity in the doped film caused by charge trapping on the dopant. The charge trapping in turn enables exciton recombination to occur by direct recombination on the dopant and thereby improves the device quantum efficiency. The propensity to trap charge allows emissive species with close metal–metal contacts to contribute a significant degree to the spectral profile of OLEDs, even when such species are present at low concentrations. The need to control the formation of such metal–metal interactions in doped films is therefore essential when square-planar complexes are employed as dopants in OLEDs, which emit either saturated monochromatic EL or broadband emission useful for white-light illumination sources.

34.4 Experimental

Materials: The syntheses, characterization, and photophysical properties of the Pt binuclear complexes **1–3** have been previously reported [30]. During our preparation of these series of compounds, only single isomers were observed in each synthesis, with the cis form being the most common. The trans form was only observed for the asymmetric pz ligand (compound **2**). While it would be very interesting to investigate different isomers of the complexes, we do not currently have a method for preparing them. Alq$_3$ and α-NPD were prepared according to literature procedures [43, 44]. BCP was purchased from Aldrich. The synthesis of mCP was based on a known literature procedure using palladium-catalyzed cross coupling of aryl halides and arylamines [45]. All of these chemicals were purified by vacuum sublimation prior to device fabrication.

Synthesis of FPt2: 2-Mercaptopyridine (0.53 g, 4.80 mmol) was added to a methanolic solution of (2-(4′,6′-difluorophenyl)pyridinato-$N,C^{2'}$)(2-(4′,6′-difluorophenyl)pyridyl-N)platinum chloride (2.00 g, 3.27 mmol). K$_2$CO$_3$ (0.50 g) was then added and the solution was heated to 60 °C for 18 h [28]. The solution was cooled and the solvent was removed under reduced pressure. The crude red product was dissolved in acetone and passed through a column of silica gel using acetone as an eluent. The solvent was removed under reduced pressure and the product was crystallized from methanol to give dark-red crystals (1.33 g, 82%). In this synthesis, only a single isomer is observed in high yield, even though multiple isomers are possible. ^{1}H NMR (DMSO-d_6): 7.67 (d, $J = 5.8$ Hz, 2H), 7.06 (dd, $J = 7.8, 7.8$ Hz, 2H), 6.78 (m, 4H), 6.58 (d, $J = 7.8$ Hz, 2H), 6.49 (ddd, $J = 8.6, 7.1, 1.7$ Hz, 2H), 6.39 (dd, $J = 6.0, 6.0$ Hz, 2H), 6.13 (ddd, $J = 6.2, 5.8, 1.4$ Hz, 2H), 5.90 (dd, $J = 9.8, 2.3$ Hz, 2H), 5.71 (ddd, $J = 11.6, 9.2, 2.3$ Hz, 2H). Absorption and emission spectra and crystallographic data for the complex are given in the Supporting Information. CCDC 614478 contains the supplementary crystallographic data for this paper. These data can be obtained free of charge from The Cambridge Crystallographic Data Centre via www.ccdc.cam.ac.uk/data_request/cif.

OLED Fabrication and Testing: Prior to device fabrication, ITO on glass was patterned as 2 mm wide stripes with a resistivity of 20 Ω per □. The substrates were cleaned by sonication in soap solution, rinsed with deionized water, boiled in trichloroethylene, acetone, and ethanol for 5 min in each solvent and dried with nitrogen. Finally, the substrates were treated with UV ozone for 10 min. Organic layers were deposited sequentially by thermal evaporation from resistively heated tantalum boats onto the substrate at a rate of 2.5 Å s^{-1}. The base pressure at room temperature was 3–5 × 10^{-6} Torr (1 Torr = 133.322 Pa). The rate for single-component layers was controlled using one crystal monitor that was located near the substrate. A second crystal monitor located near the evaporation source of the dopant was used to control the rate of dopant molecule incorporation into the host matrix. The additional monitor was screened from the host evaporation, allowing for increased precision of the dopant concentration. After organic-film deposition, the chamber was vented and a shadow mask with a 2 mm wide stripe was put onto the substrate perpendicular to the ITO stripes. A cathode consisting of 10 Å LiF followed by 1000 Å of aluminum was deposited at a rate of 0.3–0.4 Å s^{-1} for LiF and 3–4 Å s^{-1} for aluminum. OLEDs were formed at the 2 mm × 2 mm squares where the ITO (anode) and Al (cathode) stripes intersected.

The devices were tested in air within 2 h of fabrication. Device J–V and light-intensity characteristics were measured using the LabVIEW program by National Instruments with a Keithley 2400 sourcemeter/2000 multimeter coupled to a Newport 1835-C optical meter, equipped with a UV-818 Si photocathode. Only light emitting from the front face of the WOLED was collected and used in subsequent efficiency calculations. Emission spectra (PL and EL) were recorded on a PTI Model C-60SE spectrofluorometer, equipped with a 928 PMT detector and corrected for detector response.

Acknowledgement

We thank the Universal Display Corporation and the Department of Energy for financial support of this work.

Supporting Information is available online from Wiley Inter-Science or from the author.

References

1. C. W. Tang, S. A. VanSlyke, *Appl. Phys. Lett.* 1987, **51**, 913.

2. J. H. Burroughes, D. D. C. Bradley, A. R. Brown, R. N. Marks, K. Mackay, R. H. Friend, P. L. Burns, A. B. Holmes, *Nature* 1990, **347**, 539.
3. Y. Sun, N. C. Giebink, H. Kanno, B. Ma, M. E. Thompson, S. R. Forrest, *Nature* 2006, **440**, 908.
4. M. A. Baldo, D. F. O'Brien, Y. You, A. Shoustikov, S. Sibley, M. E. Thompson, S. R. Forrest, *Nature* 1998, **395**, 151.
5. Y. Ma, H. Zhang, J. Shen, C. Che, *Synth. Met.* 1998, **94**, 245.
6. M. A. Baldo, M. E. Thompson, S. R. Forrest, *Nature* 2000, **403**, 750.
7. C. Adachi, M. A. Baldo, S. R. Forrest, S. Lamansky, M. E. Thompson, R. C. Kwong, *Appl. Phys. Lett.* 2001, **78**, 1622.
8. S. Lamansky, P. Djurovich, D. Murphy, F. Abdel-Razzaq, H. E. Lee, C. Adachi, P. E. Burrows, S. R. Forrest, M. E. Thompson, *J. Am. Chem. Soc.* 2001, **123**, 4304.
9. M. Nazeeruddin, R. Humphry-Baker, D. Berner, S. Rivier, L. Zuppiroli, M. Gratzel, *J. Am. Chem. Soc.* 2003, **125**, 8790.
10. A. B. Tamayo, B. D. Alleyne, P. I. Djurovich, S. Lamansky, I. Tsyba, N. N. Ho, R. Bau, M. E. Thompson, *J. Am. Chem. Soc.* 2003, **125**, 7377.
11. Y. L. Tung, P. C. Wu, C. S. Liu, Y. Chi, J. K. Yu, Y. H. Hu, P. T. Chou, S. M. Peng, G. H. Lee, Y. Tao, A. J. Carty, C. F. Shu, F. I. Wu, *Organometallics* 2004, **23**, 3745.
12. J. Kavitha, S. Y. Chang, Y. Chi, J. K. Yu, Y. H. Hu, P. T. Chou, S. M. Peng, G. H. Lee, Y. T. Tao, C. H. Chien, A. J. Carty, *Adv. Funct. Mater.* 2005, **15**, 223.
13. Y. Kawamura, K. Goushi, J. Brooks, J. J. Brown, H. Sasabe, C. Adachi, *Appl. Phys. Lett.* 2005, **86**, 071 104.
14. Y. H. Niu, Y. L. Tung, Y. Chi, C. F. Shu, J. H. Kim, B. Chen, J. Luo, A. J. Carty, A. K. Y. Jen, *Chem. Mater.* 2005, **17**, 3532.
15. W. Sotoyama, T. Satoh, N. Sawatari, H. Inoue, *Appl. Phys. Lett.* 2005, **86**, 153 505.
16. Y. L. Tung, S. W. Lee, Y. Chi, Y. T. Tao, C. H. Chien, Y. M. Cheng, P. T. Chou, S. M. Peng, C. S. Liu, *J. Mater. Chem.* 2005, **15**, 460.
17. M. A. Baldo, S. Lamansky, P. E. Burrows, M. E. Thompson, S. R. Forrest, *Appl. Phys. Lett.* 1999, **75**, 4.
18. M. A. Baldo, D. F. O'Brien, M. E. Thompson, S. R. Forrest, *Phys. Rev. B* 1999, **60**, 14 422.
19. J. Li, P. I. Djurovich, B. D. Alleyne, M. Yousufuddin, N. N. Ho, J. C. Thomas, J. C. Peters, R. Bau, M. E. Thompson, *Inorg. Chem.* 2005, **44**, 1713.
20. W. Lu, B. X. Mi, M. C. W. Chan, Z. Hui, N. Y. Zhu, S. T. Lee, C. M. Che, *Chem. Commun.* 2002, 206.

21. C. M. Che, S. C. Chan, H. F. Xiang, M. C. W. Chan, Y. Liu, Y. Wang, *Chem. Commun.* 2004, 1484.
22. Y. Kunugi, K. R. Mann, L. L. Miller, C. L. Exstrom, *J. Am. Chem. Soc.* 1998, **120**, 589.
23. W. Lu, B. X. Mi, M. C. W. Chan, Z. Hui, C. M. Che, N. Zhu, S. T. Lee, *J. Am. Chem. Soc.* 2004, **126**, 4958.
24. V. Adamovich, J. Brooks, A. Tamayo, A. M. Alexander, P. I. Djurovich, B. W. D'Andrade, C. Adachi, S. R. Forrest, M. E. Thompson, *New J. Chem.* 2002, **26**, 1171.
25. W. Y. Wong, Z. He, S. K. So, K. L. Tong, Z. Lin, *Organometallics* 2005, **24**, 4079.
26. B. W. D'Andrade, J. Brooks, V. Adamovich, M. E. Thompson, S. R. Forrest, *Adv. Mater.* 2002, **14**, 1032.
27. B. D'Andrade, S. R. Forrest, *Chem. Phys.* 2003, **286**, 321.
28. T. Koshiyama, A. Omura, M. Kato, *Chem. Lett.* 2004, **33**, 1386.
29. K. Saito, Y. Hamada, H. Takahashi, T. Koshiyama, M. Kato, *Jpn. J. Appl. Phys. Part 2* 2005, **44**, L500.
30. B. Ma, J. Li, P. I. Djurovich, M. Yousufuddin, R. Bau, M. E. Thompson, *J. Am. Chem. Soc.* 2005, **127**, 28.
31. J. Brooks, Y. Babayan, S. Lamansky, P. I. Djurovich, I. Tsyba, R. Bau, M. E. Thompson, *Inorg. Chem.* 2002, **41**, 3055.
32. B. W. D'Andrade, S. Datta, S. R. Forrest, P. Djurovich, E. Polikarpov, M. E. Thompson, *Org. Electron.* 2005, **6**, 11.
33. X. Zhan, C. Risko, F. Amy, C. Chan, W. Zhao, S. Barlow, A. Kahn, J. L. Bredas, S. R. Marder, *J. Am. Chem. Soc.* 2005, **127**, 9021.
34. B. W. Ma, P. I. Djurovich, M. E. Thompson, *Coord. Chem. Rev.* 2005, **249**, 1501.
35. R. J. Holmes, S. R. Forrest, Y. J. Tung, R. C. Kwong, J. J. Brown, S. Garon, M. E. Thompson, *Appl. Phys. Lett.* 2003, **82**, 2422.
36. W. Lu, M. C. W. Chan, K. K. Cheung, C. M. Che, *Organometallics* 2001, **20**, 2477.
37. R. C. Kwong, S. Lamansky, M. E. Thompson, *Adv. Mater.* 2000, **12**, 1134.
38. A. Dodabalapur, L. J. Rothberg, T. M. Miller, *Appl. Phys. Lett.* 1994, **65**, 2308.
39. J. Kido, M. Kimura, K. Nagai, *Science* 1995, **267**, 1332.
40. C. H. Kim, J. Shinar, *Appl. Phys. Lett.* 2002, **80**, 2201.

41. R. S. Deshpande, V. Bulovic, S. R. Forrest, *Appl. Phys. Lett.* 1999, **75**, 888.
42. Y. Shao, Y. Yang, *Appl. Phys. Lett.* 2005, **86**, 073 510.
43. D. C. Freeman, C. E. White, *J. Am. Chem. Soc.* 1956, **78**, 2678.
44. J. P. Wolfe, S. Wagaw, S. L. Buchwald, *J. Am. Chem. Soc.* 1996, **118**, 7215.
45. T. Yamamoto, M. Nishiyama, Y. Koie, *Tetrahedron Lett.* 1998, **39**, 2367.

Chapter 35

Management of Singlet and Triplet Excitons for Efficient White Organic Light-Emitting Devices

Yiru Sun,[a] Noel C. Giebink,[a] Hiroshi Kanno,[a] Biwu Ma,[b] Mark E. Thompson,[b] and Stephen R. Forrest[a,c]

[a] *Department of Electrical Engineering,*
Princeton Institute for the Science and Technology of Materials (PRISM),
Princeton University, Princeton, New Jersey 08544, USA
[b] *Department of Chemistry, University of Southern California,*
Los Angeles, California 90089, USA
[c] *Present address: Department of Electrical Engineering and Computer Science,*
Department of Physics, and Department of Materials Science and Engineering,
University of Michigan, Ann Arbor, Michigan 48109, USA
forrest@princeton.edu

Lighting accounts for approximately 22 per cent of the electricity consumed in buildings in the United States, with 40 per cent of that amount consumed by inefficient ($\sim$15 lm W^{-1}) incandescent lamps [1, 2]. This has generated increased interest in the use of white electroluminescent organic light-emitting devices, owing to their potential for significantly improved efficiency over incandescent

---

Reprinted from *Nature*, **440**, 908–912, 2006.

Electrophosphorescent Materials and Devices
Edited by Mark E. Thompson
Text Copyright © 2006 Nature Publishing Group
Layout Copyright © 2024 Jenny Stanford Publishing Pte. Ltd.
ISBN 978-981-4877-34-3 (Hardcover), 978-1-003-08872-1 (eBook)
www.jennystanford.com

sources combined with low-cost, high-throughput manufacturability. The most impressive characteristics of such devices reported to date have been achieved in all-phosphor-doped devices, which have the potential for 100 per cent internal quantum efficiency [2]: the phosphorescent molecules harness the triplet excitons that constitute three-quarters of the bound electron–hole pairs that form during charge injection, and which (unlike the remaining singlet excitons) would otherwise recombine non-radiatively. Here we introduce a different device concept that exploits a blue fluorescent molecule in exchange for a phosphorescent dopant, in combination with green and red phosphor dopants, to yield high power efficiency and stable colour balance, while maintaining the potential for unity internal quantum efficiency. Two distinct modes of energy transfer within this device serve to channel nearly all of the triplet energy to the phosphorescent dopants, retaining the singlet energy exclusively on the blue fluorescent dopant. Additionally, eliminating the exchange energy loss to the blue fluorophore allows for roughly 20 per cent increased power efficiency compared to a fully phosphorescent device. Our device challenges incandescent sources by exhibiting total external quantum and power efficiencies that peak at 18.7 ± 0.5 per cent and 37.6 ± 0.6 lm W^{-1}, respectively, decreasing to 18.4 ± 0.5 per cent and 23.8 ± 0.5 lm W^{-1} at a high luminance of 500 cd m^{-2}.

Electrophosphorescent organic light-emitting devices (OLEDs) have been shown to harvest 100% of the excitons generated by electrical injection, corresponding to a fourfold increase in efficiency compared to that achievable in singlet-harvesting fluorescent OLEDs [3]. In this context, electrophosphorescent white OLEDs (WOLEDS) have been reported to exhibit [4–7] high quantum (5–12%) and luminous power efficiencies (6–20 lm W^{-1}) at brightnesses <100 cd m^{-2}. To date, however, blue electrophosphorescent devices have exhibited short operational lifetimes [8] that limit the colour stability of all-phosphor-doped WOLEDs. Also, compared with their fluorescent counterparts, WOLEDs employing phosphorescent blue dopants excited via the conductive host introduce an approximately 0.8 eV exchange energy loss in power efficiency. This results from the energetic relaxation following intersystem crossing into the emissive triplet state. This loss can be avoided by resonant injection from the hole transport layer (HTL) and electron transport layer (ETL) into the phosphor triplet state [6, 9], but the subsequent transfer to green and red dopants required to generate white light can reintroduce these parasitic energy losses. Here we demonstrate a new WOLED architecture that uses a fluorescent emitting dopant to harness all electrically generated high energy singlet excitons for blue emission, and phosphorescent dopants to harvest the remainder of lower-energy triplet excitons for green and red emission.

This structure takes advantage of the fortuitous connection between the proportion of singlets dictated by spin statistics (that is, one singlet versus three triplets are produced by electrical excitation [10]) and the roughly 25% contribution of blue to the perceived white light spectrum. This allows for resonant energy transfer from both the host singlet and triplet energy levels that minimizes exchange energy losses, thereby maximizing device power efficiency while maintaining the potential for unity internal quantum efficiency (IQE). This approach has the further advantages of a stable white balance with current, a high efficiency at high brightness due to reduced geminate exciton recombination [11], and an enhanced lifetime due to the combined use of a stable fluorescent blue, and long lived phosphorescent green and red, dopants in a single emissive region.

The WOLED consists of a blue fluorophore, 4,4′-bis(9-ethyl-3-carbazovinylene)-1,1′-biphenyl (BCzVBi) [12], doped in a region spatially separate from the highly efficient green and red phosphorescent dopants fac-tris(2-phenylpyridine) iridium (Ir(ppy)$_3$) and iridium(III) bis(2-phenyl quinolyl-N,C′) acetylacetonate (PQIr), respectively. All lumophores are doped into a single, common conductive host, 4,4′-bis(N-carbazolyl)biphenyl (CBP), to form the extended emissive layer (EML) which is sandwiched between the electron transporting/hole blocking layer of 2,9-dimethyl-4,7-diphenyl-1,10-phenanthroline (BCP), and the 4,4′-bis[N-(1-naphthyl)-N-phenyl-amino]-biphenyl (*a*-NPD) HTL.

The principle of device operation is illustrated in Fig. 35.1. Excitons are formed on the host with a singlet-to-triplet formation ratio of χ_s/χ_t. Singlet excitons are transferred following a resonant Förster process onto the lightly doped (5%) blue fluorophore as

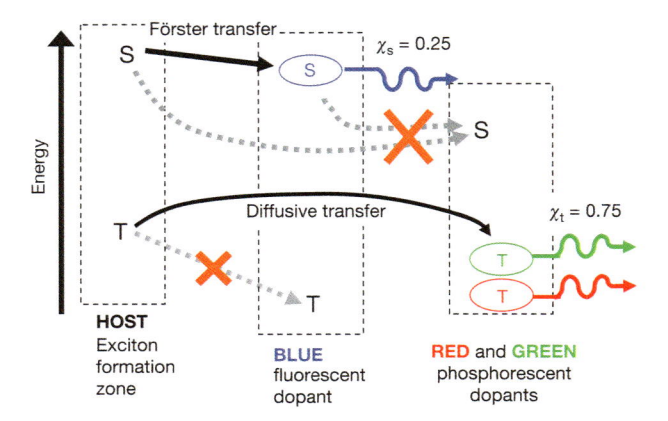

Figure 35.1 Proposed energy transfer mechanisms in the fluorescent/phosphorescent WOLED. This illustrates the separate channels for triplet (T) and singlet (S) formation and transfer directly onto their corresponding emissive dopants. The majority of excitons are formed in the host material with a singlet-to-triplet formation ratio of χ_s/χ_t. The singlet excitons in the two formation regions at each side of the light emitting layer (EML) are rapidly, and near-resonantly, transferred to the blue fluorescent dye located in these regions. The phosphor-doped region is located in the centre of the EML and separated from the exciton formation zones by spacers of undoped host material. The triplets then diffuse efficiently to the central region, where they transfer to the lower energy green or red phosphor dopants, again by a nearly resonant process to the green dopant triplet manifold, and with some energy loss to the red triplet. Diffusion of singlet excitons to the phosphor dopants is negligible due to their intrinsically short diffusion lengths [23]. Parasitic effects of charge trapping onto the phosphorescent dopants are discussed in the text.

opposed to direct trap formation [12]. The non-radiative host triplets, however, cannot efficiently transfer to the fluorophore by the Förster mechanism, or by Dexter transfer owing to the low doping concentration. On the other hand, triplets typically have long diffusion lengths [10] (~100 nm), and hence can migrate into the centre of the EML where they transfer onto the phosphors. Resonant transfer of the host triplet onto the green phosphor avoids exchange energy losses at this stage, although there are some unavoidable losses in transferring into the lowest energy red phosphor. Finally, placing an undoped host spacer with a thickness larger than the

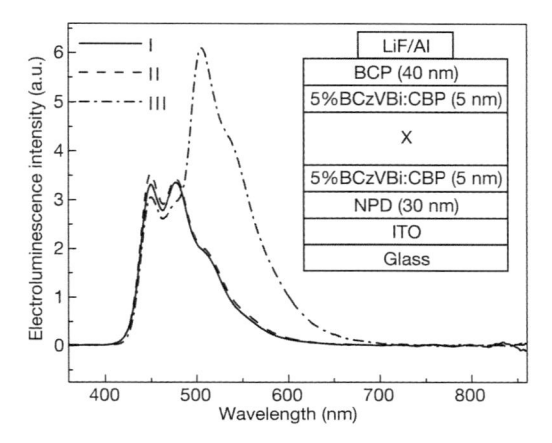

Figure 35.2 Un-normalized electroluminescence spectra of three device structures shown in the inset. The spectra were measured at a current density of 100 mA cm^{-2}. Inset, schematic cross-section of the device; see text for definitions of abbreviations used. X = CBP (16 nm) for device I; X = 5 wt% BCzVBi:CBP (15 nm) for device II; X = CBP (4 nm)/3 wt% Ir(ppy)$_3$:CBP (20 nm)/CBP (4 nm) for device III. Quantitative comparison of the three spectra suggests that excitons are primarily formed at the two interfaces, and that fluorescent doping across the entire emission layer does not increase the blue luminescence intensity (compare devices I and II). However, doping the middle of the EML with the phosphor Ir(ppy)$_3$ results in additional green emission without a corresponding decrease in blue, BCzVBi emission (devices I and III), indicating that triplets formed on CBP are efficiently transferred to Ir(ppy)$_3$, whereas the BCzVBi acts to harvest, or 'filter out', all of the singlets.

Förster radius (~3 nm) between the blue fluorophore and the phosphors prevents direct energy transfer from the blue dopant to the green and red phosphors. This device architecture is unique in that the singlet and triplet excitons are harvested along completely independent channels, and hence the transfer from host to dopant for both species can be separately optimized to be nearly resonant, thereby minimizing energy losses while maintaining a unity IQE.

Figure 35.2 provides evidence for this transfer mechanism by comparing the un-normalized electroluminescent spectra of three devices at a current density of $J = 100$ mA cm^{-2}. Device I has a 16-nm-thick CBP spacer placed between the two 5-nm-thick 5 wt% BCzVBi:CBP layers at each side of the EML, whereas device

II has a 25-nm-thick uniformly doped 5 wt% BCzVBi:CBP EML. Both devices have nearly identical emission spectra and external quantum efficiencies (EQE) of $\eta_{ext} = (2.6 \pm 0.2)\%$, indicating that ostensibly 100% of the exciton formation occurs at the edges of the EML. Furthermore, lack of short wavelength CBP emission in device I suggests that the charge density in the middle region of EML available for exciton formation directly on the host is negligible. When a 20-nm-thick 3 wt% Ir(ppy)$_3$:CBP layer is inserted in the EML and separated from the two 5 wt% BCzVBi:CBP regions by 4-nm-thick undoped CBP spacers (device III), the total efficiency is increased to $\eta_{ext} = (5.2 \pm 0.2)\%$, with the additional 2.6% emission coming from Ir(ppy)$_3$. From this we conclude that exciton diffusion from the point of origin at the edges of the EML, rather than direct charge trapping and exciton formation on the phosphor, dominates, because carriers trapped by Ir(ppy)$_3$ would result in a noticeable decrease in the BCzVBi emission. Triplet diffusion from the edges of the EML to the phosphorescent doped region is consistent with previous observations in red fluorescent/phosphorescent OLEDs, where the fluorophore 'filters' out the singlet excitons, leaving only triplets to diffuse to a spatially separated phosphor doped region [3].

To determine whether the location of the exciton formation region is predominantly at either the HTL/EML or EML/ETL interface, we compared the emission from two comparable devices with opposite symmetries, where the structure consisted of either NPD (30 nm)/5 wt% BCzVBi:CBP (10 nm)/CBP (20 nm)/BCP (40 nm), or NPD (30 nm)/CBP (20 nm)/5 wt% BCzVBi:CBP: (10 nm)/BCP (40 nm). The corresponding maximum efficiencies of these devices were $\eta_{ext} = (1.4 \pm 0.1)\%$ and $(1.8 \pm 0.1)\%$, both smaller than $\eta_{ext} = (2.6 \pm 0.2)\%$ for device II. Moreover, CBP emission at a peak wavelength of $\lambda = 390$ nm is observed in the first device, and α-NPD emission at $\lambda = 430$ nm is observed in the second structure. These observations suggest that excitons are generated at both the HTL/EML and EML/ETL interfaces, consistent with the ambipolar conductivity of CBP [13, 14]. Exciton formation at the edges, with correspondingly low generation in the bulk of the EML, can be understood as follows: large densities of holes (p) and electrons (n) pile up at the energy barriers at two EML interfaces. The exciton formation probability, which is $\sim n \times p$, is thus also significantly higher at these locations as compared with the EML bulk.

On the basis of these results, WOLEDs were fabricated by doping the middle region of the EML with both the green and red phosphorescent dopants (Fig. 35.3a, inset). As the high energy of the highest occupied molecular orbital of PQIr suggests that it can trap holes in the CBP host, a slight decrease of the blue emission intensity is observed in the WOLED. Fitting of the WOLED spectrum in Fig. 35.3b with the individual dopant spectra suggests that the ratio of emission from fluorescent to phosphorescent dopants approaches the ratio of 1/3, consistent with the singlet-to-triplet exciton formation ratio in emissive organic materials [10, 15, 16]. Furthermore, given the performance characteristics of the purely fluorescent BCzVBi device (device II in Fig. 35.2), we also find that the fraction of excitons trapped by, and formed directly on, the phosphorescent dopants in the EML is $\chi_{trap} = (18 \pm 5)\%$ (see Methods). That is, approximately 20% of the excitons are formed by direct trapping on the phosphor dopants, whereas the remaining 80% are formed at the edges of the EML, at which point the triplets subsequently diffuse into the centre where they are transferred from host to phosphor dopant before emission.

A maximum forward viewing EQE of $\eta_{ext} = (11.0 \pm 0.3)\%$ is achieved at a current density $J = (1.0 \pm 0.6)$ mA cm^{-2}, and decreases only slightly to $\eta_{ext} = (10.8 \pm 0.3)\%$ at a high forward viewing luminance of 500 cd m^{-2} (Fig. 35.3a). This device gives a maximum forward viewing power efficiency of $\eta_p = (22.1 \pm 0.3)$ lm W^{-1}. As illumination sources are generally characterized by their total emitted power, this device therefore has maximum total efficiencies [6] of $\eta_{p,t} = (37.6 \pm 0.6)$ lm W^{-1}, and $\eta_{ext,t} = (18.7 \pm 0.5)\%$. At a practical surface luminance of 500 cd m^{-2}, $\eta_{p,t} = (23.8 \pm 0.5)$ lm W^{-1}, or approximately 50% greater than for common incandescent lighting. Although the commercially available blue fluorophore has a low $\eta_{ext} = 2.7\%$ (compared with a maximum expected 5% achieved in the literature), the WOLED performance, nevertheless, represents a considerable improvement over the best all-phosphorescent devices previously reported [6, 7, 17] (see Supplementary Information).

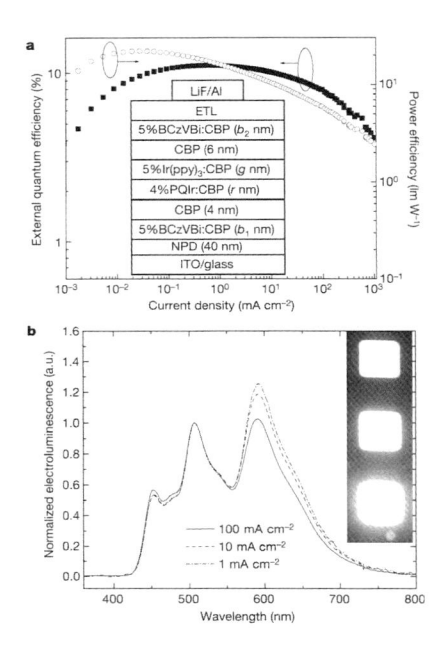

Figure 35.3 Performance characteristics of the fluorescent/phosphorescent WOLED. (a) Forward viewing external quantum efficiency (filled squares) and power efficiency (open circles) versus current density of the WOLED shown in the inset. The forward viewing external quantum efficiency peaks at $\eta_{ext} = (11.0 \pm 0.3)\%$ at $J = (1.0 \pm 0.6)$ mA cm^{-2}, and decreases slightly to $\eta_{ext} = (10.8 \pm 0.3)\%$ at a forward viewing luminance of 500 cd m^{-2}. The maximum forward viewing power efficiency is $\eta_p = (22.1 \pm 0.3)$ lm W^{-1}, with total peak and high luminance efficiencies of $\eta_{p,t} = (37.6 \pm 0.6)$ and (23.8 ± 0.5) lm W^{-1} at 500 cd m^{-2}, respectively. The forward viewing luminance at 1 A cm^{-2} is $(83{,}000 \pm 7{,}000)$ cd m^{-2}. The drive voltage for this device is (6.0 ± 0.5) V at $J = 10$ mA cm^{-2}. Inset, schematic structure of the WOLED with the following layer thicknesses: $b_1 = 15$ nm, $b_2 = 10$ nm, $r = 8$ nm, $g = 12$ nm, and with an electron transport layer (ETL) consisting of 20-nm 4,7-diphenyl-1,10-phenanthroline (BPhen) followed by 20-nm Li doped BPhen in 1:1 molar ratio. Here, BPhen is used to further reduce device drive voltage. When 40-nm BCP is used as ETL and $b_1 = 10$ nm, $b_2 = 10$ nm, $r = 12$ nm and $g = 8$ nm, η_{ext} and CRI are nearly identical to the above structure. (b) Normalized electroluminescence spectra of WOLED emission at various current densities. Note that colour dependence on current density is minimal, with CRI = 85 at all three values of current density. Inset, images of three, 4.5 mm^2 devices, each driven at four times the drive current (from 1.7 to 28 mA cm^{-2}) of the device above it in the array (equivalent to a two f-stop difference in illumination) to show the colour stability of the emission.

The intrinsic singlet-to-triplet ratio and the separation of the channels in harvesting the two excitonic species gives a well-balanced and largely current-independent colour rendition, resulting in a colour rendering index of CRI = 85 at all current densities studied, which is the highest CRI among the reported values for a WOLED. The Commission Internationale d'Eclairage (CIE) coordinates have a negligible shift from (0.40, 0.41) at 1 mA cm^{-2} to (0.38, 0.40) at 100 mA cm^{-2}. This differs from observations of an all-phosphor-doped WOLED, where blue emission becomes stronger with increasing driving voltage [6] owing to the requirement for high energy excitation of the blue phosphor. In the inset of Fig. 35.3b are images of three devices, each driven at 4 times higher drive current than the device above it in the array, to show the colour stability of the emission.

To further understand exciton diffusion, in Fig. 35.4a we plot (open circles) η_{ext} due to Ir(ppy)$_3$ emission versus the position (x) of a thin (5 nm) slab of 5 wt% Ir(ppy)$_3$:CBP located at various distances from the HTL/EML interface within a 200-nm-thick CBP EML (see Fig. 35.4a inset). Fitting (solid curve, see Methods) of the efficiency versus x yields a triplet diffusion length of $L_D = (460 \pm 30)$ Å, and predicts that $(75 \pm 5)\%$ of the phosphorescent emission results from triplet exciton diffusion from the adjacent EML interfaces, in agreement with the value calculated from analysis of the spectral content of the emission.

Compared with previous all-phosphor, high efficiency WOLEDs, the device also has a less pronounced efficiency roll-off at high current densities. For example, in Fig. 35.4b we show a comparison of η_{ext} versus J for an all-phosphor white device [6], the device of this work, and a fluorescent BCzVBi device II. The high-current decline in η_{ext} of the all-phosphor white is due to triplet–triplet annihilation [14, 18]. In contrast, there is a striking resemblance between the efficiency roll-off of the current device, and that of device II. Modelling of the roll-offs in these two structures is complicated by recombination processes such as exciton–polaron quenching [11], singlet–triplet annihilation, and field-induced exciton dissociation. Nevertheless, for both of these latter devices, the current density at the point where η_{ext} has declined by half from its peak is > 7 times that of the conventional phosphor device, while the peak

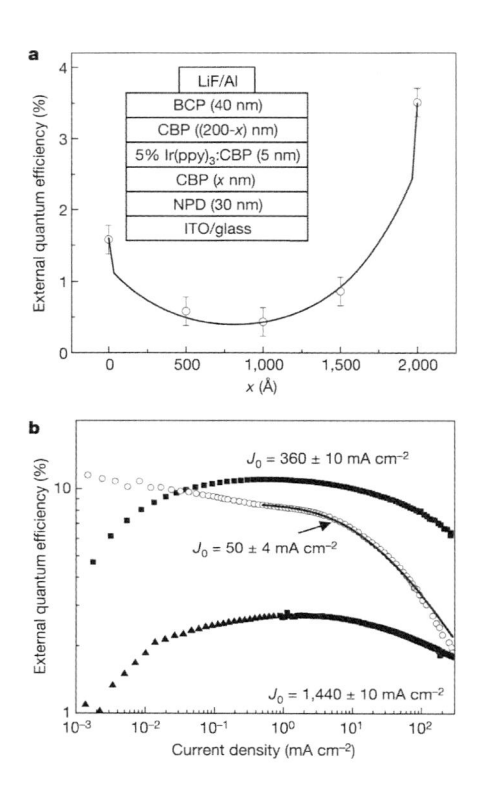

Figure 35.4 Triplet diffusion profile and reduced efficiency roll-off at high currents. (a) External quantum efficiency (open circles) from Ir(ppy)$_3$ emission at 10 mA cm^{-2} versus distance between the 50-Å slab of 5 wt% Ir(ppy)$_3$:CBP and the NPD/CBP interface in the structure shown inset. A fit following equation (2) for triplet diffusion gives the solid curve and a triplet diffusion length of $L_D = (460 \pm 30)$ Å. The error bars indicate the standard deviations in measurement. Inset, schematic cross-section of the test structures: NPD (30 nm)/CBP (x nm)/5 wt% Ir(ppy)$_3$:CBP (5 nm)/CBP ((200 − x) nm)/BCP (40 nm), with x = 0, 50, 100, 150, 200. (b) Comparison of external quantum efficiency roll-off. Open circles depict the performance of the all-phosphor white device of Ref. [6], in comparison to the white device of this work (squares), and a blue fluorescent BCzVBi device (triangles). The high current roll-off of the phosphor device is described by triplet–triplet annihilation (fit shown as solid line), yielding an onset current density $J_0 = (50 \pm 4)$ mA cm^{-2}. The device of this work clearly demonstrates a roll-off that appears qualitatively similar to that of the all-fluorescent device. For comparison, $J_0 = (360 \pm 10)$ mA cm^{-2} and $J_0 = (1{,}440 \pm 10)$ mA cm^{-2} for the WOLED of this work and the all-fluorescent device, respectively.

EQE occurs at a value of J nearly 1,000 times larger. The apparent absence of triplet–triplet annihilation suggests that the highest density of triplet excitons is at the interfaces in the fluorescent doped regions, where they subsequently diffuse towards the centre, thereby lowering the local density (Fig. 35.4a) in the region of the guest phosphors. The reduced sensitivity of η_{ext} to current density is another clear difference between the WOLED of this study and previous, high efficiency all-phosphor devices.

We note that the efficiency of the present device can be further improved by using fluorescent dopants [19] with IQE = 25% and phosphors [20, 21] with IQE = 100%, resulting in a total WOLED internal quantum efficiency of 100%. Using such 'ideal' chromophores, whose spectra are the same as the current dopants used, an approximately 34% total EQE and 60 lm W^{-1} power efficiency can in principle be achieved using this structure, corresponding to a four-fold increase over incandescent power efficiency, and even competing with high efficiency, high CRI fluorescent lighting sources. As noted above, the exchange energy difference between the host singlet and dopant triplet states can lead to a loss of luminance efficiency in all-phosphor doped WOLEDs. By applying this design concept to systems where the host singlet is resonant with the blue fluorophore singlet state, and the host triplet is resonant with the green phosphor triplet level, this structure could have a power efficiency improvement of ~20% compared to similarly ideal all-phosphor devices. The highly efficient WOLED structure reported here, with a colour rendition that is unusually independent of current density, has potential for use in the next generation of sources for solid-state indoor lighting.

Methods

Device manufacture. Devices were grown on clean glass substrates pre-coated with a 150-nm-thick layer of indium tin oxide (ITO) with a sheet resistance of 20 Ω per square. All organic layers were grown in succession without breaking vacuum (~10^{-7} torr). After organic film deposition, a shadow mask with 1-mm-diameter openings was affixed in a N_2 filled glove box before the cathode (consisting of 8-Å-thick LiF), followed by a 500-Å-thick Al cap, was deposited by high vacuum (10^{-6} torr) thermal evaporation. Current–voltage and EQE measurements were carried out using a semiconductor parameter analyser (HP 4145) and a calibrated Si photodiode (Hamamatsu S3584-08) following standard procedures [22].

Data analysis. To interpret the emission spectrum, the WOLED EQE is expressed as:

$$\eta_{\text{ext}} = (1 - \chi_{\text{trap}}) \eta_B + [(1 - \chi_{\text{trap}}) \chi_t + \chi_{\text{trap}}]\eta_{\text{GR}} \quad (35.1)$$

where η_B and η_{GR} are respectively the EQEs of a singly doped blue fluorescent device and the comparable singly doped green and red phosphorescent devices, and χ_{trap} is the fraction of excitons trapped and formed directly on the phosphorescent dopants in the EML. By fitting the WOLED spectrum in Fig. 35.3b at $J = 100$ mA cm^{-2} with the electroluminescence spectra of the three individual dopant materials, and accounting for photon energy in these power spectra, we find that $(20 \pm 2)\%$ of the total quantum efficiency is due to emission from the blue fluorescent dopant, and $(80 \pm 2)\%$ is from green and red phosphorescent dopants. Given the performance characteristics (η_B) of the purely BCzVBi device (device II in Fig. 35.2), we calculate $\chi_{\text{trap}} = (18 \pm 5)\%$ from the first term in Eq. (35.1).

Modelling. Exciton diffusion through the EML is modelled as shown in Fig. 35.4a (solid line) as follows: in steady state, and assuming that all singlet formation occurs at the HTL/EML ($x = 0$), and EML/ETL interfaces ($x = 200$ nm) (in Fig. 35.4a), a solution to the triplet diffusion equation gives:

$$n(x) = \frac{1}{\sinh\left(\frac{L}{L_D}\right)} \left[\chi_t n_R \sinh\left(\frac{x}{L_D}\right) + \chi_t n_L \sinh\left(\frac{L-x}{L_D}\right) \right]$$
$$+ \chi_s n_L \delta(x) + \chi_s n_R \delta(x - L) \quad (35.2)$$

where n is the total exciton density: $n(x = 0) = n_L$ and $n(x = L = 200 \text{ nm}) = x = L = 200 \text{ nm}) = n_R$, L_D is the triplet diffusion length, and the delta function terms account for the presence of contributing singlets at the interfaces. As the EQE from Ir(ppy)$_3$ emission is proportional to the exciton density in the Ir(ppy)$_3$-doped slab, Fig. 35.4a (solid curve) shows the fit of the efficiency at $J = 10$

mA cm^{-2} versus x using Eq. (35.2) and $\chi_t = 3\chi_s$, from which we infer $L_D = (460 \pm 30)$ Å (error bars account for the spread in fits at additional current densities). With this calculated diffusion length, integration of the total exciton density in the phosphorescent doped region in the WOLED predicts that $(75 \pm 5)\%$ of the phosphorescent emission results from triplet exciton diffusion from the adjacent EML interfaces, in agreement with the value calculated from analysis of the spectral content of the emission (Eq. (35.1)).

Acknowledgements

H.K. is currently on leave from Sanyo Electric Co., Ltd., Osaka, Japan. The authors thank the Department of Energy and Universal Display Corp. for partial support of this work, as well as R. Holmes and B. W. D'Andrade for discussions.

Supplementary Information

Supplementary Information is available at https://static-content.springer.com/esm/art%3A10.1038%2Fnature04645/MediaObjects/41586_2006_BFnature04645_MOESM1_ESM.doc.

References

1. US Department of Energy *National Lighting Inventory and Energy Consumption Estimate* Vol. 1, xii (US Govt Printing Office, Washington DC, 2001).
2. D'Andrade, B. W. & Forrest, S. R. White organic light-emitting devices for solid-state lighting. *Adv. Mater.* **16**, 1585–1595 (2004).
3. Baldo, M. A. *et al.* Highly efficient phosphorescent emission from organic electroluminescent devices. *Nature* **395**, 151–154 (1998).
4. Adamovich, V. *et al.* High efficiency single dopant white electrophosphorescent light emitting diodes. *N. J. Chem.* **26**, 1171–1178 (2002).
5. D'Andrade, B. W. & Forrest, S. R. Effects of exciton and charge confinement on the performance of white organic p-i-n electrophosphorescent emissive excimer devices. *J. Appl. Phys.* **94**, 3101–3109 (2003).

6. D'Andrade, B. W., Holmes, R. J. & Forrest, S. R. Efficient organic electrophosphorescent white-light-emitting device with a triple doped emissive layer. *Adv. Mater.* **16**, 624–627 (2004).
7. Tokito, S., Iijima, T., Tsuzuki, T. & Sato, F. High-efficiency white phosphorescent organic light-emitting devices with greenish-blue and red-emitting layers. *Appl. Phys. Lett.* **83**, 2459–2461 (2003).
8. Tung, Y. J. *et al.* A high efficiency phosphorescent white OLED for LCD backlight and display applications. *Proc. Soc. Inform. Display* **35**, 48–51 (2004).
9. Holmes, R. J. *et al.* Efficient, deep-blue organic electrophosphorescence by guest charge trapping. *Appl. Phys. Lett.* **83**, 3818–3820 (2003).
10. Baldo, M. A. Excitonic singlet-triplet ratio in a semiconducting organic thin film. *Phys. Rev. B* **66**, 14422–14428 (1999).
11. Kalinowski, J. *et al.* Triplet energy exchange between fluorescent and phosphorescent organic molecules in a solid state matrix. *Chem. Phys.* **297**, 39–48 (2004).
12. Hosokawa, C., Higashi, H., Nakamura, H. & Kusumoto, T. Highly efficient blue electroluminescence from a distyrylarylene emitting layer with a new dopant. *Appl. Phys. Lett.* **67**, 3853–3855 (1995).
13. Adachi, C., Thompson, M. E. & Forrest, S. R. Architectures for efficient electrophosphorescent organic light-emitting devices. *IEEE J. Select. Top. Quant. Electron.* **8**, 372–377 (2002).
14. Baldo, M. A. & Forrest, S. R. Transient analysis of organic electrophosphorescence: I. Transient analysis of triplet energy transfer. *Phys. Rev. B* **62**, 10958–10966 (2000).
15. Reufer, M. *et al.* Spin-conserving carrier recombination in conjugated polymers. *Nature Mater.* **4**, 340–346 (2005).
16. Segal, M., Baldo, M. A., Holmes, R. J., Forrest, S. R. & Soos, Z. G. Excitonic singlet-triplet ratios in molecular and polymeric organic materials. *Phys. Rev. B* **68**, 075211 (2003).
17. Qin, D. S. & Tao, Y. White organic light-emitting diode comprising of blue fluorescence and red phosphorescence. *Appl. Phys. Lett.* **86**, 113507 (2005).
18. Baldo, M. A., Adachi, C. & Forrest, S. R. Transient analysis of organic electrophosphorescence. II. Transient analysis of triplet-triplet annihilation. *Phys. Rev. B* **62**, 10967–10977 (2000).
19. Murata, H., Kafafi, Z. H. & Uchida, M. Efficient organic light-emitting diodes with undoped active layers based on silole derivatives. *Appl. Phys. Lett.* **80**, 189–191 (2002).

20. Kawamura, Y. *et al.* 100% phosphorescence quantum efficiency of Ir(III) complexes in organic semiconductor films. *Appl. Phys. Lett.* **86**, 071104 (2005).
21. Adachi, C., Baldo, M. A., Thompson, M. E. & Forrest, S. R. Nearly 100% internal phosphorescence efficiency in an organic light emitting device. *J. Appl. Phys.* **90**, 5048–5051 (2001).
22. Forrest, S. R., Bradley, D. D. C. & Thompson, M. E. Measuring the efficiency of organic light-emitting devices. *Adv. Mater.* **15**, 1043–1048 (2003).
23. Sokolik, I., Priestley, R., Walser, A. D., Dorsinville, R. & Tang, C. W. Bimolecular reactions of singlet excitons in tris(8-hydroxyquinoline) aluminum. *Appl. Phys. Lett.* **69**, 4168–4170 (1996).

Chapter 36

Highly Efficient, Near-Infrared Electrophosphorescence from a Pt–Metalloporphyrin Complex

Carsten Borek,[a] Kenneth Hanson,[a] Peter I. Djurovich,[a]
Mark E. Thompson,[a] Kristen Aznavour,[a] Robert Bau,[a]
Yiru Sun,[b] Stephen R. Forrest,[c] Jason Brooks,[d] Lech Michalski,[d]
and Julie Brown[d]

[a] *Department of Chemistry, University of Southern California,
Los Angeles, California 99089, USA*
[b] *Department of Electrical Engineering, Princeton University,
Princeton, New Jersey 08544, USA*
[c] *Department of Electrical Engineering and Computer Science
and Department of Physics, University of Michigan,
Ann Arbor, Michigan 48109, USA*
[d] *Universal Display Corporation, Ewing, New Jersey 08618, USA*
met@usc.edu

Organic light-emitting diodes (OLEDs) have been the subject of a significant research effort for the past two decades with a focus on devices that emit almost exclusively in the visible part of the electromagnetic spectrum [1]. Recently, there has been a growing interest in OLEDs that emit in the near-infrared (NIR) region

---

Reprinted from *Angew. Chem. Int. Ed.*, **46**, 1109–1112, 2007.

Electrophosphorescent Materials and Devices
Edited by Mark E. Thompson
Text Copyright © 2007 Wiley-VCH Verlag GmbH & Co. KGaA, Weinheim
Layout Copyright © 2024 Jenny Stanford Publishing Pte. Ltd.
ISBN 978-981-4877-34-3 (Hardcover), 978-1-003-08872-1 (eBook)
www.jennystanford.com

(700–2500 nm) [2–4]. Applications for these NIR OLEDs are particularly interesting for night-vision-readable displays [5] and sensors [6]. The efficiency of OLEDs are markedly improved when fluorescent emissive dopants are replaced with phosphorescent heavy-metal complexes that can effectively harvest both the singlet and triplet excitons formed in electroluminescence, with wavelengths (λ) ranging from the near-ultraviolet into the red (with peak emission at λ_{max} = 380–650 nm) [7–10]. Herein, we report on an efficient NIR OLED that utilizes a phosphorescent Pt–metalloporphyrin dopant, with an external quantum efficiency (EQE) greater than 6% at λ_{max} = 765 nm and a full width at half maximum of 31 nm (500 cm^{-1}).

Previously, two classes of phosphorescent complexes have been employed as dopants in NIR-emitting OLEDs. The first utilizes trivalent lanthanide cations (Ln^{3+}) as the emitting centers, for example, Er^{3+} or Nd^{3+}, chelated with chromophoric ligands to sensitize excitation-energy transfer to the lanthanide ion [11]. Schanze et al. have reported an NIR OLED utilizing Ln^{3+} in conjunction with a porphyrin/polystyrene matrix, with EQE ranging from 8.0×10^{-4} to 2.0×10^{-4}% at approximately 1 mA cm^{-2} [5]. Similarly, a Nd(phenalenone)$_3$-based OLED had an EQE of 0.007% at λ_{max} = 1065 nm [4]. The second class of NIR OLEDs is transition-metal complexes, similar to those used in the visible region. A recent report of an electrophosphorescent device that used a cyclometalated [(pyrenyl–quinolyl)$_2$Ir(acac)] complex as the phosphor gave λ_{max} = 720 nm and an EQE of 0.1% [6].

A family of complexes that have shown intense absorption and emission in the red-to-NIR region of the spectrum are the metalloporphyrins [12, 13]. There are a number of reports of OLEDs fabricated with [Pt(oep)], [Pt(tpp)] (oep = 2,3,7,8,12,13,17,18-ocatethylporphryin, tpp = 5,10,15,20-tetraphenylporphyrin), or analogues of these compounds as phosphorescent emitters, with emission maxima between 630 and 650 nm [7, 14–22], however, there has been no apparent effort to shift the Pt–porphyrin-based OLED emission into the NIR region. Porphyrin chromophores with fused aromatic moieties at the β-pyrrole positions, for example, tetrabenzoporphyrin (bp), exhibit a bathochromic shift (relative to unsubstituted porphyrin) of the absorption and emission energy, owing to the expansion of the π-electronic system of the porphyrin core [23]. The addition of bulky groups to the meso positions of the porphyrin macrocycles with β-substituted pyrroles leads to the formation of nonplanar porphyrins, and further red-shifts the absorption spectra [24]. Coordination of a heavy-metal atom increases the rate of the intersystem crossing between singlet and triplet states of the metalloporphyrins, thereby enhancing the rate of radiative decay from the triplet state. For these reasons, we focus here on Pt^{II}–tetraphenyltetrabenzoporphyrin, [Pt(tpbp)] (Fig. 36.1), as a phosphorescent dopant.

Analysis of the [Pt(tpbp)] complex by X-ray crystallography reveals a nonplanar molecular structure with a saddle-type distortion,

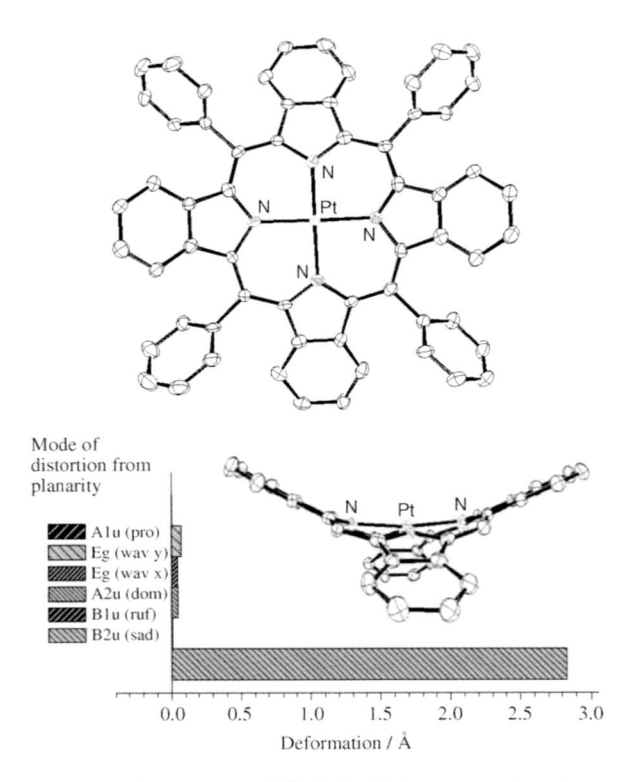

Figure 36.1 Crystal structure of [Pt(tpbp)]. Top: top view; bottom: edge view (*meso*-phenyl substituents removed for clarity) with results of NSD analysis.

similar to that found in other tpbp derivatives [25]. Crystallographic data are available in the Supporting Information and from the Cambridge Crystallographic Data Centre (CCDC-627735) [26]. An analysis of [Pt(tpbp)] using normal-coordinate structural decomposition (NSD) software [27, 28] quantifies the various distortions that accompany the macrocyclic deformation (Fig. 36.1, top). The total out-of-plane distortion ($D_{oop} = 2.83$ Å) is almost exclusively described by saddling (B2u = 2.83 Å), while doming (A2u = 0.045 Å), wave x and wave y deformations (similar to a chair conformation in cyclohexane, in either the x or y direction; Eg = 0.042 and 0.068 Å, respectively), propellering (A1u = 0.0003 Å), and ruffling (B1u = 0.006 Å) contribute only slightly to the nonplanarity. This distortion is greater than that found in [Zn(tpbp)] ($D_{oop} = 2.35$ Å) [25, yet is less than that of a [Ni(tpbp(CO$_2$Me)$_8$)] derivative ($D_{oop} = 3.43$ Å) [23].

Electrochemical analysis of [Pt(tpbp)], versus an internal ferrocene reference, shows a reversible oxidation at 0.24 V and quasireversible reduction at -1.85 V. The highest occupied (HOMO) and the lowest unoccupied (LUMO) molecular orbitals calculated from these data are 4.9 and 2.5 eV in energy relative to vacuum, respectively [29]. The absorption spectrum (Fig. 36.2) displays strong transitions for the Soret band at $\lambda_{max} = 430$ nm, with an extinction coefficient of $\varepsilon_{430nm} = 2.03 \times 10^5$ m^{-1} cm^{-1}, and the Q band with $\lambda_{max} = 611$ nm ($\varepsilon_{611nm} = 1.35 \times 10^5$ m^{-1} cm^{-1}), which differ slightly from the values for the analogous [Pd(tpbp)] complex [30].

The phosphorescence spectrum of thoroughly degassed [Pt(tpbp)] has $\lambda_{max} = 765$ nm, with a transient lifetime of $\tau = 53$ ms at room temperature (298 K); these values shift to $\lambda_{max} = 751$ nm and $\tau = 73$ μs at 77 K. The blue shift is due to a rigidochromic effect at low temperature. Assuming that the radiative rate constant does not show a temperature dependence between 298 K and 77 K, and the nonradiative decay rate is negligible at 77 K, the ratio of the lifetimes gives the photoluminescence (PL) efficiency. The photophysical properties of Pt–tetrabenzoporphrin [Pt(bp)] support these assumptions [31]. The photoluminescent quantum yield (Φ_{PL}) estimated by this method is 0.7. Radiative ($k_r = 1.3 \times 10^4$ s^{-1}) and nonradiative ($k_{nr} = 5.8 \times 10^3$ s^{-1}) decay rates were

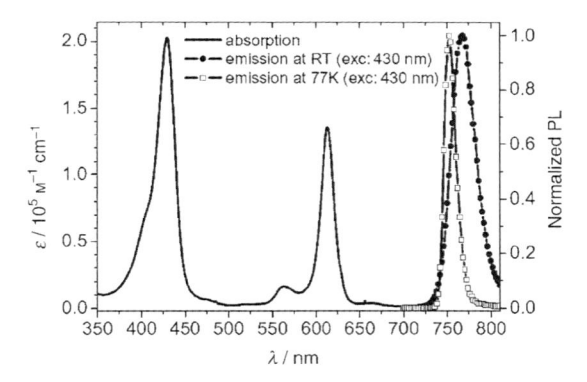

Figure 36.2 Room-temperature absorption spectrum (solid line), and normalized emission spectra at room temperature (closed circles) and at 77 K (open squares) of [Pt(tpbp)] in 2-methyl-THF.

estimated from the lifetime (τ) data ($k_r = 1/\tau_{77K}$; $k_{nr} = 1/\tau_{298K} - k_r$). The nonradiative rate is similar to that reported for [Pd(tpbp)] ($k_{nr} = 4.2 \times 10^3$ s^{-1}), but the stronger spin–orbit coupling of Pt significantly increases the radiative rate (compare [Pd(tpbp)]: $k_r = 8.8 \times 10^2$ s^{-1}) [30].

To minimize concentration quenching in [Pt(tpbp)]-based OLEDs, [Pt(tpbp)] was doped into a tris(8-hydroxyquinoline)aluminum (Alq$_3$) host [7]. The dopant singlet and triplet levels fall well below those of Alq$_3$($S_1 = 500$ nm, $T_1 = 590$ nm) [32]. The OLEDs prepared here incorporated a neat Alq$_3$ exciton-blocking layer (EBL) to prevent exciton quenching at the cathode surface. Thus, the structure used was: ITO/(400 Å) NPD/ (400 Å) Alq$_3$ + 6 wt% [Pt(tpbp)] /(500 Å) Alq$_3$/ (10 Å) LiF /(1100 Å) Al (ITO = indium tin oxide, NPD = N,N'-bis(1-naphthyl)-N,N'-diphenyl-1,1'-biphenyl-4,4'-diamine), hereafter referred to as device 1. This device has an EQE of 6.3% at 0.1 mA cm^{-2}, which gradually decreases as the current density is increased (Fig. 36.3). The intensity is roughly 0.1 µW cm^{-2} at 3 V and 750 µW cm^{-2} at 12 V. The electroluminescence (EL) spectrum for device 1 displays strong NIR emission at 769 nm, with no Alq$_3$ signal (520 nm) at any bias level (Fig. 36.3), thus indicating complete energy transfer from host to guest.

Devices were also prepared to test the operational stability of [Pt(tpbp)]-based OLEDs. For these tests, the following structure

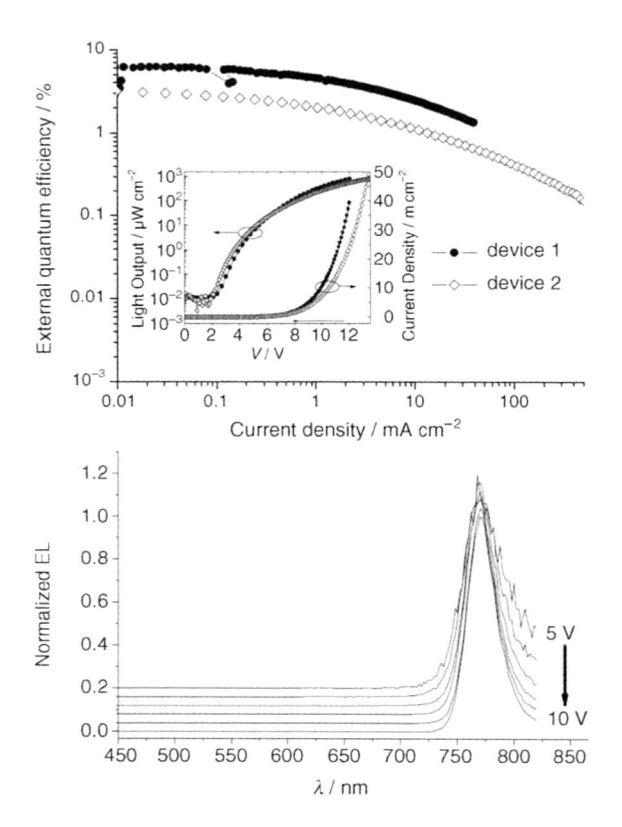

Figure 36.3 Top: external quantum efficiency versus current density. Device 1: ITO/ (400 Å) NPD/(400 Å) Alq$_3$ + 6 wt% [Pt(tpbp)]/(500 Å) Alq$_3$/(10 Å) LiF/ (1100 Å) Al; device 2: ITO/(400 Å) NPD/(300 Å) Alq$_3$ + 6 wt% [Pt(tpbp)]/ (400 Å) BAlq/(10 Å) LiF/(1000 Å) Al. Inset: Intensity and current density versus voltage for devices 1 and 2. Bottom: normalized electroluminescence spectra for device 1 at 5 V to 10 V, in 1-V steps. Spectra offset for clarity.

(device 2) was used: ITO/(400 Å) NPD /(300 Å) Alq$_3$ + 6 wt% [Pt(tpbp)] /(400 Å) BAlq/ (10 Å) LiF /(1000 Å) Al, where BAlq is aluminum(III) bis(2-methyl-8-quinolinato)4-phenylphenolate, and serves as the exciton-blocking layer. The charge transport, blocking, and host materials have previously been demonstrated to give high efficiencies and long electrophosphorescent device lifetimes [33, 34]. At a low current density of 0.1 mA cm^{-2}, device 2 has a

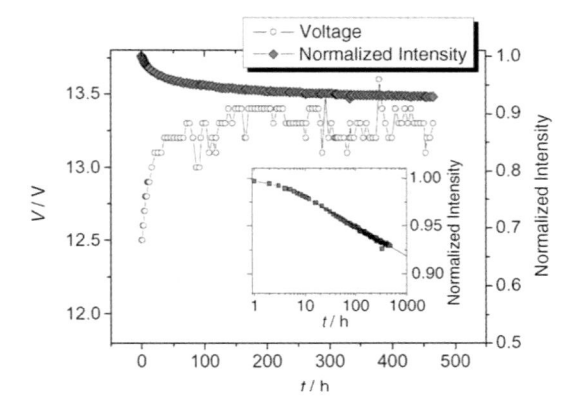

Figure 36.4 Intensity versus operating time, normalized to initial intensity (closed squares) and voltage (open circles). The initial intensity was 740 µW cm^{-2}. The inset shows the time versus radiance data plotted on a semilogarithmic scale with extrapolation to 1000 h.

maximum EQE of 3%, falling to 1.1% at 10 mA cm^{-2} (Fig. 36.3). The electroluminscence spectrum is identical to that for device 1.

The device was aged at a high constant current of 40 mA cm^{-2} corresponding to an initial intensity of 740 µW cm^{-2}. The data in Fig. 36.4 (see inset) suggest that device 2 will maintain greater than 90% of its initial intensity after 1000 h of operation. A 0.5-V rise in voltage occurring during the first 50 h corresponds to the initial fast decay in device intensity. After 100 h, the voltage stabilizes at 13.3 V. The operating voltage is sensitive to variations in temperature, which may account for some of the observed fluctuation. These initial results demonstrate that [Pt(tpbp)] devices are stable at high drive currents, although further study is needed to determine the device lifetime at radiances suitable for display use, assuming the viewer is using night-vision goggles. The long device lifetimes estimated for [Pt(tpbp)]-based OLEDs are consistent with the previously reported lifetime for [Pt(oep)]-based OLEDs (>10^6 h measured at low luminance) [35], thus illustrating the high device stability of Pt–porphyrin-based OLEDs.

In conclusion, we have demonstrated highly efficient NIR electrophosphorescent devices utilizing [Pt(tpbp)] emitters with a device peak EQE of 6.3% at 0.1 mA cm^{-2}, and lifetime of greater than 1000 h to 90% efficiency at 40 mA cm^{-2}. The very high efficiencies of the NIR devices make them suitable for many night-vision display and sensing applications. Achieving even longer-wavelength emission is possible by extending the conjugation length of the ligand structure.

Acknowledgement

The authors acknowledge financial support from the U.S. Dept. of the Army, CECOM for the phase II SBIR program (contract no. W15P7T-06-C-T201), and Universal Display Corporation.

Supporting Information

Supporting information for this article, including the procedure used to prepare and purify [Pt(tpbp)], details of the crystallographic work, NSD analysis, and the methods used to prepare and test OLEDs, is available at https://application.wiley-vch.de/contents/jc_2002/2007/z604240_s.pdf or from the author.

References

1. Z. H. Kafafi, *Organic Electroluminescence*, CRC, Boca Raton, 2005.
2. B. S. Harrison, *Appl. Phys. Lett.* 2001, **79**, 3770.
3. L. H. Slooff, A. Polman, F. Cacialli, *Appl. Phys. Lett.* 2001, **78**, 2122.
4. A. O'Riordan, E. O'Connor, S. Moynihan, P. Nockemann, P. Fias, R. Van Deun, D. Cupertino, P. Mackie, G. Redmond, *Thin Solid Films* 2006, **497**, 299.
5. K. S. Schanze, J. R. Reynolds, J. M. Boncella, B. S. Harrison, T. J. Foley, M. Bouguettaya, T.-S. Kang, *Synth. Met.* 2003, **137**, 1013.
6. E. L. Williams, J. Li, G. E. Jabbour, *Appl. Phys. Lett.* 2006, **89**, 083506.
7. M. A. Baldo, D. F. O'Brien, Y. You, A. Shoustikov, S. Sibley, M. E. Thompson, S. R. Forrest, *Nature* 1998, **395**, 151.

8. S. Lamansky, P. Djurovich, D. Murphy, F. Abdel-Razzaq, H. E. Lee, C. Adachi, P. E. Burrows, S. R. Forrest, M. E. Thompson, *J. Am. Chem. Soc.* 2001, **123**, 4304.
9. E. Holder, B. M. W. Langeveld, U. S. Schubert, *Adv. Mater.* 2005, **17**, 1109.
10. R. J. Holmes, S. R. Forrest, T. Sajoto, A. Tamayo, P. I. Djurovich, M. E. Thompson, J. Brooks, Y. J. Tung, B. W. D'Andrade, M. S. Weaver, R. C. Kwong, J. J. Brown, *Appl. Phys. Lett.* 2005, 87.
11. J. Kido, Y. Okamoto, *Chem. Rev.* 2002, **102**, 2357.
12. K. Kalyanasundaram, *Photochemistry of Polypyridine and Porphyrin Complexes*, Academic Press, London, 1992.
13. T. V. Duncan, K. Susumu, L. E. Sinks, M. J. Therien, *J. Am. Chem. Soc.* 2006, **128**, 9000.
14. D. F. O'Brien, M. A. Baldo, M. E. Thompson, S. R. Forrest, *Appl. Phys. Lett.* 1999, **74**, 442.
15. M. Colle, C. Garditz, M. Braun, *J. Appl. Phys.* 2004, **96**, 6133.
16. H. D. F. Burrows, M. , J. S. de Melo, A. P. Monkman, S. Navaratnam, *J. Am. Chem. Soc.* 2003, **125**, 15310.
17. Y. Wang, *Appl. Phys. Lett.* 2004, **85**, 4848.
18. V. A. Montes, C. Perez-Bolivar, N. Agarwal, J. Shinar, P. Anzenbacher, *J. Am. Chem. Soc.* 2006, **128**, 12436.
19. M. Ikai, F. Ishikawa, N. Aratani, A. Osuka, S. Kawabata, T. Kajioka, H. Takeuchi, H. Fujikawa, Y. Taga, *Adv. Funct. Mater.* 2006, **16**, 515.
20. L. Yanqin, R. Aurora, S. Marco, M. Marco, H. Cheng, W. Yue, L. Kechang, C. Roberto, G. Giuseppe, *Appl. Phys. Lett.* 2006, **89**, 061125.
21. Q. Hou, Y. Zhang, F. Li, J. Peng, Y. Cao, *Organometallics* 2005, **24**, 4509.
22. J. Kalinowski, W. Stampor, J. Szmytkowski, M. Cocchi, D. Virgili, V. Fattori, P. D. Marco, *J. Chem. Phys.* 2005, **122**, 154710.
23. V. V. Rozhkov, M. Khajehpour, S. A. Vinogradov, *Inorg. Chem.* 2003, **42**, 4253.
24. O. S. Finikova, S. E. Aleshchenkov, R. P. Brinas, A. V. Cheprakov, P. J. Carroll, S. A. Vinogradov, *J. Org. Chem.* 2005, **70**, 4617.
25. R.-J. Cheng, Y.-R. Chen, S. L. Wang, C. Y. Cheng, *Polyhedron* 1993, **12**, 1353.
26. http://www.ccdc.cam.ac.uk/data_request/cif.
27. W. Jentzen, X.-Z. Song, J. A. Shelnutt, *J. Phys. Chem. B* 1997, **101**, 1684.
28. http://jasheln.unm.edu/jasheln/content/nsd/nsd_welcome.htm.
29. B. W. D'Andrade, S. Datta, S. R. Forrest, P. Djurovich, E. Polikarpov, M. E. Thompson, *Org. Electron.* 2005, **6**, 11.

30. J. E. Roger, K. A. Nguyen, D. C. Hufnagle, D. G. McLean, W. Su, K. M. Gossett, A. R. Burke, S. A. Vinogradov, R. Patcher, P. A. Fleitz, *J. Phys. Chem. A* 2003, **107**, 11331.
31. T. J. Aartsma, M. Gouterman, C. Jochum, A. L. Kwiram, B. V. Pepich, L. D. Williams, *J. Am. Chem. Soc.* 1982, **104**, 6278.
32. W. Humbs, E. van Veldhoven, H. Zhang, M. Glasbeek, *Chem. Phys. Lett.* 1999, **304**, 10.
33. R. C. Kwong, M. R. Nugent, L. Michalski, T. Ngo, K. Rajan, Y.-J. Tung, M. S. Weaver, T. X. Zhou, M. Hack, M. E. Thompson, S. R. Forrest, J. J. Brown, *Appl. Phys. Lett.* 2002, **81**, 162.
34. R. C. Kwong, M. S. Weaver, M.-H. M. Lu, Y.-T. Tung, A. B. Chwang, T. X. Zhou, M. Hack, J. J. Brown, *Org. Electron.* 2003, **4**, 155.
35. P. E. Burrows, S. R. Forrest, T. X. Zhou, L. Michalski, *Appl. Phys. Lett.* 2000, **76**, 2493.

Chapter 37

Intrinsic Luminance Loss in Phosphorescent Small-Molecule Organic Light Emitting Devices due to Bimolecular Annihilation Reactions

N. C. Giebink,[a,b] B. W. D'Andrade,[c] M. S. Weaver,[c]
P. B. Mackenzie,[c] J. J. Brown,[c] M. E. Thompson,[d] and
S. R. Forrest[b]

[a]*Department of Electrical Engineering, Princeton University, Princeton, New Jersey 08544, USA*
[b]*Department of Electrical Engineering and Computer Science, and Department of Physics, University of Michigan, Ann Arbor, Michigan 48109, USA*
[c]*Universal Display Corporation, 375 Phillips Blvd., Ewing, New Jersey 08618, USA*
[d]*Department of Chemistry, University of Southern California, Los Angeles, California 90089, USA*
stevefor@umich.edu

Operational degradation of blue electrophosphorescent organic light emitting devices (OLEDs) is studied by examining the luminance loss, voltage rise, and emissive layer photoluminescence quenching that occur in electrically aged devices. Using a model where defect sites act as deep charge traps, nonradiative

---

Reprinted from *J. Appl. Phys.*, **103**, 044509, 2008.

Electrophosphorescent Materials and Devices
Edited by Mark E. Thompson
Text Copyright © 2008 American Institute of Physics
Layout Copyright © 2024 Jenny Stanford Publishing Pte. Ltd.
ISBN 978-981-4877-34-3 (Hardcover), 978-1-003-08872-1 (eBook)
www.jennystanford.com

recombination centers, and luminescence quenchers, we show that the luminance loss and voltage rise dependence on time and current density are consistent with defect formation due primarily to exciton-polaron annihilation reactions. Defect densities $\sim 10^{18}$ cm^{-3} result in $> 50\%$ luminance loss. Implications for the design of electrophosphorescent OLEDs with improved lifetime are discussed.

37.1 Introduction

The limited operational stability of organic light emitting devices (OLEDs) presents a challenge to their widespread acceptance for use in large-area displays and solid-state lighting [1–3]. While improved packaging techniques and material mechanical properties and purity [4] have lead to significant progress in eliminating extrinsic sources of degradation [1, 5], the remaining *intrinsic* luminance loss and voltage rise accompanying long term device operation are not yet well understood [1, 6, 7].

Various hypotheses have been offered to explain the basis for intrinsic degradation in device efficiency, with the most widely accepted advocating chemical degradation of a fraction of the emissive constituent molecules [1]. Presumably, bond cleavage produces radical fragments, which then participate in further radical addition reactions to form even more degradation products. These products act as nonradiative recombination centers, luminescence quenchers, and deep charge traps. For example, studies have shown that both anions and cations of tris(8-hydroxyquinoline) aluminum (Alq$_3$) are unstable [8, 9], and evidence has recently been presented that the excited states themselves may form reaction centers in the case of the common host material 4,4'-bis(9-carbazolyl)-2,2'-biphenyl (CBP) [10].

Here, we study the degradation characteristics of a typical blue phosphorescent OLED using 4, 4'-bis (3-methylcarbazol-9-yl)-2,2'-biphenyl (mCBP) as the host [11]. A model is developed that describes luminescence degradation, voltage rise, and emissive layer (EML) photoluminescence (PL) quenching in electrically aged devices. The model is based on the formation of defect states that function as luminescence quenchers, nonradiative recombination centers, and charge traps. We find that defect generation due predominantly to exciton-polaron annihilation leads to model predictions most consistent with the observed degradation effects.

This chapter is organized as follows: Section 37.2 describes the fundamental mechanisms responsible for device aging, and introduces the rate equations for defect formation, luminance loss, and voltage rise that describe long term operational degradation. Experimental methods are described in Section 37.3, and the results, including model fits to the data, are presented in Section 37.4. The physics of the processes responsible for defect generation are identified and used to analyze our experimental results in Section 37.5, with conclusions presented in Section 37.6.

37.2 Theory

Depending on their energy levels, defects can act as luminescent quenchers, nonradiative recombination centers, and deep charge traps. Luminance loss results from the first two, while voltage rise, which has been linked to the presence of fixed space charge in the emissive region [7, 12], can result from filling of the deep traps.

This situation is shown schematically in Fig. 37.1 for a single, discrete, deep defect state at energy, E_t, that lies between the highest occupied molocular orbital (HOMO) and lowest unoccupied molecular orbital (LUMO) of the host. The energetics of the phosphorescent guest are also shown in Fig. 37.1 [13, 14]. Thus, both defect and guest form discrete, deep hole traps that allow for direct exciton formation [15], although recombination is only radiative on the guest. Luminescence quenching by defects will occur if there exists an allowed transition resonant with that of the guest or host that enables Förster or Dexter energy transfer to occur [16].

Defects are assumed to act only as hole traps, with E_t representing the defect HOMO. In general, however, defects may have both their HOMO and LUMO, or a singly occupied molecular orbital [10] within the host band gap, creating both electron and hole traps. In addition, the defect state itself may not lead directly to a quenching transition, but when occupied with a trapped charge, the resulting polaron might become a quenching center [16]. The model presented below is general, however, and although it has been derived for the specific case shown in Fig. 37.1, it nevertheless remains applicable to these alternative scenarios, as discussed in Section 37.5.

To simplify the model, we assume a single recombination zone that decays exponentially [17] from one edge of the EML with characteristic length d_{rec}, as in Fig. 37.1. High efficiency electrophosphorescent OLEDs have a charge balance factor near unity [18]; hence, we assume that equal numbers of electrons and holes enter the recombination zone. Excitons either form directly on guest molecules, or they are rapidly transferred from the host due to the high doping concentration and the possibility for exothermic energy transfer. Due to the high host triplet energy, guest excitons are strongly localized, leading to a negligible possibility for and diffusion out of the recombination zone.

These considerations lead to rate equations for hole (p), electron (n), and exciton (N) densities in the recombination zone as follows:

$$\frac{dp(x,t,t')}{dt} = \frac{J}{qd_{rec}\{1-\exp[-(x_2-x_1)/d_{rec}]\}} \exp\left[-\frac{x-x_1}{d_{rec}}\right]$$
$$-\gamma \eta(x,t,t') p(x,t,t')$$
$$-\sigma v_{th}[f_D(E_t)] Q(x,t') p(xtt'), \quad (37.1)$$

$$\frac{dn(x,t,t')}{dt} = \frac{J}{qd_{rec}\{1-\exp[-(x_2-x_1)/d_{rec}]\}} \exp\left[-\frac{x-x_1}{d_{rec}}\right]$$
$$-\gamma \eta(x,t,t') p(x,t,t')$$
$$-\gamma_2 [1-f_D(E_t)] Q(x,t') n(xtt'), \quad (37.2)$$

$$\frac{dN(x,t,t')}{dt} = \gamma \eta(x,t,t') p(x,t,t')$$
$$- [1/\tau + K_{DR} Q(x,t')] N(x,t,t'). \quad (37.3)$$

The electron, hole, and exciton densities depend on the time scale of transport and energy level transitions, t (on the order of microseconds), as well as on that of degradation, t' (hours), due to formation of defects of density $Q(x,t')$. The electron and hole densities are functions of the current density, J, the elementary charge, q, and the device dimensions shown in Fig. 37.1. Excitons are formed at the Langevin rate [16, 19], $\gamma = q(\mu_n+\mu_p)/(\varepsilon\varepsilon_0)$, and

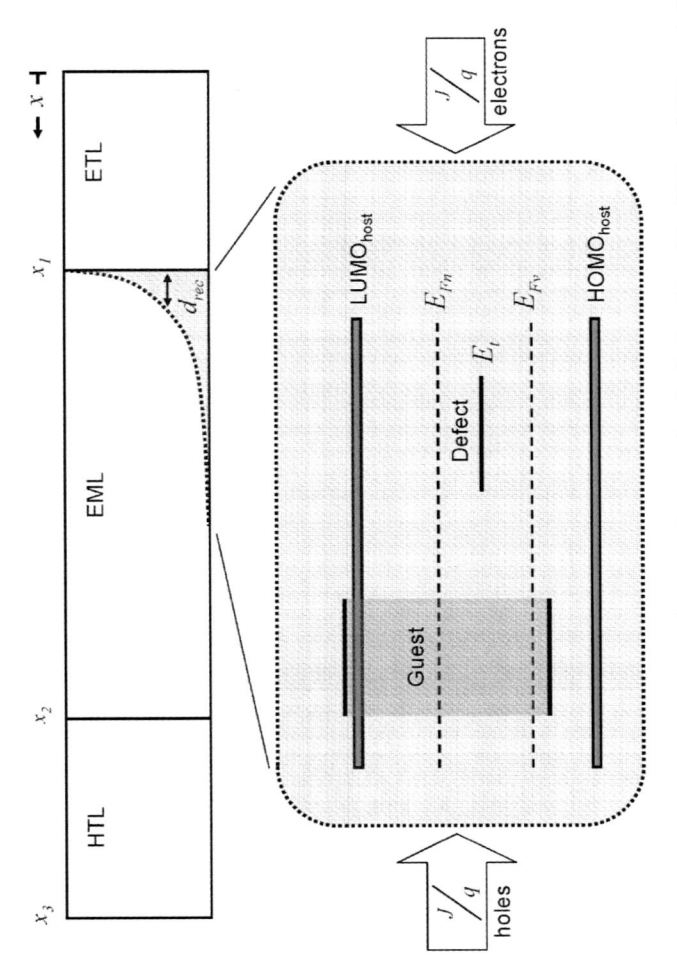

Figure 37.1 Schematic of the model geometry and assumed energy level relationships. The HTL and ETL are hole and electron transport layers respectively. The recombination zone decays exponentially from the EML/ETL interface at x_1 with characteristic length, d_{rec}. Both phosphorescent guests and defects form deep traps within the host band gap. The electron and hole quasi-Fermi levels under forward bias are E_{Fn} and E_{Fv}, respectively.

decay with natural lifetime, τ. The hole and electron mobilities in the doped EML (Refs. 13 and 20) are μ_p and μ_n, respectively, the relative dielectric constant of the EML is $\varepsilon \approx 3$, and ε_0 is the permittivity of free space.

In Eq. (37.1), holes with thermal velocity, $v_{th} \sim 10^7$ cm/s, trap at defect sites of energy, E_t, and cross section, σ. The Fermi factor, $f_D(E_t) = [\exp(E_t - E_{Fv}) + 1]^{-1}$, gives the probability that the hole trap is empty [21], where E_{fv} is the hole quasi-Fermi energy. Electrons in Eq. (37.2) nonradiatively recombine at a rate proportional to the trapped hole density, $Q(x, t')[1 - f_D(E_t)]$, and the reduced Langevin coefficient, $\gamma_2 = q(\mu_n)/(\varepsilon\varepsilon_0)$, since trapped holes are assumed to be immobile. Quenching of excitons by defects is described by the bimolecular rate coefficient, K_{DR}, in Eq. (37.3). Note that only the constant prefactors change if defects trap electrons, or both carrier types, instead of only holes, as considered above.

The defect generation mechanism has four possible routes:

$$\frac{dQ(x, t')}{dt} = \begin{cases} K_X n(x, t') \quad K_X p(x, t') & (37.4a) \\ K_X N(x, t') & (37.4b) \\ K_X N^2(x, t') & (37.4c) \\ K_X N(x, t') n(x, t'), \quad K_X N(x, t') p(x, t'), & (37.4d) \end{cases}$$

where the rate constant, K_X, is consistent in dimension with the order of the reaction. In Eq. (37.4a), the presence of an electron or hole (i.e., a polaron) leads to molecular degradation, while in Eq. (37.4b) excitons are responsible. Defect formation is a product of exciton-exciton annihilation in Eq. (37.4c), and of exciton-polaron (hole or electron) annihilation in Eq. (37.4d).

On the short time scale, t, Eqs. (37.1)–(37.3) are at steady state and can be solved to yield an expression for $N(x, t')$. The resulting, coupled differential equations containing $N(x, t')$ and $Q(x, t')$ are then solved numerically. Thus, the normalized OLED luminescence as a function of time is

$$\mathrm{EL}_{\mathrm{norm}}(t') = \frac{\int_{x_1}^{x} \frac{x}{2} N(x, t') \, dx}{\int_{x_1^{\frac{x}{2}}} N(x, 0) \, dx}, \qquad (37.5)$$

and the defect formation rate per exciton, averaged over the recombination zone is

$$F_X(t') = \frac{1}{d_{\text{rec}}} \int_{x_1}^{x_2} \frac{1}{N(x,t')} \frac{dQ(x,t')}{dt'} dx. \quad (37.6)$$

Here, the integration limits, x_1 and x_2, are defined in Fig. 37.1. The density of trapped charge increases with defect density following $\rho_T(x,t') = qQ(x,t')[1 - f_D(E_t)]$. Assuming that the growth of ρ_T is offset by an equal density of opposing charge at the cathode, and that the free charge distributions under steady-state operation are not perturbed, then the voltage rise is given by [21]

$$\Delta V(t') \approx \int_0^{x_3} x \rho_T(x,t') dx. \quad (37.7)$$

The EML PL transient will also be affected by defects. From Eq. (37.3) at time, t', the PL intensity normalized to that at $t = 0$ is

$$PL_{\text{norm}}(t)|_{t'} = \frac{\int_{x_1}^{x_2} I_0(x) \exp\{-[1/\tau + K_{\text{DR}} Q(x,t')]t\}}{\int_{x_1}^{x_2} I_0(x)}. \quad (37.8)$$

Here, $I_0(x)$ is the intensity profile of the excitation pulse in the device EML calculated by the transfer matrix method [22] for the specific device structure, incident excitation angle, wavelength, and polarization considered.

37.3 Experimental

Indium-tin-oxide (ITO) coated glass was cleaned with solvents and patterned into 2 mm^2 anode contact areas using standard photolithography techniques prior to organic film deposition. The ITO was oxygen-plasma cleaned, exposed to UV-ozone treatment, and then loaded into a vacuum chamber with a base pressure of 10^{-7} Torr. The device structure is as follows (see Fig. 37.2a, inset): a 10 nm thick hole injection layer [23], a 30 nm thick layer of the hole transporting 4,4′-bis[N-(1-naphthyl)-N-phenyl-amino]-biphenyl (NPD), a 30 nm thick EML consisting of mCBP doped with 9 wt% of the blue phosphor *fac*-tris[3-(2,6-dimethylphenyl)-7-methylimidazo[1,2-f] phenanthridine] Iridium(III), and a 5 nm thick layer of mCBP for exciton confinement within the EML. Electrons are injected into the EML through a 40 nm thick layer of Alq$_3$, capped by a cathode consisting of a 0.8 nm thick layer of LiF and a 100 nm thick

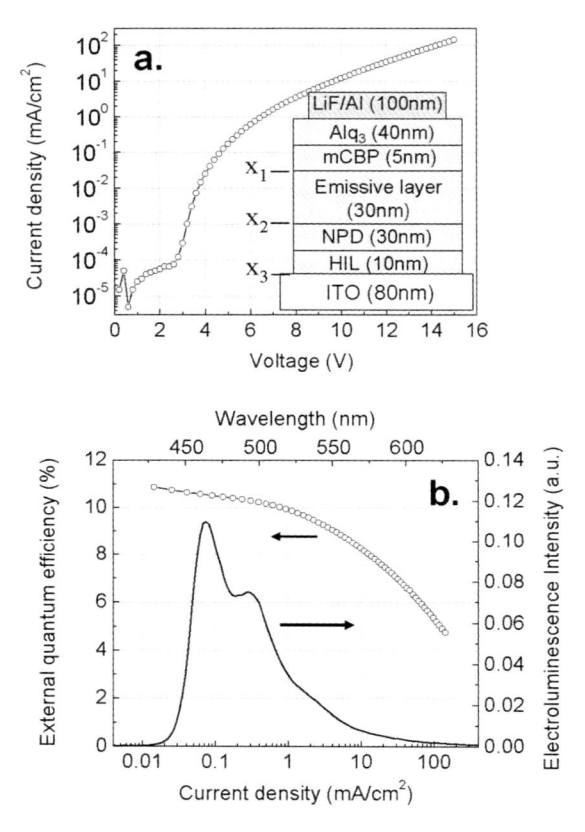

Figure 37.2 (a) Typical $J-V$ characteristics of the devices studied. Inset: schematic of the device structure, with the dimensions, $x_1 - x_3$ of Fig. 37.1, indicated as shown. (b) EQE (left scale) and emission spectrum (right scale) obtained at $J = 10$ mA /cm^2.

Al film. Following deposition, the OLEDs were transferred directly from vacuum into an oxygen and moisture-free N$_2$ glovebox, where they were encapsulated using a UV-curable epoxy, and a glass lid containing a moisture getter.

External quantum efficiency (EQE) and power efficiencies were calculated from the spectral intensity measured normal to the substrate using a SpectraScan PR705 [24]. The current versus voltage measurements were obtained using a Keithley 236 source measurement unit. Operational lifetime measurements were performed at room temperature, and devices were aged at various constant currents while monitoring their operational voltage and light output. PL transients were obtained periodically during electrical aging using a time-correlated single photon counting system from Horiba Jobin Yvon [25], with a $\lambda = 335$ nm wavelength, pulsed excitation source incident at 45° from normal. PL from the EML was obtained at a wavelength of $\lambda = 470$ nm to minimize collection of fluorescence from the transport layers.

37.4 Results

The current density–voltage (J–V) characteristics and EQE for the OLEDs taken immediately following fabrication are plotted in Figs. 37.2a and b, respectively. The device shows a peak forward viewing EQE $= (11.0 \pm 0.2)\%$. The electroluminescence emission spectrum at $J = 10$ mA/cm^2, with a peak at $\lambda = 464$ nm, is due to the dopant and remains the same at all current densities, indicating that the recombination zone remains within the EML. The recombination is highest at the EML interface adjacent to the thin mCBP blocking layer. This conclusion is supported by the lack of NPD emission, and the fact that removal of the mCBP layer results in significant Alq$_3$ emission.

Figure 37.3a provides the normalized electrophosphorescence versus time for four different drive current densities: 6.9, 15.1, 24.3, and 34.4 mA/cm^2 corresponding to initial ($t' = 0$) luminances of $L_0 = 1000, 2000, 3000$, and 4000 cd/m^2, respectively. The operational lifetime, LT$_{80}$, corresponds to the time required for the luminance to degrade to $0.8 L_0$ [6]. The rate of luminance loss increases monotonically with J; here lifetimes decrease from 110 h at 6.9 mA/cm^2, to 9 h at 34.4 mA/cm$^+$. The solid lines in Fig. 37.3a are derived from the model under the assumption that exciton localization on a dopant or host molecule leads to defect formation [Eq. (37.4b)]. The same experimental data are reproduced in Figs. 37.3b and c to compare the model predictions for exciton-exciton annihilation [Eq. (37.4c)] and exciton-polaron (electron) annihilation [Eq. (37.4d)] defect formation processes, respectively.

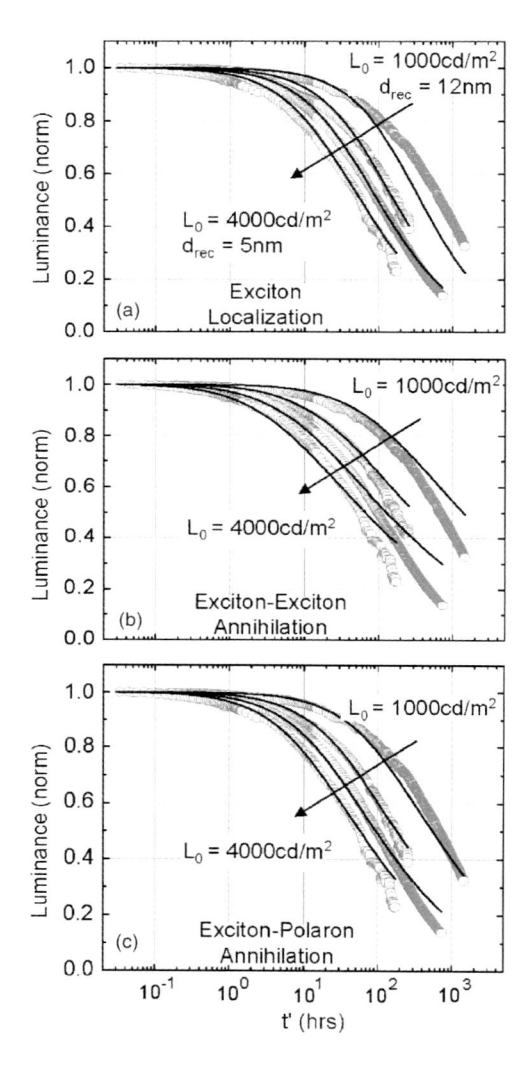

Figure 37.3 (a) Luminance degradation vs time for initial brightnesses of $L_0 = 1000, 2000, 3000,$ and 4000 cd/m^2 as indicated by the arrow. The solid black lines indicate a fit using the exciton localization degradation model discussed in the text. Note that the recombination zone width, d_{rec}, is variable in these fits. The data are reproduced for comparison with the exciton-exciton degradation model in (b) and with the exciton-polaron model in (c). All fitting parameters are given in Table 37.1.

The voltage rise corresponding to the luminance loss is plotted in each of Figs. 37.4a–c for comparison with each different degradation mechanism. The solid lines of Figs. 37.4a–c are calculated using the same exciton localization, exciton-exciton, and exciton-polaron degradation models as in Figs. 37.3a–c.

Figures 37.5a–c show PL transients obtained from an as-grown device, a device degraded to a luminance at time t' of $L(t') = 0.59L_0 (L_0 = 1000\ \text{cd/m}^2)$, and one degraded to $L(t') = 0.16L_0 (L_0 = 3000\ \text{cd/m}^2)$. The predictions from each model are again shown by solid lines in Figs. 37.5a (exciton localization), 37.5b (exciton-exciton annihilation), and 37.5c (exciton-polaron annihilation). The as-grown device shows a natural decay lifetime of $\tau = 1.10 \pm 0.08$ μs, while the degraded device transients become increasingly nonlinear, indicative of the existence of quenching. Fluorescence from NPD overlapping the $\lambda = 470$ nm detection wavelength is responsible for the sharp decrease in intensity near $t = 0$. The transients are, therefore, normalized at the onset of phosphorescence, after the fluorescence has decayed to a negligible level (i.e., at $t > 0.2$ μs).

37.5 Discussion

Configurational diagrams of the defect generation mechanisms proposed in Eqs. (37.4a)–(37.4c), are shown in Fig. 37.6. Figure 37.6a shows the exciton localization pathway, in which a direct or predissociative potential, R, crosses the exciton energy surface. In Fig. 37.6b, annihilation of two singlet (S_1) or triplet (T_1) excitons yields a ground state (S_0), and an upper excited state (S_n^* or T_n^*), which may then dissociate via a direct or predissociative reaction (route 1) along R to yield radical fragments that result in defect states. Dissociation may also occur via the hot-molecule mechanism [26] (route 2) if the upper excited state relaxes vibronically to create a hot first excited state. Similarly, Fig. 37.6c shows annihilation of an exciton (S_1 or T_1) due to collision with a polaron (D_0) to create a ground state (S_0) and an excited polaron (D_n^*), which then dissociates along routes 1 or 2, analogous to the previous case.

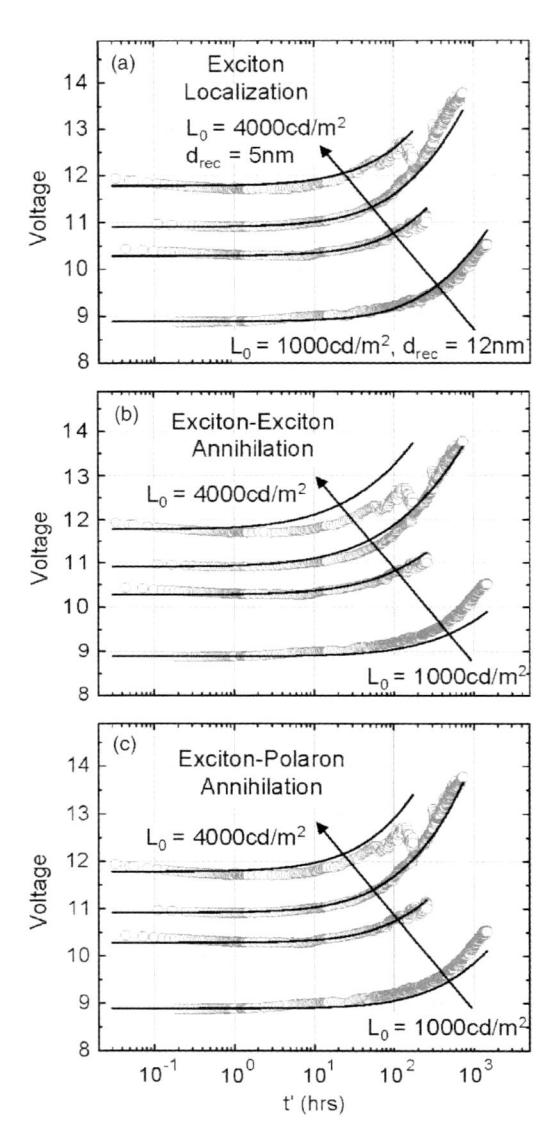

Figure 37.4 (a) Voltage rise for each of the four devices studied. The black lines are calculated using the exciton localization model. The data are reproduced in (b) and (c) for comparison with fits from the exciton-exciton and exciton-polaron models, respectively. All fitting parameters are given in Table 37.1.

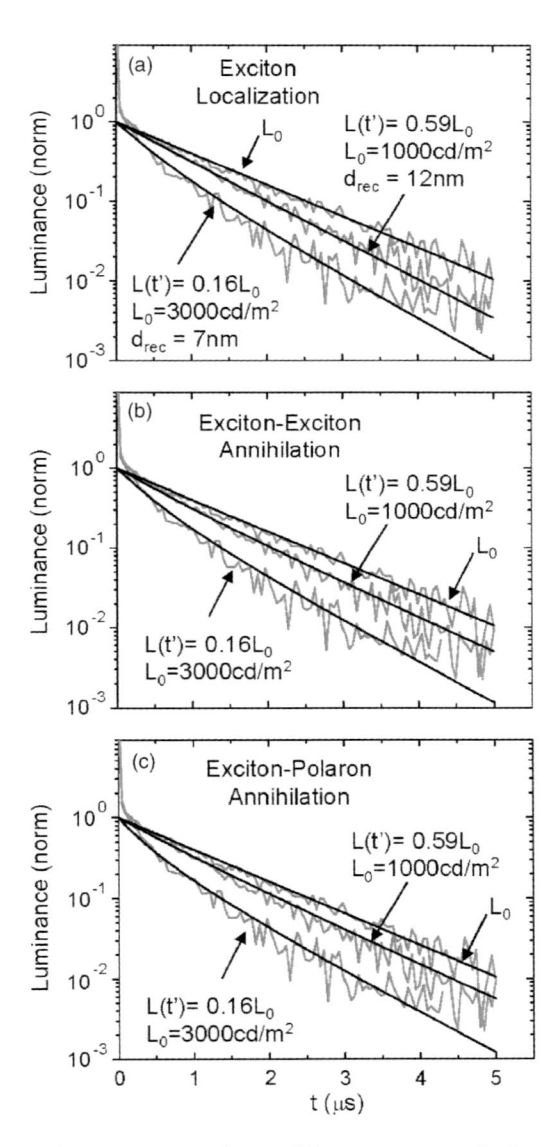

Figure 37.5 (a) PL transients obtained for an as-grown device, one aged to $L(t') = 0.59L_0$ ($L_0 = 1000$ cd/m^2) and another aged to $L(t') = 0.16L_0$ of its initial $L_0 = 3000$ cd /m^2 brightness. The solid black lines are fits from the exciton localization model. Predictions of the exciton-exciton annihilation model are shown in (b) and those for the exciton-polaron annihilation model in (c). Fitting parameters are given in Table 37.1.

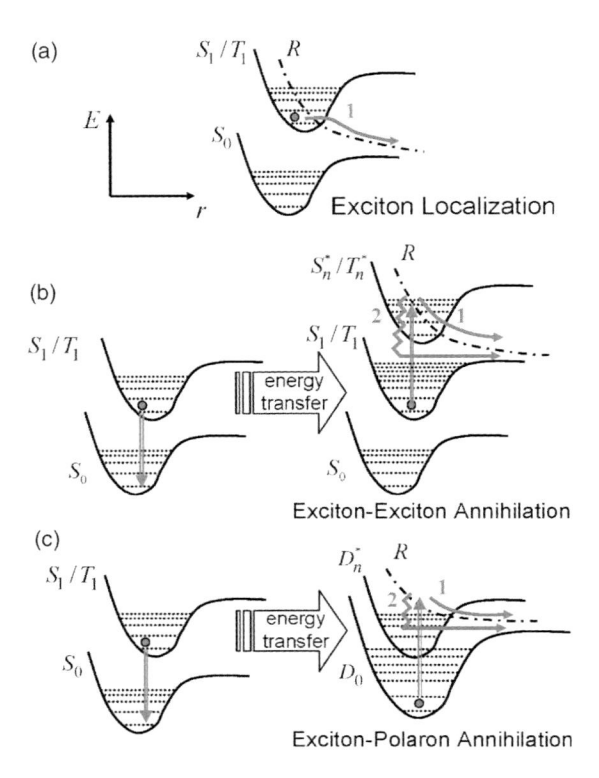

Figure 37.6 Configurational diagram showing the different dissociation mechanisms in terms of energy (E) and representative coordinate (r). In (a), a direct or predissociative potential, R, crosses the singlet or triplet first excited state energy surface. (b) shows the exciton-exciton annihilation process, which leads to a ground (S_0) and upper excited state (S_n^* or T_n^*) according to the reaction $S_1(T_1) + S_1(T_1) \rightarrow S_0 + S_n^*(T_n^*)$. Direct or predissociation may occur from the upper excited state (gray arrow, route 1), or it may relax vibronically and undergo hot-molecule dissociation (gray arrow, route 2) as discussed in the text. (c) shows the exciton-polaron mechanism, in which energy transfer from the exciton results in an excited polaron that dissociates along the analogous direct/predissociative and hot-molecule routes. The dotted lines indicate vibrational energy levels within each anharmonic electronic manifold.

To determine which of these processes are most active, we fit the data in Figs. 37.3–37.5 to the theory in Section 37.2. For each degradation model, a single set of parameters has been used to fit the luminance, voltage, and PL transient data. Both the calculated luminance degradation slope following the "knee" (i.e., onset of downward slope) of each curve (see Fig. 37.3) and the scaling of the knee position in time, t', with current density, depend primarily on the degradation mode assumed, while depending only weakly on the choice of parameter values, which determine the absolute position of the knee. Each set of fitting parameters is provided in Table 37.1. The exciton-polaron model provides the best fit to the data, as discussed below. Electron and hole mobilities representative of those found for a similar CBP host-guest combination [20] have been kept constant in all fits.

The fits to the data assuming exciton localization are shown in Fig. 37.3a. To achieve the fits shown, d_{rec} must decrease with increasing J for the position of the knee to scale correctly with current density. Despite allowing d_{rec} to vary, the fit is not accurate at the lowest initial luminance, and hence current density. It has been predicted [27] and shown experimentally for phosphorescent OLEDs (Ref. 28) that the recombination zone width is independent of J under mid to high levels of injection ($J \geq 1$ mA/cm^2), which conflicts with the wide range in d_{rec} required in Fig. 37.3a to fit the data. Note that a model in which the presence of electron or hole polarons localized on the molecules leads to aging [Eq. (37.4a)] was also examined, but poor fits resulted for all parameter combinations, and thus are not included here. The slope of the data predicted by the exciton-exciton annihilation model is too small and departs from the data in the knee region (Fig. 37.3b) at all current densities.

In contrast, the exciton-polaron annihilation model provides the best fit across the entire range of data. In Fig. 37.3c, the model deviates slightly in advanced stages of degradation [$L(t') < 0.4L_0$], where the luminance is lower than predicted. This may be due to a change in charge balance with aging, as indicated by the voltage rise observed in Fig. 37.4. This results in higher polaron densities, and thus to an increased rate of degradation, providing positive feedback not considered in the model.

Clearly, each degradation mechanism is marked by a distinct functional dependence. This is evident from approximate solution of Eqs. (37.1)–(37.3) in the limit where $Q(x, t')$ is large ($\geq 10^{17}$ cm^{-3}). Use of Eqs. (37.4a)–(37.4d) yields a polynomial of different orders in $Q(x, t')$ for each degradation mechanism: quadratic for single polaron localization [Eq. (37.4a)], fourth order for exciton localization [Eq. (37.4b)], seventh order for exciton-exciton annihilation [Eq. (37.4c)], and fifth order for exciton-polaron annihilation [Eq. (37.4d)]. The distinguishing feature of a particular degradation mode is thus the order of the polynomial used to fit the data.

The parameters in Table 37.1 are consistent with expectations for this guest-host materials combination. For example, the values suggest that defect hole traps are nearly full, lying at ~0.1 eV above the hole quasi-Fermi level. Characteristic recombination lengths are consistent with the literature reported values of between 8 and 12 nm [28, 29]. Also, the defect exciton quenching rate, $K_{DR} \sim 4 \times 10^{-12}$ cm^3 s^{-1}, is similar to that reported for other bimolecular quenching reactions in OLEDs [28]. Low capture cross sections of $\sigma \sim 10^{-17}$ cm^{-3} result from localization of large effective mass holes that are characteristic of organic molecules [16].

The relative contributions to luminance loss from defect exciton quenching and nonradiative recombination are estimated to be about 70% and 30%, respectively. This estimation is inferred from the respective decay rates. The rate of exciton decay through quenching is $-K_{DR}QN$ [see Eq. (37.3)], whereas the loss of excitons to trap formation is, from Eq. (37.2), equal to $-\gamma_2[1 - f_D]Qn$. The 70/30 split is then the ratio of these two rates. The existence of quenching is confirmed by the PL data of Fig. 37.5, and nonradiative recombination is inferred from the presence of charged defects that lead to the observed voltage rise.

The fits to the PL data in Fig. 37.5 provide further constraints to the parameter values used to model the voltage and luminance data. For example, both the PL and the luminance degradation data are strong functions of the recombination zone width, d_{rec}.

The average defect density, $Q_{av}(t') = 1/d_{rec} \int Q(x, t')dx$, calculated using the exciton-polaron model, is shown in Fig. 37.7a. The increase in defect density is linear for $t < 10$ h, and rolls off at longer times. From Figs. 37.3c and 37.7a, we infer that a defect density of

Table 37.1 Model parameter values

	Parameter	Exciton	Ex-Ex	Ex-Pol
Variable	d_{rec} (nm)	12, 9, 7, 5, ±2	10±3	8±2
	K_{DR} (cm^3 s^{-1})	$(4\pm3)\times10^{-12}$	$(4\pm3)\times10^{-12}$	$(5\pm3)\times10^{-12}$
	K_x (s^{-1} or cm^3 s^{-1})	$(6\pm3)\times10^{-6}$	$(1.7\pm0.9)\times10^{-22}$	$(7\pm2)\times10^{-24}$
	σ (cm^2)	2×10^{-17}	3×10^{-17}	10^{-17}
	$E_t - E_{Fv}$ (eV)	0.21 ± 0.05	0.15 ± 0.04	0.17 ± 0.03
Fixed, Common	J (mA/cm^2)	6.85 ($L_0=1000$ cd/m^2), 15.12 ($L_0=2000$ cd/m^2), 24.26 ($L_0=3000$ cd/m^2), 34.36 ($L_0=4000$ cd/m^2)		
	τ (μs)	1.10 ± 0.08		
	μ_p (cm^2 V^{-1} s^{-1})	2×10^{-7}		
	μ_n (cm^2 V^{-1} s^{-1})	8×10^{-8}		
	T (K)	295		
	x_1 (nm)	40		
	x_2 (nm)	70		
	x_3 (nm)	115		

10^{18} cm^{-3}, or approximately 0.1% of the molecular density leads to > 50% loss in luminescence. The rates of defect formation, $F_X(t')$, corresponding to the densities in Fig. 37.7a, are plotted in Fig. 37.7b. At 1000 cd/m², $F_X = 0.04$, or approximately one defect is formed per hour per 25 excitons created in the device. This corresponds to an absolute defect formation rate of 6×10^{11} cm^{-3} s^{-1}.

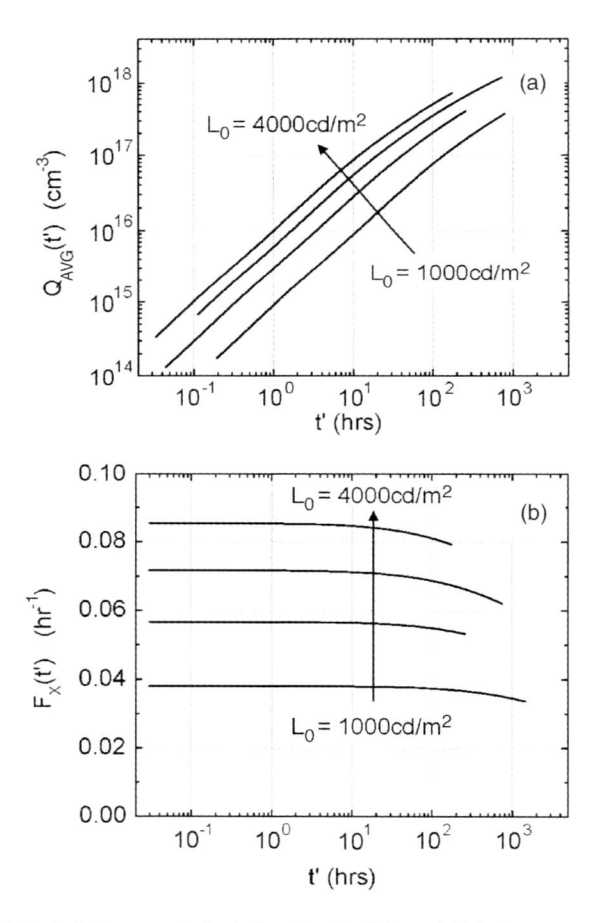

Figure 37.7 (a) Average defect density, $Q_{av}(t')$ and (b) the average defect formation rate, $F_X(t')$, per exciton, per hour, as defined in the text. The curves are calculated using the exciton-polaron model, at initial luminances of $L_0 = 1000, 2000, 3000,$ and 4000 cd/m² as indicated by the arrow.

From this result, we can estimate the probability that an exciton-polaron encounter leads to formation of a defect. Assuming an encounter radius, r, for the exciton, then in steady state, the number of encounters per unit volume is $Nn\frac{4}{3}\pi r_3$, each of which has probability, P, of resulting in a defect within the exciton lifetime, τ. Since $\Delta Q = K_x N n \tau = PNn\frac{4}{3}\pi r^3$ [see Eq. (37.4d)] defects are produced per volume during τ, the probability that an encounter leads to formation of a defect is

$$P = \frac{3K_x \tau}{4\pi r^3}, \qquad (37.9)$$

which is proportional to K_x, τ, and $1/r^3$. Using $K_x = 7 \times 10^{-24}$ cm^{-3} s^{-1} and $\tau = 1.1$ μs, and assuming nearest molecular neighbor interactions (corresponding to $r \sim 1$ nm) in Eq. (39.9), then $P \approx 2 \times 10^{-9}$, or roughly 1 in 5×10^8 encounters result in molecular dissociation.

When an annihilation event occurs, numerous pathways exist for the hot polaron to relax in addition to the dissociative routes in Fig. 37.6c. There is a near continuum of states that links the hot polaron to the ground state, allowing for extremely fast rates of internal conversion (approximately picoseconds) that compete with dissociation. Thus, in the majority of exciton-polaron annihilations, hot polarons undergo internal conversion to the ground state, bypassing the dissociative reaction.

Our conclusions conflict with that of Kondakov et al. [10] who studied electrophosphorescent OLED degradation using a CBP host. They proposed that the CBP singlet exciton is unstable, and leads to degradation products that were chemically identified in their devices. Although CBP singlets are unstasble, their density is very low due to rapid energy transfer to, as well as direct exciton formation on, the phosphorescent guests. Hence, CBP exciton instability is probably not the dominant degradation mechanism, and the exciton-polaron degradation mode identified here may have been active in that study as well.

The effects of exciton-polaron annihilation on device lifetime can be reduced by increasing d_{rec} and decreasing K_X, as shown in Figs. 37.8a and b, respectively. These results are calculated for a device with $L_0 = 1000$ cd/m^2, maintaining all other parameters at the values in Table 37.1. The device lifetime more than doubles when recombination is uniform across the EML ($d_{\text{rec}} \to \infty$), as compared to $d_{\text{rec}} = 8$ nm found for the devices studied here. Further, for $d_{\text{rec}} > 30$ nm, there is little improvement in LT_{80}. Since the voltage of an OLED is strongly dependent on layer thickness, 30 nm is near the upper limit for EML thickness in conventional, power efficient OLEDs using currently employed guest-host combinations. Also, in Fig. 37.8b, a sixfold increase of LT_{80} is calculated as K_X is reduced from 7×10^{-24} to 1×10^{-24} cm^{-3} s^{-1}.

Guest triplet excitons and host polarons are likely to be the dominant participants in the exciton-polaron defect formation reactions. The guest exciton density is much higher than that on the host, since lifetimes on the host are short (approximately nanoseconds) as compared to guest triplets with lifetimes of ∼1 µs. Since both Förster and exchange exciton-polaron annihilation mechanisms are strongly distance dependent, the physical separation of guest molecules discourages guest-guest annihilations. We therefore infer that energy exchanged by annihilation of the guest triplet exciton to the host polaron results in a dissociative process of the host molecule itself. The fragments are in close proximity to the guest molecule, and thus quench any subsequent triplets on that molecule, rendering it a permanent nonradiative center. As noted above, the fragmented molecule also acts as a deep trapping center, resulting in the observed operating voltage rise.

It is apparent that the efficiency of hot-molecule dissociation should increase with the amount of energy transferred to the polaron. Since this energy is provided by the guest, this suggests that the degradation rate (K_x) is a function of the guest exciton energy. In this case, red phosphorescent OLEDs would show the longest operational lifetimes, followed by green and then by blue devices [2, 30]. This has been observed in all OLED reliability studies reported to date in both polymer and small molecular weight systems, as well as for electrofluorescent [31] and electrophosphorescent guest-host materials combinations. For example, the longest reliability observed in the red, green, and blue Ir-based small-molecule electrophosphorescent OLEDs ranges approximately from 10^6 h for red dopants to 2×10^4 h for blue emission [2, 32]. Although there are significant differences between the device structures and test

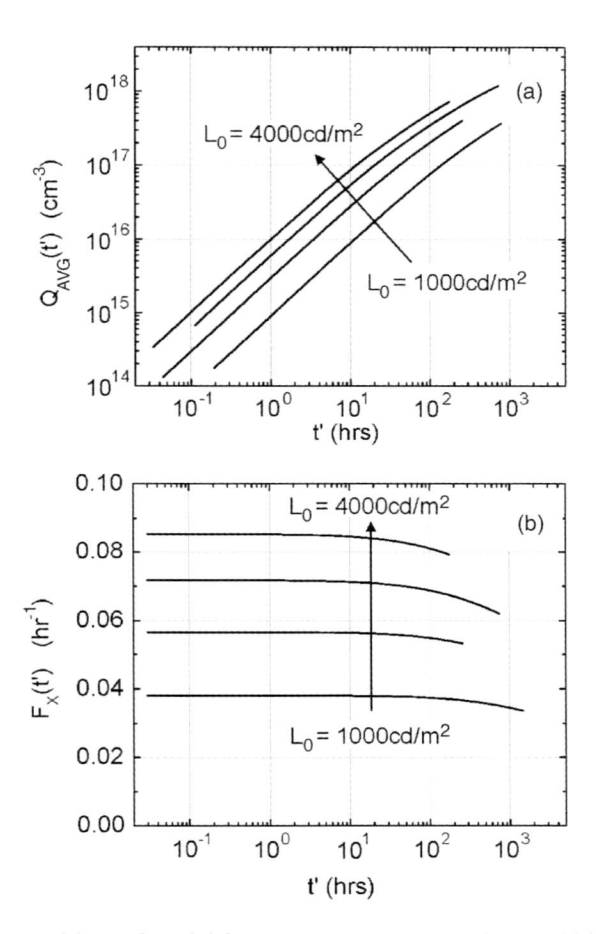

Figure 37.8 (a) Predicted lifetime improvement at $L_0 = 1000$ cd/m², obtained by increasing the recombination zone width, d_{rec}. (b) Increase in lifetime calculated for a reduction in degradation coefficient, K_X. Both (a) and (b) assume the exciton-polaron model; the filled circles indicate where the devices of this study lie on the curves.

conditions used in each of these studies, the lifetime scaling is suggestive of our energy-based model for device degradation.

Clearly, strategies for minimizing exciton-polaron annihilation must involve lowering the densities of excitons and polarons. Figure 37.8a shows the results of lowering both exciton and polaron densities by expanding the recombination zone, d_{rec}. For example, control of electron and hole mobilities in the EML as well as the strategic placement of energy barriers in the device can lead to a more uniform and distributed recombination zone [33].

Engineering host and guest molecules to lower the annihilation probability would also lead to increased operating lifetime (as in Fig. 37.8b), as well as improved efficiency at high brightness [28]. Accordingly, K_x is inversely proportional to guest exciton lifetime, τ. Hence, increasing the radiative rate would decrease K_x and result in longer-lived devices (see Fig. 37.8b). Additionally, due to the distance dependence of annihilation processes [16], increasing the intermolecular separation through addition of steric bulk to guest and host molecules might lead to decreased K_x and increased device lifetime, although this may also reduce the device efficiency by impeding exciton or charge transfer within the EML.

37.6 Conclusion

The fundamental mechanisms leading to degradation during long term operation of a typical, blue electrophosphorescent OLED have been studied. The trends in electrophosphorescence decay, voltage rise, and EML PL quenching associated with electrical aging are best fitted to a model based on the assumption that defect sites generated during operation act as exciton quenchers, deep charge traps, and nonradiative recombination centers. We find that defect generation due to exciton-polaron annihilation interactions between the dopant and host molecules leads to model predictions in good agreement with the data, although small contributions from the exciton localization and annihilation reactions cannot be discounted. A link between guest exciton energy and the annihilation induced defect formation rate was suggested, with increasing guest emission energy leading to increased defect formation rates. In addition, the defect formation rate was shown to be inversely proportional to the guest exciton lifetime. Hence, important conclusions of our model are that blue OLED operational lifetimes are less than those for green and red due to their higher energy excitations, and that increasing the exciton natural decay rate may in some cases lead to longer-lived devices. Finally, defect densities $\sim 10^{18}$ cm^{-3} are shown to result in $> 50\%$ degradation from initial luminance.

Acknowledgments

The authors thank H. Yamamoto for helpful assistance in measurements.

References

1. H. Aziz and Z. D. Popovic, *Chem. Mater.* **16**, 4522 (2004).
2. M. S. Weaver, R. C. Kwong, V. A. Adamovich, M. Hack, and J. J. Brown, *J. Soc. Inf. Disp.* **14**, 449 (2006).
3. B. D'Andrade, *Nat. Photonics* **1**, 33 (2007).
4. B. W. D'Andrade, J. Esler, and J. J. Brown, *Synth. Met.* **156**, 405 (2006).
5. P. E. Burrows, V. Bulovic, S. R. Forrest, L. S. Sapochak, D. M. McCarty, and M. E. Thompson, *Appl. Phys. Lett.* **65**, 2922 (1994).
6. R. C. Kwong, M. R. Nugent, L. Michalski, T. Ngo, K. Rajan, Y. J. Tung, M. S. Weaver, T. X. Zhou, M. Hack, M. E. Thompson, S. R. Forrest, and J. J. Brown, *Appl. Phys. Lett.* **81**, 162 (2002).
7. D. Y. Kondakov, J. R. Sandifer, C. W. Tang, and R. H. Young, *J. Appl. Phys.* **93**, 1108 (2003).
8. H. Aziz, Z. D. Popovic, N. X. Hu, A. M. Hor, and G. Xu, *Science* **283**, 1900 (1999).
9. Y. C. Luo, H. Aziz, G. Xu, and Z. D. Popovic, *Chem. Mater.* **19**, 2079 (2007).
10. D. Y. Kondakov, W. C. Lenhart, and W. F. Nichols, *J. Appl. Phys.* **101**, 024512 (2007).
11. Y. Kawamura, H. Yamamoto, K. Goushi, H. Sasabe, C. Adachi, and H. Yoshizaki, *Appl. Phys. Lett.* **84**, 2724 (2004).
12. D. Y. Kondakov, *J. Appl. Phys.* **97**, 024503 (2005).
13. N. Matsusue, S. Ikame, Y. Suzuki, and H. Naito, *Appl. Phys. Lett.* **85**, 4046 (2004).
14. A. J. Makinen, I. G. Hill, and Z. H. Kafafi, *J. Appl. Phys.* **92**, 1598 (2002).
15. C. F. Qiu, H. Y. Chen, M. Wong, and H. S. Kwok, *IEEE Trans. Electron Devices* **49**, 1540 (2002).

16. M. Pope and C. Swenberg, *Electronic Processes in Organic Crystals and Polymers* (Oxford University Press, New York, NY, 1999).
17. E. Tutis, D. Berner, and L. Zuppiroli, *J. Appl. Phys.* **93**, 4594 (2003).
18. C. Adachi, M. A. Baldo, M. E. Thompson, and S. R. Forrest, *J. Appl. Phys.* **90**, 5048 (2001).
19. J. Kalinowski, N. Camaioni, P. Di Marco, V. Fattori, and A. Martelli, *Appl. Phys. Lett.* **72**, 513 (1998).
20. S. B. Lee, T. Yasuda, M. J. Yang, K. Fujita, and T. Tsutsui, *Mol. Cryst. Liq. Cryst.* **405**, 67 (2003).
21. S. M. Sze, *Physics of Semiconductor Devices*, 2nd ed. (Wiley, New York, 2005).
22. L. A. A. Pettersson, L. S. Roman, and O. Inganas, *J. Appl. Phys.* **86**, 487 (1999).
23. Available from Universal Display Corp., Ewing, NJ 08618.
24. www.photoresearch.com
25. Horiba Jobin Yvon, Edison, NJ, USA.
26. N. Nakashima and K. Yoshihara, *J. Phys. Chem.* **93**, 7763 (1989).
27. D. Berner, H. Houili, W. Leo, and L. Zuppiroli, *Phys. Status Solidi A* **202**, 1182 (2005).
28. S. Reineke, K. Walzer, and K. Leo, *Phys. Rev. B* **75**, 125328 (2007).
29. B. W. D'Andrade and S. R. Forrest, *J. Appl. Phys.* **94**, 3101 (2003).
30. P. E. Burrows, S. R. Forrest, T. X. Zhou, and L. Michalski, *Appl. Phys. Lett.* **76**, 2493 (2000).
31. K. Fukuoka, in FPD International Forum, Session C5, 2007 (unpublished), pp. 1–1 to 1–20.
32. V. Adamovich, B. Ma, B. D'Andrade, R. C. Kwong, M. S. Weaver, and J. J. Brown, in IMID, Daegu, South Korea, 2007 (unpublished).
33. Y. R. Sun, N. C. Giebink, H. Kanno, B. W. Ma, M. E. Thompson, and S. R. Forrest, *Nature (London)* **440**, 908 (2006).

Chapter 38

Blue Light Emitting Ir(III) Compounds for OLEDs: New Insights into Ancillary Ligand Effects on the Emitting Triplet State

Andreas F. Rausch,[a] Mark E. Thompson,[b] and Hartmut Yersin[a]

[a]*Universität Regensburg, Institut für Physikalische und Theoretische Chemie, 93053 Regensburg, Germany*
[b]*University of Southern California, Department of Chemistry, Los Angeles, California 90089, USA*
hartmut.yersin@chemie.uni-regensburg.de

The sky-blue emitting phosphorescent compound Ir(4,6-dFppy)$_2$(acac) (FIracac) doped into different matrices is studied under ambient conditions and at cryogenic temperatures on the basis of broadband and high-resolution emission spectra. The emitting triplet state is found to be largely of metal-to-ligand charge transfer (MLCT) character. It is observed that different polycrystalline and amorphous hosts distinctly affect the properties of the triplet. Moreover, a comparison of FIracac with the related Ir(4,6-dFppy)$_2$(pic) (FIrpic), differing only by the ancillary ligand, reveals obvious changes of properties of the emitting state. These observations are

---

Reprinted from *J. Phys. Chem. A*, **113**(20), 5927–5932, 2009.

Electrophosphorescent Materials and Devices
Edited by Mark E. Thompson
Text Copyright © 2009 American Chemical Society
Layout Copyright © 2024 Jenny Stanford Publishing Pte. Ltd.
ISBN 978-981-4877-34-3 (Hardcover), 978-1-003-08872-1 (eBook)
www.jennystanford.com

explained by different effects of acac and pic on the Ir(III) d-orbitals. In particular, the occupied frontier orbitals, strongly involving the t_{2g}-manifold, and their splitting patterns are modified differently. This influences spin–orbit coupling (SOC) of the emitting triplet state to higher-lying ^{1,3}MLCT states. As a consequence, zero-field splittings, radiative decay rates, and phosphorescence quantum yields are changed. The important effects of SOC are discussed qualitatively and are related to the emission properties of the individual triplet substates, as determined from highly resolved spectra. The results allow us to gain a better understanding of the impact of SOC on the emission properties with the aim to develop more efficient triplet emitters for OLEDs.

38.1 Introduction

During the last several years, Ir(III) compounds have attracted much interest as highly efficient emissive dopants in organic light emitting diodes (OLEDs) [1–4]. The strong spin–orbit coupling (SOC) induced by the heavy metal center allows both singlet and triplet excitons to be used in an electroluminescence process, which leads to up to four times higher internal electroluminescence quantum yield than achievable with fluorescent emitters (triplet harvesting effect) [5, 6]. The nature of the lowest excited state and thus the emission properties can be tuned by a careful choice of the cyclometallating chromophoric ligand [2, 7–10]. Another possibility of influencing these properties is achievable by a change of the nonchromophoric ligand, the so-called ancillary or spectator ligand [11–15]. Because of its high triplet energies, the spectator ligand is not directly involved in the emission process, but it can shift the metal d-orbitals and alter their splitting patterns. Thus, the metal orbital participation in the excited state and the efficiency of spin–orbit coupling (SOC) can be influenced [11, 12]. Until now, no detailed investigations of ancillary ligand effects on the electronic properties of Ir(III) compounds based on optical high-resolution spectroscopy have been carried out.

In this study, we spectroscopically investigate the lowest excited state T_1 of the sky-blue emitting compound iridium(III)bis

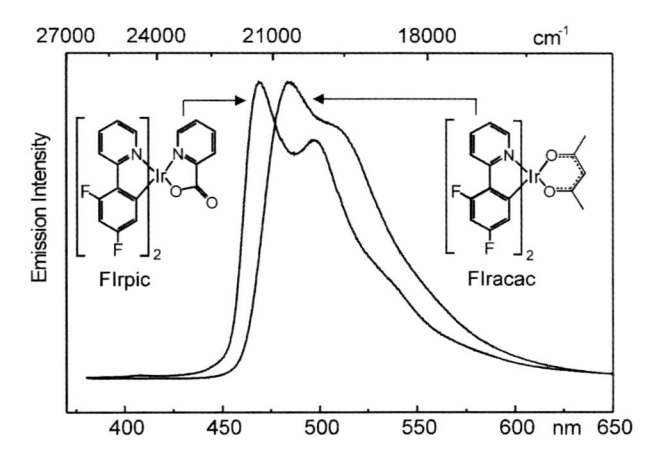

Figure 38.1 Emission of FIracac and FIrpic in CH_2Cl_2 ($\lambda_{exc} = 300$ nm) at $T = 300$ K. The insets show the structures of the compounds.

[2-(4′,6′-difluorophenyl)pyridinato-N,$C^{2'}$]-acetylacetonate (Ir(4,6-dFppy)$_2$(acac), FIracac) [16] (inset of Fig. 38.1) in different host materials, focusing on electronic properties of the lowest triplet state, such as zero-field splittings and individual decay times of the T_1 substates. The results reveal that the emitting triplet state, being largely of MLCT character, exhibits a significant dependence on the host environment. The observed behavior is comparable to the situation found recently for Ir(4,6-dFppy)$_2$(pic) (FIrpic) [17]. This latter compound, which differs only in the ancillary ligand from FIracac, represents a famous OLED emitter material [4, 16, 18, 19] and has already been spectroscopically studied in detail [17, 20]. High-resolution optical spectra at cryogenic temperatures and ambient temperature investigations allow a detailed comparison of the two emitters and reveal an insight into the influences of the ancillary ligands acac and pic on the properties of the emitting state. It is shown that the observed dissimilarities can be rationalized by different d-orbital involvements in the lowest triplet states of the respective compounds. By the exchange of acac to pic, the efficiency of spin–orbit coupling (SOC) of the T_1 substates to higher-lying singlets and triplets of MLCT character is altered and, thus, the emission properties are also changed. Further, we address shortly the differences in the nonradiative decay behavior of the two compounds.

38.2 Experimental Section

38.2.1 Synthesis

FIracac and FIrpic were synthesized according to the procedures described in Refs. 12 and 21, respectively.

38.2.2 Spectroscopy

Spectroscopic measurements were carried out with FIracac and FIrpic dissolved in CH_2Cl_2 and THF (tetrahydrofuran), respectively, at concentrations of about 10^{-5} mol/L. Doped PMMA (polymethylmethacrylate) films were prepared by dissolving the compounds (∼1 wt%) and PMMA in CH_2Cl_2. Subsequently, the solutions were spin-coated on a quartz plate. Emission spectra at 300 K were measured with a steady-state fluorescence spectrometer (Jobin Yvon Fluorolog 3). Luminescence quantum yields were determined with an integrating sphere (Labsphere, 4P-GPS-033-SL), which exhibits a highly reflective Spectralon inside coating. The estimated relative error is about ±0.10. Fluid solutions were degassed by at least three pump–freeze–thaw cycles with a final vapor pressure at 77 K of ∼10^{-5} mbar, while the PMMA films were measured under a nitrogen atmosphere. Experiments at low temperature were carried out in a He cryostat (Cryovac Konti Cryostat IT) in which He gas flow, He pressure, and heating were controlled. A pulsed Nd:YAG laser (IB Laser Inc., DiNY pQ 02) with a pulse width of about 7 ns was applied as excitation source for emission decay time measurements, using the third harmonic at 355 nm (28170 cm^{-1}). For recording site-selective emission and excitation spectra, a pulsed dye laser (Lambdaphysik Scanmate 2C) was operated, using Coumarin 102. The spectra were measured with an intensified CCD camera (Princeton PIMAX) or with a cooled photomultiplier (RCA C7164R) attached to a triple spectrograph (S&I Trivista TR 555).

Table 38.1 Emission properties of FIracac and FIrpic at ambient temperature in different solvents/matrices

	Ir(4,6-dFppy)$_2$(acac) (FIracac)			Ir(4,6-dFppy)$_2$(pic) (FIrpic)		
	CH$_2$Cl$_2$	THF	PMMA	CH$_2$Cl$_2$	THF	PMMA
λ_{max} [nm]	484	486	479	470	471	468
τ_{em} [μs]	1.0	1.2	1.2	1.9	1.8	1.7
ϕ_{PL}	0.64	0.67	0.74	0.83	0.84	0.89
k_r^a [s^{-1}]	6.4×10^5	5.6×10^5	6.2×10^5	4.4×10^5	4.6×10^5	5.2×0^5
k_{nr}^a [s^{-1}]	3.6×10^5	2.8×10^5	2.2×10^5	0.9×10^5	0.9×10^5	0.6×10^5

[a] Radiative and nonradiative rate constants are calculated from the quantum yields and emission decay times according to Eq. 38.1.

Decay times were registered using a FAST Comtec multichannel scaler PCI card with a time resolution of 250 ps.

38.3 Results and Discussion

38.3.1 Spectroscopic Introduction

Figure 38.1 shows the emission spectra of FIracac and FIrpic dissolved in CH$_2$Cl$_2$, measured at ambient conditions. The emission of FIracac has its maximum at 484 nm (20 660 cm^{-1}). In deaerated solution, the decay time constant is 1.0 μs and the phosphorescence quantum yield amounts to 64%. These values do not change remarkably when the compound is investigated in THF (Table 38.1). However, when doped into a rigid PMMA host, a blue shift of the emission is observed. This is a well-known phenomenon for compounds that exhibit a significant charge transfer with the corresponding electronic transition [22]. Further, the emission quantum yield is higher in PMMA than in fluid solution. The emission of FIrpic is slightly more structured and lies at higher energy, in CH$_2$Cl$_2$ its maximum is found at 470 nm (21 280 cm^{-1}). Further, the emission decay time (1.9 μs) is longer and the photoluminescence quantum yield (83%) is higher than that of FIracac.

The ambient temperature emission spectra of both compounds are significantly less resolved than, for example, the spectrum of

Ir(btp)$_2$(acac) (btp$^-$ = (2-benzothienyl)-pyridinate) [2, 23]. This indicates a considerably larger MLCT character in the emitting state of FIracac and FIrpic because a distinct involvement of the metal orbitals in the corresponding transitions is usually connected with pronounced metal–ligand vibrational satellites with energies below 600 cm^{-1} relative to the electronic origin. Usually, this leads to a smearing out of the emission spectrum at room temperature [24, 25].

Photophysical data of FIracac and FIrpic in different hosts measured at ambient temperature are summarized in Table 38.1. The emission quantum yields and decay times found for FIrpic closely match the values reported for other solvents and host materials [26]. A detailed comparison of the two compounds will be carried out in Section 38.3.3. The radiative and the nonradiative rate constants are calculated according to the equation

$$\phi_{PL} = k_r \tau_{em} = \frac{k_r}{k_r + k_{nr}} \quad (38.1)$$

wherein ϕ_{PL} is the photoluminescence quantum yield and τ_{em} is the emission decay time. k_r and k_{nr} represent the radiative and the nonradiative decay rates, respectively.

38.3.2 Triplet Substates and Energy Level Diagram of FIracac in CH$_2$Cl$_2$ Based on High-Resolution Spectroscopy

The ambient temperature spectra are very broad due to homogeneous and inhomogeneous broadening effects especially in fluid solution. Thus, only a crude characterization of the emitting triplet state is possible. However, detailed information can be obtained from highly resolved emission and excitation spectra at cryogenic temperatures. Therefore, FIracac is investigated in polycrystalline CH$_2$Cl$_2$ down to 1.7 K. In particular, this host has proven to be a suitable matrix for high-resolution spectroscopy of Ir(III) complexes [17, 23]. Comparable to the situation found for FIrpic [17] and Ir(btp)$_2$(acac) [23], several discrete sites (emitter molecules in specific host environments) are found. The spectra of the sites are characterized by narrow emission lines. When excited with UV light,

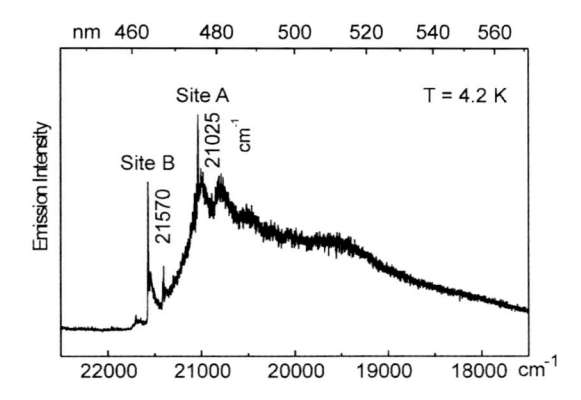

Figure 38.2 Emission spectrum of FIracac in CH_2Cl_2 at $T = 4.2$ K after excitation at 355 nm. Several discrete sites are observed together with a relatively intense inhomogeneous background. The two most intense sites are labeled as Site A and Site B.

also a broad background is observed, which is due to an additional inhomogeneous distribution of emitter molecules in the applied matrix (Fig. 38.2).

In Fig. 38.3, site-selectively excited emission spectra and a site-selectively detected excitation spectrum are displayed for the region of the electronic 0–0 transitions of the prominent site of lowest energy, which is labeled as Site A in Fig. 38.2. The emission spectrum measured at $T = 1.7$ K (part b of Fig. 38.3) shows one intense line at 21 025 cm^{-1}, which is the line of highest energy at that temperature. Therefore, it is assigned as electronic 0–0 transition from the lowest T_1 substate I to the singlet ground state S_0. With increasing temperature, an additional line appears at 21 041 cm^{-1}. This line results from the electronic 0–0 transition II → 0. With further temperature increase, line II gains intensity, but even at $T = 15$ K, line I is still the most intense one. Further, the peaks become broader. This effect is frequently observed and can be explained by the involvement of local low-energy vibrations of the dopant in its matrix cage, which exhibit slightly different energies in the electronic ground state compared to the excited state. The corresponding 1–1 transitions, which occur at higher temperature, lie near to but not exactly at the energy of the 0–0 transition. This causes a line broadening.

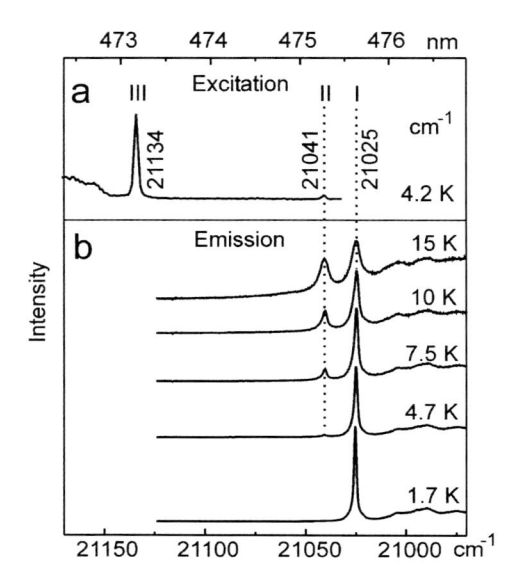

Figure 38.3 Site-selective spectra of the region of the $T_1 \leftrightarrow S_0$ 0–0 transitions of site A of FIracac in CH_2Cl_2. (a) Site-selectively detected excitation spectrum recorded at $T = 4.2$ K. The emission is detected at 21 025 cm^{-1} (electronic 0–0 transition I → 0). (b) Emission spectra recorded after selective excitation of the 0–0 transition of substate III at 21 134 cm^{-1}.

The excitation spectrum for the energy range of the electronic origins is reproduced in part a of Fig. 38.3. Two excitation lines at 21 041 and 21 134 cm^{-1} can be observed. The weak peak at 21 041 cm^{-1} is in resonance with the corresponding emission line as expected for an electronic 0–0 transition. Because no other line is observable in the relevant energy range, the high-energy line at 21 134 cm^{-1} is assigned to the 0–0 transition from the singlet ground S_0 state to the highest T_1 substate III. Interestingly, no additional line that might correspond to an electronic origin of a higher lying triplet state can be detected within the range of 1500 cm^{-1} to higher energy. Recently, Zhang, Ma, and co-workers predicted the occurrence of two close-lying triplet states T_2 and T_3 on the basis of TDDFT calculations [27]. However, their calculations do not yet include spin–orbit coupling, which might lead to significant energy shifts between the lowest triplets.

The intensity ratio obtained from the excitation spectrum shown in part a of Fig. 38.3 reveals that the transition $0 \to \text{III}$ is by a factor of 21 more allowed than the transition $0 \to \text{II}$. Thus, the transition between triplet substate III and the singlet ground state S_0 carries by far the highest oscillator strength (radiative allowedness) and mainly governs the emission properties at ambient temperature. This result is further supported by an analysis of the individual decay times of the triplet substates as obtained from the temperature dependence of the thermalized emission decay time. (Here not discussed in detail, but compare e.g. Refs. 28–30). The corresponding data and the results obtained from the high-resolution measurements are summarized in the energy level diagram as depicted in Fig. 38.4. According to an empirical ordering scheme [31–33], the T_1 state of the main site of FIracac in CH_2Cl_2 with a total zero-field splitting of 109 cm^{-1} can be assigned to be largely of MLCT character.

For the ratio of rate constants $k_{\text{III}}/k_{\text{II}}$ ($k = 1/\tau$) as determined from the temperature dependence of the thermalized emission

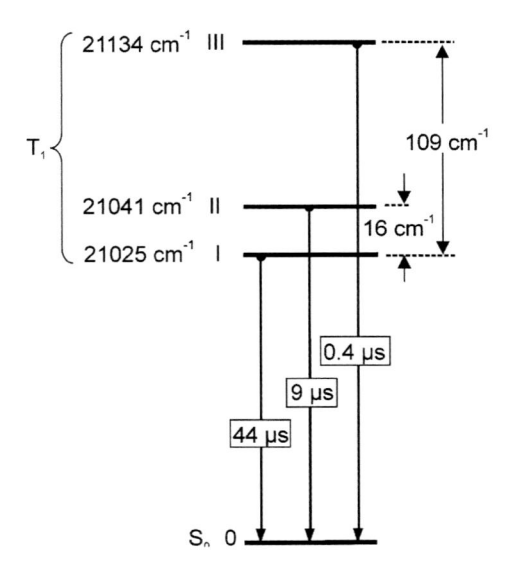

Figure 38.4 Energy level diagram and decay times for the T_1 substates I, II, and III of site A of FIracac in CH_2Cl_2.

decay time, a value of 22 is found, which is in good agreement with the ratio determined from the corresponding excitation peaks in part a of Fig. 38.3. This behavior is indicative of very similar radiative deactivation mechanisms of the substates II and III and indicates an emission quantum yield of almost 100% for both substates at cryogenic temperatures.

For completeness, it is mentioned that the other intense discrete site of FIracac in CH_2Cl_2, labeled as Site B in Fig. 38.2, exhibits different zero-field splittings with $\Delta E_{II-I} = 13$ cm^{-1} and $\Delta E_{III-I} = 92$ cm^{-1}. As explained below, such deviations between different sites are not unusual for Ir(III) compounds. Interestingly, for both compounds, FIracac and FIrpic [17], the sites of lowest energy exhibit the highest zero-field splittings. However, for Ir(btp)2(acac) in CH_2Cl_2 a manifold of sites was characterized, but a correlation between the emission energy and the zero-field splitting was not observed [23].

38.3.3 Comparison of FIracac and FIrpic: Influence of Ancillary Ligands

In Table 38.1, ambient temperature data of FIracac are compared to corresponding data of FIrpic, whereas Table 38.2 summarizes T_1 state properties of both compounds as determined at cryogenic temperatures.

The emitting triplet state of FIrpic lies about 700 cm^{-1} (~0.09 eV) higher in energy than the triplet of FIracac (Tables 38.1 and 38.2). Although this energy difference is not large, it is still important for the realization of a deeper-blue light emitting material. The observed energy shift is ascribed to different influences of the ancillary ligands acac and pic on the Ir(III) d-orbitals. In particular, the coordination of the central metal ion to nitrogen leads to a stronger bond than the coordination to oxygen [34]. DFT calculations performed on both compounds support this assumption because the Ir–N bond length to picolinate was determined to be substantially shorter than the Ir–O bond length to acetylacetonate [27]. Therefore, a higher mean ligand field strength is indicated for the pic ligand compared to the acac ligand. As consequence, the occupied d-orbitals of the t_{2g} manifold experience

Table 38.2 Energy separations and individual decay times of the triplet substates I, II, and III of FIracac in different matrices

	FIracac			FIrpic[b]
	$CH_2Cl_2^a$	THF	PMMA	CH_2Cl_2
0–0 transition $0 \leftrightarrow I$ [cm^{-1}]	21 025			21 738
ΔE_{II-I} [cm^{-1}]	16	10–15	10–15	9
ΔE_{III-I} [cm^{-1}]	109	80–125	75–120	76
τ_I [μs]	44	68	74	47
τ_{II} [μs]	9	9–19	10–20	21
τ_{III} [μs]	0.4	0.2–0.5	0.3–0.6	0.3

Note: T_1 state properties of FIrpic in CH_2Cl_2 are also displayed.
[a] Site A. [b] From Ref. 17, main site.

on average [34] a larger energy stabilization in FIrpic than in FIracac. The situation is schematically depicted in Fig. 38.5.

Interestingly, our schematic model is supported by the oxidation potentials of the compounds. They amount to 0.90 and 0.74 V (measured vs ferrocene) for FIrpic and FIracac, respectively [35, 36]. On the other hand, the π^*-orbitals of the chromophoric 4,6-dFppy ligands are—in this first order model—expected to remain almost unchanged. (Compare also Ref. 11) Thus, the d–π^* energy separations are expected to be larger for FIrpic than for FIracac. These energy differences will also be displayed in the corresponding ^{1}MLCT and ^{3}MLCT states, and indeed this trend is found experimentally for the transition energies.

In this simple consideration, we assume that the occupied π-orbitals of the chromophoric 4,6-dFppy ligands lie at the same energy for both compounds and are located below of the d-orbitals, but in energy proximity (compare for example Refs. 37–39). Thus, the close-lying d- and π-orbitals of adequate symmetry will combine to molecular orbitals. Obviously, the d–π mixing and thus the π-contribution to the occupied frontier orbitals is expected to be larger for FIrpic because of the smaller d–π energy separation compared to FIracac. Consequently, the "pureness" of MLCT character of the resulting states will become smaller with enhanced LC contributions to the emitting state. This mixing effect will also reduce the effectiveness of spin–orbit coupling, which is dominantly carried by the metal d-orbital contributions. Therefore, it is expected that the photophysical properties which are related to the effectiveness of SOC, such as ZFSs, radiative decay rates, and emission quantum yields, are strongly influenced by the discussed mixings. Indeed, this trend is observed. ZFSs (Table 38.2) and average radiative rates at ambient conditions (Table 38.1) are larger for FIracac (with less π-admixtures to the occupied d-orbitals and thus less LC admixtures to the emitting state) than for FIrpic.

The basic model presented above can be extended, if an additional effect, which can also modify zero-field splittings and radiative decay rates, is taken into account. Thus, further insights into triplet state properties are provided. We want to illustrate this issue by discussing the relevant SOC routes on the basis of interacting energy states.

A quantitative description of the triplet state properties of organometallic compounds based on quantum mechanical calculations including SOC is still very difficult (e.g., compare Ref. 40). However, the effects of SOC on the ZFS of the T_1 term and on the radiative decay rates of the triplet substates can be illustrated by formulas based on second-order perturbation theory. The energy $E(i)$ of one specific triplet substate i of T_1 (with $i =$ I, II, III) can be expressed by [20, 32, 41]

$$E(i) = E_{T_1} + \sum_{T_n} \frac{|\langle \varphi_{T_n}(j) | \hat{H}_{SO} | \varphi_{T_1}(i) \rangle|^2}{E_{T_1} - E_{T_n}} + \sum_{S_n} \frac{|\langle \varphi_{S_n} | \hat{H}_{SO} | \varphi_{T_1}(i) \rangle|^2}{E_{T_1} - E_{S_n}} \quad (38.2)$$

whereas the radiative rate constant $k_r(i)$ of T_1 substate i for the transition to the electronic ground-state S_0 is given by [20, 32, 42]

$$k_r(i) = \text{const} \times \tilde{\nu}^3 \times \left| \sum_{S_n} \frac{\langle \varphi_{S_n} | \hat{H}_{SO} | \varphi_{T_1}(i) \rangle}{E_{T_1} - E_{S_n}} \times \langle \varphi_{S_0} | e\vec{r} | \varphi_{S_n} \rangle \right|^2 \quad (38.3)$$

wherein $\hat{H}_{SO}$ is the SOC Hamiltonian, and E_{T_1}, E_{S_n}, and E_{T_n} are the unperturbed energies of the lowest triplet and of higher lying singlet and triplet states S_n and T_n, respectively. $T_n(j)$ characterizes a substate j of T_n. φ_{S_n} and $\varphi_{T_n}(j)$ represent the corresponding wave functions. In particular, they must have the same symmetry representation as the wave function $\varphi_{T_1}(i)$ of T_1 substate i, otherwise the matrix elements in Eqs. 38.2 and 38.3 vanish. $\tilde{\nu}$

represents the transition energy between the excited and the ground state and $e\vec{r}$ is the electric dipole operator.

The SOC matrix elements are usually different for the three T_1 substates. This leads to different energy stabilizations and thus to the ZFS as well as to different radiative rates. In particular, the most prominent effects of SOC are determined by the most proximate energy states because the corresponding energy denominators in Eqs. 38.2 and 38.3 are smaller than for states of higher energy. Obviously, substantial SOC in these complexes is only induced by the central metal d-orbitals. This means that the SOC matrix elements can only attain significant values, if (i) both the substates of the lowest triplet and the mixing-in states are MLCT states or contain at least some MLCT character and if (ii) the mixing MLCT states involve *different* d-orbitals [20, 32, 43]. Thus, the most prominent SOC admixtures to the substates of the lowest ^{3}MLCT state are related to couplings to the energetically most proximate ^{1,3}MLCT states that stem from different d-orbitals, whereas couplings to ^{1}LC states will be inefficient with respect to SOC. Especially, the next nearest frontier orbital with significant d-character, for example HOMO−1, will be highly important. If the corresponding d-orbital (which results from the t_{2g} manifold) experiences an energy shift, for example induced by lowering the symmetry or by exchanging the ancillary ligand, the mixings governing SOC will be altered. In particular, it is expected that the strongly asymmetric pic ligand induces a different splitting of the three occupied d-orbitals than the more symmetric acac ligand. This is schematically depicted in Fig. 38.5. From the experimental results, we can conclude that FIracac exhibits a d-orbital pattern with two more proximate frontier orbitals (HOMO and HOMO−1) than occurring for FIrpic because just this situation will result in higher SOC efficiencies and thus a larger ZFS and a higher average radiative decay rate for FIracac. Interestingly, recent calculations on both compounds support this model. A smaller energy separation between HOMO and HOMO−1 was found for FIracac than for FIrpic [27].

For completeness, it is mentioned that in pseudo-octahedrally coordinated compounds the orbitals of the t_{2g} manifold usually lie relatively close in energy [37, 38, 44]. Therefore, even small shifts of these energies/splittings will have distinct impact on the SOC

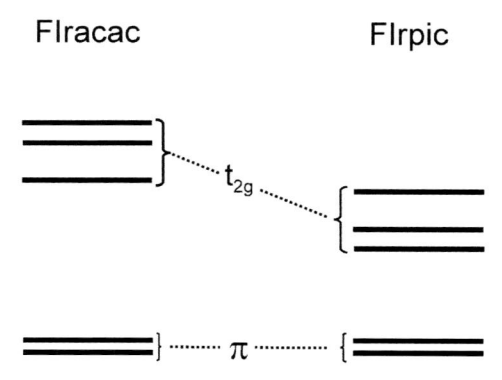

Figure 38.5 Schematic energy pattern for the occupied d-orbitals of the t_{2g} manifold and occupied π-orbitals of the chromophoric ligands. The diagram manifests trends to illustrate the discussion given in the text.

efficiencies and therefore on the emission properties due to changes of the energy denominators in Eqs. 38.2 and 38.3 (compare also next section).

In summary, both the amount of d–π mixing and the individual pattern of the occupied frontier orbitals will influence zero-field splittings and radiative decay rates. Both quantum-mechanical effects let us predict a larger ZFS and radiative decay rate for FIracac than for FIrpic - just as experimentally observed (Tables 38.1 and 38.2).

It is remarked that nonradiative processes are not included in the model presented. However, they cannot be neglected because an increase of the nonradiative deactivation rate k_{nr} can lead - despite higher radiative rates - to a lower photoluminescence quantum yield. This behavior is observed for FIracac (Table 38.1). The higher nonradiative rate for this compound compared to FIrpic can be related to differently effective deactivation processes by molecular vibrations. In particular, the higher symmetry of FIracac might be connected with a delocalization of the lowest triplet state over the two chromophoric ligands, whereas in the asymmetric FIrpic the T_1 state will probably be confined to only one chromophoric ligand (as is also indicated by DFT calculations [27]). In the latter case, the number of vibrational modes that can deactivate nonradiatively is substantially lower. Interestingly, investigations on Os(II) compounds have shown that complexes with one chromophoric ligand tend to exhibit higher photoluminescence quantum yields than corresponding complexes with two chromophoric ligands [45]. However, we cannot rule out that vibrational modes of acac and pic, respectively, might also affect the radiationless deactivation. Further, thermal population of quenching metal centered (MC, dd*) states might have an additional impact on the emission quantum yield. Because of the lower ligand field strength of acac compared to pic, the dd* states are supposed to lie at somewhat lower energy in FIracac than in FIrpic. Thus, population of those quenching dd* states at room temperature might be more pronounced in FIracac.

38.3.4 Matrix Effects on SOC and on the Triplet State Properties

In several investigations at high resolution, it was observed that the T_1 state properties of Ir(III) compounds depend on the individual site and on the host material [23, 32, 43, 46]. For example, for FIrpic in CH_2Cl_2, the T_1 states of two discrete sites have been studied in detail and were found to exhibit distinctly different zero-field splittings and substate decay times [17]. Also, for host materials in which the Ir(III) compounds give only broadband spectra (even at cryogenic temperatures), the zero-field splitting values and individual substate decay times are not discrete but are spread over specific ranges [17, 47]. Upper and lower limits for these ranges can be determined by a procedure based on temperature dependent emission decay time measurements, as described in Ref. 17. A corresponding behavior is also found for FIracac when doped into THF and PMMA, respectively. In these hosts, only broadband spectra are obtained and it is observed that the ZFS values and T_1 sublevel decay times are significantly spread (Table 38.2). Such effects occur in polycrystalline hosts [17, 23, 32] as well as in amorphous matrices [47]. However, the degree of the spread can vary (Table 38.2). Interestingly, the values observed for the discrete sites in CH_2Cl_2 (Section 38.3.2) closely correspond to the ranges observed for THF and PMMA. The observed behavior of a variation of T_1 state properties is rationalized with a sensitivity of the MLCT states of Ir(III) complexes on the host environment (matrix cage)

[17]. Changes affected by the local environment can, for example, alter the complex geometry by steric effects and influence the energies of the metal d-orbitals - in a similar manner as discussed above for the exchange of the ancillary ligand. As a consequence, the energies of the corresponding MLCT states are also altered and the energy denominators in Eqs. 38.2 and 38.3 are affected. According to these influences, the emitter molecules experience differently effective SOCs and thus exhibit different ZFSs and average radiative decay rates. It is remarked that matrix induced changes of the d–π mixtures, as discussed in Section 38.3.3, may also be of importance.

38.4 Summary and Conclusions

Photophysical properties of Ir(4,6-dFppy)$_2$(acac) (FIracac) and Ir(4,6-dFppy)$_2$(pic) (FIrpic) are investigated in detail, focusing on the emission behavior. Both compounds are highly emissive and their lowest excited electronic states are assigned to be largely of ^{3}MLCT character. The ancillary ligands acac and pic have distinct (but indirect) influence on the excited state properties by modifying the splittings and energy positions of the occupied frontier orbitals. These are largely of central metal d-character and stem from the t$_{2g}$-manifold. In a schematic and simple model, we relate especially the splitting pattern/energy separations of the frontier orbitals HOMO and HOMO–1 to the effectiveness of spin–orbit coupling of the T$_1$ substates to higher lying ^{1,3}MLCT states. Thus, relations between simple MO considerations and detailed photophysical properties such as zero-field splittings, radiative decay rates, and so forth become to survey. It is even possible to understand, qualitatively, how a purely organic host environment can modify the effectiveness of SOC. These influences are of particular importance for octahedrally coordinated complexes with emitting states of ^{3}MLCT character, especially for Ir(III) complexes, which are most suited for OLED applications.

Because of the higher MLCT character in the emitting state, FIracac should represent a better OLED material than FIrpic. However, a slightly lower energy of the emitting ^{3}MLCT state makes the compound less well suited for highly desired blue emitting

OLEDs than FIrpic. Further, the photoluminescence quantum yield of FIracac is smaller than that of FIrpic, that is FIracac is distinctly more influenced by nonradiative decay processes. This is presumably due to the involvement of both chromophoric 4,6-dFppy-ligands in the emitting state of the more symmetric FIracac. Thus, new efforts for the design of OLED triplet emitters should not only take into account a high MLCT character in the emitting state but also complex symmetry and matrix effects.

Acknowledgment

The *Bundesministerium für Bildung und Forschung (BMBF)* is gratefully acknowledged for providing the funding of this investigation. We thank the *German Academic Exchange Service (DAAD)* for financial support of the exchange program of the University of Regensburg with the University of Southern California.

References

1. *Highly Efficient OLEDs with Phosphorescent Materials*; Yersin, H., Ed., Wiley-VCH: Weinheim, 2008.
2. Tamayo, A. B., Garon, S., Sajoto, T., Djurovich, P. I., Tsyba, I. M., Bau, R., Thompson, M. E. *Inorg. Chem.* 2005, **44**, 8723.
3. Yang, C.-H., Cheng, Y.-M., Chi, Y., Hsu, C.-J., Fang, F.-C., Wong, K.-T., Chou, P.-T., Chang, C.-H., Tsai, M.-H., Wu, C.-C. *Angew. Chem., Int. Ed.* 2007, **46**, 2418.
4. Su, S.-J., Gonmori, E., Sasabe, H., Kido, J. *Adv. Mater.* 2008, **20**, 4189.
5. Adachi, C., Baldo, M. A., Thompson, M. E., Forrest, S. R. *J. Appl. Phys.* 2001, **90**, 5048.
6. Yersin, H. *Top. Curr. Chem.* 2004, **241**, 1.
7. Lamansky, S., Djurovich, P., Murphy, D., Abdel-Razzaq, F., Kwong, R., Tsyba, I., Bortz, M., Mui, B., Bau, R., Thompson, M. E. *Inorg. Chem.* 2001, **40**, 1704.
8. Tamayo, A. B., Alleyne, B. D., Djurovich, P. I., Lamansky, S., Tsyba, I., Ho, N. N., Bau, R., Thompson, M. E. *J. Am. Chem. Soc.* 2003, **125**, 7377.

9. Williams, J. A. G., Wilkinson, A. J., Whittle, V. L. *Dalton Trans.* 2008, **16**, 2081.
10. Hwang, F.-M., Chen, H.-Y., Chen, P.-S., Liu, C.-S., Chi, Y., Shu, C.-F., Wu, F.-I., Chou, P.-T., Peng, S.-M., Lee, G.-H. *Inorg. Chem.* 2005, **44**, 1344.
11. Li, J., Djurovich, P. I., Alleyne, B. D., Yousufuddin, M., Ho, N. N., Thomas, J. C., Peters, J. C., Bau, R., Thompson, M. E. *Inorg. Chem.* 2005, **44**, 1713.
12. Li, J., Djurovich, P. I., Alleyne, B. D., Tsyba, I., Ho, N. N., Bau, R., Thompson, M. E. *Polyhedron* 2004, **23**, 419.
13. Chin, C. S., Eum, M.-S., Kim, S. Y., Kim, C., Kang, S. K. *Eur. J. Inorg. Chem.* 2007, **3**, 372.
14. Di Censeo, D., Fantacci, S., de Angelis, F., Klein, C., Evans, N., Kalyanasundaram, K., Bolink, H. J., Grätzel, M., Nazeeruddin, M. K. *Inorg. Chem.* 2008, **47**, 980.
15. Chou, P.-T., Chi, Y. *Chem.–Eur. J.* 2007, **13**, 380.
16. Adachi, C., Kwong, R. C., Djurovich, P. I., Adamovich, V., Baldo, M. A., Thompson, M. E., Forrest, S. R. *Appl. Phys. Lett.* 2001, **79**, 2082.
17. Rausch, A. F., Thompson, M. E., Yersin, H. *Inorg. Chem.* 2009, **48**, 1928.
18. Su, S.-J., Sasabe, H., Takeda, T., Kido, J. *Chem. Mater.* 2008, **20**, 1691.
19. Vecchi, P. A., Padmaperuma, A. B., Qiao, H., Sapochak, L. S., Burrows, P. E. *Org. Lett.* 2006, **8**, 4211.
20. Rausch, A. F., Homeier, H. H. H., Djurovich, P. I., Thompson, M. E., Yersin, H. *Proc. SPIEsInt. Soc. Opt. Eng.* 2007, **6655**, 66550F.
21. Lamansky, S., Thompson, M. E., Adamovich, V., Djurovich, P. I., Adachi, C., Baldo, M. A., Forrest, S. R., Kwong, R. U. S. Patent 20050214576, 2005.
22. Chen, P., Meyer, T. J. *Chem. Rev.* 1998, **98**, 1439.
23. Finkenzeller, W. J., Hofbeck, T., Thompson, M. E., Yersin, H. *Inorg. Chem.* 2007, **46**, 5076.
24. Yersin, H., Huber, P., Wiedenhofer, H. *Coord. Chem. Rev.* 1994, **132**, 35.
25. Colombo, M. G., Brunold, T. C., Riedener, T., Güdel, H. U., Förtsch, M., Bürgi, H.-B. *Inorg. Chem.* 1994, **33**, 545.
26. Endo, A., Suzuki, K., Yoshihara, T., Tobita, S., Yahiro, M., Adachi, C. *Chem. Phys. Lett.* 2008, **460**, 155.
27. Gu, X., Fei, T., Zhang, H., Xu, H., Yang, B., Ma, Y., Liu, X. *J. Chem. Phys. A* 2008, **112**, 8387.
28. Finkenzeller, W. J., Yersin, H. *Chem. Phys. Lett.* 2003, **377**, 299.
29. Harrigan, R. W., Crosby, G. A. *J. Chem. Phys.* 1973, **59**, 3468.
30. Pentlehner, D., Grau, I., Yersin, H. *Chem. Phys. Lett.* 2008, **455**, 72.

31. Yersin, H., Donges, D. *Top. Curr. Chem.* 2001, **214**, 81.
32. Yersin, H., Finkenzeller, W. J. In *Highly Efficient OLEDs with Phosphorescent Materials*; Yersin, H., Ed., Wiley-VCH: Weinheim, Germany, 2008, p. 1.
33. Yersin, H., Strasser, J. *Coord. Chem. Rev.* 2000, **208**, 331.
34. Schäfer, H. L., Gliemann, G. *Einführung in die Ligandenfeldtheorie; Akad. Verlagsgesellschaft: Wiesbaden* 1967, p. 84.
35. Orselli, E., Kottas, G. S., Konradsson, A. E., Coppo, P., Fröhlich, R., De Cola, L., van Dijken, A., Büchel, M., Börner, H. *Inorg. Chem.* 2007, **46**, 11082.
36. Djurovich, P. I., private communication.
37. Ceulemans, A., Vanquickenborne, L. G. *J. Am. Chem. Soc.* 1981, **103**, 2238.
38. Kober, E. M., Meyer, T. J. *Inorg. Chem.* 1982, **21**, 3967.
39. Yersin, H., Humbs, W., Strasser, J. *Top. Curr. Chem.* 1997, **191**, 153.
40. Nozaki, K. *J. Chin. Chem. Soc.* 2006, **53**, 101.
41. Ikeda, S., Yamamoto, S., Nozaki, K., Ikeyama, T., Azumi, T., Burt, J. A., Crosby, G. A. *J. Phys. Chem.* 1991, **95**, 8538.
42. Abedin-Siddique, Z., Ohno, T., Nozaki, K., Tsubomura, T. *Inorg. Chem.* 2004, **43**, 663.
43. Rausch, A. F., Homeier, H. H. H., Yersin, H. In *Topics in Organometallic Chemistry - Photophysics of Organometallics*; Lees A. J., Ed., Springer: Berlin/Heidelberg, 2009.
44. Daul, C., Baerends, E. J., Vernooijs, P. *Inorg. Chem.* 1994, **33**, 3538.
45. Kober, E. M., Caspar, J. V., Lumpkin, R. S., Meyer, T. J. *J. Phys. Chem.* 1986, **90**, 3722.
46. Finkenzeller, W. J., Thompson, M. E., Yersin, H. *Chem. Phys. Lett.* 2007, **444**, 273.
47. Bauer, R., Finkenzeller, W. J., Bogner, U., Thompson, M. E., Yersin, H. *Org. Electron.* 2008, **9**, 641.

Chapter 39

Temperature Dependence of Blue Phosphorescent Cyclometalated Ir(III) Complexes

Tissa Sajoto,[a] Peter I. Djurovich,[a] Arnold B. Tamayo,[a] Jonas Oxgaard,[b] William A. Goddard III,[b] and Mark E. Thompson[a]

[a]*Department of Chemistry, University of Southern California, Los Angeles, California 90089, USA*
[b]*Department of Chemistry, California Institute of Technology, Pasadena, California 91125, USA*
djurovic@usc.edu, met@usc.edu

The photophysical properties for a series of facial (*fac*) cyclometalated Ir(III) complexes (*fac*-Ir(C^N)$_3$ (C^N = 2-phenylpyridyl (ppy), 2-(4,6-difluorophenyl)pyridyl (F2ppy), 1-phenylpyrazolyl (ppz), 1-(2,4-difluorophenyl)pyrazolyl (F2ppz), and 1-(2-(9,9′-dimethylfluorenyl))pyrazolyl (flz)), *fac*-Ir(C^N)2(C^N′) (C^N = ppz or F2ppz and C^N′ = ppy or F2ppy), and *fac*-Ir(C^C′)$_3$ (C^C′ = 1-phenyl-3-methylbenzimidazolyl (pmb)) have been studied in dilute 2-methyltetrahydrofuran (2-MeTHF) solution in a temperature range of 77–378 K. Photoluminescent quantum yields

---

Reprinted from *J. Am. Chem. Soc.*, **131**, 9813–9822, 2009.

Electrophosphorescent Materials and Devices
Edited by Mark E. Thompson
Text Copyright © 2009 American Chemical Society
Layout Copyright © 2024 Jenny Stanford Publishing Pte. Ltd.
ISBN 978-981-4877-34-3 (Hardcover), 978-1-003-08872-1 (eBook)
www.jennystanford.com

(Φ) for the 10 compounds at room temperature vary between near zero and unity, whereas all emit with high efficiency at low temperature (77 K). The quantum yield for *fac*-Ir(ppy)$_3$ (Φ = 0.97) is temperature-independent. For the other complexes, the temperature-dependent data indicates that the luminescent efficiency is primarily determined by thermal deactivation to a nonradiative state. Activation energies and rate constants for both radiative and nonradiative processes were obtained using a Boltzmann analysis of the temperature-dependent luminescent decay data. Activation energies to the nonradiative state are found to range between 1600 and 4800 cm^{-1}. The pre-exponential factors for deactivation are large for complexes with C$^\wedge$N ligands (10^{11}–10^{13} s^{-1}) and significantly smaller for *fac*-Ir(pmb)$_3$ (10^9 s^{-1}). The kinetic parameters for decay and results from density functional theory (DFT) calculations of the triplet state are consistent with a nonradiative process involving Ir–N (Ir–C for *fac*-Ir(pmb)$_3$) bond rupture leading to a five-coordinate species that has triplet metal-centered (^{3}MC) character. Linear correlations are observed between the activation energy and the energy difference calculated for the emissive and ^{3}MC states. The energy level for the ^{3}MC state is estimated to lie between 21 700 and 24 000 cm^{-1} for the *fac*-Ir(C$^\wedge$N)$_3$ complexes and at 28 000 cm^{-1} for *fac*-Ir(pmb)$_3$.

39.1 Introduction

Cyclometalated iridium(III) complexes have been recently shown to have phosphorescence efficiencies approaching theoretical limits (Φ = 0.8–1.0) and short radiative triplet lifetimes (τ = 1–5 μs) [1, 2]. These photophysical properties make organometallic Ir complexes excellent candidates for use in oxygen detection [3], metal ion sensing [4, 5], and luminescent labeling reagents for biological materials [6–20]; however, the most extensively investigated application of cyclometalated Ir(III) complexes is as emitters in organic light-emitting diodes (OLEDs) [21]. The high phosphorescent efficiencies, short lifetimes, and broad range of emission colors make these Ir(III) complexes ideal emitters in OLEDs, designed for flat panel displays and white light sources [22,

23]. Although a wide range of Ir(III) materials have been reported that emit from blue to the near-infrared, the number of highly efficient, blue-to-violet phosphorescent Ir(III) complexes are limited and their photophysical properties are not fully understood. In present study we have focused our attention on Ir(III) complexes with emission energies ranging from the near-UV to green in order to develop a better understanding of the processes that limit the luminescent efficiency from this class of materials.

Blue phosphorescence can be observed from Ir(III) complexes that have cyclometalated phenylpyridine (ppy) ligands modified to increase the emission energy relative to that of an efficient green phosphor, fac-Ir(ppy)$_3$ (**1**, Fig. 39.1). For example, addition of electron-withdrawing groups such as fluorine leads to a complex, fac-Ir(F$_2$ppy)$_3$ (**2**), that displays emission 40 nm higher in energy compared to **1** [24]. Similarly, a difluorinated bipyridine ligand has recently been used to form a tris-cyclometalated Ir(III) complex that displays blue phosphorescence [25]. Homoleptic complexes using non-pyridine-based heterocyclic ligands that have high triplet energies, such as phenylpyrazole [11], phenyltriazole [12], pyridylazolate [26], phenylimidazole, or phenylbenzimidazole [13], likewise phosphoresce at high energy. Other examples include heteroleptic derivatives that use electron-withdrawing ancillary ligand(s) to raise the emission energy by stabilizing the highest occupied molecular orbital (HOMO) [27–33]. Although cyclometalated Ir(III) complexes with high emission energies can be obtained using any of these approaches, the luminescent efficiencies are often well below that of **1** due to a significant increase in nonradiative decay rates. A more detailed picture of the excited-state processes is thus needed in order to optimize the phosphorescent behavior of this class of materials.

Key information about the photophysical properties of phosphorescent transition metal complexes can be obtained by studying the temperature dependence of emission from these materials. In particular, analysis of temperature-dependent transient decay from diimine (i.e., bipyridine, bpy) chelates of Ru(II) and Os(II) with metal-to-ligand charge transfer (MLCT) excited states has enabled detailed characterization the radiative and nonradiative states [34–38]. Different types of behavior are observed for these compounds in two distinct temperature regimes. At low temperatures (0–77 K), the radiative rates are found to vary due to differing thermal population of three triplet substates. The effect comes about because a significant energy gap separates the lowest and highest substate. The energy difference between the two states in the absence of an external magnetic field is defined as the zero-field splitting (zfs) [39]. The zfs values for the metal complexes are enhanced due to mixing of singlet and triplet states induced by spin–orbit coupling from the heavy metal ions. For example, the zfs values for Ru(bpy)$_3$ derivatives ($\sim$60 cm^{-1}) [37] and Os(II) analogues ($>$200 cm^{-1}) [38] are markedly larger than those found in organic molecules (typically 0.2 cm^{-1}) [39]. At higher temperatures (77–300 K), variation in the nonradiative rates for Ru(II) and Os(II) complexes reveals the existence of thermally accessible, higher-lying excited states [34, 35, 40, 41]. These higher-lying states typically reside ca. 2000 cm^{-1} above the MLCT state and provide a significant contribution to the loss of luminescent efficiency at room temperature. For polypyridyl complexes of Ru(II), nonradiative decay occurs through a triplet metal-centered (^{3}MC) ligand field state [42]. In contrast, the larger ligand field splitting of Os(II) complexes makes the ^{3}MC state relatively inaccessible at room temperature [43]. The Os(II) complexes instead suffer more severely from temperature-independent nonradiative decay [44, 45]. The temperature dependence of luminescent cyclometalated Pd(II), Rh(III), Pt(II), and Pt(IV) complexes has also been investigated, and these compounds show behavior similar to that of the Ru(II) and Os(II) complexes [46].

Temperature-dependent luminescent studies have also been carried out on cyclometalated Ir(III) complexes. Complexes examined at cryogenic temperatures show temperature-dependent variations in radiative rates similar to that seen in Ru(II) and Os(II) bisiimine complexes [47–51]. Several tris(pyridylazolate) and tris(phenyltriazolate) Ir(III) complexes have been investigated at higher temperatures, and the presence of thermally activated, nonradiative processes in these species was shown to greatly diminish the luminescent efficiency at room temperature [26, 33, 52, 53]. Of particular note is a recent report of a blue emissive Ir complex characterized by a temperature-dependent study along

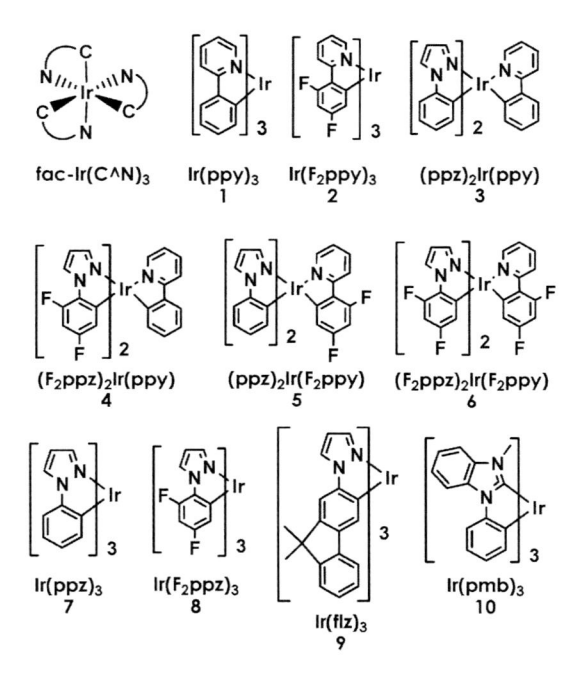

Figure 39.1 Molecular structures of the Ir complexes used in this study. All complexes are facial (*fac*) isomers.

with theoretical calculation of the potential energy surface for the excited state [54]. The calculations were used to determine the metal–ligand distortions the complex undergoes as it proceeds along the deactivation pathway. Good agreement was found between the experimental and calculated energies for deactivation. In the present paper, we utilize temperature-dependent luminescent studies in conjunction with theoretical calculations to examine the radiative and nonradiative properties for cyclometalated Ir(III) complexes, **1–10** (Fig. 39.1). We have measured the activation energies and kinetic parameters for deactivation of the excited state. Our analysis indicates that thermal population to a ^{3}MC state is the most likely deactivation processes for high-energy cyclometalated Ir(III) complexes. We use the results from both experimental and theoretical calculations to estimate the energies for the ^{3}MC state. The relative energy of the ^{3}MC state with respect to the emissive state is shown to be the principal factor that dictates the quantum efficiency for these blue cyclometalated Ir complexes.

39.2 Experimental Section

39.2.1 Synthesis

Compounds **1–10**, were prepared as previously described [24, 55].

39.2.2 Quantum Yield Measurement

The luminescent quantum yields of **1–10** were measured using an absolute method [56] more reliant than the relative method [57] typically employed in earlier studies. Measurements were carried out using a Hamamatsu C9920 system equipped with a xenon lamp, calibrated integrating sphere and model C10027 photonic multichannel analyzer. Dilute solutions ($\sim 10^{-5}$ M) of the compounds in 2-methyltetrahydrofuran (2-MeTHF) were placed in 1 cm^2 quartz cuvettes that were fitted with a Teflon stopcocks. Samples were deaerated by vigorously bubbling dry N$_2$ into the solutions using a flexible tube that was threaded through a center bore in the stopcock. It is worth noting that rubber septa were ineffective as seals for the optical cells and their use led to decreased quantum yields in less than 1 h after degassing. The quantum efficiencies were measured using either a 330 nm (for **10**) or 380 nm (for **1–6** and **9**) excitation wavelengths. The quantum efficiency data was processed using the U6039-05 software package provided by Hamamatsu. The reproducibility in the quantum efficiency measurements is ±5%.

39.2.3 Emission Intensity Measurement

Steady-state emission measurements were performed using a QuantaMaster model C-60SE spectrofluorimeter (Photon Technology International) with an excitation wavelength of 360 nm. The emission intensity measurement experiments of *fac*-Ir(ppy)$_3$ were carried out in dilute ($\sim 10^{-5}$ M), N$_2$-degassed 2-MeTHF solution using a custom Dewar. A dry ice/acetone bath was used for the measurements taken at 196 K, and liquid nitrogen was used for data recorded at 77 K.

39.2.4 Lifetime Measurement

Samples for transient luminescent decay measurements were prepared in distilled 2-MeTHF solution. The samples were deaerated by bubbling with N_2, freeze–pump–thawed (3 ×) and flame-sealed under vacuum. Measurements in the range of 77–300 K were performed using an Oxford OptistatDN-V cryostat instrument equipped with an intelligent temperature controller. For the temperature range of 300–378 K, the optical cell was placed in an insulated sample holder that was connected to a thermostat-controlled bath filled with a mixture of deionized water/ethylene glycol (1:1). All phosphorescent lifetimes were measured time-correlated single-photon counting using an IBH Fluorocube instrument equipped with a 405 nm (for **1–6**, **9**, and **10**) or a 331 nm (for **7** and **8**) LED excitation source. Fits to the temperature-dependent data were performed using the Origin (v6.1) software package.

39.2.5 Theoretical Calculations

All calculations were performed using the hybrid density functional theory (DFT) functional B3LYP as implemented by the Jaguar 7.0 program packages [58]. This DFT functional utilizes the Becke three-parameter functional [59] (B3) combined with the correlation functional of Lee, Yang, and Parr [60] (LYP) and is known to produce good descriptions of reaction profiles for transition metal containing compounds [61, 62]. The metals were described by the Wadt and Hay [63–65] core–valence (relativistic) effective core potential (treating the valence electrons explicitly) using the LACVP basis set with the valence double-ζ contraction of the basis functions, LACVP**. All electrons were used for all other elements using a modified variant of Pople's [66, 67] 6-31G** basis set, where the six d functions have been reduced to five.

All energies here are reported as $\Delta H(0\ K) = \Delta E +$ zero-point energy correction. All geometries were optimized and evaluated for the correct number of imaginary frequencies through vibrational frequency calculations using the analytic Hessian. Zero imaginary frequencies correspond to a local minimum, whereas one imaginary frequency corresponds to a transition structure.

39.3 Results and Discussion

Luminescence data for the iridium(III) complexes **1–10** recorded at room temperature and 77 K in 2-MeTHF is presented in Table 39.1. The complexes can be divided into three different groups based on their photoluminescent quantum efficiency (Φ) at room temperature. The first four complexes (**1–4**) have high Φ values (0.93–0.98), a second set (**5, 6, 9,** and **10**) has Φ ranging between 0.30 and 0.80, and a third (**7, 8**) has a very low Φ (< 0.01). The luminescent lifetimes (τ) for **1–6, 9,** and **10** fall between 1.6 and 50 µs, whereas the values for **7** and **8** are much shorter (< 10 ns). Emission from all of the complexes is highly efficient at 77 K, and the luminescent lifetimes fall in the microsecond regime. The phosphorescence displayed by these iridium complexes originates from a ligand-centered triplet state (T_1) that is strongly perturbed by admixture with a state having MLCT character (^{3}MLCT-LC) [32]. There is virtually no fluorescence as the rate of intersystem crossing for cyclometalated Ir complexes has been shown to be rapid (<100 fs) and, thus, near quantitative for triplet formation [68, 69].

The decay of the T_1 state to the ground state for these cyclometalated Ir complexes follows one of three paths, a temperature-dependent radiative process, $k_r(T)$, or one of two nonradiative (NR) processes, as illustrated in Scheme 39.1. The temperature

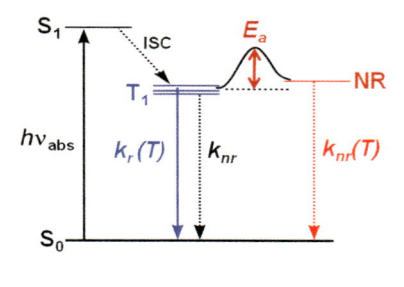

Scheme 39.1

Table 39.1 Photophysical properties of **1–10** in 2-MeTHF

No.	Complex	77 K			298 K					
		E_{0-0} (nm)	E_{0-0} (cm^{-1})	τ (μs)	λ_{max} (nm)	τ (air) (μs)[a]	τ (N$_2$) (μs)	Φ ($\pm$ 0.05)	k_r (s^{-1})	k_{nr} (s^{-1})
1	Ir(ppy)$_3$	491	20 370	4.0	508	0.036	1.6	0.97	6.1 × 10^5	1.9 × 10^4
2	Ir(F$_2$ppy)$_3$	454	22 030	2.6	466	0.037	1.7	0.98	5.8 × 10^5	1.2 × 10^4
3	Ir(ppz)$_2$(ppy)	479	20 880	4.2	500	0.038	1.7	0.95	5.6 × 10^5	2.9 × 10^4
4	Ir(F$_2$ppz)$_2$(ppy)	465	21 510	4.2	475	0.040	2.6	0.93	3.6 × 10^5	2.7 × 10^4
5	Ir(ppz)$_2$(F$_2$ppy)	463	21 600	3.4	500	0.032	1.2	0.55	4.6 × 10^5	3.8 × 10^5
6	Ir(F$_2$ppz)$_2$(F$_2$ppy)	445	22 470	3.1	457	0.044	1.3	0.66	4.6 × 10^5	3.1 × 10^5
7	Ir(ppz)$_3$	412	24 270	14			0.002	<0.01		> 10^8
8	Ir(F$_2$ppz)$_3$	388	25 770	25			0.007	<0.01		> 10^8
9	Ir(flz)$_3$	478	20 920	50	480	0.036	48	0.81	1.7 × 10^4	4.0 × 10^3
10	Ir(pmb)$_3$	378	26 455	3.1	382	0.020	1.1	0.37	3.4 × 10^5	5.7 × 10^5

[a] In toluene.

dependence of k_r comes about from thermal population of the individual triplet substates (T_I, T_{II}, and T_{III}) of T_1, each of which has a unique radiative rate [70]. Rapid thermalization between the triplet sublevels leads to emission with characteristics of a single radiative state. The nonradiative decay has a temperature-independent process, k_{nr}, that is typically associated with vibrational deactivation [71]. The temperature-dependent process, $k_{nr}(T)$, is limited by an activation barrier of E_a separating the T_1 and NR state(s) [35, 72, 73]. Similar schemes have been proposed for the radiative and nonradiative decay of triplet states in Ru and Os diimmine complexes [42, 74, 75].

The quantum efficiency for emission using Scheme 39.1 can be expressed as a function of radiative and nonradiative rates using Eq. 39.1

$$\Phi = \frac{k_r(T)}{k_r(T) + k_{nr} + k_{nr}(T)} \qquad (39.1)$$

This equation is based on unimolecular decay, assuming contributions from second-order nonradiative decay pathways, e.g., self-quenching or quenching by impurities such as oxygen, are absent. Self-quenching can be minimized in fluid solution by performing measurements at dilute concentrations ($\leq 10^{-5}$ M). Quenching by oxygen is more problematic since these Ir compounds are quenched at near diffusion-controlled rates ($k_q = 10^9 - 10^{10}$ M^{-1} s^{-1}) [76], as shown by the short lifetimes observed in aerated solution ($\tau = 0.020 - 0.044$ μs, Table 39.1). The rapid quenching is a consequence of deactivation by both energy and electron-transfer processes [76, 77]. Electron transfer is particularly effective for these complexes since their high triplet state energies and ease of oxidation make them potent reducing agents while in the excited state. Therefore, solutions need to be vigorously deoxygenated and rigorously sealed in order to obtain reproducible values for the quantum yield. For example, we observed large decreases (> 25%) within an hour during successive measurements of Φ caused by the intrusion of air when using optical cells equipped with rubber septa or poorly fitted Teflon stopcocks.

Given the prominence of *fac*-Ir(ppy)$_3$ (**1**) as the first reported tris-cyclometalated iridium complex [78], and its frequent use as a primary luminescent quantum yield standard, it is important to

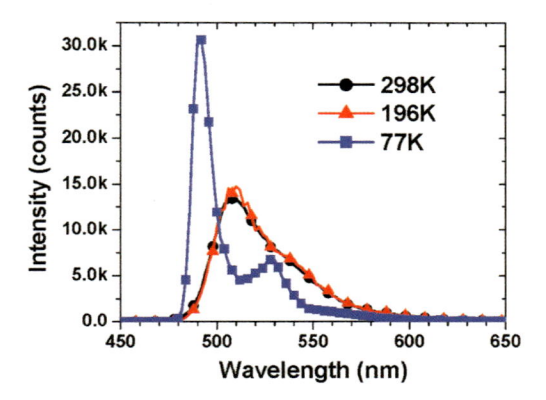

Figure 39.2 Emission intensity measurements of Ir(ppy)$_3$ (**1**) in 2-MeTHF at different temperatures.

examine the properties of this species in detail. The initial account of **1** reported its Φ as equal to 0.4 ± 0.1 [78], and this value continues to be used when the complex is employed as an emission standard. Recent measurements, either in solution or dispersed in a solid matrix, have found much higher values for Φ, ranging from 0.8 to 1.0 [1, 79–81]. We also obtained a quantum efficiency of near unity (0.97) for **1** in toluene and 2-MeTHF solutions. These high values suggest that the nonradiative channels are not available (i.e., ≪ k_{nr} ≪ k_r) and will likely not become active as the temperature is lowered. To confirm this hypothesis, the emission intensity of **1** was measured in 2-MeTHF at three different temperatures: 298, 195, and 77 K (Fig. 39.2). Although the emission line shape undergoes a blue shift and the vibronic structure is better resolved upon freezing of the solvent, there is no change in integrated emission intensity as the temperature is lowered; the peak areas are within 1.5% (6.4–6.5 × 10^5 counts). This result agrees with other studies that also show luminescence of **1** doped in a solid matrix displays no variation in intensity between room temperature and 4 K [81, 82].

Although the quantum yield for **1** remains constant below 300 K, the luminescent lifetime, and thus the radiative rate, does vary with temperature. The lifetimes of **1** at 298, 195, and 77 K are 1.6, 1.8, and 4.0 µs, respectively. It is worth noting that the 2.5 times increase in lifetime upon cooling to 77 K was used to provide support for the initial determination of $\Phi = 0.4$ [78]. Such an interpretation requires that the radiative rate be temperature-independent, an assumption commonly invoked when describing k_r values for phosphorescence. This assumption, however, is only valid for species with small values of zfs between the triplet substates. For compounds with a zfs less than $3k_BT$, thermal equilibration of the individual triplet sublevels is rapid and phosphorescence can be treated as originating from a single state, having an average $k_r = 1/3[k_r(T_I) + k_r(T_{II}) + k_r(T_{III})]$ [83]. Since most organic compounds have small values for zfs (< 1 cm^{-1}), thermal equilibration of the triplet substates occurs at temperatures well below 77 K. The values of zfs for transition metal complexes that have significant contributions of MLCT character in their excited state are markedly larger than those of organic materials, reaching values > 200 cm^{-1} [23]. Thus, thermal equilibration of the triplet sublevels for metal complexes, such as cyclometalated Ir(III) complexes, can require temperatures well above 77 K. In order to estimate the zfs for **1**, a two-level Boltzmann analysis was performed on the luminescent decay using data taken in the temperature range of 77–378 K. The analysis gives a value for k_1 of 1.8×10^{-6} s^{-1} and activation energy (E_a) of 122 cm^{-1} (Fig. 39.3). The k_1 value represents the high-temperature limit to the radiative rate, whereas E_a corresponds to the zfs between the T_I and T_{III} triplet substates, inset to Fig. 39.3. The smaller splitting between

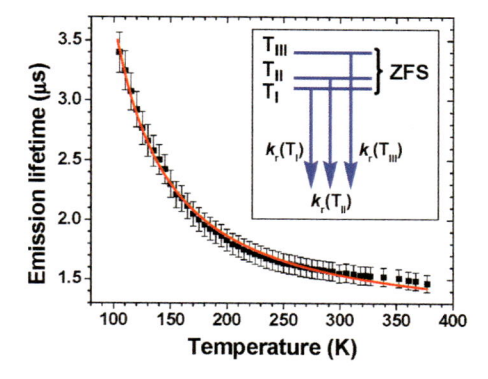

Figure 39.3 Temperature-dependent luminescent decay of **1**. The line is the fit to the data using a two-level Boltzmann analysis. Inset: schematic diagram illustrating the decay process from the triplet sublevels.

sublevels T_I and T_{II} could not be resolved in this temperature range. The k_1 and E_a values are in good agreement with results obtained by Yersin from observations of fac-Ir(ppy)$_3$ taken in THF at lower temperatures, where $k_r(T_{III}) = 1.3 \times 10^{-6}$ s^{-1} and the zfs was found to be site-dependent and range between 90 and 150 cm^{-1} [84, 85]. The decrease in lifetime observed upon warming **1** from 77 to 298 K is thus due to an increase in the radiative rate caused by thermal population of the higher triplet sublevels, particularly the T_{III} substate responsible for fast radiative decay.

The high quantum efficiency at room temperature of **1** indicates that the overall nonradiative decay rate $(k_{nr} + k_{nr}(T))$ for this complex is at least 2 orders of magnitude smaller than the radiative decay rate $(k_r = 7.5 \times 10^5$ s^{-1}, Table 39.1). The near unit value for Φ sets an upper limit for $k_{nr} + k_{nr}(T)$, dictated by the precision of the quantum yield measurement, to be less than 3.0×10^4 s^{-1}. Such a low rate implies that neither temperature-independent nor temperature-dependent nonradiative decay is effective at deactivating the excited state for **1** at temperatures of 300 K and below.

Temperature-independent nonradiative decay can occur through two different pathways: direct surface crossing from T_1 and S_0 and/or vibrational coupling to the ground state. The former process can be excluded from consideration due to the vibronic structure of the emission spectra, which indicates no major structural distortion occurs in the ligand-localized excited state relative to the ground state. Therefore, the only available temperature-independent mechanism is vibrational coupling to the ground state, which is governed by the energy gap law (EGL) [86, 87]. Compounds that follow the EGL show a linear decrease in $\ln(k_{nr})$ with increasing emission energy. Numerous studies have shown that the nonradiative rates of luminescent d^6 metal complexes follow EGL behavior [44, 45, 71]. Typically, high-frequency metal–ligand or C–C skeletal stretching vibrations provide the most effective deactivation mode to couple the ground and excited states. In particular, extensive work by Meyer and co-workers has generated enough data to allow us to predict the k_{nr} values for blue phosphors [45, 71]. A linear extrapolation to high energy ($> 20\,000$ cm^{-1}, 2.48 eV) of the EGL using slope ($-7.5 \ln(k_{nr})/$eV) and intercept (29.2) data for red emissive Os(II) bipyridyl complexes [88] leads to values of k_{nr} less than 4.0×10^4 s^{-1}. The phenomenon is exemplified by *fac*-Ir(flz)$_3$ (**9**), where the $k_{nr} = 4.0 \times 10^3$ s^{-1}. Given the high k_r values ($> 10^5$ s^{-1}) found for most of the other cyclometalated Ir complexes, simple vibrational deactivation processes should have a minor impact on the value of Φ. It follows then, that the principal mechanism that promotes nonradiative decay, lowering the quantum efficiency of complexes **3–8** and **10**, is thermal population of the NR state(s), Scheme 39.1.

In order to gain a better understanding of the thermally activated nonradiative decay processes in these materials, luminescent lifetimes for **2–10** were measured between 77 and 398 K. Kinetic parameters for the luminescent decay in 2-MeTHF solution were obtained by fitting data plotted as $1/\tau$ versus T using a Boltzmann model incorporating two temperature-dependent terms, Eq. 39.2, where τ = experimental luminescent lifetime at temperature T, k_0 = decay rate at for the lowest energy triplet substate, k_1, k_2 = decay rate constants, E_{a1}, E_{a2} = activation energies, and k_B = Boltzmann constant. Inclusion of a third term to the equation did not improve the fit. The results of the fitting of experimental data to Eq. 39.2 are summarized in Table 39.2. The fitting procedure required the value for k_0 to be fixed in order to obtain reasonable values for the kinetic parameters. An estimate for $k_0 = 10^4$ s^{-1} was used for this purpose and is based on values of $\tau = 45$–300 μs obtained for other Ir cyclometalated complexes at temperatures below 15 K [47–51, 85, 89]. Varying k_0 by ± 5000 s^{-1} did not significantly alter the results given in Table 39.2.

$$\tau = 1 \Big/ k_{\text{observed}} = \frac{1 + \exp(-E_{a1}/k_B T) + \exp(-E_{a2}/k_B T)}{k_0 + k_1 \exp(-E_{a1}/k_B T) + k_2 \exp(-E_{a2}/k_B T)}$$
(39.2)

Plots of temperature versus the decay rate display two distinct regimes, one at low temperature that corresponds to thermal redistribution between the triplet sublevels (dependent on the zfs) and the other at higher temperature that involves population of a nonradiative state. Figure 39.4 shows the data and fits for complexes **3** and **5**. The decay rate constants (k_1) obtained from the low-temperature regime are between 10^5 and 10^6 s^{-1} for **1–8** and **10**, while the values of E_{a1} fall within the range of 40–120 cm^{-1}. The

Table 39.2 Kinetic parameters for the excited-state decay of **1–10** in 2-MeTHF

No.	Complex	k_1 (s^{-1})	E_{a1} (cm^{-1})	k_2 (s^{-1})	E_{a2} (cm^{-1})	$E_{0-0} + E_{a2}$ (cm^{-1})	Φ
1	Ir(ppy)$_3$	1.8×10^6	120				0.97
2	Ir(F$_2$ppy)$_3$	1.5×10^6	50	6.2×10^{11}	4200	26 230	0.98
3	Ir(ppz)$_2$(ppy)	1.6×10^6	120	4.0×10^{14}	4800	25 680	0.95
4	Ir(F$_2$ppz)$_2$(ppy)	8.5×10^5	50	3.4×10^{13}	4600	26 110	0.93
5	Ir(ppz)$_2$(F$_2$ppy)	1.1×10^6	90	5.0×10^{12}	3300	24 900	0.55
6	Ir(F$_2$ppz)$_2$(F$_2$ppy)	1.2×10^6	60	6.1×10^{12}	3500	25 970	0.60
7	Ir(ppz)$_3$	2.5×10^5	60	1.2×10^{12}	1800	26 070	<0.01
8	Ir(F$_2$ppz)$_3$	2.1×10^5	90	2.6×10^{11}	1600	27 370	<0.01
9	Ir(flz)$_3$	4×10^4	<10				0.81
10	Ir(pmb)$_3$	8.3×10^5	40	3.7×10^9	1700	28 155	0.37

values for **9** are lower ($k_1 = 10^4$ s^{-1}, $E_{a1} < 10$ cm^{-1}). The data for E_{a1} should reflect the zfs between the T$_I$ and T$_{III}$ triplet substates and are comparable to values reported for other cyclometalated Ir(III) complexes [47, 49, 50, 89]. Large values of zfs have been correlated with an increasing degree of MLCT character in the excited state [90]. The low E_{a1} value for **9**, coupled with its low radiative rate, is consistent with a relatively small MLCT contribution to the ^{3}MLCT-LC state for this complex.

The activation energy to populate the NR state (E_{a2}, labeled E_a in Scheme 39.1) was determined from the temperature-dependent decay rate. The values of E_{a2} for **2–8** and **10** vary between 1600 and 4800 cm^{-1} (Table 39.2). The decay rate constant of the NR state (k_2) for **2–8** are all 10^{11}–10^{14} s^{-1}, whereas the value for **10** is slower (~10^9 s^{-1}). No thermal activation to an NR state is observed for either **1** or **9**. For **1**, the only thermally activated phenomenon observed within the limit of our measurement (77–398 K) is the variation in the radiative rate due to the equilibrating triplet sublevels. Likewise for **9**, an efficient blue-green phosphor ($\Phi = 0.81$), the only temperature-dependent phenomena observed between 77 and 298 K appears to originate from thermal population of the triplet substates. However, data from **9** was less reproducible than that from the other cyclometalated species as this complex has a relatively long lifetime ($\tau_{298K} = 48$ μs, Table 39.1) that makes it extremely susceptible to quenching by oxygen ($k_q = 1.5 \times 10^{10}$ M^{-1} s^{-1}; Stern–Volmer quenching constant, $k_q\tau_{298K} = 7.2 \times 10^5$ M^{-1}) and other impurities. The high sensitivity to quenching from water or other spurious impurities released on flame-sealing samples of **9** made it difficult to precisely monitor temperature-dependent variations in the luminescent decay using our experiment setup.

The luminescent quantum efficiency for complexes **2–8** and **10** is determined principally by the activation energy to the NR state. Complexes **2–4** have high Φ (>0.93) and values of E_{a2} (> 4000 cm^{-1}) that are large enough to make the NR state thermally accessible only at temperatures greater than 300 K, whereupon there is a marked decrease in the luminescent lifetime (e.g., see **3** in Fig. 39.4). Complexes **5** and **6** with moderate quantum efficiency ($\Phi = 0.55$ and 0.60, respectively) have intermediate values for E_{a2} (~3400 cm^{-1}). The temperature-dependent lifetime data for both

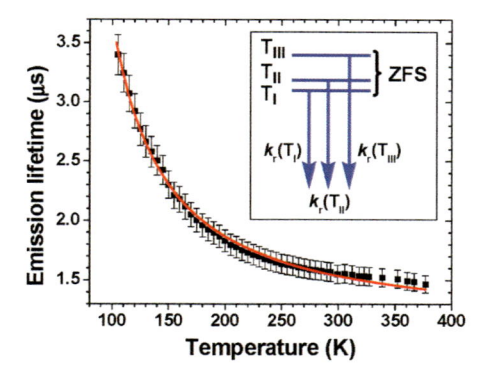

Figure 39.4 Temperature dependence of luminescence decay rate for **3** and **5**, illustrating zfs dominated decay from 100 to 250°C and $k_{nr}(T)$-dependent decay > 250°C. Lines are fits to the data using Eq. 39.2.

of these complexes show that the NR state is accessible at room temperature (e.g., **5** in Fig. 39.4).

Complexes **7** and **8** are effectively nonemissive at room temperature ($\Phi < 0.001$), with small E_{a2} values (1800 cm^{-1} for **7**, 1600 cm^{-1} for **8**). Thermally activated decay from the NR state is highly effective even below room temperature (Fig. 39.5). Both **7** and **8** are strongly emissive at 77 K, and the luminescent lifetimes show marked increase upon cooling from 298 to 77 K: from 0.002 to 14 μs for **7** and 0.007 to 25 μs for **8**. Finally, for *fac*-Ir(pmb)$_3$ (**10**), although the activation energy is small ($E_{a2} = 1700$ cm^{-1}) the quantum efficiency is relatively high ($\Phi = 0.37$). Despite having a similar value of E_{a2} to that of nonemissive **7**, the luminescent lifetime for **10** displays only a moderate dependence on temperature ($\tau_{298K} = 1.1$ μs, $\tau_{77K} = 3.1$ μs). The reason for this insensitivity is the markedly smaller pre-exponential term for **10** ($k_2 = 3.7 \times 10^9$ s^{-1}), which also allows for moderately efficient emission to occur at room temperature.

The variation in the thermally activated kinetic parameters for luminescent decay provides some clues as to the nature of the NR state. The height of the barrier to reach the NR state from the ground state for **7** ($E_{0-0} + E_{a2} = 26\,070$ cm^{-1}, 75 kcal/mol) is close to the strength of an Ir–C$_{phenyl}$ bond (80 kcal/mol) [91]. The energy barrier to the NR state for **10** ($E_{0-0} + E_{a2} = 28\,155$ cm^{-1}, 80 kcal/mol)

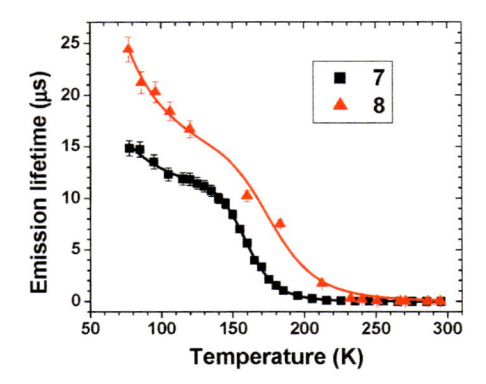

Figure 39.5 Temperature-dependent luminescent decay of **7** and **8**, illustrating zfs dominated decay from 77 to 100°C and $k_{nr}(T)$-dependent decay > 150°C. Lines are fits to data using Eq. 39.2.

is also comparable to the bond strength calculated for a Ir–C$_{phenyl}$ and Au–C$_{carbene}$ linkages (65–88 kcal/mol) [92, 93]. Given these energies, a process involving Ir–ligand bond rupture is a likely path to the NR state. Additional support for ligand dissociation comes from studies on luminescent Ru diimine complexes, where studies suggested that photosubstitution reactions in these species coincide with thermal activation to the nonradiative states [94]. A signature for a dissociative state in the Ru complexes is the magnitude of the pre-exponential term for thermal deactivation (k_2) [95]. Large values of k_2 (10^{11}–10^{14} s^{-1}), similar to those is found for **2–8**, are associated with high-frequency vibrations and consistent with bond rupture being involved in the luminescent deactivation process.

A recent publication discussing DFT calculation for **1** and **7** provides additional support for the participation of a ligand dissociation process in the excited state [96]. According to these calculations, the most favorable (lowest energy) triplet state of **7** has a five-coordinate structure, formed upon breaking an Ir–N bond of one phenylpyrazolyl ligand. On the basis of these results, we carried out DFT calculations, using a B3LYP/LACVP** model, for the triplet states of six- and five-coordinated variants of **1–10**. Selected metrical and thermodynamic data calculated for the complexes are presented in Supporting Information Tables S1 and S2, respectively. A distorted octahedral geometry around the metal

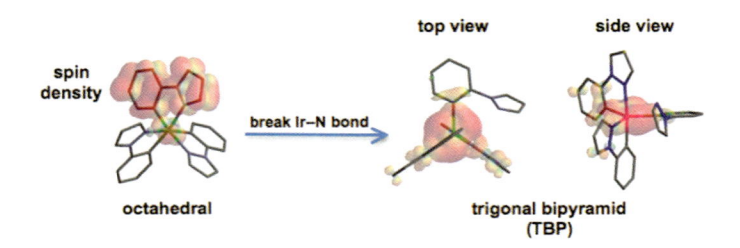

Figure 39.6 Structures and spin density surfaces calculated for the triplet states of six- and five-coordinated forms of **7**.

center, similar to that determined for the singlet state, was found for all of the six-coordinate species. One of the chelates has metal–ligand bonds that are ca. 0.03–0.05 Å shorter than those found in the other two ligands. The spin density surface is primarily localized on this distorted ligand, along with a small contribution from the metal center, and is compatible with a ^{3}MLCT-LC description of the excited state. The structures for the five-coordinate species were obtained from an unrestricted geometry optimization calculation, performed from a starting point generated by breaking an Ir–N (or Ir–C$_{carbene}$ for **10**) bond and rotating the unligated heterocyclic to a dihedral angle of 90°, relative to the phenyl group. All the five-coordinate intermediates were verified to be local minima by vibrational analysis. Structures with lowest energy were found for the rotomer with the unbound nitrogen oriented toward a hydrogen atom on the heterocycle of an adjacent chelate (Fig. 39.6).

The geometry around the metal center for the five-coordinate species closely matches that reported earlier for **1** and **7** [96] and is best described as a distorted trigonal bipyramid (TBP) with an equatorial plane containing two phenyl groups and a heterocycle of a chelating ligand, see Scheme 39.2. Previous work by Eisenstein has shown that the triplet state of d^6 ML$_5$ complexes is Jahn–Teller active and that the presence of three strong σ-donating ligands in the equatorial plane favors the TBP geometry over a square pyramidal structure [97]. For the heteroleptic compounds **3–6**, the lowest energy was found for a structure formed by breaking the Ir–N$_{pyrazole}$ bond that is trans to the phenyl of an adjacent ppz ligand (Ir–Z′, Scheme 39.2), an arrangement that places the chelated

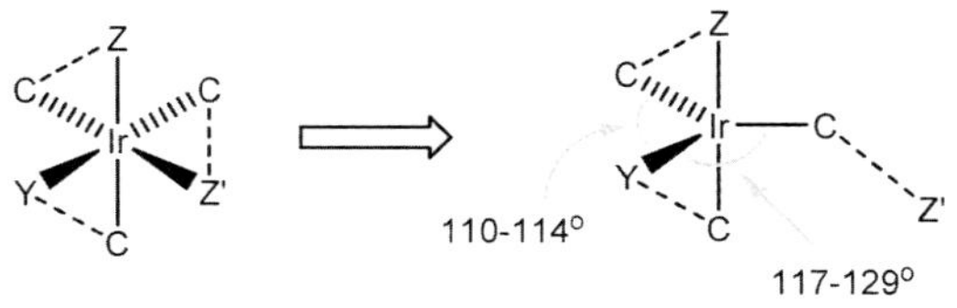

Scheme 39.2

C = phenyl, Y = pyridyl, Z, Z' = pyrazolyl

pyridyl ligand at an equatorial site. The N–Ir–C equatorial angle between the chelated ligands falls in a narrow range (110–114°) and is smaller than the C–Ir–C and N–Ir–C equatorial angles to the monodentate ligand (117–129°). The dihedral angle between the phenyl and heterocycle for the monodentate ligand is > 50° for **1–10**. The spin density is localized within the equatorial plane for the TBP geometry (Fig. 39.6), and the density at the metal center (0.81–1.11) is significantly larger than for the octahedral form (0.12–0.48), a picture consistent with a description of the TBP intermediate as a ^{3}MC state.

A key parameter to characterize the ^{3}MC state obtained from the calculations is ΔH, the difference in energy between the zero-point energies (zpe) calculated for the triplet states of the octahedral and TBP geometries of a respective complex (Table 39.3). The value of ΔH defines the relative thermodynamic stability of luminescent

Table 39.3 Calculated energy parameters for the triplet state of **1–10**

No.	Complex	ΔH(cm^{-1})	$E_{a2} - \Delta H$(cm^{-1})	$E_{0-0} + \Delta H$(cm^{-1})
1	Ir(ppy)$_3$	2117		22 487
2	Ir(F$_2$ppy)$_3$	1754	2416	23 784
3	Ir(ppz)$_2$(ppy)	1042	3788	21 922
4	Ir(F$_2$ppz)$_2$(ppy)	1655	2945	23 165
5	Ir(ppz)$_2$(F$_2$ppy)	130	3170	21 730
6	Ir(F$_2$ppz)$_2$(F$_2$ppy)	766	2744	23 236
7	Ir(ppz)$_3$	−1676	3496	22 594
8	Ir(F$_2$ppz)$_3$	−1719	3279	24 051
9	Ir(flz)$_3^a$	1137		22 057
10	Ir(pmb)$_3$	1557	143	28 012

[a] 9,9'-Dihydro analogue of flz ligand.

triplet state with respect to the product of bond rupture and originates from the differing orbital nature of the two states. The energy of the ^{3}MLCT-LC state in the octahedral form is primarily dictated by the triplet energy of cyclometalated ligand, which can vary considerably depending on the identity of the cyclometalate. On the other hand, the energy of the ^{3}MC state in the TBP intermediate is mostly metal-centered and, thus, determined mainly by the ligand field strength of the coordinating ligands, particularly the ones in the equatorial plane (two phenyls and one imine). For example, the energy of ^{3}MLCT-LC state (E_{0-0}) increases by 3900 cm^{-1} upon going from **1** (ppy ligand) to **7** (ppz ligand), whereas the calculated energy of the ^{3}MC state ($E_{0-0} + \Delta H$) for the TBP form increases only slightly, by $\sim$100 cm^{-1}. A similar difference in triplet energy occurs between the ^{3}MLCT-LC state of **2** and **8** (3740 cm^{-1}) and the ^{3}MC state (267 cm^{-1}). For compounds **1–9**, a plot of ΔH versus E_{0-0} (Fig. 39.7) displays the trend of decreasing ΔH with increasing emission energy. The variation in ΔH with respect to emission energy indicates that, for a particular compound, the TBP intermediate is destabilized to a lesser extent by the various ligands than is the octahedral form. In other words, higher emission energies thermodynamically favor the ^{3}MC state over the ^{3}MLCT-LC state. Two different trend lines can be discerned in the plot,

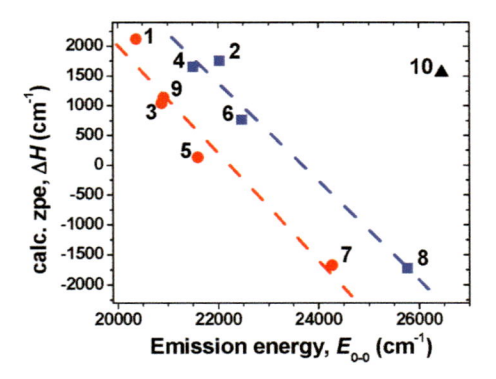

Figure 39.7 Plot of the emission energy (E_{0-0}) vs ΔH for **1–10**. Compounds with phenyl ligands in the equatorial plane of the TBP structure are shown in red, those with F$_2$phenyl are in blue. The lines are just guides for the eye.

for complexes either with difluorophenyl groups in the equatorial plane of the TBP intermediate (**2**, **4**, **6**, and **8**, blue in Fig. 39.8) or with phenyl groups (**1**, **3**, **5**, **7**, and **9** colored red in the plot). This segregation of data indicates that the difluorophenyl groups exert an equivalent destabilizing effect on both the ^{3}MLCT-LC and ^{3}MC states relative to the respective nonfluorinated complexes.

The value of ΔH can also be correlated with the activation energy (E_{a2}). Note, for example, that compound **1**, with the largest value for ΔH (2117 cm^{-1}), has an E_{a2} that is inaccessible even at 378 K. The correspondence between E_{a2} and ΔH is shown for compounds **2–8** and **10** in Fig. 39.8. The decrease in E_{a2} with decreasing ΔH is a consequence of the deviation between ^{3}MLCT-LC and ^{3}LC energies of the two different excited-state geometries of the complexes. The activation energy will decrease as the octahedral form becomes thermodynamically destabilized (less favored) relative to the TBP intermediate (smaller ΔH). The trend displayed in Fig. 39.8 also indicates that the driving force to reach the TBP intermediate is correlated with the luminescent quantum efficiency. Compounds with large ΔH (**1–4**) have high Φ, those with small ΔH (**5** and **6**) have moderate efficiency, and complexes with a negative ΔH (**7** and **8**) are nonemissive at room temperature.

For compound **10**, the relationships between E_{a2}, ΔH, and Φ do not follow the same trend as for the other complexes. The value of

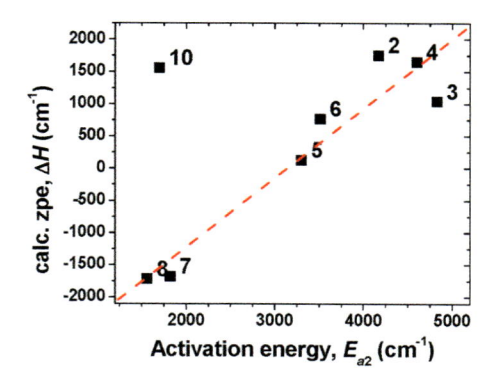

Figure 39.8 Plot of the activation energy (E_{a2}) vs ΔH for **2–8** and **10**. The line is just a guide for the eye.

E_{a2} is similar to that of complexes with low Φ (**7** and **8**), yet the value of ΔH is comparable to that of complexes with much higher Φ, such as **2** and **3**. The discrepancy can be explained using a kinetic scheme, analogous to one first proposed by Meyer for Ru(diimine) complexes [94], where the ^{3}MLCT-LC state is in equilibrium with the ^{3}MC state before undergoing irreversible return to the ground state (Eq. 39.3) [40].

$$^3\text{MLCT-LC}_{\text{octahedral}} \underset{k_b}{\overset{k_a}{\rightleftharpoons}} {}^3\text{MC}_{\text{TBP}} \overset{k_c}{\rightarrow} S_0 \quad (39.3)$$

In the case where $k_c \gg k_b$, the formation of the ^{3}MC state will be the rate-limiting step. Upon entering the ^{3}MC channel, back-reaction will not occur as the complex will undergo rapid intersystem crossing and nonradiative return to the ground state. This interpretation of the excited-state behavior is in line with the thermally activated properties exhibited by complexes **3–8**, such as the large magnitude for the pre-exponential term ($k_2 > 10^{11}$ s^{-1}). However, as the energy barrier to back-reaction ($E_{a2}-\Delta H$) is lowered, the rate of k_b will increase to the point where it will become much greater than k_c and compete effectively with nonradiative decay. The data for **10** is consistent for a case where k_b has increased such that $k_b \gg k_c$. The barrier for back-reaction for this complex is so low ($E_{a2} - \Delta H = 143$ cm^{-1}, Table 39.3) that return to the ^{3}MLCT-LC state decreases the value of $\ln(k_2)$. From the near-zero value of $E_{a2} - \Delta H$ for **10**, one can also estimate a value of $k_c \approx 10^9$ s^{-1}. This value is 10^4 times faster than the radiative rate ($k_r = 3.4 \times 10^5$ s^{-1}, Table 39.1), which still makes population of the ^{3}MC state a viable nonradiative decay process.

The data for the experimental and calculated energetic parameters for the excited states of **1–8** and **10** is summarized in Fig. 39.9. The $E_{0-0} + E_{a2}$ data represents the energy of the transition state, whereas the $E_{0-0} + \Delta H$ data provides an estimate for the energy of the ^{3}MC state (E_a and NR in Scheme 39.1, respectively). Figure 39.9 readily illustrates that the energy of the ^{3}MC state is confined to a narrower range of values (2300 cm^{-1}) for the ppy and ppz derivatives than is the ^{3}MLCT-LC state (5400 cm^{-1}). The tight energy distribution for the ^{3}MC state of complexes **1–8** implies a thermodynamic limit ($\sim$24 000 cm^{-1}, 415 nm) to the stability

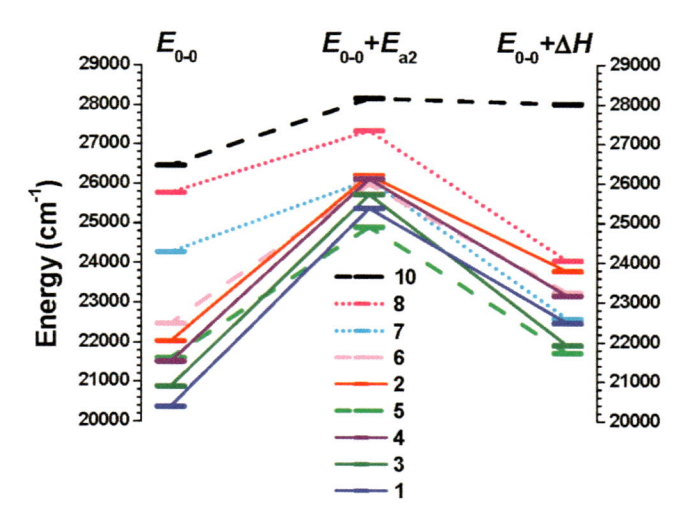

Figure 39.9 Summary of the temperature-dependent energetic parameters for **1–8** and **10**. The E_{a2} value for **1** was estimated to be 5000 cm^{-1} from data in Fig. 39.8. The lines connecting the states distinguish the compounds as having either a high (solid, $\Phi > 0.9$), intermediate (dash, $\Phi = 0.3$–0.6), or low (dot, $\Phi < 0.01$) quantum efficiency at room temperature.

of the ^{3}MLCT-LC excited state, and consequently the luminescent efficiency, that can be achieved for tris-cyclometalated complexes with these types of heterocyclic ligands. However, the high energy of $E_{0-0} + \Delta H$ for **10** suggests that heteroleptic complexes with carbene ligands can provide correspondingly high ^{3}MC state energies that could lead to deep-blue phosphorescent materials with efficiencies similar to values found for **1–4**.

39.4 Conclusion

In conclusion, analysis of the temperature dependence of luminescence from high-energy (sky-blue to near-UV) phosphorescent, tris-cyclometalated iridium complexes uncovers several important aspects regarding the triplet excited state. Once radiative rates reach values of 10^5 s^{-1}, as typically observed for these species, vibrational coupling to the ground state is not a competitive deactivation pathway for nonradiative decay. If no other unimolecular quenching process is available for these complexes, the luminescent quantum yields can approach unity, a situation that occurs for *fac*-Ir(ppy)$_3$ (**1**). One consequence of this result is the need to re-examine previous data reported for the luminescent characteristics of transition metal complexes. In the past several years the widespread use of an earlier, lower value ($\Phi = 0.4$) when using **1** as a primary quantum yield standard has led to an accumulation of errors in the literature values for the photophysical properties of a large number of related phosphorescent materials. Adoption of the higher value for the quantum yield should be employed in future work, and previous literature data that used the lower number needs to be re-evaluated with an appropriate degree of skepticism.

The temperature-dependent behavior for compounds **2–8** and **10** reveals decreases in luminescent efficiency caused by the presence of a thermally activated, nonradiative decay channel. Correlations between kinetic parameters and theoretical modeling suggest that deactivation occurs via a five-coordinate species formed by rupture of an Ir–N (Ir–C$_{carbene}$ in **10**) bond. DFT calculations for the five-coordinate intermediates are consistent with a description of the electronic structure as a ^{3}MC state. Inhibition of the bond rupture process in these types of high-energy phosphorescent complexes is expected to lead to materials with higher luminescent efficiencies. One approach to achieve this goal is to incorporate ligands that increase the energy of the ^{3}MC state and thereby raise the height of the barrier needed to thermally deactivate the complex. This can be accomplished, for example, by using cyclometalated carbenes as ancillary ligands [31]. Another method is to use rigid ligands that limit the degrees of freedom needed to dissociate a metal–ligand bond. This could involve cyclometalating ligands similar 7,8-benzoquinoline [98] or the use tripodal type chelates in order to form a hemicage structure around the complex [77]. The use of a rigid matrix is an additional method that has been shown to inhibit the thermal deactivation process in luminescent transition metal complexes [34, 99]. Any combination of these approaches could be employed to create high-performance blue phosphors in the future, by selecting for materials that inhibit a bond rupture pathway in their excited-state potential energy surface.

Acknowledgment

We would like to acknowledge Universal Display Corporation and Department of Energy for their financial support of this work. We also thank W. J. Finkenzeller and Professor H. Yersin (U. Regensburg) and Professor K. Dedeian (Delaware Valley College) for helpful discussions.

Supporting Information

Plots for the temperature dependence of emission lifetime for **2**, **4**, **6**, and **10**, tables of selected metrical parameters calculated for the triplet states of **1–10**, *xyz* coordinates for the calculated structures of **1–10**. This material is available free of charge via the Internet at https://ndownloader.figstatic.com/files/4539556.

References

1. Endo, A., Suzuki, K., Yoshihara, T., Tobita, S., Yahiro, M., Adachi, C. *Chem. Phys. Lett.* 2008, **460**, 155–157.
2. You, Y., Park, S. Y. *Dalton Trans.* 2009, 1267–1282.
3. Borisov, S. M., Klimant, I. *Anal. Chem.* 2007, **79**, 7501–7509.
4. Zhao, Q., Liu, S. J., Li, F. Y., Yi, T., Huang, C. H. *Dalton Trans.* 2008, 3836–3840.
5. Schmittel, M., Lin, H. W. *Inorg. Chem.* 2007, **46**, 9139–9145.
6. Yu, M. X., Zhao, Q., Shi, L. X., Li, F. Y., Zhou, Z. G., Yang, H., Yia, T., Huang, C. H. *Chem. Commun.* 2008, 2115–2117.
7. Lo, K. K. W., Zhang, K. Y., Leung, S. K., Tang, M. C. *Angew. Chem., Int. Ed.* 2008, **47**, 2213–2216.
8. Lo, K. K. W., Lee, P. K., Lau, J. S. Y. *Organometallics* 2008, **27**, 2998–3006.
9. Elias, B., Genereux, J. C., Barton, J. K. *Angew. Chem., Int. Ed.* 2008, **47**, 9067–9070.
10. Shao, F. W., Elias, B., Lu, W., Barton, J. K. *Inorg. Chem.* 2007, **46**, 10187–10199.
11. Shao, F. W., Barton, J. K. *J. Am. Chem. Soc.* 2007, **129**, 14733–14738.

12. Lo, K. K. W., Zhang, K. Y., Chung, C. K., Kwok, K. Y. *Chem.–Eur. J.* 2007, **13**, 7110–7120.
13. Lo, K. K. W., Tsang, K. H. K., Sze, K. S., Chung, C. K., Lee, T. K. M., Zhang, K. Y., Hui, W. K., Li, C. K., Lau, J. S. Y., Ng, D. C. M., Zhu, N. *Coord. Chem. Rev.* 2007, **251**, 2292–2310.
14. Lo, K. K. W., Lau, J. S. Y. *Inorg. Chem.* 2007, **46**, 700–709.
15. Lo, K. K. W. In *Photofunctional Transition Metals Complexes*; Yam, V. W. W., Ed., Structure and Bonding, Vol. 123; Springer: Berlin, Heidelburg, New York, 2007; pp. 205–245.
16. Lo, K. K. W., Lau, J. S. Y., Lo, D. K. K., Lo, L. T. L. *Eur. J. Inorg. Chem.* 2006, **405**, 4–4062.
17. Lo, K. K. W., Hui, W. K., Chung, C. K., Tsang, K. H. K., Lee, T. K. M., Li, C. K., Lau, J. S. Y., Ng, D. C. M. *Coord. Chem. Rev.* 2006, **250**, 1724–1736.
18. Lo, K. K. W., Chung, C. K., Zhu, N. Y. *Chem.–Eur. J.* 2006, **12**, 1500–1512.
19. Lo, K. K. W., Li, C. K., Lau, J. S. Y. *Organometallics* 2005, **24**, 4594–4601.
20. Lo, K. K. W., Hui, W. K., Chung, C. K., Tsang, K. H. K., Ng, D. C. M., Zhu, N. Y., Cheung, K. K. *Coord. Chem. Rev.* 2005, **249**, 1434–1450.
21. Thompson, M. E., Djurovich, P. I., Barlow, S., Marder, S. R. In *Comprehensive Organometallic Chemistry*; O'Hare, D., Ed., Elsevier: Oxford, 2007; Vol. 12, pp. 101–194.
22. *Highly Efficient OLEDs with Phosphorescent Materials*; Yersin, H., Ed., Wiley-VCH: Berlin, 2007.
23. Yersin, H. *Top. Curr. Chem.* 2004, **241**, 1–26.
24. Tamayo, A. B., Alleyne, B. D., Djurovich, P. I., Lamansky, S., Tsyba, I., Ho, N. N., Bau, R., Thompson, M. E. *J. Am. Chem. Soc.* 2003, **125**, 7377–7387.
25. Lee, S. J., Park, K. M., Yang, K., Kang, Y. *Inorg. Chem.* 2009, **48**, 1030–1037.
26. Yeh, Y. S., Cheng, Y. M., Chou, P. T., Lee, G. H., Yang, C. H., Chi, Y., Shu, C. F., Wang, C. H. *ChemPhysChem* 2006, **7**, 2294–2297.
27. Stagni, S., Colella, S., Palazzi, A., Valenti, G., Zacchini, S., Paolucci, F., Marcaccio, M., Albuquerque, R. O., De Cola, L. *Inorg. Chem.* 2008, **47**, 10509–10521.
28. Song, Y. H., Chiu, Y. C., Chi, Y., Cheng, Y. M., Lai, C. H., Chou, P. T., Wong, K. T., Tsai, M. H., Wu, C. C. *Chem.–Eur. J.* 2008, **14**, 5423–5434.
29. Orselli, E., Albuquerque, R. Q., Fransen, P. M., Frohlich, R., Janssen, H. M., De Cola, L. *J. Mater. Chem.* 2008, **18**, 4579–4590.
30. Di Censo, D., Fantacci, S., De Angelis, F., Klein, C., Evans, N., Kalyanasundaram, K., Bolink, H. J., Gratzel, M., Nazeeruddin, M. K. *Inorg. Chem.* 2008, **47**, 980–989.

31. Chang, C. F., Cheng, Y. M., Chi, Y., Chiu, Y. C., Lin, C. C., Lee, G. H., Chou, P. T., Chen, C. C., Chang, C. H., Wu, C. C. *Angew. Chem., Int. Ed.* 2008, **47**, 4542–4545.
32. Li, J., Djurovich, P. I., Alleyne, B. D., Yousufuddin, M., Ho, N. N., Thomas, J. C., Peters, J. C., Bau, R., Thompson, M. E. *Inorg. Chem.* 2005, **44**, 1713–1727.
33. Yang, C.-H., Li, S.-W., Chi, Y., Cheng, Y.-M., Yeh, Y.-S., Chou, P.-T., Lee, G.-H., Wang, C.-H., Shu, C.-F. *Inorg. Chem.* 2005, **44**, 7770–7780.
34. Allsopp, S. R., Cox, A., Kemp, T. J., Reed, W. J. *J. Chem. Soc., Faraday Trans. 1* 1978, **74**, 1275–1289.
35. Van Houten, J., Watts, R. J. *J. Am. Chem. Soc.* 1976, **98**, 4853–4858.
36. Hager, G. D., Watts, R. J., Crosby, G. A. *J. Am. Chem. Soc.* 1975, **97**, 7037–7042.
37. Hager, G. D., Crosby, G. A. *J. Am. Chem. Soc.* 1975, **97**, 7031–7037.
38. Yersin, H., Humbs, W., Strasser, J. *Top. Curr. Chem.* 1997, **191**, 153–249.
39. Turro, N. J. *Modern Molecular Photochemistry*; Benjamin/Cummings: Menlo Park, CA, 1978.
40. Barigelletti, F., Juris, A., Balzani, V., Belser, P., Vonzelewsky, A. *J. Phys. Chem.* 1986, **90**, 5190–5193.
41. Allsopp, S. R., Cox, A., Kemp, T. J., Reed, W. J., Carassiti, V., Traverso, O. *J. Chem. Soc., Faraday Trans. 1* 1979, **75**, 353–362.
42. Meyer, T. J. *Pure Appl. Chem.* 1986, **58**, 1193–1206.
43. Kober, E. M., Marshall, J. L., Dressick, W. J., Sullivan, B. P., Caspar, J. V., Meyer, T. J. *Inorg. Chem.* 1985, **24**, 2755–2763.
44. Caspar, J. V., Kober, E. M., Sullivan, B. P., Meyer, T. J. *J. Am. Chem. Soc.* 1982, **104**, 630–632.
45. Kober, E. M., Caspar, J. V., Lumpkin, R. S., Meyer, T. J. *J. Phys. Chem.* 1986, **90**, 3722–3734.
46. Barigelletti, F., Sandrini, D., Maestri, M., Balzani, V., von Zelewsky, A., Chassot, L., Jolliet, P., Maeder, U. *Inorg. Chem.* 1988, **27**, 3644–3647.
47. Marchetti, A. P., Deaton, J. C., Young, R. H. *J. Phys. Chem. A* 2006, **110**, 9828–9838.
48. Rausch, A. F., Thompson, M. E., Yersin, H. *Inorg. Chem.* 2009, **48**, 1928–1937.
49. Finkenzeller, W. J., Thompson, M. E., Yersin, H. *Chem. Phys. Lett.* 2007, **444**, 273–279.
50. Finkenzeller, W. J., Stossel, P., Yersin, H. *Chem. Phys. Lett.* 2004, **397**, 289–295.

51. Finkenzeller, W. J., Yersin, H. *Chem. Phys. Lett.* 2003, **377**, 299–305.
52. Harding, R. E., Lo, S. C., Burn, P. L., Samuel, I. D. W. *Org. Electron.* 2008, **9**, 377–384.
53. Lo, S. C., Shipley, C. P., Bera, R. N., Harding, R. E., Cowley, A. R., Burn, P. L., Samuel, I. D. W. *Chem. Mater.* 2006, **18**, 5119–5129.
54. Yang, L., Okuda, F., Kobayashi, K., Nozaki, K., Tanabe, Y., Ishii, Y., Haga, M. A. *Inorg. Chem.* 2008, **47**, 7154–7165.
55. Dedeian, K., Shi, J. M., Shepherd, N., Forsythe, E., Morton, D. C. *Inorg. Chem.* 2005, **44**, 4445–4447.
56. Kawamura, Y., Sasabe, H., Adachi, C. *Jpn. J. Appl. Phys., Part 1* 2004, **43**, 7729–7730.
57. Demas, J. N., Crosby, G. A. *J. Phys. Chem.* 1971, **75**, 991–1024.
58. Jaguar 7.0; Schrodinger, Inc.: Portland, OR, 2007.
59. Becke, A. D. *J. Chem. Phys.* 1993, **98**, 5648–5652.
60. Lee, C. T., Yang, W. T., Parr, R. G. *Phys. Rev. B* 1988, **37**, 785–789.
61. Niu, S. Q., Hall, M. B. *Chem. Rev.* 2000, **100**, 353–405.
62. Baker, J., Muir, M., Andzelm, J., Scheiner, A. In *Chemical Applications of Density-Functional Theory*; Laird, B. B., Ross, R. B., Ziegler, T., Eds., American Chemical Society: Washington, DC, 1996; Vol. 629, pp. 342–367.
63. Hay, P. J., Wadt, W. R. *J. Chem. Phys.* 1985, **82**, 299–310.
64. Melius, C. F., Olafson, B. D., Goddard, W. A. *Chem. Phys. Lett.* 1974, **28**, 457–462.
65. Goddard, W. A. *Phys. Rev.* 1968, **174**, 659–662.
66. Francl, M. M., Pietro, W. J., Hehre, W. J., Binkley, J. S., Gordon, M. S., Defrees, D. J., Pople, J. A. *J. Chem. Phys.* 1982, **77**, 3654–3665.
67. Hariharan, P. C., Pople, J. A. *Chem. Phys. Lett.* 1972, **16**, 217–219.
68. Hedley, G. J., Ruseckas, A., Samuel, I. D. W. *Chem. Phys. Lett.* 2008, **450**, 292–296.
69. Tang, K. C., Liu, K. L., Chen, I. C. *Chem. Phys. Lett.* 2004, **386**, 437–441.
70. Yersin, H., Donges, D. *Top. Curr. Chem.* 2001, **214**, 81–186.
71. Caspar, J. V., Meyer, T. J. *J. Phys. Chem.* 1983, **87**, 952–957.
72. Caspar, J. V., Meyer, T. J. *Inorg. Chem.* 1983, **22**, 2444–2453.
73. Forster, L. S. *Coord. Chem. Rev.* 2002, **227**, 59–92.
74. Henderson, L. J., Fronczek, F. R., Cherry, W. R. *J. Am. Chem. Soc.* 1984, **106**, 5876–5879.

75. Sauvage, J. P., Collin, J. P., Chambron, J. C., Guillerez, S., Coudret, C., Balzani, V., Barigelletti, F., De Cola, L., Flamigni, L. *Chem. Rev.* 1994, **94**, 993–1019.
76. Djurovich, P. I., Murphy, D., Thompson, M. E., Hernandez, B., Gao, R., Hunt, P. L., Selke, M. *Dalton Trans.* 2007, 3763–3770.
77. Schaffner-Hamann, C., von Zelewsky, A., Barbieri, A., Barigelletti, F., Muller, G., Riehl, J. P., Neels, A. *J. Am. Chem. Soc.* 2004, **126**, 9339–9348.
78. King, K. A., Spellane, P. J., Watts, R. J. *J. Am. Chem. Soc.* 1985, **107**, 1431–1432.
79. Kawamura, Y., Goushi, K., Brooks, J., Brown, J. J., Sasabe, H., Adachi, C. *Appl. Phys. Lett.* 2005, **86**, 071104.
80. Holzer, W., Penzkofer, A., Tsuboi, T. *Chem. Phys.* 2005, **308**, 93–102.
81. Tanaka, I., Tabata, Y., Tokito, S. *Jpn. J. Appl. Phys., Part 2* 2004, **43**, L1601–L1603.
82. Goushi, K., Kawamura, Y., Sasabe, H., Adachi, C. *Jpn. J. Appl. Phys., Part 2* 2004, **43**, L937–L939.
83. Tinti, D. S., El-Sayed, M. A. *J. Chem. Phys.* 1971, **54**, 2529–2549.
84. The value for the zfs of *fac*-Ir(ppy)$_3$ is reported to be 83 cm^{-1} in Ref. 50. Subsequent re-examination of the complex in THF has found the zfs to vary between 90 and 150 cm^{-1}, depending on the site occupied in the solvent lattice. Hofbeck, T., Yersin, H. *3rd International Symposium on Molecular Materials-MOLMAT, Book of Abstracts*; Toulouse, France, July 8–11, 2008; p. 157.
85. Goushi, K., Brooks, J., Brown, J. J., Sasabe, H., Adachi, C. *J. Photopolym. Sci. Technol.* 2006, **19**, 181–186.
86. Freed, K. F., Jortner, J. *J. Chem. Phys.* 1970, **52**, 6272–6291.
87. Englman, R., Jortner, J. *Mol. Phys.* 1970, **18**, 145–164.
88. Allen, G. H., White, R. P., Rillema, D. P., Meyer, T. J. *J. Am. Chem. Soc.* 1984, **106**, 2613–2620.
89. Rausch, A. F., Homeier, H. H. H., Djurovich, P. I., Thompson, M. E., Yersin, H. *Proc. SPIE–Int. Soc. Opt. Eng.* 2007, **6655**, 66550F.
90. Yersin, H., Humbs, W. *Inorg. Chem.* 1999, **38**, 5820–5831.
91. Nolan, S. P., Hoff, C. D., Stoutland, P. O., Newman, L. J., Buchanan, J. M., Bergman, R. G., Yang, G. K., Peters, K. S. *J. Am. Chem. Soc.* 1987, **109**, 3143–3145.
92. Scott, N. M., Dorta, R., Stevens, E. D., Correa, A., Cavallo, L., Nolan, S. P. *J. Am. Chem. Soc.* 2005, **127**, 3516–3526.

93. Hu, X. L., Castro-Rodriguez, I., Olsen, K., Meyer, K. *Organometallics* 2004, **23**, 755–764.
94. Durham, B., Caspar, J. V., Nagle, J. K., Meyer, T. J. *J. Am. Chem. Soc.* 1982, **104**, 4803–4810.
95. Barigelletti, F., Juris, A., Balzani, V., Belser, P., Vonzelewsky, A. *J. Phys. Chem.* 1987, **91**, 1095–1098.
96. Treboux, G., Mizukami, J., Yabe, M., Nakamura, S. *Chem. Lett.* 2007, **36**, 1344–1345.
97. Riehl, J. F., Jean, Y., Eisenstein, O., Pelissier, M. *Organometallics* 1992, **11**, 729–737.
98. Lamansky, S., Djurovich, P., Murphy, D., Abdel-Razzaq, F., Kwong, R., Tsyba, I., Bortz, M., Mui, B., Bau, R., Thompson, M. E. *Inorg. Chem.* 2001, **40**, 1704–1711.
99. Thompson, D. W., Fleming, C. N., Myron, B. D., Meyer, T. J. *J. Phys. Chem. B* 2007, **111**, 6930–6941.

Chapter 40

Study of Energy Transfer and Triplet Exciton Diffusion in Hole-Transporting Host Materials

Chao Wu,[a] Peter I. Djurovich,[b] and Mark E. Thompson[b]

[a]*Department of Chemical Engineering and Materials Science, University of Southern California, Los Angeles, California 90007, USA*
[b]*Department of Chemistry, University of Southern California, Los Angeles, California 90007, USA*
met@usc.edu

A device structure is used in which the hole-transporting layer (HTL) of an OLED is doped with either fluorescent or phosphorescent emitters, that is, anode/HTL-host/hole blocker/electron-transporting layer/cathode. The HTL hosts have higher HOMO energy allowing holes to be transported without being trapped by dopant molecules, avoiding direct recombination on the dopant. The unconventional mismatch of HOMO energies between host and dopant allow for the study of energy transfer in these host/guest systems and triplet exciton diffusion in the HTL-host layers of OLED devices, without the complication of charge trapping at dopants. The host materials examined here are tetraaryl-*p*-phenylenediamines. Data shows that

---

Reprinted from *Adv. Funct. Mater.*, **19**, 1–8, 2009.

Förster energy transfer between these hosts and emissive dopant in devices is inefficient. Triplet exciton diffusion in these host materials is closely related to molecular structure and the degree of intermolecular interaction. Host materials that contain naphthyl groups demonstrate longer triplet exciton diffusion lengths than those with phenyl substituents, consistent with DFT calculations and photophysical measurements.

40.1 Introduction

Shortly after the first heterojunction organic light-emitting diode (OLED) was reported [1], doped emitter systems were introduced into OLEDs and found to lead to more efficient devices with tunable emission color [2]. To obtain emission from doped emitters, the efficient harvesting of excitons by dopant molecules is crucial. Localization of excitons on dopants can occur either through direct charge trapping or by energy transfer from excitons formed in the host matrix. Both mechanisms are believed to be involved in electroluminescence, although in many cases charge trapping is found to be the principal mechanism for dopant emission [3–5]. Consequently, host materials that are used in OLEDs are often designed and optimized to promote charge trapping and subsequent carrier recombination at the dopants. Although high efficiencies have been achieved, the relative contribution these two mechanisms have on exciton formation, and the individual roles they play in the performance of OLEDs, is hard to distinguish and often missing from discussion.

Two types of excitons, singlet and triplet, can be created upon charge recombination in a host material. The exciton migrates to the dopant by intermolecular energy transfer processes. Singlet excitons can be transferred either through Förster mechanism, a process involving dipole–dipole interaction mediated energy transfer, or by hopping among neighboring molecules through the electron-exchange (Dexter) mechanism. Förster transfer involves a fairly long-range interaction, which can lead to efficient energy transfer over 8 nm [6]; however, efficient transfer typically takes place at less than 5 nm for organic materials [7]. Since Dexter transfer requires electron exchange, it is only efficient for nearest-neighbor energy transfer. The short lifetimes characteristic of singlet excitons (nanoseconds) lead to diffusion lengths of tens of nanometers. Due to the low oscillator strengths for electronic transitions in organic triplet materials, Förster energy transfer is inefficient [8]. Thus, for organic materials, triplet excitons are transferred principally through a Dexter process. Despite the short range of Dexter transfer, the longer lifetime of triplets (often greater than milliseconds) enables them to diffuse a significant distance. Triplet excitons have been reported to migrate more than 100 nm in organic amorphous films [9]; however, distances are typically closer to 100 Å [10]. Efficient Dexter transfer relies on small separations and direct orbital overlap between donor–acceptor pairs [11]. Studies have found that triplet energy transfer follows the Marcus theory for both electron and hole transfer due to the similarity to weak coupling, so that transfer is favorable in systems with small reorganization energies and large overlap integrals between the molecular orbitals of the interacting species [12, 13].

The importance of exciton diffusion to OLED device performance has been recognized since the early development of OLEDs [2, 14]. The most commonly used approach to study exciton diffusion is to insert a doped layer into devices to serve as a sensing layer at various distances away from the recombination zone in the OLED. Based on the photophysical or electroluminescence data obtained, the exciton diffusion process can be modeled and analyzed [15–20]. Exciton diffusion in OLEDs has only been investigated using a limited number of materials, such as Alq_3 and CBP. In those studies, charges were trapped or partially trapped in the doped region, since the dopant molecules have shallower (less negative) highest occupied molecular orbits (HOMOs) and deeper (more negative) lowest unoccupied molecular orbits (LUMOs) compared to the host. Therefore, the location of the recombination zone is not well confined to the interfacial region in these devices.

In this chapter, we explore a new device structure that avoids charge trapping on dopants to study energy transfer and triplet exciton diffusion in OLEDs. The selection of the host is key to the success of this study and a series of symmetric tetraaryl-*p*-phenylenediamine derivatives have been found to be good candidates for such host materials. Tetraaryl-p-phenylenediamine are known to have tunable HOMO levels, high thermal stability, good hole mobility, and high singlet and triplet energy [21, 22]. These attractive features make it possible for tetraaryl-p-phenylenediamine derivatives to simplify the device structure by being used as both hole-injection material and host. Using new host materials along with the improved device architecture, singlet and triplet excitons are independently harvested in devices and the origin of dopant emission (through charge trapping or energy transfer) is easily confirmed. The effect that the molecular structure of hosts has on triplet exciton diffusion in the solid state is also investigated.

40.2 Results and Discussion

The device architecture used for the studies reported here involves a simple structure, which is designed to avoid charge trapping at dopant molecules and provide a single recombination zone at the host–HTL/ETL interface (Fig. 40.1). This structure requires host materials to have a higher (less negative) HOMO energy than the dopants (C6 and PQIr in this study). The high HOMO energy also enables the host to serve as an effective hole-injecting layer, simplifying the structure. The injected holes will travel through host matrix, without being trapped by dopant molecules, and reach the host/ETL interface, forming a single recombination zone. It is important for both of the singlet and triplet energies of the host to be greater than those of the dopant so that the energy transfer is favorable in the host/guest system for both fluorescent and phosphorescent dopants. Tetraaryl-p-phenylenediamine derivatives fulfill all of the criteria for host materials in the architecture described above. The molecular structure of the phenylenediamine derivatives studied in this paper are given in Fig. 40.1. The unsubstituted compound, DDP, was found to crystallize rapidly after vacuum deposition as a thin film. Two approaches were adopted to modify the molecular structure and give materials that form amorphous films suitable for device studies. Adding alkyl groups to the peripheral phenyl groups at (TTP, TTTP, and TDDP) gives compounds that form glassy

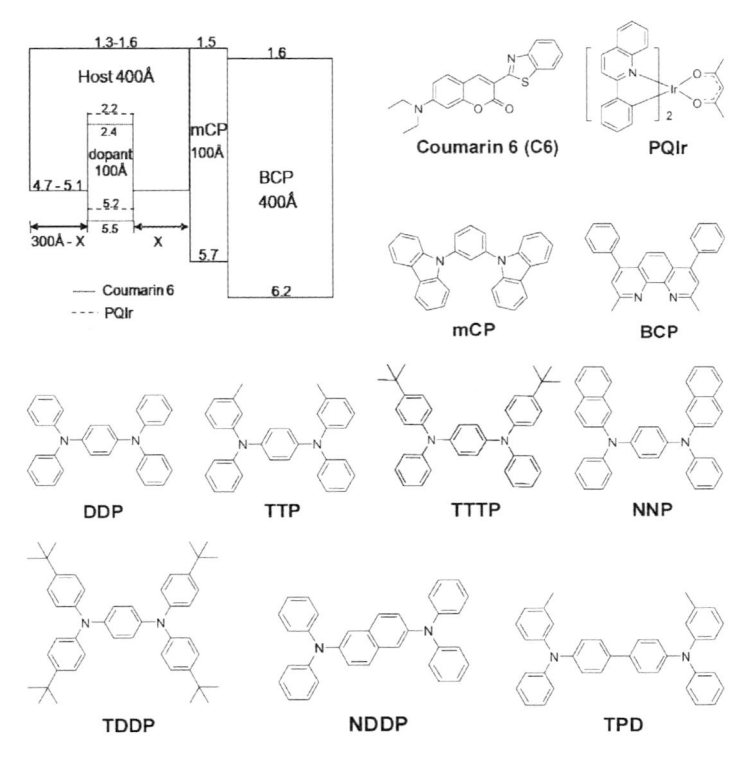

Figure 40.1 Device structure and chemical structures of the molecules used in this study. Horizontal lines and numbers on the lines in device structure represent HOMO and LUMO energy levels (eV).

thin films. Alternatively, the central phenylene was replaced with a bulkier 2,6-naphthylene, NDDP. NNP, with naphthyl groups at periphery, was also synthesized as a complementary structure. For comparative purposes, a widely used host material in OLEDs, TPD, was also chosen due to the similarity of its molecular structure and electronic properties to that of the other candidate host compounds.

40.2.1 Electrochemistry

Electrochemical data was used to estimate the HOMO energies for the materials. Measurements were carried out using cyclic voltammetry (CV) and the oxidation/reduction potentials are listed in Table 40.1. All host candidates undergo two separate, fully reversible oxidation processes. The HOMO and LUMO energies were calculated from the first oxidation and reduction potential respectively, using previously published correlations [23, 24]. All of the host candidates are more easily oxidized than the dopants chosen here, putting their HOMO levels above those of the dopants. The designed mismatch of HOMO energy between the host and dopant should prevent hole trapping by the dopant. The estimated HOMO energies are close to that of indium tin oxide (ITO;4.7 eV), which should facilitate efficient hole injection into the host layer.

40.2.2 Photophysical Properties

Photoluminescence data are also provided in Table 40.1. All of the hosts fluoresce with λ_{max} values between 390 nm and 425 nm in solution at room temperature. The alkyl groups and peripheral naphthyl groups have only a small impact on the energy of fluorescence peak position while the central naphthyl group in NDDP leads to a redshift of ~30 nm. The E_{0-0} values of the triplet state were estimated from the highest energy peak of their phosphorescence spectrum obtained in a dilute frozen solution of 2-MeTHF at 77 K. Figure 40.2 shows the emission spectra of TDDP and NNP at 77 K. Phosphorescence is dominant in the case

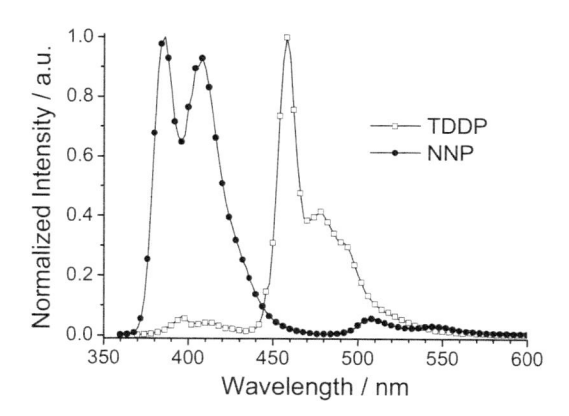

Figure 40.2 Photoluminescence (PL) spectra of TDDP and NNP in 2-MeTHF solution at 77 K.

Table 40.1 Redox, photoluminescence, energy levels and quantum yield data for each material used in devices

Molecule	$E_{1/2}^{Ox}$ [V] [a]	$E_{1/2}^{Re}$ [V] [a]	Emission [nm] Fl [b]	Ph [c]	HOMO/LUMO [ev]	Quantum Yield [%] [d]
TPD	0.36/0.47	−2.93	394/402	505	−5.1/−1.3	0.19
DDP	0.18/0.60	−2.65	390/398 [e]	452	−4.8/−1.6	0.06
TTP	0.16/0.62	−2.65	393/404	454	−4.8/−1.6	0.05
TTTP	0.13/0.56	−2.71	398/414	456	−4.8/−1.6	0.10
TDDP	0.08/0.49	−2.70	399/397	458	−4.7/−1.6	0.03
NDDP	0.24/0.55	−2.72	423/428	505	−4.9/−1.5	0.05
NNP	0.18/0.58	−2.66	393/448	508	−4.8/−1.6	0.05
C6	0.65	−1.97	487/502 [f]	–	−5.5/−2.4	0.69 [f]
PQIr	0.45	−2.19	–	598 [g]	−5.2/−2.2	0.72 [g]
mCP	0.78/0.95	−2.75	338/350	408	−5.7/−1.5	–
BCP	1.13	−2.67	359/390	473	−6.2/−1.6	

a Measured using decamethylferrocene (−0.47 V versus ferrocene) as internal reference and then converted to ferrocene. [b] In the order shown are fluorescent peak maxima in solution and for spin-coated film. [c] The phosphorescent peak with highest energy at 77 K. [d] Measured using spin-coated film. [e] The film crystallizes slowly. [f] Measured from a 2% C6-doped polystyrene film under N_2. [g] Measured from an 8% PQIr-doped polystyrene film under N_2.

of TDDP, while NNP spectrum mainly displays fluorescence. The difference comes from the energy separation between the S_1 and T_1 states, 0.4 eV for TDDP and 0.7 eV for NNP. For compounds with similar electronic configurations, which is the case here, the energy gap between states controls the relative rate of intersystem crossing [8]. The TTP, TTTP, and TDDP derivatives, consisting of a *p*-phenylenediamine core and only phenyl or alkyl–phenyl substitution at N, have triplet energies near 455 nm. Extending the π conjugation, such as with biphenyl and naphthyl cored materials, is expected to lower the triplet energy. Thus, TPD, NNP, and NDDP have triplet energies near 505 nm. The triplet energies of all the hosts are well above that of the phosphorescence dopant used in this study, PQIr ($E_{0-0} = 575$ nm or 2.16 eV).

Emission spectra of TDDP and NNP in solution and neat thin films are compared in Fig. 40.3. TDDP shows no difference between solution and film, whereas emission from NNP film displays a considerable redshift (55 nm). Of all of the hosts studied here, only NNP exhibits such marked redshift from thin film (see the Supporting Information). The most logical explanation for this observation is that the naphthyl groups lead to strong π–π interactions in solid states, since similar bathochromic shifts are typically considered to be a signature of strong intermolecular interaction [25].

40.2.3 Undoped Devices

Undoped devices with the structure ITO/host (400 Å)/BCP (400 Å)/LiF (10 Å)/Al (1200 Å) were investigated as controls. To our surprise, all the electroluminescence (EL) from devices showed redshifted emission relative to all of the components in the device. TDDP is taken as an example and spectra are shown in Fig. 40.4 where the broad, low-energy emission peaks around 500 nm. While this signal did not match any of the materials in the device, it did match the photoluminescence (PL) of the thin film composed of a 50/50 mixture of TDDP and BCP, suggesting that the origin of the low energy EL is an exciplex formed at the host/ BCP interface. Similar exciplex formation between m-MTDATA and BPhen has also been reported [26]. In order to suppress the exciplex emission,

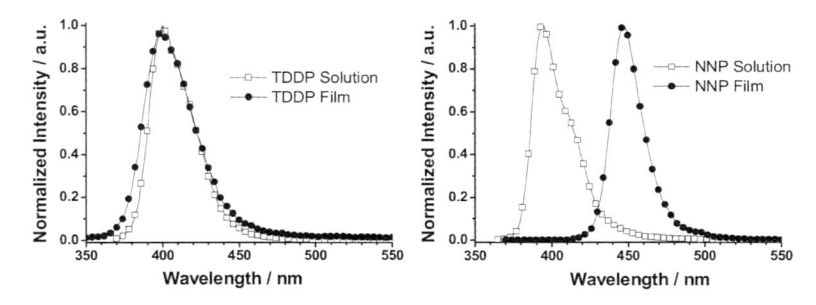

Figure 40.3 PL spectra in 2-MeTHF solution and neat film for TDDP and NNP.

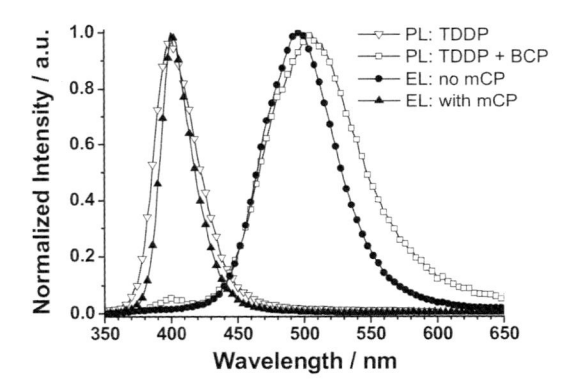

Figure 40.4 PL spectra of TDDP and TDDP:BCP (1:1) film compared with the EL spectra from undoped TDDP devices with and without an mCP interlayer.

a thin layer of N,N'–dicarbazolyl-3,5-benzene (mCP) (100 Å) was inserted between the host and BCP (ITO/host (400 Å)/mCP (100 Å)/BCP (400 Å)/LiF (10 Å)/Al (1200 Å)). These devices give only host emission (Fig. 40.4). While mCP is often employed in OLEDs as an electron blocking layer, the large energy mismatch between the HOMO of mCP and the diammine hosts (> 0.7 eV) should enable it to serve as an effective hole blocking layer. The high triplet energy of mCP (408 nm, 3.04 eV) should also prevent excitons from migrating into the BCP layer.

40.2.4 Doped Devices

In this study, OLEDs were fabricated using either 2 wt% doping of a green fluorescent emitter (C6, emission $\lambda_{max} = 490$ nm) or 8 wt% of an orange–red phosphorescent emitter (PQIr, emission $\lambda_{max} = 600$ nm) as an exciton-sensing layer in the hole-transporting layer (HTL). Doped devices had a general structure of ITO/host (300-X Å)/doped layer (100 Å)/host (X Å)/mCP(100 Å)/BCP (400 Å)/LiF (10 Å)/Al (1200 Å). Devices where X = 0, 50, 100, and 200 Å will be referred to as type I, II, III, and IV respectively in the discussion. For abbreviation purposes, Fand Pare used in front of device to indicate what kind of dopant is used, namely fluorescent (C6) or phosphorescent (PQIr). All the F and P-type devices were made using the host molecules that are listed in Fig. 40.1. The external efficiencies and voltages of the devices at 10 mA cm^{-2} for four host materials, TDDP, TPD, NDDP, and NNP, are compared in Table 40.2 (data on other current densities are given in the Supporting Information).

For the three arylenediamine-based hosts listed in Table 40.2 (TDDP, NDDP, and NNP), the voltages at 10 mA cm^{-2} are nearly constant for all of the doped devices, regardless of the identity of the dopant or its location in the device, and higher than those of their undoped counterparts. This increase in voltage may be due to either charge trapping or scattering by the dopant [27]. The data suggests charge scattering to be the main factor, because the voltage increase is independent of the different HOMO levels and doping percentages of the two dopants (PQIr and C6) [28]. The devices become more resistive, as the holes can be deflected by dopant molecules whose HOMO levels are deeper, on their way to the recombination zone. Data in Table 40.1 also suggests that holes are not trapped by dopants. According to the table, the HOMO of TPD is the closest to both dopants. This means that hole migration should be least affected in doped TPD matrices, as evidenced by the near constant voltages for all of the TPD based devices, including the undoped one. Previous work has also established that neither C6 nor Ir(ppy)$_3$ trap holes when TPD is used in such host/guest systems [14, 28]. Hole trapping by either C6 or PQIr in these new hosts is even less likely to occur, since the hosts studied in this work all have HOMO levels

Table 40.2 External quantum efficiency (EQE) and voltage data of TDDP, TPD, NDDP and NNP devices at 10 mA cm^{-2}

	Undoped	F-I	F-II	F-III	P-I	P-II	P-III	P-IV
TDDP	0.38/7.5	0.96/8.6	0.43/7.9	0.39/8.6	5.29/6.9	1.02/7.2	0.44/7.4	–
TPD	0.61/5.8	0.66/6.0	0.64/5.9	0.90/5.9	3.94/6.1	1.84/5.9	1.37/5.9	–
NDDP	0.80/4.9	0.80/4.9	0.68/6.2	0.76/5.9	0.77/6.1	2.34/6.2	2.85/5.8	–
NNP	0.59/5.0	0.67/6.2	0.53/6.2	0.46/6.1	2.49/6.1	3.23/6.0	5.48/6.0	4.20/5.8

a The data are reproducible with an error of less than 10%.

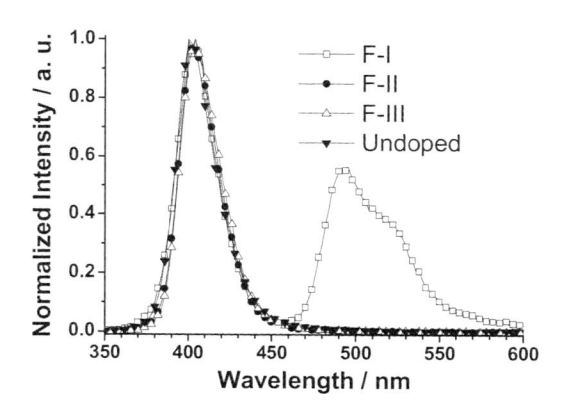

Figure 40.5 EL spectra from F-type TDDP devices at $J = 10$ mA cm^{-2}.

above that of TPD, and the HOMO energy of PQIr is similar to that of Ir(ppy)$_3$ [25].

The EL spectra from all of the undoped, mCP blocked devices only display fluorescence from the host, whereas the spectra from doped devices vary significantly depending on the dopant and position of the sensing layer. Figure 40.5 shows the EL spectra from undoped and F-type devices using a TDDP host. Emission from C6 can be observed in device F-I, whereas in the EL spectra of F-II and F-III device, only host emission is seen. The same results have also been obtained for the other three host materials. This difference in EL spectra is reflected by EQE numbers. At 10 mA cm^{-2}, the F-I device has EQE of 1%, as opposed to 0.4% for the F-II, F-III, and undoped devices. The absence of C6 emission in the F-II and F-III devices lowers the efficiency to match the undoped device. The EL spectra suggest that there is only one recombination zone near the host/mCP interface in this architecture, and that the recombination zone appears to be stationary since the EL spectra of all F-type devices are voltage-independent. A C6-doped layer that is 50 Å or more away from the mCP/host interface is ineffectively sensitized through the Förster mechanism at this distance. The TDDP emission consists of 57% of the EL spectrum of the F-I device, indicating that even shorter range energy transfer from the host to the dopant is rather inefficient. This is confirmed by the photoluminescence observed from a spin-coated TDDP film

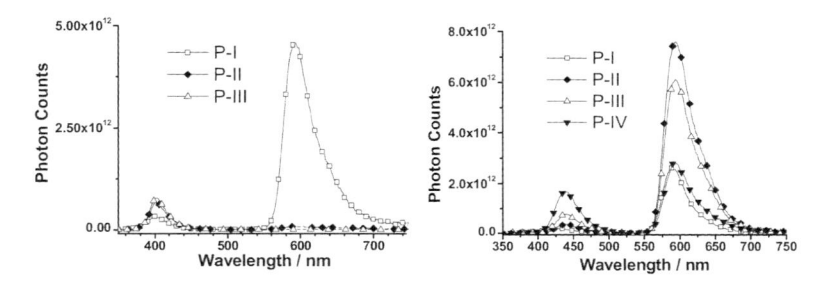

Figure 40.6 EL spectra from P-type TDDP and NNP devices at $J = 10$ mA cm^{-2}.

doped with 2% C6 (Supporting Information), which matches the EL spectrum. There are a couple of possible reasons for the poor luminescence efficiency. One is that the limited spectra overlap between the emission of hosts and the absorption of C6 results in a relatively short Förster energy transfer radius (about 25 Å). The other is C6 derivatives are known to undergo a significant degree of self-quenching when doped at concentrations greater than 1% into a host matrix [29]. The fact that the F-I device displays significant emission from TDDP provides further evidence that charge is not trapped on the C6 dopant. Previous work has shown exclusive emission from C6 when doped into a host with a deep HOMO energy such as PVK [30].

We also examined PQIr as a dopant, with the intent to collect both singlet and triplet excitons. The performance of P-type devices also depends on the position of the sensing layer. The EL spectra of P-type TDDP devices are plotted in photon counts in Fig. 40.6. These plots allow us to easily see how the contribution to EQE from fluorescence and phosphorescence changes for different types of devices. The P-I TDDP device gives nearly exclusive emission from PQIr (95.4% of the EL spectrum), while little or no PQIr emission was observed in P-II and P-III devices. It is possible that some emission in the P-I device occurs from direct recombination on the PQIr dopant. Several factors that could favor this direct recombination include a charge buildup at the host/mCP interface, the high doping percentage of PQIr, and the small difference in energy between the HOMO of TDDP and PQIr. However, a small contribution of host emission in the EL spectra suggests that recombination did take place on host molecules followed by the energy transfer from host to dopant. The decreasing phosphorescent contribution in the other TDDP devices eventually lowers the EQE of the P-III device to that of an undoped device (Table 40.2). In contrast, all the P-type NNP devices show strong PQIr emission. The contribution from PQIr to the EL spectra decreases as the dopant is placed further from the recombination zone, but it still accounts for 67% of the observed emission in the P-IV device (Fig. 40.6).

To confirm that PQIr is excited by triplet energy transfer and not by charge trapping followed by carrier recombination or singlet energy transfer, we decided to employ a triplet filter and prepared a set of P-type devices using NNP as the host. 9,10-Di(2-naphthyl)anthracene (ADN) is an ideal choice for the triplet filter, because it has a high oxidation potential and a triplet energy that is well below those of NNP and PQIr [31]. The structure of the double-doped (PQIr and ADN) P-type device is NNP (200 Å)/8%PQIr:NNP (50 Å)/NNP (50 Å)/5%ADN:NNP (100 Å)/NNP (100 Å)/mCP (100 Å)/BCP (400 Å)/LiF (10 Å)/Al (1200 Å). In this device, ADN will neither trap holes nor collect singlet energy from recombination zone, as it is too far from NNP/mCP interface. On the other hand, it will efficiently trap triplet excitons before they can reach PQIr. The EL spectrum of this double-doped device along with other control OLEDs are provided in Fig. 40.7. It is evident that the addition of the ADN-doped layer completely eliminates the PQIr emission from the double-doped device, whereas the PQIr emission is reasonably strong in the control device where the PQIr layer is at the same distance away from the NNP/mCP interface. This result is in agreement with previous data and verifies that the PQIr emission from this device structure is due to triplet exciton energy transfer, rather than charge trapping.

The behavior of the P-type TDDP devices shows that triplet excitons fail to migrate further than 50 Å in the TDDP matrix. EQEs of P-I devices follow the trend TDDP > TPD > NDDP ≈ NNP, while the order for P-III devices is reversed, NNP > NDDP > TPD > TDDP (Table 40.2). Since the impact of trapping and scattering by the dopant is similar in the various hosts, this phenomenon can be reasonably explained by the difference in transport properties of

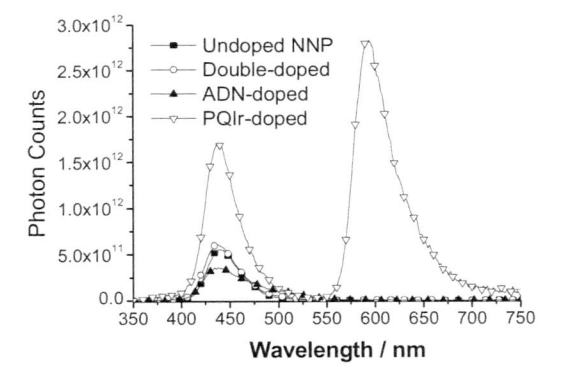

Figure 40.7 EL spectra from double-doped and control devices when $J = 10$ mA cm^{-2}. The structure of the undoped and double-doped devices is given in the text. The ADN-doped and PQIr-doped device has the structure of NNP (300 Å)/5%ADN:NNP (100 Å)/mCP (100 Å)/BCP (400 Å)/LiF (10 Å)/Al (1200 Å) and NNP (200 Å)/8%PQIr:NNP (100 Å)/NNP (200 Å)/mCP (100 Å)/BCP (400 Å)/LiF (10 Å)/Al (1200 Å), respectively.

triplet excitons in these materials. The P-I device using TDDP has the highest EQE among all the host materials, a result that could be due to the short triplet exciton diffusion length which allows the PQIr dopant to efficiently harvest triplets confined within the narrow recombination zone of the device. Using the same argument, TDDP gives solely host emission when used in the P-III architecture, since triplet excitons cannot diffuse through the spacer layer and be collected by PQIr molecules located further away. In contrast, P-III devices using the other three hosts yield much higher EQE, benefiting from the fact that more triplet excitons are able to diffuse out of recombination zone. In other words, triplet excitons can diffuse a relatively greater distance in these matrices. Figure 40.8 shows the contribution of PQIr for the four hosts studied here, from which the diffusivity of triplet excitons in different host materials can be qualitatively compared. The exciton diffusion length of TDP has been reported to be 170 Å using photocurrent spectroscopy [32]. However, based on our data, the triplet exciton diffusion length in TPD is far from ideal and could be well below the reported value. The contradiction may also come from the limitations of photocurrent spectroscopy approach itself [33].

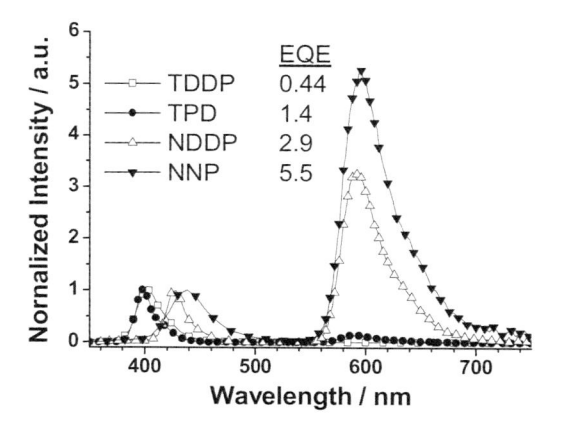

Figure 40.8 EL spectra from P-III devices using different hosts at $J = 10$ mA cm^{-2}. The spectra are normalized to the blue fluorescence peak maxima. Inset: comparison of EQE among P-III devices using different hosts.

Other DDP derivatives with various alkyl substituents were also used as hosts in P-III devices and found to have EQEs that decrease in the order of TTP > TTTP > TDDP (Fig. 40.9). The trend of EQEs can be correlated to the size and degree of alkyl substitution in the molecule, because it has been shown from a previous study that triplet states do not extend beyond alkyl groups in aromatic compounds [34]. Consequently, alkyl groups act as

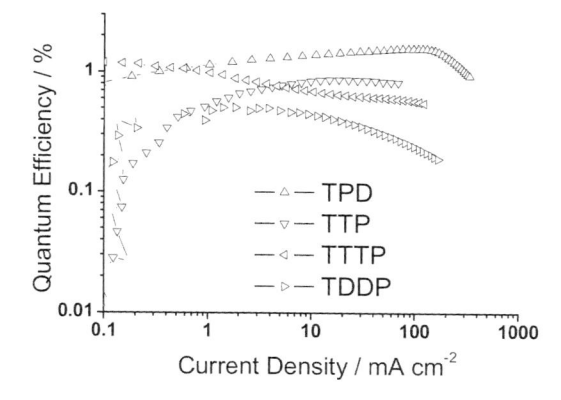

Figure 40.9 EQE of P-III devices with host materials that do not contain naphthyl groups.

a spacer to increase the separation and decrease the electronic coupling between neighboring molecules.

40.2.5 Theoretical Analysis

To rationalize why the range of diffusion lengths is so different among the various host materials, particularly those that contain naphthyl groups such as NNP, it is worthwhile to examine the details of triplet exciton diffusion. Triplet energy transfer can be modeled as a random diffusion process of triplet excitons. The diffusion length is described in Eq. 40.1.

$$L = \sqrt{D\tau} \tag{40.1}$$

Here, D is the diffusion coefficient and τ is the lifetime. Since the lifetime of phosphorescence among these host molecules are comparable, the diffusion length should depend solely on D. The diffusion coefficient is proportional to the rate constant for triplet energy transfer, k, according to Eq. 40.2 [35].

$$k = 4\pi D R_{\text{eff}} N_0 \tag{40.2}$$

Here, R_{eff} is the largest collision diameter for the two molecules undergoing triplet energy transfer and N_0 is Avogadro's number. In addition, the rate constant, k, should follow the Golden Rule given in Eq. 40.3 for a nonadiabatic process in a weakly coupled system [36].

$$k = \frac{2\pi}{h} |V|^2 \frac{\exp(-\lambda/4k_B T)}{\sqrt{4\pi \lambda k_B T}} \tag{40.3}$$

Here, λ is the reorganization energy and V is the electronic coupling term. Therefore, based on the theoretical derivation, materials with a small λ and large V should have large rate constants for energy transfer needed to support long diffusion lengths for triplet excitons.

Estimates for triplet reorganization energies can be obtained from calculations using density function theory (DFT). Following the approach described in the Experimental Section, values of λ for triplet exciton transfer for DDP (an analogue of TDD), TPD, NDDP, and NNP were computed to be 0.62, 0.85, 0.61, and 0.75 eV, respectively. The calculated reorganization energies for the compounds do not follow the order of diffusion lengths estimated from the OLED data. For example, the large reorganization energy of

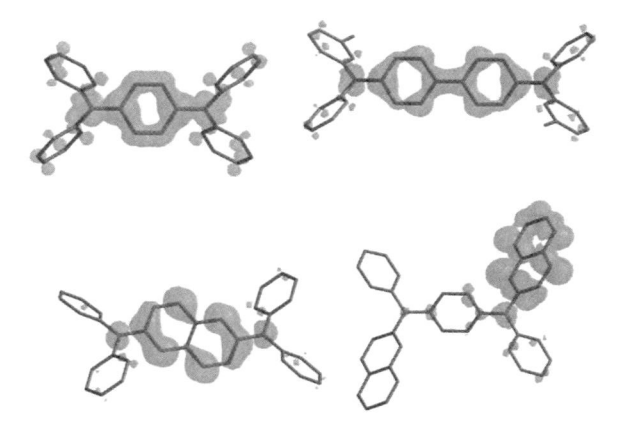

Figure 40.10 Spin density distribution of triplet states from DFT calculations. Top, DDP (left) and TPD (right); bottom, NDDP (left) and NNP (right).

NNP does not correlate with the high conductivity of triplet excitons demonstrated by the P-III devices made with NNP. The absence of any correlation between λ and the triplet diffusion length directs attention to the coupling term, V, in order to explain the trends among the host materials.

To the best of our knowledge, it has not been possible to accurately predict V in disordered materials using theoretical models. Therefore, some qualitative approaches need to be adopted to get an estimate of V. In our case, two key questions need to be addressed before we begin the analysis: where are triplet states located in the molecule, and to what extent do those states overlap intermolecularly? To answer the first question, the spin density distribution, which reflects the location of triplet states in a molecule, was determined using DFT calculations (Fig. 40.10). It can be seen that the spin density is mainly localized on the central aromatic ring(s) in DDP, TDP, and NDDP, whereas the spin resides on the outer naphthyl group in NNP. The PL spectra of these molecules provide information to the second question. As shown in Table 40.1 and Fig. 40.3, the emission peak of TDDP film is identical to that in solution, whereas a small redshift is observed for TPD and NDDP (Supporting Information). This suggests that the small aromatic π-system at the core of TDDP is unable to form effective intermolecular $\pi-\pi$ interactions, while the extended conjugated biphenyl and naphthyl systems of TPD and NDDP lead to weak but observable interactions. Additionally, the *t*-butyl groups in TDDP provide extra steric hindrance to further minimize the effects of π-stacking. In contrast, the fluorescent emission maximum of NNP undergoes a shift from 393 nm in solution to 448 nm when measured in a thin film. The large redshift from solution to film is characteristic of $\pi-\pi$ interactions between aromatic moieties caused mainly by intermolecular overlap of π-orbitals in the conjugated system. Furthermore, this noncovalent interaction is stronger when the amount of overlap is greater. This implies that the flat, electron-rich naphthyl groups in NNP molecules are likely responsible for this type of p-interaction during aggregation in the solid state. The naphthyl groups at periphery should have little difficulty to encounter adjacent peers and undergo π-stacking of naphthyl groups. As a result, the degree of $\pi-\pi$ stacking in the film should show a trend TDDP < TPD ≈ NDDP < NNP. Since the π overlap provides pathways, the value of coupling term, V, among these four molecules should generally follow the same trend. It is important to recognize that the moieties which tend to stack in these molecules are exactly where triplet states are located. With plenty of accessible naphthyl groups in NNP, this well coupled matrix offers the most effective path for diffusion of triplet excitons.

40.3 Conclusions

A simple and effective device structure is reported to study energy transfer and triplet exciton diffusion in this chapter. This device structure has been shown to have a single, stable recombination zone and is able to eliminate charge trapping by dopants, a process that is often difficult to exclude in studies on triplet exciton diffusion. A deliberate mismatch between HOMO energy of host/dopant and good hole mobility of triarylamines are key to success. Good overlap of π-system molecular orbitals is crucial to extending the triplet exciton diffusion length, with naphthyl groups providing more π-overlap in solid state than either phenyl or biphenyl groups. Despite having a larger reorganization energy, NNP displays the longest triplet exciton diffusion length as a host, which leads to the best device performance. The location of the naphthyl groups also has an effect on triplet exciton diffusion depending on specific molecular structure. NNP demonstrates that exposing naphthyl groups to the periphery of the molecule extends the diffusion length of triplet excitons. Bulky alky groups, such as *t*-butyl, on peripheral aromatic rings inhibit the rate of exciton hopping by increasing separation between adjacent molecules.

40.4 Experimental

Synthesis: The tetraaryl-*p*-phenylenediamine derivatives were synthesized by Buchwald–Hartwig coupling using a catalyst combination of Pd(OAc)$_2$ and P(*t*-Bu)$_3$ [37]. Mass spectra were recorded on an HP 5973 mass spectrometer using electron ionization 70 eV. Elemental analysis was carried out by the microanalysis laboratory at University of Illinois, Urbana-Champaign. NMR spectra were measured on a Bruker AC 250 MHz spectrometer. The glass transition temperatures (T_g) and melting points were measured using a TA Instruments 910 differential scanning calorimeter (DSC). For T_g measurements, the samples were initially heated up to 400°C at a rate of 10°C per minute, then cooled using the quench cooling accessory provided with the instrument. After that, the same samples were heated up at the rate of 5°C several times followed by natural cooling after each heating cycle. Coumarin 6 (C6) and BCP were purchased from Aldrich Chemical Co., PQIr [38] was obtained from Universal Display Corp., and mCP was synthesized according to the published literature procedure [39]. All other starting materials and solvents were purchased from commercial sources and used without further purification.

Method A: A three-neck round-bottomed flask was charged with 1,4-dibromobenzene or 2,6-dibromonaphthalene (1 equiv, 5 mmol), NaO^{t}Bu (2.5 equiv), Pd(OAc)$_2$ (0.05 equiv), and P(*t*-Bu)$_3$ (0.15 equiv). Xylene was added such that the concentration of 1,4-dibromobenzene or 2,6-dibromonaphthalene was 0.7 mol L^{-1}. The chosen diarylamines (2.5 equiv) were then added against a stream of nitrogen and the mixture was refluxed overnight. After the reaction was stopped and cooled to room temperature, the crude products were either precipitated with hexane or obtained by removing solvent under reduced pressure and then chromatographed on a column of silica gel (hexane/dichloromethane—5:1) to give the final product. Before characterization and deposition, all products were gradient sublimed using a three-zone furnace at the pressure of 10^{-6} Torr. The temperature differences between neighboring zones were 10° C and 60°C, so, for example, 180/170/120°C for the purification of DDP. The yield was calculated based on the product collected after sublimation and relative to limiting reagent, bromides.

Method B: The same as Method A except the flask was charged with N,N'-Diphenyl-p-phenylenediamine, which is limiting reagent in this case, base and catalysts first and added with 1-Bromo-4-*tert*-butylbenzene later.

1,4-Bis(diphenylamino)benzene (DDP): DDP was synthesized using method A from 1,4-dibromobenzene and diphenylamine, sublimed at 180°C with 68% yield. Anal. calcd.: C 87.35; H 5.86; N 6.79; found: C 87.33; H 5.68; N 6.92. ^{1}H NMR ($C_6D_6\delta$): 6.81 (t, 4H), 6.94 (s, 4H), 7.03 (t, 8H), 7.11 (d, 8H); MS m/z 412; T_g: N/A; T_m: 201°C.

1,4-Bis(phenyl-m-tolylamino)benzene (TTP): TTP was synthesized using method A from 1,4-dibromobenzene and 3-methyl-N-phenylaniline, sublimed at 185°C with 56% yield. Anal. calcd.: C 87.24; H 6.41; N 6.36; found: C 87.19; H 6.31; N 6.49. ^{1}H NMR ($C_6D_6\delta$): 1.97 (s, 6H), 6.69 (t, 2H), 6.81 (t, 2H), 6.98–7.07 (m, 14H), 7.17 (m, 4H); MS m/z 440; T_g: 36°C; T_m: 174°C.

1,4-Bis(phenyl-4-tert-butylphenylamino)benzene (TTTP): TTTP was synthesized using method B from 1,4-dibromobenzene and 4-*tert*-butyl-N-phenylaniline, sublimed at 215°C with 76% yield. Anal. calcd.: C 86.98; H 7.68; N 5.34; found: C 87.01; H 7.86; N 5.51. ^{1}H NMR ($C_6D_6\delta$): 1.19 (s, 18H), 6.80 (t, 2H), 6.99 (s, 4H), 7.03–7.06 (d, 4H), 7.13–7.17 (m, 12H); MS m/z 524; T_g: 72°C; T_m: N/A.

1,4-Bis(di-4-tert-butylphenylamino)benzene (TDDP): TDDP was synthesized using method A from 1,4-dibromobenzene and bis(4-*tert*-butylphenyl)amine, sublimed at 285°C with 81% yield. Anal. calcd.: C 86.74; H 8.86; N 4.40; found: C 86.88; H 8.93; N 4.21. ^{1}H NMR

($C_6D_6\delta$): 1.22 (s, 36H), 7.06 (s, 4H), 7.18–7.19 (d, 16H); MS m/z 636; T_g: N/A; T_m: 317°C.

4.8. 1,4-Bis(2-naphthylphenylamino)benzene (NNP): NNP was synthesized using method A from 1,4-dibromobenzene and 2-naphthylphenylamine, sublimed at 235°C with 90% yield. Anal. calcd.: C 89.03; H 5.51; N 5.46; found: C 89.08; H 5.29; N 5.60. ^{1}H NMR ($C_6D_6\delta$): 6.85 (t, 2H) 7.03 (s, 4H), 7.07 (t, 4H) 7.14–7.19 (m, 8H), 7.31–7.33 (d, 4H), 7.48–7.50 (d, 2H), 7.54–7.57 (t, 4H); MS m/z 512; T_g: 76°C; T_m: 200°C.

2,6-Bis(diphenylamino)naphthalene (NDDP): NDDP was synthesized using method A from diphenylamine and 2,6-dibromonaphthalene, sublimed at 275°C with 89% yield. Anal. calcd.: C 88.28; H 5.67; N 6.06; found: C 88.42; H 5.60; N 6.22. ^{1}H NMR ($C_6D_6\delta$): 6.84 (t, 4H), 7.05 (t, 8H), 7.11–7.14 (m, 12H), 7.43 (d, 2H); MS m/z 462; T_g: N/A; T_m: 277°C.

Characterization Methods: Oxidation and reduction potentials were measured by cyclic voltammetry (CV) using a EG&G Instruments model 283 potentiostat. CV scans were recorded using a Ag wire as a pseudo-reference electrode at a scan rate of 100 mV s^{-1}, in dry and degassed ethylene carbonate/dimethylcarbonate (1:1) mixture with 0.1 M tetrabutylammonium hexafluorophosphate as electrolyte. Decamethylferrocene (−0.47 V vs. ferrocene) were used as an internal reference. Both photoluminescence (PL) and electroluminescence (EL) emission spectra were obtained by a PTI QuantaMaster model C-60SE spectrofluorometer, equipped with a 928 PMT detector and corrected for detector response. 2-Methyltetrahydrofuran was used as solvent for solution PL. Films used for obtaining PL spectra and quantum yield were spin-coated on quartz substrates in air at room temperature using CH_2Cl_2 as solvent. Quantum yield was measured by a Hamamatsu PL Quantum Yield Measurement System (C9920-01). Förster radii were calculated using PhotocamCAD HD1.1.

Theoretical calculations were done using Titan version 1.0.7 to estimate the reorganization energies during the process of triplet exciton transfer. The reorganization energy of the acceptor was obtained by subtracting the T_1 energy in its optimized geometry from the T_1 energy in the geometry of the optimized S_0 state. The reorganization energy of the donor was computed in a similar way by subtracting the S_0 energy in the optimized geometry of T_1 state from the S_0 energy in optimized ground state [40]. The numbers provided in this chapter are the sum of the reorganization energies for a pair of donor and acceptor.

OLED Fabrication and Testing: The OLEDs were grown on pre-cleaned ITO-coated glass substrates with sheet resistance of 20 Ω sq^{-1}. All compounds were purified using temperature gradient vacuum sublimation prior to deposition. Organic layers were deposited by thermal evaporation from resistively heated tantalum boats at a rate of around 2 Å s^{-1}, after which a shadow mask was placed on the substrate and the cathode consisting 10 Å of LiF and 1200 Å of Al was deposited. The devices were tested in air within 1 h of fabrication. Light coming from the front surface was collected by a UV-818 Si photocathode leading to a Keithley 2400 SourceMeter/2000 multimeter coupled to a Newport 1835-C optical meter. Device current–voltage and light–intensity characteristics were measured using the LabVIEW program by National Instruments.

Acknowledgements

This work was supported by Universal Display Corp. (UDC) and Department of Energy. The authors acknowledge Dr. Jason Brooks (UDC) for helpful discussions and Thin Film Devices Inc. for ITO-coated substrates. Supporting Information is available online from Wiley InterScience or from the author.

References

1. C. W. Tang, S. A. VanSlyke, *Appl. Phys. Lett.* 1987, **51**, 913.
2. C. W. Tang, S. A. VanSlyke, C. H. Chen, *J. Appl. Phys.* 1989, **65**, 3610.
3. K. Utsugi, S. Takano, *J. Electrochem. Soc.* 1992, **139**, 3610.
4. J. Kido, H. Shionoya, K. Nagai, *Appl. Phys. Lett.* 1995, **67**, 2281.

5. R. J. Holmes, B. W. D'Andrade, S. R. Forrest, X. Ren, J. Li, M. E. Thompson, *Appl. Phys. Lett.* 2003, **83**, 3818.
6. A. L. T. Khan, P. Sreearunothai, L. M. Herz, M. J. Banach, A. Kohler, *Phys. Rev. B* 2004, 69.
7. J. R. Lacowicz, *Principles of Fluorescence Spectroscopy*, Springer, New York 2006.
8. N. J. Turro, *Modern Molecular Photochemistry*, University Science Books, Sausalito, CA 1978.
9. M. A. Baldo, D. F. O'Brien, M. E. Thompson, S. R. Forrest, *Phys. Rev. B* 1999, **60**, 14422.
10. A. Itaya, K. Okamoto, S. Kusabayashi, *Bull. Chem. Soc. Jpn.* 1976, **49**, 2037.
11. D. L. Dexter, *J. Chem. Phys.* 1953, **21**, 836.
12. G. L. Closs, M. D. Johnson, J. R. Miller, P. Piotrowiak, *J. Am. Chem. Soc.* 1989, **111**, 3751.
13. R. A. Marcus, *Angew. Chem, Int. Ed.* 1993, **32**, 1111.
14. M. A. Baldo, S. R. Forrest, *Phys. Rev. B* 2000, **62**, 10958.
15. T. A. Beierlein, B. Ruhstaller, D. J. Gundlach, H. Riel, S. Karg, C. Rost, W. Riess, *Synth. Met.* 2003, **138**, 213.
16. Y. C. Zhou, L. L. Ma, J. Zhou, X. M. Ding, X. Y. Hou, *Phys. Rev. B* 2007, **75**, 132202.
17. B. W. D'Andrade, M. E. Thompson, S. R. Forrest, *Adv. Mater.* 2002, **14**, 147.
18. Y. R. Sun, N. C. Giebink, H. Kanno, B. W. Ma, M. E. Thompson, S. R. Forrest, *Nature* 2006, **440**, 908.
19. N. C. Giebink, Y. Sun, S. R. Forrest, *Org. Electron.* 2006, **7**, 375.
20. G. Schwartz, M. Pfeiffer, S. Reineke, K. Walzer, K. Leo, *Adv. Mater.* 2007, **19**, 3672.
21. B. E. Koene, D. E. Loy, M. E. Thompson, *Chem. Mater.* 1998, **10**, 2235.
22. K. Sakanoue, M. Motoda, M. Sugimoto, S. Sakaki, *J. Phys. Chem. A* 1999, **103**, 5551.
23. B. W. D'Andrade, S. Datta, S. R. Forrest, P. Djurovich, E. Polikarpov, M. E. Thompson, *Org. Electron.* 2005, **6**, 11.
24. P. I. Djurovich, E. I. Mayo, S. R. Forrest, M. E. Thompson, *Org. Electron.* 2009, **10**, 515.
25. C. E. S. M. Pope, *Electronic Processes in Organic Crystals and Polymers*, Oxford University Press, New York 1999.
26. D. Wang, W. Li, B. Chu, Z. Su, D. Bi, D. Zhang, J. Zhu, F. Yan, Y. Chen, T. Tsuboi, *Appl. Phys. Lett.* 2008, **92**, 3.

27. K. K. Tsung, S. K. So, *Appl. Phys. Lett.* 2008, **92**, 103315.
28. M. Uchida, C. Adachi, T. Koyama, Y. Taniguchi, *J. Appl. Phys.* 1999, **86**, 1680.
29. F. Pschenitzka, J. C. Sturm, *Appl. Phys. Lett.* 2001, **79**, 4354.
30. X. Z. Jiang, R. A. Register, K. A. Killeen, M. E. Thompson, F. Pschenitzka, T. R. Hebner, J. C. Sturm, *J. Appl. Phys.* 2002, **91**, 6717.
31. J. Shi, C. W. Tang, *Appl. Phys. Lett.* 2002, **80**, 3201.
32. C. L. Yang, Z. K. Tang, W. K. Ge, J. N. Wang, Z. L. Zhang, X. Y. Jian, *Appl. Phys. Lett.* 2003, **83**, 1737.
33. S. B. Rim, P. Peumans, *J. Appl. Phys.* 2008, **103**, 124515.
34. S. C. Tse, S. K. So, M. Y. Yeung, C. F. Lo, S. W. Wen, C. H. Chen, *Chem. Phys. Lett.* 2006, **422**, 354.
35. T. Ohno, K. Nozaki, M. Nakamura, Y. Motojima, M. Tsushima, N. Ikeda, *Inorg. Chem.* 2007, **46**, 8859.
36. P. A. M. Dirac, *Proc. R. Soc. London, Ser. A* 1927, **114**, 243.
37. T. Yamamoto, M. Nishiyama, Y. Koie, *Tetrahedron Lett.* 1998, **39**, 2367.
38. S. Lamansky, P. I. Djurovich, D. Murphy, F. Abdel-Razzaq, R. Kwong, I. Tsyba, M. Botz, B. Mui, B. R. M. E. Thompson, *Inorg. Chem.* 2001, **40**, 1704.
39. V. Adamovich, J. Brooks, A. Tamayo, A. Alexander, P. I. Djurovich, B. W. D'Andrade, C. Adachi, S. R. Forrest, M. E. Thompson, *New J. Chem.* 2002, **26**, 1171.
40. I. Place, A. Farran, K. Deshayes, P. Piotrowiak, *J. Am. Chem. Soc.* 1998, **120**, 12626.

Chapter 41

Synthesis and Characterization of Phosphorescent Three-Coordinate Cu(I)–NHC Complexes[†]

Valentina A. Krylova, Peter I. Djurovich, Matthew T. Whited, and Mark E. Thompson

Department of Chemistry, University of Southern California, Los Angeles, California 90089-0744, USA
met@usc.edu

Cationic and neutral monomeric three-coordinate phosphorescent Cu(I) complexes were synthesized and characterized by XRD analysis, electrochemistry and photophysical studies in different environments. DFT calculations have aided the assignment of the electronic structure and excited state behavior of these complexes.

Luminescent Cu(I) complexes have been extensively studied for the last two decades [1]. Research to date has focused principally on four-coordinate tetrahedral homo- and heteroleptic Cu(I) complexes bearing bisimine and phosphine chelating ligands [2].

[†]Electronic supplementary information (ESI) available: Synthetic procedures, electrochemical data, absorption and emission spectra at room temperature and 77 K. CCDC 778544 and 778545. For ESI and crystallographic data in CIF or other electronic format see DOI: 10.1039/c0cc01864c
Reprinted from *Chem. Commun.*, **46**, 6696–6698, 2010.

Electrophosphorescent Materials and Devices
Edited by Mark E. Thompson
Text Copyright © 2010 The Royal Society of Chemistry
Layout Copyright © 2024 Jenny Stanford Publishing Pte. Ltd.
ISBN 978-981-4877-34-3 (Hardcover), 978-1-003-08872-1 (eBook)
www.jennystanford.com

Photophysical processes that occur in luminescent four-coordinate Cu(I) complexes have been thoroughly investigated. Strong evidence supports a Jahn–Teller based distortion (flattening) in the excited state, and consequent formation of a five-coordinate exciplex, that promotes non-radiative decay [3]. The generally accepted approach to alleviate this problem is to increase the steric bulk of ligands in order to block excited state geometrical distortion and non-emissive relaxation pathways [4]. Alternatively, Lotito and Peters have recently reported several examples of emissive three-coordinate copper(I) arylamidophosphine complexes [5]. Unlike four-coordinate tetrahedral Cu(I) compounds, the three-coordinate geometry eliminates the possibility of flattening distortion in the excited state, though exciplex formation is still possible.

We have synthesized and characterized a luminescent copper(I)–phenanthroline complex, [(IPr)Cu(phen)]OTf (**1**), bearing a monodentate N-heterocyclic carbene (NHC) ligand (IPr = 1,3-bis(2,6-diisopropylphenyl)imidazol-2-ylidene). Utilizing an anionic bisimine ligand, we were also able to prepare a neutral complex, (IPr)Cu(pybim) (**2**) (pybim = 2-(2-pyridyl)benzimidazole) (Fig. 41.1). These complexes are rare examples of emissive Cu(I)–NHC compounds. To the best of our knowledge the only other luminescent Cu(I)–NHC derivative is a binuclear copper(I) complex reported by Matsumoto and co-workers [6].

The copper(I) complexes were prepared in good yields (60%) from commercially available (IPr)Cu(Cl). Reaction of (IPr)Cu(Cl) and phenanthroline in the presence of silver triflate in THF

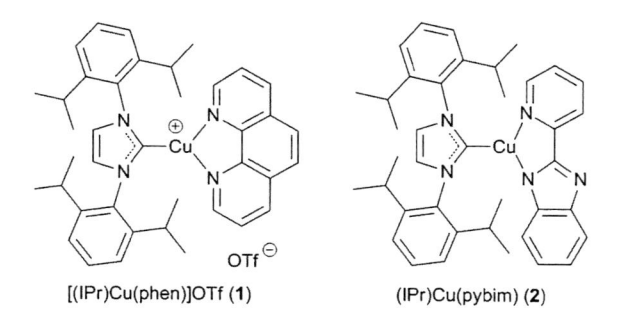

Figure 41.1 Molecular structures of Cu(I)–NHC complexes.

gave **1**. Complex **2** was obtained by deprotonation of 2-(2-pyridyl)benzimidazole with NaH in THF followed by addition of the (IPr)Cu(Cl) precursor. Pure products were isolated as yellow solids upon filtration and recrystallization. The compounds are indefinitely stable to air in the solid state and stable for several hours in solution. Complex **2** can be sublimed at 250°C.

Crystals suitable for X-ray diffraction (XRD) analysis were obtained by vapor diffusion of diethyl ether into a CH_2Cl_2 solution of each complex. Compounds **1** and **2** are monomeric and exhibit three-coordinate structures (Fig. 41.2). The geometry of both complexes is trigonal planar, and the sum of the bond angles around the copper atom are 359.38° for **1** and 360.00° for **2**. Complex

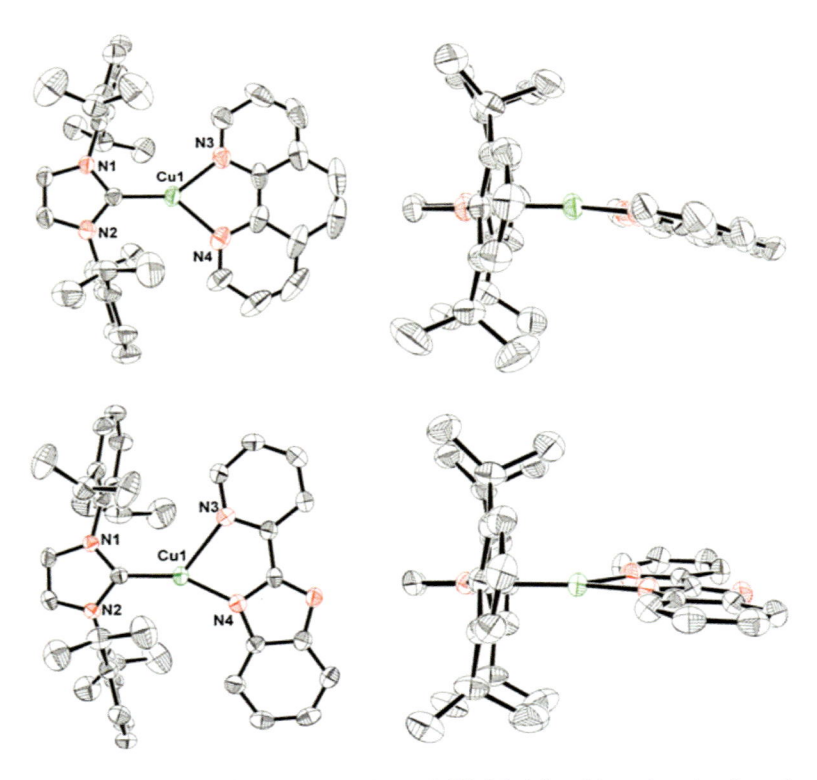

Figure 41.2 ORTEP representation of (IPr)Cu(phen)$^+$ cation (top) and (IPr)Cu(pybim) (bottom). Hydrogen atoms are omitted for clarity.

1 has an almost ideal Y-shaped copper center. The Cu–N bond distances are nearly equal (Cu(1)–N(3) = 2.039(2) Å and Cu(1)–N(4) = 2.052(2) Å), as are the C–Cu–N bond angles (C(1)–Cu(1)–N(4) = 141.49(11)° and C(1)–Cu(1)–N(3) = 136.64(11)°). The coordination environment around copper in **2** is distorted from a Y-shape due to the asymmetric anionic 2-(2-pyridyl)benzimidazole ligand. The Cu(1)–N(4) (imidazolyl) bond (1.9227(18) Å) is shorter than the Cu(1)–N(3) (pyridyl) bond (2.2907(18) Å), and the C(1)–Cu(1)–N(4) and C(1)–Cu(1)–N(3) bond angles are 154.24(8)° and 126.82(8)°, respectively. The values for the bond length between copper and the carbon atom of the NHC ligand (Cu–C = 1.877(2) Å in **1** and 1.884(3) Å in **2**) are typical for Cu(I)–NHC complexes [7]. Surprisingly, the N^N ligand and the imidazolylidene ring of the carbene ligand are coplanar. This geometry is unusual for three-coordinate copper(I)–NHC complexes, as the ancillary ligand is usually oriented perpendicular to the copper-imidazolylidene plane [8]. In complex **2** the dihedral angle between the Cu-imidazolylidene plane and the N–Cu–N chelate plane is 9°. Complex **1** has a smaller twist and the phenanthroline ligand is slightly tilted (9.5°) from the N–Cu–N plane. These distortions are likely due to steric constraints, since hydrogen atoms at the 2- and 9-positions of phenanthroline ligand are directed towards the center of phenyl rings of the NHC ligand. The importance of sterics is highlighted in our attempts to prepare an analog of **1** using bathocuproine (bcp), a 2,9-dimethyl-4,7-diphenyl substituted phenanthroline ligand. Under the reaction conditions described above only $Cu(bcp)_2^+$ and $(IPr)_2Ag^+$ were obtained as major products.

The absorption and low-temperature emission spectra of copper(I) complexes **1** and **2** are depicted in Fig. 41.3. The absorption spectrum of **1** recorded in CH_2Cl_2 shows intense bands in the region between 250–300 nm (ε = 10 000–30 000 M^{-1} cm^{-1}) that are assigned to $\pi \rightarrow \pi^*$ ligand-centered (LC) transitions. Transitions at lower energy (332 nm, ε = 3280 M^{-1} cm^{-1}; 348 nm, ε = 2280 M^{-1} cm^{-1}) can be ascribed to $d\pi \rightarrow \pi^*$ metal-to-ligand charge transfer (MLCT) absorptions. Very weak (ε < 100 M^{-1} cm^{-1}) bands at wavelengths between 400 and 500 nm are assigned to triplet MLCT states. The excitation spectrum (see ESI[†]) matches the absorption spectrum and supports the assignment. The optical spectrum of **2**

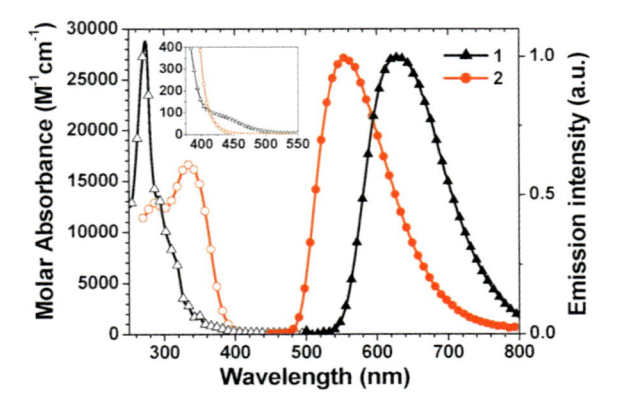

Figure 41.3 Absorption (open symbols, RT, CH_2Cl_2) and corrected low-temperature emission (closed symbols, 77 K, 2-MeTHF) spectra of complexes **1** and **2**. An expanded region of the visible absorption spectra is shown in the inset.

features a single intense absorption band at 335 nm ($\varepsilon = 16\,500$ M^{-1} cm^{-1}) that is assigned to a singlet $\pi \to \pi^*$ LC transition. No distinct triplet transitions were observed in the visible region.

Emission spectra recorded at 77 K in frozen 2-methyltetrahydrofuran (2-MeTHF) glass display a broad, featureless band centered at 630 nm for **1** and 555 nm for **2** (Fig. 41.3). The luminescence decay of **1** has two components with lifetimes of 1.8 μs and 4.6 μs (major component, 76%), whereas the excited state lifetime of **2** is 56 μs. The unstructured emission from both complexes with large Stokes shift and long lifetime is indicative of luminescence from a triplet charge transfer (CT) state.

Both copper(I) complexes are luminescent in oxygen-free CH_2Cl_2 solution at room temperature. Emission maxima are red shifted by 25 nm (**1**) and 50 nm (**2**) compared with those at 77 K (see ESI[†]). After degassing the solutions by sparging with N_2 for 15 minutes, the quantum yield (Φ) for **1** is <0.1%, whereas the value for **2** is 0.5%. Excited state lifetimes (τ) in solution are 0.08 μs (1) and 0.27 μs (**2**) (Table 41.1). The former value is comparable to the shortest lifetimes reported for Cu(I)–bisphenanthroline complexes in degassed solution [1]. The emission properties in solution are solvent-dependent. The lifetime of **2** in cyclohexane is 13.4 μs

Table 41.1 Summary of photophysical properties of **1** and **2**

	Emission lifetime/μs; quantum yield				
Complex	CH_2Cl_2 300 K	C_6H_{12} 300 K	PMMA film (2 wt%) 300 K	Crystals 300 K	2-MeTHF 77 K
1	0.08; <0.001	[a]	0.23 (11%), 1.1 (89%); 0.015	1.2; 0.026	1.8 (24%), 4.6 (76%)[b]
2	0.27; 0.005	13.4; 0.014	24.7; 0.35	33.1; 0.58	56[b]

[a] Insoluble. [b] Quantum yield not measured.

($\Phi = 1.4\%$). In more coordinating 2-MeTHF, the lifetime is 0.14 μs, whereas no emission can be observed in acetonitrile. These observations suggest that emission in solution is quenched to some extent by solvent exciplex formation.

Significant enhancement of emission occurs in rigid media. Compound **2** emits with $\Phi = 58\%$ ($\tau = 33.1$ μs) as a crystalline solid and $\Phi = 35\%$ ($\tau = 24.7$ μs) when doped (2 wt%) in a PMMA film. These data suggest that solvent coordination is not the primary quenching mechanism in fluid solution, since even in non-coordinating cyclohexane the emission is significantly quenched compared with the neat solid. This result suggests that molecular distortions in the excited state may be yet another mechanism for quenching in solution. The type of distortion is not yet clear. Barakat et al. have described a Jahn–Teller induced, Y- to T-shape distortion in the excited state of three-coordinate, trigonal planar Au(I) complexes [9]. A similar type of distortion may possibly occur in the Cu(I)-complexes reported here.

Density functional theory (DFT) calculations were performed on the complexes at the B3LYP/LACVP** level of theory using geometric parameters obtained from XRD analysis (Fig. 41.4). The HOMO (−8.45 eV) of **1** is predominantly metal-based with little contribution from the phenanthroline ligand. The LUMO (−4.75 eV) is effectively the π^* orbital of phenanthroline. The molecular orbital (MO) diagram of **1** suggests the presence of metal-to-ligand charge transfer (MLCT) transitions in this complex. In contrast, the calculated MOs of **2** have little metal character. The HOMO

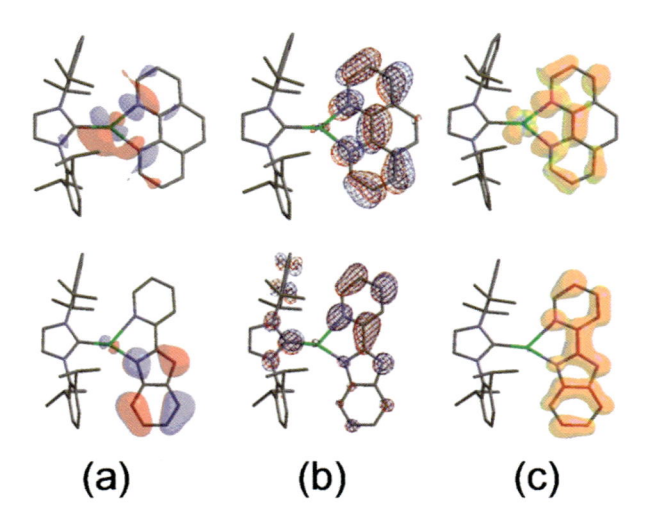

Figure 41.4 (a) HOMO, (b) LUMO and (c) triplet spin-density obtained by density functional (DFT) calculations of the [(IPr)Cu(phen)] cation (top) and (IPr)Cu(pybim) (bottom). Plot contours are shown at an isovalue of 0.032 electrons au^{-3}.

(−4.55 eV) and HOMO - 1 (−4.61 eV) of **2** are dominated by benzimidazole orbitals, and only a small contribution comes from the copper atom. Significant participation from orbitals on copper first appear in the HOMO - 2 (−5.18 eV). The LUMO (−0.79 eV) consists of orbitals on the carbene and pyridyl moieties. Thus, the observed emission of **2** originates from an intra-ligand charge-transfer (ILCT) state. The HOMO–LUMO bandgaps obtained from DFT calculations (335 nm for **1** and 330 nm for **2**) are consistent with experimental absorption wavelengths (332, 348 nm (**1**) and 335 nm (**2**)). The calculated triplet spin-density surfaces are localized on the chelating N^N ligand for both complexes with some contribution from copper in the case of **1**. This result supports the assignment of the lowest excited state as MLCT and ILCT for **1** and **2**, respectively. The nature of the frontier orbitals is consistent with the broad, structureless emission spectra obtained for both Cu(I) complexes.

Electrochemical data (see ESI†) are consistent with the results of theoretical calculations. Complex **1** undergoes irreversible oxidation and reduction at 1.23 V and -2.15 V *vs.* Fc^+/Fc, respectively. Complex **2** exhibits two irreversible oxidation peaks in CV traces at 0.67 V and 0.93 V. However, no reduction peak is observed under the applied conditions (up to -2.7 V). This result indicates that **2** is easier to oxidize than **1**, yet more difficult to reduce, and is in accord with the shallow HOMO and high LUMO energies of **2** obtained by DFT calculations.

In conclusion, new examples of luminescent, three-coordinate Cu(I)–NHC complexes were obtained in good yields. The monomeric complexes are air-stable and phosphoresce both in solution and as solids. In order to gain a deeper understanding of the photophysical properties of these, and related, three-coordinate Cu(I)–carbene complexes, we have prepared a series of derivatives and will report their properties in due course.

We gratefully acknowledge financial support of this work from Universal Display Corporation.

References

1. N. Armaroli, G. Accorsi, F. Cardinali and A. Listorti, *Top. Curr. Chem.*, 2007, **280**, 69–115; D. V. Scaltrito, D. W. Thompson, J. A. O'Callaghan and G. J. Meyer, *Coord. Chem. Rev.*, 2000, **208**, 243–266; D. R. McMillin and K. M. McNett, *Chem. Rev.*, 1998, **98**, 1201–1219.
2. N. Armaroli, *Chem. Soc. Rev.*, 2001, **30**, 113–124; A. Barbieri, G. Accorsi and N. Armaroli, *Chem. Commun.*, 2008, 2185–2193.
3. L. X. Chen, G. Jennings, T. Liu, D. J. Gosztola, J. P. Hessler, D. V. Scaltrito and G. J. Meyer, *J. Am. Chem. Soc.*, 2002, **124**, 10861–10867; L. X. Chen, G. B. Shaw, I. Novozhilova, T. Liu, G. Jennings, K. Attenkofer, G. J. Meyer and P. Coppens, *J. Am. Chem. Soc.*, 2003, **125**, 7022–7034; Z. A. Siddique, Y. Yamamoto, T. Ohno and K. Nozaki, *Inorg. Chem.*, 2003, **42**, 6366–6378; E. M. Stacy and D. R. McMillin, *Inorg. Chem.*, 1990, **29**, 393–396.
4. M. K. Eggleston, D. R. McMillin, K. S. Koenig and A. J. Pallenberg, *Inorg. Chem.*, 1997, **36**, 172–176; A. Lavie-Cambot, M. Cantuel, Y. Leydet, G. Jonusauskas, D. M. Bassani and N. D. McClenaghan, *Coord. Chem. Rev.*, 2008, **252**, 2572–2584.
5. K. J. Lotito and J. C. Peters, *Chem. Commun.*, 2010, **46**, 3690–3692.

6. K. Matsumoto, N. Matsumoto, A. Ishii, T. Tsukuda, M. Hasegawa and T. Tsubomura, *Dalton Trans.*, 2009, 6795–6801.
7. L. A. Goj, E. D. Blue, S. A. Delp, T. B. Gunnoe, T. R. Cundari, A. W. Pierpont, J. L. Petersen and P. D. Boyle, *Inorg. Chem.*, 2006, **45**, 9032–9045.
8. S. H. Hsu, C. Y. Li, Y. W. Chiu, M. C. Chiu, Y. L. Lien, P. C. Kuo, H. M. Lee, J. H. Huang and C. P. Cheng, *J. Organomet. Chem.*, 2007, 692, 5421–5428; A. Welle, S. Diez-Gonzalez, B. Tinant, S. P. Nolan and O. Riant, *Org. Lett.*, 2006, **8**, 6059–6062.
9. K. A. Barakat, T. R. Cundari and M. A. Omary, *J. Am. Chem. Soc.*, 2003, **125**, 14228–14229.

Chapter 42

A Codeposition Route to CuI–Pyridine Coordination Complexes for Organic Light-Emitting Diodes

Zhiwei Liu,[a] Munzarin F. Qayyum,[b] Chao Wu,[a]
Matthew T. Whited,[a] Peter I. Djurovich,[a] Keith O. Hodgson,[b,c]
Britt Hedman,[c] Edward I. Solomon,[b] and Mark E. Thompson[a]

[a]*Department of Chemistry, University of Southern California,
Los Angeles, California 90089, USA*
[b]*Department of Chemistry, Stanford University,
Stanford, California 94305, USA*
[c]*Stanford Synchrotron Radiation Lightsource, SLAC, Stanford University,
Stanford, California 94309, USA*
met@usc.edu

We demonstrate a new approach for utilizing CuI coordination complexes as emissive layers in organic light-emitting diodes that involves in situ codeposition of CuI and 3,5-bis(carbazol-9-yl)pyridine (mCPy). With a simple three-layer device structure, pure green electroluminescence at 530 nm from a Cu(I) complex was observed. A maximum luminance and external quantum efficiency (EQE) of 9700 cd/m^2 and 4.4%, respectively, were achieved. The luminescent species was identified as [CuI(mCPy)$_2$]$_2$ on the basis

Reprinted from *J. Am. Chem. Soc.*, **133**, 3700–3703, 2011.

Electrophosphorescent Materials and Devices
Edited by Mark E. Thompson
Text Copyright © 2011 American Chemical Society
Layout Copyright © 2024 Jenny Stanford Publishing Pte. Ltd.
ISBN 978-981-4877-34-3 (Hardcover), 978-1-003-08872-1 (eBook)
www.jennystanford.com

of photophysical studies of model complexes and X-ray absorption spectroscopy.

Tremendous improvements in organic light-emitting diodes (OLEDs) have been achieved using phosphorescent emissive Ir-based complexes [1]. Unfortunately, Ir is low in natural abundance, so there has been an increasing interest in luminescent Cu(I) complexes [2] for use in high-efficiency OLEDs. However, since most Cu(I) complexes are unstable toward sublimation [3] and hence not amenable to the vacuum deposition methods typically used to fabricate OLEDs, few devices containing such emitters have been reported [4].

Among luminescent Cu(I) complexes, those based on CuI are well-known for their structural diversity, rich photophysical behavior, and high luminance efficiency [5]. A wide range of structure motifs have been prepared by combining CuI and pyridine-based ligands in different ratios [6]. Excited states in the resulting complexes have been proposed to be halide-to-ligand charge transfer (XLCT), metal-to-ligand charge transfer (MLCT), and/or halide-to-metal charge transfer (XMCT) states on the basis of experimental and computational studies [5a, 7]. Generally, CuI-based complexes, especially with pyridine-based ligands, are highly emissive at room temperature (rt) regardless of the structure and nature of the excited state. However, the application of these complexes in OLEDs has not been demonstrated because of the aforementioned difficulties with sublimation and their poor solubility or stability in solution. In this chapter, we demonstrate that codeposition of CuI and 3,5-bis(carbazol-9-yl)pyridine (mCPy; Fig. 42.1) is an efficient way to utilize CuI complexes as emissive materials in OLEDs. Devices made using CuI:mCPy films exhibit pure green electroluminescence (EL) from a Cu(I) emitting species. The chemical composition of the luminescent species in the codeposited film has been deduced through studies of model complexes and X-ray absorption spectroscopy (XAS) and determined to be primarily $[CuI(mCPy)_2]_2$.

Figure 42.1 shows the photoluminescence (PL) spectra of a series of CuI:mCPy films made by codepositing CuI and mCPy in different molar ratios from two separate heating sources in a vacuum chamber. For comparison, the PL spectrum of a neat

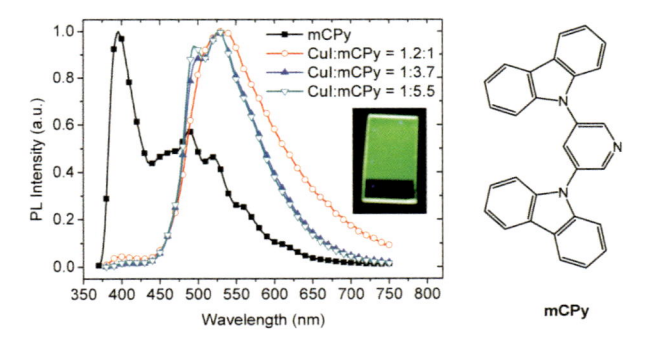

Figure 42.1 (left) PL spectra ($\lambda_{ex} = 350$ nm) of a neat mCPy film at 77 K and CuI:mCPy films at rt. Inset: photo of a CuI:mCPy film under UV light (365 nm). (right) Chemical structure of mCPy.

mCPy film at 77 K is also shown in Fig. 42.1. The spectrum for the neat mCPy film consists of an overlapping fluorescence band at 400 nm [biexponential decay, $\tau = 10.9$ (17%) and 2.0 (83%) ns] and a phosphorescence band at 500 nm (monoexponential decay, $\tau = 0.48$ s). PL was observed for codeposited CuI:mCPy films over a wide range of molar ratios, with quantum yields as high as 64%. The PL spectra from CuI:mCPy films are dominated by a band centered near 520 nm and have decay lifetimes (Table 42.1) that differ markedly from that of the mCPy film, implying the formation of phosphorescent Cu(I) complexes. The emission spectra of CuI:mCPy films sometimes show an additional feature at 495 nm (Fig. 42.1). While the shoulder at 495 nm is coincident with the phosphorescence band of neat mCPy, other data are inconsistent with emission from the neat ligand. Transient measurements at the two wavelengths gave the same biexponential lifetimes listed in Table 42.1 for each composition. In addition, the CuI:mCPy film had identical PL spectra at 298 and 77 K. If the band at 495 nm were caused by emission from neat mCPy, one would expect it to be more prominent at lower T. Thus, the 495 nm feature is likely due to a second Cu(I) complex at low concentration.

To verify that CuI:mCPy complexes are responsible for the luminescence from codeposited films, we prepared a film by codepositing CuI and 1,3-bis(carbazol-9-yl)benzene (mCP), which lacks the coordinating pyridyl group in mCPy. The emission

Table 42.1 Photophysical data for codeposited films

CuI:mCPy molar ratio	λ_{em} (nm)	PLQY (%)[a]	Lifetime (μs)[b] [percentage (%)]
1.8:1.0	—	0	—
1.2:1.0	532	8	0.5 [32], 3.0 [68]
1.0:2.3	496, 528	48	3.1 [28], 10.1 [72]
1.0:2.6	502, 528	62	4.4 [35], 12.8 [65]
1.0:3.7	500, 528	63	3.4 [21], 11.5 [79]
1.0:5.5	495, 528	64	3.5 [28], 11.6 [72]

[a]PL quantum yield. [b]Emission lifetimes were measured at 520 nm with excitation wavelength of 331 nm.

spectrum of the CuI:mCP film was identical to that of mCP (λ_{max} = 425 nm), indicating that pyridine coordination is required for formation of the emissive species in the CuI:mCPy film.

CuI:mCPy films were used to fabricate four OLEDs [ITO/NPD (25 nm)/CuI:mCPy (20 nm)/BCP (40 nm)/LiF (1 nm)/Al (100 nm), where NPD = N,N'-bis(naphthalen-1-yl)-N,N'-bis(phenyl)benzidine, BCP = bathocuproine] in which the molar ratios of the CuI:mCPy films were 0:1 (device **1**), 1:4 (device **2**), 1:6 (device **3**), and 1:10 (device **4**). Figure 42.2 shows EL spectra of the devices at 8 V. The emission from devices **2–4** was significantly different from that of device **1**, indicating that the luminescence arises

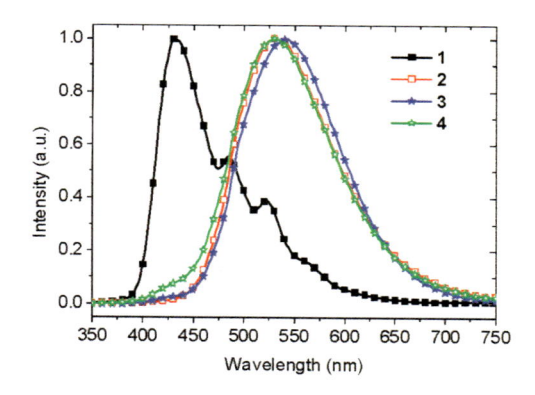

Figure 42.2 EL spectra of ITO/NPD/EML/BCP/LiF/Al devices at 8 V, where EML denotes neat mCPy (device **1**), 1:4 CuI:mCPy (device **2**), 1:6 CuI:mCPy (device **3**), or 1:10 CuI:mCPy (device **4**).

Table 42.2 Performance of devices **1–6**

Device	V_{on} (V)[a]	L_{max} (cd/m^2)	EQE_{max} (%)	PE_{max} (lm/W)	CE_{max} (cd/A)
1	6.4	2500	0.54	0.28	0.86
2	3.6	8800	2.4	3.8	7.4
3	3.5	9900	3.0	4.9	9.7
4	3.7	9100	3.2	5.3	9.8
5	3.5	9100	3.9	10.3	11.7
6	3.6	9700	4.4	10.2	13.8

[a] V_{on} is the voltage required to reach a brightness of 1 cd/m^2.

from a CuI:mCPy complex, consistent with the aforementioned PL study. EL solely from CuI:mCPy was observed in device **2** under all applied voltages. Devices **3** and **4**, with higher mCPy concentrations, exhibited a very small emissive contribution from mCPy at high applied voltages (see Supporting Information). Among OLEDs **1–4**, device **4** with the greatest amount of mCPy (CuI:mCPy = 1:10) showed the highest external quantum efficiency (EQE = 3.2%).

Two more devices were fabricated with a tris(8-hydroxyquinolinato) aluminum (Alq$_3$) electron transport layer (ETL). The structure was ITO/NPD (25 nm)/CuI:mCPy (1:5, 20 nm)/BCP (x)/Alq$_3$ (30 nm)/LiF (1 nm)/Al (100 nm), where the thickness of the BCP layer was either $x = 10$ nm (device **5**) or $x = 0$ nm (device **6**). The two devices had emission spectra similar to that of device **3**, indicating that hole–electron combination occurs within the CuI:mCPy layer and that EL arises from a CuI:mCPy complex. As shown in Table 42.2, device **6** gave the highest EQE (4.4%), power efficiency (PE = 10.2 lm/W), and current efficiency (CE = 13.8 cd/A). Device **6** with Alq$_3$ as the sole ETL gave a higher efficiency than devices containing BCP (i.e., **2–5**), likely as a result of a better balance of charge injection and transport. Moreover, the increase in efficiency suggests that further improvement in performance can be achieved by optimizing the device configuration.

Our initial lifetime measurements on CuI:mCPy-based OLEDs are encouraging. In this study, we monitored the light output of device **6** under vacuum (150 mTorr) while driving the device at a current density of 20 mA/cm^2 (brightness = 2100 cd/m^2). The half-life for the device under these conditions was 21 h, corresponding to 440 h at 100 cd/m^2. While this is well below the lifetimes reported for optimized OLEDs, it is a factor of 5 longer than that for an analogous device with the CuI:mCPy emissive layer replaced with a 20 nm layer of Ir(ppy)$_3$:mCPy [10 wt%, Ir(ppy)$_3$ = fac-tris(2-phenylpyridine)iridium]. When this Ir(ppy)$_3$-based device was driven under the same conditions, a projected half-life of 75 h at 100 cd/m^2 was observed. This indicates that the CuI:mCPy emissive layer is quite stable and promising for OLED applications.

Previous work has shown that the three main products from reactions of CuI and pyridine, namely, [CuI(py)]∞, [CuI(py)$_2$]$_2$, and [CuI(py)]$_4$ (py = pyridine), show blue, green, and orange emission, respectively, in the solid state at rt [5a]. We prepared several related complexes to model the luminescent species in the codeposited CuI:mCPy film. [CuI(mCPy)]∞ (model **A**) was synthesized via the reported method for [CuI(py)]∞ [5a] and characterized by elemental analysis and emission spectroscopy. The complex gave blue emission with λ_{max} ≈ 480 nm (Fig. 42.3) and a PL quantum yield (PLQY) of 64% in the solid state at rt. The emission exhibited a slight red shift as T decreased to 77 K (λ_{max} = 486, 508 nm). The PL lifetimes measured at 480 nm were found to be 3.0 and 16.8 μs at rt and 77 K, respectively.

Another model complex, [CuI(mCPy)]$_4$ · 3CH$_2$Cl$_2$ (model **B1**), was synthesized by mixing CuI and mCPy in CH$_2$Cl$_2$ and characterized by single crystal X-ray diffraction. The structure consists of Cu$_4$ tetrahedra with iodides capping all four faces and the mCPy pyridyl ligands at the apexes (Fig. 42.3). This structure is similar to that of [CuI(py)]$_4$ [6b]. Model **B1** gave a primary emission band at ~560 nm with a decay lifetime of 2.1 μs at rt. Exposure of crystalline **B1** to vacuum resulted in loss of CH$_2$Cl$_2$, forming [CuI(mCPy)]$_4$ (model **B2**) with dominating emission at 620 nm and a decay lifetime of 7.3 μs. Both **B1** and **B2** showed clear luminescent thermochromism with strong blue emission (λ_{max} = 500 nm) at 77 K and decay lifetimes of ~30 μs, similar to many reported (CuIL)$_4$ clusters [5a, 8]. In contrast, the codeposited films showed identical spectra at 298 and 77 K.

It was found that an amorphous material (model **C**) prepared by simple grinding of CuI with mCPy in a 1:4 molar ratio gave emission centered at 530 nm with decay lifetimes of 4.0 (37%) and 11.0

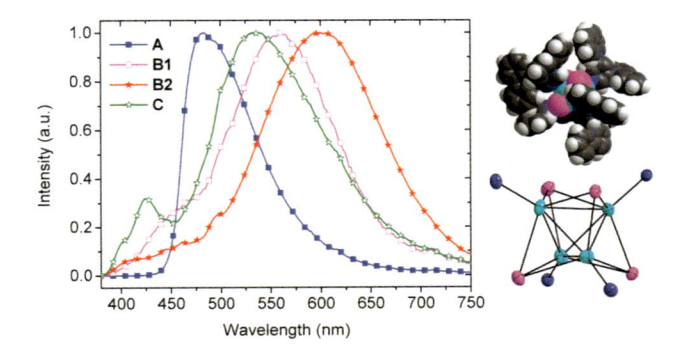

Figure 42.3 (left) PL spectra ($\lambda_{ex} = 350$ nm) of [CuI(mCPy)]∞ (model **A**), **B1**, [CuI(mCPy)]$_4$ (model **B2**), and ground CuI with mCPy (model **C**). (right top) Space-filling drawing of [CuI(mCPy)]$_4 \cdot 3$CH$_2$Cl$_2$ (model **B1**) from crystal coordinates. (right bottom) ORTEP structure (50%) of the Cu$_4$I$_4$ core of **B1** with only the coordinating N atoms of the mCPy ligands shown.

(63%) μs, quite similar to our deposited films. No complex could be isolated on dissolution of model **C**, since it transformed to polymeric [CuI(mCPy)]∞ in CH$_3$CN and tetrameric [CuI(mCPy)]$_4$ in CH$_2$Cl$_2$. Moreover, grinding either [CuI(mCPy)]∞ or [CuI(mCPy)]$_4$ with or without mCPy gave a material having photophysical properties similar to those of the codeposited CuI:mCPy film, although the most efficient emission was observed with added mCPy.

To characterize the structure of the luminescent species in codeposited films, we used XAS. In order to generate sufficient material, ~2 μm CuI:mCPy films (ratio = 1:2.5) were deposited on five four-inch Si wafers and scraped off to give a 40 mg powdered sample (model **D**). The molar ratio was confirmed by elemental analysis. Model **D** showed photophysical properties identical to those of the codeposited films used for photophysical and EL studies. Details of sample preparation, characterization, data collection, and analysis for XAS are given in the Supporting Information. Investigation of the XANES pre-edge region confirmed the presence of only monovalent Cu in the sample. The XANES spectrum was compared with those of the known two-, three-, and four-coordinate model complexes [Cu(xypz)]$_2$(BF$_4$)$_2$ [xypz = bis(3,5-dimethylpyrazoyl)-*m*-xylene], [(C$_6$H$_5$)$_4$P]$_2$[Cu(SC$_6$H$_5$)$_3$], and [Cu(py)$_4$]ClO$_4$, respectively (Fig. S12) [9]. The intensity of the peak

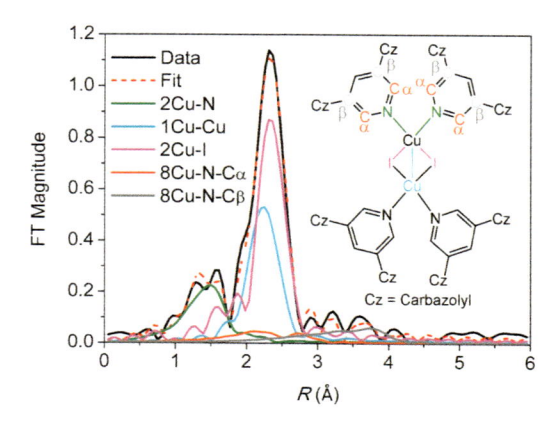

Figure 42.4 Non-phase-shift-corrected Fourier transform of codeposited CuI:mCPy film species (model **D**, CuI:mCPy = 1:2.5). Inset: chemical structure of [CuI(mCPy)$_2$]$_2$. The phase shift in the first shell was ~0.4 Å. Data, fit, and deconvoluted waves are shown.

at ~8984.5 eV, which is characteristic of 1s → 4p transitions in Cu(I) species, precluded the possibility of two-coordinate Cu(I) in the sample. The energy of the three-coordinate Cu(I) model complex closely matched that of the Cu(I) in the film. However, the energy of the 1s → 4p transition can decrease with I$^-$ ligation because of greater electron-donating ability of I$^-$ than of the N/S-based ligands of the model complexes, which results in a lower Z_{eff}. In addition, decreased antibonding of the halide orbitals with Cu(I) 4p orbitals in comparison with the N/S ligand reference complexes can also contribute to a shift to lower energy. Thus, the data are also consistent with the presence of a four-coordinate Cu(I) species in the CuI:mCPy film (model **D**).

The Fourier transform (FT) is shown in Fig. 42.4, and the EXAFS fit parameters are given in Table 42.3. Table 42.3 also gives EXAFS fit parameters for polymeric model complex **A** and related parameters (derived from crystallographic coordinates) for model **B1** and [CuI(pyCN)$_2$]$_2$ [10] (pyCN = 3-cyanopyridine) (model **E**) for comparison. The combination of one Cu–Cu and two Cu–I paths around 2.6 Å gave the best fit to the data. The distance and σ^2 parameter for the two paths were strongly correlated. On the other hand, a poor fit to the data was found when either of those paths was excluded or when any one of the two paths, in the absence of the other, was split into two. Other coordination numbers between 1 and 4 for Cu–Cu and Cu–I can be excluded on the basis of a poorer fit factor, F, and/or higher σ^2. A Cu–N path at 2.02 Å was also necessary. A slight, albeit insignificant, improvement in the fit was observed using a two Cu–N rather than a one Cu–N path.

Analysis of the EXAFS and FT data enabled us to distinguish the major Cu(I) species present in the CuI:mCPy film. The data for model **D** and the [CuI(mCPy)]∞ polymer (Figs. S13–S15) showed major differences, confirming that the CuI:mCPy film and polymer consist of different species. A complex related to model **B1** could also be eliminated as a possibility by comparison of the Cu–Cu and Cu–I coordination numbers in Table 42.3. The structural parameters

Table 42.3 EXAFS least-squares fitting results for $k = 2$–13.4 Å^{-1} for codeposited CuI:mCPy film species (model **D**) and key parameters for model complexes [CuI(mCPy)]∞ (model **A**), [CuI(mCPy)]$_4$ · 3CH$_2$Cl$_2$ (model **B1**), and [CuI(pyCN)$_2$]$_2$ (model **E**)

Model		Path	Cu-Cu	Cu-I	Cu-N	Cu-N-C$_\alpha$	Cu-N-C$_\beta$
D[a] film		Coord.	1	2	2	8	8
		R(Å)[b]	2.60	2.63	2.02	3.23	4.34
		σ^2(Å^2)[c]	755	570	1108	663	756
A[d] (CuIL)$_\infty$	---I-Cu-I--- --Cu-I-Cu--	Coord.	2	3	1		
		R(Å)	2.70	2.63	2.05		
B1[e] (CuIL)$_4$	See Figure 3 right bottom	Coord.	3	3	1		
		R(Å)	2.65	2.63	2.03		
E[e] (CuIL$_2$)$_2$	L,,I,,L Cu Cu L I L	Coord.	1	2	2		
		R(Å)	2.66	2.65	2.08		

[a] ΔE_0(eV) = -13.68 for the EXAFS fit of **D**, while the goodness of fit ($F_{normalized}$, defined by $\sum[(\chi_{obsd} - \chi_{calcd})^2 k^6]/\sum[(\chi_{obsd})^2 k^6]^{0.5}$) was 0.117. [b] The estimated standard deviation of R for each fit was ±0.02 Å. [c] The σ^2 values have been multiplied by 10^5. [d] The EXAFS fit for **A** is shown in the Supporting Information. [e] Parameters for **B1** and **E** were taken from their respective crystal structures.

for model **D** closely match those reported in the literature for model **E** [10]. Thus, on the basis of both the luminescence spectra and XAS studies, the emitting species in CuI:mCPy thin films is most likely the dimeric complex [CuI(mCPy)$_2$]$_2$ (illustrated in the Fig. 42.4 inset), with small variable amounts of [CuI(mCPy)]∞ in low enough concentration that it does not contribute significantly to the XAS.

Codeposition of mCPy with CuBr and CuCl was also examined. Both CuBr and CuCl reacted with mCPy to form luminescent films. However, neither halide material showed a luminescence efficiency comparable to the CuI-based materials. Films made using CuBr displayed an emission peak at 550 nm ($\tau = 6.7$ μs) with a maximum PLQY of 37%. The CuCl-based film had a red-shifted emission peak at 570 nm ($\tau = 3.1$ μs) and a lower PLQY of 14%. The red-shifted emission spectra and decreased PLQY are consistent with the low ligand-field strengths of the Br and Cl anions and the expected reduction in spin–orbit coupling for the lower-Z halogens.

We also examined thin films prepared by codeposition of CuI and the common OLED materials 1,3,5-tris(N-phenylbenzimidazole-2-yl)benzene (TPBi) and 2-(4-biphenylyl)-5-(4-*tert*-butylphenyl)-1,3,4-oxadiazole (PBD). Emission from both the TPBi- and PBD-based films gave substantial components for the organic materials themselves as well as a CuI complex (see Supporting Information) and only moderate PLQYs (~17 and 14%, respectively). The low efficiency and ligand contamination of the spectra are likely due to the fact that both TPBi and PBD are poor ligands for Cu, leading to incomplete formation of the emissive Cu(I) complex.

In summary, we have demonstrated that CuI coordination complexes can be formed in situ by codeposition of CuI and ligand, leading to efficient OLEDs containing Cu(I) complexes. The ligand material in this case serves a dual role as both a ligand for forming the emissive complex and as a host matrix for the formed emitter. While complex formation occurs for a number of materials, care must be taken in the design of the organic material to ensure that it is a good ligand for Cu coordination. This new approach may be easily extended to other ligands to generate functional layers for use in organic optoelectronic devices.

Acknowledgment

We gratefully acknowledge financial support of this work from Universal Display Corporation and the Center for Energy Nanoscience, an Energy Frontier Research Center funded by the U.S. Department of Energy, Office of Science, Office of Basic Energy Sciences (DE-SC0001011). Portions of this research were carried out at the Stanford Synchrotron Radiation Lightsource (SSRL). The SSRL Structural Molecular Biology Program is supported by the DOE Office of Biological and Environmental Research and by the National Institutes of Health (NIH), National Center for Research Resources (NCRR), Biomedical Technology Program (P41RR001209). This study was supported in part by the NCRR (Grant 5 P41 RR001209), a component of the NIH.

Supporting Information

Experimental details; PL and EL data; decay lifetimes; comparison of XANES data; EXAFS fit; and crystallographic data (CIF). This material is available free of charge via the Internet at http://pubs.acs.org (https://ndownloader.figstatic.com/files/4342339).

References

1. (a) Baldo, M. A., Lamansky, S., Burrows, P. E., Thompson, M. E., Forrest, S. R. *Appl. Phys. Lett.* 1999, **75**, 4. (b) Lamansky, S., Djurovich, P. I., Murphy, D., Abdel-Razzaq, F., Lee, H. E., Adachi, C., Burrows, P. E., Forrest, S. R., Thompson, M. E. *J. Am. Chem. Soc.* 2001, **123**, 4304. (c) Djurovich, P. I., Thompson, M. E. In *Highly Efficient OLEDs with Phosphorescent Materials*; Yersin, H., Ed., Wiley-VCH: Berlin, 2007; Chapter 3, pp. 131−161. (d) Thompson, M. E., Djurovich, P. I., Barlow, S., Marder, S. R. In *Comprehensive Organmetallic Chemistry*; O'Hare, D., Ed., Elsevier: Oxford, 2007; Vol. 12, pp. 101–194.
2. (a) Ford, P. C., Cariati, E., Bourassa, J. *Chem. Rev.* 1999, **99**, 3625. (b) Armaroli, N., Accorsi, G., Cardinali, F., Listorti, A. *Top. Curr. Chem.* 2007, **280**, 69. (c) Scaltrito, D. V., Thompson, D. W., O'Callaghan, J. A., Meyer,

G. J. *Coord. Chem. Rev.* 2000, **208**, 243. (d) McMillin, D. R., McNett, K. M. *Chem. Rev.* 1998, **98**, 1201.

3. Manbeck, G. F., Brennessel, W. W., Evans, C. M., Eisenberg, R. *Inorg. Chem.* 2010, **49**, 2834.

4. (a) Zhang, Q. S., Zhou, Q. G., Cheng, Y. X., Wang, L. X., Ma, D. G., Jing, X. B., Wang, F. S. *Adv. Mater.* 2004, **16**, 432. (b) Che, G. B., Su, Z. S., Li, W. L., Chu, B., Li, M. T., Hu, Z. Z., Zhang, Z. Q. *Appl. Phys. Lett.* 2006, **89**, No. 103511. (c) Su, Z. S., Che, G. B., Li, W. L., Su, W. M., Li, M. T., Chu, B., Li, B., Zhang, Z. Q., Hu, Z. Z. *Appl. Phys. Lett.* 2006, **88**, No. 213508. (d) Tsuboyama, A., Kuge, K., Furugori, M., Okada, S., Hoshino, M., Ueno, K. *Inorg. Chem.* 2007, **46**, 1992. (e) Deaton, J. C., Switalski, S. C., Kondakov, D. Y., Young, R. H., Pawlik, T. D., Giesen, D. J., Harkins, S. B., Miller, A. J. M., Mickenberg, S. F., Peters, J. C. *J. Am. Chem. Soc.* 2010, **132**, 9499. (f) Armaroli, N., Accorsi, G., Holler, M., Moudam, O., Nierengarten, J. F., Zhou, Z., Wegh, R. T., Welter, R. *Adv. Mater.* 2006, **18**, 1313.

5. (a) Kyle, K. R., Ryu, C. K., Dibenedetto, J. A., Ford, P. C. *J. Am. Chem. Soc.* 1991, **113**, 2954. (b) Kim, T. H., Shin, Y. W., Jung, J. H., Kim, J. S., Kim, J. *Angew. Chem., Int. Ed.* 2008, **47**, 685. (c) Araki, H., Tsuge, K., Sasaki, Y., Ishizaka, S., Kitamura, N. *Inorg. Chem.* 2005, **44**, 9667.

6. (a) Dyason, J. C., Healy, P. C., Pakawatchai, C., Patrick, V. A., White, A. H. *Inorg. Chem.* 1985, **24**, 1957. (b) Raston, C. L., White, A. H. *J. Chem. Soc., Dalton Trans.* 1976, 2153. (c) Rath, N. P., Maxwell, J. L., Holt, E. M. *J. Chem. Soc., Dalton Trans.* 1986, 2449. (d) Eitel, E., Oelkrug, D., Hiller, W., Strahle, J. *Z. Naturforsch., B* 1980, **35**, 1247.

7. De Angelis, F., Fantacci, S., Sgamellotti, A., Cariati, E., Ugo, R., Ford, P. C. *Inorg. Chem.* 2006, **45**, 10576.

8. (a) Perruchas, S., Le Goff, X. F., Maron, S., Maurin, I., Guillen, F., Garcia, A., Gacoin, T., Boilot, J. P. *J. Am. Chem. Soc.* 2010, **132**, 10967. (b) Tard, C., Perruchas, S., Maron, S., Le Goff, X. F., Guillen, F., Garcia, A., Vigneron, J., Etcheberry, A., Gacoin, T., Boilot, J. P. *Chem. Mater.* 2008, **20**, 7010.

9. Kau, L. S., Spirasolomon, D. J., Pennerhahn, J. E., Hodgson, K. O., Solomon, E. I. *J. Am. Chem. Soc.* 1987, **109**, 6433–6442.

10. Huang, X. C., Ng, S. W. *Acta Crystallogr.* 2004, **60**, m1055.

Chapter 43

Structural and Photophysical Studies of Phosphorescent Three-Coordinate Copper(I) Complexes Supported by an N-Heterocyclic Carbene Ligand

Valentina A. Krylova, Peter I. Djurovich, Jacob W. Aronson, Ralf Haiges, Matthew T. Whited, and Mark E. Thompson

Department of Chemistry, University of Southern California, Los Angeles, California 90089, USA
met@usc.edu

A series of four neutral luminescent three-coordinate Cu(I) complexes (IPr)Cu(N^N), where IPr is a monodentate N-heterocyclic carbene (NHC) ligand (IPr = 1,3-bis(2,6-diisopropylphenyl)imidazol-2-ylidene) and N^N denotes monoanionic pyridyl-azolate ligands, have been synthesized and characterized. A monomeric, three-coordinate geometry, best described as distorted trigonal planar, has been established by single-crystal X-ray analyses for three of the derivatives. In contrast to the previously reported (IPr)Cu(N^N) complexes, the compounds described here display a perpendicular

---

Reprinted from *Organometallics*, **31**, 7983–7993, 2012.

Electrophosphorescent Materials and Devices
Edited by Mark E. Thompson
Text Copyright © 2012 American Chemical Society
Layout Copyright © 2024 Jenny Stanford Publishing Pte. Ltd.
ISBN 978-981-4877-34-3 (Hardcover), 978-1-003-08872-1 (eBook)
www.jennystanford.com

orientation between the chelating N^N ligands and the imidazolylidene ring of the carbene ligand. The geometrical preferences revealed by X-ray crystallography correlate well with the NMR data. The conformational behavior of the complexes, investigated by variable-temperature ^{1}H NMR spectroscopy, indicate free rotation about the C_{NHC}–Cu bond in solution. The complexes display broad, featureless luminescence at both room temperature and 77 K, with emission maxima that vary between 555 and 632 nm depending on sample conditions. Luminescence quantum efficiencies of the complexes in solution ($\Phi \leq 17\%$) increase markedly in the solid state ($\Phi \leq 62\%$). On the basis of time-dependent density functional theory (TD-DFT) calculations and the experimental data, luminescence is assigned to phosphorescence from a metal-to-ligand charge-transfer (MLCT) triplet state admixed with ligand-centered (LC) character.

43.1 Introduction

Phosphorescent Cu(I) complexes have garnered a great deal of attention as inexpensive and abundant alternatives to phosphorescent materials based on heavy-metal complexes [1–3]. Since the first report of room-temperature phosphorescence from Cu(I) complexes [4], their emission properties have been considerably improved. It was shown that Cu(I) complexes can exhibit emission properties comparable to those of materials using third-row transition metals (Ir(III), Pt(II), Os(II)) [5–7]. Successful examples of applications of luminescent Cu(I) complexes in organic electronics [7–9], sensors [10], and biological systems [11, 12] have been demonstrated. Nonetheless, the copper-based materials have not been as well-developed as their third-row counterparts. It is important to obtain greater knowledge of their chemical and photophysical behavior in order to tune their properties in a predictable manner.

The most extensively studied family of luminescent Cu(I) complexes is four-coordinate cationic homo- and heteroleptic compounds, with diimine and phosphine ligands [1, 13]. However, due to their charged nature the scope of applicability for these derivatives is somewhat limited. It has been shown that stable neutral Cu(I) complexes can be isolated and often have emission properties superior to those of cationic analogues [14–16]. Moreover, studies have shown that the coordination geometry has a significant impact on the photophysical properties of copper complexes [17]. Recently a new class of neutral luminescent Cu(I) complexes having a three-coordinate geometry have been reported by us [18] and others [7, 19]. Stabilization of the three-coordinate geometry is possible through a judicious choice of ligands, in particular bulky amide, phosphine, and N-heterocyclic carbene (NHC) ligands. These complexes have good thermal stability, exhibit moderate to high phosphorescence efficiency in both solution and the solid state, and offers a new avenue to modify the emission color of Cu(I) complexes. However, there is only a limited understanding as to what controls the photophysical and intramolecular behavior of these derivatives.

A common approach to alter the emission properties of transition-metal complexes is through modification of the ligands [20–22]. The strategy can also be helpful in providing valuable insight into excited state properties. We have previously demonstrated efficient phosphorescence from air-stable, three-coordinate (NHC)Cu(N^N) complexes, where NHC is a monodentate N-heterocyclic carbene ligand and N^N is a neutral or monoanionic chelating ligand [18]. To further elucidate the photophysical behavior of this new family of Cu(I) phosphors, we have prepared a series of neutral three-coordinate NHC-Cu(I) complexes bearing chelating pyridyl-azolate ligands (Fig. 43.1, compounds **1**–**4**). These types of ligands have been extensively used in luminescent complexes with heavy metals: e.g., Ir(III) [23, 24], Pt(II) [25] and Os(II) [26, 27]. Our results show that the preferred geometric conformation of ligands in these new three-coordinate NHC-Cu(I) complexes differs from that of previously reported analogues: e.g., (IPr)Cu(pybim) (pybim = 2-(2-pyridyl)benzimidazole) (Fig. 43.1, compound **5**). Furthermore, their excited-state properties are distinct from those exhibited by third-row transition-metal complexes and four-coordinate copper(I) complexes bearing the same pyridyl-azolate ligands.

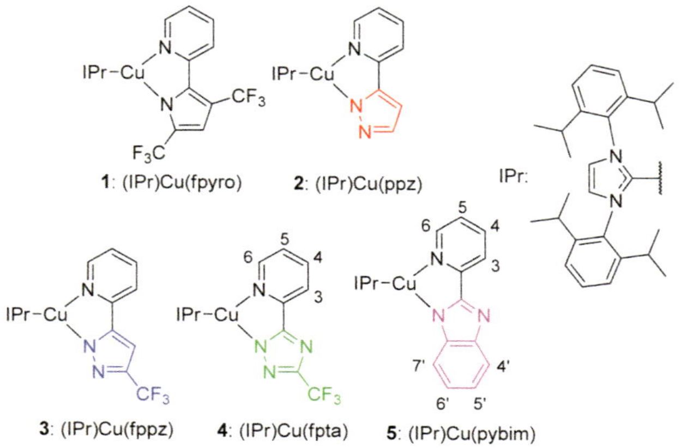

Figure 43.1 Molecular structures and numbering scheme for complexes **1**–**5** discussed in this study.

43.2 Results and Discussion

43.2.1 Synthesis and X-ray Structures

Complexes **1**–**4** were prepared following a previously published procedure for related three-coordinate NHC-Cu^I analogues [18]. Commercially available chloro[1,3-bis(2,6-diisopropylphenyl)imidazol-2-ylidene]copper(I) ((IPr)CuCl) was treated with deprotonated pyridyl-azolate ligand in tetrahydrofuran at room temperature. Products were obtained as yellow solids in good yields (40–55%), by slow recrystallization or by vacuum sublimation (225–265 °C, 10^{-6} Torr). The complexes were fully characterized by ^{1}H, ^{13}C, and ^{19}F NMR spectroscopy and elemental analysis. Complexes **1**–**4** are indefinitely stable to air in the solid state, whereas in solution they are stable for several hours in air and over several days under N_2. For example, while **1**–**4** decompose in chloroform due to trace acid, they are stable for days in acid-free solution.

Single crystals suitable for X-ray diffraction analysis were obtained either by layering a toluene solution with *n*-pentane (for **1**) or by vacuum sublimation (for **3** and **4**). Crystal structures for **1**, **3**, and **4**, along with that previously reported for **5**, are shown in Fig. 43.2, and crystallographic data are given in Table 43.1 and

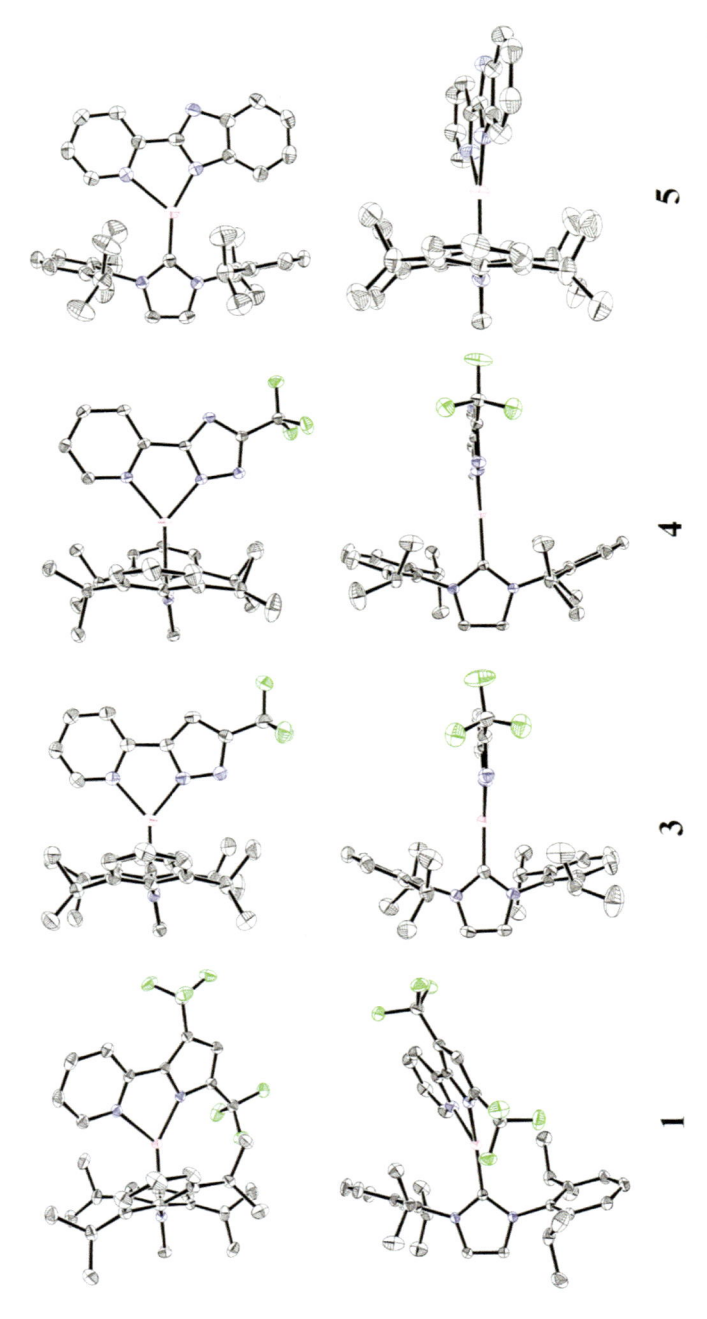

Figure 43.2 ORTEP representation of complexes **1**, **3**, **4**, and **5** [18]. Only one of the unique structures for **1** found in the unit cell is shown. Hydrogen atoms are omitted for clarity.

Table 43.1 Selected bond lengths (Å) and angles (deg) for complexes **1**, **3**, **4**, and **5** [18]

	1a	1b	3	4	5
C_{NHC}–Cu	1.8828(12)	1.8804(12)	1.862(2)	1.8838(14)	1.877(2)
Cu–N_{py}	2.1346(11)	2.1196(11)	2.0805(18)	2.0979(13)	2.2907(18)
Cu–N_{az}	1.9552(11)	1.9658(10)	1.9697(19)	1.9925(14)	1.9227(18)
C_{NHC}–Cu–N_{py}	123.93(5)	127.23(5)	139.27(9)	140.22(6)	126.82(8)
C_{NHC}–Cu–N_{az}	154.34(5)	150.17(5)	139.29(9)	138.44(6)	154.24(8)
N_{py}–Cu–N_{az}	80.14(4)	79.66(4)	81.43(8)	81.22(5)	78.94(7)

the Supporting Information. The compounds are monomeric, three-coordinate structures with the geometry around the copper center characterized as distorted trigonal planar. Surprisingly, in contrast to the previously reported three-coordinate (NHC)Cu(N^N) complexes [18], the pyridyl-azolate ligands in complexes **1**, **3**, and **4** have a perpendicular orientation with respect to the imidazolylidene ring of the carbene ligand. The dihedral angles between planes defined by the NNHC, NNHC, and CNHC atoms and Cu, N_{py}, and N_{az} atoms are 69.15° in 1, 79.98° in 3, and 81.73° in **4**. For comparison, in (IPr)Cu(pybim) (**5**) these two planes are almost coplanar (the dihedral angle is 9°) despite the steric encumbrance between the phenyl rings of the NHC ligand and hydrogen atoms at the H6 position on the pyridyl ring and H7′ position on the benzimidazolide moiety of the pybim ligand.

There are two unique molecules in the unit cell of **1**; however, both have similar geometric parameters. The N^N ligand is tilted in order to accommodate the CF_3 group at the 5′-position of the pyrrolide moiety, which sterically conflicts with the isopropyl groups of the IPr ligand and causes the geometry around copper to distort toward trigonal pyramidal. The copper atom is 0.121 Å above a plane defined by the coordinating C_{NHC}, N_{py}, and N_{az} atoms. Complex **1**, similar to complex **5**, exhibits a distorted T-shaped geometry. The C_{NHC}–Cu–N_{py} and C_{NHC}–Cu–N_{az} bond angles in **1** are 123.93(5) and 154.34(5)°, respectively. These values are close to those found in complex **5**. A similar T-shaped structure was observed by Caulton and co-workers in a three-coordinate Cu(I) (fpyro)Cu(NCMe) complex [28]. As the NCMe ligand imparts a minimal steric constraint in this complex, the T-shaped geometry was attributed to the differing σ-donor characteristics of the pyridyl and pyrrolide moieties. In **3** and **4**, however, the coordination environment around the copper atom is nominally Y-shaped. The C_{NHC}–Cu–N_{py} and C_{NHC}–Cu–N_{az} bond angles in both 3 (139.27(9) and 139.29(9)°) and 4 (140.22(6) and 138.44(6)°) are almost identical. The coordination geometry is planar, as the sum of the bond angles around the copper atom is 359.99° for 3 and 359.90° for 4. In the crystal lattice molecules 3 and 4 pack as dimers with two pyridyl-azolate ligands stacked above each other in a head-to-tail fashion (Fig. S1, Supporting Information). The intermolecular separation between the planes of pyridyl-azolate ligands in the dimer is 3.31 Å for 3 and 3.28 Å for 4, indicating the presence of weak π–π interactions. All three complexes have Cu–pyridyl distances that are longer than the corresponding formally anionic Cu–azolate bond lengths (**1**, Cu–N_{py} = 2.1346(11) Å, Cu–N_{az} = 1.9552(11) Å; 3, Cu–N_{py} = 2.0805(18) Å, Cu–N_{az} = 1.9697(19) Å; **4**, Cu–N_{py} = 2.0979(13) Å, Cu–N_{az} = 1.9925(14) Å). The Cu–N_{az} bond lengths increase in the order pyrrolide (**1**) <pyrazolide (**3**) <triazolide (**4**), which correlates with the σ-donor ability of the azolate ligand. The C_{NHC}–Cu bond lengths in **1** (1.8828(12) Å), 3 (1.862(2) Å,) and **4** (1.8838(14) Å) are comparable to the values reported for NHC-Cu^{I} complexes [29–31].

43.2.2 NMR Characterization

As revealed by crystal structure analysis, the coordination environment around copper is rather sterically crowded by the 2,6-diisopropylphenyl groups of the NHC ligand. It has been shown in other metal complexes coordinated with bulky NHC ligands that rotation about the C_{NHC}–metal bond can be severely hindered [32–36]. Rotation around the N_{NHC}–C_{aryl} and aryl–CH(CH$_3$)$_2$ bond axes is also inhibited by the steric bulk of the isopropyl groups, as evidenced by the ^{1}H NMR of (IPr)CuCl in CDCl$_3$, where the diastereotopic methyl groups appear as two distinct doublets (Fig. S2, Supporting Information). Molecular models similarly show that rotation around the N_{NHC}–C_{aryl} bond is highly unlikely in complexes **1**−**5** due to steric constraints imposed by the 2,6-isopropyl groups and the N^N ligands. Yet, despite the steric crowding, resonances for the protons on the two aromatic rings of the NHC ligand, as well as the two protons of the imidazolylidene ring, are equivalent for **1–5** in CDCl$_3$. Such equivalence with an asymmetric N^N ligand can only occur if there is either (1) no rotation, with the N^N ligand positioned in a perpendicular C_s-symmetric orientation with respect to the NHC imidazole plane, or (2) rapid rotation about the C$_{NHC}$–Cu bond axis. However, due to the asymmetry of the N^N ligand a static structure as in case 1 would lead to four diastereotopic methyl resonances for the isopropyl groups. Instead, complex **1** displays evidence consistent with case 2, as the methyl resonances appear as a simple doublet in CDCl$_3$ (Fig. S3, Supporting Information), a situation that can only come about by simultaneous rotation around the C$_{NHC}$–Cu and all the aryl–CH(CH$_3$)$_2$ bond axes. Apparently, close contact between the CF$_3$ group at the 5′-position of the fpta ligand and the methyl substituents in the isopropyl groups leads to correlated motion around the aryl–CH(CH$_3$)$_2$ bond axes during rotation about the C$_{NHC}$–Cu bond. In a related manner, the methyl resonances in complexes **2** and **3** appear as a pair of doublets (Figs. S4 and S5, Supporting Information). This pattern, similar to what is observed for (IPr)CuCl, is likewise consistent with rapid rotation around the C$_{NHC}$–Cu bond. While the ^{1}H NMR spectra for **1–3** are sharp in CDCl$_3$ at 25 °C, significant line broadening does occur at −40 °C, although no coalescence is observed before the freezing point of the solvent.

The ^{1}H NMR spectrum for (IPr)Cu(fpta) (**4**) in CDCl$_3$, in contrast, shows more complex behavior. Resonances for the fpta ligand are extremely broad at 25 °C, which indicates the presence of an additional exchange process that is slow on the NMR time scale (Fig. 43.3 and Fig. S6, Supporting Information). Upon cooling, the signals for fpta sharpen and a second set of resonances appear. At −40 °C, two sets of sharp resonances for the fpta ligand, along with a corresponding set of aliphatic resonances for the isopropyl groups, are present in a relative population of 2:1, as determined by integration of the respective signals. The same ratio is obtained when different sources of CDCl$_3$ are used and also in the presence of triethylamine (Fig. S7, Supporting Information). This indicates that protonation of the 3′-nitrogen by trace acid in CDCl3 is not responsible for the observed dynamic behavior of complex **4**. In

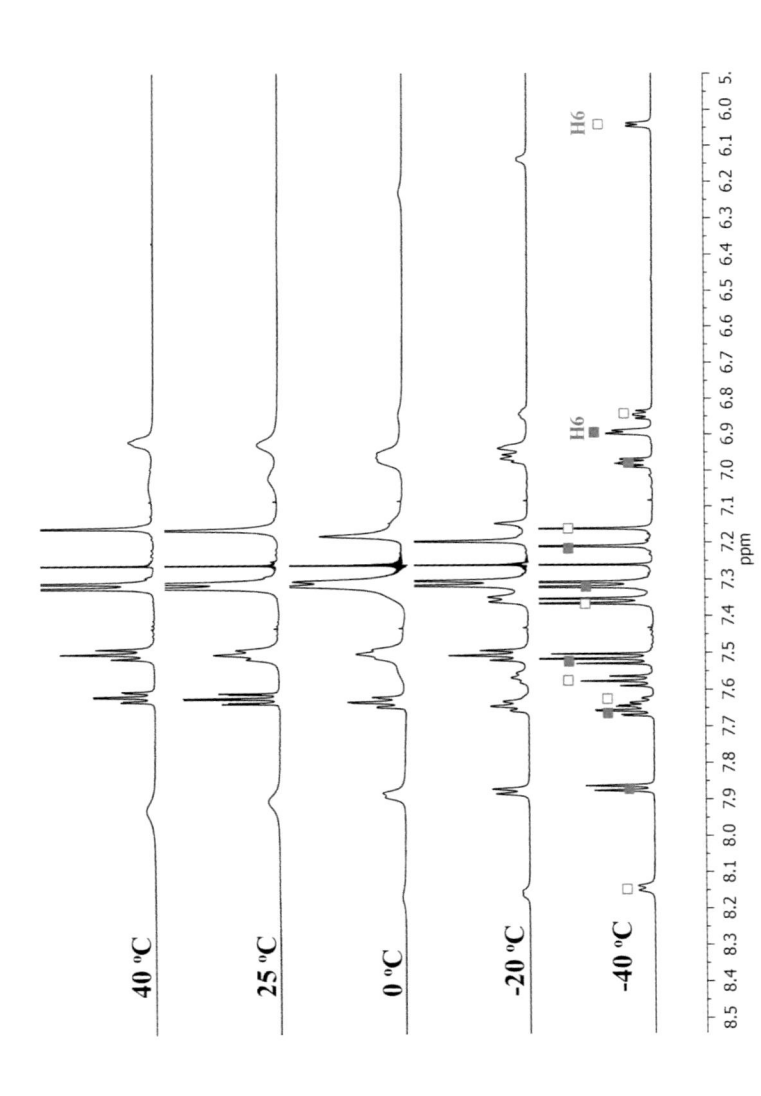

Figure 43.3 600 MHz ^{1}H NMR spectra (aromatic region) of complex **4** in CDCl$_3$ at temperatures ranging from +40 to −40 °C. At −40 °C two sets of signals correspond to the two interconverting major (closed symbols) and minor (open symbols) states.

addition, the dynamic process does not have a significant influence on the orientation of the CF_3 group, as only one resonance is observed in the ^{19}F NMR at $-40\,°C$ (Fig. S8, Supporting Information). The ^1H chemical shifts for the exchanging sites of the fpta ligand are significantly different at $-40\,°C$, particularly for the H6 hydrogen (ortho to the pyridyl ring nitrogen, Fig. 43.1), where the major and minor resonances are at δ 6.89 and 6.04 ppm, respectively. Thus, the two different sets of fpta resonances correspond to two conformers of **4** that interconvert slowly on the NMR time scale. The equilibrium between the two conformers is solvent-dependent, as the ratio of the two exchanging sites is 10:1 in CD_2Cl_2 at $-45\,°C$ (Fig. S9, Supporting Information).

The nature of the exchange process that occurs in complex **4** in $CDCl_3$ can be rationalized on the basis of the differing chemical shifts for the H6 hydrogen of the fpta ligand. In the ^1H NMR spectrum of free Hfpta, the H6 resonance is downfield at δ 8.85 ppm in $CDCl_3$ (Fig. S10, Supporting Information). Upon coordination to the copper, this hydrogen faces the NHC ligand and its resonance appears upfield near δ 6.95 ppm. Rotation about the C_{NHC}–Cu bond will bring the H6 hydrogen in and out of close proximity to the phenyl rings of the NHC ligand when the two ligands are in a respective coplanar and perpendicular orientation. In the coplanar orientation, the H6 proton is directed toward the center of one of the phenyl rings and will be shifted upfield due to shielding from the diamagnetic ring current. On the other hand, a perpendicular orientation of ligands should lead to a relative downfield shift. Hence, the chemical shift will be strongly affected by the dihedral angles between the fpta and NHC ligands. Since the H6 resonance of the minor species appears considerably upfield, it is assigned to a conformer with a predominantly coplanar orientation of the two ligands. Conversely, the major conformer is assigned to an orthogonal arrangement of ligands, in accordance with the crystal structure data of **4**, which shows that the fpta ligand is oriented nearly perpendicular to the imidazolylidene ring of the NHC ligand. The dynamic exchange process observed in $CDCl_3$ (and CD_2Cl_2) thus corresponds to a hindered rotation about the C_{NHC}–Cu bond through an asymmetric energy well (Fig. 43.4). When the temperature is raised, the equilibrium shifts and the population of the coplanar

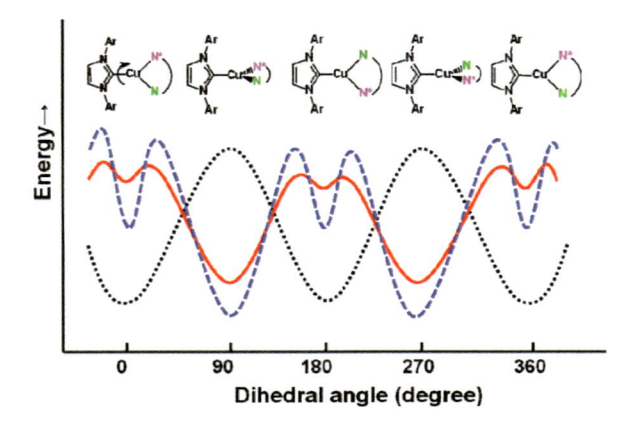

Figure 43.4 Qualitative energy diagram for rotation about the C_{NHC}–Cu bond in complexes **1–3** in CDCl$_3$ (red solid line) and **4** (blue dashed line), along with that for complex **5** (black dotted line). The dihedral angle between the planes is defined by the (N_{NHC}, N_{NHC}, C_{NHC}) and (Cu, N_{py}, N_{az}) atoms.

conformer decreases as the rotation rate about the C_{NHC}–Cu bond increases. As a result, only one set of resonances is observed in the ^{1}H NMR spectrum after the fpta resonances merge above 0°C. A similar two-state potential surface likely exists for complexes **1–3**, albeit with a much shallower energy of the coplanar conformer, and thus only one set of resonances is observed for these derivatives even at low temperature.

As mentioned earlier, compound (IPr)Cu(pybim) (**5**), in contrast to complexes **1**, **3**, and **4**, has a coplanar orientation of the two ligands in the solid state [18]. The H6 pyridyl and H7′ benzimidazolide protons of the pybim ligand in **5** are thus directed toward the centers of the aromatic rings of the carbene ligand. This conformational preference is reflected in the ^{1}H NMR spectrum recorded in CDCl$_3$ (Fig. 43.5). The resonances of the H6 and H7′ protons of the pybim ligand appear significantly upfield at 6.06 and 5.62 ppm, respectively, indicating that they are shielded by an aromatic ring current. Upon cooling to −50°C, the H6 and H7′ resonances of **5** remain sharp and are shifted further upfield. Therefore, unlike complexes **1-4**, the lowest energy conformation of **5** has the N^N ligand oriented predominantly at a dihedral

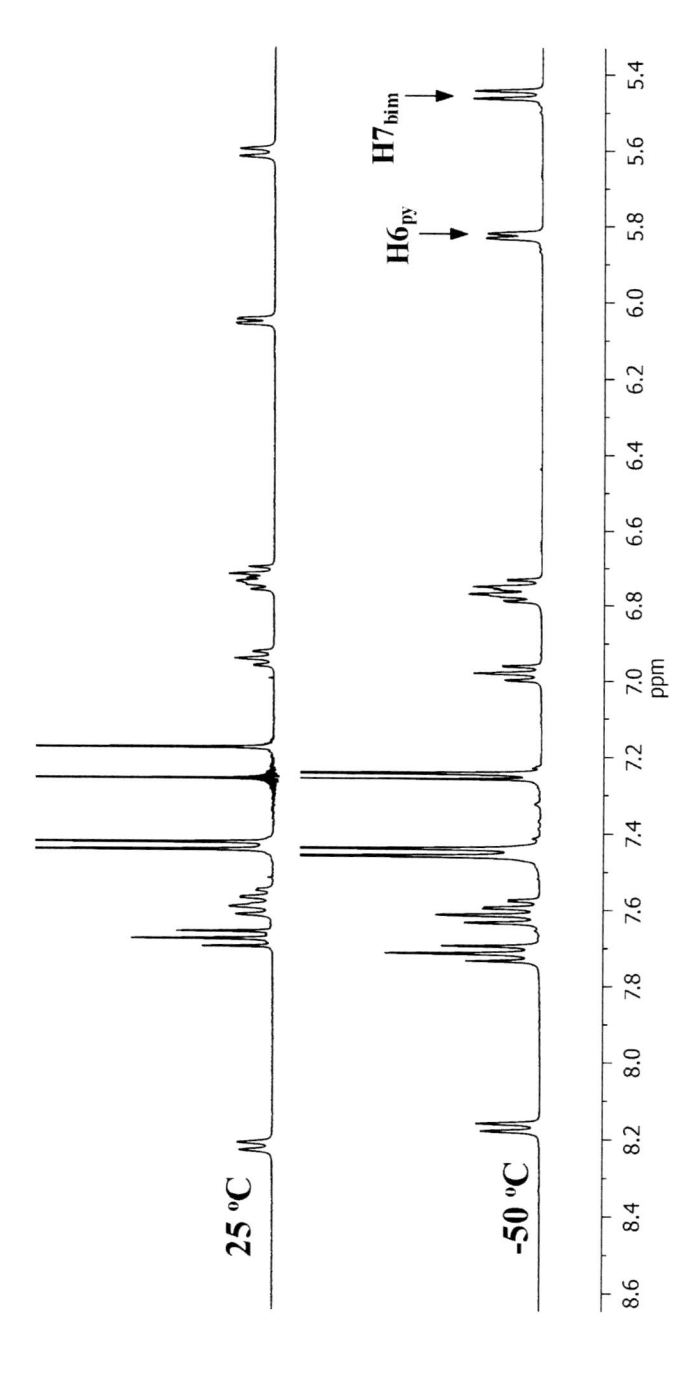

Figure 43.5 ^{1}H NMR spectra (aromatic region) of complex **5** in CDCl$_3$ at +25 and −50 °C.

angle near 0° relative to the imidazolylidene ring of the NHC ligand (Fig. 43.4). The sharp resonances in the ^{1}H NMR spectra at $-50°$C for **5** indicate that rotation about the C_{NHC}–Cu bond still persists at low temperature. The principal difference between the geometry of the (IPr)Cu(N^N) complexes in solution is thus the dihedral angle of the lowest energy conformers: orthogonal in **1–4** and coplanar in **5**.

To further probe the conformational behavior in solution, ^{1}H NMR spectra of **1–4** were also obtained in both nonpolar (benzene-d_6) and polar, coordinative (acetone-d_6) solvents (Figs. S11–S17, Supporting Information). Complexes **1**−**5** showed only one set of sharp proton signals in both benzene-d_6 (25°C) and acetone-d_6 (+25°C and −40°C); there is no evidence of solvent coordination to the copper center observed in solution. Using arguments developed above for behavior in CDCl$_3$, the ^{1}H NMR data indicate rapid rotation about the C_{NHC}–Cu bond on the NMR time scale. Downfield shifts of up to 0.45 ppm for the H6 resonances of **1–4** in acetone-d_6 are observed upon cooling to −40°C (Figs. S13–S16, Supporting Information). These shifts may in part be caused by a stabilization of the perpendicular conformation under these conditions.

43.2.3 DFT Calculations

Density functional theory (DFT) calculations were carried out for complexes **1–4** using geometric parameters obtained from X-ray analyses as starting structures. A similar approach has been used to gain insight into ground- and excited-state properties of neutral Cu(I) complexes [16]. In general, geometrical parameters for the optimized ground-state geometries are in good agreement with the corresponding values determined by X-ray crystallography. The calculated metal–ligand bonds are slightly longer than experimental values for all complexes. For example, bond lengths calculated for **1** are C_{NHC}–Cu $= 1.931$ Å, Cu–N$_{py} = 2.149$ Å, and Cu–N$_{az} = 2.009$ Å, whereas experimental values are 1.8828(12), 2.1346(11), and 1.9552(11) Å, respectively. The calculated bond angles are within 5° of the X-ray structures for **1**, **3**, and **4**.

The molecular orbital (MO) surfaces calculated for complex 1 are shown in Fig. 43.6. Complexes **2–4** have similar frontier MO compositions but variable MO energies (Fig. 43.6 and Fig. S18 in the

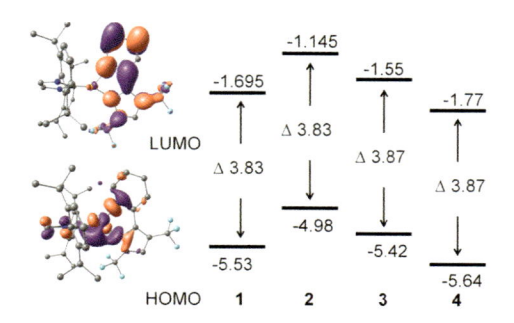

Figure 43.6 Frontier orbitals for **1** and calculated HOMO and LUMO energies (eV) for complexes **1–4**.

Supporting Information). The HOMOs have significant participation of metal d orbitals, as well as contributions from orbitals on the imidazolylidene ring of the carbene ligand and pyridyl-azolate ligand, whereas the LUMOs are localized primarily on the pyridyl-azolate ligand. Complex **2** has the highest HOMO and LUMO energies among the four complexes. Addition of an electron-withdrawing CF_3 group in complex **3** leads to equal stabilization of both the HOMO and LUMO. Introduction of a third nitrogen in the azolate moiety in **4** leads to further stabilization of both frontier orbitals. Thus, due to the presence of two electron-withdrawing CF_3 groups the HOMO and LUMO energies of complex **1** fall within the values for compounds **3** and **4**. While the energies calculated for the HOMO and LUMO are strongly influenced by the identity of the pyridyl-azolate ligand, the HOMO−LUMO gap remains effectively constant within the series. The ability to systematically vary the HOMO and LUMO energies while keeping the energy separation between them unchanged can be an advantageous feature for potential applications in organic electronics devices.

Time-dependent DFT (TD-DFT) calculations were used to determine the lowest energy electronic transitions in **1–4**; the results are summarized in Table 43.2. The lowest lying singlet excitations for all the complexes are HOMO→LUMO and can thus be ascribed to metal-to-ligand charge transfer (MLCT) transitions. The lowest triplet state for complexes **1**, **3**, and **4** is also principally HOMO→LUMO (MLCT) in character. A second triplet state (HOMO-1→LUMO),

Table 43.2 Lowest energy transitions for complexes **1-4** determined from TD-DFT calculations

Complex	States	λ (nm)	f	Major contribution
1	S_1	427	0.0002	HOMO→LUMO (97%)
	T_2	441	0	HOMO-1→LUMO (93%)
	T_1	449	0	HOMO→LUMO (92%)
2	S_1	422	0.0001	HOMO-1→LUMO (97%)
	T_2	446	0	HOMO-1→LUMO (99%)
	T_1	455	0	HOMO→LUMO (90%)
3	T_2	425	0	HOMO-1→LUMO (89%)
	S_1	427	0.0001	HOMO→LUMO (97%)
	T_1	451	0	HOMO→LUMO (99%)
4	T_2	401	0	HOMO-2→LUMO (81%)
	S_1	419	0.0001	HOMO→LUMO (97%)
	T1	442	0	HOMO→LUMO (99%)

which according to calculations lies from 400 to 2200 cm^{-1} above the lowest triplet, is mainly ligand-centered (LC) in character. In complex **2**, the positions of these two triplet states are switched, such that the lowest triplet transition is principally LC in character, while the MLCT state lies 493 cm^{-1} higher in energy. Nevertheless, on the basis of these results we assign the lowest excited state as being mixed ^{3}MLCT/LC in character.

43.2.4 Photophysical Properties

Absorption spectra for **1-4** recorded in dichloromethane solution at room temperature are shown in Fig. 43.7, and absorption and emission data in solution are summarized in Table 43.3. Intense bands ($\varepsilon = 10\,000-18\,000$ M^{-1} cm^{-1}) between 250 and 350 nm are assigned to spin-allowed $^1(\pi - \pi^*)$ ligand-centered (LC) transitions. The LC energy between 300 and 375 nm undergoes a hypsochromic shift in going from **1** to **4** that parallels trends observed in the spectra obtained from the free pyridyl-azolate ligands (Fig. S19, Supporting Information). Very weak transitions at lower energies (>375 nm) are also observed in the excitation spectra obtained in methylcyclohexane at 77 K (Fig. 43.8a). These features, assigned to triplet CT transitions, absorb down to 450 nm and are responsible

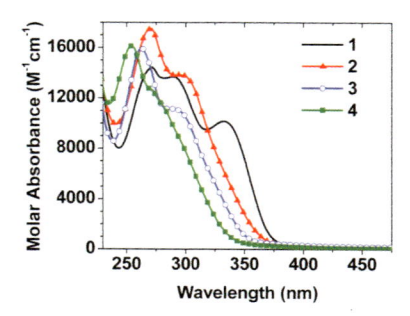

Figure 43.7 Absorption spectra of complexes **1–4** in dichloromethane at room temperature.

for the yellow color of the complexes in the solid state. The solution absorption cutoffs obtained experimentally for complexes **1–4** correlate well with the small range (442–455 nm) calculated for the lowest triplet transitions (Table 43.2).

Emission spectra recorded in frozen methylcyclohexane glass at 77 K are shown in Fig. 43.8a. Low-temperature data provide information about the nature of the lowest excited state. All four complexes have similar emission properties, regardless of differences in the pyridyl-azolate ligand. They exhibit broad featureless emission bands that are significantly red-shifted relative to the absorption. Emission maxima fall within a small range (555–570 nm), and excited-state lifetimes vary from 58 to 77 μs. These results indicate that the observed emission originates from a triplet excited state that is largely charge transfer in character. Interestingly, the emission properties of complexes **1–4** at 77 K are distinctly different from those of other transition metal complexes bearing the same pyridyl-azolate ligands. For example, the homoleptic iridium complexes mer-Ir(fppz)$_3$ and mer-Ir(fpta)$_3$ have structured emission in frozen solvent at 77 K with the first vibronic peak at 430 nm [23]. In yet another example where a pyridyl-azolate is the chromophoric ligand, the four-coordinate copper complex (POP)Cu(fpyro) (POP = bis[2-(diphenylphosphino)phenyl] ether) showed structured emission spectra at 77 K [16]. Emission from this complex, assigned to a mixed ^{3}MLCT/LC state, has a first vibronic peak (446 nm) that correlates well with the calculated lowest triplet transition (441 nm). However, all the complexes in the present study

Table 43.3 Photophysical properties of complexes **1**–**4** in solution

Complex	Absorbance λ (nm) (ε, 10^4 M^{-1} cm^{-1})[a]	Emission at room temp[b]					Emission at 77 K[c]	
		λ_{max} (nm)	τ (μs)	Φ_{PL}	k_r (s^{-1})	k_{nr} (s^{-1})	λ_{max} (nm)	τ (μs)
1	272 (1.44), 289 (1.37), 333 (1.01)	560	10	0.17	1.7×10^4	8.3×10^4	570	58
2	270 (1.75), 298 (1.38)	594	1.3	0.032	2.5×10^4	7.4×10^5	574	68
3	263 (1.59), 291 (1.11)	592	0.54	0.014	2.6×10^4	1.8×10^6	560	77
4	255 (1.61), 272 sh (1.26)	590	1.1	0.024	2.2×10^4	8.9×10^5	555	61

[a]In dichloromethane. [b]In cyclohexane deaerated with N2. [c]In methylcyclohexane.

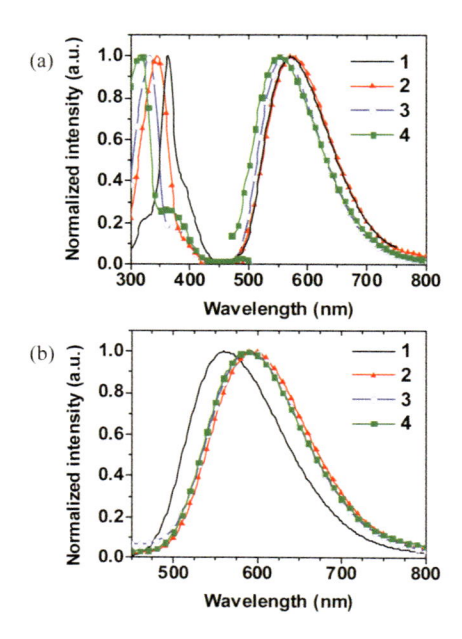

Figure 43.8 (a) Excitation and emission spectra of complexes **1–4** in methylcyclohexane at 77 K. (b) Emission spectra of complexes **1–4** in cyclohexane at room temperature.

(**1–4**) display emission at 77 K that is very broad and red-shifted in comparison to values calculated for the lowest triplet transition. A likely origin of this behavior is the large structural reorganization in the excited state. It is worth noting that the energies of the excited states obtained from TD-DFT calculations correspond to Franck–Condon (vertical) transitions. When a compound undergoes structural distortion after photoexcitation, its lowest emissive state will lie at lower energy than the Franck–Condon state. It is well-known that tetrahedral four-coordinate Cu(I) complexes are prone to a Jahn–Teller-type flattening distortion in the excited state that has significant impact on their photophysical properties [17, 37–39]. Although the three-coordinate geometry in **1–4** eliminates the possibility of a similar flattening distortion, other types of distortion may still occur in the excited state. The calculated HOMOs in these complexes have significant contributions from the copper 3d orbital that are largely antibonding in character (Fig. 43.6). Therefore, removal of an electron from the HOMO upon photoexcitation will likely induce a geometrical relaxation that can have a pronounced effect on photophysical properties, depending on the degree of metal participation in the lowest excited state. Emission from complexes **1–4** is predominantly MLCT in character (with some LC mixed in); thus, structural reorganization is likely to occur in the excited state. The type of distortion has not been confirmed experimentally; however, theoretical calculations of three-coordinate Au(I) and Cu(I) complexes postulate that a Y- to T-shaped distortion may occur in the triplet excited state for such planar species [40].

Complexes **1–4** are emissive in fluid solution at room temperature (Fig. 43.8b, Table 43.3). Measurements were performed in a noncoordinating solvent (cyclohexane) in order to avoid possible formation of exciplexes that may lead to luminescent quenching [41, 42]. The behaviors of complexes **2–4** are similar to each other in that, relative to data recorded at 77 K, spectra (centered around 592 nm) are red-shifted 20–35 nm and have short emission lifetimes ($\tau = 0.54$–1.3 µs) along with low quantum yields ($\Phi < 3\%$). The bathochromic shifts and low efficiencies are attributed to molecular distortions in the excited state that occur readily in fluid media. In contrast, emission from (IPr)Cu(fpyro) (**1**) undergoes a 10 nm blue shift relative to its 77 K spectrum and has both a longer excited state lifetime ($\tau = 10$ µs) and higher quantum yield ($\Phi = 17\%$) than complexes **2–4**. This behavior is ascribed to the smaller extent of excited-state distortion in complex **1**, which leads to an order of magnitude decrease in the nonradiative decay rate ($k_{nr} = 8.3 \times 10^4$ s^{-1}). The smaller distortion may have its origin in differences seen in the ground-state geometry of the complexes. The crystal structure for **1** shows the Cu coordinated in a T-like geometry, whereas complexes **3** and **4** (and presumably **2**) have a Y-like coordination geometry around copper. The T-like structure of **1** is closer to the optimized triplet geometry proposed for other three-coordinate d^{10} complexes [40] and may predispose **1** to undergo less distortion in the excited state than geometric changes that occur from the Y-like structures of **2–4**.

Photophysical properties from neat crystalline samples were recorded at room temperature and 77 K. Emission spectra are depicted in Fig. 43.9, and photophysical data are summarized

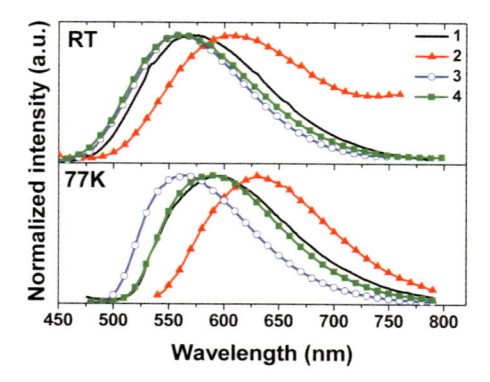

Figure 43.9 Solid-state emission of complexes **1–4** at room temperature (top) and 77 K (bottom).

in Table 43.4. At room temperature complexes **1**, **3**, and **4** are bright emitters with phosphorescence quantum yields ranging from 0.23 to 0.62. They have similar emission line shapes (λ_{max} = 560–570 nm) and excited-state lifetimes (τ = 12–13 µs). The observed enhancement of emission efficiency, relative to values from fluid solution, is consistent with the rigid, crystalline environment lowering the nonradiative decay rates by suppressing molecular vibrations and distortion in the excited state. On the other hand, emission from complex 2 is rather weak (Φ = 0.065) with a short lifetime (τ = 2 µs), and its spectrum is red-shifted almost 40 nm from those of the other derivatives. The bathochromic shift of (IPr)Cu(ppz) (**2**) and high nonradiative decay rate (and consequent low luminescence efficiency) is likely caused by a crystal-packing arrangement unlike that which occurs in derivatives with bulky CF_3 groups. The radiative rates (k_r) for all compounds at room temperature fall within a range (k_r = (1.7–5.2) × 10^4 s^{-1}) that is similar to values found in fluid solution. Upon cooling to 77 K, the emission spectra for all complexes undergo a 6–30 nm red shift accompanied by an increase in emission lifetime (τ = 17–74 µs). Such behavior is common for luminescent Cu(I) complexes and is usually ascribed to a decrease in thermally activated (E-type) delayed fluorescence [8, 43–46]. At low temperatures, thermally activated population of the higher-lying state with faster radiative rate is suppressed, leading to red-shifted emission and longer

Table 43.4 Photophysical properties of complexes **1–4** in the solid state

Complex	Emission at room temp					Emission at 77 K	
	λ_{max} (nm)	τ (μs)	Φ_{PL}	k_r (s^{-1})	k_{nr} (s^{-1})	λ_{max} (nm)	τ (μs)
1	570	13	0.23	1.7×10^4	5.9×10^4	592	28
2	605	2.0	0.065	3.3×10^4	4.7×10^5	632	17
3	560	12	0.62	5.2×10^4	3.2×10^4	566	74
4	560	12	0.48	4.0×10^4	4.3×10^4	590	35

excited-state lifetimes. Interestingly, the TD-DFT calculations for **1–4** suggest that both the S$_1$ and T$_2$ states lie close enough in energy ($\leq$2314 cm^{-1}) to be thermally accessible from the lowest triplet state, T$_1$. While our data provide strong evidence of the participation of a higher-lying state with faster radiative rate at room temperature, the nature of this state (singlet or triplet) has not been confirmed experimentally.

43.3 Conclusion

We have prepared four neutral Cu(I) complexes, (IPr)Cu(N^N), where a three-coordinate geometry is stabilized by a bulky N-heterocyclic carbene (NHC) ligand, and have systematically investigated their properties by varying the monoanionic chelating pyridyl-azolate (N^N) ligand. These complexes are easily prepared and are stable and sublimable. X-ray crystallography confirms a monomeric three-coordinate geometry and distorted-trigonal-planar copper center that is similar to those of related Cu(I) complexes reported earlier. The complexes have an orthogonal orientation between the N^N ligand and the imidazolylidene ring of the carbene ligand. ^{1}H NMR spectra indicate that rotation about the C$_{NHC}$–Cu bond occurs at room temperature, with the conformational behavior dependent on the nature of the chelating ligand as well as solvent and temperature. In case of one derivative, (IPr)Cu(fpta), a coplanar conformer could be stabilized as an intermediate in chloroform or dichloromethane at temperatures below 0°C, but this orientation is disfavored at higher temperatures and in other solvents. The geometric preference is in contrast with that of previously reported (IPr)Cu(N^N) complexes, where the lowest energy conformer in both solution and the solid state has a coplanar orientation of the two ligands. All of the new complexes reported here display yellow-orange phosphorescence in solution and the solid state at room temperature and 77 K. The emission is assigned to a mixed triplet MLCT/LC transition associated with the pyridine-azolate ligand. The photophysical properties are strongly dependent on experimental conditions. In particular, phosphorescence in fluid solution is significantly lower than in rigid media, a phenomenon attributed to quenching by excited-state distortions. In the solid state, a hypsochromic shift in emission wavelength upon going from 77 K to room temperature, along with a decrease in emission lifetime, is consistent with thermal population to a higher-lying state with a faster radiative rate. Our findings provide more insight into structure and photophysical behavior of three-coordinate NHC-Cu^I complexes and should aid in the future design of new related materials.

43.4 Experimental Section

43.4.1 Synthesis

All reactions were performed under a nitrogen atmosphere in oven-dried glassware. Chloro[1,3-bis(2,6-diisopropylphenyl)imidazol-2-ylidene]copper(I) was purchased from TCI America, and all other commercially available reagents were purchased from Sigma-Aldrich and used as received. Solvents were obtained from commercial sources and used without further purification, except for tetrahydrofuran, which was distilled over sodium/benzophenone and CDCl$_3$, which was passed through a plug of Al$_2$O$_3$. 3,5-Bis(trifluoromethyl)-2′-(2′-pyridyl)pyrrole [28, 47, 48] (Hfpyro), 2-(1H-pyrazol-5-yl)pyridine [49] (Hppz), 3-(trifluoromethyl)-5-(2-pyridyl)pyrazole [50, 51] (Hfppz), and 3-(fluoromethyl)-5-(2-pyridyl)1,2,4-triazole [52] (Hfpta) were prepared following published procedures. ^{1}H, ^{13}C, and ^{19}F NMR spectra were recorded on a Varian Mercury 400, Varian VNMRS 500, or a Varian VNMRS

600 spectrometer. The chemical shifts are given in units of ppm. All ^{1}H and ^{13}C chemical shifts were referenced to the residual solvent signals. All ^{19}F chemical shifts were referenced to CFCl$_3$ as an external standard at 0.0 ppm. Variable-temperature (VT) NMR experiments were performed as follows. Data were collected starting from the lowest temperature and then heating the sample in 10–20°C steps, allowing 15 min for temperature equilibration at each set point. ^{1}H NMR spectra of the sample at room temperature (25°C) were recorded before and after the VT NMR experiment and compared to ensure no decomposition occurred during VT studies. Elemental analyses were carried out by the Microanalysis Laboratory at the University of Illinois, Urbana-Champaign, IL.

43.4.1.1 Synthesis of (IPr)Cu(fpyro) (1)

3,5-Bis(trifluoromethyl)-2-(2′-pyridyl)pyrrole (300 mg, 1.07 mmol) was dissolved in 10 mL of THF under nitrogen, and this solution was transferred via cannula to a suspension of NaH (42.9 mg, 1.07 mmol, 60% in mineral oil) in THF (20 mL). The reaction mixture was stirred at room temperature until all the NaH had reacted and gas evolution stopped (30 min). The clear solution was then transferred to a flask charged with chloro[1,3-bis(2,6-diisopropylphenyl)imidazol-2-ylidene]copper(I) ((IPr)CuCl; 523 mg, 1.07 mmol) under nitrogen. The resulting yellow mixture was stirred at room temperature for 3 h, and then it was filtered through Celite and solvent was removed by rotary evaporation. The product was obtained as a yellow solid by crystallization from its toluene solution by layering with *n*-pentane in the freezer (430 mg, 55%). ^{1}H NMR (500 MHz, acetone-d_6, δ): 1.21 (d, $J = 6.9$ Hz, 24H, ArCHCH$_3$), 3.04 (septet, $J = 6.8$ Hz, 4H, ArCHCH$_3$), 6.61 (s, 1H, H-4 pyrrole), 7.11 (m, 1H, H-5 py), 7.34 (d, $J = 7.8$ Hz, 4H, H^{Ar}), 7.45 (t, 2H, H^{Ar}), 7.63 (s, 2H, NCH=), 7.65 (d, $J = 8.2$ Hz, 1H, H-3 py), 7.70 (d, $J = 5.0$ Hz, 1H, H-6 py), 7.77 (m, 1H, H-4 py). ^{13}C NMR (126 MHz, acetone-d_6, δ): 23.51, 25.47, 29.38, 111.02 (m), 111.65 (q, $J = 35$ Hz), 120.14 (m), 121.93, 123.33 (q, $J = 266$ Hz, CF$_3$), 124.93 (2C), 125.15, 126.04 (q, $J = 265$ Hz, CF$_3$), 129.96 (q, $J = 36$ Hz), 130.75, 137.19, 139.02, 146.53, 149.7, 152.63, 185.63. ^{19}F NMR (470

MHz, acetone-d_6, δ): −59.88, −55.07. Anal. Calcd for $C_{38}H_{41}CuF_6N_4$: C, 62.41; H, 5.65; N, 7.66. Found: C, 62.39; H, 5.64; N, 7.54.

43.4.1.2 Synthesis of (IPr)Cu(ppz) (2)

Complex **2** was prepared by following the procedure described for the preparation of **1** from 2-(1*H*-pyrazol-5-yl)pyridine (72.6 mg, 0.5 mmol) and (IPr)CuCl (243.8 mg, 0.5 mmol). The product was obtained as a yellow solid by crystallization from a toluene/*n*-pentane solvent mixture (120 mg, 40%). ^{1}H NMR (500 MHz, acetone-d_6, δ): 1.24 (d, J = 6.9 Hz, 12H, ArCHCH$_3$), 1.33 (d, J = 6.9 Hz, 12H, ArCHCH$_3$), 2.95 (septet, J = 6.9 Hz, 4H, ArCHCH$_3$), 6.33 (d, J = 1.6 Hz, 1H, H-4 pz), 6.91 (m, 1H, H-5 py), 7.23 (d, J = 1.6 Hz, 1H, H-3 pz), 7.33 (d, J = 7.7 Hz, 4H, H^{Ar}), 7.36 (d, J = 8.0 Hz, H-3 py), 7.45 (t, J = 7.8 Hz, 2H, H^{Ar}), 7.58 (m, 1H, H-4 py), 7.62−7.63 (m, 2H, NCH=, H-6 py). ^{13}C NMR (126 MHz, acetone-d_6, δ) 24.57, 24.65, 29.45, 100.93, 118.89, 120.74, 124.48, 124.75, 130.70, 136.92, 138.24, 141.02, 146.84, 149.18, 149.70, 154.76, 185.79. Anal. Calcd for $C_{35}H_{42}CuN_5$: C, 70.5; H, 7.1; N, 11.74. Found: C, 70.52; H, 7.17; N, 11.51.

43.4.1.3 Synthesis of (IPr)Cu(fppz) (3)

Complex **3** was prepared by following the procedure described for the preparation of **1** from 3-(trifluoromethyl)-5-(2-pyridyl)pyrazole (213.2 mg, 1.0 mmol) and (IPr)CuCl (487.6 mg, 1.0 mmol). The crude product after filtration and removal of solvent was purified by vacuum sublimation (225°C, 10^{-6} Torr) to give a yellow solid (349 mg, 53%). ^{1}H NMR (500 MHz, acetone-d_6, δ): 1.24 (d, J = 6.9 Hz, 12H, ArCHCH$_3$), 1.30 (d, J = 6.9 Hz, 12H, ArCHCH$_3$), 2.96 (septet, J = 6.9 Hz, 4H, ArCHCH$_3$), 6.66 (s, 1H, H-4 pz), 7.04 (m, 1H, H-5 py), 7.34 (d, J = 7.7 Hz, 4H, H+), 7.44–7.48 (m, 3H, H-3 py, H+), 7.57 (d, J = 5.0 Hz, 1H, H-6 py), 7.64 (s, 2H, NCH=), 7.70 (m, 1H, H-4 py). ^{13}C NMR (126 MHz, acetone-d_6, δ): 24.41, 24.67, 29.44, 99.77 (q, J = 1.9 Hz, C-4 pz), 119.35, 122.11, 124.51 (q, J = 267 Hz, CF$_3$ pz), 124.52, 124.77, 130.67, 136.93, 139.08, 144.35 (q, J = 34 Hz, C-3 pz), 146.81, 149.32, 150.62, 153.41, 185.72. ^{19}F NMR (470 MHz, acetone-d_6, δ): −60.57. Anal. Calcd for $C_{36}H_{41}CuF_3N_5$: C, 65.09; H, 6.22; N, 10.54. Found: C, 64.74; H, 6.21; N, 10.31.

43.4.1.4 Synthesis of (IPr)Cu(fpta) (4)

Complex **4** was prepared by following the procedure described for the preparation of **1** from 3-(fluoromethyl)-5-(2-pyridyl)-1,2,4-triazole (107 mg, 0.5 mmol) and (IPr)CuCl (243.8 mg, 0.5 mmol). The crude product after filtration and removal of solvent was purified by vacuum sublimation (265°C, 10^{-6} Torr) to give a yellow solid (145 mg, 46%). ^{1}H NMR (500 MHz, acetone-d_6, δ): 1.25 (d, J = 6.9 Hz, 12H, ArCHCH$_3$), 1.27 (d, J = 6.9 Hz, 12H, ArCHCH$_3$), 2.90 (septet, J = 6.9 Hz, 4H, ArCHCH$_3$), 7.22 (m, 1H, H-5 py), 7.37 (d, J = 7.8 Hz, 4H, H^{Ar}), 7.41 (d, J = 4.9 Hz, 1H, H-6 py), 7.50 (t, 2H, J = 7.8 Hz, H^{Ar}), 7.70 (s, 2H, NCH=), 7.80 (d, J = 7.7 Hz, 1H, H-3 py), 7.84 (m, 1H, H-4 py). ^{13}C NMR (126 MHz, acetone-d_6, δ): 24.45, 24.52, 29.48, 120.12, 122.29 (q, J = 268 Hz, CF$_3$ pz), 124.39, 124.80, 124.88, 130.89, 136.76, 139.54, 146.76, 149.68, 150.61, 156.37 (q, J = 35 Hz, C-4 pz), 163.62, 184.60. ^{19}F NMR (470 MHz, acetone-d_6, δ): −64.13. Anal. Calcd for $C_{35}H_{40}CuF_3N_6$: C, 63.19; H, 6.06; N, 12.63. Found: C, 63.24; H, 6.06; N, 12.41.

43.4.2 X-ray Crystallography

The single-crystal X-ray diffraction data for compounds **1** and **4** were collected on a Bruker SMART APEX DUO three-circle platform diffractometer with the χ axis fixed at 54.745° and using Mo Kα radiation (λ = 0.710 73 Å) monochromated by a TRIUMPH curved-crystal monochromator [53]. The diffractometer was equipped with an APEX II CCD detector and an Oxford Cryosystems Cryostream 700 apparatus for low-temperature data collection. The crystals were mounted in Cryo-Loops using Paratone oil. The frames were integrated using the SAINT algorithm [54] to give the *hkl* files corrected for *Lp*/decay. Data were corrected for absorption effects using the multiscan method (SADABS) [55]. The structures were solved by direct methods and refined on F^2 using the Bruker SHELXTL software package [56, 57]. All non-hydrogen atoms were refined anisotropically. ORTEP drawings were prepared using the ORTEP-3 for Windows V2.02 program [58]. Diffraction data for compound **3** were collected on a Bruker SMART APEX CCD diffractometer with graphite-monochromated Mo Kα radiation ($\lambda = 0.710\,73$ Å). Crystallographic data for the complexes have been deposited at the Cambridge Crystallographic Data Centre (Nos. 902418 (**3**), 902419 (**4**), 902420 (**1**)) and can be obtained free of charge via www.ccdc.cam.ac.uk.

43.4.2.1 $C_{38}H_{41}CuF_6N_4$ (1)

A clear, light yellow specimen of approximate dimensions 0.16 mm × 0.26 mm × 0.35 mm was used for the X-ray crystallographic analysis. A complete hemisphere of data was scanned on ω (0.5°) with a run time of 30 s per frame at a detector distance of 50.4 mm and a resolution of 512 × 512 pixels. A total of 4320 frames were collected. The integration of the data using a monoclinic unit cell yielded a total of 156 641 reflections to a maximum θ angle of 30.54° (0.70 Å resolution), of which 21 794 were independent (average redundancy 7.187, completeness 98.6%, $R_{\text{int}} = 3.13\%$, $R_{\text{sig}} = 2.06\%$) and 17 438 (80.01%) were greater than $2\sigma(F^2)$. The final cell constants of $a = 21.3154(8)$ Å, $b = 18.0984(7)$ Å, $c = 21.5042(8)$ Å, $\beta = 119.4410(10)°$, and $V = 7224.5(5)$ Å^3 are based upon the refinement of the *XYZ* centroids of 2097 reflections above 20 $\sigma(I)$ with $3.127° < 2\theta < 61.60°$. The ratio of minimum to maximum apparent transmission was 0.889. The calculated minimum and maximum transmission coefficients (based on crystal size) are 0.4921 and 0.7312. The structure was solved and refined using the space group $P2_1/n$, with $Z = 8$ for the formula unit, $C_{38}H_{41}CuF_6N_4$. The final anisotropic full-matrix least-squares refinement on F^2 with 899 variables converged at R1 = 3.31%, for the observed data and wR2 = 9.23% for all data. The goodness of fit was 1.011. The largest peak in the final difference electron density synthesis was 0.548 e/Å^3, and the largest hole was -0.395 e/Å^3 with an RMS deviation of 0.055 e/Å^3. On the basis of the final model, the calculated density was 1.345 g/cm^3 and $F(000)$ was 3040 e.

43.4.2.2 $C_{36}H_{41}CuF_3N_5$ (3)

A clear yellow prism of approximate dimensions 0.17 mm × 0.19 mm × 0.22 mm was used for the X-ray crystallographic analysis. A complete hemisphere of data was scanned on ω (0.3°) with a run time of 10 s per frame. The integration of the data using a monoclinic unit cell yielded a total of 20 207 reflections to a maximum θ angle of 27.51° (0.77 Å resolution), of which 7458 were independent (average redundancy 2.709, completeness 99.7%, R_{int} = 4.48%, R_{sig} = 5.00%) and 5528 (74.12%) were greater than $2\sigma(F^2)$. The final cell constants of a = 11.8209(11) Å, b = 14.4304(14) Å, c = 20.0438(18) Å, β = 104.402(2)°, and V = 3311.6(5) Å^3 are based upon the refinement of the XYZ centroids of 7208 reflections with 2.31° < 2θ < 27.47°. The ratio of minimum to maximum apparent transmission was 0.855. The calculated minimum and maximum transmission coefficients (based on crystal size) are 0.6374 and 0.7456. The structure was solved and refined using the space group $P2_1/n$, with Z = 4 for the formula unit, $C_{36}H_{41}CuF_3N_5$. The final anisotropic full-matrix least-squares refinement on F^2 with 414 variables converged at R1 = 5.00%, for the observed data and wR2 = 7.89% for all data. The goodness of fit was 1.447. The largest peak in the final difference electron density synthesis was 1.060 e/Å^3, and the largest hole was −0.545 e/Å^3 with an RMS deviation of 0.06 e/Å^3. On the basis of the final model, the calculated density was 1.332 g/cm^3 and F(000) was 1392 e.

43.4.2.3 C35H40CuF3N6 (4)

A clear yellow prismlike specimen of approximate dimensions 0.30 mm × 0.30 mm × 0.39 mm was used for the X-ray crystallographic analysis. A complete hemisphere of data was scanned on ω (0.5°) with a run time of 5 s per frame at a detector distance of 50.4 mm and a resolution of 512 × 512 pixels. A total of 4320 frames were collected. The integration of the data using a monoclinic unit cell yielded a total of 115 485 reflections to a maximum θ angle of 30.71° (0.70 Å resolution), of which 10 151 were independent (average redundancy 11.377, completeness 99.8%, R_{int} = 4.69%, R_{sig} = 2.05%) and 8352 (82.28%) were greater than $2\sigma(F^2)$. The final cell constants of $a = 11.8747(2)$ Å, $b = 14.5369(3)$ Å, $c = 19.6509(4)$ Å, $\beta = 104.9210(10)°$, and $V = 3277.79(11)$ Å^3 are based upon the refinement of the *XYZ* centroids of 132 reflections above 20σ(I) with $3.525° < 2\theta < 36.70°$. The ratio of minimum to maximum apparent transmission was 0.900. The calculated minimum and maximum transmission coefficients (based on crystal size) are 0.7672 and 0.8135. The structure was solved and refined using the space group $P2_1/n$, with $Z = 4$ for the formula unit, $C_{35}H_{40}CuF_3N_6$. The final anisotropic full-matrix least-squares refinement on F^2 with 414 variables converged at R1 $= 4.04\%$, for the observed data and wR2 $= 10.63\%$ for all data. The goodness of fit was 1.066. The largest peak in the final difference electron density synthesis was 0.667 e/Å^3 and the largest hole was -0.527 e/Å^3 with an RMS deviation of 0.074 e/ Å^3. On the basis of the final model, the calculated density was 1.348 g/cm^3 and F(000) was 1392 e.

43.4.3 Density Functional Calculations

Density functional theory (DFT) calculations were performed with the Gaussian03 [59] software package employing the B3LYP functional [60, 61] using the LANDL2DZ basis set [62–64] for Cu and 6-31G* for C, N, H, and F. Geometric parameters obtained from XRD analyses were used as a starting point for geometry optimization in the ground state. The optimized geometries were used for time-dependent density functional calculations (TD-DFT).

43.4.4 Photophysical Characterization

The UV–visible spectra were recorded on a Hewlett-Packard 4853 diode array spectrometer. Steady-state emission measurements were performed using a Photon Technology International QuantaMaster Model C-60 fluorimeter. All reported spectra are corrected for photomultiplier response. Phosphorescence lifetime measurements were performed on the same fluorimeter equipped with a microsecond Xe flash lamp or using an IBH Fluorocube instrument equipped with a 405 nm LED excitation source using the time-correlated single photon counting method. Quantum yields at room temperature were measured using a Hamamatsu C9920 system equipped with a xenon lamp, calibrated integrating sphere, and Model C10027 photonic multichannel analyzer. All solutions were prepared in air and deaerated by sparging with nitrogen for 15 min prior to performing emission, lifetime, and quantum yield measurements.

Acknowledgments

We thank Universal Display Corp. for financial support of this work. The National Science Foundation is acknowledged for the support of the X-ray diffractometer through NSF CRIF Grant 1048807. Computation for the work described in this paper was supported by the University of Southern California Center for High-Performance Computing and Communications (www.usc.edu/hpcc).

Supporting Information

CIF files giving crystallographic data and figures giving crystal-packing diagrams of 3 and 4, ^{1}H NMR spectra in acetone-d_6 (**1-5**), benzene-d_6 (**1-5**), CDCl$_3$ (**1-4**, (IPr)CuCl, and Hfpta), and CD$_2$Cl$_2$ (**4**), ^{19}F NMR of **4** in CDCl$_3$, the calculated MOs for **1-4**, and absorption spectra of free ligands and (IPr)CuCl. This material is available free of charge via the Internet at http://pubs.acs.org (https://ndownloader.figstatic.com/collections/2455828/versions/1).

Notes

The authors declare the following competing financial interest(s): one of the authors of this paper, i.e. Mark Thompson, has a financial interest in the Universal Display Corp., the principal supporter of this work. This interest has in no way influenced the content of the manuscript.

References

1. Armaroli, N., Accorsi, G., Cardinali, F., Listorti, A. *Top. Curr. Chem.* 2007, **280**, 69–115.
2. Barbieri, A., Accorsi, G., Armaroli, N. *Chem. Commun.* 2008, 2185–2193.
3. McMillin, D. R., McNett, K. M. *Chem. Rev.* 1998, **98**, 1201–1219.
4. Blaskie, M. W., McMillin, D. R. *Inorg. Chem.* 1980, **19**, 3519–3522.
5. Cuttell, D. G., Kuang, S. M., Fanwick, P. E., McMillin, D. R., Walton, R. A. *J. Am. Chem. Soc.* 2002, **124**, 6–7.
6. Miller, A. J. M., Dempsey, J. L., Peters, J. C. *Inorg. Chem.* 2007, **46**, 7244–7246.
7. Hashimoto, M., Igawa, S., Yashima, M., Kawata, I., Hoshino, M., Osawa, M. *J. Am. Chem. Soc.* 2011, **133**, 10348–10351.
8. Deaton, J. C., Switalski, S. C., Kondakov, D. Y., Young, R. H., Pawlik, T. D., Giesen, D. J., Harkins, S. B., Miller, A. J. M., Mickenberg, S. F., Peters, J. C. *J. Am. Chem. Soc.* 2010, **132**, 9499–9508.
9. Liu, Z., Qayyum, M. F., Wu, C., Whited, M. T., Djurovich, P. I., Hodgson, K. O., Hedman, B., Solomon, E. I., Thompson, M. E. *J. Am. Chem. Soc.* 2011, **133**, 3700–3703.
10. Smith, C. S., Branham, C. W., Marquardt, B. J., Mann, K. R. *J. Am. Chem. Soc.* 2010, **132**, 14079–14085.
11. Liu, F., Meadows, K. A., McMillin, D. R. *J. Am. Chem. Soc.* 1993, **115**, 6699–6704.
12. Mahadevan, S., Palaniandavar, M. *Inorg. Chem.* 1998, **37**, 693–700.
13. Scaltrito, D. V., Thompson, D. W., O'Callaghan, J. A., Meyer, G. J. *Coord. Chem. Rev.* 2000, **208**, 243–266.
14. Harkins, S. B., Peters, J. C. *J. Am. Chem. Soc.* 2005, **127**, 2030–2031.
15. Crestani, M. G., Manbeck, G. F., Brennessel, W. W., McCormick, T. M., Eisenberg, R. *Inorg. Chem.* 2011, **50**, 7172–7188.
16. Hsu, C.-W., Lin, C.-C., Chung, M.-W., Chi, Y., Lee, G.-H., Chou, P.-T., Chang, C.-H., Chen, P.-Y. *J. Am. Chem. Soc.* 2011, **133**, 12085–12099.
17. Siddique, Z. A., Yamamoto, Y., Ohno, T., Nozaki, K. *Inorg. Chem.* 2003, **42**, 6366–6378.
18. Krylova, V. A., Djurovich, P. I., Whited, M. T., Thompson, M. E. *Chem. Commun.* 2010, **46**, 6696–6698.
19. Lotito, K. J., Peters, J. C. *Chem. Commun.* 2010, **46**, 3690–3692.
20. Brooks, J., Babayan, Y., Lamansky, S., Djurovich, P. I., Tsyba, I., Bau, R., Thompson, M. E. *Inorg. Chem.* 2002, **41**, 3055–3066.

21. Lamansky, S., Djurovich, P., Murphy, D., Abdel-Razzaq, F., Lee, H. E., Adachi, C., Burrows, P. E., Forrest, S. R., Thompson, M. E. *J. Am. Chem. Soc.* 2001, **123**, 4304–4312.
22. You, Y., Park, S. Y. *Dalton Trans.* 2009, 1267–1282.
23. Yeh, Y.-S., Cheng, Y.-M., Chou, P.-T., Lee, G.-H., Yang, C.-H., Chi, Y., Shu, C.-F., Wang, C.-H. *ChemPhysChem* 2006, **7**, 2294–2297.
24. Avilov, I., Minoofar, P., Cornil, J., De Cola, L. *J. Am. Chem. Soc.* 2007, **129**, 8247–8258.
25. Chang, S. Y., Kavitha, J., Li, S. W., Hsu, C. S., Chi, Y., Yeh, Y. S., Chou, P. T., Lee, G. H., Carty, A. J., Tao, Y. T., Chien, C. H. *Inorg. Chem.* 2006, **45**, 137–146.
26. Yu, J. K., Hu, Y. H., Cheng, Y. M., Chou, P. T., Peng, S. M., Lee, G. H., Carty, A. J., Tung, Y. L., Lee, S. W., Chi, Y., Liu, C. S. *Chem. Eur. J.* 2004, **10**, 6255–6264.
27. Tung, Y. L., Wu, P. C., Liu, C. S., Chi, Y., Yu, J. K., Hu, Y. H., Chou, P. T., Peng, S. M., Lee, G. H., Tao, Y., Carty, A. J., Shu, C. F., Wu, F. I. *Organometallics* 2004, **23**, 3745–3748.
28. Flores, J. A., Andino, J. G., Tsvetkov, N. P., Pink, M., Wolfe, R. J., Head, A. R., Lichtenberger, D. L., Massa, J., Caulton, K. G. *Inorg. Chem.* 2011, **50**, 8121–8131.
29. Goj, L. A., Blue, E. D., Delp, S. A., Gunnoe, T. B., Cundari, T. R., Pierpont, A. W., Petersen, J. L., Boyle, P. D. *Inorg. Chem.* 2006, **45**, 9032–9045.
30. Diez-Gonzalez, S., Escudero-Adan, E. C., Benet-Buchholz, J., Stevens, E. D., Slawin, A. M. Z., Nolan, S. P. *Dalton Trans.* 2010, **39**, 7595–7606.
31. Teyssot, M.-L., Chevry, A., Traikia, M., El-Ghozzi, M., Avignant, D., Gautier, A. *Chem. Eur. J.* 2009, **15**, 6322–6326.
32. Shibata, T., Ito, S., Doe, M., Tanaka, R., Hashimoto, H., Kinoshita, I., Yano, S., Nishioka, T. *Dalton Trans.* 2011, **40**, 6778–6784.
33. Herrmann, W. A., Goossen, L. J., Spiegler, M. *J. Organomet. Chem.* 1997, **547**, 357–366.
34. Weskamp, T., Schattenmann, W. C., Spiegler, M., Herrmann, W. A. *Angew. Chem., Int. Ed.* 1998, **37**, 2490–2493.
35. Chianese, A. R., Li, X. W., Janzen, M. C., Faller, J. W., Crabtree, R. H. *Organometallics* 2003, **22**, 1663–1667.
36. Dible, B. R., Sigman, M. S. *Inorg. Chem.* 2006, **45**, 8430–8441.
37. Chen, L. X., Jennings, G., Liu, T., Gosztola, D. J., Hessler, J. P., Scaltrito, D. V., Meyer, G. J. *J. Am. Chem. Soc.* 2002, **124**, 10861–10867.
38. Chen, L. X., Shaw, G. B., Novozhilova, I., Liu, T., Jennings, G., Attenkofer, K., Meyer, G. J., Coppens, P. *J. Am. Chem. Soc.* 2003, **125**, 7022–7034.

39. Gothard, N. A., Mara, M. W., Huang, J., Szarko, J. M., Rolczynski, B., Lockard, J. V., Chen, L. X. *J. Phys. Chem. A* 2012, **116**, 1984–1992.
40. Barakat, K. A., Cundari, T. R., Omary, M. A. *J. Am. Chem. Soc.* 2003, **125**, 14228–14229.
41. Palmer, C. E. A., McMillin, D. R., Kirmaier, C., Holten, D. *Inorg. Chem.* 1987, **26**, 3167–3170.
42. Stacy, E. M., McMillin, D. R. *Inorg. Chem.* 1990, **29**, 393–396.
43. Czerwieniec, R., Yu, J., Yersin, H. *Inorg. Chem.* 2011, **50**, 8293–8301.
44. Blasse, G., McMillin, D. R. *Chem. Phys. Lett.* 1980, **70**, 1–3.
45. Breddels, P. A., Berdowski, P. A. M., Blasse, G. *J. Chem. Soc., Faraday Trans.* 1982, **78**, 595–601.
46. Kirchhoff, J. R., Gamache, R. E., Blaskie, M. W., Delpaggio, A. A., Lengel, R. K., McMillin, D. R. *Inorg. Chem.* 1983, **22**, 2380–2384.
47. Klappa, J. J., Rich, A. E., McNeill, K. *Org. Lett.* 2002, **4**, 435–437.
48. Pucci, D., Aiello, I., Aprea, A., Bellusci, A., Crispini, A., Ghedini, M. *Chem. Commun.* 2009, 1550–1552.
49. Uber, J. S., Vogels, Y., van den Helder, D., Mutikainen, I., Turpeinen, U., Fu, W. T., Roubeau, O., Gamez, P., Reedijk, J. *Eur. J. Inorg. Chem.* 2007, 4197–4206.
50. Singh, S. P., Kumar, D., Jones, B. G., Threadgill, M. D. *J. Fluorine Chem.* 1999, **94**, 199–203.
51. Thiel, W. R., Eppinger, J. *Chem. Eur. J.* 1997, **3**, 696–705.
52. Funabiki, K., Noma, N., Kuzuya, G., Matsui, M., Shibata, K. *J. Chem. Res., Synop.* 1999, 300–301.
53. *Bruker Instrument Service v2011.4.0.0*; Bruker AXS, Madison, WI, 2011.
54. *SAINT V7.68A*; Bruker AXS, Madison, WI, 2009.
55. SADABS V2008/1; Bruker AXS, Madison, WI, 2008.
56. *Bruker SHELXTL V2011.4-0*; Bruker AXS, Madison, WI, 2011.
57. Sheldrick, G. M. *Acta Crystallogr., Sect. A* 2008, **64**, 112–122.
58. Farrugia, L. J. *J. Appl. Crystallogr.* 1997, **30**, 565.
59. Frisch, M. J., Trucks, G. W., Schlegel, H. B., Scuseria, G. E., Robb, M. A., Cheeseman, J. R., Montgomery, J. A. Jr., Vreven, T., Kudin, K. N., Burant, J. C., Millam, J. M., Iyengar, S. S., Tomasi, J., Barone, V., Mennucci, B., Cossi, M., Scalmani, G., Rega, N., Petersson, G. A., Nakatsuji, H., Hada, M., Ehara, M., Toyota, K., Fukuda, R., Hasegawa, J., Ishida, M., Nakajima, T., Honda, Y., Kitao, O., Nakai, H., Klene, M., Li, X., Knox, J. E., Hratchian, H. P., Cross, J. B., Bakken, V., Adamo, C., Jaramillo, J., Gomperts, R., Stratmann, R. E.,

Yazyev, O., Austin, A. J., Cammi, R., Pomelli, C., Ochterski, J. W., Ayala, P. Y., Morokuma, K., Voth, G. A., Salvador, P., Dannenberg, J. J., Zakrzewski, V. G., Dapprich, S., Daniels, A. D., Strain, M. C., Farkas, O., Malick, D. K., Rabuck, A. D., Raghavachari, K., Foresman, J. B., Ortiz, J. V., Cui, Q., Baboul, A. G., Clifford, S., Cioslowski, J., Stefanov, B. B., Liu, G., Liashenko, A., Piskorz, P., Komaromi, I., Martin, R. L., Fox, D. J., Keith, T., Al-Laham, M. A., Peng, C. Y., Nanayakkara, A., Challacombe, M., Gill, P. M. W., Johnson, B., Chen, W., Wong, M. W., Gonzalez, C., Pople, J. A. *GAUSSIAN 03* (*Revision E.01*); Gaussian, Inc., Wallingford, CT, 2004.

60. Becke, A. D. *J. Chem. Phys.* 1993, **98**, 5648–5652.
61. Lee, C. T., Yang, W. T., Parr, R. G. *Phys. Rev. B* 1988, **37**, 785–789.
62. Hay, P. J., Wadt, W. R. *J. Chem. Phys.* 1985, **82**, 270–283.
63. Wadt, W. R., Hay, P. J. *J. Chem. Phys.* 1985, **82**, 284–298.
64. Hay, P. J., Wadt, W. R. *J. Chem. Phys.* 1985, **82**, 299–310.

Chapter 44

Phosphorescence versus Thermally Activated Delayed Fluorescence: Controlling Singlet–Triplet Splitting in Brightly Emitting and Sublimable Cu(I) Compounds

Markus J. Leitl,[a] Valentina A. Krylova,[b] Peter I. Djurovich,[b] Mark E. Thompson,[b] and Hartmut Yersin[a]

[a] *Institute for Physical Chemistry, University of Regensburg, 93040 Regensburg, Germany*
[b] *Department of Chemistry, University of Southern California, Los Angeles, California 90089, USA*
hartmut.yersin@ur.de, met@usc.edu

Photophysical properties of two highly emissive three-coordinate Cu(I) complexes, (IPr)Cu(py$_2$-BMe$_2$) (**1**) and (Bzl-3,5Me)Cu(py$_2$-BMe$_2$) (**2**), with two different N-heterocyclic (NHC) ligands were investigated in detail (IPr = 1,3-bis(2,6-diisopropylphenyl)imidazol-2-ylidene; Bzl-3,5Me = 1,3-bis(3,5-dimethylphenyl)-1H-benzo[d]imidazol-2-ylidene; py$_2$-BMe$_2$ = di(2-pyridyl)dimethylborate). The compounds exhibit remarkably high emission quantum

---

Reprinted from *J. Am. Chem. Soc.*, **136**, 16032–16038, 2014.

Electrophosphorescent Materials and Devices
Edited by Mark E. Thompson
Text Copyright © 2014 American Chemical Society
Layout Copyright © 2024 Jenny Stanford Publishing Pte. Ltd.
ISBN 978-981-4877-34-3 (Hardcover), 978-1-003-08872-1 (eBook)
www.jennystanford.com

yields of more than 70% in the powder phase. Despite similar chemical structures of both complexes, only compound **1** exhibits thermally activated delayed blue fluorescence (TADF), whereas compound **2** shows a pure, yellow phosphorescence. This behavior is related to the torsion angles between the two ligands. Changing this angle has a huge impact on the energy splitting between the first excited singlet state S_1 and triplet state T_1 and therefore on the TADF properties. In addition, it was found that, in both compounds, spin–orbit coupling (SOC) is particularly effective compared to other Cu(I) complexes. This is reflected in short emission decay times of the triplet states of only 34 µs (**1**) and 21 µs (**2**), respectively, as well as in the zero-field splittings of the triplet states amounting to 4 cm^{-1} (0.5 meV) for 1 and 5 cm^{-1} (0.6 meV) for **2**. Accordingly, at ambient temperature, compound **1** exhibits *two* radiative decay paths which are thermally equilibrated: one via the S_1 state as TADF path (62%) and one via the T_1 state as phosphorescence path (38%). Thus, if this material is applied in an organic light-emitting diode, the generated excitons are harvested mainly in the singlet state, but to a significant portion also in the triplet state. This novel mechanism based on two separate radiative decay paths reduces the overall emission decay time distinctly.

44.1 Introduction

In the past years, phosphorescent transition metal compounds have experienced significant research attention as they can be used as highly efficient emitter materials for organic light-emitting diodes (OLEDs) [1, 2]. Especially, complexes based on third row transition metal ions, such as Ir(III) and Pt(II), are well suited as the heavy metal center can induce significant spin–orbit coupling (SOC). This results in fast intersystem crossings [3–5], in short emission decay times for the otherwise spin-forbidden transitions from the first excited triplet state to the ground state ($T_1 \rightarrow S_0$) [1, 5–15], and to distinct zero-field splittings (ZFS) of the triplet states [1, 5–11]. In addition, when applied in an electroluminescent device, these phosphorescent materials show the *triplet harvesting effect* which allows utilizing all excitons, singlets and triplets, for the generation of light [1, 2, 10]. As a result, emitters that show the triplet harvesting effect can exhibit four times higher exciton to photon conversion efficiencies than conventional purely fluorescent emitters.

However, due to the high cost of iridium and platinum metals, more abundant central metal ions such as Cu(I) are highly attractive and have stepped into the focus of research [11, 16–40]. Such compounds might at first sight not seem to be good candidates for use in OLEDs as SOC is significantly less effective in these compounds than in Pt(II) or Ir(III) based complexes due to the smaller SOC constant of copper [41]. As a consequence, phosphorescence decay times of the order of several hundred microseconds can result. This would lead to pronounced saturation effects, if these materials were applied as emitters in OLEDs [42]. On the other hand, the energy separation between the first excited singlet S_1 and triplet T_1 state $\Delta E(S_1-T_1)$ can be relatively small in Cu(I) compounds due to a small exchange integral resulting from a pronounced metal-to-ligand charge transfer (MLCT) character [11, 16–18, 30, 34, 40, 43–45]. If this energy separation is small enough, a thermal population of the singlet state from the triplet state becomes efficient at ambient temperature. Thus, a thermally activated delayed fluorescence (TADF) from S_1 to S_0 occurs. As the $S_1 \rightarrow S_0$ transition is spin-allowed, it shows a significantly shorter emission decay time than the corresponding triplet state. For this reason, that is, the involvement of the S_1 state emission, the overall decay time of such compounds is drastically reduced and can become as short as only several microseconds at ambient temperature [11, 16–18, 30, 34, 40, 43–45]. In OLEDs using TADF emitters, the emission largely originates from the singlet state S_1. Therefore, this mechanism has been called *singlet harvesting* [11, 16–18, 30] in contrast to the *triplet harvesting* effect.

A crucial parameter that determines the effectiveness of the thermally activated delayed fluorescence is the energy separation $\Delta E(S_1 - T_1)$. If it is larger than about 3×10^3 cm^{-1} (0.37 eV), a thermal population of the singlet state S_1 is not effective. Therefore, it is important to understand how $\Delta E(S_1 - T_1)$ can be controlled by properly engineering the chemical structure of an emitter complex. For this, we investigated the two previously published, structurally related Cu(I) complexes (IPr)Cu(py$_2$-BMe$_2$) (**1**) and (Bzl-3,5Me)Cu(py$_2$-BMe$_2$) (**2**) (IPr = 1,3-bis(2,6-diisopropylphenyl)imidazol-2-ylidene; Bzl-3,5Me = 1,3-bis(3,5-dimethylphenyl)-1H-benzo-[d]imidazol-2-ylidene; py$_2$-BMe$_2$ = di(2-pyridyl)dimethylborate) [27]. The chemical structures are displayed in Table 44.1. Interestingly, compound **1** shows a highly effective TADF (with $\Delta E(S_1 - T_1) = 740$ cm^{-1} (92 meV)), whereas for compound **2** only phosphorescence but no TADF is observed at ambient temperature. In this contribution, we present detailed photophysical characterizations and discuss why despite similar chemical structures the photophysical properties of both compounds differ drastically, especially in regard of the value found for $\Delta E(S_1 - T_1)$.

44.2 Ambient Temperature Phosphorescence versus TADF

Under excitation with UV light, the powders of the studied complexes display intense blue (**1**) and yellow (**2**) luminescence at ambient temperature with short emission decay times of 11 µs (**1**) and 18 µs (**2**) and remarkably high emission quantum yields of 76% (**1**) and 73% (**2**), respectively. In Fig. 44.1, the corresponding emission spectra are displayed.

The spectra are broad and featureless with maxima at 475 nm (**1**) and 575 nm (**2**) at $T = 300$ K. The shapes of the spectra indicate that the emission originates from a charge transfer transition which, in this case, has significant metal-to-ligand charge transfer (MLCT) character. This assumption is in agreement with literature assignments of other Cu(I) compounds [11, 16–20, 27–31, 43–46] and is further substantiated by results of density functional theory (DFT) and time-dependent density functional theory (TDDFT) calculations presented in Ref. [27] and below. The emission of **2** is found at significantly lower energy than that of compound **1**. This can be rationalized by the expansion of the π-system of the IPr ligand (compound **2**) which leads to a lower lying LUMO energy than

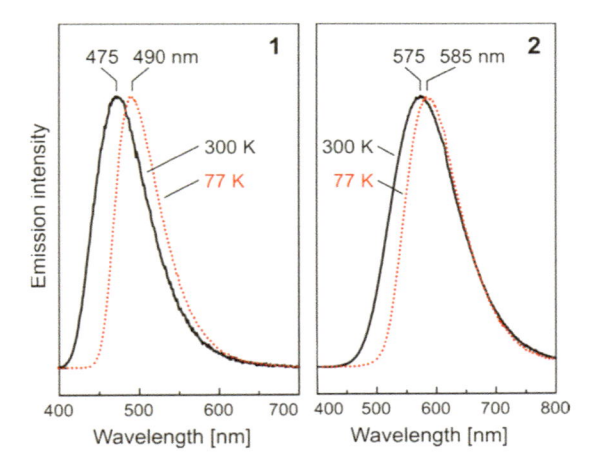

Figure 44.1 Normalized emission spectra of compound **1** and **2** as powders at ambient temperature and at 77 K. The samples were excited at $\lambda_{\text{exc}} = 350$ nm.

that of the Bzl-3,5Me ligand, whereas HOMOs of both compounds are composed of metal and py$_2$-BMe$_2$ orbitals, giving HOMO energies that are nearly unchanged. As a result, the HOMO–LUMO gap and therefore the emission energy are lower for **2** than for **1** [27].

When cooling from ambient temperature to 77 K, a red-shift of the emission from 475 to 490 nm (≈ 650 cm^{-1}) is observed for compound **1**. In addition, the emission decay time increases by a factor of about 3 from 11 to 34 µs, whereas the radiative rate $k_\text{r} = \Phi_{\text{PL}}\tau^{-1}$ decreases by about the same factor from 6.9×10^4 to 2.7×10^4 s^{-1}. The significantly longer emission decay time at 77 K of 34 µs (compared to 11 µs at 300 K) suggests that the emitting state at $T = 77$ K is the triplet state T$_1$. Further proof for this assignment is given in subsection 44.2.2. However, it is remarked that a triplet decay time of 34 µs is extraordinarily short compared to other Cu(I) compounds [11, 16–18, 22, 25, 43]. This indicates that SOC is particularly effective in compound **1**. A more detailed discussion of this aspect is given in subsections 44.2.1 and 44.2.2. The observed changes of the emission decay time or the radiative rate and the spectral shift of the emission peaks upon temperature change can be explained by the occurrence of a TADF at $T = 300$ K and are discussed in more detail in subsection 44.2.2.

In contrast, the emission decay time of compound **2** changes only slightly from 21 to 18 µs when heating from $T = 77$ K to ambient temperature. Almost no change is found for the radiative rate amounting to $k_r(77\,\text{K}) = 3.8 \times 10^4$ s+ and $k_r(300\,\text{K}) = 4.1 \times 10^4$ s^{-1}, respectively. This indicates that for compound **2** TADF is not effective and that the observed emission even at ambient temperature is phosphorescence stemming from T$_1$. The slight red-shift of the high energy flank observed on cooling may be explained (especially for this triplet emitter) by freezing out energetically higher lying emissions from an inhomogeneously broadened distribution in the powder sample (compare Ref. [46]) and is therefore *not* a result of the freezing out of the singlet emission. This is in contrast to compound **1** for which the entire spectrum is shifted (Fig. 44.1). Further support for this rationalization is given by the investigation of the emission spectra in a PMMA (poly(methyl methacrylate)) matrix at 300 and 77 K. In this situation, no such spectral change on temperature variation is observed, besides a slight narrowing on cooling.

44.2.1 Compound 2: Typical Triplet Emitter

In this subsection, we want to focus on compound **2** by investigating the emission decay time in the temperature range between 1.3 and 300 K (Fig. 44.2) which is particularly instructive.

From Fig. 44.2, it can be seen that the (thermalized) decay time for compound **2** is almost constant in the temperature range between ≈10 and 300 K and amounts to about 20 µs. Also, the radiative rate is essentially constant (compare Table 44.1) which allows us to assign the emission as phosphorescence stemming from the T$_1$ state in the entire temperature range. An emission via the TADF mechanism (compare subsection 44.2.2) is not occurring in this case. Thus, it can be concluded that the energy splitting $\Delta E(S_1 - T_1)$ between the first excited singlet and triplet state is larger than 3000 cm^{-1} (≈0.4 eV), as for such a large value no significant thermal activation is expected at $T = 300$ K.

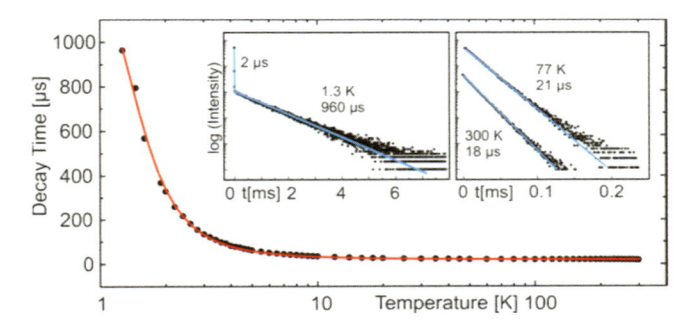

Figure 44.2 Thermalized emission decay time of compound **2** (powder) versus temperature. The sample was excited at $\lambda_{exc} = 355$ nm, and the signal was detected at $\lambda_{det} = 600$ nm. The red line represents a fit of the experimental data according to Eq. 44.1. Insets: Decay curves at $T = 1.3, 77,$ and 300 K.

Table 44.1 Structures and emission properties of the compounds (IPr)Cu(py$_2$-BMe$_2$) (**1**) and (Bzl-3,5Me)Cu(py$_2$-BMe$_2$) (**2**) as powders[a]

	1		2	
Temp. [K]	300	77	300	77
λ_{max} [nm]	475	490	575	585
Φ_{PL} [%]	76	91	73	80
τ [µs]	11	34	18	21
k_r [10^4 s^{-1}]	6.9	2.7	4.1	3.8
k_{nr} [10^4 s^{-1}]	2.2	0.3	1.5	1.0

[a] The decay time is monoexponential in the entire temperature range above ≈25 K. The radiative k_r and nonradiative k_{nr} rates were calculated according to $k_r = \Phi_{PL}\tau^{-1}$ and $k_{nr} = (1 - \Phi_{PL})\tau^{-1}$, respectively.

Interestingly, when the temperature is decreased to below ≈10 K a steep increase of the decay time from about 20 μs to 1 ms at $T = 1.3$ K is observed. A similar behavior is well-known from other transition metal compounds, such as Ir(III) and Pt(II) compounds [5–8, 11, 47, 48], and can be related to the energy splitting of the triplet state into three substates. This so-called ZFS is a consequence of SOC. Apart from the fast component with a decay time of 2 μs found at $T = 1.3$ K, the emission decay time at low temperature is governed by a Boltzmann distribution of the three substates I, II, and III. According to the monoexponential decay, these states are in a thermal equilibrium (after several μs). At low temperature, mainly emission from the energetically lowest substates I (and II) is observed. With increasing temperature, the higher lying substate III is thermally populated. Since frequently the radiative rates corresponding to the transitions from the energetically higher lying substates to the S_0 ground state are larger than the rates corresponding to the lowest substate(s), the averaged emission decay time decreases with increasing temperature [5–8, 11, 45, 47, 48]. Accordingly, the data given in Fig. 44.2 can be fitted with a modified Boltzmann function (Eq. 44.1) in order to determine the ZFS values and the decay time constants of the individual triplet substates (compare Refs. [7, 45, 49]).

$$\tau(T) = \left[1 + e^{-\frac{\Delta E(\text{II-I})}{k_B T}} + e^{-\frac{\Delta E(\text{III-I})}{k_B T}}\right]$$

$$\times \left[\tau_\text{I}^{-1} + \tau_\text{II}^{-1} e^{-\frac{\Delta E(\text{II-I})}{k_B T}} + \tau_\text{III}^{-1} e^{-\frac{\Delta E(\text{III-I})}{k_B T}}\right]^{-1} \quad (44.1)$$

In this equation, $\tau(T)$ refers to the emission decay time at a given temperature T, τ_I, τ_II, and τ_III to the individual decay times of triplet substates (I, II, and III), $\Delta E(\text{III-I})$ and $\Delta E(\text{II-I})$ to the energy splittings between the triplet substates III/I and II/I, respectively, and k_B to the Boltzmann constant.

As a result of the fitting procedure, a value of $\Delta E(\text{III-I}) = \Delta E(\text{ZFS}) = 5$ cm^{-1} (0.6 meV) was found. To the best of our knowledge, such a large ZFS has not been reported before for a Cu(I) complex. However, by this procedure, it could not be determined where substate II is energetically located with respect to substates I and III. If it is assumed that substates I and II are energetically close

(ΔE (II–I) ≈ 0 cm^{-1}; compare Refs. [11, 17, 18]), the emission decay times of the three triplet substates can be obtained. They amount to $\tau_I \approx \tau_{II} = 1.5$ ms and $\tau_{III} = 7$ μs. The results found for compound **2**, especially the value of ZFS = 5 cm^{-1} and the average emission decay time match well with an empirical ordering scheme that correlates ΔE (ZFS) with the phosphorescence decay time [1, 11]. From this perspective, it is not surprising that a ZFS of 5 cm^{-1} is found for a compound with a triplet decay time of about 20 μs.

The short emission decay component of 2 μs at $T = 1.3$ K becomes shorter and diminishes rapidly with increasing temperature and cannot be observed at temperatures higher than ≈ 25 K. Such a behavior strongly indicates the occurrence of a relatively slow spin–lattice relaxation (SLR) from the higher lying triplet substate III to the substates I and II according to the *direct* effect of SLR [5, 48]. Moreover, in a very rough estimate, one can use the measured value of τ (SLR) = 2 μs to determine the energy separation between the involved states, that is, the value of ΔE (ZFS). With the relation of ΔE (ZFS)$^3 \sim \tau$ (SLR)$^{-1}$ for the direct process of SLR and the corresponding values known from a number of other organo-transition metal compounds [5, 48], one obtains a value of about 4 cm^{-1} which nicely confirms the splitting values as determined from the fitting procedure as discussed above.

44.2.2 Compound 1: Thermally Activated Delayed Fluorescence

In Fig. 44.3, the emission decay time of compound **1** is displayed versus temperature. Similar as for compound **2**, two decay components are observed in the temperature range between 1.3 and ≈ 25 K (not displayed in Fig. 44.3). The short component can again be assigned to SLR processes, whereas the long component corresponds to the thermalized emission of the three triplet substates (compare previous subsection). As for compound **2**, a significant reduction of the decay time between $T = 1.3$ and 10 K from 110 to 40 μs is observed. Again, this can be assigned to the thermal population of higher lying triplet substates from the lowest one(s), leading to an average value of about 34 μs for all three triplet substates between ≈ 10 and ≈ 100 K ("plateau").

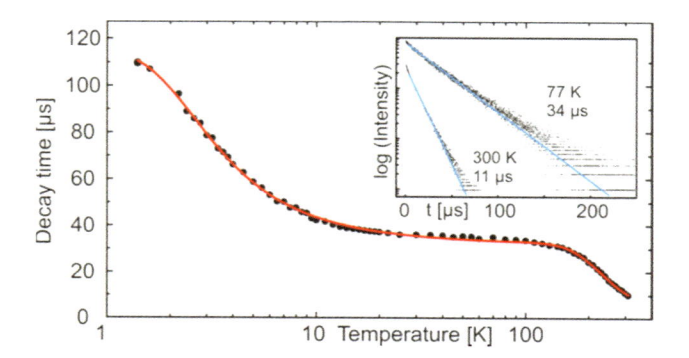

Figure 44.3 Emission decay time of compound **1** (powder) versus temperature. The sample was excited at $\lambda_{exc} = 378$ nm, and the signal was detected at $\lambda_{det} = 490$ nm. The red line represents a fit of the experimental data according to Eq. 44.2. Inset: Decay curves at $T = 77$ and 300 K.

However, in contrast to the behavior of compound **2**, the emission decay time of compound **1** is not constant up to $T = 300$ K. It decreases from about 34 to 11 μs at ambient temperature. This effect can be rationalized by the following considerations. At low temperature, only the triplet state T_1 is contributing to the emission. With increasing temperature, a thermal population of the energetically higher lying S_1 state becomes possible. As the S_1 state exhibits a significantly shorter emission decay time than that of the T_1 state, an overall reduction of the emission decay time is observed with increasing temperature. Additionally, a blue-shift of the emission occurs as the S_1 state lies energetically higher than the T_1 state. This emission mechanism corresponds to a thermally activated delayed fluorescence.

The measured data, as displayed in Fig. 44.3, can be fitted with Eq. 44.2 which represents an expansion of Eq. 44.1 by two additional terms (marked in red), which take the thermal population of the singlet state S_1 into account.

$$\tau(T) = \left[1 + e^{-\frac{\Delta E(\text{II-I})}{K_B T}} + e^{-\frac{\Delta E(\text{III-I})}{K_B T}} + e^{-\frac{\Delta E(S_1 - T_1)}{K_B T}}\right]$$
$$\times \left[\tau_I^{-1} + \tau_{II}^{-1} e^{-\frac{\Delta E(\text{II-I})}{K_B T}} + \tau_{III}^{-1} e^{-\frac{\Delta E(\text{III-I})}{K_B T}} + \tau_{S_1}^{-1} e^{-\frac{\Delta E(S_1 - T_1)}{K_B T}}\right]^{-1}$$

(44.2)

From the fitting procedure, the decay times of the three triplet substates of $\tau_I \approx \tau_{II} = 116$ μs and $\tau_{III} = 13$ μs and a value of ΔE (ZFS) $= 4$ cm^{-1} (0.5 meV) was found. The latter one is only slightly smaller than found for compound **2**. Similar as for compound **2**, the energy of triplet substate II with respect to substates I and III could not be determined. For the fitting procedure, ΔE (II–I) ≈ 0 cm^{-1} was assumed. The energy splitting between the excited triplet T$_1$ and singlet state S$_1$ is determined to ΔE (S$_1$ - T$_1$) = 740 cm^{-1} (90 meV). This value is in good agreement with the blue-shift of the emission spectrum when heating from 77 to 300 K amounting to 650 cm^{-1}. The corresponding emission decay time of the singlet state S$_1$ is found to be $\tau(S_1) = 160$ ns. Such a short decay time emphasizes the singlet nature of this state. It is remarked that in contrast to the delayed fluorescence, a prompt fluorescence is not observed for this compound as intersystem crossing (ISC) from the S$_1$ to the T$_1$ state, probably being of the order of 10 ps [50], is much faster than the prompt S$_1$ → S$_0$ emission.

Interestingly, the increase of the radiative rate and the related decrease of the emission decay time with increasing temperature is for compound **1** significantly less pronounced than that for other Cu(I) complexes. For example, the copper complexes presented in Ref. [18] show an increase of the radiative rates by the TADF process by a factor of 40–150, whereas compound **1** exhibits only an increase by a factor of 3. An explanation for this behavior can be given when the emission decay path from the triplet state to the singlet ground state is also taken into account. For the compounds in Ref. [18], the triplet state decay times are long, lying between 250 and 1200 μs, whereas compound **1** exhibits a decay time of only 34 μs. Therefore, a reduction of the decay time by involving the TADF process at higher temperatures is much less effective.

44.3 Controlling TADF by Ligand Orientation

As discussed in section 44.2, at ambient temperature, compound 1 displays an effective TADF, whereas for compound 2 thermal population of the singlet state is not observed due to the activation energy being greater than 3000 cm^{-1}. Obviously, this effect is

Table 44.2 Chemical structures of compounds **1** and **2** as well as of the modified versions **1a** and **2a**[a]

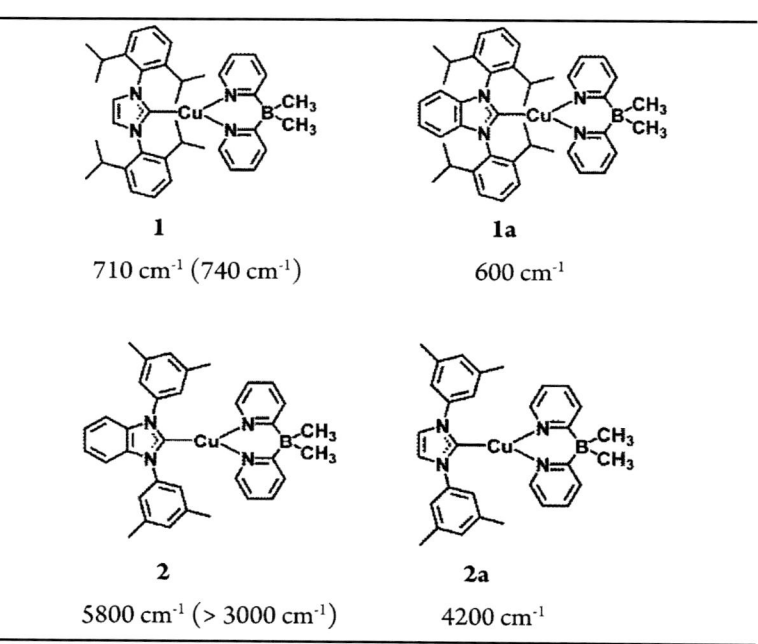

1	**1a**
710 cm^{-1} (740 cm^{-1})	600 cm^{-1}
2	**2a**
5800 cm^{-1} (> 3000 cm^{-1})	4200 cm^{-1}

[a]The values for $\Delta E(S_1 - T_1)$ obtained from TDDFT calculations and from experimental investigations (in brackets) are also displayed.

connected to differences in the chemical structures of the NHC ligands on the molecules. From Table 44.2, it can be seen that the compounds differ in two aspects. (i) The π-system of the imidazole ring in compound 1 is expanded by benzannulation in compound 2. (ii) The isopropyl groups at the 2,6-positions on the pendant phenyl rings of the NHC ligand in 1 are replaced by methyl groups at the 3,5-positions giving compound 2.

For a better understanding of the effects of these modifications on the energy gap $\Delta E(S_1 - T_1)$ between the first excited singlet and triplet state, we have performed DFT and TDDFT calculations for compounds **1** and **2** as well as for two further model compounds **1a** and **2a** as displayed in Table 44.2. Compound **1a** represents a modified version of compound **1** in which the imidazole ring is π-extended to benzimidazole, but the isopropyl groups on the phenyl rings are retained. Compound **2a** represents a modification of compound **2** where the π-system of the benzimidazole moiety is trimmed to imidazole, but the methyl groups are left unchanged. For all four structures displayed in Table 44.2, a DFT geometry optimization for the electronic ground state was performed. As starting geometry for compounds **1** and **2**, the crystal structures were used as described and provided in Ref. [27]. The starting geometries for compounds **1a** and **2a** were created by expanding and contracting the π-system, respectively, of the NHC ligand in the structures of compounds **1** and **2**. TDDFT calculations were performed on the structures obtained after geometry optimization.

It was found that compounds **1** and **1a** exhibit very similar (and small) singlet–triplet gaps of 710 and 600 cm^{-1}, respectively. This is in good agreement with the experimental value found for compound **1** amounting to $\Delta E(S_1 - T_1) = 740$ cm^{-1} (compare subsection 44.2.2). For compounds **2** and **2a**, large values of 5800 and 4200 cm^{-1}, respectively, were found (Table 44.2). These results indicate that expanding the π-system of the NHC ligand does not have a strong impact on the singlet–triplet splitting. Therefore, these modifications cannot explain the experimentally found differences with values of $\Delta E(S_1 - T_1) = 740$ cm^{-1} for **1** and of $\Delta E(S_1 - T_1) > 3000$ cm^{-1} for **2**.

Interestingly, the insensitivity of the exchange energy to benzannulation of the imidazole ring indicates that the methyl and isopropyl groups present at the IPr and Bzl-3,5Me ligands play an important role for the $\Delta E(S_1 - T_1)$ value and the occurrence of TADF. However, it seems unlikely that these groups impart a direct electronic impact on the singlet–triplet splitting. Instead, the alkyl groups can exert steric control over the orientation of the two ligands toward each other and change the electronic behavior of the compounds in this manner. In support, the X-ray structures (compare Ref. [27]) show that, for compound **1**, the IPr and py$_2$-BMe$_2$ ligands are nearly coplanar, whereas for compound **2** the Bzl-3,5Me and py$_2$-BMe$_2$ ligands are almost perpendicular to each other (Fig. 44.4).

Thus, we examined how the relative orientations of the ligands toward each other influence the singlet–triplet splitting using a model compound **1b** (Fig. 44.4). In this model compound, the

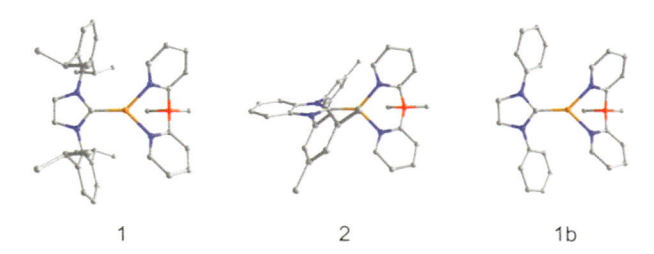

Figure 44.4 Perspective drawings of the optimized geometries of compounds **1** and **2** as well as of the model compound **1b**. Hydrogen atoms were omitted for clarity.

isopropyl groups were removed from the phenyl rings of the NHC ligand. This change allows for variation of the N–C–Cu–N torsion angle (marked green in Fig. 44.6) without encountering steric hindrance from the adjacent py$_2$-BMe$_2$ ligand. The N–C–Cu–N torsion angle of **1b** was then fixed at values between 0° and 100° in steps of 10° for a DFT geometry optimization of the singlet ground state. Interestingly, these calculations show that the spatial distribution of the HOMO changes with variation of the torsion angle. In particular, for an angle of 0°, the HOMO is localized on the copper center and py$_2$-BMe$_2$ ligand, whereas it is extended onto the imidazole ring when the angle is 90° (compare Fig. 44.5). In contrast, the LUMO remains localized on the π^*-orbitals of the NHC ligand for all torsion angles. The difference in the HOMO is due to the angular relation between the metal d and imidazole π-orbitals. When the ligands are coplanar, the two sets of orbitals are orthogonal and thus do not electronically couple to each other. However, in the perpendicular orientation, the orbitals have the appropriate symmetry to conjugate and delocalize their electronic distribution onto both ligands. Consequently, overlap between the HOMO and LUMO is small when the torsion angle is 0° (hence a small exchange energy results), whereas a significant overlap exists between the frontier orbitals with a 90° torsion thereby increasing the exchange energy. Since the lowest excited singlet and triplet states are largely comprised from transitions between these frontier orbitals ($>94\%$ for the S$_1$ and $>82\%$ for the T$_1$ state; see Table S1 in

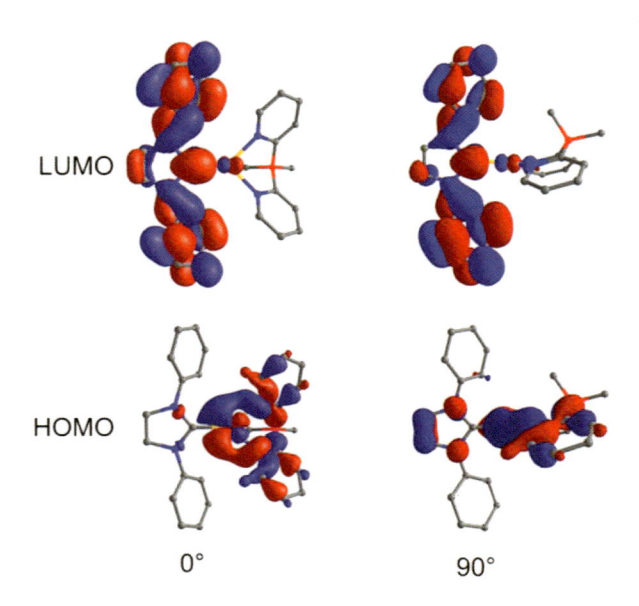

Figure 44.5 HOMOs and LUMOs of model compound **1b** displayed for a torsion angles of 0° and 90°, respectively.

the Supporting Information), variation in the degree of overlap will strongly alter the value of $\Delta E(S_1 - T_1)$.

A more accurate estimate of the dependence of the singlet–triplet splitting on the torsion angle can be made when TDDFT calculations are performed on the (torsion constrained) optimized geometries. As compound **1b** exhibits a symmetry element at all rotations (either a mirror plane or a S_2 axis), the values of $\Delta E(S_1 - T_1)$ are identical at positive and negative torsion angles. The data, displayed in Fig. 44.6, show that the singlet–triplet splitting in **1b** is lowest (540 cm^{-1}) when the N–C–Cu–N torsion angle between the two ligands is 0°. This result is in agreement with the experimental data found for compound **1** with a torsion angle of 5° and a singlet–triplet splitting of only 740 cm^{-1}. In contrast, when the torsion angle is 70° (as realized for compound **2**), a splitting of 3700 cm^{-1} is obtained from the calculations. For such a large $\Delta E(S_1 - T_1)$ energy separation, no TADF would occur. These model calculations strongly

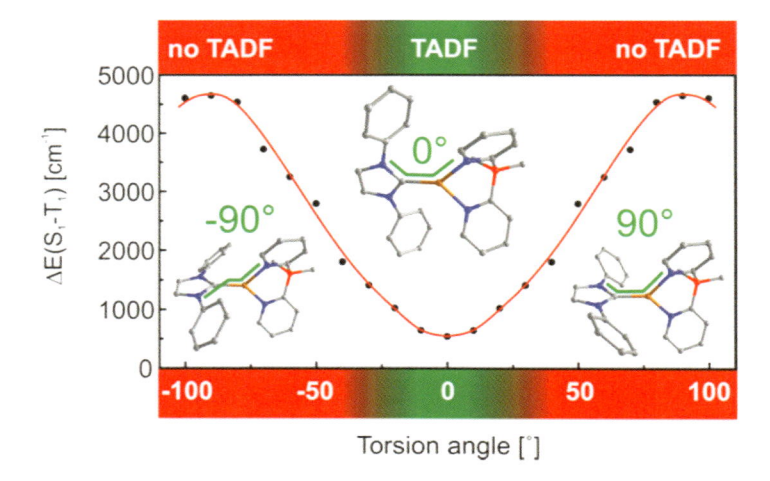

Figure 44.6 Singlet–triplet splitting $\Delta E(S_1 - T_1)$ in dependence of the torsion angle N–C–Cu–N (marked by the green line) as obtained from DFT and TDDFT calculations on the B3LYP/def2-SVP level of theory.

support the experimental results of a lower $\Delta E(S_1 - T_1)$ limit of 3000 cm^{-1} (0.37 eV) as predicted for compound **2**.

44.4 Conclusion

Materials that are applied as emitters in organic light-emitting diodes should be able to utilize all injected excitons for the generation of light. At the moment, these requirements can be met by materials that exhibit the triplet harvesting effect, typically based on high-cost Pt(II) or Ir(III) complexes, or the singlet harvesting effect, typically based on low-cost Cu(I) complexes or specific purely organic materials. Emitters showing the triplet harvesting effect hereby stand out through very effective spin–orbit coupling, whereas emitters exhibiting the singlet harvesting effect excel through a small energy splitting between the first excited triplet and singlet state resulting in a thermally activated delayed fluorescence (TADF).

Both mechanisms lead to an effective reduction of the emission decay time and enable both singlet and triplet excitons to be used for the generation of light in an OLED. One of the compounds presented in this contribution (compound **1**) combines the advantages of both the triplet and the singlet harvesting effect. (i) It exhibits relatively strong spin–orbit coupling which results in a (compared to other Cu(I) compounds known so far) very short triplet emission decay time of only 34 µs. (ii) The energy splitting between the first excited singlet and triplet state amounts to only 740 cm^{-1}. Therefore, the compound exhibits an effective TADF. The contribution of each of the two effects to the emission can be quantified according to the calculations presented in ref 18. It is found that at ambient temperature 38% of the emission intensity stems from the triplet state and 62% from the singlet state. Accordingly, the deactivation via both radiative decay paths induces a greater overall radiative decay rate. Thus, due to the combination of phosphorescence and delayed fluorescence, an effective decay time of $\tau = 11$ µs can be achieved which is shorter than the decay times of the individual processes ($\tau_{TADF}(300\ K) = 16$ µs, $\tau_{Ph} = 34$ µs) (Fig. 44.7).

Another important issue that has been revealed in this investigation is the connection between the orientation of the ligands toward each other and the value of the activation energy for a TADF process. The ligand orientation is crucial for the difference between a good OLED emitter with relatively short decay time and an emitter with too long emission decay time for good OLED performance. Therefore, the results presented here for the first time give valuable guidelines for the development of new TADF emitter materials.

44.5 Experimental Section

Compounds **1** and **2** were synthesized, purified by vacuum sublimation at 220°C, and characterized according to the procedures described in Ref. [27]. Absolute measurements of the photoluminescence quantum yields at ambient temperature and at 77 K were performed with a C9920-02 (Hamamatsu Photonics) system. Emission spectra were measured with a Fluorolog 3-22 (Horiba Jobin Yvon) spectrometer which was equipped with a cooled photomultiplier (RCA C7164R). For the measurement of the emission decay times, the same photomultiplier was used in

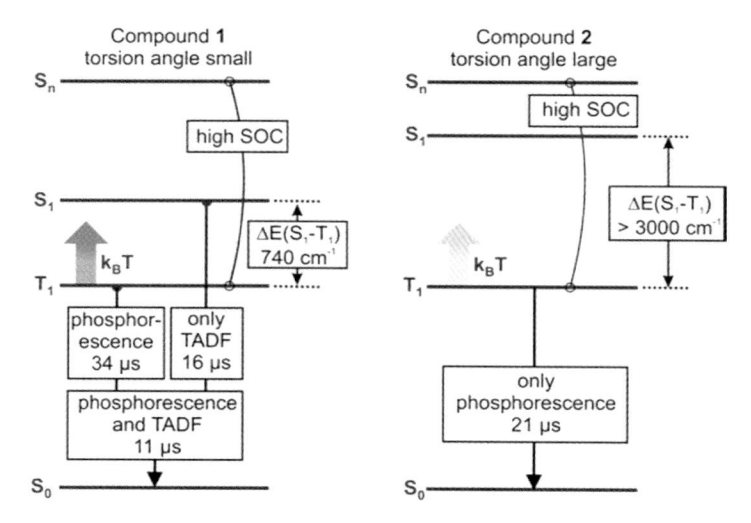

Figure 44.7 Energy level diagrams of compounds **1** and **2**. At ambient temperature, compound **2** shows only emission from the triplet state, while compound **1** additionally exhibits a TADF. The combination of TADF *and* triplet emission (phosphorescence) results in a distinct reduction of the emission decay time. The triplet states exhibit zero-field splittings of 4 and 5 cm^{-1} for compounds **1** and **2**, respectively. The TADF decay time τ (TADF) at $T = 300$ K was calculated according to τ (phosphorescence and TADF)$^{-1}$ = τ (phosphorescence)$^{-1}$ + τ (TADF)$^{-1}$. Note that the given phosphorescence decay times are the $T = 77$ K values.

combination with a FAST multichannel scaler PCI card (Comtec). As excitation source for the decay time measurements, a pulsed diode laser (Picobrite PB-375L,) with an excitation wavelength of 378 nm and a pulse width < 100 ps or a pulsed Nd:YAG laser (IB Laser Inc. DiNY pQ 02) with an excitation wavelength of 355 nm and a pulse width < 7 ns was used. For adjusting the temperature, the samples were placed into a helium cryostat (Cryovac Konti Cryostat IT) in which the helium gas flow and heating were controlled. DFT and TDDFT calculations were carried out using NWChem 6.3 on a high performance computing cluster [51]. The calculations were performed on the B3LYP/def2-SVP level of theory [52, 53], which has been shown to give good results for other Cu(I) compounds [54].

Acknowledgments

We thankfully acknowledge funding by the German Ministry of Education and Research (BMBF). We further appreciate travel expenses provided by the German Academic Exchange Service (DAAD) and the Bavaria California Technology Center (BaCaTec). Furthermore, the authors thank Randy Rückner for support regarding the high performance computation cluster on which the DFT and TDDFT calculations were performed.

Supporting Information

HOMO–LUMO coefficients at different torsion angles for compound **1b**. Emission and excitation spectra of compounds **1** and **2** (powder). Absorption spectra of compounds **1** and **2** recorded in dichloromethane. This material is available free of charge via the Internet at http://pubs.acs.org (https://ndownloader.figstatic.com/files/3871642).

References

1. Yersin, H., Ed. *Highly Efficient OLEDs with Phosphorescent Materials*; Wiley-VCH: Weinheim, Germany, 2008.
2. Baldo, M. A., O'Brien, D. F., You, Y., Shoustikov, A., Sibley, S., Thompson, M. E., Forrest, S. R. *Nature* 1998, **395**, 151–154.
3. Tang, K.-C., Liu, K. L., Chen, I. C. *Chem. Phys. Lett.* 2004, **386**, 437–441.
4. Hedley, G. J., Ruseckas, A., Samuel, I. D. W. *J. Phys. Chem. A* 2008, **113**, 2–4.
5. Yersin, H., Donges, D. *Top. Curr. Chem.* 2001, **214**, 81–186.
6. Rausch, A. F., Homeier, H. H. H., Yersin, H. *Top. Organomet. Chem.* 2010, **29**, 193–235.
7. Hofbeck, T., Yersin, H. *Inorg. Chem.* 2010, **49**, 9290–9299.
8. Bossi, A., Rausch, A. F., Leitl, M. J., Czerwieniec, R., Whited, M. T., Djurovich, P. I., Yersin, H., Thompson, M. E. *Inorg. Chem.* 2013, **52**, 12403–12415.

9. Rausch, A. F., Murphy, L., Williams, J. A. G., Yersin, H. *Inorg. Chem.* 2011, **51**, 312–319.
10. Yersin, H. *Top. Curr. Chem.* 2004, **241**, 1–26.
11. Yersin, H., Rausch, A. F., Czerwieniec, R., Hofbeck, T., Fischer, T. *Coord. Chem. Rev.* 2011, **255**, 2622–2652.
12. Cheng, G., Chow, P.-K., Kui, S. C. F., Kwok, C.-C., Che, C.-M. *Adv. Mater.* 2013, **25**, 6765–6770.
13. Sajoto, T., Djurovich, P. I., Tamayo, A. B., Oxgaard, J., Goddard, W. A., Thompson, M. E. *J. Am. Chem. Soc.* 2009, **131**, 9813–9822.
14. Fernandez-Hernandez, J. M., Beltran, J. I., Lemaur, V., Galvez-Lopez, M. D., Chien, C. H., Polo, F., Orselli, E., Fröhlich, R., Cornil, J., De Cola, L. *Inorg. Chem.* 2013, **52**, 1812–1824.
15. Tarran, W. A., Freeman, G. R., Murphy, L., Benham, A. M., Kataky, R., Williams, J. A. G. *Inorg. Chem.* 2014, **53**, 5738–49.
16. Czerwieniec, R., Kowalski, K., Yersin, H. *Dalton T.* 2013, **42**, 9826–9830.
17. Czerwieniec, R., Yu, J., Yersin, H. *Inorg. Chem.* 2011, **50**, 8293–8301.
18. Leitl, M. J., Küchle, F. R., Mayer, H. A., Wesemann, L., Yersin, H. *J. Phys. Chem. A* 2013, **117**, 11823–11836.
19. Krylova, V. A., Djurovich, P. I., Aronson, J. W., Haiges, R., Whited, M. T., Thompson, M. E. *Organometallics* 2012, **31**, 7983–7993.
20. Krylova, V. A., Djurovich, P. I., Whited, M. T., Thompson, M. E. *Chem. Commun.* 2010, **46**, 6696–6698.
21. Tsuge, K. *Chem. Lett.* 2013, **42**, 204–208.
22. Osawa, M. *Chem. Commun.* 2014, **50**, 1801–1803.
23. Kobayashi, A., Komatsu, K., Ohara, H., Kamada, W., Chishina, Y., Tsuge, K., Chang, H. C., Kato, M. *Inorg. Chem.* 2013, **52**, 13188–13198.
24. Kaeser, A., Mohankumar, M., Mohanraj, J., Monti, F., Holler, M., Cid, J. J., Moudam, O., Nierengarten, I., Karmazin-Brelot, L., Duhayon, C., Delavaux-Nicot, B., Armaroli, N., Nierengarten, J. F. *Inorg. Chem.* 2013, **52**, 12140–12151.
25. Zhang, Q., Komino, T., Huang, S., Matsunami, S., Goushi, K., Adachi, C. *Adv. Funct. Mater.* 2012, **22**, 2327–2336.
26. Wada, A., Zhang, Q., Yasuda, T., Takasu, I., Enomoto, S., Adachi, C. *Chem. Commun.* 2012, **48**, 5340–5342.
27. Krylova, V. A., Djurovich, P. I., Conley, B. L., Haiges, R., Whited, M. T., Williams, T. J., Thompson, M. E. *Chem. Commun.* 2014, **50**, 7176–7179.
28. Wallesch, M., Volz, D., Zink, D. M., Schepers, U., Nieger, M., Baumann, T., Brase, S. *Chem.—Eur. J.* 2014, **20**, 6578–6590.

29. Zink, D. M., Volz, D., Baumann, T., Mydlak, M., Flügge, H., Friedrichs, J., Nieger, M., Brase, S. *Chem. Mater.* 2013, **25**, 4471–4486.
30. Zink, D. M., Bachle, M., Baumann, T., Nieger, M., Kuhn, M., Wang, C., Klopper, W., Monkowius, U., Hofbeck, T., Yersin, H., Bräse, S. *Inorg. Chem.* 2013, **52**, 2292–2305.
31. Volz, D., Nieger, M., Friedrichs, J., Baumann, T., Bräse, S. *Langmuir* 2013, **29**, 3034–3044.
32. Cunningham, C. T., Moore, J. J., Cunningham, K. L. H., Fanwick, P. E., McMillin, D. R. *Inorg. Chem.* 2000, **39**, 3638–3644.
33. Cid, J. J., Mohanraj, J., Mohankumar, M., Holler, M., Monti, F., Accorsi, G., Karmazin-Brelot, L., Nierengarten, I., Malicka, J. M., Cocchi, M., Delavaux-Nicot, B., Armaroli, N., Nierengarten, J. F. *Polyhedron* 2014, **82**, 158–172.
34. Linfoot, C. L., Leitl, M. J., Richardson, P., Rausch, A. F., Chepelin, O., White, F. J., Yersin, H., Robertson, N. *Inorg. Chem.* 2014, **53**, 10854–10861.
35. Igawa, S., Hashimoto, M., Kawata, I., Yashima, M., Hoshino, M., Osawa, M. *J. Mater. Chem. C* 2013, **1**, 542–551.
36. Lotito, K. J., Peters, J. C. *Chem. Commun.* 2010, **46**, 3690–3692.
37. Vitale, M., Ford, P. C. *Coord. Chem. Rev.* 2001, **219–221**, 3–16.
38. Zigler, D. F., Tordin, E., Wu, G., Iretskii, A., Cariati, E., Ford, P. C. *Inorg. Chim. Acta* 2011, **374**, 261–268.
39. Kutal, C. *Coord. Chem. Rev.* 1990, **99**, 213–252.
40. Yersin, H., Leitl, M. J., Czerwieniec, R., In *Organic Light Emitting Materials and Devices XVIII*; So, F., Adachi, C., Eds., Proceedings of SPIE; SPIE: Bellingham, WA, 2014; Vol 9183, 91830N-1-91830N-11.
41. Murov, S. L., Hug, G. L., Carmichael, I. *Handbook of Photochemistry*, 2nd ed., Marcel Dekker: New York, 1993; pp. 339–341.
42. Murawski, C., Leo, K., Gather, M. C. *Adv. Mater.* 2013, **25**, 6801–6827.
43. Deaton, J. C., Switalski, S. C., Kondakov, D. Y., Young, R. H., Pawlik, T. D., Giesen, D. J., Harkins, S. B., Miller, A. J. M., Mickenberg, S. F., Peters, J. C. *J. Am. Chem. Soc.* 2010, **132**, 9499–9508.
44. Palmer, C. E. A., McMillin, D. R. *Inorg. Chem.* 1987, **26**, 3837–3840.
45. Blasse, G., McMillin, D. R. *Chem. Phys. Lett.* 1980, **70**, 1–3.
46. Coppens, P., Sokolow, J., Trzop, E., Makal, A., Chen, Y. *J. Phys. Chem. Lett.* 2013, **4**, 579–582.
47. Rausch, A. F., Thompson, M. E., Yersin, H. *Chem. Phys. Lett.* 2009, **468**, 46–51.
48. Yersin, H., Strasser, J. *Coord. Chem. Rev.* 2000, **208**, 331–364.

49. Azumi, T., O'Donnell, C. M., McGlynn, S. P. *J. Chem. Phys.* 1966, **45**, 2735–2742.
50. Iwamura, M., Watanabe, H., Ishii, K., Takeuchi, S., Tahara, T. *J. Am. Chem. Soc.* 2011, **133**, 7728–7736.
51. Valiev, M., Bylaska, E. J., Govind, N., Kowalski, K., Straatsma, T. P., Van Dam, H. J. J., Wang, D., Nieplocha, J., Apra, E., Windus, T. L., de Jong, W. A. *Comput. Phys. Commun.* 2010, **181**, 1477–1489.
52. Becke, A. D. *J. Chem. Phys.* 1993, **98**, 5648.
53. Weigend, F., Ahlrichs, R. *Phys. Chem. Chem. Phys.* 2005, **7**, 3297–3305.
54. Jesser, A., Rohrmüller, M., Schmidt, W. G., Herres-Pawlis, S. *J. Comput. Chem.* 2014, **35**, 1–17.

Chapter 45

Control of Emission Colour with N-Heterocyclic Carbene (NHC) Ligands in Phosphorescent Three-Coordinate Cu(I) Complexes[†]

Valentina A. Krylova, Peter I. Djurovich, Brian L. Conley, Ralf Haiges, Matthew T. Whited, Travis J. Williams, and Mark E. Thompson

Department of Chemistry, University of Southern California, Los Angeles, California 90089-0744, USA
met@usc.edu

A series of three phosphorescent mononuclear (NHC)–Cu(I) complexes were prepared and characterized. Photophysical properties were found to be largely controlled by the NHC ligand chromophore. Variation of the NHC ligand leads to emission colour tuning over 200 nm range from blue to red, and emission efficiencies of 0.16–0.80 in the solid state.

Phosphorescent Cu(I) complexes are an emerging class of luminescent materials based on an inexpensive and abundant metal

---

[†]The details of electronic supplementary information (ESI) are given at the end of the chapter.
Reprinted from *Chem. Commun.*, **50**, 7176–7179, 2014.

Electrophosphorescent Materials and Devices
Edited by Mark E. Thompson
Text Copyright © 2014 The Royal Society of Chemistry
Layout Copyright © 2024 Jenny Stanford Publishing Pte. Ltd.
ISBN 978-981-4877-34-3 (Hardcover), 978-1-003-08872-1 (eBook)
www.jennystanford.com

[1]. The ability to tune chemical and photophysical properties in a desirable and predictable way is highly important when considering potential applications of Cu(I)-based phosphors. The typical strategy to modulate the excited state properties of these and related luminescent materials is usually achieved through variation of the coordinating ligand(s) [2]. To date the types of ligands most commonly used to prepare phosphorescent Cu(I) complexes are di-imines or organophosphines and their derivatives [3]. Alternatively, N-heterocyclic carbenes (NHC) are an attractive class of ligands as they are electronically and sterically tunable and form strong bonds with transition metals giving robust complexes [4]. However, while NHCs have been employed as either chromophoric or ancillary ligands in luminescent Ir and Pt complexes [5] they have been rarely used as chromophoric ligands in Cu(I) complexes [6].

We have recently used NHC ligands to prepare phosphorescent 3-coordinate Cu(I) complexes (NHC)Cu(N^N), where N^N denotes a neutral diimine or monoanionic pyridyl-azolate ligand [7]. The monodentate NHC ligand, 1,3-bis(2,6-diisopropylphenyl)imidazol-2-ylidene (IPr), employed in these complexes has both a large $\pi\pi^*$ energy gap and high triplet energy, therefore the emission energy in these derivatives is controlled by variations in the N^N ligand. Herein, we report a series of luminescent (NHC)Cu(N^N) complexes **1–3** (Fig. 45.1), where the NHC ligand is principally involved in the excited state and demonstrate a wide range emission colour tunability through modification of carbene moiety. In particular, we systematically lowered the energy gap of **1** by benzannulation of imidazolylidene ring to make **2** and further introduced nitrogen atoms to form the pyrazinyl moiety in **3**. In addition, we utilize an anionic non-conjugated N^N ligand, *i.e.* di(2-pyridyl)dimethylborate (py_2BMe_2) that possesses high triplet energy to serve as an ancillary ligand [8]. To the best of our knowledge the py_2BMe_2 ligand, unlike the isoelectronic di(1-pyrazolyl)borates (pz_2BR_2, R = H, alkyl, aryl) [2c, 9], has never been used to prepare luminescent transition metal complexes. We have found that (NHC)Cu complexes with the py_2BMe_2 ligand are more robust and luminescent than the pyrazolyl-borate congeners.

The (NHC)Cu(py_2BMe_2) complexes were obtained from their respective (NHC)CuCl precursors upon addition of a stoichiometric

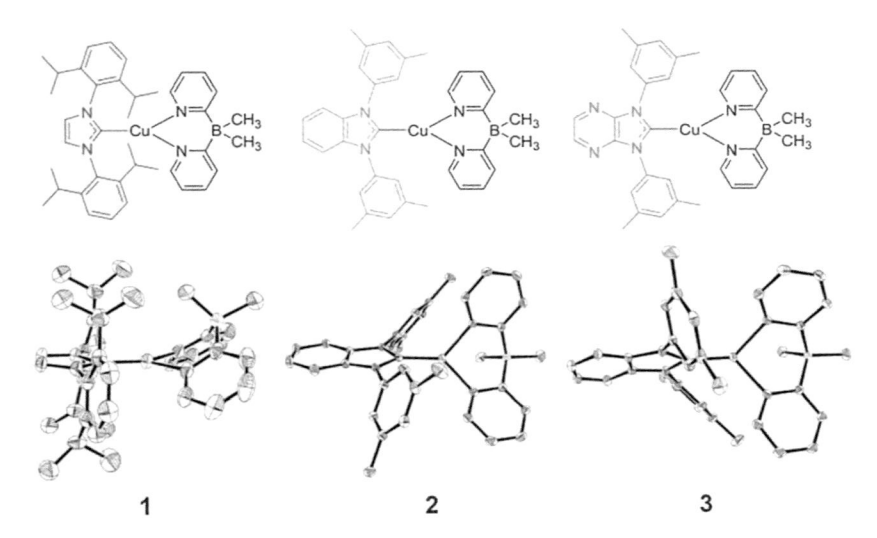

Figure 45.1 Molecular structures (top) and perspective view at 50% probability (bottom) of complexes **1–3**. Only one of the unique structures for **2** found in the unit cell is shown. Hydrogen atoms are omitted for clarity.

amount of sodium di(2-pyridyl)dimethylborate in tetrahydrofuran at RT. Complexes **1–3** are stable in solid state and in solution under anaerobic conditions. Complex **1** can be sublimed under vacuum and is stable for hours in solution, while **2** and **3** decompose slowly under aerobic conditions in solution and blacken in the solid state after 24 h exposure to air. Evidently, the isopropyl groups at *ortho* positions of phenyl groups of the NHC ligand in **1** impart greater stability of the (NHC)Cu(py$_2$BMe$_2$) complexes than in **2** and **3**. We also prepared complexes analogous to **1** and **2** using the pz$_2$BH$_2$ ligand instead of py$_2$BMe$_2$. While the analog to **1**, (IPr)Cu(pz$_2$BH$_2$), can be isolated and fully characterized (see ESI[†]), the congener to **2** decomposed rapidly upon exposure to air and was not examined further.

X-ray diffraction analyses confirmed monomeric three-coordinate structures for complexes **1–3**. Complex **2** has two unique structures in the unit cell that have similar geometric parameters. The coordination geometry in complexes **1–3** can be described as Y-shaped with the sum of bond angles around copper close to 360°

(359.98° in **1**, 359.72° in **2** and 358.64° in **3**). The Cu–N–C–B–C–N ring formed upon chelation of the py$_2$BMe$_2$ ligand adopts a boat-shaped conformation similar to that reported in metal complexes bearing related di(2-pyridyl)borate ligands [8, 10]. The relative orientations of NHC and py$_2$BMe$_2$ ligands in crystals differ within the series. In complex **1** the ligands are arranged with the pyridyl rings situated opposite the aryl rings of the NHC ligand across a crystallographic mirror plane that bisects the C$_{NHC}$, Cu and B atoms. In contrast, the py$_2$BMe$_2$ ligand in **2** and **3** is oriented about C$_{NHC}$–Cu bond so that the two pyridyl rings are situated above and below a plane defined by the N$_{NHC}$, N$_{NHC}$ and C$_{NHC}$ atoms. The Cu–N$_{py}$ bond lengths in **1** are 2.0288(15) Å and slightly shorter in **2** (1.9929(16) Å and 1.9997(16) Å) and **3** (2.010(9) Å and 2.014(9) Å). The C$_{NHC}$–Cu–N$_{py}$ angles are 132.78(4)1 in **1** and vary from 134.32(7)° and 129.27(7)° in 2 to (135.0(6)° and 128.1(6)°) in **3**. The C$_{NHC}$–Cu distances in **1–3** (1.8678(19)–1.895(2) Å) are within the range for reported NHC–Cu(I) complexes [11].

In solution ^{1}H NMR data indicates rapid boat-to-boat interconversion of the py$_2$BMe$_2$ ligand as resonances of methyl groups attached to boron atom appear as one broad singlet both at room temperature and at $-40\,°$C in acetone-d$_6$. Although the ^{1}H NMR data do not allow us to assess if there is free rotation about the C$_{NHC}$–Cu bonds in solution, the chemical shift for the protons *ortho* to the pyridyl nitrogens gives insight into the preferred molecular conformation. This resonance appears at $\delta = 8.36$ ppm in the protonated py$_2$BMe$_2^-$ ligand, whereas upon coordination to copper in **1** it is shifted markedly upfield to $\delta = 7.3$ ppm due to shielding by the diamagnetic ring current from the adjacent aryl rings of the NHC ligand. In contrast, the same resonance undergoes a much smaller shift upon coordination in **2** and **3**, appearing at $\delta = 7.97$ ppm and $\delta = 8.05$ ppm, respectively. Thus, the ^{1}H NMR data in solution correlate with the relative ligand orientation found in crystalline state; co-planar for **1**, perpendicular for **2** and **3**.

Photophysical data for complexes **1–3** are summarized in Table 45.1. The UV-visible absorption spectra for complexes **1–3** in dichloromethane are shown in Fig. 45.2. High energy bands at 290 nm in **1** ($\varepsilon \sim 7000$–14 200 M^{-1} cm^{-1}), 310 nm in **2** ($\varepsilon \sim 6500$–19 000 M^{-1} cm^{-1}) and 340 nm in **3** ($\varepsilon \sim 4000$–13 400 M^{-1} cm^{-1}) are

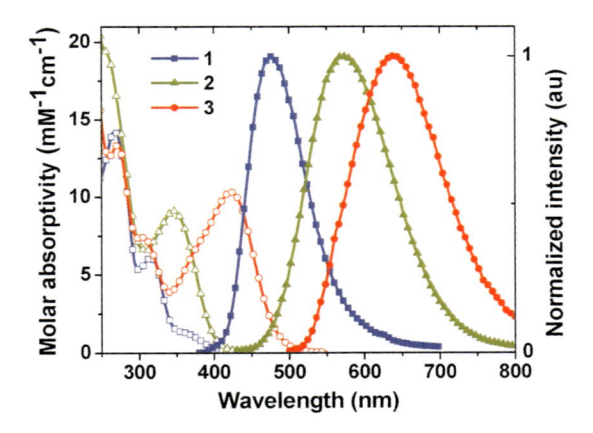

Figure 45.2 Absorption (open symbols, CH_2Cl_2) and emission (closed symbols, solid powder) spectra of complexes **1–3** at room temperature.

assigned to spin-allowed ligand centered (LC) transitions on both the NHC and py_2BMe_2 ligands. Lower energy bands, not observed in absorption spectra of precursors (see ESI†), are assigned to charge transfer (CT) transitions. In complex **1** the CT bands appear at 316 nm ($\varepsilon = 6100$ M^{-1} cm^{-1}) with a shoulder at 360 nm ($\varepsilon \sim 1300$ M^{-1} cm^{-1}). A comparison between the absorption spectrum of **1** to that of (IPr)Cu(pz$_2$BH$_2$) (see ESI†) shows the LC band is unchanged in energy in both derivatives, whereas the CT bands shift to higher energy and lower intensity ($\lambda_{max} = 305$ nm, $\varepsilon = 2000$ M^{-1} cm^{-1} and $\lambda_{max} = 330$ nm, $\varepsilon \sim 1000$ M^{-1} cm^{-1}) in the latter complex. The bathochromic shift for the low energy bands in **1** indicates that the py_2BMe_2 ligand participates in these transitions, although some CT character involving the IPr ligand may contribute as well. Upon expansion of π-system of the NHC ligand in **2** the CT band becomes more distinct and intense ($\lambda_{max} = 346$ nm, $\varepsilon = 9100$ M^{-1} cm^{-1}). Substitution of the two CH-groups with nitrogens in **3** leads to a marked red shift and increase in molar absorptivity ($\lambda_{max} = 422$ nm, $\varepsilon = 10\,300$ M^{-1} cm^{-1}). Thus, these bands in **2** and **3** are unambiguously assigned to CT transitions involving NHC ligands.

The emission spectra recorded for neat microcrystalline solids of complexes **1–3** at room temperature are broad and featureless (Fig. 45.2). Complex **1** gives sky-blue emission ($\lambda_{max} = 476$ nm), complex **2** displays yellow emission ($\lambda_{max} = 570$ nm) and **3** has orange-red emission ($\lambda_{max} = 638$ nm). The bathochromic shift in emission for complexes **2** and **3** further indicates that the lowest energy excited state is governed largely by the NHC ligand. Solid powders of **1** and **2** glow brightly upon excitation with emission quantum yields (Φ) of 0.80 and 0.70 respectively, while **3** has moderate emission efficiency ($\Phi = 0.16$). In contrast, the quantum efficiency of the (IPr)Cu(pz$_2$BH$_2$) derivative in the solid state is much lower ($\lambda = 415$ nm, $\Phi = 0.03$). The observed luminescence for **1–3** is phosphorescence as emission lifetimes (τ) are in the microsecond regime. The radiative rate constants (k_r) in the solid state vary within the small range of values ($k_r = (3.3–7.2) \times 10^4$ s^{-1}). In fluid solution the emission efficiency is substantially lower than in the solid state. In particular, complex **1** has quantum yield of 0.15 ($\tau = 2.3$ µs) in cyclohexane, while emission from **2** and **3** is almost completely quenched ($\Phi < 0.005$).

The emission spectra of neat samples of **1–3** shift to lower energies upon cooling to 77 K (Table 45.1, also see ESI†). Bathochromic shifts in emission energy at low temperature are common for Cu(I) complexes. This phenomenon is often attributed to suppression of thermally activated delayed fluorescence (TADF) and usually accompanied by a marked increase in emission lifetimes of an order of magnitude or more [9, 12]. Complexes **1–3**, however, show only a modest increase in emission lifetimes upon cooling as emission lifetimes measured at 77 K are in the range of 17–36 µs (Table 45.1). The relatively small increase in lifetime at 77 K is inconsistent with processes typically associated with TADF and suggests instead that emission measured both at room temperature and 77 K is from a state that is principally triplet in character. The behavior also implies that the radiative rate constant for the lowest triplet state is significantly enhanced in **1–3**. This unusual temperature dependence on the emission lifetime is being currently investigated in greater detail at lower temperatures.

The observed bathochromic shift of emission energy in **1–3** upon expanding the size of the π-system of a ligand chromophore and N-substitution is consistent with a decrease in separation between the highest occupied molecular orbital (HOMO) and the lowest unoccupied molecular orbital (LUMO). Computational

Table 45.1 Photophysical data for complexes **1–3**

	Absorbance,[a] λ (nm) ε (10^3 M^{-1}cm^{-1})	Emission at room temperature[b]					Emission at 77 K[b]	
		λ_{max} (nm)	τ (μs)	Φ	k_r (s^{-1})	k_{nr} (s^{-1})	λ_{max} (nm)	τ (μs)
1	268 (14.2), 316 (6.1), 360 sh (1.3)	476	11	0.8	7.2×10^4	1.8×10^4	492	36
2	257 sh (19.3), 346 (9.1)	570	15	0.7	4.7×10^4	2.0×10^4	586	17
3	271 (13.3), 305 (7.4), 422 (10.3)	638	7.5	0.16	3.3×10^4	1.0×10^5	650	21

[a]In dichloromethane.
[b]In solid state.

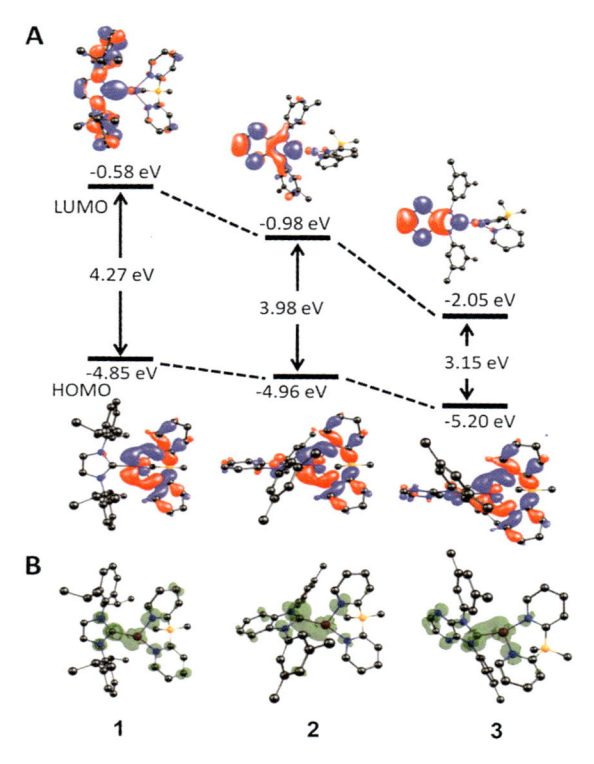

Figure 45.3 (A) HOMO and LUMO plots and energies for **1–3**. (B) Optimized triplet geometries and triplet spin density contour plots (isovalue: 0.004 e a_0^{-3}). Hydrogen atoms are omitted for clarity.

analyses of the ground and excited state properties performed using density functional theory (DFT) and time-dependent DFT (TD-DFT) calculations compare favorably with the experimental observations. The calculated wavelength and oscillator strength of the lowest singlet transitions progressively increase for **1** (λ = 381 nm, f = 0.0028), **2** (λ = 400 nm, f = 0.1440) and **3** (λ = 522 nm, f = 0.1645). This result follows the trend observed in absorption spectra, *i.e.* a decrease in energy and increase of molar absorption for the CT bands when going from **1** to **2** to **3**. The frontier molecular orbitals for **1–3** are shown in Fig. 45.3A. For all three complexes the calculated HOMOs have essentially identical spatial contours, consisting predominantly of d orbitals on copper (39–48%) mixed with orbitals on di(2-pyridyl)dimethylborate ligand

(41–45%). The LUMO in **1–3** is localized on the NHC ligand (85–94%) with minimal metal character (4–6%). A small contribution (8%) from the py$_2$BMe$_2$ orbitals appears in the LUMO of complex **1**; however, there is less (4% and 2%) in both **2** and **3**. Noteworthy is a substantial contribution (8–23%) from the carbene carbon 2p$_z$ orbital in the LUMO of all three complexes. Congruent with the orbital composition, variations of the carbene ligand have pronounced effect on LUMO energies. Complex **1** has the highest LUMO energy in the series ($E_{LUMO} = -0.58$ eV) followed by **2** ($E_{LUMO} = -0.98$ eV) and **3** ($E_{LUMO} = -2.05$ eV). The HOMO energies show a similar trend in stabilization, albeit to a lesser degree ($E_{HOMO} = -4.85$ eV, -4.96 eV, -5.20 eV, for **1–3**, respectively). The HOMO–LUMO gap is thus progressively smaller for **1** ($\Delta E_{H-L} = 4.27$ eV), **2** ($\Delta E_{H-L} = 3.98$ eV) and **3** ($\Delta E_{H-L} = 3.15$ eV).

The lowest vertical singlet and triplet excitations obtained from TD-DFT calculations are mainly HOMO→LUMO transitions (see ESI†). On the basis of the MO description given above the lowest lying transition for complex **1** can be ascribed as (M + L)LCT admixed with intraligand $\pi \rightarrow \pi^*$ (py$_2$BMe$_2$) (ILCT) character, whereas for complexes **2** and **3** the transition is principally metal–ligand to NHC–ligand charge transfer ((M + L)LCT). The calculated spin density surfaces for the triplet electronic configuration further support this assignment (Fig. 45.3B). For complexes **1–3** the spin contours are localized along the C$_{NHC}$–Cu bond axis. Both ligands also contribute to the triplet spin density; however, while complex **1** has a significant contribution from the borate ligand (31% NHC, 31% py$_2$BMe$_2$), the spin distribution is shifted toward NHC ligand for complexes **2** (52% NHC, 16% py$_2$BMe$_2$) and **3** (54% NHC, 16% py$_2$BMe$_2$).

To emphasize the importance of proper ligand design to achieve efficient room temperature phosphorescence from this family of Cu(I) compounds, we prepared complex **4** where the π-system of the NHC ligand was expanded by annulation of imidazolylidene with a peri-naphthyl moiety (Fig. 45.4, full characterization is given in ESI†). For this derivative, the intensity of the lowest lying CT absorption band centered at 450 nm is low ($\varepsilon = 450$ M^{-1} cm^{-1}) and the complex is nonemissive in the solid state at room temperature and at 77 K. Very weak, structured emission ($\Phi < 0.01$, $\tau < 10$ ns) is observed from a dilute solution of **4** in frozen 2-methyltetrahydrofuran (2-MeTHF) glass at 77 K (Fig. 45.4).

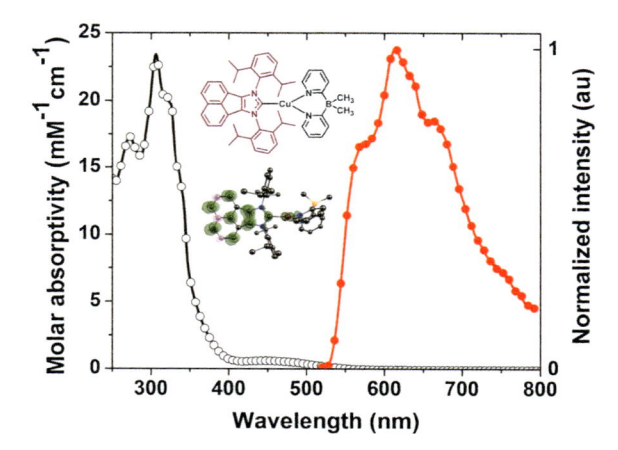

Figure 45.4 Absorption (room temperature, CH_2Cl_2) and emission (77 K, 2-MeTHF) spectra of complex **4**. (inset) Molecular structure, optimized triplet geometry and spin density contour plot (isovalue: 0.004 e a_0^{-3}).

This emission is tentatively assigned as phosphorescence since the radiative rate constant ($k_r < 10^6$ s^{-1}) and presence of vibronic features are inconsistent with fluorescence from the CT state. The HOMO for **4** calculated using DFT is essentially identical to that of complexes **1–3**. The LUMO, localized primarily on the aromatic π-system of NHC ligand, has no electron density on the 2p$_z$ orbital of CNHC atom in strong contrast to what is found in complexes **1–3**. Such an electronic distribution leads to poor overlap between frontier orbitals and thus a low oscillator strength for the lowest lying CT transitions ($\lambda = 597$ nm, $f = 0.0044$). The triplet spin density of **4** is localized on the acenaphthyl moiety and, unlike that of complexes **1–3**, has a node across the C$_{NHC}$–Cu bond axis (Fig. 45.4). The spin distribution, together with structured emission spectrum, suggests that the luminescence is ligand centered in character. Thus, extension of the π-system in this manner, while shrinking the HOMO–LUMO gap ($\Delta E_{H-L} = 2.55$ eV), decreases the energy of the ^{3}LC state on the NHC ligand and reduces electronic coupling to such a degree that it can no longer effectively interact with the MLCT states responsible for promoting fast radiative decay [3b].

In conclusion, we report a series of (NHC)–Cu(I) complexes that show phosphorescence associated primarily with NHC ligand chromophore. Judicious modification of the NHC ligand allows the emission colour to be tuned over 200 nm from blue to orange-red while retaining high emission efficiencies. The estimated triplet radiative rate constants are comparable with those of third row transition metal complexes. Taking into account electronic and steric tunability of the NHC ligands, these findings introduce a new versatile method to control the photophysical properties of luminescent Cu(I) complexes.

We are grateful to the Universal Display Corporation for financial support. B.L.C. and T.J.W. also thank National Science Foundation CHE-1054910 grant for support. Theoretical calculations were supported by the University of Southern California Center for High-Performance Computing and Communications (www.usc.edu/hpcc). The X-ray diffractometer is sponsored by National Science Foundation CRIF Grant 1048807.

Electronic Supplementary Information (ESI)

Experimental and computational details, synthesis and characterization of new compounds, absorption data for precursors and (IPr)Cu(pz$_2$BH$_2$) complex, photophysical data for **1–3** obtained at 77 K in the solid state and in methylcyclohexane glass and crystallographic data. CCDC 976577–976579 and 990891. For ESI and crystallographic data in CIF or other electronic format see DOI: 10.1039/c4cc02037e.

References

1. A. Barbieri, G. Accorsi and N. Armaroli, *Chem. Commun.*, 2008, 2185.
2. (a) J. Brooks, Y. Babayan, S. Lamansky, P. I. Djurovich, I. Tsyba, R. Bau and M. E. Thompson, *Inorg. Chem.*, 2002, **41**, 3055; (b) Y. Chi and P.-T. Chou, *Chem. Soc. Rev.*, 2010, **39**, 638; (c) J. Li, P. I. Djurovich, B. D. Alleyne, M. Yousufuddin, N. N. Ho, J. C. Thomas, J. C. Peters, R. Bau and M. E. Thompson, *Inorg. Chem.*, 2005, **44**, 1713; (d) Y. You and S. Y. Park, *Dalton Trans.*, 2009, 1267.

3. (a) G. F. Manbeck, W. W. Brennessel and R. Eisenberg, *Inorg. Chem.*, 2011, **50**, 3431; (b) C.-W. Hsu, C.-C. Lin, M.-W. Chung, Y. Chi, G.-H. Lee, P.-T. Chou, C.-H. Chang and P.-Y. Chen, *J. Am. Chem. Soc.*, 2011, **133**, 12085; (c) M. Hashimoto, S. Igawa, M. Yashima, I. Kawata, M. Hoshino and M. Osawa, *J. Am. Chem. Soc.*, 2011, **133**, 10348; (d) S. B. Harkins and J. C. Peters, *J. Am. Chem. Soc.*, 2005, **127**, 2030; (e) D. G. Cuttell, S.-M. Kuang, P. E. Fanwick, D. R. McMillin and R. A. Walton, *J. Am. Chem. Soc.*, 2002, **124**, 6; (f) M. G. Crestani, G. F. Manbeck, W. W. Brennessel, T. M. McCormick and R. Eisenberg, *Inorg. Chem.*, 2011, **50**, 7172; (g) N. Armaroli, G. Accorsi, F. Cardinali and A. Listorti, *Top. Curr. Chem.*, 2007, **280**, 69; (h) S. Igawa, M. Hashimoto, I. Kawata, M. Yashima, M. Hoshino and M. Osawa, *J. Mater. Chem. C*, 2013, **1**, 542–551; (i) M. Wallesch, D. Volz, D. M. Zink, U. Schepers, M. Nieger, T. Baumann and S. Braese, *Chem.—Eur. J.*, 2014, **20**, 6578–6590.

4. (a) D. Bourissou, O. Guerret, F. P. Gabbaie and G. Bertrand, *Chem. Rev.*, 2000, **100**, 39; (b) H. Jacobsen, A. Correa, A. Poater, C. Costabile and L. Cavallo, *Coord. Chem. Rev.*, 2009, **253**, 687.

5. (a) Y. Unger, D. Meyer, O. Molt, C. Schildknecht, I. Muenster, G. Wagenblast and T. Strassner, *Angew. Chem., Int. Ed.*, 2010, **49**, 10214; (b) K.-Y. Lu, H.-H. Chou, C.-H. Hsieh, Y.-H. O. Yang, H.-R. Tsai, H.-Y. Tsai, L.-C. Hsu, C.-Y. Chen, I. C. Chen and C.-H. Cheng, *Adv. Mater.*, 2011, **23**, 4933; (c) T. Sajoto, P. I. Djurovich, A. Tamayo, M. Yousufuddin, R. Bau, M. E. Thompson, R. J. Holmes and S. R. Forrest, *Inorg. Chem.*, 2005, **44**, 7992.

6. (a) K. Matsumoto, N. Matsumoto, A. Ishii, T. Tsukuda, M. Hasegawa and T. Tsubomura, *Dalton Trans.*, 2009, 6795; (b) R. Visbal and M. C. Gimeno, *Chem. Soc. Rev.*, 2014, **43**, 3551.

7. (a) V. A. Krylova, P. I. Djurovich, J. W. Aronson, R. Haiges, M. T. Whited and M. E. Thompson, *Organometallics*, 2012, **31**, 7983; (b) V. A. Krylova, P. I. Djurovich, M. T. Whited and M. E. Thompson, *Chem. Commun.*, 2010, **46**, 6696.

8. T. G. Hodgkins and D. R. Powell, *Inorg. Chem.*, 1996, **35**, 2140.

9. R. Czerwieniec, J.-B. Yu and H. Yersin, *Inorg. Chem.*, 2011, **50**, 8293.

10. B. L. Conley and T. J. Williams, *J. Am. Chem. Soc.*, 2010, **132**, 1764.

11. S. Diez-Gonzalez, E. C. Escudero-Adan, J. Benet-Buchholz, E. D. Stevens, A. M. Z. Slawin and S. P. Nolan, *Dalton Trans.*, 2010, **39**, 7595.

12. (a) G. Blasse and D. R. McMillin, *Chem. Phys. Lett.*, 1980, **70**, 1; (b) J. C. Deaton, S. C. Switalski, D. Y. Kondakov, R. H. Young, T. D. Pawlik, D. J. Giesen, S. B. Harkins, A. J. M. Miller, S. F. Mickenberg and J. C. Peters, *J. Am. Chem. Soc.*, 2010, **132**, 9499; (c) M. J. Leitl, F.-R. Kuechle, H. A. Mayer, L. Wesemann and H. Yersin, *J. Phys. Chem. A*, 2013, **117**, 11823.

Chapter 46

Synthesis and Characterization of Phosphorescent Platinum and Iridium Complexes with Cyclometalated Corannulene[†]

John W. Facendola,[a] Martin Seifrid,[a] Jay Siegel,[b]
Peter I. Djurovich,[a] and Mark E. Thompson[a]

[a] *Department of Chemistry, University of Southern California, Los Angeles, California 90089, USA*
[b] *School of Pharmaceutical Science and Technology, Tianjin University, 92 Weijin Road, Nankai District, Tianjin, 300072, P. R. China*
met@usc.edu

Synthesis, structural and characterization data are provided for Pt(II) and Ir(III) complexes cyclometalated with 2-(corannulene) pyridine (corpy), (corpy)Pt(dpm) and (corpy)Ir(ppz)2 (dpm = dipivaloylmethanato, ppz = 1-phenylpyrazolyl). A third compound, (phenpy)Ir(ppz)2 (phenpy = 2-(5-phenanthryl)-pyridyl), was also prepared to mimic the steric bulk of (corpy)Ir(ppz)2. X-ray analysis reveals bowl depths of 0.895 Å for (corpy)Pt(dpm) and 0.837 Å in

---

[†]The details of electronic supplementary information (ESI) are given at the end of the chapter.
Reprinted from *Dalton Trans.*, **44**, 8456–8466, 2015.

Electrophosphorescent Materials and Devices
Edited by Mark E. Thompson
Text Copyright © 2015 The Royal Society of Chemistry
Layout Copyright © 2024 Jenny Stanford Publishing Pte. Ltd.
ISBN 978-981-4877-34-3 (Hardcover), 978-1-003-08872-1 (eBook)
www.jennystanford.com

(corpy)Ir(ppz)$_2$. Neither complex displayed bowl-to-bowl stacking in the crystal lattice. A fluxional process for (corpy)Ir(ppz)$_2$ attributed to bowl inversion of corannulene is observed in solution with a barrier ($\Delta G^{\ddagger} = 13$ kcal mol^{-1}) and rate ($k = 2.5 \times 10^3$ s^{-1}) as determined using variable temperature ^{1}H NMR spectroscopy. All of the complexes display red phosphorescence at room temperature with quantum yields of 0.05 in solution and 0.2 in polymethyl methacrylate (PMMA).

46.1 Introduction

The study of fullerenes has inspired chemists to understand and probe the structural and electronic properties of these compounds, as well as spurred investigation into application of these molecules [1–4]. Corannulene, nicknamed "buckybowl", has been described as the smallest fullerene fragment that retains curvature [5–7]. As a result of its unique structure, corannulene has been proposed for use in many promising applications ranging from charge transport to end caps for nanotubes [8–10].

One interesting property exhibited by corannulene as a result of its curvature is dynamic bowl-to-bowl inversion in solution. The inversion process of corannulene has been intensively studied with various substitutions, additional fused rings, as well η^2 and η^6 ligand coordinated to a transition metal [11–16]. The energy barrier to bowl inversion has been determined using variable temperature NMR, showing that substituents on corannulene control its bowl depth and thereby affect the rate of inversion [11, 13]. The rate of inversion for the parent corannulene falls within the microsecond regime, the same time scale as radiative decay for some phosphorescent heavy transition metal complexes, particularly those of iridium and platinum [17–19].

Cyclometalated iridium and platinum complexes have been studied extensively due to their efficient phosphorescence that is, promoted by strong intersystem crossing from the heavy metal center [20–23]. The photophysical properties of these metal complexes strongly depend not only on the metal center, but also on the chemical structures of the cyclometalated (C^N) ligands [17,

18, 24]. The emission energies of such complexes are closely related to the nature of the chromophoric C^N ligand and there have been numerous in depth studies probing this relationship [20, 25, 26].

Recently, a report has appeared of 2-(corannulene)pyridine (corpy-H) ligand precursor cyclometalated onto Pd(II) [27]. Both corpy-H and a [(corpy)Pd(ACN)$_2$]$^+$ (ACN = acetonitrile) complex were structurally characterized and a columnar bowl–bowl packing arrangement of the corpy ligand was observed in the Pd species. In addition, a (corpy)Pd complex with an optically active ligand was also prepared; however, dynamic bowl inversion of corannulene could not be observed in this derivative. Moreover, aside from analysis of the UV-visible absorption spectrum of [(corpy)Pd(ACN)$_2$]$^+$, no other photophysical characterization was given for the Pd complexes.

Herein, Pt and Ir complexes containing a cyclometalated corpy ligand are synthesized and their dynamic properties probed. The Pt complex is (corpy)Pt(dpm), where dpm = η^2-dipivaloylmethane and the Ir complex is (corpy)Ir(ppz)$_2$, where ppz = 1-phenyl-1H-pyrazolyl (Fig. 46.1). A third compound, (phenpy)Ir(ppz)$_2$ (phenpy = 2-(5-phenanthryl)pyridyl)) was also synthesized to mimic the steric bulk of (corpy)Ir(ppz)$_2$ and probe the effect of atropisomerism on the dynamic behavior in solution. Variable temperature NMR is used to examine any dynamic processes that these complexes undergo in fluid solution. Both (corpy)Pt(dpm) and (corpy)Ir(ppz) are phosphorescent with the photophysical properties dictated by the corpy ligand.

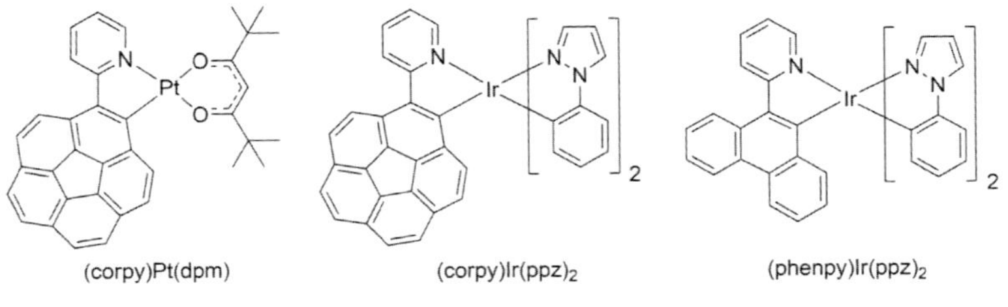

Figure 46.1 Structures of (corpy)Pt(dpm), (corpy)Ir(ppz)$_2$, and (phenpy)Ir(ppz)$_2$.

46.2 Results and Discussion

The corpy-H and phenpy-H ligand precursors were prepared by Suzuki cross coupling as described for other C^N ligands [18]. The Pt(II) and Ir(III) complexes containing corpy were prepared by routes similar to those described for other C^N complexes (Scheme 46.1) [17, 18, 31]. The (corpy)Pt(dpm) complex was

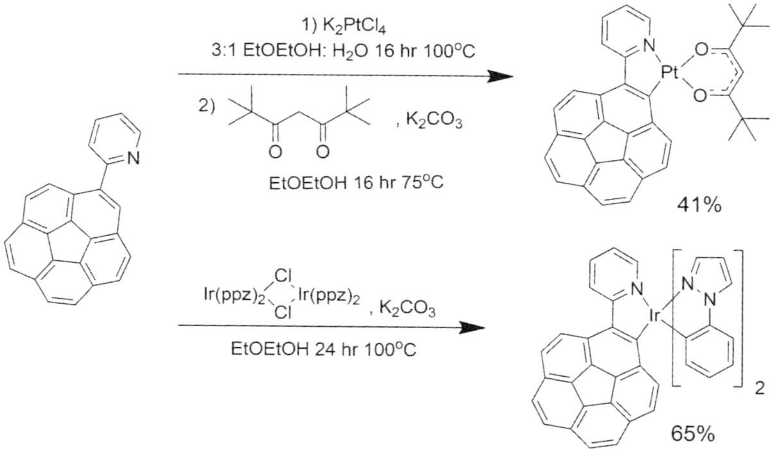

Scheme 46.1

synthesized by first preparing a cyclometalated (corpy)PtCl intermediate in a reaction between corpy-H and K_2PtCl_4, followed by addition of dipivaloylmethane and base to displace the Cl ligand. The (corpy)Ir(ppz)$_2$ and (phenpy)Ir(ppz)$_2$ complexes were synthesized by treating the [(ppz)$_2$Ir(μ-Cl)]$_2$ dimer with the correspond ligand precursor in the presence of base. The orange colored complexes were isolated as single species that are stable in air as neat solids and in fluid solution. In addition, the Ir complexes are photolytically stable as determined by monitoring the UV-visible absorption spectra in MeCN solution before and after prolonged irradiation with 254 nm light.

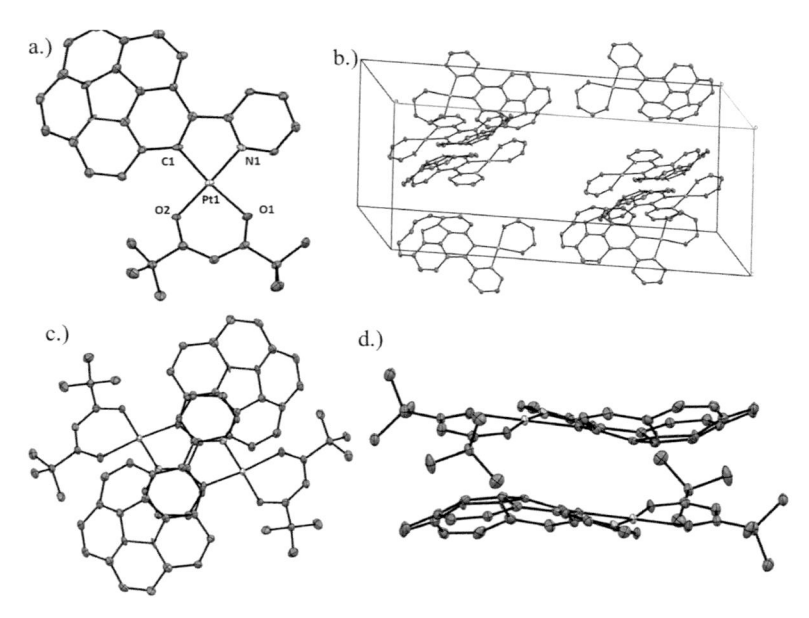

Figure 46.2 (a) Crystal structure of (corpy)Pt(dpm); (b) unit cell of (corpy)Pt(dpm) (methyl groups omitted for clarity); (c) top and (d) side view of (corpy)Pt(dpm) dimers. All hydrogens omitted for clarity. The atoms are colored blue (N) black (C), red (O) and gray (Pt).

46.2.1 Crystal Structures

X-ray diffraction analysis was performed on crystals of (corpy)Pt(dpm) and (corpy)Ir(ppz)$_2$ grown by slow diffusion of hexanes into dichloromethane of the metal complex. The structure of (corpy)Pt(dpm) is shown in Fig. 46.2a. The unit cell of (corpy)Pt(dpm) is comprised of 8 molecules in a monoclinic $C2/c$ space group (Fig. 46.2b) and contain both enantiomers of the corannulene bowl, designated P or M using the stereo-descriptor system for chiral buckybowls [28, 29]. The complex has a distorted square planar geometry around the metal center, with deviations from ideality due to chelate bite angles (C(1)–Pt–N(1) = 81.1(2)° and O(1)–Pt–O(2) = 89.85(18)°) that are comparable to values found in other cyclometalated Pt(β-diketonato) complexes [30, 31]. The corannulene and pyridyl fragments in the corpy ligand are twisted with a dihedral angle of 10.6° between the two rings. There are no metal–metal interactions as the closest Pt⋯Pt distance is 5.65 Å. The bond lengths to the metal (Pt–C(1) = 1.984(6) Å, Pt–N(1) = 1.993(5) Å, Pt–O(1) = 2.086(4) Å and Pt–O(2) = 2.021(4) Å) are also comparable to values reported for (ppy)Pt(dpm) [20] and other Pt(β-diketonato) derivatives with cyclometalated ligands [17, 30, 31]. The bowl depth of corannulene (d_{bowl} = 0.895 Å) is similar to values in unsubstituted corannulene (d_{bowl} = 0.87 Å) and corpy-H (d_{bowl} = 0.89 Å) but deeper than found in [(corpy)Pd(ACN)$_2$]$^+$ (d_{bowl} = 0.82 Å) [13, 32].

Previous crystallographic studies of corannulene derivatives have shown that the molecules can preferentially π-stack in an ordered arrangement, colloquially referred to as bowl-to-bowl packing [27, 33]. For example, the [(corpy)Pd(ACN)$_2$]$^+$ complex is stacked in a columnar arrangement displaying corannulene–corannulene π-interactions (3.3–3.4 Å) between adjacent molecules. However, no such bowl-to-bowl stacking is present in the crystal of (corpy)Pt(dpm). Instead, complexes are arranged into antiparallel pairs of enantiomers with the convex faces of the corannulene moieties pointing towards each other (Fig. 46.2c and d) [27, 33]. The corpy ligands are situated atop one another with the closest π-interaction ($\sim$3.5 Å) between the aromatic ring of one corannulene and a pyridyl group from a neighboring molecule [27].

The unit cell of (corpy)Ir(ppz)$_2$ contains two molecules of the complex as well as two dichloromethane solvate molecules in a triclinic $P\bar{1}$ space group (Fig. 46.3). The ligands are arranged in a meridional (*mer*) configuration around the pseudo-octahedral metal center with the pyrazolyls in a *trans* disposed arrangement. Both enantiomers of a single diastereomer (Λ-*P* and Δ-*M*) are present in the unit cell, where Λ and Δ describe the stereochemistry at the metal center [28, 34]. The bond angles for the atoms *trans*-disposed around the metal (C(9)–Ir(1)–C(25) = 170.90(6)°, N(1)–Ir(1)–N(3) = 174.20(5)° and N(5)–Ir(1)–C(10) = 169.71(6)°) are comparable to values in other *mer*-Ir(C°N)$_3$ complexes [18]. The dihedral angle between the two fragments of the corpy ligand is 14.2°. The bond length for Ir(1)–N(5) (2.1222(13) Å) is longer than for Ir(1)–N(1) (2.0137(13) Å) and Ir(1)–N3) (2.0154(13) Å) but comparable to distances for the Ir–N(pyridyl) *trans* to phenyl in other *mer*-Ir(C^N)$_3$ complexes [18]. Similarly, the bond length

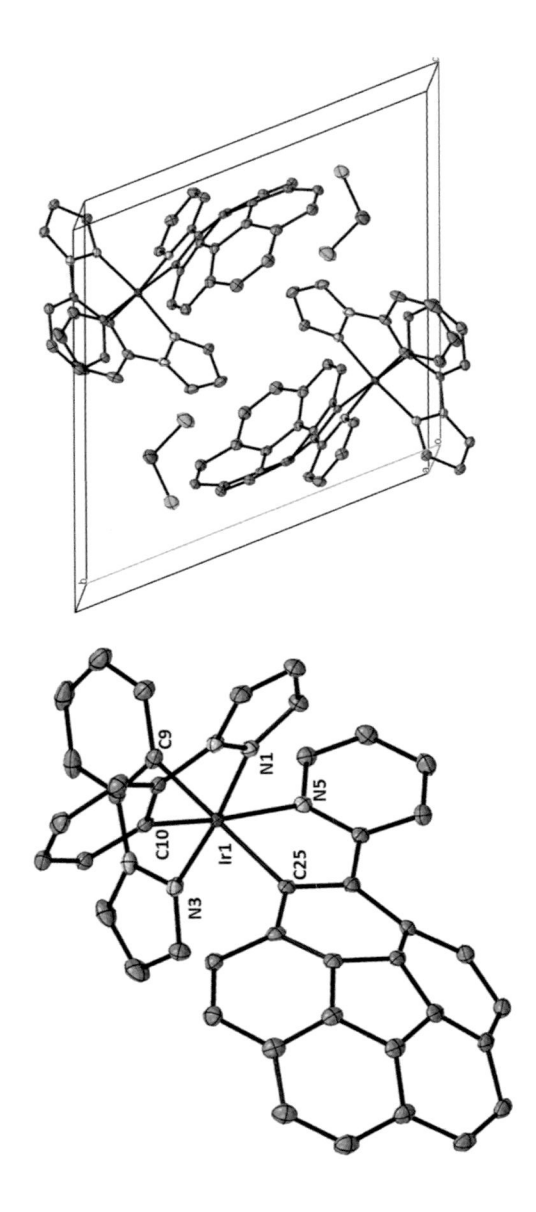

Figure 46.3 Crystal structure of (corpy)Ir(ppz)$_2$ (left) and unit cell (right). All hydrogens omitted for clarity. The atoms are colored by blue (N) black (C), and dark blue (Ir).

for Ir(1)–C(25) (2.1023(15) Å) is longer than for either Ir(1)–C(9) (2.0804(16) Å) or Ir(1)–C(10) (2.0173(16) Å) bonds of the ppz ligands. The bowl depth of corannulene ($d_{bowl} = 0.837$ Å) is shallower than in (corpy)Pt(dpm) [13, 32]. No bowl-to-bowl packing is present as the corannulenyl moiety is nested with a pyrazolyl ligand from a neighboring complex.

46.2.2 NMR and Dynamic Behavior

The Pt and Ir complexes were characterized using ^{1}H, ^{13}C, 1D and 2D NOESY, and gCOSY NMR spectroscopy (Fig. 46.4). Sharp, distinct resonances are observed in aromatic regions of the 1H NMR spectra measured at 298 K. Peak assignments for (corpy)Pt(dpm) (Fig. 46.4a), were made by analysis of the gCOSY spectrum. The two resonances furthest downfield are assigned to the proton a ($\delta =$

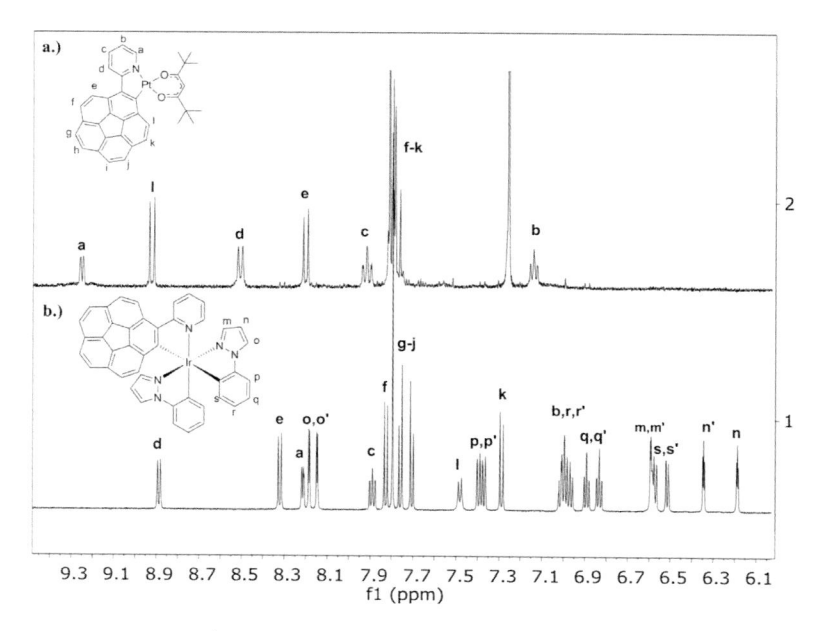

Figure 46.4 (a) ^{1}H NMR spectrum of (corpy)Pt(dpm) in CDCl$_3$, (b) ^{1}H NMR spectrum of (corpy)Ir(ppz)$_2$ in 2 : 1 CD$_2$Cl$_2$–acetone-d$_6$. The labelling scheme for the corpy ligand is the same in both spectra. Resonances marked with ′ are for the ppz ligand trans to pyridyl.

9.25 ppm) adjacent to nitrogen on the pyridyl ring and proton **l** ($\delta = 8.92$ ppm) on the corannulene ring. Both protons are deshielded due to close proximity (~2.36 Å) to the carbonyl oxygens on the dpm ligand. The remaining protons on the pyridyl ring ($\delta = 8.51$, 7.92 and 7.14 ppm) were assigned using gCOSY. The 1D-NOESY spectrum supported assignment of the resonance at $\delta = 8.20$ ppm to proton **e**; this proton is coupled to the remaining aromatic resonances on the corannulenyl ring located between $\delta = 7.70$–7.85 ppm.

The ^{1}H NMR spectrum for (corpy)Ir(ppz)$_2$ in 2:1 CD$_2$Cl$_2$–acetone-d$_6$ displays series of well-resolved resonances integrating to 26 protons that is consistent with a single species, even though multiple conformers are possible. The proton assignments shown in Fig. 46.4b were confirmed on the basis of gCOSY and 1D-NOESY spectroscopy. Resonances for protons **d** ($\delta = 8.87$ ppm) and **e** ($\delta = 8.32$ ppm) on the corpy ligand are now the furthest downfield as protons **a** ($\delta = 8.22$ ppm) and **l** ($\delta = 7.47$ ppm) are shifted upfield due to shielding by the ring currents of the adjacent ppz ligands. Resonances for protons **f–j** on corrannulene are distinct between $\delta = 7.7$–7.8 ppm as is proton **k** ($\delta = 7.27$ ppm). The remaining protons on the pyridyl and ppz ligands are likewise clearly identified and assigned, with protons on the pyrazolyl ligands being furthest upfield (**n'**, $\delta = 6.34$ ppm; *n*, $\delta = 6.19$ ppm).

The (corpy)Ir(ppz)$_2$ complex undergoes a dynamic process in fluid solution that was studied using variable temperature (VT) ^{1}H NMR spectroscopy (Fig. 46.5). A solvent mixture of 2:1 CD$_2$Cl$_2$–acetone-d$_6$ was used to provide good separation of all proton resonances and to maintain adequate solubility of the complex at low temperatures. Proton resonances of the corpy ligand, as well as specific resonances on the ppz ligands, broaden in stages when a sample is cooled below room temperature. Initially, the resonance for proton **l** ($\delta = 7.47$ ppm) broadens and merges into the baseline at 245 K. This change is concurrent with broadening of signals from protons **k, m', m, n'** and **n** that achieve coalescence at 231 K. It is readily apparent in the spectra measured at 245 K and 231 K that signals for protons **n'** and **n** broaden at different rates. Additional broadening occurs for most of the remaining protons on corpy and the ppz ligands between 231 K and 198 K. However, some resonances, specifically from the phenyl protons

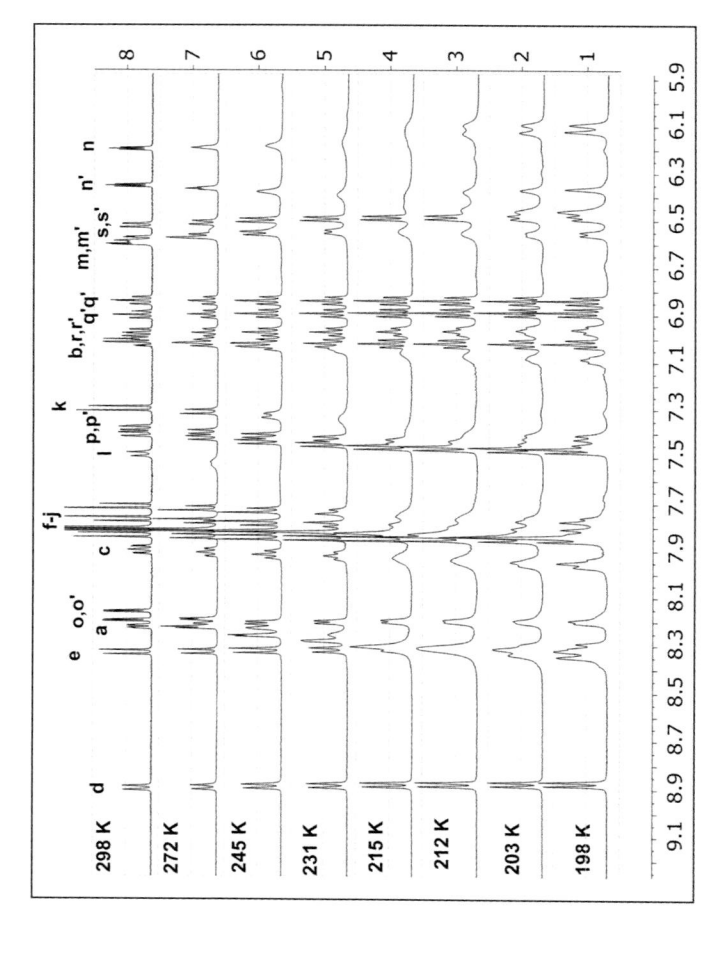

Figure 46.5 Variable temperature ^{1}H NMR spectra of (corpy)Ir(ppz)$_2$ in 2:1 CD$_2$Cl$_2$–acetone-d$_6$.

p–s on ppz, and surprisingly **q′** on ppz and **d** on pyridyl, remain sharp at all temperatures. It should also be noted that weak, broad signals grow in near the baseline at the lowest temperature reached during the experiment (198 K). These new resonances, in a molar ratio of approximately 1:4.5, are tentatively assigned to a second diastereomer of (corpy)Ir(ppz)$_2$. Unfortunately, definitive identification of this new species cannot be made as we were unable to obtain clearly resolved signals due to the inability of our spectrometer to collect data to lower temperatures.

The dynamic behavior displayed by (corpy)Ir(ppz)$_2$ in the VT ^{1}H NMR spectra is consistent a bowl-to-bowl inversion process occurring on the corannulene moiety of corpy. Corannulene has an intrinsic permanent dipole moment (2.07 D) with enhanced electron density localized in the base of the bowl [35]. The dipolar flip that accompanies inversion of the bowl should strongly affect protons closest to corpy ligand. Protons that are especially useful for interpretation of the fluxional process are labeled on molecular models of the two interconverted structures shown in Fig. 46.6. These protons (**l**, **m**, **m′**, **n** and **n′**) have anisotropic chemical shifts that depend on the orientation of the corannulene bowl. The relation of the bowl (concave or convex) to the pyrazolyl protons will dictate whether their resonances are shielded or deshielded by the direction of the dipole. The fluxional behavior is manifested as an average chemical shift at room temperature.

Kinetic analysis of the VT ^{1}H NMR spectra was carried out using the resonance at $\delta = 6.3$ ppm assigned to the pyrazolyl proton n (Fig. 46.4b and 46.6). The rate of inversion was determined through Eyring analysis of the line width ($\Delta \nu$) as a function of temperature (see ESI†) [36]. Values of $\Delta H^\ddagger$ (4.9 kcal mol^{-1}) and $\Delta S^\ddagger$ (-27 kcal mol^{-1}) were determined from curve fitting and the energy barrier, $\Delta G^\ddagger$, was found to be 13 kcal mol^{-1}, corresponding to a rate for inversion of 2.5×10^3 s^{-1} at room temperature. On the basis of studies on corannulenes substituted at the *peri*-positions, Siegel and coworkers established a correlation between the rate of inversion and bowl depth. While the substitution pattern of the cyclometalated corpy in (corpy)Ir(ppz)$_2$ is at the *ortho*-positions, the value determined for $\Delta G^\ddagger$ matches the one estimated using Siegel's data and the experimental bowl depth ($d_{\text{bowl}} = 0.837$ Å).

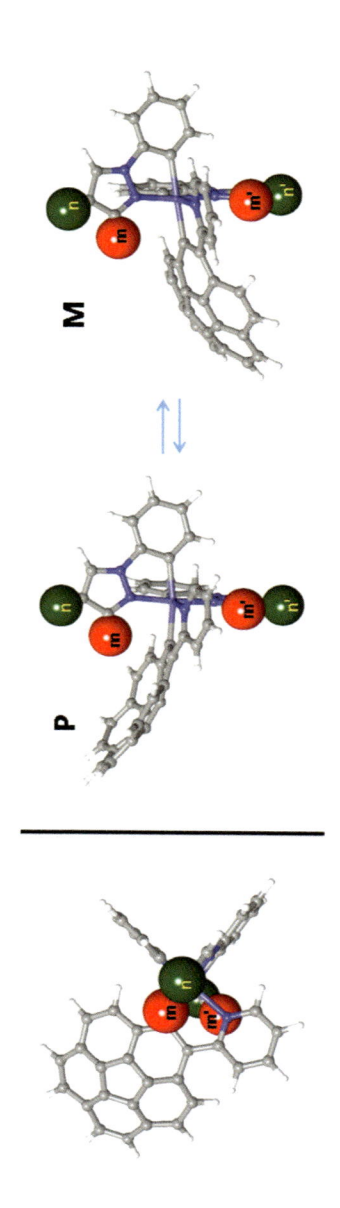

Figure 46.6 Structural representations of (corpy)Ir(ppz)$_2$. Protons **m** and **m'** are shown in red and **n** and **n'** in green. A view down the N–Ir–N axis is shown at the left. To the right is a view roughly down the N–Ir–C axis showing the Λ-*P* (left) and Λ-*M* (right) diastereomers of (corpy)Ir(ppz)$_2$.

The close correspondence of the present result to data from Siegel's study suggests that *peri*- and *ortho*-substitution patterns affect the stability of the corannulene in a similar manner.

Inversion of the corannulene bowl is likely not the only mechanism that can account for the fluxional behavior observed for (corpy)Ir(ppz)$_2$. Another process to consider is interconversion of conformers created by the twist between the corannulene and pyridyl groups of the corpy ligand. To investigate this possibility, the (phenpy)Ir(ppz)$_2$ complex was used to provide an (C^N)Ir(ppz)$_2$ analog that can still undergo conformational isomerism but not bowl inversion. Molecular models show that the torsion angle for the bonds labelled in red in Scheme 46.2 (37.7°) are comparable

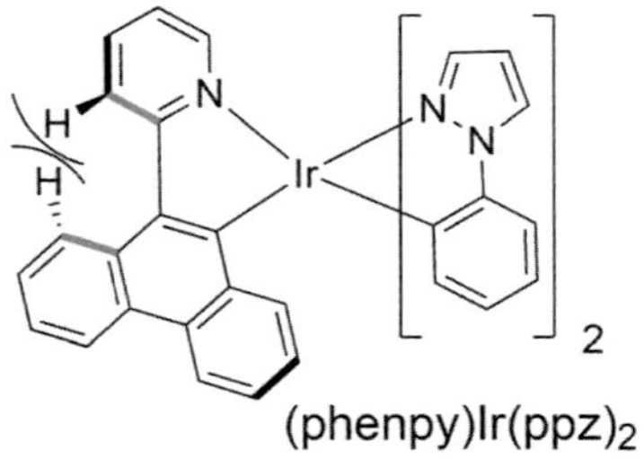

(phenpy)Ir(ppz)$_2$

Scheme 46.2

to values for the equivalent bonds in (corpy)Ir(ppz)$_2$ (34.1°). Likewise, the ^{1}H NMR spectrum of (phenpy)Ir(ppz)$_2$ at room temperature displays 26 resonances consistent with presence of a single species undergoing rapid interconversion on the NMR timescale. Proton resonances remain sharp at all temperatures up to 342 K in (CD$_3$)$_2$SO and down to 233 K in CD$_2$Cl$_2$ (see ESI†), indicating that atropisomerism is rapid even at low temperature. Therefore, the dominant process responsible for the fluxional behavior observed in (corpy)Ir(ppz)$_2$ (Fig. 46.6) is likely not due to a related atropisomerism. However, the current model assumes that a simple unimolecular inversion is the only dynamic process causing decoalescence of proton resonances in (corpy)Ir(ppz)$_2$. The fact that Eyring analysis gives a large contribution for $\Delta S^\ddagger$ suggests that the dynamic behavior is more complex. While bowl inversion is the primary mechanism, other secondary molecular distortions may participate in the fluxional process. Another feature to consider is how the difference between the dipole moments of diastereomers formed during bowl inversion affects the solvation of the complex. As (corpy)Ir(ppz)$_2$ undergoes inversion, not only does the stereochemistry of the corannulene bowl change but so does the orientation of the dipole with respect to the Ir(ppz)$_2$ fragment. The magnitude of dipole moments calculated for the two diastereomers of (corpy)Ir(ppz)$_2$ (Λ-P = 5.27 D, Λ-M = 5.23 D) are larger than that of corannulene (2.07 D). A change in the dipole of his magnitude could cause the solvent to reorganize around the complex, leading to a large $\Delta S^{\ddagger}$ for the inversion.

46.2.3 Electrochemical Properties

The redox properties of the complexes were examined by cyclic voltammetry and differential pulse voltammetry in DMF solution with 0.1 M TBAF (Table 46.1). All potential values were referenced to an internal ferrocene couple (Fc/Fc$^+$). The oxidative properties of the complexes are similar to related derivatives with cyclometalated ligands. The (corpy)Pt(dpm) complex displays an irreversible oxidative wave at E^{pa} = 0.51 V. The potential and irreversibility of this process is comparable to oxidative behavior seen in other (C$^\wedge$N)Pt(dpm) complexes with expanded π-systems [20]. The Ir complexes display reversible couples at potentials ($E^{1/2}$ = 0.30 V for (corpy)Ir(ppz)$_2$ and 0.27 V for (phenpy)Ir(ppz)$_2$) that are slightly lower than mer-(ppy)Ir(ppz)$_2$ ($E^{1/2}$ = 0.37 V) [37]. Oxidation in all of these cyclometalated complexes is typically assigned to an orbital with mixed metal-aryl character [18, 38]. A second, irreversible oxidation wave is observed in the iridium complexes at higher potentials.

Table 46.1 Redox data for the Pt and Ir complexes[a]

Compound	E_{ox1}	E_{red1}	E_{red2}	E_{red3}
(corpy)Pt(dpm)	0.51 V[b]	−2.06 V	−2.46 V	—
(corpy)Ir(ppz)$_2$	0.30 V	−2.27 V	−2.66 V[c]	−3.13 V[b]
(phenpy)Ir(ppz)$_2$	0.27 V	−2.55 V	−3.04 V[b]	—

[a]Redox potentials were recorded in 0.1 M TBAF–DMF solution and referenced to an internal Fc$^+$/Fc couple. [b]Irreversible. [c]Quasireversible.

The complexes show reversible reduction waves in DMF solution. The (corpy)Pt(dpm) complex displays two reversible waves at −2.02 V and −2.42 V. For (corpy)Ir(ppz)$_2$ reversible reduction occurs at −2.26 V while a second quasireversible wave appears at −2.66 V. The cathodic waves in both of these complexes are assigned to reduction of the corpy ligand. These potentials are markedly less negative than the first reduction potential of (phenpy)Ir(ppz)$_2$ ($E^{1/2}$ = −2.55 V). Additional irreversible reduction waves beyond −3.0 V in the Ir complexes are assigned to reduction of the ppz ligands. The lower potential for (corpy)Ir(ppz)$_2$ compared to (phenpy)Ir(ppz)$_2$ is consistent with the corannulene being a better electron acceptor than phenanthrene [39, 40]. Similarly, the second reduction for (corpy)Pt(dpm) and (corpy)Ir(ppz)$_2$ are assigned to corpy since corannulene has been shown to undergo up to three reductions in solution [24, 39, 41]. The 400 mV separation between the first and second reduction waves in the Pt and Ir complexes is smaller than that found in corrannulene (700 mV). The difference in the metal complexes indicates a decrease in coulombic repulsion in the radical anion due to the electron being delocalized onto the pyridyl ring of the corpy ligand.

46.2.4 Photophysical Properties

The absorption and emission spectra of the complexes were recorded at room temperature and 77 K as well as in a rigid PMMA matrix at room temperature (Fig. 46.7). The absorption data is listed in Table 46.2 and emission data in Table 46.3. The absorption spectra for the complexes show intense bands ($\lambda < 360$ nm, $\varepsilon > 10^4$ M^{-1} cm^{-1}) assigned to the ligand centered $\pi \to \pi^*$ transitions on the cyclometalated ligands. In particular, the bands between $\lambda = 300$–360 nm are assigned to $\pi \to \pi^*$ transition on corpy on the basis of comparison with spectra from the free corpy-H ligand. Less intense bands at lower energy ($\lambda = 350$–500 nm, $\varepsilon \approx 5 \times 10^3$ M^{-1} cm+) are assigned to allowed metal-to-ligand charge transfer (MLCT) transitions. Much weaker absorptions ($\lambda > 500$ nm, $\varepsilon < 10^2$ M^{-1} cm^{-1}) are assigned to triplet MLCT transitions that are partially allowed due to spin–orbit coupling with the singlet states by the heavy atom metal center.

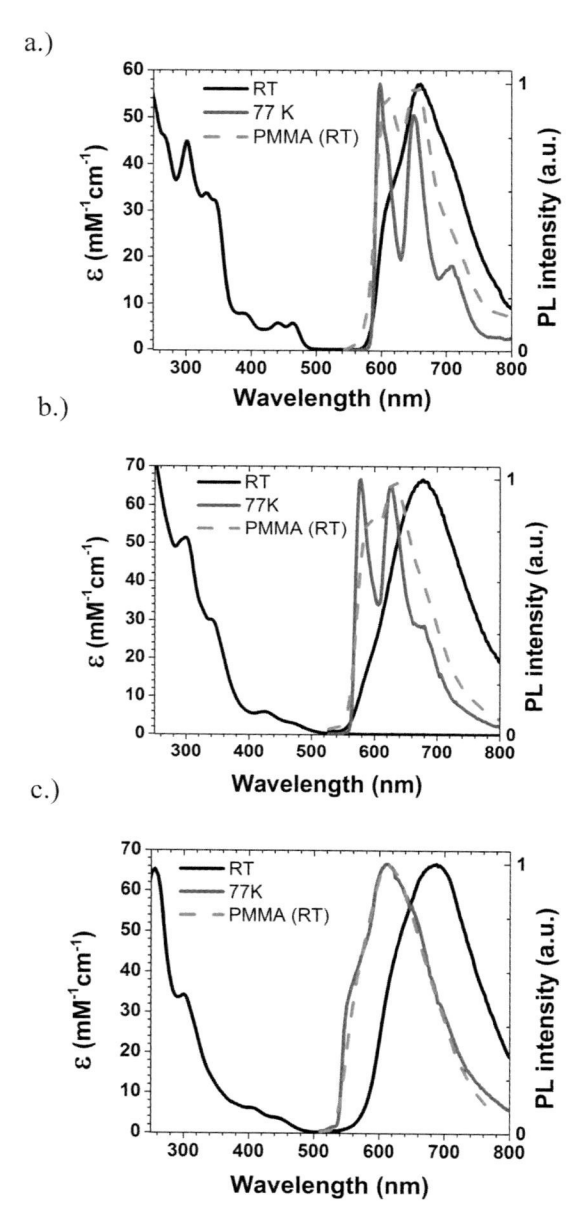

Figure 46.7 Absorption (in CH_2Cl_2) and emission (in 2-MeTHF and PMMA) spectra of (a) (corpy)Pt(dpm), (b) (corpy)Ir(ppz)$_2$ and (c) (phenpy)Ir(ppz)$_2$.

Table 46.2 Absorption data for the Pt and Ir complexes **1–3**

	λ_{max}(nm) (ε 10^3 M^{-1} cm^{-1})[a]
(corpy)Pt(dpm) (**1**)	302 (44.8), 331 (33.8), 387 (7.90), 441 (5.82), 465 (5.76)
(corpy)Ir(ppz)$_2$ (**2**)	299 (51.4), 339 (30.2), 425 (5.93), 469 (sh, 3.00)
(phenpy)Ir(ppz)**2** (**3**)	300 (34.3), 403 (6.15), 450 (sh, 3.44)

[a]Absorption spectra recorded in CH$_2$Cl$_2$.

All three complexes display broad, featureless red luminescence at room temperature in 2-MeTHF solution. At 77 K, the spectra of (corpy)Pt(dpm) and (corpy)Ir(ppz)$_2$ shows distinct vibronic structure, whereas emission from (phenpy)Ir(ppz)$_2$ remains broad and relatively featureless. The emission lifetimes at 77 K are single exponential and fall in the range τ = 9.4–15 μs consistent with phosphorescence. The vibronic structure displayed by the corpy complexes indicate that emission originates from a triplet state with significant ^{3}LC character. However, the energy of the triplet state in these complexes is over 0.2 eV lower than that of the $^3\pi$–π state in the free ligand corpy-H (E_{0-0} = 525 nm, 2.36 eV). The decrease in energy shows that a substantial stabilization of the excited state occurs upon cyclometalation of the ligand. Since the emission lifetimes of (phenpy)Ir(ppz)$_2$ and (corpy)Ir(ppz)$_2$ are comparable, as are the energies of the $^3\pi$–π states in the free ligands (see ESI[†]), the absence of distinct vibronic features in the former complex is likely due significant distortion in the excited state, as opposed to there being greater MLCT character in (phenpy)Ir(ppz)$_2$ than in (corpy)Ir(ppz)$_2$.

The photoluminescent quantum yields of the complexes in fluid solution are relatively low (Φ = 0.02–0.09). The radiative decay rate constants (k_r = 1.1–1.9 × 10^4 s^{-1}) are roughly an order of magnitude lower than values reported for highly efficient red Pt and Ir phosphors with cyclometalated ligands [17, 42], and indicates that the excited state has significant ^{3}LC character. However, the low quantum efficiency is mainly a consequence of rapid non-radiative decay (k_{nr} > 2 × 10^5 s^{-1}). These rates are nearly two orders of magnitude greater than what is found in efficient cyclometalated phosphors [18]. The luminescent spectra blue-shift upon dispersing the complexes in rigid media (polymethylmethacrylate, PMMA) and

Table 46.3 Photoluminescence (PL) data for the Pt and Ir complexes **1–3**

	Solution[a]					PMMA (1% doped)			
	298 K λ_{max} (nm) [Φ][b]	τ^c (μs)	k_r^d (10^4 s^{-1})	k_{nr}^e (10^4 s^{-1})	77 K λ_{0-0} (nm) [τ (μs)][c]	298 K λ_{max} (nm) [Φ][b]	τ^c (μs)	k_r (10^4 s^{-1})	k_{nr} (10^4 s^{-1})
1	660 [0.05]	5.1	1.0 ± 0.2	19 ± 3	592 [14.1]	654 [0.20]	12.0	1.7 ± 0.2	6.7 ± 1.5
2	678 [0.02]	1.7[f]	1.2 ± 0.2	58 ± 8	582 [9.4]	632 [0.20]	3.3 (14%), 8.2 (86%)	—	—
3	668 [0.09]	4.8	1.9 ± 0.3	19 ± 3	580 [15.0]	612 [0.32]	15	2.1 ± 0.3	4.5 ± 0.7

[a]Emission spectra recorded in 2-MeTHF.
[b]Photoluminescent quantum yield. Error is ± 10%.
[c]Error is ± 5%.
[d]Derived using Φ = $k_r \tau$.
[e]Derived using Φ = $k_r/(k_r + k_{nr})$.
[f]Measured at 800 nm. See text.

display vibronic features comparable to spectra recorded at 77 K. The quantum efficiencies also increase to $\Phi = 0.20$–0.32. The higher efficiency in PMMA is mainly due to a two to four-fold decrease in k_{nr} from values measured in fluid solution. The rigidochromic shifts and decrease in non-radiative rate constants indicate that large structural changes present in the excited state are suppressed in the rigid PMMA media. The fact that the emission lifetimes in PMMA at room temperature are comparable to values measured in 2-MeTHF at 77 K implies that vibrational deactivation remains the principal mechanism for non-radiative decay at low temperature (weak coupling limit) [43–45].

An additional feature observed in the luminescent spectrum of (corpy)Ir(ppz)$_2$ is that the emission lifetime is wavelength dependent. The emission lifetime is distinctly non-first order for wavelengths < 700 nm in spectra measured at room temperature with a longer decaying component appearing at higher energy. Spectra measured at 650 nm can be fit to a bi-exponential decay with lifetime values of $\tau = 1.6$ μs (64%), 9.4 μs (36%) in 2-MeTHF and $\tau = 3.3$ μs (14%), 8.2 μs (86%) in PMMA. In contrast, the emission lifetimes for the other two complexes remain first-order for all wavelengths under the same conditions. The lifetime data suggests that two or more different states are emitting for (corpy)Ir(ppz)$_2$. The presence of a *fac*-isomer impurity can be excluded since no evidence of photoisomerization is observed in the UV-visible spectrum of the complex after photolysis in acetonitrile for 8 hours with 254 nm light. The dynamic process observed in the ^{1}H NMR spectra for (corpy)Ir(ppz)$_2$ (Fig. 46.5) gives credence for presence of two or more different species in solution, which could lead to the non-first order decay. Note that the exchange rate found by VT ^{1}H NMR at room temperature ($k_{exchange} = 2.5 \times 10^3$ s^{-1}) is much slower than either emission decay rate ($k = 6.1$ and 1.1×10^{-5} s^{-1}). Therefore, any diastereomers that form by bowl inversion are not expected to interconvert during the lifetime of the excited state and will thus emit independently. The (corpy)Pt(dpm) complex cannot show this behavior since the bowl inversion does not lead to diastereomers, but instead to enantiomers that will decay from excited states with identical rates. The fact that excited state decay from the (phenpy)Ir(ppz)^2 complex is wavelength independent indicates atrop-isomerism is not the principal cause of non-first order decay seen from (corpy)Ir(ppz)$_2$.

A possible origin for the biexponential emission decay of (corpy)Ir(ppz)$_2$ could be related to the different ligand-orbital overlaps expected in the two diastereomers, Fig. 46.6. The excited states of organometallic Ir and Pt complexes are typically described as being mixtures of MLCT, LC and LLCT excited states (LLCT = ligand-to-ligand charge transfer) [46]. Significant spectral changes in both of the Pt and Ir complexes with corpy ligands are seen on comparing room temperature and 77 K luminescence (Fig. 46.7), involving substantial sharpening and blue shifting on cooling. These spectral changes have been termed rigidochromism [47] and are due to changes in the ratios of MLCT : LC : LLCT, favoring LC states at low temperatures for the complexes reported here. This change suggests that mixing between the two electronic configurations is sensitive to the solvation environment around the complexes. The emission lifetime is effected by the MLCT : LC : LLCT ratio, with greater MLCT character generally giving faster radiative rates [22]. We have modeled the excited state properties of the two diastereomers of (corpy)Ir(ppz)$_2$ using time-dependent density functional theory (TD-DFT). The complexes show the S_0–T_1 transitions for the two isomers are comprised of different amounts MLCT, LC and LLCT character, with the transition for the Λ-P isomer being 74% MLCT and that of the Λ-M isomer being 85% MLCT (see ESI†). Thus, since the S_0–T_1 transitions of the two isomers have different compositions it is not unreasonable to assume that the two diastereomers of (corpy)Ir(ppz)$_2$ will have unequal radiative rates, as observed. For comparison, the S_0–T_1 transition for (corpy)Pt(dpm) has > 91% MLCT character.

46.3 Conclusion

In summary, corannulene was cyclometalated onto platinum and iridium using a pendant pyridyl group to yield phosphorescent complexes. No bowl-to-bowl stacking was seen in the crystal structures of either complex. A dynamic exchange process found for (corpy)Ir(ppz)$_2$ examined using VT NMR and determined to have a rate of 2.5×10^3 s^{-1}. This process was modeled as an inversion of the corannulene bowl creating distinct diastereomers that rapidly interconvert at room temperature. The photophysics of the two compounds show that (corpy)Pt(dpm) and (corpy)Ir(ppz)$_2$ have large non-radiative rates at room temperature in solution, which decrease as the rigidity of the surrounding solvent increases. Additionally, the decay behavior of (corpy)Ir(ppz)$_2$ was non first-order at room temperature, differing from (corpy)Pt(dpm) as well as other common iridium phosphors. The absence of such irregular behavior in the decay from the (corpy)Pt(dpm) suggests that interconversion between diastereomers in (corpy)Ir(ppz)$_2$ is responsible for the unusual luminescent decay.

Future work will focus on dissecting the dynamics of the fluxional behavior observed in fluid solution. One such approach will be to change the identity of the pendant coordinating ligand with a larger, bulkier heterocycle such as a quinoline or benzoimidazole to hinder any atropisomerism between the two ligand fragments. Another approach will be to alter the ancillary phenyl pyrazole ligands, using bulky substituents in order to perturb the bowl inversion process. The inherent luminescent properties of these complexes will provide an additional spectroscopic window to help elucidate the nature of this dynamic phenomenon.

46.4 Experimental

46.4.1 Synthesis

Chemicals were received from commercial sources and used as received. All procedures were carried out in inert N$_2$ gas atmosphere despite the air stability of the complexes, the main concern being the oxidative and thermal stability of intermediates at the high temperatures of the reactions. The [(ppz)$_2$IrCl]$_2$ dimer was synthesized by the Nonoyama method which involves heating IrCl$_3$·H$_2$O to 110°C with 2–2.5 equivalents of ppz-H in a 3 : 1 mixture of 2-ethoxyethanol and deionized water [48]. Corannulene was prepared as described previously [49]. The corpy-H and phenpy-H ligand precursors were prepared by Suzuki cross coupling of their respective bromo derivatives as previously described [27, 50].

46.4.1.1 (corpy)Pt(dpm)

A 3 neck flask was charged with corpy-H (142 mg, 0.43 mmol), potassium tetrachloroplatinate(II) (75 mg, 0.18 mmol) and 18 mL of a 3 : 1 mixture of 2-ethoxy-ethanol–water. A condenser was attached to the flask and the mixture was degassed and heated to 100°C for 16 h. The reaction was cooled to ambient temperature, then water was added to the mixture and filtered and an orange-yellow precipitate was isolated. This solid was then added to a new 3 neck flask, and then charged potassium carbonate (124 mg, 0.89 mmol) and charged with 2-ethoxyethanol. A condenser was attached and the mixture was degassed after which 2,2,6,6-tetramethyl-heptane-3,5-dione (56 µL, 0.27 mmol) was added and the reaction was heated to 75°C for 16 h. The reaction was then cooled to ambient temperature and filtered and the precipitate was then washed with methanol. Column chromatography on silica gel was performed on the resultant crude mixture (100% methylene chloride) to give an orange emissive solid (52 mg, 41%). ^{1}H NMR (400 MHz, CDCl$_3$, δ) 1.33 (s, 7H), 1.43 (s, 7H) 5.99 (s, 1H) 7.14 (dd, $J = 7.44, 6.07$ Hz, 1H), 7.79 (m, 5H), 7.92 (dd, $J = 8.86, 7.45$ Hz, 1H), 8.20 (d, $J = 9.00$ Hz, 1H), 8.51 (d, $J = 8.31$ Hz, 1H), 8.92 (d, $J = 8.97$ Hz, 1H), 9.25 (dd, $J = 6.09$ Hz, 1H). ^{13}C NMR (101 MHz, CDCl$_3\delta$) 194.40, 169.26, 147.10, 138.28, 135.91, 135.83, 134.41, 131.68, 131.30, 130.74, 129.31, 127.94, 127.14, 126.39, 126.30, 125.41, 123.55, 121.46, 119.62, 93.64, 41.80, 41.43, 29.03, 28.39, 26.14. Anal. for (corpy)Pt(dpm): found: C 61.08, H 4.58, N 1.98; calcd: C 61.36, H 4.43, N 1.99.

46.4.1.2 (corpy)Ir(ppz)$_2$

A 3 neck flask was charged with corpy-H (65 mg, 0.20 mmol), [(ppz)$_2$Ir(μ-Cl)$_2$Ir(ppz)$_2$] (100 mg, 0.1 mmol), potassium carbonate (116 mg, 0.84 mmol) and 12 mL of 2-ethoxyethanol. A condenser was attached to the flask and the reaction was degassed and then heated to 100°C for 24 h. The reaction mixture was then cooled to ambient temperature and 10 mL of deionized water was added to dissolve excess potassium carbonate. The orange-red solid was vacuum filtered and washed with 10 mL of methanol and 10 mL hexanes, and then air dried. Column chromatography on silica gel was performed on the resultant crude mixture (100% methylene chloride) to give an orange-red emissive solid (36 mg, 65%). ^{1}H NMR (400 MHz, acetone-d$_6$, δ) 6.30 (dd, J = 3.01, 2.10 Hz, 1H), 6.43 (dd, J = 2.88, 2.36 Hz, 1H), 6.52 (dd, J = 7.56, 1.46 Hz, 1H), 6.57 (m, 2H), 6.68 (dd, J = 2.49, 0.71, 1H), 6.80 (ddd, J = 7.97, 7.42, 1.09 Hz, 1H), 6.87 (ddd, J = 8.01, 7.24, 1.13 Hz, 1H), 6.94 (ddd, J = 8.34, 7.46, 1.35 Hz, 1H), 6.99 (ddd, J = 8.12, 7.57, 1.67 Hz, 1H), 7.13 (ddd, J = 8.24, 6.77, 1.29 Hz 1H), 7.32 (d, J = 8.66 Hz, 1H), 7.51 (dd, J = 7.97, 1.32 Hz, 1H), 7.56 (m, 2H), 7.78 (d, J = 8.67 Hz 1H), 7.84 (d, J = 8.70 Hz, 1H), 7.89 (m, 3H), 8.03 (ddd, J = 8.19, 7.72, 1.83 Hz, 1H), 8.26 (dd, J = 5.38, 1.73 Hz, 1H), 8.43 (m, 2H), 8.49 (dd, J = 2.88, 0.63 Hz, 1H), 8.99 (d, J = 8.20 Hz, 1H). ^{13}C NMR (101 MHz, CDCl$_3\delta$) 185.45, 169.15, 154.49, 151.82, 150.01, 143.31, 142.89, 142.36, 142.24, 141.91, 141.42, 140.35, 136.68, 136.59, 136.57, 135.29, 134.62, 134.14, 133.76, 133.01, 132.53, 131.03, 130.46, 130.43, 129.15, 128.88, 127.57, 127.37, 127.32, 127.25, 127.23, 127.21, 127.17, 127.10, 126.92, 126.87, 126.73, 126.38, 126.09, 125.96, 125.78, 125.69, 125.19, 125.17, 125.03, 124.54, 124.11, 122.98, 122.29, 121.81, 121.75, 120.62, 119.91, 110.78, 110.75, 110.39, 107.19, 106.66, 106.31. Anal. for (corpy)Ir(ppz)$_2$: found: C 63.62, H 3.31, N 8.49; calcd: C 64.16, H 3.26, N 8.7.

46.4.1.3 (phenpy)Ir(ppz)$_2$

A 3 neck flask was charged with phenpy-H (105 mg, 0.41 mmol), [(ppz)$_2$Ir(μ-Cl)$_2$Ir(ppz)$_2$] (200 mg, 0.2 mmol), potassium carbonate (215 mg, 1.56 mmol) and 26 mL of dichloroethane. A condenser was attached to the flask and the reaction was degassed and then heated to 100°C for 24 h. The reaction mixture was then cooled to ambient temperature and filtered through an alumina plug. The resultant mixture was then recrystallized with dichloromethane to obtain yellow-orange emissive solid (58 mg, 20%). ^{1}H NMR (400 MHz, CD2Cl$_2$, δ) 6.29 (dd, J = 3.05, 2.54 Hz, 1H), 6.34 (dd, J = 7.29, 1.30 Hz, 1H), 6.40 (dd, J = 3.07, 2.47 Hz, 1H), 6.53 (dd, J = 7.20, 1.36 Hz, 1H), 6.73 (dd, J = 9.27, 2.30 Hz, 2H) 6.78 (ddd,

$J = 8.68, 7.58, 1.10$ Hz, 1H), 6.94 (m, 1H), 7.05 (ddd, $J = 8.39, 7.55, 1.41$ Hz, 1H), 7.29 (d, $J = 7.83$ Hz, 1H), 7.34 (d, 7.88 Hz, 1H), 7.42 (ddd, $J = 8.35, 7.60, 1.15$ Hz, 1H), 7.54 (m, 2H), 7.71 (ddd, $J = 8.87, 7.89, 1.75$ Hz, 1H), 8.02 (d, $J = 8.69$ Hz, 1H), 8.06 (dd, $J = 6.20, 2.74$ Hz, 2H), 8.28 (m, 2H), 8.50 (d, $J = 8.24$ Hz, 1H), 8.54 (d, $J = 8.24$ Hz, 1H), 8.67 (d, $J = 8.36$ Hz, 1H). ^{13}C NMR (101 MHz, CDCl$_3$, δ) 149.55, 142.88, 140.35, 132.53, 128.99, 128.44, 127.37, 127.03, 126.81, 126.77, 126.71, 126.56, 126.47, 125.78, 125.20, 122.72, 122.54, 121.75, 110.39, 106.31. Anal. for (phenpy)Ir(ppz)$_2$ + H$_2$O: found: C 59.66, H 3.57, N 9.15; calcd: C 59.18, H 3.76, N 9.33.

46.4.2 Electrochemistry

Cyclic voltammetry and differential pulsed voltammetry were performed using an VersaSTAT 3 potentiostat. Anhydrous DMF (Aldrich) was used as the solvent under inert atmosphere, and 0.1 M tetra(n-butyl)ammonium hexafluorophosphate (TBAF) was used as the supporting electrolyte. A glassy carbon rod was used as the working electrode, a platinum wire was used as the counter electrode, and a silver wire was used as a pseudo-reference electrode. The redox potentials are based on values measured from differential pulsed voltammetry and are reported relative to a ferrocene/ferrocenium (Cp$_2$Fe/Cp$_2$Fe$^+$) redox couple used as an internal reference [51], while electro-chemical reversibility was determined using cyclic voltammetry.

46.4.3 NMR Measurements

^{1}H NMR spectra were recorded on a Varian-500 and a Varian 400 NMR spectrometer. Chemical shift data for each signal are reported in ppm and measured in deuterated dichloromethane (CD$_2$Cl$_2$), deuterated chloroform (CDCl$_3$), and deuterated acetone ((CD$_3$)$_2$CO). Variable temperature NMR was measured in the range of 198–298 K. The temperature of the NMR probe was calculated using a methanol temperature standard. The rate of inversion and inversion barrier were determined by fitting the resultant data to an Eyring plot of $\ln(kh/k_b T)$ vs. $1/T$ [36].

46.4.4 X-ray Crystallography

The single-crystal X-ray diffraction data for compounds (corpy)Pt(dpm) and (corpy)Ir(ppz)$_2$ were collected on a Bruker SMART APEX DUO three-circle platform diffractometer with the χ axis fixed at 54.745° and using Mo Kα radiation ($\lambda = 0.710\,73$ Å) monochromated by a TRIUMPH curved-crystal monochromator. The crystals were mounted in Cryo-Loops using Paratone oil. Data were corrected for absorption effects using the multiscan method (SADABS). The structures were solved by direct methods and refined on F^2 using the Bruker SHELXTL software package. All non-hydrogen atoms were refined anisotropically.

46.4.5 Photophysical Measurements

Photoluminescence spectra were measured using a QuantaMaster Photon Technology International phosphorescence/fluorescence spectrofluorometer. Phosphorescent lifetimes were measured by time-correlated single-photon counting using an IBH Fluorocube instrument equipped with an LED excitation source. Quantum yield measurements were carried out using a Hamamatsu C9920 system equipped with a xenon lamp, calibrated integrating sphere and model C10027 photonic multi-channel analyzer (PMA). UV–vis spectra were recorded on a Hewlett-Packard 4853 diode array spectrometer. Samples for transient luminescent decay measurements were prepared in 2-MeTHF solution. The samples were deaerated by extensive sparging with N$_2$.

46.4.6 Computational Methods

Molecular models were created and dipole moments determined using the Jaguar 8.4 (release 17) software package on the Schrodinger Material Science Suite (v2014-2). The molecular geometries and TD-DFT calculations were performed using a B3LYP functional and a LACVP** basis set with a Poisson–Boltzmann (PBF) CH$_2$Cl$_2$ solvent dielectric continuum as implemented in Jaguar.

Acknowledgements

The authors thank Professor Travis Williams for stimulating discussions and support with VT NMR; Professor Ralf Haiges for support with single crystal X-ray crystallography. The research described here was carried out with the support of the Universal Display Corporation. The X-ray diffractometer is sponsored by National Science Foundation CRIF Grant 1048807.

Electronic Supplementary Information (ESI)

^{1}H NMR, gCOSY, 1-D NOESY, ^{13}C NMR, VT NMR of (corpy)Pt(dpm) and (phenpy)Ir(ppz)$_2$, Eyring analysis, CV of the complexes and emission spectra of the free corpy and phenpy ligand. CCDC 1034420 and 1034421. For ESI and crystallographic data in CIF or other electronic format see DOI: 10.1039/c4dt03541k.

References

1. H. W. Kroto, J. R. Heath, S. C. Obrien, R. F. Curl and R. E. Smalley, *Nature*, 1985, **318**, 162–163.
2. F. Diederich and M. Gomez-Lopez, *Chem. Soc. Rev.*, 1999, **28**, 263–277.
3. R. Taylor and D. R. M. Walton, *Nature*, 1993, **363**, 685–693.
4. M. Prato, *J. Mater. Chem.*, 1997, **7**, 1097–1109.
5. P. W. Rabideau and A. Sygula, *Acc. Chem. Res.*, 1996, **29**, 235–242.
6. L. T. Scott, *Pure Appl. Chem.*, 1996, **68**, 291–300.
7. Y.-T. Wu and J. S. Siegel, *Chem. Rev.*, 2006, **106**, 4843–4867.
8. B. M. Wong, *J. Comput. Chem.*, 2009, **30**, 51–56.
9. S. Sanyal, A. K. Manna and S. K. Pati, *ChemPhysChem*, 2014, **15**, 885–893.
10. K. Shi, T. Lei, X.-Y. Wang, J.-Y. Wang and J. Pei, *Chem. Sci.*, 2014, **5**, 1041–1045.
11. L. T. Scott, M. M. Hashemi and M. S. Bratcher, *J. Am. Chem. Soc.*, 1992, **114**, 1920–1921.
12. T. J. Seiders, K. K. Baldridge, J. M. Oconnor and J. S. Siegel, *J. Am. Chem. Soc.*, 1997, **119**, 4781–4782.

13. T. J. Seiders, K. K. Baldridge, G. H. Grube and J. S. Siegel, *J. Am. Chem. Soc.*, 2001, **123**, 517–525.
14. U. D. Priyakumar and G. N. Sastry, *J. Org. Chem.*, 2001, **66**, 6523–6530.
15. P. A. Vecchi, C. M. Alvarez, A. Ellern, R. J. Angelici, A. Sygula, R. Sygula and P. W. Rabideau, *Organometallics*, 2005, **24**, 4543–4552.
16. S. Nishida, Y. Morita, A. Ueda, T. Kobayashi, K. Fukui, K. Ogasawara, K. Sato, T. Takui and K. Nakasuji, *J. Am. Chem. Soc.*, 2008, **130**, 14954–14955.
17. J. Brooks, Y. Babayan, S. Lamansky, P. I. Djurovich, I. Tsyba, R. Bau and M. E. Thompson, *Inorg. Chem.*, 2002, **41**, 3055–3066.
18. A. B. Tamayo, B. D. Alleyne, P. I. Djurovich, S. Lamansky, I. Tsyba, N. N. Ho, R. Bau and M. E. Thompson, *J. Am. Chem. Soc.*, 2003, **125**, 7377–7387.
19. J. Li, P. I. Djurovich, B. D. Alleyne, M. Yousufuddin, N. N. Ho, J. C. Thomas, J. C. Peters, R. Bau and M. E. Thompson, *Inorg. Chem.*, 2005, **44**, 1713–1727.
20. A. Bossi, A. F. Rausch, M. J. Leitl, R. Czerwieniec, M. T. Whited, P. I. Djurovich, H. Yersin and M. E. Thompson, *Inorg. Chem.*, 2013, **52**, 12403–12415.
21. T. Sajoto, P. I. Djurovich, A. B. Tamayo, J. Oxgaard, W. A. Goddard, III and M. E. Thompson, *J. Am. Chem. Soc.*, 2009, **131**, 9813–9822.
22. H. Yersin, A. F. Rausch, R. Czerwieniec, T. Hofbeck and T. Fischer, *Coord. Chem. Rev.*, 2011, **255**, 2622–2652.
23. P.-T. Chou and Y. Chi, *Chem. – Eur. J.*, 2007, **13**, 380–395.
24. T. Sajoto, P. I. Djurovich, A. Tamayo, M. Yousufuddin, R. Bau, M. E. Thompson, R. J. Holmes and S. R. Forrest, *Inorg. Chem.*, 2005, **44**, 7992–8003.
25. Y. You and S. Y. Park, *Dalton Trans.*, 2009, 1267–1282.
26. L. Flamigni, A. Barbieri, C. Sabatini, B. Ventura and F. Barigelletti, *Photochem. Photophys. Coord. Compd. II*, 2007, **281**, 143–203.
27. M. Yamada, S. Tashiro, R. Miyake and M. Shionoya, *Dalton Trans.*, 2013, **42**, 3300–3303.
28. R. Tsuruoka, S. Higashibayashi, T. Ishikawa, S. Toyota and H. Sakurai, *Chem. Lett.*, 2010, **39**, 646–647.
29. C. Thilgen and F. Diederich, *Chem. Rev.*, 2006, **106**, 5049–5135.
30. B. Ma, P. I. Djurovich, M. Yousufuddin, R. Bau and M. E. Thompson, *J. Phys. Chem. C*, 2008, **112**, 8022–8031.
31. M. Ghedini, D. Pucci, A. Crispini and G. Barberio, *Organometallics*, 1999, **18**, 2116–2124.

32. J. C. Hanson and C. E. Nordman, *Acta Crystallogr., Sect. B: Struct. Crystallogr. Cryst. Chem.*, 1976, **32**, 1147–1153.
33. A. S. Filatov, L. T. Scott and M. A. Petrukhina, *Cryst. Growth Des.*, 2010, **10**, 4607–4621.
34. G. L. Miessler and D. A. Tarr, *Inorganic Chemistry*, Pearson Education, Upper Saddle River, N.J., 2004.
35. F. J. Lovas, R. J. McMahon, J. U. Grabow, M. Schnell, J. Mack, L. T. Scott and R. L. Kuczkowski, *J. Am. Chem. Soc.*, 2005, **127**, 4345–4349.
36. P. M. Morse, M. D. Spencer, S. R. Wilson and G. S. Girolami, *Organometallics*, 1994, **13**, 1646–1655.
37. K. Dedeian, J. M. Shi, N. Shepherd, E. Forsythe and D. C. Morton, *Inorg. Chem.*, 2005, **44**, 4445–4447.
38. P. J. Hay, *J. Phys. Chem. A*, 2002, **106**, 1634–1641.
39. C. Bruno, R. Benassi, A. Passalacqua, F. Paolucci, C. Fontanesi, M. Marcaccio, E. A. Jackson and L. T. Scott, *J. Phys. Chem. B*, 2009, **113**, 1954–1962.
40. K. Meerholz and J. Heinze, *J. Am. Chem. Soc.*, 1989, **111**, 2325–2326.
41. N. M. Shavaleev, F. Monti, R. Scopelliti, A. Baschieri, L. Sambri, N. Armaroli, M. Graetzel and M. K. Nazeeruddin, *Organometallics*, 2013, **32**, 460–467.
42. A. Endo, K. Suzuki, T. Yoshihara, S. Tobita, M. Yahiro and C. Adachi, *Chem. Phys. Lett.*, 2008, **460**, 155–157.
43. A. Ito and T. J. Meyer, *Phys. Chem. Chem. Phys.*, 2012, **14**, 13731–13745.
44. A. Ito, T. E. Knight, D. J. Stewart, M. K. Brennaman and T. J. Meyer, *J. Phys. Chem. A*, 2014, **118**, 10326–10332.
45. R. Englman and J. Jortner, *Mol. Phys.*, 1970, **18**, 145–164.
46. H. Yersin, *Highly Efficient OLEDs with Phosphorescent Materials*, Wiley-VCH, Weinheim, 2008.
47. P. Y. Chen and T. J. Meyer, *Chem. Rev.*, 1998, **98**, 1439–1477.
48. M. Nonoyama, *Bull. Chem. Soc. Jpn.*, 1974, **47**, 767–768.
49. A. M. Butterfield, B. Gilomen and J. S. Siegel, *Org. Process Res. Dev.*, 2012, **16**, 664–676.
50. O. Lohse, P. Thevenin and E. Waldvogel, *Synlett*, 1999, 45–48.
51. R. R. Gagne, C. A. Koval and G. C. Lisensky, *Inorg. Chem.*, 1980, **19**, 2854–2855.

Chapter 47

Understanding and Predicting the Orientation of Heteroleptic Phosphors in Organic Light-Emitting Materials

Matthew J. Jurow,[a] Christian Mayr,[b] Tobias D. Schmidt,[b] Thomas Lampe,[b] Peter I. Djurovich,[a] Wolfgang Brütting,[b] and Mark E. Thompson[a]

[a] Department of Chemistry, University of Southern California, Los Angeles, California 90089, USA
[b] Institute of Physics, University of Augsburg, 86135 Augsburg, Germany
met@usc.edu

Controlling the alignment of the emitting molecules used as dopants in organic light-emitting diodes is an effective strategy to improve the outcoupling efficiency of these devices. To explore the mechanism behind the orientation of dopants in films of organic host materials, we synthesized a coumarin-based ligand that was cyclometalated onto an iridium core to form three phosphorescent heteroleptic molecules, $(bppo)_2Ir(acac)$, $(bppo)_2Ir(ppy)$ and $(ppy)_2Ir(bppo)$ (bppo represents benzopyranopyridinone, ppy represents 2-phenylpyridinate, and acac represents acetylacetonate). Each emitter was doped into a $4,4'$-bis(N-carbazolyl)-$1,1'$-

---

Reprinted from *Nat. Mater.*, **15**, 85–91, 2015.

Electrophosphorescent Materials and Devices
Edited by Mark E. Thompson
Text Copyright © 2015 Nature Publishing Group
Layout Copyright © 2024 Jenny Stanford Publishing Pte. Ltd.
ISBN 978-981-4877-34-3 (Hardcover), 978-1-003-08872-1 (eBook)
www.jennystanford.com

biphenyl host layer, and the resultant orientation of their transition dipole moment vectors was measured by angle-dependent p-polarized photoluminescent emission spectroscopy. In solid films, (bppo)$_2$Ir(acac) is found to have a largely horizontal transition dipole vector orientation relative to the substrate, whereas (ppy)$_2$Ir(bppo) and (bppo)$_2$Ir(ppy) are isotropic. We propose that the inherent asymmetry at the surface of the growing film promotes dopant alignment in these otherwise amorphous films. Modelling the net orientation of the transition dipole moments of these materials yields general design rules for further improving horizontal orientation.

47.1 Introduction

Molecular orientation and solid-state morphology exert a significant impact on the performance of molecular electronic devices [1–5]. The use of phosphorescent iridium complexes with near-unity electroluminescence quantum yields as emitters in modern organic light-emitting diodes (OLEDs) has allowed for the manufacture of devices with excellent efficiencies [6, 7]. Despite high internal quantum efficiencies, approximately 80% of photons are trapped inside the thin-film structure and lost to surface plasmons and waveguide modes. To reduce the incidence of photons being dissipated to these loss channels, and enhance external quantum efficiencies (EQEs), outcoupling technologies have been successfully employed [8–10]. An alternative strategy to address this problem is to intrinsically increase the outcoupling efficiency through control of the direction of light emission. The organometallic molecules at the core of these devices emit light perpendicular to their transition dipole moment vector [4, 11] (TDV). Orienting emissive molecules with transition dipoles parallel to the substrate would eliminate the need for micro-lens arrays, gratings or other physical methods used to enhance outcoupling, and allow for large-scale manufacture of OLEDs with high EQEs (Refs. 4, 8, 9, 12–14).

Isotropic dopant orientation is observed in films of facial tris-iridium phenylpyridine (Ir(ppy)$_3$) and many other homoleptic tris-cyclometalated Ir dopants [15–17]. Heteroleptic complexes of the formula $(C^{\wedge}N)_2Ir(O^{\wedge}O)$, where $C^{\wedge}N$ is a cyclometalated ligand and $O^{\wedge}O$ is a diketonate ligand such as acetylacetonate (acac), have previously been observed to demonstrate higher EQEs than their homoleptic $Ir(C^{\wedge}N)_3$ analogues, because their average TDVs are disproportionately horizontal relative to the substrate in solid films [11, 16–25]. There are numerous reports of other dopants with ancillary ligands, especially acac and its close analogues, which feature net parallel TDV orientations to various degrees in doped films (see Supplementary Table 1 for tabulated literature data) [22–25].

Two mechanisms have been invoked to account for the disparity in the alignment properties of selected dopants in an otherwise amorphous or nearly amorphous host matrix. Past reports have speculated that large dipole moments present in the tris-cyclometalates lead to aggregation that suppresses dopant interactions with the host matrix [16, 26–28]. Other reports have proposed electrostatic interactions between electronegative regions in the $(C^{\wedge}N)_2Ir(O^{\wedge}O)$ complexes and electropositive host structures give rise to macroscopic order and thus dopant (and host) alignment [22, 29, 30]. We will show that neither the dopant dipole-based mechanism nor component electrostatics adequately describe the alignment process and propose a mechanism based on the inherent asymmetry of $(C^{\wedge}N)_2Ir(O^{\wedge}O)$ and some fac-$Ir(C^{\wedge}N)_3$ complexes that explain the many instances of observed dopant alignment in amorphous host materials.

To explore the mechanism by which these approximately spherical phosphors orient in amorphous host materials we synthesized a coumarin-based ligand from which we prepared heteroleptic iridium-based emitters: $(bppo)_2Ir(acac)$, $(bppo)_2Ir(ppy)$ and $(ppy)_2Ir(bppo)$, where bppo represents benzopyranopyridinone, and ppy represents 2-phenylpyridinate (Fig. 47.1). The coumarin functionality was employed because the carbonyl (C=O) group provides a large dipole moment [15, 17]. Each material was doped into a 4,4′-bis(N-carbazolyl)-1,1′-biphenyl (CBP) host matrix at 2, 6, 12 and 20% (v/v) and characterized by angle-dependent p-polarized emission (see Methods for details). Analysis of emission from films containing $(bppo)_2Ir(acac)$ reveals that dopant TDVs are oriented with a net horizontal alignment, whereas the emission from

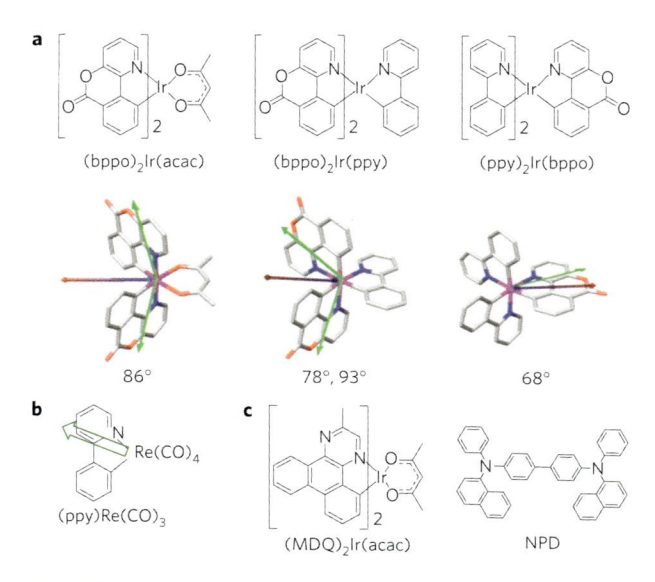

Figure 47.1 Material structure and properties. (a) Structures of phosphorescent dopants used here. Listed below each structure are the angle(s) between the permanent (μ, red arrows) and transition dipole vectors (TDVs, green arrows). The dipole moment of (bppo)$_2$Ir(acac) ($\mu = 6.18$ D) lies along the molecular C_2 axis. In (bppo)$_2$Ir(ppy) the dipole ($\mu = 8.44$ D) is tilted from this axis and lies in a plane of the bppo ligand that is trans to the C_{bppo} and N_{ppy}. In Ir(ppy)$_2$(bbpo) the dipole ($\mu = 8.25$ D) lies in the plane of the ppy ligand that is also *trans* to C_{bppo} and N_{ppy} but is tilted towards the bppo ligand. (b) TDV orientation measured in (ppy)Re(CO)$_4$ (TDV, green arrow). (c) Structures of (MDQ)$_2$Ir(acac) and NPD.

films with (ppy)$_2$Ir(bppo) and (bppo)$_2$Ir(ppy) is indicative of nearly isotropic orientation, although the dopants have substantial permanent dipole moments and similarly oriented TDVs. Understanding the causes of the alignment behaviour will allow us to develop dopant molecules that will preferentially orient in films of isotropic materials for diverse uses in photonic devices.

47.2 Results

The synthesis and photophysical properties of (bppo)$_2$Ir(ppy) and (ppy)$_2$Ir(bppo) have been reported previously [31]. A reaction between [Ir(bppo)$_2$ (μ-Cl)]$_2$ and ppy-H formed both (bppo)$_2$Ir(ppy) and (ppy)$_2$Ir(bppo) in moderate yield. Both complexes were isolated as facial isomers, with the (ppy)$_2$Ir(bppo) species being a result of ligand scrambling during the course of the reaction. See Methods for synthetic details and synthesis of (bppo)$_2$Ir(acac). The Ir complexes exhibit photophysical properties typical of phosphorescent iridium complexes. At room temperature the compounds exhibit strong green–yellow luminescence, and have quantum yields of greater than 88% and luminescence decay times in the microsecond range in solution (see Supplementary Information). Ground-state dipole moments for the complexes were calculated using density functional theory [33–35] (B3LYP-LACVP**; Fig. 47.1).

Films of the complexes doped into CBP were vapour deposited at 2, 6, 12 and 20% (v/v) to probe the impact of dipole moment and heteroleptic substitution on aggregation and concentration quenching. All species have large quantum yields in solid films, greater than 30% at any doping level tested. Concentration-related quenching of photoluminescent quantum yield, along with a bathochromic shift of emission indicative of dopant aggregation in the solid films, was observed with all species. The extent of the concentration quenching varied between the three dopants, with the (bppo)$_2$Ir(ppy) species showing the largest loss in quantum yield, followed by (bppo)$_2$Ir(acac) and finally (ppy)$_2$Ir(bppo). As concentration quenching requires an overlap of the emissive ligands, we believe that the (bppo)$_2$Ir(ppy) demonstrates the largest effect because it has two potentially emissive ligands.

Angle-dependent p-polarized emission measurements of doped films by photoluminescent excitation are used to determine the net orientation of the TDVs of emissive dopants [4, 12, 23, 25, 36–38]. We define the value Θ as the ratio of power radiated by vertical components of the contributing TDVs to the total power radiation. The details of this measurement are given in the Methods. A film with isotropic phosphors will yield a value of $\Theta = 0.33$. If the emissive TDV is aligned parallel to the substrate, $\Theta = 0$. A film with the TDV perpendicular to the substrate will give $\Theta = 1$. Vertically oriented transition dipoles couple strongly to surface plasmons in the metal electrodes of the OLED, decreasing the external efficiency of the OLED; therefore, the smallest possible value of Θ is desired [36, 39].

Emission from (ppy)$_2$Ir(bppo) and (bppo)$_2$Ir(ppy) was observed to be nearly isotropic at all measured doping levels. Films doped with (bppo)$_2$Ir(acac) exhibit Θ = 0.22 (Fig. 47.2), similar to the value observed for other heteroleptic iridium complexes with β-diketonate ligands. Interestingly, we observed that films doped with (bppo)$_2$Ir(acac) exhibit Θ = 0.22 independent of doping concentration between 6 and 20% with nearly identical line fits, despite being well into the concentration quenched regime (indicative of significant aggregation).

47.3 Discussion

Previous studies have reported that the magnitude of the ground-state dipole moment of the emitter effects the degree of aggregation and thus alignment and emission characteristics of iridium complexes in doped films [26]. Aggregated complexes were proposed to undergo a decreased interaction with the host matrix and randomly orient [40]. All tested dopants here exhibit spectral broadening and concentration quenching due to aggregation (see Supplementary Information) as expected because of their large permanent dipole moments. The concurrent observed redshifted emission in doped films indicates that these aggregates emit photons. The unchanged orientation measured in the extremely concentrated (20% v/v) (bppo)$_2$Ir(acac) film implies that the lack of aggregation is not responsible for the consistently observed net horizontal TDV orientation of heteroleptic iridium complexes containing an acac ligand.

To understand the observed alignment for the bppo and related Ir(C^N)$_3$ and heteroleptic complexes, it is essential to be able to define the orientation of the TDV relative to the molecular frame, because our optical measurement gives only the relationship of the TDV to the substrate plane. Emission from cyclometalated iridium complexes is due predominantly to a triplet metal-to-ligand-charge-transfer (^{3}MLCT) transition. In a heteroleptic Ir complex, emission is expected to originate from a ^{3}MLCT state involving the ligand(s)

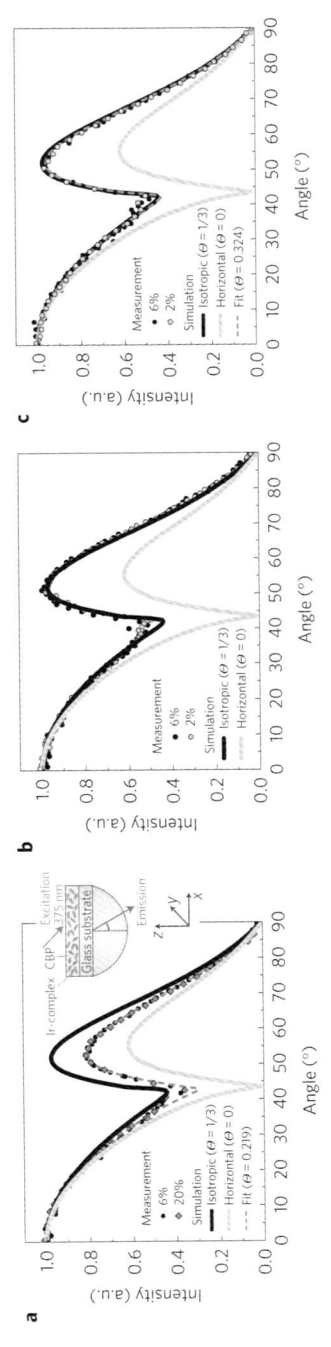

Figure 47.2 Polarized emission spectra. (a–c) Cross-sections of the measurements and simulations of the angle-dependent p-polarized photoluminescence emission spectra (considering an emission in the x–z plane) for films of 15 nm CBP doped with (bppo)$_2$Ir(acac) (at 540 nm) (a), (bppo)$_2$Ir(ppy) (at 530 nm) (b) and (ppy)$_2$Ir(bppo) (at 550 nm) (c) at different doping levels on glass substrates. The measured data have been fitted (dashed lines) to determine the degree of orientation. (bppo)$_2$Ir(acac) Θ = 0.22; (bppo)$_2$Ir(ppy) Θ = 0.33; and (ppy)$_2$Ir(bppo) Θ = 0.32. Inset image in (a) depicts experimental design.

with the lowest triplet energy [41]. Calculated triplet spin density surfaces indicate that all three bppo-based complexes considered here emit from a ^{3}MLCT state involving a single bppo ligand. The next step is to determine the orientation of the TDV for the '(bppo)Ir' fragment and ultimately to the permanent dipole moment of the molecule. Fortunately, the orientation of the TDV has been determined experimentally for the closely related cyclometalated complex (ppy)Re(CO)$_4$ by examining the polarization of emission obtained from a single crystal [42]. This Re complex emits from a ^{3}MLCT involving the '(ppy)Re' fragment. The TDV was found to lie in the plane of the ppy ligand directed by an angle of $\delta = 18.5°$ away from the Re–N bond axis (Fig. 47.1b). Considering that the molecular and photophysical properties of (bppo)$_2$Ir(acac) are similar to those of (ppy)$_2$Ir(acac), the orientation of the transition dipole for the two cyclometalated ligands can be expected to lie in a similar direction. The electron-withdrawing nature of the carbonyl functionality of the bppo ligand is expected to shift the TDV further away from the Ir–N bond than is observed for (ppy)Re(CO)$_4$; thus, we expect the bppo-based emitters to give an angle between the Ir–N bond and the TDV, defined as δ, between 20° and 40°.

The angles between the permanent and transition dipole moments for each bppo-based dopant are illustrated in Fig. 47.1a, assuming a δ value of 20°. With two identical emissive bppo ligands, (bppo)$_2$Ir(acac) and (bppo)$_2$Ir(ppy) have similar angles between their ground-state and transition dipole moment vectors, averaging 86°. (ppy)$_2$Ir(bppo) has a slightly smaller angle of 68°. The similarities of these angles, contrasted with the observation that only (bppo)$_2$Ir(acac) shows a net alignment in doped films, further indicates that the permanent molecular dipole is not responsible for dopant alignment in Ir-phosphor-based films.

Previous reports have suggested that simultaneous host and dopant interactions in aligned matrices arise from the formation of donor/host aggregates based on the component electrostatics [22, 29, 30]. The complexes studied here have very similar electrostatic surfaces (Fig. 47.3). (bppo)$_2$Ir(acac) and (bppo)$_2$Ir(ppy) are both dominated by the strongly electronegative coumarin ligand, yet demonstrate quite different orientation behaviour in CBP. Thus,

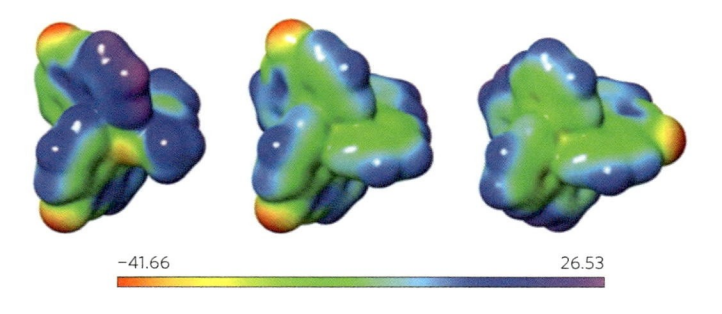

Figure 47.3 Electrostatic surfaces. Calculated electrostatic surface potentials (kcal mol^{-1}) for (bppo)$_2$Ir(acac) (left), (bppo)$_2$Ir(ppy) (middle) and (ppy)$_2$Ir(bppo) (right).

component electrostatics is unlikely to be an important factor in alignment for these dopants.

A mechanism for molecular alignment of a growing thin film of neat organic molecules during vacuum deposition has been described previously [43, 44]. These authors propose that molecules with high aspect ratios, such as CBP or NPD, preferentially lie with their long axis parallel to the surface, thus minimizing the surface free energy and increasing the film density. The CBP host used here has a comparatively low glass transition temperature ($T_g = 62\,°C$; Ref. [45]), which the previous report predicts should give an isotopic film. A separate recent study has found no dependence of the TDV orientation of two prototypical heteroleptic emitters ((ppy)$_2$Ir(acac) and (MDQ)$_2$Ir(acac)) on the T_g of the host material, with observed dopant ordering resulting from intrinsic properties of the dopant species [17].

Here we propose a mechanism, extending from previous reports on neat films [43, 44], that recognizes a surface of an amorphous (isotropic) film as inherently asymmetric during deposition, that is, organic film versus vacuum, which leads to the alignment of molecules deposited on it. The acac group presents an aliphatic region on the surface of the (C^N)$_2$Ir(acac) complex, which lies along the molecular C_2 axis. We propose that the boundary created between the organic host material on the substrate and the vacuum of the deposition chamber during fabrication causes the asymmetrical $(C^\wedge N)_2Ir(acac)$ molecules to orient before it is over coated with an amorphous layer of the host material. Molecular rearrangement and alignment on surfaces is known to occur on timescales consistent with this mechanism [43, 46–48]. Note that this mechanism does not require any alignment of the host material, and explains why $(C^\wedge N)_2Ir(O^\wedge O)$ phosphors with widely varied cyclometallating ligands all show similar degrees of dopant alignment (see Supplementary Information for tabulated values), because the shared β-diketonate ligand will give a discrete aliphatic surface 'patch' in each case.

Support for a surface-promoted alignment of dopants is seen in the recent report of an isotropic orientation for $(ppy)_2Ir(acac)$ when spin-cast in a poly-methylmethacrylate matrix (the same dopant gives $\Theta = 0.22$ in vapour deposited films) [30]. We have also examined solution-deposited films of $Ir(ppy)_3$, $(MDQ)_2Ir(acac)$ and $(bppo)_2Ir(acac)$ doped in poly-methylmethacrylate and observed emission indicative of isotropic orientation. Previous reports have shown a similar difference when comparing solution processed and thermally evaporated organic thin films, that is, isotropic and ordered films, respectively, from the two methods [44, 49]. The isotropic nature of the films processed from solution supports our proposal that an organic/vacuum interface is needed for dopant alignment. It is important to note that the vacuum and solution deposited films are in different host materials. Unfortunately, the poor solubility of CBP prevents direct comparison between vacuum and solution processing. To further test our hypothesis we compared films prepared by both vacuum deposition and spin-coating a solution of $(MDQ)_2Ir(acac)$ doped into NPD (8% v/v; see Fig. 47.1). The spin-cast films exhibited an orientation of $\Theta = 0.36$, that is, nearly isotropic (see Supplementary Information). When fabricated by vapour deposition, the same system demonstrates horizontal orientation with $\Theta = 0.24$, in good agreement with the reported values of alignment for $(C^\wedge N)_2Ir(acac)$ dopants [26]. This result indicates that the vacuum/organic boundary created during vapour deposition is critical to producing the observed alignment in the measured heteroleptic systems.

To assess consequences of dopant alignment we have developed a mathematical representation that illustrates how the orientation of a general $(C^\wedge N)_2 Ir(acac)$ and fac-$Ir(C^\wedge N)_3$ complexes affects Θ for any molecular orientation and any given TDV. The coordinate system and direction of rotation for the molecules around angles ε and φ are defined in Fig. 47.4, with the z axis orthogonal to the substrate. The model assumes that the TDV lies in the plane of the $(C^\wedge N)Ir$ ligands at an angle δ between the Ir–N bond and TDV. To probe the dependence of Θ on molecular orientation for various values of δ, the metal complex is rotated in the imposed coordinate system and Θ is calculated for the molecule using Eq. (47.1) in the Methods section. The starting point for both types of complex ($\varepsilon = \varphi = 0°$) is depicted in Fig. 47.4. The molecules were rotated stepwise around ε and Θ was calculated from the projection of the TDV of the emissive ligand(s) onto the x, y and z directions at each step in ε and φ. Figure 47.4 shows one quadrant of the possible ε and φ values; full plots for $\varepsilon, \varphi = 0° - 360°$ are presented in Supplementary Figs. 15 and 16.

For $(C^\wedge N)_2 Ir(O^\wedge O)$ complexes, values of Θ with $\delta < 10°$ are insensitive to rotation around ε but vary with changes in φ (Fig. 47.4a). As the magnitude of δ increases, a dependence of Θ on rotation around ε appears. Regions of low Θ become localized near values of $\varepsilon = 135°$, which places the C_2 axis in the x–y plane, parallel to the substrate. The mechanism we propose for dopant alignment predicts that the $(bppo)_2 Ir(acac)$ molecules will orient with the C_2 axis perpendicular to the substrate. We expect the bppo-based emitters to have values of δ between 20° and 40° (see above). Examination of the plots in Fig. 47.4a shows that if the $(bppo)_2 Ir(acac)$ dopants exhibited uniform alignment with their C_2 axes orthogonal to the substrate ($\varepsilon = 45°, 225°$ and $\varphi = 0°$), then Θ would be less than 0.22 for any value of $\delta \leq 40°$. The experimental values of Θ being higher than predicted is likely to come from two sources. First, random variation away from orthogonal order is expected for a population of molecules, which will shift the net orientation towards isotropic. Second, film roughness in much the same way contributes to a deviation towards an isotropic alignment. Examination of the CBP and doped CBP films by atomic force microscopy (AFM) shows that the organic films are not flat but have an r.m.s. roughness of 2.2 nm. We have evaluated random traces

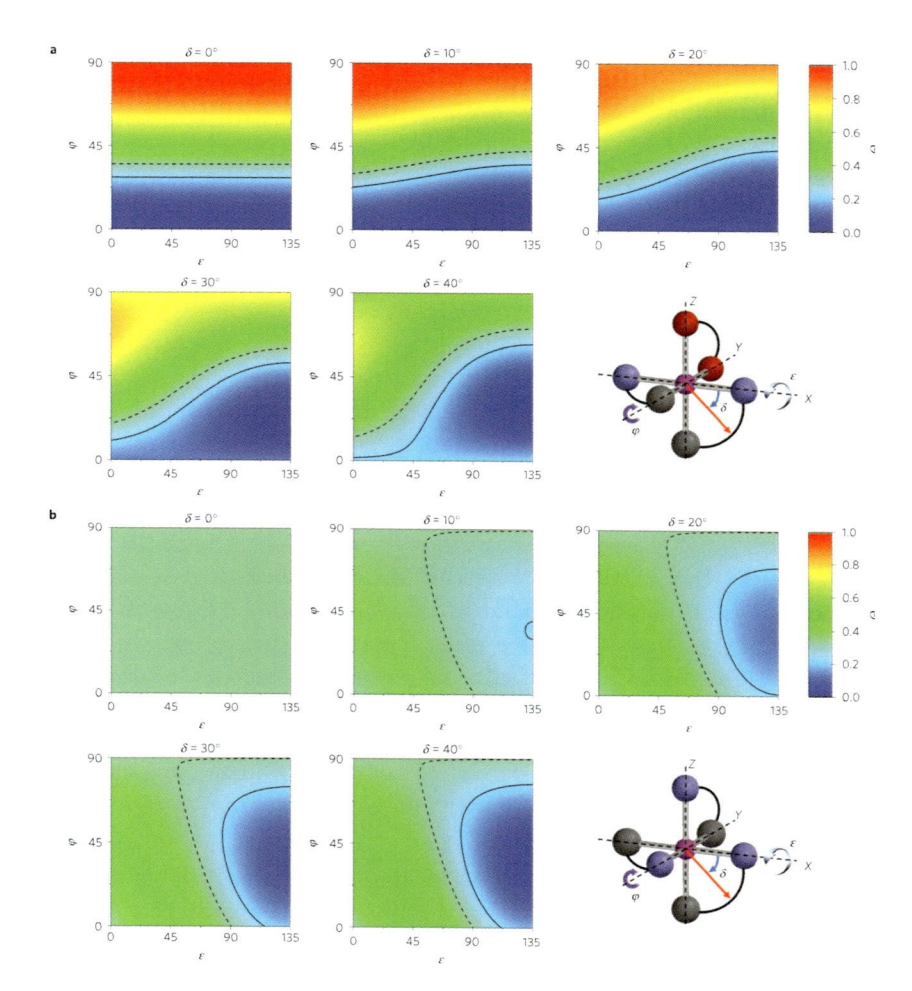

Figure 47.4 Molecular orientation. (a,b) Surface maps illustrating the dependence of the optical anisotropy parameter Θ, represented here by a colour gradient, on molecular rotation for any Ir(C^N)$_2$(O^O) molecule (a) or homoleptic fac-Ir(C^N)$_3$ molecule (b) starting from the orientation shown. The angle N–Ir–TDV is δ, where $\delta = 0°$ corresponds to a TDV oriented directly towards the nitrogen. ε is rotation about the x axis, and φ is rotation around the y axis in the direction shown in the sketch. The solid line is $\Theta = 0.22$ and the dashed line is $\Theta = 0.33$ (isotropic).

of the AFM image for an 8% doped CBP film (see Supplementary Information and Supplementary Fig. 14) and determined that the average angle the CBP surface takes relative to the substrate surface is $5.2° \pm 0.3°$. Both of these factors are expected to shift the net molecular orientation away from the ideal $\varepsilon = 45°, 225°$ and $\varphi = 0°$ towards a more isotropic value. Note that when $\varepsilon < 45°$, only a small increase in φ is needed to obtain $\Theta = 0.22$ for δ values between $30°$ and $40°$.

In contrast to (C^N)$_2$Ir(O^O) complexes, fac-Ir(C^N)$_3$ complexes with a TDV oriented along any metal–ligand bond ($\delta = 0, 90$) will yield a Θ value of 0.33 for all molecular orientations (Fig. 47.4b). In this case, emission from a perfectly aligned emitter will be experimentally indistinguishable from a randomly oriented one. As the TDV deviates from the metal–ligand bond axis ($90 > \delta > 0$), regions of low Θ appear centred at angles of $\varepsilon = 135°$ and $\varphi = 35.3°$ corresponding to a molecular geometry with the C_3 axis perpendicular to the x–y plane.

There are two examples of facial homoleptic complexes that are reported to show substantial alignment in doped films (see Ir(chpy)$_3$ and Ir(piq)$_3$ in Fig. 47.5 and Supplementary Table 1). The Θ values for the two dopants are 0.23 and 0.22, respectively [26]. The two complexes have C^N ligands that give rise to substantial structural anisotropy in the facial tris-chelated complexes, similar to heteroleptic species, as can be seen in the structural models shown in Fig. 47.5 (Refs. [22, 26]). Ir(chpy)$_3$ clearly has an aromatic and an aliphatic side, whereas the difference in Ir(piq)$_3$ is more subtle. If this anisotropy orients the molecule by our proposed mechanism, we would expect the molecular C_3 axis of these molecules to orient perpendicular to the surface, and the measured Θ value would correspond to TDV angles of $\delta \approx 10°$ (Fig. 47.6), well within the anticipated range for these C^N ligands.

Our work demonstrates the optimal design principles for Ir-based dopants to exploit the benefits of dopant orientation in OLEDs. For (C^N)$_2$Ir(O^O) it would be advantageous to shift the direction of the TDV towards the Ir–N bond axis, ideally to $\delta = 0°$ (Fig. 47.6). Note that any net orientation with $\varphi = 0° - 20°$ gives values of $\Theta < 0.1$ when $\delta = 0°$ (Fig. 47.4a). Such a broad spread of orientations is well within the range seen for the compounds reported here. The

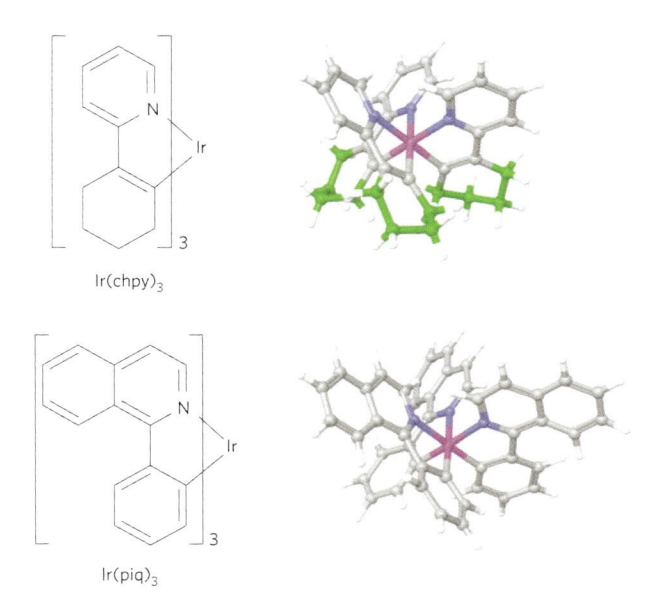

Figure 47.5 Oriented homoleptic phosphors. Molecular models of two facial Ir(C^N)$_3$ complexes that align in a CBP matrix. The aliphatic carbons of the chpy ligand have been coloured green.

outcoupling efficiency of an OLED with $\Theta = 0.1$ would increase by roughly a factor of 1.5 as compared with the isotropic case [23]. Alternatively, if one could align the dopants more uniformly with their C_2 axis perpendicular to the substrate ($\varepsilon = 45$ or $225°$ and $\varphi = 0$), Θ of ≤ 0.1 could be achieved with δ values as high as 25° (Fig. 47.6). (C^N)$_2$Ir(acac) complexes limit the operational stability of OLEDs, owing largely to the instability of the acac ligand. Heteroleptic Ir complexes, that is, (C^N)$_2$Ir(C'^N'), however, do not suffer from poor device stability. The mechanism described here, invoking structural anisotropy of the dopant to promote alignment, is equally applicable to heteroleptic complexes with aliphatic groups incorporated into either C^N or C'^N'. Figure 47.6 also shows that related fac-Ir(C^N)$_3$ complexes with δ values between 20° and 70°, if aligned with their C_3 axes perpendicular to the substrate, are also capable of generating highly anisotropic emission. Notably, δ values

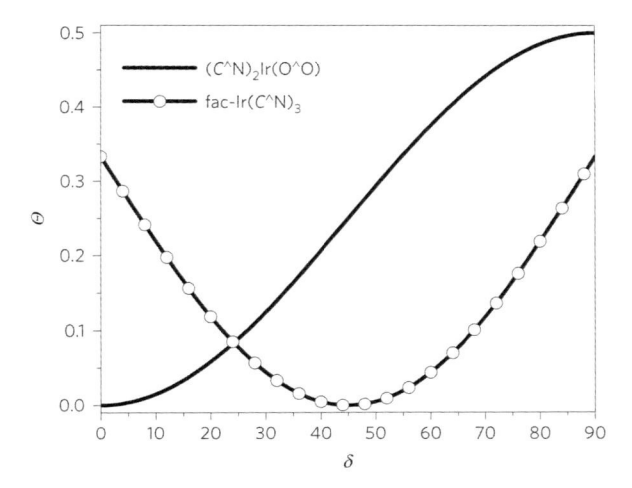

Figure 47.6 Ideal molecular orientation. Plot of δ versus Θ for $(C^\wedge N)_2$ Ir$(O^\wedge O)$ and fac-Ir$(C^\wedge N)_3$ complexes with their respective C_2 and C_3 axes oriented perpendicular to the substrate.

between 40° and 50° will give $\Theta < 0.01$ over a fairly broad range of ε and φ values.

47.4 Conclusion

We conclude that the presence of the acac group is responsible for the commonly measured value of $\Theta \sim = 0.2$ for a large variety of $(C^\wedge N)_2$Ir(acac) species. The acac ligand forms an aliphatic region on the surface of the otherwise aromatic Ir complex. We reason that interaction of this chemically anisotropic species at the boundary created between the vacuum and the organic surface during deposition is responsible for the observed net alignment of the TDVs of the dopants, wherein the phosphor's C_2 axis is largely perpendicular to the plane of the substrate. This proposed mechanism for alignment is consistent with the Θ value being unaffected by aggregation of the dopant. A similar mechanism for dopant alignment can be used to explain the low Θ values reported for facial Ir$(C^\wedge N)_3$ complexes where the $C^\wedge N$ ligand itself gives rise to significant chemical anisotropy of the dopant molecule, that is, Ir(chpy)$_3$ and Ir(piq)$_3$ (Ref. [26]).

Future work will explore the impact of the introduction of aliphatic character to different ligands. We will also explore the impact of host materials and their physical properties on the alignment behaviour of the dopants in films.

47.5 Methods

Starting materials were purchased from Sigma Aldrich and used without further purification. CBP was obtained from the Universal Display Corporation. Materials were purified by gradient sublimation before use. Chloro-bridged bis-cyclometalated iridium complexes of the formula [Ir(C^N)$_2$Cl]$_2$ were synthesized according to literature procedures [50]. 6H-[2]benzopyrano[4,3-b]pyridin-6-one (bppo) was synthesized according to the procedure in Ref. [51].

The synthesis of Iridium(III) bis(benzopyranopyridinone) (acetylacetonate) ((bppo)$_2$Ir(acac)) involved charging a flask with the chloro-bridged bppo dimer (1.4 mmol), silver triflate (3.08 mmol), bppo ligand (3.1 mmol) and potassium carbonate (13.9 mmol) under nitrogen atmosphere. The flask was charged with dry dichloroethane and stirred at reflux overnight. The solvent was removed under vacuum and the mixture was dissolved in CH$_2$Cl$_2$. The mixture was then filtered to remove silver chloride and chromatographed in 60:40 CH$_2$Cl$_2$/ethyl acetate (v/v). The collected fraction was precipitated from CH$_2$Cl$_2$ and hexanes to yield 280 mg pure (bppo)$_2$Ir(acac) (51%). ^{1}H NMR (CDCl$_3$): 1.86 (s, Me-acac, 6H), 5.32 (s, acac, 1H), 6.60 (dd, 2H), 7.04 (t, 2H), 7.42 (q, 2H), 7.66 (qd, 2H), 8.35 (dd, 2H). Electrospray ionization mass spectrum: C$_{29}$H$_{19}$IrN$_2$O$_6$ M/Z calculated 684.0872, found: 684.0851.

Iridium(III) bis(2-phenylpyridinate)(benzopyranopyridinone) ((ppy)$_2$Ir(bppo)) and iridium(III) bis(benzopyranopyridinone) (2-phenylpyridinate) ((bppo)$_2$Ir(ppy)) were synthesized by methods similar to those described in a previous report [31]. Chloro-bridged ppy dimer (0.4 mmol), silver triflate (0.88 mmol) and potassium carbonate (20 mmol) were added to a round-bottom flask under nitrogen atmosphere. The flask was charged with dry 1,2 o-dichlorobenzene. The reaction mixture was stirred at reflux overnight. The solvent was removed under vacuum and the residue was dissolved in CH_2Cl_2. The mixture was filtered to remove silver chloride and chromatographed in CH_2Cl_2 to yield 450 mg (ppy)$_2$Ir(bppo) (23%) and 325 mg (bppo)$_2$Ir(ppy) (31%). (bppo)$_2$Ir(ppy): electrospray ionization mass spectrum of $C_{35}H_{20}IrN_3O_4$, calculated M/Z: 739.1083; found: (MH$^+$) 740.1144. ^{1}H NMR (CDCL$_3$): 7.95 (d, 1H), 7.79 (m, 1H), 7.76 (d, 1H), 7.68 (m, 3H), 7.59 (m, 1H), 7.57 (m, 1H), 7.50 (dd, 1H), 7.46 (dd, 1H), 7.21 (d, 1H), 7.20 (m, 1H), 7.17 (m, 3H), 7.03 (dd, 1H), 6.97 (m, 2H), 6.87 (td, 1H), 6.74 (dd, 1H).

Optical spectra. Ultraviolet–visible absorption spectra were recorded using a Hewlett-Packard 4853 diode array spectrometer. Emission spectra were measured with a QuantaMaster Photon Technology International spectrofluorometer. Excited-state lifetimes were measured by time-correlated single-photon counting (IBH Fluorocube: LED excitation source). Quantum yields were measured using a calibrated integrating sphere (Hamamatsu C9920 system xenon lamp, and model C10027 photonic multichannel analyser).

Thin-film deposition. Film deposition and characterization methods have been performed as described in our previous report [52]. The detailed method is reproduced here for completeness. Films were made in an Angstrom Engineering EvoVac 800 VTE deposition system attached to a glove box and Inficon SQS-242 deposition software was used to control deposited material thicknesses using a 6 MHz Inficon quartz monitor gold-coated crystal sensor. All film depositions in the VTE were performed at pressures $\leq 4 \times 10^{-4}$ Pa and with deposition rates less than 1 Å s^{-1}. Organic films were stored under a nitrogen atmosphere. Doped films on glass substrates for orientation measurements were fabricated at pressures $\leq 5 \times 10^{-5}$ Pa with deposition rates of about 1.5 Å s^{-1} for CBP and various dopant concentrations (v/v). Films were encapsulated with a glass cover in a nitrogen atmosphere to prevent photodegeneration on photoluminescence excitation during orientation measurements.

Orientation measurements. To determine the molecular orientation in doped films, angular-dependent photoluminescence measurements were performed. This measurement technique has been described in our previous reports and is reproduced here for completeness [38, 53]. The sample was attached to a fused-silica half-cylinder prism by index-matching liquid and the emission angle was changed using a rotation stage. Spectra were recorded using a fibre optical spectrometer (SMS-500, Sphere Optics) and a polarizing filter to distinguish between p- and s-polarized light. The excitation of the samples was performed with a 375 nm cw laser diode with a fixed excitation angle of $45°$. The degree of orientation of the optical TDV of the emitter molecules was determined from a numerical simulation reported previously [36].

Spin-coating. Films were spin-cast from chloroform solutions. We dissolved 1 mg of each organic compound in 1 ml chloroform. We then prepared a mixture of both solutions with 4.6 ml NPD/chloroform and 0.4 ml Ir(MDQ)$_2$(acac)/chloroform resulting in a doping concentration of 8% Ir(MDQ)$_2$(acac):NPD. Variable angle spectroscopic ellipsometry was performed (Si-substrate used) to determine thickness and the optical constants of the films after spin-coating and drying for 1 h at room temperature. The thickness of the films was additionally checked by the fitting procedure of the angular-dependent photoluminescence emission spectra (p-pol for the orientation and s-pol for the actual thickness).

Theoretical calculations. Calculations were performed using the Jaguar 8.4 (release 17) software package on the Schrodinger Material Science Suite (v2014-2; Ref. [32]). Gas-phase geometry optimization was calculated using the B3LYP functional with the LACVP** basis set as implemented in Jaguar [33–35].

Electrochemistry. (bppo)$_2$Ir(acac) shows reversible reductions at -2.06 V and -2.29 V and a reversible oxidation at 0.69 V (versus Fc$^+$/Fc) by cyclic voltammetry. (bppo)$_2$Ir(ppy) shows reversible reductions at -2.12 V and -2.31 V and a reversible oxidation at 0.63 V (versus Fc$^+$/Fc). (ppy)$_2$Ir(bppo) shows a fully reversible reduction at -2.19 V and a quasi-reversible reduction at -2.77 V with a reversible oxidation at 0.48 V (versus Fc$^+$/Fc). The reversible oxidations are assigned to a metal-centred oxidation and the observed reductions are to a ligand-centred process. The more positive oxidations and more negative reductions relative to classic phosphors are proposed to be a consequence of the introduction of the extremely electron-withdrawing coumarin ligands. These strongly π electron-deficient ligands depress the highest occupied molecular orbital energy but do not change the lowest unoccupied molecular orbital as long as the emissive state remains localized on the bppo ligand. Electrochemistry of Ir(ppy)$_2$(bppo) has been reported previously [31].

Transition dipole moment, molecular orientation, and Θ value relationships. For a single emitting molecule containing only one possible TDV, $\mathbf{p} = (p_x, p_y, p_z)$, Θ is given by: $\Theta = p_z^2/(p_x^2 + p_y^2 + p_z^2)$. In molecules containing n different TDVs, or an ensemble of n differently oriented molecules, each of these TDVs must be taken into account. As an excited state is related to only one of them, each TDV must be calculated separately with respect to the contributing fraction a_i; $\sum_{i=1}^{n} a_i = 1$. Thus, the resulting Θ can be calculated as in Eq. (47.1), where $\mathbf{p}_i$ denotes the ith TDV and $p_{z,i}$ the corresponding component perpendicular to the surface.

$$\theta = \frac{\sum_{i=1}^{n} a_i p_{z,i}^2}{\sum_{i=1}^{n} a_i \mathbf{p}_i^2} \tag{47.1}$$

To explore the relationship between the anisotropy factor Θ for different orientations of a heteroleptic Ir complex comprising one acac group and two other identical ligands Ir(L)$_2$(acac) we have designed a coordinate system around an idealized molecule, depicted in Fig. 47.4a.

The biggest influence on the orientation of the molecule is achieved by the acac group. Therefore, we selected two individual angles to control the orientation of the molecule, and one to account for variations in the angle of the TDV. We define the angle N–Ir–TDV as δ, rotation around the x axis as ε and rotation around the y axis as φ.

To model the rotation of the molecule anticlockwise around the x axis we use a rotation matrix (R_x)

$$R_x = \begin{pmatrix} 1 & 0 & 0 \\ 0 & \cos(\epsilon) & -\sin(\epsilon) \\ 0 & \sin(\epsilon) & \cos(\epsilon) \end{pmatrix}$$

to represent the change in the position of the TDV of both ligands, using the coordinate system of the starting geometry (Fig. 47.4a), denoted as $\mathbf{L}_1$ (x–y plane) and $\mathbf{L}_2$ (y–z plane):

$$\mathbf{L}_1 = \begin{pmatrix} -\cos(\delta) \\ -\sin(\delta) \\ 0 \end{pmatrix}$$

$$\mathbf{L}_2 = \begin{pmatrix} \cos(\delta) \\ 0 \\ -\sin(\delta) \end{pmatrix}$$

The rotated TDV ($\mathbf{L}_1^*$ and $\mathbf{L}_2^*$) values for both ligands are obtained by multiplying the rotation matrix with the initial TDV:

$$R_x \circ \mathbf{L}_1 = \mathbf{L}_1^* = \begin{pmatrix} -\cos(\delta) \\ -\sin(\delta)\cos(\epsilon) \\ -\sin(\delta)\sin(\epsilon) \end{pmatrix}$$

$$R_x \circ \mathbf{L}_2 = \mathbf{L}_2^* = \begin{pmatrix} \cos(\delta) \\ \sin(\delta)\sin(\epsilon) \\ -\sin(\delta)\cos(\epsilon) \end{pmatrix}$$

However, as only the squared vectors are of interest to calculate the anisotropy factor, the total X-, Y- and Z-components are given by the sum of the individual contributions of the two vectors:

$$X = [-\cos(\delta)]^2 + [\cos(\delta)]^2$$

$$Y = [-\sin(\delta)\cos(\epsilon)]^2 + [\sin(\delta)\sin(\epsilon)]^2$$

$$Z = [-\sin(\delta)\sin(\epsilon)]^2 + [-\sin(\delta)\cos(\epsilon)]^2$$

The anisotropy factor Θ can then be calculated for arbitrary angles of δ and ϵ using:

$$\Theta = \frac{Z}{X + Y + Z}$$

To make the analysis more extensive, one can think about a second rotation anticlockwise around the y axis (denoted with the angle φ in the following) after the x-axis rotation. It is critical to note that the order of the rotations matters. The epsilon rotation must be applied first because the axes themselves do not change, but the molecular arrangement does (after the first rotation). Analogous to the former rotation matrix, one would achieve the following values for the three individual parameters δ, ε and φ:

$$X = [-\cos(\delta)\cos(\phi) - \sin(\delta)\sin(\epsilon)\sin(\phi)]^2$$
$$+ [\cos(\delta)\cos(\phi) - \sin(\delta)\cos(\epsilon)\sin(\phi)]^2$$

$$Y = [-\sin(\delta)\cos(\epsilon)]^2 + [\sin(\delta)\sin(\epsilon)]^2$$

$$Z = [\cos(\delta)\sin(\phi) - \sin(\delta)\sin(\epsilon)\cos(\phi)]^2$$
$$+ [-\cos(\delta)\sin(\phi) - \sin(\delta)\cos(\epsilon)\cos(\phi)]^2$$

The same procedure can also be applied to a homoleptic tris-Ir-compound such as, for example, Ir(ppy)$_3$. Here, one has to transform/rotate three different TDVs, one for each ligand. In the case that the three nitrogens are lying in the x-, (-y)- and z-direction, depicted in Fig. 47.4b, and the rotations are again performed around the x and the y axis, the TDVs for deriving Θ change to:

$$\mathbf{L}'_1 = \begin{pmatrix} -\sin(\delta)\cos(\phi) - \cos(\delta)\sin(\epsilon)\sin(\phi) \\ -\cos(\delta)\cos(\epsilon) \\ \sin(\delta)\sin(\phi) - \cos(\delta)\sin(\epsilon)\cos(\phi) \end{pmatrix}$$

$$\mathbf{L}'_2 = \begin{pmatrix} \cos(\delta)\cos(\phi) - \sin(\delta)\cos(\epsilon)\sin(\phi) \\ \sin(\delta)\sin(\epsilon) \\ -\cos(\delta)\sin(\phi) - \sin(\delta)\cos(\epsilon)\cos(\phi) \end{pmatrix}$$

$$\mathbf{L}'_3 = \begin{pmatrix} \sin(\delta)\sin(\epsilon)\sin(\phi) + \cos(\delta)\cos(\epsilon)\sin(\phi) \\ \sin(\delta)\cos(\epsilon) - \cos(\delta)\sin(\epsilon) \\ \sin(\delta)\sin(\epsilon)\cos(\phi) + \cos(\delta)\cos(\epsilon)\cos(\phi) \end{pmatrix}$$

Supplementary Information

The supplementary information is available at: https://static-content.springer.com/esm/art%3A10.1038%2Fnmat4428/MediaObjects/41563_2016_BFnmat4428_MOESM3_ESM.pdf.

Acknowledgements

The research described here was carried out with the support of the Universal Display Corporation, The Humboldt Foundation, Bavaria California Technology Center (BaCaTeC) and Deutsche Forschungsgemeinschaft (DFG Br 1728/16-1). C.M. acknowledges financial support by Bayerische Forschungsstiftung.

Author contributions

M.J.J. prepared materials and samples, designed experiments and prepared the manuscript; C.M. measured molecular orientation; T.D.S. designed and prepared mathematical models; T.L. designed and prepared mathematical models; P.I.D., W.B. and M.E.T. designed and assisted with experiments and the manuscript.

Competing financial interests

M.E.T. has a competing interest in the Universal Display Corporation.

References

1. Scharber, M. C. *et al.* Design rules for donors in bulk-heterojunction solar cells—towards 10% energy-conversion efficiency. *Adv. Mater.* **18**, 789–794 (2006).
2. Jurow, M. J. *et al.* Controlling morphology and molecular packing of alkane substituted phthalocyanine blend bulk heterojunction solar cells. *J. Mater. Chem. A* **1**, 1557–1565 (2013).

3. Tsao, H. N. *et al.* The influence of morphology on high-performance polymer field-effect transistors. *Adv. Mater.* **21**, 209–212 (2009).
4. Yokoyama, D. Molecular orientation in small-molecule organic light-emitting diodes. *J. Mater. Chem.* **21**, 19187–19202 (2011).
5. Namdas, E. B., Ruseckas, A., Samuel, I. D. W., Lo, S.-C. & Burn, P. L. Photophysics of fac-tris(2-phenylpyridine) iridium(III) cored electroluminescent dendrimers in solution and films. *J. Phys. Chem. B* **108**, 1570–1577 (2004).
6. Baldo, M. A. *et al.* Highly efficient phosphorescent emission from organic electroluminescent devices. *Nature* **395**, 151–154 (1998).
7. Baldo, M. A., Lamansky, S., Burrows, P. E., Thompson, M. E. & Forrest, S. R. Very high-efficiency green organic light-emitting devices based on electrophosphorescence. *Appl. Phys. Lett.* **75**, 4–6 (1999).
8. Mladenovski, S., Neyts, K., Pavicic, D., Werner, A. & Rothe, C. Exceptionally efficient organic light emitting devices using high refractive index substrates. *Opt. Express* **17**, 7562–7570 (2009).
9. Reineke, S. *et al.* White organic light-emitting diodes with fluorescent tube efficiency. *Nature* **459**, 234–238 (2009).
10. Brütting, W., Frischeisen, J., Schmidt, T. D., Scholz, B. J. & Mayr, C. Device efficiency of organic light-emitting diodes: Progress by improved light outcoupling. *Phys. Status Solidi A* **210**, 44–65 (2013).
11. Schmidt, T. D. *et al.* Evidence for non-isotropic emitter orientation in a red phosphorescent organic light-emitting diode and its implications for determining the emitter's radiative quantum efficiency. *Appl. Phys. Lett.* **99**, 163302 (2011).
12. Forrest, S. R. The path to ubiquitous and low-cost organic electronic appliances on plastic. *Nature* **428**, 911–918 (2004).
13. Sasabe, H. & Kido, J. Development of high performance OLEDs for general lighting. *J. Mater. Chem. C* **1**, 1699–1707 (2013).
14. Sasabe, H. & Kido, J. Recent progress in phosphorescent organic light-emitting devices. *Eur. J. Org. Chem.* **2013**, 7653–7663 (2013).
15. Mayr, C., Schmidt, T. D. & Brütting, W. High-efficiency fluorescent organic light-emitting diodes enabled by triplet–triplet annihilation and horizontal emitter orientation. *Appl. Phys. Lett.* **105**, 183304 (2014).
16. Liehm, P. *et al.* Comparing the emissive dipole orientation of two similar phosphorescent green emitter molecules in highly efficient organic light-emitting diodes. *Appl. Phys. Lett.* **101**, 253304 (2012).

17. Mayr, C. & Brütting, W. Control of molecular dye orientation in organic luminescent films by the glass transition temperature of the host material. *Chem. Mater.* **27**, 2759–2762 (2015).
18. Flämmich, M. et al. Oriented phosphorescent emitters boost OLED efficiency. *Org. Electron.* **12**, 1663–1668 (2011).
19. Helander, M. G. et al. Chlorinated indium tin oxide electrodes with high work function for organic device compatibility. *Science* **332**, 944–947 (2011).
20. Lamansky, S. et al. Highly phosphorescent bis-cyclometalated iridium complexes: Synthesis, photophysical characterization, and use in organic light emitting diodes. *J. Am. Chem. Soc.* **123**, 4304–4312 (2001).
21. Lassiter, B. E. et al. Organic photovoltaics incorporating electron conducting exciton blocking layers. *Appl. Phys. Lett.* **98**, 243307 (2011).
22. Kim, K.-H. et al. Phosphorescent dye-based supramolecules for high-efficiency organic light-emitting diodes. *Nature Commun.* **5**, 4769 (2014).
23. Kim, K.-H., Moon, C.-K., Lee, J.-H., Kim, S.-Y. & Kim, J.-J. Highly efficient organic light-emitting diodes with phosphorescent emitters having high quantum yield and horizontal orientation of transition dipole moments. *Adv. Mater.* **26**, 3844–3847 (2014).
24. Kim, S.-Y. et al. Organic light-emitting diodes with 30% external quantum efficiency based on a horizontally oriented emitter. *Adv. Funct. Mater.* **23**, 3896–3900 (2013).
25. Lee, J.-H. et al. Finely tuned blue iridium complexes with varying horizontal emission dipole ratios and quantum yields for phosphorescent organic light-emitting diodes. *Adv. Opt. Mater.* **3**, 211–220 (2014).
26. Graf, A. et al. Correlating the transition dipole moment orientation of phosphorescent emitter molecules in OLEDs with basic material properties. *J. Mater. Chem. C* **2**, 10298–10304 (2014).
27. Reineke, S., Rosenow, T. C., Lüssem, B. & Leo, K. Improved high-brightness efficiency of phosphorescent organic LEDs comprising emitter molecules with small permanent dipole moments. *Adv. Mater.* **22**, 3189–3193 (2010).
28. Reineke, S., Schwartz, G., Walzer, K., Falke, M. & Leo, K. Highly phosphorescent organic mixed films: The effect of aggregation on triplet–triplet annihilation. *Appl. Phys. Lett.* **94**, 163305 (2009).
29. Kim, K.-H. et al. Controlling emitting dipole orientation with methyl substituents on main ligand of iridium complexes for highly efficient

phosphorescent organic light-emitting diodes. *Adv. Opt. Mater.* **3**, http://dx.doi.org/10.1002/adom.201500141 (2015).

30. Moon, C.-K., Kim, K.-H., Lee, J. W. & Kim, J.-J. Influence of host molecules on emitting dipole orientation of phosphorescent iridium complexes. *Chem. Mater.* **27**, 2767–2769 (2015).

31. Ren, X. *et al.* Coumarin-based, electron-trapping iridium complexes as highly efficient and stable phosphorescent emitters for organic light-emitting diodes. *Inorg. Chem.* **49**, 1301–1303 (2010).

32. Jaguar v. Version 8.4r17 (Schrödinger LLC, 2014); www.shrodinger.com/Jaguar

33. Becke, A. D. Density-functional thermochemistry. III. The role of exact exchange. *J. Chem. Phys.* **98**, 5648–5652 (1993).

34. Hay, P. J. & Wadt, W. R. *Ab initio* effective core potentials for molecular calculations. Potentials for the transition metal atoms Sc to Hg. *J. Chem. Phys.* **82**, 270–283 (1985).

35. Stephens, P. J., Devlin, F. J., Chabalowski, C. F. & Frisch, M. J. *Ab initio* calculation of vibrational absorption and circular dichroism spectra using density functional force fields. *J. Phys. Chem.* **98**, 11623–11627 (1994).

36. Frischeisen, J., Yokoyama, D., Endo, A., Adachi, C. & Brütting, W. Increased light outcoupling efficiency in dye-doped small molecule organic light-emitting diodes with horizontally oriented emitters. *Org. Electron.* **12**, 809–817 (2011).

37. Frischeisen, J., Yokoyama, D., Adachi, C. & Brütting, W. Determination of molecular dipole orientation in doped fluorescent organic thin films by photoluminescence measurements. *Appl. Phys. Lett.* **96**, 073302 (2010).

38. Mayr, C. *et al.* Efficiency enhancement of organic light-emitting diodes incorporating a highly oriented thermally activated delayed fluorescence emitter. *Adv. Funct. Mater.* **24**, 5232–5239 (2014).

39. Weber, W. H. & Eagen, C. F. Energy transfer from an excited dye molecule to the surface plasmons of an adjacent metal. *Opt. Lett.* **4**, 236–238 (1979).

40. Kawamura, Y., Brooks, J., Brown, J. J., Sasabe, H. & Adachi, C. Intermolecular interaction and a concentration-quenching mechanism of phosphorescent Ir(III) complexes in a solid film. *Phys. Rev. Lett.* **96**, 017404 (2006).

41. Chi, Y. & Chou, P.-T. Transition-metal phosphors with cyclometalating ligands: Fundamentals and applications. *Chem. Soc. Rev.* **39**, 638–655 (2010).

42. Vanhelmont, F. W. M., Strouse, G. F., Güdel, H. U., Stückl, A. C. & Schmalle, H. W. Synthesis, crystal structure, high-resolution optical spectroscopy, and extended Hückel calculations on cyclometalated [Re(CO)$_4$ (ppy)] (ppy = 2-phenylpyridine). *J. Phys. Chem. A* **101**, 2946–2952 (1997).
43. Dalal, S. S., Walters, D. M., Lyubimov, I., de Pablo, J. J. & Ediger, M. D. Tunable molecular orientation and elevated thermal stability of vapor-deposited organic semiconductors. *Proc. Natl Acad. Sci. USA* **112**, 4227–4232 (2015).
44. Kearns, K. L. *et al.* Molecular orientation, thermal behavior and density of electron and hole transport layers and the implication on device performance for OLEDs. *Proc. SPIE* **9183**, 91830F (2014).
45. Tsai, M. H. *et al.* 3-(9-carbazolyl)carbazoles and 3,6-di(9-carbazolyl) carbazoles as effective host materials for efficient blue organic electrophosphorescence. *Adv. Mater.* **19**, 862–866 (2007).
46. Dalal, S. S., Fakhraai, Z. & Ediger, M. D. High-throughput ellipsometric characterization of vapor-deposited indomethacin glasses. *J. Phys. Chem. B* **117**, 15415–15425 (2013).
47. Brian, C. W. & Yu, L. Surface self-diffusion of organic glasses. *J. Phys. Chem. A* **117**, 13303–13309 (2013).
48. Zhu, L. *et al.* Surface self-diffusion of an organic glass. *Phys. Rev. Lett.* **106**, 256103 (2011).
49. Xing, X. *et al.* Essential differences of organic films at the molecular level via vacuum deposition and solution processes for organic light-emitting diodes. *J. Phys. Chem. C* **117**, 25405–25408 (2013).
50. Nonoyama, M. Benzo[h]quinolin-10-yl-N iridium(III) complexes. *Bull. Chem. Soc. Jpn* **47**, 767–768 (1974).
51. Zhang, W. & Pugh, G. Free radical reactions for heterocycle synthesis. Part 6: 2-bromobenzoic acids as building blocks in the construction of nitrogen heterocycles. *Tetrahedron* **59**, 3009–3018 (2003).
52. Navarro, F. F., Djurovich, P. I. & Thompson, M. E. Metal deposition for optoelectronic devices using a low vacuum vapor phase deposition (VPD) system. *Org. Electron.* **15**, 3052–3060 (2014).
53. Mayr, C., Taneda, M., Adachi, C. & Brütting, W. Different orientation of the transition dipole moments of two similar Pt(II) complexes and their potential for high efficiency organic light-emitting diodes. *Org. Electron.* **15**, 3031–3037 (2014).

Chapter 48

Deep Blue Phosphorescent Organic Light-Emitting Diodes with Very High Brightness and Efficiency

Jaesang Lee,[a] Hsiao-Fan Chen,[b] Thilini Batagoda,[b] Caleb Coburn,[c] Peter I. Djurovich,[b] Mark E. Thompson,[b] and Stephen R. Forrest[a,c,d]

[a]*Department of Electrical Engineering and Computer Science, University of Michigan, Ann Arbor, Michigan 48109, USA*
[b]*Department of Chemistry, University of Southern California, Los Angeles, California 90089, USA*
[c]*Department of Physics, University of Michigan, Ann Arbor, Michigan 48109, USA*
[d]*Department of Materials Science and Engineering, University of Michigan, Ann Arbor, Michigan 48109, USA*
stevefor@umich.edu

The combination of both very high brightness and deep blue emission from phosphorescent organic light-emitting diodes (PHOLED) is required for both display and lighting applications, yet so far has not been reported. A source of this difficulty is the absence of electron/exciton blocking layers (EBL) that are compatible with the high triplet energy of the deep blue dopant and the high frontier orbital energies of hosts needed to transport charge. Here, we show

---

Reprinted from *Nat. Mater.*, **15**, 92–98, 2015.

Electrophosphorescent Materials and Devices
Edited by Mark E. Thompson
Text Copyright © 2015 Nature Publishing Group
Layout Copyright © 2024 Jenny Stanford Publishing Pte. Ltd.
ISBN 978-981-4877-34-3 (Hardcover), 978-1-003-08872-1 (eBook)
www.jennystanford.com

that *N*-heterocyclic carbene (NHC) Ir(III) complexes can serve as both deep blue emitters and efficient hole-conducting EBLs. The NHC EBLs enable very high brightness (>7,800 cd m^{-2}) operation, while achieving deep blue emission with colour coordinates of [0.16, 0.09], suitable for most demanding display applications. We find that both the facial and the meridional isomers of the dopant have high efficiencies that arise from the unusual properties of the NHC ligand—that is, the complexes possess a strong metal–ligand bond that destabilizes the non-radiative metal-centred ligand-field states. Our results represent an advance in blue-emitting PHOLED architectures and materials combinations that meet the requirements of many critical illumination applications.

48.1 Introduction

A driving force behind the use of organic electroluminescence in displays and lighting has been the introduction of red and green electrophosphorescent devices with up to 100% internal quantum efficiencies [1, 2]. However, achieving deep blue electrophosphorescence with high efficiency and brightness, together with long-term operational stability, remains a significant challenge [3]. The design of robust and efficient blue phosphors free of electrochemically reactive moieties offers one possible solution. For example, saturated blue-emitting tris-cyclometalated iridium(III) complexes [4, 5] were introduced by using the thermodynamically stable *N*-heterocyclic carbene (NHC) ligands, or Ir(C^C:)$_3$ (Ref. [6]). This is compared with blue Ir complexes using fluorination to obtain a wide energy gap [7, 8], which has resulted in high efficiency but only an unsaturated sky-blue colour, unsuitable for displays.

Unfortunately, a significant drawback of previously demonstrated deep blue phosphorescent organic light-emitting diodes (PHOLEDs) is that they are subject to a pronounced efficiency roll-off at the high brightness required in most display and lighting applications [7–15]. For example, the current density at the half-maximum EQE, $J_{1/2}$, is typically <50 mA cm^{-2} owing to the loss of excitons and electrons from the PHOLED emission layer (EML), as well as strong bimolecular annihilation [16]. Thus, they barely achieve a brightness >3,000 cd m^{-2} at $J_{1/2}$. Up to the present work, preventing these parasitic effects has been exacerbated by the lack of exciton and charge blocking layers that are compatible with the high-energy triplet emitters required for deep blue emission.

Based on our understanding of the unique energetics of the Ir(C^C:)$_3$ complexes, we introduce a device design that significantly improves the efficiency of the deep blue PHOLEDs, especially at high brightness. Here, Ir(C^C:)$_3$ complexes are used as the phosphorescent emitting molecules, electron/exciton blocking layers (EBL) and dopants that conduct holes across the EML. The combined effects of these multiple uses lead to a marked improvement in EQE at high current densities. Specifically, $J_{1/2}$ is increased in devices with an EBL by more than a factor of 280 compared to those without it, and is improved by a further 50% (resulting in a cumulative improvement by a factor of 420) by grading the dopant across the EML, thereby reducing triplet annihilation losses at very high brightness [17, 18].

Facial (*fac*-) and meridional (*mer*-)Ir(C^C:)$_3$-based PHOLEDs exhibit Commission Internationale d'Eclairage (CIE) coordinates of [0.16, 0.09] and [0.16, 0.15], with maximum external quantum efficiencies of EQE = 10.1 ± 0.2 and 14.4 ± 0.4% at low luminance, decreasing slightly to 9.0 ± 0.1 and 13.3 ± 0.1% at $L = 1,000$ cd m^{-2}. Surprisingly, the device efficiencies are decreased by 50% at unusually high brightness, giving $L = 7,800 \pm 400$ and $22,000 \pm 1,000$ cd m^{-2} (corresponding to $J_{1/2} = 160 \pm 10$ and 210 ± 1 mA cm^{-2}), respectively. The *fac*- or *mer*-Ir(C^C:)$_3$-based devices produce unparalleled brightness at $J_{1/2}$ compared with the PHOLEDs with similar colour coordinates [15, 19]. To our knowledge, the *fac*-isomer-based device achieves the brightest deep blue emission among the PHOLEDs reported so far, which nearly meets the most stringent National Television System Committee (NTSC) requirements.

An additional finding is that the *mer*-Ir(C^C:)$_3$ is equally or more efficiently luminescent than the *fac*-isomer in solution and the solid state [20], whereas conventional red- and green-emitting Ir(C^N)$_3$ complexes follow the opposite trend—that is, *fac* is more efficient than *mer* [21, 22]. We find that the strong Ir–NHC ligand bonds [4] in Ir(C^C:)$_3$ result in reduced non-radiative decay via the metal-centred ligand-field states for both isomers [23]. Our studies of the

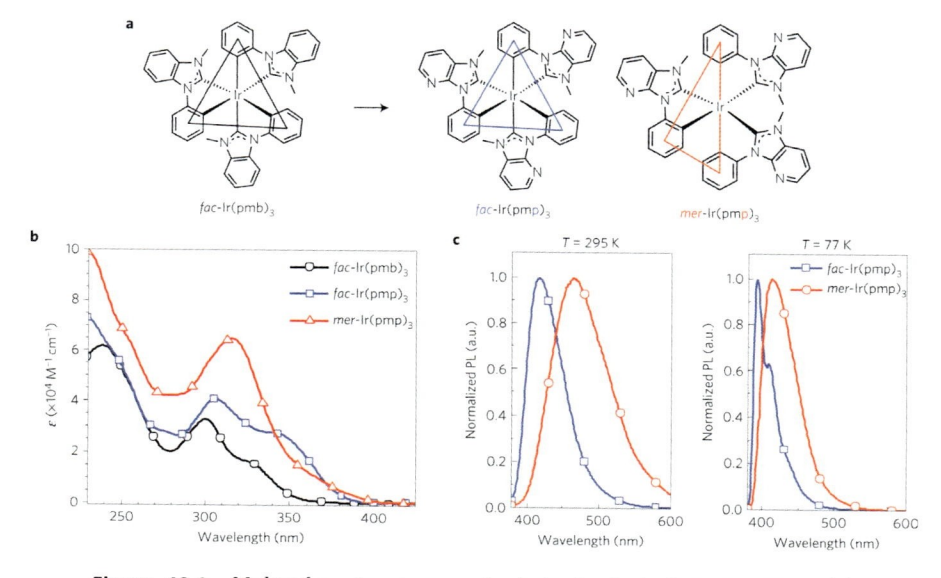

Figure 48.1 Molecular structure and photophysical characteristics. (a) Molecular structural formulae of *fac*-Ir(pmb)$_3$ and the *fac*- and *mer*-isomers of Ir(pmp)$_3$. Here, *fac*- and *mer*-Ir(pmp)$_3$ have the C_3 and C_1 molecular symmetries, respectively, in pseudo-octahedral coordinates. (b) Absorption spectra of *fac*-Ir(pmb)$_3$ (circles), *fac*-Ir(pmp)$_3$ (squares) and *mer*-Ir(pmp)$_3$ (triangles) in dichloromethane (CH$_2$Cl$_2$) solution. (c) Photoluminescence spectra of *fac*- and *mer*-Ir(pmp)$_3$ in degassed 2-methyltetrahydrofuran (2-MeTHF) at temperatures of $T = 295$ K and 77 K.

photophysics of Ir(C^C:)$_3$ complexes, along with their employment in device architectures designed for their optimum performance, provides a solution for achieving efficient deep blue emission at very high brightness.

48.2 Results

The structure of our newly synthesized NHC Ir(III) complex, tris-(N-phenyl, N-methyl-pyridoimidazol-2-yl)iridium(III), Ir(pmp)$_3$, is based on the near-ultraviolet-emitting tris-(N-phenyl, N-methyl-benzimidazol-2-yl) Ir(III), or Ir(pmb)$_3$, whose benzannulated component in the NHC ligands is replaced with a fused pyridyl ring, as shown in Fig. 48.1a. The greater electronegativity of the nitrogen atom versus methine (CH) lowers the reduction potential of *fac*-Ir(pmp)$_3$ to $E_{red} = -2.77 \pm 0.05$ V relative to *fac*-Ir(pmb)$_3$ ($E_{red} = -3.19 \pm 0.05$ V), although their oxidation potentials are nearly identical ($E_{ox} = 0.47 \pm 0.05$ and 0.45 ± 0.05 V, respectively). The absorption spectra of *fac*-Ir(pmb)$_3$ and *fac*-Ir(pmp)$_3$ in Fig. 48.1b show that their spin-allowed metal-to-ligand charge-transfer (^{1}MLCT) transitions have high-energy onsets, at $\lambda = 350$ and 380 nm, respectively. The observed redshift for the absorption spectrum of *fac*-Ir(pmp)$_3$ results from its smaller energy gap compared to *fac*-Ir(pmb)$_3$ inferred from their E_{red} and E_{ox}. Accordingly, the photoluminescence (PL) spectrum of the *fac*-Ir(pmp)$_3$ in Fig. 48.1c has a redshifted peak wavelength of $\lambda_{max} = 418$ nm in the deep blue compared to the near-ultraviolet emission of *fac*-Ir(pmb)$_3$ ($\lambda_{max} = 380$ nm; Ref. [4]). Meanwhile, the PL spectrum of *mer*-Ir(pmp)$_3$ is broad and exhibits a large room-temperature bathochromic shift ($\lambda_{max} = 465$ nm) relative to the *fac*-isomer. The lower emission energy of the *mer*-Ir(pmp)$_3$ compared to the *fac*-isomer is due to its lower oxidation potential ($E_{ox} = 0.23 \pm 0.05$ V) and nearly identical reduction potentials ($E_{red} = -2.80 \pm 0.05$ V), which result in a correspondingly reduced energy gap. The emission from both Ir(pmp)$_3$ isomers undergoes a pronounced rigidochromic shift at $T = 77$ K, with the *fac*-isomer exhibiting a vibronically structured line shape.

Figure 48.2a shows the temperature-dependent transient PL characteristics of *fac*- and *mer*-Ir(pmp)$_3$ in de-aerated 2-MeTHF, with quantum yields of $\Phi_{PL} = 76 \pm 5$ and $78 \pm 5\%$, respectively, at $T = 295$ K (compare with Φ_{PL} of *fac*-Ir(pmb)$_3 = 37 \pm 5\%$; Ref. [23]) and $\Phi_{PL} = 95 \pm 5\%$ at 77 K for both isomers. The triplet lifetimes, τ, were obtained from mono-exponential fits to the data at room temperature. Radiative (k_r) and non-radiative (k_{nr}) rate constants are calculated using the relationship [22] $k_r = \Phi_{PL}/\tau$, where $\Phi_{PL} = k_r/(k_r + k_{nr})$. The *mer*-isomer has a shorter triplet lifetime of $\tau = 0.8 \pm 0.1$, versus 1.2 ± 0.1 µs for the *fac*-isomer, which results in its higher $k_r = (1.0 \pm 0.2) \times 10^6$, versus $(6.4 \pm 1.3) \times 10^5$ s^{-1}, and $k_{nr} = (2.7 \pm 0.4) \times 10^5$, versus $(2.0 \pm 0.4) \times 10^5$ s^{-1}. At $T = 77$ K, triplet lifetimes of *fac*-Ir(pmp)$_3$ were extracted from multi-exponential fits. Accordingly, *fac*-Ir(pmp)$_3$ has two relatively well-resolved lifetimes of $\tau_1 = 3.9 \pm 0.2$ µs (weighting: 45%) and $\tau_2 = 9.2 \pm 0.2$ µs (55%). In

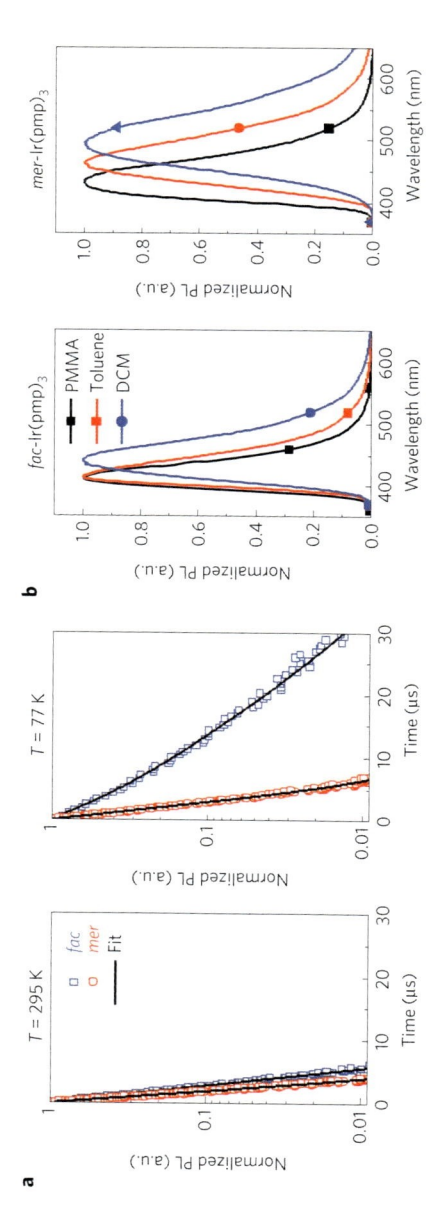

Figure 48.2 Temperature and solution dependence of the phosphors. (a) Transient phosphorescence decay of diluted *fac*- and *mer*-Ir(pmp)$_3$ in 2-MeTHF obtained at $T = 295$ and 77 K, with the fits based on mono- or multi-exponential decays. (b) Photoluminescence spectra of diluted *fac*- and *mer*-Ir(pmp)$_3$ in different polarity media, poly(methyl methacrylate) (PMMA), toluene and dichloromethane (DCM).

Table 48.1 Photophysical properties of *fac*- and *mer*-Ir(pmp)$_3$ dispersed in degassed 2-methyltetrahydrofuran (2-MeTHF) solution at room temperature (295 K) and 77 K

Temperature	295 K				77 K	
	Φ_{PL} (%)*	τ (μs)†	k_r (10^5 s^{-1})	k_{nr} (10^5 s^{-1})	Φ (%)‡	τ (μs)†
fac-Ir(pmp)$_3$	76 ± 5	1.2 ± 0.1	6.4 ± 1.3	2.0 ± 0.4	95 ± 5	3.9 ± 0.2, 9.2 ± 0.2
mer-Ir(pmp)$_3$	78 ± 5	0.8 ± 0.1	10 ± 2	2.7 ± 0.4	95 ± 5	1.0 ± 0.1

*Photoluminescence quantum yield (Φ_{PL}). †Mono- and multi-exponential fits are used for extracting triplet lifetimes (τ) at temperatures of $T = 295$ and 77 K, respectively. ‡Calculated by referencing the integrated emission intensity to that of *fac*-Ir(ppy)$_3$ ($\Phi_{PL} \approx 100\%$). Errors for the model parameters (k_r and k_{nr}) are the 95% confidence interval.

contrast, the *mer*-isomer still shows a mono-exponential decay, with an only slightly increased $\tau = 1.0 \pm 0.1$ μs at $T = 77$ K. Table 48.1 summarizes the photophysical parameters of both isomers.

Figure 48.2b shows the PL spectra of diluted *fac*- and *mer*-Ir(pmp)$_3$ in media of different polarity, poly(methyl methacrylate) (PMMA), toluene and dichloromethane (DCM), having dipole moments of 0, 0.42 and 1.6 D (Ref. [24]), respectively. Both complexes exhibit stronger positive solvatochromism in a more polar medium. Further, the PL spectrum of the *mer*-isomer is more redshifted and broadened than that of the *fac*-isomer, because the excited states of the former complex are more stabilized in a similar polar solvent.

Figure 48.3 shows the structures of PHOLEDs using *fac*- or *mer*-Ir(pmp)$_3$ (denoted by devices D$_{fac}$ or D$_{mer}$) along with the highest occupied molecular orbital (HOMO) and lowest unoccupied molecular orbital (LUMO) energies for all organic materials studied [25–27]. The LUMO energies are calculated from the reported or measured E_{red} (following Ref. [27]). The EBLs consist of the Ir(C^C:)$_3$ themselves (that is, *fac*-Ir(pmp)$_3$ and *fac*-Ir(pmb)$_3$ for D$_{fac}$ and D$_{mer}$, respectively), which have equal or shallower LUMO energies than that of the host, as well as equal or larger triplet energy levels than the dopants. This enables efficient hole injection into the hole-conducting dopants, while blocking electrons transported via the host. The doping concentration in the EML is linearly graded from 20 vol% at the EBL interface to 8 vol% at

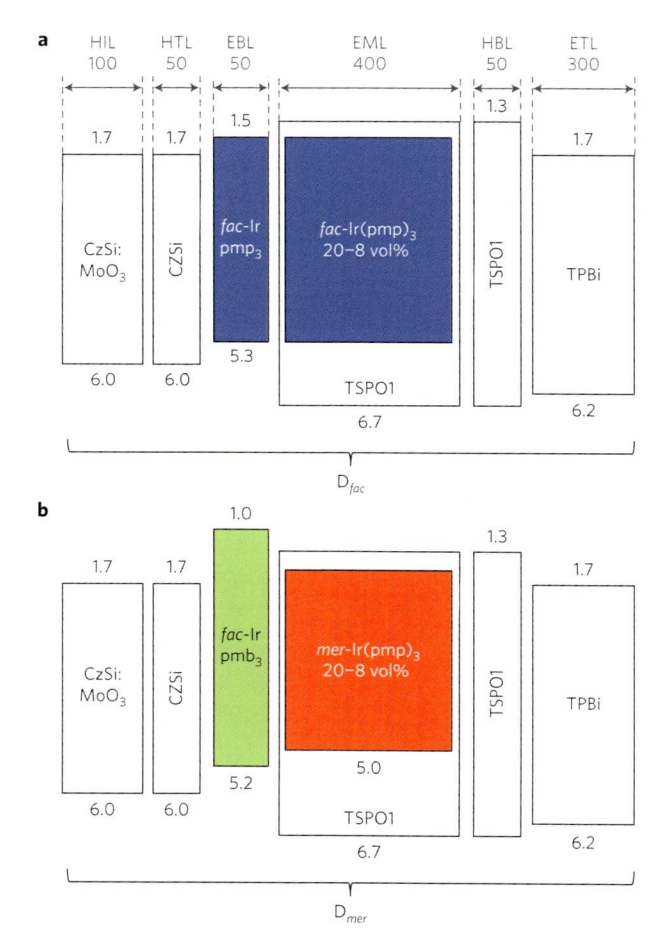

Figure 48.3 Device structure and its frontier orbital energies. (a,b) Energy level diagrams of phosphorescent organic light-emitting devices (PHOLEDs) based on *fac*- (a) and *mer*-Ir(pmp)$_3$ (b), denoted by devices D$_{fac}$ and D$_{mer}$, respectively. Highest occupied molecular orbital (HOMO) and lowest unoccupied molecular orbital (LUMO) energies of comprising materials in eV are either calculated or obtained from the literature [26–28]. HIL, hole injection layer; HTL, hole blocking layer; EBL, electron/exciton blocking layer; EML, emissive layer; HBL, hole/exciton blocking layer; ETL, electron-transport layer; the corresponding number below each abbreviation is the thickness of the layer in nanometres. CzSi stands for 9-(4-tert-butylphenyl)-3,6-bis(triphenylsilyl)-9H-carbazole and TPBi stands for 1,3,5-tris(1-phenyl-1H-benzimidazol-2-yl)benzene.

the hole blocking layer (HBL) interface to create a uniform triplet distribution across the EML (Ref. [17]).

Holes and electrons injected into the EML are mainly transported via the dopant and the host (diphenyl-4-triphenylsilylphenyl-phosphine oxide, or TSPO1), respectively (see Supplementary Section 1). Owing to the nested HOMO and LUMO energies of Ir(pmp)$_3$ in TSPO1 (Fig. 48.3), the majority of electrons transported via the host are trapped by the dopant and then radiatively recombine with the holes on the dopant. As TSPO1 is preferably electron-transporting owing to its diphenylphosphine oxide group [28], triplets are primarily formed at the EBL/EML interface (top, Fig. 48.4a). This necessitates the use of EBLs with high triplet and shallow LUMO energies, as shown in Fig. 48.3. In the graded EML, an initially high doping concentration (20 vol%) near the EBL/EML interface facilitates hole injection and transport, which gradually reduces owing to the decreasing dopant fraction (8 vol%) at the EML/HBL interface (bottom, Fig. 48.4a). The resulting triplet densities [17] (or recombination profiles) of both types of EML are shown in Fig. 48.4b (see Methods). In the uniformly doped EML, ~47% of triplets are concentrated near the EBL/EML interface ($x = 0$–10 nm), which decreases to approximately 33% in the graded EML owing to the deeper hole penetration. The distributed recombination profile reduces the probability of bimolecular annihilation quenching [16]. In addition, radiative recombination in the graded-EML device occurs farther from the EBL/EML interface compared to the uniformly doped EML PHOLED, resulting in enhanced light outcoupling (see Supplementary Section 2). These combined effects lead to the increased EQE of D$_{mer}$, as shown in the inset of Fig. 48.4b.

Figure 48.5a shows the PL spectra of *fac*- and *mer*-Ir(pmp)$_3$ uniformly doped at 14 vol% into TSPO1. Figure 48.5b shows the electroluminescence (EL) spectra of D$_{fac}$ and D$_{mer}$ measured at a current density of $J = 10$ mA cm^{-2}, which result in deep blue CIE coordinates of [0.16, 0.09] and [0.16, 0.15], respectively. In the inset we show images of the packaged devices along with their characteristic emission colour. The EL spectra have nearly identical CIE coordinates to the PL, confirming the emission is solely from the dopants in the PHOLEDs. Figure 48.5c shows the current density–voltage–luminance ($J-V-L$) characteristics of D$_{fac}$ and D$_{mer}$. D$_{mer}$

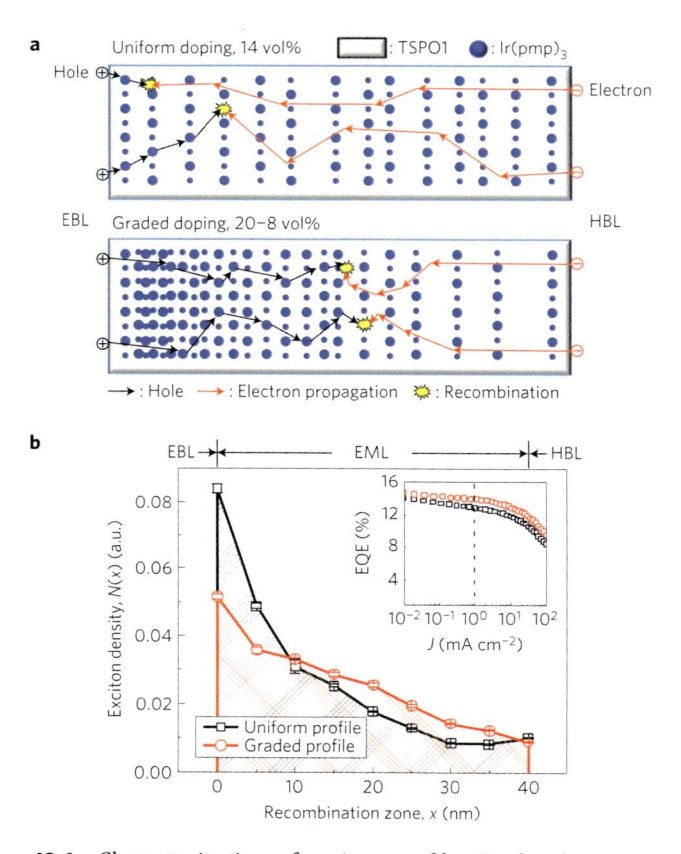

Figure 48.4 Characterization of exciton profiles in the device emission layer. (a) Charge-transport mechanisms in the emission layers (EML) of uniformly doped Ir(pmp)$_3$ (blue circle) at 14 vol% (top) and linearly graded doping at 20–8 vol% (bottom) into TSPO1 (white rectangular background), from the EBL to the HBL boundaries. Black and red arrows describe the hole- and electron-transport trajectories, respectively, which then recombine radiatively, as illustrated by a yellow starburst. This conceptually demonstrates that recombination occurs relatively closer to the HBL boundary for the graded versus uniformly doped EML owing to the improved hole injection and transport of the former. (b) Triplet density distributions of the uniformly doped (squares) and graded doped (circle) EMLs in D$_{mer}$ measured at $J = 10$ mA cm^{-2}. The external quantum efficiency (EQE) versus J of D$_{mer}$ for both types of EML structure is compared in the inset.

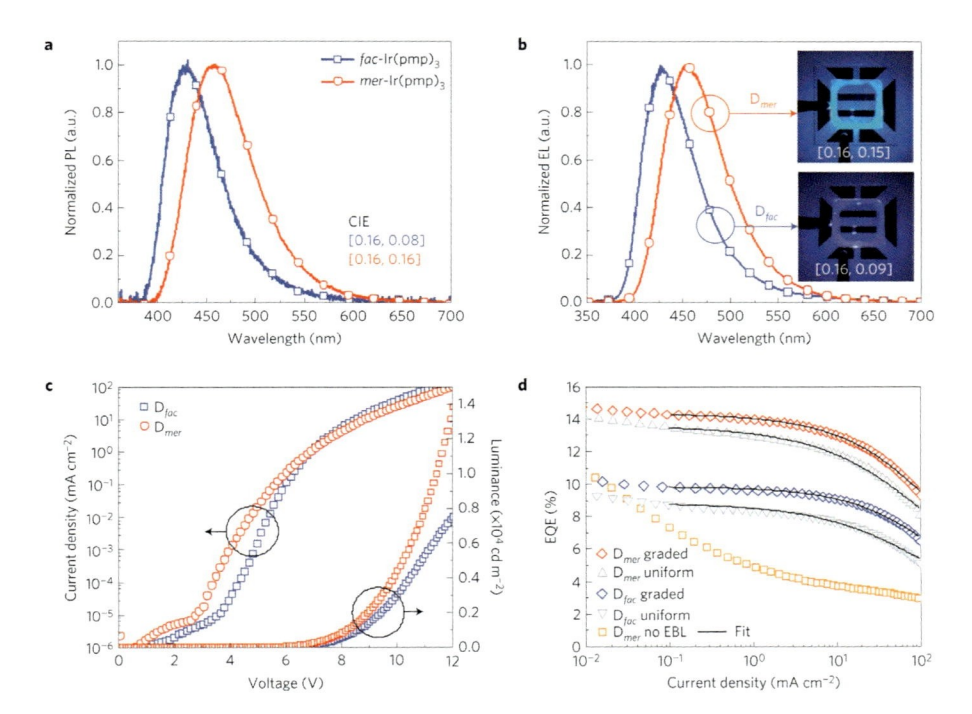

Figure 48.5 Electrophosphorescent device characteristics. (a) Photoluminescence (PL) spectra of *fac*- and *mer*-Ir(pmp)$_3$ doped at 14 vol% into TSPO1 excited by a HeCd laser (wavelength, $\lambda = 325$ nm). (b) Electroluminescence (EL) spectra of D$_{fac}$ and D$_{mer}$ measured at a current density of $J = 10$ mA cm^{-2}. The insets are photographs of 2 mm^2 packaged PHOLEDs whose illumination is reflected from a white background to avoid saturation of the camera sensor. The bright irregular square shape in each image is due to light scattered from the epoxy package seal. The package is a sandwich of two glass slides, one containing electrodes (dark regions) and the other serving as a lid. (c) Current density–voltage–luminance (J–V–L) characteristics of D$_{fac}$ and D$_{mer}$. (d) EQE versus J for PHOLEDs of either linearly graded or uniformly doped emission layers consisting of either *fac*- or *mer*-Ir(pmp)$_3$. Fits (solid lines) are based on the model in the Methods. EQE versus J for the *mer*-based PHOLED without an electron/exciton blocking layer (EBL) is also plotted.

turns on at a lower voltage than D_{fac} (3 V versus 4 V), which is presumably due to different charge injection, transport and trapping characteristics [20], and the lower HOMO energy of mer-Ir(pmp)$_3$ than the fac-isomer. Although the current densities of D_{fac} at high voltage (>7 V) are greater than those of D_{mer}, the latter device still achieves a higher luminance owing to its redshifted emission and higher EQE at all current densities.

Figure 48.5d shows the EQE–J characteristics of D_{fac}, D_{mer} and analogous devices whose EMLs are uniformly doped at 14 vol% by fac and mer-Ir(pmp)$_3$, and a device without an Ir(C^C:)$_3$-based EBL. By employing fac-Ir(pmb)$_3$ as the EBL, the uniformly doped EML PHOLED has a markedly higher EQE and reduced efficiency roll-off at high J compared to the PHOLED lacking an EBL. Therefore, $J_{1/2}$ increases by almost 280 times, from 0.5 ± 0.1 to 140 ± 10 mA cm^{-2}, and EQE increases by at least 40% at all current densities. The EQEs of the graded-EML devices employing an EBL are further improved by ~10% at all current densities compared to uniformly doped EML PHOLEDs, and $J_{1/2}$ is increased by a further 50%, leading to a cumulative improvement by a significant factor of 420. Thus, D_{fac} and D_{mer} attain EQE = 10.1 ± 0.2 and 14.4 ± 0.4% at low luminance, decreasing only slightly to 9.0 ± 0.1 and 13.3 ± 0.1% at L = 1,000 cd m^{-2}, and by 50% at L = 7,800 ± 400 and 22,000 ± 1,000 cd m^{-2} (corresponding to $J_{1/2}$ = 160 ± 10 and 210 ± 10 mA cm^{-2}), respectively. The difference in EQE of D_{fac} versus D_{mer} is consistent with the trend found in the solid-state PL quantum yields for fac- versus mer-Ir(pmp)$_3$ when doped in TSPO1 at the same concentrations (2–30 vol%, see Supplementary Section 3).

The EQE–J plot of graded and uniformly doped fac- and mer-Ir(pmp)$_3$ PHOLEDs in Fig. 48.5d was modelled to analyse the effect of the distributed recombination profile on device performance (see Methods and Table 48.2). As the doping concentration changes from uniform (14 vol%) to graded (20–8 vol%), the triplet–triplet annihilation rate (k_{TT}) increases whereas the triplet–polaron annihilation rate (k_{TP}) decreases for both fac- and mer-Ir(pmp)$_3$-based devices. The increased k_{TT} is due to a high triplet concentration in the dopant-rich (>14 vol%) region of the EML (Fig. 48.5b). However, the reduced local density of triplets in the graded EML compensates for the higher k_{TT}. Also, the significantly lower k_{TP} for

Table 48.2 Parameters for triplet–triplet (k_{TT}) and triplet–polaron annihilation (k_{TP})

	Doping	Fixed parameter τ (μs)*	Fitting parameters	
			k_{TT} (10^{-12} cm^{-3} s^{-1})	k_{TP} (10^{-13} cm^{-3} s^{-1})
D_{fac}	Uniform	1.0 ± 0.1	4.6 ± 0.8	8.6 ± 0.1
	Graded		5.6 ± 0.4	3.2 ± 0.5
D_{mer}	Uniform	0.85 ± 0.03	5.8 ± 0.8	10 ± 1
	Graded		7.3 ± 0.4	6.4 ± 0.4

*Triplet lifetime (τ) was obtained from the transient, solid-state phosphorescence decay fit [16]. Errors for the model parameters (k_{TT} and k_{TP}) are the 95% confidence interval.

the graded devices, which is proportional to the charge density [29], is consistent with the measured profiles.

48.3 Discussion

48.3.1 Implication of the Employed Device Design

The graded deep-blue-emitting phosphor Ir(pmp)$_3$ across the EML serves as a wide-energy-gap hole transporter that evenly distributes the exciton formation zone, thereby reducing bimolecular triplet annihilation. Note that light-blue-emitting materials such as iridium(III)bis [(4,6-difluorophenyl)-pyridinato-N,C_2']picolinate (FIrpic) achieve higher emission efficiency (>20%; Refs. [30, 31]) than reported here. This is a result of their relatively low emission energies and the large choice of host materials. Given the difficulties of synthesizing hole-transporting hosts for deep blue PHOLEDs, using hole-transporting gradient doping via the phosphor itself is clearly an effective strategy for improving efficiency.

Although co-host systems using two different compounds to separately transport holes and electrons have previously been shown to improve charge balance in the EML (Ref. [31]), they do not generally eliminate the need for an EBL and/or HBL, as charge carriers and excitons can still leak out from the emission zone without them. To prevent leakage, the comparatively high electron mobility in the recent deep blue OLED EML (Refs. [15,

32]) necessitates using an EBL, which has minimal impact on the conventional hole-transport dominated structures [33]. The significantly improved EQE and $J_{1/2}$ of our PHOLED primarily results from the Ir(C^C:)$_3$ EBL (Fig. 48.5d). As noted, effective blocking by the dopant itself is a unique property of the Ir(C^C:)$_3$ family of phosphors, with their very shallow LUMO energies and wide HOMO–LUMO gaps. Indeed, previously reported deep blue PHOLEDs lacking the EBL are similar to that of our unblocked device with its severe efficiency roll-off, whereas the blocked PHOLED characteristics are similar to those of conventional red- and green-emitting devices. This indicates the achievement of both charge and exciton confinement in the EML.

48.3.2 Origin of Both Highly Emissive *fac*- and *mer*-Ir(pmp)$_3$

The *mer*-isomers of the conventional red- and green-emitting Ir(C^N)$_3$ complexes [21, 22] typically have a non-radiative decay rate (k_{nr}) at least an order of magnitude larger than their *fac*-isomers. This difference is attributed to a more efficient thermal population of non-radiative triplet metal-centred (^{3}MC) ligand-field states that comprise Ir–ligand antibonding orbitals in the *mer*-isomer [23, 24]. The asymmetric molecular structure (C_1) of *mer*-isomers leads to *trans*-disposed Ir–N linkages that are more labile compared to the three equivalent Ir–N bonds in the C_3-symmetric *fac*-isomers [35]. Therefore, the ^{3}MC states of the *mer*-isomer are stabilized and thermally accessible compared to the *fac*-isomer. However, for Ir(pmp)$_3$, the difference in k_{nr} between *fac*- and *mer*-isomers is less than a factor of two as a result of the strong Ir–carbene bonds destabilizing the ^{3}MC states for both isomers. The lack of *mer*-to-*fac* isomerization for Ir(pmp)$_3$ substantiates its strong metal–ligand bond nature, whereas the Ir(C^N)$_3$ complexes, having weaker Ir–N bonds, allow such conversion [23]. At $T = 77$ K, the quantum yields of *fac*- and *mer*-Ir(pmp)$_3$ increase to near unity owing to suppressed non-radiative decay via a thermal population to ^{3}MC states.

The relative dominance of ligand-centred (^{3}LC) over ^{3}MLCT excited states in *fac*-Ir(pmp)$_3$ compared to the *mer*-isomer is reflected in the more pronounced temperature dependence of the transient PL response (Fig. 48.2a). The broader PL spectrum, both at T = 295 and 77 K, more pronounced bathochromic shift in a polar medium (Fig. 48.2b), and rigidochromic shift in a frozen media (see Fig. 48.1c) confirm that emission from *mer*-Ir(pmp)$_3$ originates more from a polar excited state (^{3}MLCT), rather than the relatively nonpolar ^{3}LC-dominant states of the *fac*-isomer [36] (Supplementary Section 4).

Compared to the structurally analogous violet-emitting *fac*-Ir(pmb)$_3$ complex, *fac*- and *mer*-Ir(pmp)$_3$ achieve a higher phosphorescence efficiency of $\Phi_{PL} = 76 \pm 5$ and $78 \pm 5\%$, respectively, relative to $37 \pm 5\%$ for *fac*-Ir(pmb)$_3$ (Ref. [23]); the difference once more is due to the increased stabilization of the triplet states in Ir(pmp)$_3$. Another possible explanation for the enhanced Φ_{PL} is a decrease in the torsion angle between the phenyl and pyridoimidazole groups in Ir(pmp)$_3$ relative to Ir(pmb)$_3$, caused by a steric interference between the H-atoms at the 1,7 phenyl and benzimidazole group positions. Substitution of the methine (CH) for N in Ir(pmp)$_3$ eliminates this conflict.

48.3.3 Differences between D$_{fac}$ and D$_{mer}$

The difference in the solid-state PL efficiencies of the *fac*- versus *mer*-Ir(pmp)$_3$, which leads to the difference in EQE of D$_{fac}$ versus D$_{mer}$, is probably due to different degrees of the emitter aggregation quenching. This may be induced by their different static dipole moments (17.2 D versus 10.8 D for *fac*- and *mer*-Ir(pmp)$_3$, respectively), as high dipole moments can promote aggregation [37]. Hence, when *fac*- or *mer*-Ir(pmp)$_3$ is doped at the same concentration of 11 vol% into wide-energy-gap hosts, the latter complex achieves higher PL quantum efficiencies (Supplementary Section 3).

A polar host such as TSPO1 is comprised of phosphine oxide moieties (P=O), which may lead to further phosphorescence losses owing to crystallite formation [38]. That is, strong interactions between TSPO1 molecules lead to crystalline domains within an amorphous film that result in structured, lower-energy emission bands [28] superposed with those obtained in solution [39]

(Supplementary Section 3). We speculate, therefore, that quenching by TSPO1 crystalline domains results in two phenomena: the lower PL efficiencies of Ir(pmp)$_3$ in TSPO1 versus p-bis(triphenylsilyly) benzene (UGH2; $\Phi_{PL} = 66 \pm 3$ versus $85 \pm 4\%$ for *mer*-Ir(pmp)$_3$ at 11 vol%) despite similarly high triplet energies [26] ($E_T = 3.37$ and 3.5 eV, respectively) and reduced PL quantum yields of Ir(pmp)$_3$ in TSPO1 at lower concentrations than those obtained at the optimal concentration of 14 vol%. The latter is in contrast to the results found with the conventional red, green and light-blue dopants into nonpolar hosts having maximum efficiency at very diluted concentrations (<2 vol%; Ref. [40]).

On the other hand, the polar TSPO1 enables a high solid-state solubility of the Ir(pmp)$_3$ by preventing interactions between dopants, leading to their physical separation [28, 41]. This enables the comparatively high optimal doping concentration of 14 vol% of Ir(pmp)$_3$, as well as a narrower and less bathochromically shifted Ir(pmp)$_3$ emission, resulting in a more saturated blue emission compared with that in other nonpolar hosts. Finally, given that the PL quantum yields at 14 vol% in TSPO1 are higher than at either 20 vol% or 8 vol% of Ir(pmp)$_3$ (the initial and terminal concentrations of the graded doping in the EML), the optimal doping concentrations for EL and PL are different because the dopant also serves as a hole transporter in the PHOLED, affecting the charge balance, and hence the EQE.

48.4 Conclusions

We find that deep-blue-emitting Ir(C^C:)$_3$ complexes can be simultaneously employed as triplet-emitting dopants, hole transporters and EBLs. This combination of uses significantly reduces electron and exciton losses. In particular, *fac*-Ir(pmp)$_3$-based PHOLEDs achieved remarkably reduced efficiency roll-off at high current density, resulting in very high brightness (>7,800 cd m^{-2}) with CIE coordinates of [0.16, 0.09], closest to the NTSC requirement among reported Ir-based PHOLEDs. The highly emissive *mer*-isomer of Ir(pmp)$_3$, which is due to the strong Ir–NHC ligand bond, enables even brighter PHOLED (>22,000 cd m^{-2}) operation in the blue. Our advances in materials design and device architectures provide the guidelines for designing efficient and, more importantly, long-lived deep blue PHOLEDs.

48.5 Methods

Synthesis of fac- and mer-isomers of Ir(pmp)$_3$. A mixture of N^3-methyl-N^2-phenylpyridine-2,3-diamine [42] (4.11 g, 20.63 mmol), triethyl orthoformate (172 ml), concentrated HCl (0.9 ml), and 5 drops of formic acid was heated at reflux for 12 h under N^2. The solution was cooled to room temperature, and the resulting solid (imidazolium chloride hydrochloric salt) was collected by filtration (2.81 g, 46%). A mixture of [IrCl(COD)]$_2$ (700 mg, 1.04 mmol), ligand (1.76 g, 6.24 mmol), silver oxide (1.45 g, 6.24 mmol), and triethylamine (635 mg, 6.24 mmol) in chlorobenzene (100 ml) was bubbled with N_2 for 10 min. The solution was heated to reflux for 12 h under N_2. The reaction mixture was cooled and chlorobenzene was evaporated. The crude solid was coated on silica and twice purified by column chromatography [dichloromethane (DCM) to DCM/acetone = 95/5] to yield *fac*-Ir(pmp)$_3$ (310 mg, 18%) and *mer*-Ir(pmp)$_3$ (890 mg, 53%). CHN analyses were carried out on both compounds; theoretical: C 57.34, H 3.70, N 15.43; *mer*-Ir(pmp)$_3$: C 57.22, H 3.52, N 15.52; *fac*-Ir(pmp)$_3$: C 56.95, H 3.73, N 15.50. ^{1}H and ^{13}C nuclear magnetic resonance (NMR) data for the two complexes are given as follows; *fac*-Ir(pmp)$_3$: ^{1}H NMR (CDCl$_3$, 400 MHz) d 8.84 (dd, $J = 8.0$, 1.2 Hz, 3H), 8.37 (dd, $J = 8.0$, 1.2 Hz, 3H), 7.39 (dd, $J = 8.0$, 1.2 Hz, 3H), 7.14–7.07 (m, 6H), 6.76 (td, $J = 8.0$, 1.2 Hz, 3H), 6.61 (dd, $J = 8.0$, 1.2 Hz, 3H), 3.28 (s, 9H); ^{13}C NMR (CDCl$_3$, 101 MHz) d 191.74, 147.61, 146.59, 146.25, 142.83, 136.44, 128.69, 125.36, 121.49, 116.89, 115.97, 114.68, 33.67; HRMS (*m/z*, ESI$^+$) Calcd for C$_{39}$H$_{31}$IrN$_9$ 818.2332 [M + H$^+$], found: 818.2347.

mer-Ir(pmp)$_3$: ^{1}H NMR (CDCl$_3$, 400 MHz) d 8.88 (dd, $J = 8.0$, 1.2Hz, 1H), 8.82 (dd, $J = 8.0$, 1.2 Hz, 1H), 8.78 (dd, $J = 8.0$, 1.2 Hz, 1H), 8.42 (dd, $J = 4.8$, 1.2 Hz, 2H), 8.38 (dd, $J = 4.8$, 1.2 Hz, 1H), 7.52 (dd, $J = 8.0$, 1.2 Hz, 1H), 7.49 (dd, $J = 8.0$, 1.2 Hz, 1H), 7.45 (dd, $J = 8.0$, 1.2 Hz, 1H), 7.22–7.13 (m, 3H), 7.07–6.97 (m, 3H), 6.93 (dd, $J = 8.0$, 1.2 Hz, 1H), 6.87 (dd, $J = 8.0$, 1.2 Hz, 1H), 6.76 (td, J

= 8.0, 1.2 Hz, 1H), 6.70 (td, J = 8.0, 1.2 Hz, 1H), 6.66 (td, J = 8.0, 1.2 Hz, 1H), 6.56 (dd, J = 8.0, 1.2 Hz, 1H), 3.32 (s, 3H), 3.27 (s, 3H), 3.21 (s, 3H); ^{13}C NMR (CDCl$_3$, 101 MHz) d 190.48, 187.76, 186.63, 148.74, 148.62, 147.84, 147.15, 146.74, 146.65, 146.54, 146.50, 145.36, 143.06, 143.03, 142.79, 138.90, 138.51, 135.82, 129.18, 128.80, 125.38, 125.27, 125.15, 121.32, 120.91, 120.86, 117.12, 117.09, 116.77, 116.35, 116.18, 115.90, 115.22, 115.12, 114.56, 110.01, 33.66, 33.61, 32.97; HRMS (m/z, ESI$^+$) Calcd for C$_{39}$H$_{31}$IrN$_9$ 818.2332 [M + H$^+$], found: 818.2313.

Device fabrication and characterization. PHOLEDs were grown on pre-cleaned glass substrates coated with 80-nm-thick indium tin oxide (ITO) by vacuum thermal evaporation in a chamber with a base pressure 6 × 10^{-7} torr. The devices consist of 10 nm MoO$_3$ doped at 15 vol% in 9-(4-tert-butylphenyl)-3,6-bis(triphenylsilyl)-9H-carbazole (CzSi) as a hole injection layer (HIL)/5 nm CzSi hole-transport layer (HTL)/5 nm Ir(C^C:)$_3$-based electron and exciton blocking layer (EBL)/40 nm Ir(pmp)$_3$ doped in TSPO1 to form the EML/5 nm TSPO1 hole and exciton blocking layer (HBL)/30 nm 1,3,5-tris(1-phenyl-1H-benzimidazol-2-yl)benzene (TPBi) electron-transport layer (ETL)/1.5 nm 8-hydroxyquinolinolato-Li electron injection layer (EIL)/100 nm Al (cathode). The devices were patterned using a shadow mask with an array of circular openings, resulting in contacts with a measured diameter of d = 430 μm. The standard deviation for a population of more than 20 devices leads to a variation in area of ∼2%. The EBLs used for the *fac*- and *mer*-Ir(pmp)$_3$-based devices were *fac*-Ir(pmp)$_3$ and *fac*-Ir(pmb)$_3$, respectively (see Fig. 48.1a). The current density–voltage–luminance (J–V–L) characteristics were measured using a parameter analyser (HP4145, Hewlett–Packard) and a calibrated photodiode (FDS1010-CAL, Thorlab) according to standard procedures [43]. The emission spectra at J = 10 mA cm^{-2} were recorded using a calibrated spectrometer (USB4000, Ocean Optics) coupled to the device with an optical fibre.

Probing the recombination zone. The exciton density, $N(x)$, in the EML as a function of distance, x, from the EBL/EML interface was determined by measuring the relative emission intensity from a 1.5-nm-thick 'sensing' layer comprised of 5 vol% doped red-emitting phosphor—that is, iridium (III) bis (2-phenyl quinolyl-N,C$^{2'}$) acetylacetonate (PQIr), inserted at different positions within the EML, as previously [17].

EQE modelling. Based on triplet–triplet and triplet–polaron annihilation dynamics, EQE versus J is modelled using:

$$\frac{dn(x, t)}{dt} = G(x) - \gamma n^2(xt) \quad (48.1)$$

$$\frac{dn(x, t)}{dt} = G(x) - \frac{N(x, t)}{\tau} - k_{TP} n(x, t) - \frac{1}{2} k_{TT} N^2(xt) \quad (48.2)$$

where n is the electron density and N is the triplet density, $\gamma = q(\mu_n + \mu_p)/\varepsilon\varepsilon_0$ is the Langevin recombination rate, and μ_n and μ_p are the respective electron and hole mobilities in the doped EML. $\varepsilon = 3$ is the dielectric constant, q is the charge, x is position, t is time, G is the charge generation rate, and k_{TT} and k_{TP} are the triplet–triplet and triplet–polaron annihilation rates, respectively. Here, $G(x) = J/q \cdot R(x)$, where R is the measured recombination profile in Fig. 48.4b, different from previous assumptions based on the constant recombination zone width [16, 18]. Equation (48.1) assumes that $p = n$ in the EML, and τ is obtained from the transient phosphorescence decay of Ir(pmp)$_3$ doped into TSPO1. $N(x)$ in Eq. (48.2) is obtained in steady state and integrated over the EML to obtain N. Then, N/J is normalized by the EQE at $J = 0.1$ mA cm^{-2} and fitted to $J = 100$ mA cm^{-2} (Fig. 48.5d, lines), where bimolecular quenching is active.

Supplementary Information

The supplementary information is available at: https://static-content.springer.com/esm/art%3A10.1038%2Fnmat4446/MediaObjects/41563_2016_BFnmat4446_MOESM6_ESM.pdf.

Acknowledgements

This work was supported by the Air Force Office of Scientific Research (AFOSR) and Universal Display Corporation.

Author contributions

J.L. designed, fabricated and optimized the PHOLEDs, and analysed the optical and electrical properties of materials with S.R.F., M.E.T. and P.I.D. H.-F.C., T.B. and M.E.T. synthesized and measured the photophysical and electrochemical properties of materials. C.C. provided EQE roll-off theory and modelling. S.R.F. supervised the project, analysed data, and wrote the manuscript with J.L.

Competing financial interests

M.E.T. and S.R.F. have an equity interest in one of the sponsors of this work (Universal Display Corp.).

References

1. Baldo, M. A. *et al.* Highly efficient phosphorescent emission from organic electroluminescent devices. *Nature* **395,** 151–154 (1998).
2. Baldo, M. A., Lamansky, S., Burrows, P. E., Thompson, M. E. & Forrest, S. R. Very high-efficiency green organic light-emitting devices based on electrophosphorescence. *Appl. Phys. Lett.* **75,** 4–6 (1999).
3. Giebink, N. C. *et al.* Intrinsic luminance loss in phosphorescent small-molecule organic light emitting devices due to bimolecular annihilation reactions. *J. Appl. Phys.* **103,** 044509 (2008).
4. Sajoto, T. *et al.* Blue and near-UV phosphorescence from iridium complexes with cyclometalated pyrazolyl or N-heterocyclic carbene ligands. *Inorg. Chem.* **44,** 7992–8003 (2005).
5. Schildknecht, C. *et al.* Novel deep-blue emitting phosphorescent emitter. *Proc. SPIE Int. Soc. Opt. Eng.* **5937,** 59370E (2005).
6. Hopkinson, M. N., Richter, C., Schedler, M. & Glorius, F. An overview of N-heterocyclic carbenes. *Nature* **510,** 485–496 (2014).

7. Chang, C.-F. *et al.* Highly efficient blue-emitting iridium(III) carbene complexes and phosphorescent OLEDs. *Angew. Chem. Int. Ed.* **47,** 4542–4545 (2008).
8. Chiu, Y.-C. *et al.* En route to high external quantum efficiency (∼12%), organic true-blue-light-emitting diodes employing novel design of iridium(III) phosphors. *Adv. Mater.* **21,** 2221–2225 (2009).
9. Sasabe, H. *et al.* High-efficiency blue and white organic light-emitting devices incorporating a blue iridium carbene complex. *Adv. Mater.* **22,** 5003–5007 (2010).
10. Hsieh, C.-H. *et al.* Design and synthesis of iridium bis(carbene) complexes for efficient blue electrophosphorescence. *Chem. Eur. J.* **17,** 9180–9187 (2011).
11. Lu, K.-Y. *et al.* Wide-Range color tuning of iridium biscarbene complexes from blue to red by different N^N ligands: An alternative route for adjusting the emission colors. *Adv. Mater.* **23,** 4933–4937 (2011).
12. Fleetham, T., Wang, Z. & Li, J. Efficient deep blue electrophosphorescent devices based on platinum(II) bis(n-methyl-imidazolyl)benzene chloride. *Org. Electron.* **13,** 1430–1435 (2012).
13. Lee, S. *et al.* Deep-blue phosphorescence from perfluoro carbonyl-substituted iridium complexes. *J. Am. Chem. Soc.* **135,** 14321–14328 (2013).
14. Hang, X.-C., Fleetham, T., Turner, E., Brooks, J. & Li, J. Highly efficient blue-emitting cyclometalated platinum(II) complexes by judicious molecular design. *Angew. Chem. Int. Ed.* **52,** 6753–6756 (2013).
15. Fleetham, T., Li, G., Wen, L. & Li, J. Efficient 'Pure' blue OLEDs employing tetradentate Pt complexes with a narrow spectral bandwidth. *Adv. Mater.* **26,** 7116–7121 (2014).
16. Baldo, M. A., Adachi, C. & Forrest, S. R. Transient analysis of organic electrophosphorescence. II. Transient analysis of triplet–triplet annihilation. *Phys. Rev. B* **62,** 10967–10977 (2000).
17. Zhang, Y., Lee, J. & Forrest, S. R. Tenfold increase in the lifetime of blue phosphorescent organic light-emitting diodes. *Nature Commun.* **5,** 5008 (2014).
18. Erickson, N. C. & Holmes, R. J. Engineering efficiency roll-off in organic light-emitting devices. *Adv. Funct. Mater.* **24,** 6074–6080 (2014).
19. Park, M. S. & Lee, J. Y. Indolo acridine-based hole-transport materials for phosphorescent OLEDs with over 20% external quantum efficiency in deep blue and green. *Chem. Mater.* **23,** 4338–4343 (2011).

20. Holmes, R. J. et al. Saturated deep blue organic electrophosphorescence using a fluorine-free emitter. *Appl. Phys. Lett.* **87,** 243507 (2005).
21. Deaton, J. C., Young, R. H., Lenhard, J. R., Rajeswaran, M. & Huo, S. Photophysical properties of the series *fac*- and mer-(1-Phenylisoquinolinato-N^C$^{2'}$)$_x$(2-phenylpyridinato-N^C$^{2'}$)$_{3-x}$ iridium(III) ($x = 1$–3). *Inorg. Chem.* **49,** 9151–9161 (2010).
22. Tamayo, A. B. et al. Synthesis and characterization of facial and meridional tris-cyclometalated iridium(III) complexes. *J. Am. Chem. Soc.* **125,** 7377–7387 (2003).
23. Sajoto, T. et al. Temperature dependence of blue phosphorescent cyclometalated Ir(III) complexes. *J. Am. Chem. Soc.* **131,** 9813–9822 (2009).
24. Lauerhaas, J. M. et al. Reversible luminescence quenching of porous silicon by solvents. *J. Am. Chem. Soc.* **114,** 1911–1912 (1992).
25. Xiao, L. et al. Recent progresses on materials for electrophosphorescent organic light-emitting devices. *Adv. Mater.* **23,** 926–952 (2011).
26. Yook, K. S. & Lee, J. Y. Organic materials for deep blue phosphorescent organic light-emitting diodes. *Adv. Mater.* **24,** 3169–3190 (2012).
27. Djurovich, P. I., Mayo, E. I., Forrest, S. R. & Thompson, M. E. Measurement of the lowest unoccupied molecular orbital energies of molecular organic semiconductors. *Org. Electron.* **10,** 515–520 (2009).
28. Jeon, S. O., Jang, S. E., Son, H. S. & Lee, J. Y. External quantum efficiency above 20% in deep blue phosphorescent organic light-emitting diodes. *Adv. Mater.* **23,** 1436–1441 (2011).
29. Reineke, S., Walzer, K. & Leo, K. Triplet-exciton quenching in organic phosphorescent light-emitting diodes with Ir-based emitters. *Phys. Rev. B* **75,** 125328 (2007).
30. Su, S.-J., Gonmori, E., Sasabe, H. & Kido, J. Highly efficient organic blue- and white-light-emitting devices having a carrier- and exciton-confining structure for reduced efficiency roll-off. *Adv. Mater.* **20,** 4189–4194 (2008).
31. Seino, Y., Sasabe, H., Pu, Y.-J. & Kido, J. High-performance blue phosphorescent OLEDs using energy transfer from exciplex. *Adv. Mater.* **26,** 1612–1616 (2014).
32. Zhang, Q. et al. Efficient blue organic light-emitting diodes employing thermally activated delayed fluorescence. *Nature Photon.* **8,** 326–332 (2014).

33. Yang, C.-H. *et al.* Deep-blue-emitting heteroleptic iridium(III) complexes suited for highly efficient phosphorescent OLEDs. *Chem. Mater.* **24,** 3684–3695 (2012).
34. You, Y. & Nam, W. Photofunctional triplet excited states of cyclometalated Ir(III) complexes: Beyond electroluminescence. *Chem. Soc. Rev.* **41,** 7061–7084 (2012).
35. Tsuchiya, K., Ito, E., Yagai, S., Kitamura, A. & Karatsu, T. Chirality in the photochemical mer→fac geometrical isomerization of tris(1-phenylpyrazolato,$N,C^{2\prime}$)iridium(III). *Eur. J. Inorg. Chem.* **2009,** 2104–2109 (2009).
36. Kober, E. M., Sullivan, B. P. & Meyer, T. J. Solvent dependence of metal-to-ligand charge-transfer transitions. Evidence for initial electron localization in MLCT excited states of 2,21-bipyridine complexes of ruthenium(II) and osmium(II). *Inorg. Chem.* **23,** 2098–2104 (1984).
37. Reineke, S., Rosenow, T. C., Lüssem, B. & Leo, K. Improved high-brightness efficiency of phosphorescent organic LEDs comprising emitter molecules with small permanent dipole moments. *Adv. Mater.* **22,** 3189–3193 (2010).
38. Padmaperuma, A. B., Sapochak, L. S. & Burrows, P. E. New charge transporting host material for short wavelength organic electrophosphorescence: 2,7-bis(diphenylphosphine oxide)-9,9-dimethylfluorene. *Chem. Mater.* **18,** 2389–2396 (2006).
39. Calcagno, P. *et al.* Understanding the structural properties of a homologous series of bis-diphenylphosphine oxides. *Chem. Eur. J.* **6,** 2338–2349 (2000).
40. Kawamura, Y. *et al.* 100% phosphorescence quantum efficiency of Ir(III) complexes in organic semiconductor films. *Appl. Phys. Lett.* **86,** 071104 (2005).
41. Erk, P. *et al.* 11.2: Efficient deep blue triplet emitters for OLEDs. *SID Symp. Dig. Tech. Pap.* **37,** 131–134 (2006).
42. Shi, F., Xu, X., Zheng, L., Dang, Q. & Bai, X. Method development for a pyridobenzodiazepine library with multiple diversification points. *J. Comb. Chem.* **10,** 158–161 (2008).
43. Forrest, S. R., Bradley, D. D. C. & Thompson, M. E. Measuring the efficiency of organic light-emitting devices. *Adv. Mater.* **15,** 1043–1048 (2003).

Chapter 49

Hot Excited State Management for Long-Lived Blue Phosphorescent Organic Light-Emitting Diodes

Jaesang Lee,[a] Changyeong Jeong,[a] Thilini Batagoda,[b] Caleb Coburn,[c] Mark E. Thompson,[b] and Stephen R. Forrest[a,c,d]

[a]*Department of Electrical and Computer Engineering, University of Michigan, Ann Arbor, Michigan 48109, USA*
[b]*Department of Chemistry, University of Southern California, Los Angeles, California 90089, USA*
[c]*Department of Physics, University of Michigan, Ann Arbor, Michigan 48109, USA*
[d]*Department of Materials Science and Engineering, University of Michigan, Ann Arbor, Michigan 48109, USA*
stevefor@umich.edu

Since their introduction over 15 years ago, the operational lifetime of blue phosphorescent organic light-emitting diodes (PHOLEDs) has remained insufficient for their practical use in displays and lighting. Their short lifetime results from annihilation between high-energy excited states, producing energetically hot states (>6.0 eV) that lead to molecular dissociation. Here we introduce a strategy to avoid dissociative reactions by including a molecular hot excited

---

Reprinted from *Nat. Commun.*, **8**, 15566, 2017.

Electrophosphorescent Materials and Devices
Edited by Mark E. Thompson
Text Copyright © 2017, The Author(s)
Layout Copyright © 2024 Jenny Stanford Publishing Pte. Ltd.
ISBN 978-981-4877-34-3 (Hardcover), 978-1-003-08872-1 (eBook)
www.jennystanford.com

state manager within the device emission layer. Hot excited states transfer to the manager and rapidly thermalize before damage is induced on the dopant or host. As a consequence, the managed blue PHOLED attains T80 = 334 ± 5 h (time to 80% of the 1,000 cd m^{-2} initial luminance) with a chromaticity coordinate of (0.16, 0.31), corresponding to 3.6 ± 0.1 times improvement in a lifetime compared to conventional, unmanaged devices. To our knowledge, this significant improvement results in the longest lifetime for such a blue PHOLED.

49.1 Introduction

Organic light-emitting diodes (OLEDs) are an important technology for attractive, high efficiency displays and lighting. New applications enabled by OLEDs include flexible [1], wearable [2], transparent [3] and high-resolution displays [4], as well as efficient and high intensity illumination [5]. The primary impediment to large-scale commercialization of OLEDs, however, is the short operational lifetime of blue-emitting devices [6]. Red and green OLEDs are almost universally based on electrophosphorescent emission due to their 100% internal quantum efficiency (IQE) [7, 8] and operational lifetimes of T95 >10,000 h which are sufficient for most display and lighting applications [6]. (Here, TX is the time elapsed for the luminance to decrease to X % of its initial value of L_0 = 1,000 cd m^{-2} under constant current operation.)

In contrast, the realization of long-lived blue electrophosphorescent OLEDs (PHOLEDs) has not been achieved since its first demonstration in 2001 (Ref. [9]). Surprisingly, T80 of the blue PHOLEDs with 1931 Commission Internationale de l'Eclairage (CIE) chromaticity coordinates of $y < 0.4$ are <10 h [10, 11]. Even for greenish-blue devices with $y \geq 0.4$, T80 is <160 h, which is still too small for practical use [12–16]. This has led to the use of significantly less efficient fluorescent OLEDs for blue emission. Even so, the lifetime of blue fluorescent OLEDs is insufficient for many applications [17] and are at least ten times less than state-of-the-art red and green PHOLEDs [18]. In the same vein, the lifetime of green thermally assisted delayed fluorescent OLEDs is only T95 = 1,300 h [19], and considerably less for blue.

The short operational lifetime of blue PHOLEDs has been convincingly attributed to annihilation between excited states (that is, exciton–exciton or exciton–polaron) in the device emission layer (EML) [20–22] that result in a hot (that is, multiply excited) exciton or polaron while the remaining state nonradiatively transitions to the ground state. This process is analogous to Auger recombination in inorganic semiconductor light-emitting diodes and lasers that also has been found to adversely affect device performance [23]. The hot state in the PHOLED EML can attain up to double the energy of the initial excited state (≥ 6.0 eV). Thus, there is a possibility that their dissipation on dopant or host molecules can induce chemical bond dissociation [24]. Indeed, there are no bonds in organic molecules used in OLEDs that can tolerate the concentration of such a high energy without inducing molecular dissociation. The probability of this reaction increases with twice the excited state energy, and hence is particularly dominant for blue PHOLEDs compared with red and green-emitting analogues.

The key to realizing long-lived blue PHOLEDs is, therefore, to prevent the hot state energies from ever leading to molecular dissociation reactions in the first place. This can be accomplished by reducing bimolecular annihilations, or by bypassing the dissociative processes altogether. In this work, we demonstrate a strategy to thermalize the hot states without damaging the dopant or host molecules. For that purpose, we add an ancillary, protective dopant called an excited state manager into the EML whose triplet exciton energy is higher than the emitting triplets on the dopant. Thus, the manager does not trap excited states, but rather it efficiently returns them to the dopant where they can emit light. Further, by providing exothermic energy transfer pathways from the hot states to the manager, the probability of direct dissociation of the active materials comprising the EML are reduced, or possibly eliminated.

By locating the manager dopant in the region where the triplet excitons have the highest density and thus bimolecular annihilation is most probable, the longest-lived managed blue PHOLEDs achieve T80 = 334 $\pm$ 5h at L_0 = 1,000 cd m^{-2} with CIE coordinates of (0.16, 0.30), which is a 3.6 $\pm$ 0.1 and 1.9 $\pm$ 0.1 times improved lifetime compared to conventional and graded-EML devices [25] of T80 = 93 ± 9 and 173 ± 3 h, respectively. Indeed, the lifetime of managed blue PHOLEDs is at least 30 times longer than previously reported blue PHOLEDs with similar colour coordinates [10, 11]. Our strategy contrasts with previous methods that have employed third components [19, 26, 27], but none of which directly address the siphoning of energy from the most vulnerable constituents of blue PHOLEDs; that is, the dopant and host molecules in the EML. Based on our results, we provide the selection criteria for ideal manager molecules that can enable further improvement in the stability of blue PHOLEDs.

49.2 Results

49.2.1 Hot Excited State Management to Extend PHOLED Lifetime

Figure 49.1a shows the Jablonski diagram of an EML containing an excited state manager and the possible relaxation pathways for excitons. The manager can enable the transfer of the hot singlet/triplet state (S_n^*/T_n^*, where $n \gg 1$) resulting from triplet–triplet annihilation (TTA, process 2) to the lowest excited state of the manager (S_M/T_M) via process 3'. The hot state can be either an exciton or polaron state resulting from either TTA or triplet–polaron annihilation, respectively [20].

Figure 49.1b shows the calculated energy levels of S_n^*/T_n^* for EML molecules used in this work (see below). When TTA occurs between one or more molecular species in the EML [28], either the singlet or triplet state is promoted to S_n^*/T_n^* >5.4 eV. While most hot states rapidly relax to the lowest excited states (S_1/T_1, process 2'), those that have sufficient energy can lead to the chemical bond dissociation via $S_n^*/T_n^* \rightarrow D$ (process 3), where D represents dissociative states. Dissociation requires energy in excess of the bond dissociation energy of the excited molecule. For example, bond dissociation energies of weak bonds in the host ($D_{1,\text{mCBP}}$ and $D_{2,\text{mCBP}}$, Fig. 49.1b and Supplementary Note 1) are at 3.5–5 eV above the

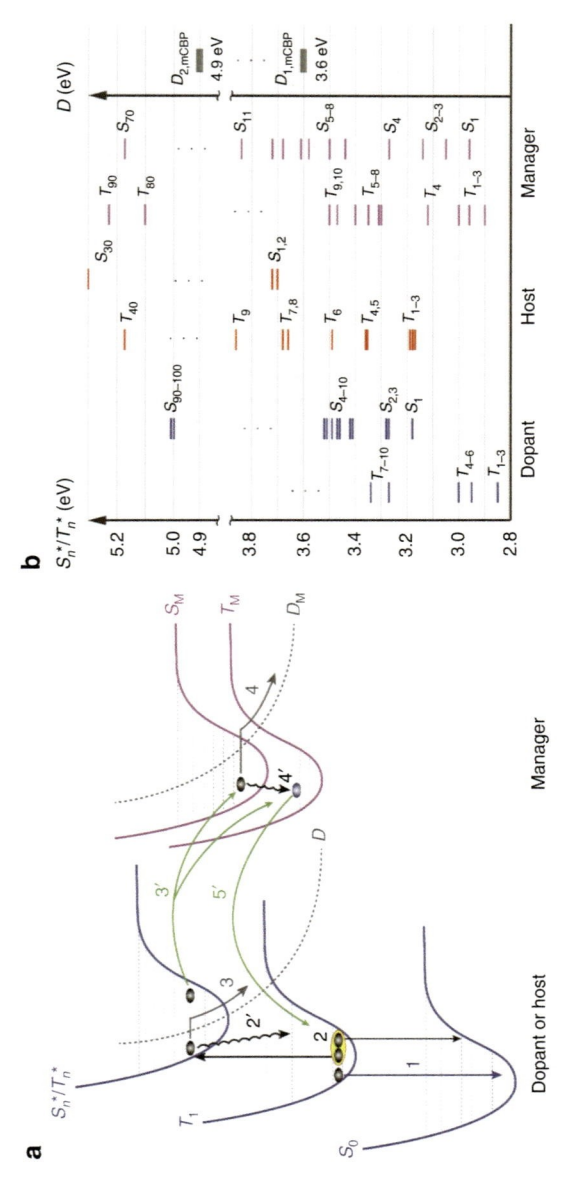

Figure 49.1 Energetics of the excited states in the PHOLED EML. (a) Jablonski diagram of the EML containing the manager. Here, S_0 is the ground state, T_1 is the lowest energy triplet state and S^*/T^* is a hot singlet/triplet manifold of the dopant or host. D represents the dissociative state via the predissociative potential of the manager. S_M/T_M is the lowest singlet/triplet state of the manager. Possible energy-transfer pathways are numbered as follows: (1) radiative recombination, (2) TTA resulting in excitation to S^*/T^*, (2′) internal conversion and vibrational relaxation, (3) and (4) dissociative reactions leading to molecular dissociation, (3′) exothermic Förster energy transfer for singlet-to-singlet transitions, and (3)′ and (5′) Dexter energy transfer for triplet-to-triplet transitions, and (4′) intersystem crossing and vibrational relaxation. (b) Calculated energies of exciton states for the molecules in the EML (dopant, host and manager) and a few of dissociative states for mCBP used as a host.

ground state. TTA can readily supply this energy, while that of the lowest triplet (T_1) is insufficient to induce the dissociation reactions.

By introducing a manager whose energy S_M/T_M is greater than that of the dopant, excitons formed on, or transferred to the manager can be returned to the dopant for emission. Also, exothermic transfer from S_n^*/T_n^* to S_M/T_M is allowed, and damage to these molecules via dissociative reactions (process 3) is reduced provided that the rate for $S_n^*/T_n^* \rightarrow S_M/T_M$ is comparable or higher than $S_n^*/T_n^* \rightarrow D$. Since TTA can yield both hot singlets and triplets [29], the hot state resonantly transfers via a Förster or Dexter process to the manager via process 3. A transferred singlet undergoes vibrational relaxation and Förster transfer back to the lowest dopant singlet state, provided that the manager molecule has a high photoluminescence quantum yield [30]. Alternatively, the thermalized singlet state intersystem crosses to the triplet state ($S_M \rightarrow T_M$ via process 4'), which subsequently transfers back to the dopant or host ($T_M \rightarrow T_1$) via process 5'. This leads to radiative recombination (process 1), or is recycled back to S_n^*/T_n^* by a repeat process. It is also possible that the high energy S_M/T_M state can result in dissociation of the manager itself via $S_M/T_M \rightarrow D_M$ (process 4), that is, where the manager serves as a sacrificial additive to the EML. Process 4 is not optimal since the number of effective managers decreases over time, providing less protection for the host and dopant as the device ages. Even in this case, however, the manager can still increase device stability.

From the foregoing discussion, three primary criteria must be met for effective molecular design of the manager: First, the exciton energy of the manager should be higher than lowest exciton states (S_1/T_1) of the dopant; second, the rate of transfer to the manager (process 3') must be comparable to or higher than that for dissociation (process 3); and third, the manager should be sufficiently stable such that it does not degrade on a time scale short compared to that of the unmanaged device (process 4).

We introduce meridional-tris-(N-phenyl, N-methyl-pyridoimidazol-2-yl)iridium (III) [mer-Ir(pmp)3] as the manager in the PHOLED EML. The EML also consists of the blue dopant, iridium (III) tris[3-(2,6-dimethylphenyl)-7-methylimidazo[1,2-f] phenanthridine] [Ir(dmp)3] and the host, 4,4'-bis[N-(1-naphthyl)-N-phenyl-amino]-biphenyl (mCBP). Figure 49.2a shows molecular formulae of Ir(dmp)$_3$ and *mer*-Ir(pmp)$_3$. The manager is characterized by a relatively strong metal–ligand bond [30] and a glass transition temperature of 136°C. The triplet energy of *mer*-Ir(pmp)$_3$ is 2.8 eV calculated from its peak phosphorescence spectrum ($\lambda = 454$ nm), while its onset is at $E_{T1} \approx 3.1$ eV, higher than that of the dopant of $E_{T1} \approx 2.8$ eV (Fig. 49.3a). Thus, *mer*-Ir(pmp)$_3$ fulfills criterion (i), although both criteria (ii) and (iii) are possibly not met by this molecule. Hence, these complexes have not been optimized for rapid transfer via process 3'. This is a function of the intimate orbital overlap between manager and dopant or host; a property controlled by the steric and orbital characteristics of all molecules involved. Nor is *mer*-Ir(pmp)$_3$ particularly stable, which can lead to manager depletion with time (process 4). In spite of these shortcomings, we find significant lifetime improvements for blue PHOLEDs using this manager molecule.

49.2.2 Performance of Managed PHOLEDs

Figure 49.2b shows the energy level diagram of the managed devices. The lower energy (>1 eV) of the HOMO of the dopant compared with that of the host suggests that hole transfer is predominantly supported by the dopant molecules and only slightly by the manager, while electrons are transported by both the host and the manager having nearly identical lowest unoccupied molecular orbital (LUMO) energies (Supplementary Fig. 2). The EML doping schemes of the control and managed PHOLEDs are given in Fig. 49.2c (denoted as GRAD and M0, respectively; see Methods). For GRAD, the concentration of the dopant is linearly graded from 18 to 8 vol% from the hole transport layer (HTL) to the electron transport layer interfaces to enable a uniform distribution of excitons and polarons throughout the EML. This structure was previously shown [25] to reduce bimolecular annihilation, and thereby achieve an extended lifetime compared to conventional, non-graded-EML devices (denoted CONV; see Methods). In device M0, 3 vol% of the manager is uniformly doped across the EML, and the concentration of the dopant is graded from 15 to 5 vol%. To investigate the lifetime dependence on the manager position,

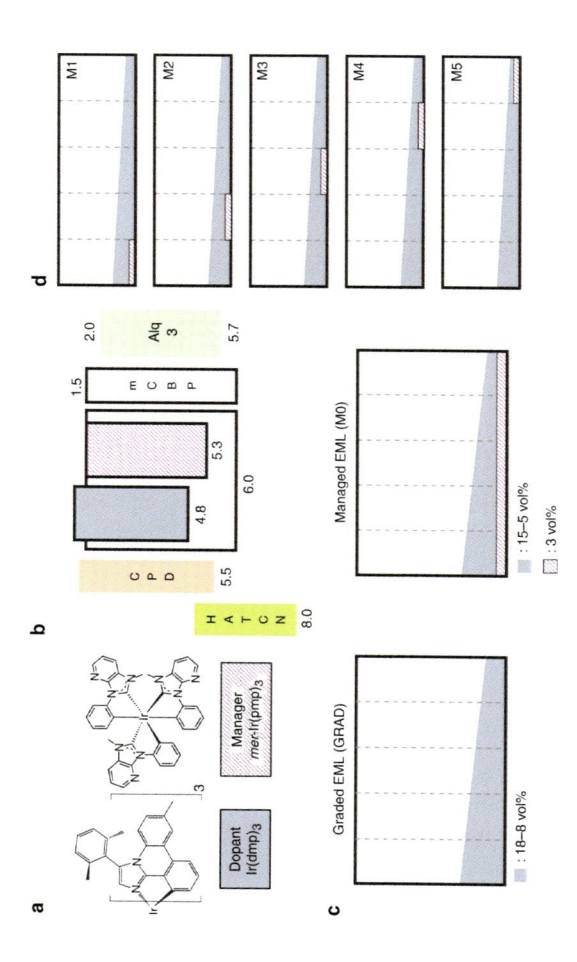

Figure 49.2 Energy and doping schemes of the PHOLEDs. (a) Molecular formulae of Ir(dmp)$_3$ and *mer*-Ir(pmp)$_3$, used for the dopant and the manager, respectively. (b) Energy level diagram of the PHOLED with the manager, denoted 'managed PHOLED'. Numbers in the figure are energies referred to the vacuum level. (c) Doping scheme of the 50 nm-thick EML for the graded-EML and managed PHOLEDs, denoted as GRAD and M0, respectively. GRAD has the dopant graded from 18 to 8 vol% in the mCBP host, while M0 is a similarly graded device but with the 3 vol% of the manager replacing the dopant of the same amount, compared to GRAD, to keep the total doping concentration the same for both devices. (d) Managed PHOLEDs M1–M5 have selectively doped 10 nm-thick zones of the EML. The zones have a manager doping of 3 vol% substituting the dopant of the same amount. The other details of the EML are identical to that of GRAD.

the manager is doped at 3 vol% into 10 nm-thick zones at various locations within the 50 nm-thick EML of devices M1–M5, shown in Fig. 49.2d. Except for the zone with the manager, the remainder of the EMLs for M1–M5 are identical to that of GRAD, keeping the total doping concentrations of all devices the same.

Figure 49.3a shows the electroluminescence (EL) spectra of GRAD, M0, M3 and M5 measured at a current density of $J = 5$ mA cm^{-2}. The GRAD and managed PHOLEDs exhibit nearly identical EL spectra with CIE chromaticity coordinates of (0.16, 0.30). This confirms that radiative recombination in managed devices occurs solely on the dopant, while triplets formed on the manager efficiently transfer back to the dopant via process 5′ in Fig. 49.1.

Figure 49.3b,c shows the current density–voltage (J–V) and external quantum efficiency (EQE)–J characteristics of GRAD, M0, M3 and M5. Table 49.1 summarizes properties of their EL characteristics at $L_0 = 1{,}000$ cd m^{-2}. The initial operating voltages (V_0) of the managed PHOLEDs (M0–M5) are higher than GRAD by ∼1 V and the voltage at $J = 5$ mA cm^{-2} shows a similar trend. This is due to a reduced fraction of the dopant in managed PHOLED EMLs compared to that of GRAD, and due to the manager acting as a hole trap with its HOMO energy of 5.3 ± 0.1 versus 4.8 ± 0.1 eV for the dopant. For example, when a small concentration (<5 vol%) of the manager is added as a substitute of the same amount for the dopant, the device resistance marginally increases (Supplementary Note 2). The EQE for all devices is 9–10%, consistent with the PLQY of the dopant of $44 \pm 1\%$ when doped in mCBP at 13 vol%. The EQE of the managed PHOLEDs at $L_0 = 1{,}000$ cd m^{-2} is slightly (<1.0%) higher than that of GRAD, leading to the maximum difference in drive current density of $J_0 < 0.6$ mA cm^{-2} needed to achieve the same L_0.

Figure 49.4a shows the time evolution of the increase in operating voltage, $\Delta V(t) = V(t) - V_0$, and normalized luminance loss, $L(t)/L_0$ ($L_0 = 1{,}000$ cd m^{-2}) of CONV, GRAD, M0 and M3 under constant current. Table 49.1 includes the lifetime characteristics (T90, T80 and $\Delta V(t)$) for all the managed PHOLEDs. Managed PHOLEDs M0–M5 have increased T90 and T80 relative to those of GRAD. For example, the longest-lived device M3 attains T90 = 141 $\pm$ 11 h and T80 = 334 $\pm$ 5 h, corresponding to a 3.0 $\pm$ 0.1 and 1.9 $\pm$ 0.1 times improvement from those of GRAD and a 5.2 $\pm$ 0.2 and

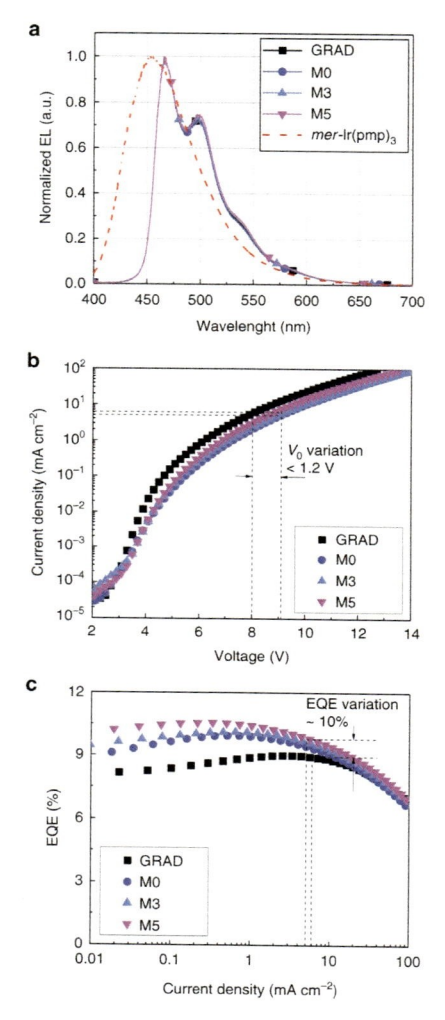

Figure 49.3 Performance of the PHOLEDs. (a), Normalized electroluminescent (EL) spectra of the GRAD and managed PHOLEDs, M0, M3 and M5, measured at a current density of $J_0 = 5$ mA cm^{-2}. For comparison, the PL spectrum of the manager [mer-Ir(pmp)$_3$] is also shown. (b) Current density–voltage. (c) External quantum efficiency (EQE)–current density characteristics of GRAD and selected managed PHOLEDs. Note that between GRAD and the managed PHOLEDs, the absolute difference of the operating voltages (V_0) and EQE at an initial luminance of $L_0 = 1,000$ cd m^{-2} for the lifetime test are <1.2 V and 1.0%, respectively.

Table 49.1 Electroluminescent and lifetime characteristics for CONV, GRAD and managed PHOLEDs (M0-M5) at $L_0 = 1{,}000$ cd m^{-2}

Device	J_0 (mA cm^{-2})	EQE (%)	V_0 (V)	CIE*	T90 (h)	T80 (h)	ΔV(T90) (V)	ΔV(T80) (V)
CONV	6.7 ± 0.1	8.0 ± 0.1	6.6 ± 0.0	[0.15, 0.28]	27 ± 4	93 ± 9	0.3 ± 0.1	0.4 ± 0.1
GRAD	5.7 ± 0.1	8.9 ± 0.1	8.0 ± 0.0	[0.16, 0.30]	47 ± 1	173 ± 3	0.6 ± 0.1	0.9 ± 0.1
M0	5.5 ± 0.1	9.4 ± 0.1	9.2 ± 0.0	[0.16, 0.30]	71 ± 1	226 ± 9	0.9 ± 0.1	1.2 ± 0.1
M1	5.4 ± 0.1	9.5 ± 0.1	8.8 ± 0.1	[0.16, 0.29]	99 ± 3	260 ± 15	1.2 ± 0.1	1.6 ± 0.1
M2	5.4 ± 0.1	9.3 ± 0.0	8.9 ± 0.1	[0.16, 0.31]	103 ± 0	285 ± 8	0.7 ± 0.1	1.0 ± 0.1
M3	5.3 ± 0.1	9.6 ± 0.0	9.0 ± 0.1	[0.16, 0.30]	141 ± 11	334 ± 5	1.1 ± 0.1	1.5 ± 0.2
M4	5.2 ± 0.1	9.6 ± 0.2	8.6 ± 0.0	[0.16, 0.31]	126 ± 7	294 ± 16	1.0 ± 0.1	1.3 ± 0.1
M5	5.1 ± 0.1	9.9 ± 0.1	8.6 ± 0.0	[0.16, 0.31]	119 ± 6	306 ± 3	0.9 ± 0.1	1.2 ± 0.1

EQE, external quantum efficiency.
*Measured at current density of $J = 5$ mA cm^{-2}.
Errors for the measured values are s.d. from at least three devices.

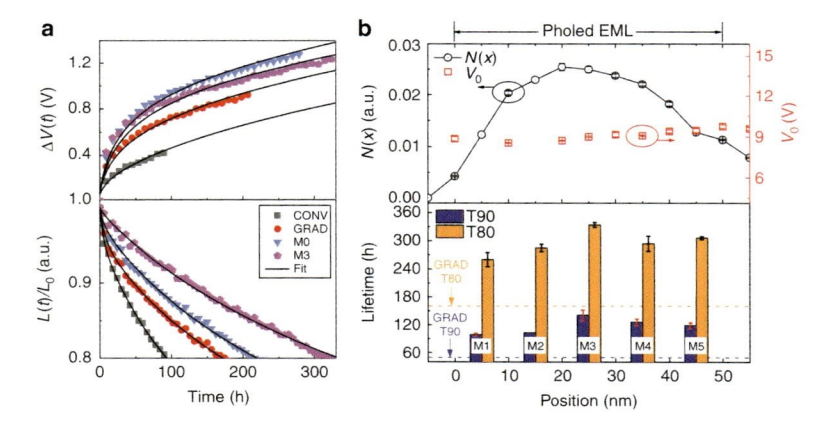

Figure 49.4 Lifetime and modelling of the PHOLEDs. (a) Lifetime characteristics of CONV, GRAD, managed PHOLEDs M0 and M3. Top and bottom show the time evolution of the operating voltage change, $\Delta V(t) = V(t) - V0$, and the normalized luminance degradation, $L(t)/L_0$, respectively. Solid lines are fits based on the model in Methods (see fitting parameters in Table 49.2). (b) (Top) Exciton density profile, $N(x)$, of the PHOLED emission layer (EML) as a function of position, x, and operating voltages of the devices using delta-doped sensing layer at $J = 5$ mA cm^{-2} (Supplementary Note 3). The origin of the x-axis is at the HTL/EML interface. The operating current density results in a luminance of $L_0 = 1{,}000$ cd m^{-2}. (Bottom) Lifetimes (T90 and T80) of managed devices (M1–M5) as functions of the position of the managed EML zones. T90 and T80 of the managed devices are compared with those of the GRAD (dotted lines). Note that the variation in lifetime qualitatively follows the exciton density profile, suggesting that placing the manager at the point of highest exciton density results in the longest device lifetime. Error bars represent 1 s.d. for at least three devices.

3.6 ± 0.1 times improvement compared with CONV, respectively. Here, T90 and T80 are used to determine the short- and long-term effectiveness of the excited state management.

The upper panel of Fig. 49.4b shows the measured triplet density profile, $N(x)$, in the GRAD EML at $J = 5$ mA cm^{-2}, where x is the distance from the EML/HTL interface (see Methods, Supplementary Note 3). The T90 and T80 of M1–M5 versus manager position in the EML are given in the lower panel of Fig. 49.4b. Note that the variation in lifetime qualitatively follows the exciton density profile. For example, M3 includes the manager at 20 nm $< x <$ 30 nm, which is at the point of highest exciton density relative to those of other managed devices. Hence, the effectiveness of the manager at this position should be largest, as is indeed observed. Finally, the change in operating voltage, $\Delta V(t)$, required to maintain a constant current is larger for M0–M5 than that of GRAD, while their rate of luminance degradation is reduced. This suggests the formation of polaron traps that have no effect on the luminance.

49.3 Discussion

The degraded molecular products (or defects) can be formed in any and all layers of aged PHOLEDs, but those located in the EML play a dominant role in affecting the device luminance. On the other hand, changes in the operating voltage can arise from defects generated both within and outside the EML. To model the time evolution of the device performance, we consider that two types of charge traps, A and B, with volume densities of Q_A and Q_B, respectively, are generated by the hot states within the EML. Using thermally stimulated current measurements, we observe an increase in the rate of generation of charge traps and a decreased density of the original transport sites compared to unmanaged devices (Supplementary Note 4). When hot states are generated in blue-emitting devices, all molecular bonds are potentially vulnerable to dissociation by high energy ($E_{S^*/T^*} \sim 5.4$–6 eV) focused momentarily on a single bond. Dissociated molecular fragments either become neutral species by disproportionation, or they participate in radical addition reactions with neighbouring molecules to form high-molecular-mass products [31].

To detect degraded molecular products in the aged device, we use laser desorption (LDI)/ionization mass spectroscopy (MS) on fresh and photo-degraded materials. *Mer*-Ir(*pmp*)$_3$ shows lower mass defects compared to the parent Ir complex, which are found even in the fresh sample. In degraded *mer*-Ir(*pmp*)$_3$, additional higher mass defects are also observed. Similar high and low mass species have also been reported for degraded mCBP. High mass defects have a smaller energy gap than the parent molecule, while the small mass defects show the opposite trend [32]. Details of these investigations will be reported elsewhere.

The small- and large-energy gap defects (relative to the dopant) are identified as the hole traps, Q_A and Q_B, in Fig. 49.5a. Both traps are charged when filled, leading to an increase in voltage, $\Delta V(t)$. Shockley–Read–Hall (SRH) nonradiative recombination occurs for holes trapped on Q_A. Likewise, exciton quenching via triplet states at Q_A results in a decrease in luminance (Fig. 49.5a). On the other hand, large-energy-gap Q_B defects can capture excited states that are subsequently transferred to the dopant, and thus do not affect the PHOLED luminance. Note that triplets on the dopant (at energy $E_{T,dop}$) are transferred from exciplex states originally formed between the hole on the dopant and the electron on the host ($E_{T,ex}$) [25], as well as from excitons directly formed on the manager ($E_{T,M}$).

Based on these considerations, we developed a lifetime model [20] for fitting both $L(t)/L_0$ and $\Delta V(t)$ of CONV, GRAD, and managed PHOLEDs ('Methods' section). The best fit is provided by assuming that defects generated in the EML are the result of TTA in the devices studied here (Supplementary Note 5). A comparison of lifetime among devices tested at $L_0 = 1{,}000$ cd m^{-2} results in nearly identical initial and steady-state exciton populations, provided that their natural triplet lifetimes and bimolecular annihilation rates are also similar. When GRAD and M3 are driven at $J_0 = 5.3 \pm 0.1$ mA cm^{-2}, the initial luminance levels are 1,000 cd m^{-2} versus 930 cd m^{-2} for M3 and GRAD, respectively. These conditions lead to a slight overestimation of <40 h for T80 = 173 $\pm$ 3 h for GRAD at $L_0 = 1{,}000$ cd m^{-2}.

The model also includes polaron traps generated outside of the EML with a density of Q_{ext}, resulting in the increase of the operating voltage without affecting luminance ('Methods' section). These traps originate from the degradation of charge transport and blocking layers, all of which are commonly observed in aged devices [24, 33, 34].

Table 49.2 summarizes the parameters used for fitting the lifetime data for CONV, GRAD and the managed PHOLEDs. The defect generation rates k_{QA} and k_{QB} are similar for most devices, yielding nearly similar Q_A and Q_B in managed PHOLEDs, which are smaller than those in the GRAD and CONV over the same operational period,

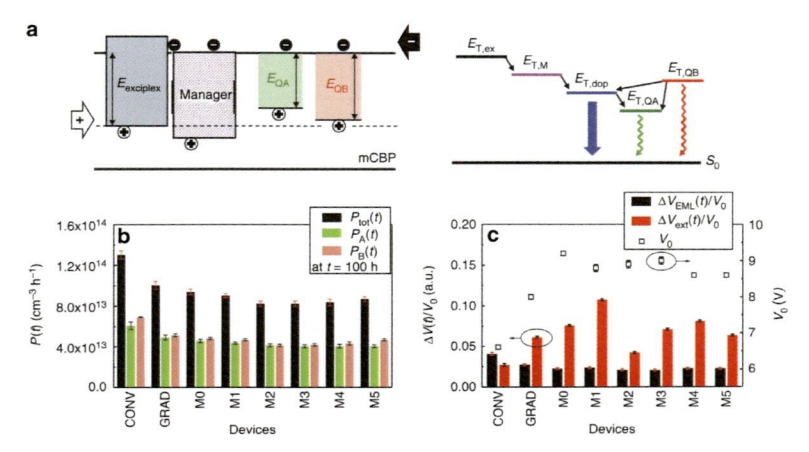

Figure 49.5 Analysis of the effectiveness of the manager. (a) (Left) Energy level diagram of the doped EML along with proposed positions of Q_A and Q_B. Here, Q_A and Q_B are assumed to be hole traps, with Q_A deeper in the energy gap than Q_B. Holes are transported by the dopant and the manager, and are potentially trapped by Q_A and Q_B. Electrons are transported by the host and the manager. (Right) Energy diagram of the triplet exciton states in the EML. The sources of triplet excitons in the as-grown device due to charge recombination are twofold: triplet exciplexes ($E_{T,ex}$) generated between the host and the dopant, and triplet excitons directly formed on the manager ($E_{T,M}$). Both can exothermically transfer to the dopant ($E_{T,dop}$). Q_A, the deep hole trap, has a low-energy triplet state that results in exciton quenching ($E_{T,QA}$), while Q_B, the shallow trap ($E_{T,QB}$), transfers excitons to the lower energy sites. (b) Average Q_A and Q_B generation rates in the EML, $P_A(t)$ and $P_B(t)$, from hot states in CONV, GRAD, and managed PHOLEDs. The total defect generation rate is $P_{tot}(t)$ where t = 100 h. (c) Relative contributions to the voltage rise with respect to V_0 induced by defects within and outside the EML (that is, $Q_A + Q_B$ and Q_{ext}, respectively) at $t = 100$ h. The separate contributions to the voltage rise, $\Delta V_{EML}(t)/V_0$ and $\Delta V_{ext}(t)/V_0$, along with V_0 are shown. Error bars represent one s.d. for at least three devices.

t. For example, Q_A and Q_B in M3 at $t = 100$ h are (4.9 ± 0.1) and $(5.0 \pm 0.1) \times 10^{15}$ cm^{-3}, while those in GRAD are (5.5 ± 0.2) and $(5.7 \pm 0.1) \times 10^{15}$ cm^{-3}, and those in CONV are (6.6 ± 0.2) and $(7.5 \pm 0.1) \times 10^{15}$ cm^{-3}, respectively. Compared to CONV and GRAD, the reduction in SRH recombination ($k_{Qn}Q_A n$) and direct exciton quenching ($k_{QN}Q_A N$) leading to a reduced rate of luminance loss in managed PHOLEDs that is attributed to their lower Q_A. Here, k_{Qn}

Table 49.2 Model parameters for the lifetime model for CONV, GRAD and managed PHOLEDs

Device	k_{QN} (10^{-11} cm^3 s^{-1})	k_{Qp} (10^{-7} cm^3 s^{-1})	k_{QA} (10^{-21} cm^3 s^{-1})	k_{QB} (10^{-21} cm^3 s^{-1})	k_{Qext} (10^{-21} cm^3 s^{-1})
CONV	3.3 ± 0.4	0.7 ± 0.2	0.9 ± 0.1	1.0 ± 0.1	0.06 ± 0.01
GRAD	2.3 ± 0.2	0.9 ± 0.2	0.9 ± 0.1	1.0 ± 0.1	0.2 ± 0.01
M0	2.3 ± 0.1	1.3 ± 0.2	1.0 ± 0.1	1.0 ± 0.1	0.5 ± 0.1
M1	2.1 ± 0.1	1.6 ± 0.2	0.9 ± 0.1	1.0 ± 0.1	0.8 ± 0.1
M2	1.9 ± 0.1	3.0 ± 0.7	0.9 ± 0.1	0.9 ± 0.1	0.5 ± 0.1
M3	1.9 ± 0.1	3.0 ± 0.8	0.9 ± 0.1	0.9 ± 0.1	1.0 ± 0.3
M4	2.1 ± 0.1	2.1 ± 0.5	0.9 ± 0.1	1.0 ± 0.1	0.8 ± 0.2
M5	2.0 ± 0.1	0.9 ± 0.1	0.9 ± 0.1	1.0 ± 0.1	0.3 ± 0.1

Errors for the model parameters are the 95% confidence interval for fit.

and k_{QN} are the reduced Langevin and defect-exciton recombination rates, respectively, and n and N are the steady-state densities of electrons and excitons, respectively.

The rate of defect formation within the EML is given by $P(t) = \frac{1}{d_{EML}} \int_{EML} \frac{dQ(x,t)}{dt} dx$. Figure 49.5b shows the rates for generating Q_A and Q_B, and $Q_A + Q_B$ ($P_A(t)$, $P_B(t)$, and $P_{tot}(t)$, respectively) at $t = 100$ h. For example, for CONV, $P_{tot} = (1.3 \pm 0.1) \times 10^{14}$ cm^{-3} h^{-1} is reduced to $P_{tot} = (1.0 \pm 0.1) \times 10^{14}$ cm^{-3} h^{-1} for GRAD, and decreases further to $P_{tot} = (0.8 \pm 0.1) \times 10^{14}$ cm^{-3} h^{-1} for M3. It is remarkable that only a 15% decrease in the defect formation rate for managed versus graded doping devices leads to a nearly twofold improvement in T80. This result suggests that even a small change in the probability of dissipation of excess energy and the resulting defect density can have large effects on device lifetime, consistent with previous work [20, 31, 35, 36].

Note that since the luminance loss is primarily due to Q_A, the high P_A of CONV and GRAD of (6.1 ± 0.4) and (4.9 ± 0.3) × 10^{13} cm^{-3} h^{-1} leads to a luminance of <800 and 850 ± 10 cd m^{-2}, respectively, as opposed to that of M3 = 920 ± 10 cd m^{-2} with $P_A = (4.0 \pm 0.1) \times 10^{13}$ cm^{-3} h^{-1} at $t = 100$ h. On the other hand, M3, M4 and M5 have similar P_A, yielding a luminance of 915 ± 5 cdm^{-2}, while P_B are (4.2 ± 0.1), (4.3 ± 0.2) and (4.7 ± 0.1) × 10^{13} cm^{-3} h^{-1}, respectively. This larger variation in P_B is because Q_B can return excitons to the dopants where they have a renewed opportunity to luminesce, and thus its effect is small compared to P_A.

The percentage contributions of $k_{Qn}Q_A n$ to the luminance degradation (that is, $k_{Qn}Q_A n + k_{QN}Q_A N$) is $90 \pm 2\%$ for most devices. This indicates that SRH recombination is the dominant mechanism due to the large density of injected polarons that are lost prior to exciton formation.

The diverse defects with different, distributed energetic characteristics can lead to somewhat larger uncertainties in the hole trapping rate (k_{Qp}) compared with other parameters extracted from the model (Table 49.2). Nevertheless, we note that k_{Qp} is generally higher for the managed PHOLEDs than that for CONV or GRAD, resulting from energy levels arising from multiple species (Supplementary Note 4). This is offset by the relatively small density of Q_A in the managed PHOLEDs, additional exciton generation via Q_B, and reduced exciton loss due to the smaller k_{QN}.

Compared to CONV and GRAD, the managed PHOLEDs have a lower rate of exciton-defect interactions (k_{QN}), indicating that fewer excitons are eliminated due to the quenching by Q_A (Fig. 49.5b). Now, $k_{QN} \cong 2.0 \times 10^{-11}$ cm^3 s^{-1} of the aged PHOLEDs is larger by nearly two orders of magnitude than the TTA rate of $k_{TT} \cong 1.0 \times 10^{-13}$ cm^3 s^{-1} obtained from the transient PL of the as-grown PHOLED EML. Thus, the reduction of luminance is severely impacted by defect-related exciton loss compared to increased TTA, while the latter process still plays a critical role in triggering molecular dissociation reactions.

Figure 49.5c shows $\Delta V_{EML}(t)/V_0$ and $\Delta V_{ext}(t)/V_0$ for CONV, GRAD and managed PHOLEDs. These are the relative contributions to the total voltage rise induced by defects within and outside of the EML (that is, $Q_A + Q_B$ and Q_{ext}, respectively) at $t = 100$ h with respect to V_0. CONV and GRAD have relatively high $\Delta V_{EML}(t)$ compared to the managed devices due to the higher defect densities in the EML. The generation rate of Q_{ext} that produces $\Delta V_{ext}(t)$ is k_{Qext}, which is generally higher for the managed PHOLEDs than CONV and GRAD. This results from the higher resistivity of the devices due to thick EML, as well as the low hole conductivity in the managed EML. Using an approximation based on space-charge-limited transport [37], the mobility in the managed EML is reduced by ~20% compared to that of the GRAD EML. Polaron-induced degradation in the transport layers is accelerated in the managed devices due to the increased polaron density arising from lower hole mobilities [24, 38]. Thus, while the EML defects (Q_A and Q_B) are sufficient to accurately model $L(t)/L_0$, those formed in other non-luminescent layers of the PHOLEDs (Q_{ext}) were also included to fully account for $\Delta V(t)$.

The reduced lifetime improvement from 3.0 ± 0.1 to 1.9 ± 0.1 times increases in T90 and T80, respectively, for M3 versus GRAD is due to the degradation of the manager molecules themselves via process 5. Thus, to achieve further increased efficiency, reduced luminance degradation and smaller voltage increase of the devices, manager molecules with improved stability and hole mobility compared with *mer*-Ir(pmp)$_3$ are required.

We demonstrate a strategy to dissipate the energy of hot excited states that otherwise lead to dissociative reactions and deteriorate the operational stability of blue PHOLEDs. By introducing excited state manager molecules into the PHOLED EML, we achieve to our knowledge the longest lifetime reported thus far for blue-emitting devices (Supplementary Table 1). We also developed a phenomenological model that establishes the roles and characteristics of defects present in the device. Our findings emphasize the importance of excited state management or similar approaches to further improve the lifetime of blue PHOLEDs. While such approaches based on an understanding of the fundamental underlying processes leading to device failure are essential, they must be accompanied by the development of highly stable dopants, managers, hosts and transport materials; a challenge made all the more difficult by the very wide energy gaps required for blue PHOLEDs.

49.4 Methods

Device fabrication and characterization. PHOLEDs were grown by vacuum sublimation in a chamber with a base pressure of 4×10^{-7} Torr on pre-patterned indium-tin-oxide (ITO) glass substrates (VisionTek Systems Ltd., United Kingdom). The device and the structures of GRAD and managed PHOLEDs are as follows: 70 nm ITO anode/5 nm dipyrazino[2,3,-f:2′,3′-h]quinoxaline 2,3,6,7,10,11-

hexacarbonitrile (HATCN) hole injection layer/10 nm N, N'-Di (phenyl-carbazole)-N, N'-bis-phenyl-(1,1'-biphenyl)-4,4'-diamine (CPD) HTL/50 nm EML/5 nm mCBP:Ir(dmp)$_3$ 8 vol% exciton blocking layer/5 nm mCBP hole blocking layer/25 nm tris-(8-hydroxyquinoline)aluminium (Alq$_3$) electron transport layer/1.5 nm hydroxyquinolato-Li (Liq) electron injection layer/100 nm Al cathode. The conventional PHOLED (CONV) has the following structure [20, 25]: 5 nm HATCN/30 nm CPD/35 nm 13 vol% Ir(dmp)$_3$ uniformly doped in mCBP/5 nm mCBP/25 nm Alq$_3$/1.5 nm Liq/100 nm Al. The device area is 2 mm^2 defined by the intersection of a 1 mm wide ITO strip and an orthogonally positioned 2 mm wide metal cathode patterned by deposition through a shadow mask. HATCN and Alq$_3$ were purchased from Luminescence Technology Corporation (Taiwan), CPD was from P&H Technology (South Korea), mCBP and Ir(dmp)$_3$ were provided by Universal Display Corporation (Ewing, NJ, USA) and *mer*-Ir(pmp)$_3$ was synthesized following previous methods [30]. The J–V–L characteristics of the PHOLEDs were measured [39] using a parameter analyzer (Hewlett-Packard, HP4145) and a calibrated Si-photodiode (Thorlab, FDS1010-CAL). The PHOLED emission spectra were recorded using a calibrated spectrometer (OceanOptics, USB4000). For lifetime tests, PHOLEDs were operated at constant current (Agilent, U2722) and the luminance and voltage data were automatically collected (Agilent, 34972A). Errors quoted for the measured electroluminescent and lifetime characteristics (J_0, V_0, EQE, T90, T80 and $\Delta V(t)$) are s.d.'s taken from a population of from three devices.

49.4.1 Exciton Profile Measurement

The exciton density profile, $N(x)$, was measured across the EML by inserting ultrathin ($\sim$1 Å) red phosphorescent (iridium (III) bis (2-phenylquinolyl-N, C$^{2'}$) acetylacetonate (PQIr)) sensing layers at different locations within the EML in a series of blue PHOLEDS [40, 41]. The integrated emission intensities of PQIr and Ir(dmp)$_3$ at J_0 are converted into the number of excitons at x via:

$$I_{\text{sens}}(\lambda, x) = a_{\text{PQIr}}(x) I_{\text{PQIr}}(\lambda) + a_{\text{Ir(dmp)}_3}(x) I_{\text{Ir(dmp)}_3}(\lambda) \quad (49.1)$$

where $I_{\text{sens}}(\lambda, x)$ is the emission intensity consisting of the combined spectra of Ir(dmp)$_3$ ($I_{\text{Ir(dmp)3}}(\lambda)$) and PQIr ($I_{\text{PQIr}}(\lambda)$). The relative weights of $a_{\text{PQIr}}(x)$ and $a_{\text{Ir(dmp)3}}(x)$, respectively, were used. Then, the outcoupled exciton density, $\eta_{\text{out}}(x)N(x)$, is equal to the relative number of excitons emitting on the PQIr at x as:

$$\eta_{\text{out}}(x) N(x) = A \cdot \text{EQE}(x) \cdot \frac{a_{\text{PQIr}}(x) \int I_{\text{PQIr}}(\lambda)/\lambda d\lambda}{a_{\text{PQIr}}(x) \int I_{\text{PQIr}}(\lambda)/\lambda d\lambda + a_{\text{Ir(dmp)}_3}(x) \int I_{\text{Ir(dmp)}_3}(\lambda)} \quad (49.2)$$

Here, EQE(x) is external quantum efficiency of the device with the sensing layer at x; thus the right-hand side of Eq. (49.2) gives the relative number of excitons at position x. Also, $\int_{\text{EML}} N(x)dx = 1$, and $\eta_{\text{out}}(x)$ is the outcoupling efficiency calculated as the fraction of outcoupled light emitted at x based on Green's function analysis [42]. The Förster transfer length of $\sim$3 nm [25] limits the spatial resolution of the measurement.

Since the thickness of delta-doped PQIr is less than a monolayer, PQIr molecules are spatially dispersed to avoid emission loss by concentration quenching. A delta-doped sensing layer only slightly affects the charge transport as opposed to previously used 1–2 nm-thick, doped layers [25]. This leads to a variation in operating voltages at J_0 of <0.5 V among all sensing devices (see also the upper panel of Fig. 49.4b and Supplementary Note 3).

49.4.2 Mass Spectrometry Measurement

Materials used in the PHOLEDs were prepared in N2-filled encapsulated vials. They were photodegraded by the laser irradiation at $\lambda =$ 442 nm for >5 h. For the LDI–MS measurement, the material was dissolved in toluene/THF, and the solution is placed onto the target plate and subsequently evaporated. The Bruker Autoflex Speed mass spectrometer is run in reflection mode. The spectrometer was calibrated with a series of known peptides and matrix peaks. Mass spectra of degraded materials were compared to those of their pristine counterparts.

49.4.3 Lifetime Degradation Model

The rate equations for holes (p), electrons (n) and excitons (N) are:

$$\frac{dp(x, t, t')}{dt'} = G(x) - \gamma n(x, t, t') p(x, t, t')$$
$$- k_{Qp} [Q_A(x, t) + Q_B(x, t)] p(xtt'),$$
$$\frac{dn(x, t, t')}{dt'} = G(x) - \gamma n(x, t, t') p(x, t, t')$$
$$- k_{Qn} [Q_A(x, t) + Q_B(x, t)] n(xtt'),$$
$$\frac{dN(x, t, t')}{dt'} = \gamma n(x, t, t') p(x, t, t') + k_{Qn} Q_B(x, t) n(x, t, t')$$
$$- \frac{1}{\tau_N} + k_{QN} Q_A(x, t) N(xtt'), \tag{49.3}$$

There are two different time scales: t' is the duration of charge transport and energy transfer ($\sim$ µs) and t is the device degradation time ($\sim$h) due to the formation of defects, $Q_A(x,t)$ and $Q_B(x,t)$. The triplet decay lifetime is $\tau_N = 1.4 \pm 0.1$ µs, obtained from the transient PL decay of thin-film EMLs of the GRAD and managed PHOLEDs. Also, $G(x) = J_0 \cdot N(x)/e \cdot \int_{EML} N(x) dx$ is the generation rate of excitons due to charge injection at current J_0, $\gamma = e(\mu_p + \mu_n)/\varepsilon_r \varepsilon_0$ is the Langevin recombination rate, where e is the elementary charge, μ_n and μ_p are the electron and hole mobilities in the EML, respectively, and ε_0 and $\varepsilon_r \sim 3$ are the vacuum and relative permittivities, respectively. It follows that $k_{Qn} = e\mu_n/\varepsilon_r \varepsilon_0$ is the reduced Langevin recombination rate describing the recombination of immobile trapped holes and mobile electrons.

The trap densities, Q_A and Q_B, resulting from the TTA increase at rates k_{QA} and k_{QB} are given by:

$$\frac{dQ_A(x, t)}{dt} = k_{QA} \{N(x, t)\}^2,$$
$$\frac{dQ_B(x, t)}{dt} = k_{QB} \{N(x, t)\}^2. \tag{49.4}$$

Equation (49.3) is solved in steady state ($t' \to \infty$), yielding $n(x, t, t')$, $p(x, t, t')$ and $N(x, t, t') = n(x, t)$, $p(x, t)$ and $N(x, t)$, respectively. This set of equations is numerically solved with $Q_A(x,t)$ and $Q_B(x,t)$ to fit both the luminance loss and voltage rise as a function of t using:

$$L(t)/L(0) = \int N(x, t) \eta_B(x) dx \tag{49.5}$$

and

$$\Delta V(t) = \frac{e}{\varepsilon \varepsilon_0} \left(\int_{\text{EML}} x Q(x,t)\, dx + \int_{\text{ext}} x' Q_{\text{ext}}(x',t)\, dx' \right) \quad (49.6)$$

Here, $\eta_B(x)$ is the outcoupling efficiency of the excitons emitted at x and $Q_{\text{ext}}(x',t)$ is introduced to account for the voltage rise caused by traps present outside the EML. The uniqueness of the fit that yields parameters, k_{QN}, k_{Qp}, k_{QA}, k_{QB} and k_{Qext}, has been tested, with results in Supplementary Note 5.

Note that when extracting k_{Qext} and thus $\Delta V_{\text{ext}}(t)$ from the fits, the polaron densities in the EML at J_0 are used. However, k_{Qext} should more accurately reflect the polaron densities in the transport layers due to charge trapping by Q_{ext}, and thus, a reduction in layer conductivity. This simplifying assumption leads to its large variation among devices compared with other parameters. Initial values of Q_A, Q_B and Q_{ext} are set at 10^{15} (cm^{-3}), which accurately traces the time evolution of $\Delta V(t)$ and converge to their final values after the iteration of the least-square algorithm.

Supplementary Information

Supplementary Information accompanies this paper at http://www.nature.com/naturecommunications

Acknowledgements

This work was supported by grant DE-EE0007077 of the US Department of Energy, FA9550-14-1-0245 of the US Air Force Office of Scientific Research, and Universal Display Corporation (UDC). The authors thank UDC for providing the host and dopant materials, Mr Xiao Liu for transient photoluminescence measurements, Mr Quinn Burlingame for insightful discussions, and James Windak in the chemistry department at the Univ. of Michigan for the LDI-TOF mass spectroscopy measurement.

Additional Information

Data availability. The data that support the findings of this study are available from the authors upon request.

Author contributions: J.L. designed, fabricated and characterized the PHOLEDs with C.J. J.L. developed the model for lifetime characteristics. C.C. and J.L. obtained thermally stimulated current measurements. T.B. and M.E.T. synthesized the manager material and performed DFT calculations. S.R.F. supervised the project, analysed data and wrote the manuscript with J.L.

Competing interests: S.R.F. and M.E.T. declare an equity interest in one of the sponsors of this work (UDC). The remaining authors declare no competing financial interests.

References

1. Kim, S. *et al.* Low-power flexible organic light-emitting diode display device. *Adv. Mater.* **23**, 3511–3516 (2011).
2. Kim, S.-W. Organic light emitting diode display. US patent 2014/0312319 A1 (2014).
3. Lee, C. & Park, G. Transparent display panel and transparent organic light emitting diode diplay device including the same. US patent 2016/0055794 A1 (2014).
4. Jung, Y. K. *et al.* 52-3: distinguished paper: 3 stacked top emitting white OLED for high resolution OLED TV. *SID Symp. Dig. Tech. Pap.* **47**, 707–710 (2016).
5. Lee, J., Slootsky, M., Lee, K., Zhang, Y. & Forrest, S. R. An electrophosphorescent organic light emitting concentrator. *Light Sci. Appl.* **3**, e181 (2014).
6. Hack, M., Brown, J. J., Weaver, M. S. & Premutico, M. Lifetime OLED display. US patent 9257665 B2 (2016).
7. Baldo, M. A. *et al.* Highly efficient phosphorescent emission from organic electroluminescent devices. *Nature* **395**, 151–154 (1998).
8. Baldo, M. A., Lamansky, S., Burrows, P. E., Thompson, M. E. & Forrest, S. R. Very high-efficiency green organic light-emitting devices based on electrophosphorescence. *Appl. Phys. Lett.* **75**, 4–6 (1999).

9. Adachi, C. et al. Endothermic energy transfer: A mechanism for generating very efficient high-energy phosphorescent emission in organic materials. *Appl. Phys. Lett.* **79**, 2082–2084 (2001).
10. Zhuang, J. et al. Highly efficient phosphorescent organic light-emitting diodes using a homoleptic iridium(III) complex as a sky-blue dopant. *Org. Electron.* **14**, 2596–2601 (2013).
11. Klubek, K. P., Dong, S.-C., Liao, L.-S., Tang, C. W. & Rothberg, L. J. Investigating blue phosphorescent iridium cyclometalated dopant with phenyl-imidazole ligands. *Org. Electron.* **15**, 3127–3136 (2014).
12. Oh, C. S., Choi, J. M. & Lee, J. Y. Chemical bond stabilization and exciton management by CN modified host material for improved efficiency and lifetime in blue phosphorescent organic light-emitting diodes. *Adv. Opt. Mater.* **4**, 1281–1287 (2016).
13. Kang, Y. J. & Lee, J. Y. High triplet energy electron transport type exciton blocking materials for stable blue phosphorescent organic light-emitting diodes. *Org. Electron.* **32**, 109–114 (2016).
14. Zhang, L. et al. Highly efficient blue phosphorescent organic light-emitting diodes employing a host material with small bandgap. *ACS Appl. Mater. Interfaces* **8**, 16186–16191 (2016).
15. Seo, J.-A. et al. Long lifetime blue phosphorescent organic light-emitting diodes with an exciton blocking layer. *J. Mater. Chem. C* **3**, 4640–4645 (2015).
16. Jeon, S. K. & Lee, J. Y. Four times lifetime improvement of blue phosphorescent organic light-emitting diodes by managing recombination zone. *Org. Electron.* **27**, 202–206 (2015).
17. Huang, H.-L., Balaganesan, B., Fu, Y.-H., Lin, H.-Y. & Chao, T.-C. P-159: electron transporting materials for highly efficient and long lifetime blue OLED devices for display and lighting applications. *SID Symp. Dig. Tech. Pap.* **47**, 1722–1724 (2016).
18. D'Andrade, B., Esler, J., Lin, C., Weaver, M. & Brown, J. 61.5L: late-news paper: extremely long lived white phosphorescent organic light emitting device with minimum organic materials. *SID Symp. Dig. Tech. Pap.* **39**, 940–942 (2008).
19. Tsang, D. P.-K. & Adachi, C. Operational stability enhancement in organic light-emitting diodes with ultrathin Liq interlayers. *Sci. Rep.* **6**, 22463 (2016).
20. Giebink, N. C. et al. Intrinsic luminance loss in phosphorescent small-molecule organic light emitting devices due to bimolecular annihilation reactions. *J. Appl. Phys.* **103**, 44509 (2008).

21. Giebink, N. C., D'Andrade, B. W., Weaver, M. S., Brown, J. J. & Forrest, S. R. Direct evidence for degradation of polaron excited states in organic light emitting diodes. *J. Appl. Phys.* **105**, 124514 (2009).
22. Wang, Q. & Aziz, H. Degradation of organic/organic interfaces in organic light-emitting devices due to polaron–exciton interactions. *ACS Appl. Mater. Interfaces* **5**, 8733–8739 (2013).
23. Sadi, T., Kivisaari, P., Oksanen, J. & Tulkki, J. On the correlation of the Auger generated hot electron emission and efficiency droop in III-N light-emitting diodes. *Appl. Phys. Lett.* **105**, 91106 (2014).
24. Schmidbauer, S., Hohenleutner, A. & König, B. Chemical degradation in organic light-emitting devices: mechanisms and implications for the design of new materials. *Adv. Mater.* **25**, 2114–2129 (2013).
25. Zhang, Y., Lee, J. & Forrest, S. R. Tenfold increase in the lifetime of blue phosphorescent organic light-emitting diodes. *Nat. Commun.* **5**, 5008 (2014).
26. Hong, S., Kim, J. W. & Lee, S. Lifetime enhanced phosphorescent organic light emitting diode using an electron scavenger layer. *Appl. Phys. Lett.* **107**, 41117 (2015).
27. Hsin, M.-H. *et al.* P-161: 89.3% lifetime elongation of blue TTA-OLED with assistant host. *SID Symp. Dig. Tech. Pap.* **47**, 1727–1729 (2016).
28. Baldo, M. A., Adachi, C. & Forrest, S. R. Transient analysis of organic electrophosphorescence. II. Transient analysis of triplet-triplet annihilation. *Phys. Rev. B* **62**, 10967–10977 (2000).
29. Bachilo, S. M. & Weisman, R. B. Determination of triplet quantum yields from triplet-triplet annihilation fluorescence. *J. Phys. Chem. A* **104**, 7711–7714 (2000).
30. Lee, J. *et al.* Deep blue phosphorescent organic light-emitting diodes with very high brightness and efficiency. *Nat. Mater.* **15**, 92–98 (2016).
31. Kondakov, D. Y., Lenhart, W. C. & Nichols, W. F. Operational degradation of organic light-emitting diodes: mechanism and identification of chemical products. *J. Appl. Phys.* **101**, 24512 (2007).
32. Sandanayaka, A. S. D., Matsushima, T. & Adachi, C. Degradation mechanisms of organic light-emitting diodes based on thermally activated delayed fluorescence molecules. *J. Phys. Chem. C* **119**, 23845–23851 (2015).
33. Adamovich, V. I., Weaver, M. S. & D'Andrade, B. W. Long lifetime phosphorescent organic light emitting device (OLED) structures. US patent 8866377 B2 (2014).

34. Kwong, R. C. et al. High operational stability of electrophosphorescent devices. *Appl. Phys. Lett.* **81**, 162–164 (2002).
35. Winter, S., Reineke, S., Walzer, K. & Leo, K. Photoluminescence degradation of blue OLED emitters. *Proc. SPIE* **6999**, 69992N (2008).
36. Fujimoto, H. et al. Influence of material impurities in the hole-blocking layer on the lifetime of organic light-emitting diodes. *Appl. Phys. Lett.* **109**, 243302 (2016).
37. Lampert, M. A. Simplified theory of space-charge-limited currents in an insulator with traps. *Phys. Rev.* **103**, 1648–1656 (1956).
38. Xia, S. C., Kwong, R. C., Adamovich, V. I., Weaver, M. S. & Brown, J. J. in *Proceedings of the 45th IEEE International Annual Reliability Physics Symposium*, 253–257 (2007).
39. Forrest, S. R., Bradley, D. D. C. & Thompson, M. E. Measuring the efficiency of organic light-emitting devices. *Adv. Mater.* **15**, 1043–1048 (2003).
40. Erickson, N. C. & Holmes, R. J. Investigating the role of emissive layer architecture on the exciton recombination zone in organic light-emitting devices. *Adv. Funct. Mater.* **23**, 5190–5198 (2013).
41. Coburn, C., Lee, J. & Forrest, S. R. Charge balance and exciton confinement in phosphorescent organic light emitting diodes. *Adv. Opt. Mater.* **4**, 889–895 (2016).
42. Celebi, K., Heidel, T. D. & Baldo, M. A. Simplified calculation of dipole energy transport in a multilayer stack using dyadic Green's functions. *Opt. Express* **15**, 1762–1772 (2007).

Chapter 50

Eliminating Nonradiative Decay in Cu(I) Emitters: > 99% Quantum Efficiency and Microsecond Lifetime

Rasha Hamze,[a] Jesse L. Peltier,[b] Daniel Sylvinson,[a] Moonchul Jung,[a] Jose Cardenas,[a] Ralf Haiges,[a] Michele Soleilhavoup,[b] Rodolphe Jazzar,[b] Peter I. Djurovich,[a] Guy Bertrand,[b] and Mark E. Thompson[a]

[a] *Department of Chemistry, University of Southern California, Los Angeles, California, USA*
[b] *UCSD-CNRS Joint Research Laboratory (UMI 3555), Department of Chemistry and Biochemistry, University of California, San Diego, La Jolla, California 92093-0358, USA*
met@usc.edu

Luminescent complexes of heavy metals such as iridium, platinum, and ruthenium play an important role in photocatalysis and energy conversion applications as well as organic light-emitting diodes (OLEDs). Achieving comparable performance from more–earth-abundant copper requires overcoming the weak spin-orbit coupling of the light metal as well as limiting the high reorganization energies

---

Reprinted from *Science*, **363**(6427), 601–606, 2019.

Electrophosphorescent Materials and Devices
Edited by Mark E. Thompson
Text Copyright © 2017, The Author(s)
Layout Copyright © 2024 Jenny Stanford Publishing Pte. Ltd.
ISBN 978-981-4877-34-3 (Hardcover), 978-1-003-08872-1 (eBook)
www.jennystanford.com

typical in copper(I) [Cu(I)] complexes. Here we report that two-coordinate Cu(I) complexes with redox active ligands in coplanar conformation manifest suppressed nonradiative decay, reduced structural reorganization, and sufficient orbital overlap for efficient charge transfer. We achieve photoluminescence efficiencies >99% and microsecond lifetimes, which lead to an efficient blue-emitting OLED. Photophysical analysis and simulations reveal a temperature-dependent interplay between emissive singlet and triplet charge-transfer states and amide-localized triplet states.

Organometallic complexes of heavy metals often phosphoresce from high-energy, long-lived triplet states with high luminance efficiency, enabling applications ranging from photocatalysis [1] and chemo- and biosensing [2, 3] to dye-sensitized solar cells [4] and organic electronics [5, 6]. By contrast, phosphorescence of typical organocopper complexes is inefficient [7, 8] compared to organo-Ir and organo-Pt phosphors [9]. This difference largely stems from the rates of two intersystem crossing (ISC) processes. The spin-orbit coupling (SOC) parameter (ξ) that facilitates ISC is smaller for the Cu nucleus ($\xi = 857$ cm^{-1}) than for heavier metals such as Ir and Pt ($\xi = 3909$ and 4481 cm^{-1}, respectively) [10]. Therefore, the rate of ISC from the lowest excited singlet state (S_1) to the lowest triplet excited state (T_1) typically is on the order of 10^{10} to 10^{11} s^{-1} in Cu complexes [11, 12] as opposed to 10^{13} to 10^{14} s^{-1} in heavy metal complexes [13, 14]. The rate of radiative ISC from T_1 to the ground state (S_0) is markedly slower in organocopper phosphors ($k = 10^3$ to 10^4 s^{-1}) compared to Ir and Pt emitters ($k > 10^5$ s^{-1}) [5, 15]. Moreover, the lowest-energy optical transitions in most Cu(I) complexes are typically metal-to-ligand charge transfer (MLCT) events, which are associated with large reorganization energies. In these MLCT transitions, formal oxidation at the d^{10} metal center induces Jahn-Teller distortion [16] that not only increases nonradiative decay rates but also leads to ISC rates even slower than those expected based only on SOC considerations [17].

In the past decade, thermally assisted delayed fluorescence (TADF) has emerged as a useful alternative for harvesting both singlet and triplet excitons generated in organic light-emitting diodes (OLEDs) [18–21]. This process is accomplished by bringing the S$_1$ and T$_1$ manifolds close enough in energy to give high rates for exo- and endothermic ISC between them. (Fig. 50.1A). The resulting equilibrium between S$_1$ and T$_1$ typically favors the longer-lived, weakly emissive triplet; however, a high radiative rate from S$_1$ can lead to a high radiative efficiency for the TADF process. A conundrum in purely organic, donor-acceptor–type TADF systems is raised by the opposing requirements for minimizing the energy gap between the two lowest excited states (ΔE_{S1-T1}) through poor highest occupied molecular orbital (HOMO)–lowest unoccupied molecular orbital (LUMO) overlap and achieving high radiative rates for the S$_1$ state via large orbital overlap [22]. TADF also occurs in Cu(I) complexes and was first observed by McMillin and co-workers [23]. In such Cu complexes, the HOMO is typically metal-based, and the LUMO is composed of ligand π^* orbitals. However, Cu TADF systems present a major departure from their pure organic counterparts: Forward ISC in the Cu(I) compounds is fast enough to quantitatively depopulate the S1 state, thereby completely outpacing prompt fluorescence and resulting in monoexponential emission decay at all temperatures. To illustrate, a highly emissive Cu complex with the smallest recorded $\Delta E_{S1-T1} = 33$ meV has a radiative rate (k_r) of 2×10^5 s^{-1} in the solid state at room temperature (it is only weakly emissive in fluid solution) [20]. Fitting the temperature-dependent photoluminescent decay of this complex to the modified Boltzmann Eq. 50.1, where k_{S1} and k_{T1} are the rate constants for radiative decay from the S$_1$ and T$_1$ states, respectively, k_B is Boltzmann's constant, and T is temperature, gives a derived rate of fluorescence (prompt emission from S$_1$) of 2×10^6 s^{-1} [24], which is a full order of magnitude faster than the recorded rate of emission (TADF) at room temperature. It is therefore reasonable to conclude that the limiting factor in determining the rate of TADF in Cu systems is the endothermic ISC from T$_1$ to S$_1$, which is tied to SOC as shown in Eq. 50.2, where ρ_{FC} denotes the Frank-Condon density of states and $|<S_1|\hat{H}_{SO}|T_1>|$ denotes the SOC matrix element [25]. Nevertheless, even the most efficient TADF-based Cu emitters have low photoluminescence (PL) quantum yields (Φ_{PL}) in fluid or polymeric matrices. High rates of nonradiative decay are observed in nonrigid environments owing to substantial distortions in the excited state of frequently studied tetrahedral motifs [26, 27]:

$$\tau_{\text{TADF}} = \frac{3 + \exp\left(\frac{\Delta E_{S_1-T_1}}{k_B T}\right)}{3k_{T_1} + k_{S_1}\exp\left(\frac{\Delta E_{S_1-T_1}}{k_B T}\right)} \quad (50.1)$$

$$k_{\text{ISC}} = \frac{2\pi}{\hbar}\rho_{\text{FC}}|<S_1|\hat{H}_{\text{SO}}|T_1>|^2 \quad (50.2)$$

Reports of linear Cu complexes with high Φ_{PL} in nonrigid matrices highlight the appeal of low coordination in limiting excited-state reorganization [28–30]. Unfortunately, the MLCT nature of the radiative transitions in these derivatives leads to excited-state lifetimes that are relatively long ($\tau \sim 20$ μs), thus limiting their luminescent efficiency. However, a recent paper by Di et al. has reported efficient green electroluminescence (EL) in OLEDs using two-coordinate carbene-gold and carbene-copper complexes [maximum external efficiency (EQE_{max}) = 26.3% for the former and 9.7% for the latter] [19]. The complexes discussed by Di et al. consist of Au(I) or Cu(I) ions coordinated to a cyclic (alkyl)(amino)carbene (CAAC) [31, 32] and an N-bound amide. Here we examine a closely related family of two-coordinate, neutral CAAC-Cu(I)-amide complexes with notable photophysical properties (see Fig. 50.1; compound **1b** was examined by Di et al.). Optimizing the steric encumbrance of substituents on the carbene, we achieve complexes with $\Phi_{\text{PL}} > 99\%$ and short emission lifetimes ($\tau = 2$ to 3 μs) in fluid and polymeric media. Electrochemical, photophysical, and computational analyses reveal a picture of ligand-based frontier orbitals with minimal metal contribution. Coplanar ligand conformation in these complexes is critical to maintaining high Φ_{PL}. Finally, one of the complexes is used in blue OLEDs.

Structural and electronic properties of CAAC-Cu-amide complexes

The two-coordinate CAAC-Cu-amide complexes were prepared in high yields (80 to 90%) following literature procedures (Fig. 50.1B) [19]. Complexes **1** to **5** were isolated as white to yellow powders and display varying degrees of sensitivity to O_2 and moisture depending on the steric bulk around the carbene and the nature of the amide. Complex **1b**, bearing 1,8-dimethyl carbazolide (CzMe), was found to

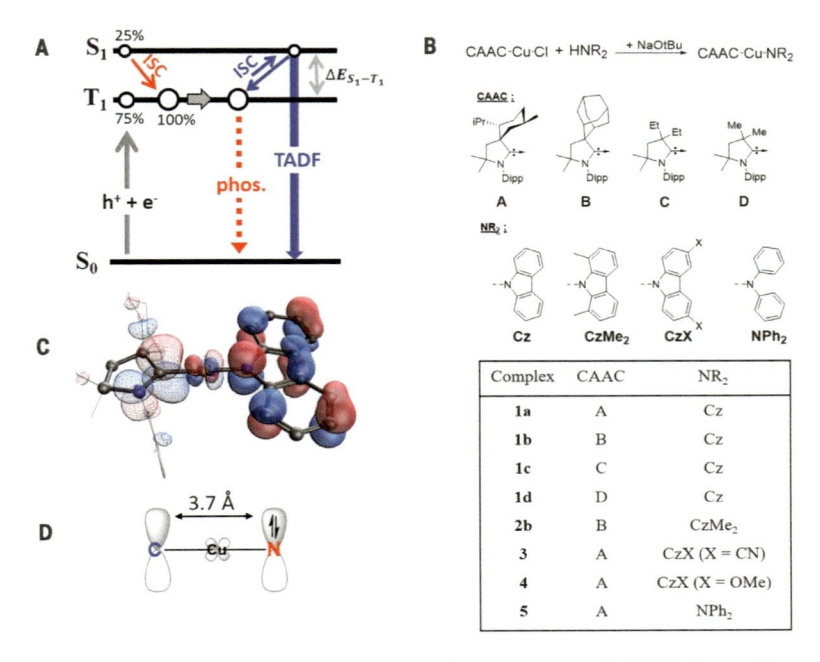

Figure 50.1 Background and chemical structures of CAAC-Cu-amides. (A) Scheme depicting the radiative processes [phosphorescence (phos.) and TADF] in an OLED. (B) Complexes studied in this work. Dipp, 2,6-diisopropylphenyl; tBu, tert-butyl; iPr, isopropyl; Et, ethyl; Me, methyl. (C) HOMO (solid) and LUMO (mesh) surfaces of complex **1a**. (D) Simplified picture of the HOMO and LUMO of complex **1a**.

be highly sensitive to H_2O and CO_2 in ambient air as a solid and in solution (see the supplementary materials).

X-ray diffraction analysis was performed on single crystals of the complexes **1a**, **1c**, **1d**, **2b**, and **3** to **5** (see Figs. S1 to S8 and Tables S1 to S8). The structures all display a near-linear coordination geometry around the Cu center (174° to 179°), with a 3.73 to 3.77 Å separation between the carbene carbon and the amide nitrogen owing to similar bond lengths (C–Cu = 1.88 to 1.89 Å; Cu–N = 1.85 to 1.87 Å). Dihedral angles between the ligands are <9° in complexes **1a**, **1c**, **1d**, and **3** to **5**, whereas complex **2b** has a near orthogonal conformation (dihedral angle = 83°). Despite the asymmetry of the CAAC, only one set of ^{1}H NMR (nuclear magnetic resonance) spectroscopic resonances for the carbazolide is observed in all complexes. This observation indicates rapid exchange on the NMR time scale, likely caused by rotation around the $C_{carbene}$–N_{amide} axis. Variable temperature NMR experiments performed on complex **1a** down to −60°C show no signs of coalescence, which suggests a low energy barrier for the rotation in question. Complex **2b** also shows one set of carbazolide resonances in its ^{1}H NMR spectrum, consistent with an orthogonal conformation for the ligands.

The electrochemical properties of the Cu complexes **1a** and **3** to **5**, precursors (CAAC^{Men}-CuCl, where CAAC^{Men} is ligand A in Fig. 50.1B, and the free amines), and potassium carbazolide (KCz) were examined (see Fig. S11 and table S11 for details). The Cu complexes undergo irreversible oxidation at potentials that vary over a 1-V range, depending on the donor strength of the amide ligand. Relative to their parent amines, the oxidation potentials of the Cu-amides decrease by 0.6 to 0.7 V. All potentials fall well below the Cu(I) oxidation potential of CAAC^{Men}-CuCl. The oxidation potential of **1a** is anodically shifted by 0.73 V compared to that of KCz, consistent with metalation. Reduction potentials are quasi-reversible, with values that are unchanged from the parent CAAC^{Men}-CuCl. The data show that the redox potentials are independently controlled by the ligands: Oxidation is primarily at the amide, and reduction at the π-accepting carbene.

The redox noninnocent nature of the ligands is also captured by density functional theory (DFT; B3LYP/LACVP**). As shown in Fig. 50.1C, the HOMO is principally amide-based, with substantial electron density residing in the filled p orbital of N_{amide}. The LUMO is localized largely on the unfilled p orbital of $C_{carbene}$. The nature of the frontier molecular orbitals and coplanar orientation of the ligands allow for a simplified representation of the valence structure (Fig. 50.1D) as a donor-bridge-acceptor linear system, wherein the metal d orbitals act as a weak electronic bridge between the parallel donor (N_{amide} 2p$_z$) and acceptor ($C_{carbene}$ 2p$_z$) orbitals, thereby illustrating the potential for long-range π interaction [33, 34]. The ground state of these complexes is marked by a large permanent dipole, $\mu_g \sim 11.3$ D, in close agreement with the report by Föller and Marian for an isoelectronic Au complex [35].

Absorption spectra of complexes **1a**, **1b**, **2b**, and **3** to **5** in tetrahydrofuran (THF) (Fig. 50.2A) show high-energy bands ($\lambda < 350$ nm) corresponding to $\pi - \pi^*$ transitions of the ligands. Broad, low-energy bands apparent in these complexes are assigned to singlet interligand charge transfer (^{1}ICT) from the electron-rich amide to the electron-accepting carbene. The onset of the ICT bands for **1a** and **3** to **5** falls in the order expected based on the oxidation potentials of their amide ligands (inset of Fig. 50.2A). A notable feature of the ICT transitions in these complexes is their high extinction coefficients ($\varepsilon > 10^3$ M^{-1} cm^{-1}), which is surprising considering the ~3.7 Å separation between the HOMO and LUMO. These values are a factor of 10 greater than what is typically observed for MLCT transitions in organocopper complexes. We tentatively attribute the strongly allowed nature of the charge-transfer (CT) transitions in these complexes to the small but nonnegligible contribution of the Cu d orbitals acting as an effective electronic bridge between the donor and the acceptor components. In addition, the coplanarity of the ligands leads to a parallel orientation of the filled $2p_z$ orbital on the amide N and the empty $2p_z$ orbital on the carbene C, maximizing long-range orbital overlap.

A characteristic property of the ICT band in these complexes is the pronounced hypsochromic shift as solvent polarity increases (Fig. 50.2B). The absorption onset of the ICT band undergoes a blue shift of 2400 cm^{-1} in **1a** and 2600 cm^{-1} in **5** upon increasing solvent polarity. The magnitude of the shift reflects a strong change in the electronic dipole moment upon excitation. The direction of the shift is a consequence of the ground-state dipole being much larger in magnitude and opposite in orientation relative to its excited-state counterpart [36]. Similar hypsochromic shifts of the ICT absorption band are observed upon freezing the solvent matrix and are more pronounced in methylcyclohexane (MeCy) than in 2-methyltetrahydrofuran (2-MeTHF) (Fig. 50.2C). The blue shift in 2-MeTHF at 77 K is likely due to the solvent dipoles being frozen around the large solute dipole, stabilizing the ground state (relaxing the potential energy surface) and destabilizing the ICT excited state. The blue shift recorded in MeCy at 77 K, where the solubility of **1a** is reduced, can be brought about by long-range dipole-dipole

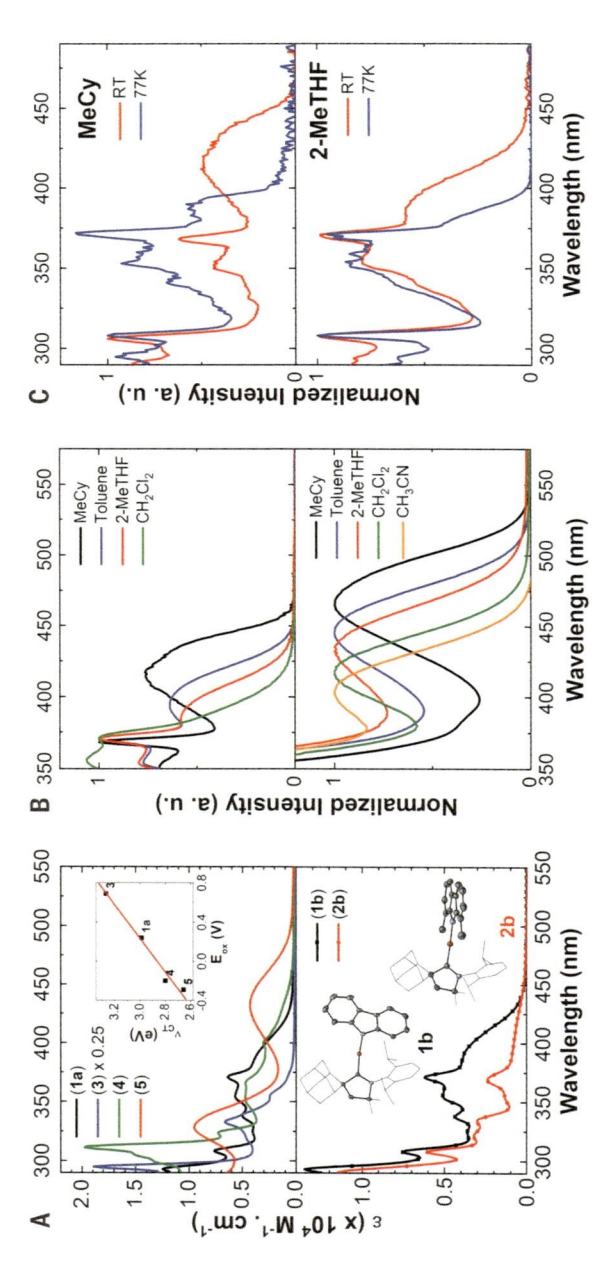

Figure 50.2 Electronic spectra of CAAC-Cu-amides. (A) Absorption spectra of complexes 1a, **1b**, **2b**, and **3** to **5** in THF. The inset in the top spectrum shows a linear relation between the energy of the CT absorption band (v_{CT}) in MeCy and the oxidation potential (E_{ox}) of the complexes. The crystal structures of complexes **1b** and **2b** are shown as insets in the bottom spectrum. The pendant adamantyl and Dipp groups are depicted in wireframe for clarity. (B) Blue shift of the ICT absorption band with increasing solvent polarity observed in complexes **1a** and **5**. (C) Absorption spectra of complex **1a** at room temperature (RT) and 77 K, showing a blue shift in the ICT band at low temperature. a.u., arbitrary units.

interactions between the solute molecules [37]. In both instances, the hypsochromic shift is absent when the solvent glass is thawed.

Efficient luminescence from allowed interligand charge transfer transitions

The Cu complexes all luminesce with high efficiency in fluid solution, as well as when doped in polystyrene (PS) matrices, manifesting microsecond radiative lifetimes (Fig. 50.3A and Table 50.1). Powdered samples of the carbazolide complexes **1a** to **1d** and **3** are poorly emissive, whereas **2b**, **4**, and **5** exhibit stronger luminescence in their microcrystalline forms than in solution (see Figs. S22, S23, and S26 and Table S15). Emission spectra of **1a** to **1d**, **4**, and **5** in 2-MeTHF solutions are broad and featureless at room temperature, consistent with the ICT origin of these transitions. The PL efficiency improves as the steric encumbrance on the carbene increases in the series **1d** < **1c** < **1b** < **1a** (Φ_{PL} = 0.1, 0.6, 0.7, and 1.0, respectively). Because **1a** to **1d** have similar radiative rate constants (k_r = 2.0 × 10^5 to 4.3 × 10^5 s^{-1}), the principal effect of increasing steric bulk is to decrease the rates of nonradiative decay. Complexes **4** and **5** show red-shifted emission relative to complex **1a**, with radiative rates comparable to that of 1a (Table 50.1). The sparingly soluble complex **3** shows narrow, structured emission centered at 426 nm in solution and has a radiative rate constant that is lower than the other complexes, features we attribute to emission from a state with predominant triplet-carbazolide (^{3}Cz) character, vide infra (Fig. 50.3D). In addition, complex **3** displays a low energy (λ ~600 nm) concentration-dependent emission band in solution, characterized by a rise time in its PL decay traces (see Fig. S34 and Table S17), which is consistent with the diffusional process required to form a luminescent excimer.

To eliminate the complications of aggregation and excimer formation in photophysical studies, we doped the complexes into thin films [1 weight % in PS], where excimer formation is suppressed. At room temperature, samples in the rigid matrix exhibit a blue shift in their emission relative to spectra recorded in solution (i.e., rigidochromism) and suppressed rates of nonradiative decay (Fig. 50.3A). Complexes **1a** and **4** display broad emission

Table 50.1 Photophysical data for complexes **1a**, **1b**, **2b**, and **3** to **5** doped 1% by weight into PS films or dissolved at 10^{-5} M concentration in 2-MeTHF. n.d, not determined

Complex	Emission at room temperature					Emission at 77 K	
	λ_{max} (nm)	τ (μs)	Φ_{PL}	k_r (s^{-1})	k_{nr} (s^{-1})	λ_{max} (nm)	τ (μs)
			1% weight PS films				
1a	474	2.8	1.0	3.5×10^5	3.6×10^3	480	64
3	426	240 (70%)	0.82	1.5×10^3	3.2×10^2	424	6900
		1300 (30%)					
4	518	2.3	1.0	4.3×10^5	$<4.4 \times 10^3$	490	550
5	532	2.6	0.78	3.0×10^5	8.5×10^4	536	264
			2-MeTHF				
1a	492	2.5	1.0	3.9×10^5	$<8.0 \times 10^3$	430	7300
1b	510	2.3	0.68	3.0×10^5	1.7×10^5	430	430 nm: 3000
							500 nm: 48
1c	500	1.8	0.56	3.1×10^5	2.4×10^5	430	7000
1d	510	0.54	0.11	2.0×10^5	1.6×10^6	430	5000
2b	542	0.86 (79%)	0.12	$1.1 \times 10^{5*}$	$8.0 \times 10^{5*}$	438	430 nm: 2400
		2.3 (21%)					500 nm: 37
3	428	450 nm: 8.3	0.11	n.d.	n.d.	422	12,000
	590†	600 nm: 8.0					
4	558	0.28	0.25	8.9×10^5	2.7×10^6	470	Seconds
5	580	0.87	0.16	1.8×10^5	9.7×10^6	500	215

*Calculated from a weighted average of the two contributions to τ.
† Excimer peak.

with near unity quantum efficiency in PS ($\Phi_{PL} = 1.0$), whereas complex **5** is less efficient ($\Phi_{PL} = 0.78$) (Table 50.1). Thin films of complex **3** give narrow, structured emission at room temperature with biexponential decay lifetimes of 240 μs and 1.3 ms. The slow radiative rates in **3** ($k_r = 1.5 \times 10^3$ s^{-1}) are consistent with our initial ^{3}Cz assignment, owing to the highly destabilized ICT in this complex (Fig. 50.3D). Notably, the high Φ_{PL} and k_r values for complexes **1a**, **4**, and **5** in solution and thin films are comparable to phosphors containing heavy metals, such as Ir and Pt.

The photoluminescent properties are dramatically altered on cooling to 77 K. A vibronically structured, long-lived emission (τ of ms to s, $k_r < 10^3$ s^{-1}) is observed for **1a** to **1d**, **3**, and **4** in frozen glasses of 2-MeTHF and MeCy (Table 50.1 and Fig. 50.3B). The emission at 77 K is assigned to a low-lying triplet state localized on the carbazolide ligand (^{3}Cz), as the phosphorescence spectrum of KCz in frozen 2-MeTHF replicates the same profile as **1a** (Fig. 50.3B). The blue shift in emission observed in MeCy at 77 K corresponds with the hypsochromic shift observed in ICT absorption at that temperature, as destabilizing the ICT transition leaves ^{3}Cz as the lowest-lying emissive state (Fig. 50.3E). Luminescence from **5** is broad and featureless in frozen glassy matrices, with long excited-state lifetimes ($\tau = 215$ μs in 2-MeTHF), consistent with emission from a ^{3}ICT state (Fig. 50.3D).

The luminescent properties of **1a** and **5** in thin PS films were examined as a function of temperature to probe the ICT manifold while avoiding complications from ^{3}Cz-dominated emission in frozen solvents. Temperature-dependent emission of thin PS films of both complexes display broad, featureless ICT spectra at all temperatures between 10 and 300 K (see Figs. S30 and S31) and have excited state lifetimes that increase with decreasing temperatures (Fig. 50.3C). Fits of the temperature-dependent PL decay curve to a Boltzmann distribution (Eq. 50.1) give $\Delta E_{^1CT-^3CT}$ = 0.063 eV (510 cm^{-1}) for **1a** and 0.071 eV (570 cm^{-1}) for **5** [coefficient of determination (R^2) values for the fits of **1a** and **5** in Fig. 50.3C are 0.996 and 0.997, respectively]. The radiative lifetimes of ^{1}ICT (**1a**, $\tau = 150 \pm 15$ ns; **5**, $\tau = 210 \pm 21$ ns) and ^{3}ICT (**1a**, $\tau = 64 \pm 6$ μs; **5**, $\tau = 250 \pm 15$ μs), derived from the fits to Eq. 50.1, are comparable to values reported by Yersin and co-workers and Bräse

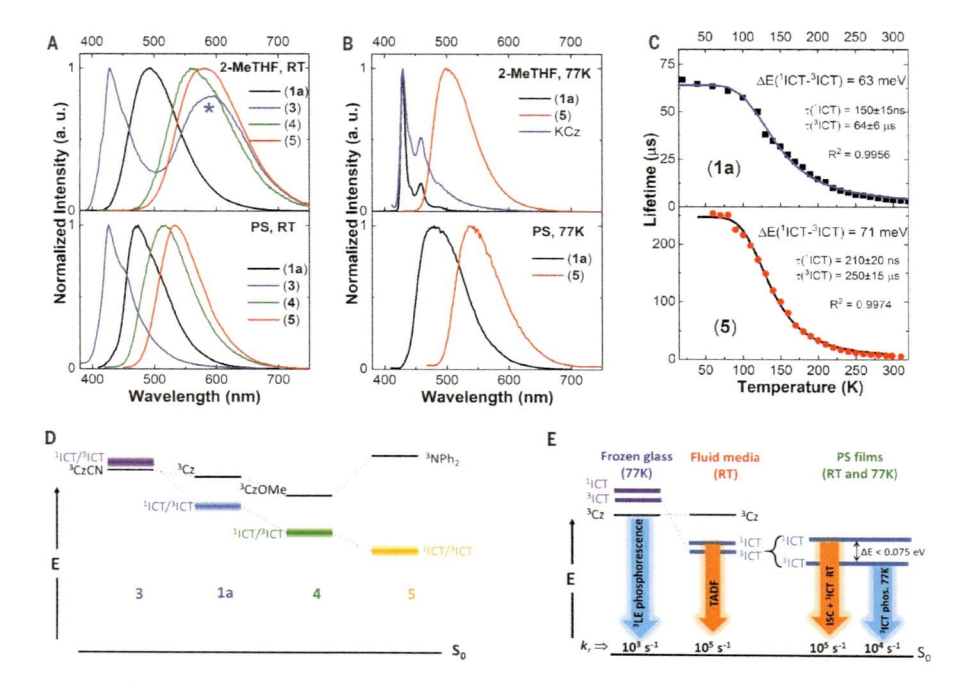

Figure 50.3 Luminescence behavior of CAAC-Cu-amides. (A) Room-temperature emission spectra of complexes **1a** and **3** to **5** at 10^{-5} M in 2-MeTHF and doped 1% by weight into PS films. The asterisk indicates excimer emission. (B) 77 K emission spectra of complexes **1a** and **5** in 2-MeTHF (10^{-5} M) and 1% PS films. Also shown is the gated phosphorescence spectrum of KCz at 77 K (top). (C) Temperature-dependent PL decays of complexes **1a** (top) and **5** (bottom) as well as the data fit to Eq. 50.1. The parameters obtained from the fit, $\Delta E_{1_{CT}-3_{CT}}$, $\tau_{1_{ICT}}$, and $\tau_{3_{ICT}}$ are shown in each plot. (D) State diagram depicting ^{1,3}CT/^{3}LE ordering in the reported complexes. The relative energies of the states are based on emission spectra. (E) Jablonski diagram depicting the different processes operating in various media at room temperature and 77 K.

and co-workers in the fastest Cu(I) TADF emitters reported to date [22, 38]. The rate of emission at low temperature attributed to ^{3}ICT-based phosphorescence is faster for **1a** than for **5**, owing to the close-lying ^{3}Cz in **1a**, which can enhance SOC by mixing with ^{3}ICT through configuration interaction [9, 12].

We also investigated the role of ligand conformation in the photophysical properties using complexes **1b** and **2b**: The former has coplanar orientation of carbene and carbazole ligands, whereas the ligands are nearly orthogonal in the latter. The ICT absorption band in **2b** has an extinction coefficient reduced threefold relative to **1b**: $\varepsilon_{ICT} = 1300$ and 4000 M^{-1} cm^{-1}, respectively (Fig. 50.2A). Similarly, the efficiency and radiative rate constant for **2b** are reduced nearly fourfold relative to **1b** ($\Phi_{PL} = 0.12$ and 0.68, and $k_r = 1.1 \times 10^5$ and 3×10^5 s^{-1}, respectively). The marked decrease in k_r observed for **2b** is important in light of the decrease in HOMO-LUMO overlap and the expected decrease in $\Delta E_{1_{CT}-3_{CT}}$ [35]. These observations highlight the impact of a coplanar ligand conformation on maintaining the strongly allowed nature of the ICT transitions in absorption and emission and are therefore incompatible with the rotationally accessed spin-state inversion (RASI) mechanism described by Di et al. [19], as previously noted by Föller and Marian and Penfold and co-workers [35, 39]. Moreover, the suggestion by Di et al. that the S_1 state lies below the T_1 state in energy when the carbene and carbazole ligands are orthogonal is not supported by these experimental results and is counter to what is expected on the basis of fundamental quantum mechanical considerations [40].

Time-dependent DFT (TDDFT) was used to model the main electronic transitions of the excited states in these complexes (see the supplementary materials for details). The ^{3}ICT state shares the same orbital parentage as ^{1}ICT and lies within 0.25 eV of the latter (Table S22), in agreement with recent reports from Föller and Marian [35] and Tafett et al. [41]. In addition, the ^{3}Cz state is only 0.03 to 0.1 eV higher in energy than the ^{3}ICT state in all the carbazolide-based complexes except for **3**, where it is the lowest triplet state. The triplet state localized on the diphenylamide in **5**, ^{3}LE (i.e., ^{3}NPh$_2$), is destabilized relative to ^{3}ICT by 0.5 eV.

We have further modeled the effects of solvation on the excited states of complex **1a** at 77 and K using a multiscale hybrid approach that 300 K using a multiscale hybrid approach that used molecular dynamics simulations in conjunction with TDDFT, as detailed in the supplementary materials. At room temperature, it was found that the ^{3}ICT state is the lowest triplet state in all cases owing to stabilization by the solvent dipoles (Fig. S41). A similar procedure was followed to study the effect of solvation at 77 K, where single-point TDDFT calculations were performed on a frozen equilibrated cell using the hybrid scheme described in the supplementary materials. Here it was found that the ^{3}Cz state is the lowest-lying triplet, in accordance with the experimental observation of ^{3}Cz emission in 2-MeTHF at 77 K. The destabilization of the ^{3}ICT state can be attributed to its associated dipole (4.25 D), which is opposite in direction to the large dipole in the ground ^{1}ICT (11.8 D) and ^{3}Cz states (11.27 D). Hence, solvent molecules in a frozen matrix are expected to be ordered so as to stabilize the ground state, whereas the ^{3}CT state would be destabilized. Negative solvatochromic effects observed in absorption can be explained by the same rationale.

Exploration of CAAC-Cu-amides as emitters in blue OLEDs

OLEDs incorporating **1a** as an emitter were fabricated by vapor deposition (see the supplementary materials for details), following the general architecture outlined in Fig. 50.4A and only changing the emissive layer to screen different wide bandgap host materials. Commonly used hosts in blue OLEDs [1,3-bis(triphenylsilyl)benzene (UGH3); 3,3′-bis(carbazol-9-yl)biphenyl (mCBP); and dicarbazolyl-3,5-benzene (mCP)] were examined. In addition, we also tried a Cu-based host, that is, (CAAC)Cu-C$_6$F$_5$, with a high triplet energy; however, these devices degraded rapidly during operation (see the supplementary materials). Thin films doped into the various established hosts show similar trends with Φ_{PL} in UGH3 > mCBP > mCP (0.9, 0.6, and 0.3 respectively). OLEDs prepared with **1a** doped into UGH3 at 20 volume % give EQE$_{max}$ = 9.0% and 16 cd/A at 2 mA/cm^2 (Fig. 50.4B), consistent with the high triplet energy of the UGH3 host (T$_1$ = 3.5 eV). Although the EQE values for green- and yellow- or orange-emitting Cu-based OLEDs have been reported to be > 20% [42, 43], the highest efficiencies previously reported for blue-emitting (EL λ_{max} < 500 nm) Cu-based OLEDs are < 6% [44, 45]. The low EQE$_{max}$ of the mCBP and mCP devices can be explained by a low triplet energy for the hosts (mCBP, T$_1$ = 2.8 eV; mCP, T$_1$ = 2.9 eV), which do not confine triplet excitons on the Cu emitter as efficiently as UGH3. The roll-off in device efficiency as the current is raised for the UGH3-based device (Fig. 50.4B) is comparable to the roll-off observed for UGH3-based OLEDs with an iridium-based

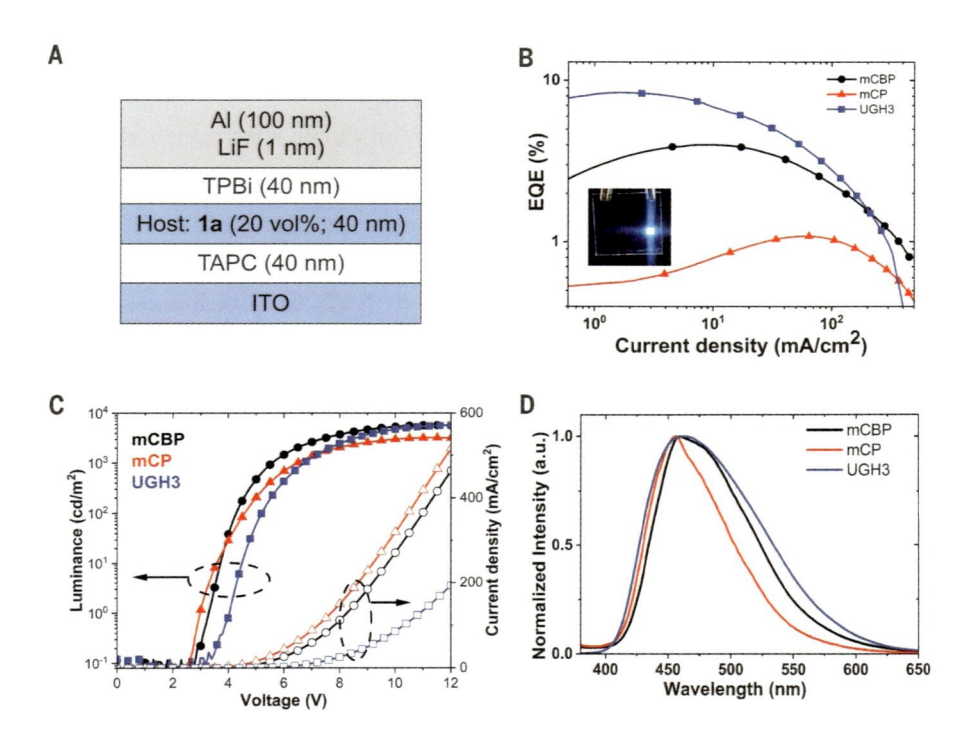

Figure 50.4 OLED demonstration. (A) OLED device architecture energy scheme for **1a**-based OLEDs. TPBi, 2,2′,2″-(1,3,5-benzinetriyl)-tris(1-phenyl-1-H-benzimidazole); TAPC, 4,4′-cyclohexylidenebis[N,N-bis(4-methylphenyl)benzenamine]; ITO, indium tin oxide. (B) EQE traces of devices using different hosts. The inset is a photograph of a **1a**-based device. (C and D) Current (J)–voltage (V)–luminance (L) traces (C) and EL spectra (D) of devices using different hosts.

emitter [46], which has been attributed to increased triplet-triplet and triplet-polaron annihilation at higher current density [47, 48]. All devices exhibit blue EL, similar to the PL for **1a** in a PS film, with minor shifts due to "solvent" effects of the different host matrices. This is in contrast to the green-emissive OLEDs reported by Di et al. [19] for an OLED with **1b** doped into polyvinylcarbazole. We expect that stable, high–triplet energy host materials for blue OLEDs using emitter **1a** should positively affect both efficiency and device stability.

Outlook

We have prepared a series of two-coordinate CAAC-Cu-amide complexes with emission tunable across the visible spectrum, high Φ_{PL} (up to >99%) in nonrigid media, and $k_r > 10^5$ s^{-1}. Emission stems from a strongly allowed amide-to-carbene ICT transition, with the coplanar ligand conformation and coupling through the metal d orbitals ensuring strong ε_{ICT} and resultant k_r. By contrast, a near-orthogonal arrangement of ligands leads to both low ε_{ICT} and a decrease in k_r owing to poor orbital overlap. In the Cu-carbazolide complexes, there exists a closely lying ^{3}Cz-centered state that dominates emission in frozen solvent glasses, owing to the destabilization of the ICT manifold in such media. Within the ICT manifold, efficient TADF is observed with $\Delta E_{1_{CT}-3_{CT}} < 75$ meV, among the smallest values recorded for mononuclear Cu(I)-based TADF systems [49].

The $\tau_{1_{ICT}}$ values we obtain are typical in Cu(I) TADF-based emitters; however, they are far longer than the prompt fluorescence rates recorded in pure organic TADF systems, with τ_{S_1} values on the order of 10 ns [50]. Another distinction from organic TADF systems is the much longer-lived (millisecond to second) phosphorescence at low temperature for organics. The discrepancy between organometallic and organic TADF suggests that a spin-pure treatment of the CT manifold is inadequate in organometallic emitters, owing to stronger SOC effects [51]. The slow ^{1}ICT decay in Cu-based systems is likely due to considerable mixing with ^{3}ICT via SOC. The opposite is true for ^{3}ICT, where considerable singlet character in the nominally triplet state leads to markedly faster decay than expected for a spin-pure triplet. The acronym TADF is an imprecise description for what is observed in Cu-based complexes of the type described here, because neither the singlet nor triplet states are spin-pure. Lifetimes as high as 0.2 μs for the ^{1}ICT state suggest that this state has substantial triplet contribution, thus fluorescence is not the best description for this type of emission [51, 52]. The process observed here is better described as thermally enhanced luminescence, where both the lower-energy and higher-energy states are highly emissive, albeit with the lower state having a much longer radiative lifetime. Thermal enhancement here manifests in a shortening of the radiative lifetime but does not markedly increase the already high luminance efficiency.

The extent of Cu involvement in the electronic properties of these complexes appears to be the answer to the Cu-TADF conundrum we posited: The metal contribution is large enough to induce high exo- and endothermic k_{ISC}, yet low enough to ensure small reorganization energies. This work outlines the design parameters for attaining Cu(I) complexes with photophysical properties akin to their heavy metal counterparts: maintaining a two-coordinate geometry around the metal center, with redox-active ligands in a coplanar orientation. These results therefore open the door to investigation of these complexes in fields where traditional heavy metal–based phosphors have been used, for example, optical sensing, photoredox catalysis, and solar fuel generation, to name a few.

Supplementary Information

The supplementary information is available at www.sciencemag.org/content/363/6427/601/suppl/DC1

Materials and Methods

Figs. S1 to S61

Tables S1 to S22

References [53–69].

Acknowledgments

Funding: This work was supported by the Universal Display Corporation for the University of Southern California authors; a National Science Foundation (NSF) Graduate Research Fellowship (grant no. DGE-1650112) supported J.L.P.; and the NSF (CHE-1661518) supported all UCSD authors. Computation for the work described in this paper was supported by the

University of Southern California's Center for High-Performance Computing (https://hpcc.usc.edu/). Author contributions: R.Ham., J.L.P., J.C., M.S., and R.J. synthesized and characterized the compounds. R.Ham. collected and analyzed the spectroscopic data. M.J. and R.Ham. fabricated and tested the OLEDs. R.Ham., J.L.P., R.J., and R.Hai. determined the x-ray structures. D.S. performed the theoretical calculations. R.Ham., P.I.D., and M.E.T. wrote the manuscript. G.B and M.E.T. directed the project.

Competing interests: R.Ham., M.J., P.I.D., and M.E.T. are inventors on a patent application submitted by the University of Southern California based partly on the intellectual property

in this report. One of the authors (M.E.T.) has a financial interest in the Universal Display Corporation. Data and materials availability: Structural data for all of the copper complexes reported here are freely available from the Cambridge Crystallographic Data Centre (CCDC nos. 1865272 to 1865277 for **1a**, **1c**, **1d**, and **3** to **5**; CCDC no. 1857183 for **2b**; and CCDC no. 1861278 for **Cu Host**). The supplementary materials for this paper include synthetic and characterization data (^{1}H and ^{13}C NMR spectra, mass spectrometry, and elemental analysis) for each copper complex; electrochemical data, ORTEP drawings, and crystallographic information on the compounds listed above; photophysical data; and details of DFT, TDDFT, and molecular dynamics calculations.

References

1. K. Kalyanasundaram, *Coord. Chem. Rev.* **46**, 159–244 (1982).
2. M. H. Keefe, K. D. Benkstein, J. T. Hupp, *Coord. Chem. Rev.* **205**, 201–228 (2000).
3. K. K.-W. Lo, M.-W. Louie, K. Y. Zhang, *Coord. Chem. Rev.* **254**, 2603–2622 (2010).
4. M. Grätzel, *Inorg. Chem.* **44**, 6841–6851 (2005).
5. S. Lamansky et al., *J. Am. Chem. Soc.* **123**, 4304–4312 (2001).
6. H. J. Bolink, E. Coronado, R. D. Costa, N. Lardiés, E. Ortí, *Inorg. Chem.* **47**, 9149–9151 (2008).
7. P. C. Ford, A. Vogler, *Acc. Chem. Res.* **26**, 220–226 (1993).

8. N. Armaroli, G. Accorsi, F. Cardinali, A. Listorti, in *Photochemistry and Photophysics of Coordination Compounds I*, V. Balzani, S. Campagna, Eds. (Springer, 2007), pp. 69–115.
9. H. Yersin, A. F. Rausch, R. Czerwieniec, T. Hofbeck, T. Fischer, *Coord. Chem. Rev.* **255**, 2622–2652 (2011).
10. S. Fraga, K. M. S. Saxena, J. Karwowski, *Handbook of Atomic Data.* Physical Science Data (Physical Science Data Series, vol. 5, Elsevier, 1976).
11. L. Bergmann, G. J. Hedley, T. Baumann, S. Bräse, I. D. W. Samuel, *Sci. Adv.* e1500889 (2016).
12. T. J. Penfold, E. Gindensperger, C. Daniel, C. M. Marian, *Chem. Rev.* **118**, 6975–7025 (2018).
13. G. J. Hedley, A. Ruseckas, I. D. W. Samuel, *Chem. Phys. Lett.* **450**, 292–296 (2008).
14. M. Kleinschmidt, C. van Wüllen, C. M. Marian, *J. Chem. Phys.* **142**, 094301 (2015).
15. J. Brooks et al., *Inorg. Chem.* **41**, 3055–3066 (2002).
16. S. Lamansky, R. C. Kwong, M. Nugent, P. I. Djurovich, M. E. Thompson, *Org. Electron.* **2**, 53–62 (2001).
17. Z. A. Siddique, Y. Yamamoto, T. Ohno, K. Nozaki, *Inorg. Chem.* **42**, 6366–6378 (2003).
18. H. Uoyama, K. Goushi, K. Shizu, H. Nomura, C. Adachi, *Nature* **492**, 234–238 (2012).
19. D. Di et al., *Science* **356**, 159–163 (2017).
20. H. Yersin, R. Czerwieniec, M. Z. Shafikov, A. F. Suleymanova, *ChemPhysChem* **18**, 3508–3535 (2017).
21. H. Yersin, Ed., *Highly Efficient OLEDs: Materials Based on Thermally Activated Delayed Fluorescence* (Wiley, 2019).
22. R. Czerwieniec, M. J. Leitl, H. H. H. Homeier, H. Yersin, *Coord. Chem. Rev.* **325**, 2–28 (2016).
23. J. R. Kirchhoff et al., *Inorg. Chem.* **22**, 2380–2384 (1983).
24. R. Czerwieniec, J. Yu, H. Yersin, *Inorg. Chem.* **50**, 8293–8301 (2011).
25. P. K. Samanta, D. Kim, V. Coropceanu, J.-L. Brédas, *J. Am. Chem. Soc.* **139**, 4042–4051 (2017).
26. M. Iwamura, H. Watanabe, K. Ishii, S. Takeuchi, T. Tahara, *J. Am. Chem. Soc.* **133**, 7728–7736 (2011).
27. C. E. McCusker, F. N. Castellano, *Inorg. Chem.* **52**, 8114–8120 (2013).
28. M. Gernert, U. Müller, M. Haehnel, J. Pflaum, A. Steffen, *Chemistry* **23**, 2206–2216 (2017).

29. S. Shi et al., *Dalton Trans.* **46**, 745–752 (2017).
30. R. Hamze et al., *Chem. Commun. (Camb.)* **53**, 9008–9011 (2017).
31. M. Melaimi, R. Jazzar, M. Soleilhavoup, G. Bertrand, *Angew. Chem. Int. Ed.* **56**, 10046–10068 (2017).
32. O. Back, M. Henry-Ellinger, C. D. Martin, D. Martin, G. Bertrand, *Angew. Chem. Int. Ed.* **52**, 2939–2943 (2013).
33. M. M. Hansmann, M. Melaimi, D. Munz, G. Bertrand, *J. Am. Chem. Soc.* **140**, 2546–2554 (2018).
34. D. C. Rosenfeld, P. T. Wolczanski, K. A. Barakat, C. Buda, T. R. Cundari, *J. Am. Chem. Soc.* **127**, 8262–8263 (2005).
35. J. Föller, C. M. Marian, *J. Phys. Chem. Lett.* **8**, 5643–5647 (2017).
36. C. Reichardt, *Chem. Rev.* **94**, 2319–2358 (1994).
37. V. Bulović, R. Deshpande, M. E. Thompson, S. R. Forrest, *Chem. Phys. Lett.* **308**, 317–322 (1999).
38. D. Volz et al., *Chem. Mater.* **25**, 3414–3426 (2013).
39. S. Thompson, J. Eng, T. J. Penfold, *J. Chem. Phys.* **149**, 014304 (2018).
40. S. P. McGlynn, T. Azumi, M. Kinoshita, *Molecular Spectroscopy of the Triplet State* (Prentice-Hall, Inc., 1969).
41. E. J. Taffet, Y. Olivier, F. Lam, D. Beljonne, G. D. Scholes, *J. Phys. Chem. Lett.* **9**, 1620–1626 (2018).
42. Y. Liu, S.-C. Yiu, C.-L. Ho, W.-Y. Wong, *Coord. Chem. Rev.* **375**, 514–557 (2018).
43. M. Wallesch et al., in *Proceedings SPIE 9183, Organic Light Emitting Materials and Devices XVIII*, F. So, C. Adachi, Eds. (SPIE, 2014), 918309.
44. D. Liang et al., *Inorg. Chem.* **55**, 7467–7475 (2016).
45. X.-L. Chen et al., *Chem. Mater.* **25**, 3910–3920 (2013).
46. X. Ren et al., *Chem. Mater.* **16**, 4743–4747 (2004).
47. S. Reineke, K. Walzer, K. Leo, *Phys. Rev. B* **75**, 125328 (2007).
48. L. Zhang, H. van Eersel, P. A. Bobbert, R. Coehoorn, *Chem. Phys. Lett.* **652**, 142–147 (2016).
49. M. Osawa et al., *J. Mater. Chem. C Mater. Opt. Electron. Devices* **1**, 4375–4383 (2013).
50. P. L. Santos et al., *J. Mater. Chem. C Mater. Opt. Electron. Devices* **4**, 3815–3824 (2016).
51. R. Baková, M. Chergui, C. Daniel, A. Vlček Jr., S. Záliš, *Coord. Chem. Rev.* **255**, 975–989 (2011).
52. N. J. Turro, *Modern Molecular Photochemistry* (University Science Books, 1991).

Chapter 51

Rapid Multiscale Computational Screening for OLED Host Materials

Daniel Sylvinson M. R., Hsiao-Fan Chen, Lauren M. Martin, Patrick J. G. Saris, and Mark E. Thompson

Department of Chemistry, University of Southern California, Los Angeles, California 90089, USA
met@usc.edu

The design of new host materials for phosphorescent organic light emitting diodes (OLEDs) is challenging because several physical property requirements must be met simultaneously. A triplet energy (E_T) higher than that of the chosen emitting dopant, appropriate highest occupied molecular orbital/lowest unoccupied molecular orbital energy levels, good charge carrier transport, and high stability are all required. Here, computational methods were used to screen structures to find the most promising candidates for OLED hosts. The screening was carried out in three *Tiers*. The *Tier* 1 selection, based on density functional theory calculations, identified a set of eight molecular structures with $E_T > 2.9$ eV, suitable for hosting blue phosphorescent

---

Reprinted from *ACS Appl. Mater. Interfaces*, **11**, 5276–5288, 2019.

Electrophosphorescent Materials and Devices
Edited by Mark E. Thompson
Text Copyright © 2019 American Chemical Society
Layout Copyright © 2024 Jenny Stanford Publishing Pte. Ltd.
ISBN 978-981-4877-34-3 (Hardcover), 978-1-003-08872-1 (eBook)
www.jennystanford.com

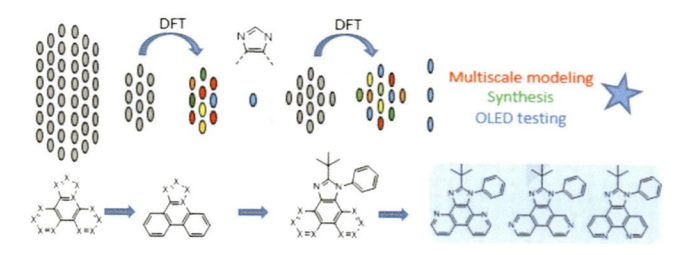

dopants such as iridium(III)bis((4,6-di-fluorophenyl)-pyridinato-N,C2′)picolinate. Phenanthro[9,10-d]imidazole was chosen as the starting point for the *Tier* 2 selection. Thirty-seven unique molecular structures were enumerated by isoelectronic nitrogen transmutation of up to two CH fragments of the phenanthrene. Three molecules, that is, imidazo[4,5-f]-phenanthrolines with nitrogens at the 1,10-, 3,8-, and 4,7-positions, were selected for *Tier* 3, which involved the use of molecular dynamics simulations and electron coupling calculations to predict differences in charge transport between the three materials. The three were explored experimentally through synthesis and device fabrication. The singlet, triplet, and frontier orbital energies computed using single-molecule density functional theory calculations (*Tiers* 1 and 2) were consistent with the experimental values in a fluid solution, and the multiscale modeling scheme (*Tier* 3) correctly predicted the poor device performance of one material. We conclude that screening host materials using only single-molecule quantum mechanical data was not sufficient to predict whether a given material would make a good OLED host with certainty; however, they can be used to screen out materials that are destined to fail due to low singlet/triplet energies or a poor match of the frontier orbital energies to the dopant or transport materials.

51.1 Introduction

Improvements in organic light emitting diode (OLED) technology for both display and white lighting applications have been realized through the development of both new organic materials and device architectures. Since the first report of heterojunction and doped OLEDs in the late 1980s [1], significant strides have been made to improve the overall performance of red, green, blue, and white OLEDs. The development of phosphorescent materials capable of harvesting both singlet and triplet electrogenerated excitons has yielded maximum internal quantum efficiencies of 100% [2–6]. Devices with effective host–guest emissive layers and supporting transport materials have enabled stable and efficient phosphorescent OLEDs (PHOLEDs) [5–10].

Among the important design criteria for blue OLED host materials are high triplet energy, appropriate highest occupied molecular orbital (HOMO)/lowest unoccupied molecular orbital (LUMO) energy-level alignment with respect to the emissive dopant, balanced carrier transport, and robust molecular structure. Finding host materials that satisfy all of these criteria for a given phosphorescent dopant is complex because these properties are not usually independently tunable with a molecular structure. The common approach of synthesizing and characterizing a large number of materials in the hope of uncovering satisfactory candidates is both time consuming and challenging. Here, we discuss exploring and screening a chemical space using high-throughput computing as an alternative to the restrictive and time-consuming Edisonian approach. High-throughput computing can accelerate the materials discovery process by allowing for selection of promising structural fragments and elimination of structural moieties that do not meet minimum thresholds for key host parameters before time is spent in the laboratory to prepare and test them. Such an approach has been used extensively in biology, medicine, and catalysis but has only recently been reported for molecular optoelectronic materials [11–16]. A recent study reported by the Aspuru-Guzik group describes computationally screening a library of thermally assisted delayed fluorescence (TADF)-emitting dopants for OLEDs with an initial size of 1.6 million candidates [15]. This study led to a set of 20 TADF-based emitters that gave high electroluminescent efficiency. Single-molecule modeling studies are well suited to probe emissive dopants for OLEDs since they are dispersed in the host matrix in the OLED, largely eliminating dopant–dopant interactions. The study presented here is focused on using computational methods to screen host materials for OLEDs, where intermolecular interactions may play a significant role in the properties of the material. Our strategy is to explore a diverse set of candidate structures through a *Tiered*, multiscale computational approach by understanding structure–property relationships at each Tier.

The conceptual starting point of our search for high triplet energy host materials is triphenylene, which has a triplet energy in the deep-blue range (E_T = 425 nm, 2.9 eV) as well as a large aromatic system, leading to efficient charge transport [17, 18]. Additionally, the rigid tetracyclic structure inhibits one potential OLED degradation pathway involving exocyclic bond cleavage of the host material, as is widely suspected for the standard phenylcarbazole derivatives [19, 20]. Due to the synthetic challenges and structural limitations inherent to the synthesis of regiospecifically substituted heterotriphenylenes [21], this report explores a related but more synthetically accessible class of molecular structures in which one of the 6-membered rings of triphenylene is replaced with a 5-membered heterocyclic ring, which can readily be formed from orthoquinone precursors (Scheme 51.1). This phenanthroheterole-based structure is referred to as H2P due to its two peripheral hexagonal "H-rings" and one pentagonal "P-ring".

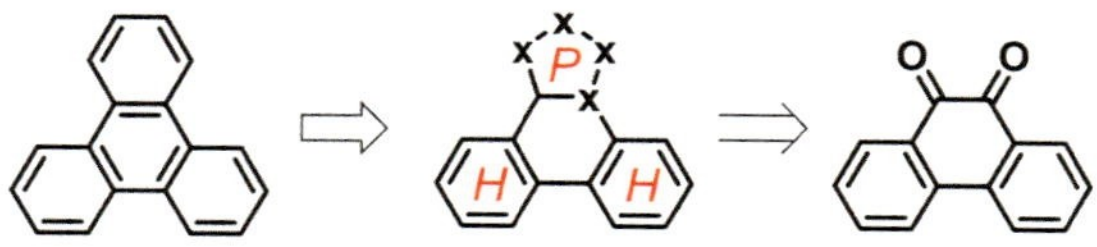

Scheme 51.1

The initial screening of candidate materials was carried out with density functional theory (DFT) modeling of the electronic properties of a number of structures with different heterocyclic P-rings (Scheme 51.1) to identify those with high triplet energies and frontier orbital energies suitable for hosting our desired sky-blue dopant phosphor, i.e., iridium(III)bis((4,6-di-fluorophenyl)-pyridinato-N,C2′)picolinate (FIrpic). The primary requirement for hosting FIrpic is a triplet energy greater than 2.7 eV [22] to prevent luminescence quenching by triplet energy transfer. The first *Tier* of our screening process examined a range of P-ring substitutions to find the heteroatom substitutions that gave the highest triplet energy. A second *Tier* of screening started with imidazole for the P-ring and involved nitrogen incorporation into the H-rings, leading to the choice of three promising organic host materials to be prepared and tested in FIrpic-based phosphorescent OLEDs. A third *Tier* of screening searched for high charge carrier mobilities by modeling the solid-state properties of the three materials through molecular dynamics (MD) simulations. Computationally inexpensive electron coupling dimer-splitting calculations were used to predict charge transport properties from MD simulations, revealing poor electron mobility predicted for one of the three materials. Although our gas-phase DFT predictions (*Tiers* 1 and 2) and experimental measurements of the physical properties of the three materials showed them to be very similar, OLEDs prepared with each as a host for a blue phosphorescent dopant gave different performances, as predicted by *Tier* 3, with external efficiencies ranging from 0.6 to 9%.

51.2 Experimental Section

51.2.1 Quantum Mechanics

All QM calculations reported in this work were performed using the Materials Science Suite [23] of programs developed by Schrödinger LLC. on a 64-core workstation. Density functional theory was used for ground-state optimization and estimation of HOMO/LUMO energies, T_1 (triplet) state energy, and hole/electron reorganization energies (λ^+, λ^-) for the molecules. Hole/electron reorganization energies were calculated using Eqs. 51.1 and 51.2.

$$\lambda^+ = (E^0_+ - E^0_0) + (E^+_0 - E^+_+) \qquad (51.1)$$

$$\lambda^- = (E^0_- - E^0_0) + (E^-_0 - E^-_-) \qquad (51.2)$$

In the above expressions, the superscripts represent the state of the molecule (0, +, and − for neutral, cationic, and anionic states, respectively), whereas subscripts indicate the state at which the structure is optimized. For instance, according to this terminology,

E_+^0 represents the energy of the neutral ground state at the cation-optimized geometry.

Calculations involving cationic, anionic, and triplet states were performed using the unrestricted Kohn–Sham DFT scheme. Excited singlet-state calculations were performed using time-dependent DFT (TDDFT).

DFT calculations in *Tier* 1 and *Tier* 2 were performed using B3LYP/MIDIX and B3LYP/LACV3P** levels for screening, respectively. All DFT calculations were implemented using the Schrödinger's Jaguar [24, 25] program. Canvas [26] was used to visualize and sort results between *Tiers*.

51.2.2 Molecular Dynamics/Electron Coupling Calculations

All MD simulations reported in this work were carried out using the Desmond [27] module in the Materials Science Suite on a GPU workstation. The following multistage MD simulation protocol was implemented for each of the molecular systems reported in this work. A disordered system of 125 molecules in the simulation box (with periodic boundary conditions) was prepared using the disordered system builder facility available within Schrodinger's Materials Science Suite [23], following which a series of six short MD simulations of 1.2 ns each were performed in the canonical (NVT) ensemble with the temperature between two consecutive stages stepped up by 100 K increments starting from 300 to 800 K for the final stage. Following the NVT stages, a 30 ns-long MD simulation was carried out in the isothermal–isobaric (NPT) ensemble at a fixed temperature and pressure of 300 K and 1.01 atm, respectively, to expedite the simulation process. Postsimulation trajectory analysis on the last 6 ns of the NPT stimulation was performed to log density. The standard deviation of the calculated density was found to be less than 0.5% of the averaged densities calculated for all the systems ensuring convergence. The OPLS-2005 force field [28] available in the Materials Science Suite was used for all of the simulations.

To calculate the hole/electron hopping rates, charge coupling calculations were performed on the MD equilibrated structures. The hopping rates (κ) for charge transfer between two molecules can be calculated using the Marcus theory, Eq. 51.3.

$$\kappa = \frac{4\pi^2}{h} \frac{H_{ab}^2}{\sqrt{4\pi \lambda k_B T}} \exp\left(-\frac{(\lambda + \Delta G)^2}{4\lambda k_B T}\right) \quad (51.3)$$

Here, λ, H_{ab}, k_B, and T denote the hole/electron reorganization energy, intermolecular coupling parameter, Boltzmann constant, and temperature, respectively. ΔG is the free energy difference for the charge-transfer process, which is equal to the enthalpy difference for the process in this case. It should be noted that the reorganization energy λ is made up of two terms: inner-sphere reorganization energy (λ_{in}) accounting for the change in energy caused by geometric relaxation of a molecule upon addition/removal of an electron and outer-sphere reorganization energy (λ_{out}) accounting for relaxation/polarization of the surrounding medium, which is expected to be much smaller in rigid solids and is hence neglected in the computation of mobility in most cases. Additionally, λ_{out} is more tedious to compute than λ_{in}, which can be easily derived from gas-phase DFT calculations. Therefore, in this work, λ_{out} is neglected and λ in Eq. 51.3 becomes λ_{in}. The hopping rates are computed by two different approaches in this study. In the first approach, H_{ab} is approximated as half the difference between the LUMO and LUMO + 1 energies of the neutral dimer for electron transfer, $H_{ab}^- \approx \frac{1}{2}(E_{LUMO+1} - E_{LUMO})$ and similarly half the difference between HOMO and HOMO − 1 energies of the dimer for hole transfer, $H_{ab}^+ \approx \frac{1}{2}(E_{HOMO} - E_{HOMO-1})$. These dimer frontier orbital splittings are a measure of the intermolecular coupling for the electron and hole transfer processes and are obtained from single-point DFT calculations on all dimer pairs in the MD equilibrated structure within a closest approach distance of 4 Å between them. This approach is commonly referred to as the energy-splitting-in-dimer method in the literature [29–40]. Using the dimer frontier orbital splitting as a surrogate for the coupling parameter vastly reduces computational cost compared to more rigorous treatments, such as constrained-DFT (CDFT)-based approaches. It should be noted that in this approach, the free energy change for the process is not computed and set to zero for all of the dimer pairs. It has been shown that this approach can lead to an overestimation of the coupling parameter for nonsymmetrical cases and does not account for variation in onsite energies, so although being computationally very inexpensive, it is clearly not the best method to estimate coupling [41]. To confirm the validity of this rather crude approach in describing the charge transport properties of the systems under study, a more sophisticated treatment based on CDFT was implemented [42–44]. In this approach, H_{ab} is taken as the coupling matrix element between the wave function of the initial state where the electron/hole is localized on one of the monomers and the final state where the electron/hole is localized on the other monomer. ΔG is approximated as just the computed energy difference between the two states. CDFT is used to localize the electron/hole onto just one monomeric unit. To include electrostatic effects on the charge-transfer process from the neighboring molecules, the CDFT calculations were performed with each dimer embedded in a field of atomic point charges obtained from the partial charges (charges from the OPLS-2005 Force Field) of the atoms in the neighboring molecules within 4 Å from the dimer. The CDFT coupling calculations were performed on 80 randomly selected dimer pairs that are within a distance of 4 Å apart. Both the CDFT and dimer frontier orbital splitting coupling calculations were carried out at the B3LYP/LACV3P** level.

51.2.3 Synthesis

All starting materials were purchased from Sigma Aldrich and used as received, with the exception of 3,8-phenanthroline-5,6-dione [45] and 4,7-phenanthroline-5,6-dione [46], which were prepared by known procedures. Yields are reported as recovered yield after sublimation.

51.2.3.1 General procedure for synthesis of 1H-imidazo[4,5-*f*]-phenanthrolines

A 50 mL round-bottom flask was charged with dione (1 equiv), pivaldehyde (1 equiv), aniline (1.2 equiv), ammonium acetate (10 equiv), and acetic acid. The flask was fitted with a reflux condenser and placed in an oil bath at 130°C. The reaction mixture was allowed to reflux for 3 h with magnetic stirring under ambient atmosphere.

The mixture was allowed to cool to room temperature and poured into 100 mL 1 M aqueous NaOH. The resulting precipitate was collected by filtration. The precipitate was purified by silica column chromatography (hexane/ethyl acetate 1:3) followed by sublimation at 10^{-6} torr.

51.2.3.2 2-(Tert-butyl)-1-phenyl-1H-imidazo[4,5-*f*][1,10] phenanthroline (10d)

1,10-Phenanthroline-5,6-dione (630 mg, 3 mmol), pivaldehyde (258 mg, 3 mmol), aniline (335 mg, 3.6 mmol), ammonium acetate (2.31 g, 30 mmol), and acetic acid (15 mL) were used. Yield 755 mg, 71%. ^{1}H NMR (400 MHz, CDCl$_3$)δ 9.12 (dd, J = 4.1 Hz, J = 1.8 Hz, 1H), 9.05 (dd, J = 7.9 Hz, J = 2.0 Hz, 1H), 8.95 (dd, J = 4.5 Hz, J = 1.8 Hz, 1H), 7.72–7.61 (m, 4H), 7.56–7.61 (m, 2H), 7.61 (dd, J = 8.5 Hz, J = 4.4 Hz, 1H), 6.90 (dd, J = 8.5 Hz, J = 1.8 Hz, 1H), 1.39 (s, 9H); ^{13}C NMR (100 MHz, CDCl$_3$) δ 160.50, 148.55, 147.35, 144.63, 144.08, 139.19, 134.17, 130.39, 130.37, 130.06, 129.74, 127.52, 127.32, 123.95, 123.23, 121.92, 119.95, 35.52, 30.53. CHNS: calculated [C, 78.38; H, 5.72; N, 15.90] found: [C, 78.94; H, 5.83; N, 15.25].

51.2.3.3 2-(Tert-butyl)-1-phenyl-1H-imidazo[4,5-*f*][3,8] phenanthroline (10c)

3,8-Phenanthroline-5,6-dione (1.5 g, 7.14 mmol), pivaldehyde (615 mg, 7.14 mmol), aniline (798 mg, 8.56 mmol), ammonium acetate (5.5 g, 71.4 mmol), and acetic acid (36 mL) were used. ^{1}H NMR (400 MHz, CDCl$_3$)δ 10.13 (d, J = 0.8 Hz, 1H), 8.81 (d, J = 5.8 Hz, 1H), 8.62 (d, J = 5.8 Hz, 1H), 8.42 (d, J = 5.8 Hz, 1H), 8.35 (d, J = 5.8 Hz, 1H), 8.08 (d, J = 0.8 Hz, 1H), 7.74–7.64 (m, 3H), 7.59–7.54 (m, 2H), 1.43 (s, 9H); ^{13}C NMR (100 MHz, CDCl$_3$) 160.87, 147.08, 144.57, 143.96, 143.59, 139.27, 135.00, 131.49, 130.64, 130.54, 130.32, 129.51, 128.33, 123.13, 119.98, 117.35, 116.28, 35.58, 30.53. CHNS: calculated [C, 78.38; H, 5.72; N, 15.90] found: [C, 77.49; H, 5.86; N, 15.01].

51.2.3.4 2-(Tert-butyl)-1-phenyl-1H-imidazo[4,5-*f*][4,7] phenanthroline (10b)

4,7-Phenanthroline-5,6-dione (1.05 g, 5 mmol), pivaldehyde (438 mg, 5 mmol), aniline (558 mg, 6 mmol), ammonium acetate (3.85 g, 50 mmol), and acetic acid (25 mL) were used. Yield 640 mg, 36%. ^{1}H NMR (400 MHz, CDCl$_3$)δ 9.14 (dd, $J = 4.5$ Hz, $J = 1.6$ Hz, 1H), 8.87 (dd, $J = 8.2$ Hz, $J = 1.8$ Hz, 1H), 8.83 (dd, $J = 8.2$ Hz, $J = 1.8$ Hz, 1H), 8.43 (dd, $J = 4.4$ Hz, $J = 1.6$ Hz, 1H), 7.60–7.51 (m, 4H), 7.4–7.43 (m, 2H), 7.35 (dd,$J = 8.3$ Hz, $J = 4.3$ Hz, 1H), 1.45 (s, 9H); ^{13}C NMR (100 MHz, CDCl$_3$) 161.36, 150.59, 148.70, 143.15, 140.22, 140.03, 138.11, 131.84, 130.91, 130.81, 129.76, 128.89, 128.48, 122.74, 122.16, 120.18, 119.54, 35.66, 30.67. CHNS: calculated [C, 78.38; H, 5.72; N, 15.90] found: [C, 78.39; H, 5.74; N, 15.36].

51.2.4 Equipment

UV–visible spectra were recorded on a Hewlett-Packard 4853 diode array spectrometer. Photoluminescence spectra were measured using a QuantaMaster Photon Technology International phosphorescence/fluorescence spectrofluorometer. Quantum yield measurements were carried out using a Hamamatsu C9920 system equipped with a xenon lamp, calibrated integrating sphere, and model C10027 photonic multichannel analyzer. Photoluminescence lifetimes were measured by time-correlated single-photon counting using an IBH Fluorocube instrument equipped with an LED excitation source. UV–visible spectra were recorded in dichloromethane, and all the other photophysical measurements were carried out in 2-methyltetrahydrofuran (2-MeTHF). NMR spectra were recorded on Varian 500 and Varian 400 NMR spectrometers and referenced to the residual solvent resonance.

51.2.5 Electrochemistry

Cyclic voltammetry and differential pulse voltammetry were performed using a VersaSTAT 3 potentiostat. Anhydrous dimethylformamide (DMF) (Aldrich) was used as the solvent under inert atmosphere, and 0.1 M tetra(*n*-butyl)-ammonium hexafluorophosphate was used as the supporting electrolyte. A silver wire was used as the pseudo-reference electrode, and a platinum wire was used as the counter electrode. A glassy carbon rod was used as the working electrode. The redox potentials are based on the values measured using differential pulsed voltammetry and are reported relative to the ferrocenium/ferrocene (Cp_2Fe^+/Cp_2Fe) redox couple used as an internal reference while electrochemical reversibility was confirmed by cyclic voltammetry.

51.2.6 OLED Fabrication and Testing

OLEDs were fabricated on prepatterned indium tin oxide (ITO)-coated glass substrates (20 ± 5 Ω/cm^2, Thin Film Devices, Inc.). Prior to organic deposition, the substrates were cleaned by subsequent washes in tergitol solution, water, and acetone followed by a 10 min UV-ozone treatment. All organic materials were sublimed by thermal gradient sublimation in a 3-zone furnace. Organic layers were deposited by vacuum thermal evaporation at deposition rates of 1–2 Å/s at chamber pressures of 10^{-7}–10^{-6} Torr in an EVO Vac 800 deposition system from Angstrom Engineering. Aluminum cathodes were deposited through a shadow mask in a crossbar structure defining device areas of 4 mm^2.

OLED current–power and current–voltage curves, under applied forward bias of 0–12 V, were measured using a Keithley power source meter model 2400, a Newport multifunction optical meter model 1835-C, a low-power Newport silicon photodiode sensor model 818-UV, and a fiber bundle (used to direct the light into the photodiode). Electroluminescence spectra of the OLEDs were collected with a Photon Technology International QuantaMaster model C-60 fluorimeter at several voltages, between 3 and 11 V, to ensure emission characteristics remained constant.

51.3 Results and Discussion

51.3.1 Molecular Search Strategy

For our study, we chose to search for host materials for the well-studied sky-blue dopant FIrpic. The choice of this phosphorescent dopant sets limits on the electronic properties for the host material. To achieve a high luminance efficiency for a FIrpic dopant, the triplet energy of the host must be > 2.65 eV. To use the FIrpic-doped host material in an efficient OLED stack, the LUMO of the host needs to be near or above that of FIrpic, so the desired LUMO energies of the host are > -2.1 eV. Thus, we will use these two criteria to identify the most promising compounds in our computational search of the H2P structure space to carry into the experimental study. Isoelectronic, heteroatomic trans-mutations of CH fragments of the parent cyclopentaphenanthrene structure (Scheme 51.1) have qualitatively different effects whether they be in the P- or H-rings. For neutral aromatic structures, the H-rings may include pnictogens (only N was considered), whereas the P-ring may also include chalcogens (O and S were considered). Because the chalcogens, which provide a large chemical space with potentially desirable properties, are only available to the P-ring, our structure search was broken into two *Tiers*: *Tier* 1 selection (Fig. 51.1) focused on the heteroatoms of the P-ring heterocycle, whereas *Tier* 2 focused on aza substitution in the H-ring using the best candidates identified in *Tier* 1.

Tier 1 identified the candidates with triplet energies over 2.7 eV, based on a FIrpic dopant. Furthermore, the selection strategy in *Tier* 1 involved maximizing the LUMO energy, rather than optimizing it based on the LUMO of FIrpic. This strategy anticipates the aza substitution in *Tier* 2, which will categorically stabilize LUMO energies, so a destabilized LUMO energy from *Tier* 1 allows room for tunability through the desired range of LUMO energies in *Tier* 2 [11].

The *Tier* 1 selection involved a survey of 15 P-ring heterocycles, and the *Tier* 2 selection involved 37 unique H-ring aza substitution patterns. By carrying out the selection processes serially, choosing the best candidate in *Tier* 1 to carry into *Tier* 2, the number of structures that needed to be modeled dropped from 555 (15 × 37), corresponding to all H-ring substitutions for each P-ring structure, to 52 (15 + 37) calculations necessary to survey the optimal chemical space of H2P structures.

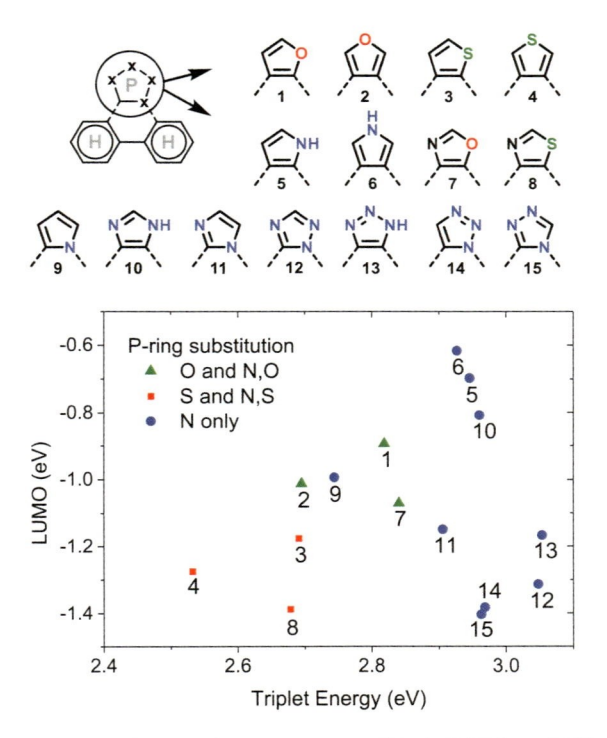

Figure 51.1 (Top) Elemental enumeration of O, S, N/NH, and CH in the P-ring of the H2P structure. (Bottom) Lowest unoccupied molecular orbital energy vs triplet energy for iterative cyclo-pentaphenanthrene structures.

51.3.2 *Tier* 1 Selection

The first *Tier* selection involved incorporation of oxygen, sulfur, and nitrogen for X into the P-ring of the H2P framework to produce the library of chemically relevant structures shown in Fig. 51.1. DFT calculations were performed at the B3LYP/MIDIX level for screening, which is a low level of theory suited for rapid screening. The triplet energies of S- and O-substituted P-rings were lower than that of N-substituted compounds, with nonchalcogen-containing compounds **10–15** having the highest triplet energies. The highest triplet energies were predicted for triazoles **12** and **13**, which, although are promising candidates, fail to meet the *Tier* 1 requirement of maximized LUMO energy. *Tier* 2 N-substitutions of

12 and **13** would likely result in LUMO energies below the desired range. The target structures were therefore **5**, **6**, and **10** since they present both high triplet and less negative LUMO energies. Compound **10** was selected because it and substituted versions of it can be readily prepared from phenanthrene-9,10-dione or its aza-substituted analogs, as illustrated in Scheme 51.2. It is noteworthy that whereas 6,6,6,5-membered tetracyclic imidazo[4,5-*f*]1,10-phenanthroline derivatives have been investigated as ligands in phosphorescent emitters for OLEDs [47–49], they have never been used as host materials for either fluorescent or phosphorescent OLEDs. Anticipating a crystalline rather than glassy morphology of vapor-deposited films of planar **10**, as well as potentially recalcitrant synthetic preparations, substituent functional groups were chosen for the imidazole before *Tier* 2 selection. When R_1 = phenyl, the singlet and triplet energies are markedly red-shifted due to conjugation of the phenyl group with the imidazole ring. For example, when $R_1 = R_2 = H$, the S_1/T_1 energies are predicted by DFT to be 3.54/2.96 eV, whereas when $R_1 = Ph$, $R_2 = H$, the S_1/T_1 energies are predicted to be 3.33/2.56 eV. Thus, $R_1 = H$ or alkyl is preferred for high triplet energy; however, in synthetic trials,

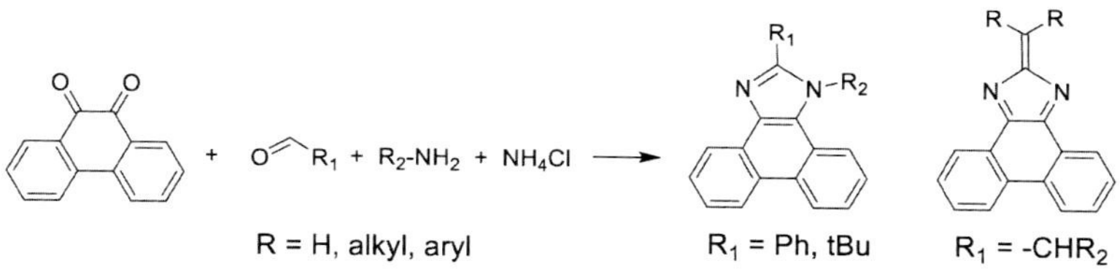

Scheme 51.2

we found that when R_1 is methyl, ethyl, or isopropyl, the reaction does not give **10** but the corresponding alkylidine 2H-imidazole compound instead (Scheme 51.2 far right). The derivative with R_2 = *tert*-butyl is well behaved due to the lack of α-protons, giving exclusively the desired imidazophenanthrene and is therefore the best choice in terms of S_1/T_1 energies as well as stability and solubility of the compound. In contrast to R_1 = phenyl, when R_2 is phenyl, steric interactions force the aromatic ring out of conjugation, so the orbital and excited-state energies are largely unaffected, for example, $R_1 = H$, $R_2 = Ph$ give S_1/T_1 energies of 3.30/2.89 eV (based on DFT calculations).

51.3.3 *Tier* 2 Selection

The second-*Tier* selection explored nitrogen substitution in the H-rings of **10**. Incorporating 0, 1, or 2 nitrogens into the H-rings of the parent phenanthro[4,5-*f*]imidazole gives a total of 37 unique aza-substituted compounds (see the Supporting Information (SI)). For the *Tier* 2 screening, we chose to carry out DFT calculations on the 37 aza-substitutions of **10** with $R_1 = tert$-butyl and $R_2 =$ phenyl using the B3LYP hybrid functional and LACV3P** basis set. In *Tier* 2, we chose a markedly larger basis set than was used in *Tier* 1, so direct comparisons of theoretically predicted properties to experimentally determined values will be meaningful. In these calculations, we predicted a number of parameters, including singlet and triplet energies, as well as molecular orbital energies and surfaces for each of the 37 molecules. The number of nitrogens and the site of nitrogen substitution in the H-rings markedly affect the electronic properties of the H2P molecules. For a complete listing of the parameters predicted in the *Tier* 2 calculations, see the SI. Figure 51.2 plots two of the calculated *Tier* 2 properties: LUMO energy vs triplet energy. As expected [11, 50], the LUMO is stabilized with nitrogen substitution in the H-ring, shifting from −1.0 eV for **10a** to a range of −1.2 to −1.4 for a single nitrogen to −1.4 to −1.9 for two nitrogens, with the site of nitrogen substitution markedly affecting the degree of stabilization of the LUMO. The greatest stabilization is seen for substitution in the 3- and 8-positions and the least for substitution at the 2- and 9-positions. The larger range of LUMO energies seen for the molecules with two nitrogens is due to an additive effect when both nitrogens are in a single H-ring. The ortho, meta, and para derivatives together give LUMO energies within a range of roughly 0.2 eV and an average of −1.7 eV, whereas the materials with a single nitrogen in each H-ring give the same range of LUMO energies but with an average LUMO energy of −1.55 eV. The site of nitrogen substitution also affects the triplet

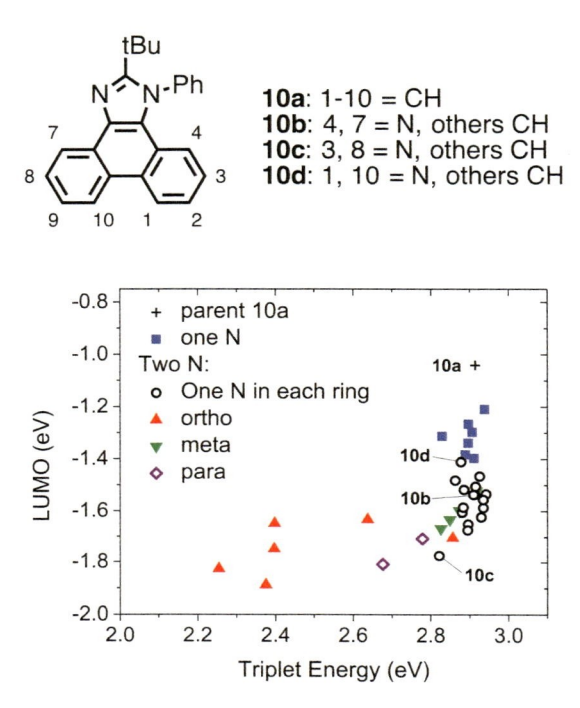

10a: 1-10 = CH
10b: 4, 7 = N, others CH
10c: 3, 8 = N, others CH
10d: 1, 10 = N, others CH

Figure 51.2 LUMO vs triplet energy for second-*Tier* iteration of aza substitution in the phenanthrene section of the H-rings of the parent phenantho[4,5-*f*]imidazole. Compounds **10a–d** are illustrated by colored circles. The identities of the other compounds in the screen are given in the SI.

energy but not in the same manner as it does the LUMO energies. Substitution of a single nitrogen into the two H-rings, two nitrogens meta in a single H-ring, or a single nitrogen in both H-rings gives a minimal change in the triplet energy of the molecule, relative to the unsubstituted compound **10a**. In contrast, substitution of two nitrogens into a single H-ring in either an ortho or para disposition lowers the triplet energy substantially below 2.8 eV in nearly every case (the exception has N in the 2,3-positions). The reason for the marked red shift in the ortho- and para-substituted derivatives is a filled–filled interaction [51] of the two nitrogen lone pairs in these compounds, leading to a marked destabilization of the out-of-phase combination of the lone pair orbitals and a resultant narrowing of the HOMO–LUMO gap. This interaction of filled nonbonding orbitals is not seen for the meta-substituted derivative or those with a single nitrogen in each H-ring.

The *Tier* 2 screen also included estimation of the electron and hole reorganization energies for each of the H2P molecules. Reorganization energies are useful parameters to evaluate the kinetics of intermolecular hole and electron hopping and assess charge carrier conduction. A lower reorganization energy reduces the barrier to carrier hopping between molecules in the thin film, and efficient carrier conduction is more favorable for a given material. The compounds in the *Tier* 2 screen ranged from low to moderate reorganization energies for holes (0.24–0.30 eV) and electrons (0.18–0.40 eV), suggestive of efficient carrier transport in the H2P family (see the SI for a full listing).

We chose to focus on three imidazo[4,5-*f*]phenanthroline derivatives **10b**, **10c**, and **10d** for characterization and study in blue OLEDs due to their high triplet energies and favorable reduction potentials. Additionally, these three materials have symmetric nitrogen substitution patterns in the H-rings, so only a single regioisomer will be formed for each derivative by our synthetic approach.

Figure 51.3 shows the DFT-optimized ground-state geometries along with the HOMO and LUMO orbitals for **10a–d**. The orbital density distributions of the HOMOs and LUMOs for **10b–d** are very similar. The LUMOs are localized on the phenanthroline, whereas the HOMOs involve both the phenanthroline and imidazole fragments. The addition of *tert*-butyl and phenyl groups has a negligible effect on the composition of the HOMO or LUMO of these compounds. For comparison, the HOMO and LUMO are shown for **10a** as well. The HOMO matches that seen for **10b–d**, but the LUMO of **10a** is localized on the phenyl group, rather than the phenanthroline, due to the relative difficulty of reducing phenanthrene compared to phenanthroline. The LUMO + 1 orbital of **10a** is a good match to the LUMO orbital of **10b–d**. Nitrogen substitution in the H-rings stabilizes the phenanthrene system, such that the phenanthroline orbitals fall below the phenyl-based π-orbitals for **10b–d**.

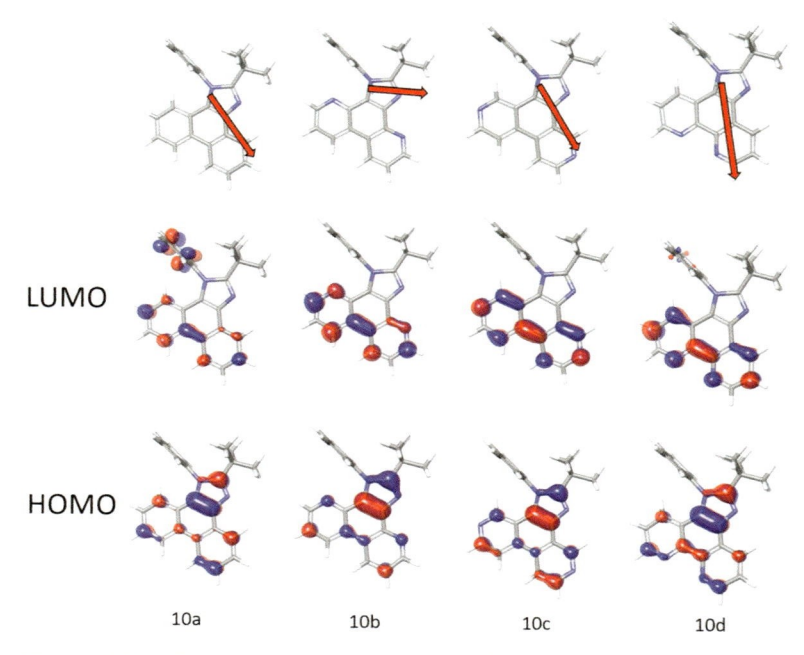

Figure 51.3 Highest occupied molecular orbital and lowest unoccupied molecular orbital diagrams calculated at the B3LYP/LACV3P** level of theory. The permanent dipole moment for each molecule is illustrated in the images at the top.

51.3.4 Synthesis and Physical Characterization of H2P Host Materials

Compounds **10b–d** were synthesized in 60–80% yields as illustrated in Scheme 51.2, where the phenanthrenedione precursor is replaced with the appropriate phenanthroline-dione. The H2P compounds exhibit intense $\pi - \pi^*$ absorptions between 250 and 290 nm. A weak tail from 320 to 400 nm was observed for **10b** and **10d**, whereas a more intense and distinct absorption is seen for **10c** in the same spectral region. The absorption features at longer wavelengths suggest a degree of charge-transfer character, which is the strongest for **10c**. Solutions of **10b**, **10c**, and **10d** have emission maxima at 392, 402, and 396 nm, respectively, at room temperature. Singlet excited-state energies corresponding to the energy of the short wavelength edge of the emission band (intensity $= 0.1 \times \lambda_{max}$) for the three compounds are 3.47, 3.20, and 3.42 eV, respectively. The triplet energies (E_T) for **10b**, **10c**, and **10d** at 2.91, 2.76, and 2.83 eV, respectively, were determined from the high energy edge of the phosphorescence spectra measured in 2-MeTHF at 77 K. There is a good correspondence between the calculated and experimental triplet energies. The calculated triplet energies are within 0.05 eV for **10a**, **10b**, and **10d** and within 0.13 eV for **10c**. Triplet energies obtained from neat solids are lower than those in solution. Triplets in the solid state were found to be $E_T = 2.65$ eV for **10b**, 2.47 eV for **10c**, and 2.62 eV for **10d** (Fig. 51.4). The triplet energy depression in the solid state, of approximately 250 meV, for all three materials could not have been predicted from the modeling in *Tiers* 1 and 2, prompting us to continue with a third *Tier* (vide infra).

The photoluminescent quantum yields (Φ_{PL}) of 10% FIrpic-doped films are 0.61, 0.40, and 0.20 for **10d**, **10b**, and **10c**, respectively. The photoluminescence efficiency of an FIrpic-doped film of **10b–d** will limit the internal quantum efficiency for OLEDs made with each host material. The low Φ_{PL} for doped films of **10c** is due to the low triplet energy of that host, such that significant quenching of FIrpic emission is expected. However, **10b** with the highest triplet energy did not provide the highest Φ_{PL}, likely due to a lower average triplet energy in the solid state than that of **10d**, leading to partial quenching of FIrpic emission in **10b**.

Cyclic voltammograms of **10a–d** showed quasi-reversible reduction waves; however, only **10d** gives a quasi-reversible oxidation, whereas oxidation of **10b** and **10c** is irreversible (see Table 51.1 for potentials). DFT calculations suggest that reduction occurs predominantly on the phenanthroline fragment and oxidation involves the imidazole ring (Fig. 51.3). The electrochemical potentials are consistent with these predictions. The imidazole groups are not significantly affected by the nitrogen position in the H-rings, leading to a closely spaced set of oxidation potentials for **10b–d** (1.11–1.17 V). In contrast, the phenanthroline fragment is significantly affected by the nitrogen position, leading to a more disparate set of reduction potentials (-2.25 to -2.54 V). Placement of nitrogens in the 3-, 8-positions of the phenanthroline gives **10c** the lowest reduction potential. HOMO and LUMO levels in Table 51.1

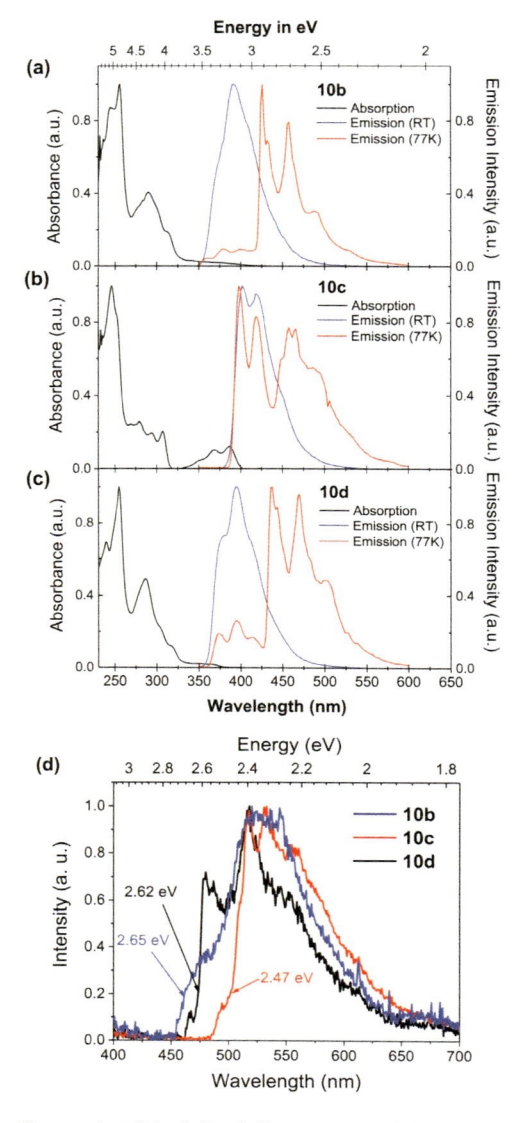

Figure 51.4 Absorption (black line), fluorescence (blue line), and cryogenic (77 K) spectra (red line) of (a) **10b**, (b) **10d**, and (c) **10c**. In each subfigure, the bottom axis refers to wavelength and the top axis denotes energy in eV. (d) Solid-state phosphorescence spectra: collected during the 0.1–1.0 ms period after the excitation pulse to remove the contribution from fluorescence.

were calculated from redox potentials using previously published correlations [52, 53].

The HOMO and LUMO values for **10a–d** were computed using the adiabatic scheme (implemented in the Materials Science Suite) at the B3LYP/LACV3P** level where single-point energies were computed for the cation, anion, and neutral species using the finite-element Poisson–Boltzmann solver continuum solvent model [54, 55] with DMF as the solvent on the corresponding gas-phase optimized structures. The HOMO/LUMO energies calculated using this procedure are in reasonable agreement with experimental values as seen in Table 51.1, differing by 0.2–0.3 eV; however, the trends in the values between the four compounds experimentally were reproduced in the theoretical values.

51.3.5 *Tier* 3 Selection

The gas-phase calculations in *Tiers* 1 and 2 agree well with the properties of the three host compounds in fluid solution. However, these single-molecule calculations cannot predict the solid-state properties of the bulk materials. The decrease in triplet energy from solution to neat solid (vide supra) prompted us to include a third *Tier* of modeling to address solid-state properties. Because modeling excited-state energies in amorphous solids is challenging, and because triplet energy depressions were similar for the three materials, attention was turned to charge transport, the critical role of a host material. To screen for the performance of **10b–d** as host materials, we carried out theoretical studies of the three materials as amorphous solids using MD and electron coupling DFT calculations. The search criterion for this *Tier* 3 screen is maximization of charge mobility. MD simulations were performed on a cell of 125 molecules of each host material with periodic boundary conditions to model solid-state morphologies. The final density for each simulation was between 1.11 and 1.14 g/cm^3. As a first approximation of charge mobility, the number of $\pi - \pi$ interactions for **10b**, **10c**, and **10d** was counted, with more interactions presumably being favorable for charge transport. Here, the $\pi - \pi$ interactions are classified into two types: (1) face–face interactions, where two aromatic rings are within a distance of 4.4 Å and a maximum angle of 30° with respect

Table 51.1 Summary of calculated and experimental physical properties for **10b**, **10c**, and **10c**

	E^{ox}/E^{red} (V/V)[a]	HOMO (eV) Calcd	HOMO (eV) Exp.[d]	LUMO (eV) Calcd	LUMO (eV) Exp.[d]	$\lambda(+)^e$ (eV)	$\lambda(-)^f$ (eV)	E_T (eV) Calcd	E_T (eV) Solution	E_T (eV) Solid	Δ(sol–sol)	Φ_{PL}^g
10a	0.89[b]/−2.97[b]	−5.47	−5.85	−1.04	−1.27	0.24	0.21	2.90	2.86	2.70	0.16	
10b	1.15[c]/−2.54[b]	−5.91	−6.21	−1.54	−1.76	0.27	0.40	2.91	2.92	2.65	0.27	0.40
10c	1.11[c]/−2.25[b]	−5.88	−6.15	−1.77	−2.10	0.25	0.30	2.84	2.71	2.47	0.24	0.20
10d	1.17[b]/−2.51[b]	−5.82	−6.24	−1.41	−1.79	0.28	0.22	2.90	2.85	2.62	0.23	0.61

[a] Oxidation and reduction potentials vs ferrocene/ferrocenium redox couple.
[b] Quasi-reversible.
[c] Irreversible.
[d] HOMO and LUMO energy levels were calculated from the redox potential with published correlation [51, 52].
[e] Calculated hole reorganization energy.
[f] Calculated electron reorganization energy.
[g] Photoluminescent quantum yield of vacuum-deposited films containing 10% FIrpic-doped H2P.

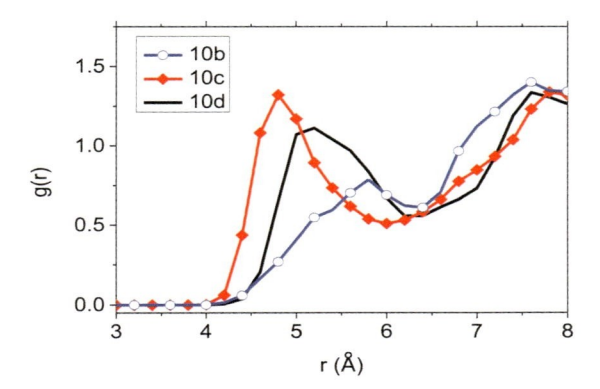

Figure 51.5 Center-of-mass radial distributions [$g(r)$] from MD simulations of three host materials, obtained by averaging over 30 ns.

to each other; and (2) edge–face interactions, where two aromatic rings are within a distance of 5.5 Å with a minimum subtended angle of 60° between them. The number of $\pi-\pi$ contacts includes the total number of face–face and edge–face interactions, giving a total of 324, 409, and 364 for **10b**, **10c**, and **10d**, respectively. Compound **10c** has substantially more face-to-face π-contacts than either **10b** or **10d**, whereas **10d** has more face-to-edge π-contacts than the other two compounds. This preliminary analysis suggests a trend in charge transport as **10c** > **10d** > **10b**. A similar trend is seen for the center-of-mass (COM) radial distribution functions (RDFs), as shown in Fig. 51.5. The COM RDFs reflect differences in solid-state center-to-center distances, which could have an effect on charge transport in the host material. The **10b** RDF clearly shows differences in distances in the range of 4–6 Å. There is approximately a 35 and 50% difference in height between the nearest-neighbor peak maximum of the **10b** RDF and the nearest-neighbor peak maximum of **10d** and **10c** RDFs, respectively. This indicates a lower proportion of short center-to-center distances in bulk **10b** with respect to the other H2Ps.

A plausible explanation for the dissimilar morphologies of **10b** from **10c** and **10d** involves the dipole moments of each molecule. In all three cases, the permanent dipole moment for each molecule lies in the H2P plane; however, in **10c** and **10d** the dipole extends from the imidazole ring into the phenanthroline, whereas in **10b** the dipole is largely within the imidazole ring (Fig. 51.3). In the amorphous solid, the molecules tend to form dimer pairs with adjacent dipoles in a roughly antiparallel orientation. For **10c** and **10d**, closely spaced dimers can be formed with antiparallel dipoles; however, for **10b**, overlapping the dipoles of adjacent molecules is sterically hindered by the *tert*-butyl and phenyl groups on the imidazoles.

Complimentary to structural analysis of MD simulations, a quantum mechanical analysis was performed. Dimer frontier orbital splitting electron coupling calculations (see the Methods section) were performed on the MD equilibrated structure for all dimer pairs within a contact distance of 4 Å from each other (note that this is the closest atom–atom distance for any pair, not the center-to-center distance or radius) amounting to a total of 838, 857, and 880 dimer pairs for **10b**, **10c**, and **10d**, respectively. These calculations were used to estimate the charge carrier hopping rates according to Eq. 51.3 to assess variations in charge transport between neat host materials. Figure 51.6 shows histograms of the calculated hole and electron hopping rates. Notably, the distribution of electron hopping rates for **10b** is significantly narrower, and the rates are slower than those of **10c** and **10d**. There were no significant differences in hole hopping rates between the three host materials. This suggests that on average the dimer pairs formed for **10b** give comparable HOMO–HOMO overlap to those of **10c** and **10d**, but poorer LUMO–LUMO overlap compared to **10c** and **10d**.

Although feasible for this study, transport calculations carried out for a large number of dimer pairs of molecules is too time consuming and impractical to be repeated for a large number of different materials. To develop a more rapid, high-throughput *Tier 3* materials screen, we sought to test the validity of a smaller scale estimate of the hopping rates for each of the materials using a truncated random subset of dimer pairs. Calculations were repeated for twenty random dimers from the exhaustive set of dimer pairs with a spacing of 4 Å or less. The distribution of hopping rates for the smaller sets mirrors what we observed for the exhaustive calculations (Fig. 51.6), suggesting that a less rigorous calculation

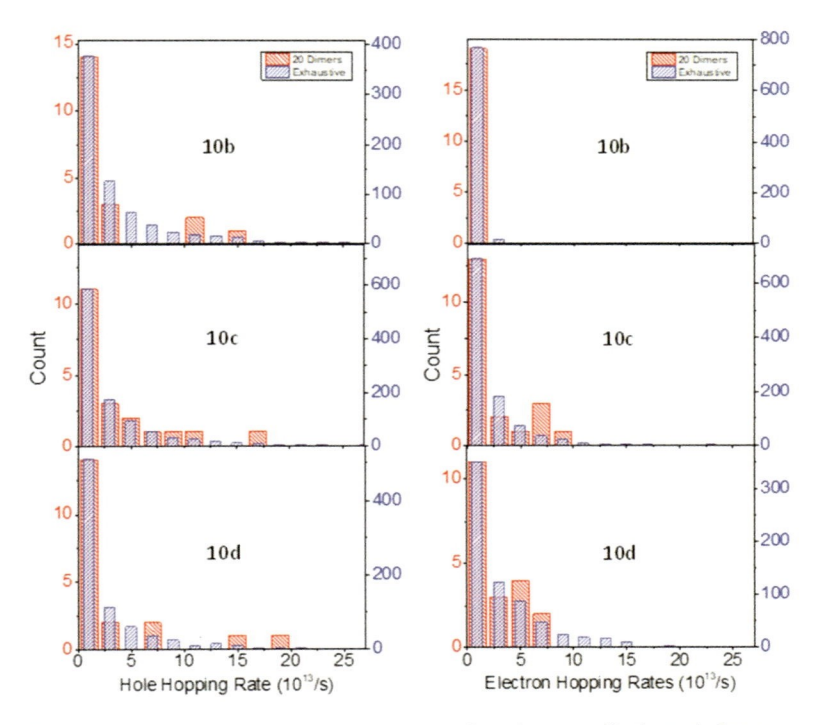

Figure 51.6 Histogram plots showing the distribution of hole and electron hopping rates extracted from the frontier dimer orbital splitting coupling calculations of both the exhaustive dimer set and the smaller 20 dimer subset for the three host materials (top, **10b**; middle, **10c**; bottom, **10d**).

could be used in the future to compare the range of hopping rates expected for different materials.

The calculated hopping rates of 500 randomly selected dimer pairs from the exhaustive dimer-splitting coupling calculations of each system were then used to compute charge carrier mobilities (Table 51.2). The mobilities were also computed for the hopping rates (80 dimers) calculated from the CDFT approach to validate the simple dimer-splitting method. A simple transport model proposed by Goddard et al. [56] was used to compute the mobilities. In this model, the carrier mobility is described by the Einstein relation (Eq. 51.4)

$$\mu_{h/e} = \frac{eD}{k_B T} \quad (51.4)$$

Table 51.2 Number of $\pi-\pi$ contacts and calculated mobilities for **10b**, **10c**, and **10c**

| | π-contacts | | | | Isotropic mobility ($\times 10^{-4}$ m^2/(V s)) | | | |
| | | | | | Dimer splitting | | CDFT | |
	Total	π–ff (face-face)	π–ef (edge-face)	π–ff/ π–ef	μ_h	μ_e	μ_h	μ_e
10b	324	77	247	0.31	4.45	0.67	20.3	23.4
10c	409	142	267	0.53	6.47	3.46	27.5	53.8
10d	364	72	292	0.25	6.16	5.34	14.9	41.7

where D, the charge diffusivity, is calculated from the hopping rates (κ_i) obtained from the coupling calculations, Eqs. 51.5 and 51.6.

$$D = \frac{1}{6}\sum_i r_i^2 \kappa_i P_i \qquad (51.5)$$

$$P_i = \frac{k_i}{\sum_i k_i} \qquad (51.6)$$

The index i runs over all dimer pairs for which coupling calculations were done.

Both coupling methods predict that the electron mobility of **10b** would be significantly lower than that of **10c** and **10d**. The dimer-splitting method was found to predict a greater disparity in the electron mobility of **10c/10d** versus **10b**. Also, the hole mobility of **10c** as predicted by the CDFT approach is much larger than that of **10b** and **10d** compared to the dimer-splitting method. The reason for this disparity between the two methods may be attributed to the fact that the dimer-splitting approach does not account for the Gibbs free energy change (ΔG in Eq. 51.3) associated with the charge-transfer process in different dimer pairs, which can be substantial in disordered systems [57, 58]. It is important to stress that the dimer-splitting method is only useful in qualitative estimates of the coupling, which are used in calculating carrier mobilities to rank-order the members of the structurally similar molecules of a *Tier 3* set. The average coupling parameters computed using the two methods are in reasonably good agreement with each other (see

SI). The low electron mobility for **10b** is consistent with its greater center-of-mass intermolecular spacing leading to a lower average hopping rate for the 500 dimer pairs examined. *Tier* 3, performed at various levels of theory, predicts that **10b** may display poor electron mobility in the solid state and therefore fails to meet the selection criterion. It should be noted that the crude mobility calculations presented above are used here to qualitatively gauge the trends in the transport properties of the candidates and are not expected to be quantitatively accurate. More sophisticated multiscale methods like Kinetic Monte Carlo-based approaches among others have been developed to adequately compute mobilities of amorphous organic materials but tend to be more computationally expensive [57–63]. After three *Tiers* of screening on a 555-membered structure space, two materials were selected: **10c** and **10d**. For the benefit of validating the above theoretical methods, **10b** was also investigated experimentally as an example of a poorly performing material.

51.3.6 OLED Fabrication and Testing

We employed a relatively simple configuration for OLED devices with H2P host materials: ITO/NPD (20 nm)/mCP (5 nm)/H2P: FIrpic (10%, 30 nm)/BCP (50 nm)/LiF (1 nm)/Al (NPD = N,N'-di(1-naphthyl)-N,N'-diphenyl-(1,1'-biphenyl)-4,4'-diamine, mCP = 1,3-di(9H-carbazol-9-yl)benzene, BCP = 2,9-dimethyl-4,7-diphenyl-1,10-phenanthroline). NPD/mCP functioned as the hole-transporting layers, and BCP was introduced as both the hole-blocking and electron-transporting layer (energy-level diagram is given in Fig. 51.7a). Figure 51.7b shows the current density–voltage–brightness (J–V–L) characteristics of these devices. The highest current is observed for the devices based on 10c, with comparable light output at a given voltage observed for **10c** and **10d**, leading to higher external quantum efficiency (EQE) for **10d** than **10c**. The maximum EQE values are 0.7, 5.0, and 9.0% for devices with **10b**, **10c**, and **10d**, respectively. It is useful to compare these OLEDs to FIrpic-based OLEDs with a similar architecture but a conventional host material. N,N'-Dicarbazolyl-4,4'-biphenyl is a common OLED host with a triplet energy comparable to **10c** (2.6 eV). It was used as the host in a device architecture very similar to the one used

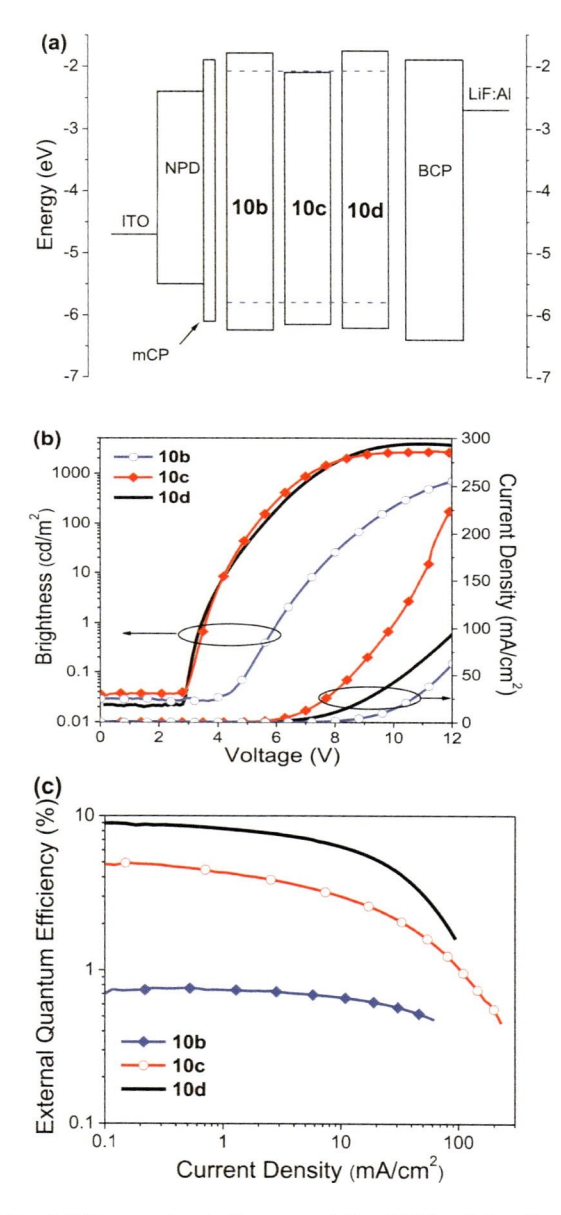

Figure 51.7 (a) Energy-level diagram of the OLEDs (blue line is FIrpic). The energies of the three different hosts are illustrated next to each other, but each device contained only a single host material. (b) shows $J-V-L$ characteristics and (c) is a plot of EQE versus current density.

here. The peak EQE was reported as 6.1% [64]. Shifting to a host material with a triplet energy of 2.9 eV, i.e., mCP, puts the host triplet close to those of **10b** and **10d**. The FIrpic-doped mCP OLED gave a peak efficiency of 7.5% [64]. The efficiencies of OLEDs based on **10c** and **10d** are comparable to those made with similar triplet energy carbazole-based hosts, whereas the efficiencies of **10b**-based devices fall well short of comparable carbazole-based OLEDs.

To understand the source of the differences in device efficiencies, we roughly deconvoluted the EQE into four limiting factors as shown in Eq. 51.7. The photoluminescent quantum yield (Φ_{PL}) was measured for phosphor-doped host films and is assumed to be unchanged in the device. The usable exciton fraction (χ) takes into account the statistical branching ratio of electrogenerated singlets to triplets with respect to the emissive species, which is unity for a phosphorescent dopant such as Firpic. The outcoupling factor (η_e) accounts for losses due to wave-guiding and plasmon absorption and is usually 0.2–0.3 [65]. Lastly, the charge recombination factor (η_r) is the ratio of excitons formed to injected charge carrier pairs, sometimes referred to as charge balance. Charge recombination factors that are less than unity are due to differential charge injection or transport between electrons and holes.

Table 51.3 lists the parameters from Eq. 51.7 for each host. The EQE and Φ_{PL} were measured from devices and thin films, respectively. The χ and η_e are assumed to be 1 and 0.2, respectively. The η_r is calculated from the other four parameters.

$$\Phi_{EL} = \Phi_{PL} \chi \eta_r \eta_e \tag{51.7}$$

The highest efficiency is seen for the device with a **10d**-based emissive layer. The device performance with **10c** is noteworthy since the device exhibited an EQE at its theoretical limit. Considering a

Table 51.3 Efficiency parameters of the OLEDs for **10b**, **10c**, and **10d**[a]

Host	EQE (%)	Φ_{PL}	Charge recombination factor η_e
10b	0.7	0.4	0.09
10c	4	0.2	1
10d	9	0.6	0.75

[a] η_r was computed assuming $\chi = 1$ and $\eta_e = 0.2$.

20% Φ_{PL} and outcoupling of 0.2, the maximum achievable efficiency for the OLEDs is no greater than 4%, assuming that the dopant is isotropically dispersed in the **10c** host [66]. This suggests near-unit efficiency for carrier recombination in these devices. The **10b**-based device exhibited the lowest device efficiency among the three hosts despite a 40% quantum yield in a 10% FIrpic-doped film due to a very low charge recombination factor. The trend in $J-V$ characteristics shows current densities of **10c** >**10d** > **10b** at a given voltage, agreeing well with the *Tier* 3 screening. Furthermore, the turn-on voltage (voltage at a brightness of 0.1 cd/m^2) for devices utilizing **10b** is ca. 2 V higher than those for devices with **10c** and **10d**. The turn-on voltage is tied to both injection barriers and carrier mobilities. The dependence on carrier mobility is due to the need for both carriers to diffuse into the EML prior to recombination to avoid exciplex formation at the interfaces between the EML and the transport layers. We expect the barriers to charge injection of the three host materials to be the same, so the principal factor controlling the turn-on voltage is likely the carrier mobilities of the devices. The trend in turn-on voltages is consistent with charge carrier mobilities within the devices of **10c, 10d** > **10b**.

By inspection of the energy-level diagram (Fig. 51.7a) and the nature of charge transport in other blue OLEDs [67], the dopant (FIrpic) mediates hole transport, whereas the host is expected to mediate electron transport. To test the predictions of the MD study

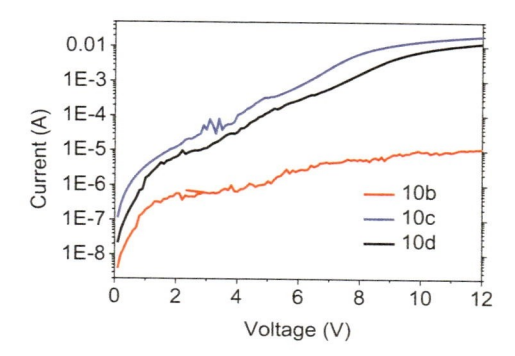

Figure 51.8 Current–voltage plots for electron-only devices, that is, ITO/BCP (10 nm)/H2P (40 nm)/LiF/Al.

with regard to electron transport characteristics of the three hosts, we fabricated electron-only devices with a structure of ITO/BCP (10 nm)/H2P (40 nm)/LiF/Al. In this device architecture, holes are blocked so the current passing through the organic layers is purely an electron current. Figure 51.8 shows the current versus voltage characteristics of the electron-only devices for the three materials. As anticipated, the trend in the $J-V$ characteristics shows current densities of **10c** > **10d** > **10b** at a given voltage, and the device with **10b** gave a current that is roughly 3 orders of magnitude lower than those of devices with either **10c** or **10d**.

51.4 Summary

In this chapter, we have described a computational approach to screen a family of 6,6,6,5-membered tetracyclic materials for use as host materials in blue-emitting phosphorescent OLEDs (PHOLEDs). Using computational methods to accelerate the materials discovery process, we identified the best candidates from a large library of different host structures before preparing them. Synthesizing and physically testing only those candidate materials with a high likelihood of success minimize the time taken to discover useful materials. The screening of the materials was based on DFT and TDDFT calculations for the candidate molecules in the gas phase and was used to identify the materials that gave the most promising triplet and LUMO energies for blue PHOLEDs. The criterion used to choose the host material also depends on the other materials used to fabricate the blue OLEDs. Here, we chose to search for host materials that are optimally designed for an FIrpic dopant, an NPD/mCP hole-transporting stack, and a BCP electron-transporting layer. If it were desirable to change the other materials in the PHOLED, that is, the emissive dopant or transport materials, the screening process could be reassessed to find the best host material for the new set of PHOLED materials. It is important to stress that the DFT/TDDFT calculations would not need to be run again as the new set of PHOLED materials would simply change the parameters used to screen the database of calculated molecular properties. One could take this process further and consider the materials in this library for another application all together. If one were interested in finding materials for photovoltaic applications for example, a different set of search criteria could be used to screen the same database of calculated properties to find the best candidate materials for that application. In the context of the *Tier* 1 and *Tier* 2 screening approaches reported here, the *Tier* 2 screen might need to be repeated with a different candidate or candidates advanced from *Tier* 1.

The screening methods based on gas-phase computational modeling allowed us to identify the materials with suitable triplet and LUMO energies to serve as host materials for blue PHOLEDs. The calculations gave excellent agreement with the experimental parameters measured in a fluid solution. Note that *Tiers* 1 and 2 were not sufficient to predict device performance. Extending our modeling studies of the three synthesized materials into the solid state in *Tier* 3, using MD simulations and modeling of the resultant equilibrated amorphous solids, we predicted the differences in the carrier transport we observed in the materials. Although the two host materials identified after *Tier* 3 screening (**10c** and **10d**) appeared to be well suited as PHOLED host materials, the gas-phase modeling studies did not include condensed-phase polarization or packing effects. We observed a marked red shift in the triplet energy between the gas-/solution-phase triplet energies and those in the solid state to the extent that emission from Firpic was either partially (**10d**) or mostly (**10c**) quenched. We are currently exploring further multiscale modeling methods to develop an expanded *Tier* 3 screen that will predict triplet energies of organic materials in the solid state. With such a multiscale "solid-state" method in hand, we can use a computational approach to accelerate materials discovery of compounds with triplet energies high enough to act as host materials in deep-blue PHOLEDs.

The total chemical space of materials considered in our *Tier* 1/*Tier* 2 screening process was 555 structures; however, it could be expanded to much larger chemical space. Recent reports from other groups have shown a similar computational screening approach used with libraries that are several orders of magnitude larger than the one considered here [12, 14, 15]. In a study from the Aspuru-Guzik group, the researchers identified 20 candidate TADF emitters for PHOLEDs from a 1.6 million member library, prepared them, and demonstrated external efficiencies as high as 20% for one of them, comparable to some of the best Ir-phosphor-based PHOLEDs. It is important to stress that this study involved dopants, which will be present in the OLED at low concentration and thus have no interactions between dopants. This is an important distinction from the thin-film host materials considered here, where intermolecular interactions mediate energy and electron transport. Gas-phase modeling studies are sufficient to predict PHOLED performance for an emissive dopant, such as a TADF material, but condensed-phase studies are needed for materials that will transport charge or excitons in the device. In the present study, we used a multiscale approach involving MD in conjunction with DFT calculations to model the solid-state properties of the three materials that we prepared and studied; however, this would not be practical for libraries with thousands of members or more. We are currently working to develop methods to streamline the condensed-phase modeling studies, with the goal to rapidly identify new host and transport materials for optoelectronic applications.

Supporting Information

The Supporting Information is available free of charge on the ACS Publications website at DOI: 10.1021/acsami.8b16225.

Cyclic voltammetry, list of molecules in *Tier* 2 library, FMO plots for unsubstituted H2Ps, list of calculated properties for molecules in *Tier* 2 library, average H_{ab} values computed using the dimer-splitting and CDFT approaches (PDF)

Author Information

Funding: One of the authors (Thompson) has a financial interest in the Universal Display Corporation.

Notes: The authors declare the following competing financial interest(s): Mark Thompson has a financial interest in the Universal

Display Corporation. Hsiao Fan Chen was a postdoc in Thompson's lab at the time this work was done, but he is now an employee of the Universal Display Corporation.

Acknowledgments

This work was supported by grant DE-EE0007077 of the US Department of Energy, and Universal Display Corporation. The authors would like to thank Schrödinger Inc. for access to their Materials Suite family of software tools, used extensively here, and Dr Mathew Halls for helpful discussions.

References

1. Tang, C. W., VanSlyke, S. A. Organic electroluminescent diodes. *Appl. Phys. Lett.* 1987, **51**, 913–915.
2. Baldo, M. A., O'Brien, D., You, Y., Shoustikov, A., Sibley, S., Thompson, M., Forrest, S. Highly efficient phosphorescent emission from organic electroluminescent devices. *Nature* 1998, **395**, 151–154.
3. Baldo, M., Lamansky, S., Burrows, P., Thompson, M., Forrest, S. Very high-efficiency green organic light-emitting devices based on electrophosphorescence. *Appl. Phys. Lett.* 1999, **75**, 4–6.
4. O'Brien, D., Baldo, M., Thompson, M., Forrest, S. Improved energy transfer in electrophosphorescent devices. *Appl. Phys. Lett.* 1999, **74**, 442–444.
5. Adachi, C., Baldo, M. A., Thompson, M. E., Forrest, S. R. Nearly 100% internal phosphorescence efficiency in an organic light-emitting device. *J. Appl. Phys.* 2001, **90**, 5048–5051.
6. Kwong, R. C., Nugent, M. R., Michalski, L., Ngo, T., Rajan, K., Tung, Y.-J., Weaver, M. S., Zhou, T. X., Hack, M., Thompson, M. E., et al. High operational stability of electrophosphorescent devices. *Appl. Phys. Lett.* 2002, **81**, 162–164.
7. Tsutsui, T., Yang, M.-J., Yahiro, M., Nakamura, K., Watanabe, T., Tsuji, T., Fukuda, Y., Wakimoto, T., Miyaguchi, S. High quantum efficiency in organic light-emitting devices with iridium-complex as a triplet emissive center. *Jpn. J. Appl. Phys.* 1999, **38**, no. L1502.

8. Adachi, C., Baldo, M. A., Forrest, S. R., Thompson, M. E. High-efficiency organic electrophosphorescent devices with tris (2-phenylpyridine) iridium doped into electron-transporting materials. *Appl. Phys. Lett.* 2000, **77**, 904–906.
9. Adachi, C., Baldo, M. A., Forrest, S. R., Lamansky, S., Thompson, M. E., Kwong, R. C. High-efficiency red electrophosphorescence devices. *Appl. Phys. Lett.* 2001, **78**, 1622–1624.
10. Adachi, C., Kwong, R., Forrest, S. R. Efficient electrophosphorescence using a doped ambipolar conductive molecular organic thin film. *Org. Electron.* 2001, **2**, 37–43.
11. Halls, M. D., Djurovich, P. J., Giesen, D. J., Goldberg, A., Sommer, J., McAnally, E., Thompson, M. E. Virtual screening of electron acceptor materials for organic photovoltaic applications. *New J. Phys.* 2013, **15**, no. 105029.
12. Aspuru-Guzik, A., Adams, R., Baldo, M., Aguilera-Iparraguirre, J., Gómez-Bombarelli, R. *34.4: Invited Paper: Combinatorial Design of OLED-Emitting Materials*, SID Symposium Digest of Technical Papers; Wiley Online Library, 2015; pp. 505–506.
13. Pyzer-Knapp, E. O., Suh, C., Gómez-Bombarelli, R., Aguilera-Iparraguirre, J., Aspuru-Guzik, A. What is high-throughput virtual screening? A perspective from organic materials discovery. *Annu. Rev. Mater. Res.* 2015, **45**, 195–216.
14. Hachmann, J., Olivares-Amaya, R., Atahan-Evrenk, S., Amador-Bedolla, C., Sánchez-Carrera, R. S., Gold-Parker, A., Vogt, L., Brockway, A. M., Aspuru-Guzik, A. The Harvard clean energy project: large-scale computational screening and design of organic photovoltaics on the World Community Grid. *J. Phys. Chem. Lett.* 2011, **2**, 2241–2251.
15. Gómez-Bombarelli, R., Aguilera-Iparraguirre, J., Hirzel, T. D., Duvenaud, D., Maclaurin, D., Blood-Forsythe, M. A., Chae, H. S., Einzinger, M., Ha, D.-G., Wu, T., Markopoulos, G., Jeon, S., Kang, H., Miyazaki, H., Numata, M., Kim, S., Huang, W., Hong, S. I., Baldo, M., Adams, R. P., Aspuru-Guzik, A. Design of efficient molecular organic light-emitting diodes by a high-throughput virtual screening and experimental approach. *Nat Mater* 2016, **15**, 1120–1127.
16. Shin, Y., Liu, J., Quigley, J. J., Luo, H., Lin, X. Combinatorial design of copolymer donor materials for bulk heterojunction solar cells. *ACS Nano* 2014, **8**, 6089–6096.
17. Clark, W., Litt, A. D., Steel, C. Triplet lifetimes of benzophenone, acetophenone, and triphenylene in hydrocarbons. *J. Am. Chem. Soc.* 1969, **91**, 5413–5415.

18. Togashi, K., Nomura, S., Yokoyama, N., Yasuda, T., Adachi, C. Low driving voltage characteristics of triphenylene derivatives as electron transport materials in organic light-emitting diodes. *J. Mater. Chem.* 2012, **22**, 20689–20695.
19. Kondakov, D., Lenhart, W., Nichols, W. Operational degradation of organic light-emitting diodes: mechanism and identification of chemical products. *J. Appl. Phys.* 2007, **101**, no. 024512.
20. Schmidbauer, S., Hohenleutner, A., König, B. Chemical degradation in organic light-emitting devices: mechanisms and implications for the design of new materials. *Adv. Mater.* 2013, **25**, 2114–2129.
21. Saris, P. J. G., Thompson, M. E. Gram scale synthesis of benzophenanthroline and its blue phosphorescent platinum complex. *Org. Lett.* 2016, **18**, 3960–3963.
22. Adachi, C., Kwong, R. C., Djurovich, P., Adamovich, V., Baldo, M. A., Thompson, M. E., Forrest, S. R. Endothermic energy transfer: a mechanism for generating very efficient high-energy phosphorescent emission in organic materials. *Appl. Phys. Lett.* 2001, **79**, 2082–2084.
23. *Materials Science Suite 2016-4*; Schrödinger, LLC: NY, 2016.
24. Bochevarov, A. D., Harder, E., Hughes, T. F., Greenwood, J. R., Braden, D. A., Philipp, D. M., Rinaldo, D., Halls, M. D., Zhang, J., Friesner, R. A. Jaguar: a high-performance quantum chemistry software program with strengths in life and materials sciences. *Int. J. Quantum Chem.* 2013, **113**, 2110–2142.
25. *Jaguar*; Schrödinger, LLC: NY, 2016.
26. *Canvas*; Schrödinger, LLC: NY, 2016.
27. *Desmond Molecular Dynamics System*; D. E. Shaw Research: NY, 2016.
28. Banks, J. L., Beard, H. S., Cao, Y. X., Cho, A. E., Damm, W., Farid, R., Felts, A. K., Halgren, T. A., Mainz, D. T., Maple, J. R., Murphy, R., Philipp, D. M., Repasky, M. P., Zhang, L. Y., Berne, B. J., Friesner, R. A., Gallicchio, E., Levy, R. M. Integrated modeling program, applied chemical theory (impact). *J. Comput. Chem.* 2005, **26**, 1752–1780.
29. Huang, J. S., Kertesz, M. Intermolecular transfer integrals for organic molecular materials: can basis set convergence be achieved? *Chem. Phys. Lett.* 2004, **390**, 110–115.
30. Lemaur, V., Da Silva Filho, D. A., Coropceanu, V., Lehmann, M., Geerts, Y., Piris, J., Debije, M. G., Van de Craats, A. M., Senthilkumar, K., Siebbeles, L. D. A., Warman, J. M., Bredas, J. L., Cornil, J. Charge transport properties in discotic liquid crystals: a quantum-chemical insight into structure-property relationships. *J. Am. Chem. Soc.* 2004, **126**, 3271–3279.

31. Hutchison, G. R., Ratner, M. A., Marks, T. J. Intermolecular charge transfer between heterocyclic oligomers. Effects of heteroatom and molecular packing on hopping transport in organic semiconductors. *J. Am. Chem. Soc.* 2005, **127**, 16866–16881.
32. Kwon, O., Coropceanu, V., Gruhn, N. E., Durivage, J. C., Laquindanum, J. G., Katz, H. E., Cornil, J., Bredas, J. L. Characterization of the molecular parameters determining charge transport in anthradithiophene. *J. Chem. Phys.* 2004, **120**, 8186–8194.
33. Newton, M. D. Quantum chemical probes of electron-transfer kinetics - the nature of donor-acceptor interactions. *Chem. Rev.* 1991, **91**, 767–792.
34. Cornil, J., Calbert, J. P., Bredas, J. L. Electronic structure of the pentacene single crystal: relation to transport properties. *J. Am. Chem. Soc.* 2001, **123**, 1250–1251.
35. Cheng, Y. C., Silbey, R. J., da Silva, D. A., Calbert, J. P., Cornil, J., Bredas, J. L. Three-dimensional band structure and bandlike mobility in oligoacene single crystals: a theoretical investigation. *J. Chem. Phys.* 2003, **118**, 3764–3774.
36. da Silva, D. A., Kim, E. G., Bredas, J. L. Transport properties in the rubrene crystal: electronic coupling and vibrational reorganization energy. *Adv. Mater.* 2005, **17**, 1072–1076.
37. Jordan, K. D., Paddonrow, M. N. Long-range interactions in a series of rigid nonconjugated dienes. 1. distance dependence of the Pi+,Pi- and Pi+Star,Pi-Star splittings determined by ab initio calculations. *J. Phys. Chem.* 1992, **96**, 1188–1196.
38. Paddon-Row, M. N., Jordan, K. D. Analysis of the distance dependence and magnitude of the Pi+, Pi- and Pi+asterisk, Pi-asterisk splittings in a series of diethynyl[N]staffanes - an ab initio molecular-orbital study. *J. Am. Chem. Soc.* 1993, **115**, 2952–2960.
39. Liang, C. X., Newton, M. D. Ab initio studies of electron-transfer - pathway analysis of effective transfer integrals. *J. Phys. Chem.* 1992, **96**, 2855–2866.
40. Coropceanu, V., Cornil, J., da Silva Filho, D. A., Olivier, Y., Silbey, R., Bredas, J.-L. Charge transport in organic semiconductors. *Chem. Rev.* 2007, **107**, 926–952.
41. Valeev, E. F., Coropceanu, V., da Silva, D. A., Salman, S., Bredas, J. L. Effect of electronic polarization on charge-transport parameters in molecular organic semiconductors. *J. Am. Chem. Soc.* 2006, **128**, 9882–9886.

42. Wu, Q., Voorhis, T. V. Extracting electron transfer coupling elements from constrained density functional theory. *J. Chem. Phys.* 2006, **125**, no. 164105.
43. Wu, Q., Van Voorhis, T. Constrained density functional theory and its application in long-range electron transfer. *J. Chem. Theory Comput.* 2006, **2**, 765–774.
44. Ratcliff, L. E., Grisanti, L., Genovese, L., Deutsch, T., Neumann, T., Danilov, D., Wenzel, W., Beljonne, D., Cornil, J. Toward fast and accurate evaluation of charge on-site energies and transfer integrals in supramolecular architectures using linear constrained density functional theory (Cdft)-based methods. *J. Chem. Theory Comput.* 2015, **11**, 2077–2086.
45. Botana, E., Da Silva, E., Benet-Buchholz, J., Ballester, P., de Mendoza, J. Inclusion of cavitands and calix[4]arenes into a metallobridged para-(1h-imidazo[4,5-f][3,8]phenanthrolin-2-yl)-expanded calix[4]arene. *Angew. Chem., Int. Ed.* 2007, **46**, 198–201.
46. Imor, S., Morgan, R. J., Wang, S., Morgan, O., Baker, A. D. An improved preparation of 4,7-phenanthrolino-5,6:5′,6′-pyrazine. *Synth. Commun* 1996, **26**, 2197–2203.
47. Tordera, D., Pertegas, A., Shavaleev, N. M., Scopelliti, R., Orti, E., Bolink, H. J., Baranoff, E., Graetzel, M., Nazeeruddin, M. K. Efficient orange light-emitting electrochemical cells. *J. Mater. Chem.* 2012, **22**, 19264–19268.
48. Zhang, F., Si, C., Wei, D., Wang, S., Zhang, D., Li, S., Li, Z., Zhang, F., Wei, B., Cao, G., Zhai, B. Solution-processed organic light-emitting diodes based on yellow-emitting cationic iridium(iii) complexes bearing cyclometalated carbene ligands. *Dyes Pigm.* 2016, **134**, 465–471.
49. Zhao, G.-W., Hu, Y.-X., Chi, H.-J., Dong, Y., Xiao, G.-Y., Li, X., Zhang, D.-Y. High efficient oleds based on novel Re(I) complexes with phenanthroimidazole derivatives. *Opt. Mater.* 2015, **47**, 173–179.
50. Winkler, M., Houk, K. N. Nitrogen-rich oligoacenes: candidates for N-channel organic semiconductors. *J. Am. Chem. Soc.* 2007, **129**, 1805–1815.
51. Knight, E. T., Myers, L. K., Thompson, M. E. Structure and bonding in group 4 metallocene acetylide and metallacyclopentadiene complexes. *Organometallics* 1992, **11**, 3691–3696.
52. D'Andrade, B. W., Datta, S., Forrest, S. R., Djurovich, P., Polikarpov, E., Thompson, M. E. Relationship between the ionization and oxidation potentials of molecular organic semiconductors. *Org. Electron.* 2005, **6**, 11–20.

53. Djurovich, P. I., Mayo, E. I., Forrest, S. R., Thompson, M. E. Measurement of the lowest unoccupied molecular orbital energies of molecular organic semiconductors. *Org. Electron.* 2009, **10**, 515–520.
54. Tannor, D. J., Marten, B., Murphy, R., Friesner, R. A., Sitkoff, D., Nicholls, A., Ringnalda, M., Goddard, W. A., Honig, B. Accurate first principles calculation of molecular charge-distributions and solvation energies from ab-initio quantum-mechanics and continuum dielectric theory. *J. Am. Chem. Soc.* 1994, **116**, 11875–11882.
55. Marten, B., Kim, K., Cortis, C., Friesner, R. A., Murphy, R. B., Ringnalda, M. N., Sitkoff, D., Honig, B. New model for calculation of solvation free energies: correction of self-consistent reaction field continuum dielectric theory for short-range hydrogen-bonding effects. *J. Phys. Chem.* 1996, **100**, 11775–11788.
56. Wen, S.-H., Li, A., Song, J., Deng, W.-Q., Han, K.-L., Goddard, W. A. First-principles investigation of anistropic hole mobilities in organic semiconductors. *J. Phys. Chem. B* 2009, **113**, 8813–8819.
57. Bässler, H. Charge transport in disordered organic photoconductors - a Monte-Carlo simulation study. *Phys. Status Solidi B* 1993, **175**, 15–56.
58. Friederich, P., Symalla, F., Meded, V., Neumann, T., Wenzel, W. Ab initio treatment of disorder effects in amorphous organic materials: toward parameter free materials simulation. *J. Chem. Theory Comput.* 2014, **10**, 3720–3725.
59. Massé, A., Coehoorn, R., Bobbert, P. A. Universal size-dependent conductance fluctuations in disordered organic semiconductors. *Phys. Rev. Lett.* 2014, **113**, no. 116604.
60. Friederich, P., Gomez, V., Sprau, C., Meded, V., Strunk, T., Jenne, M., Magri, A., Symalla, F., Colsmann, A., Ruben, M., Wenzel, W. Rational in silico design of an organic semiconductor with improved electron mobility. *Adv. Mater.* 2017, **29**, no. 1703505.
61. Friederich, P., Meded, V., Poschlad, A., Neumann, T., Rodin, V., Stehr, V., Symalla, F., Danilov, D., Ludemann, G., Fink, R. F., Kondov, I., von Wrochem, F., Wenzel, W. Molecular origin of the charge carrier mobility in small molecule organic semiconductors. *Adv. Funct. Mater.* 2016, **26**, 5757–5763.
62. Kordt, P., van der Holst, J. J. M., Al Helwi, M., Kowalsky, W., May, F., Badinski, A., Lennartz, C., Andrienko, D. Modeling of organic light emitting diodes: from molecular to device properties. *Adv. Funct. Mater.* 2015, **25**, 1955–1971.

63. Kwiatkowski, J. J., Nelson, J., Li, H., Bredas, J. L., Wenzel, W., Lennartz, C. Simulating charge transport in tris(8-hydroxyquinoline) aluminium (Alq3). *Phys. Chem. Chem. Phys.* 2008, **10**, 1852–1858.

64. Holmes, R. J., Forrest, S. R., Tung, Y.-J., Kwong, R. C., Brown, J. J., Garon, S., Thompson, M. E. Blue organic electrophosphorescence using exothermic host–guest energy transfer. *Appl. Phys. Lett.* 2003, **82**, 2422–2424.

65. Gather, M. C., Reineke, S. Recent advances in light outcoupling from white organic light-emitting diodes. *J. Photonics Energy* 2015, **5**, no. 057607.

66. Flämmich, M., Gather, M. C., Danz, N., Michaelis, D., Bräuer, A. H., Meerholz, K., Tünnermann, A. Orientation of emissive dipoles in oleds: quantitative in situ analysis. *Org. Electron.* 2010, **11**, 1039–1046.

67. Zhang, Y., Lee, J., Forrest, S. R. 10-Fold increase in the lifetime of blue phosphorescent organic light-emitting diodes. *Nat. Commun.* 2014, **5**, no. 5008.

Chapter 52

"Quick-Silver" from a Systematic Study of Highly Luminescent, Two-Coordinate, d^{10} Coinage Metal Complexes

Rasha Hamze,[a,*] Shuyang Shi,[a,*] Savannah C. Kapper,[a]
Daniel Sylvinson Muthiah Ravinson,[a] Laura Estergreen,[a]
Moon-Chul Jung,[a] Abegail C. Tadle,[a] Ralf Haiges,[a]
Peter I. Djurovich,[a] Jesse L. Peltier,[b] Rodolphe Jazzar,[b]
Guy Bertrand,[b] Stephen E. Bradforth,[a] and Mark E. Thompson[a]

[a]*Department of Chemistry, University of Southern California, Los Angeles, California 90089, USA*
[b]*UCSD-CNRS Joint Research Laboratory (UMI 3555), Department of Chemistry and Biochemistry, University of California, San Diego, La Jolla, California 92093-0358, USA*
met@usc.edu

A systematic study is presented on the physical and photophysical properties of isoelectronic and isostructural Cu, Ag, and Au complexes with a common amide (N-carbazolyl) and two different carbene ligands (i.e., CAAC = (5R,6S)-2-(2,6-diisopropylphenyl)-6-isopropyl-3,3,9-trimethyl-2-azaspiro[4.5]decan-2-ylidene, MAC = 1,3-bis(2,6-diisopropylphenyl)-5,5-dimethyl-4-keto-tetrahydropyri

---

*These authors contributed equally.
Reprinted from *J. Am. Chem. Soc.*, **141**, 8616–8626, 2019.

Electrophosphorescent Materials and Devices
Edited by Mark E. Thompson
Text Copyright © 2019 American Chemical Society
Layout Copyright © 2024 Jenny Stanford Publishing Pte. Ltd.
ISBN 978-981-4877-34-3 (Hardcover), 978-1-003-08872-1 (eBook)
www.jennystanford.com

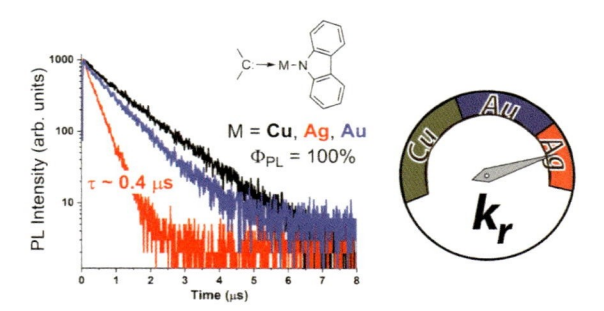

dylidene). The crystal structures of the (carbene)M(I)(N-carbazolyl) (M^{CAAC}) and (MAC)M(I)(N-carbazolyl) (M^{MAC}) complexes show coplanar carbene and carbzole ligands and C–M–N bond angles of ~180°. The electrochemical properties and energies for charge transfer (CT) absorption and emission compounds are not significantly affected by the choice of metal ion. All six of the (carbene)M(Cz) complexes examined here display high photoluminescence quantum yields of 0.8–1.0. The compounds have short emission lifetimes ($\tau = 0.33-2.8$ μs) that fall in the order Ag < Au < Cu, with the lifetimes of (carbene)Ag(Cz) roughly a factor of 10 shorter than for (carbene)Cu(Cz) complexes. Detailed temperature-dependent photophysical measurements (5–325 K) were carried out to determine the singlet and triplet emission lifetimes (τ_{fl} and τ_{ph}, respectively) and the energy difference between the singlet and triplet excited state, ΔE_{S1-T1}. The τ_{fl} values range between 20 and 85 ns, and the τ_{ph} values are in the 50–200 μs regime. The emission at room temperature is due exclusively to E-type delayed fluorescence or TADF (i.e., $T_1 \xrightarrow{\Delta} S_1 \rightarrow S_0 + h\nu$). The emission rate at room temperature is fully governed by ΔE_{S1-T1}, with the silver complexes giving ΔE_{S1-T1} values of 150–180 cm^{-1} (18–23 meV), whereas the gold and copper complexes give values of 570–590 cm^{-1} (70–73 meV).

52.1 Introduction

There have been several recent reports of highly luminescent, two-coordinate, d^{10} metal complexes of coinage metals, i.e., Cu,

Ag, Au [1]. These compounds have the general formula of (carbene)M[(I)](amide), such that the metal complex is neutral, with the metal ion in a linear coordinate geometry (C–M–N bond angle ≈180°). This combination of ligands leads to fascinating chromophores characterized by high luminescence efficiencies in solution and the solid state (close to unity in many cases) with phosphorescent lifetimes in the 1–3 µs range. The emission of the copper-based compounds can be tuned in color from violet to red through appropriate choice of carbene and amide ligands [1b, c]. These photophysical properties make the (carbene)M(amide) complexes attractive candidates for a range of applications including photocatalysis [2] and chemo- and biosensing [3] to dye-sensitized solar cells [4] and organic electronics [5]. In particular, these emissive materials are promising dopants in organic LEDs where efficient microsecond phosphorescence is requisite to achieving high electroluminescence efficiency. Indeed, OLEDs have been reported with external quantum efficiencies (EQEs) ranging from 10% to 25% that utilize (carbene)M[(I)](amide) emitters with M = Cu, Ag, Au [1a–c, 6]. These coinage-metal-based phosphors emit via E-type fluorescence or thermally assisted delayed fluorescence (TADF), unlike traditional Ir-based emitters that rely on strong spin–orbit coupling to induce emission from what is principally a triplet metal-to-ligand charge transfer (^{3}MLCT) excited state. The fast radiative lifetimes for these coinage metal complexes are due largely to the small energy separation between the lowest singlet and triplet excited states (ΔE_{S1-T1}), which facilitates intersystem crossing (ISC) from the long-lived triplet to the faster radiating singlet state. Herein, we systematically examine the photophysical properties of isoelectronic and isostructural Cu, Ag, and Au complexes with a common amide (N-carbazole) and two different carbene ligands (Fig. 52.1, CAAC = (5R,6S)-2-(2,6-diisopropylphenyl)-6-isopropyl-3,3,9-trimethyl-2-azaspiro[4.5]-decan-2-ylidene [7], MAC = 1,3-bis(2,6-diisopropylphenyl)-5,5-dimethyl-4-keto-tetrahydropyridyli-dene [1c]), all of which have high luminescence efficiency and short lifetimes.

The photophysical properties of d^{10} metal complexes have been widely studied since the initial discovery of their luminescent characteristics [8]. A few systematic photophysical studies

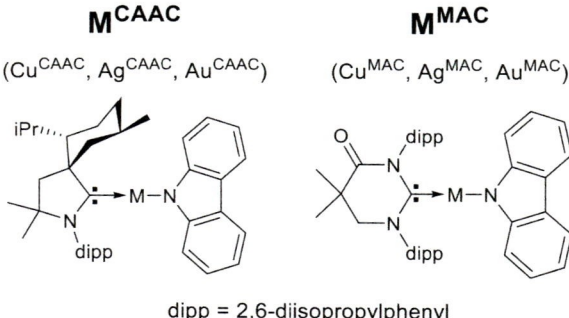

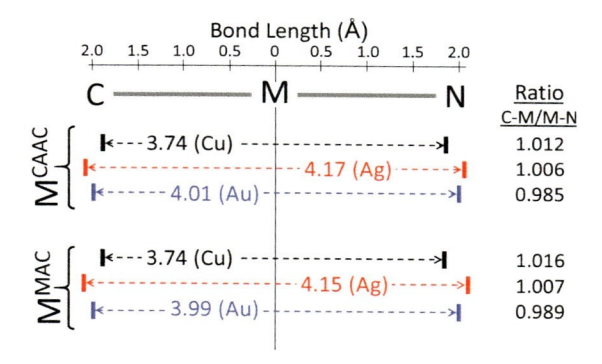

Figure 52.1 (Carbene)M(Cz) complexes studied here. M^{CAAC} and M^{MAC} are used to refer to the CAAC- and MAC-based complexes of Cu, Ag, and Au collectively. X-ray crystallographic studies of the six compounds show a linear coordination geometry (C–M–N = 174–180°) and coplanar ligands (dihedral angles = 0.1–3°). The C–M and M–N bond lengths are illustrated for the six compounds above. The C⋯N distances for both M^{CAAC} and M^{MAC} are Ag > Au > Cu.

have been reported for isostructural monovalent Cu, Ag, and Au complexes, including clusters [8q], and three- or four-coordinate mononuclear complexes [9]. However, emission from mononuclear complexes is often quenched in fluid solution owing to excited-state distortion and/or exciplex formation, which has limited the breadth of data analysis [10]. Copper-based complexes typically emit from predominantly MLCT excited states [8q, 11], or metal-halide-to-ligand charge transfer states for metal halide complexes [8q, 12]

In contrast to Cu(I) complexes, the Ag(I) and Au(I) analogues often show weak MLCT contributions to their excited states [13]. Hsu et al. reported Ag(I) complexes that displayed both fluorescence and weak ligand-centered phosphorescence owing to slower ISC than for the Cu(I) and Au(I) derivatives [9a]. The Au(I) derivatives showed faster ISC but still gave largely ligand-centered phosphorescence [9a].

The two-coordinate (carbene)M(amide) derivatives studied here behave quite differently from three- and four-coordinated complexes that emit from MLCT states. Whereas related (carbene)Cu(aryl) compounds also emit from MLCT excited states [14], pairing an electron accepting carbene with an electron donating amide leads to a lowest excited state that is principally an LLCT transition [1c, 6, 15]. A coplanar orientation of the ligands leads to parallel alignment of the donor (amide) and acceptor (carbene) p-orbitals that enhances interaction between the donor and acceptor ligands. Emission from the (carbene)Cu(Cz) complexes (Cz = carbazolyl) was shown to be E-type delayed fluorescence and is similarly observed in Ag and Au derivatives [1c, 6, 15]. In this process, the singlet (S_1) and triplet (T_1) manifolds are close enough in energy to give high rates of ISC between the long-lived T_1 state and the fluorescent S_1 state. Thus, luminescence entails thermally promoted ISC from the T_1 to the S_1 state followed by rapid emission from the S_1 state (i.e., $T_1 \xrightarrow{\Delta} S_1 \rightarrow S_0 + h\nu$). The process relies on an equilibrium between S_1 and T_1. Organic TADF materials typically have a triplet state whose lifetime is long enough to become essentially nonemissive at room temperature. Therefore, a small ΔE_{S1-T1} can greatly accelerate radiative decay via TADF. In our recent papers, we have suggested that thermally enhanced luminescence (TEL) is a better description for the emission process in these metal complexes since both the singlet and triplet states can have high emission efficiencies and the triplet has a lifetime in the microsecond regime [1b, c].

Here we extend our earlier photophysical studies of Cu^{CAAC} [1b] and Cu^{MAC} [1c] to include the isostructural Ag(I) and Au(I) complexes. We have investigated the structural and photo-physical properties of M^{CAAC} and M^{MAC} derivatives to investigate the role of the metal ion in the excited-state properties. Complexes with a given carbene ligand have similar structures, redox potentials, excited-state energies, and photoluminescent efficiencies ($\Phi^{PL} = 0.8–1.0$), but each has distinct excited-state dynamics. We have characterized the excited-state energies and temperature dependence of these materials and show that the smallest ΔE_{S1-T1} gaps, and markedly the fastest radiative rates ($k_r > 10^6$ s^{-1}), are found in the silver complexes.

52.2 Experimental Section

The copper-based compounds, Cu^{CAAC} and Cu^{MAC}, were prepared following literature procedures [1b, c]. The syntheses and characterization of M^{CAAC} and M^{MAC} for M = Ag and Au are given in the Supporting Information for this chapter.

52.2.1 Electrochemical Measurements

Cyclic voltammetry and differential pulsed voltammetry were performed using a VersaSTAT 3 potentiostat. Anhydrous acetonitrile (DriSolv) was used as the solvent under inert atmosphere, and 0.1 M tetra(n-butyl)ammonium hexafluorophosphate (TBAF) was used as the supporting electrolyte. A glassy carbon rod was used as the working electrode; a platinum wire was used as the counter electrode, and a silver wire was used as a pseudoreference electrode. The redox potentials are based on values measured from differential pulsed voltammetry and are reported relative to a ferrocene/ferrocenium (Cp$_2$Fe/Cp$_2$Fe$^+$) redox couple used as an internal reference, while electrochemical reversibility was determined using cyclic voltammetry.

52.2.2 X-ray Crystallography

The structures of (CAAC)CuCl, Ag^{CAAC}, Au^{CAAC}, Ag^{MAC}, and Au^{MAC} were determined by single-crystal X-ray crystallography. The X-ray intensity data were measured on a Bruker APEX DUO system equipped with a TRIUMPH curved-crystal monochromator and a Mo Kα fine-focus tube ($\lambda = 0.71073$ Å). The frames were integrated and corrected for Lp/decay with the Bruker SAINT software package using a SAINT V8.38A (Bruker AXS, 2013) algorithm. Data were corrected for absorption effects using the multiscan method (SADABS). The structures were solved by intrinsic phasing and refined on F^2 using the Bruker SHELXTL Software Package. The details of the data collection, structure solution, and metric data for the crystal and molecular structure are given for each compound in the SI.

52.2.3 Photophysical Measurements

All samples in fluid solution were degassed by extensive sparging with N_2. Doped polystyrene thin films were prepared from a solution of polystyrene (PS). PS pellets (100 mg) were mixed with 2 mL of toluene and sonicated for 1 h until all PS pellets were dissolved. The chosen metal complex (1 mg) was then dissolved in this PS solution. A sample of 0.5 mL was dropcast on a quartz substrate (2 cm × 2 cm) using a pipet to achieve an even surface. The film was left to air-dry for 30 min and then placed in the vacuum of the glovebox antichamber for another 30 min. The resulting film is approximately 200 μm thick.

The UV–vis spectra were recorded on a HewlettPackard 4853 diode array spectrometer. Steady state emission measurements were performed using a QuantaMaster Photon Technology International spectrofluorometer. All reported spectra are corrected for photomultiplier response. Phosphorescence lifetime measurements were performed using an IBH Fluorocube instrument equipped with 331 nm LED and 405 nm laser excitation sources using a time-correlated single photon counting method. Long-lived phosphorescence decays (> 30 μs) were measured using a QuantaMaster Photon Technology International spectrofluoremeter equipped with a Xe flash lamp. Quantum yields at room temperature were measured using a Hamamatsu C9920 system equipped with a xenon lamp, calibrated integrating sphere, and model C10027 photonic multichannel analyzer (PMA).

Temperature-dependent measurements in the range 5–320 K were performed using a JANIS ST-100 Standard Optical Cryostat instrument equipped with an intelligent temperature controller.

Emission lifetimes were measured using the Photon Technology International QuantaMaster model C-60 fluorimeter in the range 5–150 K and the IBH Fluorocube instrument in the range 150–320 K.

Ultrafast time-resolved emission was used to determine the intersystem crossing rates for $S_1 \rightarrow T_1$. Data were collected using a time-correlated single photon counting (TCSPC) system (Becker and Hickl SPC 630) operating in tandem with a 250 kHz Ti:sapphire regenerative amplifier (Coherent RegA 9050). Excitation pulses were centered at 400 nm produced by doubling 800 nm amplified output onto a type I BBO. The resulting fluorescence emission was collected at a right angle to the sample, passed through a 0.125 m double monochromator (Digikröm CM112) set to transmit at wavelengths between 490 and 570 nm, depending on the maximum emission of the sample. The emission was detected using a Hamamatsu R3809U-50 photomultiplier tube with 20 ps instrument response time. The samples used for these measurements were toluene solutions of M^{CAAC} or M^{MAC}.

52.2.4 Molecular Modeling

All calculations reported in this work were performed using the Q-Chem 5.1 program [16]. Ground-state optimization calculations were performed using the B3LYP functional along with the LACVP** basis set. Time-dependent density functional theory (TDDFT) calculations on the ground-state optimized geometries were performed at the CAM-B3LYP/LACVP** level for a balanced description CT and LE states. Atomic contributions to the NTOs of the excited states were computed using the Hirshfeld method [17] available in the Multiwfn package [18]. Transition densities were plotted by taking the product of the hole and electron NTOs.

52.3 Results and Discussion

The two copper compounds, Cu^{CAAC} and Cu^{MAC}, were prepared by literature procedures [1b, c]. The silver and gold MCAAC and

Results and Discussion

$$\begin{array}{c}\text{CAAC}\\ \text{MAC}\end{array}\!\!\overset{\!\!\diagdown}{\text{C}}\!\!-\!\!\text{H}\Bigg]^+ \text{Cl}^- \;+\; \text{MLX} \longrightarrow \;\overset{\!\!\diagdown}{\text{C}}\!\!\to\!\!\text{M}\!-\!\text{Cl} \;\xrightarrow{\text{KCz}}\; \overset{\!\!\diagdown}{\text{C}}\!\!\to\!\!\text{M}\!-\!\text{N}\!\!\begin{array}{c}\diagup\\ \diagdown\end{array}$$

M = Ag: MLX = Ag_2O or AgCl
M = Au: MLX = $(Me_2S)AuCl$

(52.1)

M^{MAC} complexes were prepared by first isolating the (carbene)MCl complex followed by treatment with potassium carbazolide (KCz), Eq. 52.1 (see SI for synthetic procedures and characterization of the complexes).

An X-ray diffraction study of the six (carbene)MCz complexes revealed a similar coordination and interligand conformation environment around the metal center. The C–M–N bond angles are all close to 180° (range = 174–180°), and the carbene and carbazole ligands are nearly coplanar (dihedral angles between the ligands = 0.1–6°). The individual C–M and M–N bond lengths, which vary between the metals, are illustrated in Fig. 52.1. The $C_{C:}\cdots N_{Cz}$ distance ($C_{C:}$ = the carbene carbon bound to the metal and N_{Cz} is the carbazolyl nitrogen), which will become important in our analysis of the photophysical data, falls in the order Cu (~3.7 Å) < Au (~4.0 Å) < Ag (~4.15 Å). The trend follows the order expected on the basis of the atomic radii of the three metals [19]. The C–M and M–N bond lengths are not measurably different between the CAAC and MAC complexes of a given metal ion. However, the ratio of C–M/M–N bond lengths (Fig. 52.1) decreases in the order Cu > Ag > Au. This trend is consistent with progressively stronger bonding of the metal favoring the carbene over the carbazolyl ligand in the order Au > Ag > Cu.

To delineate the effect of the metal ion on the frontier orbital energies, the oxidation (E_{ox}) and reduction (E_{red}) potentials of the (carbene)M(Cz) complexes were determined using cyclic and differential pulsed voltammetry. All M^{CAAC} and M^{MAC} complexes show irreversible oxidation and reversible reduction, with the identity of the metal having only a minor effect on the potentials. A near common value in oxidation potential (E_{ox} = 0.26 ± 0.06 V for M^{CAAC} and E_{ox} = 0.23 ± 0.06 V for M^{MAC}, both versus ferrocene/ferrocenium) indicates a process localized on the carbazolyl ligand. The reduction potentials are dependent on the electrophilicity of the two carbene ligands (E_{red} = -2.78 ± 0.06 V for M^{CAAC} and $E_{red} = -2.45 \pm 0.06$ V for M^{MAC}), thereby denoting a process centered on the carbene ligand. Thus, the ΔE_{ox-red} values are greater for M^{CAAC} than the M^{MAC} series by roughly 200 mV. This shift in the ΔE_{ox-red} for the two families of compounds is reflected in the red shift observed for the absorption and emission energies of M^{MAC} relative to M^{CAAC} complexes, *vide infra*.

The absorption spectra of the M^{CAAC} and M^{MAC} complexes are shown in Fig. 52.2. Spectra for the silver and gold complexes are similar to those of the copper analogues, with the principal difference being in the absorptivity of the lowest energy transition. Absorption bands between 300 and 375 nm are assigned to transitions localized on the carbazolyl ligand, whereas transitions on the carbenes fall below 300 nm based on comparison with the (carbene)M(Cl) precursors. The lowest energy band is assigned to a carbazolyl → carbene charge transfer (CT) transition [1c, 15]. The absorption bands for the M^{MAC} compounds are red-shifted from their M^{CAAC} counterparts by roughly 50 nm (ca. 150 mV), as expected from the ΔE_{ox-red} values. The carbazolyl-centered bands show minor shifts depending on the metal ion, with their energies falling in the order Au > Cu > Ag.

The M^{CAAC} and M^{MAC} (M = Ag, Au) complexes display negative solvatochromism in their absorption spectra and positive solvatochromism in their emission spectra (*vide infra*) analogous to what was observed for Cu^{CAAC} and Cu^{MAC} congeners (see SI for solvent-dependent spectra). In the copper complexes, the solvatochromism was attributed to the excited-state transition being oriented in the direction opposite to the ground-state dipole [1b, c]. The opposing direction of ground and excited-state dipole moments makes the complexes highly polarizable. The same explanation is valid for the silver and gold analogues (see SI for computed ground and excited-state dipole moments for M^{CAAC} and M^{MAC}). Both the ground- and excited-state dipoles lie essentially along the C–M–N bond axis, i.e., the metal's d_{z^2} axis. Thus, a polar solvent interacting with the large ground-state dipole stabilizes the ground state and destabilizes the excited state giving a blue shift in absorption. Solvent reorganization around the excited complex leads to stabilization of the excited state and destabilization of the ground state, resulting in the observed red-shifted emission in polar solvents. The shift in absorption is

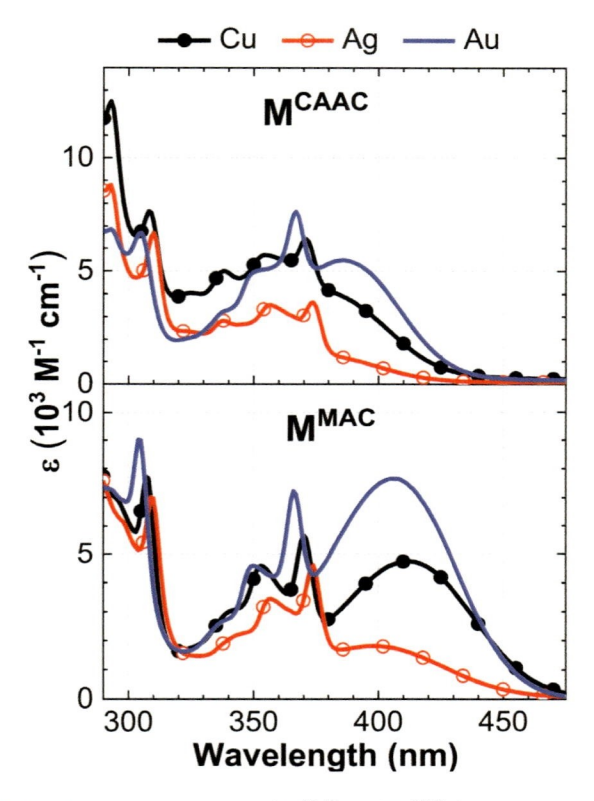

Figure 52.2 Absorption spectra of M^{CAAC} and M^{MAC} (M = Cu, Ag, or Au) in tetrahydrofuran.

larger than the shift in emission for the same solvent systems, because the excited-state dipole moment is of a smaller magnitude than that of the ground state. This difference leads to weaker interactions with the polar solvent for the excited state than for the ground state.

Modeling the complexes with time-dependent density functional theory (TDDFT) allows us to estimate the excited-state energies and oscillator strengths of the lowest energy absorptions. TDDFT calculations were carried out using CAM-B3LYP/LACVP** for the (carbene)MCz complexes and are given in Table 52.1. The computational data agrees qualitatively with the experimental results, predicting a higher energy for the CT transition on M^{CAAC} than that

Table 52.1 Excited-state energies and $S_0 \rightarrow S_1$ oscillator strengths (f) from TDDFT calculations

	^{3}CT energy (eV)	^{1}CT energy (eV)	$\Delta E_{S_1-T_1}$ (eV)	f
Cu^{CAAC}	3.03	3.29	0.26	0.114
Ag^{CAAC}	3.09	3.29	0.20	0.076
Au^{CAAC}	3.06	3.36	0.30	0.163
Cu^{MAC}	2.79	2.99	0.20	0.101
Ag^{MAC}	2.91	3.02	0.11	0.069
Au^{MAC}	2.90	3.14	0.24	0.152

of M^{MAC}, as well as the observed ordering of the oscillator strengths, i.e., Au > Cu > Ag. The lowest energy singlet (^{1}CT) and triplet (^{3}CT) states for all the complexes are N$_{Cz}$ → C$_{C:}$ transitions with little, but not insignificant, metal character. The metal contributes less than 15% to the NTOs associated with the S_1 and T_1 states in all cases, and the contribution decreases in the order Au > Cu > Ag for both M^{CAAC} and M^{MAC} (see SI for details). Figure 52.3 shows the natural transition orbitals (NTOs) calculated for Ag^{CAAC} and Ag^{MAC}. It is important to stress that the metal still serves as an effective bridge to electronically couple the carbazolyl donor to the carbene acceptor despite providing only a small percentage of electron density to the hole and electron NTOs (< 5% in some cases). This small contribution from the metal to the NTOs is the principal reason that the LLCT transitions in M^{CAAC} and M^{MAC} have such large extinction coefficients ($\varepsilon = 1000$–8000 M^{-1} cm^{-1}). The magnitude of the extinction coefficient is correlated with the degree of metal character in the NTOs, with silver having the lowest metal contribution and thus the weakest absorption, whereas gold has the largest contribution, and consequently the strongest absorption. Also shown in Fig. 52.3 are the transition densities (the product of the hole and electron NTOs) that define the region of overlap between the NTOs for absorption to the ^{1}CT state. The contribution of the metal to the transition densities is greater than that for the NTOs, which illustrates the importance of the metal to modulating the absorptivity.

The emission spectra of M^{CAAC} and M^{MAC} compounds recorded in 2-MeTHF and 1 wt% doped polystyrene (PS) thin films at room

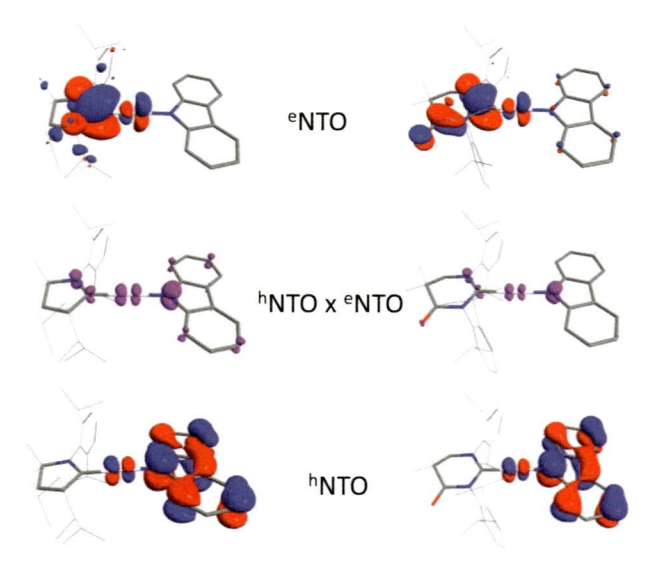

Figure 52.3 Hole (bottom) and electron (top) NTOs and transition densities (center) for the ^{1}CT state of Ag^{CAAC} (left) and Ag^{MAC} (right). The isovalues for the transition densities are increased relative to the NTOs by 20 times to improve visualization.

temperature (RT) and 77 K are shown in Fig. 52.4. The M^{MAC} compounds are red-shifted from their M^{CAAC} counterparts, but to a lesser degree than seen in the absorption spectra. The silver compounds show a small red shift relative to the copper and gold analogues. The spectra of the PS thin films are similar at RT and 77 K; however, the spectra in 2-MeTHF solutions show a drastic change in their spectral profile at 77 K. The reason for this change in line shape at 77 K has the same origins as the solvatochromism [1b, c]. Emission at RT is from a CT state whereas the emission at 77 K originates from a triplet-state localized on the carbazolide ligand (^{3}Cz). In fluid solution the polar 2-MeTHF solvent molecules organize around the large dipole moment of the complexes to stabilize the ground state. At 77 K, the solvent molecules are held rigidly in this ordered structure, which is destabilizing with respect to the CT excited state, shifting it to energies higher than that of the ^{3}Cz state. (Fig. S4 and Ref. [1b]). In contrast, the energy of the

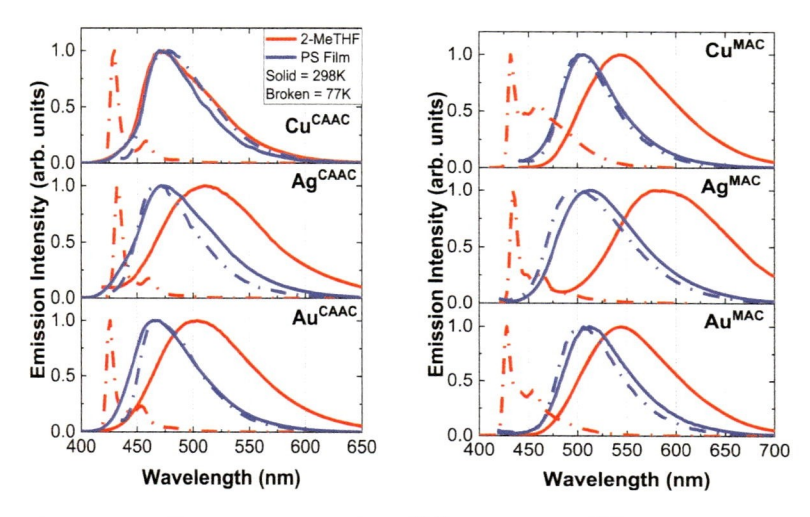

Figure 52.4 Emission spectra for M^{CAAC} (left) and M^{MAC} (right) at room temperature (RT) and 77 K in 2-MeTHF and at a 1 wt% doping level in polystyrene (excitation at 400 nm).

CT excited state does not change at low temperature in PS because the polymer chains are immobile and have a small permanent dipole moment. Thus, whereas minor changes in line shape do take place on cooling doped PS films to 77 K, the CT transition remains the lowest excited state. The temperature independence of spectral line shape in PS is also apparent in the excitation spectra of doped polystyrene thin films at RT and 77 K (Fig. S5). Therefore, polystyrene films doped with M^{CAAC} and M^{MAC} complexes were used for temperature-dependent photophysical studies.

Emission efficiencies and kinetic parameters for the (carbene)M(Cz) compounds are given in Table 52.2. Most of the compounds give high photoluminescence quantum yields (Φ_{PL}) with short lifetimes ($\tau = 0.3$–3 µs). The luminescence efficiency of the Ag and Au complexes doped into PS ($\Phi_{PL} = 0.8$–1.0) are comparable to the values previously reported for the copper analogues [1b, c]. The M^{CAAC} compounds give similarly high Φ_{PL} in 2-MeTHF and methylcyclohexane (see SI), whereas values for the M^{MAC} derivatives in 2-MeTHF are lower than those in polystyrene (Cu^{MAC} and Au^{MAC}, $\Phi_{PL} \approx 0.5$; Ag^{MAC}, $\Phi_{PL} = 0.06$). The low efficiency for Ag^{MAC} is due

Table 52.2 Emission properties of M^{CAAC} and M^{MAC} (M = Cu, Ag, or Au)

	λ_{max} (nm)	τ (μs)	Φ_{PL}	k_r (10^5 s^{-1})	k_{nr} (10^5 s^{-1})	$\lambda_{max,77K}$ (nm)	τ_{77K} (μs)	$k_{ISC}^{S1 \to T1}$ (10^{10} s^{-1})[a]
				2-MeTHF Solution				
Cu^{CAAC}	474	2.5	1.0	3.9	<0.08	430	7300	2.1
Ag^{CAAC}	512	0.37	0.71	19	7.8	432	21000	0.71
Au^{CAAC}	502	1.20	0.95	7.9	0.42	426	530	>4
Cu^{MAC}	542	1.1	0.55	5.0	4.1	432	[b]	0.45
Ag^{MAC}	568	0.04	0.06	15	270	434	9900	0.63
Au^{MAC}	544	0.79	0.50	6.3	6.3	428	260	~3.4
				1% in PS Film				
Cu^{CAAC}	470	2.8	1.0	3.5	<0.04	479	61	[c]
Ag^{CAAC}	472	0.50	1.0	20	<0.20	472	11	[c]
Au^{CAAC}	472	1.14	1.0	8.8	<0.9	472	45	[c]
Cu^{MAC}	506	1.4	0.9	6.4	0.71	502	140	[c]
Ag^{MAC}	512	0.33	0.79	24	6.3	500	7.7	[c]
Au^{MAC}	512	0.83	0.85	10	1.8	506	43	[c]

[a]Intersystem crossing rates were measured in toluene solution. [b]Cu^{MAC} gives a biexponential decay with lifetimes of 2.1 ms (60%) and 340 μs (40%). The former is due to ^{3}Cz emission and the latter to the ^{3}CT emission. [c]Not measured.

to having a nonradiative decay rate ($k_{nr} = 2.7 \times 10^7$ s^{-1}) that is nearly 2 orders of magnitude larger than any of the other complexes. This difference is possibly due to the fact that the Dipp groups on the MAC ligand are less constrained than the menthyl group on the CAAC ligand, making Ag^{MAC} susceptible to formation of exciplexes with the solvent. Moreover, the six-membered ring in the MAC ligand has fewer constraints to torsional distortion than the five-membered ring in the CAAC ligand. The long separation between the carbene and carbazolyl for the silver complex (C$_{C:}$ ⋯ N$_{Cz}$ = 4.15 Å) and the weaker backbonding for silver [20] may also increase the flexibility of the molecule, which will further contribute to the high k_{nr} values for Ag^{MAC} in fluid solution.

Our previous studies showed that the Cu^{CAAC} and Cu^{MAC} complexes emit via a TADF (or TEL) process, and the data recorded here suggests that the Ag and Au derivatives behave similarly. A hallmark of a thermally promoted emission process is a marked increase in the luminescence lifetime on cooling since emission is from the long-lived triplet state at 77 K. The measured lifetimes of the complexes doped into PS thin films increase 10- to 100-fold upon cooling to 77 K. The emission lifetime undergoes a greater increase in frozen 2-MeTHF solution; however, this change is due to emission originating from the ^{3}Cz state.

The measured luminescence lifetimes of the (carbene)M(Cz) complexes fall in the order Cu > Au > Ag, Table 52.2. This trend does not match the sequence expected on the basis of the spin–orbit coupling constants for the metal ions [21], but it matches the order predicted by the calculated ΔE_{S1-T1} values; i.e., a small ΔE_{S1-T1} will give a high rate of thermal promotion to the S$_1$ and larger K_{eq} for T$_1$ S$_1$. Remarkably, the silver compounds give lifetimes substantially less than 1 μs (Ag^{CAAC}, $\tau = 0.5$ μs; Ag^{MAC}, $\tau = 0.33$ μs). The radiative rates for these two emitters are > 10^6 s^{-1}, more than a factor of 2 greater than the analogous gold compounds and 4 times larger than the copper analogues. To put this in context, the radiative rates for the silver compounds are roughly a factor of 10 faster than state-of-the-art iridium-based emitters used in OLEDs [22]. To probe the origin of the high radiative rates for Ag^{CAAC} and Ag^{MAC}, we turned to temperature-dependent photophysical measurements to determine the ΔE_{S1-T1} and kinetic parameters for the M^{CAAC} and M^{MAC} compounds.

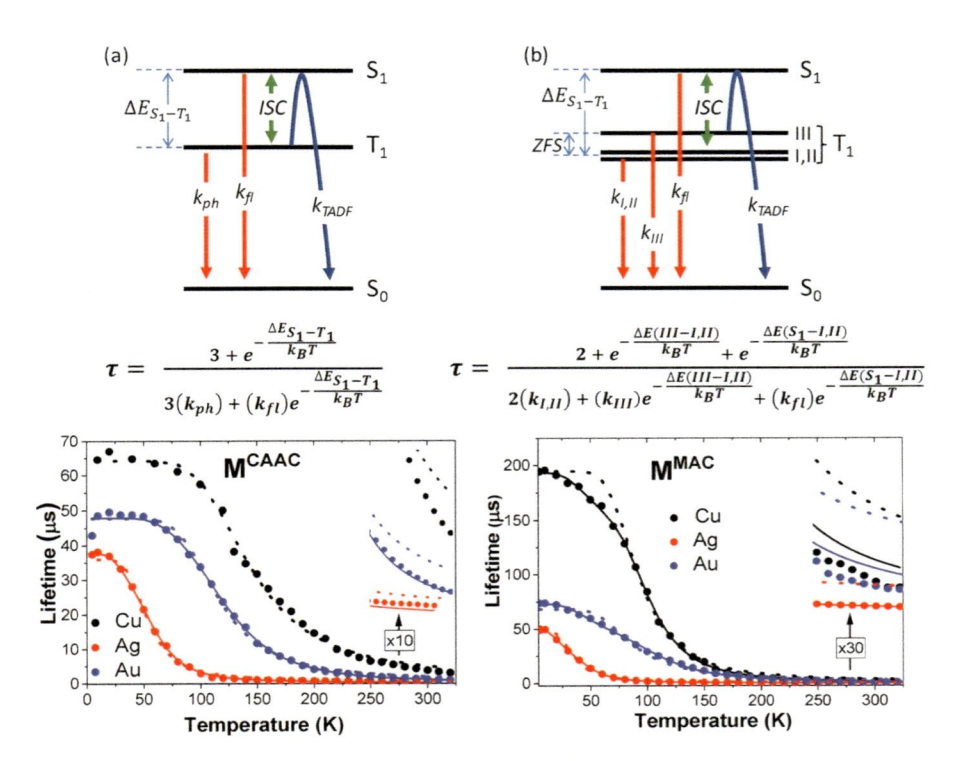

$$\tau = \frac{3 + e^{-\frac{\Delta E_{S_1-T_1}}{k_B T}}}{3(k_{ph}) + (k_{fl})e^{-\frac{\Delta E_{S_1-T_1}}{k_B T}}}$$

$$\tau = \frac{2 + e^{-\frac{\Delta E(III-I,II)}{k_B T}} + e^{-\frac{\Delta E(S_1-I,II)}{k_B T}}}{2(k_{I,II}) + (k_{III})e^{-\frac{\Delta E(III-I,II)}{k_B T}} + (k_{fl})e^{-\frac{\Delta E(S_1-I,II)}{k_B T}}}$$

Figure 52.5 Two-level (a) and three-level (b) models for emission from M^{CAAC} and M^{MAC}. The equations giving the measured lifetimes for the two models are given below each scheme. Temperature-dependent lifetimes of M^{CAAC} (c) and M^{MAC} (d) doped at 1 wt% in polystyrene. Data (symbols) and fits (dashed lines and solid lines represent fits to two-level and three-level models, respectively).

Luminescence decay lifetimes were measured for the (carbene)M(Cz) complexes between 5 and 300 K. The data were evaluated with two different kinetic models, illustrated in Fig. 52.5a,b. One model assumes two emitting states (S_1 and T_1), with an energy spacing of ΔE_{S1-T1}. The other model, similar to analysis of other phosphorescent d^{10} metal complexes [8a], assumes three emitting states (S_1, T_{III}, and $T_{I/II}$) in which the triplet state is split into a single and a degenerate pair of triplet sublevels. There are thus two energy spacings, ΔE_{S1-T1}, and the energy between the T_{III} and $T_{I,II}$ states, i.e., the zero-field splitting (ZFS). The data were fit using Boltzmann models for both schemes (equations given in Fig. 52.5a,b) assuming that the rates of intersystem crossing (S_1 T_1) and internal conversion between triplet sublevels are faster than either fluorescence (k_{fl}) or phosphorescence (k_{ph} or k_{III}, $k_{I,II}$). In addition, it is assumed that nonradiative decay is markedly slower than either k_{fl}, k_{ph} or (k_{III}, $k_{I,II}$), consistent with the near unit Φ_{PL} for these compounds in polystyrene thin films. Note here that both singlet and triplet states are CT in origin. The kinetic and energy splitting parameters derived from these fits are given in Table 52.3.

The two- and three-level fits give the same trends for ΔE_{S1-T1} upon comparing Cu, Ag, and Au CAAC and MAC complexes. The photoluminescent lifetimes at room temperature are well-correlated with the singlet–triplet splitting; i.e., ΔE_{S1-T1} and τ_{meas} give the order Ag < Au < Cu for a given carbene ligand. This trend is inversely related to the metal-ligand bond distances that fall in the order Ag > Au > Cu (Fig. 52.1). A longer bond between the metal and ligands likely weakens the interaction between p-orbitals on N_{Cz} and $C_{C:}$, thus decreasing ΔE_{S1-T1} and concomitantly shortening the lifetime for the silver complexes. In addition to ΔE_{S1-T1}, the three-level fit gives the zero-field splitting (ZFS) energy. These ZFS values are markedly greater than values reported previously for three- and four-coordinated Cu, Ag or Au complexes [8a, 23]. We have proposed that the near collinear orientation between the d_{z^2} orbital on the metal and the transition dipole moment is likely the origin of the large ZFS values [1c]. Recent calculations on related two-coordinated d^{10} metal complexes have described large spin–orbit coupling matrix elements between the T_1 and higher lying singlet and triplet states [24]. Interactions promoted by the magnetically anisotropic d-orbitals and higher lying states could be responsible for the large values of ZFS for the ^{3}CT state. Although the three-level fits give good R^2 values, the fact that the ZFS energy is a significant fraction of ΔE_{S1-T1} for the same compounds suggests that the two energies may not be unique fits to the data.

Both the two- and three-level fits to the temperature-dependent emission decay data match the data collected below 150 K but give inaccurate values for the emission decay at higher temperatures. The reason for this discrepancy is that the fits are dominated by the large changes in lifetime at low temperature, which come about as a result of a significant contribution for emission from the triplet sublevels.

Table 52.3 Photophysical parameters obtained from Boltzmann fits to temperature-dependent lifetimes[a]

	Two-level model 5–300 K		Three-level model 5–300 K				Full kinetic scheme 200–325 K	
	ΔE_{S1-T1} (cm^{-1})	τ_{ph}/τ_{fl} (μs/ns)	ΔE_{S1-T1} (cm^{-1})	ZFS (cm^{-1})	$\tau_{III}/\tau_{I,II}$ (μs)	τ_{fl} (ns)	ΔE_{S1-T1} (cm^{-1})	τ_{fl} (ns)
Cu^{CAAC}	500[a]	64/150[b]	c	c	c	c	590	72
Ag^{CAAC}	180	36/95	280	d	3.4/38	30	150	83
Au^{CAAC}	430	50/81	650	220	2.4/49	23	570	25
Cu^{MAC}	360[e]	180/230[e]	500[e]	85[e]	44/190[e]	70[e]	570	28 [36]
Ag^{MAC}	110	48/200	200	d	4.3/51	60	180	46 [80]
Au^{MAC}	260	68/330	470	91	9.1/73	70	570	24 [23]

[a]Fitting equations are given in Fig. 52.5 for the two-level and three-level data and using Eq. 52.3 for the full kinetic scheme. Values of τ_{fl} for the M^{MAC} complexes derived from the Strickler-Berg analysis are given in brackets. [b]From Ref. [1b]. [c]Fits to the three-level model did not give reasonable values for the parameters. [d]An insufficient amount of data was collected between 5 and 15 K to accurately determine the ZFS. [e]From Ref. [1c].

The fits are relatively insensitive to the radiative rate of ^{1}CT (k_{fl}) such that large deviations in k_{fl} gives acceptable fits. Therefore, to better fit the data, and thus k_{fl}, we considered only the temperatures above 200 K. At these temperatures variations in the lifetime can be treated as being controlled principally by ΔE_{S1-T1}, and not by either ZFS or direct phosphorescence from the triplet level(s). This assumption is supported by the fact that at temperatures above 200 K emission from M^{CAAC} and M^{MAC} due to TADF is greater than 90% (M = Au) and 95% (M = Cu and Ag) (see Fig. S7). Thus, in this higher temperature regime we can ignore effects from ZFS and use a full kinetic scheme for a two-level system shown in Fig. 52.6, where emission is controlled simply by ΔE_{S1-T1}, and the fluorescence and phosphorescence rates. Dias, Penfold, and Monkman employed these conditions to show that the rate constant for emission from S$_1$ (k_{TADF}) is given by the relationship in Eq. 52.2, where τ_{PF} is the lifetime for prompt fluorescence and $\Phi_{ISC}^{S_1 \to T_1}$ is the quantum yield for intersystem crossing (ISC) [25]. For these metal complexes, τ_{PF} corresponds to rate constant for ISC from the singlet excited state (i.e., S$_1 \to$ T$_1$) since this ISC rate constant ($k_{ISC}^{S_1}$) is $\gg k_{fl}$. To verify this inequality, the ISC rate was estimated using ultrafast time-correlated single photon counting measurements of the luminescence decay of each complex in toluene solution (Table 52.2). Both gold compounds show an initial decay rate comparable to the instrument response function of 20 ps, consistent with S$_1 \to$ T$_1$ ISC rate constants ($k_{ISC}^{S_1}$) greater than 10^{10} s^{-1}. The rate constants for the silver compounds are slower ($k_{ISC}^{S_1} \sim 10^{10}$ s^{-1}), as are values for Cu^{CAAC} and Cu^{MAC} ($k_{ISC}^{S_1} = 10^{10}$ and 5×10^9 s^{-1}, respectively). Considering the difference in energy between S$_1$ and T$_1$ states (ΔE_{S1-T1}), we can estimate the rate constants for return to the lowest excited singlet state (T$_1 \to$ S$_1$, $k_{ISC}^{T_1}$) to be $\sim 10^8$ s^{-1} for the copper and gold complexes, and $\sim 10^9$ s^{-1} for the silver compounds (see SI for details). These ISC rates validate the following inequalities: k_{ph}, $k_{nr}^{T_1} \ll k_{ISC}^{T_1}$, k_{fl} and $k_{nr}^{S_1} \ll k_{fl}$, $k_{ISC}^{S_1}$. This allows Eq. 52.2 to be simplified to Eq. 52.3 (see SI for derivation). Thus, a plot of ln(k_{TADF}) vs $1/T$ will be linear with a slope of $\Delta E_{S1-T1}/k_B$ and an intercept [ln(b)] that is a function of $k_{ISC}^{S_1}$ and k_{fl}. These plots are in fact linear in the 200–325 K temperature regime (Fig. 52.6), and the ΔE_{S1-T1} values calculated accordingly are given in Table 52.3. Solving for k_{fl} from the intercept of this plot gives

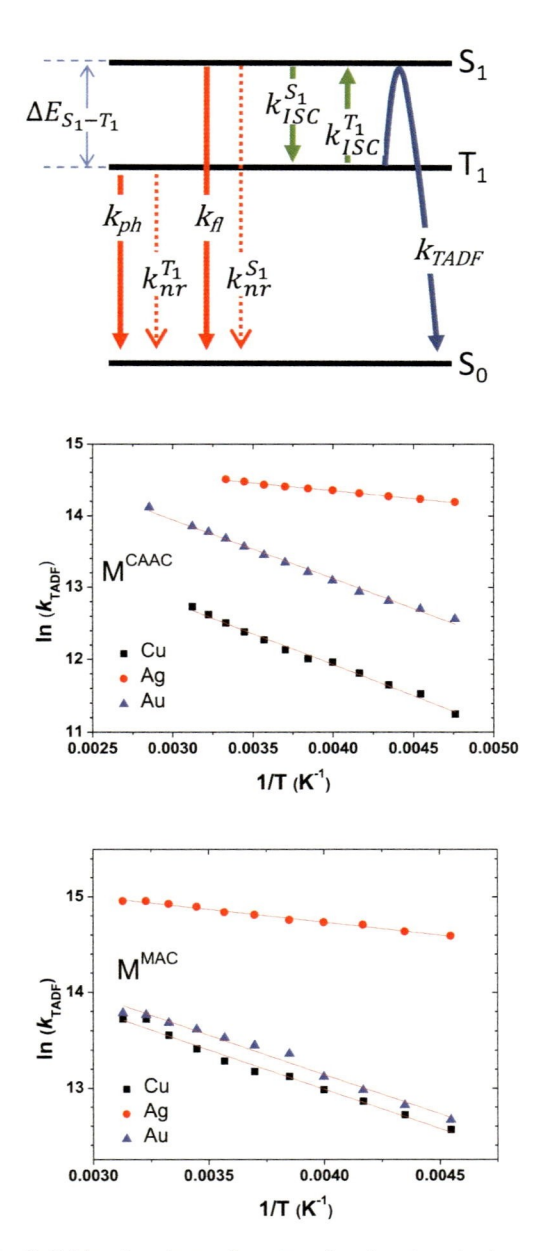

Figure 52.6 Full kinetic scheme for a two-level system is shown at the top, and the fits to the temperature-dependent lifetime data from 200 to 325 K (symbols) to Eq. 52.3 (solid lines) are shown for M^{CAAC} and M^{MAC}.

Eq. 52.4 and, since $k_{ISC}^{S_1} \gg b$, this reduces to $k_f = 3b$.

$$k_{TADF} = \frac{k_{ph} + k_{nr}^{T_1} + k_{ISC}^{T_1}(1 - \Phi_{ISC}^{S_1 \to T_1})}{1 + k_{ISC}^{T_1}\tau_{PF}} \quad (52.2)$$

$$\ln(k_{TDAF}) = \ln\left(\frac{k_{ISC}^{S_1}}{3}\left(1 - \frac{k_{ISC}^{S_1}}{k_{fl} + k_{ISC}^{S_1}}\right)\right) - \frac{\Delta E_{S_1-T_1}}{k_B T}$$

$$= \ln(b) - \frac{1}{T}\left(\frac{\Delta E_{S_1-T_1}}{k_B}\right) \quad (52.3)$$

$$k_f = \frac{3bk_{ISC}^{S_1}}{k_{ISC}^{S_1} - 3b} \quad (52.4)$$

The ΔE_{S1-T1} energies derived from the full kinetic scheme (Table 52.3, Fig. 52.5) are similar to values derived from the Boltzmann two- and three-level fits. The measured emission rates also clearly follow the values for ΔE_{S1-T1} in all three models. However, radiative rate constants for emission from the S_1 state are better estimated in the full kinetic scheme since k_{fl} is not convoluted with k_{ph} as in the case for fits to the two- and three-level models. The radiative lifetimes for the S_1 CT state (τ_{fl}) range between 24 and 83 ns and fall in the order Ag > Cu > Au.

The τ_{fl} values observed for the M^{MAC} compounds are very close to those predicted by a Strickler–Berg analysis of the absorption spectra. Strickler–Berg analysis extends the Einstein equation for predicting atomic spectra to molecular emitters and uses the intensity of the CT absorption band to estimate the oscillator strength for the $S_0 \to S_1$ transition [26]. Although we cannot accurately measure the area of the absorption bands for the M^{CAAC} compounds owing to overlap with transitions on Cz, the integrated intensities can be reliably determined for the red-shifted CT bands of the M^{MAC} complexes in THF (Fig. S8). The radiative rate constants predicted from Strickler–Berg analysis (k_{fl} = 36, 80, and 23 ns for Cu^{MAC}, Ag^{MAC}, and Au^{MAC}, respectively, Table 52.3) compare favorably to k_{fl} values determined from fits to Eqs. 52.2–52.4. Although it would be preferable to determine the extinction coefficients for the M^{MAC} complexes in polystyrene, using spectra of the three M^{MAC} complexes doped in polystyrene that are scaled to extinction coefficients measured in THF gives k_{fl} values within 10% of the values provided in Table 52.3. The agreement between k_fl values determined using the two methods supports the assumptions used in the full kinetic scheme and provides a more reliable estimate than the Boltzmann fits to the lifetime data between 5 and 325 K. The measured emission rates, k_fl, k_ph, and ΔE_{S1-T1} values illustrate that the rate of emission is not restricted by the rate of ISC between S_1 and T_1, nor by k_fl or k_ph, but instead by ΔE_{S1-T1}. The value of K_eq for the T_1 S_1 equilibrium [$K_\text{eq} = \frac{1}{3}\exp(-\frac{\Delta E_{S1-T1}}{k_B T})$] can then be determined from the singlet–triplet splitting energy. At room temperature the K_eq values are 0.13 and 0.03 for a ΔE_{S1-T1} of 200 and 500 cm^{-1}, respectively. Since emission primarily comes from the singlet state, a larger equilibrium constant increases the amount of time the molecule spends in this state and shortens the luminescence lifetime. The analysis establishes that the key parameter for achieving fast emission from these complexes is a small separation in energy between the lowest singlet and triplet CT states.

High efficiency OLEDs prepared by thermal evaporation have been reported for both Cu$^\text{CAAC}$ and Cu$^\text{MAC}$ [1b, c]. The gold (carbene)M(Cz) complexes have good thermal stability and were thus used to fabricate high efficiency OLEDs (see SI for details). OLEDs using Au$^\text{MAC}$ as a dopant emitted green light ($\lambda_\text{max} = 510$ nm) with an external quantum efficiency (EQE) of 18% (at a brightness of 50 cd/m^2, which drops to 15% at 1000 cd/m^2), comparable to that of the Cu$^\text{MAC}$-based OLED [1c] and a closely related (carbene)Au(Cz)-based OLED [1a]. Unfortunately, neither Ag$^\text{CAAC}$ nor Ag$^\text{MAC}$ sublime cleanly making it impossible to properly compare analogous OLEDs made using silver-based emitters. However, an OLED doped with a related (carbene)-Ag(Cz) complex has been reported and gave a similar emission color with an EQE of 14% [1d].

52.4 Conclusion

In exploring the properties of the isoelectronic (carbene)M(Cz) compounds for Cu, Ag, and Au, we found many similarities and some stark differences for the different metals. The structures of the CAAC and MAC complexes are nearly identical for the three metals, all giving coplanar carbene and carbazole ligands, and $C_{C:}-M-N_{Cz}$ bond angles close to 180°. The electrochemical properties of the M^{CAAC} and M^{MAC} compounds are not significantly affected by the choice of metal ion, nor are the energies for absorption and emission. All six of the (carbene)-M(Cz) complexes examined here display high Φ_{PL} values in a polystyrene matrix, and most give similarly high Φ_{PL} in fluid solution as well. In contrast to these nearly metal-independent properties, the extinction coefficient for the CT bands in the absorption spectra is markedly affected by the metal ion and fall in the order Au > Cu ≫ Ag, which parallels the strengths of π-interactions between the different metal ions and the π-acidic or π-basic ligands. The compounds have short emission lifetimes ($\tau = 0.33–2.8$ µs) that fall in the order Ag < Au < Cu for both M^{CAAC} and M^{MAC}, with the lifetimes of (carbene)AgCz roughly a factor of 10 shorter than for (carbene)CuCz complexes.

Temperature-dependent lifetime data was used to determine the energy gap between the singlet and triplet excited states (ΔE_{S1-T1}), the zero-field splitting of the triplet sublevels, and the radiative rates for emission from the singlet and triplet excited states. Emission at room temperature is due to thermal equilibration between the lower energy triplet and higher lying singlet and subsequent emission from the singlet. The emission decay rates for these complexes are determined by ΔE_{S1-T1}. The silver complexes have the shortest radiative lifetimes ($\tau = 0.4–0.5$ µs at RT) and also the smallest values for ΔE_{S1-T1} (150–180 cm^{-1}, 18–23 meV). The radiative lifetimes for fluoresence ($S_1 \rightarrow S_0$) for the M^{CAAC} and M^{MAC} compounds fall between 24 and 84 ns. Our previous studies on Cu^{CAAC} and Cu^{MAC} compounds gave much longer singlet lifetimes, which suggested that the S_1 state had substantial triplet character [1c, 15]. The radiative lifetimes reported here for the singlet state were measured without the errors to Boltzmann fits of temperature-dependent data between 5 and 325 K that were inherent in our previous study. The radiative lifetimes derived from the 200–325 K fits match the values predicted from Strickler–Berg analysis using absorption spectra of the M^{MAC} complexes.

Although the radiative lifetimes for the triplet excited states are much shorter than purely organic TADF materials, the M^{CAAC} and M^{MAC} compounds have T_1 phosphorescence lifetimes that are markedly longer than the observed emission lifetimes at room temperature. We have found that for the M^{CAAC} and M^{MAC} compounds the rate of emission via thermally enhanced luminescence, i.e., $T_1 \to S_1 \to S_0 + h\nu$, is controlled not by the ISC rates between T_1 and S_1, nor by k_{fl} and k_{ph} from the respective S_1 and T_1 states, but is fully governed by ΔE_{S1-T1}. The energy separation affects the rate of upconverting $T_1 \to S_1$ ISC; however, this rate is still on the order of 10^3–10^4 times faster than the emission decay rate. More importantly, ΔE_{S1-T1} alters the values for the $T_1 S_1$ equilibrium constants; K_{eq} is roughly 0.1 for the Ag complexes and < 0.03 for the Cu and Au complexes. These equilibria determine the amount of time the complex spends in the S_1 state, and thus the overall emission rate. Decreasing ΔE_{S1-T1} to 100 cm^{-1} or less is expected to further increase the equilibrium constant (maximum $K_{eq} = 0.33$) and decrease the emission lifetime, but at some point the fluorescence rate (k_{fl}) will become rate limiting. Such a small ΔE_{S1-T1} can be achieved by decoupling a donor and acceptor that comprise the CT state, but the subsequent decrease in oscillator strength for the CT transition will increase the fluorescence radiative lifetime, a limitation present in organic TADF molecules. Fortunately, electronic coupling through the metal ion in the M^{CAAC} and M^{MAC} complexes is sufficiently strong to confer a high oscillator strength for the singlet CT transition, especially for the gold-based materials ($\varepsilon = 6000$–8000 M^{-1} cm^{-1}).

The metal center in these two-coordinated complexes behaves unlike the metal in other photoactive coordination complexes. In complexes that emit from MLCT excited states using d^6, d^8, or d^{10} metal centers, the metal is typically considered to be a redox active donor to the ground state, whose electrochemical potential is modified by coordinating ligands. In contrast, the metal ions in the two-coordinate complexes described here are redox "innocent" and do not greatly alter the electrochemical potentials in the ground state. Instead, the metal serves as a monatomic electrical conduit that modulates the electron coupling between the two redox active ligands. The role of the metal is simply to affect the kinetics of electron transfer from one ligand to the other, altering the probability of the absorption transition and the rate of return to the ground state. The metal accomplishes these functions by the strength of coupling in the former process, and the ability to promote electron spin flips in the excited state in the latter case. Consequently, weak electronic coupling combined with rapid intersystem crossing (< 1 ns) can lead to submicrosecond radiative decay in what are nominally triplet emitters.

Supporting Information

The Supporting Information is available free of charge on the ACS Publications website at DOI: 10.1021/jacs.9b03657.

Detailed synthetic procedures and characterization for (CAAC)MCl, M^{CAAC}, (MAC)MCl, and M^{MAC} for M = Ag and Au; NMR spectra; single-crystal X-ray diffraction data; cyclic and differential pulse voltammetry data; photophysical data including temperature-dependent and ultrafast lifetimes measurements; derivation for Eqs. 52.1–52.4; as well as method descriptions for molecular modeling studies and calculated dipole moments for M^{CAAC} and M^{MAC} compounds

(http://pubs.acs.org/doi/suppl/10.1021/jacs.9b03657/suppl_file/ja9b03657_si_001.pdf)

X-ray data for (CAAC)AgCl

(http://pubs.acs.org/doi/suppl/10.1021/jacs.9b03657/suppl_file/ja9b03657_si_002.cif)

X-ray data for Ag^{CAAC}

(http://pubs.acs.org/doi/suppl/10.1021/jacs.9b03657/suppl_file/ja9b03657_si_003.cif)

X-ray data for Au^{CAAC}

(http://pubs.acs.org/doi/suppl/10.1021/jacs.9b03657/suppl_file/ja9b03657_si_004.cif)

X-ray data for Ag^{MAC}

(http://pubs.acs.org/doi/suppl/10.1021/jacs.9b03657/suppl_file/ja9b03657_si_005.cif)

X-ray data for Au^{MAC}

(http://pubs.acs.org/doi/suppl/10.1021/jacs.9b03657/suppl_file/ja9b03657_si_006.cif)

Acknowledgments

This project was supported by several funding sources. The Universal Display Corporation supported Hamze, Shi, Kapper, Sylvinson, Jung, Haiges, Djurovich, and Thompson, whose work involved synthesis and characterization of (carbene)MCz compounds, steady state and temperature-dependent photophysical measurements, theoretical modeling, X-ray structures for M^{CAAC}, OLED studies, and data analysis/fitting. A grant from the Department of Energy, Basic Energy Sciences (grant DE-SC0016450), supported Tadle, Estergreen, Bradforth, and Thompson, whose work involved the X-ray structures for M^{MAC} compounds, ultrafast (ISC) photophysical measurements, and data analysis. The team from the University of California at San Diego was supported by a National Science Foundation (NSF) Graduate Research Fellowship grant DGE-1650112 and NSF grant CHE-1661518 (Peltier) (Jazzar, Bertrand), whose work involved syntheses of the CAAC ligands and the (CAAC)MCl complexes for Cu, Ag, and Au. Computation for the work described in this paper was supported by the University of Southern California's Center for High-Performance Computing (https://hpcc.usc.edu/).

References

1. (a) Di, D. W., Romanov, A. S., Yang, L., Richter, J. M., Rivett, J. P. H., Jones, S., Thomas, T. H., Abdi Jalebi, M., Friend, R. H., Linnolahti, M., Bochmann, M., Credgington, D. High-performance light-emitting diodes based on carbene-metal-amides. *Science* 2017, **356** (6334), 159–163. (b) Hamze, R., Peltier, J. L., Sylvinson, D., Jung, M., Cardenas, J., Haiges, R., Soleilhavoup, M., Jazzar, R., Djurovich, P. I., Bertrand, G., Thompson, M. E. Eliminating nonradiative decay in Cu(I) emitters: >99% quantum efficiency and microsecond lifetime. *Science* 2019, **363** (6427), 601–606. (c) Shi, S., Jung, M. C., Coburn, C., Tadle, A., Sylvinson M. R., D., Djurovich, P. I., Forrest, S. R., Thompson, M. E. Highly efficient photo- and electroluminescence from two-coordinate Cu(I) complexes

featuring nonconventional N-heterocyclic carbenes. *J. Am. Chem. Soc.* 2019, **141**(8), 3576–3588. (d) Romanov, A. S., Jones, S. T. E., Yang, L., Conaghan, P. J., Di, D., Linnolahti, M., Credgington, D., Bochmann, M. Mononuclear silver complexes for efficient solution and vacuum-processed OLEDs. *Adv. Opt. Mater.* 2018, **6**(24), 1801347.

2. Kalyanasundaram, K. Photophysics, photochemistry and solar energy conversion with tris(bipyridyl)ruthenium(II) and its analogues. *Coord. Chem. Rev.* 1982, **46**, 159–244.

3. (a) Keefe, M. H., Benkstein, K. D., Hupp, J. T. Luminescent sensor molecules based on coordinated metals: a review of recent developments. *Coord. Chem. Rev.* 2000, **205**(1), 201–228. (b) Lo, K. K.-W., Louie, M.-W., Zhang, K. Y. Design of luminescent iridium(III) and rhenium(I) polypyridine complexes as in vitro and in vivo ion, molecular and biological probes. *Coord. Chem. Rev.* 2010, **254**(21), 2603–2622.

4. Grätzel, M. Solar energy conversion by dye-sensitized photovoltaic cells. *Inorg. Chem.* 2005, **44**(20), 6841–6851.

5. (a) Lamansky, S., Djurovich, P., Murphy, D., Abdel-Razzaq, F., Lee, H. E., Adachi, C., Burrows, P. E., Forrest, S. R., Thompson, M. E. Highly phosphorescent bis-cyclometalated iridium complexes: Synthesis, photophysical characterization, and use in organic light emitting diodes. *J. Am. Chem. Soc.* 2001, **123**(18), 4304–4312. (b) Bolink, H. J., Coronado, E., Costa, R. D., Lardiés, N., Ortí, E. Near-quantitative internal quantum efficiency in a light-emitting electrochemical cell. *Inorg. Chem.* 2008, **47**(20), 9149–9151.

6. (a) Romanov, A. S., Jones, S. T. E., Yang, L., Conaghan, P., Di, D. W., Linnolahti, M., Credgington, D., Bochmann, M. Mononuclear silver complexes for efficient solution and vacuum-processed OLEDs. *Adv. Opt. Mater.* 2018, **6**(24), 1801347. (b) Romanov, A. S., Becker, C. R., James, C. E., Di, D. W., Credgington, D., Linnolahti, M., Bochmann, M. Copper and gold cyclic (alkyl)(amino)carbene complexes with sub-microsecond photoemissions: structure and substituent effects on redox and luminescent properties. *Chem. - Eur. J.* 2017, **23**(19), 4625–4637.

7. (a) Frey, G. D., Donnadieu, B., Soleilhavoup, M., Bertrand, G. Synthesis of a room-temperature-stable dimeric copper(I) hydride. *Chem. - Asian J.* 2011, **6**(2), 402–405. (b) Melaimi, M., Jazzar, R., Soleilhavoup, M., Bertrand, G. Cyclic (alkyl)(amino)carbenes (CAACs): recent developments. *Angew. Chem., Int. Ed.* 2017, **56**(34), 10046–10068.

8. (a) Yersin, H., Czerwieniec, R., Shafikov, M. Z., Suleymanova, A. F. TADF material design: photophysical background and case studies focusing

on Cu-I and Ag-I complexes. *ChemPhysChem* 2017, **18**(24), 3508–3535. (b) Leitl, M. J., Zink, D. M., Schinabeck, A., Baumann, T., Volz, D., Yersin, H. Copper(I) complexes for thermally activated delayed fluorescence: from photophysical to device properties. *Top. Curr. Chem.* 2016, **374**(3), 25. (c) Yam, V. W. W., Au, V. K. M., Leung, S. Y. L. Light-emitting self-assembled materials based on d(8) and d(10) transition metal complexes. *Chem. Rev.* 2015, **115**(15), 7589– 7728. (d) Ni, W. X., Li, M., Zheng, J., Zhan, S. Z., Qiu, Y. M., Ng, S. W., Li, D. Approaching white-light emission from a phosphorescent trinuclear gold(I) cluster by modulating its aggregation behavior. *Angew. Chem., Int. Ed.* 2013, **52**(50), 13472–13476. (e) Partyka, D. V., Teets, T. S., Zeller, M., Updegraff, J. B., Hunter, A. D., Gray, T. G. Constrained digold(I) diaryls: syntheses, crystal structures, and photophysics. *Chem. - Eur. J.* 2012, **18**(7), 2100–2112. (f) Zhan, S. Z., Li, M., Zhou, X. P., Wang, J. H., Yang, J. R., Li, D. When Cu4I4 cubane meets Cu-3(pyrazolate)(3) triangle: dynamic interplay between two classical luminophores functioning in a reversibly thermochromic coordination polymer. *Chem. Commun.* 2011, **47**(46), 12441–12443. (g) Zhan, S. Z., Li, M., Zhou, X. P., Ni, J., Huang, X. C., Li, D. From simple to complex: topological evolution and luminescence variation in a copper(I) pyridylpyrazolate system tuned via second ligating spacers. *Inorg. Chem.* 2011, **50**(18), 8879–8892. (h) Czerwieniec, R., Hofbeck, T., Crespo, O., Laguna, A., Concepcion Gimeno, M., Yersin, H. The lowest excited state of brightly emitting gold(I) triphosphine complexes. *Inorg. Chem.* 2010, **49**(8), 3764–3767. (i) Lotito, K. J., Peters, J. C. Efficient luminescence from easily prepared three-coordinate copper(I) arylamidophosphines. *Chem. Commun.* 2010, **46**(21), 3690–3692. (j) Deaton, J. C., Switalski, S. C., Kondakov, D. Y., Young, R. H., Pawlik, T. D., Giesen, D. J., Harkins, S. B., Miller, A. J. M., Mickenberg, S. F., Peters, J. C. E-Type delayed fluorescence of a phosphine-supported Cu-2(mu-NAr2)(2) diamond core: harvesting singlet and triplet excitons in OLEDs. *J. Am. Chem. Soc.* 2010, **132**(27), 9499–9508. (k) Miller, A. J. M., Dempsey, J. L., Peters, J. C. Long-lived and efficient emission from mononuclear amidophosphine complexes of copper. *Inorg. Chem.* 2007, **46**(18), 7244–7246. (l) Yang, C., Messerschmidt, M., Coppens, P., Omary, M. A. Trinuclear gold(I) triazolates: A new class of wide-band phosphors and sensors. *Inorg. Chem.* 2006, **45**(17), 6592–6594. (m) Rawashdeh-Omary, M. A., Omary, M. A., Patterson, H. H. Oligomerization of Au(CN)(2)(−) and Ag(CN)(2)(−) ions in solution via ground-state aurophilic and argentophilic bonding. *J. Am. Chem. Soc.* 2000, **122**(42), 10371–10380. (n) Yam, V. W. W., Lo, K. K. W. Luminescent polynuclear d(10) metal complexes. *Chem. Soc. Rev.* 1999, **28**(5), 323–334.

(o) Omary, M. A., Hall, D. R., Shankle, G. E., Siemiarczuk, A., Patterson, H. H. Luminescent homoatomic exciplexes in dicyanoargentate(I) ions doped in alkali halide crystals. 2. "Exciplex tuning" by varying the dopant concentration. *J. Phys. Chem. B* 1999, **103**(19), 3845–3853. (p) Omary, M. A., Patterson, H. H. Luminescent homoatomic exciplexes in dicyanoargentate(I) ions doped in alkali halide crystals. 1. "Exciplex Tuning" by site-selective excitation. *J. Am. Chem. Soc.* 1998, **120**(31), 7696–7705. (q) Ford, P. C., Vogler, A. Photochemical and photophysical properties of tetranuclear and hexanuclear clusters of metals with d10 and s2 electronic configurations. *Acc. Chem. Res.* 1993, **26**(4), 220–226. (r) King, C., Khan, M. N. I., Staples, R. J., Fackler, J. P. Luminescent mononuclear gold(I) phosphines. *Inorg. Chem.* 1992, **31**(15), 3236–3238. (s) Schaefer, W. P., Marsh, R. E., McCleskey, T. M., Gray, H. B. A Luminescent gold complex - bis-Mu-bis(dicyclohexylphosphino)ethane-P,P′-digold bis(hexafluorophosphate). *Acta Crystallogr., Sect. C: Cryst. Struct. Commun.* 1991, **47**, 2553–2556. (t) Henary, M., Zink, J. I. Luminescence from the chair and cube isomers of $Ag_4I_4(PPH_3)_4$. *Inorg. Chem.* 1991, **30**(15), 3111–3112. (u) Yam, V. W. W., Lai, T. F., Che, C. M. Novel luminescent polynuclear gold(I) phosphine complexes - synthesis, spectroscopy, and X-ray crystal-structure of $Au_3(Dmmp)_2^{3+}$ Dmmp = bis(dimethylphosphinomethyl)-methylphosphine. *J. Chem. Soc., Dalton Trans.* 1990, no. 12, 3747–3752. (v) Kutal, C. Spectroscopic and photochemical properties of d^{10} metal-complexes. *Coord. Chem. Rev.* 1990, **99**, 213–252. (w) King, C., Wang, J. C., Khan, M. N. I., Fackler, J. P. Luminescence and metal-metal interactions in binuclear gold(I) compounds. *Inorg. Chem.* 1989, **28**(11), 2145–2149. (x) Nagasundaram, N., Roper, G., Biscoe, J., Chai, J. W., Patterson, H. H., Blom, N., Ludi, A. Single-crystal luminescence study of the layered compound $KAu(CN)_2$. *Inorg. Chem.* 1986, **25**(17), 2947–2951. (y) Segers, D. P., Dearmond, M. K., Grutsch, P. A., Kutal, C. Multiple luminescence from borohydridobis-(triphenylphosphine)copper(I). *Inorg. Chem.* 1984, **23**(18), 2874–2878. (z) Ziolo, R. F., Lipton, S., Dori, Z. Photoluminescence of phosphine complexes of d^{10} metals. *J. Chem. Soc. D* 1970, **17**, 1124.

9. (a) Hsu, C.-W., Lin, C.-C., Chung, M.-W., Chi, Y., Lee, G.-H., Chou, P.-T., Chang, C.-H., Chen, P.-Y. Systematic investigation of the metal-structure–photophysics relationship of emissive d10-complexes of group 11 elements: the prospect of application in organic light emitting devices. *J. Am. Chem. Soc.* 2011, **133**(31), 12085–12099. (b) Osawa, M., Kawata, I., Ishii, R., Igawa, S., Hashimoto, M., Hoshino, M. Application of neutral d(10) coinage metal complexes with an anionic bidentate ligand in delayed fluorescence-type organic light-emitting diodes. *J. Mater. Chem.*

C 2013, **1**(28), 4375–4383. (c) Crespo, O., Diez-Gil, C., Gimeno, M. C., Jones, P. G., Laguna, A., Ospino, I., Tapias, J., Villacampa, M. D., Visbal, R. Influence of the group 11 metal on the emissive properties of complexes M{(PR2) (2)C2B9H10}L. *Dalton Transactions* 2013, **42**(23), 8298–8306. (d) Teets, T. S., Partyka, D. V., Esswein, A. J., Updegraff, J. B., Zeller, M., Hunter, A. D., Gray, T. G. Luminescent, three-coordinate azadipyrromethene complexes of d10 copper, silver, and gold. *Inorg. Chem.* 2007, **46**(16), 6218–6220.

10. (a) Penfold, T. J., Karlsson, S., Capano, G., Lima, F. A., Rittmann, J., Reinhard, M., Rittmann-Frank, M. H., Braem, O., Baranoff, E., Abela, R., Tavernelli, I., Rothlisberger, U., Milne, C. J., Chergui, M. Solvent-induced luminescence quenching: static and time-resolved X-ray absorption spectroscopy of a copper(I) phenanthroline complex. *J. Phys. Chem. A* 2013, **117**(22), 4591–4601. (b) Stacy, E. M., McMillin, D. R. Inorganic exciplexes revealed by temperature-dependent quenching studies. *Inorg. Chem.* 1990, **29**(3), 393–396. (c) Siddique, Z. A., Yamamoto, Y., Ohno, T., Nozaki, K. Structure-dependent photophysical properties of singlet and triplet metal-to-ligand charge transfer states in copper(I) bis(diimine) compounds. *Inorg. Chem.* 2003, **42**(20), 6366–6378.

11. (a) Iwamura, M., Takeuchi, S., Tahara, T. Ultrafast excited-state dynamics of copper(I) complexes. *Acc. Chem. Res.* 2015, **48**(3), 782–791. (b) Iwamura, M., Takeuchi, S., Tahara, T. Real-time observation of the photoinduced structural change of bis(2,9-dimethyl-1,10-phenanthroline)copper(I) by femtosecond fluorescence spectroscopy: A realistic potential curve of the Jahn-Teller distortion. *J. Am. Chem. Soc.* 2007, **129**(16), 5248–5256. (c) Armaroli, N., Accorsi, G., Cardinali, F., Listorti, A., Photochemistry and photophysics of coordination compounds: copper. In *Photochemistry and Photophysics of Coordination Compounds I*; Balzani, V., Campagna, S., Eds., Springer, 2007; Vol. *280*, pp. 69–115.

12. Liu, Z., Djurovich, P. I., Whited, M. T., Thompson, M. E. Cu4I4 clusters supported by P^N-type ligands: new structures with tunable emission colors. *Inorg. Chem.* 2012, **51**(1), 230–236.

13. (a) Kriechbaum, M., Winterleitner, G., Gerisch, A., List, M., Monkowius, U. Synthesis, Characterization and luminescence of gold complexes bearing an NHC ligand based on the imidazo 1,5-a quinolinol scaffold. *Eur. J. Inorg. Chem.* 2013, **2013**(32), 5567–5575. (b) Shafikov, M. Z., Czerwieniec, R., Yersin, H. Ag(I) complex design affording intense phosphorescence with a landmark lifetime of over 100 ms. *Dalton Transactions* 2019, **48**(8), 2802–2806.

14. Shi, S., Collins, L. R., Mahon, M. F., Djurovich, P. I., Thompson, M. E., Whittlesey, M. K. Synthesis and characterization of phosphorescent two-coordinate copper(i) complexes bearing diamidocarbene ligands. *Dalton Transactions* 2017, **46**(3), 745–752.
15. Hamze, R., Peltier, J. L., Sylvinson, D., Jung, M., Cardenas, J., Haiges, R., Soleilhavoup, M., Jazzar, R., Djurovich, P. I., Bertrand, G., Thompson, M. E. Eliminating nonradiative decay in Cu(I) emitters: 99% quantum efficiency and microsecond lifetime. *Science* 2019, **363**(6427), 601.
16. Shao, Y. H., Gan, Z. T., Epifanovsky, E., Gilbert, A. T. B., Wormit, M., Kussmann, J., et al. Advances in molecular quantum chemistry contained in the Q-Chem 4 program package. *Mol. Phys.* 2015, **113**(2), 184–215.
17. Hirshfeld, F. L. Bonded-atom fragments for describing molecular charge-densities. *Theor. Chim. Acta* 1977, **44**(2), 129–138.
18. Lu, T., Chen, F. W. Multiwfn: A multifunctional wavefunction analyzer. *J. Comput. Chem.* 2012, **33**(5), 580–592.
19. Slater, J. C. Atomic Radii in Crystals. *J. Chem. Phys.* 1964, **41**(10), 3199–3204.
20. Comas-Vives, A., Harvey, J. N. How important is backbonding in metal complexes containing N-heterocyclic carbenes? Structural and NBO analysis. *Eur. J. Inorg. Chem.* 2011, **2011**(32), 5025–5035.
21. Murov, S. L., Carmichael, I., Hug, G. L. *Handbook of Photochemistry*, 2nd ed., Marcel Dekker: New York, 1993.
22. Yersin, H., Rausch, A. F., Czerwieniec, R., Hofbeck, T., Fischer, T. The triplet state of organo-transition metal compounds. Triplet harvesting and singlet harvesting for efficient OLEDs. *Coord. Chem. Rev.* 2011, **255**(21), 2622–2652.
23. Monkowius, U., Zabel, M., Fleck, M., Yersin, H. Gold(I) complexes bearing P Boolean and N-ligands: an unprecedented twelve-membered ring structure stabilized by aurophilic interactions. *Z. Naturforsch., B: J. Chem. Sci.* 2009, **64**(11–12), 1513–1524.
24. (a) Föller, J., Ganter, C., Steffen, A., Marian, C. M. Computer-aided design of luminescent linear N-heterocyclic carbene copper(I) pyridine complexes. *Inorg. Chem.* 2019, **58**, 5446. (b) Thompson, S., Eng, J., Penfold, T. J. The intersystem crossing of a cyclic (alkyl)(amino) carbene gold (I) complex. *J. Chem. Phys.* 2018, **149**(1), 014304.
25. Dias, F. B., Penfold, T. J., Monkman, A. P. Photophysics of thermally activated delayed fluorescence molecules. *Methods Appl. Fluoresc.* 2017, **5**(1), 012001.
26. Strickler, S. J., Berg, R. A. Relationship between absorption intensity and fluorescence lifetime of molecules. *J. Chem. Phys.* 1962, **37**(4), 814–822.

Chapter 53

Highly Efficient Photo- and Electroluminescence from Two-Coordinate Cu(I) Complexes Featuring Nonconventional N-Heterocyclic Carbenes

Shuyang Shi,[a] Moon Chul Jung,[b] Caleb Coburn,[c] Abegail Tadle,[a] Daniel Sylvinson M. R.,[a] Peter I. Djurovich,[a] Stephen R. Forrest,[c,d,e] and Mark E. Thompson[a,b]

[a]*Department of Chemistry, University of Southern California, Los Angeles, California 90089, USA*
[b]*Mork Family Department of Chemical Engineering and Materials Science, University of Southern California, Los Angeles, California 90089, USA*
[c]*Department of Physics, University of Michigan, Ann Arbor, Michigan 48109, USA*
[d]*Department of Electrical and Computer Engineering, University of Michigan, Ann Arbor, Michigan 48109, USA*
[e]*Department of Materials Science and Engineering, University of Michigan, Ann Arbor, Michigan 48109, USA*
met@usc.edu

A series of six luminescent two-coordinate Cu(I) complexes were investigated bearing nonconventional N-heterocyclic carbene ligands, monoamido-aminocarbene (MAC*) and diamidocarbene (DAC*),

---

Reprinted from *J. Am. Chem. Soc.*, **141**, 3576–3588, 2019.

Electrophosphorescent Materials and Devices
Edited by Mark E. Thompson
Text Copyright © 2019 American Chemical Society
Layout Copyright © 2024 Jenny Stanford Publishing Pte. Ltd.
ISBN 978-981-4877-34-3 (Hardcover), 978-1-003-08872-1 (eBook)
www.jennystanford.com

along with carbazolyl (Cz) as well as mono- and dicyano-substituted Cz derivatives. The emission color can be systematically varied over 270 nm, from violet to red, through proper choice of the acceptor (carbene) and donor (carbazolyl) groups. The compounds exhibit photoluminescent quantum efficiencies up to 100% in fluid solution and polystyrene films with short decay lifetimes ($\tau \approx 1\ \mu s$). The radiative rate constants for the Cu(I) complexes ($k_r = 10^5$–$10^6\ s^{-1}$) are comparable to state of the art phosphorescent emitters with noble metals such as Ir and Pt. All complexes show strong solvatochromism due to the large dipole moment of the ground states and the transition dipole moment that is in the opposite direction. Temperature-dependent studies of (MAC*)Cu(Cz) reveal a small energy separation between the lowest singlet and triplet states ($\Delta E_{S_1} - T_1 = 500\ cm^{-1}$) and an exceptionally large zero-field splitting (ZFS = 85 cm^{-1}). Organic light-emitting diodes (OLEDs) fabricated with (MAC*)Cu(Cz) as a green emissive dopant have high external quantum efficiencies (EQE = 19.4%) and brightness of 54 000 cd/m^2 with modest roll-off at high currents. The complex can also serve as a neat emissive layer to make highly efficient OLEDs (EQE = 16.3%).

53.1 Introduction

Organic light-emitting diode (OLED) technology is widely used in commercial applications such as displays and lighting. For luminescent materials to be applied in OLEDs, it is essential that both singlet (S_1) and triplet (T_1) excitons be harvested in the emission layer to achieve high efficiency. Organometallic complexes with noble metals such as Ir(III), Pt(II), Re(I), and Ru(II) can induce efficient spin–orbit coupling (SOC) between T_1 and S_1, and can achieve high radiative rate constants from T_1 to S_0 [1–8]. Thermally activated delayed fluorescence (TADF) is an alternate mechanism to harvest both singlet and triplet excitons that does not require heavy metals for efficient SOC. Cu complexes have been reported to undergo efficient phosphorescence from charge transfer (CT) states via TADF [9–19]. Most of the reported luminescent mononuclear Cu(I) complexes are four- [13, 15, 20–23] or three-coordinate [24–30], and typically undergo structural distortion in the excited state. The geometric change leads to large Franck–Condon factors and an increase in nonradiative decay rates due to strong vibronic coupling between the ground and excited states [31]. Therefore, although photoluminescence quantum yields (Φ_{PL}) of copper complexes can be high in the crystalline solid, the efficiency markedly decreases in fluid solution. Interestingly, we recently reported a two-coordinate cationic Cu carbene complex [(DAC)$_2$Cu][BF$_4$] (DAC = 1,3-bis(mesityl)-5,5-dimethyl-4,6-diketopyrimidinyl-2-ylidene) that is brightly emissive in fluid solution with Φ_{PL} = 65% [32], which is among the highest values reported for mononuclear Cu(I) complexes [21, 29, 33]. The high Φ_{PL} of this complex demonstrates that a two-coordinated Cu center can achieve high efficiency in fluid media if the ligands provide sufficient steric hindrance to prevent structural distortion in the excited state. Although the nonradiative rate constant (k_{nr} = 10^3–10^4 s^{-1}) of [(DAC)$_2$]Cu(BF$_4$) in solution is low as compared to most of the other luminescent Cu complexes [20, 25], the radiative rate constant ($k_r = \sim 10^4$ s^{-1}) is at least an order of magnitude smaller than those of Ir- and Pt-based emitters ($k_r = 10^5$–10^6 s^{-1}) [2]. Recent studies have also appeared describing highly efficient emission from two-coordinated Cu(I) complexes bearing cyclic alkyl-amino carbene (CAAC) ligands [27, 34–37]. These complexes have fast radiative rates ($k_r > 10^5$ s^{-1}) and highlight two-coordinate Cu(I) structures, particularly (CAAC)Cu(amide) complexes [35, 37], as promising luminescent complexes for application to OLEDs and other optoelectronic applications, provided that the appropriate carbene is chosen as a ligand.

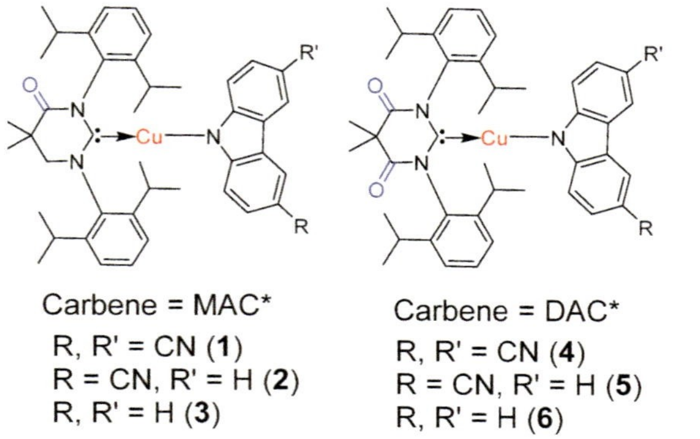

Scheme 53.1 Molecular structures of 1–6.

Carbene = MAC*
R, R' = CN (**1**)
R = CN, R' = H (**2**)
R, R' = H (**3**)

Carbene = DAC*
R, R' = CN (**4**)
R = CN, R' = H (**5**)
R, R' = H (**6**)

In this work, we characterize a series of two-coordinate neutral Cu complexes (Scheme 53.1) bearing nonconventional cyclic amido-carbenes [38–41] including (MAC*)Cu(CzCN$_2$) (**1**), (MAC*)Cu(CzCN) (**2**), (MAC*)Cu(Cz) (**3**), (DAC*)Cu(CzCN$_2$) (**4**), (DAC*)Cu(CzCN) (**5**), and (DAC*)Cu(Cz) (**6**) (MAC = cyclic monoamido-aminocarbene, DAC = cyclic diamidocarbene, "*" indicates that the aryl group bound to N is 2,6-diisopropylphenyl, Cz = carbazole). The carbonyl groups incorporated into MACs and DACs increase the π-accepting properties of the carbene, and concomitantly lower the energy of the LUMO. The diisopropylphenyl group on the carbene ligands provides sufficient steric hindrance to suppress the formation of an exciplex between the complexes and the solvent. All of the complexes show efficient TADF with short decay lifetime in polystyrene films and solution. The radiative rates ($k_r = 10^5$–10^6 s^{-1}) are comparable to those of (CAAC)Cu(amide) compounds [37], and, more surprising, to state-of-the-art emitters bearing noble metal such as Ir and Pt [2]. OLEDs using complex **3** as dopants and neat emitters were grown using vapor deposition, and show an external quantum efficiency up to 19.4%, with only modest efficiency roll-off at high brightness.

53.2 Experimental Section

53.2.1 Synthesis

All reactions were carried out using standard Schlenk and glovebox techniques using dried and degassed solvents. The DAC*Cl [42] precursor, 3-cyanocarbazole [43], and 3,6-dicyanocarbazole [44] were synthesized by following the literature procedure. CuCl and carbazole were purchased from Sigma-Aldrich. 3-Chloropivaloyl chloride was purchased from Arctom Chemicals. NMR spectra were recorded on a Varian 400 NMR spectrometer.

53.2.1.1 Synthesis of (MAC*)CuCl (1a): 3-chloro-N-(2,6-diisopropylphenyl)-N'-((2,6-diisopropylphenylimino)methyl)-2,2-dimethylpropanamide (1c)

N,N'-Bis(2,6-diisopropylphenyl) formamidine (500 mg, 1.37 mmol) and triethylamine (287 mL, 2.06 mmol) were dissolved in dichloromethane (20 mL) and stirred at 0 °C for 10 min, after which 3-chloropivaloyl chloride (0.195 mL, 1.51 mmol) was added dropwise. The solution mixture was stirred for 3 h at 0 °C. The solvent was removed under reduced pressure to afford a white powder, which was extracted with toluene and filtered through Celite. Removal of the residual solvent afforded the product as a white solid. Yield: 650 mg (98%). ^{1}H NMR δ_H (CDCl$_3$, 400 MHz, 298 K): 1.14 (m, 18H, CH(CH$_3$)$_2$), 1.30 (d, $J = 6.8$ Hz, 6H, CH(CH$_3$)$_2$), 1.54 (s, 6H, C(CH$_3$)$_2$), 2.95 (sept, $J = 6.5$ Hz, 4H, CH(CH$_3$)$_2$), 3.77 (s, 2H, CCH$_2$Cl), 7.08 (m, 3H, Ar–H), 7.27 (m, 2H, Ar–H), 7.39 (m, 1H, Ar–H), 8.59 (s, 1H, NCHN). ^{13}C NMR δ_C (CDCl$_3$, 101 MHz, 298 K): 23.17 (s, CH(CH$_3$)$_2$), 24,11 (s, CH(CH$_3$)$_2$), 24.22 (s, C(CH$_3$)$_2$), 24.83 (s, CH(CH$_3$)$_2$), 27.35 (s, CH(CH$_3$)$_2$), 28.94 (s, CH(CH$_3$)$_2$), 46.25 (s, C(CH$_3$)$_2$), 53.25 (s, CCH$_2$Cl), 123.05 (s, m-ArH), 124.01 (s, m-ArH), 124.20 (s, p-ArH), 129.18 (s, p-ArH), 132.95 (s, o-NAr), 138.91 (s, o-NAr), 145.21 (ipso-NAr), 145.51 (ipso-NAr), 148.51 (s, NCN), 174.74 (s, C=O). MALDI-TOF: m/z calculated, 447.34 [M–Cl$^-$]$^+$; found, 447.64 [M]$^+$.

53.2.1.2 1,3-bis(2,6-diisopropylphenyl)-5,5-dimethyl-4-keto-tetrahydropyrimidin-1-ium Chloride (1b)

1c (650 mg, 1.34 mmol) was dissolved in toluene (20 mL), and the solution was refluxed for 16 h at 110 °C during which a white precipitate formed. The reaction mixture was cooled to rt, and the white precipitate was collected by vacuum filtration and washed with cold toluene. Yield: 450 mg (69%). ^{1}H NMR δ_H (CDCl$_3$, 400 MHz, 298 K): 1.20 (d, J = 6.8 Hz, 6H, CH(CH$_3$)$_2$), 1.30 (dd, J = 9.9, 6.8 Hz, 12H, CH(CH$_3$)$_2$), 1.40 (d, J = 6.6 Hz, 6H, CH(CH$_3$)$_2$), 1.76 (s, 6H, C(CH$_3$)$_2$), 3.04 (sept, J = 6.8 Hz, 2H, CH(CH$_3$)$_2$), 3.25 (sept, J = 6.8 Hz, 2H, CH(CH$_3$)$_2$), 4.66 (s, 2H, CCH$_2$N), 7.20–7.30 (m, 4H, Ar–H), 7.45 (td, J = 7.9, 4.0 Hz, 2H, Ar–H), 9.82 (s, 1H, N=CH–N). ^{13}C NMR δ_C (CDCl$_3$, 101 MHz, 298 K): 24.00 (s, CH(CH$_3$)$_2$), 24.41 (s, CH(CH$_3$)$_2$), 24.59 (s, C(CH$_3$)$_2$), 24.77 (s, CH(CH$_3$)$_2$), 24.87 (s, CH(CH$_3$)$_2$), 28.98 (s, CH(CH$_3$)$_2$), 29.23 (s, CH(CH$_3$)$_2$), 39.17 (s, C(CH$_3$)$_2$), 61.76 (s, CCH$_2$N), 124.86 (s, m-ArH), 125.59 (s, m-ArH), 131.66 (s, p-ArH), 131.97 (s, p-ArH), 129.56 (s, o-NAr), 134.48 (s, o-NAr), 144.35 (ipso-NAr), 145.98 (ipso-NAr), 159.29 (s, NCN), 169.67 (s, C=O). Anal. Calcd for C$_{30}$H$_{43}$ClN$_2$O: C, 74.58; N, 5.80; H, 8.97. Found: C, 74.31; N, 6.17; H, 8.96.

53.2.1.3 (N,N′-bis(diisopropylphenyl)-5,5-dimethyl-4-keto-tetrahydropyrimidin-2-ylidene)-Cu(I) chloride (MAC*CuCl) (1a)

KHMDS (136 mg, 0.68 mmol) was added to a THF solution (20 mL) of **1b** (300 mg, 0.62 mmol) at rt, and the solution was stirred for 1 h before CuCl (67 mg, 0.68 mmol) was added. The reaction mixture was stirred at rt for 16 h, filtered through Celite, and the solvent was concentrated to 3 mL under reduced pressure. Hexane (20 mL) was added to the solution, and a white precipitate formed. Yield: 300 mg (88%). ^{1}H NMR δ_H (acetone-d_6, 400 MHz, 298 K): 1.17 (d, $J = 6.8$ Hz, 6H, CH(CH$_3$)$_2$), 1.35 (m, 18H, CH(CH$_3$)$_2$), 1.59 (s, 6H, C(CH$_3$)$_2$), 3.13 (sept, $J = 6.8$ Hz, 2H, CH(CH$_3$)$_2$), 3.38 (sept, $J = 6.8$ Hz, 2H, CH(CH$_3$)$_2$), 4.14 (s, 2H, CCH$_2$N), 7.30 (d, $J = 7.6$ Hz, 2H, m-ArH), 7.37 (d, $J = 7.7$ Hz, 2H, m-ArH), 7.42 (t, $J = 7.7$ Hz, 1H, p-ArH), 7.47 (t, $J = 7.7$ Hz, 1H, p-ArH). ^{13}C NMR δ_C (acetone-d_6, 101 MHz, 298 K): 23.25 (s, CH(CH$_3$)$_2$), 23,67 (s, CH(CH$_3$)$_2$), 23.69 (s, CH(CH$_3$)$_2$), 23.75 (s, C(CH$_3$)$_2$), 24.25 (s, CH(CH$_3$)$_2$), 28.27 (s, CH(CH$_3$)$_2$), 28.58 (s, CH(CH$_3$)$_2$), 37.87 (s, C(CH$_3$)$_2$), 60.75 (s, CCH$_2$N), 124.19 (s, m-ArH), 125.11 (s, m-ArH), 129.65 (s, p-ArH), 130.05 (s, p-ArH), 136.19 (s, o-NAr), 140.14 (s, o-NAr), 144.55 (ipso-NAr), 145.62 (ipso-NAr), 171.16 (s, C=O), 208.89 (s, NCN). Anal. Calcd for C$_{30}$H$_{42}$ClCuN$_2$O: C, 66.03; N, 5.13; H, 7.76. Found: C, 66.02; N, 5.43; H, 7.73.

53.2.1.4 Synthesis of [(DAC*)Cu]$_2$Cl$_2$ (2a)

KHMDS (450 mg, 2.28 mmol) was added to a THF solution (20 mL) of **2b** (1.39 g, 2.28 mmol) at rt, and the solution was stirred for 1 h before CuCl (230 mg, 2.28 mmol) was added. The reaction mixture was stirred at rt for 16 h. The solvent was evaporated under reduced pressure, and the obtained red solid was redissolved in toluene (20 mL) and filtered through Celite. The filtrate was concentrated to 3 mL under reduced pressure. Hexane (20 mL) was added to the solution, and a red precipitate formed. Yield: 400 mg (31%). ^{1}H NMR δ_H (acetone-d_6, 400 MHz, 298 K): 1.18 (d, $J = 6.8$ Hz, 24H, CH(CH$_3$)$_2$), 1.33 (d, $J = 6.8$ Hz, 24H, CH(CH$_3$)$_2$), 1.86 (s, 12H, C(CH$_3$)$_2$), 3.03 (sept, $J = 6.8$ Hz, 8H, CH(CH$_3$)$_2$)), 7.38 (d, $J = 7.8$ Hz, 8H, m-ArH), 7.53 (d, $J = 7.8$ Hz, 4H, p-ArH). ^{13}C NMR δ_C (acetone-d_6, 101 MHz, 298 K): 23.35 (s, CH(CH$_3$)$_2$), 23.58 (s, CH(CH$_3$)$_2$), 24.37 (s, C(CH$_3$)$_2$), 28.70 (s, CH(CH$_3$)$_2$), 51.92 (s, C(CH$_3$)$_2$), 124.78 (s, m-ArH), 130.69 (s, p-ArH), 135.02 (s, o-Ar), 145.42 (ipso-N–Ar), 172.23 (s, C=O), 213.64 (s, NCN). Anal. Calcd for C$_{60}$H$_{80}$Cl$_2$Cu$_2$N$_4$O$_4$: C, 64.38; N, 5.01; H, 7.19. Found: C, 64.16; N, 5.29; H, 7.19.

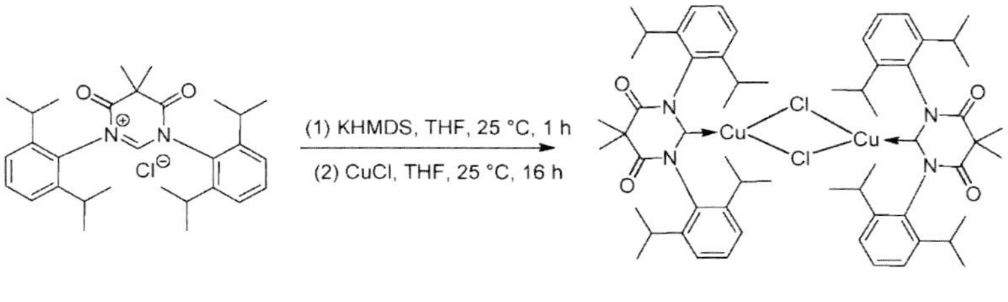

(1) KHMDS, THF, 25 °C, 1 h
(2) CuCl, THF, 25 °C, 16 h

(DAC*)OTf (**2b**) $[(DAC^*)Cu]_2Cl_2$ (**2a**)

53.2.1.5 Synthesis of Complexes 1–6: General Procedure

The carbazole ligand and NaO^{t}Bu were dissolved in THF and stirred for 3 h at rt. (carbene)CuCl was added to the reaction mixture and stirred for 16 h. The resulting mixture was filtered through Celite, and the solvent was removed under reduced pressure to afford a solid. The solid was redissolved in dichloromethane, and hexane was added to precipitate the desired product.

(1) NaO^{t}Bu, THF, RT, 3 h
(2) (Carbene)CuCl, THF, RT, 16 h
Carbene
Cu

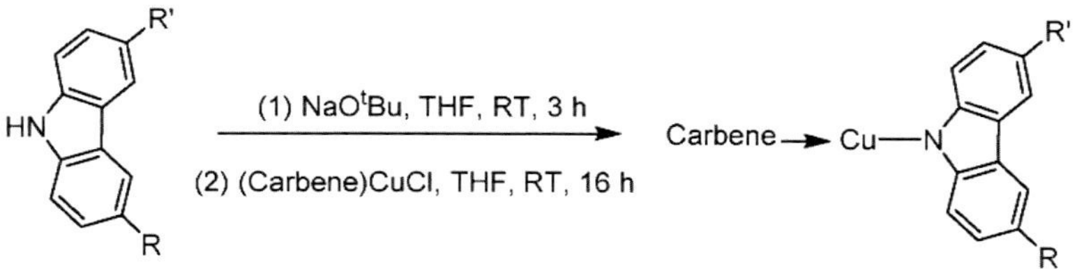

Carbene = MAC*

R, R' = CN (**1**)
R = CN, R' = H (**2**)
R, R' = H (**3**)

Carbene = DAC*

R, R' = CN (**4**)
R = CN, R' = H (**5**)
R, R' = H (**6**)

53.2.1.6 (MAC*)Cu(CzCN$_2$) (1)

The complex was made from (MAC*)CuCl (160 mg, 0.29 mmol), [Cz(CN)$_2$] (64 mg, 0.29 mmol), and NaO^{t}Bu (29 mg, 0.30 mmol) as a white solid. Yield: 168 mg (85%). ^{1}H NMR δ_H (acetone-d_6, 400 MHz, 298 K): 1.20–1.28 (m, 18H, CH(CH$_3$)$_2$), 1.43 (d, J = 6.8 Hz, 6H, CH(CH$_3$)$_2$), 1.68 (s, 6H, C(CH$_3$)$_2$), 3.30 (sept, J = 6.9 Hz, 2H, CH(CH$_3$)$_2$), 3.55 (sept, J = 6.9 Hz, 2H, CH(CH$_3$)$_2$), 4.34 (s, 2H, CCH$_2$N), 5.59 (d, J = 8.5 Hz, 2H, CH^1(Cz)), 7.22 (d, J = 10.8

Hz, 2H, CH^2(Cz)), 7.58 (d, J = 7.8 Hz, 2H, m-ArH), 7.64 (d, J = 7.8 Hz, 2H, m-ArH), 7.84 (m, 2H, p-ArH), 8.34 (s, 2H, CH^3(Cz)). ^{13}C NMR δ_C (acetone-d_6, 101 MHz, 298 K): 23.43 (s, CH(CH$_3$)$_2$), 23.58 (s, CH(CH$_3$)$_2$), 23.63 (s, CH(CH$_3$)$_2$), 23.81 (s, C(CH$_3$)$_2$), 24.40 (s, CH(CH$_3$)$_2$), 28.48 (s, CH(CH$_3$)$_2$), 28.70 (s, CH(CH$_3$)$_2$), 38.06 (s, C(CH$_3$)$_2$), 61.08 (s, CCH$_2$N), 99.09 (s, CN-Cz), 115.76 (s, CH^1(Cz)), 120.61 (s, ipso-CN(Cz)), 123.43 (s, ipso-C(Cz)), 124.74 (s, CH^3(Cz)), 125.02 (s, m-ArH), 125.91 (s, m-ArH), 127.16 (s, CH^2(Cz), 130.31 (s, p-ArH), 130.69 (s, p-ArH), 136.51 (s, o-Ar), 140.44 (s, o-Ar), 145.53 (ipso-N-Ar), 146.66 (ipso-N–Ar), 152.27 (s, ipso-N(Cz)), 171.18 (s, C=O), 208.74 (s, NCN). Anal. Calcd for C$_{44}$H$_{48}$CuN$_5$O: C, 72.75; N, 9.64; H, 6.66. Found: C, 72.70; N, 9.36; H, 6.83.

53.2.1.7 (MAC*)Cu(CzCN) (2)

The complex was made from (MAC*)CuCl (200 mg, 0.37 mmol), CzCN (71 mg, 0.37 mmol), and NaO^{t}Bu (36 mg, 0.37 mmol) as a white solid. Yield: 200 mg (78%). ^{1}H NMR δ_H (acetone-d_6, 400 MHz, 298 K): 1.21–1.31 (m, 18H, CH(CH$_3$)$_2$), 1.43 (d, J = 6.8 Hz, 6H, CH(CH$_3$)$_2$), 1.67 (s, 6H, C(CH$_3$)$_2$), 3.30 (sept, J = 6.8 Hz, 2H, CH(CH$_3$)$_2$), 3.55 (sept, J = 6.8 Hz, 2H, CH(CH$_3$)$_2$), 4.31 (s, 2H,CCH$_2$N), 5.51 (d, J = 8.5 Hz, 1H,CH^7(Cz)), 5.63 (d, J = 7.9 Hz, 1H, CH^1(Cz)), 6.87 (t, J = 7.4 Hz, 1H, CH^5(Cz)), 6.95 (t, J = 7.6 Hz, 1H, CH^6(Cz)), 7.09 (d, J = 8.5 Hz, 1H, CH^2(Cz)), 7.55 (d, J = 7.8 Hz, 2H, m-ArH), 7.61 (d, J = 7.8 Hz, 2H, m-ArH), 7.79–7.84 (m, 2H, p-ArH), 7.87 (d, 1H, CH^4(Cz)), 8.15 (s, 1H, CH^3(Cz)). ^{13}C NMR δ_C (acetone-d_6, 101 MHz, 298 K): 23.44 (s, CH(CH$_3$)$_2$), 23.59 (s, CH(CH$_3$)$_2$), 23.60 (s, CH(CH$_3$)$_2$), 23.82 (s, C(CH$_3$)$_2$), 24.38 (s, CH(CH$_3$)$_2$), 28.49 (s, CH(CH$_3$)$_2$), 28.71 (s, CH(CH$_3$)$_2$), 38.01 (s, C(CH$_3$)$_2$), 61.12 (s, CCH$_2$N), 96.68 (s, CN–Cz), 115.07 (s, CH^7(Cz)), 115.16 (s, CH^1(Cz)), 117.01 (s, CH^5(Cz)), 119.21 (s, CH^4(Cz)), 121.42 (s, ipso-CN(Cz)), 123.21 (s, ipso-C(Cz)), 123.71 (s, CH^3(Cz)), 124.15 (s, ipso-C(Cz)), 124.48 (s, CH^6(Cz)), 124.90 (s, m-ArH), 125.59 (s, CH^2(Cz)), 125.81 (s, m-ArH), 130.13 (s, p-ArH), 130.50 (s, p-ArH), 136.48 (s, o-Ar), 140.47 (s, o-Ar), 145.45 (ipso-N–Ar), 146.56 (ipso-N–Ar), 150.50 (s, ipso-N(Cz)), 151.59 (s, ipso-N(Cz)), 171.23 (s, C=O), 209.25 (s, NCN). Anal. Calcd for C$_{43}$H$_{49}$CuN$_4$O: C, 73.63; N, 7.99; H, 7.04. Found: C, 73.45; N, 8.16; H, 6.88.

53.2.1.8 (MAC*)Cu(Cz) (3)

The complex was made from (MAC*)CuCl (2.0 g, 3.67 mmol), Cz (613 mg, 3.67 mmol), and NaO^{t}Bu (353 mg, 3.67 mmol) as a yellow solid. Yield: 2.1 g (85%). ^{1}H NMR δ_H (acetone-d_6, 400 MHz, 298 K): 1.14–1.36 (m, 18H, CH(CH3)2), 1.43 (d, $J = 6.8$ Hz, 6H, CH(CH$_3$)$_2$), 1.66 (s, 6H, C(CH$_3$)$_2$), 3.30 (sept, $J = 6.8$ Hz, 2H, CH(CH$_3$)$_2$), 3.55 (sept, $J = 6.8$ Hz, 2H, CH(CH$_3$)$_2$), 4.26 (s, 2H, CCH$_2$N), 5.56 (d, $J = 8.1$ Hz, 2H, CH^1(Cz)), 6.71 (t, $J = 7.3$ Hz, 2H, CH^3(Cz)), 6.81 (t, $J = 6.9$ Hz, 2H, CH^2(Cz)), 7.53 (d, $J = 7.8$ Hz, 2H, m-ArH), 7.58 (d, $J = 7.8$ Hz, 2H, m-ArH), 7.71 (d, $J = 6.9$ Hz, 2H, CH^4(Cz)), 7.73–7.79 (m, 2H, ArH). ^{13}C NMR δ_C (acetone-d_6, 101 MHz, 298 K): 23.46 (s, CH(CH$_3$)$_2$), 23,58 (s, CH(CH$_3$)$_2$), 23.61 (s, CH(CH$_3$)$_2$), 23.82 (s, C(CH$_3$)$_2$), 24.37 (s, CH(CH$_3$)$_2$), 28.50 (s, CH(CH$_3$)$_2$), 28.72 (s, CH(CH$_3$)$_2$), 37.97 (s, C(CH$_3$)$_2$), 61.16 (s, CCH$_2$N), 114.60 (s, CH^1(Cz)), 114.95 (s, CH^3(Cz)), 118.39 (s, CH^4(Cz)), 122.73 (s, CH^2(Cz)), 123.91 (s, ipso-C(Cz)), 124.78 (s, m-ArH), 125.70 (s, m-ArH), 129.93 (s, p-ArH), 130.30 (s, p-ArH), 136.47 (s, o-Ar), 140.51 (s, o-Ar), 145.37 (ipso-N–Ar), 146.46 (ipso-N–Ar), 149.85 (s, ipso-N(Cz)), 171.29 (s, C=O), 209.83 (s, NCN). Anal. Calcd for C$_{42}$H$_{50}$CuN$_3$O: C, 74.58; N, 6.21; H, 7.45. Found: C, 74.21; N, 6.01; H, 7.47.

53.2.1.9 (DAC*)Cu(CzCN$_2$) (4)

The complex was made from (DAC*)CuCl (200 mg, 0.36 mmol), [Cz(CN)$_2$] (78 mg, 0.36 mmol), and NaO^{t}Bu (35 mg, 0.36 mmol) as a yellow solid. Yield: 180 mg (68%). ^{1}H NMR δ_H (acetone-d_6, 400 MHz, 298 K): 1.25 (dd, $J = 11.3$ Hz, 6.8 Hz, 24H, CH(CH$_3$)$_2$), 1.95 (s, 6H, C(CH$_3$)$_2$), 3.23 (sept, $J = 6.8$ Hz, 4H, CH(CH$_3$)$_2$), 5.59 (d, $J = 8.0$ Hz, 2H, CH^1(Cz)), 7.28 (d, $J = 8.5$ Hz, 2H, CH^2(Cz)), 7.68 (d, $J = 7.8$ Hz, 4H, m-ArH), 7.97 (t, $J = 7.8$ Hz, 2H, p-ArH), 8.36 (s, 2H, CH^3(Cz)). ^{13}C NMR δ_C (acetone-d_6, 101 MHz, 298 K): 23.34 (s, CH(CH$_3$)$_2$), 23.72 (s, CH(CH$_3$)$_2$), 24.36 (s, C(CH$_3$)$_2$), 28.80 (s, CH(CH$_3$)$_2$), 52.37 (s, C(CH$_3$)$_2$), 99.53 (s, CN–Cz), 115.72 (s, CH^1(Cz)), 120.49 (s, ipso-CN(Cz)), 123.56 (s, ipso-C(Cz)), 124.81 (s, CH^3(Cz)), 125.62 (s, m-ArH), 127.39 (s, CH^2(Cz)), 131.41 (s, p-ArH), 135.20 (s, o-Ar), 146.59 (ipso-N–Ar), 152.14 (s, ipso-N(Cz)), 172.11 (s, C=O), 213.90 (s, NCN). Anal. Calcd for C$_{44}$H$_{46}$CuN$_5$O$_2$ + 0.5H$_2$O: C, 70.52; N, 9.34; H, 6.32. Found: C, 70.81; N, 9.26; H, 6.29.

53.2.1.10 (DAC*)Cu(CzCN) (5)

The complex was made from (DAC*)CuCl (200 mg, 0.36 mmol), CzCN (69 mg, 0.36 mmol), and NaO^{t}Bu (35 mg, 0.36 mmol) as an orange solid. Yield: 200 mg (76%). ^{1}H NMR δ_H (acetone-d_6, 400 MHz, 298 K): 1.25 (dd, J = 8.3 Hz, 6.8 Hz, 24H, CH(CH$_3$)$_2$), 1.94 (s, 6H, C(CH$_3$)$_2$), 3.22 (sept, J = 6.8 Hz, 4H, CH(CH$_3$)$_2$), 5.51 (d, J = 8.5 Hz, 1H, CH^7(Cz)), 5.61 (d, J = 8.1 Hz, 1H, CH^1(Cz)), 6.91 (t, J = 7.8 Hz, 1H, CH^5(Cz)), 7.00 (t, J = 7.6 Hz, 1H, CH^6(Cz)), 7.13 (d, J = 8.5 Hz, 1H, CH^2(Cz)), 7.65 (d, J = 7.8 Hz, 4H, m-ArH), 7.89 (d, J = 7.8 Hz, 1H, CH^4(Cz)), 7.91 (t, J = 7.8 Hz, 2H, p-ArH), 8.17 (s, 1H, CH^3(Cz)). ^{13}C NMR δ_C (acetone-d_6, 101 MHz, 298 K): 23.33 (s, CH(CH$_3$)$_2$), 23.69 (s, CH(CH$_3$)$_2$), 24.37 (s, C(CH$_3$)$_2$), 28.80 (s, CH(CH$_3$)$_2$), 52.20 (s, C(CH$_3$)$_2$), 97.27 (s, CN–Cz), 115.06 (s, CH^7(Cz)), 115.09 (s, CH^1(Cz)), 117.40 (s, CH^5(Cz)), 119.31 (s, CH^4(Cz)), 121.24 (s, ipso-CN(Cz)), 123.33 (s, ipso-C(Cz)), 123.77 (s, CH^3(Cz)), 124.34 (s, ipso-C(Cz)), 124.67 (s, CH^6(Cz)), 125.51 (s, m-ArH), 125.84 (s, CH^2(Cz)), 131.22 (s, p-ArH), 135.15 (s, o-Ar), 146.49 (ipso-N–Ar), 150.33 (s, ipso-N(Cz)), 151.49 (s, ipso-N(Cz)), 172.17 (s, C=O), 214.12 (s, NCN). Anal. Calcd for C$_{43}$H$_{47}$CuN$_4$O$_2$: C, 72.19; N, 7.83; H, 6.62. Found: C, 72.04; N, 7.88; H, 6.62.

53.2.1.11 (DAC*)Cu(Cz) (6)

The complex was made from (DAC*)CuCl (100 mg, 0.18 mmol), Cz (30 mg, 0.18 mmol), and NaO^{t}Bu (18 mg, 0.19 mmol) as a purple solid. After the precipitation, the compound was purified by sublimation. Yield: 86 mg (70%). ^{1}H NMR δ_H (acetone-d_6, 400 MHz, 298 K): 1.25 (dd, J = 6.8, 5.8 Hz, 24H, CH(CH$_3$)$_2$), 1.93 (s, 6H, C(CH$_3$)$_2$), 3.20 (sept, J = 6.8 Hz, 4H, CH(CH$_3$)$_2$)), 5.55 (d, J = 8.1 Hz, 2H, CH^1(Cz)), 6.75 (t, J = 7.7 Hz, 2H, CH^3(Cz)), 6.86 (t, J = 7.5 Hz, 2H, CH^2(Cz)), 7.62 (d, J = 7.9 Hz, 4H, m-ArH), 7.74 (d, J = 7.6 Hz, 2H, CH^4(Cz)), 7.86 (d, J = 7.8 Hz, 2H, p-ArH). ^{13}C NMR δ_C (acetone-d_6, 101 MHz, 298 K): 23.34 (s, CH(CH$_3$)$_2$), 23.65 (s, CH(CH$_3$)$_2$), 24.38 (s, C(CH$_3$)$_2$), 28.84 (s, CH(CH$_3$)$_2$), 51.98 (s, C(CH$_3$)$_2$), 114.59 (s, CH^1(Cz)), 115.50 (s, CH^3(Cz)), 118.40 (s, CH^4(Cz)), 122.97 (s, CH^2(Cz)), 124.13 (s, ipso-C(Cz)), 125.37 (s, m-ArH), 131.01 (s, p-ArH), 135.12 (s, o-Ar), 146.39 (ipso-N–Ar), 149.71 (s, ipso-N(Cz)),

172.24 (s, C=O), 214.46 (s, NCN). MALDI-TOF: m/z calcd, 689.30 [M]$^+$; found, 690.54 [M]$^+$.

53.2.2 X-ray Crystallography

The X-ray intensity data were measured on a Bruker APEX DUO system equipped with a TRIUMPH curved-crystal monochromator and a Mo Kα fine-focus tube (λ = 0.71073 Å). The frames were integrated with the Bruker SAINT software package using a SAINT V8.37A (Bruker AXS, 2013) algorithm. Data were corrected for absorption effects using the multiscan method (SADABS). All non-hydrogen atoms were refined anisotropically. CCDC 1873677, 1873680, 1873678, and 1873679 contain the supplementary crystallographic data for **1–4**, respectively. These data can be obtained free of charge at http://www.ccdc.cam.ac.uk/conts/retrieving.html, or from the Cambridge Crystallographic Data Centre, CCDC, 12 Union Road, Cambridge CB2 1EZ, UK (fax, 44-1223-336-033; or e-mail, deposit@ccdc.cam.ac.uk).

53.2.3 Electrochemical Measurements

Cyclic voltammetry and differential pulsed voltammetry were performed using a VersaSTAT 3 potentiostat. Anhydrous acetonitrile (DriSolv) was used as the solvent under inert atmosphere, and 0.1 M tetra(n-butyl)ammonium hexafluorophosphate (TBAF) was used as the supporting electrolyte. A glassy carbon rod was used as the working electrode, a platinum wire was used as the counter electrode, and a silver wire was used as a pseudoreference electrode. The redox potentials are based on values measured from differential pulsed voltammetry and are reported relative to a ferrocene/ferrocenium (Cp$_2$Fe/Cp$_2$Fe$^+$) redox couple used as an internal reference, while electrochemical reversibility was determined using cyclic voltammetry.

53.2.4 Photophysical Characterization

The UV–visible spectra were recorded on a Hewlett-Packard 8453 diode array spectrometer. Photoluminescent emission measurements were performed using a Photon Technology International QuantaMaster model C-60 fluorimeter. Emission lifetimes at room temperature were measured by time-correlated single-photon counting using an IBH Fluorocube instrument equipped with an LED ($\lambda = 405$ nm) excitation source. Quantum yield measurements were carried out using a Hamamatsu C9920 system equipped with a xenon lamp, calibrated integrating sphere, and model C10027 photonic multichannel analyzer (PMA). Temperature-dependent measurements in the range of 5–320 K were performed using a JANIS ST-100 Standard Optical Cryostat instrument equipped with an intelligent temperature controller. Emission lifetimes were measured using the Photon Technology International QuantaMaster model C-60 fluorimeter in the range of 5–150 K and the IBH Fluorocube instrument in the range of 150–320 K. All samples in fluid solution were deaerated by extensive sparging with N_2.

53.2.5 Density Functional Theory (DFT) Calculations

Ground-state geometries of all complexes reported were optimized at the B3LYP/LACVP** level. Time-dependent DFT (TD-DFT) calculations were performed on the optimized structures at the CAM-B3LYP/LACVP** level. These DFT and TD-DFT calculations were performed using Q-Chem 5.1 [45]. TD-DFT calculations with solvent effects were performed using the IEF-PCM solvation model in the nonequilibrium limit using the ptSS (perturbative state-specific) approach as implemented in Q-Chem 5.1.

53.2.6 OLED Fabrication and Characterization

Glass substrates with 1 mm wide indium tin oxide (ITO) strips were cleaned by sequential sonication in tergitol, deionized water, acetone, and isopropanol, followed by 15 min UV ozone exposure. Organic materials and metals were deposited in a vacuum thermal evaporator with a base pressure of 10^{-7} Torr at rates of 0.2–1 Å/s. A 2 mm² device area was defined by deposition through shadow masks to define 1 mm wide cathodes consisting of 100 nm Al that were oriented perpendicular to the ITO strips. The device structure was: glass substrate/70 nm ITO/5 nm hexaazatriphenylene hexacarbonitrile (HATCN)/40 nm 4,4′-cyclohexylidenebis [N,N-bis(4-methylphenyl)benzenamine] (TAPC)/10 nm N,N′-dicarbazolyl-3,5-benzene (mCP)/EML/45 nm 2,2′,2″-(l,3,5-benzenetriyl)-tris(L-phenyl-l-H-benzimidazole) (TPBi)/1.5 nm 8-hydroxyquinolinato lithium (LiQ)/100 nm Al. Here, the EML is 25 nm of compound 3, either neat or doped into 3,3′-di(9H-carbazol-9-yl)-1,1′-biphenyl (mCBP) at 10 or 40 vol %.

A HP4156A semiconductor parameter analyzer was used to source voltage while measuring the photocurrent using a calibrated large area Thorlabs FDS1010-CAL photodiode that collected all light exiting the bottom of the flat glass substrate. A fiber-coupled OceanOptics USB4000-VIS-NIR spectrometer was used to measure the output spectra.

53.3 Results and Discussion

53.3.1 Synthesis and Characterization

The pyrimidinium salt precursor to MAC* was synthesized by condensation of N,N′-bis(2,6-diisopropylphenyl) formamidine and 3-chloropivaloyl chloride in the presence of excess triethylamine at 0°C, followed by intramolecular cyclization in refluxing toluene for 16 h. The overall yield of MAC*Cl was ca. 70%. The DAC*Cl salt was synthesized following the literature procedure [42]. The (carbene)-CuCl precursors were synthesized from the free MAC* and DAC* carbenes generated in situ, followed by addition of CuCl. MAC*CuCl was isolated as a white solid, whereas the DAC* analogue is a deep red solid. The red color for the DAC* complex is likely due to the formation of a dimeric species [(DAC*)Cu(μCl)$_2$Cu(DAC*)], as reported for the mesityl-substituted DAC analogue [46]. The (carbene)CuCz complexes were synthesized from reaction of the (carbene)CuCl with sodium carbazolide and isolated in 68–85% yield. Complexes 1 and 2 are white solids, whereas complexes 3–6 are yellow, orange-yellow, orange, and purple, respectively. Complexes 1–5 are air- and moisture-stable and can be sublimed under vacuum, whereas complex 6 reacts with ambient moisture to generate free CzH and a yellow solid, presumably DAC*CuOH.

Single crystals of 1–4 suitable for X-ray diffraction studies were grown by slow evaporation of CH_2Cl_2/pentane solutions. The complexes show a near linear geometry at the copper center (C_{NHC}–Cu–N_{Cz} = 173–180°) (Fig. S1). Bond lengths to the ligands (Cu–C_{NHC} = 1.868(2)–1.902(2) Å, Cu–N_{Cz} = 1.852(1)–1.855(2) Å) are comparable to the values reported for mononuclear Cu–C_{NHC} and Cu–Cz complexes [25–27, 32, 47]. Dihedral angles between ligand planes are in the range between 1–16° and lead to a near parallel orientation of 2p orbitals on C_{NHC} and N_{Cz}. This geometric arrangement, in light of the close distance between these 2p orbitals on the ligated atoms ($C_{NHC} \cdots N_{Cz}$ = 3.719(2)–3.762(2) Å), suggests that a long-range π interaction could be present across the metal center [48, 49].

The electrochemical properties of the complexes and their precursors (carbene–CuCl) were examined by cyclic voltammetry and differential pulse voltammetry in acetonitrile solution (Table 53.1). The redox potentials are referenced to an internal ferrocene (Fc^+/Fc) couple, and converted to HOMO and LUMO energies [50]. All of the complexes display quasi-reversible reduction and irreversible oxidation waves. The reduction potentials are similar in complexes with the same carbene ligand and are largely unchanged from the precursors MAC*CuCl and DAC*CuCl. Reduction potentials for the DAC* analogues are anodically shifted relative to the MAC* analogues by ca. 0.90 V due to the second carbonyl group in DAC*. The oxidation potentials increase from complexes **3** to **1** (and **6** to **4**) consistent with the donor strength of the Cz ligands decreasing with addition of the nitrile groups. Oxidation potentials of chloride precursors (E_{ox} = 0.91 and 1.03 V for MAC*CuCl and DAC*CuCl) are markedly larger than the Cu–Cz complexes. The redox behavior suggests that the reduction potential is determined by the identity of the carbene ligand, whereas the oxidation potential is controlled by carbazolyl ligand.

Theoretical calculations support the assignment of the electrochemical results. The contours for the LUMO and HOMO obtained from the Density Functional Theory (DFT) calculations of the (carbene)Cu(Cz) complexes (for example, see Fig. 53.1) show the LUMO is primarily localized on the carbene ligand, whereas the HOMO is predominantly on the Cz ligand. Both LUMO and HOMO

Table 53.1 Redox data for complexes **1–6** and carbene–CuCl and dipole moments calculated for **1–6** in the ground state (S_0), lowest ligand-localized triplet state (^{3}Cz), and charge transfer singlet state (^{1}ICT)

Complex	Potentials				Dipole moment μ (μ_Z)[c]		
	E^a_{ox} (V)	E^a_{red} (V)	HOMO[b] (eV)	LUMO[b] (eV)	μ_{gs} S_0	μ_{es} ^{3}Cz	μ_{es} ^{1}ICT
1	0.60	−2.45	−5.48	−1.94	19.1 (18.9)	19.1 (18.8)	5.5 (−4.3)
2	0.38	−2.45	−5.23	−1.94	14.9 (14.8)	12.5 (12.4)	8.6 (−8.6)
3	0.17	−2.50	−4.98	−1.88	10.6 (10.2)	9.9 (9.4)	13.7 (−13.2)
4	0.60	−1.60	−5.48	−2.94	17.2 (17.2)	16.7 (16.7)	8.9 (−8.9)
5	0.38	−1.56	−5.23	−2.99	13.7 (13.0)	11.4 (10.3)	13.5 (−12.8)
6	0.17	−1.60	−4.98	−2.94	8.3 (8.3)	0.60 (0.5)	17.0 (−17.0)
(MAC*)CuCl	0.91	−2.50	−5.84	−1.88			
(DAC*)CuCl	1.03	−1.60	−5.97	−2.94			

[a]Redox potentials obtained in acetonitrile with 0.1 M TBAPF$_6$ versus internal Fc$^+$/Fc. [b]HOMO = 1.15(E_{ox}) + 4.79; LUMO = 1.18(E_{red})−4.83 [50]. [c]Obtained from TD-DFT calculations (CAM-B3LYP/LACVP**) using geometry optimized structures. μ_Z is the projection of μ along the Cu–N bond axis. Negative values indicate dipole moments are opposite in direction from the dipole moments in the ground state. All dipole moment values are reported in Debye.

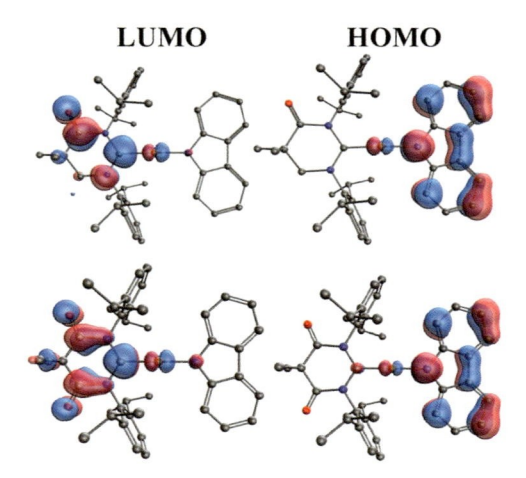

Figure 53.1 Frontier orbitals. LUMO (left) and HOMO (right) of complexes **3** (top) and **6** (bottom).

have little, but not insignificant, metal character. DFT and TD-DFT calculations were carried out at the CAM-B3LYP/LACVP** level to predict the lowest energy excited states for these complexes [51]. The CAM-B3LYP functional was found to provide an accurate description of both the charge transfer (CT) and the ligand-localized (Cz) states (vide infra). One state corresponds to an intramolecular charge transfer (ICT) involving the carbazolyl (donor) and the carbene (acceptor). In compounds **1** and **2**, the ICT state lies higher in energy than the ^{3}Cz state, whereas in **3–6** the ICT state is lowest in energy. The singlet and triplet (^{1}ICT and ^{3}ICT) levels for these configurations are predicted to be close in energy (< 0.3 eV). See the Supporting Information for further discussion of the modeling studies.

Ground (μ_{gs}) and excited state (μ_{es}) dipole moments for compounds **1–6** were estimated using DFT and TD-DFT calculations and are given in Table 53.1. Dipole moments were calculated for the ^{3}Cz, ^{1}ICT, and ^{3}ICT states using optimized structures of the ground state. The combination of linear geometry and large separation between the donor and acceptor imparts a large μ_{gs} to all of the complexes. Addition of a carbonyl group decreases μ_{gs} by ∼2 D upon going from MAC* to DAC* analogues. The ICT transitions redistribute electron density away from carbazolyl toward the carbene, which effectively leads to a μ_{es} that is in the direction opposite to μ_{gs}. In Table 53.1 we give both the magnitude of the μ vector and its projection scalar onto the axis defined by the Cu–N bond (μ_Z). The μ_{es} values for the ^{3}Cz states of compounds **1–5** are similar to their μ_{gs} values, because the transition involves only the redistribution of electron density within the carbazolyl ligand. However, in **6**, the ^{3}Cz state was found to contain a significant contribution from the HOMO–LUMO ICT transition (27%) leading to a marked reduction in μ_{es}.

53.3.2 Photophysical Properties

Absorption spectra of complexes 1–6 in 2-methyltetrahydrofuran (2-MeTHF) are shown in Fig. 53.2a. Structured absorption bands between 300–380 nm are assigned to allowed transitions localized on the carbazolyl ligand (^{1}Cz) by comparison to similar absorption features found in potassium carbazolide (Fig. S2) [37]. Strong (ε = 4000–6000 M^{-1} cm^{-1}), featureless absorption bands at lower energy are distinct in 2–6. A similar band present in 1 is obscured by the structured ^{1}Cz transitions and has an absorption edge at 400 nm. The bands red-shift in the series 1 → 6 as the donor strength of the carbazolyl ligand increases (from Cz(CN)$_2$ to Cz) and the electrophilicity of the carbene increases (from MAC* to DAC*). A plot of redox gaps versus the onset energy of the low energy bands shows a linear trend (inset in Fig. 53.2a). These bands are therefore assigned to allowed ICT transitions on the basis of this correlation and TD-DFT calculations. A hypsochromic shift of 71 nm (4100 cm^{-1}) is observed in the onset of the ^{1}ICT band upon varying the solvent from nonpolar methylcyclohexane to polar acetonitrile (Fig. 53.2b). In contrast, the absorption bands for the localized ^{1}Cz transitions show only slight spectral shifts in the same solvents. The ^{1}ICT bands display pronounced negative solvatochromism due to the excited-state transition that is oriented in the direction opposite to the ground state (Table 53.1).

Emission spectra of complexes **1–5** in 2-MeTHF at room temperature are shown in Fig. 53.3a, and their photophysical properties are summarized in Table 53.2. Complex **6** is nonemissive in

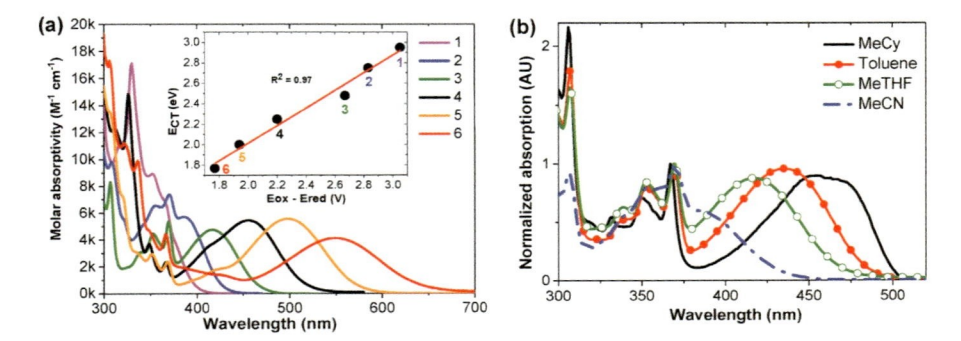

Figure 53.2 Absorption spectra of complexes **1–6** at room temperature. (a) Spectra of complexes **1–6** in 2-MeTHF. Inset: The energy of the low energy band (obtained from onset) versus the redox gaps. (b) Spectra of complex **3** in methylcyclohexane (MeCy), toluene, 2-MeTHF, and acetonitrile.

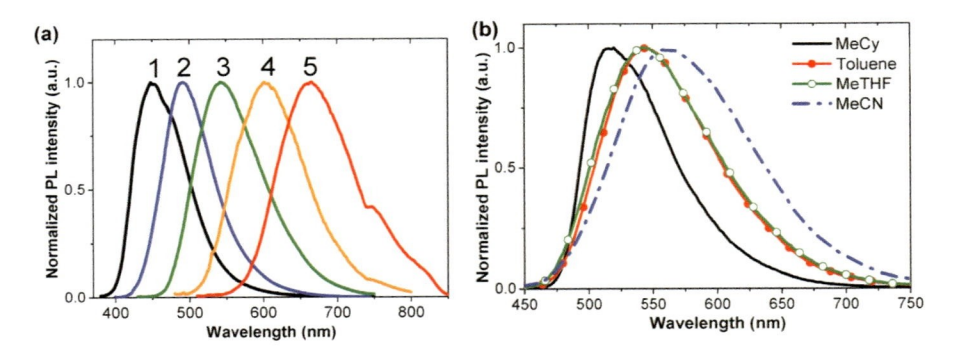

Figure 53.3 Emission spectra of complexes **1–5** in fluid solution at room temperature. (a) Spectra of complexes **1–5** in 2-MeTHF. (b) Spectra of complex **3** in methylcyclohexane, toluene, 2-MeTHF, and acetonitrile.

2-MeTHF. Complexes **1–5** show broad and featureless bands indicative of emission from the ICT states. The emission color ranges over 220 nm, from deep blue to red, through variation of either the Cz donor or the carbene acceptor ligands. The quantum efficiency (Φ_{PL}) ranges from 2% to 100% with short decay lifetimes ($\tau = 0.052$–2.3 μs). The radiative rate constants of complexes **1-5** are high ($k_r = 10^5$–10^6 s^{-1}) and comparable to values for efficient phosphorescent emitters with noble metals such as Ir(III) and Pt(II). Emission spectra of the complexes are also solvatochromic. For example, the

Table 53.2 Luminescent properties of complexes 1–6 in different media

Complex	$\lambda_{max,RT}$ (nm)	Φ_{RT}	τ_{RT} (μs)	$k_{r,RT}$ (10^5 s^{-1})	$k_{nr,RT}$ (10^5 s^{-1})	$\lambda_{max,77K}$ (nm)	τ_{77K} (μs)
				Solution in 2-MeTHF			
1	448	0.24	2.3	1.0	3.3	424	9300
2	492	1.00	1.2	8.3	<0.083	428	2200
3	542	0.55	1.1	5.0	4.1	432	430 nm, 2100 (62%); 342 (38%); 520 nm, 181
4	602	0.05	0.080	6.2	119	492	18
5	666	0.02	0.052	3.8	180	536	210
				1 wt% in Polystyrene Film			
1	432	0.80	2.6 (47%), 14 (53%)	0.9[a]	0.23[a]	424	3200
2	468	1.00	1.3	7.7	<0.077	464	100
3	506	0.90	1.4	6.4	0.71	502	140
4	548	0.78	1.2	6.5	1.8	544	150
5	616	0.30	0.75	4.0	9.3	612	410
6	704	0.03	0.19	1.6	51	682	150
				Neat Solid			
1	438	0.05	0.37 (33%), 1.8 (67%)	0.38[a]	7.2[a]	438	7100
2	474	0.76	0.75	10.0	3.2	468	90
3	492	0.53	0.84	6.3	5.5	482	160
4	550	0.68	1.0	6.8	3.2	558	280
5	616	0.15	0.33	4.5	26	598	180
6	658	0.12	0.39	3.1	22	634	310

[a]Calculated from the weighted average of the two contributions to τ.

peak maximum of complex **3** undergoes a bathochromic shift of 47 nm (1620 cm^{-1}) going from nonpolar (MeCy) to polar (MeCN) solution (Fig. 53.3b). The Φ_{PL} and τ decrease as well across a series of solvents (Φ_{PL} = 0.90, 0.73, 0.55, and 0.18 and τ = 1.55, 1.1, 1.1, and 0.41 µs for MeCy, toluene, MeTHF, and MeCN, respectively). The decrease in luminescent efficiency is caused by an increase in the rate of nonradiative decay (k_{nr}) as values for k_r remain near constant in the various solvents. The increase in nonradiative decay rates in polar solvents indicates deactivation of the excited state induced by formation of exciplexes [52–55] and/or effects of outer-sphere quenching on the ICT state [56, 57].

The emission spectra of the complexes undergo marked hypsochromic shifts in frozen solvents, as shown for complexes **1–4** in 2-MeTHF at rt and 77 K in Fig. 53.4. The spectral shifts upon freezing the fluid solutions are a result of destabilization of the ICT states in the rigid environment, which locks the solvent dipoles organized around the dipole moment of the ground state. Thus, a solvation shell that stabilizes the ground state will destabilize the excited state, thereby shifting the ICT state to higher energy. Varying ratios of ^{3}Cz/ICT emission are observed from complexes with different carbene and/or Cz ligands. Complexes **1** and **2** display vibronic fine structure from the lower lying ^{3}Cz state at 77 K in 2-MeTHF and have decay lifetimes at 430 nm in the range of milliseconds. The ICT state in these derivatives is destabilized to the point that emission is exclusively from the ^{3}Cz state. Both ^{3}Cz and ICT emission are observed in complex **3** in 2-MeTHF at 77 K as both states are similar in energy. Complex **4** (and **5**, see the Supporting Information) shows only broad ICT emission in 2-MeTHF at 77 K. The ICT emission in **4** blue-shifts by 125 nm (4300 cm^{-1}) upon cooling to 77 K; however, this change is insufficient to raise the energy of the ICT state above that for the ^{3}Cz state.

To further probe the nature of the hypsochromic shift, emission spectra of **3** in 2-MeTHF were obtained between 77 and 295 K (Fig. 53.5a). The emission spectrum at 77 K displays features from both the ^{3}Cz (τ = 2.1 ms at 430 nm) and the ^{3}ICT (τ = 180 µs at 520 nm) states. Negligible change in the emission profile is observed until warming to 93 K, which is close to the glass transition temperature of 2-MeTHF (T_g = 90–91 K) [58–60]. At temperatures above T_g,

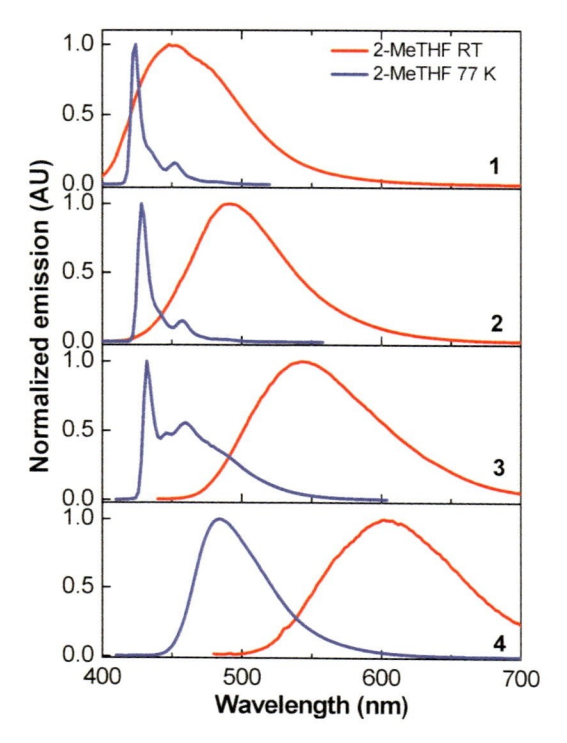

Figure 53.4 Emission spectra of complexes **1–4** (top to bottom) at room temperature (RT) and 77 K in 2-MeTHF.

the dielectric relaxation rate of the solvent increases and the ICT state is stabilized through solvation effects. The softening of the matrix causes the ICT band to red-shift and increase in the intensity between 93 and 105 K. The change in lowest state energy from ^{3}Cz to ICT as a function of temperature is schematically illustrated in Fig. 53.5b. Emission from the ^{3}Cz state disappears at temperatures above 105 K, and luminescence is due exclusively to the ICT state. The ICT state is fully stabilized even before the melting point of 2-MeTHF ($T_m = 137$ K) as the emission profile is nearly unchanged from 105 K to room temperature.

It is common for luminescent Cu(I) complexes to exhibit low quantum efficiencies in fluid solution due to high nonradiative decay rates caused either by unimolecular distortion of the complexes

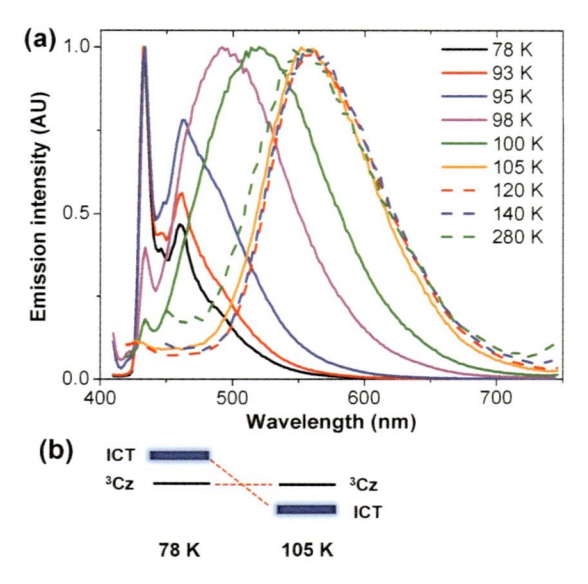

Figure 53.5 (a) Emission spectra of complex **3** at 78–280 K in 2-MeTHF. (b) Schematic energy diagram depicting the effect of rigidochromism on the ordering of the ICT/^{3}Cz states.

in the excited state or by bimolecular formation of excimers or exciplexes with the solvent [52, 55]. The effect that these processes have on the Φ_{PL} can be reduced in the solid state and in rigid media. Indeed, complexes **2–6** display comparable or even higher Φ_{PL} as neat solids than in 2-MeTHF solution (Table 53.2). High values of Φ_{PL} are also observed for the complexes when doped into polystyrene (PS) films. Complexes **1–6** doped at 1 wt% in PS show broad featureless ICT emission at room temperature (Fig. 53.6a). The k_r values of the complexes in the doped films are similar to those in 2-MeTHF, whereas values for k_{nr} are much smaller. The magnitude of k_{nr} increases across the series **1 → 6**, a trend that is consistent with the energy gap law [61]. In a series of materials with similar excited-state character, the energy of the transition (ΔE) is exponentially dependent on the nonradiative rate constant. For complexes **2–6**, a plot of log(k_{nr}) versus ΔE gives a linear fit (Fig. 53.6b). However, complex **1** falls well off the line, suggesting a different orbital composition in the excited state for this complex.

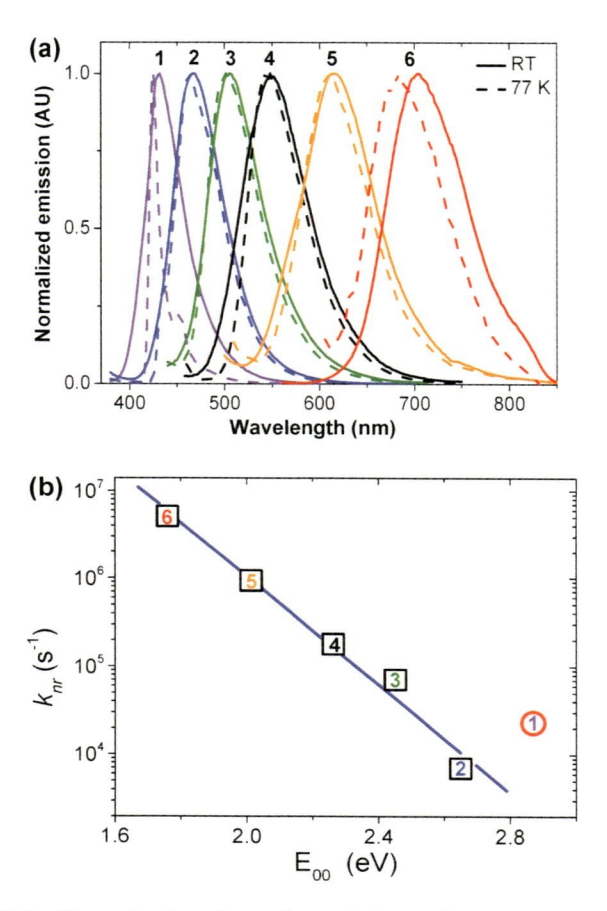

Figure 53.6 Photophysics of complexes **1–6** in polystyrene films. (a) The emission spectra at room temperature (RT) and 77 K. (b) Energy gap law plot at room temperature.

It is also worth noting that the value of k_r is considerably smaller for 1 than for complexes **2–6** in both 2-MeTHF and PS films. The deviations in k_r and k_{nr} from values found for complexes **2–6** are consistent with an increased contribution from the ^{3}Cz state in emission from complex **1**.

The emission properties of complexes 1–6 in PS films were measured at 77 K (Fig. 53.6a), and decay lifetimes were found to be strongly temperature-dependent. For complex 1, the emission is well-resolved and shows a characteristic vibronic structure of Cz at 77 K (Fig. S2). The lifetime increases from a few microseconds at room temperature to 3.2 ms at 77 K. The vibronic structure along with the long lifetime at 77 K is consistent with emission from the locally excited ^{3}Cz state. For complexes 2–6, the emission spectra remain broad and featureless at 77 K, indicative of emission from ICT states. The emission spectra show a minimal blue-shift at 77 K; however, the decay lifetimes markedly increase (by 2 orders of magnitude). The large increase in emission lifetime on cooling to 77 K suggests that the luminescence at room temperature is due to thermally activated delayed fluorescence (TADF), which requires a small energy separation between the ^{1}ICT and ^{3}ICT states.

To obtain parameters governing the temperature-dependent luminescent properties, the emission lifetime of complex **3** in PS film was recorded at intervals between 5 and 320 K (Fig. 53.7a). The lifetime steadily increases upon cooling until near 150 K, whereupon the increase becomes more pronounced, until 50 K where the rate of increase slows. The luminance decayed monoexponentially between 130 K and room temperature and was better fit as a biexponential decay between 5 and 120 K (see the Supporting Information for representative traces between 5 and 120 K and a table of lifetimes). The data shown in Fig. 53.7a between 5 and 120 K are weighted averages of the two measured lifetimes. A similar situation was observed for iridium(III)bis[2-(4′,6′-difluorophenyl)pyridinato-$N,C^{2\prime}$]-picolinate (FIrpic), where the change to biexponential decay at low temperature was attributed to the effects of site inhomogeneity of the host matrix on the dopant triplet sublevels [62]. The increase in lifetime with decreasing temperature is attributed to successive depopulation of states at high energy that have radiative rate constants faster than the lowest lying state. At temperatures above 50 K, the emission is dominated by a higher-lying S_1 state, whose rate of thermal population increases with heating. At temperatures below 50 K, thermal activation between triplet substates is observed. Under the assumption of fast thermalization between all states, the temperature-dependent decay curve can be fit to a Boltzmann distribution equation (Eq. 53.1) [62].

$$\tau = \left(2 + e^{-(\Delta E(\text{III}-\text{I})/k_B T)} + e^{-(\Delta E(S_1-\text{I})/k_B T)}\right) \big/ \left(2/\tau_{\text{I,II}} + e^{-(\Delta E(\text{III}-\text{I})/k_B T)}/\tau_{\text{III}} + e^{-(\Delta E(S_1-\text{I})/k_B T)}/\tau_{S_1}\right)$$

(53.1)

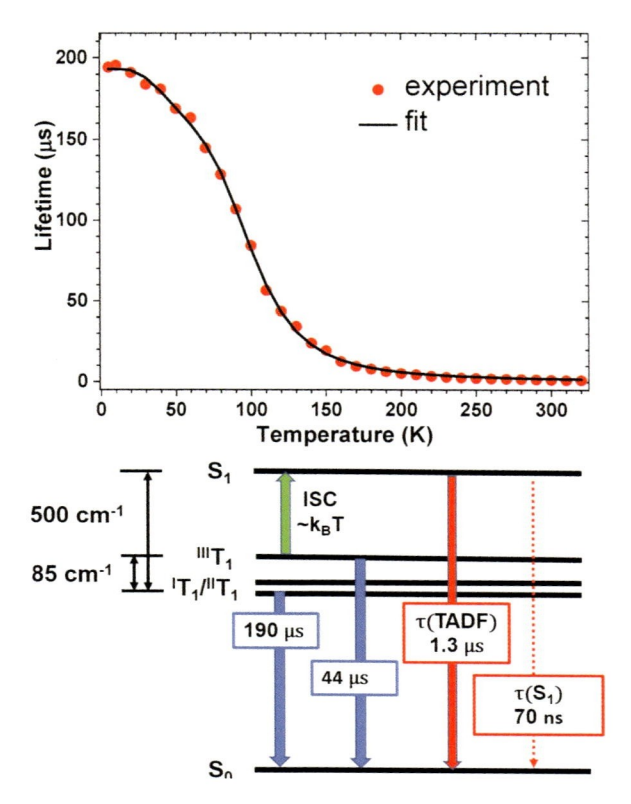

Figure 53.7 Emission lifetime versus temperature of complex **3** in the PS film (top). Energy level diagram for complex **3** derived from fit to Eq. 53.1 (bottom).

Here, S_1 represents the lowest singlet state, whereas I, II, and III represent the triplet substates $^{I}T_1$, $^{II}T_1$, and $^{III}T_1$, and k_B is the Boltzmann constant. Substates I and II are treated as degenerate because the energy splitting between these two states is normally much smaller than that between substates II and III in molecules with a prolate geometry such as complex **3** [63]. Moreover, small splittings ($\Delta E < 10$ cm^{-1}) between substates I and II are typically found in copper complexes [2]. Fits of the experimental lifetime data to Eq. 53.1 reveal the decay rate of each state and the energy separation between them (Fig. 53.7b). Fits to a simple Boltzmann expression that does not include a term to account for splitting of the triplet substates are poor, with a plateau at $T < 50$ K (Fig. S8). The exchange energy, $\Delta E(S_1-{}^lT_1)$, is 500 cm^{-1} (62 meV). This energy separation is among the smallest reported for Cu-based TADF emitters [9–11, 16, 33]. The decay lifetime of the S_1 state ($\tau_{S_1} = 70$ ns) is among the fastest values of τ_{S_1} for Cu complexes [9] and consistent with the high k_r as mentioned above. The decay lifetimes for the $^{III}T_1$ and $^lT_1/^{II}T_1$ substates are 44 and 190 μs, respectively. The zero-field splitting (ZFS), which corresponds to ΔE(III–II/I), is 85 ± 20 cm^{-1} (error is derived from the sensitivity of the fit to the ZFS value; see Fig. S7). The ZFS in **3**, induced by the effective spin orbit coupling (SOC) with the metal center, is exceptionally large for a Cu(I) complex [9, 64, 65]. Note that because the two lowest states have lifetimes in the microsecond regime, we cannot experimentally rule out the possibility that these are two closely spaced triplet states instead of substates associated with ZFS. However, our TD-DFT calculations, which correspond well with the experimental ICT energies, find a T_2 state localized on Cz that is 0.29 eV (2340 cm^{-1}) higher in energy than the T_1 state (see Table S2). Therefore, we consider it highly unlikely that the energy separation is between T_1 and a second, faster radiating T_2 state.

The values obtained for $k_r(S_1-S_0)$, $\Delta E(S_1-{}^lT_1)$, and ZFS from the temperature dependence analysis of complex **3** are unique in comparison to data reported for luminescent four-coordinate Cu(I) complexes [9, 65, 66]. To a first approximation, the radiative rate constant for a $S_0 \rightarrow S_1$ transition is assumed to be proportional to the magnitude of the exchange energy in a one-electron transition between the HOMO and LUMO [65]. This relationship is due to the fact that both parameters increase in concert with the degree of spatial overlap between the orbitals involved in the S_1 and T_1 states. Therefore, a large $k_r(S_1-S_0)$ and small $\Delta E(S_1-{}^lT_1)$ are considered to be mutually exclusive properties. Yet in complex **3**, the values for $k_r(S_1-S_0) = 1.2 \times 10^7$ s^{-1} and $\Delta E(S_1-{}^lT_1) = 500$ cm^{-1} (Fig. 53.7) are large and small, respectively, as compared to values reported for other Cu(I) luminophores [65]. For example, Cu(pop)(pz$_4$B) [pop = bis(2-(diphenylphosphanyl)phenyl) ether, pz = pyrazol-1-yl] is reported to have $k_r(S_1-S_0) = 1.1 \times 10^7$ s^{-1} and $\Delta E(S_1-{}^lT_1) = 1000$ cm^{-1}, whereas the values are 3.9×10^6 s^{-1} and 370 cm^{-1} for Cu(dppb)(pz$_2$Bph$_2$) [dppb = 1,2-bis-(diphenylphosphino)-benzene, ph = phenyl] [67–69]. The deviation of **3** from trends observed in other Cu(I) emitters is likely due to two factors: the linear, coplanar arrangement of ligands and the minimal, but essential participation of the metal center to the ^{1}ICT transition. The geometry of the complex maximizes any potential π-overlap between the 2p orbitals on the C_{NHC} and N_{Cz} atoms ligated to Cu, whereas the highly polarizable 3d orbitals of the metal provide enough electron density to impart high oscillator strength to the ^{1}ICT transition. On the other hand, the 3.7–3.8 Å spacing between the carbene C-2p and carbazole N-2p orbitals minimizes the overlap between the HOMO and LUMO, as well as stabilization of the unpaired electrons in the ^{3}ICT state, thereby minimizing the exchange energy.

The large ZFS in **3** is likewise a consequence of the linear geometry of the complex. In the simple approximation of a one-electron transition between states, SOC can only occur between T_1 and singlet transitions at higher energy that use a d orbital different than that in the HOMO, but have the same LUMO as the ^{3}ICT state [2, 70]. TD-DFT calculations find the d orbital participating in the HOMO of **3** to be nominally $3d_{xz}$ (or $3d_{yz}$) on Cu, whereas the S_3 (E = 3.77 eV, f = 0.007) and T4 (3.37 eV, f = 0.000) states are exclusively transitions from HOMO−3 to LUMO, where the HOMO−3 is principally comprised of the $3d_{z^2}$ orbital on Cu (see Tables S1 and S2). The S_0–S_3 and S_0–T_4 transitions can thus be described as having MLCT character. The large ZFS found in **3** presumably involves SOC between these ^{1,3}MLCT and ^{3}ICT states. Note that because of the anisotropy of the molecule, the dipole moments for the ^{1,3}MLCT and ^{3}ICT transitions, as well as $3d_{z^2}$ orbital, are all directed along the C_{NHC}–Cu–N_{Cz} axis. This fortuitous arrangement likely maximizes the strength of the SOC interaction between the ^{1,3}MLCT and ^{3}ICT states and leads to both the large ZFS and the high radiative rate for the T_1–S_0 transition.

53.3.3 Electroluminescence

OLEDs using complex **3** as the dopant were fabricated by vapor deposition. The device structure is glass substrate/70 nm ITO/5 nm hexaazatriphenylene hexacarbonitrile (HATCN)/40 nm 4,4′-cyclohexylidene-bis [N,N-bis(4-methylphenyl)benzenamine]

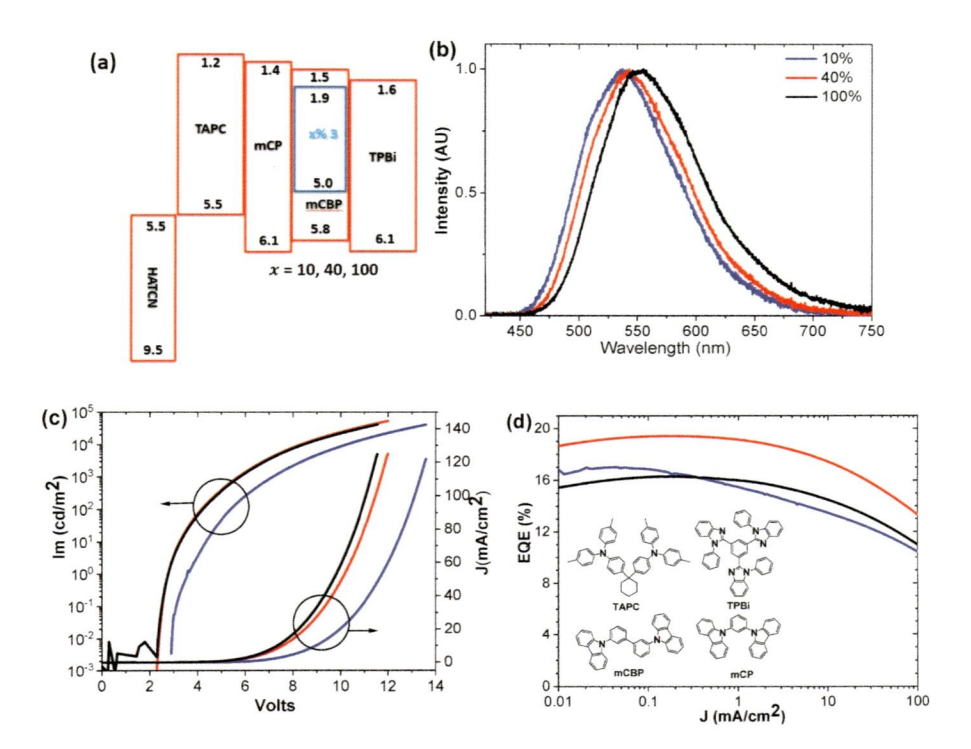

Figure 53.8 Electroluminescent device characteristics containing complex 3 at doping concentrations of 10% (blue), 40% (red), and 100% (black). (a) Frontier orbital energies in eV. (b) Electroluminescent spectra. (c) Current density–voltage–luminance $(J-V-L)$. (d) External quantum efficiency. Inset: Molecular structures of materials used in the devices.

(TAPC)/10 nm N,N′-dicarbazolyl-3,5-benzene (mCP)/25 nm EML/ 65 nm 2,2,2″-(l,3,5-benzenetriyl)-tris(N-phenyl-l-H-benzimidazole) (TPBi)/1.5 nm 8-hydroxyquinolinato lithium (LiQ)/100 nm Al. Here, the EML is compound 3 doped into 3,3′-di(9H-carbazol-9-yl)-1,1′-biphenyl (mCBP) at 10 or 40 vol%, or used as a neat material (100%). The frontier orbital energies for the materials used in this OLED are given in Fig. 53.8a, and the molecular structures of the materials are shown in the inset of Fig. 53.8d.

The device characteristics are shown in Fig. 53.8, and the data are summarized in Table S5. The devices emit bright green light with an emission peak maximum between 534 and 555 nm (Fig. 53.8b), consistent with the photoluminescence spectra of the dopant. The slight red shift in the emission peak maximum with higher doping concentration is likely due to a combination of aggregation and solvation effects. The J–V–L (J = current density, L = luminance) characteristics of the devices indicate that the turn-on voltage (defined at brightness of 0.1 cd/m^2) decreases from 3.5 to 2.7 V as the doping concentration increases from 10% to 100% (Fig. 53.8c). The device resistance also decreases with increasing dopant concentration. The low resistance at high doping concentration suggests that complex **3** is an efficient charge carrier. The maximum EQE of the devices ranges from 15.4% to 19.4% with the highest value achieved at a doping concentration of 40% (Fig. 53.8d). The device efficiency is among the highest reported for OLEDs based on Cu(I) dopants [16, 19, 22, 71, 72]. Interestingly, the device using a neat emissive layer also demonstrates a high efficiency (EQE = 16.3%). Indeed, the photoluminescent efficiencies of both the vapor-deposited 10% doped (Φ_{PL} = 100%) and the neat (Φ_{PL} = 74%) films of complex **3** are high, leading to the high EQE of the devices. In addition, the roll-off in efficiency at high drive current for all devices is much smaller than reported for OLEDs using four- and three-coordinate Cu dopants where the EQE dropped to < 2% at 100 mA/cm^2 in most cases [16, 21, 27, 72, 73] and to 7% in one exceptional case [71]. The small roll-off here is likely due to the short emission decay lifetime in complex **3** that limits triplet–triplet and triplet–polaron annihilation at high brightness/current [74, 75].

53.4 Conclusion

A series of two-coordinate Cu complexes bearing nonconventional NHCs were investigated. X-ray analysis reveals a linear geometry at the copper center with the ligands in a coplanar conformation. Efficient absorption and emission occurs from an ICT transition between the carbazolyl and carbene ligands. The energy of the ICT emission is tied directly to the donor strength of the carbazole, and the acceptor strength of the carbene. The emission energy can be systematically tuned across the visible spectrum, ranging from violet (λ_{max} = 432 nm) to deep red (λ_{max} = 704 nm). The ICT transitions are solvatochromic and rigidochromic due to the large dipole moment of the ground states and the relatively small dipole in the excited states. The compounds display efficient TADF with Φ_{PL} up to 100%. The quantum yield is lowest for two compounds with absorption and emission in the red end of the spectrum, due to energy gap law considerations. The radiative rate constants (k_r = 10^5–10^6 s^{-1}) are comparable to those of efficient Ir and Pt phosphorescent emitters.

Vapor-deposited OLEDs fabricated using complex 3 as an emissive dopant show both high EQE and extremely low roll-off in efficiency at high current, qualities attributed to the high quantum efficiency and the short exciton decay time of the Cu(I) dopant. These two-coordinate Cu(I) complexes present new opportunities for use as dopants in OLEDs. Lifetime studies will need to be carried out to determine if these copper-based emitters could serve as alternatives to the state-of-the-art Ir(III) complexes commonly used in OLEDs. However, the performance of these unoptimized OLEDs is already comparable to metrics found for devices using cyclometalated Ir(III) dopants. Moreover, the host-free devices based on 3 demonstrate the potential of these complexes to be used as neat emitters in OLEDs.

The linear Cu(I) compounds explored here meet several of the crucial requirements to achieve efficient thermally enhanced luminescence, for example, TADF [9]. First, the small metal character in the lowest singlet and triplet transitions of these complexes minimizes Jahn–Teller distortion in the excited state. Therefore, geometric changes in the excited state are suppressed that inhibit nonradiative deactivation. Second, the complexes show strong ICT emission at ambient temperature, and a $\Delta E(S_1$–$T_1)$ of 500 cm^{-1} was determined for complex 3. The small $\Delta E(S_1$–$T_1)$ in the ICT transitions is due to the spatially separated HOMO and LUMO. Small values for $\Delta E(S_1$–$T_1)$ can also be achieved by organic TADF materials [76, 77]. However, the intersystem crossing rate for the $T_1 \rightarrow S_0$ transition in purely organic TADF compounds is relatively slow, resulting in a "dark" triplet state. In contrast, the high rate of ISC in the Cu(I) compounds makes $T_1 \rightarrow S_0$ transitions efficient, leading to both the S_1 and the T_1 being bright states. Such "dual channel" emission has been proposed as a means to improve efficiency of TADF compounds [64]. For the Cu(I) compounds, fast ISC also occurs for thermal activation to the S_1 state. Thus, no long-lived delayed component is present in the transient decay lifetime. Finally, the large k_r of the $S_1 \rightarrow S_0$ transition should be as high as possible to obtain short emission decay lifetime. The short prompt lifetime ($\tau_{S_1} = 70$ ns) determined from temperature-dependent studies of **3** is consistent with the high radiative rate ($k_r = 6.4 \times 10^5$ s^{-1} in polystyrene film).

The linear geometry of these complexes imparts extraordinary photophysical properties as compared to the more thoroughly investigated four-coordinate Cu(I) counterparts. The efficient luminescence of our Cu(I) complexes is a consequence of the small $\Delta E(S_1-T_1)$ and high k_r to the $S_1 \rightarrow S_0$ transition, which are normally considered mutual exclusive properties. For these linear complexes, strong coupling between the HOMO and LUMO through the d orbitals of the Cu center induces a high k_r ($S_1 \rightarrow S_0$) despite the large spatial separation between the valence orbitals. The zero-field splitting found for complex **3** (ZFS = 85 cm^{-1}) is also much larger than the highest value reported for the more common four-coordinate Cu(I) complexes (ZFS = 15 cm^{-1}) [64]. The ZFS is presumably due to efficient spin orbit coupling between the ^{1,3}MLCT and ^{3}ICT states. The ^{1,3}MLCT state involves the $3d_z^2$ orbital on Cu, which is oriented along the $C_{NHC}-Cu-N_{Cz}$ axis and is near collinear with the transition dipole moments. All of these combined properties maximize the strength of the SOC and lead to the large ZFS.

In conclusion, the linear Cu(I) complexes present new possibilities to design highly efficient luminescent Cu(I) materials, not only for use in OLEDs, but also for other applications that rely on luminescent metal complexes. Their optoelectronic properties can be extensively modified to tune the electronic energy states with respect to emission color, energy separation, quantum yield, decay lifetimes, and oscillator strength [37]. Linear Cu(I) complexes, presently the least investigated luminescent Cu(I) complexes, might be the most extraordinary examples among these materials in all respects.

Supporting Information

The Supporting Information is available free of charge on the ACS Publications website at DOI: 10.1021/jacs.8b12397.

Absorption and emission of potassium carbazolide (KCz) in 2-MeTHF, details of theoretical studies, temperature-dependent photophysical measurements of **3**, summary of the OLED characteristics, and NMR spectra of all of the precursors and metal complexes

(http://pubs.acs.org/doi/suppl/10.1021/jacs.8b12397/suppl_file/ja8b12397_si_001.pdf)

X-ray crystallographic data for compound **1**

(http://pubs.acs.org/doi/suppl/10.1021/jacs.8b12397/suppl_file/ja8b12397_si_002.cif)

X-ray crystallographic data for compound **2**

(http://pubs.acs.org/doi/suppl/10.1021/jacs.8b12397/suppl_file/ja8b12397_si_003.cif)

X-ray crystallographic data for compound **3**

(http://pubs.acs.org/doi/suppl/10.1021/jacs.8b12397/suppl_file/ja8b12397_si_004.cif)

X-ray crystallographic data for compound **4**

(http://pubs.acs.org/doi/suppl/10.1021/jacs.8b12397/suppl_file/ja8b12397_si_005.cif)

Acknowledgments

We would like to thank Rasha Hamze for helpful discussions and providing the spectra of potassium carbazolide, shown in the Supporting Information. This work was supported by the Universal Display Corp. and the Air Force Office of Scientific Research (contract no. FA9550-18-1-016). Computation for the work described in this paper was supported by the University of Southern California's Center for High-Performance Computing (hpc.usc.edu).

References

1. Lamansky, S., Djurovich, P., Murphy, D., Abdel-Razzaq, F., Lee, H.-E., Adachi, C., Burrows, P. E., Forrest, S. R., Thompson, M. E. Highly phosphorescent bis-cyclometalated iridium complexes: synthesis, photophysical characterization, and use in organic light emitting diodes. *J. Am. Chem. Soc.* 2001, **123**(18), 4304–4312.
2. Yersin, H., Rausch, A. F., Czerwieniec, R., Hofbeck, T., Fischer, T. The triplet state of organo-transition metal compounds. Triplet harvesting and singlet harvesting for efficient OLEDs. *Coord. Chem. Rev.* 2011, **255**(21), 2622–2652.
3. Tsuzuki, T., Shirasawa, N., Suzuki, T., Tokito, S. Color tunable organic light-emitting diodes using pentafluorophenyl-substituted iridium complexes. *Adv. Mater.* 2003, **15**(17), 1455–1458.
4. Kalinowski, J., Fattori, V., Cocchi, M., Williams, J. A. G. Light-emitting devices based on organometallic platinum complexes as emitters. *Coord. Chem. Rev.* 2011, **255**(21), 2401–2425.
5. Fleetham, T., Ecton, J., Wang, Z., Bakken, N., Li, J. Single-doped white organic light-emitting device with an external quantum efficiency over 20%. *Adv. Mater.* 2013, **25**(18), 2573–2576.
6. Brooks, J., Babayan, Y., Lamansky, S., Djurovich, P. I., Tsyba, I., Bau, R., Thompson, M. E. Synthesis and characterization of phosphorescent cyclometalated platinum complexes. *Inorg. Chem.* 2002, **41**(12), 3055–3066.
7. Li, X., Zhao, G.-W., Hu, Y.-X., Zhao, J.-H., Dong, Y., Zhou, L., Lv, Y.-L., Chi, H.-J., Su, Z. Rational design and characterization of novel phosphorescent rhenium(I) complexes for extremely high-efficiency organic light-emitting diodes. *J. Mater. Chem. C* 2017, **5**(30), 7629–7636.
8. Tokel, N. E., Bard, A. J. Electrogenerated chemiluminescence. IX. Electrochemistry and emission from systems containing tris(2,2′-bipyridine)ruthenium(II) dichloride. *J. Am. Chem. Soc.* 1972, **94**(8), 2862–2863.
9. Yersin, H., Czerwieniec, R., Shafikov Marsel, Z., Suleymanova Alfiya, F. TADF material design: photophysical background and case studies focusing on CuI and AgI complexes. *ChemPhysChem* 2017, **18**(24), 3508–3535.
10. Wallesch, M., Volz, D., Zink Daniel, M., Schepers, U., Nieger, M., Baumann, T., Bräse, S. Bright coppertunities: multinuclear CuI complexes with N–P ligands and their applications. *Chem. - Eur. J.* 2014, **20**(22), 6578–6590.

11. Leitl, M. J., Krylova, V. A., Djurovich, P. I., Thompson, M. E., Yersin, H. Phosphorescence versus thermally activated delayed fluorescence. controlling singlet–triplet splitting in brightly emitting and sublimable Cu(I) compounds. *J. Am. Chem. Soc.* 2014, **136**(45), 16032–16038.
12. Ohara, H., Kobayashi, A., Kato, M. Simple and extremely efficient blue emitters based on mononuclear Cu(I)-halide complexes with delayed fluorescence. *Dalton Trans.* 2014, **43**(46), 17317–17323.
13. Bergmann, L., Friedrichs, J., Mydlak, M., Baumann, T., Nieger, M., Brase, S. Outstanding luminescence from neutral copper(I) complexes with pyridyl-tetrazolate and phosphine ligands. *Chem. Commun.* 2013, **49**(58), 6501–6503.
14. Tsuboyama, A., Kuge, K., Furugori, M., Okada, S., Hoshino, M., Ueno, K. Photophysical properties of highly luminescent copper(I) halide complexes chelated with 1,2-bis(diphenylphosphino)benzene. *Inorg. Chem.* 2007, **46**(6), 1992–2001.
15. Cuttell, D. G., Kuang, S.-M., Fanwick, P. E., McMillin, D. R., Walton, R. A. Simple Cu(I) complexes with unprecedented excited-state lifetimes. *J. Am. Chem. Soc.* 2002, **124**(1), 6–7.
16. Zhang, F., Guan, Y., Chen, X., Wang, S., Liang, D., Feng, Y., Chen, S., Li, S., Li, Z., Zhang, F., Lu, C., Cao, G., Zhai, B. Syntheses, photoluminescence, and electroluminescence of a series of sublimable bipolar cationic cuprous complexes with thermally activated delayed fluorescence. *Inorg. Chem.* 2017, **56**(7), 3742–3753.
17. Wang, Z., Zheng, C., Wang, W., Xu, C., Ji, B., Zhang, X. Synthesis, structure, and photophysical properties of two four-coordinate CuI–NHC Complexes with Efficient Delayed Fluorescence. *Inorg. Chem.* 2016, **55**(5), 2157–2164.
18. Mohankumar, M., Holler, M., Meichsner, E., Nierengarten, J.-F., Niess, F., Sauvage, J.-P., Delavaux-Nicot, B., Leoni, E., Monti, F., Malicka, J. M., Cocchi, M., Bandini, E., Armaroli, N. Heteroleptic copper(I) pseudorotaxanes incorporating macrocyclic phenanthroline ligands of different sizes. *J. Am. Chem. Soc.* 2018, **140**(6), 2336–2347.
19. Deaton, J. C., Switalski, S. C., Kondakov, D. Y., Young, R. H., Pawlik, T. D., Giesen, D. J., Harkins, S. B., Miller, A. J. M., Mickenberg, S. F., Peters, J. C. E-Type delayed fluorescence of a phosphine-supported $Cu_2(\mu\text{-}NAr_2)_2$ diamond core: harvesting singlet and triplet excitons in OLEDs. *J. Am. Chem. Soc.* 2010, **132**(27), 9499–9508.
20. Felder, D., Nierengarten, J.-F., Barigelletti, F., Ventura, B., Armaroli, N. Highly luminescent Cu(I)–phenanthroline complexes in rigid matrix and

temperature dependence of the photophysical properties. *J. Am. Chem. Soc.* 2001, **123**(26), 6291–6299.

21. Miller, A. J. M., Dempsey, J. L., Peters, J. C. Long-lived and efficient emission from mononuclear amidophosphine complexes of copper. *Inorg. Chem.* 2007, **46**(18), 7244–7246.

22. Igawa, S., Hashimoto, M., Kawata, I., Yashima, M., Hoshino, M., Osawa, M. Highly efficient green organic light-emitting diodes containing luminescent tetrahedral copper(I) complexes. *J. Mater. Chem. C* 2013, **1**(3), 542–551.

23. Harkins, S. B., Peters, J. C. A highly emissive Cu_2N_2 diamond core complex supported by a [PNP]-ligand. *J. Am. Chem. Soc.* 2005, **127**(7), 2030–2031.

24. Elie, M., Weber Michael, D., Di Meo, F., Sguerra, F., Lohier, J. F., Pansu Robert, B., Renaud, J. L., Hamel, M., Linares, M., Costa Ruben, D., Gaillard, S. Role of the bridging group in bis-pyridyl ligands: enhancing both the photo- and electroluminescent features of cationic (IPr)CuI complexes. *Chem. - Eur. J.* 2017, **23**(64), 16328–16337.

25. Krylova, V. A., Djurovich, P. I., Conley, B. L., Haiges, R., Whited, M. T., Williams, T. J., Thompson, M. E. Control of emission colour with N-heterocyclic carbene (NHC) ligands in phosphorescent three-coordinate Cu(I) complexes. *Chem. Commun.* 2014, **50**(54), 7176–7179.

26. Marion, R., Sguerra, F., Di Meo, F., Sauvageot, E., Lohier, J.-F., Daniellou, R., Renaud, J.-L., Linares, M., Hamel, M., Gaillard, S. NHC Copper(I) complexes bearing dipyridylamine ligands: synthesis, structural, and photoluminescent studies. *Inorg. Chem.* 2014, **53**(17), 9181–9191.

27. Gernert, M., Müller, U., Haehnel, M., Pflaum, J., Steffen, A. A cyclic alkyl(amino)carbene as two-atom π-chromophore leading to the first phosphorescent linear CuI complexes. *Chem. - Eur. J.* 2017, **23**(9), 2206–2216.

28. Shi, S., Djurovich, P. I., Thompson, M. E. Synthesis and characterization of phosphorescent three-coordinate copper(I) complexes bearing bis(amino)cyclopropenylidene carbene (BAC). *Inorg. Chim. Acta* 2018, **482**, 246–251.

29. Hashimoto, M., Igawa, S., Yashima, M., Kawata, I., Hoshino, M., Osawa, M. Highly efficient green organic light-emitting diodes containing luminescent three-coordinate copper(I) complexes. *J. Am. Chem. Soc.* 2011, **133**(27), 10348–10351.

30. Lotito, K. J., Peters, J. C. Efficient luminescence from easily prepared three-coordinate copper(i) arylamidophosphines. *Chem. Commun.* 2010, **46**(21), 3690–3692.

31. Robinson, G. W., Frosch, R. P. Electronic excitation transfer and relaxation. *J. Chem. Phys.* 1963, **38**(5), 1187–1203.
32. Shi, S., Collins, L. R., Mahon, M. F., Djurovich, P. I., Thompson, M. E., Whittlesey, M. K. Synthesis and characterization of phosphorescent two-coordinate copper(I) complexes bearing diamidocarbene ligands. *Dalton Trans.* 2017, **46**(3), 745–752.
33. Czerwieniec, R., Kowalski, K., Yersin, H. Highly efficient thermally activated fluorescence of a new rigid Cu(I) complex [Cu(dmp)(phanephos)]. *Dalton Trans.* 2013, **42**(27), 9826–9830.
34. Hamze, R., Jazzar, R., Soleilhavoup, M., Djurovich, P. I., Bertrand, G., Thompson, M. E. Phosphorescent 2-, 3-and 4-coordinate cyclic (alkyl)(amino) carbene (CAAC) Cu(I) complexes. *Chem. Commun.* 2017, **53**(64), 9008–9011.
35. Di, D. W., Romanov, A. S., Yang, L., Richter, J. M., Rivett, J. P. H., Jones, S., Thomas, T. H., Jalebi, M. A., Friend, R. H., Linnolahti, M., Bochmann, M., Credgington, D. High-performance light-emitting diodes based on carbene-metal-amides. *Science* 2017, **356**(6334), 159–163.
36. Romanov, A. S., Becker, C. R., James, C. E., Di, D. W., Credgington, D., Linnolahti, M., Bochmann, M. Copper and gold cyclic (alkyl)(amino)carbene complexes with sub-microsecond photoemissions: structure and substituent effects on redox and luminescent properties. *Chem. - Eur. J.* 2017, **23**(19), 4625–4637.
37. Hamze, R., Peltier, J. L., Sylvinson, D., Jung, M., Cardenas, J., Haiges, R., Soleilhavoup, M., Jazzar, R., Djurovich, P. I., Bertrand, G., Thompson, M. E. Eliminating nonradiative decay in Cu(I) emitters: >99% quantum efficiency and microsecond lifetime. *Science* 2019, **363**, 601.
38. Blake, G. A., Moerdyk, J. P., Bielawski, C. W. Tuning the electronic properties of carbenes: a systematic comparison of neighboring amino versus amido groups. *Organometallics* 2012, **31**(8), 3373–3378.
39. Braun, M., Frank, W., Reiss, G. J., Ganter, C. An N-heterocyclic carbene ligand with an oxalamide backbone. *Organometallics* 2010, **29**(20), 4418–4420.
40. Hudnall, T. W., Bielawski, C. W. An N,N′-diamidocarbene: studies in C–H insertion, reversible carbonylation, and transition-metal coordination chemistry. *J. Am. Chem. Soc.* 2009, **131**(44), 16039–16041.
41. Cesar, V., Lugan, N., Lavigne, G. Reprogramming of a malonic N-heterocyclic carbene: a simple backbone modification with dramatic consequences on the ligand's donor properties. *Eur. J. Inorg. Chem.* 2010, **3**, 361–365.

42. Hudnall, T. W., Moerdyk, J. P., Bielawski, C. W. Ammonia N-H activation by a N,N[prime or minute]-diamidocarbene. *Chem. Commun.* 2010, **46**(24), 4288–4290.
43. Skuodis, E., Tomkeviciene, A., Reghu, R., Peciulyte, L., Ivaniuk, K., Volyniuk, D., Bezvikonnyi, O., Bagdziunas, G., Gudeika, D., Grazulevicius, J. V. OLEDs based on the emission of interface and bulk exciplexes formed by cyano-substituted carbazole derivative. *Dyes Pigm.* 2017, **139**, 795–807.
44. Wu, S., Liu, Y., Yu, G., Guan, J., Pan, C., Du, Y., Xiong, X., Wang, Z. Facile preparation of dibenzoheterocycle-functional nanoporous polymeric networks with high gas uptake capacities. *Macromolecules* 2014, **47**(9), 2875–2882.
45. Shao, Y., Gan, Z., Epifanovsky, E., Gilbert, A. T. B., Wormit, M., et al. Advances in molecular quantum chemistry contained in the Q-Chem 4 program package. *Mol. Phys.* 2015, **113**(2), 184–215.
46. Collins, L. R., Lowe, J. P., Mahon, M. F., Poulten, R. C., Whittlesey, M. K. Copper diamidocarbene complexes: characterization of monomeric to tetrameric species. *Inorg. Chem.* 2014, **53**(5), 2699–2707.
47. Díez-González, S., Stevens, E. D., Scott, N. M., Petersen, J. L., Nolan, S. P. Synthesis and characterization of [Cu(NHC)2]X complexes: catalytic and mechanistic studies of hydrosilylation reactions. *Chem. - Eur. J.* 2008, **14**(1), 158–168.
48. Rosenfeld, D. C., Wolczanski, P. T., Barakat, K. A., Buda, C., Cundari, T. R. 3-Center-4-electron bonding in [(silox)2MoNtBu]2(μ-Hg) controls reactivity while frontier orbitals permit a dimolybdenum π-bond energy estimate. *J. Am. Chem. Soc.* 2005, **127**(23), 8262–8263.
49. Hansmann, M. M., Melaimi, M., Munz, D., Bertrand, G. Modular approach to kekulé diradicaloids derived from cyclic (alkyl)-(amino)carbenes. *J. Am. Chem. Soc.* 2018, **140**(7), 2546–2554.
50. Sworakowski, J., Lipiński, J., Janus, K. On the reliability of determination of energies of HOMO and LUMO levels in organic semiconductors from electrochemical measurements. A simple picture based on the electrostatic model. *Org. Electron.* 2016, **33**, 300–310.
51. Yanai, T., Tew, D. P., Handy, N. C. A new hybrid exchange–correlation functional using the Coulomb-attenuating method (CAM-B3LYP). *Chem. Phys. Lett.* 2004, **393**(1), 51–57.
52. Eggleston, M. K., McMillin, D. R., Koenig, K. S., Pallenberg, A. J. Steric effects in the ground and excited states of Cu(NN)2+ systems. *Inorg. Chem.* 1997, **36**(2), 172–176.

53. Gothard, N. A., Mara, M. W., Huang, J., Szarko, J. M., Rolczynski, B., Lockard, J. V., Chen, L. X. Strong steric hindrance effect on excited state structural dynamics of Cu(I) diimine complexes. *J. Phys. Chem. A* 2012, **116**(9), 1984–1992.
54. Palmer, C. E. A., McMillin, D. R., Kirmaier, C., Holten, D. Flash photolysis and quenching studies of copper(I) systems in the presence of Lewis bases: inorganic exciplexes? *Inorg. Chem.* 1987, **26**(19), 3167–3170.
55. Shaw, G. B., Grant, C. D., Shirota, H., Castner, E. W., Meyer, G. J., Chen, L. X. Ultrafast structural rearrangements in the mlct excited state for copper(I) bis-phenanthrolines in solution. *J. Am. Chem. Soc.* 2007, **129**(7), 2147–2160.
56. Capano, G., Rothlisberger, U., Tavernelli, I., Penfold, T. J. Theoretical rationalization of the emission properties of prototypical Cu(I)–phenanthroline complexes. *J. Phys. Chem. A* 2015, **119**(27), 7026–7037.
57. Penfold, T. J., Karlsson, S., Capano, G., Lima, F. A., Rittmann, J., Reinhard, M., Rittmann-Frank, M. H., Braem, O., Baranoff, E., Abela, R., Tavernelli, I., Rothlisberger, U., Milne, C. J., Chergui, M. Solvent-induced luminescence quenching: static and time-resolved X-ray absorption spectroscopy of a copper(I) phenanthroline complex. *J. Phys. Chem. A* 2013, **117**(22), 4591–4601.
58. Mizukami, M., Fujimori, H., Oguni, M. Glass transitions and the responsible molecular motions in 2-methyltetrahydrofuran. *Prog. Theor. Phys. Suppl.* 1997, **126**, 79–82.
59. Qi, F., El Goresy, T., Böhmer, R., Döß, A., Diezemann, G., Hinze, G., Sillescu, H., Blochowicz, T., Gainaru, C., Rössler, E., Zimmermann, H. Nuclear magnetic resonance and dielectric spectroscopy of a simple supercooled liquid: 2-methyl tetrahydrofuran. *J. Chem. Phys.* 2003, **118**(16), 7431–7438.
60. Richert, R. Molecular probing of dielectric relaxation in the glass-transition region. *Chem. Phys. Lett.* 1992, **199**(3), 355–359.
61. Caspar, J. V., Meyer, T. J. Application of the energy gap law to nonradiative, excited-state decay. *J. Phys. Chem.* 1983, **87**(6), 952–957.
62. Rausch, A. F., Thompson, M. E., Yersin, H. Matrix effects on the triplet state of the OLED emitter Ir(4,6-dFppy)(2)(pic) (FIrpic): investigations by high-resolution optical spectroscopy. *Inorg. Chem.* 2009, **48**(5), 1928–1937.
63. Richert, S., Tait, C. E., Timmel, C. R. Delocalisation of photoexcited triplet states probed by transient EPR and hyperfine spectroscopy. *J. Magn. Reson.* 2017, **280**, 103–116.

64. Hofbeck, T., Monkowius, U., Yersin, H. Highly efficient luminescence of Cu(I) compounds: thermally activated delayed fluorescence combined with short-lived phosphorescence. *J. Am. Chem. Soc.* 2015, **137**(1), 399–404.
65. Czerwieniec, R., Leitl, M. J., Homeier, H. H. H., Yersin, H. Cu(I) complexes - Thermally activated delayed fluorescence. Photophysical approach and material design. *Coord. Chem. Rev.* 2016, 325, 2–28.
66. Leitl, M. J., Zink, D. M., Schinabeck, A., Baumann, T., Volz, D., Yersin, H. Copper(I) complexes for thermally activated delayed fluorescence: from photophysical to device properties. *Topics Curr Chem* 2016, **374**(3), 1 DOI: 10.1007/s41061-016-0019-1.
67. Czerwieniec, R., Yersin, H. Diversity of copper(I) complexes showing thermally activated delayed fluorescence: basic photophysical analysis. *Inorg. Chem.* 2015, **54**(9), 4322–4327.
68. Czerwieniec, R., Yu, J. B., Yersin, H. Blue-light emission of Cu(I) complexes and singlet harvesting. *Inorg. Chem.* 2011, **50**(17), 8293–8301.
69. Czerwieniec, R., Yu, J. B., Yersin, H. Correction to blue-light emission of Cu(I) complexes and singlet harvesting (vol 17, pg 8293, 2011). *Inorg. Chem.* 2012, **51**(3), 1975–1975.
70. Li, E. Y. T., Jiang, T. Y., Chi, Y., Chou, P. T. Semi-quantitative assessment of the intersystem crossing rate: an extension of the El-Sayed rule to the emissive transition metal complexes. *Phys. Chem. Chem. Phys.* 2014, **16**(47), 26184–26192.
71. Zhang, J., Duan, C., Han, C., Yang, H., Wei, Y., Xu, H. Balanced dual emissions from tridentate phosphine-coordinate copper(I) complexes toward highly efficient yellow OLEDs. *Adv. Mater.* 2016, **28**(28), 5975–5979.
72. Zhang, Q., Komino, T., Huang, S., Matsunami, S., Goushi, K., Adachi, C. Triplet exciton confinement in green organic light-emitting diodes containing luminescent charge-transfer Cu(I) complexes. *Adv. Funct. Mater.* 2012, **22**(11), 2327–2336.
73. Liang, D., Chen, X.-L., Liao, J.-Z., Hu, J.-Y., Jia, J.-H., Lu, C.-Z. Highly efficient cuprous complexes with thermally activated delayed fluorescence for solution-processed organic light-emitting devices. *Inorg. Chem.* 2016, **55**(15), 7467–7475.
74. Giebink, N. C., D'Andrade, B. W., Weaver, M. S., Mackenzie, P. B., Brown, J. J., Thompson, M. E., Forrest, S. R. Intrinsic luminance loss in

phosphorescent small-molecule organic light emitting devices due to bimolecular annihilation reactions. *J. Appl. Phys.* 2008, **103**(4), 044509.

75. Wang, Q., Aziz, H. Degradation of organic/organic interfaces in organic light-emitting devices due to polaron–exciton interactions. *ACS Appl. Mater. Interfaces* 2013, **5**(17), 8733–8739.

76. Uoyama, H., Goushi, K., Shizu, K., Nomura, H., Adachi, C. Highly efficient organic light-emitting diodes from delayed fluorescence. *Nature* 2012, **492**, 234.

77. Kaji, H., Suzuki, H., Fukushima, T., Shizu, K., Suzuki, K., Kubo, S., Komino, T., Oiwa, H., Suzuki, F., Wakamiya, A., Murata, Y., Adachi, C. Purely organic electroluminescent material realizing 100% conversion from electricity to light. *Nat. Commun.* 2015, **6**, 8476.

Chapter 54

Platinum-Functionalized Random Copolymers for Use in Solution-Processible, Efficient, Near-White Organic Light-Emitting Diodes

Paul T. Furuta,[a] Lan Deng,[a] Simona Garon,[b] Mark E. Thompson,[b] and Jean M. J. Fréchet[a]

[a] *Department of Chemistry, University of California, Berkeley, Berkeley, California 94720-1460, USA*
[b] *Department of Chemistry, University of Southern California, Los Angeles, California 90089-0744, USA*
frechet@cchem.berkeley.edu, mthompso@chem1.usc.edu

Electroluminescent devices based on cyclometalated metal complexes of Pt [1] and Ir [2] are of considerable interest due to their high quantum efficiencies (QE). The strong spin–orbit coupling of the heavy metal ions results in intersystem crossing from singlet to triplet exited state, allowing the complexes to utilize both singlet and triplet excitons, theoretically approaching 100% internal QE [3]. Due to ligand effects and excimer emission, complexes based on Pt can emit over a wide range of wavelengths [1c, 4], thus enabling the fabrication of organic light-emitting devices (OLEDs) in a range of colors, including white [5].

Efficient white light devices are of particular interest, as they may be useful in a wide range of applications from backlight for displays in portable devices to broad area illumination sources for room lighting. Such white OLEDs have been prepared by both solution [6] and vacuum [5, 7] deposition methods with best efficiencies recorded for devices obtained by vacuum deposition of small molecules.

We now report on near-white phosphorescent polymer OLEDs based on a single luminescent dopant, anchored to a polymer chain, which emits simultaneously from monomer (blue) and aggregate (orange) states. White polymer OLEDs have involved luminescence from different species, either uniformly dispersed in a single layer [6a, d] or in separate layers [6b], resulting in a current-dependent color balance. Unfortunately, white light-emitting devices based on multiple independent emitters are expected to change color throughout the life of the device, due to differential aging of the different components. We have previously reported that vacuum-deposited OLEDs utilizing (2-(4′,6′-difluorophenyl)pyridinato-$N,C^{2'}$)(2,4-pentanedionato)Pt(II) dopant (FPt) emit white light from simultaneous blue FPt "monomer" emission and red emission from the excimer or aggregate FPt emission, with the ratio of blue to red emission controlled by the total FPt concentration (high concentration favors red emission). The excimer lacks a bound ground state, limiting energy transfer between dopants. Since the stability of both emitting species depends on only a single complex, the age-dependent color shifts should decrease.

In addressing the challenge of making a solution-processed device, which emits white light from a combination of monomer and excimer/aggregate states, we now report the synthesis of a multifunctional polymer-containing electron- (ET) and hole-transporting (HT) moieties as well as emitter complexes. First, a series of random terpolymers **5** containing triphenylamine (TPA), oxadiazole, and β-diketonate units was prepared by polymerization of various ratios of monomers **2**, **3** [8], and **4** [9] using alkoxyamine initiator [10] **1** at 125°C. This living polymerization system [11] was chosen due to its compatibility with the various functional monomers and its ability to form predictable molecular weight polymers with low polydispersity. It should be noted that other living polymerizations such as ATRP involving metal catalysts might not be well suited for the preparation of **5**, as the presence of catalyst residues might impair device performance. Conjugated polymer systems containing phosphor complexes can suffer from quenching of phosphorescence for wavelengths lower than red [12], which is the reason for choosing polymers with a photoinert backbone. The diketonate moieties of terpolymers **5a–d** were then metalated (Scheme 54.1) using [Pt-2-(4′,6′-difluorophenyl)pyridinato-(μ-Cl)]$_2^4$ and Na$_2$CO$_3$ in a 2-ethoxyethanol/anisole mixture at 80°C to give the final Pt-containing polymers **6a–d**. These air-stable metalated terpolymers have good solubility, enabling solution processing, and can be stored at room temperature.

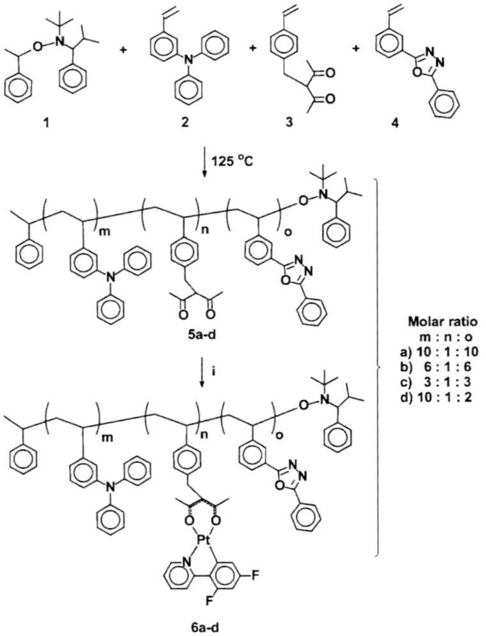

Scheme 54.1 (Conditions: (i) 2-ethoxyethanol, anisole, [Pt-2-(4′,6′-difluorophenyl)pyridinato(μ-Cl)]$_2$, Na$_2$CO$_3$, 80°C.)

All solution process steps were performed in air. OLEDs were prepared on indium-tin oxide (ITO)-coated glass substrates, upon which a 450 Å layer of poly(3,4-ethylenedioxythiophene)/poly(styrenesulfonate) (PEDOT/PSS) was spin-coated and baked (90 °C for 45 min) prior to addition of polymers **6a–d**. The polymer layers (**6a–d**, 1000 Å) were spin coated from a chloroform solution, followed by thermal vacuum evaporation of 150 Å of 2,9-dimethyl-4,7-diphenyl-1,10-phenanthroline (BCP), 250 Å aluminum-tris(8-hydroxyquinolate) (Alq$_3$), and finally 1000 Å of LiF/Al for the cathode. The combined BCP/Alq$_3$ layers act as an efficient electron injecting contact [1a, 13].

To achieve the goal of white light emission [14], two differents forms of the Pt complex must be present within the polymer film: the monomeric form for blue monomeric emission and the aggregate form for orange excimer emission. Therefore, in addition to intrinsic structural features of the polymer, the concentration of the Pt complex [5] will largely determine the extent of FPt-FPt interaction within the polymer. Figure 54.1 shows the electroluminescence (EL) spectra of devices made from polymers **6a–d**. Polymers **6a**, **6b**, and **6c** have the same HT/ET balance of 1:1, but have a Pt complex loading that increases from 6 to 10 and 18 wt%. A comparison of the EL of **6a** and **6b** shows that the latter has a broader spectrum beyond 550 nm. This is due to the higher concentration of complex, leading to an increased proportion of FPt with strong FPt–FPt interaction, hence the stronger emission in the orange to red part of the spectrum. This effect is much greater in **6c**, where the blue and green bands of the complex are nearly absent compared to the orange-red excimer/aggreagate emission. Interestingly, while the loading of the Pt complex in **6b** and **6d** are identical (10 wt% each), the spectrum of **6d** with its large proportion of HT units has a noticeable contribution of the blue and blue-green monomer emission and less of the excimer emission than both **6b** and **6a**, even though the concentration of Pt complex is higher in **6d** than in **6a**. This appears to be an effect of the polarity of the environment, where the higher ratio of nonpolar TPA to polar oxadiazole favors the monomer complex emitting species over the excimer, compared to **6a–c**.

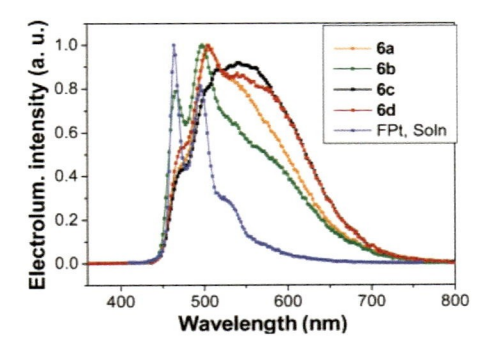

Figure 54.1 EL spectra of devices **6a–d** and the PL spectrum of FPt.

Table 54.1 Devices **6a–d** performance, color, and composition

Polymer	Turn on voltage, V	EQE %	CIE coordinates	Ratio of m:n:o
6a	7.8	4.6	$x = 0.33; y = 0.50$	10:1:10
6b	8.0	3.2	$x = 0.36; y = 0.48$	6:1:6
6c	8.1	3.0	$x = 0.38; y = 0.50$	3:1:3
6d	8.1	3.5	$x = 0.30; y = 0.43$	10:1:2
6d	8.1	3.5	$x = 0.30; y = 0.43$	10:1:2

As mentioned previously, the colors emitted by devices made from **6a**, **6b**, and **6d** differ slightly, and this is reflected in their CIE coordinates (Table 54.1). However, all devices appear nearly white to the eye (pure white is $x = 0.33$, $y = 0.33$). Device **6c**, having greater excimer emission, was orange-white in color. The spectra were reproducible for each device made with the same polymer. The devices give EL spectra consistent with only dopant emission; there is no copolymer backbone, HT, or ET emission observed at any bias level. The ratio of monomer/aggregate contributions in the EL spectra is also invariant with the applied voltage, leading to a voltage-independent, good quality, near-white broad emission. The performance of the devices also depended on the concentration of Pt complex, with **6a** having the best efficiency (4.6%) as well as the lowest turn on voltage (7.8 V); the other devices likely suffer Förster losses from high complex loading.

In conclusion, we have demonstrated that a random terpolymer, containing HT, ET, and phosphorescent Pt complex moieties, can be spin cast from solution to fabricate near-white light-emitting diodes with up to 4.6% external quantum efficiency (EQE), which is among the highest values reported for a solution-processed white-emitting device. Further changes in the composition ratio could potentially tune the emission closer to pure white.

Supporting Information

Experimental and characterization data of polymers **5a–d** and **6a–d** and OLED structure, characterization, and performance data for all devices (PDF). This material is available free of charge via the Internet at http://pubs.acs.org (https://ndownloader.figstatic.com/files/5152384).

Acknowledgment

Financial support of this research by the Air Force Office of Scientific Research, Universal Display Corporation, and by the U.S. Department of Energy under Contracts DE-AC03-76SF00098 and DE-FG02-02ER83565 are acknowledged with thanks.

References

1. (a) O'Brien, D. F., Baldo, M. A., Thompson, M. E., Forrest, S. R. *Appl. Phys. Lett.* 1999, **74**, 442–444 (b) Chan, S. C., Chan, M. C. W., Wang, Y., Che, C. M., Cheung, K. K., Zhu, N. *Chem. Eur. J.* 2001, **7**, 4180–4189. (c) Lu, W., Mi, B. X., Chan, M. C. W., Hui, Z., Che, C. M., Zhu, N., Lee, S. T. *J. Am. Chem. Soc.* 2004, **126**, 4958–4971.
2. (a) Baldo, M. A., Lamansky, S., Burrows, P. E., Thompson, M. E., Forrest, S. R. *Appl. Phys. Lett.* 1999, **75**, 4–6. (b) Ikai, M., Tokito, S., Sakamoto, Y., Suzuki, T., Taga, Y. *Appl. Phys. Lett.* 2001, **79**, 156–158.
3. Adachi, C., Baldo, M. A., Thompson, M. E., Forrest, S. R. *J. Appl. Phys.* 2001, **90**, 5048–5051.

4. Brooks, J., Babayan, Y., Lamansky, S., Djurovich, P. I., Tsyba, I., Bau, R., Thompson, M. E *Inorg. Chem.* 2002, **41**, 3055–3066.
5. (a) Andrade, B. W. D., Brooks, J., Adamovich, V., Thompson, M. E., Forrest, S. R. *Adv. Mater.* 2002, **14**, 1034–1036. (b) Adamovich, V., Brooks, J., Tamayo, A., Alexander, A. M., Djurovich, P. I., D'Andrade, B. W., Adachi, C., Forrest, S. R., Thompson, M. E. *New J. Chem.* 2002, **26**, 1171–1178.
6. (a) Kido, J., Shionoya, H., Nagai, K. *Appl. Phys. Lett.* 1995, **67**, 2281–2283. (b) Wang, Y. Z., Sun, R. G., Meghdadi, F., Leising, G., Epstein, A. J. *Appl. Phys. Lett.* 1999, **74**, 3613–3615. (c) Paik, K. L., Baek, N. S., Kim, H. K. *Macromolecules* 2002, **35**, 6782–6791. (d) Gong, X., Moses, D., Heeger, A. J., Xiao, S. *J. Phys. B* 2004, **108**, 8601–8605.
7. (a) Andrade, B. W. D., Thompson, M. E., Forrest, S. R. *Adv. Mater.* 2002, **14**, 147–151. (b) Feng, J., Li, F., Gao, W., Liu, S., Liu, Y., Wang, Y. *Appl. Phys. Lett.* 2001, **78**, 3947–3949. (c) Jiang, X., Zhang, Z., Zhao, W., Zhu, W., Zhang, B., Xu, S. *J. Phys. D: Appl. Phys.* 2000, **33**, 473–476.
8. Dailey, S., Feast, W. J., Peace, R. J., Sage, I. C., Till, S., Wood, E. L. *J. Mater. Chem.* 2001, **11**, 2238–2243.
9. Dailey, S., Feast, W. J., Peace, R. J., Sage, I. C., Till, S., Wood, E. L. *J. Mater. Chem.* 2001, **11**, 2238–2243.
10. Dao, J., Benoit, D., Hawker, C. J. *J. Polym. Sci., Polym. Chem.* 1998, **36**, 2161–2167.
11. Hawker, C. J., Bosman, A. W., Harth, E. *Chem. Rev.* 2001, **101**, 3661–3688.
12. (a) Sudhakar, M., Djurovich, P. I., Hogen-Esch, T. E., Thompson, M. E. *J. Am. Chem. Soc.* 2003, **125**, 7796–7797. (b) Chen, X., Liao, J. L., Liang, Y., Ahmed, M. O., Tseng, H. E., Chen, S. A. *J. Am. Chem. Soc.* 2003, **125**, 636–637.
13. Yang, J., Gorden, K. C. *Chem. Phys. Lett.* 2003, **372**, 577–582.
14. Furuta, P., Brooks, J., Thompson, M. E., Fréchet, J. M. J. *J. Am. Chem. Soc.* 2003, **125**, 13165–13172.

Index

1-phenyl-3-methylimidazolium iodide 559, 561–562, 565
2-ethoxyethanol 149, 175, 205, 211, 355, 357, 360, 511, 559–560, 562, 843–844, 1063
2-MeTHF 201, 222–223, 373, 440, 443–444, 446, 453, 465, 483, 486, 492, 500, 521, 523–526, 556, 574, 675, 680, 682, 685, 712, 737–738, 819–820, 838, 841, 880, 883, 933, 936–938, 940, 956, 965, 998, 1000, 1036–1037, 1039–1042, 1051
2-phenylpyridinate 851, 853, 866
4,4′-bis(9-ethyl-3-carbazovinylene)-1,1′-biphenyl (BCzVBi) 607, 609–610
9,10-Di(2-naphthyl)anthracene (ADN) 720–721

absorption bands 155–156, 396, 484, 494, 523, 996, 1008, 1036
absorption energies 295, 484, 487–488, 495, 523, 526
absorption peaks 45, 87, 236
absorption spectra 2, 97, 151, 155–158, 161, 184–186, 220, 225, 283, 285, 297, 372–374, 422, 443, 449, 455, 483–484, 500, 522–523, 539, 542, 548, 553, 572, 574, 623–624, 736–737, 769–770, 783, 807, 815, 818, 837, 867, 881, 933, 996, 999, 1008, 1010, 1036–1037
absorption transitions
  lowest energy triplet 443
  strong metal-to-ligand-charge-transfer 538
ADN, *see* 9,10-Di(2-naphthyl)anthracene
aggregation 292, 725, 853, 855–856, 865, 891, 935, 1048
alkyl groups 28, 287, 290, 710, 712, 722, 801
all-phosphor 613–615
Alq_3 4–9, 14–19, 39–56, 61–63, 70–73, 75–77, 83–84, 86, 97–98, 104, 106–107, 127, 137, 141, 152, 164–167, 243–244, 264, 271–273, 285, 294–295, 297–299, 306, 309, 311, 403, 418–423, 583–585, 593, 599, 625–626, 632, 637, 709, 747, 919, 1064
Alq_3 devices 5–6, 8, 17, 19, 48, 50, 52–54, 63, 140
  undoped 40, 44
aluminum 23–24, 61, 70, 82, 127, 137, 152, 164, 264, 286, 305–308, 583, 600, 625–626, 632, 747
ambient temperature phosphorescence 792–793, 795, 797

amorphous host materials 853
analogues, mononuclear 572, 574

β-diketonate ligand 201, 241, 287–288
bathocuproine (BCP) 15–20, 44–45, 48, 50–52, 55–56, 61–63, 71, 75–77, 82, 84–88, 96, 127, 152, 164–165, 243–244, 264, 270–274, 285, 294–295, 297–299, 306, 308, 319, 324, 339, 341–343, 403, 418–423, 429, 432–433, 583, 585, 593, 599, 607, 610, 614, 714–716, 720–721, 726, 736, 746–747, 973, 1064
BCP, see bathocuproine
BCP layer 15, 17–19, 44, 51–52, 63, 75, 88, 270, 272–273, 585, 592, 715, 747
BCzVBi, see 4,4'-bis(9-ethyl-3-carbazovinylene)-1,1'-biphenyl
benzopyranopyridinone 851, 853, 866
binuclear platinum complexes 572–576, 579
bis-cyclometalated iridium diimine complexes 521
blue phosphorescent cyclometalated Ir(III) complexes 675–700

CAAC, see cyclic alkyl-amino carbene
cyclic alkyl-amino carbene (CAAC) 930–931, 940, 987, 989, 992, 1012–1013, 1021–1022
cyclohexane 624, 737–738, 772–773, 816

cyclometalated carbene ligands for near-UV phosphorescence 550
cyclometalated complexes 173, 184, 200, 218, 375, 464, 496, 688, 836
cyclometalated corannulene 823–848
cyclometalated platinum complex 437–456
cyclometalates 171, 201, 214–215, 226, 372, 374–375, 378, 464, 520, 695
cyclometalating ligands 148–149, 172, 174–175, 192, 201, 204, 212, 350, 355–356, 359–361, 363, 371, 373, 378–379, 462–464, 477, 479, 490–491, 499, 511, 517, 539, 552–553, 699
  monoanionic 197, 347, 349
  phenylpyridine-based 510
  single 200

DAC, see diamidocarbene
diamidocarbene (DAC) 1019, 1021–1023, 1025, 1028–1029, 1032–1036
differential pulsed voltammetry 202, 351–352, 441, 468–469, 477, 564–565, 846, 957, 992, 995, 1030
dimers 180, 182, 205, 211, 213, 236, 283, 287, 355–356, 359, 473, 517, 761, 826–827, 843, 953–954, 971
  chloride-bridged 152–153, 155, 171, 174, 180, 182, 197
dipole, excited-state 996–997
dipole moments, excited-state 996
dopant aggregation 292, 581, 588, 855

dopant alignment 852, 858,
    860–861, 865
dopant concentrations 7, 9, 47,
    112, 151, 167, 236, 241, 286,
    290, 294, 297, 318, 323, 340,
    419, 589, 600, 867, 1048
dopant energies 301, 415, 583
dopant energy gap 301
dopant interactions 853, 858, 949
dopant materials 616, 922
dopants 70–71, 127–128,
    131–132, 146, 148, 163, 165,
    233–236, 263, 269, 271–273,
    279–280, 282, 286–288, 290,
    292, 295, 297, 301, 319–320,
    414–417, 419–420, 432–433,
    580, 585–593, 596–598, 600,
    607, 609, 639, 707–710, 712,
    716, 718–720, 851–855,
    859–861, 863–866, 878–879,
    883, 885, 890, 892, 902–907,
    909, 914–915, 948–949, 976,
    979, 1048–1049, 1062
  blue 593, 609, 906
  blue fluorescent 606, 616
  complex 292
  fluorescent 94, 273, 615
  lower energy 71
  phosphor 42, 148, 234, 272,
    422, 428, 431, 608, 611
  red 581, 597, 607, 650
  red phosphor 606, 608

EBL, see electron/exciton blocking
    layer
EGL, see energy gap law
EIL, see electron injection layer
electrochemical gap 487–488,
    510, 521, 532
electroluminescence spectra 4,
    73–74, 77, 118, 128, 238, 242,
    286, 310, 318, 337, 616, 626,
    957

electroluminescent devices 9, 44,
    98–99, 510, 790, 1061
  doped 2
  efficient 198, 580
  single-layer 508, 510–511, 528
electroluminescent spectra 33–34,
    65, 75, 135, 141, 308, 321,
    1047
electron-donating 200–201, 522
electron/exciton blocking layer
    (EBL) 280, 295, 297,
    429–430, 625, 877, 879,
    883–890, 892, 894
electron hopping rates 970–971
electron injection layer (EIL) 894,
    919
electron transfer quenching 442,
    451–452
  oxidative 443, 451–454
  reductive 443, 451–453
electron transfer quenching
    studies 437–456
electron transport layer (ETL) 73,
    81–82, 137, 152, 250, 295,
    324, 413, 585, 593, 607, 612,
    747, 884, 894
electronic spectroscopy 219, 372,
    480, 522
electrophosphorescence 14,
    59–66, 89, 105, 234, 253, 263,
    311, 313
  blue 116, 122, 327–328, 334,
    338, 344, 414
  blue organic 327–334, 428
  high-efficiency 136, 312
  near-infrared 621–628
electrophosphorescent devices
    13–20, 60, 62, 70, 83, 126,
    300, 305–312, 622
  blue 328, 414, 606
  efficient 415
  efficient deep blue/near
    ultraviolet 428
  efficient NIR 627

green 878
green-emitting 65
high operational stability 305–312
orange-red 305
red 13, 310
red organic 116, 414
saturated red 20
electrophosphorescent organic light-emitting devices 233
elemental analysis 26, 467, 517, 726, 748–749, 758, 944
emission
  aggregate 280, 290, 297
  AlQ$_3$ 4–5, 7–8, 18–19, 33–36, 46, 54, 98, 639
  balanced 288, 597
  balanced monomer/aggregate 288, 290, 292
  BCzVBi 609–610
  blue 65, 111, 121, 140, 161, 238, 320, 510, 594, 607, 613, 650, 748, 902
  blue host 83
  blue monomer 235, 280
  deep blue 107, 877–879
  dopant 297, 429, 708, 710, 1065
  electrophosphorescent 902
  excimer 191, 234, 236–238, 241–243, 444, 448, 938, 1061, 1064–1065
  exciplex 237, 281, 714
  fluorescent 1, 111
  green 18, 136, 167, 186, 306, 308, 575, 609
  host 61, 70, 76, 292, 715, 718–719, 721
  infrared 8, 20
  ligand-based 161
  light 17, 88, 248, 508–509, 852
  low-energy 580, 590
  near-UV 537, 557
  photoluminescence 236–237
  red 24, 148, 186, 386, 510, 575, 589–590, 594, 596, 607, 1062
  red-shifted 349, 351, 581, 714, 774, 888, 935
  saturated red 2, 8, 24–25, 36, 140, 146, 166
  solid-state 774
  thermalized 663, 797
  triplet 40, 65, 74, 86, 148, 329, 806
  weak 34, 439, 543
  white 241–242, 279, 281–282, 295, 297, 594
  white light 233–244, 315, 1064
  yellow 186, 322, 816
emission color 24, 70, 136, 148, 174, 201, 317, 463, 508, 510, 538, 540, 581, 586, 598, 676, 757, 1009, 1020, 1037, 1050
emission data 522, 769, 837
emission decay 151, 444, 545, 1004
emission decay rates 841, 1010–1011
emission decay time 659–660, 790–791, 793–794, 796–799, 804, 806
emission efficiencies 161, 199, 438, 774, 811, 816, 821, 889, 991, 1000
emission energies 190, 295, 348, 375, 384, 462, 465, 485–487, 489, 491–493, 496, 499–500, 510, 527, 539–541, 543, 550, 554, 584, 587, 622, 652, 664, 677, 687, 695, 793, 812, 816, 825, 881, 889, 996, 1048
emission intensity measurement 680, 685
emission lifetimes 151, 378, 437, 441, 526, 558, 700, 773–774, 776, 816, 839, 841–842, 930, 988, 994, 1002, 1010–1011, 1031, 1043–1044

emission peaks 98, 320, 487, 594, 724, 752, 793
  highest energy 451
  low-energy 714
  sharp 17, 31
emission spectra 28, 31, 61, 64, 70, 73, 97–99, 109–110, 155, 158–162, 172, 185–187, 198, 200, 220–225, 280, 283, 287, 302, 372–373, 402–403, 418, 456, 483–485, 487, 491–492, 499–500, 526, 528, 538, 540, 542, 554, 556, 565–566, 574–575, 599–600, 658–662, 745, 747, 770, 772–774, 792, 794, 815–816, 938, 996, 998, 1036–1037, 1039–1043
  ambient temperature 659
  angular-dependent photoluminescence 868
  featureless 526, 574
  low-temperature 219, 736
  mononuclear 571
  PHOLED 919
  red-shifted 172, 752
  single-crystal 574
  structured 770
  structureless 739
  time-resolved 120
emission spectroscopy 748
  angle-dependent *p*-polarized photoluminescent 852
emissive dopants 192, 266, 296, 301, 413–414, 463, 499–500, 538, 608, 708, 855, 949, 977, 979, 1049
  efficient 656
  fluorescent 622
  green 1020
  red 581
  single 242, 279
emissive layers 72–73, 105, 163–164, 239, 254, 270, 280, 282, 286, 295, 307–308, 310, 317, 319, 323, 329, 331, 339, 341, 413, 415, 420, 429–431, 438, 580, 592–595, 632, 743, 748, 884, 940, 949, 975, 1020, 1048
  single 282, 580, 593, 596–597
emissive materials 61, 330, 340, 528, 744, 989
emissive sites 7, 14, 24, 60, 71, 76–77
  phosphorescent 33, 36
EML, *see* light-emitting layer
energy barriers 419, 610, 652, 691, 697, 824, 833
energy gap 75, 112, 199–200, 222, 263, 295, 338, 415, 678, 714, 800, 812, 878, 881, 913, 915, 918, 929, 1010
  optical 285, 418, 488, 499
energy gap law (EGL) 492–493, 499, 527, 687, 1041
energy-level alignments 116, 949
energy separation 374, 465, 549, 558, 665, 667, 714, 768, 791, 797, 803, 989, 1011, 1020, 1043–1045, 1050
energy transfer 1–3, 7–8, 13–18, 23, 40, 50, 53, 56, 61, 66, 73, 88, 93–94, 98–100, 116, 120–122, 233–234, 239–240, 288, 292, 329–330, 332, 334, 338–339, 384–387, 393, 396, 400, 402, 407–408, 442, 450–452, 454, 606, 644, 707–710, 720, 723, 725, 921
  backward 119, 122, 136
  direct 97, 609
  direct triplet 71
  efficient 2, 8, 40, 82, 93, 129, 235, 292, 387, 592, 597, 708
  efficient exothermic 332, 338
  endothermic 115–122, 328, 332–333

exothermic 119–120, 327–328, 334, 344, 634
Förster 93, 708–709
host 414
host-to-guest 422
resonant 126, 239, 607
triplet exciton 720
energy transfer efficiency 93, 100
energy transfer pathway, endothermic triplet 239
EQE, *see* external quantum efficiencies
ETL, *see* electron transport layer
excimer 233, 235–236, 238, 240–241, 287, 437–439, 444–445, 447–448, 454, 571, 1041, 1062, 1064
  red shifted emissive phosphorescent triplet 455
excimer emission intensity 445, 447–448
excimer formation 191, 237, 241, 393, 445–449, 455, 935
excimer lifetimes 448, 455
excimer states 241–242
excitation spectra 157, 223, 237, 288, 339–340, 443–444, 658, 660–663, 736, 769, 807, 1000
excitation wavelengths 223, 332, 680, 746, 806
excited state energetics 450
excited-state lifetimes 173, 349, 441, 737, 770, 773–775, 867, 930, 937
excited state management 912, 918
excited state manager molecules 918
excited state monomers 455
excited-state oxidation 438, 455
excited state redox properties 439, 450
excited states 2–4, 29, 59–60, 93–94, 115–116, 126, 136–137, 146–148, 155, 159, 162–163, 166–167, 172–173, 199–200, 221, 225–227, 286–287, 375–378, 438–439, 455, 464–465, 485, 492, 508–509, 548–549, 554, 556, 572, 575–576, 679, 686–687, 692–693, 697–698, 734, 738, 744, 772–774, 839, 841–842, 890–891, 928–929, 939, 988–991, 996–997, 999–1000, 1010–1012, 1021, 1039, 1041, 1049
  complex 451
  emissive 167, 235
  emitting 199, 222
  high-energy 901
  ligand-based 159–160
  ligand-centered 495
  ligand-localized 687
  molecular 92, 492
  nonemissive 544, 550, 557–558
  phosphorescent 157, 296
exciton confinement 105, 112, 269–270, 637, 890
exciton diffusion 14, 52, 316–324, 332, 448, 610, 613, 616, 709
exciton formation 17, 40, 47–48, 54–56, 82, 88, 99, 105, 111–112, 166, 319–320, 337–338, 342–343, 404, 413, 588, 597, 608, 610, 633, 649, 708, 889, 917
exciton-polaron annihilation 633, 641, 646, 649
exciton-polaron model 640, 642, 645–646, 648, 651
exciton recombination 282, 403, 598
excitons 1, 3, 14, 17–18, 39–41, 43–44, 51, 53–55, 60–61, 63, 75, 91–92, 94, 104–105, 127–128, 131, 148, 253–254, 262, 264, 268, 270–273,

305–306, 317, 319–320, 323–324, 332, 334, 338–339, 341, 414–415, 420, 422, 606–611, 613, 634, 636, 641, 646, 648–649, 651–652, 708, 790–791, 878–879, 894, 903–904, 906–907, 914, 916–917, 919–922

exothermic host 327–334
  emissive 334

external quantum efficiencies (EQE) 2, 4–5, 13–15, 60, 62–63, 81–82, 84, 86, 97, 103–104, 127, 129–130, 136, 163–164, 250–254, 261, 263, 268, 270–271, 280, 306, 308, 310, 383–384, 420–421, 508–511, 528, 530–533, 587–588, 590–592, 610, 614–616, 622, 625–626, 638–639, 717–722, 743, 852–853, 879, 886–888, 891–892, 895–896, 909–911, 919–920, 973–975, 989, 1009, 1020, 1047–1048, 1065–1066

extinction coefficients 155–156, 184, 186, 219, 239, 374, 395, 462, 483–484, 490, 495–496, 548, 574, 624, 939, 998, 1008, 1010

flat-panel displays, very-high-efficiency 247–256

fluorescence 3, 14, 34–35, 39, 41, 43, 60, 64, 73, 91–92, 100, 104, 148, 157, 262, 290, 639, 641, 682, 714, 719, 799, 820, 929, 942, 966, 991, 1004, 1006, 1011
  delayed 100, 774, 799, 805, 928, 949, 988–989, 991
  phosphor-sensitized 125–132

fluorescent devices 60, 119, 311–312
  phosphor-sensitized 130

fluorescent dyes 1, 7, 40–42, 69, 92–94, 98, 100–101, 126, 130, 136, 306, 395

fluorescent materials 92, 94, 104, 126–127, 148

fluorescent OLEDs 126–127, 132, 248, 310–311, 313, 606

fluorine-free emitter 427–434

glycerol 177, 284, 356, 358–359, 376, 560

HBL, see hole blocking layer
heavy-metal complexes 136, 756
  phosphorescent 622
heterocyclic compounds, electron-deficient 308
heteroleptic phosphors 851–872
heterostructure devices 271
high-efficiency fluorescent organic light-emitting devices 91
high-efficiency green OLEDs 60–66
high-efficiency organic electrophosphorescent devices 81–88
high-efficiency red electrophosphorescence devices 135–142
high-efficiency single dopant white electrophosphorescent light emitting diodes 279–302
high-energy ligand field states 553
highest occupied molecular orbital (HOMO) 62, 74, 86, 88, 111–112, 214–215, 221–222, 224, 238–239, 253, 272, 285, 296, 338, 368, 371–372, 374, 415–416, 432–433, 449, 455,

462, 465, 477, 480, 487–488, 515, 520, 539, 582, 584, 633, 667, 677, 738–739, 768–769, 793, 802–803, 816, 818–820, 883–884, 929, 931–933, 949, 953, 963–965, 967–968, 970, 1033–1036, 1045–1046
highly phosphorescent bis-cyclometalated iridium complexes 145–166
HIL, see hole injection layer
hole blocking layer (HBL) 17, 152, 295, 306–307, 318, 339, 429, 585, 593, 607, 715, 884–885, 889, 894, 919
hole injection layer (HIL) 117, 306, 329, 884, 894
hole-transport layer (HTL) 16–17, 82, 85, 88, 117, 127, 137, 151, 241, 248, 250, 257, 295–296, 301, 319, 585, 593, 607, 707, 716, 884, 894, 907
hole-transporting host materials 707–708, 710, 712, 716–728
holes 17–18, 44, 52, 74, 88, 101, 104, 111, 137, 148, 152, 204, 239, 250, 256, 262, 270, 272–273, 295, 305–308, 312, 320, 339, 344, 383, 403, 406, 416, 419, 428–429, 433, 438, 585, 588, 594, 596, 634, 636, 646, 707, 716, 884–886, 914–915, 921, 963, 971, 975, 977, 994, 998–999
HOMO, see highest occupied molecular orbital
host
  fluorescent 70–71, 338–339
  high-energy 328
  nonpolar 98, 892
  wide-energy-gap 337, 891
  wide-gap 338, 344
host evaporation 286, 600
host exciton 31, 585

host exciton lifetime 76
hot excited state management for long-lived blue PHOLEDs 902–922
HTL, see hole-transport layer

ICT, see intramolecular charge transfer
ILCT, see intra-ligand charge-transfer
imidazole 308, 800–802, 948, 951, 960–962, 965, 970
indium tin oxide (ITO) 4, 43, 47, 71, 77, 82, 105, 109, 117, 127, 137, 151, 243, 248, 257, 264, 285–286, 317, 319, 323, 338, 419, 423, 429, 508, 516, 529, 585, 599–600, 615, 625–626, 637, 712, 894, 918, 941, 957, 1031, 1064
internal quantum efficiency (IQE) 19–20, 25, 103–105, 108, 112, 126–127, 136, 148, 606–607, 615, 902, 965
intra-ligand charge-transfer (ILCT) 739, 819
intramolecular charge transfer (ICT) 933, 1035, 1040
IQE, see internal quantum efficiency
iridium 59–60, 70, 72, 81–82, 96, 103, 105, 116–117, 125–126, 135–137, 142, 146, 150–151, 167, 175–177, 250, 257, 261, 266, 280, 282, 285, 296, 305–306, 310, 318–320, 324, 328, 337, 357–358, 362, 414, 417, 427–429, 469–472, 512–515, 537, 551, 561, 563, 607, 866, 880, 889, 895, 906, 919, 927, 948, 950

iridium complexes 118, 145, 384, 517, 537–566, 682, 823, 836, 856
  cyclometalated 538, 856
  facial tris-cyclometalated 379
  heteroleptic 856
  phosphorescent 852, 855
  phosphorescent organometallic 306
  tris-cyclometalated 222, 371, 698
iridium trichloride hydrate 559–560, 562
isomers 161, 284, 288, 350, 363, 368, 376, 428–430, 433, 538, 551–553, 599, 679, 842, 879, 881, 883, 890
  facial 347–348, 350, 360, 363–364, 367, 372, 374–378, 431, 551–552, 855
  meridional 347–351, 356, 360–364, 366–368, 372–373, 375–376, 378, 431–432, 538, 551, 553, 878
ITO, *see* indium tin oxide

lanthanide complexes 2, 25
laser dyes 157, 225–226, 394
LECs, *see* light-emitting electrochemical cells
ligands 136–137, 142, 145–147, 152–153, 155, 159–163, 167, 172, 180–182, 184, 186–187, 190–191, 198–200, 211, 218–219, 224–225, 350, 360–363, 463–465, 473–474, 507–508, 510–511, 538–539, 541, 543–544, 548–550, 552–553, 571–573, 676–677, 693, 695, 699, 734, 752, 755–757, 762, 764–765, 775–776, 801–803, 812–815, 819–820, 824–826, 837, 855–856, 863, 865–866, 870–871, 931–933, 1011, 1021–1022
  2-phenylpyridyl 201, 218, 497
  acetylacetonate 117, 366
  amide 932–933, 989
  ancillary 145, 174, 182, 199–200, 461–462, 464–465, 477, 479–480, 484–485, 487–490, 492, 495–497, 499, 541, 655, 657, 664, 667, 670, 677, 699, 736, 812, 853
  aromatic 66, 199, 286
  bipyridyl 510, 520, 522, 527
  biquinolyl 522, 527
  borate 461, 473, 500, 814, 819
  bpy 510
  btp 137, 159
  carbazole 939, 995, 1010, 1026
  carbazolyl 995–996, 1033, 1036
  coumarin-based 851, 853
  cyclometalated 147, 171–174, 180, 182, 349, 443, 449, 489, 695, 828, 836–837, 839, 853, 858
  diamine 172, 532
  diimine 173, 201, 349, 517, 522–523, 532
  diketonate 284, 853
  emissive 855, 861
  isocyanide 477, 485
  monodentate 694
  N-heterocyclic carbene 537, 755, 757
  NHC 538, 550, 736, 761–762, 764, 767, 800–802, 811–816, 819–821, 878, 880
  nonchromophoric 464–465, 656
  phenylpyrazolyl 520, 692
  pyrazolyl 348, 371, 474, 480, 493, 552, 830–831
  pyridine-based 744

light-emitting devices 104, 248
light-emitting electrochemical cells (LECs) 508–510, 528–532
  lifetimes of 532
  polymer-based 509, 532
  solid-state 509
light-emitting layer (EML) 82–86, 88, 104, 106, 108, 117, 119, 127, 131, 137, 250, 253–254, 339, 341, 343–344, 415–416, 418–420, 422, 429, 593–595, 597, 607–611, 616, 632, 634, 636–637, 639, 650, 652, 746, 878–879, 883–886, 888–890, 892, 894–895, 903–904, 906–907, 909, 912–917, 919–922, 976, 1032, 1047
lowest unoccupied molecular orbital (LUMO) 17, 62, 72, 86, 111, 198, 215, 238–239, 296, 338, 348, 368, 372, 374, 415, 432–433, 462, 465, 477, 479, 484, 488, 499, 507, 515, 520, 539, 541, 583, 624, 633, 738–739, 768–769, 802–803, 816, 819–820, 883–884, 907, 929, 931–933, 947, 949, 953, 958–959, 961–964, 968, 970, 1033–1035, 1045–1046, 1049–1050
luminance loss 631–633, 639, 641, 646, 915–916, 921
LUMO, *see* lowest unoccupied molecular orbital
LUMO energies 203, 222, 224, 285, 295, 342, 344, 355, 372, 407, 415, 417–418, 433, 467, 477, 487–488, 520, 522, 532–533, 541, 543, 548, 583–584, 592, 712, 768, 792, 819, 883, 885, 958, 960–962, 978, 1033

meridional complexes 351, 377
metal complexes 136, 349, 371, 439, 464, 490, 509–510, 678, 686–687, 761, 814, 824, 837, 988–989, 991, 1003–1004, 1006, 1051
  luminescent transition 544, 699, 812
  phosphorescent heavy transition 824
  phosphorescent transition 677

N-heterocyclic carbene (NHC) 537–538, 541, 550, 734, 755, 757, 760, 775, 789, 811–815, 819–820, 878, 1019–1052
natural transition orbitals (NTOs) 994, 998–999
NHC, *see N*-heterocyclic carbene
NHC complexes 733–740
NMR, *see* nuclear magnetic resonance
NMR spectra 149, 152, 174, 181, 202, 283–284, 351, 362–363, 377, 379, 467, 516, 565, 726, 762, 767, 775–777, 783, 830, 833, 841, 846, 944, 956, 1012, 1023, 1051
NMR spectroscopy 223, 357, 359, 361–363, 473, 517, 756, 758, 824, 831
non-Lambertian spatial emission patterns 83, 106
nonradiative decay 17, 84, 121, 485, 557, 678, 684, 688, 697, 699, 928–944, 1004, 1039
NTOs, *see* natural transition orbitals
nuclear magnetic resonance (NMR) 26, 150–151, 175–177, 205–210, 284–285, 288, 357, 379, 470–472, 512–514, 560–564, 599, 727–728, 761,

764, 777–779, 783, 841, 844–846, 848, 866–867, 893–894, 931, 955–956

OLED host materials 280, 947–980
OLEDs, see organic light-emitting devices
organic electroluminescent devices 1–10
organic light-emitting devices (OLEDs) 14–15, 17, 23–25, 34, 59–61, 69–73, 82–84, 97–98, 100, 103–112, 125–127, 135–137, 146, 148, 163–167, 233–234, 247–248, 262–263, 294–296, 306–307, 328–329, 338–343, 383–384, 413–417, 419, 429, 431–432, 438, 531–532, 580–583, 585–589, 598, 605–606, 621–622, 631–632, 655–671, 707–709, 744–752, 790–791, 863–864, 902–903, 927–928, 930–931, 940–941, 947–951, 957, 973–976, 1009, 1020–1022, 1046–1050
  based on phosphorescent hosts 69–78
  high-efficiency 81–82
  phosphor-doped 384, 451
  phosphorescent small-molecule 632–652
  for very-high-efficiency flat-panel displays 248–256
  white 234, 239, 315
organometallic complexes 181, 928, 1021
oxidation
  irreversible 198, 218, 407, 438, 480, 489, 739, 932, 995
  reversible 297, 348, 371, 407, 521, 553, 624, 868

phenanthroline 172–173, 199, 349, 734, 738, 948, 954–956, 963, 965, 970
phenanthroline ligand 736, 738
phenyl ligand 364, 695
phenylpyrazole-based iridium diimine complexes 532
PHOLED emission layer 878, 912
PHOLEDs, see phosphorescent organic light-emitting diodes
phosphor molecules 122, 126
phosphor sensitization 126–127
phosphorescence
  blue 329, 339, 537–566, 677
  delayed 120–121, 333–334
  near-UV 537–566
  short emissive lifetime 414
  weak 162–163, 182, 329
phosphorescence efficiencies 84, 103–112, 162, 182, 192, 676, 891
phosphorescence emission rate 89, 110
phosphorescence energy 161–162, 375, 452
phosphorescence lifetime 3, 20, 27, 33, 174, 202, 283, 351, 466, 516, 565, 782, 993, 1010
phosphorescence quenching 383–388
phosphorescence spectra 159–161, 166–167, 330, 388, 965
phosphorescent bis-cyclometalated iridium complexes 146–168
phosphorescent complexes 263, 622, 842
  high-energy 699
phosphorescent compounds 66, 192, 383
phosphorescent cyclometalated iridium complexes 171–192

phosphorescent cyclometalated
    platinum complexes 197–228
phosphorescent dopants 70, 73,
    75, 136, 151, 243, 261–264,
    266, 269, 271–273, 275, 288,
    295, 350, 387, 419, 433, 558,
    579, 581–582, 588–589,
    606–608, 611, 616, 623, 710,
    854, 949, 975
  blue 292, 951
  efficient 148, 438
  emissive 70
  red 592, 611, 616
phosphorescent dyes 1–3, 10, 14,
    25, 40, 42–43, 69–70, 92
phosphorescent organic
    light-emitting diodes
    (PHOLEDs) 248, 250–252,
    254, 256, 310–312, 877–879,
    883–884, 886–888, 890, 892,
    894, 896, 901–902, 910, 912,
    918–920, 949, 977, 979
  blue 902–904, 907, 918–919,
    977–978
  doped EML 885, 888
  long-lived Blue 902–922
  managed 907, 909–910,
    914–918, 921
  multilayer white 315–324
phosphors 42–43, 60, 94, 116,
    120, 122, 126–127, 130,
    136–137, 139–140, 148, 155,
    163, 166–167, 173, 234–235,
    237, 266, 274, 317–319, 350,
    383, 385–388, 413, 415, 429,
    432–433, 540, 582, 608–610,
    615, 622, 757, 839, 860,
    889–890, 937
  red 14, 61, 135–136, 316, 324,
    387, 608–609
photoelectron spectroscopy 285,
    423
photoluminescence spectra 31,
    118, 156, 160, 167, 185,
    291–292, 294, 423, 454, 508,
    540, 847, 956, 1048
photonic multichannel analyzer
    (PMA) 680, 783, 847, 956,
    993, 1031
platinum 3, 24–26, 31, 33, 69–70,
    73, 137, 139, 198–199,
    203–210, 212–215, 218, 235,
    279, 283–284, 287–288, 292,
    371–372, 374, 414, 437–439,
    442, 448, 451, 495, 520,
    573–576, 579–582, 584,
    586–588, 590, 592, 597–598,
    621–628, 678, 756–757,
    790–791, 796, 823–828, 830,
    836–839, 841–844, 847–848,
    927–928, 937, 1020–1022,
    1037, 1049, 1061–1062, 1064
platinum binuclear complexes
    579–600
platinum-functionalized random
    copolymers 1061–1066
platinum porphyrins 23–24,
    30–31, 33–34, 60
PMA, *see* photonic multichannel
    analyzer
polarons 55–56, 633, 636, 641,
    650–651, 903, 907
polymer organic light-emitting
    devices 261–272
polystyrene 7, 15, 30, 402, 415,
    557, 574, 585, 587, 935, 993,
    1000, 1003–1004, 1008, 1041
porphyrins 23–36, 60, 69

QE, *see* quantum efficiency
quantum efficiency (QE) 6–9, 18,
    20, 24–26, 28–32, 36, 50–51,
    56, 60, 62–65, 69, 73–75, 83,
    100, 106, 108, 129, 146, 157,
    162, 172, 227, 253, 262, 269,
    271–272, 274, 297–301, 323,
    330–331, 340, 351, 378, 384,

447–448, 516, 529, 531–532, 538, 541, 565, 580, 588, 592, 598, 680, 684–685, 688, 691, 698

radial distribution functions (RDFs) 969
RDFs, *see* radial distribution functions
red electroluminescent devices 507
reduction, reversible 218, 371, 407, 455–456, 837, 868, 995

saturated calomel electrode (SCE) 72, 202, 371, 438, 441, 449–450, 452, 454–456, 469, 521
saturated deep blue organic electrophosphorescence 427–434
SCE, *see* saturated calomel electrode
sensitizers 94, 97–98, 100, 132, 349, 439, 463, 538
  phosphorescent 91–92, 97, 100, 127
silver 44, 551, 560–563, 994, 996, 998, 1002
silver complexes 988, 992, 1004, 1010
singlet-triplet splitting 540, 801, 803–804, 1004
solid-state luminescence spectra 189
solid-state phosphorescence spectra 966
solid-state photoluminescence spectra 191
states
  bound ground 233, 235, 1062
  dopant triplet 615

electronic ground 661, 801
emissive triplet 544, 606
emitting triplet 655–657, 660, 664
excitonic 13, 15–16, 243
guest triplet 61, 328
high-energy triplet 116, 415
ligand-based 159, 200
ligand-centered 462
ligand-centered triplet 226, 682
ligand-localized LUMO 218
long-lived triplet 3, 64, 928, 1002
low-energy triplet 157, 387
metal-centered ligand-field 879
metal-halide-to-ligand charge transfer 990
metal-to-ligand charge transfer 105
Stern–Volmer analysis 385, 388, 451
Strickler–Berg analysis 1008, 1010

TADF, *see* thermally activated delayed fluorescence
TCSPC, *see* time-correlated single photon counting
TD-DFT, *see* time-dependent density functional theory
thermal equilibration 686, 1010
thermally activated delayed fluorescence (TADF) 789–806, 816, 928–929, 931, 949, 988–989, 991, 1002, 1006, 1021, 1043, 1049
thin films, doped 282, 287, 386–387, 419, 448, 454
time-correlated single photon counting (TCSPC) 440, 994
time-dependent density functional theory (TD-DFT) 756, 792, 842, 939, 944, 952, 994, 997

transition-metal complexes 66, 510, 622, 686, 699, 757, 770
transition metal compounds, phosphorescent 790
transitions
  excited-state 996, 1036
  nonemissive 87, 110
  one-electron 1045–1046
  vibronic 485, 526, 589, 594
triplet emission lifetimes 988
triplet energies 116, 215, 253, 273, 319, 329–330, 368, 375, 384–387, 414–418, 428, 432, 437, 450–452, 454–455, 464, 488, 499, 510, 539–541, 543, 545, 548, 558, 606, 634, 695, 710, 714, 720, 858, 907, 940, 947, 950–951, 958–962, 965, 967, 973, 975, 978
triplet energy level alignment 82
triplet energy transfer
  quenchers 450
triplet energy traps 263
triplet excimers 233–244
triplet exciton diffusion 61, 613, 617, 707–710, 712–728
triplet excitons 3, 9, 19, 40, 43, 69, 77, 82, 85–86, 88–89, 92, 105, 110, 112, 116, 126, 146, 151, 199, 242, 254, 262, 317, 333, 383, 414, 428, 580, 605–616, 622, 656, 709–710, 719–721, 723–726, 804, 903, 915, 928, 940, 1021, 1061
  emissive 126
  high-energy 428
triplet metal-to-ligand-charge-transfer 856
triplet-state energies 387, 423
tris-chelate complexes 162, 172, 182
  neutral metal 173

tris-cyclometalates 348, 350, 375, 378, 853
tris-ligand complexes 174
  neutral 147

ultrahigh energy gap
  hosts 413–424
ultrahigh energy gap organosilicon compounds 413
ultraviolet photoelectron spectroscopy (UPS) 297, 417–418, 432, 583–584
UPS, see ultraviolet photoelectron spectroscopy
UV photoelectron spectroscopy 583

white emission, balanced 237, 282, 595
white organic light emitting diodes (WOLEDs) 235–236, 238, 240–241, 243, 279–282, 286, 294–295, 298–299, 301, 315–316, 320, 323, 581, 593, 595–598, 600, 606–607, 611–615, 617
WOLEDs, see white organic light emitting diodes

X-ray absorption spectroscopy (XAS) 744, 749, 752
X-ray crystallographic analysis 780–781
XAS, see X-ray absorption spectroscopy

yellow double-doped OLEDs 126–132